PATTY'S INDUSTRIAL HYGIENE

Fifth Edition

Volume 4

VII SPECIALTY AREAS AND ALLIED PROFESSIONS

PATTY'S INDUSTRIAL HYGIENE

Fifth Edition

Volume 4

VII SPECIALTY AREAS AND ALLIED PROFESSIONS

ROBERT L. HARRIS

Editor

CONTRIBUTORS

J. D. Bachmann
G. M. Barron
L. R. Birkner
M. S. Black
D. S. Bloswick
K. A. Busch
F. I. Cooper
K. J. Donham
B. L. Epstien
J. F. Gamble
M. A. Golembiewski
M. Hertzberg
T. A. Hethmon
W. E. Horner
R. G. Lieckfield Jr.
N. A. Leidel
R. McIntyre-Birkner
D. J. McKee
P. R. Morey
H. J. Muranko
J. S. Ostendorf
R. C. Poore
W. J. Popendorf
B. Rogers
G. R. Rosenblum
R. Sesek
M. A. Shapiro
L. K. Simkins
T.J. Villnave
A. G. Worthan

A Wiley-Interscience Publication

JOHN WILEY & SONS, INC.

New York / Chichester / Weinheim / Brisbane / Singapore / Toronto

This book is printed on acid-free paper. ♾

Published simultaneously in Canada.

For ordering and customer service, call 1-800-CALL-WILEY.

Library of Congress Cataloging in Publication Data:

Patty's industrial hygiene / [edited by] Robert L. Harris. — 5th ed.
[rev.]
v. ⟨ ⟩ cm.
Fourth ed. published as: Patty's industrial hygiene and toxicology.
Includes bibliographical references and index.
ISBN 0-471-29749-6 (Vol. 4) (cloth : alk. paper); 0-471-29784-4 (set)
1. Industrial hygiene. I. Harris, Robert L., 1924– .
II. Patty, F. A. (Frank Arthur), 1897– Industrial hygiene and toxicology.
RC967.P37 2000
613.6′2—dc21 99-32462

Printed in the United States of America.

10 9 8 7 6 5 4 3 2 1

Contributors

John D. Bachmann
U.S. Environmental Protection Agency
Research Triangle Park, North Carolina

Gerald M. Barron
Pittsburgh, Pennsylvania

Lawrence R. Birkner, MA, MBA, CIH
McIntyre, Birkner and Associates
Thousand Oaks, California

Marilyn S. Black, Ph.D
Air Quality Services, Inc.
Atlanta, Georgia

Donald S. Bloswick, Ph.D, PE
Department of Mechanical Engineering
College of Engineering, University of
Utah, Salt Lake City, Utah

Kenneth A. Busch, MS
Cincinnati, Ohio

Frederick I. Cooper
Cooper Environmental
Moraga, California

Kelley J. Donham, DVM, University of
Iowa, Institute of Agricultural Medicine
and Occupational Health, Oakdale, Iowa

Barbara L. Epstien, MPH, CIH
Air Quality Services, Inc.
Atlanta, Georgia

John F. Gamble, Ph.D, Occupational
Health and Epidemiology Division,
Exxon Biomedical Sciences Inc.
East Millstone, New Jersey

Mark A. Golembiewski, CIH
Daly City, California

Martin Hertzberg, Ph.D
Copper Mountain, California

W. Elliot Horner, Ph.D
Air Quality Services, Inc.
Atlanta, Georgia

Thomas A. Hethmon, CIH, ROH
Phelps Dodge Corp.
Phoenix, Arizona

Robert G. Lieckfield, Jr., CIH
Clayton Environmental Consultants
Novi, Michigan

Nelson A. Leidel, Sc.D
Punta Gorda, Florida

Ruth McIntyre-Birkner, MBA
McIntyre, Birkner and Associates
Thousand Oaks, California

David J. McKee, Ph.D
U.S. Environmental Protection Agency
Research Triangle Park, North Carolina

Philip R. Morey, Ph.D, CIH
Air Quality Services, Inc.
Gettysburg, Pennsylvania

Henry J. Muranko, MPH, CIH, CSP
Muranko and Associates
Scottsdale, Arizona

Judith S. Ostendorf, MPH, CCM, COHN-S, Public Health Nursing
The University of North Carolina
Chapel Hill, North Carolina.

Ronald C. Poore, CIH
Clayton Environmental Consultants
Novi, Michigan

William J. Popendorf, Ph.D, CIH
Department of Biology, Utah State University, Logan, Utah

Bonnie Rogers, Dr.PH., COHN-S, FAAN, Public Health Nursing
The University of North Carolina
Chapel Hill, North Carolina

Gary R. Rosenblum, MS, CIH, ARCO
Los Angeles, California

Richard Sesek, Ph.D, Department of Mechanical Engineering, College of Engineering, University of Utah
Salt Lake City, Utah

Maurice A. Shapiro, Professor of Environmental Health Engineering, Graduate School of Public Health,
Pittsburgh, Pennsylvania

Lisa K. Simkins, PE, CIH
Alamo, California

Timothy J. Villnave, DC, MSPH
Salt Lake Worksite Ergonomics
Salt Lake City Utah

Anthony G. Worthan, MPH
Air Quality Services, Inc.
Atlanta, Georgia

Preface

Industrial hygiene is an applied science and a profession. Like other applied sciences such as medicine and engineering, it is founded on basic sciences such as biology, chemistry, mathematics, and physics. In a sense it is a hybrid profession because within its ranks are members of other professions—chemists, engineers, biologists, physicists, physicians, nurses, and lawyers. In their professional practice all are dedicated in one way or another to the purposes of industrial hygiene, to the anticipation, recognition, evaluation, and control of work-related health hazards. All are represented among the authors of chapters in these volumes.

Although the term "industrial hygiene" used to describe our profession is probably of twentieth century origin, we must go further back in history for the origin of its words. The word "industry," which has a dictionary meaning, "systematic labor for some useful purpose or the creation of something of value," has its English origin in the fifteenth century. For "hygiene" we must look even earlier. Hygieia, a daughter of Aesklepios who is god of medicine in Greek mythology, was responsible for the preservation of health and prevention of disease. Thus, Hygieia, when she was dealing with people who were engaged in systematic labor for some useful purpose, was practicing our profession, industrial hygiene.

Industrial Hygiene and Toxicology was originated by Frank A. Patty with publication of the first single volume in 1948. In 1958 an updated and expanded Second Edition was published with his guidance. A second volume, Toxicology, was published in 1963. Frank Patty was a pioneer in industrial hygiene; he was a teacher, practitioner, and manager. He served in 1946 as eighth President of the American Industrial Hygiene Association. To cap his professional career he served as Director of the Division of Industrial Hygiene for the General Motors Corporation.

At the request of Frank Patty, George and Florence Clayton took over editorship of the ever-expanding *Industrial Hygiene and Toxicology* series for the Third Edition of Volume I, *General Principles*; published in 1978, and Volume II, *Toxicology*, published in

1981–1982. The First Edition of Volume III, *Theory and Rationale of Industrial Hygiene Practice,* edited by Lewis and Lester Cralley, was published in 1979 with its Second Edition published in 1984. The ten-book, two-volume Fourth Edition of *Patty's Industrial Hygiene and Toxicology,* edited by George and Florence Clayton, was published in 1991–1994, and the Third Edition of Volume III, *Theory and Rationale of Industrial Hygiene Practice,* edited by Robert Harris, Lewis Cralley, and Lester Cralley, was published in 1994. With the agreement and support of George and Florence Clayton, and Lewis and Lester Cralley, it is a signal honor for me to follow them and Frank A. Patty as editor of the Industrial Hygiene volumes of *Patty's Industrial Hygiene and Toxicology.*

Industrial hygiene has been dealt with very broadly in past editions of *Patty's Industrial Hygiene and Toxicology.* Chapters have been offered on sampling and analysis, exposure measurement and interpretation, absorption and elimination of toxic materials, occupational dermatoses, instrument calibration, odors, industrial noise, ionizing and nonionizing radiation, heat stress, pressure, lighting, control of exposures, safety and health law, health surveillance, occupational health nursing, ergonomics and safety, agricultural hygiene, hazardous wastes, occupational epidemiology, and other vital areas of practice. These traditional areas continue to be covered in this new edition. Consistent with the past history of *Patty's,* new areas of industrial hygiene concerns and practices have been addressed as well: aerosol science, computed tomography, multiple chemical sensitivity, potential endocrine disruptors, biological monitoring of exposures, health and safety management systems, industrial hygiene education, and other areas not covered in earlier editions.

Although industrial hygiene has been practiced in one guise or another for centuries, the most systematic approaches and the most esoteric accomplishments have been made in the past fifty or sixty years—generally in the years since Frank Patty published his first book. This accelerated progress is due primarily to increased public awareness of occupational health and safety issues and need for environmental control as is evidenced by Occupational Safety and Health, Clean Air, and Clean Water legislation at both federal and state levels.

Industrial hygienists know that variability is the key to measurement and interpretation of workers' exposures. If exposures did not vary, exposure assessment could be limited to a single measurement, the results of which could be acted upon, then the matter filed away as something of no further concern. We know, however, that exposures change. But not only do exposures change—change is characteristic of the science and practice of our profession as well. We must be alert to recognize new hazards, we must continue to evaluate new and changing stresses, we must evaluate performance of exposure controls and from time to time upgrade them. These volumes represent the theory and practice of industrial hygiene as they are understood by their chapter authors at the time of writing. But, as observed by the Greek philosopher Heracleitus about 2500 years ago, "There is nothing permanent except change." Improvements and changes in theory and practice of industrial hygiene take place continuously and are generally reported in the professional literature. Industrial hygienists, the practitioners, the teachers, and the managers, must stay abreast of the professional literature. Furthermore, when an industrial hygienist develops new knowledge, he/she has what almost amounts to an ethical obligation to share it in our journals.

One cannot ponder the rapid changes and advancements made in recent decades in science and technology, and in our own profession as well, without wondering at what the next two or three decades will bring. Developments in computer technology and information processing and exchange have greatly influenced manufacturing (robotics, computer controlled machining) and the general conduct of commerce and business in the past one or two decades. This change will only accelerate with computer speeds and capacities doubling every 18 months or so, and processing units approaching microsize. The possibility for continuously monitoring and computer storage of exposures of individual workers may become reality within the next decade. The human genome project holds promise for prevention and cure of many diseases, including some associated with conditions of work. World population continues to increase geometrically and is expected to be about eight billion in the year 2020; with improvements in preventive health care this will be an increasingly older population. Genetic engineering and highly effective pesticides are already improving yields of agricultural commodities; if all goes well in this area, and if we can avoid set-backs as might be associated with potential endocrine disruptors, feeding the expanding human population may not be a limiting factor. Globalization of manufacturing and commerce has already begun to reduce manufacturing employment in the United States and in Europe, and to expand opportunities for expanding populations in some developing nations. The United States and other developed nations are on their way to becoming world centers of information and innovation.

How will all of this affect the future practice of industrial hygiene? In the Preface to the Fourth Edition of *Patty's,* George and Florence Clayton suggested that the future of industrial hygiene is limited only by the narrowness of vision of its practitioners. More recently, Lawrence Birkner, past president of the American Academy of Industrial Hygiene, and his co-worker and spouse, Ruth McIntyre Birkner, in writing about the future of the occupational and environmental hygiene profession, say much the same thing. (See "The Future of the Occupational and Environmental Hygiene Profession" in *A.I.H.A.* Journal, pp. 370–374, 1997) Larry and Ruth report that we must be aware of the changes likely to take place in the next couple of decades, and must develop strategies now to assure the profession's full participation in protecting the health and safety of workers, and the environment, of tomorrow.

ROBERT L. HARRIS

Raleigh, North Carolina

Contents

USEFUL EQUIVALENTS AND CONVERSION FACTORS

1 kilometer = 0.6214 mile
1 meter = 3.281 feet
1 centimeter = 0.3937 inch
1 micrometer = 1/25,4000 inch = 40 microinches = 10,000 Angstrom units
1 foot = 30.48 centimeters
1 inch = 25.40 millimeters
1 square kilometer = 0.3861 square mile (U.S.)
1 square foot = 0.0929 square meter
1 square inch = 6.452 square centimeters
1 square mile (U.S.) = 2,589,998 square meters = 640 acres
1 acre = 43,560 square feet = 4047 square meters
1 cubic meter = 35.315 cubic feet
1 cubic centimeter = 0.0610 cubic inch
1 cubic foot = 28.32 liters = 0.0283 cubic meter = 7.481 gallons (U.S.)
1 cubic inch = 16.39 cubic centimeters
1 U.S. gallon = 3,7853 liters = 231 cubic inches = 0.13368 cubic foot
1 liter = 0.9081 quart (dry), 1.057 quarts (U.S., liquid)
1 cubic foot of water = 62.43 pounds (4°C)
1 U.S. gallon of water = 8.345 pounds (4°C)
1 kilogram = 2.205 pounds
1 gram = 15.43 grains
1 pound = 453.59 grams
1 ounce (avoir.) = 28.35 grams
1 gram mole of a perfect gas ≎ 24.45 liters (at 25°C and 760 mm Hg barometric pressure)
1 atmosphere = 14.7 pounds per square inch
1 foot of water pressure = 0.4335 pound per square inch
1 inch of mercury pressure = 0.4912 pound per square inch
1 dyne per square centimeter = 0.0021 pound per square foot
1 gram-calorie = 0.00397 Btu
1 Btu = 778 foot-pounds
1 Btu per minute = 12.96 foot-pounds per second
1 hp = 0.707 Btu per second = 550 foot-pounds per second
1 centimeter per second = 1.97 feet per minute = 0.0224 mile per hour
1 footcandle = 1 lumen incident per square foot = 10.764 lumens incident per square meter
1 grain per cubic foot = 2.29 grams per cubic meter
1 milligram per cubic meter = 0.000437 grain per cubic foot

To convert degrees Celsius to degrees Fahrenheit: °C (9/5) + 32 = °F
To convert degrees Fahrenheit to degrees Celsius: (5/9) (°F − 32) = °C
For solutes in water: 1 mg/liter ≎ 1 ppm (by weight)
Atmospheric contamination: 1 mg/liter ≎ 1 oz/1000 cu ft (approx)
For gases or vapors in air at 25°C and 760 mm Hg pressure:
To convert mg/liter to ppm (by volume): mg/liter (24,450/mol. wt.) = ppm
To convert ppm to mg/liter: ppm (mol. wt./24,450) = mg/liter

CONVERSION TABLE FOR GASES AND VAPORS[a]

(Milligrams per liter to parts per million, and vice versa; 25°C and 760 mm Hg barometric pressure)

Molecular Weight	1 mg/liter ppm	1 ppm mg/liter	Molecular Weight	1 mg/liter ppm	1 ppm mg/liter	Molecular Weight	1 mg/liter ppm	1 ppm mg/liter
1	24,450	0.0000409	39	627	0.001595	77	318	0.00315
2	12,230	0.0000818	40	611	0.001636	78	313	0.00319
3	8,150	0.0001227	41	596	0.001677	79	309	0.00323
4	6,113	0.0001636	42	582	0.001718	80	306	0.00327
5	4,890	0.0002045	43	569	0.001759	81	302	0.00331
6	4,075	0.0002454	44	556	0.001800	82	298	0.00335
7	3,493	0.0002863	45	543	0.001840	83	295	0.00339
8	3,056	0.000327	46	532	0.001881	84	291	0.00344
9	2,717	0.000368	47	520	0.001922	85	288	0.00348
10	2,445	0.000409	48	509	0.001963	86	284	0.00352
11	2,223	0.000450	49	499	0.002004	87	281	0.00356
12	2,038	0.000491	50	489	0.002045	88	278	0.00360
13	1,881	0.000532	51	479	0.002086	89	275	0.00364
14	1,746	0.000573	52	470	0.002127	90	272	0.00368
15	1,630	0.000614	53	461	0.002168	91	269	0.00372
16	1,528	0.000654	54	453	0.002209	92	266	0.00376
17	1,438	0.000695	55	445	0.002250	93	263	0.00380
18	1,358	0.000736	56	437	0.002290	94	260	0.00384
19	1,287	0.000777	57	429	0.002331	95	257	0.00389
20	1,223	0.000818	58	422	0.002372	96	255	0.00393
21	1,164	0.000859	59	414	0.002413	97	252	0.00397
22	1,111	0.000900	60	408	0.002554	98	249.5	0.00401
23	1,063	0.000941	61	401	0.002495	99	247.0	0.00405
24	1,019	0.000982	62	394	0.00254	100	244.5	0.00409
25	978	0.001022	63	388	0.00258	101	242.1	0.00413
26	940	0.001063	64	382	0.00262	102	239.7	0.00417
27	906	0.001104	65	376	0.00266	103	237.4	0.00421
28	873	0.001145	66	370	0.00270	104	235.1	0.00425
29	843	0.001186	67	365	0.00274	105	232.9	0.00429
30	815	0.001227	68	360	0.00278	106	230.7	0.00434
31	789	0.001268	69	354	0.00282	107	228.5	0.00438
32	764	0.001309	70	349	0.00286	108	226.4	0.00442
33	741	0.001350	71	344	0.00290	109	224.3	0.00446
34	719	0.001391	72	340	0.00294	110	222.3	0.00450
35	699	0.001432	73	335	0.00299	111	220.3	0.00454
36	679	0.001472	74	330	0.00303	112	218.3	0.00458
37	661	0.001513	75	326	0.00307	113	216.4	0.00462
38	643	0.001554	76	322	0.00311	114	214.5	0.00466

CONVERSION TABLE FOR GASES AND VAPORS (*Continued*)

(*Milligrams per liter to parts per million, and vice versa; 25°C and 760 mm Hg barometric pressure*)

Molecular Weight	1 mg/liter ppm	1 ppm mg/liter	Molecular Weight	1 mg/liter ppm	1 ppm mg/liter	Molecular Weight	1 mg/liter ppm	1 ppm mg/liter
115	212.6	0.00470	153	159.8	0.00626	191	128.0	0.00781
116	210.8	0.00474	154	158.8	0.00630	192	127.3	0.00785
117	209.0	0.00479	155	157.7	0.00634	193	126.7	0.00789
118	207.2	0.00483	156	156.7	0.00638	194	126.0	0.00793
119	205.5	0.00487	157	155.7	0.00642	195	125.4	0.00798
120	203.8	0.00491	158	154.7	0.00646	196	124.7	0.00802
121	202.1	0.00495	159	153.7	0.00650	197	124.1	0.00806
122	200.4	0.00499	160	152.8	0.00654	198	123.5	0.00810
123	198.8	0.00503	161	151.9	0.00658	199	122.9	0.00814
124	197.2	0.00507	162	150.9	0.00663	200	122.3	0.00818
125	195.6	0.00511	163	150.0	0.00667	201	121.6	0.00822
126	194.0	0.00515	164	149.1	0.00671	202	121.0	0.00826
127	192.5	0.00519	165	148.2	0.00675	203	120.4	0.00830
128	191.0	0.00524	166	147.3	0.00679	204	119.9	0.00834
129	189.5	0.00528	167	146.4	0.00683	205	119.3	0.00838
130	188.1	0.00532	168	145.5	0.00687	206	118.7	0.00843
131	186.6	0.00536	169	144.7	0.00691	207	118.1	0.00847
132	185.2	0.00540	170	143.8	0.00695	208	117.5	0.00851
133	183.8	0.00544	171	143.0	0.00699	209	117.0	0.00855
134	182.5	0.00548	172	142.2	0.00703	210	116.4	0.00859
135	181.1	0.00552	173	141.3	0.00708	211	115.9	0.00863
136	179.8	0.00556	174	140.5	0.00712	212	115.3	0.00867
137	178.5	0.00560	175	139.7	0.00716	213	114.8	0.00871
138	177.2	0.00564	176	138.9	0.00720	214	114.3	0.00875
139	175.9	0.00569	177	138.1	0.00724	215	113.7	0.00879
140	174.6	0.00573	178	137.4	0.00728	216	113.2	0.00883
141	173.4	0.00577	179	136.6	0.00732	217	112.7	0.00888
142	172.2	0.00581	180	135.8	0.00736	218	112.2	0.00892
143	171.0	0.00585	181	135.1	0.00740	219	111.6	0.00896
144	169.8	0.00589	182	134.3	0.00744	220	111.1	0.00900
145	168.6	0.00593	183	133.6	0.00748	221	110.6	0.00904
146	167.5	0.00597	184	132.9	0.00753	222	110.1	0.00908
147	166.3	0.00601	185	132.2	0.00757	223	109.6	0.00912
148	165.2	0.00605	186	131.5	0.00761	224	109.2	0.00916
149	164.1	0.00609	187	130.7	0.00765	225	108.7	0.00920
150	163.0	0.00613	188	130.1	0.00769	226	108.2	0.00924
151	161.9	0.00618	189	129.4	0.00773	227	107.7	0.00928
152	160.9	0.00622	190	128.7	0.00777	228	107.2	0.00933

CONVERSION TABLE FOR GASES AND VAPORS (*Continued*)
(*Milligrams per liter to parts per million, and vice versa; 25°C and 760 mm Hg barometric pressure*)

Molecular Weight	1 mg/liter ppm	1 ppm mg/liter	Molecular Weight	1 mg/liter ppm	1 ppm mg/liter	Molecular Weight	1 mg/liter ppm	1 ppm mg/liter
229	106.8	0.00937	253	96.6	0.01035	277	88.3	0.01133
230	106.3	0.00941	254	96.3	0.01039	278	87.9	0.01137
231	105.8	0.00945	255	95.9	0.01043	279	87.6	0.01141
232	105.4	0.00949	256	95.5	0.01047	280	87.3	0.01145
233	104.9	0.00953	257	95.1	0.01051	281	87.0	0.01149
234	104.5	0.00957	258	94.8	0.01055	282	86.7	0.01153
235	104.0	0.00961	259	94.4	0.01059	283	86.4	0.01157
236	103.6	0.00965	260	94.0	0.01063	284	86.1	0.01162
237	103.2	0.00969	261	93.7	0.01067	285	85.8	0.01166
238	102.7	0.00973	262	93.3	0.01072	286	85.5	0.01170
239	102.3	0.00978	263	93.0	0.01076	287	85.2	0.01174
240	101.9	0.00982	264	92.6	0.01080	288	84.9	0.01178
241	101.5	0.00986	265	92.3	0.01084	289	84.6	0.01182
242	101.0	0.00990	266	91.9	0.01088	290	84.3	0.01186
243	100.6	0.00994	267	91.6	0.01092	291	84.0	0.01190
244	100.2	0.00998	268	91.2	0.01096	292	83.7	0.01194
245	99.8	0.01002	269	90.9	0.01100	293	83.4	0.01198
246	99.4	0.01006	270	90.6	0.01104	294	83.2	0.01202
247	99.0	0.01010	271	90.2	0.01108	295	82.9	0.01207
248	98.6	0.01014	272	89.9	0.01112	296	82.6	0.01211
249	98.2	0.01018	273	89.6	0.01117	297	82.3	0.01215
250	97.8	0.01022	274	89.2	0.01121	298	82.0	0.01219
251	97.4	0.01027	275	88.9	0.01125	299	81.8	0.01223
252	97.0	0.01031	276	88.6	0.01129	300	81.5	0.01227

[a]A. C. Fieldner, S. H. Katz, and S. P. Kinney, "Gas Masks for Gases Met in Fighting Fires," *U.S. Bureau of Mines, Technical Paper No. 248,* 1921.

PATTY'S INDUSTRIAL HYGIENE

Fifth Edition

Volume 4

VII SPECIALTY AREAS AND ALLIED PROFESSIONS

CHAPTER FIFTY-TWO

Statistical Design and Data Analysis

Nelson A. Leidel, Sc.D and Kenneth A. Busch, MS

1 INTRODUCTION

The work of professionals in industrial hygiene and allied disciplines such as environmental health can substantially benefit from use of airborne exposure study design and data analysis methodologies that are based in mathematical statistics and probability theory. It has been said that "the science of statistics deals with making decisions based on observed data in the face of uncertainty" (1).

Section 2 of this chapter discusses some major areas of industrial hygiene practice where statistical methods perform an important role. Section 2.1 discusses the need for statistically sound study designs for both experimental and observational studies. Brief discussions appear in Section 2.2 concerning statistical methods used for occupational epidemiological studies and in Section 2.3 concerning estimating possible threshold levels and low-risk levels for occupational exposures. Finally, Section 2.4 introduces the area of application for statistics that is of primary interest and receives most attention in this chapter. This is the estimation of occupational exposures to airborne contaminants and calculation of error limits for such estimates. Nine possible objectives of occupational exposure estimation are discussed, which have their own special requirements for study design strategies.

Occupational exposure study designs and related data analysis methods have come to be broadly called sampling strategies (2). These sampling strategies are plans of action, based on statistical theory used to determine a logical, efficient framework for application

Patty's Industrial Hygiene, Fifth Edition, Volume 4. Edited by Robert L. Harris.
ISBN 0-471-29749-6

of general scientific methodology and professional judgment. These are broadly outlined in Section 2.5.

Section 3 presents some basic statistical theory relevant to occupational exposure data. Distributional models are given that identify the contributions of various sources of variation to the overall (net) random error in occupational exposure estimates. The National Institute for Occupational Safety and Health (NIOSH) nomenclature for exposure data is first given in Section 3.1. Then in Section 3.2 a model is given for the contributions of the various components of variation to the net random error in occupational exposure measurements (due to the measurement procedure used). Section 3.3 extends the model for total error to include random and systematic variations in true exposure levels (over times, locations, or workers doing similar work).

Section 3 also includes information on the mathematical characteristics of basic distributional models. This is the starting point for deriving sampling distributions of industrial hygiene exposure data taken by various sampling strategies. General properties of the normal distribution model are given in Section 3.4, and of the lognormal distribution model (both two-parameter and three-parameter) in Section 3.5. Section 3.6 discusses the adequacy of normal and lognormal distribution models for certain general types of continuous variable data (specifically occupational exposure measurements). These data models are then used to apply special-interest applications of statistical theory to occupational exposure study designs (Section 5) and specialized data analyses (Section 6).

Section 4 discusses basic principles of statistically sound study design and data analysis that apply to all industrial hygiene surveys, evaluations, or studies. Section 4.1 presents general study design principles for experimental or observational studies intended to estimate means, variances, quantiles, tolerance limits, or proportions below a designated value for a single population and for those studies seeking to compare any of these parameters between two study groups [e.g., between an exposed ("treatment") group and an unexposed ("control") group]. Section 4.2 then discusses general principles of data analysis. Particular emphasis is given to the necessity for appropriate selection, estimation, and verification of a distributional model (or models) for the study data.

Section 5 gives particular study designs to be used for collecting data to estimate individual occupational exposures and exposure distributions. The section first discusses the cornerstone concept of worker target populations. Section 5.1 discusses in detail another important concept, the determinant variables affecting occupational exposure levels experienced by a target population. Section 5.2 discusses exposure measurement strategies selected to measure a short- or long-period, time-weighted average (TWA) exposure of an individual worker on a given day. Both practical and statistical considerations are discussed for long-term and short-term exposure estimates. Lastly, exposure monitoring strategies are presented in Section 5.3 for measuring multiple exposures (e.g., multiple workers on a single day, a single worker on multiple days, or multiple workers on multiple days). Eight possible elements of monitoring programs are discussed in detail, with examples of both exposure screening and exposure distribution monitoring programs.

Section 6 presents specialized applied methods for formal statistical analysis of occupational exposure data generated by the study designs discussed in Section 4 and Section 5. The first portion of Section 6 (Section 6.1, Section 6.2, Section 6.3, Section 6.4, Section 6.5) covers methods for computing confidence intervals for true exposures of individual workers that have been estimated using exposure measurement strategies. In addition, statistical significance tests of hypotheses are presented for classifying individual exposure estimates relative to an exposure control limit. The middle portion of Section 6 (Section 6.6, Section 6.7, Section 6.8, Section 6.9, Section 6.10) covers inferential methods for computing tolerance limits, tolerance intervals, and point estimates of exposure distribution fractiles. The last portion of Section 6 (Section 6.11, Section 6.12, Section 6.13) covers graphical techniques for presenting lognormally distributed data and their associated tolerance limits.

2 GENERAL AREAS OF APPLICATION FOR STATISTICS IN INDUSTRIAL HYGIENE

In the practice of industrial hygiene and allied professions, statistical theory is applicable to both:

1. *Experimental* studies (i.e., studies with planned intervention on determinant factors suspected of altering the phenomenon under study).
2. *Observational* surveys and studies (i.e., studies with no deliberate human intervention).

With statistically valid study design and data analysis, both experimental and observational studies can validly and reliably identify causes of occupational health problems, screen workplaces for excessive exposure conditions, estimate worker exposure levels, evaluate the effectiveness of engineering controls, and determine the protection levels afforded by personal protective equipment. Of course, depending on factors such as relative cost, conditions amenable to experimentation, availability of relevant laboratory models, and interpretability of available field data, one or the other of these two general types of studies is usually a clear choice for any given research objective. However, experimentation generally has several fundamental advantages over observational surveys and studies. These are discussed in Section 2.1 in the context of health effects studies related to workplace contaminant exposures. An advantage of observational studies is that subjects are humans; epidemiologic studies are discussed in Section 2.2.

2.1 Need for Statistically Valid Study Design

Typical industrial hygiene *experimental* studies include estimating distributions of worker exposures for selected conditions, determining the efficacy of exposure control systems, and estimating the accuracy and variability of measurement methods. Experiments may be done to study the effects of determinant variables on worker exposure levels, on the efficacy of control measures, or on exposure measurement procedures. In these experiments, the "treatments" are usually a set of controlled conditions under which an exposure level (or other physical or biological response) is measured as the response variable. Thus, "treatments" are usually exposure groups in this type of experiment. If a factor is "controlled" through deliberate selection of particular existing conditions, one has a *quasi-*

experimental study in a field setting. In contrast to experimental studies, one conducts an *observational* study when conditions must be taken as found in a field setting (i.e., without opportunity to experiment with preselected, controlled levels for the determinant variables).

An inherent problem for industrial hygienists is that "safe" exposure levels for many substances are unknown and must be determined using the best available evidence (3). To do this, pertinent exposure level and health effects data are collected and statistically analyzed. One approach is to expose suitable animal species and observe biological effects that may occur. A second important approach to estimation of acceptable exposure levels, the epidemiological study on exposed workers, is discussed in a limited manner in this section, and references to more comprehensive presentations are supplied.

For either type of industrial hygiene study, experimental or observational, the resulting data will have stochastic components (i.e., random or chance variations) that cannot be ignored when evaluating the data. Different types of biological data may have fundamentally different statistical properties. Population health effects may be measurable in:

1. The average amount of change in a quantitative biological parameter measured on a continuous scale. Examples of continuous variable measurements are lung volume, heart rate, and body weight.
2. The presence or absence of a qualitative biological abnormality. An example would be a pathological condition such as a tumor.
3. Values that exist only at a limited number of discontinuous (i.e., discrete) points on an ordered scale. For example, severity of lung histopathology has been graded on a 6-point rating scale.

The first and second types of data are the most frequently encountered, and they will be discussed in Section 4 in relation to principles of study design and data analysis. The results from any industrial hygiene study should be evaluated both in terms of internal validity and external validity. With regard to *internal validity*, suppose a researcher were to conclude that the performance of control method B is "significantly better" than the performance of method A. If the conclusion was erroneous because of random errors that had affected the exposure measurements used to evaluate performances of both methods, then the study conclusion would not have internal validity. That is, chance would have led to an incorrect conclusion that there was a true difference in performance. As will be discussed later, variability exists in exposure results for numerous reasons. This variability can then cast doubt on the internal validity of any conclusions drawn from the exposure results.

It is also essential to examine the *external validity* of exposure results from any given industrial hygiene study. That is, how valid are the study results *outside of the study sample*? As before, suppose a researcher was to conclude that the performance of control method B is "significantly better" than the performance of method A. If the conclusion were based on worker activities or control method conditions that are irrelevant to tasks and circumstances in the real world, then the conclusion and research findings would have no external validity. External validity also includes topics such as possible *nonrandom sampling errors* (i.e., "mistakes") and *biases* in exposure results. Biases are systematic errors that deprive a statistical result of representativeness by distorting its expected value. Unlike internal validity, for which there are objective statistical computations to justify conclusions, evaluating external validity must sometimes be a subjective matter that relies largely on professional judgment.

For those researchers who wish to generalize their exposure results to larger populations or other settings (workplaces), a two-stage process is involved during which external validity problems can arise (4, 5). First, researchers must define a *target population* of persons, settings, and times (see Section 5). Second, researchers must draw a sample of exposed workers to represent the target population. However, these samples usually cannot be drawn in a formal randomized manner. Instead, sampled workers are usually selected merely in a convenient manner that gives an intuitive impression of representativeness. However, the sample of workers selected, as well as the settings, and conditions of any given workplace study, particularly observational studies, may severely hamper the *generalizability* of subsequent exposure estimates.

Cook and Campbell (6) have suggested that it is useful to distinguish between (*1*) target populations, (*2*) formally representative samples that correspond to known populations, (*3*) samples actually achieved in field research, and (*4*) achieved populations. They have noted:

> To criticize the study because the achieved sample of settings was not formally representative of the target population may appear unduly harsh in light of the fact that financial and logistical resources were limited, and so sampling was conducted for convenience rather than formal representativeness. . . . *it is worth noting that accidental samples of convenience do not make it easy to infer the target population*, nor is it clear what population is actually achieved.

The objectives of an occupational health study should be the primary determinant of its statistical design, *not* expedient considerations of "available" specialized experimental facilities or presence of "experts" on a given professional staff. Availability of specialized research staff and personal interests of these researchers can be powerful incentives for inappropriately designing a research study.

Finney (7) discusses a professional ethics code for professional statisticians that includes maintenance of an attitude of scientific objectivity that is not biased by the interests of an employer or client, preservation of confidentiality of data furnished for analysis and of the results of that analysis, an obligation to screen raw data for possible mistakes or outliers, an obligation to perform data analyses using best available methods (e.g., most powerful significance tests) unless simpler methods are believed to produce equivalent results, and an obligation not to use a "canned" data analysis protocol (e.g., personal computer or mainframe software) unless its underlying assumptions and correct interpretation of results are understood. Finney also discusses the responsibility of a client to furnish the consulting statistician with complete and accurate raw data, along with adequate background information on the sampling strategy and environmental circumstances surrounding collection of the data, so that appropriate data analyses and valid interpretation of results can be performed.

It is the responsibility of the researcher, not the statistician, to assure those determinant factors such as the species used as animal models, exposure techniques, biological parameters that can be accurately measured, and exposure measurement procedures are given consideration and are appropriately incorporated into the study protocol. The statistician then addresses such tactical problems as sample size, allocation of experimental units (subjects) to treatments (exposure groups), schedules for sampling, and data analysis techniques. Proper consideration and selection of study design and data analysis parameters and methodologies will assure the researcher that the study will have adequate (but not excessive) statistical power and technical capability to detect the anticipated effects that are the focus of the research.

Sometimes technically appropriate experimental facilities or situations are available, but the feasible sample sizes are only marginal insofar as the production of definitive results is concerned. Nevertheless, an expedient decision may be made to proceed with the study under the misguided rationale that "some information is better than none." This incorrect and wasteful practice often results in studies that are predestined to be inconclusive and possibly misleading to subsequent investigators. The potential waste of time and resources can be minimized by first securing a statistician's evaluation of (or better yet, assistance with) the study design and data analysis plan. The statistician can usually warn the researcher if the planned study has low statistical power or for other reasons (e.g., bias due to confounded factors) lacks the statistical capability to detect the desired size of effect of a determinant factor, if indeed the effect exists.

If some species of experimental animal could serve as a perfect biological model for health effects on humans caused by workplace exposure, then there would be no question that controlled, animal exposure studies would yield better information than observational studies of humans. The deficiencies of human observational studies are analogous to those that would occur in an animal study for which subjects were constrained within a measured but uncontrolled laboratory environment for 8 hr each day, but allowed to roam freely through the streets and alleys outside their laboratory, eating and breathing whatever they encountered, during the other 16 hr.

Unfortunately, the perfect biological model for human health effects does not exist, but nevertheless the advantages of the carefully controlled, animal exposure experiment are several. First, in an experimental animal study one can control the primary study factor(s) (e.g., exposure levels, exposure duration, exposure schedule) and determine its direct effect, at differing levels of interest, on one or more response variables. In observational studies of humans, the primary study factor(s) can be observed, but not controlled, and toxicological and mathematical extrapolation of observed results to other exposure levels of interest must be performed.

Second, with animal studies there is more freedom to design experimental studies for complete elimination of bias (we shall put aside for now the interspecies extrapolation bias that may exist) and for high sensitivity to small effects. Given large enough sample sizes, an experimental study can be designed to have suitably high statistical power (probability) of detecting a small effect (if detecting this small effect is worth the required time and resources). Even a much larger observational study may not be able to detect a statistically significant, moderate-sized exposure effect, since there are unknown, or uncontrolled and unmeasured (even unmeasurable), secondary factors that operate within the workplaces and home environments to modify or distort the workers' biological responses to workplace exposures. Here, making a simple comparison of response variable values between exposed and unexposed groups will yield an imprecise estimate of the exposure effect (i.e., a wide confidence interval for the true magnitude of the effect) and probably will also yield a biased (substantially inaccurate) estimate of the effect. Bias can occur because of unequal levels of secondary factors in the control and exposed groups, so that one observes the joint (net) effect due to both the primary factor and to unintended differences in levels of the secondary factors. This joint effect is said to consist of *confounded effects* (i.e., joint effect of two or more factors confused with each other).

In observational studies for which levels of some secondary factors are known, methods of formal statistical analysis can be useful to adjust for the part of the bias due to confounded effects of these factors. For secondary factors that have effects that are additive to each other and to the effect of the primary factor, several methods for bias adjustment are available. For continuous response variables, these include *covariance analysis* (ANCOVA), and *analysis of variance* (ANOVA) for randomized blocks. For discrete response variables, there are procedures such as the *chi-square test for matched data* (Cochran's Q test) and *two-way analysis of variance of ranks.*

In observational studies no statistical adjustment can usually be made for any bias due to unknown secondary factors or due to covariates for which the levels were not measured. However, in experimental studies this dilemma of unknown or uncorrectable bias need never occur if the experiments are statistically well designed. Bias can be prevented in experiments by using statistical design features, such as selection of subjects at random from the same pool for assignment to control and test groups, or restricted random selection with matching on a secondary factor or within blocks (e.g., age groups, genders, weight ranges). Randomizing within blocks not only prevents bias but also improves precision. Similar blocking can be employed in observational studies during the data analysis, but only regarding the known secondary factors. It is *random assignment* of subjects to the exposure groups that is the best protection against bias due to effects of *unknown* secondary factors. Random assignment can be used in experimental studies, but not in observational studies for which there will always be some bias (one hopes it will be small) due to unequal representation of secondary factors (i.e., those not taken account of in the data analysis) in the exposed and unexposed groups.

In spite of the inherent limitations and deficiencies of observational studies on workers, they are an invaluable source of information on health effects due to work-related diseases. To some research scientists, the fact that the subjects of observational studies are workplace-exposed humans is an overriding consideration that transcends the statistical advantages and experimental flexibility afforded by animal exposure studies. Effects observed in workplace observational studies can be strongly suggestive of effects due to the primary factor(s) studied, but the results cannot be considered definitive by themselves. Replication of the result found for a given toxic agent, preferably in other studies of different worker target populations, for several occupational settings and several cultural, ethnic, or socioeconomic groups, affords much higher credibility and substantiation to the initial findings. This is true because the universal presence of a similar observed effect in different occupational settings (that involve different secondary factors) would tend to implicate the primary factor as the cause, not the variety of secondary factors.

On the other hand, a single controlled animal experiment that has been toxicologically and statistically well-designed (with balance and randomization used in all phases of the work) and competently conducted, theoretically can yield a definitive result (within the limits of random error, which can be governed by its experimental design). Even so, an experimental scientist feels better after the initial findings have been replicated in an independently conducted study. Also, the criticism can be made, "What good is a 'definitive' result for animals if it does not apply to humans?" An extrapolation from animals to humans must only be defended on the basis of the similarities of their appropriate biological mechanisms and structures; supported presumptively by empirical species similarities previously observed. A good statistical experimental design does not assure validity of the interspecies extrapolation; however, it does assure validity of the findings for the animal species used.

In the U.K., an Occupational Exposure Standard (OES) is sometimes determined by applying an "Uncertainty Factor" to a "No Observed Adverse Effect Level" (NOAEL), where the *NOAEL* has been determined in experimental animal exposure studies (8). As stated above, a major problem in using this approach has been difficulty in choosing the *uncertainty factor*, since the factor may have to be based largely on biomedical professional judgment in the context of a human risk assessment that requires interspecies extrapolation of an experimentally determined *animal* dose–response curve.

Finally, observational studies in workplaces can be an economical source of estimates of chronic effects due to long-term, low-level workplace exposures (although the estimates are often somewhat imprecise and ill-defined). Observational studies use existing data (although these data sometimes are not readily accessible), whereas a chronic exposure animal study requires at least several months and sometimes several years to conduct.

2.2 Epidemiological Studies

In the occupational epidemiological study, an attempt is made to associate an observed incidence or severity of adverse health effects in groups of human workers with factors such as industry, job type, or some measure of exposure to potentially toxic materials. The last type of study may attempt to estimate a dose–response relationship between level of exposure to the material and prevalence or incidence of health effect being studied. However, the type of dose–response relationships utilized for occupational settings generally are more correctly referred to as exposure-response functions since the effective dose at the critical site(s) within each individual worker's body is never really known (9). Ulfvarson has noted (10):

> The concept of exposure of an employee to a substance in the work environment may denote at least two things. It may indicate the dose of the substance absorbed in the body. It may also merely indicate the presence of the employee in an environment in which there is a more or less-well determined concentration of the substance, from which an uptake of the substance is deduced.

For practical reasons, the latter concept of exposure is currently used by industrial hygienists.

Additionally, the *exposure variable* utilized may be either an average exposure level over some long period (such as a worker's total working lifetime) or a time-integrated (cumulative) exposure [e.g., ppm-years, (fibers/cm^3) years, (mg/m^3)-years]. Ford et al. (11) have defined a cumulative exposure (CE) estimate and a lifetime, weighted-average, exposure (LWAE) estimate. The CE estimate is computed by quantitative integration of the terms: (*1*) exposure duration multiplied by (*2*) matching exposure level (concentration) measured in a subject's various working zones. The CE estimate assumes that a given working zone has an exposure level that is uniform over time, and the exposure level estimate for that working zone is taken to be the geometric mean (GM) of (a supposed random sample of) exposure levels for that zone. The LWAE estimate is then defined to be LWAE = CE/(total duration of exposure). It is interesting that the Ford et al. study (11) included neurobehavioral test data that had a better interworker correlation with average exposure level (LWAE) than with cumulative exposure (CE) or with total duration of exposure.

Estimation of these or similar exposure–response relationships is desirable since they can be used in the process of selecting exposure-control limits for workers in occupational environments (12, 13). Recognizing the statistical nature of the problem, Roach (14) was the first to suggest that an occupational hazard such as silica dust exposure leading to silicosis should be analyzed as an exposure–response function similar to the dose–response function utilized in toxicological research.

A major obstacle in estimation of exposure–response relationships is accurate and sufficiently precise estimation of worker exposure levels and their relation to doses delivered to the body. Gamble and Spirtas (15) have presented a systematic approach for using occupational titles to classify the "effect" and "exposure" of workers in retrospective, exposure–response estimation. Esmen (16) has proposed a process for reconstruction of the integrated exposure of one or more agents over a reasonably long period of time. He presents the basis for a model and a simplified procedure for what he calls a *retrospective industrial hygiene survey*. Roach (17, 18) has proposed sampling strategies that consider the biological half-times of the measured substances.

Combined with the substantial advantage of having human workers as subjects, occupational observational studies generally carry the frustrating disadvantages of lack of control of the exposure conditions, uncontrolled effects of the exposures outside the occupational environment, and interactions between exposure effects and demographic and socioeconomic factors. These design problems for observational studies were discussed at length in Section 2.1. Statistically it should be possible to estimate a set of exposure-response relationships for any work-related disease given that a reasonably homogeneous and large enough group of workers is available. As a practical matter, it is difficult to obtain homogeneous and unbiased worker populations of adequate size for which one has adequate, accurate exposure information (19). As a result, there have been a minimal number of exposure–response functions estimated in the last 50 years for occupational populations. These include functions such as those estimated by Hatch (20) for silicosis in miners and other dusty trades, Lundin et al. (21) for respiratory cancer in uranium miners, Berry et al. (22) for asbestosis in asbestos textile workers, and Dement et al. (23) for lung cancer and other nonmalignant respiratory diseases in chrysotile asbestos textile workers. Berry et al. (22) contains an appendix with a particularly good discussion of measures of exposure and occupational dose–response relationships.

Hornung (24) discusses use of exposure matrices to determine exposure levels for the various combinations of job factors and personal factors that are relevant in an epidemiological study. In comparison to the larger uncertainties due to job and personal factors, and measurement method biases, sampling and analytical error contribute relatively little to the uncertainty of exposure level estimates for missing job matrix cells (i.e., cells for which there are health effects data but for which no direct exposure-measurements data exist). Approaches used to "fill in" the missing cells include:

1. Interpolation between surrounding cells.
2. Use of mathematical prediction models based on principles of physics, data on production levels, and environmental factors.
3. Use of statistical models fitted to measurement data in surrounding cells.

In experimental clinical studies of volunteer subjects, better control over, and accurate measurement of, short-term human exposures can be obtained. Of course, an overriding necessity is safety of the human subjects, and this usually precludes use of exposure levels high enough, or exposures long enough, to produce the chronic toxicity that can occur in the workplace.

Ulfvarson (10) has reviewed and extensively discussed the limitations to the use of typical worker exposure data in epidemiologic studies. He concluded that there are a considerable number of possible biases in the estimation of uptake of a substance into the bodies of a group of workers, when body uptake is uncritically derived from typical sources of airborne exposure levels. He defined a positive bias as where the uptake is overestimated in comparison to the true uptake. He also rated the validity of his conclusions regarding the probable sign of the bias as follows: {3} = self evident, {2} = a conclusion with some reservation, and {1} = an educated guess. Some of the possible biases listed by Ulfvarson (with his own ratings {in braces} of the validity of their sign) follow:

1. Positive bias {2} due to the use of a measurement strategy intended to collect a sufficient mass of contaminant for the purpose of exceeding the minimum detectable level for the analytical method.
2. Positive bias {3} due to the use of a "worst-case," biased measurement strategy.
3. Positive or negative bias due to the use of daily TWA exposure results that do not represent the unsampled workers to whom they are applied.
4. Positive bias {2} due to the use of data from area (static) sampling devices that were deliberately located to yield high results such as would be obtained with a source sampling strategy.
5. Negative bias {2} due to failure to obtain repeat measurements when the first result demonstrates compliance, but may have been unusually low.
6. Positive bias {2} due to use of exposure data from establishments that do not represent the unsampled establishments to which they are applied.

7. Positive and negative biases due to seasonal variations in exposure levels.
8. Positive bias {3} due to rotation of workers to unexposed work areas, when the exposure measurements are only taken during the work operation.
9. Positive bias {3} due to the use of effective respirators by workers.
10. Negative bias {2} due to existence of unfavorable exposure patterns creating a "resonance" between intermittent airborne levels and levels in the body.
11. Negative bias {2} due to increased lung ventilation resulting from hard physical labor.

Ulfvarson (10) recommends that an epidemiologist use the list of possible biases as a checklist and try to find out the premises of the available data and thus the most probable sign of bias.

In connection with human risk assessment, Kimbrough (25) makes similar severe criticisms of the exposure assessment methods used in both animal exposure studies and human epidemiological studies:

1. Indirect exposure assessments are said to have doubtful accuracy; if possible, direct measurements of body burdens are preferable. In assessments of inhalation or ingestion exposures, physical and chemical properties that affect "bioavailability" should not be ignored (e.g., particle size and presence of trace elements to which the toxin may be bound).

2. Kimbrough also criticizes the validity and accuracy of some types of extrapolations of toxicologic data from animal exposure studies to humans. An expedient basis for extrapolation has sometimes been to make an assumption that humans are at least as sensitive as the most sensitive animal species tested; this is not necessarily true.

3. Multiple causes exist for many chronic diseases, which can lead to confounding (of the primary deterministic variable or environmental factor that is under study) with correlated causative agents that are not part of the study.

4. Biological half-lives of toxins are sometimes shorter at high dosages; therefore, making the usual assumption of a constant excretion rate can lead to underestimation of exposures that (a) were sustained at an earlier time, (b) resulted in presently observed body burdens, and (c) caused chronic biological effects.

5. Animal dose–response data, calculated from dosages on the basis of body weight or body surface area, are not necessarily predictive of biological effects in all humans receiving similar dosages. In an epidemiological study, the average exposure level and correlated response rate assigned to a cohort group may differ from rates for selectively exposed or specially sensitive cohort subgroups (e.g., smokers and nonsmokers working under the same conditions).

General statistical methodology for epidemiological studies is given in texts such as those by MacMahon and Pugh (26), Mauser and Bahn (27), Friedman (28), Lilienfeld and Lilienfeld (29), Schlesselman (30), and Rothman and Boice (31). Only since about 1980 have specialized epidemiologic texts appeared that deal specifically with studies of occupational groups. One is Monson (32), a second is Chiazze et al. (33), and a third is Karvonen and

Mikheev (34). In addition, state-of-the-art general methodology for epidemiological studies in occupational populations is reported in journals such as *American Journal of Epidemiology, American Journal of Industrial Medicine, Archives of Environmental Health, British Journal of Industrial Medicine, International Archives of Occupational and Environmental Health*, and *Journal of Occupational Health*. Both theoretical and practical problems of performing occupational health field studies are being solved, thanks to the extensive field experience that investigators have accrued in occupational observational studies. The March 1976 issue of *Journal of Occupational Medicine* deals entirely with occupational epidemiology, including an article by Enterline (35) warning of the pitfalls in epidemiological research. Potential problems in occupational studies include inaccurate or uninterpretable cause-of-death statements on death certificates, improper control groups, lack of reliable and accurate quantitative exposure estimates, overlapping exposure and follow-up periods, and competing (but sometimes unknown) causes of death.

The research needs in epidemiology have been summarized by an authoritative Second Task Force appointed by Department of Health, Education, and Welfare. In Chapter 15 of their 1977 report (36), a subtask force chaired by Brian MacMahon made an assessment of the state-of-the-art of epidemiology and other statistical methods used in environmental health studies. Among the specific recommendations were the following:

1. For clinical environmental research, guidelines are needed for the protection of human subjects and special attention should be given to statistical design and analysis of such studies.
2. For epidemiological studies, better exposure data are needed. Health professionals should help make decisions about what environmental data are collected by governmental agencies. Also, routine surveillance is needed of disease incidence in occupational groups along with surveillance of exposure levels.
3. Animal studies should be used to identify biochemical or physiological early-indicator effects of serious chronic disease in worker groups believed to be exposed to potentially hazardous agents.
4. In all areas of environmental health research, more powerful statistical techniques are needed in areas of multivariate analysis, time-series and sequential analysis, and nonparametric methods. Better dose–response models for mixtures of toxic agents are needed, as are better models for animal-to-human extrapolation (particularly for carcinogenesis).

Another useful reference to more specific methodological techniques is the "Steelworker Series" of 10 epidemiological reports by J. W. Lloyd and C. K. Redmond, published between 1969 and 1978 in the *Journal of Occupational Medicine* (37, 38). These reports are considered by some to constitute the evolution to the present state-of-the-art of modern methodology for epidemiological studies of an occupational population. A paper by Kupper et al. (39) discusses methods for selecting suitable samples of industrial worker groups and valid control groups.

2.3 Threshold and Low-Risk Exposure Levels

Ideally, the industrial hygienist would like to be able to assess the net risk of an adverse health effect as a function of a worker's past, present, and future exposures and in relation to age, race, gender, and other individual susceptibility factors. If such comprehensive toxicological knowledge were available in relation to past exposures, appropriate limits on future exposures of an individual could be recommended to control the chances of the worker experiencing adverse chronic health effects.

However, achieving such a high level of toxicological understanding is not realistic—one would have to know all the "exposure–time–response" relationships for the toxic material of interest (i.e., the relationships between level of exposure, length of exposure, pattern of intermittent exposures, and the incidence and severity of health effects that occur to some or all of an exposed population). In most cases the industrial hygienist has only an approximate estimate of some "low-risk" exposure level at which the incidence of an adverse health effect *appears* to be low or absent in one or more working populations.

Better yet, if possible, one would like to know a "safe" level of exposure below which all adverse health effects other than the most minor and transient ones are absent in *almost all* workers. This type of level is often called an exposure threshold level, or simply *threshold level*. The existence of such thresholds is a controversial question, which we will not attempt to answer here. This concept has been used as the basis of *threshold limit values* (TLV) recommended by the American Conference of Governmental Industrial Hygienists in their annually updated publication (40). Excellent discussions on the question of the existence of thresholds have been given by Elkins (41), in the *American Industrial Hygiene Association Quarterly* (42), by Stokinger (43–45), Smyth (46, 47), Hatch (48), Bingham (49), Hermann (50), and Thomas (51).

Statistical models have been developed that can aid us in extrapolating to "acceptably low-risk" exposure levels from higher-risk, dose–response data. Since a low- or minimal-risk exposure level always exists (regardless of whether a true threshold exists) a low-risk level of exposure can usually be selected that is "sufficiently safe." Note that the determination of "sufficiently safe" involves cost-benefit considerations of a socioeconomic and political nature and is not solely a scientific determination. Hartley and Sielken (52) have reviewed the technical aspects of statistically estimating "safe doses" in carcinogenesis experiments. They leave the definition of an "acceptable" increase in the risk of carcinogenesis to the regulatory agencies. Acceptable risk levels that have been suggested are 10^{-8} (1 in 100 million) and 10^{-6} (1 in 1 million). Either of these low-risk exposure levels is effectively impossible to determine by direct experimentation with animals because of the enormous sample sizes that would be required to distinguish such minuscule tumor incidence differences. The problem becomes particularly acute when there is a "normal" or "background" (control) tumor incidence.

Therefore, various procedures have been proposed for extrapolation to low-risk exposure levels using a mathematical model for an exposure-response curve that has been fitted to higher-level, exposure–response data. Among these mathematical modeling procedures are the Mantel-Bryan procedure (53, 54) that is based on extrapolation using a conservative slope on logarithmic-normal probability graph paper. Other models are reviewed by Armitage (55). Some of the models are more flexible than others. For example, Hartley and Sielken (56, 57) use a polynomial instead of a straight line to extrapolate. Other models are derived from assumed biological mechanisms. Crump et al. (58) assume a multistage biological mechanism for development of cancer. The Crump model is fitted to simple incidence versus dose data, whereas the Hartley-Sielken model can be fitted to time-to-tumor data.

In the context of setting an exposure limit for coal mine respirable dust, it has not been possible to clearly distinguish (on the basis of goodness of fit) between threshold-limit-value dose–response models and a nonthreshold (logistic) dose–response model (59). In order to be used as aids in the process of selecting an exposure limit, models must be extrapolated to low dose levels beyond the range of data to which they were fitted. Biological arguments usually tend to favor the nonthreshold type of model, but care must be taken that choice of a particular model is based on evidence of validity of the model, rather than on a preference for the recommended exposure limit that it yields compared to other limits yielded by alternative models. Krewski and Brown (60) have provided a comprehensive list of references for carcinogenic risk assessment, which are grouped by carcinogen bioassay, carcinogenicity screening, quantitative risk assessment, and regulatory considerations.

It is difficult to determine exposure–time–response relationships, from limited and unstructured epidemiological data such as are usually available for humans, that reflect a disease (adverse health effect) state that can be attributed to working environments. Nonetheless, exposure control limits can be estimated from carefully conducted, extensive epidemiological studies of workers. For example, the lower limit of a range of average personal exposure levels for a large group of workers in a given plant could be taken as a conservative control limit (i.e., more worker protective), if the "health profile" of these workers is found to be similar to that of a cohort group of unexposed but otherwise similar individuals. The condition that the unexposed cohort group be "otherwise similar" is often difficult to attain, relevance of this necessary condition to validity of the estimated health effects has been previously discussed in Section 2.1. But in any case, this approach is considerably less desirable than use of the sought after, but generally unattainable, exposure–response curves. With the latter approach the exposure experience of the entire exposed work population related to the occurrence of some adverse health effect is used to estimate an appropriate low-risk, exposure control limit. With the former approach that usually is based on only a small fraction of the total exposed group, an exposure control limit is usually set on the basis of the estimated lower range of exposures (see above) in the highest exposure group that did not show the health effect. However, such failure to find the health effect in a small sample of workers may have limited statistical significance. That is, it is possible to fail to observe a health effect in a small random sample of workers even though they had been selected from a larger group containing an unacceptable proportion who would show the adverse health effect. Thus this procedure may yield an estimated control limit that is set too high for the workforce. Also, the susceptibility characteristics of a small sample of workers, who work and live in close proximity, may be different from the rest of the workforce and different from other similarly exposed groups.

2.4 Objectives of Occupational Exposure Estimation

Statistical methodologies can make substantial contributions toward achieving the three goals vital to the objective of effective worker protection: recognition, evaluation, and control of chemical and physical stresses to workers. Industrial hygienists are called on to examine the work environment and recognize workplace stresses and factors that have the potential for adversely affecting worker health. Then they must evaluate the magnitude of the stresses and interpret the results to be able to give an expert opinion regarding the general healthfulness of the workplace, either for short periods or for a lifetime of worker exposure. Finally, there must be a determination of the need for, or effectiveness of, control measures that minimize adverse health effects of workplace exposures.

The practical aspects of achieving these three goals require answers to two initial questions:

1. Are exposure measurements necessary?
2. If so, what type of measurements are needed in relation to our reasons for taking the exposure measurements?

Attempting to answer these two broad questions will require answering additional questions such as the following that will influence the choice of study design strategy:

1. Are rough estimates needed only or are precise estimates of exposure levels needed?
2. Are only worst-case estimates of the higher exposure levels needed or are estimates of exposure distributions needed?
3. What agents will be measured?
4. Which workers will be sampled?
5. How many samples will be taken?
6. When will the samples be taken?
7. At what locations will the samples be taken?
8. What actions can be taken based on results of the data analysis?

Regarding our data analysis strategies one must ask: How will the data be analyzed and how were decisions be reached regarding research hypotheses in order to decide if the exposure levels are acceptable?

Answers to the preceding questions must be arrived at by first clearly defining the objective(s) of estimating worker exposures. Clear definition of the objective(s) will facilitate the formulation of appropriate study designs and data analyses. Typically the objective(s) of estimating worker exposures will be one or more of the following nine.

2.4.1 Hazard Recognition

Hazard recognition is the identification of hazardous agents present in the workplace that are used in such a fashion that exposures create a possible health hazard. It involves the (*1*) identification of the hazardous agents present in the workplace, (*2*) identification of the job activities and work operations that involve their use, and (*3*) determination of whether the hazardous agents are intermittently or routinely released into the workplace air. The existing control measures that might reduce the hazardous exposures must be identified. It must also be determined if hazardous conditions could occur during an irregular episode or during an accident. These include hazardous agents, explosive concentrations, or insufficient oxygen.

2.4.2 Hazard Evaluation

Hazard evaluation is similar to, but wider in scope, than hazard recognition. Where hazardous agents arise from industrial operations, it must be determined if it is possible for a hazardous exposure to occur. This can involve an evaluation of the severity of health hazard(s) from hazardous agents, explosive concentrations, or insufficient oxygen due to industrial operations. It may involve determining if worker complaints or health problems could be due to hazardous exposure levels in the workplace.

Hazard evaluation can include estimating the distribution(s) of worker exposure levels. Specifically, the shape, location, and dispersion of the exposure distribution could be estimated. Percentiles of exposure distributions (along with associated tolerance limits) could be estimated as a function of average-exposure level. The important determinant factors affecting worker exposure levels might be identified. These are the qualitative or quantitative variables or factors that influence or affect worker exposure levels. Periods of abnormally high exposure during the work shift should be looked for.

2.4.3 Control Method Evaluation

The adequacy of existing control measures to substantially reduce the probability of unacceptable worker exposures would be determined. The adequacy of newly installed controls and their control effectiveness could be compared to the precontrol exposure situation. The important design and operating parameters affecting the efficacy of control methods could be identified and evaluated.

2.4.4 Exposure Screening Program

This is a limited exposure monitoring program that is designed to identify target populations of workers (see Section 5) with other-than-acceptable, exposure level distributions (i.e., those worker groups with substantial proportions of exposures exceeding an exposure control limit or regulatory standard) Such groups will then receive additional exposure monitoring. The program uses an Action Level as a screening cut-off to select appropriate target populations for inclusion in a limited exposure surveillance program that provides for either: (*1*) minimal periodic exposure measurements to be made on only a few workers or (2) a more extensive exposure distribution monitoring program. An exposure screening program uses the least resources but provides for reasonable protection of the worker target population. This type of a limited program, as given in Leidel et al. (2), was recommended by NIOSH for use in regulatory monitoring programs (see Section 2.4.6).

2.4.5 Exposure Distribution Monitoring Program

This is a more extensive exposure monitoring program designed to quantify target population exposure distributions over an initial temporal base period (e.g., one day, several weeks, months, or years). Initial exposure distribution estimates would be periodically updated with routine exposure monitoring of the target population(s). Accomplishing this objective requires the appropriate definition of target populations and the identification of those determinant variables that affect the exposure distributions (see Section 5).

2.4.6 Regulatory Monitoring Program

Federal or state regulations may require the establishment of exposure monitoring programs [e.g., Occupational Safety and Health Administration (OSHA), Mine Safety and Health Administration (MSHA)]. The primary goal of this objective is "acceptable quality maintenance" regarding worker exposure levels. That is, the goal is to ensure that all worker exposure levels meet applicable permissible exposure limits (PELs) set by the regulatory agency. This program may involve elements from both an exposure screening program and an exposure distribution monitoring program. However, most OSHA-type monitoring programs are basically exposure screening programs.

2.4.7 Epidemiological Studies

When appropriate data are available, epidemiologists generally attempt to associate an observed incidence or severity of adverse health effects in target populations with exposure levels of hazardous agents. Demonstrating that an increasing response is associated with increasing exposure gives strong support to a hypothesis of disease causation due to exposure to the agent investigated. Using occupational exposure estimation for this purpose generally requires valid, reasonably precise, retrospective estimation of exposure levels for individual workers during all periods of their working lives. In exploratory studies, sometimes population average or population minimum exposures must be used as rough approximations to exposure levels of individual workers. However, this can lead to serious misclassification errors.

Monson (32) has suggested seven types of criteria to follow when interpreting occupational epidemiologic data:

1. Consistency.
2. Specificity.
3. Strength of association.
4. Dose–response relationships.
5. Coherence.
6. Temporal relationship.
7. Statistical significance.

However, when using the criterion of dose-response relationships, Monson (61) has cautioned:

> The lack of a dose–response relationship is fairly weak evidence against causality. The measure of exposure may be misclassified, there may be a threshold necessary for the exposure to cause the disease, there may be bias in the measure of exposure. The presence of a dose–response relationship is relatively strong evidence for causality.

2.4.8 *Measurement Method Comparison*

Often in industrial hygiene one desires to demonstrate the equivalency of a new exposure measurement method with currently used "standard" method. This approach is less desirable than defining a performance standard for monitoring a workplace contaminant, and then performing a measurement method validation for a proposed new method (see Section 2.4.9). However, a methods comparison may be necessary if the "control" method has been used in past epidemiological studies or is the basis of a regulatory exposure level [e.g., Mine Research Establishment (MRE) horizontal elutriator for coal dust, vertical elutriator for cotton dust, USPHS/NIOSH method for asbestos fibers]. This objective involves determining if the new method can be used as an adequately precise predictor of results that would be obtained if the standard method were used.

2.4.9 *Measurement Method Validation*

To properly evaluate workplace exposure results, it is highly advisable to estimate the accuracy of the selected measurement method and also identify the causes and magnitude of the random error components in the method. Then one can estimate the following for the method: (*1*) the repeatability (total precision error measured as variability in replicate measurements on the same sample by the same analyst) and (*2*) the reproducibility (variability between laboratories).

It is also highly advisable to estimate two important "detection limits" for the measurement method. The first of these is limit of detection (LOD), which is the minimum detectable level that can be reliably differentiated from "background noise." The LOD value should be reported for all observed values at or below the LOD (i.e., "zero" values should not be reported).

The second important "detection limit" is the limit of quantitation (LOQ), which is the minimum value for which the method yields quantitative estimates that have acceptably low uncertainty. The LOQ is greater than the LOD. The LOQ is the minimum measurement level for which an interval estimate should be reported (e.g., confidence interval, confidence limits). Sometimes the accuracy and precision estimates for the method will be compared to a performance standard to determine the acceptability of the measurement procedure (e.g., the "$\pm 25\%$ accuracy at a 95% confidence level" criteria often used by NIOSH and OSHA as discussed in Leidel et al. (2).

For most practical applications in which an assessment is made of the precision of analytical methods, the Bonferroni approach is adequate to calculate an upper confidence limit for the coefficient of variation (62). The measured bias is treated as known and each measurement is corrected for bias before analyzing the corrected data for estimation of precision.

To determine with reasonable confidence that the specified accuracy standard is met, the upper confidence limit for the CV is then compared to the accuracy standard. For use in marginal cases, a sophisticated significance test for agreement of the measured total accuracy (bias included) has been developed (63, 64).

2.5 Sampling Strategies for Occupational Exposure Estimation

It is important to recognize that sampling strategies describe a major portion of a series of exposure assessment steps used to achieve the goal of worker protection from adverse health effects due to hazardous workplace exposures. Figure 52.1 presents an overview of exposure sampling strategies as they relate to personal and group exposure assessments. Most exposure estimation sampling strategies have the following four common elements indicated in Figure 52.1.

1. Establishing the purposes and objectives for estimating occupational exposures (e.g., see Section 2.4).
2. Determining the study design for both the preliminary and full-scale studies (the principles of study design and data analysis are discussed in Section 4 and specific study designs are given in Section 5).
3. Collecting the experimental or observational data and other necessary information.
4. Performing the data analysis (see Section 6).

Figure 52.1 also indicates that sampling strategy results are then combined with other information (e.g., medical information, worker symptoms) and other constraints, and these

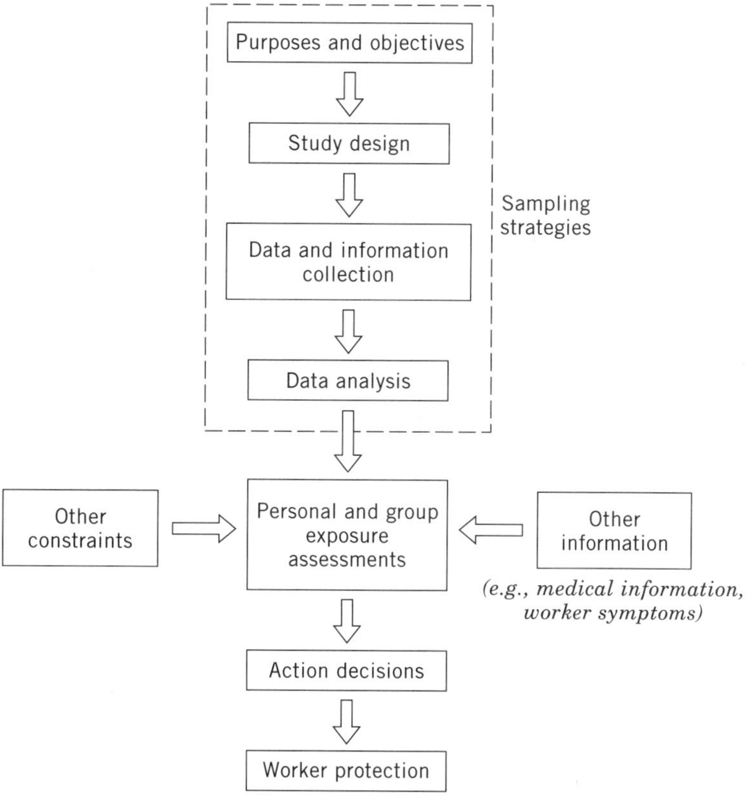

Figure 52.1. Overview of exposure sampling strategies.

are the three elements of personal and group exposure assessments. These assessments then lead to action decisions necessary for worker protection from hazardous exposures.

General types of occupational exposure sampling strategies suitable for the nine objectives of exposure estimation discussed in Section 2.4 are summarized in Table 52.1. Solid bullets indicate primary strategies and hollow bullets indicate secondary strategies in Table 52.1. Study design strategies can include:

1. Making workplace observations.
2. Estimating possible airborne levels from material usage rates.
3. Measuring worst-case exposures.
4. Conducting an exposure screening monitoring program to identify higher-risk workers or worker target populations for additional monitoring (see Section 5.3).
5. Conducting an exposure distribution monitoring program to quantify target population exposure distributions (see Section 5.3).
6. Identifying and quantifying determinant variables for exposure level distributions (see Section 5.1).

Data analysis strategies can include:

1. Using an action level as a screening cut-off value to identify medium- to high-risk workers or target populations for additional monitoring.
2. Performing hypothesis testing for classifying individual TWA exposure point estimates (see Section 6.1.2, Section 6.2.2, Section 6.3.2, Section 6.4.2, and Section 6.5.2).
3. Calculating interval estimates for individual TWA exposures (see Section 6.1.1, Section 6.2.1, Section 6.3.1, Section 6.4.1, and Section 6.5.1).
4. Calculating point and interval estimates for proportions and fractiles of exposure distributions (see Section 6.7, Section 6.8, and Section 6.10).
5. Using lognormal probability paper for analyzing worker or target population exposure distributions (see Section 6.12 and Section 6.13).
6. Using control charts or semilogarithmic paper for analyzing daily TWA exposure levels and distribution parameters (see Section 6.11).

3 STATISTICAL THEORY: COMPONENTS OF VARIATION, DISTRIBUTIONAL MODELS FOR OCCUPATIONAL EXPOSURE DATA, AND TESTS OF SIGNIFICANCE USED FOR COMPLIANCE/NONCOMPLIANCE DECISION MAKING

To plan an assessment study for measuring occupational exposures and properly analyzing the resulting worker exposure data, one must first understand how various types of errors can affect individual exposure measurements. Evaluating the errors that affect an individual exposure measurement is analogous to evaluating how the sizes of individual trees of the

Table 52.1. Sampling Strategies for Exposure Estimation Objectives[a]

● Primary strategy
○ Secondary strategy

		1. Hazard recognition	2. Hazard evaluation	3. Control-method evaluation	4. Exposure-screening monitoring program	5. Exposure-distribution monitoring program	6. Regulatory monitoring program	7. Epidemiological study	8. Measurement-methods comparison	9. Measurement-method validation
Study-design strategies	1. Workplace observations	●	○							
	2. Estimate exposure levels from material-usage rates	○								
	3. Measuring worst-case exposures		●	○						
	4. Exposure-screening monitoring program to identify higher-risk workers or target populations for additional monitoring		○		●		●			
	5. Exposure-distribution monitoring program to quantify target-population exposure distributions		○			●	○	●	○	
	6. Identifying and quantifying determinant variables for exposure-level distributions		○	●			○	●	○	
Data-analysis strategies	1. Using an Action Level as a screening cut-off value to identify medium- to high-risk workers or target populations for additional monitoring				●		●			
	2. Performing hypothesis testing for classifying individual TWA-exposure point estimates		○		○	○	○			
	3. Calculating interval estimates for individual TWA-exposure estimates (e.g., confidence limits)		●	●	○	●	●	●	●	●
	4. Calculating point and interval estimates for proportions and fractiles of exposure distributions (e.g., tolerance limits)			●	○	●	●	●	●	●
	5. Using lognormal probability paper for analyzing and summarizing worker or target-population exposure distributions		○	●		●	○	●	○	○
	6. Using control charts or semilogarithmic paper for analyzing and summarizing daily TWA-exposure levels and distribution parameters				○	●	●			

same type and age vary randomly within sections of a forest. Adequate distributional models for errors in occupational exposure results, and the sources of variation affecting a single-exposure measurement, will be discussed in Section 3.2 and Section 3.3. Next, one must understand the patterns in which groups of data will occur. For example, values of the location parameter (mean values) of the within-group distributions may differ between groups due to assignable causes. This is analogous to interpreting differences between average tree sizes obtained from different forests as being due to such factors as soil fertility, rainfall, pollution, and so on. Models for describing the behavior of such data families will be discussed in Section 3.4, Section 3.5, and Section 3.6.

3.1 Nomenclature for Exposure Concentration Data

Exposure concentration data from occupational settings can be reported as single measurements or as other estimates (e.g., time-weighted average concentrations). An exposure measurement is defined as the measured airborne concentration of a material in a single air sample taken near a worker such that it is a valid measurement of that worker's exposure. For the sake of brevity, phrases such as exposure, exposure data, and exposure measurement(s) will be used for the longer and more accurate phrases exposure concentration, exposure intensity, exposure concentration data, exposure intensity data, exposure concentration measurement(s), and exposure intensity measurement(s). Use of the single word exposure will not refer to the dose of contaminant or agent received by a worker's body; rather, the term biological dose will have that meaning.

Technical Note: The use of the word sample in both the statistical sense and the industrial hygiene physical sense in this chapter can present a source of confusion. Conceptually, a statistical sample consists of one or more items selected from a parent population, each of which has some characteristic measured. However, in the physical sense an industrial hygiene sample or exposure sample is an exposure measurement determined, for example, from a measured amount of an airborne material collected on a physical device (e.g., filter, charcoal tube, passive dosimeter). Such industrial hygiene sampling is usually performed by drawing a measured volume of air through a filter, sorbent tube, impingement device, or other instrument to trap and collect the airborne contaminant. Passive dosimeters rely on diffusion to move the contaminant to the collecting media. In the sense of this chapter, an occupational exposure sample and accompanying act of sampling combine both the concept of a statistical sample (i.e., one result among many that could occur under the same conditions) and the physical sample, which is a physical agent that is measured or material sample that is chemically analyzed or interpreted.

An exposure estimate is an estimate of a workplace exposure concentration or intensity over a specified time period that is calculated from one or more exposure measurements. An exposure estimate may be obtained for a period as short as a few seconds or represent a period from minutes to years. In the latter case it is known as a TWA exposure, which is the time-integrated instantaneous exposure (e.g., the cumulative concentration) divided by the length of time for the exposure period. If cumulative exposure could be estimated from a single air sample, for example, it would be the quotient of the weight of the material in the air sample divided by volume of air sampled during the measurement period.

Typically, the duration of TWA exposure estimates for chemical agents will be about 15 min. or less to estimate short-term, acute-exposure risk; 8-hr to estimate a workday exposure risk; and 40-hr or longer to estimate prolonged, chronic exposure risk. If it is appropriate from a toxicological standpoint, longer time-averaging exposure periods (a month or longer) can be used to estimate chronic exposure risk. Certain nomenclature has been developed by Leidel et al. (2) to describe several different types of TWA exposure estimates. These are illustrated in Figure 52.2 and include:

1. *Full-period single-sample estimate*: A single exposure measurement taken for the full duration of the desired time-averaging period (e.g., 40 hr for a workweek TWA, 8 hr for an 8-hr workday TWA, or 15 min. for a 15-min. short-term TWA).

2. *Full-period consecutive samples estimate*: The TWA of a continuous series of exposure measurements (equal or unequal, nonoverlapping time intervals) covering the full duration of the desired time-averaging period.

3. *Partial-period consecutive samples estimate*: The TWA of a series (continuous or noncontinuous) of exposure measurements (equal or unequal, nonoverlapping time intervals) obtained for a total duration less than the desired time-averaging period. For an 8-hr

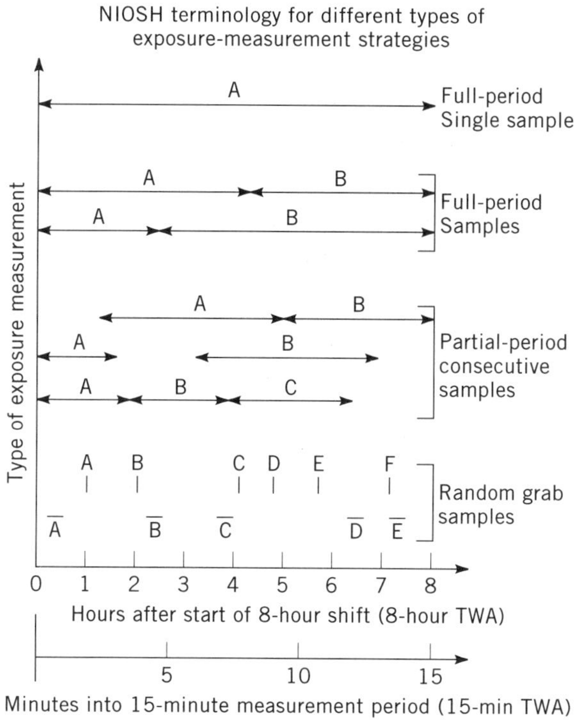

Figure 52.2. Types of exposure measurements that could be taken for an 8-hr TWA or 15-min short-term average exposure standard (from Leidel et al., 1977).

TWA exposure estimate, this would mean that exposure samples assumed to represent the entire 8-hr exposure would be selected to cover about 4 to less than 8-hr. Several samples totaling less than 4 hr (e.g., eight 15-min samples) can usually be better treated as grab samples for the purposes of statistical analysis.

4. *Grab-samples estimate*: The average of several short-period samples taken during random intervals of the time-averaging period. Sometimes it is not feasible, due to technical limitations in measurement methods (e.g., direct reading instruments, some colorimetric detector tubes), to obtain a type 1 or type 2 exposure estimate. In such situations, grab samples may be taken during several short intervals (e.g., seconds or several minutes up to less than about 30 min) within the desired longer time-averaging period such as 15 min. or 8 hr.

Adequate distributional models for describing variability of the preceding four types of exposure estimates will be presented in Section 3.6. Selection of particular intervals to be sampled is discussed in Section 5.2.2.

3.2 Net Error Model for Exposure Measurements

Suppose a worker's exposure to a workplace contaminant is to be measured. Assume that an appropriate measurement method is available that on the average can give valid (i.e., representative and accurate) determinations of airborne concentrations. The sampling equipment and laboratory instruments must be properly calibrated to reduce systematic errors or biases. This does not imply that every sample will give an exactly correct answer (i.e., the true value of the airborne concentration at the time and place where the sample is taken). To the contrary, random analytical error causes every sample result to differ somewhat from the respective true average exposure that existed during the time period of the sample. "Exposure" is used here synonymously with "exposure concentration" because it is assumed that no difference exists between the true concentration measured and the true exposure concentration intended to be measured (e.g., as the concentration in a worker's "breathing zone"). The discrepancies between the reported results and the unknown true exposures are termed random errors because they are assumed to vary in magnitude and direction in a random manner from sample to sample. Random errors, within limits, are inherent to any measurement method and equipment. The presence of random error does not imply that the method has been improperly used (i.e., that mistakes have been made). Of course, a discrepancy outside the usual range of variability (an excessively imprecise result) for the method could indicate that a mistake has been made and such data might be discarded, especially if the suspect result can be associated with an identifiable irregularity that occurred during the sampling procedure or in the sample analysis.

To systematically approach the statistical treatment of random errors, one uses a mathematical statistical model. The true average concentration at the spatial location and temporal period of the exposure sample is denoted by the symbol μ and the particular reported result from the sample is denoted by the symbol X. Thus the total error of a single sample ϵ_T is given by

$$\epsilon_T = X - \mu \tag{1}$$

The total (net) error ϵ_T is the algebraic sum of independent measurement errors, which typically result from the component sampling and analytical steps in the measurement procedure. For example,

$$\epsilon_T = \epsilon_S + \epsilon_A \tag{2}$$

where ϵ_S is a positive or negative random sampling error and ϵ_A is a positive or negative random analytical error. For independent analytical errors, the size and sign of the error does not depend on the size or sign of the sampling error. All ϵ's have the same concentration units as the reported result X (e.g., ppm, mg/m^3).

Thus any exposure estimate X that is both an industrial hygiene and statistical sample can be represented as the following algebraic sum of the true concentration μ and the net sampling and analytical error for the particular sample:

$$X = \mu + \epsilon_S + \epsilon_A \tag{3}$$

If multiple samples could be taken at exactly the same point in space and over the same time period and if the true value μ were identical for all samples, they would be true replicate samples. Note that in actual industrial hygiene sampling it may be difficult to obtain duplicate samples that are true replicates. Nevertheless, a given sample must be thought of as a random sample from the hypothetical population of all replicate samples that might have been obtained under exactly the same exposure conditions. Seim and Dickeson (65) have suggested a device for collecting actual replicate samples in the workplace.

3.3 Sources of Variation in Exposure Results

Routinely a population of exposure sample measurements from a given sampling strategy (e.g., grab samples for a worker during a work shift, 8-hr TWA exposures for several workers on a work shift, a series of 8-hr TWA exposures for a worker on several work shifts) will exhibit variability (i.e., scatter or dispersion). An important part of the interpretation of such results is an analysis of the pattern of variation, or distribution, of the data. From the sample results one usually attempts to draw inferences about the population distribution of exposure levels (i.e., about the pattern of all levels occurring for the same conditions under which the sample results were obtained). When analyzing sample data, it is important to understand the sources of variation that combine to create observed total (net) variability (due to total errors measured as differences between measurement results and true exposure levels). The sizes of these variations are a function of both the exposure levels and the measurement method. Both random and systematic errors can be affected by both the exposure levels being measured (Section 3.3.1 and Section 3.3.3) and the measurement procedure used to obtain the sample results (Section 3.3.2 and Section 3.3.4).

3.3.1 Random Variation in Workplace Exposure Levels

An elementary mathematical model for component random errors and related net random variation in replicate sample results was discussed in Section 3.2. Recognition of random influences on the measurement process can be extended to a recognition of other random influences that affect the true value of what is measured. This extension of the model enables a better understanding of the sources of variation in exposure results at different sites and times. Random changes in the determinant variables affecting the workplace exposure levels can lead to random variation in the exposure results. Exposure determinant variables are qualitative or quantitative variables, factors, or parameters that influence or affect true airborne exposure levels. The types of determinant variables are discussed in detail in Section 5. For now, note that random variation in determinant variables can result in any or all of the following types of variability in real exposure levels in the workplace:

1. Intraday (within a day) exposure level fluctuations within the same workplace.
2. Interday (between days) exposure level fluctuations within the same workplace.
3. Interworker (between workers) exposure level differences between different workers within a job group or occupational category.

It is important to realize that random variation in exposure levels and in subsequent exposure measurements can be accounted for (but not prevented) by appropriate statistical procedures. Generally the magnitude of random exposure variations over space, time, or workers cannot be quantified or predicted before making numerous exposure measurements, since often it is not possible to predict how the many determinant variables will affect workplace exposure levels.

3.3.2 Random Variation in the Measurement Procedure

In the previous section, there is a discussion of sources of random variation in true workplace exposure levels. Random errors also occur in the exposure level measurement procedure, and these physical variations lead to random variations in corresponding exposure measurements data. Examples of possible sources of random physical errors in the process of exposure measurement are

1. Random changes in pump flow rate (or mass flow rate with passive dosimeters) during sample collection.
2. Random changes in collection efficiency of the sampling device.
3. Random changes in desorption efficiency of the samples during analysis.

It is important to realize that variations in multiple-measurement results obtained during an hour, a day, or over many days that are due to fluctuations in the actual workplace exposure levels will usually exceed measurement procedure variation by a substantial amount (often by factors of 10 or 20). Thus the predominant component of variation in these results will be due to the considerable variation in what is being measured and not due to the measurement method itself. This is illustrated in Figure 52.3.

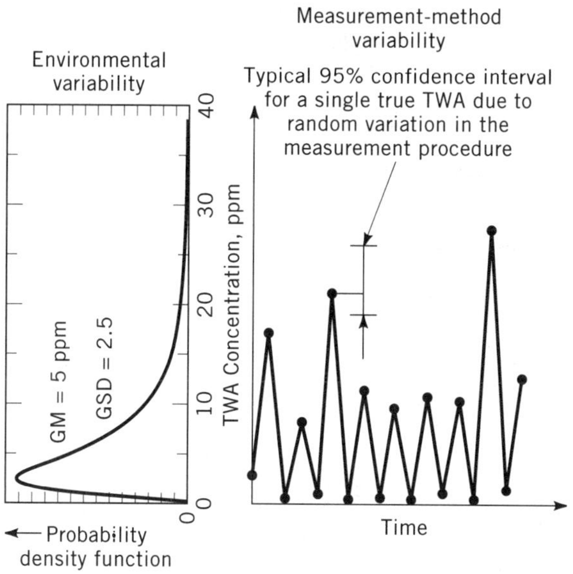

Figure 52.3. Variation components in multiple measurement results. Actual Workplace TWA Exposures: Comparison of variability from: 1. Measurement method (sampling and analytical steps) 2. Environmental fluctuations in actual workplace-exposure levels. Random errors in measurement method modelled with a normal-distribution. Environmental fluctuations in values being measured modelled with a 2-parameter lognormal distribution.

However, in contrast to random variation in the true exposure levels, the relatively smaller random measurement errors can generally be quantified before making the measurements (i.e., ranges of error can be estimated probabilistically from methods evaluation experiments performed before the exposure measurements). Then, the effects of the known distribution of random measurement errors on forthcoming exposure results can be minimized by the application of sampling strategies and programs based on statistical principles.

3.3.3 Systematic Variation in Worker Exposure Levels

In contrast to random variations, systematic variations in workplace exposure levels or systematic errors in the measurement procedure cannot be predicted with statistical methodologies based in probability theory. Instead, the study design must anticipate and make provision for systematic errors. During a data analysis performed to compare exposure results between two groups, the comparison should be made within blocks (i.e., within appropriate subgroups that are homogeneous in all other respects), or the measurements should be corrected for possible systematic errors due to extraneous factors before any statistical analyses are performed.

Systematic biases or shifts in the determinant variables affecting the workplace exposure levels will lead to systematic shifts in the exposure results. For example, some systematic shifts in determinant variables, and their consequences, are

1. Changes in a worker's exposure situation (such as several different jobs or operations during a work shift or over several days) can result in intraday or interday shifts in worker exposure.
2. Production or process changes can cause shifts in worker exposure levels (intraday or interday).
3. Control procedure or control system changes can cause shifts in worker exposure levels (intraday or interday).

3.3.4 Systematic Errors in the Measurement Procedure

Besides the random errors in a measurement procedure, there can also be systematic errors that occur during the measurement procedure and lead to systematic errors (biases) in exposure results. Examples are

1. Mistakes in pump calibration and drops in pump battery voltage leading to systematic errors in air flow rate.
2. Use of the sampling device at temperature or altitude conditions substantially different than the calibration conditions (see Ref. 66).
3. Physical or chemical interferences during sample collection.
4. Sample degradation during storage before analysis.
5. Intralaboratory errors (similar errors in groups of analyses performed within a laboratory) due to chemical or optical interferences, improper procedures, mistakes in analytical instrument calibration, or failure to properly follow the steps of an analytical procedure.
6. Interlaboratory differences due to use of different methods, different equipment, or different training of personnel.

Systematic measurement errors may be identified and their effects minimized with the use of quality assurance programs.

In the statistical sense, a substantial systematic shift or error in either the measurement process or the exposure levels being measured creates a different population with another location (central tendency) on the exposure level scale. If the systematic shift(s) goes undetected, the resulting two (or several) "side-by-side" sample populations can be mistakenly analyzed as a single distribution. The inferential statistical procedures presented in this chapter will not detect and do not allow for the analysis of highly inaccurate results caused by systematic errors or shifts. Unfortunately, systematic errors in a measurement procedure or systematic shifts in exposure levels sometimes go undetected and introduce considerably larger errors into the exposure results than would be caused by the usual random variations. This can lead to reporting inferences from the sample results that have erroneously higher uncertainty (less precision) than is stated, if precision is calculated from the known amount of random variability that has previously existed when no biases are present. Since vague or uncertain inferences generally have little value, it becomes critical to identify potential systematic errors and take steps in the study design to eliminate them, or correct for them, if possible, when the results are statistically analyzed.

For example, for organic solvent exposure measurements, Olsen et al. (67) found larger random variability in analyses of replicate field samples than in analyses of replicate laboratory samples of the same pure compound. A large (5- to 10-fold) systematic error (bias) was also detected that was associated with the use of two different sampling strategies.

3.3.5 Location of the Measurement Device in Relation to the Worker

A most important goal of personal exposure measuring for chemical agents is to obtain valid estimates of the concentrations breathed by workers. A valid exposure estimate is one that measures what it is purported to measure. More specifically, criteria for evaluating the validity of worker exposure estimates include:

1. *Relevance.* Is the air concentration in the sample equivalent to the concentration breathed by the worker(s) of interest?
2. *Calibration.* Are the exposure measurements unbiased estimates of the true concentration sampled (i.e., are measurements accurate on the average)?
3. *Precision.* Was adequate exposure information obtained to derive a sufficiently precise exposure estimate for the worker(s) of interest (either on a given day or over some longer period such as several years of employment)?

Concerning relevance of an exposure estimate to a worker's actual exposure, exposure measurements should be taken in the worker's breathing zone (i.e., air that would most nearly represent that inhaled by the worker). There are three basic classes of exposure measurement techniques:

1. *Personal.* The measurement device is directly attached to a worker and worn continuously during all work and rest operations. Thus the device collects air from the breathing zone of the worker.
2. *Breathing Zone.* The measurement device is held by a second person who attempts to sample the air in the breathing zone of a worker.
3. *General Air.* The measurement device is placed in a fixed location in a work area. This technique is also called area sampling.

If measurements taken by the general air technique are to be used, then it is necessary to demonstrate that they are valid estimates of personal exposures. Normally this is difficult to do. Refer to Ref. 68 for a discussion of this subject.

3.4 The Normal Distribution Model

3.4.1 Descriptive Parameters

A utilitarian mathematical model for the frequency distribution of some types of continuous variable occupational health data is the normal distribution. This model has a mathematical formula that can be used to describe and compute the normal distribution curve. In routine practice the formula is rarely directly applied since tables for the distribution are readily available in technical handbooks. This distribution is also available in inexpensive hand-held calculators or personal computer spreadsheet programs. The distributional formula relates ordinates $f(X)$ of the curve to values of the variable X. It is called a distribution function or simply distribution.

Technical Note: This terminology is not universally used. The definition of the term distribution function is according to Hald (69), but others refer to $f(X)$ as the frequency function or probability density function.

The distribution function $f(X)$ for a normal distribution is represented by the special notation $N(X; \mu, \sigma^2)$. That is, for the normal distribution model (a.k.a. normal curve), the ordinate or height of the probability density curve is given by the formula:

$$f(X) = N(X; \mu, \sigma^2) = \frac{1}{\sigma\sqrt{2\pi}} \exp\left(\frac{-\frac{1}{2}(X - \mu)^2}{\sigma^2}\right) \tag{4}$$

Note that two constants (parameters), μ, the mean, and σ, the standard deviation completely characterize the normal distribution. Thus the notation $N(X; \mu, \sigma^2)$ is statistical shorthand for "the normal distribution of a variable X that has the true mean μ and variance σ^2." All normal curves have the same general appearance—a bell-shaped curve that is symmetrical about its mean. The true mean μ, also called the expected value of the random variable X, denoted $E(X)$, is the weighted average value of all values of the random variable X. The weighting function is the distribution function [i.e., each X is weighted by its probability density $f(X)$].

Mathematically, the mean $E(X)$ of any distribution function, say $f(X)$, is the corresponding distribution curve's center of gravity, which is defined by

$$E(X) = \int_{-\infty}^{+\infty} Xf(X)dX \tag{5}$$

For the mean of the normal distribution the general $f(X)$ in equation 5 is replaced by $N(X; \mu, \sigma^2)$ so that

$$E(X) = \int_{-\infty}^{+\infty} \frac{X}{\sigma\sqrt{2\pi}} \exp\left[\frac{-\frac{1}{2}(X - \mu)^2}{\sigma^2}\right]dX = \mu \tag{6}$$

The integration in equation 6 is not obvious and the details of its evaluation are not presented here. The point to note is that for the normal distribution, the parameter μ in its formula is the mean of the distribution.

Similarly, it can be shown that for the normal distribution, the weighted average value of squared deviations $(X - \mu)^2$ is σ^2, that is,

$$E(X - \mu)^2 = \int_{-\infty}^{+\infty} \frac{(X - \mu)^2}{\sigma\sqrt{2\pi}} \exp\left[\frac{-\frac{1}{2}(X - \mu)^2}{\sigma^2}\right] dX = \sigma^2 \tag{7}$$

For any distribution the mean square of deviations from the mean, denoted by $E[X - E(X)]^2$ is known as the variance of X. The variance is the square of the standard deviation. For the normal distribution, the variance is equal to its second parameter σ^2, so that the standard deviation is σ. The mode of any distribution is the point on the X scale at which the maximum of the distribution function occurs. The median is the middle X value, that is, the value exceeded by 50% of the area under the distribution curve. Hereafter the "proportion of the distribution between two values of X," may be referred to also which should be understood to mean "proportion of the area under the distribution curve between two values of X." Since the total area under any distribution curve is unity (1.0) the "proportion of the area between two X values" can also be called the "area between two X values." For the normal distribution the mode, median, and mean are all equal to μ.

The two parameters μ and σ of a normal distribution completely determine its location (central tendency) and shape (dispersion or variability). The location parameter is the mean μ, which is the center point of the curve. The variability parameter (or measure of dispersion) is the standard deviation σ, which indicates how much dispersion there is of the X values about their mean. Table 52.2 gives some examples of relationships between the mean, standard deviation, and proportions of the total distribution that lie within various intervals containing the mean. See Section 6.9 for a procedure to calculate intervals of a normally distributed variable X (with known parameters) which contain designated proportions of the distribution. Such values can be expressed by fractile terminology [this is Hald's (62) terminology, some others use quantile]. The fractile X_P is the value of X that has proportion P of the area of the distribution $f(X)$ at or below it.

Values of X (e.g., concentration results for replicate samples) occur within intervals above and below m with predictable relative frequencies (or probabilities) that are equal to corresponding areas under the curve (the sample distribution curve). Ordinates of the sample distribution curve do not give probabilities of corresponding sample results. Rather, it is the area under the sample distribution curve between two values of X that is equal to

Table 52.2. Areas under the Normal Curve

X Interval	z Interval[a]	Percent of Data within Interval
$\mu - \sigma$ to $\mu + \sigma$	-1.0 to 1.0	68.4
$\mu - 1.645\sigma$ to $\mu + 1.645\sigma$	-1.645 to 1.645	90.0
$\mu - 1.960\sigma$ to $\mu + 1.960\sigma$	-1.960 to 1.960	95.0
$-\infty$ to $\mu + 1.645\sigma$	$-\infty$ to 1.645	95.0
$\mu - 2\sigma$ to $\mu + 2\sigma$	-2 to 2	95.4
$\mu - 1.960\sigma$ to $+\infty$	-1.960 to ∞	97.5
$\mu - 2.576\sigma$ to $\mu + 2.576\sigma$	-2.576 to 2.576	99.0
$\mu - 3\sigma$ to $\mu + 3\sigma$	-3 to 3	99.7

[a]Where $z = (X - \mu)/\sigma$ = standard normal deviate.

the relative frequency (proportion) of replicate samples that would occur in that interval. For this area to represent a probability that is a proportion between 0 (impossibility) and 1 (certainty) a distribution curve is standardized such that the total area under the curve is exactly unity (1.0).

3.4.2 *Coefficient of Variation*

On the average, random errors in exposure concentration measurements, due to errors in exposure measurement procedures during sampling (e.g., elapsed time, air flow rate) and during subsequent chemical analyses, are generally proportional to the level of airborne concentration measured. Therefore, it is appropriate to express the magnitudes of these errors as fractions of the concentration levels. In this way, the measurement variability of an exposure measurement procedure can be expressed as a constant value that is independent of the concentration measured. A measure of this proportional variability called the coefficient of variation (*CV*) is defined by: $CV = \{E[X - E(X)]^2\}^{1/2}/[E(X)]$ [i.e., $CV =$ (standard deviation)/(expected measurement)]. Some chemists know it as the relative standard deviation (RSD) or (s_r). For the normal distribution, $CV = \sigma/\mu$.

If a measurement procedure is composed of two or more independent steps (e.g., obtaining the sample, subsequent laboratory analysis) it can be shown (see equation 2) that the net error ($\epsilon_T = \epsilon_S + \epsilon_A$) for the combined steps of the procedure has the following total coefficient of variation:

$$\text{CV}_T = [\text{CV}_S^2 + \text{CV}_A^2]^{1/2} = \sigma_T/\mu \qquad (8)$$

where the subscript *S* denotes the sampling step and the subscript *A* denotes the analytical step. It is important to realize that the *CV*s are not directly additive; instead, the CV_T increases as the square root of the sum of the squares of the component *CV*s.

The total coefficient of variation CV_T generally can be treated as constant within the range of concentrations at which the measurement method is routinely applied. At a given concentration μ within the application range, the standard deviation of the measurement error is given by

$$\sigma_T = (\mu)(\text{CV}_T) \qquad (9)$$

where

$$\sigma_T = [\sigma_S^2 + \sigma_A^2]^{1/2} \qquad (10)$$

Note that a random variable that is assumed to be normally distributed can theoretically attain negative values, whereas airborne concentration measurements cannot attain negative values. Nevertheless, the normal distribution model is usually an adequate approximation to the sampling distribution of replicate sample chemical analyses. Most measurement methods used in industrial hygiene have net random errors whose standard deviation is small compared to the true mean airborne concentration. Thus the portion of the normal distribution model lying left of zero has negligible area and the model adequately predicts the distribution of (positive) replicate measurements. This is illustrated in Figure 52.4 which shows the relative frequency of the many possible results one might obtain with one measurement of an 80-ppm true exposure using a $CV_T = 0.10$ measurement method. As indicated in Table 52.2 about 68% of the possible results are within the region centered about μ, from 72 ppm (μ − σ) to 88 ppm (μ + σ). Additionally, there is only a 2.5% likelihood that a measurement of the actual 80 ppm will lie less than 64.3 ppm (μ − 1.960σ) or greater then 95.7 ppm (μ + 1.960σ) [Note: σ = (0.10) (80 ppm) = 8.0 ppm.]

3.5 The Lognormal Distribution Model

A second mathematical model of great utility for several types of industrial hygiene data is the logarithmic-normal, or lognormal distribution. The general properties of the several types of lognormal distributions have been extensively discussed by Aitchison and Brown (70). Section 3.5.1 will discuss a variate whose logarithm is distributed according to normal law (i.e., the case of a two-parameter lognormal). This is the simplest case because it involves primarily an interplay of the mathematical properties of the logarithmic function (used as a data transformation) and the well-known statistical properties of the normal distribution. These were discussed in the previous section. In Section 3.5.2, the definition and scope of the lognormal distribution will be extended with the use of a third parameter to shift the origin of the distribution's measurement scale.

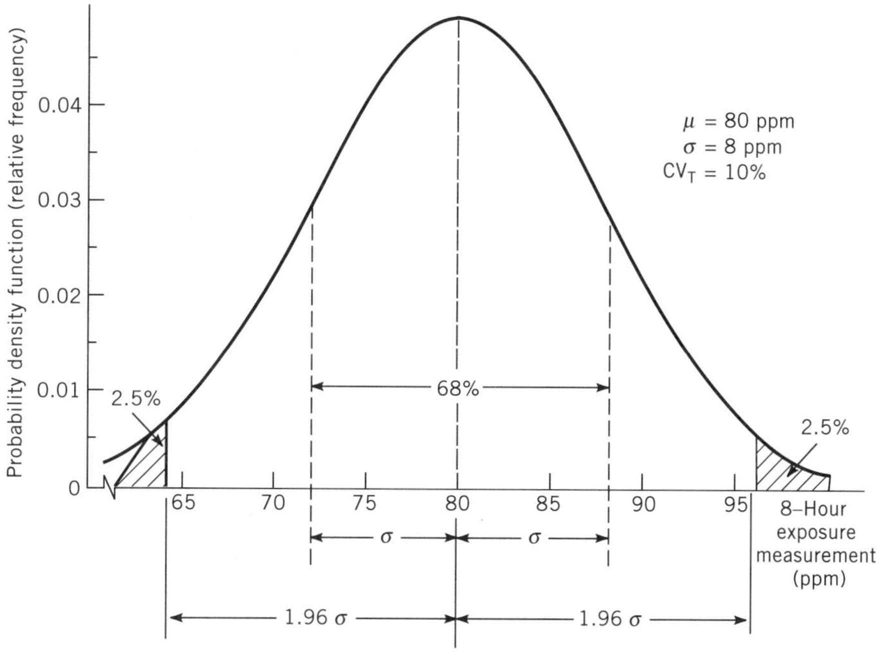

Figure 52.4. Predicted sampling distribution of simultaneous, single 8-hr samples from an employee with a true exposure average μ of 80 ppm. μ = 80 ppm, σ = 8 ppm CV_T = 10%.

3.5.1 Two-Parameter Lognormal

A two-parameter lognormal curve has the general formula

$$f(X) = \frac{1}{X(\ln \sigma_g)\sqrt{2\pi}} \exp\left(\frac{-\tfrac{1}{2}(\ln X - \ln \mu_g)^2}{\ln^2 \sigma_g}\right) \qquad (11)$$

where $0 < X < \infty$. Equation 11 is called the lognormal probability density function, which is generally shortened to lognormal distribution function. Its general structure is similar to equation 4 for the normal distribution function. The relationship between these two distributions is that for a random variable X, which is lognormally distributed, the values ln X or log X will be normally distributed. As with the normal distribution, the basic two-parameter lognormal distribution is fully characterized by two parameters. However, the lognormal parameters are known as the geometric mean (GM) and geometric standard deviation (GSD). The true geometric mean μ_g is defined by

$$E(\ln X) = \int_{-\infty}^{+\infty} (\ln X)N(\ln X; \ln \mu_g, \ln^2 \sigma_g)d \ln X = \ln \mu_g \qquad (12)$$

and the true geometric standard deviation σ_g is defined by

$$E(\ln X - \mu_g)^2 = \int_{-\infty}^{+\infty} (\ln X - \ln \mu_g)^2 N(\ln X; \ln \mu_g, \ln^2 \sigma_g)d \ln X = \ln^2 \sigma_g \qquad (13)$$

In these equations the notation used is analogous to the $N(X; \mu, \sigma^2)$ notation defined in equation 4 for a normally distributed variable X with mean μ and variance σ^2. Thus μ and σ_g are antilogs to the base e of the mean and standard deviation, respectively, of the natural logarithmic transform of X. The interpretation of μ_g and σ_g parameters for a two-parameter lognormal distribution differs somewhat from the interpretation of μ and σ for a normal distribution. The similarity is that the geometric mean is the location parameter and the geometric standard deviation is the variability, or dispersion, parameter. Note that the distribution function for the basic two-parameter lognormal model originates at zero and does not exist in the region below zero.

In Table 52.2 for a normal distribution, multiples of σ were added to and subtracted from μ to obtain intervals of X that contain specified proportions of the distribution. The factors that multiply s are standard normal deviates (i.e., values of a normally distributed variable with mean zero and variance one), which are known as Z values. Values of Z are listed in tables of the standard normal distribution available in statistical texts and other scientific reference books. A given Z value, denoted Z_P, corresponds to a "two-sided" probability $(2P - 1)$ that a randomly selected value of X will be within the interval $(\mu - Z_P\sigma)$ to $(\mu + Z_P\sigma)$. For example, Table 52.2 shows that $Z_{0.8413} = 1.000$, for a two-sided probability of 0.684; $Z_{0.95} = 1.645$, for a two-sided probability of 0.90; and $Z_{0.975} = 1.960$, for a two-sided probability of 0.950. Corresponding intervals for a lognormal distribution are of the form μ_g/σ_X to $\mu_g\sigma_X$. Table 52.3 gives examples of two-parameter

Table 52.3. Areas under the Lognormal Curve

X Interval	Percent of Data within Interval
μ_g/σ_g to $\mu_g\sigma_g$	68.4
$\mu_g/\sigma_g^{1.645}$ to $\mu_g\sigma_g^{1.645}$	90.0
$\mu/\sigma_g^{1.960}$ to $\mu_g\sigma_g^{1.960}$	95.0
0 to $\mu_g\sigma_g^{1.645}$	95.0
μ_g/σ_g^2 to $\mu_g\sigma_g^2$	95.4
$\mu_g/\sigma_g^{1.960}$ to $(+\infty)$	97.5
$\mu_g/\sigma_g^{2.576}$ to $\mu_g\sigma_g^{2.576}$	99.0
μ_g/σ_g^3 to $\mu_g\sigma_g^3$	99.7

lognormal intervals corresponding to the intervals for a normal distribution in Table 52.2. See Section 6.10 for a procedure to calculate intervals of a lognormally distributed variable X (with known parameters) that contain designated proportions of the distribution. Such intervals are bounded by fractiles X_P (see the definition given in Section 3.4.1).

3.5.2 Three-Parameter Lognormal

The utility of the basic two-parameter lognormal model can be considerably expanded by the introduction of a third parameter that is a change-of-origin parameter. A simple displacement of a random variate X that is not lognormally distributed can sometimes be made to define a transformed variate $X_T = (X - k)$ that is lognormally distributed. If the range of X is $k < X < \infty$, the range of X_T will be $0 < X_T < \infty$. The two-parameter model can be thought of as a special case of a three-parameter lognormal model, for which $k = 0$. The third parameter k for the lognormal model is sometimes selected to be the lower bound of the known range of values of the original variate X and can be thought of as the threshold of the three-parameter lognormal distribution, just as zero is the threshold for the basic two-parameter model.

The three-parameter lognormal model is useful when the data, such as exposure results, exhibit more skewness to the right than would be expected for a lognormal distribution that is not located close to the zero origin. Such a distribution could result if there were lognormal random additive variations in exposure levels combined with a fixed-background exposure level. An appropriate constant k is subtracted from each data value to create transformed data values that are then analyzed using techniques appropriate for two-parameter lognormally distributed data. Estimated parameters of the transformed distribution, along with appropriate confidence limits, are calculated. Examples are the GM of X_T and its confidence limits, or tolerance limits for the random variable X_T. The constant k is then added to all calculated values of X_T to estimate corresponding values for X relevant to the original data distribution. Details for appropriate estimation of k are presented in Section 6.12, Step 5 of the Solution.

3.6 Adequate Distributional Models for Exposure Results

To design efficient and statistically powerful studies, make rational decisions in hypothesis tests, and make valid inferences regarding expected limits on true occupational exposures, it is necessary to use adequate distributional models of exposures for the target populations the samples represent. Adequate distributional models are the keystones for valid use of the parametric statistical methods that will be presented in Section 5 and Section 6. The rationale and mathematical statistical properties of models given in this section are also discussed in depth in Busch and Leidel (71).

The adequacy of a model is dependent on its ability to serve as a workable forecaster of parameters (e.g., mean) or functions of parameters (e.g., CV or fractiles) of the parent population. Moroney (72) has noted:

> Probably there never is a mathematical function that fits a practical case absolutely perfectly. Nor is it at all necessary that there should be. What we seek is not a perfect description of a distribution but an adequate one; that is to say, one that is good enough for the purpose we have in view.

Wilkins (73) has remarked that statistical tests for goodness-of-fit have the characteristic that the test will reject any hypothetical model for a practical data set if there is a sufficient number of samples. The critical value for the test statistic (based on the permissible departure of the observed data distribution from the chosen distributional model) can even be smaller than the precision provided by the measurement methodology used to obtain the data. Even though one might observe a "statistically significant lack of fit," the proposed distributional model may still be adequate for our needs. One must also be cautious regarding hypothesis test outcomes of "no statistically significant lack of fit." These outcomes may be due to low statistical power for the tests resulting from small sample sizes.

3.6.1 Applications for the Normal Distribution

The normal distribution is usually an adequate model for the following populations of industrial hygiene results.

3.6.1.1 Populations of Replicate Analyses Performed on an Industrial Hygiene Sample (e.g., Aerosol Filters or Charcoal Tubes). Replicate analyses of a given sample are defined to be repeated analyses with variability equal to that which would exist in analyses of physically different samples, if these could have been obtained without sampling errors, of exactly the same concentration in exactly the same setting. In other words, replicate analyses are those with variability that reflects the total random error of the analytical procedure, not just components of error due to some (but not all) steps in the analysis. For an unbiased analytical method (i.e., without systematic error), the expected value of truly replicate analyses of a given physical sample is not the true time-integrated concentration the sample was obtained from. Rather, the expected value of replicate analyses is an air concentration equivalent to the amount actually present in the sample (i.e., the expected value includes the random sampling error for that sample). The sample coefficient of variation ($\widehat{CV}_A$) computed from replicate analyses is a measure of dispersion for the analytical procedure and is usually taken to be an approximation to the true CV_A measuring the proportional error due to the analytical step of the exposure measurement procedure. Note that the total (net) error of the measurement procedure is $CV_T = (CV_A^2 + CV_S^2)^{1/2}$, where CV_S is the coefficient of variation for sampling errors (i.e., those introduced by the physical sampling portion of the measurement procedure during which the contaminated air is moved onto or through the sampling media).

Section 6.6 presents the computation of tolerance limits for a variable that is normally distributed with unknown parameters. Based on results for n replicate samples, this procedure can be used to compute a tolerance interval that we can be 95% confident will contain at least 95% of analytical results for the same concentration by the same method under the same conditions. Procedures are given in Section 6.12 and Section 6.13 that detail the use of logarithmic probability paper to compute and display tolerance limits for lognormally distributed data. These procedures can be modified, where a normal distribution is expected, by substituting normal probability paper (where a linear scale is used for the original data instead of a logarithmic scale).

3.6.1.2 Populations of Replicate Measurements of Calibrated Test Concentrations. The arithmetic mean of the replicate measurements is the best estimate of a calibrated test concentration. The sample coefficient of variation calculated from a set of replicate measurements is an estimate of the measurement procedure's total CV_T (combined CV for the sampling and analytical portions of the method). Refer to Section 3.4.2 and Step 1 of this section. In this context the term sampling error refers primarily to physical random error in the volume of air sampled. Here, sampling error does not include intersubject variability, nor does it include the real variations in air concentrations of the contaminant over space and time. Such spatial and temporal variations are indeed appropriate parts of the total error of some special types of exposure estimates (and will be so included, e.g., for an 8-hr TWA estimated from grab samples, see Section 3.6.2 and Section 6.4.1).

Industrial hygiene researchers will usually obtain multiple measurements taken simultaneously at sampling locations in a small spatial volume (e.g., a sphere less than 30 cm in diameter), when attempting to estimate the variability (CV) of a measurement method. Unfortunately, unless the sampled workplace atmosphere is truly homogeneous, the sample results may lead to a variability estimate for the measurement method that is erroneously high. It is usually assumed that each of the measurements is a sample of the same true concentration, but this may not be a valid assumption, and the researcher must demonstrate that the sample environment is truly homogeneous.

Figure 52.5 illustrates the use of a normal distribution to model a sampling distribution of replicate measurements of a 5-ppm calibrated test concentration. The figure assumes a measurement method with a $CV_T = 0.10$ method was used. The top portion of the figure also indicates what a typical histogram might look like for relatively few replicate measurements. Note that the histogram is not exactly symmetrical about $\mu = 5.0$ ppm due to random sampling variation. The bottom portion of Figure 52.5 illustrates how a normal distribution for this type of sample data can be estimated by fitting a straight line to a plot of the sample cumulative distribution on normal-probability graph paper. On this type of graph paper, the sample distribution should yield a linear plot, aside from the expected deviations accountable to random sampling variations.

It is important to note that normality of replicate exposure measurements (or at least approximate normality that is sufficient to meet the requirements of any inferential statistical methodology one desires to use) is to be expected from theoretical considerations.

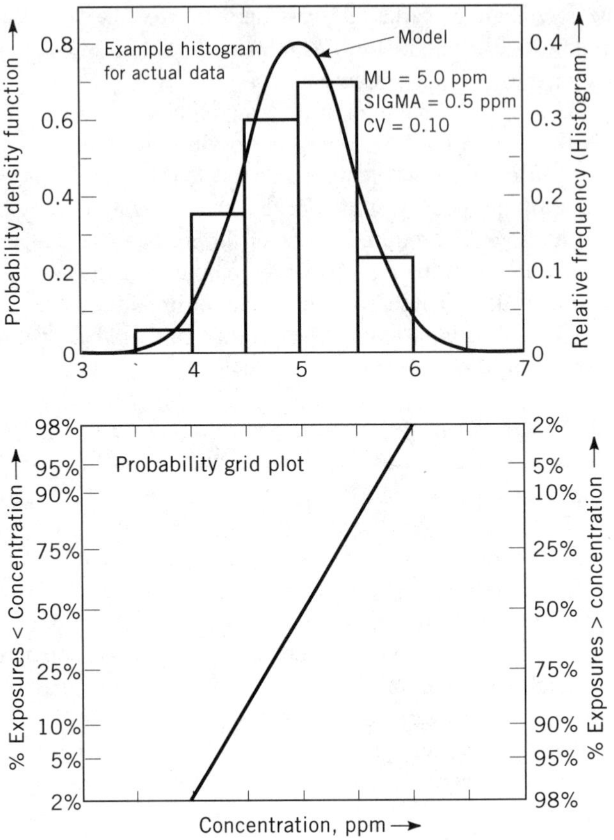

Figure 52.5. Normal distribution model for sampling distribution of replicate measurements of a 5-ppm calibrated test concentration with a $CV_T = 0.10$ method.

The total net error of any particular exposure measurement is the net error resulting from many random incremental positive and negative physical influences during the various sampling and analytical stages of a measurement. Determinant variables that can lead to positive and negative errors in measurements include unavoidable technician variations, small environmental variations (e.g., humidity, temperature), and functional variations in component parts of the sampling and analytical equipment (e.g., voltage, pump flow rate, operating temperature). Insofar as these various sources of random additive errors operate independently, their net influences tend to make the net error follow a normal probability density curve. The proof that this is true would be similar to the proof of the central limit theorem from mathematical statistics.

3.6.1.3 Populations of Replicate Full-Period Single-Sample Estimates of Exposure. The justification for using the normal distribution to model the random errors of replicate exposure measurements was discussed in Section 3.6.1.2. A single full-period sample estimate can be considered a sample of one from a hypothetical parent population of all replicate exposure measurements that could have been obtained at exactly the same point in space and over the same time period. Even though the sample is continuously collecting airborne material from constantly varying levels (i.e., from a distribution of true levels that generally vary substantially), it is assumed that on the average such samples faithfully and accurately integrate all the instantaneous concentration levels. The result of this integration is the mass of formerly airborne material collected in the sample. Thus the single full-period measurement is a time-integrated exposure estimate for the duration of the sample. A previously well-determined total coefficient of variation CV_T for the measurement procedure is used as the measure of dispersion for the population of possible replicate samples from which the one at hand is considered to be a random sample. This CV_T is treated here as known and will be used for inferential decision making and confidence interval calculations for the full-period type of exposure estimate. Section 6.1 presents applied statistical procedures for computing confidence limits (Section 6.1.1) and classifying exposures relative to an exposure control limit (Section 6.1.2) for the case of exposure estimates based on full-period single samples.

3.6.1.4 Populations of Full-Period Consecutive-Sample Estimates of Exposure. The justification for using the normal distribution to model the random errors of averages of sets of consecutive samples taken during the time-averaging period of the exposure estimate is an extension of the preceding discussion for the full-period single-sample estimate. It is assumed that the random proportional errors of the set of consecutive measurements are independent and all have the same known total coefficient of variation for the measurement method.

Section 6.2 and Section 6.3 present applied statistical procedures for computing confidence limits on the true exposure (Section 6.2.1 and Section 6.3.1) and for classifying exposures relative to an exposure control limit (Section 6.2.2 and Section 6.3.2) when exposure estimates are based on the average of full-period consecutive samples. The confidence interval and decision-making computations are presented for two different types of situations. Uniform exposure methods are given in Section 6.2 to be used when one believes that all consecutively sampled periods had equal true average exposure concentrations. Conservative nonuniform exposure methods are given in Section 6.3, to be used if one believes that the sampled periods had substantially different exposure concentrations.

3.6.2 Applications for a Lognormal Distribution

A lognormal distribution (either two- or three-parameter) is usually an adequate model for four general types of populations discussed in this section. However, be alert that a population of industrial hygiene data may be a composite of several different distributions. For example, a substantial portion of the data may occur at zero concentration with the remainder occurring in a two-parameter lognormal distribution and one or more three-parameter lognormal distributions.

3.6.2.1 Populations of True Exposure Levels at Different Times during Periods of Hours to Years. The justification for using the lognormal distribution to model workplace exposure levels at different times for a given worker, or averages for a target population of workers at different times (see Section 5) has been presented in Ref. 74. According to Hahn and Shapiro (75), the choice of a distributional form for many types of data can be based on theoretical considerations. However, for other types of data where not enough is known about the underlying physical mechanisms to be able to do this, they suggest basing choice of the distributional form on goodness-of-fit tests for observed data. They suggest use of the versatile Johnson and Pearson family of distributions for empirical fitting of approximate distributional models. They say this empirical modeling approach is usually adequate within the range of the sample data used to fit the model.

Stoline (76) compared goodness-of-fit of lognormal distributional models to the fit of more general Box-Cox distributional models. He found that for most of the data sets reasonable goodness of fit could be achieved with either the lognormal or Box-Cox distributional forms. The Box-Cox model fit slightly better, but not statistically significantly better, for a few samples. These empirical results for environmental monitoring data relate strictly only to the types of water contamination data that Stoline examined, but the statistical properties of air monitoring data are likely to be similar. Therefore, given that there is likely to be a choice between nearly equivalent Box-Cox and lognormal distributional alternatives, the lognormal model would be preferable because existing normal distribution theory could be easily applied (to a logarithmic transform).

Conditions conducive to (but not all necessary for) the occurrence of lognormal distributions are found in populations of workplace exposure levels. These conditions include:

1. Physical causes of variability tend to cause the same proportional changes in concentration, irrespective of whatever concentration is present.
2. The true exposure levels cover a wide range of values, often several orders of magnitude. The variation of the true exposure levels is of the order of the size of the exposure levels.
3. The true exposure levels lie close to a physical limit (zero concentration).
4. A finite probability exists of unusually large values (or data "spikes") occurring.

Section 6.7 presents an applied statistical procedure for the computation of tolerance limits for a lognormally distributed population with unknown parameters. This procedure is useful for computing, from a sample of exposure levels, a tolerance interval that one can be 95% confident will contain at least 95% of the population of true exposure levels. Section 6.8 details the computation of a point estimate and confidence limits for the proportion of a lognormally distributed population that exceeds a specified value (such as an exposure control limit). Section 6.11 suggests the use of semilogarithmic graph paper for plotting variables that are lognormally distributed in time, which frequently is the case for true exposure levels at different times during a period. Section 6.12 and Section 6.13 present procedures for using logarithmic probability paper to estimate the parameters of logarithmic normal distributions and for displaying tolerance limits for estimated lognormal distributions.

3.6.2.2 Grab-Sample Populations of Intraday Exposure Measurement Populations Consisting of Short-Term Samples. The duration of each grab-sample is short compared to the total interval from which the samples were obtained (e.g., less than about 5% of the interval). Grab-sample exposure estimates reflect the lognormal intraday distribution of true exposure levels that is sampled (i.e., environmental variability). Grab-sample populations also have a component of normally distributed measurement process (sampling and analysis) error (i.e., measurement method variability), but this component generally is negligible compared to the larger amount of lognormal variation of the true concentrations. These two variability contributors are graphically illustrated in Figure 52.6.

For evaluation of a worker's individual health risk, an inference must be made concerning the relation between the TWA exposure control limit and the true arithmetic mean of the entire population of grab samples from which the few samples at hand were selected at random. If the true exposures occurring during the interval could be considered uniform (effectively equal), then the sample arithmetic mean of the grab samples would be an adequate estimate of the TWA exposure. But, if there are subintervals with respectively stable but different exposures, then a sample arithmetic mean for each period of equal exposure should be computed from the grab samples of each period and a TWA estimate then computed from the series of arithmetic means. Confidence limits for such estimates

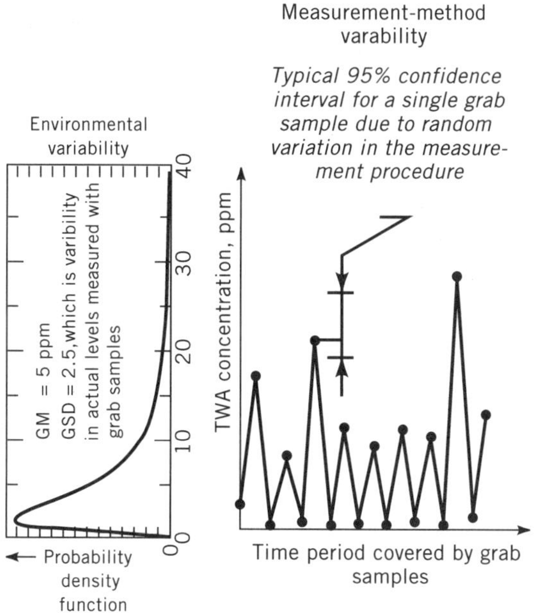

Figure 52.6. Comparison of variability contributions in a grab-sample estimate. Grab-Samples Estimate: Comparison of variability from: 1. Measurement method (sampling and analytical steps) 2. Environmental fluctuations in actual values measured with grab samples. Random errors in measurement method used for grab samples are modelled with a normal distribution. Environmental fluctuations in actual values being measured with grab samples are modelled with a 2-parameter lognormal distribution.

of TWA exposures can be based on the normal distribution, since the only errors are those of the measurement procedure.

However, if there were general lognormal variability among the total set of grab-sample intervals making up the period of the standard, then an estimate of the TWA exposure based on grab samples would require computing a sample geometric mean and converting it to an estimate of the TWA exposure. A procedure for doing this is given in Section 6.4, but this is not a recommended procedure because the correction factor for converting a GM estimate to a TWA exposure estimate is a function of the GSD. When the sample geometric standard deviation for each interval of uniform exposure has to be computed (to estimate the variability of the worker's exposure during each interval), this complex estimation technique introduces considerable sampling error into the TWA exposure estimate.

Section 6.7 presents an applied statistical procedure for the computation of tolerance limits for a lognormally distributed population with unknown parameters. This procedure is useful for computing tolerance intervals that one can be 95% confident will contain 95% of the true short-term exposure levels during a given day for which grab samples were taken. Also, Section 6.11 suggests the use of semilogarithmic graph paper for plotting variables that are lognormally distributed in time, which typically is the case for grab-sample measurements on a given day. Section 6.12 and Section 6.13 present procedures for using logarithmic probability paper to estimate the parameters of lognormal distributions and for displaying corresponding tolerance limits.

3.6.2.3 Populations of Daily 8-hr TWA Exposure Estimates for a Worker. For evaluation of a worker's individual health risk due to a chronic exposure, the long-term average of daily exposures could be estimated by the arithmetic mean of a random sample of daily TWA estimates, if the daily exposures could be considered uniform. If not, then a sample arithmetic mean for each group of daily TWAs from each multiday period of equal exposure would have to be computed and a long-term TWA estimate computed from the series of arithmetic means. In either of these cases (where uniform exposures on each day can be assumed), confidence limits for the long-term TWA exposure would be based on the normal distribution, since the only errors are in the measurement procedure.

The CE (cumulative exposure) and LWAE (lifetime, weighted-average exposure) methods of Ford et al. (11), described earlier in Section 2.2, have the desirable property of being practical exposure evaluation tools in epidemiological studies because they permit cumulative or weighted-average personal exposure estimates to be made even when individual, personal exposure concentration measurements had not been made. However, for this purpose it is questionable to use a geometric mean as the exposure estimate for a given working zone. Exposure concentrations in a given working zone usually vary over time, at least to some extent; therefore, an arithmetic mean concentration would be theoretically preferable to a geometric mean concentration because an arithmetic mean is directly proportional to the cumulative mass of contaminant to which a worker was exposed while in that zone.

Admittedly, if the true exposure level within a working zone were really constant over the exposure period of interest [as assumed by Ford et al. (11)] then either the arithmetic mean or the geometric mean would be a consistent estimator of the true exposure concentration for a working zone. (A consistent estimator is one for which, as the sample size becomes very large, the probability approaches 1.0 that an estimate of that type will be infinitesimally close to the true value of the parameter.) But for smaller sample sizes, the arithmetic mean would usually be preferable to the geometric mean as an estimate of the true exposure level. This is so because the arithmetic mean would usually be closer to being an unbiased estimate due to the fact that average net errors of sampling and analysis are usually more nearly normally distributed than they are lognormally distributed. A more detailed discussion of the rationale for choice of the normal distribution model for estimates of average exposure is given in Busch and Leidel (71).

Thus, for the case of day-to-day lognormal random variability (without multiday periods of equal exposures), the sample geometric mean of a random sample of daily exposures would need to be computed and converted to an estimate of the long-term arithmetic mean exposure using a function of the sample geometric standard deviation as a correction factor. This complex estimation procedure for the arithmetic mean of a lognormal distribution has a relative large variance and should be avoided if possible. An alternative procedure, for sufficiently large samples (e.g., at least 30 days), is to compute the sample arithmetic mean and compute its confidence limits under normal distribution assumptions. The justification for this is that the sample means of many random samples are approximately normally distributed even though the single samples are not normally distributed. A detailed discussion of this point is given as a technical note after the next paragraph.

With populations of daily 8-hr TWA exposure estimates, measures of central tendency can be misleading regarding occupational health risk. Analogously, one may report that a river has an average depth of 2 ft, thus inferring that it is safe to wade in, but people can drown in those parts of the river that are more than 5 ft deep. It is important to consider the upper tail (higher values) of any exposure distribution. Tolerance limits provide an indication of the potential upper levels of exposure distributions or an indication of the potential widths of exposure distributions. Often, tolerance limits are the relevant statistical values instead of estimates and confidence limits for central tendency parameters (arithmetic means and medians). Frequently, investigators compute the latter values merely because elementary statistical texts contain equations for estimates of, and confidence limits on, central tendency parameters, while these texts fail to discuss the concept and use of tolerance limits.

Technical Note: The reader should be cautioned that it is possible to erroneously apply the central limit theorem of mathematical statistics and conclude that daily TWA exposure estimates that are calculated from lognormally distributed, intraday exposure levels (e.g., from grab samples) should have an (interday) normal distribution. To the contrary, the theorem merely implies that the means of n identically distributed (e.g., lognormally) independent random variables will be approximately normally distributed regardless of the distribution of the individual variables. But note that the sample means (daily TWAs) of the individual variables (all possible instantaneous exposure values over some multiday period) are not the means of independent random samples obtained from the same lognormal distribution. Different lognormal distributions of intraday exposure levels exist for the various days, and another lognormal distribution exists for interday variability of the daily TWAs. Each "mean" (TWA) is then merely a single sample from the interday lognormal distribution of daily TWAs.

An appropriate application of the central limit theorem would be to a multiday exposure average computed from a lognormal population of daily TWA exposure estimates. Each daily TWA would constitute a single sample. If n randomly selected TWAs were drawn from the lognormal interday population of TWAs and a sample mean calculated, then the distribution of such multiday exposure averages (each estimated from n samples) would be approximately normally distributed. The approximation improves as the sample size n increases.

3.6.2.4 Populations of Daily 8-hr TWA Exposures for a Group of Workers Having Similar Expected Exposures (e.g., in the Same Exposure Environment, from the Same Job Type or Occupational Group). For evaluation of individual worker exposures and attendant health risk, neither the arithmetic mean nor the geometric mean of any such population of multiple exposures of a group of workers is an appropriate parameter, unless the distribution has negligible variation. The difficulty is that a particular worker's individual distribution of exposures may consistently lie in the high (or low) tail of a multiday, multiworker distribution because the true multiday exposure average of that worker may be substantially different from the central tendency of the multiworker exposure distribution. Geometric standard deviations would be the appropriate measure of dispersion for the separate interday and interworker components of the total variation of daily exposures.

Both types of exposure distributions (multiple work shifts for a given worker and multiple workers for a given work shift) are usually approximately lognormal. However, to be able to use a single lognormal distribution for exposures of different workers on different work shifts (or days), the following conditions must apply. The between days for a given worker and between workers on a given day random variations must be independent. Such independence can be assured by randomly selecting the worker–day combinations for which exposures are to be measured. A suitable procedure to do this would be to randomly select the days to be sampled and then randomly select a different worker for measurement on each selected day. This selection procedure is required in order that the resulting data will follow a single lognormal distribution. If the same group of workers were measured on each of the same several days, exposures would be intercorrelated and would not constitute a simple random sample from the same lognormal distribution. The method of analysis of variance (of a logarithmic transform of exposure results) would then have to be used to separate the total variation into components due to worker-to-worker (on the same day) and day-to-day (for the same worker) lognormal variations. This technique is so complex that its complete exposition is inappropriate in this chapter.

Busch and Leidel (71) present normal and lognormal distributional models for the "components of variance" that are associated with the sources of variability that affect various types of industrial hygiene sampling data. They also give formulas for mean and variance parameters of related distributional models for total (net) error. These formulas are expressed in terms of the distributional parameters for the component errors.

If cross-classified exposure data must be analyzed to determine tolerance limits, a professional statistician's assistance most likely will be needed. Additional discussion is given in Section 6.8 concerning an example computation of lognormal tolerance limits determined from exposures for randomly selected worker-day combinations.

3.7 Decision Values for the Unknown Mean of a Normal Distribution with Known Coefficient of Variation

Frequently, n independently collected consecutive samples $X_1, X_2, \cdots, X_n$ are obtained that collectively span the period of a worker's time-weighted average exposure (e.g., 40-hr, 8-hr, 15-min. TWA). Such samples are termed full-period consecutive samples and the average of the n measurements is used to estimate the TWA exposure. Assume that these measurements have net random errors that are normally and independently distributed with the same total coefficient of variation CV_T. Then the mean $\bar{X}$ of the n measurements will also be normally distributed. This normal distribution of $\bar{X}$ values has the following mean and variance:

$$\mu_{\bar{X}} = E(\bar{X}) = \left(\frac{1}{n}\right)[E(X_1) + E(X_2) + \cdots + E(X_n)]$$

$$= \left(\frac{1}{n}\right)(\mu_1 + \mu_2 + \cdots + \mu_n) \tag{14}$$

$$\sigma^2_{\bar{X}} = \left(\frac{1}{n^2}\right)(\mathrm{CV}^2_T)(\mu^2_1 + \mu^2_2 + \cdots + \mu^2_n) \tag{15}$$

If a worker's work shift exposure were uniform (i.e., all consecutively sampled periods having effectively equal true average concentrations) there would be $\mu_i = \mu$ for $i = 1, 2, \cdots, n$, and the TWA measurement mean exposure $\bar{X}$ could then be treated as a random sample from a normal distribution with mean μ and variance $\sigma^2/n = (1/n)(CV^2_T)(\mu^2)$. (*Note*! Substantially different exposure situations during a workshift generally result in a nonuniform 8-hr TWA exposure.). In 95% of such uniform mean measurements, the average $\bar{X}$ would be within an interval $\mu \pm [(1.96)(CV_T)(\mu)/n^{1/2}]$. Equivalently, two-sided intervals $\bar{X} \pm [(1.96)(CV_T)(\mu)/n^{1/2}]$ would contain μ for 95% of the $\bar{X}$ values. Note that the latter probability intervals are centered about the randomly varying TWA measurement means $\bar{X}$.

To create a decision-making test (i.e., statistical significance test of a null hypothesis of compliance of a reported TWA exposure mean for a worker with an exposure control limit or exposure standard, denoted ECL), assume that the worker's true TWA exposure level μ is equal to the value ECL. Under this null hypothesis, decision intervals surrounding the ECL can be computed that would contain the TWA exposure mean $\bar{X}$ in at least 95% of similar cases. Such a decision interval, for the case of uniform exposures, would be of the form

$$\mathrm{ECL} \pm \frac{(1.96)(\mathrm{CV}_T)(\mathrm{ECL})}{\sqrt{n}} \tag{16}$$

The lower bound will be termed the lower decision value (LDV) and the upper bound will be termed the upper decision value (UDV). Similar open decision intervals that are upper-bounded only (by UDV, i.e., one-sided decision intervals) can be computed that would contain the TWA measurement mean $\bar{X}$ for at least 95% of similar cases, given that $\mu \leq$

ECL. Such a one-sided decision interval, $\bar{X} \leq [\text{ECL} + (1.645)(CV_T)(\text{ECL})/n^{1/2}] = \text{UDV}$, is open on the lower side and its upper bound will be denoted as the UDV for $\bar{X}$. This is because the decision interval will be exceeded only by 5% or less of $\bar{X}$ values, when the null hypothesis, H_0: $\mu \leq$ ECL, is true. Hence the interval's upper bound will be denoted as $\text{UDV}_{5\%}$. Therefore, in case $\bar{X} > \text{UDV}_{5\%}$, the hypothesis H_0 is considered unlikely to be true, since this occurrence would be infrequent under H_0, and an alternative hypothesis, H_1: $\mu >$ ECL, is accepted because H_1 gives the observed $\bar{X}$ a more reasonable probability of having occurred by chance. Specific applied methods with examples will be presented in Section 6.1, Section 6.2 and Section 6.3 that apply the one-sided $\text{LDV}_{5\%}$ and similar $\text{UDV}_{5\%}$ concepts to decision making regarding compliance or noncompliance of a particular TWA exposure estimate with an exposure control level or standard.

Instead of decision values, if one desires to calculate a one-sided, 95% lower confidence limit ($\text{LCL}_{1,.95}$) for μ, one could rearrange the following probability statement:

$$P\left\{\bar{X} \leq \left[\mu + \frac{(1.645)(\mu)(\text{CV}_T)}{\sqrt{n}}\right]\right\} = 0.95$$

and obtain the $\text{LCL}_{1,.95}$ as

$$\mu \geq \frac{\bar{X}}{1 + (1.645)(CV_T)/n^{1/2}}$$

The analogous one-sided 95% upper confidence limit ($\text{UCL}_{1,.95}$) for μ is

$$\mu \leq \frac{\bar{X}}{1 - (1.645)(CV_T)/n^{1/2}}$$

To compute two-sided, 95% confidence limits ($\text{LCL}_{2,.95}$ and $\text{UCL}_{2,.95}$) for an interval estimate, use the above formulas with 1.96 substituted in place of 1.645. In summary, for normally distributed and independent measurement errors, these formulas give exact 95% confidence limits in cases where CV_T is known and there are uniform exposures in n equal-duration, consecutive, sampling periods. For the general case of unequal sampling durations and nonuniform exposure, approximate confidence limits can be obtained using methods given in Section 6.2.1 and Section 6.3.1. Specific cases of confidence limit applications are also discussed with examples in Section 6.1, Section 6.2, and Section 6.3.

4 PRINCIPLES OF STUDY DESIGN AND DATA ANALYSIS FOR EVALUATING OCCUPATIONAL EXPOSURES

One important goal of research is to make inferences about some population or draw other general conclusions, based on sample survey results or experimental study data. Often an investigator will seek the assistance of a statistician in analyzing nondefinitive results from a research study. Unfortunately, sometimes the results presented for analysis are not only fragmentary, but incoherent, so that next to nothing can be done with them except perhaps compute some trivial descriptive statistics. This does not have to happen. Research dollars do not have to be wasted on unproductive studies. Adherence to statistical principles of study design and related data analysis can produce substantially better results.

A study should be initiated, conducted, and the results evaluated only if the investigator has a clear purpose in mind and a clear idea about the precise way the results will be analyzed to yield the desired information. Far too often studies are conducted in the blithe and uncritical belief that a subsequent "statistical analysis" will yield something useful, especially when a statistician is engaged to "juggle the data."

The methodological tools of statistics cannot extract information or inferences that are not inherent to the data. The use of statistical technique to analyze study results requires asking certain questions concerning the hypotheses to be tested and the parameters to be estimated. The ability to test an hypothesis depends on the study design used and the circumstances in which the data were collected (77). Statistical distributional assumptions are required for application of most methods of statistical analysis and these assumptions depend partly on the circumstances and pattern of the experimentation or data collection. Besides using the right research tools, one must also use valid study designs and adequate sample sizes in order to have a good chance of detecting changes or effects of practical significance (i.e., those effects large enough to be of interest). The effects of interest are often small enough so that they might be obscured by random errors or hidden by other confounding effects, unless special attention were given to designing a study with sufficient statistical power.

To solve these problems, the investigator must know enough of statistical principles and techniques to be able to recognize when advice is needed from a statistician before the study is initiated. It is hoped that review and adherence to the principles of study design and data analysis discussed in this section will provide this basic level of awareness of statistical principles. Many of these principles are common sense, but unfortunately, common sense is frequently uncommon. Altman (78) believes that the general standard of statistics in medical journals is poor, and the situation is not any better for industrial hygiene journals. Of course, uniform guidelines for statistical design cannot be precisely applied in every study. Section 2 identified observational and experimental studies as the two major classes of research studies. Special considerations may be involved in subclasses of these study types such as exploratory, methodological, and pilot or preliminary studies, where the primary purpose is to test feasibility or to evaluate alternative approaches or techniques.

4.1 Study Design Principles and Implementation Guidelines

The following guidelines are based on those suggested by Crow et al. (79), Soule (80), and Green (81).

4.1.1 Establish the Study Objectives

Clearly establish the purpose and scope of the study. By whom, for what purpose, and how will the results be used? There should be a complete, clear, and concise statement of the objectives for the study. State how the anticipated results will specifically be used to meet the objectives. Are the results intended to be definitive or will this be a pilot test or feasibility study? Classify the study as descriptive (i.e., principal objective is to estimate only basic statistical parameters such as means) or analytic or inferential (i.e., designed to test a specific research hypothesis involving some process of inference in probability). Any statistical review should, in part, concern itself with how well the study design can meet the stated objectives and study hypotheses, if any. The study results can only be as relevant and productive as is permitted by the initial conception of the research problem.

State the study conditions and parameters used to represent the conditions to which the results will relate. Provide a clear and complete definition of the study target population to which inferences will be made or hypotheses investigated, based on results obtained from a sample. The target population is a subset of the general population that is both subject to the study exposure(s) and at risk of the development of the occupational disease or adverse health effect(s). Any given sample is a member of the sample population that consists of all samples that could have been selected from the target population. Also, include ranges of the determinant variables and define the reporting units (e.g., individuals, job types, establishments, industries).

Chapter 2 of the American Industrial Hygiene Association (AIHA) manual edited by Hawkins et al. (82) discusses the concept of a homogeneous exposure group (HEG) and outlines several methods for defining HEGs: (*1*) task-based approach, (*2*) job-description-based approach, (*3*) chemical-based approach, (*4*) process/job/agent/task-based approach, and (*5*) data-analysis-based approach. HEGs are recommended for use in determining exposure assessment priorities and as an element in the selection of related exposure monitoring sampling strategies.

4.1.2 Formulate the Design for a Preliminary Study

Examine the precision afforded by different study sizes, with consideration for the benefits of a statistically powerful study versus the disadvantages of a less powerful study. Weak or equivocal results carry appreciable risk of wrong decisions and resultant limited conclusions. Plan to take replicate samples within each "treatment" (i.e., within each combination of time, location, and any other determinate variable). Statistically significant differences between "treatments" can only be demonstrated by comparison of the measured differences to variability within treatments. Attempt to obtain an equal number of randomly selected replicate samples for each combination of determinant variables. Taking measurements of "representative" or "typical" experimental units (i.e., reporting units) does not constitute random sampling, although random sampling would usually tend to be both of these. The element of random sampling is essential to methods of statistical inference that are based in the mathematical theory of probability. To test whether a treatment (exposure) has an effect, collect samples both where the condition is present and where the treatment condition is absent, but all other determinant variables are the same. Usually, an effect can only be demonstrated by comparison with such a proper control. However, in some cases the only "controls" also differ from the exposed group with respect to another variable. If this extra-experimental variable has been measured for individual subjects, it may be possible to correct for the extra-exposure part of the total exposure group difference by means of such statistical techniques as covariance analysis or comparison within blocks.

4.1.3 Review the Design with All Collaborators

Discuss the design with collaborators, reach an understanding, and keep notes about what decisions hinge on each outcome. Collaborators should anticipate and discuss all determinant variables that might affect the results. Review the study design in sufficient detail to discover any procedures that might lead to bias in the results. Obtain a pertinent peer review of the study design and use the comments as if the reviewers were collaborators in the study. Review and discuss the robustness of the chosen statistical methods (i.e., the tendency for inferences based on the methods to remain valid despite a violation of one or more assumptions underlying the theoretical development of the method). Provide for examining the study results to detect errors caused by serious violations of the methodology assumption. Discuss the objectives, goals, and study design with representatives of management and labor, as appropriate to their respective interests. Review how the results should be reported and to whom.

4.1.4 Conduct the Preliminary Study

Those who skip a preliminary study due to "not enough time" usually end up wasting time by attempting to analyze a study with trivial or equivocal results. This is the opportunity to test the feasibility of the study design before substantial resources are committed to the research. Obtain sufficient data to provide adequate estimates of the variance components that will be encountered; these can be used as a basis to develop an efficient design for the more definitive main study to follow. Obtain adequate information to evaluate the adequacy of the measurement equipment, personnel training and readiness, facilities that will be needed for statistically valid data analysis, and to determine appropriate ranges for the study variables. Verify that the chosen exposure measurement method is adequate and appropriate for the entire range of study conditions and determinant variables anticipated. Experience from past similar studies can sometimes be used as a substitute for a preliminary study or pretest. However, a careful critique of the relevancy of past studies should be made before using the data to design the new study.

4.1.5 Complete the Study Design

Use the variability estimates from the preliminary study to estimate the power of the study to detect the size of effect to which the study must be sensitive. If possible, specify the basis for the sample size calculation in a statement such as:

> To detect a true difference of _____ (units or percent) between group A and group B, with _____ percent statistical power and a _____ percent probability of making a type-I error, _____ trials are needed.

Present the design in clear terms to assure that its provisions can be followed without confusion. Include the intended data analysis methods as part of the design, including validity checks on the governing statistical assumptions. Review the principles of data analysis given in Section 4.2. Consider the necessity for a data transformation (see Section 4.2.1).

4.1.6 Conduct the Study

During the study, maintain communication among all collaborators, so that problems and intermediate results may be evaluated and dealt with, in keeping with the study objectives and the related study design previously agreed on. If measurements are to be taken in the plant, advise representatives of management and labor in advance and during your sampling.

4.1.7 Conduct the Data Analysis

Using the principles presented in Section 4.2, follow the data analysis methods that were previously selected for the study. Stay with the results from these methods because they were previously chosen as the most appropriate statistical methods to test the desired hypotheses. In a well-designed study, with proper attention to randomization and balance, unexpected or undesirable results are not valid reasons for subsequent rejection of the statistical methodology and then hunting for a "better" one. Remember that overinterpretation is an attempt to compensate for underplanning.

4.1.8 Prepare a Report

Report results of the study in relation to its original objectives and goals. Discuss both statistical and practical significance of the results. Present related conclusions based on inferences indicated and supported by the results for test data. State necessary limitations on the inferences in your discussion and conclusions. Present summary data, statistics, and results in clear graphs and tables. In general, graphs are superior to tables for portraying trends, correlations, scatter, outliers, and so forth. If the results suggest a need for further studies, outline the course that such studies should take.

Just as exposure measurements should be accurate and precise estimates of worker exposure, so should reports be accurate and precise communications of your study objectives, results, and conclusions. Study and use of guides such as those by Bates (83), Crews (84), and the CBE Style Manual Committee (85) can contribute greatly to the quality of written reports.

4.1.9 Implement Appropriate Follow-up

Discuss study results with appropriate representatives of management and labor so that corrective action to reduce health hazards can be implemented, if necessary, along with additional exposure monitoring, biological monitoring, and medical surveillance programs as required. Are such evaluations needed for air pollution, water pollution, hazardous waste disposal, or safety?

4.2 Principles of Data Analysis

4.2.1 Choose a Distributional Model

Choose a distributional model (see Section 3.6) for the target population that the sample data represent. Note that a data transformation may be necessary (i.e., converting the data into such form that they follow a common distribution with known properties and readily available analytical methodology). Data transformations may be needed for other reasons than just giving the transformed results a convenient distributional form. Murphy (86) has discussed six objectives of transformation:

1. Normalization.
2. Stabilization of the variance.
3. To make the effects linear and additive.
4. To make the mean a good measurement of "the typical value".
5. To linearize a relationship between two or more variables.
6. To remodel the distribution into a more familiar one.

4.2.2 Review Statistical Assumptions

Review assumptions that underlie the statistical methods to be used. If a specific distributional model can be assumed, proceed to Section 4.2.6. The parametric methods outlined therein have the benefits of (*1*) lower sample sizes (hence lower costs) for the same statistical power or (*2*) moderate gains in power for the same sample size, when compared to the more robust nonparametric methods. (The "distribution free" methods do not depend on a specific probability model with one or more parameters for the distributional form of the parent population.) However, if the sample data are not adequately fitted by the assumed parametric distribution (see Section 3.6) or if other assumptions (e.g., independence) of the inferential methods are suspect, then one has unknown risks of incorrect hypothesis testing decisions, inaccurate confidence intervals, or inaccurate tolerance limits. The sizes of such inaccuracies depend on the robustness of the methods used. If sample size is insufficient (less than about 10) to examine the pattern of the data (see next section) and test the distributional model, proceed to Section 4.2.6. Then one may be in the position of having to use parametric methods without even weak verification of assumptions. A larger sample size (e.g., 30 or greater) is needed to empirically estimate distributional fractiles from ranked sample data without assuming any mathematical model for the distributional form (see Section 4.2.5).

4.2.3 Qualitatively Examine the Data

Plot the grouped sample data as a histogram or individually on appropriate probability paper (e.g., normal, lognormal, Weibull) to qualitatively examine data regarding the distributional assumptions and to investigate unusual patterns in the data (see Section 6.12). Data that are lognormally distributed in time can be plotted on semilogarithmic graph paper so that the data plot is symmetrical about the distribution's geometric mean. This is a qualitative way of looking for trends in the data over time (see Section 6.11).

Gross errors due to mistakes in sampling or chemical analytical procedures can contaminate a data sample. Barnett and Lewis (87) have surveyed most of the extensive statistical theory that has been developed concerning detection of data "outliers" and concerning methods for performing analyses of data sets that may be contaminated with outliers. In particular, they give "robust" methods for estimating the mean of an underlying distribution when a sample supposedly taken entirely from that distribution contains one or more outliers. Methods are given for several types of data models (e.g., normal and lognormal distributions), for various numbers and types of outliers (e.g., outliers in left tail, in right tail, or in both tails), for known and unknown variances, and for different objectives of the data analysis.

4.2.4 Estimate Sample Distributional Parameters

Use the sample data to calculate estimates of the parameters of the assumed distributional model [e.g., arithmetic mean and standard deviation for a normal distribution (see Section 3.4.1), geometric mean and geometric standard deviation for a two-parameter lognormal distribution (see Section 3.5.1), and if necessary the third parameter (for origin translocation) for a three-parameter lognormal distribution (see Section 3.5.2)].

When the data sample contains data values that are too low (or too high) to be quantitatively measured, tables of factors given in Cohen (88) can be used to compute an optimal maximum-likelihood estimate (MLE) of the mean of a normal (or lognormal) distribution. The Cohen tables include factors for both censored samples and truncated samples. Formulas for the MLE are given in Hald (62), based on theory for sampling from a censored normal distribution.

Hornung and Reed (89) advocate use of an easier-to-compute estimator of the geometric mean when there are nondetectable concentration values present in a sample taken from an assumed lognormal distribution. They merely replace values below the limit of detection L by $L/\sqrt{2}$ and then calculate the estimated geometric mean. In a practical sense, they say that this estimate of central tendency is sufficiently accurate for the usual sample sizes used in industrial hygiene sampling. The maximum-likelihood estimate is admittedly more precise, but Hornung and Reed (89) judge the Hald (62) technique to be not usually worth the effort of its complex calculation.

4.2.5 Verify Distributional Model

To test a distributional model qualitatively, use the sample estimates of parameters from the previous step to plot the estimated target population distribution on the appropriate probability paper and compare to the actual sample data distribution. Mage (90) has suggested an objective method for testing normal distributional assumptions using probability paper.

With larger sample sizes (at least 30), a histogram of the sample data can be compared to the shape of the fitted distributional function. To quantitatively verify the distributional model quantitatively, the classic chi-square test can be applied to the histogram's interval frequencies. This goodness-of-fit test usually is the first to come to mind, but it generally requires substantially larger sample sizes than occur in industrial hygiene. Lilliefors (91) has adapted the Kolmogorov-Smirnov test when the mean and variance of the distributional model are unknown (as is typically the case with occupational exposure data). This modified goodness-of-fit test can be used with small sample sizes (e.g., 10 to 20), for which the accuracy of the chi-square test would be questionable. Also, it is claimed to be a more powerful test than the chi-square test for any sample size. Iman (92) has provided graphs for use with the Lilliefors test.

Waters et al. (93) stress that although a probability plot may be quite useful to interpret the nature of a confirmed nonlognormal distributional condition, a preliminary formal test of significance is needed to confirm the fact that any seeming nonlinearity viewed in a probability plot is not in fact merely a random sampling variation. They propose use of an approximate test based on use of their "ratio metric" goodness-of-fit test statistic, which is the ratio of the arithmetic sample mean divided by the maximum-likelihood estimate of the arithmetic mean. The ratio metric test is presented as a simpler to compute, but nearly as accurate, alternative to the established and powerful Shapiro-Wilk W test (94). Our view is that the savings in computational effort for the Waters et al. (93) test for nonlognormality seems not important enough to warrant use of an approximate test instead of the powerful and more rigorous Shapiro-Wilk W test.

If the fit is judged inadequate, return to Section 4.2.1 and choose another data transform or distributional model.

4.2.6 Apply Chosen Statistical Methodology to the Sample Data Analysis

If the fit of the distributional model is judged adequate (or assumed to be adequate in which case one has been able to delete the steps following Section 4.2.2) or if the statistical methodology is sufficiently robust for its intended application, proceed with the data analysis. Calculate appropriate confidence intervals and tolerance limits, perform hypothesis tests, and so on based on the assumed distributional model. If the data or data transforms are not adequately fitted by available continuous-variable, parametric distribution models, certain nonparametric (distribution-free) methods are available. However, nonparametric methods generally require sample sizes larger than occur with sample sets used for exposure estimation; therefore, for these types of data, the distributional forms have been investigated using prior data collected for that purpose, and these forms can usually be assumed to apply to similar samples taken thereafter.

4.2.7 Report Results and Interpretation

If a report or journal article is written, it should include statements covering (*1*) the statistical rationale used to develop the study design and (*2*) the related protocol used for the statistically valid data analysis. These statements should contain sufficient detail so that a reader with statistical expertise would be able to duplicate the results if supplied with the investigator's raw data. If a personal computer or mainframe statistical program was used, a precise reference to it would usually preclude the need for presenting additional computational details. If the statistical analysis had been performed by a consultant or collaborator, their assistance would usually be needed in the writing, or at least in the editing, of the final report.

5 STUDY DESIGNS FOR ESTIMATING OCCUPATIONAL EXPOSURES AND EXPOSURE DISTRIBUTIONS

Study design considerations, including sample size estimation techniques, for obtaining occupational exposure estimates and their distributions will be discussed from two perspectives in this section. The first perspective, presented in Section 5.2, discusses exposure measurement strategies for making individual exposure estimates (see Section 3.1). The discussion examines different approaches to measuring exposures for individual workers on a single day.

The second perspective covers monitoring strategies for obtaining exposure estimates for worker target populations. Monitoring strategies are built on appropriate measurement strategies. Exposure monitoring will be defined as a series of steps necessary for estimating multiple exposures of a target population. This material is presented in Section 5.3. Fundamental to exposure assessments and exposure monitoring programs is the concept of a worker target population. A target population is usually defined as a function of:

1. Numbers of workers exposed under the conditions that are being studied (e.g., process factors and determinant variables).
2. Exposure measurement averaging times.
3. Time period to be represented by the target population exposure distribution.
4. Ranges of determinant variables affecting the exposure levels to which the target population is exposed.

The number of workers can be as few as one worker, measurement averaging times typically are either 15 min. or 8 hr, and the time period can be as short as a few hours on a given day. Determinant variables will be discussed in Section 5.1. Typical examples of target populations for several combinations of temporal periods and numbers of workers include the following:

1. Daily 8-hr exposure estimates for a single worker over several days to years.
2. Eight-hour exposure estimates for several workers on a single day.
3. Daily 8-hr exposure estimates for many workers over several days to years.

The elementary case of a single worker on a single day is governed by exposure measurement considerations, which are discussed in Section 5.2. For this situation there is but one exposure, but the parent population is considered to be all measurements (with their errors) that could have been made of that one exposure on the same day.

The other two types of target populations are considered under Section 5.3, Exposure Monitoring Strategies. In these cases the parent populations consist of multiple daily exposures and/or multiple workers exposed, as well as the exposure measurement populations for these workers at these times.

The relationship between target populations and determinant variables is graphically shown in Figure 52.7. This figure is a flowchart for exposure assessments, which are less comprehensive than exposure monitoring programs. Exposure assessments generally include the following elements:

1. Defining an appropriate target population of workers.
2. Selecting a sample population from the target population for exposure measuring.
3. Selecting one or more exposure measurement strategies.

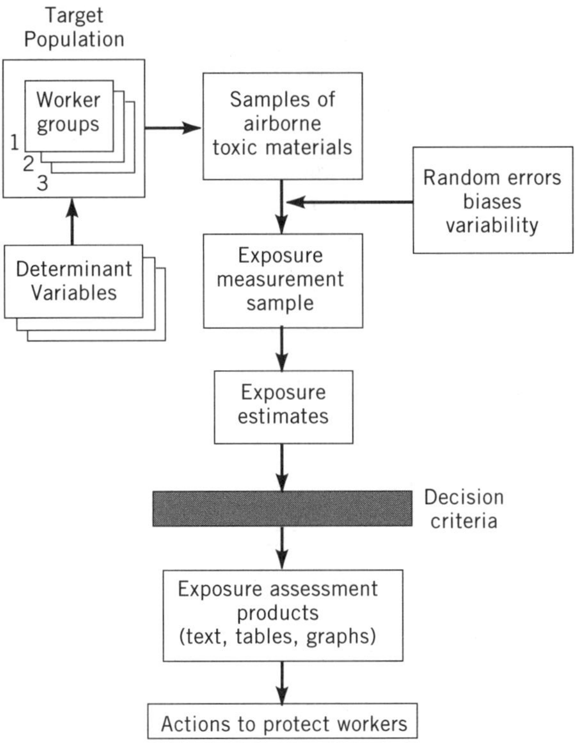

Figure 52.7. Exposure assessment flow chart.

4. Obtaining exposure measurement samples for the sample population.
5. Calculating exposure point estimates and interval estimates.
6. Possibly estimating an exposure distribution for the target population.
7. Comparing the exposure estimates (and possibly the exposure distribution) to decision criteria for acceptability of exposure levels.
8. Reporting the exposure assessment products for the target population (e.g., text, tables, graphs).
9. Taking any actions necessary to protect workers in the target population.

5.1 Determinant Variables Affecting Occupational Exposure Levels

Both exposure measurement strategies and exposure monitoring strategies must consider the determinant variables that affect the true exposure levels that are to be estimated with worker exposure measurements. Determinant variables are the qualitative or quantitative factors that determine, or at least are associated with and affect, the actual worker exposure levels. They generally can be classified as process, environmental, temporally associated, behavioral, and incidental. Note that a failure to identify and consider significant determinant variables generally has more deleterious consequences for a study than identifying and investigating too many determinant variables, including some that have little effect on exposure levels. One cannot make reliable inferences from exposure results beyond the number and range of determinant variables represented by the sampling distribution.

The following groups of typical determinant variables are presented as examples only and are not inclusive. The significant determinant variables for any given group of workers and time period must be identified, from experience with similar situations if possible, or using a research study if necessary.

5.1.1 Process Factors for Chemical Agents

These are factors related to primary contaminant levels or to the control of emissions and/or exposure levels.

1. Process type and operation (see Refs. 95–97).
2. Chemical composition of material used in operation.
3. Physical state and properties of material used (e.g., vapor pressure, size distributions of particulates and aerosols).
4. Rate of operation (e.g., mass or volumetric rate, revolutions per minute, linear rate, items in a given time period).
5. Energy conditions of operation (e.g., temperature, pressure).
6. Degree of process automation.
7. Emissions from adjacent operations.
8. Airflow patterns around workers (e.g., from exhaust ventilation, from adjacent operations).
9. Heating and ventilation airflows
10. Exposure control methods (e.g., local exhaust ventilation, respirators)

The scope of this chapter does not include mathematical modeling of exposure levels as a function of process factors (e.g., using multiple regression analysis). The assumptions underlying the usual regression analysis methods include an assumption that the independent variables are measured without error, but this assumption would not be valid for most quantitative process factors. The textbook by Fuller (98) brings together previously published rigorous statistical theory that is needed for the correct application of regression analysis to industrial hygiene problems in which both the dependent and independent variables are measured with error.

5.1.2 Environmental Determinant Variables

These are environmental factors, variations of which can modify exposure levels.

1. Meteorological conditions.
2. Age, size, and physical layout of plant.

3. Job category (e.g., responsibilities, work operations, work areas, time spent at each) (see Ref. 99).

5.1.3 Temporally Associated Determinant Variables

Time is an independent variable correlated with worker exposure levels. Time cannot be a direct cause of changes in workplace exposure levels but may be useful to predict exposure levels that follow time cycles, have systematic time trends, or have autocorrelation between present and past levels. Time series models are useful when the causative determinant variables are either unknown or unmeasured.

1. Contaminant buildup in the workplace air from morning to afternoon.
2. Exponential clearance due to air flushing and dilution during nonworking hours.
3. Cyclical or trending process operations with respect to work shift, season of year, and year or decade.

5.1.4 Behavioral Determinant Variables

Behavioral factors affect work habits that in turn affect exposure levels. Effects of these factors cannot be modeled mathematically, but the factors can be helpful in explaining (or at least rationalizing) observed exposure differences and effects.

1. Worker job practices, movements, and habits.
2. Worker training.
3. Worker attitudes.
4. Management and supervisory attitudes.
5. Presence of exposure measurement equipment, industrial hygiene personnel, or supervisory personnel.

5.1.5 Incidental Determinant Variables

Irregular changes in exposure levels can occur due to episodes, incidents, accidents, and otherwise unintended happenings such as described by Crocker (100):

1. Spills due to falls, punctures, tears, corrosion, etc.
2. Equipment maintenance or lack thereof.
3. Failure of process equipment prone to corrosion and leakage (e.g., pump packings, tank vents).
4. Interruption of utilities to process equipment or exposure control systems.
5. Interaction due to accidental mixing or simultaneous release of two vessels' contents.
6. Interruption or increase in flow of one or more process streams.
7. Vessel failure.
8. Accidental overpressurization, overheating, or overcooling of process equipment.

9. Sudden plant flooding, violent storms, or earthquakes.
10. Operator errors or instrument failure.

5.2 Exposure Measurement Strategies

Exposure measurement strategies deal with the considerations necessary to measure individual worker exposures on a given day to obtain short- or long-period TWA exposure estimates. Monitoring strategies for measuring multiple exposures (e.g., multiple workers on a single day or a single worker on multiple days) are presented in Section 5.3.

5.2.1 Practical Considerations

The adequate and preferable strategies for a given measurement situation are governed by both practical and statistical considerations. An adequate measurement strategy is one that is good enough for the purpose in mind. Thus one should clearly identify the objective of the intended exposure estimation (see Section 2.4) and the required precision of the estimates. A measurement strategy can then be selected that meets this precision requirement.

Most of the following discussion will concern measurements obtained for estimating individual worker exposures to chemical agents; most of the concepts presented are applicable to estimating individual workers' exposures to physical agents as well. Generally these exposure estimates are then compared to exposure control limits to determine the relative hazard for individual workers or the acceptability of the workplace exposure levels (e.g., in relation to OSHA PELs or ACGIH TLVs). It should be noted that there are other specialized purposes for airborne contaminant measurements. These include:

Source sampling at potentially hazardous operations. Hubiak et al. (101) have discussed the utility of short-term source sampling for identifying work operations that need additional engineering effort to reduce contaminant emissions (and related worker exposures). Sometimes these can be considered worst-case exposure levels, which are used in screening strategies for hazard evaluations.

Evaluating work practices to determine exposure variability and detect hazardous levels due to inappropriate work practices. This technique can also assist in differentiating between exposure levels due to work practices and those due to inadequate engineering controls. With direct-reading measurements it is possible to make on-the-spot recommendations for improving work practices and to recognize the appropriate direction for engineering control research.

Worker training to improve the effectiveness of work practices and engineering controls. Selected workers can perform the work operation while others observe the work practices and note the resulting exposure levels. The goal is to have all workers approach the results of the best or "cleanest" worker. Showing workers their exposure profile recorded on a strip chart will help them to better understand what a change in work practices or use of engineering controls can achieve.

Evaluating individual job tasks for their relative contribution to a worker's 8-hr TWA exposure. This may lead to developing appropriate administrative controls, improved work practices, engineering controls, or at least to identifying the need for personal respiratory controls.

Continuous monitoring to detect extraordinary exposure levels, so that a warning can be provided before a serious hazard develops. Generally, fixed sampling systems are used with a central analyzer. These systems may sample from multiple points in the workplace or in the return air of recirculation systems. Holcomb and Scholz (102) have reported an evaluation of typical continuous monitoring equipment.

Screening measurements for qualitative detection of airborne hazards during emergencies and uncontrolled releases. A screening strategy may also be used for quantitative determination of airborne hazards for hazard evaluation and spot checking of exposure levels. Typically, detector tubes and direct-reading meters are used for molecular size contaminants and aerosol monitors are used for particulates. Schneider (103) and Leichnitz (101) have discussed the use of colorimetric detector tubes for screening. King et al. (105) have described a simultaneous direct-reading indicator tube system for rapid qualitative measurements.

Some nonstatistical considerations affecting the selection of measurement strategies follow:

1. Amount of information available regarding the nature and concentration of airborne contaminants to be measured and possible interfering chemicals.
2. Availability and cost of sampling equipment (e.g., pumps, filters, detector tubes, direct-reading meters, passive dosimeters).
3. Availability and cost of sample analytical facilities (e.g., for filters, charcoal tubes, dosimeters).
4. Availability and cost of personnel to take the measurements.
5. Location of work operations and workers to be sampled.
6. The need for obtaining results immediately, within a day or two, or after several weeks.

5.2.2 Statistical Considerations for Long-Term Exposure Estimates

Generally the statistical design of exposure measurement strategies is concerned with reducing the limit of random error in an exposure estimate calculated from several exposure measurements, which also has the effect of increasing the power of a related hypothesis test for compliance or noncompliance with an exposure control limit. The imprecision (uncertainty) of a long-term (e.g., 8 hr) exposure estimate is governed by three classes of factors:

1. Random variation in the measurement procedure (i.e., the precision of the air sampling/chemical analysis method, see Section 3.3.2, Section 3.4.2, and Section 3.6.1) or the random variation in the true exposure levels during the estimation period (see Section 3.3.1 and Section 3.6.2).
2. Sample size (i.e., number of measurements obtained during the time-averaging period for the TWA estimate) and the duration of each measurement.
3. Sampling period selection (i.e., whether random or systematic sampling is to be used to select subintervals for sampling during the TWA period of interest, or a cumulative sample is to be taken over the entire period).

It should be realized that there is no "best" measurement strategy for all situations. However, some strategies are more desirable than others. The following discussion points out the statistical considerations for the four different types of TWA exposure estimates (see Section 3.1). Remember that the word period refers to the duration of the desired time-averaging period (e.g., 15 min. for a 15-min. TWA estimate, 8 hr for an 8-hr TWA estimate).

The full-period consecutive-samples estimate is the "best" strategy in that it yields an estimate with the least uncertainty (narrowest confidence interval). This is because the uncertainty of this type of exposure estimate is a function only of the measurement method error (see Section 3.2) and is independent of the substantial variability in the actual exposure levels measured (see Section 3.6.1). There are only moderate statistical benefits to be gained (106) from increased sample sizes (e.g., eight 1-hr samples versus four 2-hr samples), but with the substantially increased analytical costs (per 8-hr TWA exposure estimate) the practical benefits are negligible. Figure 52.8 illustrates the effects of increased sample size. Generally two consecutive samples for the full time-averaging period provide sufficient precision for most exposure estimation purposes. For a method with very large CV_T, more samples may be warranted.

The full-period single-sample estimate (e.g., one 8-hr sample) is the "second-best" strategy, if an appropriate measurement method is available. An exposure estimate calculated from one 8-hr measurement is almost as precise as an estimate computed from two 4-hr measurements, since both strategies employ full-period sampling. The disadvantage of a single measurement is that a bias or mistake in the measurement is difficult to detect. Also, substantial differences in exposure levels during the measurement period are not revealed, but this feature is also an advantage in that temporal variability is "integrated out" by the cumulative, physical sampling procedure itself.

The partial-period consecutive-samples estimate is substantially less desirable than the preceding two estimates. The major problem created by this strategy is that exposure levels are unknown during the unsampled portion of the TWA period. Strictly speaking, the measurement samples are representative only of the periods that are actually sampled. For example, if one desires to estimate an 8-hr TWA from a sample or samples spanning only 5 hr, then a problem is created of assuring that the 5-hr period results represent the entire 8-hr period. Reliable knowledge or professional judgment may sometimes be used to extrapolate the 5-hr TWA to an 8-hr estimate. This should be done only after considering the effect on the real 8-hr TWA of any substantially higher or lower actual exposure levels during the unsampled period. Figure 52.9 illustrates the effect that a liberal assumption of zero exposure (i.e., less worker protective), for the unsampled portion of the TWA period has on the value required to demonstrate noncompliance at a statistical test power of 50%.

The grab-samples estimate is the least preferable strategy for estimating an 8-hr TWA exposure. This exposure estimate has substantially larger uncertainty than the first two types of estimates. This is because the uncertainty of a grab-samples estimate is dominated by the considerable variability of the exposure levels measured, which is substantially larger than the measurement method variability. This is illustrated in Figure 52.6 which compares the relative contributions to variability in a grab-samples estimate. Regarding sample size, Figure 52.10 shows that a reasonable number of grab samples for an exposure estimate is between 8 and 11, taken at random intervals during the TWA period. However,

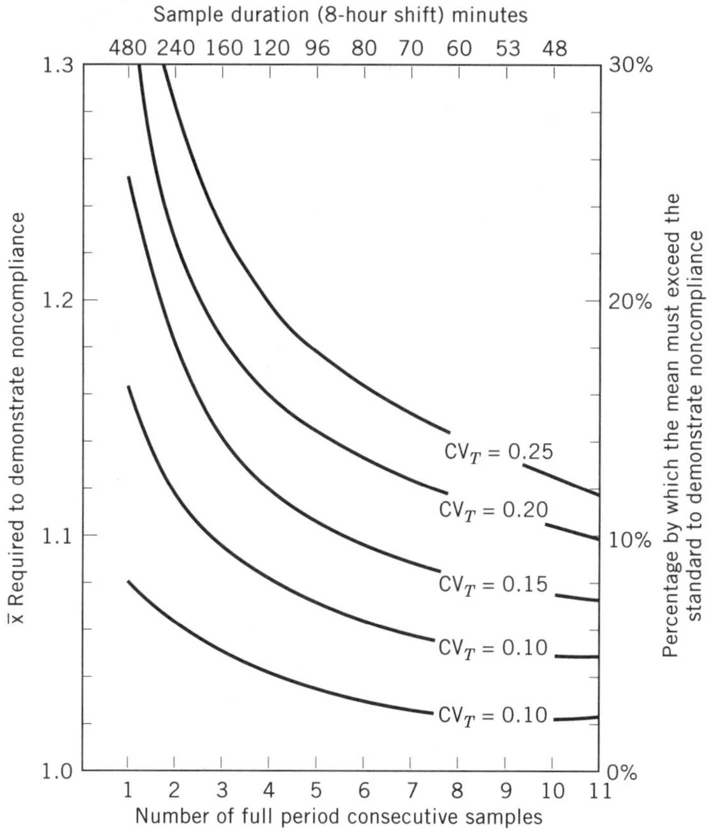

Figure 52.8. Effect of sample size for a full-period, consecutive-samples estimate on noncompliance demonstration when test power is 50 percent and test confidence level is 95%; CV_T is coefficient of variation of exposure measurement method.

this applies only if the worker's exposure levels are adequately uniform during the TWA period. If the worker is at several work locations or operations during the TWA period, then at least 8 to 11 grab-sample measurements should be obtained during each period of anticipated uniform exposure that substantially contributes to the TWA exposure. If one has to take fewer than 8 to 11 measurements during each uniform exposure period, then allocate the total number of measurements in proportion to the duration of each period. That is, take more measurements during the longer periods of anticipated uniform exposure.

If grab samples are taken, the duration of each measurement need be only long enough to collect sufficient mass of contaminant to reach the minimum level of detection for the analytical method. That is, any increase in sample duration beyond the minimum time to collect a sufficient mass of contaminant is unnecessary and unproductive.

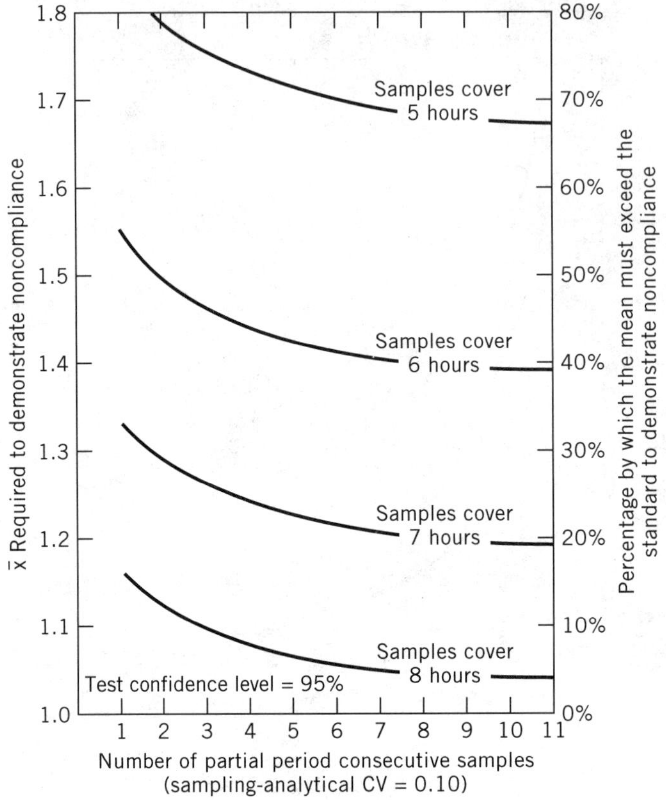

Figure 52.9. Effect of (1) total time covered by all samples and (2) sample size for a partial-period, consecutive-samples estimate on noncompliance demonstration when test power is 50% and test confidence level is 95%.

For grab samples it is desirable to choose the sampling periods in a statistically random fashion. The accuracy of the probability levels for the statistical methodologies presented in Section 6 for testing hypotheses of compliance or noncompliance depends on implied assumptions regarding the lognormality and independence of the grab-sample results that are averaged. These assumptions are not unduly restrictive if precautions are taken to avoid bias when selecting the sampling times during the period of the exposure estimate.

For grab sampling, a TWA estimate represents a period longer than the total interval measured, but an unbiased estimate of the true average is obtained when samples are taken at random intervals. It is valid to sample instead at equal intervals if the series is known to be stationary with contaminant levels varying randomly about a constant mean, and if exposure fluctuations are of short duration compared to the length of the sampled interval. However, if the average exposure level and its confidence interval were calculated from samples taken at equally spaced intervals, biased results could occur if there were exposure

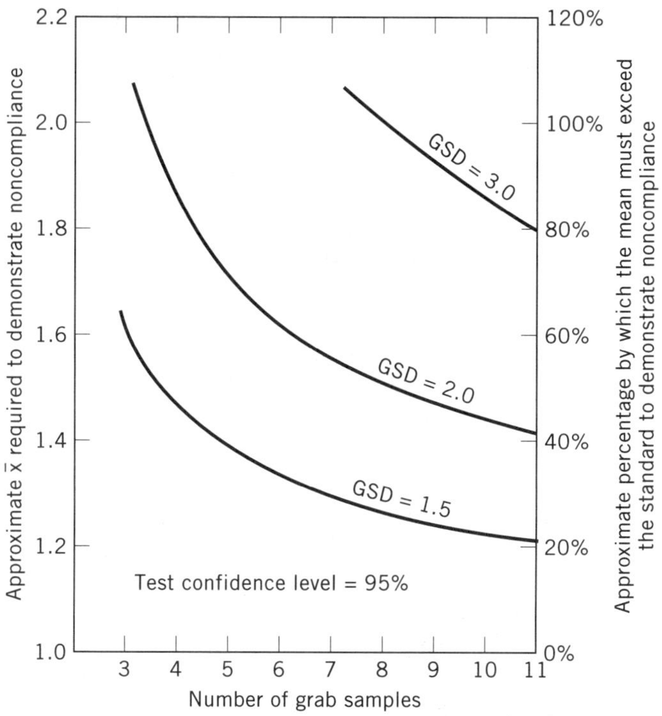

Figure 52.10. Effect of sample size for a grab-samples estimate on noncompliance demonstration when test power is 50% and test confidence level is 95%; GSD is the geometric standard deviation for intraday variation in the airborne environmental concentrations.

cycles in the operation that were in phase with the intervals sampled. The important benefits of random sampling are that subsequent results are unbiased even if cycles and trends occurred during the period of the exposure estimate.

The word random refers to the method used for selecting the sample. A random sample is one chosen in a manner such that each possible sample has a fixed and determinate probability of selection. A practical way of defining a random sample for an exposure measurement is one obtained such that any portion of the TWA exposure estimate period has the same chance of being sampled as any other. Ordinary haphazard or seemingly purposeless choices generally do not guarantee true randomness. Devices such as random number tables, or random numbers generated by inexpensive, hand-held calculators or personal computer spreadsheet software, can be used to prevent subjective biases that tend to be inherent in arbitrary personal choices. Reference 107 details a formal statistical method for choosing random sampling periods. Rather than using the random number table discussed in the method, it would be more efficient to generate the random numbers for the procedure with a hand-held calculator or personal computer software.

5.2.3 Statistical Considerations for Short-Term Exposure Estimates

Short-term exposure estimates (e.g., 15 min. or less) generally are obtained only for determination of peak exposures during short periods. These estimates may be used for comparison with ceiling exposure control limits designed to prevent acute health effects. Short-term samples taken to measure short-term exposures are statistically analyzed in a manner similar to short-term samples taken to measure long-term exposures. However, two important differences should be noted.

The first difference is that the measurements taken for estimation of peak exposures are best taken in a nonrandom fashion. That is, all available knowledge relating to the exposure level determinant variables such as work area, worker, work practices, and type of operation should be utilized to obtain samples during periods of maximum expected exposure.

The second difference is that measurements obtained for short-term estimates are generally taken to represent a much shorter period than short-term samples taken for estimating 8-hr TWA exposures. Each short-term measurement usually consists of a single instrument reading (if a direct-reading device is available) or a 5- to 15-min. sample if a minimum mass of material needs to be collected (e.g., on a charcoal tube or filter). A series of samples spanning 15 min. could also be taken and the measurements averaged.

Leidel et al. (2) recommend that a minimum of three short-term exposure estimates be obtained on any given work shift for a worker and the highest of all estimates be used as an estimate of the worker's peak exposure for that work shift. This recommended minimum number of estimates is not based on statistical considerations, but on practicality. Taking at least three samples increases the probability of detecting an exposure close to the highest exposure. Three samples also facilitate the detection of gross mistakes or biased measurements. However, usually only the highest value (not the average of the three or more) would be compared to a ceiling exposure limit. If measurements are obtained for evaluation with a short-term TWA limit, such as a threshold limit values–short-term exposure limit (TLV–STEL) of the American Conference of Governmental Industrial Hygienists [e.g., ACGIH (40)], the total sampling time should equal the time-averaging period for the limit, which is typically 15 min. Thus, for some colorimetric detector tubes, it might be necessary to take several consecutive samples and average the results.

Although short-term measurements taken for estimation of peak exposures are usually best taken in a nonrandom (biased) fashion, random sampling may be useful in some work situations where the exposure levels appear uniform during a work shift. Professional judgment may be unable to identify particular periods with a risk of higher-than-usual exposure. For this case, a statistical procedure is given below that can be used as a peak exposure detection strategy. The sample size recommendations given are based on combinatorial probability formulas detailed in Ref. 108.

PURPOSE. Provide a sample size to assure (i.e., have a probability of at least 90 or at least 95%) that at least one randomly sampled period will be from the higher exposures present during the work shift or other total duration examined (i.e., from the highest 10% or highest 20% of the work shift exposure distribution).

ASSUMPTIONS. No limiting assumptions are required. The derivation of this method is based on the hypergeometric sampling distribution (i.e., sampling from a finite population without replacement). No mathematical model is assumed for the distribution of exposure levels present in the finite-size population of possible measured periods; therefore, these sample sizes may be larger than would be needed if a parametric distributional form of the exposures were confidently known. However, this nonparametric procedure is useful when the form of the exposure distribution is irregular or unknown. Tables 52.4, 52.5, and 52.6 give the required sample sizes for 32, 48, 96 possible sampling periods in a total time period whose highest short-term exposure levels are to be estimated.

EXAMPLE. For a target population of 32 consecutive 15-min. periods in an 8-hr work shift for a worker, determine the necessary sample size such that there is at least 90% confidence that at least one sampled period will be from those periods with the highest 20 percent of exposures occurring during the work shift.

SOLUTION. There are 32 discrete, nonoverlapping, 15-min. periods in an 8-hr work shift. Table 52.4 indicates that a random sample of 10 of the 32 15-min periods will have

Table 52.4. Required Sample Size for Detecting at Least One of the Higher Exposures Among 32 Periods in the Total Duration (15-min Sample Periods in an 8-hr Work Shift)

To Obtain at Least One Sample Period from the	At a Confidence Level of	Random Sample from At Least
Top 20%	0.90	10 periods
Top 20%	0.95	12 periods
Top 10%	0.90	17 periods
Top 10%	0.95	20 periods

Table 52.5. Required Sample Size for Detecting at Least One of the Higher Exposures Among 48 Periods in the Total Duration (10-min Sample Periods in an 8-hr Work Shift)

To Obtain at Least One Sample Period from the	At a Confidence Level of	Random Sample from at Least
Top 20%	0.90	10 periods
Top 20%	0.95	13 periods
Top 10%	0.90	21 periods
Top 10%	0.95	25 periods

Table 52.6. Required Sample Size for Detecting at Least One of the Higher Exposures Among 96 or More Sample Periods Less Than 5 min Each from an 8-hr Work Shift

To Obtain at Least One Sample Period from the	At a Confidence Level of	Random Sample From at Least
Top 20%	0.90	10 periods
Top 20%	0.95	13 periods
Top 10%	0.90	21 periods
Top 10%	0.95	27 periods

at least 90 percent probability of containing one or more of the 6 periods during which the 20% highest exposures will occur. The number 6 is the largest integer representing 20% or less of 32.

Where the short-term time-averaging period is 10 min, there would be 48 such periods in an 8-hr work shift and the sample sizes in Table 52.5 would be appropriate. For example, to have at least 90% probability that the sampled periods include at least one of the four periods that have 10% or less of the highest exposures, 21 of the 48 periods should be selected at random and sampled.

Less than 10-min time-averaged measurements may sometimes be obtained, as with a 3-minute colorimetric tube or spot readings with a direct-reading meter. Then the sample sizes in Table 52.6 are appropriate.

5.3 Exposure Monitoring Strategies

The exposure monitoring strategies presented in this section are guidelines for measuring multiple exposures (e.g., multiple workers on a single day, a single worker on multiple days, multiple workers on multiple days). These strategies should be used with the exposure measurement strategies (for measuring an individual's exposure on a single occasion) treated in the previous section. Monitoring strategies for monitoring programs should always consider eight elements, which will be presented in further detail in the following sections:

5.3.1. Need for exposure estimates.
5.3.2. Airborne chemical(s) to be measured.
5.3.3. Strategy for initial monitoring.
5.3.4. Criteria for decision making.
5.3.5. Strategy for periodic monitoring for continuing hazard evaluation.
5.3.6. Occasions requiring extraordinary monitoring.
5.3.7. Criteria for termination of monitoring.
5.3.8. Procedures for follow-up.

A review report by Rappaport (109) supports a school of thought for certain industrial hygienists who believe that OSHA PELs should not be formulated as upper bounds on 8-hr TWA exposures of single employees. Rather, the report says that the entire exposure distribution should be controlled for the purpose of controlling both the long-term average exposure and the proportion of high 8-hr TWA exposures in the right tail of the temporal exposure distribution. The report says that in practice the tail area of the exposure distribution can also be controlled by merely controlling the long-term mean of the distribution.

Admittedly, it is primarily the long-term cumulative exposure that must be controlled because it is believed to be correlated with risk of chronic toxicity. The usual OSHA 8-hr TWA permissible exposure limits have in fact been used as a means to control corresponding long-term arithmetic mean exposures. This type of OSHA PEL usually has its basis either:

1. In dose-response relationships seen in chronic, animal exposure experimental studies (i.e., different groups of animals exposed for a long time to a range of carefully controlled, constant, repeated daily doses) or
2. In effects in observational studies of human workers whose corresponding "working lifetime" cumulative exposures have been estimated in connection with epidemiological studies.

Therefore, the argument is sometimes made that it would be "logical" to protect human workers against anticipated biological effects of chronic exposure by formulating OSHA PELs that are upper limits on long-term averages of daily TWA exposure levels.

But an argument against use of long-term average PELs is that it would not be feasible to enforce such exposure limits. This is true because it would be necessary to sample a worker's full range of exposures using multiday random sampling strategies and a sufficiently large sample size to determine a reasonably precise estimate of the long-term mean. This could be done validly (but perhaps not feasibly) only for those types of exposure measurements data that do not have patterns of temporal autocorrelation or systematic (average) interworker differences. For those cases the following comments concerning lack of validity of limited period multiday exposure averages as legal exposure limits would not apply. In general, unless random variability (with no time trends or cycles) exists, a group of consecutive sampling days cannot be treated as a random sample (or even as a representative sample) of a given employee's long-term distribution of daily exposures. Any number of working days, whether sampled randomly or consecutively during a limited period of time (e.g., one year), would not be representative of the entire distribution of past and future working days during a working lifetime. Production levels and operations change in practice, and this produces related changes in work tasks and exposure levels. Resulting exposure levels may vary systematically (e.g., cyclically), rather than randomly,during extended periods of time. Thus, arguments against use of long-term average exposure limits for individual workers relate mainly to infeasibility of the sampling strategies that would be needed for meaningful enforcement.

There are even stronger arguments against taking an average of multiple worker exposure levels to compare to an exposure control limit. An OSHA PEL is intended to apply to every individual worker, and interworker exposure variability typically exceeds variability of physical sampling and chemical analysis. Simple indiscriminate (unbalanced) mixing of interworker and interday variability is invalid because the pattern of interworker differences may be similar at different times. That is, interworker differences at different points in time may be autocorrelated, rather than fully random, and a multiworker average would be biased as an estimate of the average exposure of any individual worker. It would be biased high for some and low for others. Of course, if it can be validly assumed that there are no consistent interworker exposure differences, then any worker's exposure measurement would apply to any other worker in the target population.

On the other hand, current OSHA PELs based on single-day, 8-hr TWAs do permit valid enforcement decisions (compliance, no decision, or noncompliance) to be made. With 8-hr TWA PELs, random sampling plans are possible that properly support related statistical decision theory that is given in this report. Of course, it would be pointless to make a statistically valid enforcement decision about compliance with a PEL that is biologically meaningless. However, this is not the case for current 8-hr TWA PELs. When it is determined that compliance with an 8-hr TWA PEL exists on several occasions (or on a deliberately selected "worst occasion"), it is usually possible to conclude that the related long-term mean exposure is also at a compliance level (unless operations change).

The preference expressed here for 8-hr TWA PELs does not imply that there is no need to monitor exposures of multiple employees at different times. The remainder of this section deals with strategies for such exposure monitoring.

Section 2.4 listed two major types of monitoring programs as possible objectives of exposure estimation. The first type is an exposure screening program, which is a limited exposure monitoring program designed to identify target populations of workers with other-than-acceptable exposure distributions for follow-up periodic monitoring. The program uses minimal resources consistent with reasonable protection for workers. This type of program uses an action level as a screening cut-off to identify appropriate target populations for inclusion in a limited exposure surveillance program or a more extensive exposure distribution monitoring program. The latter program is a more extensive one intended to quantify exposure distributions of target populations. Generally this is first done for an initial base period; then the initial estimates of the exposure distributions are periodically updated with more current estimates from routine exposure monitoring.

There are several commonalities between exposure screening and exposure distribution monitoring programs. Guidelines noting the similarities and differences in the two types of programs will be presented in the following sections, which detail eight possible elements for these two types of monitoring programs.

5.3.1 Need for Exposure Measurements

Both exposure screening and exposure distribution monitoring programs need to begin by determining the need for exposure estimates. Desirable predecessors to these programs, which may negate the need for exposure measurements, include:

1. Conducting a workplace materials survey to determine if potentially harmful materials are being used in the workplace (see Ref. 110).

2. Conducting a walkthrough survey to identify process operations that may be potentially hazardous and to determine if workers may be exposed to hazardous airborne concentrations of materials released into the workplace (see Ref. 111).

3. Estimating airborne concentrations based on the amount of material released into the workplace air. This may be useful for contaminants that are low to moderate hazards (see Ref. 112). However, this technique usually requires a substantial safety factor to account for uncertainty, which may limit its usefulness.

4. Source sampling at potentially hazardous operations. This may be useful for screening strategies, such as may be used for hazard evaluations. The usual assumption is that the resulting exposure estimates are worst-case exposure levels and all worker exposures will be less than the values found at the source(s) of the contaminant (see Section 5.2.1 above Refs. 111, 113). This assumption may be invalid if there are more than one or two contaminant sources.

5. Preparing a written determination of the need (or lack of need) for exposure measurements. The written determination would consider:

a. Any information, observation, or calculation that might indicate worker exposures.

b. Any measurement taken.

c. Any worker remarks of symptoms that may be due to exposure to workplace materials or operations.

d. Any possible changes in production, process, or controls that could result in hazardous increases in airborne levels of contaminants or render control procedures inadequate.

5.3.2 Airborne Chemical(s) to Be Measured

The next element, in both exposure screening and exposure distribution monitoring programs, is determination of the airborne materials to be measured. This is best done by considering the information acquired from the various steps in the previous program element. It may be necessary to first do screening measurements (see Section 5.2.1, no. 6), if prior information is unavailable and qualitative detection is promptly required (e.g., emergencies and uncontrolled releases).

5.3.3 Strategy for Initial Monitoring

Initial monitoring is the first monitoring program element where an exposure screening program differs considerably from an exposure distribution monitoring program. For an exposure screening program, the objective of initial monitoring is to selectively obtain exposure estimates only for "maximum-risk" workers. These can be defined as those workers "believed to have the greatest exposure."

For exposure screening monitoring, the most efficient approach to sampling is a nonrandom selection of the highest-risk workers. The selection process must use competent professional judgment that relies on experience and knowledge of the exposure level determinant factors pertinent to the target population. Some factors to consider in selecting the maximum-risk worker are given in Ref. 114. Related determinant variables affecting the exposure levels of the target populations are discussed above in Section 5.1. The important point with this approach is to sample only those workers whose exposures are believed to represent the higher exposures of the target population. However, this approach is subject to frailties of professional judgment that could lead to erroneous conclusions regarding the exposure levels for the highest-risk workers.

For an exposure screening program, if maximum-risk workers cannot be identified for each operation or target population with reasonable confidence, a second approach is random sampling. The same sample size theory applies here as was discussed in paragraph 5.2.3 for random sampling of a homogeneous-risk target population, based on the combinatorial probability formulas detailed in Ref. 108. This approach is less efficient than the first, which relies on professional judgment to select a nonrandom sample. However, the results are independent of any mistakes that might occur in professional judgment.

PURPOSE. Provide a large enough sample size to assure (i.e., have a probability of at least 90 or at least 95%) that at least one randomly sampled worker will have a high exposure relative to most other workers in the target population (i.e., be from the highest 10% or highest 20% of the specified target population exposure distribution).

ASSUMPTIONS. The sample of workers to be measured is assumed to be randomly chosen from the specified target population. The derivation of this method is based on the theory of random sampling without replacement (see Section 5.2.3). No mathematical form is assumed for the exposure distribution of the target population. Thus these sample sizes may be larger than necessary for some situations where the exposure distribution is confidently known, such as when a two- or three-parameter lognormal distribution can be reliably used as a model. However, these sample size recommendations are robust.

EXAMPLE. Estimate an appropriate sample size for a 26-worker target population such that there is at least 90% probability that the sample will include at least one higher-risk worker from the top 10 or lesser percent of the 26-worker exposure distribution.

SOLUTION. Table 52.7 indicates that for the N = 26 workers, a random sample of size $n = (N - 8) = 18$ workers will have at least 90% probability of containing one or more of the $N_0 = 2$ workers in 26 who represent no more than 10% of the highest exposures (7.7% in this example). The value 2 for N_0, the number of high values in the population, is the largest integer representing no more than 10% of $N = 26$ workers.

COMMENTS. At least 18 of the 26 workers in the target population need to be randomly sampled to be at least 90% sure to "catch" one worker from the two that constitute no more than the upper 10% of the target population. With this small target population (and extremely small number of highest-exposure workers) it is necessary to measure almost

Table 52.7. Necessary Sample Size n for $P \geq 90\%$ that One or More Items Will Be from the Highest 10% or Less (i.e., Will Be from the N_0 Highest Items in N)

N—Size of Target Population	n—Necessary Sample Size	N_0—Number of "Highest 10% or Less" Values
10–19	$(N - 1) = 9$ to 18	1
20	$(N - 6) = 14$	2
21–24	$(N - 7) = 14$ to 17	2
25–27	$(N - 8) = 17$ to 19	2
28–29	$(N - 9) = 19$ to 20	2
30–31	$(N - 14) = 16$ to 17	3
32–33	$(N - 15) = 17$ to 18	3
34–35	$(N - 16) = 18$ to 19	3
36–37	$(N - 17) = 19$ to 20	3
38–39	$(N - 18) = 20$ to 21	3
40–41	$(N - 23) = 17$ to 18	4
42–43	$(N - 24) = 18$ to 19	4
44–45	$(N - 25) = 19$ to 20	4
46	$(N - 26) = 20$	4
47–48	$(N - 27) = 20$ to 21	4
49	$(N - 28) = 21$	4
50	$(N - 32) = 18$	5

70% of the total. This method is inefficient for small target populations but works considerably better with larger ones. For a target population of 100 workers, the necessary sample would be about 20 (or only 20% of the total) to have at least 90% probability of sampling one worker from the 10 workers that constitute the highest 10% of the exposure distribution. Necessary sample size tables for other upper tail percentages of exposure distributions or other confidence levels can be computed from the material in Ref. 108.

Since the target population is finite (26 in this example) only discrete probability levels P are attainable with the sampling plan. The sample size n is chosen to have at least 90 percent or greater probability ($P \geq 0.90$) that the n items in the sample will include one or more of the N_0 highest items. Since integer numbers N_0 of highest values cannot be chosen to exactly represent 10% of the population size N (unless $N = 10, 20, 30$, etc.), N_0 is chosen to be the largest integer for which $N_0 \leq [(0.10)(N)]$.

Compared to an exposure screening program, the sampling procedure for an exposure distribution monitoring program is considerably different. An objective of the latter program is to estimate the exposure distribution of the target population over an initial base period, which may range from one day to several years. The necessary sample size for initial monitoring is influenced by the desired precision of the exposure distribution estimate for the target population.

One approach to exposure distribution estimation is to plot a sample of exposure estimates on lognormal probability graph paper. This technique is presented in Section 6.12. An expanded discussion is given in Ref. 115. An example of this approach is given in Paik et al. (116) and the *American Industrial Hygiene Association Journal* (117). For this approach, a minimum sample of about 6–10 TWA exposure estimates generally is most cost effective. That is, one should randomly sample at least 6–10 8-hr TWA exposure estimates for the defined target population over the desired temporal base period. The rationale underlying the recommendation for 6–10 TWA estimates is illustrated in Figure 52.11. If one is considering the computation of one-sided tolerance limits for a lognormally distributed variable, such as exposure estimates from a target population of workers (see Section 3.6.2.4) as will be discussed in Section 6.7, this figure indicates that the K coefficient in the tolerance limit algorithm decreases relatively little after n values in the range of 6–10. That is, there will be small decreases in K after $n = 10$ compared to the costs incurred for the additional samples.

Note that 6 to 10 exposure samples will only provide the roughest estimate of the shape of the actual exposure distribution. It will tell one almost nothing about the goodness-of-fit of a particular distributional model, such as lognormal. Data presented by Daniel and Wood (118), and discussed in Ref. 115, indicate that considerably larger samples (such as 30–60) are necessary to obtain an estimate with low uncertainty of the central 80% of the target population exposure distribution. Even with these sample sizes, unusual behavior in the 10% upper and lower tails still cannot be confidently determined. However, as mentioned in Section 3.6, an exposure distribution estimate with moderate uncertainty may suffice because it is an adequate estimate. The adequacy of any distributional estimate is dependent on its ability to serve as a workable forecaster of the unsampled portions of the exposure distribution for the target population. That is, is the estimate good enough for our purpose? This reinforces the point that the purposes of exposure measurements and monitoring need to be clearly defined before the measurements are obtained.

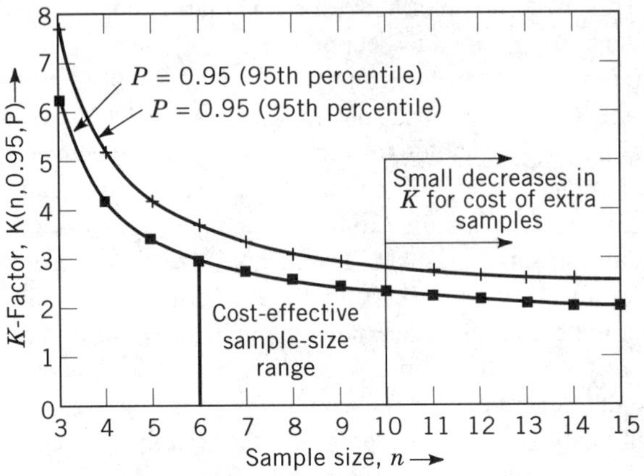

Figure 52.11. Cost-effective sample sizes for estimating an exposure distribution, where for a lognormal distribution the one-sided upper tolerance limit for the 100Pth percentile concentration (at the 95% confidence level) is: $\text{UTL}_{(1,0.95,P)}$ = antilog[(mean of log concs) + (K-factor)(std dev of log concs)]. Thus the cost-effective sample sizes for reasonably-precise tolerance limits are in the range of 6 to 10 TWA estimates to estimate an exposure distribution.

Besides providing information on the range and frequency of exposures, the exposure distribution monitoring strategy has another important advantage. The initial monitoring can also yield an estimate of the variability parameter of the distribution, to be used as an indicator to judge if the target population has been adequately defined. Remember that a target population of exposures generally is defined in terms of (*1*) the numbers and types of workers, (*2*) the temporal period to be covered, and (*3*) ranges of determinant variables that affect the exposure levels to which the population is exposed. If the variability of the sample exposure distribution exceeds a geometric standard deviation (GSD) of about 3.0, then the suitability of the factors used to define the target population should be reexamined (this approximate criterion is based on the professional judgment of one of the authors, NAL). Other factors may be needed that define more limited population groupings in order to achieve less variability for the exposure distribution. Such a process may need to be repeated several times until one achieves an exposure distribution variability parameter that is small enough to permit meaningful exposure monitoring. Lyles and Kupper (119) and Rappaport et al. (120) have also pointed out that interworker exposure variability often exists, and it must not be underestimated by taking repeated measurements of the same worker when the objective is to assess the exposure distribution of a worker population.

If too many workers from populations with several different exposure distributions having substantially different median-exposure levels are pooled into one target population, then a pooled exposure distribution of excessive variability can result. A highly disperse exposure distribution generally is unsatisfactory because exposure estimates for individual workers will have a considerable amount of uncertainty. For example, for a two-parameter lognormal distribution with an arithmetic mean of 100 ppm and a GSD of 2.5 (GM = 65.7 ppm), the 5th-percentile worker exposure is about 15 ppm and the 95th-percentile worker exposure would be about 300 ppm! Note that 1 in 10 of the worker exposures would lie either below 15 ppm or above 300 ppm. For this exposure distribution, the best one could say for any individual worker is that there is 90% confidence that during the sampled temporal period the worker's exposure was between 15 and 300 ppm. However, this highly imprecise estimate might be adequate if the exposure control level was substantially higher than the 95th-percentile exposure level, such as 1000 ppm. For a lognormal distribution, Lyles and Kupper (119) give the steps of a SAS computer program that computes the exact sample size needed to assure that a tolerance limits test will have specified power to produce a significant result (at a chosen significance level) when there is a specified population percentage (of individual TWA's) exceeding a specified percentile of a lognormal exposure distribution that has a specified GSD. Examples are given that show exact sample sizes from two to three times larger than approximate sample sizes previously published by Selvin et al. (121).

5.3.4 Criteria for Decision Making

Decision making is a monitoring program element where the objectives of an exposure screening program and an exposure distribution monitoring program are similar, but the techniques used differ considerably because the available information is different. The two types are discussed under separate headings below.

In this section, criteria for decision making about exposure distributions cannot be discussed in a definitive framework of "legal" and "illegal" (or other black and white definitions of acceptable or unacceptable) parameters of exposure distributions. Generally, well-accepted toxicological interpretations are unavailable regarding parameters such as percentages of days workers are (or possibly are) exposed above an 8-hr, TWA exposure control limit, when the long-term (multiple day) mean is below the control limit. Therefore, provision of explicit definitions of acceptable and unacceptable exposure distributions will not be attempted. These definitions must be derived for individual work exposure situations in the context of the particular toxicological, regulatory, and administrative considerations of a given occupational exposure environment. The technical discussions of this section are intended to assist the competent professional in estimating and interpreting worker exposure distributions. The statistical tools for making controlled-risk decisions and judgments about the acceptability of an exposure distribution, based on an estimate of that distribution, are to be found in Section 6. In particular, procedures are presented in Section 6.7 and Section 6.13 for computing and displaying tolerance limits for a lognormal distribution of exposures, and in Section 6.8 for computing a point estimate and confidence limits for the percentage of lognormally distributed values exceeding an exposure control limit.

Decision Making in an Exposure Screening Program. The main objective of exposure screening is to identify target populations with unacceptable exposure distributions for follow-up exposure monitoring or other actions (e.g., worker medical surveillance, worker training programs, engineering exposure controls, personal exposure controls, or other specialized measurement actions such as noted in Section 5.2.1). The decision-making technique used is to compare the exposure estimate with an appropriate action level that serves as a first screening cut-off value. A secondary objective would be to subclassify the unacceptable exposure distributions into subgroups that require either (*1*) follow-up monitoring at a normal frequency (e.g., at least every 2 months) or (*2*) follow-up monitoring at an increased frequency (e.g., at least monthly). The second screening cut-off value for these decisions is the exposure control limit chosen for adequate worker protection. For target populations judged to have acceptable exposure distributions, generally the indicated action for an exposure screening program would be to redetermine the need for exposure measurements (see Section 5.3.1) each time there is a change in production, process, or control measures that could result in a substantial increase in exposure levels.

The original action level was developed by NIOSH for regulations promulgated by the Occupational Safety and Health Administration (OSHA) as discussed by Leidel et al. (122). Also refer to Ref. 123. For that action level, only single-day exposure estimates for the maximum-risk workers from each examined target population are compared to an action level set at 50% of the regulatory exposure standard for the particular substance. The NIOSH action level concept uses single-day exposure estimates to reach decisions regarding the acceptability of possible exposure levels on unmeasured days. Single-day estimates are the sole basis for deciding whether further exposure monitoring or other actions should be performed for the target population's workers.

Note that a required assumption for the application of an action level is that the companion exposure control limit (ECL) is sufficiently protective of workers. That is, the use of an action level to reduce employer costs of exposure monitoring, medical surveillance, worker training, and so forth presumes that exposures below the ECL create minimal or acceptable risks to workers. The rationale of using an action level is to screen worker exposures so that there is a low probability that a minimal proportion of daily exposures will exceed the ECL linked to the action level. Thus it is implicit that a small proportion of exposures below the ECL lead to minimal or acceptable risks to workers. If this is not the case, then it is not appropriate to use an action level as a justification or screening value to reduce or eliminate exposure monitoring, medical surveillance, and so forth.

It is also important to note that the 50% action level should be thought of as a regulatory action level, since:

1. It is an approximate value developed for simplicity of application.
2. It was a value intended to reduce employer exposure monitoring burdens.
3. It incorporated feasibility considerations for regulatory monitoring requirements.

During the development of the 50% action level for regulatory purposes, the premise used was that "the employer should try to limit to 5% probability, that no more than 5% (or greater) of an employee's actual (true) daily-exposure averages exceed the standard" (124). This is not meant to imply that a work situation with as many as 5% of overexposure days is acceptable to OSHA. The stated premise is part of the "risk tradeoff" that is characteristic of statistical decision theory. Measurements subject to random errors are used to make decisions about overexposure. The limit of 5% overexposure days cited above is merely the minimum of hypothetical overexposure incidences that, if they occurred, would be discovered at least 95% of the time they were measured when the GSD of the exposure distribution is 1.22 or less.

The Leidel et al. (122) report demonstrated that an action level appropriately computed to achieve the stated goal of a low 5% probability level is a variable value that is a function of the interday variability of the true daily TWA exposure levels. Regarding the interday variability measured by a geometric standard deviation (GSD) of an assumed two-parameter lognormal distribution, Leidel et al. (124) noted: "Higher GSDs require lower fractional action levels. A GSD of 2.0 requires an action level as low as 0.115 of the standard!" However, NIOSH decided to recommend to OSHA a single-value action level computed as 50% of an exposure standard. This pragmatic decision was intended (*1*) to increase the probability of employer compliance with regulatory monitoring requirements as a result of the simplicity of applying a single action level, (*2*) to lower implementation costs, and (*3*) to substantially increase feasibility. The alternative would have been recommending a procedure that required an employer to obtain a GSD estimate of the interday exposure level variability for each exposure situation before computing an appropriate action level.

The decision to recommend a single-value action level (50%) keyed to a single GSD of 1.22 did lead to a regulatory screening strategy with some limitations. The potential problems and limited performance characteristics of the "bare bones" regulatory screening strategy in Leidel et al. (2) have been comprehensively analyzed by Tuggle (125) who noted: "The intent of the NIOSH decision scheme is to give indications about periodic monitoring: whether to terminate (TERM), initiate/increase (INCR), or continue (CONT) exposure measurements." Tuggle (125) concluded the following:

> For environments with a very low fraction of exposures above the PEL, the NIOSH scheme renders the correct, TERM decision with a high probability—for all variabilities. For low variability environments, the relative probability of a TERM decision decreases sharply (the relative probability of an INCR decision increases sharply) as the fraction of exposures above the PEL increases. In other words, at low variability, the NIOSH scheme is very "sensitive" to an increasing fraction of overexposures. However, as exposure variability increases, this sensitivity falls off, and the NIOSH decision scheme assumes an increasing probability of incorrect, TERM decisions for high-exposure risk environments for which monitoring should definitely be continued or even increased. For example, consider environments with moderate-to-high variability and with 25 percent of all exposures above the PEL—one exposure in four an overexposure: Figure 3 shows that the NIOSH decision scheme produces incorrect, TERM decisions for sun environments, from about 20% to well over 50% of the time.

Tuggle (125) noted the following limitations of a simplistic regulatory screening strategy and suggested some desirable characteristics for an augmented exposure screening program:

1. The regulatory screening strategy bases a decision on, at most, only two measurements: the last one and possibly the one preceding. An augmented strategy should use all the data collected.

2. The regulatory screening strategy considers only a qualitative aspect of the exposure estimate, that is, whether the exposure is below the 50% action level or above the exposure standard. An augmented strategy should also consider the quantitative aspect of how much an exposure estimate is below the action level or above the exposure standard.

3. The regulatory screening strategy does not determine or account for the range of exposure level variabilities actually encountered. An augmented strategy should, since exposure variability (GSD) is a factor equally as important as median exposure (GM) in evaluating the occurrence of overexposures.

4. Lastly, the regulatory screening strategy produces incorrect decisions to terminate exposure sampling, particularly with substantial proportions of overexposures and moderate- to high-variability exposure level environments. An augmented strategy should limit such incorrect decisions to an acceptably low probability that is independent of exposure variability.

Tuggle (125) suggested a simple modification to the regulatory screening strategy given by Leidel et al. (2) to reduce incorrect decisions to terminate exposure monitoring that result from analyzing initial monitoring results. He recommended that the first decision criterion require, in all instances, two (instead of one) consecutive exposure estimates below the 50% action level to terminate routine exposure monitoring. He demonstrated that this would substantially reduce the probability of incorrect decisions to terminate monitoring. Tuggle (126) subsequently recommended an augmented monitoring strategy based on one-sided tolerance limits for a distribution of daily exposure estimates assumed to follow a two-parameter lognormal distribution.

It should be noted that the NIOSH "employee exposure determination and decision strategy" recommended by NIOSH to OSHA for incorporation in regulations (127) does maintain the objective of a 5% limit on the risk of making incorrect decisions of noncompliance during the actual periods and for the actual workers that were measured for TWA exposures. However, the NIOSH regulatory screening strategy does not achieve 95 percent probability of terminating a process that has 5% or more overexposure days, unless the GSD of the process is at or below the value 1.22 to which the 50% action level is keyed. The limitations of a constant-value action level that is keyed to a single GSD were also discussed by NIOSH in Leidel et al. (122). However, a constant value was selected for practical reasons stated earlier. Tuggle's mathematical statistical analysis of the NIOSH-recommended action level of 50% is correct, but practical remedies are still not at hand.

Decision Making in an Exposure Distribution Monitoring Program. Compared to an exposure screening program, the decision-making process for evaluating initial monitoring results from an exposure distribution monitoring program is considerably different. However, the objectives of the two monitoring programs are similar.

For an exposure distribution monitoring program, the objective of the decision-making element is to classify the exposure distributions or target populations into one of the following three classes: acceptable, marginal, or unacceptable. An acceptable exposure distribution is one for which we are confident that essentially all exposures are less than the desired exposure control limit (ECL). A marginal exposure distribution is one for which we are confident that at least a moderate proportion of exposures exceed the desired exposure control limit. An unacceptable exposure distribution is one for which only confident that at least a substantial proportion of worker exposures exceed the desired exposure control limit.

If the exposure distribution for a target population is judged acceptable, generally the follow-up action would be limited to routine exposure monitoring at an appropriate future time. The objective of follow-up action would be to monitor and revise the initial or previous estimate of the exposure distribution for the target population. If the judgment for the exposure distribution is marginal, then the indicated action generally would be more definitive exposure monitoring, which could include the specialized measurement actions noted in Section 5.2.1. This additional monitoring should identify the marginally and unacceptably exposed workers and could assist in implementing overexposure prevention measures. Lastly, if the exposure distribution is judged unacceptable, the indicated action would follow the same course specified for a target population identified with an action level screening approach as unacceptably exposed.

Two statistical techniques that are useful in an evaluation of exposure distribution estimates for target populations of workers are (*1*) probability plotting, including computation and graphical display of tolerance limits, and (*2*) distribution fitting to estimate one or more quantiles (a formal technique based in statistical estimation theory).

The first technique utilizes the type of lognormal probability plot presented in Section 6.12 and Section 6.13 and Ref. 115. Exposure estimates are plotted on lognormal probability paper along with lines representing the desired ECL exposure limit and tolerance limits for the true exposure distribution. Rough estimates of overexposure "risks" (stated in percentages of overexposed workers or workdays with their attendant uncertainties) can then be obtained from such exposure probability plots. The estimates are given by the indicated probabilities at the intersection of the exposure estimate distribution plots and the ECL exposure limit line and the exposure values above and below the ECL.

This analytical approach is illustrated on lognormal probability paper in Figure 52.12. The two dashed lines indicate the estimated exposure distributions (see Section 6.12) and the two thick solid lines bounding the shaded areas represent upper tolerance limits (one-sided upper confidence limits for differing percentiles *P* of the parent distribution of ex-

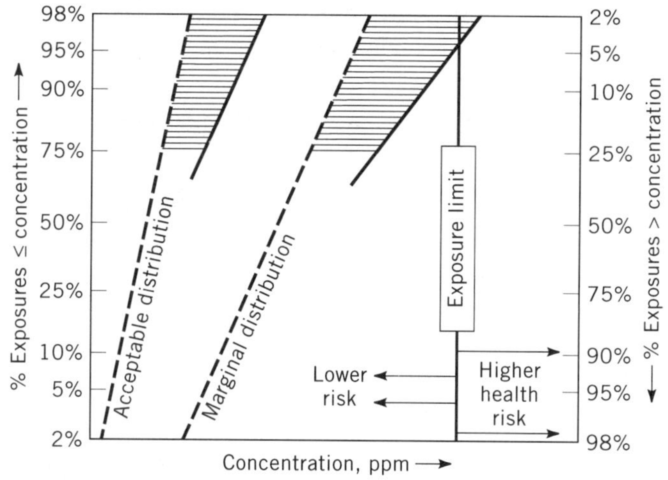

Figure 52.12. Graphical criteria for acceptable, marginal, or unacceptable exposure distributions.

posures). The shaded regions represent the uncertainty in the estimated exposure distribution, since these are areas in which the true exposure distribution could lie.

The calculations and plotting required for this approach can be quite tedious and time consuming. However, one of the authors (NAL) has created a customized program to do the necessary calculations and plotting by using the unique transform-programming language available in SigmaPlot® 5.0 from SPSS Inc. Both *lognormal probability* and *semi-logarithmic time-series* plots can be created rapidly from existing data bases.

Note that the qualitative definitions acceptable, marginal, or unacceptable exposure distribution involve both objective quantitative and subjective qualitative aspects of decision making using competent professional judgment based on relevant training and experience. The professional judgment exercised should consider at least the following information: (*1*) individual aspects of the target population, (*2*) chemical(*s*) creating the exposures and their health effects at different exposure levels, (*3*) possible exposure levels (both routine and episodic) (*4*) time patterns of exposure, (*5*) knowledge of exposure level determinant factors pertinent to the target population, (*6*) the overexposure risk estimates, (*7*) slope (variability) of the exposure distribution, and (*8*) experience with similar exposure situations. Note that this technique should be used with caution and decisions made with some degree of conservatism that favors protection of worker health in the absence of definitive information (i.e., err in the direction of marginal or unacceptable exposure distribution decisions by placing the "burden of proof" on acceptable decisions).

The second evaluation technique, formal point estimation, complements the other approach, which is graphical. The two techniques can be used independently, but the authors believe that both techniques are useful for evaluation of any exposure distribution data set. The quantitative inferential approach is to calculate the best estimate (point estimate) for the proportion of the exposure distribution exceeding some desired value (e.g., exposure control limit, exposure standard). In addition, confidence limits for the true value of the parameter whose point estimate was obtained should also be calculated to indicate the uncertainty of the estimate due to limited sample sizes. Approximate confidence limits could be either one or two sided and typically will be calculated at the 95% confidence level. The details for this technique are presented in Section 6.8.

5.3.5 Strategy for Periodic Monitoring for Continuing Hazard Evaluation

Periodic follow-up exposure monitoring applies both to an exposure screening program and to an exposure distribution monitoring program and has a similar objective for both programs. The objective is that individual workers or target populations of workers with potentially other-than-acceptable (i.e., marginal or unacceptable, see Section 5.3.4) exposure distributions be periodically monitored, often enough to promptly detect unacceptable exposure situations. In addition, an exposure distribution monitoring program should periodically monitor at a frequency adequate to identify substantial changes in the target populations' exposure distributions.

Control charts are commonly used in quality control operations, but Hawkins and Landenberger (128) discuss some limitations of conventional control charting techniques when they are applied to industrial hygiene data. They point out that the usual industrial-hygiene sample sizes are adequate only for the most sophisticated types of control chart analysis, such as moving average and moving range methods. However, if one does desire to apply quality assurance procedures to lognormally distributed data, review the techniques given by Morrison (129).

Remember that for an exposure screening program the set of possible decisions suggested (Section 5.3.4) regarding the target population's exposure distribution are acceptable, unacceptable with "normal" frequency follow-up monitoring, and unacceptable with "increased frequency" follow-up monitoring.

For regulatory purposes, NIOSH recommended to OSHA that minimum requirements for "normal" frequency of monitoring for an exposure screening program be set at 2–3 months (2). It should be noted that, in addition to basic health and safety protection considerations, feasibility and simplicity of application considerations played an important role in this recommendation. Lynch et al. (130) have commented for NIOSH:

> Four such exposure measurements during the year are considered minimal to detect any significant fluctuations in the average level of environmental contaminants. Employee–exposure measurements four times per year per exposed employee are a reasonable minimal burden on the employer. The employer should not interpret this to be an absolute minimum that would always be appropriate to determine each employee's exposure. There is information that a waiting time between sampling events (for typical data) of over one month results in a relatively low level of confidence in the exposure estimates. However, a maximum waiting time of three months between measurements should protect each employee and give some idea of variation without putting an inordinate burden on the employer.

For an "increased frequency" of regulatory follow-up monitoring, NIOSH recommended a minimum frequency of once a month (2). When unexpectedly or unusually high exposure results are obtained by either initial monitoring or periodic monitoring, more intensive monitoring is needed to quickly identify increasing exposure trends that may lead to hazardous exposures.

For other than regulatory monitoring programs, the selection of an adequate frequency for periodic exposure monitoring should be based on experience with similar exposure situations, knowledge of the exposure level determinant factors pertinent to the target population, and use of competent professional judgment. Guidelines for the factors to be considered include:

1. Nature and degree of health hazard from the exposure situation, evaluated at all possible exposure levels, even those that might be unexpected.

2. Ratio of previous exposure levels to the desired exposure-control limit, and possible trends in this ratio.

3. Degree of variability seen in previous exposure distributions, variability between the distributions, and trends over time. These can be studied by plotting the sample data sets from several longitudinal points in time on semilogarithmic paper (see Section 6.11). Another complementary approach would be the plotting of several sample data distributions on the same lognormal probability graph.

4. Determinant variables affecting the target population's exposure levels (see Section 5.1).

5. Reliability of decision-making techniques used and quantity and quality of exposure estimates available. Exposure screening program decision techniques yield decisions of less reliability than the quantitative techniques available for exposure distribution estimates.

5.3.6 Occasions Requiring Extraordinary Monitoring

With either an exposure screening program or an exposure distribution monitoring program, additional exposure monitoring is indicated whenever a change occurs in production, process, personnel, exposure controls, or any other determinant variable that could lead to substantial increases in worker exposure levels.

5.3.7 Criteria for Termination of Monitoring

This program element concerns a specialized type of decision making. As with monitoring of exposures for the purpose of decision making about exposure distributions, the objectives of termination of routine periodic monitoring are similar in an exposure screening program and in an exposure distribution monitoring program, but the techniques used and available information differ considerably. In some situations it may be desirable to conserve exposure measurement resources by terminating routine monitoring of target populations for which one is confident that the exposure distributions have been consistently acceptable. As with selecting an adequate frequency of periodic monitoring (Section 5.3.5), decisions to terminate monitoring should be based on experience with similar exposure situations, knowledge of the exposure level determinant factors pertinent to the target population, availability of monitoring resources, and availability of competent professional judgment. Use the factors given in Section 5.3.5 when considering termination of routine monitoring.

For regulatory purposes in an exposure-screening program, NIOSH recommended to OSHA that regulations allow cessation of monitoring for a worker if two consecutive exposure estimates taken at least one week apart were both less than the 50-% action level. See Section 5.3.4 for a discussion of an analysis by Tuggle (125) of this regulatory termination rule and his suggested alternative approach for situations where a two-parameter lognormal distribution of exposures can be assumed.

For exposure distribution monitoring programs, decisions to terminate monitoring should also be based on experience with similar exposure situations, knowledge of the exposure level determinant factors pertinent to the target population, availability of monitoring resources, and use of competent professional judgment. Also use the factors given in Section 5.3.5 when considering termination of routine monitoring.

There is an additional quantitative tool available when considering termination of an exposure distribution monitoring program. Tuggle (126) has recommended the use of one-sided tolerance limits if one can assume that all available exposure estimates are from a single two-parameter lognormal distribution. The computed one-sided upper tolerance limit for the target population would be compared to the selected exposure control limit (ECL) and monitoring would be continued if the tolerance limit exceeded the ECL. All available exposure estimates would be considered sample data and utilized for computation of a tolerance limit.

However, if the exposure data are not stable over time (i.e., if differing distributions of exposures exist over time for the same target population of workers), competent professional judgment may be used to censor some of the earlier exposure estimates. If this is not done, the computed tolerance limit will probably be larger than it should be, which would cause errors in the decision making that lead toward unjustified continuation of monitoring. Thus if all the data are used, instead of a smaller sample based on the more recent exposure data, the decision making would tend to be conservative (i.e., favor continuation of possibly unnecessary monitoring, but reduce possible risks to workers).

5.3.8 Procedures for Follow-up

This program element should always be used for all types of monitoring programs. One should discuss the planning, conduct, and results of each program element with appropriate representatives of management and labor. If necessary, implement corrective action to reduce potential or identified health hazards. Write periodic evaluation reports as appropriate. Determine if additional exposure monitoring, biological monitoring, or medical surveillance programs are necessary. Determine if areas outside the workplace should be evaluated regarding possible environmental air or water pollution, hazardous waste disposal practices, or safety practices.

6 APPLIED METHODS FOR ANALYZING OCCUPATIONAL EXPOSURE DATA

This sixth and last section of this chapter is devoted to the applied statistical methods that the authors have suggested be used for analysis of occupational exposure data. Data to be analyzed by these methods can be generated by the study designs discussed in Section 4 and Section 5. The first portion of this section (Section 6.1, Section 6.2, Section 6.3, Section 6.4, and Section 6.5) covers methods for computing one- and two-sided confidence limits and confidence intervals for the true exposures of individual workers during the same periods as their exposure measurements. In addition, classification tests for individual exposure estimates relative to an exposure control limit are presented. The second portion of Section 6 (Section 6.6, Section 6.7, Section 6.8, Section 6.9, and Section 6.10) covers inferential methods for exposure distributions such as computing tolerance limits, tolerance intervals, and point estimates of the proportions of values in distributions that exceed a chosen limit, along with associated confidence limits. The third portion of the section (Section 6.11, Section 6.12, and Section 6.13) presents graphical techniques for plotting lognormally distributed data and associated tolerance limits. Appropriate sample size recommendations for specialized objectives were given earlier in Section 5.2.3 and Section 5.3.3.

It is strongly recommended that the calculations presented in these methods be performed on a hand-held, programmable calculator or personal computer spreadsheet software. By programming the algorithms given in this chapter and then running the numerical examples, one can ensure that the correct equations are being used. Additionally, it is very useful and instructive if one subsequently graphs the results to obtain a better understanding of the results. Piele (131) presents series of examples demonstrating how to set up and solve statistical problems with either Lotus 1-2-3, Release 2.01 or higher, or Microsoft Excel, Version 2.0 or higher. His emphasis is on spreadsheet concepts and functions that can be applied to a wide range of quantitative problems.

The methodologies for the exposure classification tests (statistical hypothesis tests) given in Section 6.1, Section 6.2, Section 6.3, Section 6.4, and Section 6.5 were originally developed by NIOSH for OSHA in support of their regulatory enforcement procedures. The purpose of the methodologies is to limit the risk of making unjustified noncompliance decisions (i.e., decisions not supported by sufficient accuracy in the exposure measurement data) regarding regulatory exposure limits. The necessity for development of such procedures is a reflection of the nature of the legal system used in the United States.

The presentation of statistical hypothesis tests in this chapter may unfortunately give the erroneous impression that industrial hygienists should view any ECL, ACGIH TLV, or legal standard as a definitive boundary between "safe" and "dangerous" exposure levels for all workers, which is definitely not the case. The hypothesis tests for individual worker exposure estimates presented in this section should receive routine application only in legal proceedings concerning regulatory issues. That is, an employer should not attempt to judge the acceptability of a workplace environment by statistically comparing only a few imprecise exposure measurements with some exposure limit.

It is strongly recommended that the computation of confidence limits be the preferred procedure for calculating the uncertainty of individual worker exposure measurements. The primary purpose of computing confidence limits should be the estimation of the uncertainty of individual worker exposure estimates due to the random errors of the exposure measurement procedure used. It is important to realize that the confidence limits do not reflect the substantial interday and interworker exposure level variability. When attempting to judge the acceptability of exposures in a workplace, an industrial hygienist should if possible evaluate the multiday and multiworker exposure distributions (see Section 5.3.4). Of course, an employer continually has the legal obligation to maintain exposure levels below applicable regulatory exposure limits on all workdays for all workers.

6.1 Full-Period Single-Sample Estimate

The methods in this section are applicable to exposure estimates calculated from a single exposure measurement taken for the full duration of the desired time-averaging period (e.g., 40 hr for a workweek TWA, 8 hr for an 8-hr workday TWA, 15 min for a 15-min TWA). It is suggested that these computations be programmed into a PC spreadsheet program or a program such as SigmaPlot® 5.0, which could also plot the TWA estimates and the computed confidence limits on a *semilogarithmic time-series* plot as discussed in Section 6.11 of this chapter.

6.1.1 Confidence Limits for a True TWA Exposure

PURPOSE. Compute one- or two-sided confidence limits for a true worker exposure that is estimated from a single exposure measurement when the total coefficient of variation for the exposure measurement procedure is known.

ASSUMPTIONS. The sample result is assumed to be a valid measurement of the worker's exposure at the time and place of the sampling (refer to discussions in Section 3.1 and

Section 3.3.5). The random errors in full-period single-sample estimates X are assumed to be independent and normally distributed with zero mean (see Section 3.6.1.3). This assumption implies that the arithmetic mean of many replicate exposure measurements (i.e., taken at exactly the same point in space and over the same time period) would be equal to the true exposure average. This true mean (or "expected value") is unknown, but the total coefficient of variation for the exposure measurement procedure (denoted by CV_T, see Section 3.4.2) is assumed known. Adequate confidence limits can be computed for the true concentration based on an assumption that the standard deviation is equal to the product of CV_T and the true full-period exposure μ, where μ is estimated by measurement X. The accuracy of the computed confidence limit(s) will be a function of the accuracy of the coefficient of variation used.

EXAMPLE. An exposure measurement procedure with a CV_T of 0.09 was used to obtain an 8-hr TWA exposure estimate for a worker. A single 8-hr measurement yielded an estimate $X = 0.20$ mg/m^3. Compute two-sided 95 percent confidence limits for the worker's true exposure based on the exposure estimate X.

SOLUTION.

1. Compute the two-sided 95% upper and lower confidence limits ($UCL_{2,\ 0.95}$ and $LCL_{2,\ 0.95}$) for the worker's true exposure μ from:

$$UCL_{2,\ 0.95} = \frac{X}{1 - (1.96)(CV_T)}$$

$$LCL_{2,\ 0.95} = \frac{X}{1 + (1.96)(CV_T)}$$

In this example

$$\begin{aligned} UCL_{2,\ 0.95} &= \frac{0.20}{1 - (1.96)(0.09)} \\ &= \frac{0.20}{0.824} \\ &= 0.24 \text{ mg/m}^3 \end{aligned}$$

$$\begin{aligned} LCL_{2,\ 0.95} &= \frac{0.20}{1 + (1.96)(0.09)} \\ &= \frac{0.20}{1.176} \\ &= 0.17 \text{ mg/m}^3 \end{aligned}$$

Thus, in this example, the two-sided confidence interval at the 95% confidence level for the true worker exposure that was estimated by the single-sample exposure measurement of 0.20 mg/m^3 is from $LCL_{2,\ 0.95} = 0.17$ mg/m^3 to $UCL_{2,\ 0.95} = 0.24$ mg/m^3. That is, one can be 95% confident that the true exposure μ is between 0.17 and 0.24 mg/m^3.

2. To compute either one-sided 95% confidence limit ($UCL_{1,\,0.95}$ or $LCL_{1,\,0.95}$) for the true exposure, use the formula for the corresponding two-sided limit with a Z value of 1.645 substituted for the multiplier 1.96. (The multiplier 1.96 corresponds to 2.5% of the area in a standard normal distribution allocated to each tail, whereas 1.645 corresponds to the total 5% allocated to one tail.)

COMMENTS. Note that, for any given estimate and associated confidence interval, the interval either does or does not contain the true value. However, one can say, in a special sense, that there is a 95% probability of the true exposure being bounded by the computed confidence interval. That is, if repeated samples were taken, and confidence limits calculated for each sample, 95% of the confidence limits would enclose the true worker exposure.

Also, the indicated uncertainty in the exposure estimate reflected in the width of the asymmetrical confidence interval (−15 to +21% about the point estimate in this example) does not incorporate uncertainty due to interday exposure level variability (see Section 3.3.1 and Section 3.6.2.3). The confidence interval pertains only to-the worker's true exposure on the particular day and during the period of that day that was actually measured.

6.1.2 *Classification of Exposure (Hypothesis Testing)*

PURPOSE. Classify an exposure estimate based on a single-day exposure measurement as noncompliance exposure, possible overexposure, or compliance exposure relative to an exposure control limit (ECL).

ASSUMPTIONS. The same as used for Section 6.1.1.

EXAMPLE. A charcoal tube and personal pump were used to sample a worker's TWA exposure to α-chloroacetophenone. The analytical laboratory reported a TWA exposure estimate X of 0.040 ppm and stated that the measurement procedure had a CV_T of 9%. The applicable ECL was 0.05 ppm. With 95% confidence, classify the TWA exposure estimate X relative to this ECL.

SOLUTION

1. For an employer, in this example $X \leq$ ECL. However, if it had been that $X >$ ECL, no statistical significance test (of the null hypothesis of noncompliance against an alternative of compliance) would need to be made because the estimate itself would exceed both the ECL and LDV (lower decision value). In that case it would be a noncompliance exposure for the employer.

However, since $X \leq$ ECL, it is necessary to compute the one-sided LDV to do a hypothesis test for compliance. The decision is to be made with at least 95% confidence; that is, with 5% or less probability of being wrong (in statistical terminology, a 0.05 or less probability of a type 1 error). Thus, a subscript of 5% for LDV will be used to identify the decision test's level of significance (also called the size of the significance test).

$$\begin{aligned} \mathrm{LDV}_{5\%} &= \mathrm{ECL} - (1.645)(CV_T)(ECL) \\ &= (0.05 \text{ ppm}) - (1.645)(0.09)(0.05 \text{ ppm}) \\ &= (0.05 \text{ ppm})(1 - 0.148) = (0.05)(0.852) \\ &= 0.043 \text{ ppm} \end{aligned}$$

Then the exposure estimate X is classified according to the following criteria:

a. If $X \leq \text{LDV}_{5\%}$, classify as compliance exposure, (i.e., reject the null hypothesis and accept the alternative hypothesis) or

b. If $X > \text{LDV}_{5\%}$, classify as possible overexposure, (i.e., do not reject the null hypothesis).

Since the exposure measurement $X = 0.040$ ppm is less than the $\text{LDV}_{5\%}$ of 0.043 ppm, the employer can decide that the measured worker had a compliance exposure on the day and during the period of the measurement.

2. For a compliance officer, because $X \leq \text{ECL}$, no statistical test of the null hypothesis of compliance against an alternative of noncompliance need be made because the estimate is already less than the ECL and therefore would be less than the UDV (upper decision value). Hence it would be a compliance exposure for the compliance officer.

However, if the result had been $X > \text{ECL}$, it would have been necessary to compute the one-sided UDV test statistic for noncompliance:

$$\text{UDV}_{5\%} = ECL + (1.645)(CV_T)(ECL)$$

Then the exposure estimate is classified according to the following criteria:

a. If $X > \text{UDV}_{5\%}$, classify as noncompliance exposure, (i.e., reject the null hypothesis and accept the alternative hypothesis), or

b. If $X \leq \text{UDV}_{5\%}$ classify as possible overexposure, (i.e., do not reject the null hypothesis).

COMMENTS. Figure 52.13 is a graphical interpretation of decision regions bounding an *exposure standard* (i.e., ECL) using LDV values for employer decisions and UDV values for compliance officer decisions. The observed value X for the TWA exposure estimate is denoted by TWA inside a diamond in Figure 52.13.

Note the limited applicability of these tests. The outcomes of these hypothesis tests apply only to the particular worker's exposure on the day and during the period of the exposure measurement.

The use of CV_T in the classification formulas is equivalent to calculating the standard deviation of X (the single exposure measurement) as $[(\text{CV}_T)(\text{ECL})]$ instead of $[(\text{CV}_T)(\mu)]$, where μ is the true exposure for which X is the estimate. The use of a one-sided, 5% significance level, lower decision value or a one-sided, 5% significance level, upper decision value ($\text{LDV}_{5\%}$ or $\text{UDV}_{5\%}$, see Section 3.7) computed in this manner is correct, since the classification rule selected for use as a compliance officer's test for noncompliance is algebraically equivalent to a significance test of the null hypothesis of compliance (i.e., of H_0: $\mu \leq \text{ECL}$). Similarly, the classification rule selected for use as the employer's test for compliance is equivalent to a significance test of the null hypothesis of noncompliance (i.e., of H_0: $\mu > \text{ECL}$). In both cases, setting $\mu = \text{ECL}$ to compute the decision value serves to keep the size of the test at or below 0.05 ($\leq 5\%$ probability of incorrectly rejecting the null hypothesis).

The rationale for the hypothesis test is

1. For a test at the 5% level of significance, calculate a one-sided upper decision value ($\text{UDV}_{5\%}$) or lower decision value ($\text{LDV}_{5\%}$). These decision values are in fact critical values for the measurement X, under the null hypothesis that the true exposure is equal to the

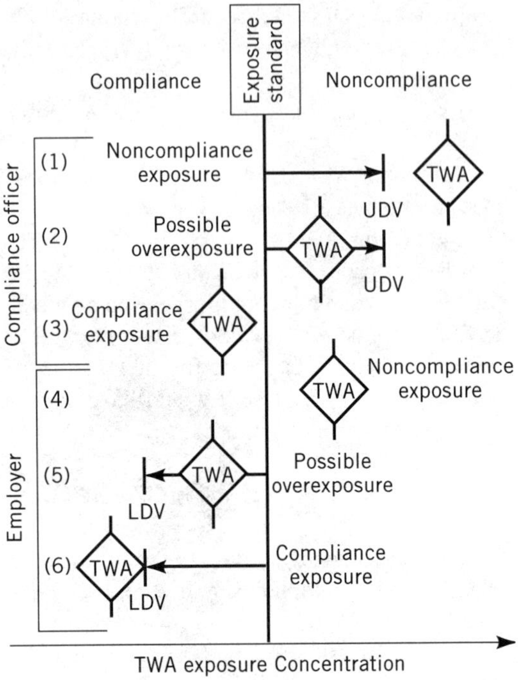

Figure 52.13. Classification of exposure with hypothesis testing (one-sided decision values).

ECL. If the null hypothesis were true, the exposure estimate would not exceed the UDV (or be less than the LDV for the complementary hypothesis test) more than 5% of the time for the same true exposure and measurement conditions.

2. For a test of noncompliance, if the exposure estimate exceeds the one-sided upper decision value ($UDV_{5\%}$), reject the null hypothesis of compliance and decide for noncompliance. Similarly, for a test for compliance, if the exposure estimate is less than the one-sided lower decision value ($UDV_{5\%}$), reject the null hypothesis of noncompliance and decide for compliance.

6.2 Full-Period Consecutive-Samples Estimate for Uniform Exposure

The methods in this section are for exposure estimates calculated from a series of consecutive exposure measurements (equal or unequal time duration) that collectively span the full duration of the desired time-averaging period.

6.2.1 *Confidence Limits for a True TWA Exposure*

PURPOSE. Compute one- or two-sided confidence limits for a true TWA exposure, based on multiple exposure measurements in a uniform exposure situation.

ASSUMPTIONS. The random errors for measurements X_i are assumed to be normally and independently distributed (see Section 3.6.1.3). The true worker exposure average for a sampled period during the total period of the exposure estimate TWA (i.e., true arithmetic mean of the hypothetical parent population of replicate exposure measurements at exactly the same point in space during the same time period of length T_i is unknown, but is assumed to be about the same for all periods sampled (uniform exposure environment). The coefficient of variation of replicate measurements made by the exposure measurement procedure is assumed known and constant (see Section 3.4.2). The accuracy of the computed confidence limit(s) will be a function of the accuracy of the coefficient of variation used.

It is assumed here that all sampled periods have equal true average concentrations; if it is expected that the samples have significantly different values because of different exposure situations during the workshift, then the conservative procedure in Section 6.3.1 can be used. Where exposures are highly variable over the different sampling periods during the duration of the estimated TWA, the use of the formulas in this section will underestimate the random sampling error in TWA, thus underestimating the width of the confidence interval.

Example. An exposure measurement procedure with a CV_T of 0.08 was used to obtain an 8-hr TWA exposure estimate denoted TWA. A personal pump and three charcoal tubes were used to consecutively measure a worker's approximately uniform exposure to isoamyl alcohol. The analytical lab reported the following exposure estimates for the three tubes: $X_1 = 90$ ppm over $T_1 = 150$ min, $X_2 = 140$ ppm over $T_2 = 100$ min, and $X_3 = 110$ ppm over $T_3 = 230$ min. The ECL chosen by the industrial hygienist was 100 ppm. The CV_T for the measurement procedure is 0.08 (8%). Compute the one-sided 95% upper confidence limit for the true 8-hr TWA exposure. The TWA estimate is 110 ppm.

Solution

1. If the sample durations ($T_1, T_2, \cdots, T_n$) are approximately equal, the one-sided 95% upper confidence limit for the true TWA exposure can be computed from the short equation:

$$UCL_{1,\ 0.95} = TWA/\{1 - [(1.645)(CV_T)]/(n)^{1/2}\}$$

In this example the sample durations are not approximately equal, but for illustrative purposes and as a first approximation:

$$\begin{aligned} UCL_{1,\ 0.95} &= \frac{110 \text{ ppm}}{1 - [(1.645)(0.08)]/(3)^{1/2}} \\ &= \frac{110 \text{ ppm}}{1 - 0.0760} \\ &= \frac{110 \text{ ppm}}{0.924} \\ &= 119 \text{ ppm} \end{aligned}$$

Thus an approximate one-sided 95% $UCL_{1,\ 0.95}$ for the true TWA exposure is 119 ppm based on the measured 110 ppm TWA estimate.

2. Since the sample durations are not equal, compute the more exact one-sided 95 percent upper confidence limit for the true TWA from the longer equation:

$$\mathrm{UCL}_{1,\ 0.95} = \frac{(\mathrm{TWA})(T_1 + \cdots + T_n)}{(T_1 + \cdots + T_n) - [(1.645)(\mathrm{CV}_T)(T_1^2 + \cdots + T_n^2)^{1/2}]}$$

In the example:

$$\begin{aligned}\mathrm{UCL}_{1,\ 0.95} &= \frac{(110\ \mathrm{ppm})(150 + 100 + 230)}{(480) - [(1.645)(0.08)(150^2 + 100^2 + 230^2)^{1/2}]}\\ &= \frac{(110\ \mathrm{ppm})(480)}{(480) - (38.5)}\\ &= 120\ \mathrm{ppm}\end{aligned}$$

In this case the short equation gave a confidence limit that was only slightly lower than the more accurate estimate. The $\mathrm{UCL}_{1,0.95}$ is 120 ppm based on the 110 ppm TWA estimate. That is, one can be 95% confident that the true TWA exposure is less than 120 ppm.

3. To compute a one-sided 95% lower confidence limit ($\mathrm{LCL}_{1,\ 0.95}$) substitute a plus sign for the minus sign in the denominator of the step 1 or 2 formulas. To compute a two-sided 95 percent confidence interval ($\mathrm{LCL}_{2,\ 0.95}$ to $\mathrm{UCL}_{2,\ 0.95}$) for the true TWA exposure, substitute a Z value of 1.96 for 1.645.

COMMENTS. This method is an extension of the procedure presented in Section 6.1.1 for a TWA estimate based on a single exposure measurement. Review the comments of that section.

The indicated uncertainty in the TWA estimate reflected in the 10-ppm difference between the TWA of 110 ppm and the $\mathrm{UCL}_{1,\ 095}$ of 120 ppm (9% higher than the TWA estimate in this example) does not incorporate uncertainty due to interday exposure level variability (see Section 3.3.1 and Section 3.6.2.3). The one-sided confidence limit pertains only to the worker's true exposure on the day actually sampled and during the total period of the three measurements.

6.2.2 *Classification of Exposure (Hypothesis Testing)*

PURPOSE. Classify a TWA exposure estimate that is based on multiple exposure measurements from a uniform exposure situation, as noncompliance exposure, possible overexposure, or compliance exposure, regarding an exposure control limit (ECL).

ASSUMPTIONS. The same as used for Section 6.2.1. The formulas of this section strictly apply only to the case of uniform exposure. Where true exposures are highly variable between the sampling periods over the duration of the TWA estimate (nonuniform exposure situation), the use of the formulas in this section will underestimate the random sampling error in the TWA estimate. This misapplication would therefore increase the chance of deciding a noncompliance exposure (with the test for the compliance officer) or deciding a compliance exposure (with the test for the employer). For a nonuniform exposure situation, use the conservative methods given below in Section 6.3.2.

EXAMPLE. The same as used for Section 6.2.1, except classify with 95% confidence the 110-ppm TWA exposure estimate relative to the ECL of 100 ppm.

SOLUTION

1. If the sample periods (T_1, T_2, $\cdots$, T_n) are equal (or, for a somewhat inexact test, approximately equal), classify the TWA estimate using the short equations in step 2 (for an employer) or step 3 (for a compliance officer). For unequal sample periods, classify using the longer equations in step 4 (for an employer) or step 5 (for a compliance officer).

2. For an employer, if TWA $\geq$ ECL, a possible overexposure is indicated by the TWA itself and no statistical test for compliance need be made. This is the case for this example. However, when there are approximately equal sample periods and TWA $<$ ECL, compute the one-sided LDV (lower decision value) to do a test for compliance:

$$\mathrm{LDV}_{5\%} = ECL - \frac{(1.645)(\mathrm{CV}_T)(\mathrm{ECL})}{(n)^{1/2}}$$

Then classify the TWA exposure estimate based on the multiple measurements according to:

a. If TWA $\leq$ $\mathrm{LDV}_{5\%}$, classify as compliance exposure.

b. If TWA $>$ $\mathrm{LDV}_{5\%}$, classify as possible overexposure.

3. For a compliance officer, if TWA $\leq$ ECL, it is superfluous to make a statistical test for noncompliance, since the TWA estimate obviously would also be less than the UDV because by definition the ECL is always less than the UDV. However, for TWA $>$ ECL, and where there are approximately equal sampling periods, the compliance officer would compute the one-sided $\mathrm{UDV}_{5\%}$ critical value of the TWA:

$$\mathrm{UDV}_{5\%} = \mathrm{ECL} + \frac{(1.645)(\mathrm{CV}_T)(\mathrm{ECL})}{(n)^{1/2}}$$

Then the exposure estimate would be classified according to:

a. If TWA $>$ $\mathrm{UDV}_{5\%}$, classify as noncompliance exposure.

b. If TWA $\leq$ $\mathrm{UDV}_{5\%}$, classify as possible overexposure.

4. For an employer, when the periods of the measurements are not approximately equal and TWA $<$ ECL, compute the one-sided LDV critical value of the TWA exposure estimate:

$$\mathrm{LDV}_{5\%} = \mathrm{ECL} + \frac{(1.645)(CV_T)(\mathrm{ECL})(T_1^2 + \cdots + T_n^2)^{1/2}}{T_1 + \cdots + T_n}$$

Then use the classification criteria of step 2 to perform a hypothesis test for compliance.

5. For a compliance officer, when the periods of the measurements are not approximately equal and TWA $>$ ECL, compute the one-sided $\mathrm{UDV}_{5\%}$ to be used as the critical value of the TWA exposure estimate:

$$\mathrm{UDV}_{5\%} = \mathrm{ECL} + \frac{(1.645)(\mathrm{CV}_T)(\mathrm{ECL})(T_1^2 + \cdots + T_n^2)1/2}{T_1 + \cdots + T_n}$$

Then use the classification criteria of step 3 to perform a hypothesis test for noncompliance. For this example:

$$\mathrm{UDV}_{5\%} = 100 + \frac{(1.645)(0.08)(100)(150^2 + 100^2 + 230^2)1/2}{150 + 100 + 230}$$
$$= 100 + 8.0 = 108 \text{ ppm}$$

Since the TWA = 110 ppm exceeds the $\mathrm{UDV}_{5\%}$ of 108 ppm, the TWA exposure estimate of 110 ppm is classified as a statistically significant noncompliance exposure. The compliance officer can state that, for the measured worker on the day and during the period of the three measurements, the true 8-hr TWA exposure was a noncompliance exposure. The statistical decision of noncompliance is made with 5% or less probability of being wrong.

COMMENTS. The same as for Section 6.1.2, which should be reviewed before the use of this section.

The outcomes of these hypothesis tests are valid only for the worker's exposure on the day and during the period of the exposure measurements.

In the example, the measurement results indicate a sufficiently uniform exposure over the duration of the TWA estimation period, so that use of the formulas in this section is appropriate.

6.3 Full-Period Consecutive-Samples Estimate for Nonuniform Exposure

The methods in this section are a generalization of those in Section 6.2. They are for exposure estimates calculated from a series of consecutive exposure measurements (equal or unequal time duration) that collectively span the full duration of the desired TWA exposure period. The following methods are longer than those of Section 6.2 but are not limited by the assumption of nearly equal arithmetic mean exposures during the TWA period (uniform exposure environment). For highly nonuniform exposure situations, use of the less complex methods of Section 6.2 (intended for uniform exposure situations) may slightly underestimate the sampling error in the TWA estimate. However, the conservative methods of this section will usually slightly overestimate the sampling error in the TWA estimate. The $\mathrm{LDV}_{5\%}$ of Section 6.3.2 will be slightly lower than that from Section 6.2.2 and the $\mathrm{UDV}_{5\%}$ from Section 6.3.2 will be slightly higher than that of Section 6.2.2.

6.3.1 *Confidence Limits for a True TWA Exposure*

PURPOSE. Compute one- or two-sided confidence limits for a true TWA exposure, based on an exposure estimate calculated from multiple exposure measurements from a nonuniform exposure situation.

ASSUMPTIONS. The same as used for Section 6.2.1, except that there is no assumption that arithmetic mean exposures are all essentially equal (for the n sample periods that make up the TWA period).

EXAMPLE. Two charcoal tubes and a personal pump were used to obtain an 8-hr TWA exposure estimate, denoted TWA, for isoamyl alcohol. The results for the two tubes were reported as $X_1 = 30$ ppm over $T_1 = 300$ min. and $X_2 = 140$ ppm over $T_2 = 180$ min.

with a CVT of 0.08 for the measurement procedure. The ECL chosen by the industrial hygienist was 100 ppm. For this nonuniform exposure situation, compute the two-sided 95% confidence interval for the true 8-hr TWA exposure, based on the following estimated TWA = 71 ppm:

$$\text{TWA} = \frac{(300\ \text{min})(30\ \text{ppm}) + (180\ \text{min})(140\ \text{ppm})}{480\ \text{min}} = 71.2\ \text{ppm}$$

SOLUTION

1. If the sample durations (T_1, T_2, $\cdots$, T_n) were approximately equal, the two-sided 95% confidence interval for the true TWA exposure could be computed from the short equation:

$$\text{LCL}_{2,\ 0.95} \text{ and } \text{UCL}_{2,\ 0.95} = \text{TWA} \pm \frac{(1.96)(\text{CV}_T)(X_1^2 + \cdots + X_n^2)^{1/2}}{(n)(1 + \text{CV}_T^2)^{1/2}}$$

In this example the durations are not approximately equal, but for illustrative purposes and as a first approximation:

$$\text{LCL}_{2,\ 0.95} \text{ and } \text{UCL}_{2,\ 0.95} = 71.2 \pm \frac{(1.96)(0.08)(30^2 + 140^2)^{1/2}}{(2)(1 + 0.08^2)^{1/2}}$$
$$= 71.2 \pm 11.2 = 60.0 \text{ and } 82.4 \text{ ppm}$$

which is reported as 60 and 82 ppm. Thus an approximate two-sided 95% confidence interval for the true TWA exposure is 60 to 82 ppm. The best point estimate for the true exposure is the TWA estimate of 71 ppm.

2. Since the sample durations are not equal, compute the more accurate two-sided 95% confidence interval for the true TWA exposure from the longer equation:

$$\text{LCL}_{2,\ 0.95} \text{ and } \text{UCL}_{2,\ 0.95} = \text{TWA} \pm \frac{(1.96)(\text{CV}_T)(T_1^2X_1^2 + \cdots + T_n^2X_n^2)^{1/2}}{(T_1 + \cdots + T_n)(1 + \text{CV}_T^2)^{1/2}}$$

In the example:

$$\text{LCL}_{2,\ 0.95} \text{ and } \text{UCL}_{2,\ 0.95} = 71.2 \pm \frac{(1.96)(0.08)(300^2 \times 30^2 + 180^2 \times 140^2)^{1/2}}{(300 + 180)(1 + 0.08^2)^{1/2}}$$
$$= 71.2 \pm 8.7 = 62.5 \text{ and } 79.9 \text{ ppm}$$

which is reported as 62 and 80 ppm. Thus an improved, more accurate value for the two-sided 95% confidence interval for the true average exposure is 62 to 80 ppm with a point estimate TWA of 71 ppm. That is, one can be 95% confident that the true TWA exposure is between 62 and 80 ppm.

3. To compute either a lower or upper one-sided 95% confidence limit for the true average exposure, substitute a Z value of 1.645 for 1.960 in step 1 or 2.

COMMENTS. Similar to those given in Section 6.1.1 and Section 6.2.1.

6.3.2 *Classification of Exposure (Hypothesis Testing)*

PURPOSE. Classify a TWA exposure estimate that is based on multiple exposure measurements from a nonuniform exposure situation, as noncompliance exposure, possible overexposure, or compliance exposure, relative to an exposure control limit (ECL).

ASSUMPTIONS. The same as used for Section 6.2.1, except that there is no assumption that arithmetic mean exposures are essentially equal during n sample periods within the TWA period.

EXAMPLE. The same as used for Section 6.3.1, except classify the TWA estimate of 71 ppm relative to the 100-ppm ECL.

SOLUTION

1. If the sample periods (T_1, T_2, $\cdots$, T_n) are approximately equal, classify the TWA estimate using the short equation in step 2 (for an employer) or the equation in step 3 (for a compliance officer). For unequal sample periods, use the longer equations in step 4 (for an employer) or step 5 (for a compliance officer).

2. For an employer, if TWA $\geq$ ECL, no statistical test for compliance need be made. Since by definition the ECL always exceeds the LDV, therefore the TWA must also exceed the LDV. However, when TWA $<$ ECL and if there are approximately equal sample periods, compute the following LDV for use in a one-sided hypothesis test for compliance:

$$\text{LDV}_{5\%} = \text{ECL} - \frac{[(1.645)(\text{CV}_T)(X_1^2 + \cdots + X_n^2)^{1/2}](\text{ECL/TWA})}{(n)(1 + \text{CV}_T^2)^{1/2}}$$

For this example the two measurement periods are not approximately equal, but as an illustration (and as a first approximation of $\text{LDV}_{5\%}$):

$$\begin{aligned}\text{LDV}_{5\%} &\approx 100 - \frac{[(1.645)(0.08)(30^2 + 140^2)^{1/2}](100/71)}{(2)(1 + 0.08^2)^{1/2}} \\ &= 100 - 13.2 = 86.8 \text{ ppm}\end{aligned}$$

which is reported as 87 ppm after appropriate rounding. Then classify the TWA exposure estimate based on the multiple measurements according to:

a. If TWA $\leq \text{LDV}_{5\%}$, classify as compliance exposure.

b. If TWA $> \text{LDV}_{5\%}$, classify as possible overexposure.

Since the TWA point estimate of 71 ppm is substantially less than the approximate $\text{LDV}_{5\%} = 87$ ppm, the employer can state that, for the measured worker on the day and during the period of the measurement, the exposure estimate was a compliance exposure. The probability of this decision being incorrect is considerably less than 5%, since TWA is considerably less than its critical value ($\text{LDV}_{5\%}$) for the test of compliance. However, since the example has unequal sampling periods, the equation in step 4 below was used to obtain a more accurate value of $\text{LDV}_{5\%}$. The formula in step 4 yields a higher critical value ($\text{LDV}_{5\%} = 90$ ppm, see below) for TWA, but it does not change the classification decision. The step 4 procedure yields a more highly statistically significant decision of compliance exposure than is provided by the more conservative step 2 test.

3. For a compliance officer, if TWA $\leq$ ECL (as in this example), it is superfluous to compute a statistical test for noncompliance (see discussion in Section 6.2.2, part 3 of the solution). However, for TWA > ECL, and where there are approximately equal sample periods, the compliance officer would compute the following noncompliance one-sided test statistic:

$$\mathrm{UDV}_{5\%} = \mathrm{ECL} + \frac{[(1.645)(\mathrm{CV}_T)(X_1^2 + \cdots + X_n^2)^{1/2}](\mathrm{ECL/TWA})}{(n)(1 + \mathrm{CV}_T^2)^{1/2}}$$

Then the exposure estimate TWA, which is based on the multiple measurements, would be classified according to:

a. If TWA > $\mathrm{UDV}_{5\%}$ classify as noncompliance exposure.

b. If TWA $\leq$ $\mathrm{UDV}_{5\%}$, classify as possible overexposure.

4. For an employer, when the periods of the measurements are not approximately equal and TWA < ECL, compute the one-sided LDV critical value of the TWA:

$$\mathrm{LDV}_{5\%} = \mathrm{ECL} - \frac{[(1.645)(\mathrm{CV}_T)(T_1^2X_1^2 + \cdots + T_n^2X_n^2)^{1/2}](\mathrm{ECL/TWA})}{(T_1 + \cdots + T_n)(1 + \mathrm{CV}_T^2)^{1/2}}$$

Then use the classification criteria of step 2. In this example:

$$\begin{aligned}\mathrm{LDV}_{5\%} &= 100 - \frac{[(1.645)(0.08)(300^2 \times 30^2 + 180^2 \times 40^2)^{1/2}](100/71)}{(300 + 180)(1 + 0.08^2)^{1/2}} \\ &= 100 - 10.3 = 89.7 \text{ ppm}\end{aligned}$$

which is reported as 90 ppm.

5. For a compliance officer, when the periods of the measurements are not approximately equal and TWA > ECL, compute the noncompliance one-sided test statistic:

$$\mathrm{UDV}_{5\%} = \mathrm{ECL} + \frac{[(1.645)(\mathrm{CV}_T)(T_1^2X_1^2 + \cdots + T_n^2X_n^2)^{1/2}](\mathrm{ECL/TWA})}{(T_1 + \cdots + T_n)(1 + \mathrm{CV}_T^2)^{1/2}}$$

Then use the classification criteria of step 3.

COMMENTS. The same as for Section 6.1.2, plus the following statistical note applicable to hypothesis testing for nonuniform exposures.

Statistical Note: The following notation is used. Let

μ_i = true exposure during ith exposure interval of the time-averaging period,
$i = 1, 2, \cdots, n$
$\mu = (1/T)(T_1\mu_1 + T_2\mu_2 + \cdots + T_n\mu_n)$ = true TWA exposure
T_i = length of ith sampling interval
$T = T_1 + T_2 + \cdots + T_n$ = time-averaging period of TWA and associated ECL
X_i = Exposure estimate for the ith exposure interval

To test H_0: $\mu \leq$ ECL (i.e., compliance exposure) against the alternative hypothesis H_1: μ > ECL (i.e., noncompliance exposure), a critical value must be selected for the measured exposure TWA. The critical value for a test of noncompliance is intended to be the TWA that would be exceeded at most only 5% of the time when H_0 is true. The correct critical value (denoted by $\text{UDV}_{5\%}$ for the upper decision value) is

$$\text{UDV}_{5\%} = \text{ECL} + [(1.645)(\text{CV}_T)]\left[\frac{\left(\sum_{i=1}^{n} T_i^2\mu_i^2\right)}{(T^2)(1 + \text{CV}_T^2)}\right]^{1/2}$$

where

$$\left(\frac{1}{T}\right)\sum_{i=1}^{n} T_i\mu_i = \text{ECL}$$

is a restriction on the μ_i's. A problem of this nonuniform case is that many combinations of μ_i's exist that all have the desired time-weighted average (ECL) under the null hypothesis H_0. The range of choices for μ_i's is discussed below, and our preferred selection is given for use in the classification decision formulas given elsewhere in Section 6.3.2.

1. Maximum statistical power (probability of rejecting H_0 when H_1 is true) would be obtained by choosing

$$\mu_i = \left(\frac{T}{T_i}\right)\left(\frac{\text{ECL}}{n}\right) \qquad i = 1, \cdots, n$$

However, if this choice is made, the probability of falsely rejecting the null hypothesis [i.e., of falsely deciding for noncompliance (H_1) when compliance (H_0) actually exists] would be higher than the intended 5%. This procedure might be acceptable for the type of "screening monitoring" that is followed by confirmatory sampling. But usually this choice of μ_i's under H_0 would be unsatisfactory because it gives a liberal test of noncompliance. It could also be shown to give a liberal test of compliance. It would usually not be a good choice for use by either the compliance officer or the employer.

2. Minimum statistical power to reject H_0 would be obtained by choosing

$$\mu_i = \frac{T}{T_1}(\text{ECL}) \quad \text{and} \quad \mu_i = 0$$

$$i = 2, \cdots, n$$

The index value $i = 1$ could be assigned to any of the n intervals. All the exposure is put into this single interval and none into the other intervals. This choice provides a conservative test with less than a 5% chance of making a type 1 decision error (falsely rejecting H_0). The advantage to making a noncompliance decision with this procedure is that the choice of μ_is would be incontrovertible. Also, if an employer could show compliance using this procedure, there would be high assurance that the ECL was not exceeded on the day of the exposure measurement. However, the conservatism of this procedure would make it undesirable for routine use by a compliance officer because it provides less statistical power to detect true overexposure situations, and this could compromise worker protection.

3. On balance, it seems best to try to control the size of the hypothesis test as close to the intended 5% level as one can. To attempt this, a substitution to follow μ_i estimates into the UDV (or LDV) formulas given earlier in this statistical note:

$$\hat{\mu}_i = \left(\frac{\text{ECL}}{\text{TWA}}\right)(X_i)$$

$$i = 1, 2, \cdots, n$$

This procedure provides μ_i values that have ECL as their time-weighted average (under H_0), and the $\hat{\mu}_i$ values are in the same ratios to each other as are their sample estimates. The resulting formulas for UDV and LDV have been presented in Section 6.3.2.

6.4 Grab-Sample Estimate, Small-Sample Size (Less than 30 Measurements During the Time-Averaging Period)

Grab samples are samples taken during intervals that are short compared to the duration of the time-averaging period they are drawn from (e.g., intervals of seconds to about 2 min. for a 15-min. TWA period and up to about 30 min for TWA periods of 8-hr). Grab-sample measurements should be taken at random intervals during the desired time-averaging period (see Ref. 107). It is suggested that these computations be programmed into a PC spreadsheet program or a program such as SigmaPlot® 5.0, which could also plot the TWA estimates and the computed confidence limits on a *semilogarithmic time-series* plot as discussed in Section 6.11 of this chapter.

6.4.1 Confidence Limits for a True TWA Exposure, Small Number of Grab Samples

PURPOSE. Compute one- or two-sided confidence limits for an exposure estimate based on less than 30 grab samples.

ASSUMPTIONS. Unfortunately, there are no simple statistical methods available to determine exact one- or two-sided confidence limits for an arithmetic mean that is estimated from a small sample size (less than 30) of grab samples as an exposure estimate. This is because grab-sample data generally reflect a lognormal distribution of true exposure levels among the short intervals that are sampled (see Section 3.6.2.2). Thus the statistical problem is to estimate confidence limits for the arithmetic mean of a lognormal distribution. It is possible to easily compute one- or two-sided exact confidence limits for the geometric mean of a lognormal distribution of grab-sample results. But it is important to note that a geometric mean exposure is not the most suitable parameter for estimation of worker exposure risk. Therefore, this section will present a well-known approximate procedure for confidence limits on the true arithmetic mean exposure. The geometric mean is also an inappropriate parameter for comparison to an exposure control limit (ECL). Fortunately, an exact exposure classification procedure for grab-sample estimates was made available by NIOSH (2, 132), which is presented in Section 6.4.2. Also, more recently, Armstrong (133) has provided tables of factors that facilitate the calculation of exact confidence limits for the arithmetic mean of a lognormal distribution. Refer to the notes at the end of Section 6.4.2.

For small sample sizes, Aitchison and Brown (70) state that statistical theory fails to provide a means of obtaining exact confidence limits or intervals for the arithmetic mean (μ_X) of a lognormally distributed variable X. For large sample sizes, their suggestion is to compute the limits:

$$\bar{X} \pm \frac{Z_P s_X}{n^{1/2}}$$

where $Z_P = 1.96$ for two-sided confidence limits at the $P = 0.95$ confidence level and s_X is the sample standard deviation computed from a random sample of n lognormally distributed X's (e.g., concentrations in n grab samples taken at random intervals). See Section 6.5.1 for our recommended slight modification to the Aitchison and Brown (70) large-sample ($n \geq 30$) confidence limits. The authors have modified the formula only slightly by using Student's t statistic, $t_{P,n-1}$, in place of the standard normal deviate, Z_P.

For the small-sample case ($n < 30$), which is the subject of this section, no simple theory for exact confidence limits on μ_X is available. However, such confidence limits are not needed to perform statistical tests of significance under null hypotheses of compliance ($\mu_X \leq$ ECL) and noncompliance ($\mu_X >$ ECL). The hypothesis tests concerning μ_X, the arithmetic mean concentration, can be carried out with the sample mean ($\bar{y}$) and sample standard deviation (s_y) of the logarithmic transformation ($y = \ln X$). Detailed explanations of two special methods for testing hypotheses about μ_X are given in Section 6.4.2. In case confidence limits on μ_X are desired in the small-sample case, an often-used approximate method is outlined here.

SOLUTION

1. First, for the ln-transformed data, compute exact confidence limits on μ_y, the true mean of $y = \ln X$. These exact limits are given by

$$\bar{y} - \frac{t_{P,n-1} s_y}{\sqrt{n}} < \mu_y < \bar{y} + \frac{t_{P,n-1} s_y}{\sqrt{n}}$$

where $t_{P,n-1}$, is the Student's t statistic for $(n - 1)$ degrees of freedom for the P level of significance. For example, for two-sided 95% confidence limits and a sample size of 10, $t_{0.95,\,9} = 2.262$.

2. Then, detransform each logarithmic-value confidence limit on μ_y to a corresponding value of μ_X using the relationship:

$$\mu_x = \exp\,(\mu_y)\,\exp\left(\frac{\sigma_y^2}{2}\right) \approx \exp\,(\mu_y)\,\exp\left(\frac{s_y^2}{2}\right)$$

The approximation is due to the necessary use of the sample variances s_y^2 instead of the true variance σ_y^2 to perform the detransformation (i.e., to convert the geometric mean of

$X[GM_X = \exp(\mu_y)]$ to the corresponding arithmetic mean of $X(\mu_X)$). This approximate method is adequate for most practical industrial hygiene applications.

COMMENTS. This approach is graphically illustrated in Figure 52.14.

6.4.2 *Classification of Exposure (Hypothesis Testing), Small Number of Grab Samples*

PURPOSE. Based on multiple grab samples, classify an exposure estimate as noncompliance exposure, possible overexposure, or compliance exposure relative to an exposure control limit (ECL).

ASSUMPTIONS. Note: This method cannot process zero data values. It is assumed that none of the sample results are zero. Refer to Ref. 115 for a discussion of how to treat a sample population that includes zero values.

The grab-sample measurements are drawn from a single two-parameter lognormal distribution of true exposure levels. If it is suspected that the parent population of exposure levels is more accurately described by a three-parameter lognormal distribution (see a Section 3.5.2), then the location constant *k* should be estimated (see Section 6.12, Step 5 of the Solution) and subtracted from all measurements and the ECL before the method in this section is used.

Grab-sample populations also have a component of approximately normally distributed, measurement-procedure error, but this component is assumed negligible compared to the lognormally distributed environmental variations in the true exposure levels sampled. The net effect of measurement errors is to increase slightly the variability of the lognormal distribution without changing its shape appreciably (71).

The true arithmetic mean TWA exposure (i.e., true arithmetic mean of a 2-parameter lognormal distribution of all exposure levels that occurred at the same sampling position over the duration of the TWA period) is assumed unknown, and the geometric standard

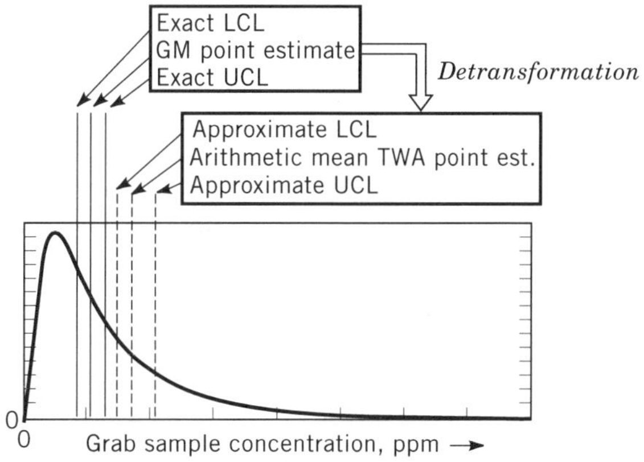

Figure 52.14. Two-sided confidence limits for a true TWA exposure estimated with a grab-samples exposure measurement.

deviation parameter of the parent lognormal exposure distribution is also assumed unknown. Both are estimated from the sample population of grab-sample results.

The grab-sample measurements are assumed to be a random sample of all exposure levels that occurred at the sampling position during the TWA period of interest. One should not attempt to estimate an 8-hr TWA-exposure based on short samples selected at random from only a small portion of the TWA period (e.g., the last 2 hr). The sample periods from the TWA period should be chosen as a random sample from the entire TWA period.

The statistical theory for the method in this section is contained in Bar-Shalom et al. (132).

EXAMPLE. A personal pump and eight charcoal tubes were used to estimate a worker's 8-hour TWA-exposure to ethyl alcohol. Each tube was placed on the worker for 20 min. The 20-min. periods were randomly selected from the 24 possible nonoverlapping periods during the workshift. The 8-hr ECL used by the employer is 1000 ppm. The following results were reported by the company laboratory: $X_1 = 1225$ ppm, $X_2 = 800$ ppm, $X_3 = 1120$ ppm, $X_4 = 1460$ ppm, $X_5 = 975$ ppm, $X_6 = 980$ ppm, $X_7 = 525$ ppm, and $X_8 = 1290$ ppm. With 95% confidence, classify the exposure estimate, TWA $= 1058$ ppm, relative to the ECL of 1000 ppm.

SOLUTION

1. Compute a standardized value (x_i) for each X_i sample result by dividing by the ECL, and then compute the base 10 logarithm (y_i) of each x_i. That is, $y_i = \log_{10}(X_i/\text{ECL})$.

2. In this example:

Original Sample Results X_i (ppm)	Standardized Results X_i/ECL	$y_i = \log_{10}$ (X_i/ECL)
1225	1.225	0.0881
800	0.800	−0.0969
1120	1.120	0.0492
1460	1.460	0.1644
975	0.975	−0.0110
980	0.980	−0.0088
525	0.525	−0.2798
1290	1.290	0.1106

3. Compute the classification variables, $\bar{y}$, s, and n. First compute the arithmetic mean $\bar{y}$ of the logarithmic values y_i, then compute the standard deviation s of the y_i values. The number of grab samples used in the computations is n.

In this example $\bar{y} = 0.0020$, $s = 0.1400$, and $n = 8$. These three variables (sample estimates of two parent population parameters and the sample size) will be used in the classification procedure.

4. Using the classification chart in Figure 52.15 for the average of several grab-sample measurements, plot the classification point that has coordinates $\bar{y}$ and s. A family of curves form the boundaries of three classification regions. Each pair of boundaries is for a given sample size n. Pairs of classification boundaries for odd values of n ranging from $n = 3$ to 25 are provided (all even sample sizes except for 4 will have to be interpolated).

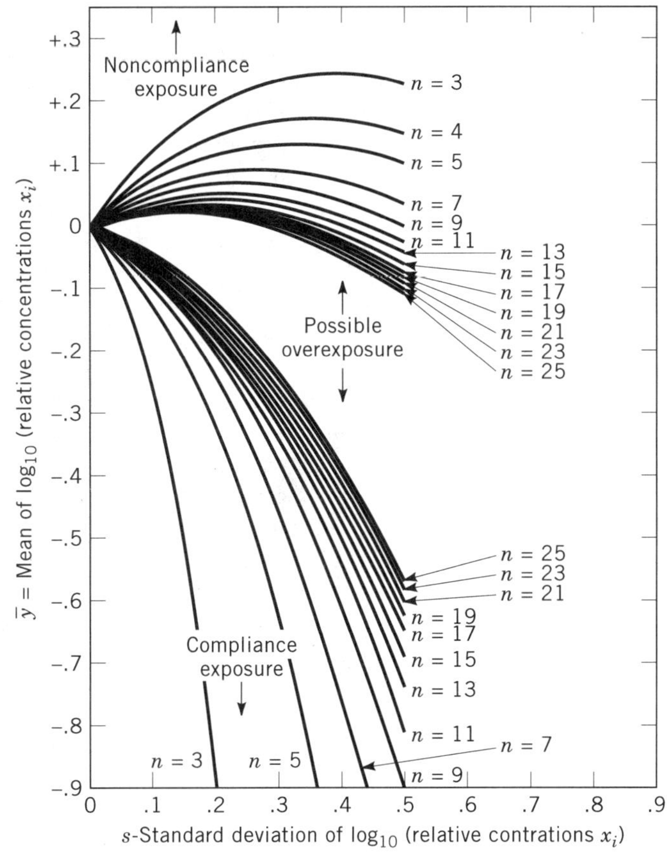

Figure 52.15. Grab-sample, measurement-average classification chart.

5. Classify the exposure estimate:

a. If the classification point lies on or above the upper curve of the pair corresponding to the number of measurements n (i.e., the n-curve pair), then classify as noncompliance exposure.

b. If the classification point lies below the lower curve of the n-curve pair, then classify as compliance exposure.

c. If the classification point is between the two curves for sample size n, then classify as possible overexposure.

d. If s exceeds 0.5 (greater than the range of the classification chart), a possible cause is that one or more of the measurements is relatively distant from the main portion of the sample distribution (i.e., an outlier that could be a mistake). Another explanation is that two or more substantially different exposure level distributions are being mistakenly analyzed as a single distribution (see Section 3.3.4).

6. In this example the classification point lies between the upper interpolated $n = 8$ curve and the lower interpolated $n = 8$ curve. Thus the TWA exposure estimate of 1058 ppm, estimated from the eight grab-sample measurements, is classified as a possible overexposure relative to the ECL of 1000 ppm. That is, the estimate is not precise enough to be able to confidently say that the true exposure was not in compliance.

COMMENTS. The x_i ratios (X_n/ECL) computed in step 1 of the solution are standardized exposure values that make the concentrations of contaminant independent of the ECL (in concentration units). This enables us to use the same decision chart for any concentration level. All values $x_i = (X_i/\text{ECL})$ are comparable to a single scale of compliance with an ECL of unity for x_i. That is, the ECL for the transformed variable x will always be unity, which corresponds to zero on the $y = \ln(x)$ ordinate scale of the classification chart in Figure 52.15.

Rappaport and Selvin (134) emphasize that most 8-hr TWA PELs were formulated for the purpose of indirectly limiting corresponding long-term average exposures. They present a formal statistical significance test for noncompliance that compares an estimate of the long-term arithmetic average exposure to a permissible exposure level. The test implicitly assumes lognormal distributions of both interday and interworker 8-hr TWA exposures, and the test statistic (denoted by T) is calculated from 8-hr TWA exposure measurements made on a sample of workdays selected at random during a period of one year or so. They state that different employees should be measured on the different sampling days, selected at random from within a "uniform exposure group."

Hewett (135) agrees with these authors that a "compliance" test should not be made on a multiworker-average exposure level, even when the sampled workers are within a "uniform exposure group". Controlling a population average exposure would not necessarily protect a high percentage of the individual workers. For this purpose, the rationale for a tolerance limits test is preferable. The rationale for a tolerance limits test is already well understood and accepted.

Although it is not agreed that a long-term average PEL is a feasible regulatory concept, it is noted that the T statistic can also be adapted to the present purpose of testing for a significant difference between an 8-hr permissible exposure level (PEL) and the arithmetic mean (μ_c) of a lognormal distribution of grab-sample exposure levels. The proposed test statistic T is given by

$$T = \frac{\bar{x}_c - \text{PEL}}{s_{\bar{x}_c}}$$

where

$$\bar{x}_c = e^{[\bar{x}_{\ln(c)} + 1/2(s_{\ln(c)})^2]} \qquad s_{\bar{x}_c} = \sqrt{\frac{(\text{PEL})^2[(s_{\ln(c)})^2 + \tfrac{1}{2}(s_{\ln(c)})^4]}{n - 2}}$$

$\bar{x}_c$ is the maximum-likelihood estimator of true mean μ_c, $s_{\bar{x}_c}$ is the estimated standard deviation of $\bar{x}_c$, and $\bar{x}_{\ln(c)}$ and $s_{\ln(c)}$ are the sample mean and sample standard deviation of a natural logarithmic transform [ln(c)], calculated from n concentration measurements (denoted by c).

With respect to the size of the test (i.e., the probability a of a type I error), for $n > 5$ and GSD ≤ 3.0, an adequate approximation to the distribution of the T statistic is said to be Student's t distribution with $(n - 1)$ degrees of freedom.

The power of this T test [i.e., the probability $(1 - \beta)$ of obtaining a statistically significant difference when a hypothetical true difference exists] was evaluated in Rappaport and Selvin (134), both for an employer's test and for a compliance officer's test. For a test with 5% size ($\alpha = 0.05$), Table 52.8 provides approximate minimum sample sizes (n) that are needed to attain at least 90 percent power [i.e., $(1 - \beta) \geq 0.90$]. The formula below for n is the basis for computed values of n that correspond to central tendency and variability parameters of various assumed lognormal distributions of concentrations.

$$n \cong \frac{[Z_{(1-\alpha)} + Z_{(1-\beta)}]^2\{[\ln(\text{GSD})]^2 + \tfrac{1}{2}[\ln(\text{GSD})]^4\}}{\left(1 - \dfrac{\mu_c}{\text{PEL}}\right)^2}$$

where $\ln(\text{GSD}) = \sigma_{\ln(c)}$ = standard deviation of ln (c)
μ_c/PEL = arithmetic mean of c/permissible exposure limit

For a small sample ($n < 20$) of lognormally distributed air concentrations, Selvin and Rappaport (136) state that the simple arithmetic mean is a less-variable estimator of the population mean than the maximum-likelihood estimator for purposes of estimation of average occupational exposure. For larger sample sizes, even without use of the bias correction factor given in Aitchison and Brown (70), Selvin and Rappaport (136) state that the maximum-likelihood estimate (MLE) provides a more precise estimator of the arithmetic mean. Additional theory is given in Aitchison and Brown (70), which shows that for any sample size the MLE, combined with a correction for bias, provides a minimum variance unbiased estimate of the arithmetic mean of a lognormal distribution.

However, for *all sample sizes*, it is the *MLE* that must be used in order for available statistical decision theory to be applicable to making *exact* controlled-risk decisions of

Table 52.8. Minimum Values of n Needed for $\alpha = 0.05$ and $(1 - \beta) \geq 0.90$

	μ_0/PEL	GSD = 1.5	GSD = 2.0	GSD = 2.5	GSD = 3.0	GSD = 3.5
Compliance	0.10	2	6	13	21	30
	0.25	3	10	19	30	43
	0.50	7	21	41	67	96
	0.75	25	82	164	266	384
Noncompliance	1.25	25	82	164	266	384
	1.50	7	21	41	67	96
	2.00	2	6	11	17	24
	3.00	1	2	3	5	6

compliance and noncompliance with exposure control limits. Lyles and Kupper (119) present the exact (MLE) test in an algebraic (i.e., nongraphical) computational format, and they note that their algebraic format gives results equivalent to the graphical exact test procedure that was developed earlier, under contract, by NIOSH (132). The algebraic format for the exact test can accommodate any sample size, whereas the NIOSH decision chart is applicable only for sample sizes in the range from $n = 3$ to $n = 25$. However, the NIOSH graphical format has the intended advantage of being easier to implement in a field setting, and its sample size range was deliberately selected to be sufficient for field problem applications. The algebraic exact test statistic could be computed by means of a specialized computer program, but it would require inputting critical values of both the chi-squared and Student's *t* distributions.

Recently, Lyles and Kupper (119) have shown that the Rappaport and Selvin (134) test method requires sample sizes that are too large, and that their test also forfeits statistical power. Therefore, that test should only be used as a "conservative" test procedure. Lyles and Kupper (119) give an alternative approximate test statistic that is easy to compute, which they claim gives reasonably accurate test results except when variability is high (e.g., GSD > 5.7) or sample size is low (e.g., $n < 10$). Their paper contains a SAS computer program for an approximate sample size which is recommended by them for use when either the exact MLE test procedure, or their alternative approximate test procedure, is to be used. Sample sizes are keyed to: (*1*) desired statistical power of the test, (*2*) probability of a type I error, (*3*) ratio of the true average exposure to the exposure limit, and (*4*) GSD of day-to-day exposure levels in a *uniform setting*.

Therefore, although it is less convenient to apply than the approximate *T* test proposed by Rappaport and Selvin (134), or the alternative approximate test by Lyles and Kupper (119), the authors recommend uniform use of the bias-corrected MLE estimate of μ_c for making statistical decisions of compliance or noncompliance in a legal context. The graphical decision chart procedure given in Leidel et al. (2) (and presented earlier in this section) eases the clerical task of using the bias-corrected MLE estimate of the mean of lognormally distributed grab samples to make tests of compliance or noncompliance.

It is important to note that statistical tests for the mean of a lognormal distribution have been applied here to a single-shift average of grab samples taken during the same shift. The same statistical tests have also been proposed by others (134, 137–139) to test a *multishift average* of TWA exposures for compliance with a (hypothetical) lifetime-average occupational exposure limit. But Hewett correctly points out that no legal limit exists for any long-term average exposure level (i.e., for average exposures over periods longer than a single shift) (135, 140). Therefore, even though the proposed *statistical* test for a long-term mean is theoretically valid, such a test has no regulatory relevance at this time and is of academic interest only.

Armstrong (133) has published a useful table of factors that facilitate computation of exact confidence limits for the arithmetic mean of a lognormal distribution. Confidence limits for the true arithmetic mean are obtained by multiplying the appropriate factors, which are listed as function of geometric standard deviation (GSD) and sample size (*n*), times the estimated geometric mean (GM). The factors were determined by Armstrong from theory and tables published by Land (141, 142). Armstrong then used the resulting confidence limits to perform a significance test that is equivalent to the graphical exposure classification method given for small numbers of grab samples in Leidel et al. (2). Individual readers may or may not prefer using an exposure classification method that is based on confidence limits in preference to our graphical decision charts.

Hewett and Ganser (143) give yet another procedure for calculating Land's exact confidence limits for the mean of lognormally distributed concentration data. They present two graphs which give respective *C*-factors for use in calculating the 95% LCL or 95% UCL. The desired confidence limit is calculated as:

$$\mathrm{CL} = \exp\{\ln(\mu_y) + \mathrm{C}[s_y/(n-1)^{1/2}]\}$$

where $\mu = \exp(\bar{y} + (\frac{1}{2})s_y^2)$

CL is either the 95% UCL for the true mean μ_y, or the 95% LCL, depending on which chart is used to find the C-factor.

Armstrong (133) compares examples of exact confidence limits with corresponding approximate confidence limits computed by four different methods, including methods presented earlier in this chapter. He concludes that adequate accuracy for practical applications requires calculation of the exact limits, rather than any of the approximate limits, whenever GSD > 1.5 or $n < 25$.

6.5 Grab-Sample Estimates of the TWA Exposure, Large-Sample Size (30 or More Measurements During the Time-Averaging Period)

Usually one collects far fewer than 30 measurements during an 8-hr TWA period, or 15-min. TWA period, because of the cost of each measurement (even for inexpensive ones such as with colorimetric tubes) and limited availability of personnel to take the measurements. However, if one has a direct-reading instrument available for the contaminant of interest, then it is feasible to obtain more than 30 samples during the desired TWA period. If the larger number of samples can be taken at random intervals, this strategy is preferable to the small-sample size approach (less than 30) discussed in the previous section, since for larger sample sizes the uncertainty regarding the true TWA exposure is considerably less. Additionally, for sample sizes of 30 or more, the statistical analysis is less complex because the distribution of the average of the exposure measurements is adequately described by a normal distribution. This section will present methods for computing both confidence limits and hypothesis tests for TWA exposure estimates that are calculated from large samples. One does not have to calculate the logarithms of the standardized measurements (as was necessary in Section 6.4 for TWA estimates based on small sample sizes). The hypothesis tests are less complex, and this method can process zero data values, unlike the procedure in Section 6.4. It is suggested that these computations be programmed into a PC spreadsheet program or a program such as SigmaPlot® 5.0, which could also plot the TWA estimates and the computed confidence limits on a *semilogarithmic time-series* plot as discussed in Section 6.11 of this chapter.

6.5.1 Confidence Limits for a True TWA Exposure, ≥30 Grab Samples

PURPOSE. Compute one-or two-sided confidence limits for an exposure estimate based on 30 or more grab samples.

ASSUMPTIONS. This procedure is robust regarding the actual distribution of exposure levels occurring during the TWA period of the grab-sample measurements. Both the true exposure level (i.e., arithmetic mean of the distribution of all exposure levels that occurred at the sampling position over the duration of the TWA period) and the standard deviation of the parent exposure distribution are assumed unknown. Both are estimated from the results obtained from grab samples.

The grab-sample measurements are assumed to be a random sample of all exposure levels that occurred at the sampling position during the TWA period of interest. One should not attempt to estimate an 8-hr TWA exposure average based on short samples selected at random from only a small portion of the TWA period (e.g., the last 2 hr). The sample periods from the TWA period should be chosen as a random sample from the entire TWA period.

EXAMPLE. A direct-reading ozone meter with strip chart recorder was used to continually measure a worker's exposure to ozone. The following 35 measurement values were read off the strip chart record at 35 times randomly selected within the 8-hr period. All values are given in ppm.

0.084	0.062	0.127	0.057	0.101	0.072	0.077
0.0145	0.084	0.101	0.105	0.125	0.076	0.043
0.079	0.078	0.067	0.073	0.069	0.084	0.061
0.066	0.085	0.080	0.071	0.103	0.075	0.070
0.048	0.092	0.066	0.109	0.110	0.057	0.107

Compute the two-sided 95% confidence limits for the true TWA exposure based on the sample mean estimate of 0.0794 ppm ozone.

SOLUTION

1. Compute the arithmetic mean $\bar{X}$ and standard deviation s of the $n = 35$ measurements X_i. Here, $\bar{X} = 0.0794$ ppm and $s = 0.0233$ ppm.

2. Compute the two-sided 95% confidence interval (limits $\mathrm{LCL}_{2,\,0.95}$ and $\mathrm{UCL}_{2,\,0.95}$) for the true TWA exposure, based on the TWA exposure estimate $\bar{X}$, from the equations:

$$\mathrm{LCL}_{2,\ 0.95} = \bar{X} - \frac{(t_{0.95,n-1})(s)}{\sqrt{n}}$$

$$\mathrm{UCL}_{2,\ 0.95} = \bar{X} + \frac{(t_{0.95,n-1})(s)}{\sqrt{n}}$$

Note that the equation factor $t_{0.95,\ 34} = 2.032$ will differ for sample sizes other than 35. See the comments section that follows the solution.

3. In this example:

$$\begin{aligned}\mathrm{LCL}_{2,\ 0.95} &= 0.0794 - \frac{(2.032)(0.0233)}{(35)^{1/2}}\\ &= 0.0794 - 0.0080 = 0.071 \text{ ppm}\end{aligned}$$

$$\begin{aligned}\mathrm{UCL}_{2,\ 0.95} &= 0.0794 + \frac{(2.032)(0.0233)}{(35)^{1/2}}\\ &= 0.0794 + 0.0080 = 0.087 \text{ ppm}\end{aligned}$$

Thus, the two-sided 95% confidence interval for the true TWA exposure, based on an exposure estimate of 0.079 ppm obtained from 35 samples, is 0.071 to 0.087 ppm. That is, one can be 95% confident that the true TWA exposure is between 0.071 and 0.087 ppm.

COMMENTS. The factor 2.032 in the equations for $\mathrm{LCL}_{2,\ 0.95}$ and $\mathrm{UCL}_{2,\ 0.95}$ is taken from the Student's t table for $(n - 1) = 34$ degrees of freedom. For the large-sample ($\geq$30) n values appropriate for this procedure, the normal distribution Z value of 1.960 can be used as an approximation to Student's t.

The indicated uncertainty in the TWA exposure estimate, as reflected in the width of the confidence interval ($\pm$10% of the TWA estimate in this example), does not incorporate uncertainty due to interday exposure level variability (see Section 3.3.1 and Section 3.6.2.3). The confidence interval width reflects only the intraday variability of the exposure levels occurring over the 8-hr work shift period of the grab-sample measurements. That is, the computed confidence interval is valid only for the worker's TWA exposure on the day and during the period of the grab samples, at the same sampling position.

6.5.2 Classification of Exposure (Hypothesis Testing), ≥30 Grab Samples

PURPOSE. Classify a TWA exposure estimate based on 30 or more grab-sample measurements as noncompliance exposure, possible overexposure, or compliance exposure relative to an exposure control limit (ECL).

ASSUMPTIONS. The same as used for Section 6.5.1.

EXAMPLE. The same data as used for Section 6.5.1, except classify with 95 percent confidence the TWA exposure estimate regarding an ECL of 0.1 ppm.

SOLUTION

1. As in Section 6.5.1, compute the arithmetic mean $\bar{X}$ (the TWA estimate) and standard deviation s of the $n = 35$ measurements X_1. The results are $\bar{X} = 0.0794$ ppm, $s = 0.0233$ ppm.

2. For an employer, if $\bar{X} > \mathrm{ECL}$, no statistical test for compliance need be made, since the exposure estimate exceeds the ECL. If $\bar{X} \leq \mathrm{ECL}$, compute the one-sided lower decision value for comparison with the sample mean:

$$\mathrm{LDV}_{1-P} = \mathrm{ECL} - \frac{(t_{P,n-1})(s)}{(n)^{1/2}}$$

$$\text{For this example, } \mathrm{LDV}_{5\%} = 0.1000 - \frac{(1.691)(0.0233)}{(35)^{1/2}} = 0.093 \text{ ppm}$$

Note that the factor 1.645 from the normal distribution's Z values could have been used as an approximation to the correct Student's t value of 1.691 [for $(n - 1) = 34$ degrees of freedom and $1 - P = 0.05$ significance level for a one-sided test].

Then classify the exposure estimate based on the TWA estimate X of a single worker's TWA exposure on a single day according to:

a. If $\bar{X} \leq \text{LDV}_{5\%}$, classify as compliance exposure.

b. If $\bar{X} > \text{LDV}_{5\%}$, classify as possible overexposure.

Since the $\bar{X} = 0.079$ ppm is less than the $\text{LDV}_{5\%}$ of 0.093 ppm, the employer can state that for the measured worker, on the day and during the period of the TWA exposure estimate, the exposure estimate was a statistically significant compliance exposure at the 5% significance level. To check for even higher degrees of statistical significance, $\text{LDV}_{1\%}$ and $\text{LDV}_{0.1\%}$ could also be computed by using $t_{0.99,n-1}$, and $t_{0.999,n-1}$, respectively, instead of $t_{0.95,n-1}$ in the equation for LDV.

3. For a compliance officer, in this example TWA < ECL, so no statistical test for noncompliance need be made since the TWA estimate $\bar{X}$ is less than the ECL and hence certainly less than the UDV. For $\bar{X} >$ ECL, the compliance officer would compute the noncompliance one-sided upper decision value for comparison with the sample mean:

$$\text{UDV}_{5\%} = \text{ECL} + \frac{(t_{0.95,n-1})(s)}{(n)^{1/2}}$$

where $t_{0.95,n-1}$ is the 95th percentile (one-sided) of Student's t distribution with $(n - 1)$ degrees of freedom. Note that 1.645 can be used as an approximation for $t_{0.95,n-1}$, since $n \geq 30$ when this procedure is used. Then classify the TWA exposure estimate $\bar{X}$ according to:

a. If $\bar{X} > \text{UDV}_{5\%}$, classify as noncompliance exposure.

b. If $\bar{X} \leq \text{UDV}_{5\%}$, classify as possible overexposure.

COMMENTS. Figure 52.13 is a graphical interpretation of decision regions using LDVs for employer decisions and UDVs for compliance officer decisions. The observed value $\bar{X}$ for the TWA exposure estimate is denoted by TWA inside a diamond in Figure 52.13.

Note the limited validity of these tests. The outcomes of these hypotheses tests are valid only for the worker's exposure on the day and during the period of the grab-sample exposure measurements, at the same sampling position.

6.6 Tolerance Limits for a Normally Distributed Variable with Unknown Parameters

PURPOSE. Based on a random sample of n observations, compute one- or two-sided tolerance limits that have a desired γ level of confidence (probability) that the tolerance interval they bound will contain a specified proportion P of the normal distribution represented by the random sample.

ASSUMPTIONS. The parent population for which the tolerance limit(s) statement will be made is assumed to be adequately described by a normal distribution.

Both the true mean and true standard deviation parameters of the parent normal distribution are assumed to be unknown. These two distributional parameters are estimated by sample values, $\bar{X}$ and s, computed from a random sample of n observations.

A tolerance limit can be thought of as a confidence limit for a designated percentile of the parent distribution of single observations. However, note that this procedure for estimating tolerance limits is less "robust" (against lack of normality of the parent population) than procedures for estimating confidence limits on the arithmetic mean distributional parameter. This is true because sample means tend to have a "bell-shaped" (nearly normal) distribution even if the single observations were not from a normal distribution (at least for moderate-to-large sample sizes). Thus, one must be cautious and avoid overinterpretation of tolerance limits. That is, tolerance limits should not be used for "fine-line" decision making. A useful rule-of-thumb is that they probably should not be reported to more than two significant figures, which generally is the usual level of precision for industrial hygiene measurements.

The primary use of these methods should be to obtain an indication of the potential upper values of a parent distribution (e.g., higher-valued exposure measurements by computation of a one-sided upper tolerance limit for the upper-95th-percentile exposure measurement) or an indication of the potential width of the parent distribution (e.g., potential range of exposure measurements by computation of a two-sided tolerance interval for the central 90% of exposure measurements). It is suggested that these computations be programmed into a PC spreadsheet program or a program such as SigmaPlot® 5.0, which could also plot the TWA estimates and the computed tolerance limits on a *semilogarithmic time-series* plot as discussed in Section 6.11 of this chapter.

EXAMPLE. An exposure measurement procedure was used to repeatedly measure a calibrated reference concentration of 100 ppm. The six replicate measurements of the 100 ppm were reported as 95.7, 90.9, 109.4, 107.6, 101.1, and 84.7 ppm. Compute a two-sided, $100\gamma = 95\%$ confidence level, tolerance interval for the central $P = 95\%$ of the parent distribution of possible measurements from a "true" concentration of 100 ppm.

SOLUTION

1. The mean will not be assumed known since any measurement method systematic error could cause the true mean of measurements to differ from the true concentration of 100 ppm. Compute the arithmetic mean $\bar{X}$ and standard deviation s of the $n = 6$ measurements X_i. Here, $\bar{X} = 98.2$ ppm and $s = 9.63$ ppm.

2. Compute the two-sided, 95% confidence level, upper and lower tolerance limits (UTL, LTL) for the central 95% of the parent population from the equations:

$$\text{UTL}_{2,\gamma,P} = \bar{X} + (K)(s) \quad \text{and} \quad \text{LTL}_{2,\gamma,P} = \bar{X} - (K)(s)$$

where K is a factor obtained from appropriate statistical tables for two-sided tolerance limits for a normally distributed variable (e.g., Ref. 144), which is given as a function of the following three independent variables:

a. Sample size ($n = 6$ in this example)
b. Confidence level (e.g., $\gamma = 0.95$)
c. Proportion of the population (e.g., $P = 0.95$)

3. For this example $K = 4.414$ and the $\text{UTL}_{2,\,0.95,\,0.95}$ calculations are

$$\begin{aligned}\text{UTL}_{2,\,0.95,\,0.95} &= 98.2 + (4.414)(9.63)\\ &= 98.2 + 42.5\\ &= 140.7\ \text{ppm}\end{aligned}$$

which is reported as 141 ppm after appropriate rounding. Then compute

$$
\begin{aligned}
\mathrm{LTL}_{2,\ 0.95,\ 0.95} &= 98.2 - (4.414)(9.63) \\
&= 98.2 - 42.5 \\
&= 55.7 \text{ ppm}
\end{aligned}
$$

which is reported as 56 ppm after rounding.

4. Thus the two-sided tolerance interval (95% confidence level, central 95% of the parent population) is 56 to 141 ppm. That is, based on the frugal sample of six replicate measurements of a 100-ppm reference concentration, the best one can say about the measurement procedure is that we can be 95% confident that 95% of future results with this procedure at 100 ppm will lie between 56 and 141 ppm (under conditions similar to those during which the replicate samples were obtained).

COMMENTS. Tolerance limits (and tolerance interval) can be thought of as γ-level specialized confidence limits (and confidence interval) for a fractile interval containing the designated proportion P of the parent distribution.

Note that we cannot say that there is a 95% probability of 95% of the parent distribution lying in the tolerance interval 56 to 141 ppm. For any given tolerance interval such as this one, the computed interval either does or does not contain the stated percentage of the parent population. However, since 95% (100γ) of such tolerance intervals would each contain at least 95% ($100P$) of the parent population, the chances are 19 to 1 that this particular tolerance interval does in fact have 95% or more of the distribution.

Frequently researchers or writers will erroneously state that 95% of their population of sample results will lie in the interval $(\bar{X} - 2s)$ to $(\bar{X} + 2s)$. (Generally 2.000 is used as an approximation to the more exact value of 1.960 for the standard normal deviate.) This type of inferential statement is true only for large (i.e., in the hundreds) sample sizes. Typically, researchers only have small sample sizes (e.g., less than 30) and the K factors for the appropriate two-sided tolerance limit computations are substantially greater than 2. Thus it is possible to substantially underestimate the width of a parent distribution from small sample sizes if the proper tolerance interval computations are not performed. If one had naively estimated the central 95 percent of the results as bounded by $(\bar{X} \pm 2s)$, then one would have substantially underpredicted the possible range of future measurement method results as about 79 to 117 ppm.

The range of the six results was about 91 to 109 ppm, yet the two-sided tolerance interval for 95% of the results was 56 to 141 ppm. This wide tolerance interval is partially due to the frugal sample size used for its computation. However, it also demonstrates how poor an indicator the range of a small sample can be for the span of a normal distribution.

6.7 Tolerance Limits for a Lognormally Distributed Variable with Unknown Parameters

PURPOSE. Based on n samples taken at random, compute one- or two-sided tolerance limits that have a desired γ level of confidence that the tolerance interval they bound will contain a desired proportion P of the lognormal distribution from which the sample was taken.

Assumptions. There is a value of the response variable for each member of a large parent population of "sampling units" to which the tolerance limits statement will apply.

Examples of sampling units could be replicate workers (same sampling period and same type of work) or repeated exposure periods for a given worker. The corresponding population of exposure measurements is assumed to be adequately described by a single 2-parameter lognormal distribution. If it is suspected that the parent population's exposure measurements are more accurately described by a three-parameter lognormal distribution (see Section 3.5.2), then the location constant k for the parent distribution should be estimated (see Section 6.12, step 5 of the solution) and subtracted from all sample values before the method in this section is used.

Both the geometric mean and geometric standard deviation of the parent population distribution are unspecified and assumed unknown. Both distributional parameters are to be estimated from response variable measurements made on a random sample of n sampling units selected from the parent population. It is suggested that these computations be programmmed into a PC spreadsheet program or a program such as SigmaPlot® 5.0, which could also plot the estimates and the computed tolerance limits on a *semilogarithmic time-series* plot as discussed in Section 6.11 of this chapter.

This method cannot process zero data values. It is assumed that none of the sample results are zero (or less than zero) Refer to Ref. 115 for a discussion of how to treat a sample population that includes zero values.

Note that this procedure for estimating tolerance limits is less robust than procedures for estimating confidence limits for the geometric mean distributional parameter. The reason for this is explained in Section 6.6 relevant to tolerance limits for a normally distributed variable. Similar comments would apply here, since tolerance limits for a lognormally distributed variable X are merely a detransformation of tolerance limits for the normally distributed variable $y = \ln X$. The robustness that exists for the sample mean ($\bar{y}$) of the log transformation also exists for its detransformation [GM_X = antilog ($\bar{y}$)].

Example. Five workers were randomly selected from a target population of workers. Each selected worker's 8-hr TWA exposure was measured on a different workday, randomly selected within a 6-month period. The five results were reported as 11.1, 10.6, 21.4, 3.9, and 4.9 ppm. Compute the one-sided, 95% confidence level, upper tolerance limit for the lower 95% of a parent distribution of TWA exposures that is defined by randomly selected workers measured on randomly selected days.

Solution. Assume that general (i.e., group average) exposure levels (X) for the work environment are lognormally distributed among workdays and that percentage differences among individual workers' exposures are similar on each workday. Workers' mean exposure levels are also lognormally distributed. Then, the net variability due to both sources of variability (i.e., due to days and workers) is also lognormally distributed. Under this model, the following analysis is appropriate.

1. Compute the base 10 logarithm, $y_i = \log X_i$ of each sample value. Here the five logarithmic values y_i are 1.045, 1.025, 1.330, 0.591, and 0.690.

2. Compute the sample mean $\bar{y}$ and standard deviation s of the $n = 5$ logarithmic values. Here, $\bar{y} = 0.936$ and $s = 0.298$. As supplementary information, one can compute the sample geometric mean GM of X by taking the antilog$_{10}$ of $\bar{y}$ (i.e., $10^{0.936} = 8.63$ ppm)

and the sample geometric standard deviation GSD by taking the antilog_{10} of s (i.e., $10^{0.298} = 1.986$).

3. Compute the one-sided, 95% confidence level, upper tolerance limit ($UTL_{1,\,0.95,\,0.95}$) for the lower 95% of the defined population of worker-day combination exposure levels (or compute $LTL_{1,\,0.95,\,0.95}$, the lower 95% tolerance limit for the upper 95% of the distribution). Use the following equations:

$$UTL_{1,\ 0.95,\ 0.95} = \text{antilog}_{10}[\bar{y} + K)(s)]$$

$$LTL_{1,\ 0.95,\ 0.95} = \text{antilog}_{10}[\bar{y} - K)(s)]$$

where K is a factor obtained from appropriate statistical tables for one-sided tolerance limits for a normally distributed variable.

4. For the values of this example $K = 4.202$ and the calculations are

$$\begin{aligned} UTL_{1,\ 0.95,\ 0.95} &= \text{antilog}_{10}[0.936 + (4.202)(0.298)] \\ &= \text{antilog}_{10}[2.188] = 10^{2.188} \\ &= 154 \text{ ppm} \end{aligned}$$

5. Thus the one-sided upper tolerance limit (95% confidence level, lower 95% of the defined population) is 154 ppm. That is, based on the frugal sample of only five exposure estimates over a 6-month period from the target population of workers, the best one can say is that there is a 95% confidence that 95% of the daily exposure averages over the 6-month period were below 154 ppm.

COMMENTS. This type of one-sided upper tolerance limit can be thought of as a one-sided upper confidence limit for the true value of a lognormally distributed random variable that exceeds a specified proportion (percentile) of the population of values. This concept is graphically illustrated in Figure 52.16 for a lognormally distributed exposure population.

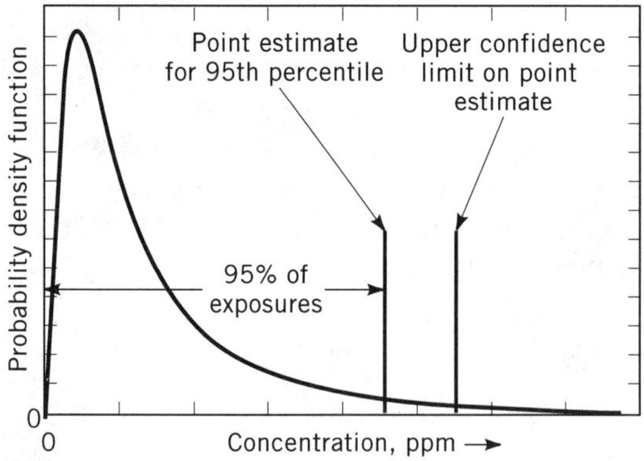

Figure 52.16. One-sided, 95 confidence level, upper tolerance limit for the lower 95 percent (95th percentile) of a lognormally distributed variable with unknown parameters.

A companion technique is presented in the following section for computing a point estimate (and associated confidence limits) for the true proportion of a lognormal distribution exceeding a specified value of the random variable (e.g., proportion of all exposures exceeding a specified exposure level such as a PEL or TLV).

Note that we cannot say that there is a 95% probability that 95% of the parent distribution was below 154 ppm. For any given computed one-sided upper tolerance limit such as this one, the limit either does or does not exceed the stated percentage of the parent population. However, since 95% of such upper tolerance limits would each exceed at least 95% of the parent population, chances are at least 19 to 1 that this particular upper tolerance limit does in fact exceed 95% or more of the distribution.

The range of the five results was 4 to 21 ppm, yet the one-sided upper tolerance limit for the lower 95% of the results was 154 ppm. The high $UTL_{1,\,095,\,0.95}$ is partially due to the frugal sample size used for its computation. Note that the tolerance limit K factor, which is 4.202 in this example, would be only about half as large if n were 50 instead of the 5 in this example. This demonstrates how poor an estimator the range of a small sample is for the span of a lognormal distribution, particularly since the distribution is skewed toward higher values.

To be able to partition the total exposure variability GSD into components due to exposure variability between workers and exposure variability between days, it would be necessary to measure exposures of several workers on each of several days and examine the data by an appropriate analysis of variance method (ANOVA), using a logarithmic transformation of the exposure concentrations as the response variable.

Finally, note that in some real exposure settings, there may be "interaction" (i.e., lack of independence) between interworker exposure variations and interday exposure variations. In such cases, the interworker exposure distributions would have to be determined separately for each workday. For the solution to the example, it was assumed that equal percentage differences among individual workers' true exposures exist on each workday, which is equivalent to assuming "no interaction" on the scale of the logarithmic transformation.

6.8 Point Estimate and Confidence Limits for the Proportion of a Lognormally Distributed Population (Unknown Parameters) Exceeding a Specific Value

PURPOSE. Compute a point estimate P and confidence limits for the true proportion p of a two-parameter lognormal distribution exceeding some specified value (e.g., exposure control limit), when given a random sample from the distribution.

ASSUMPTIONS. Values of the response variable (e.g., single or multiple worker 8-hr TWA exposure estimates) for the parent population are assumed to be adequately described by a single two-parameter lognormal distribution. If it is suspected that the parent population is more accurately described by a three-parameter lognormal (see Section 3.5.2), then the location constant k for the parent distribution should be estimated (see Section 6.12, step 5 of the solution) and subtracted from all sample values before the method in this section is used.

Both the geometric mean (GM) and geometric standard deviation (GSD) of the parent population distribution are unspecified and assumed unknown. Both distributional parameters are to be estimated from response variable measurements made on a random sample of n sampling units selected from the parent population.

Note: This procedure cannot process zero data values. It is assumed that none of the sample results are zero (or less than zero). Refer to Ref. 115 for a discussion of how to treat a sample population that includes zero values. Note that this procedure for computing a point estimate P of the actual proportion p of a lognormal distribution exceeding a specified value is less robust than procedures for estimating the geometric mean distributional parameter. One must be cautious and avoid overinterpretation of these point estimates and their confidence limits. A point estimate of distributional tail area is sensitive to departures from the assumption of lognormality, and these estimates should not be used for "fine-line" decision making. A useful rule-of-thumb is that estimated tail areas (probabilities) should not be reported to more than two decimal places. The primary use of these methods should be to obtain an indication of the potential frequency of high-exposure levels from a parent lognormal distribution.

The procedure assumes that the exposure estimates are from a "stable" unvarying parent distribution. Experience, professional judgment, and knowledge of the exposure level determinant factors must be relied on here for assurance of the validity of this assumption. Only current sample data that represent a stable, unvarying exposure situation should be used in the following computations. One way of assuring that this condition is met is to plot the sample data on semilogarithmic paper (see Section 6.11). If the data are judged as trending upward (or downward) with time, then this procedure should not be used because an erroneous point estimate could result. Only if the long-term exposure plot appears "level" (i.e., constant mean after measurement errors have been smoothed out) should one use this procedure.

EXAMPLE. Initial monitoring of a target population of 35 workers exposed to chromic acid mist and chromates was done by sampling one worker on each of 6 days. A different worker was selected at random each day for exposure measurement. The monitoring yielded the following six 8-hr TWA exposure estimates for six different workers: 0.105, 0.052, 0.082, 0.051, 0.180, and 0.062 mg/m^3. Compute a point estimate P of the true proportion p of 8-hr TWA exposures experienced by the 35 workers that have exceeded an exposure control limit ECL of 0.10 mg/m^3 and the associated one-sided 95% upper confidence limit $\text{UCL}_{1,\,0.95}$ for the true proportion p.

SOLUTION. As in the tolerance limit example of Section 6.7, assume that true exposures for randomly selected, worker-day combinations are affected by two types of independently distributed, lognormally distributed, random variations:

a. Lognormally distributed, daily geometric means (over all workers in the target population)

b. Lognormally distributed, multiplicative (proportional) factors for differences between workers; ratios between exposures of individual workers are constant from day to day.

To be able to use a single lognormal distribution as an appropriate model for the type of exposure monitoring data collected in these two examples, it is essential to do the exposure monitoring according to a particular scheme whereby each randomly selected worker is sampled on only a single randomly selected day, and only one worker is measured each day. If a given worker were sampled repeatedly, the exposure measurements would be intercorrelated (as opposed to independent) samples and a more complex approach (i.e.,

ANOVA) would be needed in order to properly identify the day-to-day and worker-to-worker components of total variability. This problem had to be dealt with in a complex effort by Rappaport et al. (120) to use ANOVA in order to estimate intraworker and interworker components of variance. In this ANOVA study, any day-to-day variability of the general exposure environment would have been corrupted by unaccounted-for inter-correlations among interworker variations. To prevent this from happening, the different workers had to be measured on different days. Considerable statistical care and sophistication must be taken to interpret and use such variance components correctly in applications such as compliance testing, or for calculation of the required sample size for determination of tolerance limits on TWA exposures for a designated worker population.

1. Compute the sample geometric mean (GM) and geometric standard deviation (GSD) for the sample of n results. For the n = six 8-hr TWA exposure estimates, the sample GM = 0.080 mg/m^3 and the sample GSD = 1.627.

2. Compute $g(\text{ECL}) = [\log(\text{ECL}) - \log(\text{GM})]/\log(\text{GSD}) = 0.458$.

3. From a table of areas P under the standard normal curve from Z to $+\infty$, obtain the P value for $Z = g(\text{ECL})$. For $g(\text{ECL}) = 0.458$, $P = 0.32$, which is the integral of the standard normal curve from g(ECL) to plus infinity. Note that this computed P estimate should be about the same as a point estimate for p obtained with an approximate graphical plotting technique such as that described in Section 6.12 (i.e., a P point estimate for p that is the indicated probability at the intersection of an exposure distribution line and an exposure control limit line).

4. To compute confidence limit(s) on the actual proportion p, first decide whether one- or two-sided limits are desired and select the confidence level for the computation. Then from a table of Z values, obtain the related Z value for the desired confidence level. For example, to compute a one-sided 95% $\text{UCL}_{1,\,0.95}$ (or one-sided 95% lower confidence limit, $\text{LCL}_{1,\,0.95}$) for p, use a Z value of 1.645 (or -1.645). For two-sided 95% limits on p, use ± 1.960, and for two-sided 99% limits use ± 2.576.

5. Computing confidence limit(s) for the actual proportion p involves the solution to a quadratic equation. The two necessary quadratic roots are given by

$$U = \frac{-b \pm (b^2 - 4ac)^{1/2}}{2a}$$

where

$$a = \frac{1}{2n - 3} - \frac{1}{Z^2} \qquad b = \frac{2g}{Z^2}\left(\frac{2n - 3}{2n - 2}\right)^{1/2}$$

and

$$c = \frac{1}{n} - \frac{g^2}{Z^2}\frac{2n - 3}{2n - 2}$$

The two U values are standard normal deviates (i.e., Z values) corresponding to the lower and upper confidence limits for the true proportion p of exposures exceeding ECL. Use of the larger value from the U equation leads to a one- or two-sided (depending on the Z value selected) LCL on the area to the right of ECL (i.e., an LCL on p). Use of the smaller value from the U equation leads to a UCL for p.

In this example, $Z = +1.645$ to obtain a one-sided 95% upper confidence limit. Since $n = 6$ and $g = 0.458$ in this example, the three intermediate functional variables for U are $a = -0.258$, $b = 0.321$, and $c = 0.0969$. The resulting two values for U are -0.251 and $+1.494$. The smaller value (-0.251) corresponds to a probability of 0.599 or about 0.60, which is the $UCL_{1,\,0.95}$ for p.

6. The point estimate P of 0.32 for p (the actual proportion of overexposures) has been computed and the one-sided 95% upper confidence limit of 0.60 for p. Given the assumptions of this inferential method, our best estimate is that 32% of the 8-hr TWA exposures experienced by the 35 workers (on the 45 days of exposure that were sampled at random) exceeded the exposure control limit of 0.10 mg/m^3. However, this point estimate P for the actual proportion p of overexposure worker-days was based on a frugal sample of only six exposure estimates (one worker on each of six days). Note that the one-sided, 95% upper confidence limit for p is 60%, which indicates the large amount of uncertainty in the estimated proportion P. Thus it should be stated that the true (actual) percentage of worker-days exceeding the exposure limit could have been as high as 60%.

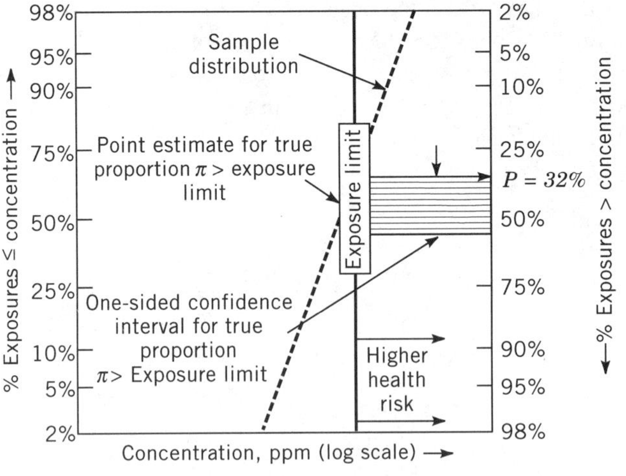

Figure 52.17. Point estimate P and one-sided, upper confidence limit for the actual proportion π of an exposure distribution exceeding an exposure control limit (ECL).

COMMENTS. This approach is graphically illustrated in Figure 52.17. The thick, angled, dashed line represents the estimated exposure distribution (a.k.a. sample distribution) plotted on lognormal probability paper (see Section 6.12). The bold, horizontal, solid line at the top of the shaded region points to the point estimate P of 0.32 from the intersection of the estimated exposure distribution line and the exposure limit ECL line. The bold, horizontal solid line at the bottom of the shaded region below the P value line represents the one-sided, upper confidence limit for the true proportion p of exposures exceeding ECL. The shaded region represents the uncertainty in P, since this area indicates the region in which the true exposure distribution and resulting p could lie.

If one had good reason to assume that a particular worker's exposure distribution over some long period (e.g., two months) is the same as the target population's distribution, then one could infer that any given worker was overexposed on 32% of the exposure days, but could have been overexposed as many as 60% of the exposure days. However, one must recognize that individual worker overexposure risks (resulting from individual exposure distributions) can be "masked" by the exposure distribution for the target population. For example, the lower portion of the target population exposure distribution might be created by workers with consistently low exposures, perhaps due to better work practices. Then these "lower-exposure tail" workers would have individual overexposure risks substantially lower than the 32% estimated for the group as a whole.

Hewett and Ganser (143) have stated that the approximate method presented here for determining confidence limits on percent exceedance is "inaccurate." The authors' method was derived from related theory given in Hald (62), who approximate the distribution of $(\bar{y} + ts_y)$ by a normal distribution. The approximation is good, and gives adequate accuracy except for very small sample sizes such as the one used in an example by Hewett and Ganser to compare results by their approximate graphical method to results obtained by our approximate method. Their example used the following hypothetical data: $n = 5$, GM $= 1$, GSD $= 2$, and ECL $= 18.4$. The value chosen for the ECL in the example was chosen to be equal to the 95% UTL for 95% of a parent lognormal distribution, and they make their comparison under the assumption that the expected 95% UCL for exceedance is 0.05 (5%) in this example.

However, the following data indicate that exceedance above an ECL that equals a sample UTL (as for these calculations) would be *less* than 0.05, even for larger sample sizes.

n	$f = (n - 1)$	$E(UTL)$	$P[y > E(UTL)]$
5	4	$\mu + 4.807\sigma$	0.000 000 77
6	5	$\mu + 4.213\sigma$	0.000 012 6
8	7	$\mu + 3.609\sigma$	0.000 153
10	9	$\mu + 3.291\sigma$	0.000 499
31	30	$\mu + 2.516\sigma$	0.005 94
61	60	$\mu + 2.319\sigma$	0.010 2
121	120	$\mu + 2.199\sigma$	0.013 9

The smallness of these probabilities is to be expected because it is the 95th percentile of the distribution of sample UTLs which exceeds 95% of the parent distribution of concentration data. Most (95%) of the sample UTLs will have true exceedances that are *less* than 0.05.

A modification of our approximate method will now be presented that is easier to calculate. It removes the correction for bias in the sample standard deviation, but the small negative bias is less than 2% for sample sizes larger than 10. Except for modifications to coefficients a, b, and c, as given below, our two methods are the same and should produce nearly identical results for large sample sizes.

Leidel and Busch (4th ed. chapter) Method w/ Bias Correction	Leidel and Busch (present chapter) Simplified Method w/o Bias Correction
$a = [1/(2n - 3)] - (1/Z^2)$	$a = 1$
$b = (2g/Z^2)[(2n - 3)/(2n - 20)]^{1/2}$	$b = -2g$
$c = (1/n) - [g^2(2n - 3)]/[Z^2(2n - 2)]$	$c = g^2[(1 - (Z^2/(2n - 2))] - Z^2/n$

An interesting comparison is given below between results obtained with the two versions of the authors' method and the method proposed by Hewett and Ganser (143), which is based on theory and tables from Odeh and Owen (145). The two examples for small $n = 5$ are from Hewett and Ganser (143).

Example #1: $n = 5$, ECL = 5, GM = 2.575 ($\bar{y} = 0.9458$), GSD = 1 ($s_y = 0.4209$)

	Leidel/Busch (4th ed. chapter)	Leidel/Busch (present chapter)	Hewett/Ganser (143)
Est'd π	0.057	0.057	0.057
95% LCL	0.000015	0.0025	0.005
95% UCL	0.262	0.344	0.645

Example #2: $n = 5$, ECL = 18.4, GM = 1 ($\bar{y} = 0$), GSD = 2 ($s_y = 0.6931$)

	Leidel/Busch (4th ed. chapter)	Leidel/Busch (present chapter)	Hewett/Ganser (143)
Est'd π	0.000013	0.000013	0.000013
LCL	0 ($<10^{-10}$)	0 ($<10^{-10}$)	0. (off graph)
UCL	0.0103	0.00048	0.05 (off graph)

Correspondingly, the upper portion of the group's exposure distribution might be due to workers with consistently high exposures, perhaps due to "dirtier" job locations or poorer work practices such as might result from inadequate training. These "upper-exposure tail" workers would have individual overexposure risks substantially higher than the 32% estimated for the group as a whole.

A companion technique for computing a one-sided tolerance limit (e.g., the exposure value that has 95% confidence of exceeding at least the lower 95% of the exposure distribution) was presented in the previous Section 6.7.

6.9 Fractile Intervals of a Normally Distributed Population with Known Parameters

PURPOSE. Compute fractile intervals for a population adequately described by a specified normal distribution.

ASSUMPTIONS. The term fractile refers to the value of the random variate below which a stated proportion of the distribution lies. The population for which the fractile interval statement will be made is assumed to be adequately described by a normal distribution. Both the arithmetic mean and standard deviation of the population distribution are assumed known so that neither distributional parameter will be estimated from sample data. Note that these assumptions are rarely met in practice, but this procedure is presented for illustrative purposes to assist one to better understand the normal distribution model. The normal tolerance limits procedure of Section 6.6 should be used where the distributional parameters are estimated from typical industrial hygiene sample sizes (e.g., less than 50). If there is any question whether one has a sufficient sample size to compute fractiles with this procedure, use the normal tolerance limit procedure.

EXAMPLE. Suppose that on a particular day, a worker's true 8-hr TWA exposure is "known" to be 25 ppm. On that day, if the worker's exposure were measured using a procedure with a total coefficient of variation of 10 percent ($CV_T = 0.10$), one would expect the value of the measurement to follow a normal distribution with mean $\mu = 25$ ppm and standard deviation $\sigma = (\mu)(CV_T) = 2.5$ ppm (see Section 3.4 and Section 3.6.1). Compute the lower and upper values that bound the central 95% of the normal distribution of possible measurement values. These boundaries are known as the 2.5% and 97.5% fractiles of the normal distribution. There is a 95% probability that any single random measurement will lie within such a fractile interval.

SOLUTION

1. Table 52.2 indicates that the X interval for the central 95% of a normal distribution is from $(\mu - 1.96\sigma)$ to $(\mu + 1.96\sigma)$.

2. In this example the central 95% interval is bounded by 25 ppm $\pm$ (1.96)(2.5 ppm) = 25 ppm $\pm$ 4.9 ppm, which is 20.1 to 29.9 ppm.

3. Thus the central fractile interval enclosing 95% of the possible measurements, assuming a normal distribution with mean value equal to the true TWA exposure of 25 ppm and CV_T of 10% is 20.1 to 29.9 ppm. It can also be said that there is a 95% probability that any single random measurement will lie within the interval 20.1 to 29.9 ppm.

4. If one is interested in the 95th percentile, Table 52.2 indicates that 95% of the measurements will be below $(\mu + 1.645\sigma)$. In this example this 95% fractile is bounded by [25 ppm + (1.645)(2.5 ppm)] = [25 + 4.1] ppm = 29.1 ppm.

5. Thus the lower 95% of a normal distribution of possible measurements of a true TWA exposure of 25 ppm consists of all measurements at or below 29.1 ppm. One can also say that there is a 95% probability that a single random measurement of a true TWA of 25 ppm will be less than 29.1 ppm.

6.10 Fractile Intervals of a Lognormally Distributed Population with Known Parameters

PURPOSE. Compute fractile intervals for a population adequately described by a specified lognormal distribution.

ASSUMPTIONS. The population for which the fractile interval statement will be made is assumed to be adequately described by a known two-parameter lognormal distribution.

Both the population geometric mean and population geometric standard deviation of the population distribution are assumed known so that neither distributional parameter will need to be estimated from sample data. Note that these assumptions are rarely met in practice, but this procedure is presented for illustrative purposes to assist one to better understand the lognormal distribution model. The two-parameter lognormal tolerance limits procedure of Section 6.7 for samples should be used where the distributional parameters are estimated from typical industrial hygiene sample sizes (e.g., less than 50).

EXAMPLE. Suppose that the interday variability of a worker's 8-hr TWA exposures is adequately described by a known two-parameter lognormal distribution with a population geometric mean μ_g of 20 ppm and population geometric standard deviation σ_g of 1.9 (see Section 3.5 and Section 3.6.2). Note that the σ_g is dimensionless (does not have concentration units attached to it) unlike an arithmetic standard deviation σ, which has the same units as the corresponding arithmetic mean μ. Compute the lower and upper daily exposure values bounding the central 90% of the daily exposure distribution. These boundaries are the 5 and 95% fractiles of the two-parameter lognormal distribution.

SOLUTION

1. Table 52.3 indicates that the X interval for the central 90% of a two-parameter lognormal distribution lies in the fractile interval $(\mu_g)/(\sigma_g^{1.645})$ to $(\mu_g)(\sigma_g^{1.645})$.

2. In this example the central 90% fractile interval lies between (20 ppm)/($1.9^{1.645}$) and (20 ppm)($1.9^{1.645}$) which is 7.0 to 57.5 ppm.

3. Thus the central 90% fractile interval, for the two-parameter lognormal distribution of 8-hr TWA exposures with a true geometric mean μ_g of 20 ppm and true geometric standard deviation σ_g of 1.9, is about 7 to 57 ppm. That is, 90% of the daily TWAs lie within the interval 7 to 57 ppm. One can also say that there is a 90% probability that a single random daily exposure will lie within the fractile interval 7 to 57 ppm. Note that this central fractile interval of the two-parameter lognormal distribution was computed to be balanced in the sense that it has equal percentages (5%) in each of the left and right tails of the distribution that lie outside the central fractile interval.

4. If one is interested in the 95th percentile, Table 52.3 indicates that 95% of daily TWA exposures will be below $(\mu_g)(\sigma_g^{1.645})$. In this example, this 95th percentile is equal to (20 ppm)($1.9^{1.645}$), which is 57.5 ppm.

5. Thus the lower 95% portion of the two-parameter lognormal distribution consists of daily 8-hr TWA exposures at or below 57.5 ppm. That is, 95% of the daily TWAs will be at or below 57.5 ppm. One can also say there also is a 95% probability that a single random daily exposure will be less than 57.5 ppm.

6.11 Use of Semilogarithmic Graph Paper to Make Time Plots of Variables Believed to be Lognormally Distributed in Time

PURPOSE. To plot data that is believed to be lognormally distributed over time on semilog graph paper to check for possible cycles or trends with time.

ASSUMPTIONS. This procedure is useful to qualitatively look for exposure trends or cycles in time, when the distribution of available data is skewed to the right (toward higher values). No quantitative inferences are made with this procedure. To produce a nearly symmetrical data plot for lognormally distributed data, the mid-point of the logarithmic scale should be given a value close to the geometric mean of the sample data. Data that are approximately described by a two-parameter lognormal model (see Section 3.6.2) should show symmetrical random variability around the geometric mean (i.e., with no apparent trend or other systematic pattern in time). It is suggested that this plotting be done with a personal computer program such as SigmaPlot® 5.0, which can plot the TWA estimates on a *semilogarithmic time-series* graph with "true" date and time scaling (i.e., dates are plotted as a continuous variable, not as labels).

Note that zero data values cannot be displayed with this procedure.

EXAMPLE. A worker's exposure to chromic acid mist and chromates was measured on 7 February. The 8-hr TWA exposure estimate of 0.105 mg/m^3 was judged a possible overexposure relative to a company's exposure control limit (ECL) of 0.1 mg/m^3.

Before making major capital improvements to the local exhaust ventilation, the company industrial hygienist decided to explore the interday variability of 8-hr TWA exposures for the worker after the initial measurement, on 5 other days in February and March. The measurement results in mg/m3 in chronological order are: 7 Feb., 0.105; 15 Feb., 0.052; 20 Feb., 0.082; 28 Feb., 0.051; 14 Mar., 0.180; and 25 Mar., 0.062. Plot the data to yield a symmetrical distribution over time.

SOLUTION

1. Use semilogarithmic graph paper with enough cycles and with scaling that is appropriate for the sample data. Semilog paper utilizes a logarithmic scale for one variable and a linear scale for the second variable. In this procedure, the exposure data is plotted on the logarithmic scale, and time is plotted on the linear scale. Almost 60 types of semilog graph papers are presented by Craver (144) including a 2-cycle by 36 divisions (3 years divided into months), 2-cycle by 52 divisions (1 year divided into weeks), 1-cycle by 60 divisions (5 years divided into months), and 1-cycle by 366 divisions (1 year divided into days).

2. In this example the range of the exposure measurements covers two decades (0.051 to 0.180 mg/m^3 falls within the interval 0.01 to 1.0 mg/m^3), so that 2-cycle semilog paper would be appropriate. Also, the time variable covers 46 days (almost 8 weeks), but the time values are not evenly spaced so that about 50 or more divisions would be appropriate on the linear scale (for time in days). If the time values were spaced exactly one or more weeks apart (i.e., at multiples of seven days), then a linear scale with about eight or more major divisions would be appropriate.

COMMENTS. If one desires to apply quality assurance procedures to lognormally distributed data, review the article by Morrison (129).

6.12 Use of Logarithmic Probability Graph Paper to Fit a Lognormal Distribution to Data

PURPOSE. Graphically estimate a two- or three-parameter lognormal distribution by fitting a straight line to a plot of the sample cumulative distribution on lognormal probability graph paper.

ASSUMPTIONS. It is assumed that the sample data are drawn from a single two-parameter lognormal distribution. The distribution will then yield a linear plot, aside from the expected deviations accountable to random sampling variations. The procedure will also explain how to qualitatively detect three-parameter lognormal distributions as a characteristic type of nonlinearity in the plot and transform the sample results so they plot linearly (see part 5 of the solution). The plotting procedure will explain how to detect qualitatively multimodal distributions, such as mixtures of two or more lognormal distributions (see part 6 of the solution). Neither the true geometric mean μ_g nor true geometric standard deviation σ_g need be known for this plotting procedure.

Note: This method cannot process zero or negative data values. It is assumed that all sample results are nonzero. Refer to Ref. 115 for a discussion of how to treat a sample population that includes zero values.

The sample results are assumed to adequately represent the parent distribution that is to be estimated. The calculations and plotting required for this approach can be quite tedious and time consuming. However, one of the authors (NAL) has created a customized program to do the necessary calculations and plotting by using the unique transform-programming language available in SigmaPlot® 5.0 from SPSS Inc. Both *lognormal probability* and *semilogarithmic time-series* plots (Section 6.11 of this chapter) can be created rapidly from existing data bases.

EXAMPLE. The same as used in the preceding Section 6.11. Plot the data to estimate the long-term lognormal distribution of daily exposures that would exist if the conditions of the 2-month representative period of the measurements persisted and the exposure distribution was stable over the long run (i.e., unchanging μ_g and σ_g).

SOLUTION

1. Only the basic aspects of probability plotting will be presented in this solution. Extensive details for this procedure are presented in Ref. 115. Use logarithmic probability paper with enough cycles for the range of sample data. Logarithmic probability paper utilizes one logarithmic scale for the measured variable and the other scale is a (nonlinear) cumulative probability scale. The same configuration of plotted points would exist if the probability scale were replaced by one that is linear in either the Z value or the probit of cumulative probability. The probit is a transformed variable equal to 5 plus the Z value (standard normal deviate) corresponding to the cumulative area under a normal distribution that is equal to the probability to be plotted. If the cumulative percentage of a normal distribution were plotted on a linear scale instead of on the special cumulative probability scale, it would form an S-shaped curve called an *ogive*. However, when the ogive curve is plotted against a cumulative probability scale, a linear function (straight line) results. Therefore, in this procedure the values of the exposure are plotted against the logarithmic scale and the expected cumulative percentages determined from positions (ranks) in the ordered (ranked) data are plotted against the probability scale. The latter expected values are given as plotting positions in Ref. 146. Craver (147) presents four types of normal probability paper (with different systems for numbering the linear scale for the random variable such as exposure concentration) and three types of lognormal probability paper (1-, 2-, and 3-cycle logarithmic scales for a random variable such as exposure concentration).

2. In the example data given in Section 6.11, the range of the exposure measurements covers two decades (0.01 to 1.0 mg/m^3), so that 2-cycle logarithmic probability paper would be appropriate for fitting a lognormal distribution as a model for the exposure data distribution.

3. Rank the sample data from lowest-exposure result to the highest-exposure value and obtain the expected cumulative percentage (plotting position) for each of the $n = 6$ values from Ref. 146. For this example the plotting coordinates for each of the 6 coordinate pairs (i.e., measurement in mg/m^3 as ordinate vs. cumulative percentage as the abscissa plotting location on the cumulative probability scale) are (0.051, 10.3%), (0.052, 26.0%), (0.062, 42.0%), (0.082, 58.0%), (0.105, 74.0%), and (0.180, 89.7%).

4. Using these plotting coordinates, plot a point for each sample value. One can also plot the individual uncertainties for each measurement by first calculating individual confidence limits for each measurement with a procedure selected from Section 6.1.1, Section 6.2.1, Section 6.3.1, or Section 6.5.1. This will qualitatively aid one in comparing the amount of measurement procedure uncertainty to the uncertainty due to environmental variability of the true exposure levels (see Section 3.3).

5. If the plotted distribution has a substantial "hockey stick" appearance, with a flattening of the curve (approaching zero slope) at substantial portions of lower cumulative probability (the lowest 20 or more percent of the sample) then it might be indicative of a three-parameter lognormal distribution (see Section 3.5.2) Such a distribution can result if there are lognormal random variations that are added to a constant background level of the same contaminant. Such a data plot can bek and subtracting this constant ("background level") from each measurement value before plotting. An adequate k can be estimated from the initial "hockey stick" plot by noting the value that is asymptotically approached by the "blade" of the hockey stick. That is, estimate k from the concentration value that the measurements appear to converge to at the lowest cumulative probabilities. A detailed example is given in Ref. 148.

6. If the data plot appears to have one or more "dog legs" or kinks in the central region of the plot, then it may be that the plotted distribution is a mixture of two or more individual lognormal distributions (multimodal data) One should then attempt to classify the sample data into two or more appropriate lognormal distributions by individually examining the determinant variables for each of the sample data. Additional qualitative interpretations of lognormal probability plots are given in Ref. 149.

6.13 Use of Logarithmic Probability Graph Paper to Display Tolerance Limits for an Estimated Lognormal Distribution

PURPOSE. Graphically display one- or two-sided tolerance limits, at the 95% confidence level, for the 75th, 90th, and 95th percentiles of an estimated lognormal distribution. Model the parent distribution of daily TWA exposures by plotting the estimated distribution and associated tolerance limits on lognormal probability paper.

ASSUMPTIONS. The same as used for Section 6.7. The calculations and plotting required for this approach can be quite tedious and time consuming. However, one of the authors (NAL) has created a customized program to do the necessary calculations and plotting by using the unique transform-programming language available in SigmaPlot® 5.0 from SPSS

Inc. Both *lognormal probability* and *semilogarithmic time-series* plots (Section 6.11 of this chapter) can be created rapidly from existing databases.

EXAMPLE. The same as used for Section 6.12. Compute the one-sided, 95% confidence level, upper tolerance limits for the 75th, 90th, and 95th percentiles of the parent distribution of daily TWA exposures for the worker exposed to chromic acid mist and chromates over the 2-month period. Then plot the three UTL points on the same lognormal probability plot used to display the lognormal distribution estimate obtained in the previous Section 6.12.

SOLUTION

1. Compute the base 10 logarithm of each sample value. Here, the 6 logarithmic values in chronological order from 7 February to 25 March are: -0.979, -1.284, -1.086, -1.292, -0.745, and -1.208.

2. Compute the arithmetic mean $\bar{y}$ and standard deviation s of the $n = 6$ logarithmic values. Here, $\bar{y} = -1.099$ and $s = 0.211$. As supplementary information, one can compute the sample geometric mean (GM) by taking the antilog_{10} of $\bar{y}$ ($10 - 1.099 = 0.080$ mg/m^3) and the sample geometric standard deviation (GSD) by taking the antilog_{10} of s ($10^{0.211} = 1.63$)

3. Compute the three one-sided upper tolerance limits(UTL_1, confidence level, percentile) for the parent population from the three equations given below, for the particular case of $n = 6$:

$$\text{UTL}_{1,\ 0.95,\ 0.75} = \text{antilog}_{10}[\bar{y} + (1.895)(s)]$$

$$\text{UTL}_{1,\ 0.95,\ 0.90} = \text{antilog}_{10}[\bar{y} + (3.006)(s)]$$

$$\text{UTL}_{1,\ 0.95,\ 0.95} = \text{antilog}_{10}[\bar{y} + (3.707)(s)]$$

where each tolerance limit factor K was obtained from appropriate tables for one-sided tolerance limits, such as Table 7 on page T-15 of Natrella (144).

4. In this example the calculations are (all with 95% confidence):

$$\begin{aligned}\text{UTL}_{1,\ 0.95,\ 0.75} &= \text{antilog}_{10}[-1.099 + (1.895)(0.211)] \\ &= 0.20\ \text{mg/m}^3 \text{ for the 75th percentile}\end{aligned}$$

$$\begin{aligned}\text{UTL}_{1,\ 0.95,\ 0.90} &= \text{antilog}_{10}[-1.099 + (3.006)(0.211)] \\ &= 0.34\ \text{mg/m}^3 \text{ for the 90th percentile}\end{aligned}$$

$$\begin{aligned}\text{UTL}_{1,\ 0.95,\ 0.95} &= \text{antilog}_{10}[-1.099 + (3.707)(0.211)] \\ &= 0.48\ \text{mg/m}^3 \text{ for the 95th percentile}\end{aligned}$$

5. Thus the (concentration, percentile) coordinates on the (logarithmic, cumulative probability) axes for the three one-sided upper tolerance limits (95% confidence level) are (0.20 mg/m^3, 75%), (0.34 mg/m^3, 90%), and (0.48 mg/m^3, 95%). That is, based on the frugal sample of six daily exposures over a 2-month representative period for the worker, the best one can conclude is that one is 95% confident that under these conditions 75% of daily exposures would be below 0.20 mg/m^3, 90% would be below 0.34 mg/m^3, and 95% would be below 0.48 mg/m^3. The three tolerance limit values could then be plotted on the

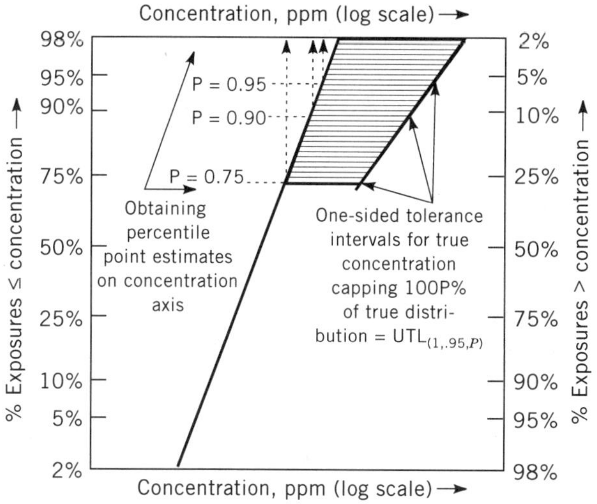

Figure 52.18. One-sided upper tolerance limits for specified concentration percentiles.

same lognormal probability plot used to graphically display the estimated lognormal distribution obtained in Section 6.12.

COMMENTS. A tolerance limit can be thought of as a confidence limit for a specified fractile of a population. Thus a "tolerance line" created by connecting the three tolerance limit points can be thought of as an approximate confidence line for the estimated lognormal population distribution. This approach for graphical analysis and presentation is illustrated in Figure 52.18. The thick, angled, solid line across the full vertical span of the figure represents the estimated exposure distribution plotted on lognormal probability paper (see Section 6.12). The shorter, thick, angled, solid line at the right of the shaded region represents the many possible one-sided, upper tolerance limits for the actual exposure concentrations exceeding selected proportions P of the actual exposure distribution. The shaded region represents the uncertainty in estimating actual exposure concentrations exceeding selected proportions P, since this area indicates the region in which the true exposure distribution could lie. Percentile point estimates (i.e., those exposure concentrations exceeding specified proportions P of the exposure distribution) are estimated by: (*1*) selecting a P value on the left ordinate; (*2*) moving horizontally to the thick, dashed exposure distribution line; and (*3*) moving vertically to the logarithmic exposure concentration scale. One-sided, upper tolerance limits (95% confidence limits) for selected percentile point estimates are similarly obtained by moving horizontally to the thick, solid line to the right of the shaded region and then moving vertically to the logarithmic exposure concentration scale.

BIBLIOGRAPHY

1. A. H. Bowker and G. J. Lieberman, *Engineering Statistics*, Prentice-Hall, Englewood Cliffs, NJ, 1972, p. 1.

2. N. A. Leidel, K. A. Busch, and J. R. Lynch, *NIOSH Occupational Exposure Sampling Strategy Manual*, U.S. Department of Health, Education, and Welfare (NIOSH) Publication 77-173, Cincinnati, OH, 1977.
3. W. W. Lowrance, *Of Acceptable Risk—Science and the Determination of Safety*, William Kaufmann, Los Altos, CA, 1979.
4. G. H. Bracht and G. V. Glass, *Am. Educ. Res. J.* **5**, 437–474 (1968).
5. T. D. Cook and D. T. Campbell, *Quasi-Experimentation—Design and Analysis Issues for Field Settings*, Houghton Mifflin, Boston, 1979, pp. 70–80.
6. *Ibid.*, p. 71.
7. D. J. Finney, *Biometrics* **47**, 331–339 (1991).
8. S. Fairhurst, *Ann. Occup. Hyg.* **39**(3), 375–385 (1995).
9. T. F. Hatch, *Arch. Environ. Health* **16**, 571–578 (1968).
10. U. Ulfvarson, *Int. Arch. Occup. Environ. Health* **52**, 285–300 (1983).
11. D. P. Ford, B. S. Schwartz, S. Powell, T. Nelson, L. Keller, S. Sides, J. Agnew, K. Bolla, and M. Bleecker *Am. Ind. Hyg. Assoc. J.* **52**, 226–234 (1991).
12. T. F. Hatch, *J. Occup. Med.* **14**, 134–137 (1972).
13. T. F. Hatch, *Arch. Environ. Health* **27**, 231–235 (1973).
14. S. A. Roach, *Br. J. Ind. Med.* **10**, 220 (1953).
15. J. Gamble and R. Spirtas, *J. Occup. Med.* **18**, 399–404 (1976).
16. N. Esmen, *Am. Ind. Hyg. Assoc. J.* **40**, 58–65 (1979).
17. S. A. Roach, *Am. Ind. Hyg. Assoc. J.* **27**, 1–12 (1966).
18. S. A. Roach, *Ann. Occup. Hyg.* **20**, 65–84 (1977).
19. H. Buchwald, *Ann. Occup. Hyg.* **15**, 379–391 (1972).
20. T. H. Hatch, *Am. Ind. Hyg. Assoc. Quar.* **16**, 30–35 (1955).
21. F. E. Lundin, J. K. Wagoner, and V. E. Archer, *Radon Daughter Exposure and Respiratory Cancer Quantitative and Temporal Aspects*, NIOSH-NIEHS Joint (1962). Monograph No. 1, U.S. Department of Health, Education, and Welfare (1977).
22. G. Berry, J. C. Gilson, S. Holmes, H. C. Lewinsohn, and S. A. Roach, *Br. J. Ind. Med.* **36**, 98–112 (1979).
23. J. M. Dement, R. L. Harris, M. J. Symons, and C. Shy, *Ann. Occup. Hyg.* **26**, 869–887 (1982).
24. R. W. Hornung, *Appl. Occup. Environ. Hyg.* **6**, 516–520 (1991).
25. R. D. Kimbrough, *Appl. Occup. Environ. Hyg.* **6**, 759–763 (1991).
26. B. MacMahon and T. F. Pugh, *Epidemiology Principles and Methods*, Little Brown, Boston, 1970.
27. J. S. Mausner and A. K. Bahn, *Epidemiology, An Introductory Text*, Saunders, Philadelphia, 1974.
28. G. D. Friedman, *Primer of Epidemiology*, McGraw-Hill, New York, 1974.
29. A. M. Lilienfeld and D. E. Lilienfeld, *Foundations of Epidemiology*, 2nd ed., Oxford University Press, New York, 1980.
30. J. J. Schlesselman, *Case-Control Studies: Design, Conduct, Analysis*, Oxford University Press, New York, 1982.

31. K. J. Rothman and J. D. Boice, Jr., *Epidemiologic Analysis with a Programmable Calculator*, Epidemiology Resources, Boston, 1982.
32. R. R. Monson, *Occupational Epidemiology*, 2nd ed., CRC Press, Boca Raton, FL, 1990.
33. L. Chiazze, Jr., F. E. Lundin, and D. Watkins, eds., *Methods and Issues in Occupational and Environmental Epidemiology*, Ann Arbor Science Publishers, Ann Arbor, Michigan, 1983.
34. M. Karvonen and M. I. Mikheev, *Epidemiology of Occupational Health*, World Health Organization, Regional Office for Europe, WHO Regional Publications, European Series No. 20, Copenhagen, 1986.
35. P. E. Enterline, *J. Occup. Med.* 18, 150–156 (1976).
36. U.S. Department of Health Education and Welfare, *Advances in Health Survey Research Methods: Proceedings of a National Invitational Conference*, sponsored by the National Center for Health Services Research, U.S. Department of Health, Education and Welfare (HRA) Publication No. 77-3154, 1977.
37. J. W. Lloyd and A. Ciocca, *J. Occup. Med.* **11**, 299–310 (1969).
38. J. F. Collins and C. K. Redmond, *J. Occup. Med.* **20**, 260–266 (1978).
39. L. L. Kupper, A. J. McMichael, and R. Spirtas, *J. Am. Stat. Assoc.* **70**, 524–528 (1975).
40. ACGIH, *1998–1999 Threshold Limit Values for Chemical Substances and Physical Agents and Biological Exposure Indices*, American Conference of Governmental Industrial Hygienists, Cincinnati, Ohio, 1998.
41. H. B. Elkins, *Ind. Hyg. Quar.* **9**, 22–25 (1948).
42. *Am. Ind. Hyg. Assoc. Quar.* **16**, 27–39 (1955).
43. H. E. Stokinger, *Pub. Health Rep.* **70**, 1–11 (1955).
44. H. E. Stokinger, *Am. Ind. Hyg. Assoc. J.* **23**, 45–47 (1962).
45. H. E. Stokinger, *Arch. Environ. Health* **25**, 153–157 (1972).
46. H. F. Smyth, Jr., *Am. Ind. Hyg. Assoc. Quar.* **17**, 129–185 (1956).
47. H. F. Smyth, Jr., *Am. Ind. Hyg. Assoc. J.* **23**, 37–44 (1962).
48. T. F. Hatch, *Arch. Environ. Health* **22**, 687–689 (1971).
49. E. Bingham, *Arch. Environ. Health* **22**, 692–695 (1971).
50. E. R. Hermann, *Arch. Environ. Health* **22**, 699–706 (1971).
51. H. F. Thomas, *Ann. Occup. Hyg.* **22**, 389–397 (1979).
52. H. O. Hartley and R. L. Sielken, Jr., *Biometrics* **33**, 1–30 (1977).
53. N. Mantel and W. R. Bryan, *J. Nat. Cancer Inst.* **27**, 455–470 (1961).
54. N. Mantel, N. R. Bohidar, C. C. Brown, J. L. Ciminera, and J. W. Tukey, *Cancer Res.* **35**, 865–872 (1975).
55. P. Armitage, *Biometrics Supplement: Current Topics in Biostatistics and Epidemiology*, 119–129 (1982).
56. H. O. Hartley and R. L. Sielken, Jr., *A Non-Parametric for "Safety" Testing of Carcinogenic Agent*, Food and Drug Administration Technical Report 1, Institute of Statistics, Texas A & M University, College Station, TX, 1975.
57. H. O. Hartley and R. L. Sielken, Jr., *A Non-Parametric for "Safety" Testing of Carcinogenic Agent*, Food and Drug Administration Technical Report 2, Institute of Statistics, Texas A & M University, College Station, TX, 1975.
58. K. S. Crump, H. A. Guess, and K. L. Deal, *Biometrics* **33**, 437–451 (1977).

59. P. Morfeld, H. J. Vautrin, A. Kosters, K. Lampert, and C. Piekarski, *Appl. Occup. Environ. Hyg.* **12**(12), 840–848 (1997).
60. D. Krewski and C. Brown, *Biometrics* **37**, 353–366 (1981).
61. Ref. 32, p. 100.
62. A. Hald, *Statistical Theory with Engineering Applications*, Wiley, New York, 1952.
63. D. L. Bartley, T. J. Fischbach, S. A. Shulman, and R. Song. *Generalized Confidence Limits: Application to the Estimation of Accuracy*, unpublished NIOSH paper, 1993, 27 pages.
64. T. Fischbach, E. Kennedy, S. Shulman, K. Busch, P. Eller, R. Song, and L. Doemeny, *Am. Ind. Hyg. Assoc. J.* **57**, 452–455 (1996).
65. H. J. Seim and J. A. Dickeson, *Am. Ind. Hyg. Assoc. J.* **44**, 562–566 (1983).
66. Ref. 2, Technical Appendix A.
67. E. Olsen, B. Laursen, and P. S. Vinzents, *Am. Ind. Hyg. Assoc. J.* **52**, 204–211 (1991).
68. Ref. 2, Technical Appendix C.
69. Ref. 62, p. 91.
70. J. Aitchison and J. A. C. Brown, *The Lognormal Distribution*, Cambridge University Press, Cambridge, England, 1957.
71. K. A. Busch and N. A. Leidel, "Statistical Models for Occupational Exposure Measurements and Decision Making," in *Advances in Air Sampling*, ACGIH Industrial Hygiene Science Series, Lewis Publishers, Chelsea, MI, 1988, pp. 319–336.
72. M. J. Moroney, *Facts from Figures*, Penguin Books, Baltimore 1951, p. 261.
73. P. E. J. Wilkins, *Air Pollution Control Assoc.* **26**, 935 (1976).
74. Ref. 2, Technical Appendix M.
75. G. J. Hahn and S. S. Shapiro, *Statistical Models in Engineering*, Wiley, New York, 1968.
76. M. R. Stoline, *Environmetrics* **2**, 85–106 (1991).
77. D. Finney, *J. Statist. Med.* **1**, 5–13 (1982).
78. D. G. Altman, *Statist. Med.* **1**, 59–71 (1982).
79. E. L. Crow, F. A. Davis, and M. W. Maxfield, *Statistics Manual*, Dover Publications, New York, 1960.
80. R. D. Soule, "An Industrial Hygiene Survey Checklist," in National Institute for Occupational Safety and Health, *The Industrial Environment—Its Evaluation and Control*, Department of Health, Education, and Welfare, Cincinnati, OH, 1973.
81. R. H. Green, *Sampling Design and Statistical Methods for Environmental Biologists*, Wiley, New York, 1979.
82. N. C. Hawkins, S. K. Norwood, and J. C. Rock, Eds., *A Strategy for Occupational Exposure Assessment*, American Industrial Hygiene Association, Akron, OH, 1991, Chapt. 2.
83. J. D. Bates, *Writing with Precision—How to Write so That You Cannot Possibly Be Misunderstood*, Acropolis Books, Washington, DC, 1978.
84. F. Crews, *The Random House Handbook*, Random House, New York (1974).
85. CBE Style Manual Committee, *CBE Style Manual*, 5th ed., Council of Biological Editors, Bethesda, MD, 1983.
86. E. A. Murphy, *Biostatistics in Medicine*, Johns Hopkins University Press, Baltimore, 1982, Chapt. 4.
87. V. Barnett and T. Lewis, *Outliers in Statistical Data*, Wiley, Chichester, England 1978.
88. A. C. Cohen, *Technometrics* **3**, 535 (1961).

89. R. W. Hornung and L. D. Reed, *Appl. Occup. Environ. Hyg.* **5**, 46–51 (1990).
90. D. T. Mage, *Am. Statistician.* **36**, 116–120 (1982).
91. H. W. J. Lilliefors, *Am. Stat. Assoc.* **62**, 399–402 (1967).
92. R. I. Iman, *Am. Statistician* **36**, 109–112 (1982).
93. M. A. Waters, S. Selvin, and S. M. Rappaport, *Am. Ind. Hyg. Assoc. J.* **52**, 493–502 (1991).
94. S. S. Shapiro and M. B. Wilk, *Biometrika* **52**, 591–611 (1965).
95. Ref. 2, p. 27.
96. W. A. Burgess, *Recognition of Health Hazards in Industry*, Wiley, New York 1981.
97. L. V. Cralley and L. J. Cralley, eds. *Industrial Hygiene Aspects of Plant Operations*, Vol. 1, *Process Flows*, Macmillan, New York, 1982.
98. W. A. Fuller, *Measurement Error Models*, Wiley, New York, 1987.
99. M. Corn and N. A. Esmen, *Am. Ind. Hyg. Assoc. J.* **40**, 47–57 (1979).
100. B. B. Crocker, *Chem. Engr.*, 97 (Mayer 1970).
101. R. J. Hubiak, F. H. Fuller, G. N. VanderWerff, and M. Ott, *Occ. Health Safety* **50**, 10–18 (1981).
102. M. L. Holcomb and R. C. Scholz, *Evaluation of Air Cleaning and Monitoring Equipment Used in Recirculation Systems*, National Institute for Occupational Safety and Health, Publication DHHS (NIOSH) 81–113, Cincinnati, OH, 1981.
103. D. Schneider, *Draeger Rev.* **46**, 5–12 (1980).
104. K. Leichnitz, *Draeger Rev.* **46**, 13–21 (1980).
105. M. V. King, P. M. Eller, and R. J. Costello, *Am. Ind. Hyg. Assoc. J.* **44**, 615–618 (1983).
106. Ref. 2, Technical Appendix E.
107. Ref. 2, Technical Appendix F.
108. Ref. 2, Technical Appendix A.
109. S. M. Rappaport, *Ann. Occup. Hyg.* **35**, 61–121 (1991).
110. Ref. 2, p. 21.
111. Ref. 2, p. 24.
112. Ref. 2, pp. 28–30.
113. Ref. 2, p. 27.
114. Ref. 2, pp. 33–34.
115. Ref. 2, Technical appendix I.
116. N. W. Paik, R. J. Walcott, and P. A. Brogan, *Am. Ind. Hyg. Assoc. J.* **44**, 428–432 (1983).
117. *Am. Ind. Hyg. Assoc. J.* **44**, 697 (1983).
118. C. Daniel and F. S. Wood, *Fitting Equations to Data*, Wiley-Interscience, New York, Appendix 3A, 1971.
119. R. H. Lyles and L. L. Kupper, *Am. Ind. Hyg. Assoc. J.* **57**, 6–15 (1996).
120. S. M. Rappaport, R. H. Lyles, and L. L. Kupper, *Ann. Occup. Hyg.* **39**(4), 469–495 (1995).
121. S. Selvin, S. Rappaport, R. Spear, J. Schulman, et al., *Am. Ind. Hyg. Assoc. J.* **48**, 89–93 (1987).
122. N. A. Leidel, K. A. Busch, and W. E. Crouse, *Exposure Measurement Action Level and Occupational Environmental Variability*, National Institute for Occupational Safety and Health, U.S. Department of Health, Education, and Welfare (NIOSH) Publication 76–131, Cincinnati, OH, 1975.
123. Ref. 2, pp. 10–11 and Technical Appendix L.

124. Ref. 122, p. 29.
125. R. M. Tuggle, *Am. Ind. Hyg. Assoc. J.* **42**, 493–498 (1981).
126. R. M. Tuggle, *Am. Ind. Hyg. Assoc. J.* **43**, 338–346 (1982).
127. Ref. 2, Figure 1.1.
128. N. C. Hawkins and B. D. Landenberger, *Appl. Occup. Environ. Hyg.* **6**, 689–695 (1991).
129. J. Morrison, *Appl. Stat.* **7**, 160 (1958).
130. J. R. Lynch, N. A. Leidel, R. A. Nelson, and R. F. Boggs, *The Standards Completion Program Draft Technical Standards Analysis and Decision Logics*, National Institute for Occupational Safety and Health, National Technical Information Service Publication PB 282 989, Springfield, VA, 1978, p. 13.
131. D. T. Piele, *Introductory Statistics with Spreadsheets*, Addison-Wesley, Reading, MA, 1990.
132. Y., D. Bar-Shalom, Budenaers, R. Schainker, and A. Segall, *Handbook of Statistical Tests for Evaluating Employee Exposure to Air Contaminants*, National Institute for Occupational Safety and Health, U.S. Department of Health, Education, and Welfare (NIOSH) Publication 75-147, Cincinnati, OH, 1975.
133. B. G. Armstrong, *Am. Ind. Hyg. Assoc. J.* **53**, 481–485 (1992).
134. S. M. Rappaport and S. Selvin, *Am. Ind. Hyg. Assoc. J.* **48**, 374–379 (1987).
135. P. Hewett, *Ann. Occup. Hyg.* **42**(6), 413–417 (1998).
136. S. Selvin, and S. M. Rappaport, *Am. Ind. Hyg. Assoc. J.* **50**, 627–630 (1989).
137. R. H. Lyles, L. L. Kupper, and S. M. Rappaport, *Ann. Occup. Hyg.* **41**(1), 63–76 (1997).
138. S. M. Rappaport, L. L. Kupper, R. H. Lyles, *Ann. Occup. Hyg.* **42**(6), 417–420 (1998).
139. S. M. Rappaport, L. L. Kupper, R. H. Lyles, *Ann. Occup. Hyg.* **42**(6), 421–422 (1998).
140. P. Hewett, *Ann. Occup. Hyg.* **42**(6), 420–421 (1998).
141. C. J. Land, *Am. Stat. Assoc.* **68**, 960–963 (1973).
142. C. Land, "Hypothesis Tests and Interval Estimates," in E. Crow and K. Shimizu, eds., *Lognormal Distributions. Theory and Applications* Marcel Dekker, New York, (1988), pp. 87–112.
143. P. Hewett and G. H. Ganser, *Appl. Occup. Environ. Hyg.* **12**(2), 132–142 (1997).
144. M. G. Natrella, *Experimental Statistics*, National Bureau of Standards Handbook 91, Superintendent of Documents, U.S. Government Printing Office, Washington, D.C., 1963.
145. R. E. Odeh and D. B. Owen, *Statistics: Textbooks and Monographs Series*, Vol. 32—*Tables for Normal Tolerance Limits, Sampling Plans, and Screening*. 1980.
146. Ref. 2, Table 1-1.
147. J. S. Craver, *Graph Paper from Your Copier*, H. P. Books, Tucson, AZ, 1980.
148. Ref. 2, pp. 103, 104.
149. Ref. 2, p. 102, Table 3.

CHAPTER FIFTY-THREE

Occupational Health Nursing

Bonnie Rogers, DrPH, COHN-S, FAAN
and Judith Ostendorf, MPH, RN, COHN-S, CCM

1 INTRODUCTION

Occupational health nursing, then called industrial nursing, began in the late 1800s in the northeastern part of the United States. Betty Moulder worked for a group of coal mining companies in Pennsylvania, and Ada Stuart was employed by the Vermont Marble Company, providing health services for ill and injured workers and their families (1). Since that time, the scope of practice in occupational and environmental health nursing has greatly expanded with increased emphasis on health promotion and health protection services. Many factors have influenced the evolution of occupational health nursing practice. Among them are the changing population and workforce, the introduction of new chemicals and work processes into the work environment, increased work demands, technological advances and regulatory mandates, increased focus on illness/injury prevention, and a rise in health care costs and workers' compensation claims.

Because of the dynamic nature of occupational health nursing and the complexities of work-related health problems, it is important that the occupational health nurse utilize an interdisciplinary approach to address the health needs of the workforce. While individual care is provided to all ill and injured workers, occupational health nurses provide programs and services to maintain, monitor, and enhance the health of the aggregate workforce. Occupational health nurses practice within the context of a prevention framework in order to maintain, protect, and promote worker health and improve the health and safety of the work environment.

Patty's Industrial Hygiene, Fifth Edition, Volume 4. Edited by Robert L. Harris.
ISBN 0-471-29749-6

2 SCOPE OF PRACTICE

The definition of occupational and environmental health nursing has recently been revised by the American Association of Occupational Health Nurses and states: "Occupational and environmental health nursing is the specialty practice that focuses on the promotion, prevention, and restoration of health within the context of a safe and healthy environment. It includes the prevention of adverse health effects from occupational and environmental hazards. It provides for and delivers occupational and environmental health and safety services to workers, worker populations, and community groups. Occupational and environmental health nursing is an autonomous specialty and nurses make independent nursing judgements in providing health care services." In addition, AAOHN has established Standards of Occupational and Environmental Health Nursing (Table 53.1) which set a framework for practice in occupational and environmental health nursing for which Nurses are accountable (2).

As part of the occupational health team, the occupational health nurse utilizes many skills to improve and foster worker health and safety. The scope of occupational health nursing practice, as shown in Figure 53.1, is broad and comprehensive and includes the following (3).

2.1 Worker/Workplace Assessment and Surveillance

Worker health and hazard assessment and surveillance activities are designed to identify worker and workplace health problems and the state of worker's health in order to match the job with the employee and to protect workers from work-related health hazards. Knowledge of job demands and analysis of job tasks are essential for an accurate assessment. Various assessments/examinations can be performed such as occupational history taking, preplacement, periodic, and return-to-work assessments and examinations. Preplacement examinations also help to establish baseline data for comparison with future health monitoring results.

Table 53.1. Standards of Occupational and Environmental Health Nursing

Standard I:	Assessment
Standard II:	Diagnosis
Standard III:	Outcome identification
Standard IV:	Planning
Standard V:	Implementation
Standard VI:	Evaluation
Standard VII:	Resource management
Standard VIII:	Professional development
Standard IX:	Collaboration
Standard X:	Research
Standard XI:	Ethics

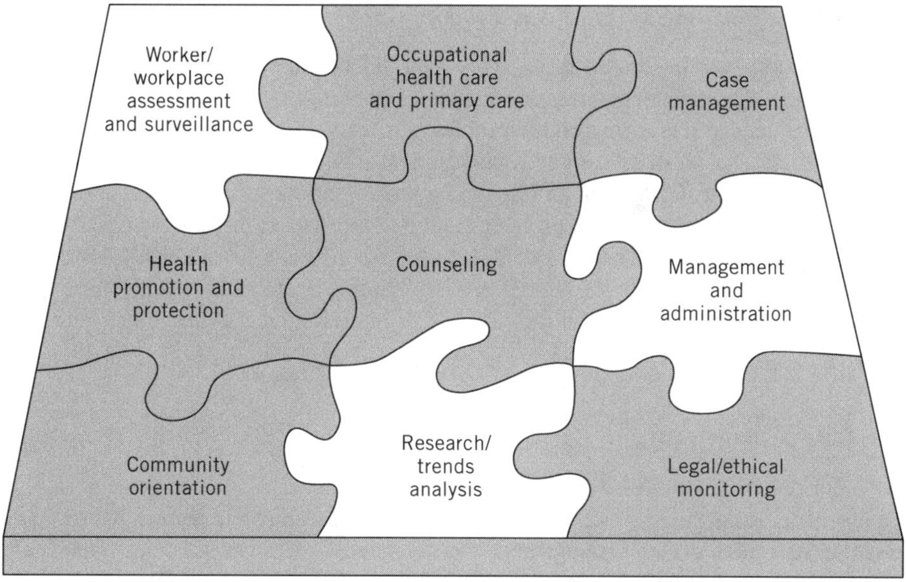

Figure 53.1. Occupational health nursing practice expertise (© B. Rogers, 1998).

Workplace hazard detection and surveillance activities are designed to identify potential and actual hazards that are harmful to workers' health. Knowledge about the nature of the work and work processes, control strategies including engineering, work practice, administrative and personal protective equipment, and the conduct of comprehensive walk-through assessment surveys, including observation of work practices and use of personal protective equipment, are essential to effective hazard detection.

If a hazard is identified, services of an industrial hygienist may be needed to measure levels of exposure of specific substances. In collaboration with a physician and other health care professionals, the occupational health nurse will need to review data obtained in the hazard assessment in order to recommend health surveillance activities and to participate in the implementation of control strategies. Knowing the magnitude and distribution of occupational health illness and injury events will provide a picture of the overall health of the workforce and potential for linkage to morbidity and premature mortality. This can lead to targeting high risk jobs and exposures which is integral to any assessment and surveillance program in specifically focusing risk reduction control mechanisms. Working with other health care professionals such as industrial hygienists and physicians, the occupational health nurse will want to design programs to identify vulnerable workers who are symptomatic, remove them from exposure to prevent further insult, observe and sample the work environment to determine the exposure source(s), and reduce or eliminate the exposure agent. Utilizing a multidisciplinary approach can increase alternatives for problem solving, thus adding to both the effectiveness and efficiency of programmatic interventions.

2.2 Occupational Health Care and Primary Care

In many occupational health settings, the occupational health nurse is the primary provider of service to the worker population; however, a collaborative multidisciplinary approach will often be needed depending on the problem. Primary care activities incorporate direct health care services to ill and injured workers and include diagnosis, treatment, referral for medical care and follow-up, and emergency care. Nonoccupational health care may also be provided for minor health problems and chronic disease monitoring for employees with stable conditions. The occupational health nurse must work collaboratively with the physician to develop appropriate health care guidelines where appropriate. It is imperative that the nurse be thoroughly familiar with her or his state nurse practice legislation that defines the legal scope and limitations of practice.

2.3 Case Management

The occupational health nurse acts to coordinate and manage quality health care services and resources from the onset of illness or injury to help return the worker to work or to an optimal alternative. Case management is often focused on high-cost, catastrophic cases; however, it is also beneficial to apply case management practices to monitor the outcomes of every case. Early intervention is a key component of case management services as it provides for immediate problem identification and engages the worker in care planning from the beginning of the illness/injury to recovery.

The occupational health nurse assesses an employee's needs, plans and implements services required, and evaluates the outcome of the case management process. The nurse identifies a network of health care providers for service delivery that is outcome directed, effective, and cost efficient. As part of the case management process it is essential to determine if the intervention was effective, if the outcomes anticipated were achieved, if cost savings were realized, and if the worker, health care provider, and employer were satisfied with the service and outcome.

2.4 Health Promotion/Protection

Health promotion and health protection activities are designed to improve employees' general health and well-being and to increase employees' awareness and knowledge about toxic exposures in the workplace, life-style risk factors related to health and illness, and strategies to altering behaviors that contribute to health hazard risk. In addition, organizational strategies to enhance workplace health must be emphasized. Occupational and environmental health nurses practice at all levels of prevention with an emphasis on cost containment while preserving and improving quality health services. Leavell and Clark (4) classified levels of prevention as primary, secondary, and tertiary. Primary prevention may be accomplished in the prepathogenesis period by measures designed to promote general optimum health or by specific protection of human beings against disease agents or the establishment of barriers against agents in the environment. As soon as the disease is detectable, early in pathogenesis, secondary prevention may be accomplished by early diagnosis and prompt and adequate treatment. When the process of pathogenesis has progressed and the disease has advanced beyond its early stages, secondary prevention may also be accomplished by means of adequate treatment to prevent sequelae and limit disability. Later, when the disability have been stabilized, tertiary prevention may be accomplished by rehabilitation.

Using the Leavell and Clark framework, occupational health programmatic activities should emphasize preventive strategies (5). Primary prevention strategies are directed toward health promotion and risk reduction. Health promotion is aimed at maintaining and improving the physical and mental well-being of workers or groups of workers, and improving the concept of health in the workplace. Immunization programs, health education classes, and exercise programs are good examples of health promotion activities. Risk reduction measures are designed to eliminate or avert health hazards in order to reduce the risk of disease and prevent injuries. For example, occupational health team members can conduct walk-through surveys to identify workplace hazards and to observe all work practices so that appropriate control strategies can be instituted.

Secondary prevention is aimed at early disease detection, prompt treatment, and prevention of further limitations. In the occupational health setting, early detection includes preplacement and periodic screening examinations for health monitoring so that illnesses and injuries can be identified as soon as possible, appropriate referrals for management of the problem can be made, and the hazard situation can be eliminated. Screening programs such as mammography and analysis of injury trend data are additional examples of early detection strategies.

Tertiary prevention is intended to restore the injured individual's health as fully as possible. Rehabilitation strategies for substance-abusing employees, return to work/work hardening programs, and chronic disease monitoring are examples of tertiary preventive measures.

2.5 Counseling

Health counseling is an integral component of occupational health nursing practice. The occupational health nurse is in the best position to provide counseling services to workers being the health care provider most available to the employee. Counseling activities are intended to help employees clarify health problems and to provide for strategic interventions to deal with crisis situations and appropriate referrals. Counseling activities can relate to such areas as stress, behavioral, marital, social, and interpersonal situations just to name a few. Issues that may interfere with the worker's ability to work or perform the job and the employee will probably benefit from some form of intervention such as listening, supporting, or referral. Behavioral indicators such as increased absenteeism, changes in mood or appearance, or social withdrawal at the workplace may be triggers suggestive of emotional distress. The occupational health nurse needs to build a supportive, trusting, and confidential relationship with the employee, have a knowledge of community referral resources, and recognize when to refer an employee for additional assistance so he or she can get appropriate help and can also remain productive on the job.

2.6 Management and Administration

Increasingly, the occupational health nurse is assuming a major role in the management and administration of the occupational health unit and in policy making decisions to ensure effective occupational health and safety programs and services for workers. This means that the nurse manager must be cognizant of occupational health and safety laws and regulations and develop and support strategies to meet compliance parameters. The nurse must be fully aware of the corporate culture and its impact on the workforce. The occupational health unit goals must support the corporate mission so as to avoid service fragmentation and conflicts in programming and needed resources.

The occupational health nurse is often the health care manager at the work site with responsibilities for program planning and goal development; budget planning and management; organizing, staffing and coordinating activities of the unit, including development of policy, procedures, and protocol manuals; and evaluating unit performance based on achievement of goals and objectives. Cost effectiveness and cost containment of health care services are an integral part of the scope of management responsibilities.

2.7 Community Organization

Community organization activities involve developing a network of resources to provide services efficiently and effectively to workers and employers. Company collaboration and partnerships can be a vital and mutually satisfying experience enabling the occupational health nurse to develop a support system to meet the health and safety needs of employees. Services provided by voluntary or governmental agencies such as parenting programs, cardiac or drug rehabilitation services, or home health can be cost beneficial to both the employee and employer. The occupational health nurse can help the industry to create a health partnership with the community by collaboratively working on programs such as Healthy Babies which can impact both the company and community at large by promoting sound parental care to effect healthy birth outcomes. Providing or sponsoring health fairs for workers, their families, and the community is another example of successful partnerships. Providing service in times of distress such as disaster preparedness of the workforce and providing valuable service during community disasters creates a natural alliance and fosters good will.

2.8 Research and Trend Analysis

Research activities are directed toward identifying practice-related health problems and participating in research activities to identify contributing factors and ultimately recommend corrective actions. The importance of research in occupational health cannot be overstated and the occupational health nurse plays a large part in identifying problems for research. The conduct of research is needed to support and expand the knowledge base in the specialty practice (6). Research and practice go hand in hand with the mission to improve and foster the health and well-being of the worker and workforce, and improve working conditions, eliminating or minimizing potential or actual hazards. For example, understanding the effects of toxic exposures, designing strategies to prevent work-related accident/injuries or illnesses, evaluating the cost-effectiveness of health interventions, or understanding human behavior and motivation related to health promotion activities are important occupational health nursing investigations. Through data collection and a review of records, the occupational health nurse can identify trends in illness and injury patterns that may be related to the workplace exposures. It is important that research findings be disseminated so that information can be utilized to improve clinical outcomes and practice procedures. Knowledge used can then be built on to advance the profession and the practice. As part of a research team, the occupational health nurse can participate in the design, data collection, analysis, and report phases of research studies and ultimately contribute to problem resolution.

2.9 Legal–Ethical Monitoring

Legal and ethical monitoring activities involve knowledge and integration of laws and regulations governing nursing practice and occupational health (e.g., Occupational Safety and Health Act; Hazard Communication), and recognizing and resolving ethical problems that affect workers with regard to occupational health and safety. Ethical issues abound in the work environment and the occupational health nurse is faced with many challenges in ethical-decision making. Issues related to confidentiality of employee health records, hazardous exposures, truth telling, inappropriate screening of employees, discrimination, and professional incompetence or illegal practice are but a few of these ethical challenges. The occupational health nurse is guided by a Code of Ethics that is founded on ethical theories and principles and provides a framework regarding acts of care. The nurse needs to recognize and understand both personal and corporate values related to occupational health and safety and that these values may sometimes compete. The nurse is obligated to act in the best interest of the worker and provide effective leadership skills in ethical health care. The nurse as a moral agent is concerned with values, choices, and duties related to the "good" of individuals and larger societies, and with upholding and advancing the standards of the profession. In this role, the occupational health nurse not only brings a special expertise to occupational health dilemmas but is also able to structure the issues so that sound and deliberate decisions are made using a reasoned approach.

The legal scope of practice in occupational health nursing covers a broad spectrum and will be discussed in detail as implementation of laws and regulations require planning, communicating, and monitoring. Knowledge of occupational safety and health laws and regulations places the OHN in a key position to interact with the injured and/or ill employee to promote employee safety and health, and help the employer comply with health and safety standards and guidelines. It is important for the OHN to be knowledgeable not only about nursing regulations but also about OSHA, occupational safety and health standards, and other legislative and regulatory mandates that impact the workers.

2.9.1 Occupational Safety and Health Act

The major law governing occupational safety and health is the Occupational Safety and Health Act (7). The Occupational Safety and Health Act (OSH Act), signed into law on December 29, 1970 by President Richard Nixon, was the first comprehensive national policy to protect American workers against workplace safety and health hazards. The OSH Act was passed " . . .to assure so far as possible every working man and woman in the Nation safe and healthful working conditions and to preserve our human resources" (OSHA Act, 1970). In addition, OSHA's General Duty Clause states that each employer "shall furnish . . .a place of employment which is free from recognized hazards that are causing or likely to cause death or serious physical harm to his employees" (OSH Act, 1990).

Under the OSH Act the Occupational Safety Health and Administration (OSHA) was created within the Department of Labor to:

- Encourage employers and employees to reduce workplace hazards and to implement new or improve existing safety and health programs.
- Provide for research in occupational safety and health to develop innovative ways of dealing with occupational safety and health problems.
- Establish "separate but dependent responsibilities and rights" for employers and employees for the achievement of better safety and health conditions.
- Maintain a reporting and recordkeeping system to monitor job-related injuries and illnesses.
- Establish training programs to increase the number and competence of occupational safety and health personnel.
- Develop mandatory job safety and health standards and enforce them effectively.
- Provide for the development, analysis, evaluation and approval of state occupational safety and health programs.

OSHA meets one of its purposes by developing mandatory job safety and health standards in order to reduce on the job injuries and to limit workers' risk of developing occupational disease. Standards are developed through a rulemaking process which usually requires three to five years to complete and provides for extensive input from industry and other interested groups or individuals. Standards are enforced through workplace inspections conducted by OSHA compliance safety and health officers who identify work-related hazards, recommend corrections for hazard abatement, and issue citations.

2.9.2 OHNs Role in Occupational Safety and Health

Occupational health nursing incorporates many roles, but basically focuses on promotion, protection and restoration of workers' health within the context of a safe and healthy work environment; including preventing injury and illness and reducing health hazards. In order to be successful in preventing injury and illness and reducing health hazards, OHNs must be aware of pertinent legislation and regulations developed specifically to protect workers' safety and health. They must also stay current regarding changes in legislation and regulations and discuss these changes with employers and employees, particularly with respect to health mandates, health surveillance, and appropriate education and training for employees.

Some of the key standards and laws relating to employee safety and health include the American with Disabilities Act (ADA), Family Medical Leave Act (FMLA), Recordkeeping Standard, Occupational Noise Exposure Standard, Access to Medical Records Standard, Hazard Communication Standard (HazCom), Control of Hazardous Energy Standard (Lockout/Tagout), Bloodborne Pathogen Standard (BBP), Respiratory Protection Standard, and DOT's Drug and Alcohol Testing Regulation (Table 53.2). The occupational health nurse's role with regard to selected laws/related acts identified in the table will be discussed in more detail.

The Americans with Disabilities Act (ADA) prohibits discrimination against a qualified individual with a disability who is able to perform the essential job functions with or without reasonable accommodations. The Equal Employment Opportunity Commission (EEOC) enforces the Act. OHNs may be actively involved in developing and implementing the company's ADA policy. This is done by following ADA guidelines in conducting medical examinations, reviewing job descriptions to ascertain physical requirements for essential functions and determining job suitability, identifying need for and recommending reasonable accommodations, and maintaining confidential medical records in order to protect the privacy of the employee with a disability, thus protecting the disabled individual from discrimination.

The Family Medical Leave Act requires employers with 50 or more employees to provide a maximum of 12 weeks unpaid, job-protected leave in any 12-month period for employees whose circumstances meet certain criteria for eligibility. Examples include the birth of the employee's child, adoption or foster placement of a child with the employee, employee's need to care for a parent, spouse, or child with a serious health condition, or when the employee is unable to perform his or her job functions because of a serious health condition.

The OHN guides the employer in creating policy about and complying with the Family Medical Leave Act (FMLA) through her/his knowledge and understanding of this Act and its requirements, worker's compensation law and its interaction with the FMLA as well as the employee's health and personal issues. He or she also has a responsibility to educate employees who meet the criteria for eligibility under the FMLA about the Act's provisions and requirements for requesting leave and to facilitate the process. If employee health records are involved, they must be treated as confidential medical records and maintained in separate files.

Recordkeeping Regulations within the OSH Act of 1970 requires employers having 11 or more employees to prepare and maintain records of occupational injuries and illnesses. Employers with 10 or less are exempt, as are employers in retail trade, finance, insurance, real estate, and service industries. Specific recording and reporting requirements are provided in the recordkeeping regulation (**29** *CFR* 1904). The purpose of this regulation is to provide valid statistical data about the occurrence of occupational injuries and illnesses that will allow OSHA to identify high hazard industries and to inform employees of the status of their employer's safety record. This recordkeeping system is the responsibility of the Bureau of Labor Statistics (BLS) and has very complex guidelines. Primarily the employer is required to keep two forms:

- OSHA 200 Log, an annual summary of occupational injuries and illnesses.
- OSHA 101, a supplementary record of occupational injuries and illnesses.

OHNs often have the responsibility of OHSA recordkeeping and typically maintain the OSHA 200 Log and OSHA 101 required forms. The OSHA 200 Log represents one calendar year and has each recordable injury and illness logged on it within six working

Table 53.2. Laws and Related Acts

Laws/Related Acts	Date	Agency	Purpose
American with Disabilities Act (ADA)	1990	EEOC	To protect disabled individuals from discrimination.
Family Medical Leave Act (FMLA) (**29** *CFR* 825.118)	1993	EEOC	To provide employees, under specific circumstances, up to 12 weeks of unpaid leave annually
Recordkeeping Standard (**29** *CFR* 1904)	1970	OSHA	To permit BLS survey material to be compiled, to help define high hazard industries, and to inform employees of the status of their employer's record.
Occupational Noise Exposure Standard (**29** *CFR* 1910.95)	1972, rev. 1983	OSHA	To prevent the hearing loss of workers by identifying the progression of hearing loss so that preventive measures can be taken.
Access to Medical and Exposure Records Standard (**29** *CFR* 1910.1020)	1980, rev. 1988	OSHA	To permit direct access by employees or their designated representatives and by OSHA to employer-maintained exposure and medical records in order to provide employees with information to assist in the management of their own safety and health.
Hazard Communication Standard (Haz Com) (**29** *CFR* 1910.1200)	1983	OSHA	To ensure that the physical and health hazards of all chemicals produced or imported are evaluated, and that this information is transmitted to employers and employees.
The Control of Hazardous Energy Standard (lockout/tagout) (**29** *CFR* 1910.147)	1989	OSHA	To establish requirements for the prevention of equipment startup to protect the employee from injury when he or she is performing maintenance on equipment.
Bloodborne Pathogen Standard (BBP) (**29** *CFR* 1910.1030)	1991	OSHA	To reduce morbidity and mortality associated with diseases such as HBV and HIV.
Respiratory Protection Standard (**20** *CFR* 1910.134)	1994	OSHA	To protect the health and safety of workers who wear respirators by reducing their exposure to toxic substances.
Drug and Alcohol Testing Program (**49** *CFR* Part 40)	1994	DOT	To prevent drug and alcohol use in safety-sensitive employees in the aviation, motor carrier, railroad, and mass transit industries.

days from the time the employer learns of the occurrence. The employee's name, department, job title, description of injury or illness, and number of lost work days and limited work days are documented on the 200 Log. The OSHA 101, Supplementary Record of Occupational Injuries and Illnesses, contains much more detail about each injury and illness and also must be completed within six working days from the time the employer learns of the work-related injury or illness. Annually employers are selected to participate in the annual statistical survey and are provided with OSHA Form 200S that they must complete. Information for Form 200S is copied from the 200 Log and returned to BLS. If the employers selected have fewer than 11 employees they are notified and sent the survey at the beginning of the year. The OHN assists OSHA and the employer by providing the Bureau of Labor Statistics with accurate records so they can better differentiate the high hazard industries.

In 1980, OSHA issued a standard requiring employers to provide employees with information to assist in the management of their own safety and health. This standard, Access to Employee Exposure and Medical Records (**29** *CFR* 1910.1020), permits direct access by employees, their designated representatives, and OSHA to employer-maintained exposure and medical records in order to yield both direct and indirect improvements in the detection, treatment, and prevention of occupational disease. If no records exist, the employer must provide records of other employees with job duties similar to those of the employee.

Exposure and related medical records are maintained by the OHN so it is her/his responsibility to respond to request to examine and copy exposure and medical records. Examples of exposure and medical records which might be requested are environmental monitoring, biological monitoring results, material safety data sheets, or any other record that reveals the identify of a toxic substance or health harmful physical agent. The OHN protects the employee's privacy by developing and implementing a clearly written policy addressing where and how the records are stored, where to store records after employees leave the company, and the protocol pertaining to release of information and access to medical and exposure records, including consent forms for employees to sign. By providing the correct, accurate records to employees who request access, the OHN assists the employees in deriving the information needed to assist in the management of their own safety and health.

One of the OSHA standards, the Hazard Communication Standard (HazCom) (**29** *CFR* 1910.12,) was promulgated in 1983 in order to establish uniform requirements to ensure that the hazards of all chemicals produced, imported, or used within the United States are evaluated in order to prevent illness, injury, or death from chemical exposure in the worksite. The evaluation is the responsibility of the chemical manufacturers and importers. The Standard requires the employer to

- Develop a written hazard communication program.
- Identify and assess all chemical substances in the workplace for potential physical or health hazards to the workers.
- Label all chemical containers with warning notices regarding handling and disposal procedures of hazardous substances.

- Ensure that Material Safety Data Sheets (MSDS) are available to employees.
- Provide employee training about hazards of chemicals and measures that can be taken to protect workers from dangerous exposures.

The role of the OHN may encompass coordinating the HazCom Program: including obtaining pertinent MSDS sheets from manufacturers or suppliers, keeping an inventory of hazardous chemicals, ensuring that all chemicals are properly labeled and stored, providing surveillance of exposed employees and maintaining medical records of surveillance activities, and developing, presenting, and documenting the employee education and training program. Teaching employees, as a group during the education and training program and individually during surveillance activities, about the health hazards of chemicals is a natural role for OHNs. The OHN's involvement with the Hazard Communication Standard ensures that information about the physical and health hazards of chemicals at the worksite is disseminated to the employees, enabling the employees to take an active part in their own health protection, and the employer to comply with the standard.

The Bloodborne Pathogens Standard (**29** *CFR* 1910.1030) was established by OSHA in 1992 in order to eliminate or minimize exposure to blood and other body fluids. The Standard requires the employer to develop and implement an exposure control plan which must:

- Be written and accessible to employees.
- Include documentation of exposure determination.
- Provide a method for implementing the exposure control plan, including methods of compliance, employee education and training, description of the hepatitis B vaccination program and postexposure follow up procedures, and recordkeeping and communication procedures.
- A procedure for evaluating exposure incidents.

OHNs play a key role in assisting management in complying with the Bloodborne Pathogen Standard that protects employees from potential exposure to BBP. They assist management in developing an effective program consistent with the Standard elements, including exposure determination, employee education and training, development and implementation of the hepatitis B vaccination program, exposure reporting and post-exposure evaluation and follow-up, as well as recordkeeping and communication procedures related to the BBP Standard. OHNs also serve as consultants to assist in establishing methods of compliance, such as engineering and work practice controls (8).

In developing an effective program, engineering and work practice controls designed to eliminate or minimize employee exposure, and use of universal precautions and personal protective equipment are the methods used to prevent occupational transmission of bloodborne pathogens. Examples of engineering controls are self-sheathing needles and puncture-resistant disposal containers for contaminated sharp instruments. Examples of work practice controls are use of universal precautions and restriction of eating, drinking, smoking, etc., where blood or other potentially infectious materials are kept. Universal precautions treat all human blood and other body fluids as if they are contaminated with bloodborne pathogens. Potentially exposed employees would use appropriate personal protective equipment, including gloves, gowns, laboratory coats, face shields or masks and eye protection.

The OHN should be responsible for the development and implementation of the hepatitis B vaccination program and post-exposure follow-up. The Standard requires that the hepatitis B vaccine be offered, at no expense, to all exposed employees. If the employee chooses not to accept the offer of the hepatitis B vaccination, s/he must sign a mandatory declination statement. The post-exposure evaluation and follow-up is critical to employee safety and health; therefore, the most important component of effective post-exposure evaluation is the method for reporting exposures. Exposures must be reported and acted upon immediately. Counseling the employee about the risk of infection, importance of early testing to determine if transmission has occurred, and recommendations for post-exposure prophylaxis and prevention of transmission of disease is essential and should occur over several sessions.

Employee education and training is also provided by the OHN. Information about the hazards of bloodborne pathogens and methods to prevent transmission to workers may be introduced at employee orientation. These programs are an ideal time to introduce the method for and importance of appropriate reporting of exposures.

In addition to maintaining employee health records, including test results for all exposed employees, documentation of the training sessions may also be the responsibility of the OHN. The records must be complete and accurate and include training dates, contents of sessions, name and qualification of trainer, and names and job titles of all who attend. Employee health records are required to be maintained for the length of employment plus 30 years and training records must be maintained.

OHNs participate in all aspects of compliance with the BBP Standard beginning with developing and implementing an exposure control plan and ending with maintaining the health and training records for the appropriate number of years. The OHN plays an active role in partnering with OSHA and the employer to reduce morbidity and mortality associated with exposures such as HB and HIV.

2.10 Interdisciplinary Relationships

Typically the occupational safety and health team includes the occupational health nurse, the industrial hygienist, occupational health physician, ergonomist, and safety manager. Collaboration occurs between professional disciplines in order to assess, plan, implement, and evaluate care and occupational health services (6, 9). Thus, the OHN should have an understanding of each discipline's contribution.

- Industrial hygienists are trained in recognition, evaluation, control and investigation of occupation health hazards. They focus on the identification and control of occupational health hazards by using analytical methods to detect the extent of workplace chemical, biological, or physical exposures and implement engineering controls and work practices to correct, reduce, or eliminate them,

- Physicians trained in occupational medicine provide treatment of occupational disease and injury. Large corporations may employ a medical director but smaller companies generally use physician services on a part-time, on-call, or consulting basis (10).
- Ergonomists study the physical and behavioral interaction between humans and their environments and strive to design a system in which the workplace layout, work methods, machines and equipment, and work environment are compatible with the physical and behavioral limitations of the workers. The better the system, the higher the level of safety and work efficiency (10).
- Safety managers administer the safety program and advise and guide management and others on all matters pertaining to safety. They must be knowledgeable about the equipment, facilities, manufacturing process, and workers' compensation (10).

Depending on the industry and related occupational health problems, disciplines that also may be included as part of an interdisciplinary team, include toxicologists and epidemiologists. The toxicologist evaluates human health effects due to chemical exposure and studies the chemical's harmful actions on the living systems. An epidemiologist studies work-related illness and injury trends, applying epidemiologic methods to determine causal relationships by analyzing and interpreting risk data. The epidemiologist is usually able to identify the cause of a particular occupational disease. Interdisciplinary collaboration is not limited to the company itself, but includes the community. OHNs may partner with a physical therapist to have her/him on site to work with employees experiencing musculoskeletal problems and provide collaborative opportunities with the ergonomist to design and implement engineering controls to help prevent the problem from recurring.

OHNs use their knowledge and skills to work with benefits managers to reduce healthcare costs for current employees as well as retirees through integrating health systems including health promotion, health care, employee assistance programs, case management, short and long term disability management, and worker's compensation. OHNs can also plan and coordinate health fairs for the corporation, which may involve community groups, insurance providers, physical equipment businesses, and donations of health foods, such as bagels by local grocery stores.

Occupational health nurses are important team members in the occupational safety and health arena with thorough knowledge of, and skills in, occupational health nursing and related knowledge in areas of environmental health, occupational safety and ergonomics, industrial hygiene, toxicology, epidemiology and the social and behavioral sciences. This broad knowledge base enables the OHN to communicate with, assemble, coordinate, and participate in interdisciplinary activities. An example of interdisciplinary functioning is ergonomics.

Typically employees report symptoms of repetitive motion discomfort to the OHN. OHNs are knowledgeable about the jobs performed by workers, risk factors associated with cumulative trauma disorders, (CTD), and particular jobs with increased numbers of discomfort complaints. After a problem has been identified by performing a worksite analysis, OHNs may collaborate with safety and engineering professionals and assist in developing a solution by applying knowledge of human factors and ergonomics. An important program component, medical or occupational health management, is key to the success of the ergonomics program. Early reporting of symptoms allows early identification and treatment of CTDs and ideally will eliminate and proactively reduce the risk of CTD development. Communicating with health care providers (HCP) about modified duty opportunities, accurate recordkeeping, and employee education and training about early signs and symptoms of CTDs in order to facilitate early detection and improved recovery are important to the success of the program.

In order to prevent and reduce injuries and illnesses, the occupational health nurse may consult with the industrial hygienist, occupational health physician, ergonomist, and safety manager in order to formulate a plan to assess the complaints, investigate to see if a problem exists, evaluate the environment including the workstation, plan and implement a solution, and evaluate the outcomes. Preparation might include meeting to review the problem, conducting a worksite assessment, preparing to videotape the job being performed, and discussing questions to review with employees performing the job. This type of interdisciplinary collaboration is essential for a dynamic health-oriented occupational health service (11).

3 SUMMARY

The practice of occupational health nursing is guided by public health principles in places of work among populations of workers. Occupational health nurses are professional health care providers who are specifically trained in recognition of the associations between conditions of work and human health. They serve with physicians, industrial hygienists, safety specialists, and others as vital members of the team of professionals who deal with problems in occupational health and safety.

Occupational health nursing is a dynamic and evolving professional field. Continuing study by its practitioners is necessary to stay abreast of research findings and new developments pertinent to its practice.

Occupational health nurses are very often in the forefront of any effort to protect and enhance the health of workers. Detection of the first hint of a work-related health problem may depend on the astute observations of the occupational health nurse. For many workers, the occupational health nurse is the primary source of guidance for health maintenance, behavior modification, and promotion of health.

Occupational health nurses are health professionals with clearly definable professional obligations. The American Association of Occupational Health Nurses has adopted a Code of Ethics that provides ethical guidance for occupational health nurses in their professional practice. The occupational health nurse must be familiar with laws and regulations governing occupational health and safety, and the nurse practice act for the state in which the nurse practices.

BIBLIOGRAPHY

1. American Association of Occupational Health Nurses. Atlanta, GA, 1976.
2. *Standards of Occupational Health Nursing Practice.* American Association of Occupational Health Nurses, Atlanta, GA, 1999.

3. B. Rogers, "Occupational Health Nursing Expertise," *J. UEOEH* **20**(Suppl.) (1998).
4. H. R. Leavell and E. G. Clark, *Preventative Medicine for the Doctor and his Community*, McGraw Hill, New York 1965.
5. J. Wachs, and J. Parker-Conrad, "Occupational Health Nursing in 1990 and the Coming Decade," *Appl. Occup. Environ. Hyg.* **5**, 200–203 (1990).
6. B. Rogers, *Occupational Health Nursing: Concepts and Practice*, W. B. Saunders, Philadelphia, 1994.
7. Occupational Safety and Health Act, *Public law 91-596*. 91st Congress. S. 2193. Dec. 29, 1970; **29** *USC* 654.
8. L. Goldstein and S. Johnson, "OSHA Bloodborne Pathogens Standard: Implications for the Occupational Health Nurse. *AAOHN J.* **39**(4), 82–188 (1991).
9. B. Rogers and E. Lawhorn, "Occupational Health Nursing Specialty Practice," in M. K. Salazar, ed., *AAOHN Core Curriculum for Occupational Health Nursing*, 1997.
10. National Safety Council, "Occupational Health Programs," in *Accident Prevention Manual for Business and Industry*, 10th ed., Philadelphia, W. B. Saunders, 1992, pp. 3–6, 93–105.
11. U.S. Department of Labor, Occupational Safety and Health Administration. *Ergonomics Program Management Guidelines for Meatpacking Plants*. OSHA 3123. Government Printing Office, Washington, DC, 1990.

CHAPTER FIFTY-FOUR

Ergonomics

Donald S. Bloswick, Ph.D., PE and Timothy Villnave, DC, MSPH

1 INTRODUCTION

1.1 Definition

The word "ergonomics" is derived from Greek words "ergon" meaning work and "nomos" meaning laws (1). Ergonomics can therefore be defined as the study of the laws of work, or the relationship between the worker and the work environment. In its broadest sense, ergonomics deals with the relationship between the user and the products and systems in the working, home, recreational, and general living environment. In this chapter, however, discussion will be limited to the workplace.

In the UAW-Ford Ergonomics Process (2) ergonomics is defined as:

> The integration and application of scientific disciplines such as physiology, psychology, and engineering in designing and improving the workplace to minimize errors, reduce the risk of accidents or chronic injuries while increasing employee well-being.

The Board of Certification in Professional Ergonomics (3) has defined ergonomics as:

> A body of knowledge about human abilities, human limitations, and other human characteristics that are relevant to design. Ergonomic design is the application of this body of knowledge to the design of tools, machines, systems, tasks, jobs, and environments for safe, comfortable, and effective human use.

The Occupational Safety and Health Administration (4) has defined ergonomics to be:

Patty's Industrial Hygiene, Fifth Edition, Volume 4. Edited by Robert L. Harris.
ISBN 0-471-29749-6

The field of study that seeks to fit the job to the person rather than the person to the job. This is achieved by evaluation and design of workplaces, environments, jobs, tasks, equipment, and processes in relationship to human capabilities and interactions in the workplace.

1.2 History

Although not formally defined as such, the field of ergonomics dates back nearly 300 years. In 1713 Ramazzini, in discussing the stress of hand intensive work, noted that: The maladies that afflict the clerks aforesaid arise from three causes: First, constant sitting; secondly, the incessant movement of the hand and always in the same direction; thirdly, the strain on the mind from the effort not to disfigure the books by errors or cause loss to their employers when they add, subtract, or do other sums in arithmetic. . . . In a word, they lack the benefits of moderate exercise, for even if they wanted to take exercise, they have no time for it; they are working for wages and must stick to their writing all day long. Furthermore, incessant driving of the pen over paper causes intense fatigue of the hand and the whole arm because of the continuous and almost tonic strain on the muscles and tendons, which in the course of time results in failure of power in the right hand (5).

Ramazzini also wrote "Manifold is the harvest of diseases reaped by Craftsmen . . . as the . . . cause I assign certain violent and irregular motions and unnatural postures . . . by which . . . the natural structure of the living machine is so impaired that serious diseases gradually develop" (6).

In 1949 the Ergonomics Research Society was organized to bring together professionals from a variety of disciplines concerned with the relationship between the human and the work environment. This initial group was composed primarily of physical, biological, and psychological professionals from the United Kingdom. One of the primary concerns of this founding group was to address the issues associated with the interactions between humans and machines in complex military systems. As names for the society they suggested "Society for Human Ecology" and "Society for the Study of Human Environment". They finally adopted "Ergonomics" as a term which recognized the importance of the behavioral sciences, physiology, and anatomy (7). In the United States, at about this same time, engineers and behavioral scientists were studying similar issues, and called this field of study "Human Factors" or "Human Factors Engineering". In time, the distinction between ergonomics and human factors became based more on content area than geography. Budnick (8) notes that those concentrating on the physical aspects of work (anthropometry, occupational biomechanics, work physiology) became known as "ergonomists". Professionals concerned more with the psychological, the relationship between humans and their environment (sensations/perception and the decision making process), were called human factors professionals. It must be emphasized, however, that these distinctions are not clear cut and considerable overlap exists between the two professional groups.

1.3 Overview

As noted earlier, although ergonomics is concerned with the overall relationship between the human and his/her surroundings in the general living environment, in this chapter discussion is limited to the relationship between the worker and the work environment, and the need to optimize the "fit" between the worker and the job. Poor fit may result in stress to the operator with resulting job related injuries or illnesses. In addition, this poor fit may adversely affect product quality or production efficiency. This fit can be presented as the overlap between the individuals capabilities and the task requirements, as noted in Figure 54.1

This fit is highest when the task requirements match the capabilities of the individual. Fit can be maximized by decreasing the physical task requirements through task analysis and redesign or by increasing the capabilities of the worker through, for example, physical conditioning or training.

Some general workplace factors which must be considered in the optimization of the relationship between the human and the working environment include the following issues which will be discussed in more detail later:

1. Forces required to perform the task.
2. Postures involved in the task.
3. Energy required to perform the task.
4. Frequency at which the task is performed.
5. Rest between exertions.
6. Environmental conditions.

1.4 Impact of Ergonomics in the Work Setting

The implementation of an effective ergonomics program in the work setting have been shown to (*1*) reduce injury and illness costs and (*2*) increase productivity.

1.4.1 Reduction of Injury/Illness Cost

The implementation of an effective ergonomics program has been shown to reduce the incidence and severity of ergonomic related disorders and overall costs associated with these disorders. Implementation of ergonomics controls in one auto parts manufacturing plant is credited with much of the 67% decline in injury rates over a one year period (9) and two Goodyear facilities experienced an approximate 80% decrease in lost time injuries

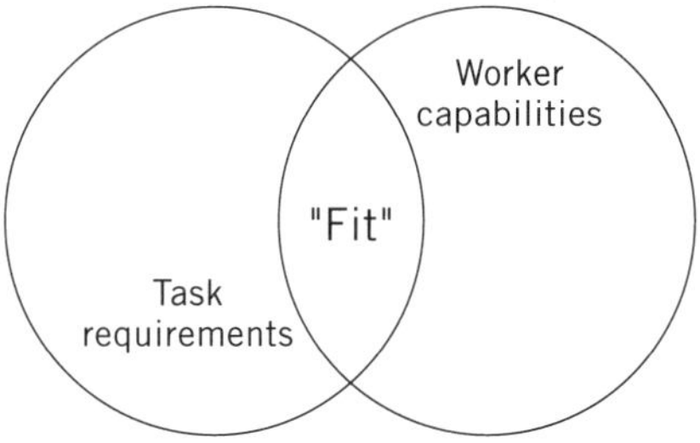

Figure 54.1. Ergonomics is the fit between worker capabilities and task requirements.

during the two years subsequent to the implementation of an overall ergonomics program (10). An early intervention/medical surveillance program at Beech Aircraft is estimated to have saved approximately $200,000 per year by reducing upper extremity cumulative trauma disorders (11). In the early 1990s, 43 out of 100 employees at the Siemens home office in Auburn Hills, Michigan complained of pain in their shoulders, back, elbows, and fingers. This was reduced to zero after the implementation of an ergonomics program including back cushions, lumbar supports, keyboard/mouse wrist rests, document holders, training in stretching, and the use of frequent short rest breaks (12). Between 1992 and 1995, Silicon Graphics realized a 40–50% drop in reportable upper extremity cumulative trauma disorder (UECTD) cases through the establishment of a self-directed ergonomics resource center (13) Sikorsky noted a 75% drop in lost work day incident and severity rates and a 25% drop in OSHA recordable incidents in its helicopter assembly plant after the implementation of an ergonomics program which included training, the establishment of work teams, and the use of engineering controls (14). The investment of approximately $375,000 in equipment, equipment modification, and ergonomic chairs in one AT&T facility resulted in an 80% decrease in its $400,000 workers' compensation cost expense (15). In its Dubuque Works, John Deere implemented a comprehensive ergonomics program which reduced repetitive motion injuries by 60% and nearly eliminated back injuries (16). Pepsico implemented an ergonomics program in Hawaii which included training, exercise, task and equipment analysis/redesign, and the hiring of a consultant and experienced a reduction in back injury reports from 15 cases in 1991 to 4 cases in 1992 (17). An in-depth training program involving tool selection, work methods, and scaffold placement at The Newport News Shipbuilding welding department, reduced wrist injuries from approximately 40 per year to 6 per year. At the same facility, a lifting training program in the maintenance department reduced the frequency of back injuries from approximately one per month to zero. In 1996 the ergonomic case rate was reduced by 30%, the lost time ergonomic case rate reduced by 55%, and over $1 million saved in workers' compensation costs (18). In one large poultry processor a corporate wide ergonomics program was established at 13 plants in 1990. Between 1990 and 1995, upper extremity and lifting related claims decreased approximately 50%, the upper extremity severity rate decreased 20%, and the lifting severity rate decreased 36% (19).

Table 54.1 indicates what this payoff can be in terms of dollars. For example, if a company is operating at a 4% profit margin and incurs one carpal tunnel syndrome release surgery at a total cost of $20,000 (surgical costs, time off from work, training of replacements, etc.), the sales force would have to generate an *additional* $500,000 in sales to offset the cost of the injury ($\$500,000 \times 0.04 = \$20,000$).

It should be noted that in some facilities there is also an increase in ergonomic related cases reported on the OSHA 200 log during the first year or so after the implementation of a comprehensive ergonomics program. In 1990, one manufacturing plant of approximately 1,100 employees, 700 involved in direct labor, had 10 UECTD cases, costing approximately $100,000. In 1991, after the implementation of an aggressive medical management and ergonomics program, the number of cases *increased* to 45 but the cost *decreased* to approximately $40,000. This trend continued in 1992 (20). In another facility of approximately 4,500 workers, the implementation of an overall ergonomics program contributed to the reduction of lost workdays associated with UECTDs. Lost workdays

Table 54.1. Total Sales to Pay Back Accident Cost for Different Profit Margins

	Company Profit Margin				
Accident Costs (U.S. dollars)	2%	4%	6%	8%	10%
10,000	500,000	250,000	167,000	125,000	100,000
20,000	1,000,000	500,000	333,000	250,000	200,000
50,000	2,500,000	1,250,000	833,000	625,000	500,000
75,000	3,750,000	1,875,000	1,250,000	938,000	750,000
100,000	5,000,000	2,500,000	1,667,000	1,250,000	1,000,000
500,000	25,000,000	12,500,000	8,333,000	6,250,000	5,000,000
1 million	50,000,000	25,000,000	16,667,000	12,250,000	10,000,000

went from 613 in 1990 to 149 in 1991, and approximately 20 during the first half of 1992, while the actual *incidence* of ergonomic disorders remained relatively constant (**20**). This phenomenon is the sign of an environment where workers are encouraged to report minor disorders early, before they become serious enough to require extensive time off work and expensive medical intervention.

1.4.2 Increased Productivity

Implementation of an ergonomics program may not only reduce injury and illness costs but may also increase productivity. For example, Fleming Companies Incorporated experienced an increase in productivity as well as a 50% drop in back injury incidence and a decrease in workers' compensation costs as a result of a back program and ergonomic job analysis program (21). Ethicon, a Johnson and Johnson company, reported that as a result of an ergonomics effort, production at that particular facility was increased by over 10% (22). Implementation of an ergonomics program including training, conditioning, stretching, adjustable chairs, equipment modification, and the use of an ergonomist at Red Wing Shoes reduced manufacturing time as well as reducing workers' compensation costs from \$4.4 million in 1990 to about \$1.3 million in 1995 (23). The implementation of an ergonomics program including training, job rotation, and task/workstation redesign at General Seating increased productivity as well as reduced lost work days by 70% (24).

2 DISORDERS ASSOCIATED WITH PHYSICAL WORK FACTORS

The physical design of the workplace, tools, and materials influence how tasks are performed. For example:

1. To assemble a unit, a seated line worker reaches forward 20 in. to a height 17 in. above her work surface to grasp a 2 lb part in a bin.
2. To repair a truck, a mechanic bends at the waist for 30 sec while operating a 10 lb pneumatic wrench.

3. To respond to a request, an office worker cradles a telephone handpiece between her head and shoulder while keying into her computer.

Muscles contracted, tendons transmitted muscle forces, ligaments guided bone motion at joints, and the heart pumped blood to accomplish these tasks. Usually, the worker is able to perform required job activities without injury. On occasion, however, the physical demands of the task can exceed the physiologic capabilities of the person which produces tissue necrosis, inflammation, or chronic adaptive changes.

2.1 Epidemiologic Association between Physical Work Factors and Occupational Injury/Illness

Musculoskeletal disorders (MSDs) are frequently associated with physical work factors. These disorders involve nerves, tendons, muscles, and supporting structures and generally affect the low back and upper extremity.

The National Institute for Occupational Safety and Health (25) evaluated over 600 studies from an epidemiologic perspective and saw strong evidence of a causal relationship between several body motions/positions/responses and MSDs. These associations include:

1. Static or extreme neck posture and neck or neck/shoulder.
2. Repeated, forceful elbow motions and elbow MSDs.
3. Repeated, forceful hand/wrist exertion with non-neutral wrist posture and carpal tunnel syndrome.
4. Repeated, forceful hand/wrist exertion with non-neutral wrist posture and wrist tendinitis.
5. High vibration intensity/duration from hand tools and hand–arm vibration syndrome.
6. Lifting with forceful low back movements and low back disorders.
7. Whole body vibration exposure and back disorders.

Hagberg et al. (26) reviewed over 500 publications and concluded that there was strong evidence of a relationship between select work activities and specific MSDs such as shoulder tendinitis, hand-wrist tendinitis, carpal tunnel syndrome, and nonspecific musculoskeletal disorders (occupational cervicobrachial disorder, cumulative trauma disorder, occupational overuse syndrome, repetitive strain injury). Evidence was also found for an association between work activities and tension neck syndrome and osteoarthrosis. Physical work conditions were classified as:

1. Fit, reach, see.
2. Cold, vibration and local mechanical stress.
3. Awkward postures.
4. Musculoskeletal load.
5. Static load.
6. Task invariability.

A steering committee of The National Academy of Sciences/National Research Council determined that there is a relationship between high level, biomechanical stressor exposure (i.e., exertion, posture, contact stress, vibration, and cold) of the upper extremity, neck, and back and the occurrence of MSDs (27). The committee noted that it was difficult to assess the literature because studies were frequently characterized by poor temporal contiguity, multiple types of exposure/health outcome measurements, and lack of control of confounders. These shortcomings, according to the committee, were countered by relationship consistency among the studies, biological plausibility, and the reduction of MSDs incidence when physical work factors were controlled. Other epidemiologic reviews cite similar concerns but conclude a causal relationship exists (25, 28, 29). It should also be noted, however, that some skeptics contend that research weakness makes it premature to associate physical work factors with MSDs (30–32).

The core issue of association controversy centers on the multifactorial nature of work related musculoskeletal disorders as illustrated in Figure 54.2 (26, 27, 33). Features of the physical work environment such as heavy tools/parts can require the generation of high force within body tissues and lead to injury/illness (26). Organizational work design characterized by high work pace or boring work may be a significant contributor to MSDs (34,

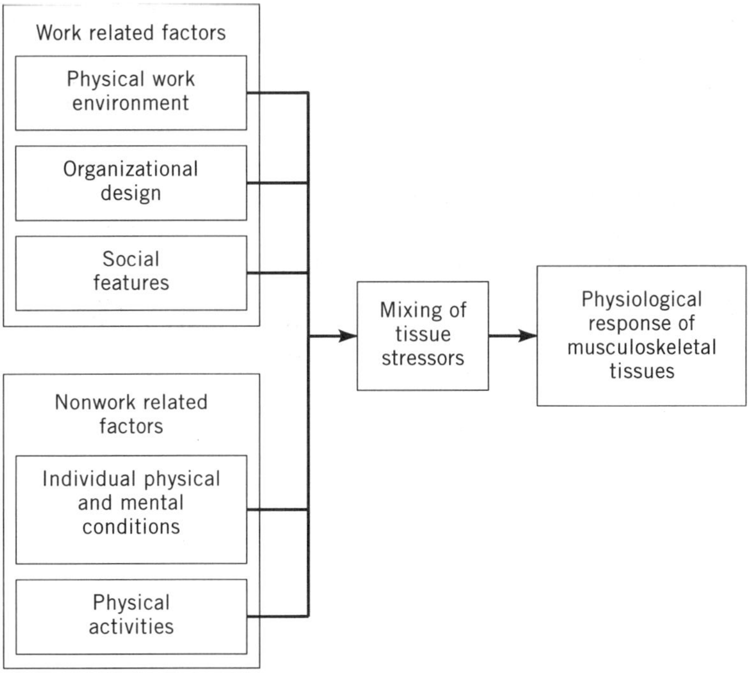

Figure 54.2. Musculoskeletal disorders can have multifactorial causes involving both work and nonwork factors. A review of these factors can determine the level of contribution made by the physical work environment to the incidence or exacerbation of an injury/illness.

35). Social work features such as job dissatisfaction or poor employee appraisal rating have been associated with the incidence of back injury (36, 37).

Nonwork related individual physical conditions have been associated with musculoskeletal injury/illness. Loss of fitness (38, 39), familial history (38–40), and gender (41) are examples. Nonwork physical activities may expose the individual to risk conditions related to the development of MSDs. Recreational sports such as golf (42), volleyball (43), and triathlon training (44) can produce soft/hard tissue trauma. Cigarette smoking, particularly if there is a chronic cough, has been associated with low back pain (45).

In summary, the multifactorial nature of MSDs has lead to much confusion and controversy as to the role of physical work factors. At this time, epidemiologic reviews tend to conclude that there is good evidence for a causal relationship between several physical work factors and MSDs. Studies involving MSDs and work related social features such as monotonous work, low social support, and limited job control have been evaluated as having some evidence of association and reviews of personal factors research have concluded that there is some evidence of association between age, smoking, inflammatory disorders, and diabetes and work related MSDs. Disagreement as to the relative importance of various factors will probably continue for some time. Like other multifactorial pathologies such as certain cancers and lung disorders, epidemiologic research is needed to clarify the situation.

2.2 Survey Reports of Physical Work Factors and Occupational Injury/Illness

There are no *direct* statistics relating the incidence of occupational disorders with physical work factors. The Bureau of Labor Statistics Annual Survey of Occupational Injuries and Illnesses (ASOII), a Department of Labor publication, contains work related injury/illness statistics derived from a random sample of 165,000 private sector establishments. Excluded are self-employed workers, farms with fewer than 11 employees, private households, and employees in federal, state, and local government. The 1996 survey (46) involving disabling claims (cases that had lost work days) revealed that:

1. Overexertion and repetitive motion contributed to 600,400 injury/illnesses, 33% of all disabling cases.
2. 311,900 cases were due to overexertion from lifting (nearly 17%).
3. Repetitive motion cases produced a median value of 17 lost work days, the highest of all event/exposures.

Although these are significant numbers, overexertion/repetitive motion disabling cases have diminished in incidence over the last several years (see Table 54.2). Comparing 1996 with 1992 data, the number of injury/illnesses requiring days away from work declined over 20% for overexertion and nearly 18% for repetitive motion. NIOSH (25) attributed these improvements to:

1. More intensive efforts to prevent this style of event/exposure.
2. More effective prevention and treatment programs which are reducing days away from work.

Table 54.2. A Yearly Comparison of the Number of Nonfatal Occupational Injury/Illnesses Involving Days away from Work for Events of Overexertion and Repetitive Motion Shows a Reduction of Cases for Both Categories[a]

	Year				
Event of Exposure	1992	1993	1994	1995	1996
Overexertion	659.1	635.8	613.3	559.9	526.6
Repetitive motion	89.9	94.3	92.6	82.6	73.8

[a]Case numbers are derived from the Annual Survey of Injuries and Illnesses conducted by the Bureau of Labor Statistics (number in thousands).

3. Reluctance of employers/employees to report or record disorders.
4. Changes in criteria used by health care providers to diagnose conditions.

Despite lower incidence, overexertion and repetitive motion continue to be associated with approximately 28% and 4% respectively of injuries and illnesses involving lost work days (see Table 54.3).

The accuracy of the ASOII is criticized from several perspectives. Under reporting is suggested by Leigh et al. (47) who concluded that the 1992 Annual Survey underestimated the incidence of disabling injury/illness by 50% to 70%. Over-reporting is claimed by some clinicians who feel disorders are improperly attributed to occupational activities on a frequent basis (48, 49) and medical diagnosis categorization in the Survey is disputed due to vagueness (50).

There are several other national injury/illness surveys such as the National Health Interview Survey, the National Electronic Injury Surveillance System, and the National Hospital Discharge Survey. However, data from these reviews are weakened by limited information, sample bias, and lack of association with work activities (51). Deriving information from occupational injury/illness surveys is further complicated by the lack of consistent guidelines for defining, classifying, and reporting occupational injury/illness (52, 53). The ASOII remains the only routinely published, national source of information about occupational injury/illness among U.S. workers (25).

Table 54.3. A Yearly Comparison of the Percent of Nonfatal Occupational Injury/Illnesses Involving Days away from Work for Events of Overexertion and Repetitive Motion Reveals a Relatively Static Situation[a]

	Year				
Event or Exposure	1992	1993	1994	1995	1996
Overexertion	28.3%	28.2%	27.4%	27.4%	28.0%
Repetitive motion	3.8%	4.2%	4.1%	4.0%	3.9%

[a]These percent values are derived from the Annual Survey of Occupational Injuries and Illnesses conducted by the Bureau of Labor Statistics.

Low back MSDs comprised 16% of all workers' compensation claims during 1989 for Liberty Mutual Insurance Company, yet accounted for 33% of all claims costs (54). The mean direct cost per case was $8,321, up from $6,807 in 1986. At the Boeing Company, 19% of all workers' compensation claims among the 31,200 employees were low back MSDs related (37). Low back claims were responsible for 41% of the total direct claims costs over the 15 month study period. Andersson et al. (55) cited nationwide direct cost estimates of from 4.6 billion to 13.4 billion for work related low back MSDs.

Upper extremity cumulative trauma disorders accounted for 0.83% of all 1989 Liberty Mutual workers' compensation claims and 1.64% of all claims costs (56). The mean direct cost per case was $8,070. It was noted that injury/illness classification was based on insurance claims coding which may not have always been sensitive enough to identify a UECTD. Surveys and research have not defined the prevalence of work-related upper extremity MSDs but it is estimated that this group of conditions is the second largest, with low back disorders first (57).

2.3 Mechanism of Disorders

Disorders associated with physical work factors are generally not a one-time event. A 22 yr-old worker with no history of low back pain might lift a 75-lb canister from the ground to a shelf and strain a lumbar muscle. The more common history is long-term exposure to physical work factors such as repetitive motion with high force and extreme posture at the wrist while assembling a part, leading to a musculoskeletal disorder.

Armstrong et al. (58) proposed a dose–response model that reflected the various work related factors and the physiological response of body tissues (see Figure 54.3). Work

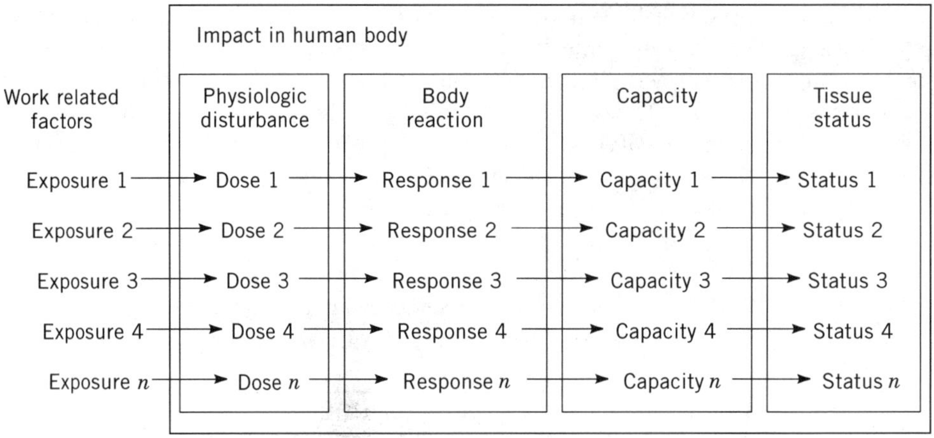

Figure 54.3. The proposed dose–response model by Armstrong et al. illustrates the impact that exposure to various work related factors can have through a dose/response/capacity circuit. A task that requires a tube to be squeezed with the hand (exposure) creates a mechanical disturbance to a hand flexor tendon (dose), deformation of the tendon (response), reformation of normal tendon length (capacity), and good integrity for the next activity (status).

requirements such as holding a vibrating hand tool, keyboarding data into a computer, or lifting a 30-lb box produce an exposure external to the body that leads to a dose internal to the body. Dose may be a mechanical disturbance to the body such as a knee ligament stretch/deformation when a squat postural position is assumed. Dose could involve a physiologic disturbance such as the accumulation of metabolic waste in soft tissues during a sustained muscle contraction. Dose may also be a psychological disturbance such as anxiety over a work decision or workload. Response is the change that occurs as a result of the dose. Tensile force through a tendon from muscle contraction can produce a stretch of tendon tissue. As repeat muscle contraction occurs, there may be a series of responses such as elastic deformation of tendon tissues, collagen fiber separation/tearing, local inflammatory response, and scar tissue formation. Capacity is the body's ability to respond to the dose. If the tensile forces across a ligament are within normal limits, the ligament stretches slightly and then returns to its normal length. If the tensile force exceeds physiologic capability, ligament fibers tear. A tissue status (i.e., torn ligament fibers) follows the dose/response/capacity sequence and sets the stage for the next exposure.

2.4 Common Disorders Related to Physical Work Factors

Hagberg et al. (26) applied accepted criteria for causality (strength of association, temporal relationship, consistency, predictive performance, and coherence) to epidemiologic studies that associated MSDs with physical work factors. Upper extremity conditions were supported by evidence and commonly accepted low back conditions are reviewed.

2.4.1 Hand/Wrist/Forearm Region

2.4.1.1 Wrist Tendinitis, Tenosynovitis, Peritendinitis. Tendons course over all sides of the wrist and are primarily involved with wrist extension and flexion motions. A sheath surrounds each tendon to facilitate sliding and provide protection from mechanical irritation as the tendon passes over bony structures. Tendinitis occurs when everyday microdamage from repeat tensile loading exceeds the body's ability to respond/repair (59). A low-grade inflammation reaction persists as the tendon attempts to heal. The key symptom is wrist pain that may be an intermittent dull ache or sharp/stabbing in nature. It is commonly worsened by hand grasping or wrist motion. Hand weakness is usually described along with an ache or tightness over the forearm to the elbow (5).

Tenosynovitis involves inflammation between the tendon and synovial membrane of the surrounding sheath. A mechanism described by Hegmann and Moore (31) involves a narrowing of the space between the tendon and the sheath leading to a friction irritation as the tendon slides. Symptoms include the gradual development of pain characterized by varying intensity and duration. The wrist pain is usually worsened by wrist motion and an ache may be described over the forearm to the elbow. Hand weakness is described and there may be slight swelling/warmth over the affected area (5, 31). When this condition affects tendons along the radial side of the wrist that control thumb motion, it is called DeQuervain's tenosynovitis.

Peritendinitis is an inflammation at the junction of the tendon and muscle. At this zone, the force created by a contracting muscle cell is transmitted to tendon fibers through numerous fingerlike projections. Pain is described over the affected area that may be several inches up the forearm. There may be a "creaking noise" near the involved tissue during motion of the thumb/wrist. Hand weakness and local swelling/warmth is common (31).

2.4.1.2 Carpal Tunnel Syndrome. Carpal tunnel syndrome has received more attention than any other upper extremity MSD. The wrist bones are shaped in the form of an arch, the ends of which are enclosed by a fibrous band called the transverse carpal ligament. Nine tendons and the median nerve pass through the tunnel of the arch (see Fig. 54.4). Carpal tunnel syndrome is a compression/irritation of the median nerve at the wrist that could be generated from a variety of sources including thickening of tendon sheaths, direct pressure of the tendons on the median nerve, thickening of the transverse carpal ligament, or contact trauma resulting from the use of the hand/wrist to pound on parts (31). Hagberg et al. (26) describe the potential mechanism of the disorder:

> While entrapment is occurring, the nerve is injured because there is increased pressure on it. In the early stages, this increased pressure will impair blood flow, oxygenation and axonal transport system. If the pressure is high, mechanical blocking of the depolarization process in the nerve will occur. Increased pressure on the nerve can be caused by edema of the surrounding tissues (i.e. swelling) . . . Edema may result from mechanical irritation of surrounding structures such as tendons or muscles. If a nerve is entrapped at one point there may be increased susceptibility to mechanical pressure both distal and proximal to the entrapment point. This increased distal and proximal susceptibility is caused by the impairment in the retrograde and anterograde axonal flows. Thus multiple entrapments along the same nerve are somewhat common.

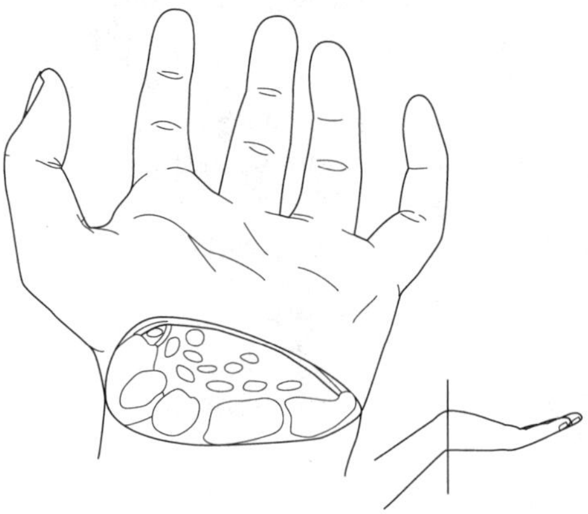

Figure 54.4. A cut through the wrist reveals the carpal tunnel formed by the carpal bones and the transverse carpal ligament. Nine wrist flexor tendons travel through the wrist along with the median nerve.

Symptoms of carpal tunnel syndrome include tingling/numbness along the distribution of the median nerve in the hand (thumb, second, third, and half of the fourth finger). There may be shooting pain in the hand, hand weakness, and atrophy of the thenar muscle. Symptoms are usually worse at night and can make sleeping difficult.

2.4.1.3 Raynaud's Phenomenon. Vibration induced into the upper extremity from operating compressed air and electrical tools is associated with a set of symptoms called Raynaud's Phenomenon (also known as Hand-Arm Vibration Syndrome or White Finger). Although the exact mechanism is unknown, it is felt that the vibration damages small blood vessels primarily in the hand, prompting episodes of arterial spasm (5, 60). There are complaints of hand tingling/numbness, cold, swollen fingers, pallor, and stiffness that are prompted by exposure to cold and vibration. The vascular changes may create damage to local nerve tissue and lead to further hand paresthesia (26).

2.4.2 Neck/Shoulder Region

2.4.2.1 Shoulder Tendinitis. Four muscles originate from the scapulae (infraspinatus, supraspinatus, teres minor, and subscapularis) and have tendons that broaden as they insert into the head of the humerus. This tendon group is known as the rotator cuff. Along the front of the humerus, the biceps brachii muscle forms a tendon that courses over the top of the rotator cuff (see Fig. 54.5). The tissue most commonly affected in this group is the supraspinatus tendon. One theory suggests that vascular insufficiency predisposes this tendon to degeneration. Impingement of the tendon between the humerus and the acromion is also proposed as a mechanism for this disorder (31).

Symptoms include localized, deep pain of gradual onset with a generalized ache in shoulder muscle tissue. It is usually painful to lie on the shoulder during sleep. Abduction and flexion shoulder motions can be painful, particularly near the 90 degree position. It is usually difficult to work with the hands over the head with this condition (5, 26, 31).

2.4.2.2 Tension Neck Syndrome. Several muscles course over the back of the cervical/upper thoracic spine and insert into the shoulder region. Their function includes moving and stabilizing the head/neck/shoulder region. When an overhead reach is performed, this muscle group rotates the scapulae so that the humerus is in a position to successfully perform the action. When the head is flexed, for example to look down on a work surface, these muscles stabilize the head/neck position.

This muscle group is subject to a cycle of pain/spasm that typically evolves into a chronic condition. Herington and Morse (5) describe the development of a small muscle spasm that leads to pain. The pain prompts more muscle spasms that interfere with local blood flow. Nutrients are blocked from tissues and metabolic waste products collect leading to more pain/muscle spasm reactions. Typical symptoms include pain over the neck/shoulder region of varying intensity/duration (see Fig. 54.6) and loss of shoulder strength. This condition has also been called myofascial pain syndrome and occupational cervicobrachial disorder.

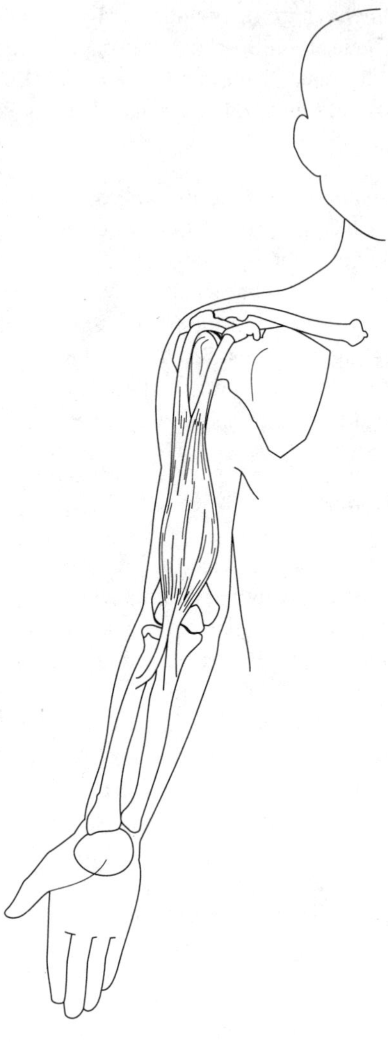

Figure 54.5. The biceps brachii muscle is illustrated along the front of the shoulder. This muscle has two tendons in the shoulder region: the most lateral tendon is the one of chief clinical concerns.

2.4.3 Low Back

2.4.3.1 Regional Low Back Pain. The spine is composed of 24 stacked bones called vertebra that form a flexible column. The low back typically refers to the lumbar region which makes up the lower third of the spine. This spinal area is formed by five vertebra separated by cartilagenous soft tissue called discs. The center of the disc is a soft gelatinous material, called the nucleus pulposus. The outside of the disc is made up of angled collagen fibers called the annulus fibrosus. The vertebra articulate at two facet joints located along the posterior aspect of each segment. Numerous ligaments and muscles stabilize the region.

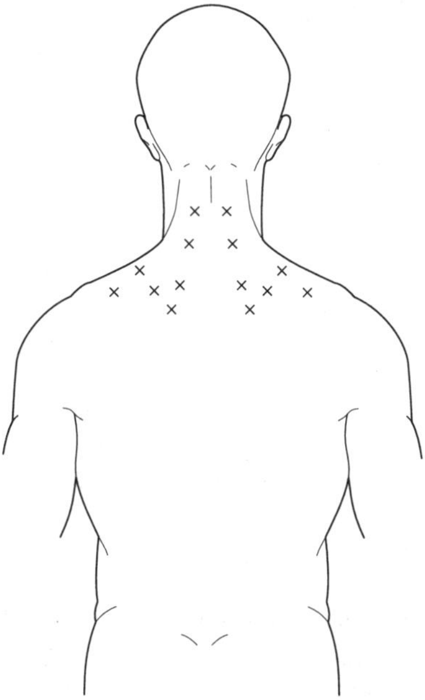

Figure 54.6. The Xs in this figure mark the common locations people with tension neck syndrome describe as painful.

The low back has three major functions (61):

1. Mobility: trunk bending and rotation is promoted.
2. Stability: transmission of body weight, attachment of internal organs, guidance/control of motion, and support for external load moments occurs.
3. Protection: boney structures of the spine surround the spinal cord and nerves.

The low back is subject to four types of forces that deform spinal tissue and can lead to injury (61, 62). A compression force perpendicular to the surface of the vertebra is created by upper body weight and load handling. Compression forces cause bulging of the disc (See Section 3.2 "Body Biomechanics" for additional explanation of this compressive force). Tension forces can occur across tendon tissue connected to low back muscle. When a load is lifted, low back muscles contract and exert force through tendon which slightly stretches/deforms. If the tendon tissue is over-stretched, fibers may become torn. When the spine rotates the torsional forces create stress on the disc and may increase the intradiscal pressure. Shear forces (parallel with the face of the disc) also occur during load handling. This shear is accentuated by the lordotic curve of the low back and increases strain on disc material.

The exact etiology of low back pain is often not known (57). Amid this uncertainty, suggested causes of pain include disc derangement, ligament tears, muscle/tendon tears, compression fractures of the vertebral body, and facet joint tissue changes (31, 63).

The worker will commonly provide a history of a lift, twist, carry, push/pull, or slip/fall event although the reason for pain may be unknown. After the event, pain may be felt immediately or appear gradually over several days. The pain can vary in intensity, be one sided/both sided over the low back, and restrict worker posture/motion. Symptoms sometimes radiate into the buttock (31, 52, 63).

2.4.3.2 ***Lumbar Radiculopathy.*** Nerve roots exit the spinal canal through openings between adjacent vertebral segments. These roots may become irritated from pressure created by a herniated/bulging disc tissue or by bony changes that narrow the vertebral opening. As with regional low back pain, the worker will commonly provide history of an activity that places a strain across low back tissues, however, symptoms may develop without a specific precipitating event. This differs from regional low back pain in that the dominant patient complaint often involves pain radiating into the lower extremity. There may be tingling or numbness in the body sites innervated by the involved nerve (31, 57, 63).

2.4.4 Generalized Conditions

Some work-related disorders can occur at several different locations in the body. Two conditions that have multisite capability and a strong association with occupational activities include osteoarthritis (degenerative joint disease) and cumulative trauma disorders.

2.4.4.1 ***Degenerative Joint Disease.*** Freely moving joints are composed of hard subchondral bone covered by a layer of hyaline cartilage along the articulating surface. A fibrous capsule attaches to each bone and encloses the joint. The inner lining of the capsule secretes a fluid, which, along with the articular cartilage, promotes a low coefficient of friction. Ligaments reinforce the joint capsule and guide/restrict motion.

Norkin suggests that tissue injury/illness may produce a wobbling of the joint or deviation from normal alignment. Abnormal forces occur across the joint which promote a breakdown of articular cartilage. Friction is increased, which leads to more articular cartilage erosion. A chain of events becomes established that perpetuates and affects all the joint tissues. Brisson et al. (63) consider high peak loading a potential instigator of the process. The force may induce microfractures to subchondral bone which decreases bone give/shock absorption capability. The articular cartilage covering the affected bone sustains greater compression forces and may become damaged.

Symptoms of degenerative joint disease include local joint pain, particularly after an activity that strains the involved joint. Stiffness and swelling commonly occur in the morning after prolonged inactivity. Changes on radiographs may be observed.

2.4.4.2 ***Cumulative Trauma Syndrome.*** Cumulative trauma syndrome is a classification of MSDs that have a causal mechanism of long term exposure to a stressor (i.e., vibration, awkward posture) (50). An assembly worker who makes 800 widgets a day by squeezing two parts together with a high force, pinch grip may sustain wrist tendinitis. A laboratory employee who looks down through a microscope for several hours a day to view slides may strain neck/shoulder muscle tissue. A data entry clerk who keys numbers on a computer keyboard for eight hours a day may experience carpal tunnel syndrome. These three conditions involve different tissues (tendon, muscle, and nerve) in different parts of the body, but all can be classified as cumulative trauma syndromes due to the duration of exposure to the injury source. Other terms used to similarly classify MSDs include occupational overuse syndrome, repetitive motion disorder, and repetitive strain injury.

3 SELECT BIOMECHANICAL CONCEPTS

3.1 The Body as a Lever System

From a biomechanical standpoint, the body is comprised of a complex array of levers in which external forces generate moments about a joint, or fulcrum, which are resisted by the forces produced by muscles.

These lever systems are classified as first, second, or third class levers. First class levers have the fulcrum placed between the load and the force. Figure 54.8 shows how the weight of the head is resisted by the force in the neck muscles with the atlantooccipital joint acting as a fulcrum. Another example of a first class lever system is in the low back where the erector spinae muscles act to support the weight of the upper body with the vertebrae and disc of the low back acting as a fulcrum.

Second class levers have the fulcrum located at one end and the force at the opposite end. An example of this is the force exerted by the gastrocnemius muscle by way of the achilles tendon, which lifts the body with the toes acting as the fulcrum. This is illustrated in Figure 54.9.

Third class levers have the fulcrum at one end and the weight on the other end with the force acting at any point between the weight and the fulcrum. An example of this is in the arm where the activity of the brachialis muscles provides the force to lift a load in the hand; the interface between the ulna and the humerus serving as the fulcrum. This is shown in Figure 54.10.

3.2 Body Biomechanics

As noted in the previous section, the body is a biomechanical lever system. The weight, or force, of the external load generates a moment about the fulcrum (resultant moment)

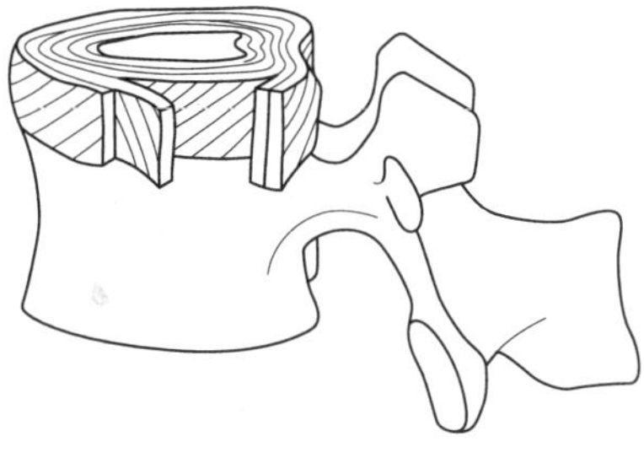

Figure 54.7. A lumbar vertebra is illustrated with a disc. The center of the disc is composed of a soft, gelatinous material (nucleus pulposus) that is surrounded by angled collagen fibers along the outside of the disc structure (annulus fibrosus).

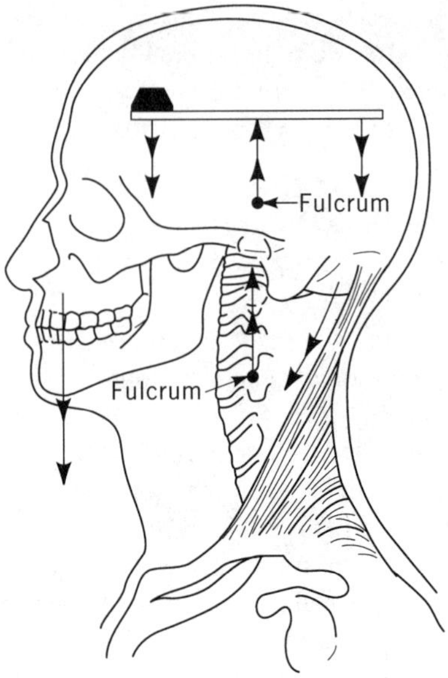

Figure 54.8. The action of the muscles of the neck against the weight of the head is an example of a first-class lever formed by anatomical structures. The atlantooccipital joint acts as a fulcrum.

which must be resisted by an equal and opposite moment generated by the muscles of the body (reactive moment). Moment may be defined as the affect of a force acting a given distance away from a fulcrum or point of rotation. Moment is calculated as the product of force times distance which represents the magnitude of the "tendency" to rotate about the fulcrum. The external force, or weight, and the internal muscle force combine to cause a compressive force on the fulcrum.

As an example, the moment at the elbow can be calculated for a lower arm holding a 10-lb weight in the hand (see Figure 54.11) (64). If the length to center of mass of the elbow is 7 in., the weight of the forearm/hand unit is 3.5 lb, and the length of the forearm/hand is 13 in., then the moment at the elbow can be calculated as follows:

$$\text{MOMENT} = (\text{Dist from Center of Object Wt to Fulcrum}) \times (\text{Wt of Object})$$

$$\text{Moment due to weight of forearm/hand } (M_{\text{fh}}) = 7 \text{ in.} \times 3.5 \text{ lb}$$

$$M_{\text{fh}} = 24.5 \text{ in.-lb}$$

$$\text{Moment due to 10-lb weight in the hand } (M_{\text{W}}) = 13 \text{ in.} \times 10 \text{ lb}$$

$$M_{\text{W}} = 130 \text{ in.-lb}$$

$$\text{Total moment at the elbow for lifting a 10-lb weight}$$
$$(M_t) = 24.5 + 130 = 154.5 \text{ in.-lb}$$

This resultant moment must be countered by the muscles of the body. For example, in Figure 54.12 the force of the biceps muscle is represented schematically by a force vector

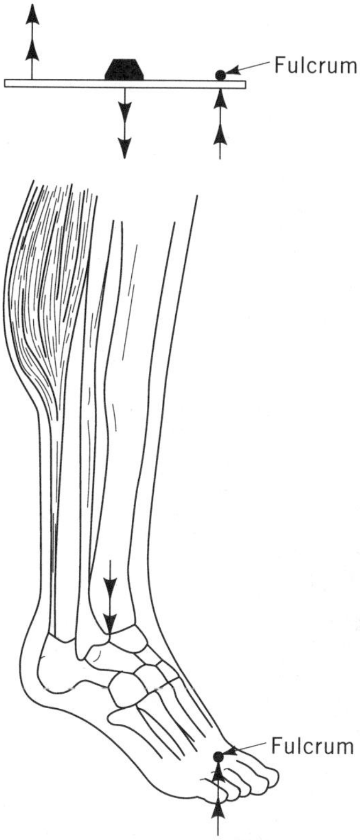

Figure 54.9. The ankle joint, as an example of an anatomical second-class lever system. The fulcrum is located at the base of the big toe.

acting 2 in. away from the elbow (64). (This is a third class lever.) The reactive moment generated by the biceps muscle acting through its moment arm must be equal to the resultant moment of 154.5 in.-lb calculated for Figure 54.11.

$$\begin{aligned} \text{If } M(\text{resultant}) &= M(\text{reactive}) \\ \text{then } 154.5 \text{ in.-lb} &= 2 \text{ in} \times \text{Muscle Force} \\ \text{Muscle Force} &= 154.5/2 = 77.25 \text{ lb} \end{aligned}$$

In the case of manual material handling tasks the moment about the low back is caused by the weight of the load, the distance that the load is held out from the body, the weight of the torso, and the torso flexion angle. These moments must be resisted by an equal and opposite moment caused by the muscles in the back.

Figure 54.13 represents a situation (first class lever) which would result if a 160-lb person were to hold a 50-lb weight out in front of the body with the torso flexed approx-

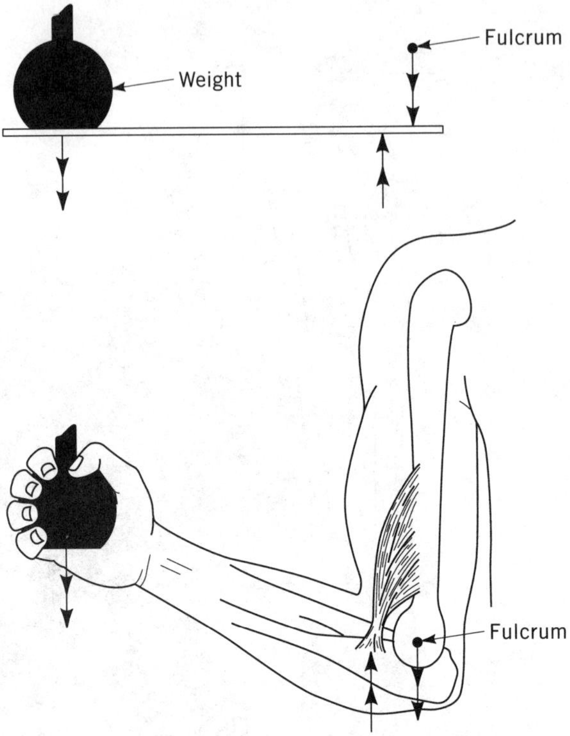

Figure 54.10. An anatomical third-class lever is formed between ulna and humerus. The brachialis muscle provides the activating force; the fulcrum is forced by the trochlea of the humerus.

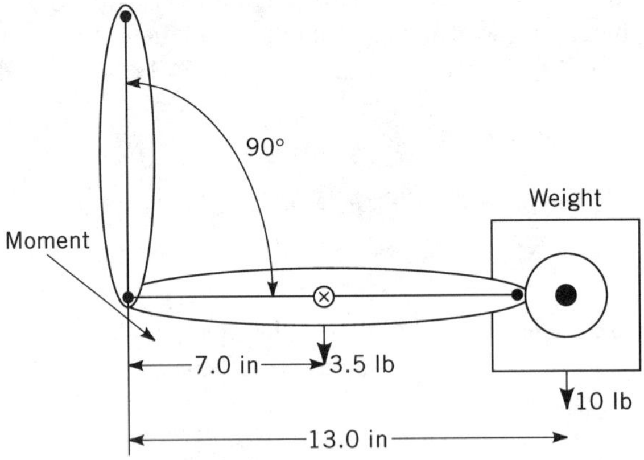

Figure 54.11. Example of moment calculation at the elbow. Adapted from Ref. 64.

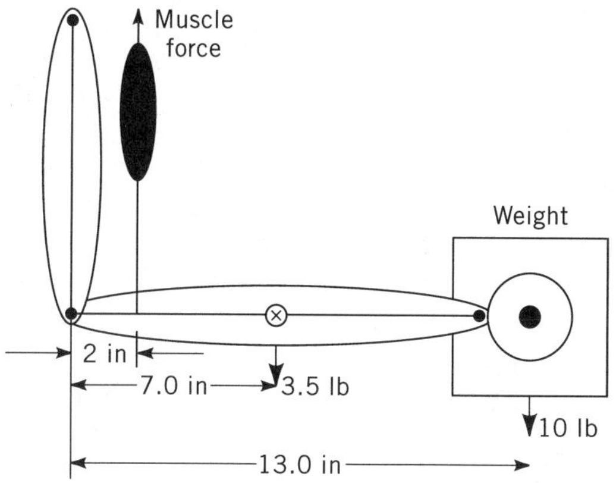

Figure 54.12. Example of muscle force calculation at the elbow. Adapted from Ref. 64.

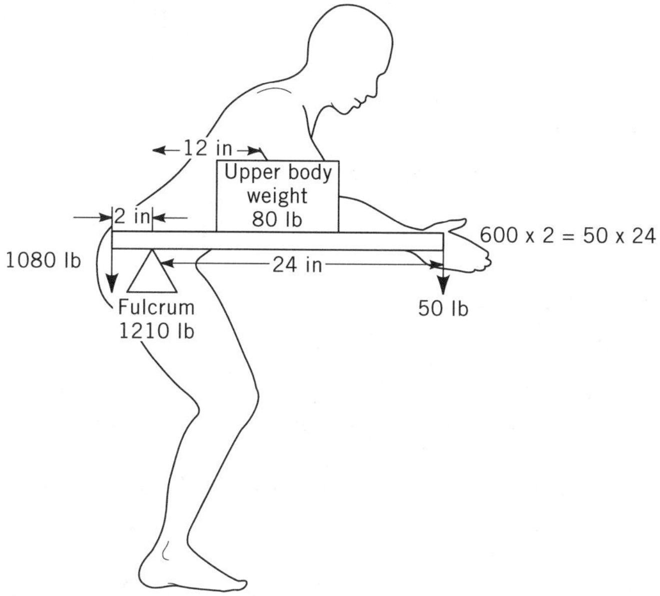

Figure 54.13. Example of back compressive force estimation.

imately half way (45° from vertical). It is assumed that the upper body weighs approximately 50% of total body weight or 80 lb. The clockwise moment caused by the load is 50 lb × 24 in. (= 1200 in.-lb) and the clockwise moment caused by the upper body weight is 80 lb × 12 in. (= 960 in.-lb) for a total of 2160 in.-lb. To balance this, the back muscle must generate the same 2160 in.-lb moment with a 2-in.-moment arm. This requires a force of 1080 lb since 1080 lb sincc 1080 lb × 2 in. = 2160 in.-lb. The fulcrum (low back) must accept all of the forces for a total compressive force of 1210 lb (1080 + 80 + 50). NIOSH (65) has established that back compressive forces in excess of 770 lb " . . . can be tolerated by most young, healthy workers" and back compressive forces in excess of 1430 lb " . . . are not tolerable . . . in most workers." Here we see that a load of only 50 lb can cause back compressive forces which approach the NIOSH upper limit.

While compressive force is of primary interest in the estimation of low back stresses during manual material handling, shear forces also exist. These shear forces try to translate or "slide" the vertebrae and are resisted by the muscles, ligaments, and the end-plate between the vertebral bodies and the intervertebral discs.

4 ANTHROPOMETRY

Anthropometry deals with the dimension of the human body. Designing a workstation to accommodate people of different sizes presents a tremendous challenge. Poor workplace design results in an increased risk of acute as well as chronic injury. Designing the job to reduce or eliminate bending and reaching can help reduce the risk of injury and make a more comfortable and productive workstation.

4.1 Anthropometric Data

Anthropometric data is usually given in terms of "percentiles". For example, a 50th percentile or average male is approximately 69.1 in. tall. This 50th percentile male (by height) is one who is taller than 50% of all males from a specific population. A 25th percentile (by height) female would be taller than 25% of other females. In the industrial workplace, tools or other systems are generally developed to accommodate from the 5th to the 95th percentile of the industrial population. When the population is comprised of both males and females this means the design should include from the 5th percentile female up to the 95th percentile male.

One common source of anthropometric data is tabular information such as that presented in Figures 54.14 and 15, which provide functional anthropometric dimensions for 5th, 50th, and 95th percentile females and males in a variety of postures (66).

Another source of anthropometric data is based on statistical procedures to predict the length of a link. It has been noted that the links of our body are in approximate proportion to our overall height. This relationship is shown in Figure 54.16 (67).

It is assumed that the length of any link can be approximated by multiplying the individual's height by the factor associated with that link or:

$$\text{Link Length} = f_i \times \text{stature}$$

where f_i is the proportionality factor associated with link i.

If one knows the factor associated with the link and the height of an operator or population of operators, one can estimate workstation design parameters. For example, one could estimate the height of a table which would allow a 50% male (average male) to put

Measurement (inches)	Males			Females		
	95th	50th	5th	95th	50th	5th
1-Forward grip reach	33.3	30.9	28.5	30.1	28.0	25.8
1a-Bust to grip reach	21.9	20.9	19.8	18.3	18	17.5
2-Elbow height	46.9	43.5	40.2	43.1	40.2	37.2
3-Hip height	39.2	36.0	32.9	35.8	32.9	29.9
4-Fingertip height	28.5	26.0	23.4	27.4	24.8	22.2
5-Shoulder height	61.0	56.7	52.4	56.1	52.2	48.2
6-Eye height	71.9	67.3	62.8	64.2	60.0	55.9
7-Stature	73.6	69.1	64.6	68.1	64.0	59.8
8-Vertical grip reach	87.0	81.9	76.8	80.5	75.8	71.1
9-Thigh thickness	7.3	6.3	5.3	7.3	6.1	4.9
10-Sitting eye height	33.9	31.5	29.1	31.9	29.5	27.2
11-Sitting height	38.4	36.0	33.7	36.2	33.9	31.5
12-Vertical grip reach (seated)	53.4	49.4	45.5	49.2	45.7	42.1
13-Elbow-fingertip length	20.3	18.9	17.5	18.5	17.1	15.8
14-Knee height	23.8	21.7	19.5	21.7	19.9	18.1
15-Popliteal height	19.5	17.5	15.6	17.7	15.9	14.2
16-Buttock-popliteal length	21.9	19.7	17.5	21.3	19.3	17.3
17-Buttock-knee length	25.6	23.6	21.7	24.6	22.6	20.7

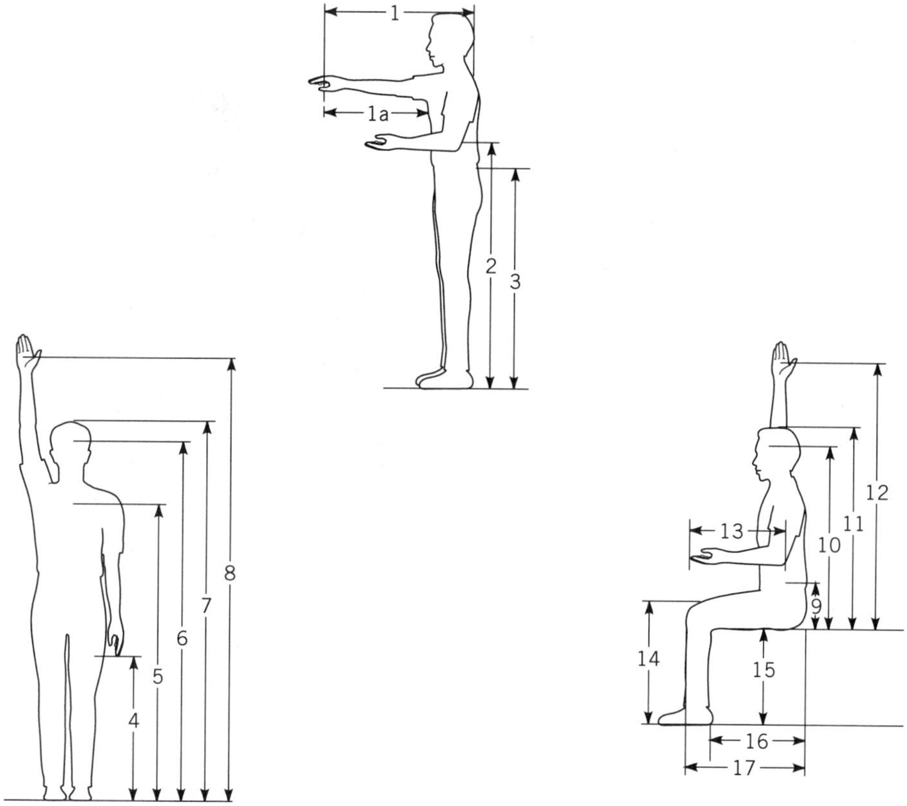

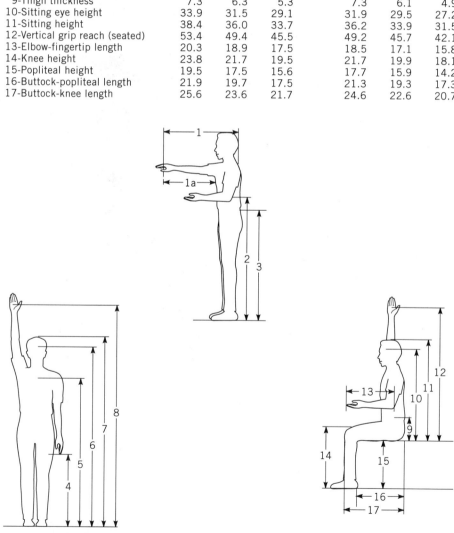

Figure 54.14. Functional anthropometric dimensions. Adapted from Ref. 66.

Measurement (inches)	Males 95th	Males 50th	Males 5th	Females 95th	Females 50th	Females 5th
19-Shoulder breadth (bideltoid)	20.3	18.5	16.7	17.3	15.8	14.2
20-Shoulder -elbow length	15.8	14.4	13.0	14.4	13.2	12.0
21-Sitting shoulder height	25.8	23.6	21.5	24.4	22.2	20.1
22-Sitting elbow height	11.6	9.7	7.7	11.2	9.3	7.3
23-Hip breadth	16.1	14.2	12.2	17.3	14.8	12.2
24-Chest (bust) depth	11.4	10.0	8.7	11.8	10.0	8.3
25-Abdominal depth	13.0	10.8	8.7	12.2	10.2	8.3
26-Upper limb length	33.5	31.1	28.7	30.5	28.1	25.8
27-Shoulder-grip length	28.5	26.4	24.2	26.0	24.0	22.1
28-Span	76.8	71.3	65.8	68.7	64.0	59.3
29-Hand length	8.1	7.5	6.9	7.5	6.9	6.3
30-Hand breadth	3.9	3.5	3.2	3.4	3.0	2.6

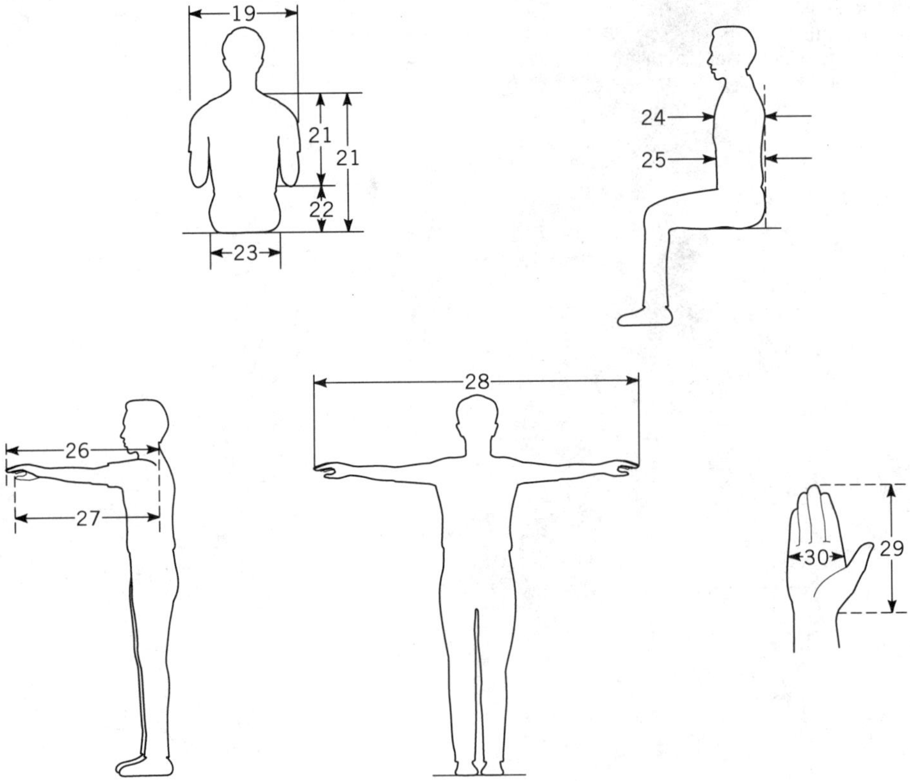

Figure 54.15. Functional anthropometric dimensions. Adapted from Ref. 66.

his hands on the table with the upper arm vertical and the lower arm horizontal. This would require an estimation of elbow height. As noted above, a 50 percentile male is 69.1 in. tall. From the figure we can see that the factor associated with elbow height is 0.63. Elbow height for a 50% male can be calculated as:

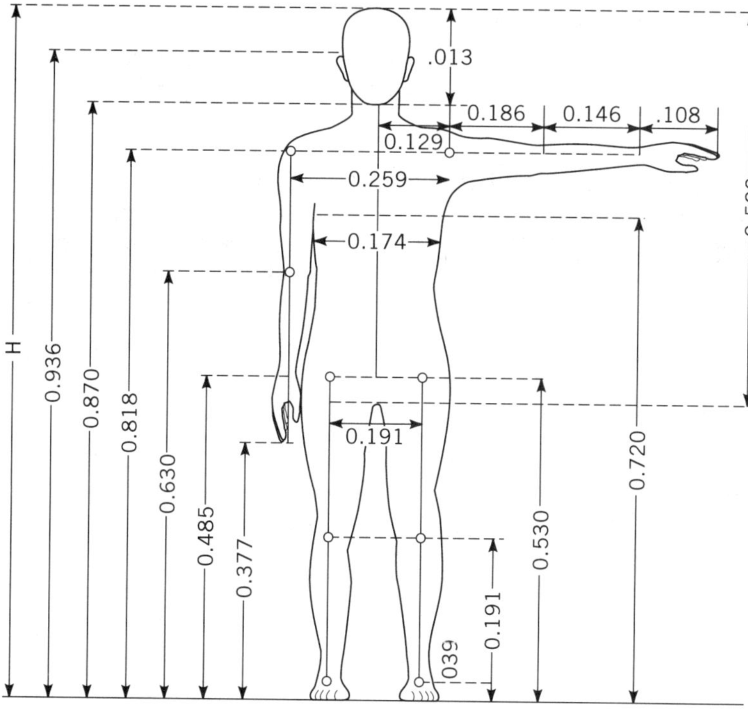

Figure 54.16. Body link lengths as a function of height.

$$0.63 \times 69.1 \text{ in.} = 43.5 \text{ in.}$$

If the table were constructed so that the top was approximately 44.5 in. from the floor (add 1 in. for shoes) the average male would be in a relatively neutral posture with the upper arms vertical at his side and his lower arms horizontal.

One could also use a combination of link lengths to estimate the maximum overhead fingertip reach of this same person by adding the factors associated with the floor-shoulder (0.818), shoulder–elbow (0.186), elbow–wrist (0.146), and wrist–fingertip (0.108) and multiplying by the worker height (69.1 in.).

$$0.818 + 0.186 + 0.146 + 0.108 = 1.258$$

This would result in a maximum overhead (*fingertip*) reach of:

$$1.258 \times 69.1 \text{ in.} = 86.93 \text{ in.}$$

Note that this is fingertip reach and would not allow the worker to exert any force or perform manipulative activities at this overhead distance.

4.2 Using Anthropometric Data

When applying anthropometric data it is important to keep several factors in mind. Below is a list of steps that are useful in applying the information properly:

1. Analyze the job.
   a. *Determine what is involved in doing the job properly.* A step by step analysis of the job attributes will assist in the determination of the type of anthropometric data needed. For example, are there pinch points to guard? Does the operator need access to the inside of the machinery for maintenance? Are controls and displays used regularly? Is the job done predominantly standing, sitting, or both?
   b. *Determine stressful elements of the job.* Does the job have elements that can contribute to musculoskeletal disorders? Are these elements caused by poor job design because of improper use of anthropometric data?
2. Find the relevant body dimensions.
   a. *What body dimensions are needed in designing the workplace*? An analysis of the job to determine what data is necessary to correct the problem is an important step in the efficient use of anthropometric data.
   b. *Determine what percentage of the population has to be accommodated.* In the typical industry setting, a person is eligible for a job as long as he/she meets certain physical requirements. The range of values and the amount of adjustability that should be designed into these jobs must accommodate a large proportion of the population. As noted earlier, 5th percentile to 95th percentile is normally used as the design guideline.
   c. Obtain the data from either the tables or from the statistical prediction methods and add appropriate allowances for clothing, etc. Depending on the physical requirements of the job, the surrounding environment, or many other factors, the operator may have to wear different types of protective clothing. Since most anthropometric data reflects unclad or lightly clad individuals, allowances should be added for jobs requiring protective clothing.
3. Apply the data to design/redesign the workplace.
   a. Make design recommendations with operator input and consent.
   b. Use analysis procedures to check design or design changes and determine if they have minimized/corrected the problem or introduced new stresses.
   c. Follow-up at a later date to recheck.

The three general principles for applying anthropometric data in the workplace are to (*1*) design for the extreme, (*2*) design for the average, and (*3*) design for adjustability.

Design for the Extreme. As noted above, in the industrial workplace one usually designs tools, workplaces, or other systems to accommodate the 5th to the 95th percentile of the population. In some cases an even larger range must be accommodated in order to protect the worker from serious hazards.

Design for the Average. Using this principle is frequently a mistake. Depending on the application, designing for the average will often lead to discomfort for a large portion of the population. In some cases, however, designing for the average is necessary. Doorknobs, light switches, urinals, etc., are all designed to accommodate the "average" person.

Design for Adjustability. If the designer can build adjustability into the layout, then this should be done. Adjustability allows for many people to use the system or equipment. This assumes, however, that the worker will perform proper adjustments. To ensure that workers will adjust the system, it must be designed to facilitate quick, easy adjustment with a minimum of tools or other devices. In addition, workers must be taught the basics of correct posture and how to adjust the equipment correctly.

5 RISK FACTORS

5.1 Background

Risk factors are human activities and responses related to the physical attributes of tasks that increase the likelihood of the occurrence of MSDs (67, 68). Risk factors may directly (i.e., high force) or indirectly (i.e., recovery time) influence the beginning or development (i.e., duration) of a disorder (26). Derived principally from epidemiologic studies, risk factors reflect a detailed analysis of the work conditions associated with MSDs.

Interacting with the physical features of work such as the workstation, parts, tools and equipment, can require the body to generate/absorb high forces, assume awkward postures, and perform fast motions. These human physical stresses have been associated with MSDs (25, 26, 68). Although research is continuing to develop, it is generally considered that (68):

1. The greater the exposure to a risk factor, the greater the risk of MSDs.
2. The more the number of risk factors, the greater the risk of MSDs.
3. The greater the recovery time between successive risk factor exposures, the less the risk of MSDs.

Human force, posture, motion, and vibration are risk factors. They are quantified by magnitude, repetition, duration, and recovery that are also considered to be risk factors. Identification of the presence/absence of risk factors and their quantifying characteristics is done through job analysis techniques.

5.2 Specific Risk Factors

5.2.1 Force

Force is the strain that occurs across body tissues during an exertion. When a box is lifted off the ground, a compression force is created through lumbar vertebra and discs. When a part is grasped, a tension force is created in the muscle/tendon units of hand muscles. In general, the greater the force, the greater the risk of MSDs. Several investigations associate forceful hand exertions (i.e., grasping heavy fabrics in a sewing operation, various occupations handling tools/parts/equipment requiring hand forces greater than 9 lb, forceful gripping among grinders, butchers, grocery workers and frozen food factory workers) with carpal tunnel syndrome (69–71). Jobs in an investment casting plant that required hand forces of greater than 13 pounds (72) and intensive hand tasks associated with food processing (73) were related to hand/wrist cumulative trauma disorders. Forceful jabbing motions performed by the upper extremity with a lance to clean furnace openings were associated with elbow arthritis (74). Mail sack weight has been implicated as the reason letter carriers have greater incidence of shoulder disability than other groups (75).

Heavy physical work is frequently related to low back pain (45, 76–80). Hildebrandt (76) surveyed nearly 9,000 workers from 33 trades and 34 professions and found relatively higher prevalence rates of back pain among workers in the construction industry, supervisory production employees, plumbers, drivers, and cleaners. Low prevalence was found among sedentary professions such as chemists, scientists, bookkeepers, secretaries, and administrators. Marras et al. (81) determined that load moment was a predictor for low back injury for medium and high risk jobs when he assessed 400 industrial lifting jobs in 48 industries.

A slight variation of the force risk factor involves body contact with surfaces. The contact puts pressure on soft tissues just under the skin such as the nerves and tendons and has been associated with symptoms of MSDs. Contact stress at the thumb (82), wrist (83), and palm of the hand (84) has been associated with nerve irritation.

5.2.2 Posture

Posture is the positioning of the body as a result of joint motion. When a load is held far away the resulting body posture can increase the moment about a joint and increase the muscle and tendon forces at the joint. In addition, when joints are positioned near their end of range of motion, greater strains tend to occur across muscles/tendons and ligaments to stabilize the body part. In general, the more a joint deviates from its neutral position, the greater the risk of injury.

Specific postures have been associated with MSDs or MSDs symptoms. For example:

at the wrist:

- Activities with a flexed or extended wrist were related to carpal tunnel syndrome (48).
- Extension position was related to carpal tunnel syndrome (69).
- Tasks that required ulnar deviation had a higher prevalence of wrist symptoms (85).

At the shoulder:

- Elevating the shoulder greater than 90° was associated with stiffness/pain in the neck and shoulder (86).
- Abduction greater than 30 degrees was related to shoulder disorders (74).
- Abduction or flexion exceeding 60 degrees for a duration of greater than one hour/day was associated with acute shoulder and neck pain (87).

At the cervical spine:

- a mean load moment of 58.4 in.-lb in a flexion position (near full flexion) produced pain in muscles along the back side of the neck within 15 minutes (88).
- 60 degrees of neck flexion took 120 minutes to generate severe pain symptoms while it took over 300 minutes at 30 degrees of neck flexion to produced severe pain symptoms (89).

At the low back:

- rotation of the low back has been causally linked to low back pain (45, 77, 78).
- low back flexion was predictive of medium and high risk jobs (81).

5.2.3 Motion

Motion is the displacement of one body part relative to another around a joint. Velocity, the speed of body part motion, and acceleration, the rate of change of speed of body part motion, have both been investigated. Marras (90) monitored wrist position, angular velocity, and angular acceleration in 40 workers at eight industrial facilities who were divided into low risk and high risk groups. The mean values of both angular velocity and acceleration velocity were significantly different between the low and high risk jobs. In another study, wrist angular velocity was higher among those performing high risk tasks but not statistically significant (85). A low back, tri-axial electrogoniometer was used to assess low back motions while industrial tasks were performed in positions considered low, medium, and high risk for low back disorders (81). Trunk lateral velocity and trunk twisting velocity were good predictors of the medium and high risk jobs.

5.2.4 Vibration

Vibration is oscillatory motion of the body. It can be created regionally in the upper extremity, known as segmental or hand–arm vibration, by using a hand tool such as a pneumatic wrench or performing select tasks (i.e., finishing operations using a grinder). Platers exposed to a 4.6–4.7 m/s^2 four hour, frequency-weighted energy equivalent acceleration had a 42 percent prevalence for white fingers. The platers also had a higher odds ratio for a positive Allen test compared to a nonexposed group (91). Use of vibratory hand tools has been related to carpal tunnel syndrome (92, 93). NIOSH felt disorders related to hand–arm vibration were significant enough to develop Criteria for a Recommended Standard (94). There may be an association between segmental vibration and grip muscle force (95). Hand–arm vibration stimulates the tonic muscle reflex, which promotes muscle contraction, and reduces tactility, which also prompts greater muscle contraction to hold/manipulate the vibrating object. The higher grip muscle forces increase the magnitude of the force risk factor.

Vibration can also be generated in the whole body when standing or sitting on a vibrating surface. Low back pain has been associated with driving tractors (96, 97), forklifts (97–99), urban buses (100), crane operators (101), and cars (102). Degenerative changes in the lumbar spine were reported among operators of earth-moving machines who had been exposed to whole-body vibration for at least ten years (103). Bovenzi and Zadini (100) found higher occurrence of low back symptoms with increased magnitude, duration, and dose of whole body vibration, while Bongers et al. (101) indicated that the risk ratio for disability due to intervertebral disc disorders increased with duration of exposure to whole body vibration.

5.2.5 Cold Temperature

Cold temperature is an environmental hazard that can freeze body tissues and cause frostbite and trench foot. Cold can also influence physiologic responses such as grip strength, manual dexterity, and cutaneous sensory sensitivity (68). Although force was considered the major risk factor, frozen food workers were among groups with the highest prevalence of carpal tunnel syndrome (71). In developing a risk factor based checklist, Keyserling et al. (104) included cold as an item to be considered when evaluating stress in relation to handling tools.

5.2.6 Repetition

Repetition is the repeat performance of a similar force, posture, motion, or vibration absorption over the course of accomplishing a task. Repetition is a quantifier of the risk factors that are physical stressors. Generally, the greater the repetition of the risk factor, the greater the degree of risk. Ranney et al. (105) reported that 54% of 146 female workers in highly repetitive jobs had clinical findings of upper extremity MSDs. Microscopic changes in fibrous tissue density were described by Armstrong et al. (106) in synovial, subsynovial, and adjacent connective tissues on cadaver specimens from 30–40 mm proximal to 20–40 mm distal to the wrist crease. Maximum changes occurred 0 to 20 mm distal to the wrist crease suggesting that mechanical stress at the wrist from repeated hand exertions promotes fibrous changes in local tissues and may contribute to increased pressure in the carpal tunnel, a proposed mechanism of carpal tunnel syndrome. Kuorinka and Koskinen (107) report that the number of pieces handled per time unit was related to wrist muscle-tendon syndrome among light mechanical industry workers. Although both high force and high repetitiveness were predictors of carpal tunnel syndrome, high repetitiveness was more significant among 652 industrial workers (70). Those in high repetitive/high force jobs tend to transfer out of those positions more frequently than those in low force/low repetitive jobs (72). Lift frequency was related to medium and high risk jobs for low back injury (81). Low back muscle fatigue was shown to occur sooner for light load/high frequency tasks than for heavy load/low frequency tasks (108).

5.2.7 Duration

Duration is the amount of time that there is exposure to a risk factor. It may involve the length of time during a task, during a work day, or over the occupational experience of the worker. As with repetition, duration is a quantifier of the risk factors that are physical stressors. Generally, the greater the duration of the risk factor, the greater the degree of risk. Duration of employment was influential in the prevalence of severe disability among female garment workers who did piecework. Those who worked 10 to 19 years had over twice the incidence of severe disability compared to those who worked from 0 to 4 years (107). Ohlsson et al. (108) reported higher rates of hand pain among assembly workers with longer employment duration. Over the course of a day, video display terminal operators reported greatest discomfort at the end of the work shift (109). Length of employment was significant for incidence of low back trouble among fork-lift truck drivers (98). Operators of earth-moving machines who had worked at least ten years with exposure to whole body vibration demonstrated greater low back degenerative changes when compared to controls (103).

5.2.8 Recovery

Recovery is a time period of rest, low stress activity, or an activity that allows a strained body part to rest. Recovery may occur within a task or between tasks that involve the same force, posture, motion, or vibration exposure. As with repetition, recovery is a quantifier of the risk factors that are physical stressors. Generally, the greater the recovery from the risk factor, the less the degree of risk. Taking 15-sec rest pauses every 6 min. reduced discomfort levels among video display terminal operators (112). The occurrence of sick leave due to MSDs was reduced by 0.5 years when comparing sewing machine operators who work 5-hr days versus those who work 8-hr days (113).

5.3 NIOSH Review of Work-Related Musculoskeletal Disorders and Risk Factors

NIOSH analyzed over 600 epidemiologic studies to assess the relationship between select MSDs and physical work, psychosocial, and individual factors (25). Evidence of causality included strength of association, consistency, temporality, exposure-response relationship, and coherence of evidence. Evidence of work relatedness was divided into four categories:

1. *Strong evidence of work-relatedness.* A causal relationship was shown to be very likely between intense or long-duration exposure to the specific risk factor(s) and MSDs when the epidemiologic criteria of causality was used. A positive relationship was observed between exposure to the specific risk factor and MSDs in studies in which chance, bias, and confounding factors could be ruled out with reasonable confidence in at least several studies.

2. *Evidence of Work-Relatedness.* Some convincing epidemiologic evidence showed a causal relationship when the epidemiologic criteria of causality for intense or long-duration exposure to the specific risk factor and MSDs in studies in which chance, bias, and confounding factors were not the likely explanation.

3. *Insufficient Evidence of Work-Relatedness.* The available studies were of insufficient number, quality, consistency, or statistical power to permit a conclusion regarding the presence or absence of a causal association. Some studies suggested a relationship to specific risk factors, but chance, bias, or confounding may explain the association.

4. *Evidence of No Effect of Work Factors.* Adequate studies consistently show that the specific workplace risk factor(s) was not related to the development of MSDs.

The review concluded that there was strong evidence for increased risk of some work-related MSDs when exposed to select physical work factors when exposures are intense, prolonged, and particularly when workers are exposed to several risk factors simultaneously (see Table 54.4). Despite limited detailed quantitative information about exposure–disorder relationships (dose–response), there were a large number of cross-sectional studies, reinforced by a limited number of prospective studies. Investigations involving psychosocial factors were evaluated as modest in strength. Personal factors had some support in influencing the occurrence of MSDs but were not seen as synergistic with physical work factors.

5.4 Interaction of Physical Risk Factors

Analyzing the workplace from a risk factor perspective is complicated by the combination of parameters in a task (see Table 54.5). When performing data entry at a video display terminal workstation, there may be repeat finger flexion motion with wrist extension (awkward posture) while striking the keyboard (force) at a rate of 150 keys per minute (repetition) for two hours (duration). A warehouse task may be characterized by bending and twisting at the waist (awkward posture) to (acceleration/velocity) lift a heavy case (force) quickly and place on a pallet at a rate of 100 cases an hour (repetition) for four hours (duration). It is a challenge to determine the degree of injury/illness risk accurately when a task is characterized by multiple physical risk factors.

Several references list tasks as risk factors as opposed to specific human stressors. The incidence of injury/illness relative to those tasks is well founded. However, those tasks are usually composed of a combination of human stressor risk factors. Tasks cited as high risk include:

- Lifting/lowering.
- Pushing/pulling.
- Carrying.
- Gripping.
- Heavy dynamic exertion (combination of lifting/pushing/carrying).
- Reaching.
- Overhead work.
- Static exertion (bending at the waist for three minutes while performing a task).
- Wearing bulky clothes (usually requires greater generation of muscle forces to move).
- Using gloves (usually requires greater generation of muscle forces to grasp).

6 WORKSTATION DESIGN

Workstation design should consider the purpose and processes of a job and match those criteria with worker characteristics/capabilities in a manner that enhances the well being of the operator and the company. A successful design follows principles from several

Table 54.4. A Summary of Findings is Presented from the NIOSH Epidemiologic Review of Studies that Examine the Relationship Between Physical Risk Factors and MSDs[a]

		Level of Evidence			
Body Part	Risk Factor	Strong Evidence	Evidence	Insufficient Evidence	Evidence of No Effect
Neck and neck/shoulder	Force		X		
	Posture	X			
	Vibration			X	
	Repetition		X		
Shoulder	Force			X	
	Posture		X		
	Vibration			X	
	Repetition		X		
Elbow	Force		X		
	Posture			X	
	Repetition			X	
	Combination	X			
Hand/wrist: Carpal Tunnel Syndrome	Force		X		
	Posture			X	
	Repetition		X		
	Vibration		X		
	Combination	X			
Hand/wrist: tendinitis	Force		X		
	Posture		X		
	Repetition		X		
	Combination	X			
Hand/wrist: Hand-arm vibration syndrome	Vibration	X			
Back	Heavy physical work		X		
	Lifting/forceful movement	X			
	Awkward posture		X		
	Static work posture			X	
	Whole body vibration	X			

[a]Ref. 25.

Table 54.6. A Framework to Evaluate a Task Relative to Risk Factors[a]

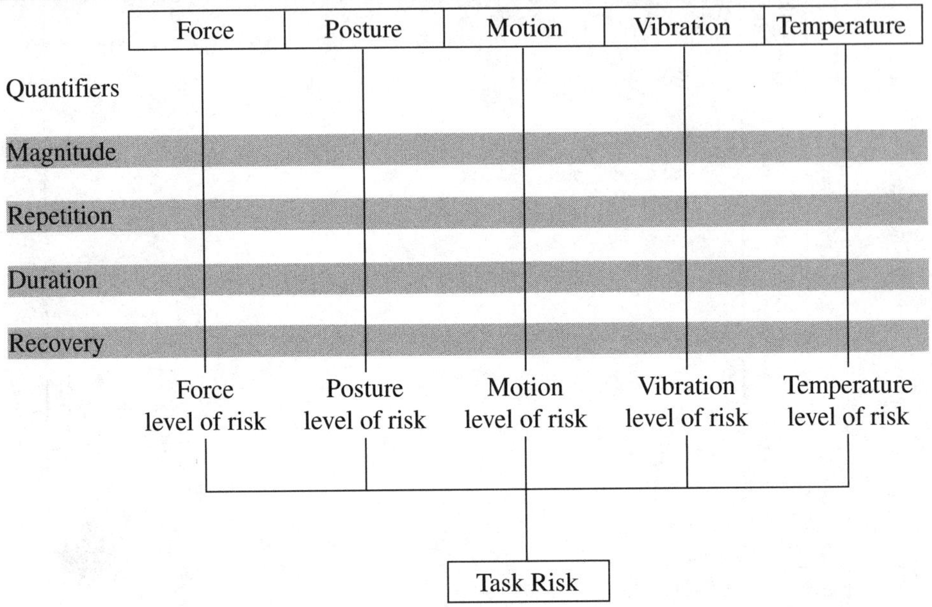

[a]Each physical stressor risk factor (i.e., force) is qualified by magnitude, repetition, duration, and recovery to define its level of risk. Once each physical stressor risk factor is considered and quantified, consider them as an aggregate to get a sense of task risk.

specialties that have been grouped into human factors engineering. This section concentrates on features related to the physical design of the workplace. Anthropometry, a fundamental topic of workstation design, is addressed earlier in this chapter and will not be discussed.

6.1 Standing Workstations

Standing workstations work well when tasks involve frequent reaches, operator mobility, the use of heavy tools/equipment, or the need to generate large forces. General guidelines include:

Work surface height: the nature of the work and elbow height are critical features; generally, for light and heavy work, the work surface should be below the elbow: specific recommended heights are (114):

- For heavy work, 75–90 cm (males) and 70–85 cm (females).
- For light work, 90–95 cm (males) and 85–90 cm (females).
- For precision work, 100–110 cm (males) and 95–105 cm (females).

If the work surface is nonadjustable and the design is intended for a general population, Sanders and McCormick (115) suggest placing the work surface at the high end of the height range; it is assumed that shorter workers will stand on an elevated platform.

Other concerns include ensuring that there is room for toe space and knee/thigh space below the work surface; floor padding and a stool for foot placement enhance worker comfort (116).

6.2 Sitting Workstations

Sitting workstations are best applied when tasks involve precise hand movement, low hand forces, and few needs to reach/move to obtain work items. General guidelines include:

- Work surface height: work activities greatly influence the suggested height of the work surface—lower for keyboarding and higher for reading/writing. Sanders and McCormick (115) recommend:
  - Acquiring equipment with height adjustability to allow for various worker sizes and multiple tasks.
  - Placing the work surface such that the elbows are at the working height (i.e., a computer keyboard frequently places the keyboard home row two inches above the work surface; the working height is the keyboard home row, hence, the work surface should be placed such that the elbows are in the same plane as the computer home row).
  - Ensuring adequate thigh/knee/foot clearance under the work surface.
- Other concerns: reaches above the shoulder and behind the back should be avoided; padding may be needed if contact is made with the forearm along the work surface edge; details involving chairs and specialty concerns with computer workstations are addressed in another chapter.

6.3 Man–Machine Interface

Interaction with equipment happens at many workplaces as part of the work process. Machines have control devices such as buttons and levers that the operator activates to start/stop/modify work events. Frequently, machines also have displays that give the operator feedback regarding the work process/work condition and prompt an operator action. Select control/display design characteristics reduce the chance for operator judgment error or confusion in responding to a work process.

6.3.1 Display/Control Guidelines

Use visual style displays when:

- The message is complex, long, and may be referred to later (as in a monitoring process).
- The message does not need immediate attention.
- The operator remains in a stationary location.

Use auditory style displays when:

- The message is simple, short, and will not be returned to later.
- The message requires immediate action.
- The operator is not confined to one location.

Displays of numbers and letters should:

- Use simple, uncluttered lettering fonts. Avoid the use of slanted fonts since they are more difficult to read.
- Avoid using all capital letters for instructions. The readability of instructions can be further enhanced by listing the instructions rather than presenting the information in a paragraph format.
- Orient letters and numbers in an upright position. Do not orient them radially on circular or semicircular dials.
- Avoid unusual numerical progressions (i.e., 3, 6, 9, 12 . . .)
- Follow minimum dimensions for easy reading:
  - height (in in.) = viewing distance in in./200
  - width/height ratio = $\frac{3}{5}$
  - stroke width/height ratio = $\frac{1}{6}$

There is no one "correct" method for the placement of controls and displays. The layout is intimately dependent upon the relative interaction of these two components. Grandjean (114) suggests the following:

- Controls should be located close to the corresponding display. Controls should be below, or if necessary, to the right of the display.
- If controls and displays must be located on separate panels, they should be presented in the same order and arrangement.
- Identification labels should be identical for controls and displays and should be placed above or to the right of the control or display.
- Controls which are normally operated in sequence should be arranged in sequential order from left to right.
- If controls are not operated in a particular sequence, the controls and corresponding displays should be arranged in functional groups emphasized by color, shape, labeling, knob size/shape, border, or physical arrangement.
- The most often used controls and displays should be located directly in front of the operator.
- General United States stereotypes should be followed:
  - When a control is moved or turned to the right, the pointer should move to the right or up.
  - When a control is moved up or forward, the pointer should move up or to the right.

- Clockwise movements should be associated with increases in the display.
- When a lever is moved up or to the right, the display readings should increase or a switch should move to the "on" position.

7 WORKSITE ANALYSIS

7.1 Surveillance Methods

Surveillance methods included here include the use of OSHA and Workers' Compensation data and information which may be gathered directly from workers or supervisors.

7.1.1 Records Review

Employers with 11 or more employees at any time during the present or prior calendar year must maintain a log and summary of all recordable occupational injuries and illnesses. At present this record is kept on OSHA Form 200. OSHA recordable injuries/illnesses are those which cause any of the following:

- Fatality.
- Lost work days.
- Transfer to another job.
- Termination of employment.
- Medical treatment (other than first aid).
- Loss of consciousness or restriction of work or motion.

The OSHA Form 200 includes a brief description of the injury or illness including days lost or restricted. It also indicates if the disorder is associated with repeated trauma. Review of the OSHA 200 compilation of recordable injuries or illnesses can provide substantial information relating to the frequency and severity of ergonomic related injuries and illnesses in the facility or department (47). Statistics relating to injuries or illnesses are often reported in terms of incidence rate (IR). This is the number of *new* cases per 100 worker years and may be computed for all musculoskeletal disorders or by particular type of musculoskeletal disorder. The incidence rate is calculated as follows:

$$IR = \frac{\text{Number of new cases during a time period} \times 200{,}000 \text{ hr}}{\text{Total hours worked by all workers for the time period}}$$

One measure of incident *severity* is the number of musculoskeletal cases involving lost work days, or sometimes the total of lost work days per 100 worker years. For example, this might be calculated as:

$$SR = \text{Number of new cases during a time period which involve lost work days} \times \frac{200{,}000\ \text{hr}}{\text{Total hours worked by all workers for the time period}}$$

Another measure of severity would be based on total lost work days (LWD). This would be represented as:

$$SR\ (LWD) = \frac{\text{Number of days lost during a time period} \times 200{,}000\ \text{hr}}{\text{Total hours worked by all workers for the time period}}$$

Remember that the OSHA recordable incidence rate (IR), lost work day incidence severity rate (SR), or lost work days severity rate [SR(LWD)] noted above may represent (*1*) total OSHA recordables, (*2*) OSHA recordables for all musculoskeletal disorders, or (*3*) OSHA recordables for a particular *type* of musculoskeletal disorder. This rate can therefore be used to identify the frequency/severity of general musculoskeletal disorders or a specific type of musculoskeletal disorder within the facility.

Workers' Compensation records can be used as a surveillance method for musculoskeletal disorders associated with ergonomic hazards. Workers' Compensation records can also be used to identify jobs or departments in which a significant number of musculoskeletal disorders have occurred. One advantage of Workers' Compensation records is that they also identify jobs or departments where musculoskeletal disorders have been associated with a high dollar cost in terms of medical and/or rehabilitation expenses.

Another way to identify jobs or work areas which present significant stress to the operator is to determine which jobs cause high absenteeism, high worker turnover, or have quality problems. These types of production related measures are sometimes associated with jobs that have significant musculoskeletal stress.

7.1.2 Worker Perceptions

Interviews of workers and supervisors are often a valuable tool to identify jobs with high musculoskeletal stress. For example, the following broad questions can point to the most difficult jobs which are frequently characterized by musculoskeletal stresses:

- What are the hardest tasks?
- What are the least favorite tasks?
- What tasks generate the most worker complaints?
- What tasks pose the highest risk of injury?
- What tasks are performed by the workers with the least seniority?
- What tasks seem to cause more worker injuries or symptoms?

Date ______/______/______

______ Plant ______ Dept# Job Name ______________________

Shift ______________ Hours worked/week ______ years ______ months Time on THIS Job

Other jobs you have done in the last year (for more than 2 weeks)

______ Plant ______ Dept# ______________ Job Name ______ months ______ weeks Time on THIS Job

______ Plant ______ Dept# ______________ Job Name ______ months ______ weeks Time on THIS Job

(If more than 2 jobs, include those you worked on the most)

Have you had any pain or discomfort during the last year?

☐ Yes ☐ No (If NO, stop here)

If YES, carefully shade in area of the drawing which bothers you the MOST.

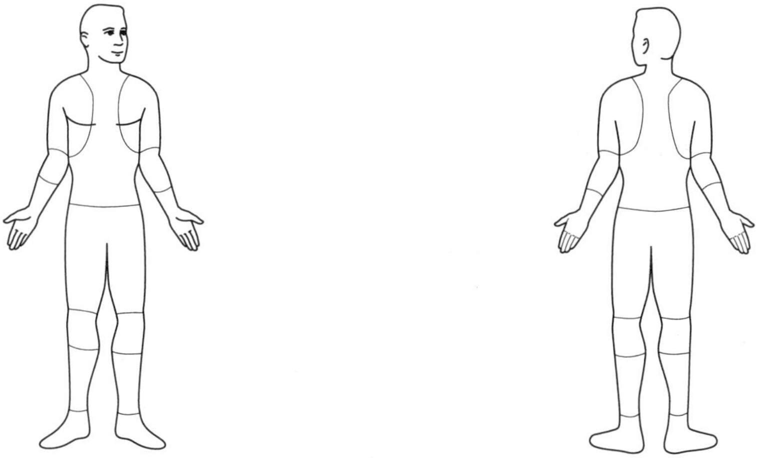

Front Back

Figure 54.17. Symptoms Survey Form.

These are just examples of questions which might identify stressful jobs. Others which reflect the unique feature of a worksite or industry can be added. Valuable worker input is also available through the use of symptoms surveys such as that shown in Figure 54.17 (118). NIOSH suggests that the following procedures be followed for best results when using this form:

(Complete a separate page for each area that bothers you)

Check Area: ☐ Neck ☐ Shoulder ☐ Elbow/Forearm ☐ Hand/Wrist ☐ Fingers
☐ Upper Back ☐ Low Back ☐ Thigh/Knee ☐ Low Leg ☐ Ankle/Foot

1. Please put a check by the word(s) that best describe your problem

☐ Aching	☐ Numbness (asleep)	☐ Tingling
☐ Burning	☐ Pain	☐ Weakness
☐ Cramping	☐ Swelling	☐ Other
☐ Loss of Color	☐ Stiffness	

2. When did you first notice the problem? ________ (month) ________ (year)
3. How long does each episode last? (Mark an X along the line) ________/

________/ ________/ ________/ ________/ ________/

1 hour 1 day 1 week 1 month 6 months

4. How many separate episodes have you had in the last year? ________
5. What do you think caused the problem? ________

____

6. Have you had this problem in the last 7 days? ☐ Yes ☐ No
7. How would you rate this problem? (mark an X on the line)

NOW

____

None Unbearable

When is it the WORST

____

None Unbearable

8. Have you had medical treatment for this problem? ☐ Yes ☐ No

8a. If NO, why not? ________

8a. If YES, where did you receive treatment?

☐ 1. Company Medical Times in past year ________

☐ 2. Personal doctor Times in past year ________

☐ 3. Other Times in past year ________

Did treatment help? ☐ Yes ☐ No ________

9. How much time have you lost in the last year because of this problem? ______ days
10. How many days in the last year were you on restricted or light duty because of this problem?

______ days

11. Please comment on what you think would improve your symptoms

____

____

____

____

Figure 54.17. Continued.

- No name should be used on the forms; the collection process should ensure anonymity.
- Survey participation should be voluntary.
- Workers should fill out the form on their own. If needed, the survey may be administered to groups by a trained person who offers explanation of how the form should be filled out.
- The survey should be conducted on work time.

One way to use the results of this type of survey is to rank-order the number and severity of complaints by body part, from the highest to lowest. Once ergonomic changes have been made to correct problem jobs, a second survey can be used to indicate if the musculoskeletal stresses have been reduced.

7.2 Checklists

The qualification and quantification of musculoskeletal stress in industrial tasks can be accomplished by a variety of checklists and analytical methods. In general, checklists point out the existence of ergonomic hazards and more detailed analytical tools quantify the hazard level and direct the analyst toward abatement measures which are likely to reduce the hazard to an acceptable level. The following checklists are from Cohen, Gjessing, Fine, Bernard, and McGlothlin (118).

Checklists can provide information ranging from the most general overview of a complete work area to a detailed review of specific hand tools or a computer workstation. In this section, five checklists are presented which allow the user to identify most ergonomic hazards in the workplace. Figure 54.18, Workstation Checklist, illustrates a general workstation checklist which can be used to identify potential stressors in the workplace which deserve further attention. Figure 54.19, Task Analysis Checklist, is also a general checklist and facilitates the identification of ergonomic related stressors which might exist during task performance. Figure 54.20, Handtool Analysis Checklist, allows the identification of ergonomic risk factors associated with hand tools. Figure 54.21, Material Handling Checklist, allows the user to identify potential ergonomic problems which might arise during the performance of manual material handling tasks. Figure 54.22, Computer Workstation Checklist, allows the user to identify potential ergonomic risk factors during seated tasks, particularly those associated with the use of computer workstations. Figure 54.23 provides a guide to be used when video-taping jobs for further analysis, or to be used as a record of workplace or task conditions before or after ergonomic abatements have been performed.

7.3 Analytical Tools

7.3.1 Rodger's Analysis

Rodgers (99) has developed a job survey tool which emphasizes upper extremity cumulative trauma issues.

Workstation Checklist

"No" responses indicate potential problem areas which should receive further investigation.

1. Does the work space allow for full range of movement?	☐ yes	☐ no
2. Are mechanical aids and equipment available?	☐ yes	☐ no
3. Is the height of the work surface adjustable?	☐ yes	☐ no
4. Can the work surface be tilted or angled?	☐ yes	☐ no
5. Is the workstation designed to reduce or eliminate		
bending or twisting at the wrist?	☐ yes	☐ no
reaching above the shoulder?	☐ yes	☐ no
static muscle loading?	☐ yes	☐ no
full extension of the arms?	☐ yes	☐ no
raised elbows?	☐ yes	☐ no
6. Are the workers able to vary posture?	☐ yes	☐ no
7. Are the hands and arms free from sharp edges on work surfaces?	☐ yes	☐ no
8. Is an armrest provided where needed?	☐ yes	☐ no
9. Is a footrest provided where needed?	☐ yes	☐ no
10. Is the floor surface free of obstacles and flat?	☐ yes	☐ no
11. Are cushioned floor mats provided for employees required to stand for long periods?	☐ yes	☐ no
12. Are chairs or stools easily adjustable and suited to the task?	☐ yes	☐ no
13. Are all task elements visible from comfortable positions?	☐ yes	☐ no
14. Is there a preventative maintenance program for mechanical aids, tools, and other equipment?	☐ yes	☐ no

Figure 54.18. Workstation Checklist.

This analysis method requires that the analyst measure effort level, continuous effort time, and efforts per minute in one of three categories. Effort Level is identified as either light, moderate, or heavy. Continuous Effort Time is less than 6 sec., 6–20 sec., or greater than 20 sec. Efforts Per Minute is less than 1/min., 1–5/min., or greater than 5/min. These three categories are each given a value of 1, 2, or 3 respectively. Each combination of effort level, continuous effort time, and efforts per minute is identified by a three-digit code ranging from 111 to 332. The three-digit codes identify the job as low, moderate, high, or very high change priority. A form for this analysis is shown in Figure 54.24.

Effort level actually relates to the percent of maximum voluntary contraction exerted by the operator. This, however, is difficult to determine, so the Rodger's Analysis provides a table with observable cues with which one can estimate effort level. This is shown in Table 54.6.

Task Analysis Checklist

"No" responses indicate potential problem areas which should receive further investigation.

1. Does the design of the primary task reduce or eliminate		
bending or twisting of the back or trunk?	☐ yes	☐ no
crouching?	☐ yes	☐ no
bending or twisting the wrist?	☐ yes	☐ no
extending the arms?	☐ yes	☐ no
raised elbows?	☐ yes	☐ no
static muscle loading?	☐ yes	☐ no
clothes wringing motion?	☐ yes	☐ no
finger pinch grip?	☐ yes	☐ no
2. Are mechanical devices used when necessary?	☐ yes	☐ no
3. Can the task be done with either hand?	☐ yes	☐ no
4. Can the task be done with two hands?	☐ yes	☐ no
5. Are pushing or pulling forces kept minimal?	☐ yes	☐ no
6. Are required forces judged acceptable by the workers?	☐ yes	☐ no
7. Are the materials		
able to be held without slipping?	☐ yes	☐ no
easy to grasp?	☐ yes	☐ no
free from sharp edges and corners?	☐ yes	☐ no
8. Do containers have good handholds?	☐ yes	☐ no
9. Are jigs, fixtures, and vises used where needed?	☐ yes	☐ no
10. As needed, do gloves fit properly and are they made of the proper fabric?	☐ yes	☐ no
11. Does the worker avoid contact with sharp edges when performing the task?	☐ yes	☐ no
12. When needed, are push buttons designed properly?	☐ yes	☐ no
13. Do the job tasks allow for ready use of personal equipment that may be required?	☐ yes	☐ no
14. Are high rates of repetitive motion avoided by		
job rotation?	☐ yes	☐ no
self-pacing?	☐ yes	☐ no
sufficient pauses?	☐ yes	☐ no
adjusting the job skill level of the worker?	☐ yes	☐ no
15. Is the employee trained to		
proper work practices?	☐ yes	☐ no
when and how to make adjustments?	☐ yes	☐ no
recognizing signs and symptoms of potential problems?	☐ yes	☐ no

Figure 54.19. Task Analysis Checklist.

Handtool Analysis Checklist

"No" responses indicate potential areas which should receive further investigation.

1. Are tools selected to limit or minimize		
exposure to excessive vibration?	☐ yes	☐ no
use of excessive force?	☐ yes	☐ no
bending or twisting the wrist?	☐ yes	☐ no
finger pinch grip?	☐ yes	☐ no
problems associated with trigger finger?	☐ yes	☐ no
2. Are tools powered where necessary and feasible?	☐ yes	☐ no
3. Are tools evenly balanced?	☐ yes	☐ no
4. Are heavy tools suspended or counterbalanced in ways to facilitate use?	☐ yes	☐ no
5. Does the tool allow adequate visibility of the work?	☐ yes	☐ no
6. Does the tool grip/handle prevent slipping during use?	☐ yes	☐ no
7. Are tools equipped with handles of textured, non-conductive material?	☐ yes	☐ no
8. Are different handle sizes available to fit a wide range of hand sizes?	☐ yes	☐ no
9. Is the tool handle designed not to dig into the palm of the hand?	☐ yes	☐ no
10. Can the tool be used safely with gloves?	☐ yes	☐ no
11. Can the tool be used by either hand?	☐ yes	☐ no
12. Is there a preventative maintenance program to keep tools operating as designed?	☐ yes	☐ no
13. Have employees been trained		
in the proper use of tools?	☐ yes	☐ no
when and how to report problems with tools?	☐ yes	☐ no
in proper tool maintenance?	☐ yes	☐ no

Figure 54.20. Handtool Analysis Checklist.

A good example is a job where an operator is standing and working on a conveyor belt which is approximately waist height (Fig. 54.25). Once each minute the conveyor belt presents the operator with a square packing box with four compartments. Every 15 seconds the operator is required to spend 10 seconds wrapping (severe wrist flexions and pinch grip) a ceramic figurine (weight is 5 lb) which is placed into one of the four compartments in the packing box. While the operator is placing the first two figurines in the closest compartments he/she assumes a reasonably upright posture. When he/she places figurines in the two outside compartments he/she must reach out with the arms and bend forward. In this case, the most stressful activity for the neck, shoulders, and back is the 30 sec when the operator bends forward and extends his/her arms to place the two figurines in the two outside compartments. The continuous effort time for the neck, shoulder, and back is 30 sec, for a Continuous Effort category of 3. There is one effort per minute so the Effort per Minute category is 2. The neck is forward about 20° for an Effort Level category of 2.

Materials Handling Checklist

"No" responses indicate potential problem areas which should receive further investigation.

1. Are the weights of loads to be lifted judged acceptable by the workforce?	☐ yes	☐ no
2. Are materials moved over minimum distances?	☐ yes	☐ no
3. Is the distance between the object load and the body minimized?	☐ yes	☐ no
4. Are walking surfaces		
level?	☐ yes	☐ no
wide enough?	☐ yes	☐ no
clean and dry?	☐ yes	☐ no
5. Are objects		
easy to grasp?	☐ yes	☐ no
stable?	☐ yes	☐ no
able to be held without slipping?	☐ yes	☐ no
6. Are there handholds on these objects?	☐ yes	☐ no
7. When required, do gloves fit properly?	☐ yes	☐ no
8. Is the proper footwear worn?	☐ yes	☐ no
9. Is there enough room to maneuver?	☐ yes	☐ no
10. Are mechanical aids used whenever possible?	☐ yes	☐ no
11. Are working surfaces adjustable to the best handling heights?	☐ yes	☐ no
12. Does material handling avoid		
movements below knuckle height and above shoulder height?	☐ yes	☐ no
static muscle loading?	☐ yes	☐ no
sudden movements during handling?	☐ yes	☐ no
twisting at the waist?	☐ yes	☐ no
extended reaching?	☐ yes	☐ no
13. Is help available for heavy or awkward lifts?	☐ yes	☐ no
14. Are high rates of repetition avoided by		
job rotation?	☐ yes	☐ no
self-pacing?	☐ yes	☐ no
sufficient pauses?	☐ yes	☐ no
15. Are pushing or pulling forces reduced or eliminated?	☐ yes	☐ no
16. Does the employee have an unobstructed view of handling the task?	☐ yes	☐ no
17. Is there a preventive maintenance program for equipment?	☐ yes	☐ no
18. Are workers trained in correct handling and lifting procedures?	☐ yes	☐ no

Figure 54.21. Material Handling Checklist.

Computer Workstation Checklist

"No" responses indicate potential problem ares which should receive further investigation.

1. Does the workstation ensure proper worker posture, such as		
horizontal thighs?	☐ yes	☐ no
vertical lower legs?	☐ yes	☐ no
feet flat on the floor or footrest?	☐ yes	☐ no
neutral wrists?	☐ yes	☐ no
2. Does the chair		
adjust easily?	☐ yes	☐ no
have a padded seat with a rounded front?	☐ yes	☐ no
have an adjustable backrest?	☐ yes	☐ no
provide lumbar support?	☐ yes	☐ no
have casters?	☐ yes	☐ no
3. Are the height and tilt of the work surface on which the keyboard is located adjustable?	☐ yes	☐ no
4. Is the keyboard detachable?	☐ yes	☐ no
5. Do keying actions require minimal force?	☐ yes	☐ no
6. Is there an adjustable document holder?	☐ yes	☐ no
7. Are arm rests provided where needed?	☐ yes	☐ no
8. Are glare and reflections avoided?	☐ yes	☐ no
9. Does the monitor have brightness and contrast controls?	☐ yes	☐ no
10. Do the operators judge the distance between eyes and work to be satisfactory for their viewing needs?	☐ yes	☐ no
11. Is there sufficient space for knees and feet?	☐ yes	☐ no
12. Can the workstation be used for either right- or left-handed activity?	☐ yes	☐ no
13. Are adequate rest breaks provided for task demands?	☐ yes	☐ no
14. Are high stroke rates avoided by		
job rotation?	☐ yes	☐ no
self-pacing?	☐ yes	☐ no
adjusting the job to the skill of the worker?	☐ yes	☐ no
15. Are employees trained in		
proper posture?	☐ yes	☐ no
proper work methods?	☐ yes	☐ no
when and how to adjust their workstations?	☐ yes	☐ no
how to seek assistance for their concerns?	☐ yes	☐ no

Figure 54.22. Computer Workstation Checklist.

Protocol for Videotaping Jobs for Risk Factors

The following is a guide to preparing a videotape and related task information for facilitating job analyses and assessments of risk factors for work-related musculoskeletal disorders.

Materials needed:

Videocamera and blank tapes
Spare batteries (at least 2) and battery charger
Clipboard, pens, paper, blank checklists
Stopwatch, strain gauge (optional) for weighing objects

Videotaping Procedures:

1. To verify the accuracy of the video camera to record in real time, videotape a worker or job with a stopwatch running in the field of view for at least 1 min. The play-back of the tape should correspond to the lapsed time on the stopwatch.
2. Announce the name of the job on the voice channel of the video camera befor ethe taping of any job. Restrict running time comments to the facts. Make no editorial comments.
3. Tape each job long enough to observe all aspects of the task. Tape 5 to 10 min for all jobs, including at least 10 complete cycles. Fewer cycles may be needed if all aspects of the job are recorded at least 3 to 4 times.
4. Hold the camera still, using a tripod if availabe. Don't walk unless absolutely necessary.
5. Begin taping each task with a whole-body shot of the worker. Include the seat/chair and the surface the worker is standing on. Hold this for 2 to 3 cycles, then zoom in on the hands/arms or other body parts which may be under stress due to the job task.
6. It is best to tape several workers to determne if workers of varying body size adopt different postures or are affected in other ways. If possible, try to tape the best and worst case situations in terms of worker "fit" to the job.
   The following suspected upper body problems suggest focusing on the parts indicated:

   ______ wrist problems/complaints .... hand/wrist/forearms
   ______ elbow problems/complaints .... arms/elbows
   ______ shoulder problems/complaints .... arms/shoulders

   For back and lower limb problems, the focus would be on movements of the trunk of the body and leg, knee, and foot areas under stress due to task loads or other requirements.
7. Video from whatever angles are needed to capture the body part(s) under stress.
8. Briefly tape the jobs performed before and after the one under actual study to see how the targeted job fits into the total department process
9. For each taped task, obtain the following information to the maximum extent possible:

   ______ if the task is continuous or sporadic
   ______ if the worker performs the work for the entire shift, or if there is rotation with other
   ______ workers
   ______ measures of work surface heights and chair heights and whether adjustable
   ______ weight, size and shape of handles and textures for tools in use; indications of
   ______ vibration in power tool usage
   ______ use of handwear
   ______ weight of objects lifted, pushed, pulled, or carried
   ______ nature of environment in which work is performed-(too cold or too hot?)

Figure 54.23. Protocol for Videotaping Jobs for Risk Factors.

The arms are away from the body so the shoulder effort level category is also a 2. The back is bent forward with no, or low, load for 30 sec so this is also an effort level category of 2. For the arms/elbows and the wrist/hands/fingers the continuous effort time is the 10 sec of severe wrist flexion and pinch grip during the 15 sec of wrapping each figurine, so the continuous effort time category is 2. Four figurines are placed per minute so the efforts

Body part	Effort level 1=Light 2-Moderate 3=Heavy	Continuous effort time 1-< 6 sec 2=6-20 sec 3=> 20 sec	Efforts/Minute 1=<1/min 2=1-5/min 3=>5/min	Change priority
Neck				
Shoulder				
Back				
Arms/Elbows				
Wrists/Hands/ Fingers				
Legs/Knees/ Ankles/Feet/Toes				

Change priority					
Moderate (M)	1,2,3 1,3,2 2,1,3 2,2,2 2,3,1 2,3,2 3,1,2	High (H)	2,2,3 3,1,3 3,2,1 3,2,2	Very High (VH)	3,2,3 3,3,1 3,3,2

Not noted = low
(X33 can not exist)

Figure 54.24. Rodgers' Ergonomic Job Analysis Form

per minute category is 2. The arms rotate while exerting moderate force, so the effort level category is 2 and the wrist/hands/fingers have severe wrist flexion and pinch grip for an effort level category of heavy or 3. The legs, knees, ankles, feet, and toes are continuously standing so the continuous effort time is greater than 20 sec. for a continuous effort category of 3. Since there are no postural changes the efforts per minute is assigned a category of 1. Since the body is bending forward the effort level category for the legs, knees, ankles, feet, and toes is defined as moderate, or a 2. The change priorities, as defined by the categories below the table, indicate that the shoulder and wrist/hands/fingers have a high priority for change and the other body parts have a moderate priority for change.

The Rodger's Analysis provides a general overview of the ergonomic stresses in a job and assigns a change priority as low, moderate, high, or very high. This analysis method is often used as a screening tool to determine which jobs need further, more detailed

Table 54.6. Method of Estimating Effort Level After Rodgers

Body Part	Light	Moderate	Heavy
Neck	Head turned partly to side or back/forward slightly	Head turned to side; head fully back; forward about 20°	Same as moderate but with force or weight; head stretched forward
Shoulders	Arms slightly away from sides; arms extended with some support	Arms away from body, no support; working overhead	Exerting forces or holding weight with arms away from body or overhead
Back	Leaning to side or bending; arching back	Bending forward, no load; lifting moderately heavy loads near body; working overhead	Lifting or exerting force while twisting; high force or load while bending
Arms, elbows	Arms away from body, no load; light forces/lifting near body	Rotating arm while exerting moderate force	High forces exerted with rotation; lifting with arms extended
Hands, fingers, wrists	Light forces or weights handled close to body; straight wrists; comfortable power grips	Grips with wide or narrow span; moderate wrist angles, especially flexion; use of gloves with moderate force	Pinch grips; strong wrist angles; slippery surfaces
Legs, knees, ankles, feet, toes	Standing, walking without bending or leaning; weight on both feet	Bending forward, leaning on table; weight on one side; pivoting while exerting force	Exerting high forces while pulling or lifting; crouching while exerting force

Body part	Effort level 1=Light 2-Moderate 3=Heavy	Continuous effort time 1-< 6 sec 2=6-20 sec 3=> 20 sec	Efforts/Minute 1=<1/min 2=1-5/min 3=>5/min	Change priority
Neck	2	3	2	M
Shoulder	2	3	2	M
Back	2	3	2	M
Arms/Elbows	2	2	2	M
Wrists/Hands/ Fingers	3	2	2	H
Legs/Knees/ Ankles/Feet/Toes	2	3	1	M

Change priority

Moderate (M)		High (H)		Very High (VH)	
	1,2,3 1,3,2 2,1,3 2,2,2 2,3,1 2,3,2 3,1,2		2,2,3 3,1,3 3,2,1 3,2,2		3,2,3 3,3,1 3,3,2

Not noted = low
(X33 can not exist)

Figure 54.25. Example Rodgers' Ergonomic Analysis

analysis. Another advantage of the Rodger's Analysis is that simply observing and categorizing the effort levels is a good start in identifying risk factors which must be abated.

7.3.2 Rapid Upper Limb Assessment (RULA)

McAtamney and Corlett (120) developed rapid upper limb assessment (RULA) as a survey method for use in ergonomics investigations of the workplace. This analysis method is based primarily on a posture analysis of the upper limb (upper arm, lower arm, wrist) and the torso (neck, trunk, legs). It also includes some consideration of whether the posture is static or repeated more than four times per minute and a consideration of the force, or load, handled by the operator.

Figures 54.26 and 54.27 indicate the posture scores assigned to the upper extremity and torso respectively. Tables 54.7 and 54.8 are then used to translate the individual posture

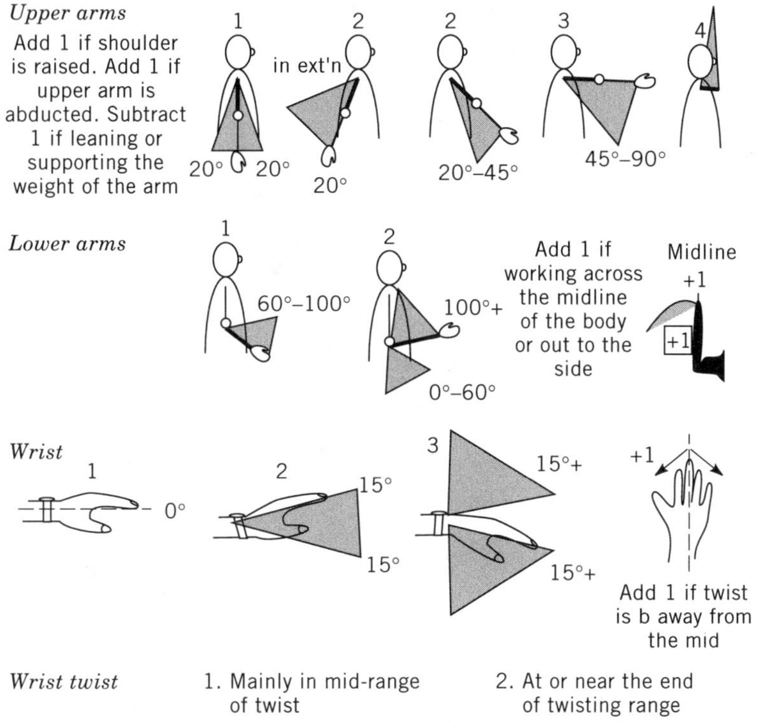

Figure 54.26. RULA Posture Score Form for Upper Extremity. From Ref. 120.

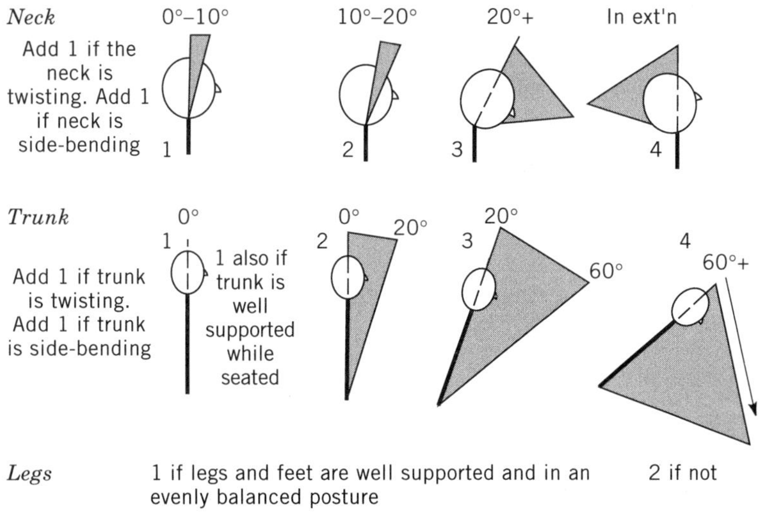

Figure 54.27. RULA Posture Score Form for Neck, Trunk, and Legs. From Ref. 120.

Table 54.7. RULA Upper Extremity Posture Score Combination[a]

		Wrist Posture Score							
		1		2		3		4	
		W. Twist		W. Twist		W. Twist		W. Twist	
Upper Arm	Lower Arm	1	2	1	2	1	2	1	2
1	1	1	2	2	2	2	3	3	3
	2	2	2	2	2	3	3	3	3
	3	2	3	3	3	3	3	4	4
2	1	2	3	3	3	3	4	4	4
	2	3	3	3	3	3	4	4	4
	3	3	4	4	4	4	4	5	5
3	1	3	3	4	4	4	4	5	5
	2	3	4	4	4	4	4	5	5
	3	4	4	4	4	4	5	5	5
4	1	4	4	4	4	4	5	5	5
	2	4	4	4	4	4	5	5	5
	3	4	4	4	5	5	5	6	6
5	1	5	5	5	5	5	6	6	7
	2	5	6	6	6	6	7	7	7
	3	6	6	6	7	7	7	7	8
6	1	7	7	7	7	7	8	8	9
	2	8	8	8	8	8	9	9	9
	3	9	9	9	9	9	9	9	9

[a]Table A into which the individual posture scores for the upper limbs are entered to find posture score A

Table 54.8. RULA Neck, Trunk, Legs Posture Score Combination[a]

	Trunk Posture Score											
	1		2		3		4		5		6	
	Legs		Legs		Legs		Legs		Legs		Legs	
Neck Posture Score	1	2	1	2	1	2	1	2	1	2	1	2
1	1	3	2	3	3	4	5	5	6	6	7	7
2	2	3	2	3	4	5	5	5	6	7	7	7
3	3	3	3	4	4	5	5	6	6	7	7	7
4	5	5	5	6	6	7	7	7	7	7	8	8
5	7	7	7	7	7	8	8	8	8	8	8	8
6	8	8	8	8	8	8	8	9	9	9	9	9

[a]Table B into which the individual posture scores for the neck, trunk and legs are entered to find posture score B.

scores for the components of the upper extremity and torso into overall posture scores for the upper extremity and torso. The information in Figure 54.28 is used to modify the posture score based on the muscle use and force, or load. Figure 54.29 is a RULA scoring sheet, which can be used to generate the overall score for the upper extremity and torso

Give a score of 1 if the posture is
- mainly static, eg held for longer than 1 min
- repeated more than 4 times/min

(**a**)

0	1	2	3
• No resistance or less than 2 kg intermittent load or force	• 2–10 kg intermittent load or force	• 2–10 kg static load • 2–10 kg repeated load or force	• 10 kg or more load • 10 kg or more repeated load or force • Shock or forces with a rapid build-up

(**b**)

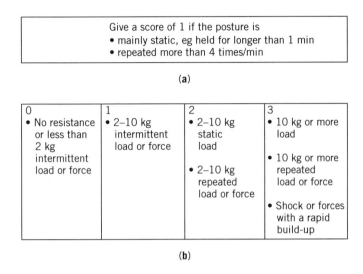

Figure 54.28. RULA Muscle Use and Force/Load Score Form. From Ref. 120. (**a**) The muscle use scores which are added to posture scores A and B. (**b**) The force or load score which is added to posture scores A and B.

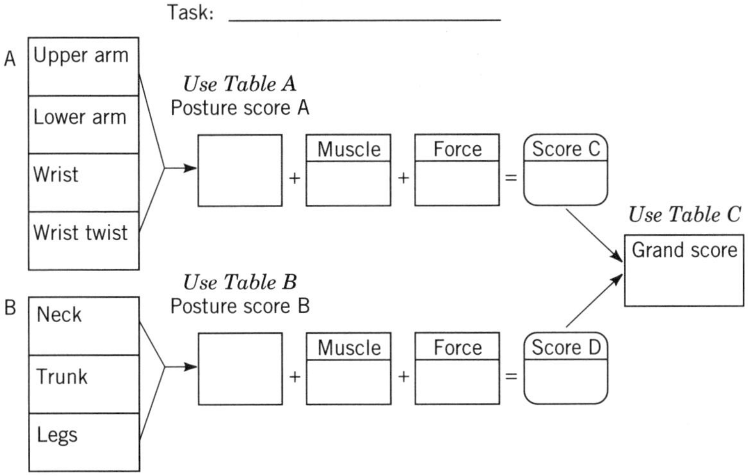

Figure 54.29. RULA Scoring Sheet (Grand Score). From Ref. 120.

based on the posture scores and the muscle use and/or load modifier. These scores (scores C and D) are used to find the grand score, which indicates the overall risk to the body due to muscloskeletal loading (Fig. 54.30). A grand score of 1 or 2 indicates a posture that is acceptable if not maintained or repeated for long periods. A grand score of 3 or 4 indicates that further investigation is needed and that job modifications may be necessary. A grand score of 5 or 6 indicates that further investigation and job modification have a high priority and a grand score of 7 indicates that the investigation and job modification should be accomplished immediately.

For example, consider the job described in Section 7.2.2 (Rodger's Analysis). The scores for this analysis are illustrated in Figure 54.31. From Figure 54.26 it can be seen that the upper arm posture would be a 2, the lower arm posture would be a 1, the wrist posture would be a 3, and the wrist twist posture would be a 2. Using Table 54.7 a posture score of 4 is calculated for the upper extremity. The muscle use score would be 0 and the force/load score would be 1. So the total score for the upper extremity would be 5.

Based on Figure 54.27 the posture scores for the neck, trunk, and legs are assumed to be 3, 3, and 1 respectively. Using Table 54.8 the overall posture score for the torso is a 4. As was the case with the upper extremity, the muscle use score is a 0, and the force/load score is a 1. This results in a total score for the torso of 5. Using Figure 54.30 the upper extremity score of 5 and the torso score of 5 are translated into a grand score of 6 for this job. This suggests that a relatively high priority should be given to further investigation and possible modification of this job.

Score D (neck, trunk, leg)

Score C (upper limb)	1	2	3	4	5	6	7+
1	1	2	3	3	4	5	5
2	2	2	3	4	4	5	5
3	3	3	3	4	4	5	6
4	3	3	3	4	5	6	6
5	4	4	4	5	6	7	7
6	4	4	5	6	6	7	7
7	5	5	6	6	7	7	7
8	5	6	6	7	7	7	7

Table C into which score C (posture score A plus the muscle use score and the force or load score) and score D (posture score B plus the muscle use score and the force or load score) are entered to find the grand score.

Figure 54.30. RULA Priority Sheet. From Ref. 120.

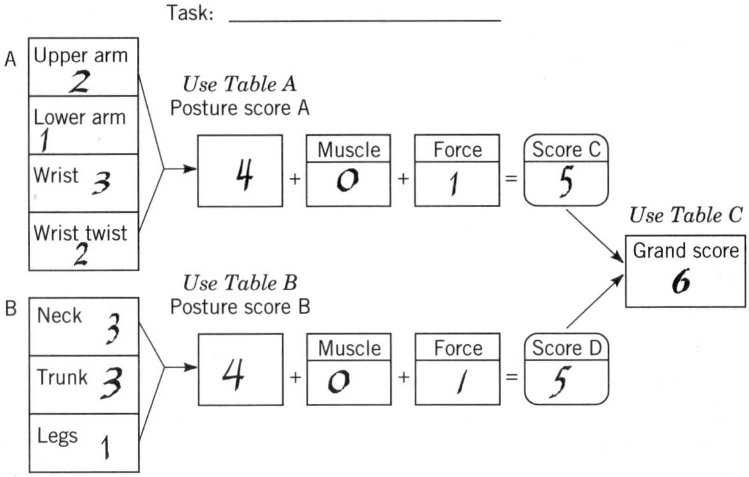

Figure 54.31. Example RULA analysis

7.3.3 1994 Revised NIOSH Equation

Jobs that require the manual lifting and handling of objects have been associated with increased rates of low back pain and other related musculoskeletal disorders. Because of the hazards associated with lifting, the National Institute for Occupational Safety and Health (NIOSH) has developed a method to quantify the risk in manual material handling jobs (121). The NIOSH Revised Lifting Equation or RLE discussed here is based on a combination of biomechanical, psychophysical, and physiological data (122). It establishes acceptable lifting limits based on selected task parameters and facilitates the development of effective controls.

Biomechanical, psychophysical, and physiological data are combined into *one equation* which provides a Recommended Weight Limit or RWL. If the load which is being lifted exceeds the RWL some fraction of the workforce will be at an increased risk of lifting-related low back pain. If the load exceeds three times the RWL nearly all workers will be at an increased risk of a work-related injury. The NIOSH RLE can be used to analyze both lifts and lowers but for simplicity only the word "lift" will be used in the following discussion.

To use the RWL it is necessary to observe the job and record the object weight, posture at the beginning (origin) and end (destination) of the lift, the lift frequency, lift duration, and grip. These variables should be measured as accurately as possible while observing the job, however, it is often necessary to estimate some variables and it is wise to record the job on videotape and/or with still photographs for future reference. Definitions of these variables are noted below:

1. Object weight (L)—measured in pounds.
2. Horizontal Location (H)—the location of the hands, measured out from the center of a line connecting the ankles, in inches. If the object is not held at the center of mass, the distance to the object's center of mass should be used.

3. Vertical Location (V)—the location of the center of the hands measured from the floor (or standing surface), in inches.
4. Vertical Displacement (D)—the vertical distance moved by the object during the lift (origin to destination), in inches. This can also be calculated as the difference between the vertical location (V) of the hands at the origin and destination of the lift.
5. Asymmetric Angle (A)—the location of the object from the worker's mid-sagittal plane, in degrees. Note that this angle defines the object *location* or worker *posture* at the beginning and end of the lift and not the rotational movement between the origin and destination of the lift.
6. Frequency of Lifting (F)—the average number of lifts per minute.
7. Duration of Lifting—classified as short, moderate, or long. A lift is defined as a short duration if it has a work duration of one hour or less and is followed by a recovery time (period without lifting activities) equal to 1.2 times the lift duration. For example, a lift duration of 50 min. would have to be followed by a recovery period of 60 min. (50 min. × 1.2) for the 50 min. lift period to be categorized as short. If a subsequent lift duration started less than 60 min. after the end of the 50 min. lift period the two lift durations must be combined. A lift is defined as a moderate duration if it has a work duration of more than one hour but less than two hours and is followed by a recovery time equal to 0.3 times the lift duration. For example a lift duration of 90 min. would have to be followed by a recovery period of 27 min. (90 min. × 0.3) for the 90 min. lift period to be categorized as short. If a subsequent lift duration started less than 27 min. after the end of the 90 min. lift period the two lift durations must be combined. A long duration lift is one that has a duration between two and eight hours.
8. Object Coupling—classified as good, fair, or poor. A good coupling is one in which the hands can grasp the load at handles or another easy-to-grip location. A fair coupling is one where the hands can be flexed about 90°, such as when lifting a box while holding the bottom. A poor coupling is one where the hands must grasp an object such as a watermelon or nonrigid bag.

This information is entered into the STEP 1 section of the worksheet noted as Figure 54.32. Multipliers for the horizontal location, vertical location, displacement, asymmetry, frequency, and coupling are determined from the tables shown in Figure 54.33 and entered as STEP 2 of the worksheet. The recommended weight limit or RWL is then calculated for the origin and destination of the lift by multiplying the 51-lb load constant by each of the task multipliers. The 51-lb load constant represents an approximate load that, when lifted in the optimum posture, would be acceptable to 75% of female workers and 90% of male workers and results in a disc compression force of less than 770 lb (122).

A lifting index or LI is calculated for the origin and destination of the lift by dividing the object weight by the RWL. If the LI is less than 1.0 the load is less than the RWL and can be lifted by most members of the workforce without an increased risk of lifting-related low back pain. If the LI exceeds 3.0 the load exceeds three times the RWL and nearly all workers will be at an increased risk of a work-related injury. If it is determined that the lift is stressful, a review of the relative magnitudes of the multipliers provides information

JOB ANALYSIS WORKSHEET

DEPARTMENT ______________________

JOB TITLE ______________________

ANALYST'S NAME ______________________

DATE ______________________

JOB DESCRIPTION ______________________

STEP 1. Measure and record task varibles

Object Weight (lbs)		Hand Location (in)				Vertical Distance (in)	Asymmetric Angle (degrees)		Frequency Rate	Duration	Object Coupling
		Origin		Dest.			Origin	Dest	lifts/min	(HRS)	
L (AVG)	L (Max)	H	V	H	V	D	A	A	F		C

STEP 2. Determine the multipliers and compute the RWL's

	RWL =	**LC**	×	**HM**	×	**VM**	×	**DM**	×	**AM**	×	**FM**	×	**CM**			
ORIGIN	**RWL =**	51	×		×		×		×		×		×		=		**lbs**
DESTINATION	**RWL =**	51	×		×		×		×		×		×		=		**lbs**

STEP 3. Compute the LIFTING INDEX

ORIGIN LIFTING INDEX = $\frac{\text{OBJECT WEIGHT (I)}}{\text{RWL}}$ = ________ = ☐

DESTINATION LIFTING INDEX = $\frac{\text{OBJECT WEIGHT (I)}}{\text{RWL}}$ = ________ = ☐

Figure 54.32. Single-Task Job Analysis Worksheet for NIOSH RLE.

	Object Weight	Hand Location				Vertical Distance	Twisting		Freq. Rate	Dur.	Coupling
		V		V			A				
Task #	lbs	O	D	O	D	D	O	D	F	Hrs	C
											G C P
											G C P
											G C P
											G C P
											G C P

H	HM
in	
≤10	1.00
11	0.91
12	0.83
13	0.77
14	0.71
15	0.67
16	0.63
17	0.59
18	0.56
19	0.53
20	0.50
21	0.48
22	0.46
23	0.44
24	0.42
25	0.40
>25	0.00

F	V<30"					
lifts/	< 1 hour		1–2 hours		2–8 hours	
min	V<30"	V≥30"	V<30"	V≥30"	V<30"	V≥30"
0.2	1.00	1.00	0.95	0.95	0.85	0.85
0.5	0.97	0.97	0.92	0.92	0.81	0.81
1	0.94	0.94	0.88	0.88	0.75	0.75
2	0.91	0.91	0.84	0.84	0.65	0.65
3	0.88	0.88	0.79	0.79	0.55	0.55
4	0.84	0.84	0.72	0.72	0.45	0.45
5	0.80	0.80	0.60	0.60	0.35	0.35
6	0.75	0.75	0.50	0.50	0.27	0.27
7	0.70	0.70	0.42	0.42	0.22	0.22
8	0.60	0.60	0.35	0.35	0.18	0.18
9	0.52	0.52	0.30	0.30	0.00	0.15
10	0.45	0.45	0.26	0.26	0.00	0.13
11	0.41	0.41	0.00	0.23	0.00	0.00
12	0.37	0.37	0.00	0.21	0.00	0.00
13	0.00	0.34	0.00	0.00	0.00	0.00
14	0.00	0.31	0.00	0.00	0.00	0.00
15	0.00	0.28	0.00	0.00	0.00	0.00
>15	0.00	0.00	0.00	0.00	0.00	0.00

H	HM
in	
0	0.78
5	0.81
10	0.85
15	0.89
20	0.93
25	0.96
30	1.00
35	0.96
40	0.93
45	0.89
50	0.85
55	0.81
60	0.78
65	0.74
70	0.70
>70	0.00

D	DM
in	
≤10	1.00
15	0.94
20	0.91
25	0.89
30	0.88
35	0.87
40	0.87
45	0.86
50	0.86
55	0.85
60	0.85
70	0.85
>70	0.00

A	AM
deg	
0	1.00
15	0.95
30	0.90
45	0.86
60	0.81
75	0.76
90	0.71
105	0.66
120	0.62
135	0.57
>135	0.00

Coupling Type	CM	
	V<30 in	V≥30 in
GOOD	1.00	1.00
FAIR	0.95	1.00
POOR	0.90	0.90

NOTES:

Figure 54.33. Task variables and multipliers for NIOSH RLE.

which can be used in task redesign. The lowest multipliers have the most impact (reduction) on the load constant of 51 lb and indicate the task parameters which should first be considered for redesign. For example, if the horizontal multiplier is the smallest, an attempt should be made to redesign or modify the task to reduce the horizontal distance during the lift/lower. This will increase the horizontal multiplier, increase the RWL, and reduce the LI.

As an example of the use of the left index, a lift performed for 30 min. with a 60 min. recovery time (Figure 54.34). The lift involves lifting 40-lb boxes (without handles so the box must be lifted from the bottom) from the floor (V at origin = 0 in.) to a 40-in.-high conveyor (V at destination = 40 in.) five times per minute. The horizontal location of where the hands contact the load is measured to be 20 in. from the center of the ankles at the origin and 15 in. at the destination of the lift. The load is located 30° to the side of the sagittal plane at the origin and 45° to the side of the sagittal plane at the destination of the lift. The lift duration category is short since the lift duration is 30 min. and the recovery time of 60 min. exceeds $1.2 \times 30 = 45$ min. The coupling is fair since the box is lifted from the bottom. This information is recorded in the STEP 1 section of the worksheet shown as Figure 54.34. This gathered data is used to determine the task multipliers from Figure 54.33 which are then noted in "STEP 2" of Figure 54.34. It can be seen that the RWL or recommended weight limit is 10.6 at the origin and 19.0 at the destination of the lift. In the STEP 3 section of the worksheet the lifting index or LI is calculated to be 3.77 for the origin and 2.10 for the destination of the lift. While the lift is stressful both at the origin and destination of the lift, the LI at the origin is above 3.0 and merits attention first. The LI can be reduced by reducing the load or by redesigning the task. Since the horizontal multiplier (0.5) and the vertical multiplier (0.70) at the origin of the lift are the lowest multipliers, the load should be located closer to the body and held higher to increase these multipliers. If the load could be moved closer at the origin of the lift (so the horizontal distance is 10 in.) and elevated to 30 in. the HM and the VM would both be 1.0. The RWL can be calculated to be 30.3 and the LI calculated to be 1.32. This is still somewhat stressful, but much better than the LI of 3.77 for the original lift. The same process could be used to reduce the LI for the destination of the lift.

The NIOSH RLE does not apply in the following situations:

- Lifting/lowering with one hand.
- Lifting/lowering for over eight hours.
- Lifting/lowering while seated or kneeling.
- Lifting/lowering in a restricted work space.
- Lifting/lowering unstable objects.
- Lifting/lowering while carrying, pushing, or pulling.
- Lifting/lowering with wheelbarrows or shovels.
- Lifting/lowering with high speed motion (faster than about 30 in./sec)
- Lifting/lowering with unreasonable foot/floor coupling (<0.4 coefficient of friction between the sole and the floor).
- Lifting/lowering in an unfavorable environment (i.e., temperature significantly outside 66–79°F (19–26° C) range; relative humidity outside 35–50% range)

JOB ANALYSIS WORKSHEET

DEPARTMENT ____________
JOB TITLE ____________
ANALYST'S NAME ____________
DATE ____________

JOB DESCRIPTION ____________

STEP 1. Measure and record task variables

Object Weight (lbs)		Hand Location (in)				Vertical Distance (in)	Asymmetric Angle (degrees)		Frequency Rate	Duration	Object Coupling
		Origin		Dest.			Origin	Dest	lifts/min	(HRS)	
L (AVG)	L (Max)	H	V	H	V	D	A	A	F		C
40	40	20	0	15	40	40	30	45	5	SHORT	Fair

STEP 2. Determine the multipliers and compute the RWL's

	RWL =	LC	×	HM	×	VM	×	DM	×	AM	×	FM	×	CM		
ORIGIN	RWL =	51	×	.5	×	.70	×	.87	×	.90	×	.8	×	.95	=	10.6 lbs
DESTINATION	RWL =	51	×	.67	×	.93	×	.87	×	.86	×	.8	×	1.0	=	19.0 lbs

STEP 3. Compute the LIFTING INDEX

ORIGIN LIFTING INDEX = $\frac{\text{OBJECT WEIGHT (I)}}{\text{RWL}} = \frac{40}{10.6} = 3.77$

DESTINATION LIFTING INDEX = $\frac{\text{OBJECT WEIGHT (I)}}{\text{RWL}} = \frac{40}{19} = 2.10$

Figure 54.34. Example NIOSH RLE single-task analysis.

The NIOSH RLE can also be used to analyze manual material handling jobs with varying weights and/or postures for different tasks within the job. The reader is referred to the Applications Manual for the Revised NIOSH Lifting Equation (121) for further information on this and other details of this analysis method.

7.3.4 Biomechanical Models

There are several commercially available biomechanical models which allow the estimation of back compressive force and joint moments as a function of load, posture and anthropometry (123, 124). It is sometimes, however, useful to be able to estimate back compressive force and shoulder moment using a simpler paper and pencil calculation.

7.3.4.1 Low Back Compressive Force Estimation. One can estimate the back compressive force through the use of load weight, body weight, torso posture, and the distance that the load is held out from the body if some simplifying assumptions are made. A worksheet to facilitate data recording and analysis is shown in Figure 54.35.

Back compressive forces are often of interest when the material handling job requires lifting from below mid-thigh or when the torso is flexed (bent forward). If one uses the NIOSH Revised Lifting Equation as an initial analysis the Horizontal Multiplier (HM) and Vertical Multiplier (VM) are indicators of the potential for low back stress. If HM or VM are low a follow-up back compressive force analysis is suggested.

The estimated back compressive force may be compared to some generally accepted limits established by The National Institute for Occupational Safety and Health (NIOSH). NIOSH indicates that compressive forces in excess of 770 lb will put some portion of the work force at risk and compressive forces in excess of 1430 lb will put most members of the workforce at risk (65).

In Figure 54.35, the largest of the terms A, B, or C contribute most to the lifting hazard so task redesign priorities may be established by comparing the relative values of these three terms. Note that: (*1*) term A is the back muscle force reacting to upper body weight (to lower this, change the upper body angle with the horizontal); (*2*) term B is the back muscle force reacting to load moment (to lower this, change the magnitude of the load or the distance that the load is held out from the body); and (*3*) term C is the direct compressive component of upper body weight and load (to lower this, change the magnitude of the load or body weight). Actually only terms A and B must be considered since they are task dependent and can be modified to reduce the back compressive force. Term C is seldom, if ever, the largest term and when it is the largest, the back compressive force will be low.

In summary:

1. Measure task parameters and body weight and record on the top of the back compressive force worksheet.
2. Calculate the values of terms A, B, and C and add them together to get an estimate of back compressive force.
3. Compare the total back compressive force to 770 lb (stressful for some members of the workforce) and 1430 lb (stressful for most members of the workforce).

COMPRESSIVE FORCE WORKSHEET

BW =	BODY WEIGHT =	______
L =	LOAD IN HANDS =	______
HB =	HORIZ DISTANCE FROM HANDS TO LOW BACK =	______
TFF =	TORSO FLEXION FACTOR = ______	

Note:	if torso is vertical use TFF	=	0
	if torso is bent 1/4 of the way use TFF	=	0.38
	if torso is bent 1/2 of the way use TFF	=	0.71
	if torso is bent 3/4 of the way use TFF	=	0.92
	if torso is horizontal use TFF	=	1.00

Fc = A + B + C

Where	**A** = 3(BW)TFF =	3(______) × (______) =	______
	B = 0.5(L * HB) =	0.5(______) × (______) =	______
	C = 0.8[(BW)/2 + L] =	0.8[(______)/2 + ______] =	______

A + B + C = TOTAL COMPRESSIVE FORCE ESTIMATE (lbs)

========

NIOSH indicates that compressive forces excess of 770 lb will put some portion of the work force at risk and compressive forces in excess of 1430 lb will put most members of the workforce at risk. The largest of the terms A, B, or C contribute most to the lifting hazard.

A = 3(BW) TFF = Back muscle force reacting to upper body weight. To lower this, change the upper body angle with the horizontal.

B = 0.5(L × HB) = Back muscle force reacting to load moment. To lower this, change the magnitude of the load or the distance that the load is held out from the body.

C = 0.8[(BW)/2 + L] = Direct compressive component of the upper body weight and load. To lower this, change the magnitude of the load (or the body weight).

NOTE: This is just an estimate and its accuracy varies depending on posture, especially as the hands move out in front of the body. If precise compressive force data is required, a computer model such as that developed at the University of Michigan should be used!

Figure 54.35. Worksheet for back compressive force estimation.

4. If necessary redesign the job to reduce back compressive force starting with the largest of terms A or B.

This simplified model tends to underpredict low back compressive force for two lifting conditions: (*1*) with low weights (5–10 lb.) when the worker is standing straight up with the arms extended way out in front of the body, and (*2*) with all weights (highest percentage underprediction with low weights) when the worker is in an extreme squat position with the knees bent. If precise compressive force data is required, a computer model such as that developed at the University of Michigan (123) or ErgoWeb (124) should be used.

As an example (Fig. 54.36) a job in which 160-lb males are required to lift 40-lb boxes (without handles so the box must be lifted from the bottom) from the floor (V = 0 inches) to a 40-in. high conveyor (V = 40 in.) can be analyzed. The back compressive force can be estimated at the origin and destination of the lift. Since the back compressive force is generally higher when loads are lifted from a low level, the example will analyze the origin (V = 0) of this lift. Assume that the horizontal distance from the hands to the low back (location of the low back can be estimated by the hips) is 15 in. and the torso is horizontal so the torso flexion factor (TFF) is 1.0. As shown on Figure 54.36 the low-back compressive force is estimated to be 876 lb which is above the 770 lb that NIOSH has indicated will put some portion of the work force at risk. Term "A" is the highest so primary attention should be paid to task redesigns which will allow a reduction in torso flexion.

7.3.4.2 Shoulder Moment Estimation The shoulder moment resulting from a particular two handed lift can also be estimated if some simplifying assumptions are made. The moment at the shoulder depends on the weight of the load, body weight (arm weight), and the distance that these two weights are located in front of the point of rotation (shoulder). A work sheet to facilitate these calculations is shown on Figure 54.37.

Shoulder moment is often of interest when the material handling job requires that the load be held out from the body. If one uses the NIOSH Revised Lifting Equation as an initial analysis, the Horizontal Multiplier (HM) is an indicator of the potential for shoulder stress. If HM is low a follow-up shoulder moment analysis is suggested.

There are no generally accepted limits with which the estimated shoulder moment may be compared. Two of the variables which determine the shoulder moment which individuals may be able to generate on a task are gender and arm posture. Tables have been developed which provide an estimate of the maximum shoulder moment capability for an average male and female as a function of the included angle of the forearm and upper arm. These tables and a figure indicating these angles are shown Figure 54.38. The metric which is proposed as a measure of the stress at the shoulder is the ratio of the shoulder moment required by the task as calculated by the worksheet (Mtask) and the maximum strength of an average male/female in that posture (Mcap). While there are no empirically determined acceptable limits for this ratio, it is proposed that ratios below 0.5 (task required shoulder moment is less than half of the maximum for the average male/female) will not present a hazard for most workers unless the frequency is quite high, and ratios above 1.0 (task required shoulder moment exceeds the maximum for the average male/female) will present a hazard for many members of the workforce. The relative contribution of the arm weight to the moment (Mb) and load weight to the moment (Mf) do not provide much meaningful information.

In summary:

1. Measure task parameters and body weight and record on the top of the shoulder moment worksheet.
2. Calculate the values of terms Mb, and Mf and add them up to get an estimate of total shoulder moment.
3. Compare the shoulder moment resulting from the task (Mtask) to the maximum shoulder moment of an average male/female (Mcap) by using the ratio Mtask/Mcap.

COMPRESSIVE FORCE WORKSHEET

BW =	BODY WEIGHT =	160
L =	LOAD IN HANDS =	40
HB =	HORIZ DISTANCE FROM HANDS TO LOW BACK =	15
TFF =	TORSO FLEXION FACTOR =	1.0

Note:	if torso is vertical use TFF	=	0
	if torso is bent 1/4 of the way use TFF	=	0.38
	if torso is bent 1/2 of the way use TFF	=	0.71
	if torso is bent 3/4 of the way use TFF	=	0.92
	if torso is horizontal use TFF	=	1.00

Fc = A + B + C

Where **A** = 3(BW)TFF =	3(160) × (1.0)	=	480
B = 0.5(L * HB) =	0.5(40) × (15)	=	300
C = 0.8[(BW)/2 + L] =	0.8[(160)/2 + 40]	=	96

A + B + C = TOTAL COMPRESSIVE FORCE ESTIMATE (lb) 876

NIOSH indicates that compressive forces excess of 770 lb will put some portion of the work force at risk and compressive forces in excess of 1430 lb will put most members of the workforce at risk. The largest of the terms A, B, or C contribute most to the lifting hazard.

A = 3(BW) TFF = Back muscle force reacting to upper body weight. To lower this, change the upper body angle with the horizontal.

B = 0.5(L × HB) = Back muscle force reacting to load moment. To lower this, change the magnitude of the load or the distance that the load is held out from the body.

C = 0.8[(BW)/2 + L] = Direct compressive component of the upper body weight and load. To lower this, change the magnitude of the load (or the body weight).

NOTE: This is just an estimate and its accuracy varies depending on posture, especially as the hands move out in front of the body. If precise compressive force data is required, a computer model such as that developed at the University of Michigan should be used!

Figure 54.36. Example for back compressive force estimation.

Ratios below 0.5 will not present a hazard for most workers unless the frequency is quite high, and ratios above 1.0 will present a hazard for many members of the workforce.

4. If necessary redesign the job to reduce shoulder moment.

As was the case with the estimate of compressive force, it must be emphasized that this is just an estimate and it's accuracy varies depending on posture. If a one-handed lift is analyzed the shoulder moment capability (M_{cap}) can be estimated by halving the values in Figure 54.36 such as that developed at the University of Michigan (123) or ErgoWeb (124) should be used.

SHOULDER MOMENT WORKSHEET

BW =	BODY WEIGHT (lb)	______
D =	HORIZONTAL DISTANCE FROM LOAD TO SHOULDER JOINTS (in)	______
L =	LOAD WEIGHT (lb)	______
A =	Forearm angle in degrees	______
B =	Upper arm angle in degrees	______

$\boldsymbol{M_t = M_b + M_f}$

Where $M_b = 0.0115 \times D \times BW = 0.0115 \times$ ______ $\times$ ______ $=$ ______

$M_f = 0.5 \times D \times L =$ $0.5 \times$ ______ $\times$ ______ $=$ ______

$\boldsymbol{M_t = M_b + M_f}$ = (in lb) --------

Note that:

M_b = Moment at the shoulder due to the weight of the arm
M_f = Moment at the shoulder due to the weight of the load in the hands
M_t = Total moment at the shoulder = M_{task}

Substitute BW, D, L, Into the above equation to estimate the total moment required at the shoulder (M_{task} expressed as in lb). The tables on the next page indicate the maximum strength of an average male/female in that posture (upper arm angle, lower arm angle). Record the value from the tables based on angles A and B (M_{cap}). The ratio of M_{task}/M_{cap} Represents the required shoulder moment as percent of the maximum for the average male/female.

M_{task} = ______ (from above)

M_{cap} = ______ (from above)

M_{task}/M_{cap} = ______ = ______ (×100.0 = percent maximum)

Ratios (M_{task}/M_{cap}) below 0.5 will not present a hazard for most workers unless the frequency is quite high, and ratios above 1.0 will present a hazard for many members of the workforce.

Figure 54.37. Worksheet for shoulder moment estimation.

As an example (Fig. 54.39) the job noted above in which 160-lb males are required to lift 40-lb boxes from the floor (V = 0 inches) to a 40-in. high conveyor (V = 40 in.). The shoulder moment can be estimated at the origin and destination of the lift. Since the shoulder moment is generally higher when loads are above waist level the example will analyze the origin (V = 40) of this lift. Assume that the horizontal distance from the hands to the low back is 24 in., the lower arm (elbow) included angle is 135°, and the upper arm (shoulder) angle is 45°. The estimated shoulder moment of 524 in.-lb represents 73% of the average male maximum in that posture. This is above 50% of the average male maximum so consideration should be given to task redesign which will lower the 73% value.

Proactive analysis can be used to propose and justify initial job design or job redesigns. For example, an analysis was run using the biomechanical model of the shoulder discussed earlier for five different loads at three different locations. The shoulder moment was de-

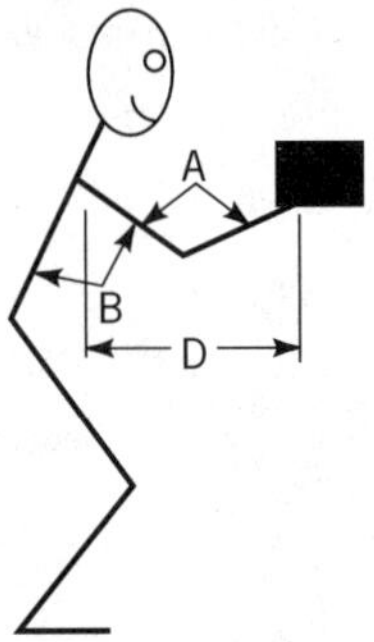

Forearm angle (A)	45	90	135	180
Upper arm angle (B)		50% Male		
0	632	691	751	810
45	598	658	717	777
90	565	624	684	743
135	531	591	650	710
180	498	557	617	676
		50% Female		
0	332	363	395	426
45	314	346	377	408
90	297	328	359	391
135	279	310	342	373
180	626	293	324	355

Figure 54.38. Arm angles and shoulder moment capabilities for 50 percentile males and females.

termined for a 50% male (height = 70 in., weight = 166 lb) for loads of 10, 20, 30, 40 and 50 lb held directly out from the shoulder. The results are noted in Tables 54.9 and 54.10 in terms of the actual moment (in.-lb) and as a percent of the average maximum moment for a 50% male.

The numbers in bold type in Figure 54.21 indicate when the shoulder moment required by the task (or planned task) exceeds 50% of the average maximum for a 50 percentile (average) male. It can be seen that 50 lb loads held at 12, 20 and 28 in. out from the shoulder, 40 lb loads held 20 and 28 in. out from the shoulder, and 30-lb loads held 20 and 28 in. out from the shoulder have the potential to cause shoulder stress. Note that these calculations do not recognize the fatigue caused by repetitive lifting.

7.3.5 Metabolic Models

The recognition of metabolic stresses in task design/redesign requires (*1*) the estimation of the aerobic capacity, or ability to do work, for the individual or population, (*2*) the estimation of the energy requirements of the job and (*3*) specification of administrative

SHOULDER MOMENT WORKSHEET

BW =	BODY WEIGHT (lb)	160
D =	HORIZONTAL DISTANCE FROM LOAD TO SHOULDER JOINTS (in)	24
L =	LOAD WEIGHT (lb)	40
A =	forearm angle in degrees	135
B =	Upper arm angle in degrees	45

$\boldsymbol{M_t = M_b + M_f}$

Where $M_b = 0.0115 \times D \times BW = 0.0115 \times$ 24 $\times$ 160 = 44

$M_f = 0.5 \times D \times L =$ 0.5 $\times$ 24 $\times$ 40 = 480

$\boldsymbol{M_t = M_b + M_f}$ = (in lb) 524

Note that:

M_b = Moment at the shoulder due to the weight of the arm
M_f = Moment at the shoulder due to the weight of the load in the hands
M_t = Total moment at the shoulder = M_{task}

Substitute BW, D, L, Into the above equation to estimate the total moment required at the shoulder (M_{task} expressed as in lb). The tables on the next page indicate the maximum strength of an average male/female in that posture (upper arm angle, lower arm angle). Record the value from the tables based on angles A and B (M_{cap}). The ratio of $M_{\text{task}}/M_{\text{cap}}$ Represents the required shoulder moment as percent of the maximum for the average male/female.

M_{task} = 524 (from above)

M_{cap} = 717 (from above)

$M_{\text{task}}/M_{\text{cap}} = \frac{524}{717} =$ .73 (×100.0 = percent maximum)

Ratios ($M_{\text{task}}/M_{\text{cap}}$) below 0.5 will not present a hazard for most workers unless the frequency is quite high, and ratios above 1.0 will present a hazard for many members of the workforce.

Figure 54.39. Example for shoulder moment estimation.

Table 54.9. Shoulder Moment (in.-lb) as a Function of Load and Posture For a 50% Male

	Horizontal Location (in. from Shoulder Joint)		
Load (lb)	12	20	28
10	83	138	194
20	143	238	334
30	203	338	474
40	263	438	614
50	323	538	754

Table 54.10. Shoulder Moment (as a Percent of Average Maximum) as a Function of Load and Posture for a 50% Male

	Horizontal Location (in. from Shoulder Joint)		
Load (lbs)	12	20	28
10	13.2	19.6	26.1
20	22.7	33.7	45.0
30	32.3	47.9	63.8
40	41.8	62.0	82.6
50	51.4	76.2	101.5

controls to reduce whole-body fatigue. Metabolic stresses are often of interest when the material handling job has a high lift/lower frequency or contains considerable walking, carrying, and/or pushing/pulling. If one uses the NIOSH Revised Lifting Equation as an initial analysis the Frequency Multiplier (FM) is an indicator of the potential for metabolic stress. If FM is low a follow-up metabolic analysis is suggested.

7.3.5.1 Estimation of Aerobic Capacity. The metabolic load for a particular worker depends on the aerobic capacity of the worker and the job energy rate requirements. Since maximal aerobic capacity testing is often performed until exhaustion, an alternative to this direct measurement is an assignment of maximum physical work capacity (MPWC) based on an individual's age. This relationship is MPWC $=$ 16 kcal $\times$ PFI for males and MPWC $=$ 12 kcal $\times$ PFI for females where PFI indicates the Physical Fitness Index. An individuals physical fitness index (PFI) varies with age as shown in Table 54.11 (125, 126).

A workers capability to work is also dependent on the MPWC and the task duration. This total capacity to do work for a given time may be called physical work capacity or PWC. This can be estimated as a function of age as:

Table 54.11. Physical Fitness Index as a Function of Age

Age	PFI[a]
20	1.16
25	1.13
30	1.09
35	1.00 (average)
40	0.95
45	0.93
50	0.91
55	0.88
60	0.83
65	0.79

$$PWC = [(\log 4400 - \log t) * IAC]/3.0$$

where PWC = Physical work capacity in kcal/min.
IAC = Individual aerobic capacity assumed to be 16 kcal/min. for males or 12 kcal/min. for females. IAC = PFI × 12 for women and PFI × 16 for men, where PFI is the Physical Fitness Index and is based on age (see above).
t = time duration of a work activity in minutes. (The equation is based on work durations of 8 hrs or less.)

Table 54.12 and 54.13 are based on this equation and summarize physical work capacities in kcal/min for men and women ages 20 through 65 for some common work durations.

7.3.5.2 Estimating Job Energy Expenditure Rate Requirements A relatively simple method of estimating job energy requirements by breaking the job down into components was developed by Thomas Bernard under contract with the Motor Vehicle Manufacturers Association (MVMA) (127). This method uses measures of arm use, walk distance, lift frequency, lift weight and push/pull force and distance. The standard error of the estimate for this method is 62 kcal/hr. A worksheet for this method is shown on Figure 54.40.

As an example, assume that the assembly of a differential requires the picking up of parts weighing 4.5 lb. There is extensive hand movement beyond an envelope of 20 in. and the job is performed by two workers who walk over a path of about 20 ft three times per minute. A completed worksheet for this job is shown as Figure 54.41. The energy requirements for this job are estimated to be 244.2 kcal/hr or 4.07 kcal/min. (The actual measured values were 240–258 kcal/hr.)

Depending on the physical work capacity or PWC of the worker, the metabolic load for the worker, or worker population, can be estimated. For example, a 510-min. work day is assumed (8 hr × 60 min./hr plus 30 min. a lunch if the worker's activities at lunch are

Table 54.12. Physical Work Capacity in kcal/minute for Females as a Function of Age and Work Duration[a]

AGE	PFI[a]	120 min.	240 min.	480 min.	510 min.
20	1.16	7.26	5.86	4.46	4.34
25	1.13	7.07	5.71	4.35	4.23
30	1.09	6.82	5.51	4.20	4.08
35	1.00	6.26	5.05	3.85	3.74
40	0.95	5.94	4.80	3.66	3.56
45	0.93	5.82	4.70	3.58	3.48
50	0.91	5.69	4.60	3.50	3.41
55	0.88	5.51	4.45	3.39	3.29
60	0.83	5.19	4.19	3.19	3.11
65	0.79	4.94	3.99	3.04	2.96

[a]PFI = Physical fitness index.

Table 54.13. Physical Work Capacity in kcal/minute for Males as a Function of Age and Work Duration

AGE	PFI[a]	120 min.	240 min.	480 min.	510 min.
20	1.16	9.68	7.82	5.95	5.79
25	1.13	9.43	7.61	5.80	5.64
30	1.09	9.09	7.34	5.59	5.44
35	1.00	8.34	6.74	5.13	4.99
40	0.95	7.93	6.40	4.88	4.74
45	0.93	7.76	6.27	4.77	4.64
50	0.91	7.59	6.13	4.67	4.54
55	0.88	7.34	5.93	4.52	4.39
60	0.83	6.92	5.59	4.26	4.14
65	0.79	6.59	5.32	4.05	3.94

[a]PFI = Physical fitness index.

unknown). If this job were to be performed by females 35 years and older or males 65 years or older the task energy requirement of 4.07 kcal/min. would exceed the worker PWC noted in Tables 54.12 and 54.13. In these cases the task may be redesigned to reduce the task energy requirement so that they are below the worker's PWC for the work duration.

7.3.5.3 Administrative Controls If it is not possible to redesign a job to reduce its overall energy requirements to acceptable levels, administrative controls should be applied to reduce the risk of whole body fatigue for workers. These controls may take the form of careful work/rest cycling in the form of rest breaks to avoid whole-body fatigue. The equation noted below gives a general scheme for allocating rest allowances. It can be used to determine the percentage of rest time recommended for each hour of work (128).

$$\%\ Rest\ Period = \frac{(PWC - E_{task})}{(E_{rest} - E_{task})} \times 100$$

where $\%\ Rest\ Period$ = percent each hour for rest.
PWC = physical work capacity.
E_{task} = average energy rate of the job (kcal/min).
E_{rest} = average energy during rest (1.5 kcal/min. often used).

For example an estimate of a work/rest program for the previous job which requires 4.07 kcal/minute if it is performed for an 8-hr shift (510 min. including lunch) by a 50-year old female fellow. Assume that the energy at rest is 1.5 kcal/min. PWC = 3.41 kcal/min. (from Table 54.7). Using equation above gives:

$$\%\ Rest\ Period = \frac{(3.41\text{–}4.07)}{(1.5\text{–}4.07)} \times 100 = (0.68/2.57 \times 100) = 26\%$$

This analysis indicates that this worker would require 26% rest or over 15 min. of rest/hr. This suggests that job redesign to reduce the job energy requirements is appropriate. If

BASAL METABOLIC RATE (in kcal/hour) 117

ARMS

DESCRIPTION	VALUE
Little hand/arm movement.	0
Most hand movement within 20 in.	1
Frequent hand movement more than 20 in.	2
Bend, stoop, extend reach, etc.	3

_____ × 25 = _____
arm value

WALK DESCRIPTION
Distance (in feet) covered during walking or carrying in one minute.
(Do not include the push/pull distance.)

_____ × .7 = _____
walk value

LIFT

DESCRIPTION FOR WEIGHT	VALUE
Most parts and tools less than 4 lb.	1
Most parts and tools between 4 lb. and 11 lb.	2
Most parts and tools greater than 11lb.	3

DESCRIPTION FOR FREQUENCY	VALUE
Less than 2 complete work cycles per min.	1
Between 2 and 5 complete work cycles per min.	2
More than 5 complete work cycles per min.	3

_____ × _____ × _____ = _____ × 4.4 = _____
arm value (from above) weight val freq val lift value

PUSH/PULL

DESCRIPTION FOR FORCE
Average force (lbs) exerted during push/pull.

DESCRIPTION FOR DISTANCE
Average (in feet) covered during push/pull in one minute.

5.2 + 1.1 × _____) × (_____)/3 = _____
push/pull force distance

TOTAL METABOLISM (kcal/hour) = _____
TOTAL METABOLISM (kcal/min) = _____

Figure 54.40. Worksheet for Bernard Method to estimate task metabolic load (25).

PWC exceeds E_{task} the result of the above calculations will be negative which means additional rest is not needed.

7.3.6 Liberty Mutual Psychophysical Data

Snook and Ciriello (129) performed several experiments in which the following task variables were controlled:

BASAL METABOLIC RATE (in kcal/hour) 117

ARMS DESCRIPTION — VALUE

Description	Value
Little hand/arm movement.	0
Most hand movement within 20 in.	1
Frequent hand movement more than 20 in.	2
Bend, stoop, extend reach, etc.	3

2 (arm value) × 25 = 50

WALK DESCRIPTION
Distance (in feet) covered during walking or carrying in one minute.
(Do not include the push/pull distance.)

60 (walk value) × .7 = 42

LIFT DESCRIPTION FOR WEIGHT — VALUE

Description for Weight	Value
Most parts and tools less than 4 lb.	1
Most parts and tools between 4 lb. and 11 lb.	2
Most parts and tools greater than 11lb.	3

Description for Frequency	Value
Less than 2 complete work cycles per min.	1
Between 2 and 5 complete work cycles per min.	2
More than 5 complete work cycles per min.	3

2 (arm value (from above)) × 2 (weight val) × 2 (freq val) = 8 (lift value) × 4.4 = 35.2

PUSH/PULL

DESCRIPTION FOR FORCE
Average force (lbs) exerted during push/pull.

DESCRIPTION FOR DISTANCE
Average (in feet) covered during push/pull in one minute.

5.2 + 1.1 × 0 (push/pull force)) × (____ (distance))/3 = 0

TOTAL METABOLISM (kcal/hour) = 244.2
TOTAL METABOLISM (kcal/min) = 4.07

Figure 54.41. Example of task metabolic load estimation (25).

- Frequency of lift, lower, carry, push, pull.
- Load size.
- Vertical height of load or handle.
- Lift, lower, push, pull, carry distance.

Based on their feelings of exertion or fatigue, subjects adjusted the weight or force of the object to a level where they could work "... as hard as they could without straining themselves, or without becoming unusually tired, weakened, overheated, or out of breath."

This is generally called "psychophysical" data. Tables 54.14, 54.15, and 54.16 are extracted from this research and provide generally acceptable force or load levels for lifting, lowering, pushing, pulling, and carrying which would be acceptable for 50% and 90% of males and females for different task parameters.

For example, from Table 54.14 it can be seen that 50% of females could be expected to lift a 20-in. wide load weighing 20 lb. from knuckle height to shoulder height at a frequency of one lift every 14 sec. Ninety percent could lift 15 lb. From Table 54.15 it can be seen that with a handle height of 56.7 in. 50% of males could be expected to exert a sustained push force of 51 lb over a distance of 49.9 ft one time every 5 min. Ninety percent could push 29 lb.

7.3.7 Summary of Analytical Tools

Table 54.17 presents a summary of the analytical tools discussed in Section 7.2 with the relevant Applicable Work Condition and the Calculation Output.

8 ERGONOMICS STANDARDS AND GUIDELINE DOCUMENTS

There are several ergonomics related standards and guideline documents that have been written by governmental bodies and voluntary organizations. The reader should realize that this arena is in a state of flux. As this chapter is being written, the federal Occupational Safety and Health Administration (OSHA) *is promulgating* an ergonomics program standard while concurrently, *two voluntary organizations are writing ergonomics guideline documents*. An ergonomics standard has been issued by one state OSHA agency and two state OSHAs are proposing standards.

8.1 Federal OSHA Ergonomics Standard

OSHA has taken gradual, measured steps in the development of an ergonomics standard. The federal agency hired its first ergonomist during 1979 and in the early 1980s, held ergonomics training classes. In 1986, OSHA issued a request for comments and information on reducing back injuries from manual lifting in general industry. The first OSHA citations for ergonomic hazards were issued in 1987 against four Chrysler plants. While investigating cumulative trauma disorders within the meatpacking industry in the late 1980s, OSHA issued Ergonomics Program Management Guidelines For Meatpacking Plants (a document that outlines how to organize an ergonomics program) in 1990. This document is discussed later in this chapter. A settlement agreement with 13 employers in the early 1990s produced the establishment of ergonomics programs covering nearly 500,000 workers. An advanced notice of proposed rulemaking for an ergonomics standard was published in 1992 that lead to the production of a draft proposed ergonomic protection standard in 1995. Following confrontational stakeholder meetings, the draft was abandoned. Congress prohibited OSHA from advancing the standard for nearly two years but relaxed its opposition in 1998. A draft standard was published in late 1999 that may cover

Table 54.14. Approximate Maximum Acceptable Weight of Lift and Lower (lb) (Load = 20″ Wide, Lift Distance = 30″)[a]

	Floor Level to Knuckle Height One lift/Lower Every								Knuckle Height to Shoulder Height One Lift/Lower Every								Shoulder Height to Arm Reach One Lift/Lower Every							
Parental Capable	5	9	14	1	2	5	30	8	5	9	14	1	2	5	30	8	5	9	14	1	2	5	30	8
		s				min		h		s				min		h		s				min		h
Approximate Maximum Acceptable Weight of lift for Males (lb)																								
90	*15*	18	22	29	33	35	37	44	18	22	26	29	31	31	35	37	15	20	22	26	26	29	31	35
50	*31*	35	42	57	64	71	73	84	29	37	44	49	51	53	57	64	24	33	37	44	46	46	53	57
Approximate Maximum Acceptable Weight of lift for Females (lb)																								
90	11	13	15	18	18	18	20	29	11	13	15	20	20	20	22	26	9	11	11	15	15	15	18	20
50	*18*	22	22	26	26	29	31	42	15	18	20	24	26	26	29	35	13	15	15	20	22	22	24	26
Approximate Maximum Acceptable Weight of lower for Males (lb)																								
90	18	22	24	33	37	40	42	53	22	24	31	31	33	33	35	42	15	18	22	24	26	26	26	33
50	*33*	42	46	60	68	75	77	99	40	44	53	53	60	60	62	75	29	33	37	42	49	49	49	60
Approximate Maximum Acceptable Weight of lower for Females (lb)																								
90	11	13	15	18	18	20	22	29	13	13	15	18	20	22	22	29	11	11	11	13	15	18	18	22
50	18	20	22	24	29	29	31	42	18	20	22	26	29	31	31	40	15	18	18	20	22	24	24	31

[a]Italicized values exceed 8 hour physiological criteria.

over 1.5 million employers and 18–25 million employees in select office, retail, and manual handling and manufacturing operations.

Without the standard, OSHA has cited businesses for ergonomics hazards through the General Duty Clause of the Occupational Safety and Health Act of 1970. Section 5 of the Act, termed the General Duty Clause, states that "A. Each Employer: 1) shall furnish to each of his employees employment and a place of employment which are free from recognized hazards that are causing or likely to cause death or serious physical harm to his employees" (130). A general duty clause violation exists when it is established that:

- A condition or activity in the employer's workplace presents a hazard to employees.
- The cited employer or the employer's industry recognizes that hazard.
- The hazard is causing or likely to cause death or serious physical harm.
- Feasible means exist to eliminate or materially reduce the hazard.

OSHA has had mixed success in using the General Duty Clause. In a recent case, a plastics plant avoided fines for citations that included ergonomics hazards by implementing a proactive ergonomics program (131). An administrative law judge ruled against OSHA in another case that lasted nearly five years involving a tire manufacturer cited for 13 ergonomics citations. The judge determined that OSHA failed to establish the first element of a general duty clause violation, that the conditions or activities in respondent's workplace, as alleged in the citation, presented a hazard to employees (132).

8.2 State of California Ergonomics Standard

Since 1983, there has been organized activity within the State of California to develop ergonomics related standards to prevent work injuries. Initial efforts focused on video display terminals and were characterized by the adversarial perspectives of labor and business. In 1993, the State passed Workers' Compensation Reform legislation that contained a mandate for the Occupational Safety and Health Standards Board to develop an ergonomics standard by January 1, 1995. The State of California OSHA wrote a proposed standard, however, it was not approved by the OSH Standards Board. When January 1, 1995, arrived with no standard in place, the California Labor Federation sued the Standards Board for failing to meet the deadline. The judge ruled against the Standards Board and issued a court order for the Board to adopt a standard on ergonomics by December 1996. Through 1995 and 1996, several versions of the ergonomics standards were written by the Standards Board. In July 1997, the standard went into effect but was subject to two court order changes in its first year. Further legal challenges are anticipated.

Table 54.15. Approximate Maximum Acceptable Forces of Push/Pull (lb) ("handle" Height = 56.7″)[a]

Percent Capable	8.0 ft One Push/Pull Every							24.9 ft One Push/Pull Every							49.9 ft One Push/Pull Every							100 ft One Push/Pull Every					149.9 ft One Push/Pull Every					200.1 ft Push One Push/ Pull Every			
	6	12	1	2	5	30	8	15	22	1	2	5	30	8	25	35	1	2	5	30	6	1	2	5	30	6	1	2	5	30	8	2	5	30	6
		s			min		h		s			min		h		s			min		h			min		h			min		h		min		h
Approximate Maximum Acceptable Forces of push for Males (lb)																																			
Initial Forces																																			
90	44	49	55	55	57	57	68	31	35	48	48	49	49	57	35	40	42	42	44	46	55	33	35	42	42	53	29	31	35	35	44	26	31	31	40
50	71	79	88	88	93	93	112	51	55	73	73	77	77	93	57	64	66	68	73	73	66	53	60	66	68	84	44	51	57	57	73	44	49	49	62
Sustained Forces																																			
90	22	29	33	35	40	40	49	*18*	*20*	29	29	33	35	40	*18*	20	24	26	29	31	35	*18*	22	26	29	35	*15*	18	22	24	29	15	16	20	24
50	*37*	49	60	62	68	71	84	*29*	*35*	49	51	57	60	71	*31*	*37*	*44*	44	51	53	62	*33*	*37*	44	51	62	*26*	*31*	37	42	51	*26*	31	35	42
Approximate Maximum Acceptable Forces of push for Females (lb)																																			
Initial Forces																																			
90	31	33	37	40	44	46	49	33	35	35	35	40	42	44	26	31	31	31	33	35	37	26	29	31	33	37	26	29	31	33	37	26	29	31	33
50	44	49	55	57	64	66	71	46	51	51	53	57	60	64	40	44	44	44	49	51	55	40	42	46	49	55	40	42	46	49	55	37	40	44	49

Sustained Forces																																			
90	*13*	18	22	22	24	26	31	*13*	15	15	15	18	20	24	*11*	*13*	*13*	13	15	15	20	*11*	*13*	13	13	16	*11*	*11*	11	13	18	*9*	9	9	13
50	*26*	*35*	42	44	46	51	62	*26*	*31*	*31*	33	35	37	46	*22*	*24*	*26*	*26*	31	31	40	*22*	*24*	26	26	35	*20*	*22*	*24*	24	33	*18*	*18*	20	26
Approximate Maximum Acceptable Forces of pull for Males (lb)																																			
Initial Forces																																			
90	31	35	40	40	42	42	51	24	29	35	35	37	40	46	29	33	33	33	35	37	44	26	29	33	33	42	22	24	29	29	35	22	24	24	31
50	44	51	57	57	82	82	73	35	40	53	53	55	57	68	42	46	49	49	53	53	64	37	42	49	49	60	33	35	42	42	53	31	35	35	44
Sustained Forces																																			
90	18	22	26	29	33	33	40	13	18	22	24	26	26	33	*15*	18	20	20	22	24	29	15	16	20	24	29	*13*	15	18	20	22	13	13	15	20
50	*29*	35	44	46	51	53	62	*22*	*29*	35	37	42	44	51	*24*	*29*	31	33	37	37	44	*24*	29	33	37	44	*20*	*24*	26	31	37	*20*	22	26	31
Approximate Maximum Acceptable Forces of pull for Females (lb)																																			
Initial Forces																																			
90	29	35	37	40	44	46	49	29	31	35	35	40	42	44	22	26	29	31	33	35	37	26	29	31	33	37	26	29	31	33	37	26	29	31	33
50	42	49	53	55	62	64	68	42	44	49	51	55	57	62	31	35	42	42	46	49	53	37	40	44	46	53	37	40	44	46	53	35	40	42	46
Sustained Forces																																			
90	*13*	*20*	22	22	24	26	33	*15*	*18*	20	20	22	24	29	*13*	*15*	*15*	18	18	20	24	*13*	*15*	15	18	22	*13*	*13*	15	15	20	*11*	11	11	15
50	*22*	*35*	37	40	42	46	55	*26*	*29*	*33*	35	37	40	49	*20*	*24*	*29*	*29*	31	33	42	*24*	*26*	26	29	37	*22*	*24*	*24*	26	35	*18*	*20*	20	26

[a]Italicized values exceed 8 hour physiological criteria.

Table 54.16. Approximate Maximum Acceptable Weight of Carry at Waist (lb) (load = 20″ Wide, Lift Distance = 30″)[a]

Percent Capable	8.0 ft One Carry Every							14.1 ft One Carry Every							27.9 ft One Carry Every						
	6 s	12 s	1 min	2 min	5 min	30 min	8 h	10 s	16 s	1 min	2 min	5 min	30 min	8 h	18 s	24 s	1 min	2 min	5 min	30 min	8 h
Males																					
90	22	31	37	37	42	46	55	20	24	33	33	37	42	49	22	24	29	29	33	37	44
50	*42*	55	66	66	73	84	97	37	44	60	60	66	75	86	37	42	51	53	57	64	77
Females																					
90	24	26	29	29	29	29	40	20	22	29	29	29	29	40	22	24	26	26	26	26	35
50	*33*	35	40	40	40	40	55	*26*	29	40	40	40	40	53	*31*	33	35	35	35	35	49

Table 54.17. Summary of Ergonomic Analysis Tools Described in this Chapter[a]

Analytical Tool	Applicable Work Condition	Calculation Output
Rodger's Analysis	Exposure to job risk factor effort level, continuous effort time, and efforts/minute	Priority for change = low, moderate, high, very high
RULA: Rapid Upper Limb Assessment Survey	Exposure to job risk factors of posture, force, muscle use, and motion	A grand score is generated that reflects the need for further investigation and job changes
Revised NIOSH Lifting Equation	Lifting/lowering tasks	A recommended weight limit is calculated based on task characteristics and a lifting index is generated that indicates relative risk by comparing the recommended weight limit to actual task load weight
Back Compressive Force Model	Lifting/lowering tasks performed in the sagittal plane	The L5/S1 spinal compressive force is calculated and compared to benchmark risk values established by NIOSH
Shoulder Moment Estimation Model	Lifting/lowering tasks performed in the sagittal plane	The task shoulder moment is calculated and compared to the maximum shoulder moment capabilities of the 50th percentile male or female
AAMA/Bernard Predictive Metabolic Model	Physically demanding tasks characterized by activities such as lifting/lowering, pushing/pulling, walking, reaching, bending	An estimated metabolic rate required to perform the job/task is calculated and compared to the metabolic rate at which the worker(s) can work
Liberty Mutual Tables of Acceptable Weights for Lift/Lowering Tasks	Lifting/lowering tasks Carrying tasks Pushing/pulling tasks	A load weight limit is generated from task characteristics

[a]This table is a brief overview intended to give the analyst a general sense of the analytical tools contained in this chapter. It is extremely important to understand the development, proper application, assumptions, limitations, and interpretation of the calculation in order to appropriately use the analytical tool. These points will be discussed for each of the tools.

The relatively brief California ergonomics standard is presented below:

Article 106. Ergonomics

Section 5110. Repetitive Motion Injuries.

(a) Scope and application. This section shall apply to a job, process, or operation where a repetitive motion injury (RMI) has occurred to more than one employee under the following conditions:

(1) Work related causation. The repetitive motion injuries (RMIs) were predominantly caused (i.e., 50% or more) by a repetitive job, process, or operation;

(2) Relationship between RMIs at the workplace. The employees incurring the RMIs were performing a job process, or operation of identical work activity. Identical work activity means that the employees were performing the same repetitive motion task, such as but not limited to word processing, assembly, or loading;

(3) Medical requirements. The RMIs were musculoskeletal injuries that a licensed physician objectively identified and diagnosed; and

(4) Time requirements. Thc RMIs were reported by the employees to the employer in the last 12 months but not before July 3, 1997.

Exemption: Employers with 9 or fewer employees.

(b) Program designed to minimize RMIs. Every employer subject to this section shall establish and implement a program designed to minimize RMIs. The program shall include a worksite evaluation, control of exposures which have caused RMIs and training of employees.

(1) Worksite evaluation. Each job, process, or operation of identical work activity covered by this section or a representative number of such jobs, processes, or operations of identical work activities shall be evaluated for exposures which have caused RMIs.

(2) Control of exposures which have caused RMIs. Any exposures that caused RMIs shall, in a timely manner, be corrected or if not capable of being corrected have the exposures minimized to the extent feasible. The employer shall consider engineering controls, such as work station redesign, adjustable fixtures or tool redesign, and administrative controls, such as job rotation, work pacing or work breaks.

(3) Training. Employees shall be provided training that includes an explanation of:

(A) The employer's program;

(B) The exposures which have been associated with RMIs;

(C) The symptoms and consequences of injuries caused by repetitive motion;

(D) The importance of reporting symptoms and injuries to the employer; and

(E) Methods used by the employer to minimize RMIs.

(c) Satisfaction of an employer's obligation. Measures implemented by an employer under subsection (b)(1), (b)(2), or (b)(3) shall satisfy the employer's obligations under that respective subsection, unless it is shown that a measure known to but not taken by the employer is substantially certain to cause a greater reduction in such injuries and that this alternative measure would not impose additional unreasonable costs.

8.3 Federal OSHA's Ergonomics Program Management Guidelines For Meatpacking Plants

In the 1980s', OSHA field inspectors found a high incidence and severity of cumulative trauma disorders (CTDs) and other work-related disorders due to ergonomic hazards within the meatpacking industry. To protect workers, health and safety program management guidelines were written and accepted as part of settlement agreements with large meatpacking firms.

Realizing the whole industry would benefit from the recommendations of the guidelines, OSHA published Ergonomics Program Management Guidelines For Meatpacking Plants in 1991 (133). Although OSHA developed the Meatpackers' Guidelines for use within that specific industry, other employers have used the document as a reference. OSHA has stated that the guidelines are not a new standard or regulation and failure to implement the guidelines is not in itself a violation of the OSHAct. However, OSHA expects employers to implement effective ergonomics programs adapted to their particular workplaces. Details of the guideline are outlined later in this chapter.

8.4 ANSI Z-365: Control of Cumulative Trauma Disorders Proposed Standard

The American National Standards Institute (ANSI) Accredited Standards Committee on Control of Cumulative Trauma Disorders has released a proposed standard that describes a process involving management responsibilities, training, employee involvement, surveillance, case evaluation, and job analysis. The draft provides a good background on ergonomic concerns and gives a blueprint for injury prevention program development. Although the writing of this document began in 1990, it will likely see further revisions.

8.5 ANSI B-11: Ergonomic Guidelines for the Design, Installation, and Use of Machine Tools

The ANSI Machine Tool Safety Standards Committee (B11) took the initiative in the early 1990s to compose a voluntary standard involving the design, installation, and use of tools/machines from an ergonomics perspective to reduce workplace injury and improve product production (134). The Machine Tool Safety Standards Committee (B11) formed a subcommittee composed of representatives from manufacturing, higher education, safety, design, and ergonomics to generate consensual ergonomic guidelines to assist the design, installation, and use of individual and integrated machine tools and auxiliary components in manufacturing systems. The standard was approved in 1994 and is characterized by three themes:

- Necessity of communication among all individuals (users, installers, manufacturers, designers) involved with the machine tools and auxiliary components to ensure their effective design, installation, and use.
- Knowledge of ergonomics concepts and principles among all individuals.
- Ability to apply ergonomics concept/principle knowledge effectively to machine tools and auxiliary components.

The standard is divided into sections that include (*1*) basic ergonomic factors, (*2*) ergonomic consideration for the design of machine tools, (*3*) machine tool installation considerations, (*4*) ergonomics considerations for the use of machine tools, (*5*) environmental considerations, and (*6*) training. General concepts as well as specific recommendations are provided.

8.6 ANSI/HFS 100-1988: American National Standard for Human Factors Engineering of Visual Display Terminal Workstations

This 90 page document was produced as a cooperative effort between ANSI and The Human Factors Society (HFS), a professional organization. The standard specifies acceptable conditions for visual display terminals, associated furniture, and the office environment. Another chapter in this text has computer workstations as its focus.

8.7 NIOSH Criteria For A Recommended Standard: Occupational Exposure to Hand–Arm Vibration

The National Institute for Occupational Safety and Health (NIOSH) concluded that Hand–Arm Vibration Syndrome is a chronic disorder with a latency period varying from a few months to several years (94). The syndrome is dependent on many interacting factors including:

- Vibration level produced by the tool.
- Hours of tool use per day.
- Environmental conditions.
- Type and design of the tool.
- Manner in which the tool is held.
- Vibration spectrum produced by the tool.
- Vibration "tolerance" of the worker.
- Tobacco and drug use by the worker.

NIOSH felt that due to the complex interactions among these and other factors and the general lack of objective data, it was not possible at that time to determine meaningful dose–response relationships. The agency declined to establish specific Recommended Exposure Limits (REL). However, NIOSH recognized that hand–arm vibration is a significant workplace hazard and recommended medical monitoring, engineering controls, good work practices, use of protective clothing and equipment, worker training programs, and administrative controls such as limited daily use time.

NIOSH has not re-addressed this issue since writing the criteria document in 1989. OSHA does not have and is not currently promulgating a standard for hand-arm vibration. The OSHA Technical Manual does not consider this physical agent to be a health hazard. Threshold limit values have been published by the American Conference of Governmental Industrial Hygienists (135).

9 ERGONOMICS AND THE AMERICANS WITH DISABILITIES ACT

At first glance, it does not appear that ergonomics concepts would have any application to legal civil rights. Ergonomics is concerned with the physical features of the workplace as they relate to physical capabilities of workers. Passage of the Americans with Disabilities Act (ADA) brought ergonomics into the civil rights arena. Under this law, the physical features of the workplace can not be a barrier to employment opportunities for a qualified individual with a disability. Ergonomists quantify task physical demands and can help define job essential functions that assist decision making. The purpose of this section is to provide a basic understanding of the law. Books and numerous newsletters offer further information on ADA and are important references as court cases further define the statue. The reference for all material presented in this section is *A Technical Assistance Manual on the Employment Provisions (Title I) of the Americans with Disabilities Act* (136).

9.1 ADA Overview

The Act became effective July 26, 1992, for employers with 25 or more workers and July 26, 1994, for employers with 15 or more workers. It protects qualified individuals from employment discrimination and is enforced by the U.S. Department of Justice, Civil Rights Division, Disability Rights Section. Remedies that may be required from a charge filed by a worker with the agency include compensatory and punitive damages, back pay, front pay, and attorney's fees as well as reasonable accommodation, reinstatement, and job offers. The Act requires the affected employer to consider the following issues:

- Who is a disabled person?
- If a disabled person applies for a position within the company, does the person have to be hired?
- What process changes and work accommodations are required?
- What are company obligations to an employee who becomes disabled due to an on the job injury?

The Act is being further defined as complaints are made to the EEOC and the courts make decisions.

9.2 ADA Law Definitions

The Americans With Disabilities Act is a federal law which prohibits an employer to discriminate against qualified individuals with disabilities in all aspects of the employment process. A qualified individual is a person who satisfies the requisite skill, experience, education and other job-related requirements of the employment position such individual holds or desires; and who with or without reasonable accommodation, can perform the essential functions of such position.

Reasonable accommodation is any modification or adjustment to a job, an employment practice, or the work environment that makes it possible for an individual with a disability to enjoy an equal employment opportunity. Examples of reasonable accommodations include making facilities accessible and usable, job restructuring, modified work schedules, flexible leave policies, acquisition or modification of equipment and devices, and adjusting and modifying of examinations, training materials, and policies.

Legal obligation to reasonable accommodation is limited by the imposition of undue hardship on business operation. This requires significant difficulty or expense in relation to the size of the employer, the resources available, and the nature of the operation. The concept of undue hardship includes any action that is unduly costly, extensive, substantial, disruptive, or that would fundamentally alter the nature or operations of the business.

The essential functions are the basic activities that are required to perform a job. Criteria for a function to be considered essential includes:

- The position exists to perform the function (i.e., a person is hired to proofread documents; the ability to proofread accurately is an essential function, because it is the reason that this position exists).
- There are a limited number of other employees available to perform the function, or among whom the function can be distributed (i.e., it may be an essential function for a file clerk to answer the telephone if there are only three employees in a very busy office and each employee has to perform many different tasks).
- A function is highly specialized, and the person in position is hired for special expertise or ability to perform it (i.e., a sales position for a company which is doing business primarily in Japan; knowing the Japanese language is an essential function of the job)

An individual with a disability is a person who:

- Has a physical or mental impairment (i.e., a physiological or mental disorder which compromises the normal working of the body such as leg amputee, epilepsy, dyslexia, or contagious disease (HIV), impairment does not include height or weight within normal range, pregnancy, or inability to read due to dropping out of school), or
- Has a record of such an impairment (an individual may have a history of a impairment which currently may or may not affect them such as heart disease, cancer, or rehabilitated drug user), or

- Is regarded as having such an impairment (an individual who is perceived as having an impairment such as high blood pressure which is under control or a facial scar on a person with high public contact)

that

- Substantially limits (i.e., an individual must be unable to perform, or be significantly limited in the ability to perform, an activity compared to an average person in the general population),
- One or more major life activities that an average person can perform with little or no difficulty such as wailing, speaking, breathing, seeing, performing manual tasks, learning, sitting, standing, lifting, and reading.

The employer may require a qualification standard that an individual not pose a direct threat to the health or safety to the individual or others. The employer must show significant risk of substantial harm, the specific risk must be identified, it must be a current risk (not one that is speculative or remote), the assessment of the risk must be based on objective medical or other factual evidence regarding a particular individual and reasonable accommodation can not eliminate or reduce the concerns below the level of a direct threat.

9.3 Application of ADA

Employers cannot discriminate against individuals with disabilities in regard to any employment practices or terms, conditions, and privileges of employment. This includes application, testing, hiring, assignments, evaluation, disciplinary actions, training, promotion, medical examinations, layoff/recall, termination, compensation, leave, or benefits.

The following sequence outlines regulation logic. This is not legal advice. If you have a specific situation with a worker or potential worker, seek appropriate counsel.

Question One: Does the individual have requisite skills, experience, education, and other job related requirements?
If NO—No job offer (be ready to defend decision by ADA rules).
If YES—Individual is classified as "otherwise qualified".

Question Two: Can the individual perform essential job functions with or without reasonable accommodation?
If NO—or I Do NOT KNOW—Post offer/pre-employment capability evaluation.
If YES—Job offer can be made without concern of accommodation.

Question Three: Does the individual have a disability?
If NO—Job accommodation is not required. No job offer.
If YES—Accommodation evaluation.

Question Four: Can reasonable accommodations be made so the individual can perform job essential functions?
If NO—This has to be due to undue hardship or lack of effective controls. No job offer.
If YES—"Direct threat" analysis.

Question Five: Is there a "direct threat"?
If NO—Job offer can be made. The individual is a qualified individual with a disability who can perform the essential functions of the job with reasonable accommodation without creating a direct threat.
If YES—No job offer.

9.4 Job Analysis

The ADA does not require that an employer conduct a job analysis to identify the essential functions of a job, however, to comply with ADA regulations, it is wise to know some specific job requirements. These include:

1. Requisite skills (i.e., the necessary education, experience, and licenses required to hold the position).
2. Essential job functions [i.e., the purpose of the job (results or outcome) and the importance of actual job functions in achieving the purpose].
   - Purpose Example: move heavy packages from the dock to a storage room as opposed to lift and carry heavy packages from the dock to the storage room.
   - Importance Example: the frequency, amount of time and the consequences of not performing a function.
3. The manner by which a job is performed (the manner by which a job is performed can be described, however, it is erroneous to conclude it is the only way a job can be done.

An ergonomist's skills are helpful in identifying the purpose of a job and quantifying job functions such as amount of force required to push a pedal/lever, weight of tool/load, reach distance, frequency of widget production, or length of time task is performed.

9.5 Actions which Constitute Discrimination

The following situations have been defined by EEOC as examples of violations of the ADA law:

1. Refusing to make reasonable accommodation to the known physical or mental limitations of a qualified applicant or employee with a disability, unless the accommodation would pose an undue hardship on the business.

2. Using qualification standards, employment tests, or other selection criteria that screen out or tend to screen out an individual with a disability unless they are job-related and necessary for the business.
3. Failing to use employment tests in the most effective manner to measure actual abilities. Tests must accurately reflect the skills, aptitude, or other factors being measured, and not the impaired sensory, manual, or speaking skills of an employee or applicant with a disability (unless those are the skills the test is designed to measure).

10 CONTROLS

Once ergonomics hazards have been identified and measured, the analyst needs to compose an appropriate control strategy. As mentioned earlier in the chapter, the goal is to increase the overlap between worker capabilities and task demands to improve the fit of the job with the worker. The creation of control options depends on the experience and imagination of the analyst. Although specific solutions vary, there is a standard thought process that can be applied.

As with controlling other hazards, a three level hierarchy is followed: engineering solutions, administrative controls, and personal protective equipment.

10.1 Engineering Controls

The preferred method for controlling ergonomics hazards is through engineering techniques. When the design of the workplace reduces the magnitude of risk factors, the likelihood of injury/illness is lessened. When considering engineering controls, the analyst should think in terms of:

- The complete production system: what the work process is and why it is organized as it is.
- The local workstation: the arrangement and dimensions of the worker's immediate workspace.
- Tools/equipment: devices used by the worker in the work process.

10.1.1 Work System Design

A workplace is frequently composed of a series of interrelated work areas. A part is created at one work station, modified at a second workstation, and altered at a third workstation before being packed for shipment at a fourth workstation. What happens in the part production at one workstation effects activities at workstations after it. The work process should be set up to minimize the magnitude of risk factors through the manufacturing plan.

Example:

Shipment cases arrive to an inspection work station on a floor level, roller conveyer from a packing work station. The worker lifts the case to a 30-in. high work surface for a content check once every two minutes for eight hours. The risk condition is repeat lifting at a moderate frequency for an extended duration.

A work system solution would involve reviewing the overall manufacturing process to recommend a control that would allow the cases to arrive at the inspection station on a conveyor at the level of the work surface to avoid the lift.

10.1.2 Workstation Design

Some risk conditions are created solely by the design of the work station itself. Three engineering principles should be followed:

1. The workstation should be designed for the task that is to be accomplished.
   Example:
   A common finding in many offices is the conversion of a desk/secretary return or credenza into a computer workstation. An office worker is usually forced into awkward postures to accomplish tasks.
2. The work station should be adjustable to accommodate workers of various size.
   Example:
   As mentioned in the Anthropometry Section, designing for the average size worker usually forces those workers not between the 5th and 95th percentile to assume awkward postures. The reasonable reach to a part bin for a six foot two worker on day shift may be an unreasonable reach for a five foot two worker on the swing shift.
3. The work station should allow adequate space for required task motions.
   Example:
   Reduced space causes the worker to assume awkward postures while performing tasks. A kitchen was modified by adding three ovens due to increased business. They were placed in what had been a small storage area. The cramped space made it difficult for the cooks to load/unload the ovens. This space restriction caused one cook to sustain a significant burn while unloading a tray of fresh-baked cookies.

10.1.3 Tool and Equipment Design

Tools and equipment should be selected to perform a specific task while simultaneously minimizing the magnitude of risk factors. Common risk conditions associated with use of tools/equipment include: long term, steady force generated by muscle contraction, awkward postural position, fast motion, vibration, and cool temperatures.

Example:

A bar code scanner was used to read labels on three by five cards. The scanner had a "gun handle" design which required the worker to repeatedly bend at the wrist to perform the task. The most cost effective solution involved placing the three by five cards on a slant board which eliminated the bending of the wrist.

10.2 Administrative Controls

Administrative controls are workplace policy, procedures, and practices that minimize the exposure of workers to risk conditions. They are considered less effective than engineering controls in that they do not usually eliminate the hazard. Rather, they lessen the duration and frequency of exposure to the risk condition. Administrative controls are applied when the cost or practicality of engineering controls are prohibitive.

Common administrative controls include:

- Providing rest breaks to relieve fatigued muscle-tendon groups.
- Increasing the number of employees assigned to a task to alleviate severe conditions such as lifting heavy objects.
- Rotating workers through different jobs.
- Providing sufficient numbers of standby/relief personnel.
- Job enlargement—having the worker do several different tasks as opposed to one task the whole shift.
- An effective tool maintenance program to ensure equipment is in proper working order.
- An effective housekeeping program to minimize ergonomic/safety hazards.
- Training employees in proper work techniques that reduce the risk of injury/illness such as correct use of the work station, maintenance/use of tools, and lifting technique.
- Monitoring work practices to ensure proper work techniques are followed.

10.3 Personal Protective Equipment

A device that a worker wears to prevent full exposure to the ergonomic hazard is considered personal protective equipment. This type of control is applied when it is not possible to contain the risk condition through engineering or administrative measures. It is considered the least effective control method.

Examples

- Knee pads to protect the knee when performing long term kneeling tasks.
- Gel shoe insoles to cushion the foot/lower extremity when standing long term on a hard surface.
- Gloves that absorb vibration from tools.
- Gloves that protect the hands from knife cuts.
- Gloves/clothes that protect against the cold.
- Proper fit of personal protective equipment is important to avoid other potential injury. If gloves are too tight, they restrict hand circulation. If gloves have a slippery grip surface, more force is required from hand muscles to hold objects.

10.3.1 Back Belts and Wrist Braces

Back belts and wrist braces are controversial items. Although widely used in industry, they are not considered to be personal protective equipment by most ergonomics professionals and OSHA (131). In a NIOSH back belt study (137), it was concluded that:

> "The effectiveness of using back belts to lessen the risk of back injury among uninjured workers remains unproven. The Working Group does not recommend the use of back belts to prevent injuries among uninjured workers, and does not consider back belts to be personal protective equipment. The Working Group further emphasizes that back belts do not mitigate the hazards to workers posed by repeated lifting, pushing, pulling, twisting, or bending."

A different conclusion was reached by Kraus (138) in an epidemiological study of low back injuries within a national chain of home improvement stores. After a back support policy was instituted, the incidence of low back injury dropped from 31 to 20 back injuries per one million employee work hours. When combined with sound workplace design and training in proper lifting techniques, back belts can contribute to a safer work environment. Back belts and hand/wrist braces are an adjunct to injury/illness treatment. It is recommended that their use in the workplace be determined by a health care professional (133).

11 COMPONENTS OF ERGONOMICS PROGRAMS

This section is based extensively on OSHA publication 3123, "Ergonomics Program Management Guidelines for Meatpacking Plants" (133) which presents the primary components of an effective ergonomics program. While these Program Management Guidelines were prepared for meatpacking plants they can be used as the foundation for an overall ergonomics program in nearly any facility. The overall program elements include:

1. Worksite Analysis
   - Review of Available Data
   - Screening Survey
   - Ergonomic Job Hazard Analysis
   - Periodic Ergonomic Survey
2. Hazard Prevention and Control
   - Engineering Controls
   - Work Practice Controls
   - Personal Protective Equipment
   - Administrative Controls
3. Medical Management
4. Training and Education
   - General Training
   - Job-Specific Training

Training for Supervisors

Training for Managers

Training for Engineers and Maintenance Personnel

Worksite analysis identifies existing hazards and conditions, operations that create hazards and close scrutiny and tracking of injury and illness records to identify patterns of traumas or strains that may indicate the development of CTDs or manual material handling strains.

The worksite analysis uses a systematic process to identify and quantify ergonomic risk factors. The process is divided into four parts:

A. Gathering information from available sources.

The first step is to analyze existing records of illness and injury. This process involves a review of the injury/illness records to locate departments, work areas or positions where the rate or cost of ergonomic related disorders are high or are increasing. This review may include the OSHA 200 log and medical, safety, and insurance records (refer to 29 *Code of Federal Regulations* 1904 for the complete requirement on maintaining the OSHA 200 (117).

As noted earlier in this chapter, the incidence rate is presented in terms of ergonomic related disorders per 200,000 work hours (100 full-time equivalent employees for one year). The incidence rate should be analyzed to determine the overall ergonomic related incidence and to see if specific types of ergonomic disorders are concentrated in specific departments, work areas, shifts, tasks, or if trends exist.

B. Conducting baseline screening surveys to determine which jobs need a closer analysis.

The second component of this process involves an analysis of the workplace to determine the existence of ergonomic hazards. This includes an analysis of the job, workstation, or process which contributes to the risk of developing an ergonomic related disorder. This analysis frequently focuses on risk factors for upper extremity cumulative trauma disorders (UECTD) and back disorders.

C. Ergonomic Job Hazard Analysis

Jobs identified as being potentially hazardous via the screening surveys will need further detailed analysis to determine proper abatement strategies. Also, positions being considered in a rotation scheme or for light duty should receive this job hazard analysis. This analysis should be conducted by individuals with training and experience in these areas.

D. Periodic Ergonomic Surveys

Periodic surveys should be conducted to identify previously unnoticed risk factors or failures or deficiencies in work practice or engineering controls. This can be done using the methods used in the initial screening survey.

11.2 Hazard Prevention and Control

Once ergonomic hazards are identified through the systematic worksite analysis, the next step is to design measures to prevent or control these hazards. Ergonomic hazards are prevented primarily by effective design of the workstation, tools, and job. To be effective, a program should use appropriate engineering and work practice controls, personal protective equipment, and administrative controls to correct or control ergonomic hazards.

A. Engineering Controls

Engineering solutions, where feasible, are the preferred method of control for ergonomic hazards. The focus of an ergonomics program is to make the job fit the person, not to make the person fit the job. This is accomplished by redesigning the workstation, work method, or tool to reduce the demands of the job such as high force, repetitive motion, and awkward postures.

The following are examples of engineering controls that have been found to be effective and achievable in the industry.

(a) *Workstation Design* Work stations should be designed to accommodate the persons who actually work on a given job; it is not adequate to design for the "average" or typical worker. Workstations should be easily adjustable and either designed or selected to fit a specific task, so that they are comfortable for the workers using them. The work space should be large enough to allow for the full range of required movements.

(b) *Design of Work Methods*. Traditional work method analysis considers static postures and repetition rates. This should be supplemented by addressing the force levels and the hand and arm postures involved. The tasks should be altered to reduce these and the other stresses identified with CTDs. The results of such analyses should be shared with the health care providers; e.g., to assist in compiling lists of "light-duty" and "high-risk" jobs.

(c) *Tool Design and Handles*. Attention should be paid to the selection and design of tools to minimize the risks of upper extremity CTDs and back injuries. In any tool design, a variety of sizes should be available and the tool matched properly to the task. As an example, an in-line style tool should be used on a horizontal surface that is between elbow and waist height and a pistol-grip style tool should be used on vertical surfaces of the same height.

B. Work Practice Controls

An effective program for hazard prevention and control also includes procedures for safe and proper work that are understood and followed by managers, supervisors and workers. Key elements of a good work practice program for ergonomics include proper work techniques, employee conditioning, regular monitoring, feedback, maintenance, adjustments and modifications, and enforcement.

(a) *Proper Work Techniques*. A program for proper work techniques, such as proper tool maintenance, correct lifting techniques and proper work techniques.

(b) *New Employee Conditioning Period*. Strenuous tasks will usually require conditioning, or break-in periods, which may last several weeks. New and returning employees should be gradually integrated into a full workload as appropriate for specific jobs and individuals. Employees should be assigned to an experienced trainer for job training and evaluation during the break-in period. Employees reassigned to new jobs should also have a break-in period.

(c) *Monitoring*. Regular monitoring at all levels of operation helps to ensure that employees continue to use proper work practices. This monitoring should include a periodic review of the techniques in use and their effectiveness, including a determination of whether the procedures in use are those specified; if not, then it should be determined why changes have occurred and whether corrective action is necessary.

(d) *Adjustments and Modifications*. Modify work practice controls when the dynamics of the work place change. Such adjustments include changes in expected production rates and staffing for tasks.

C. Personal Protective Equipment (PPE)

PPE should be selected with ergonomic stressors in mind. Appropriate PPE should be provided in a variety of sizes, should accommodate the physical requirements of workers and the job, and should not contribute to extreme postures and excessive forces. The following factors need to be considered when selecting PPE in the meat industry:

(a) Proper fit is essential. For example, gloves that are too thick or that fit improperly can reduce blood circulation and sensory feedback, contribute to slippage, and require excessive grip strength. The gloves in use should facilitate the grasping of the tools and knives needed for a particular job while protecting the worker from injury.

(b) Protection against extreme cold is necessary to minimize stress on joints.

(c) Braces, splints, back belts, or other similar devices are not PPE.

(d) Other types of PPE that may be selected for use (e.g., arm guards) should not increase other ergonomic stressors.

D. Administrative Controls

Administrative controls include job rotation and job enhancement, adequate mandatory rest breaks, and the implementation of an exercise/conditioning program.

The purpose of job rotation is to alleviate physical fatigue and stress of a particular set of muscles–tendon–nerve groups by rotating employees among one or two other jobs that use different muscle–tendon–nerve groups. Caution must be used in deciding which jobs are used; although different jobs may appear to have different stressors they may actually pose the same physical demands.

Rest pauses are needed to relieve fatigued muscle-tendon groups. The length of time needed depends on the task's overall effort and total cycle time.

11.3 Medical Management

A medical management program for ergonomic related disorders should include documentation of the following:

1. Health care providers who understand the prevention, early recognition, evaluation, treatment and rehabilitation of UECTDs, ergonomic principles, employee physical assessment, and OSHA recordkeeping requirements.
2. Periodic workplace walk-throughs by health care providers to gain an understanding of the operations and work practices, identify potential light duty jobs, and maintain contact with the employees.
3. Periodic surveys of the work population for symptoms of work related ergonomic disorders.
4. Compilation of a list of light duty jobs.
5. Health surveillance program including the determination of a baseline for new or transferred workers, implementation of a conditioning period for new or transferred workers, and a periodic health survey of all workers exposed to ergonomic stresses.

6. Training of employees in the causes, preventions, early symptoms, and means of prevention of ergonomic related disorders.
7. Encouragement to report early symptoms of ergonomic related disorders.
8. Standard protocols for the evaluation and treatment of ergonomic related disorders.
9. Maintenance of appropriate illness/injury records such as the OSHA 200 log, OSHA Form 101, and OSHA Form 200-s.
10. Periodic monitoring of surveys, OSHA forms, or health surveillance program records for trends in ergonomic related disorders.

11.4 Training and Education

It is essential that all employees are informed about ergonomic hazards and abatement methods at a level consistent with their need and ability to use the information. The program should be designed and implemented by qualified persons. Appropriate special training should be provided for personnel responsible for administering the program.

The program should be presented in language and at a level of understanding appropriate for the individuals being trained. It should provide an overview of the potential risk of illnesses and injuries, their causes and early symptoms, the means of prevention, and treatment.

The program should also include a means for adequately evaluating its effectiveness. This might be achieved by using employee interviews, testing, and observing work practices, to determine if those who received the training understand the material and the work practices to be followed.

Training for affected employees should consist of both general and specific job training:

11.4.1 General Training

Employees who are potentially exposed to ergonomic hazards should be given formal instruction on the hazards associated with their jobs and with their equipment. This includes information on the varieties of CTDs, what risk factors cause or contribute to them, and how to recognize and report symptoms, and how to prevent these disorders. This instruction should be repeated for each employee as necessary. OSHA's experience indicates that, at a minimum, annual retraining is advisable.

11.4.2 Job Specific Training

New employees and reassigned workers should receive an initial orientation and hands-on training prior to being placed in a full-production job. Training lines may be used for this purpose. Each new hire should receive a demonstration of the proper use of and procedures for all tools and equipment. The initial training program should include the following:

Use of special tools and devices associated with individual workstations.

Use of appropriate guards and safety equipment, including personal protective equipment.

Use of proper lifting techniques and devices

Training and education should be provided to employees, supervisors, managers, engineers and maintenance personnel, and health care providers.

1. Employees should receive general training on the types and risk factors for ergonomic disorders and appropriate abatement measures.
2. Supervisors should receive the same training as the employees plus additional training so that they can recognize and correct hazardous work practices and reinforce employee training.
3. Manager training should provide a basic understanding of the ergonomics of production processes and the overall cost-effectiveness of ergonomic hazard abatement measures.
4. Plant engineers and maintenance personnel should be trained so that they can prevent ergonomic hazards through proper work place, tool, and process design and abate these hazards when discovered.
5. Health care providers should be knowledgeable in the prevention, early recognition, evaluation, treatment and rehabilitation of CTDs, and in the principles of ergonomics, physical assessment of employees, and OSHA recordkeeping requirements.

Additional information on the establishment and operation of an ergonomics program can be found in Ref. 114.

12 ERGONOMICS CERTIFICATION, ORGANIZATIONS, AND PUBLICATIONS

12.1 Ergonomics Certification

The Board of Certification in Professional Ergonomics (BCPE) a nonprofit corporation (3) which provides procedures for examining and certifying qualified practitioners of ergonomics. Applicants may be certified as either a Certified Professional Ergonomist (CPE) or Certified Human Factors Professional (CHFP). The BCPE certifies practitioners of ergonomics, or a person who has:

1. Superior knowledge of available ergonomics information.
2. Command of the methodologies used by ergonomists and is applying that knowledge to the design of a product, system, job, or environment.
3. Applied this knowledge to the analysis, design, testing, and evaluation of products, systems, and environments.

Applicants must meet all of the following requirements to be certified:

1. Masters degree, or equivalent, in one of the correlative fields of ergonomics, such as biomechanics, human factors, ergonomics, industrial engineering, industrial hygiene, kinesiology, psychology, or systems engineering.

2. Four years of demonstrable experience in ergonomics practice.
3. Evidence of experience in applying ergonomics to "design" by submission of a work product to the BCPE.
4. Receiving a passing score on the BCPE examination.

The BCPE recently created a new level of ergonomics certification, Certified Ergonomics Technologist (CET), that will encourage certification for health care providers (119). The CET certificate has requirements in the same areas as the CPE but with less depth of knowledge and experience requirement.

For more information contact:
Board of Certification in Professional Ergonomics
P.O. Box 2811
Bellingham, WA 98227-2811
USA
Phone: (360) 671-7601
Fax (360) 671-7681
E-mail: BCPEHQ@aol.com
Internet: http://www.bcpe.org/

The Oxford Research Institute (ORI) has established a process for certification in industrial ergonomics or human factors engineering which uses a controlled peer review process. Qualified applicants may be certified as a Certified Industrial Ergonomist (CIE) or Certified Human Factors Engineering Professional (CHFEP). Specialty certification may be requested in a number of different areas.

To be certified applicants must provide evidence of the following:

1. Résumé which details BS plus six years, masters degree plus five years or doctoral degree plus four years of experience in related field of employment.
2. Specialized training or formal education in fields related to ergonomics or human factors engineering.
3. Duplicate copies of two or three work samples such as books, articles, technical reports, inventions, patents, awards, honors, demonstrations, video tapes or other media where the applicant is the principle author.
4. Two letters of recommendation from sponsors who are familiar with the applicants' work in the specialty field.
5. Receiving a passing score on the certification examination is not presently a requirement but may be implemented. As of March 1997 and "for the next several months" this requirement will be waived for applicants who are "practicing ergonomists."

For more information contact:
Oxford Research Institute
10153 Vantage Point Court

New Market, MD 21774
USA
Phone: (301) 865-4506

Roy Matheson and Associates (RMA) provides certification as a Certified Ergonomic Evaluator Specialist (CEES). Applicants must be certified in a related safety, health, or engineering field. Applicants must complete two 15–20 page papers on worksite evaluations, one of which must deal with risk of back injury and one with risk of an upper extremity cumulative trauma disorder. Applicants must submit recommendations from three people who have received ergonomic evaluations from the applicant and submit proof of 25 workstation assessments.

For more information contact:
Roy Matheson and Associates
P.O. Box 492
Keene, NH 03431
USA
Phone: (800) 443-7690
Fax (603) 358-0116
E-mail: info@roymatheson.com
Internet: http://www.roymatheson.com

The Board of Certified Safety Professionals (BCSP) was organized in 1969 as a peer certification board to certify practitioners in the safety profession. Qualified applicants may be certified as a Certified Safety Professional (CSP). Individuals who hold the CSP designation may, by examination, seek specialty designation in ergonomics.

For more information contact:
Board of Certified Safety Professionals
208 Burwash Avenue
Savoy, IL 61874
USA
Phone: (217) 359-9263
Fax (217) 359-0055
E-mail: bcsp@bscp.com
Internet: http://www.bcsp.com

12.2 Professional Organizations in Ergonomics

P.O. Box 1369
Santa Monica, CA 90406-1369
USA
Phone (310) 394-1811
Fax (310) 394-2410
E-mail: 70732.2420@compuserve.com
Internet: http://www.hfes.org

The International Ergonomics Association (IEA)
Secretary General

Prof. ir. Pieter Rookmaaker
Netherlands Railways
SE ARBO/Ergonomics
P.O. Box 2286
3500 GG UTRECHT
The Netherlands
Phone +31-30-354455
Fax +31-30-2399456
E-mail: seaergo@knoware.nl
Internet: http://www.spd.louisville.edu/~ergonomics/international_ergonomics_association.html

The Ergonomics Society
Devonshire House
Devonshire Square
Loughborough LE11 3DW
United Kingdom
Phone +44 1509 234904
Fax +44 1509 234904
E-mail: ergosoc@ergonomics.org.uk
Internet: http://www.ergonomics.org.uk/

In addition to the above organizations, several professional societies have technical interest groups relating to ergonomics and include ergonomic issues at regional and national conferences. Probably the largest of these are the Institute of Industrial Engineers (IIE), The American Industrial Hygiene Association (AIHA), and The American Society of Safety Engineers (ASSE).

12.3 Ergonomics Related Publications

Table 54.18 lists related publications.

GLOSSARY

MSD. Musculoskeletal disorders

UECTD. Upper extremity cumulative trauma disorders

CTD. Cumulative trauma disorders

PWC. Physical work capacity

RMI. Repetitive motion injuries

M_{task}. Moment at a joint caused by external load

M_{cap}. Moment capability at a joint

E_{rest}. Energy consumption at rest

E_{task}. Energy consumption during task performance

Table 54.18. Ergonomics Related Publications

Publication	Publisher
Journals	
Applied Ergonomics	Elsevier Science Inc. 660 White Plains Road Tarrytown, NY 10591-5153 (212)989-5800
Applied Occupational and Environmental Hygiene	Elsevier Science Inc. 660 White Plains Road Tarrytown, NY 10591-5153 (212)989-5800
Clinical Biomechanics	Elsevier The Boulevard Langford Lane Kidlington, Oxford OX5 1GB UK
Ergonomics	Taylor & Francis 1900 Fronst Road Suite 101 Bristol, PA 19007 (215)785-5800 (215)785-5515 fax
Ergonomics in Design	Human Factors and Ergonomics Society PO Box 1369 Santa Monica, CA 90406-1369 (310)394-1811
Human Factors	Human Factors and Ergonomics Society 1124 Montana Ave, Suite B Santa Monica, CA 90403 (310)394-1811
International Journal of Human Factors in Manufacturing	John Wiley & Sons 605 Third Avenue New York, NY 10158-0012 (212)850-6645 (212)850-6021 fax subinfo@jwiley.com
International Journal of Industrial Ergonomics	Elsevier Science PO Box 211 1000 AE Amsterdam The Netherlands 0 20-485 36 42 0 20-485 35 98 fax Fulfilment-F@Elsevier.NL NY Office (212)633-3750

Table 54.18. (continued)

Publication	Publisher
Journal of Back and Musculoskeletal Rehabilitation	Elsevier Science Ireland Customer Relations Manager Bay 15K Shannon Industrial Estate Shannon, Co. Clare, Ireland (+353-61)471944 (+353-61)472144 fax Eirmailjournal@elsevier.ie
Journal of Biomechanical Engineering	ASME Department B99 22 Law Drive Box 2900 Fairfield, MJ 07007-2300 (800)843-2763 (201)882-1717 fax (201)882-5155 fax infocentral@asme.org
Journal of Biomechanics	Elsevier Science Inc. 660 White Plains Road Tarrytown, NY 10591-5153 (212)989-5800
Journal of Occupational Safety and Ergonomics	Ablex Publishing 355 Chestnut Street Norwood, NJ 07648 (201)767-8450 (201)767-6717 fax
Occupational Ergonomics	Chapman & Hall 115 5th Avenue New York, NY 10003 (212)780-6233 (212)260-1363 fax fogarty@chaphall.com
Work	Elsevier Science Ireland Customer Relations Manager Bay 15K Shannon Industrial Estate Shannon, Co. Clare, Ireland (+353-61)471944 (+353-61)472144 fax Eirmailjournal@elsevier.ie

Table 54.18. (continued)

Publication	Publisher
	Trade Publications
Material Handling Engineering	Penton Publishing 1100 Superior Avenue Cleveland, OH 44114-2543 (216)696-7000
Occupational Hazards	Penton Publishing 1100 Superior Avenue Cleveland, OH 44114-2543 (216)696-7000
Occupational Health & Safety	Stevens Publishing 222 Rosewood Drive Danvers, MA 01923 (508)750-8500
Professional Safety	ASSE 1800 E. Oakton St Des Plaines, IL 60018-2187 (708)692-4121
Safety & Health	National Safety Council 1121 Spring Lake Drive Itasca, IL 60143-3201
Workplace Ergonomics	Medical Publications Stevens Publishing 3630 I-35 Waco, TX 76706

ACKNOWLEDGMENTS

Copyright 1999, University of Utah. All rights reserved. This chapter based in part on software and intellectual property licensed from the University of Utah exclusive to ErgoWeb Inc., http://wwwergoweb.com.

BIBLIOGRAPHY

1. P. B. Gove, ed. *Webster's 3rd New Twentieth Century Dictionary Unabridged*, 3rd ed., Merriam-Webster, Inc., Springfield, 1993.
2. UAW-Ford National Joint Committee on Health and Safety, *Fitting Jobs to People: The UAW-Ford Ergonomics Process Implementation Guide*, University of Michigan, Ann Arbor, 1988.
3. Board of Certification in Professional Ergonomics, *Board of Certification in Professional Ergonomics Home Page*, 1996.
4. ErgoWeb, *OSHA Draft Ergonomic Protection Standard: Summary of Key Provisions*, 1995–1998.

5. T. N. Herington, and L. H. Morse, "Cumulative Trauma/Repetitive Motion Injuries," in T. N. Herington and L. H. Morse, eds., *Occupational Injuries: Evaluation, Management, and Prevention*, Mosby, St. Louis. 1995 pp. 333–345
6. B. Ramazzini, "Essai sur les Maladies de Artisans," *De Morbis Artificum*, Chapts. 1 and 52, 1777.
7. O. G. Edholm and K. F. H. Murrell, *The Ergonomics Research Society: A History, 1949–1970*. Wykeham Press, Winchester, 1973
8. P. M. Budnick, "Ergonomics," in G. L. Key, ed. *Industrial Therapy* Mosby, St. Louis, 1994, pp. 75–84.
9. Bureau of National Affairs, "Costs, Rewards of Redesign," in *Back Injuries: Costs, Causes, Cases & Prevention: A BNA Special Report*, Bureau of National Affairs: Washington, D.C. 1988 pp. 49–50.
10. D. T. Geras, C. D. Pepper, and S. H. Rodgers, "An Integrated Ergonomics Program at the Goodyear Tire and Rubber Company," in A. Mital, ed., *Advances in Industrial Ergonomics and Safety*, Taylor and Francis, Bristol, PA, 1989, pp. 21–28.
11. "Early intervention reduces injuries at Beech Aircraft company hygienist says," *Occupati. Safety and Health Reporter*, **22**(2), 61–62 (1992).
12. "Communication Drives Process at Siemens," *CTD News*, **6**(1), 5–6 (Jan. 1997).
13. "Silicon Graphics Melds High-and Low-Tech," *CTD News*, **6**(1), 7–8 (Jan. 1997).
14. J. Thaler, "The Sikorsky Success Story," *Workplace Ergonomics*, 22–25 (March/April 1996).
15. M. Gauf, ed. "AT&T Uses Cost-Conscious Program to Fight CTDs," *Ergonomics That Work: Case Studies of Companies Cutting Costs Through Ergonomics*, 1995, *CTD News*, 53–62 (1995).
16. J. L. Jegerlehner, "Workers' Participation Helps Reduce Cumulative Trauma Disorder Injuries," in A. C. Bittner and P. C. Champney, eds. *Advances in Industrial Ergonomics and Safety*, Taylor & Francis, Bristol, PA 1995, pp. 339–342.
17. D. Fehrenbacher, and J. L. Wick, "A Successful Back Injury Reduction Program," in A. C. Bittner and P. C. Champney, eds., *Advances in Industrial Ergonomics and Safety*, Taylor & Francis, Bristol, PA, 1995, pp. 347–350.
18. "Training a 'Limbsaver' at Newport News," *CTD News* **6**(1), 9–10 (Jan. 1997)
19. R. J. Jones, "Corporate Ergonomics Program of a Large Poultry Processor," *Am. Indust. Hyg. Assoc. J.*, **28** 132–137 (Feb. 1997).
20. D. S. Bloswick, *Technical Reports File*, University of Utah, Department of Mechanical Engineering: Salt Lake City, UT, 1992.
21. Bureau of National Affairs, "Warehousing & Distribution: Fleming Cos. Inc.," in *Back Injuries: Costs, Causes, Cases & Prevention: A BNA Special Report*, Bureau of National Affairs: Washington, DC 1988, pp. 80–82.
22. A. R. Longmate, and T. J. Hayes, "Making a Difference at Johnson & Johnson: Some Ergonomic Intervention Case Studies," *Ind. Management*, **32**(2), 27–30 (1990).
23. M. Gauf, ed., "Red Wing Shoes' Early Warning System," Ergonomics That Work: Case Studies of Companies Cutting Costs Through Ergonomics, *CTD News*, 79–83 (1995).
24. M. Gauf, ed., "Problem-Solving by Committee at General Seating, Ergonomics That Work: Case Studies of Companies Cutting Costs Through Ergonomics," *CTD News* 63–68 (1995).
25. B. P. Bernard, *DHHS (NIOSH) Publication No. 97-141, Musculoskeletal Disorders and Workplace Factors*, July, 1997.

26. M. Hagberg et al., *Work Related Musculoskeletal Disorders (WMSDs): A Reference Book for Prevention.* Taylor & Francis, Briston, PA, 1995.
27. National Academy of Sciences, *Work Related Musculoskeletal Disorders: A Review of the Evidence.* National Academy Press, Washington, DC, 1998.
28. F. Gerr, R. Letz, and P. J. Landrigan, "Upper-Extremity Musculoskeletal Disorders of Occupational Origin," *Annual Rev of Public Health,* **12** 543–566 (1991).
29. S. R. Stock, "Workplace Ergonomic Factors and the Development of Musculoskeletal Disorders of the Neck and Upper Limbs: A Meta-Analysis," *Am. J. Industr. Med.,* **19** 87–107 (1991).
30. E. Scalia, *Ergonomics: OSHA's Strange Campaign to Run American Business,* August 1994.
31. K. T. Hegmann and J. S. Moore, "Common Neuromusculoskeletal Disorders," in P. M. King, ed., *Sourcebook of Occupational Rehabilitation,* Plenum Press; New York, 1998.
32. M. I. Vender, M. L. Kasdan, and K. L. Trupper, "Upper Extremity Disorders: A Literature Review to Determine Work-Relatedness," *J. Hand Surgery,* **20**(A), 534–541 (1995).
33. G. D. Wright, "Points of View: The Failure of the "RSI" Concept," *Med. J. Australia,* 147 (Sept. 1987).
34. K. Ekberg et al., "Cross-Sectional Study of Risk Factors for Symptoms in the Neck and Shoulder Area," *Ergonomics,* **38**(5), 971–980 (1995)
35. P. Rey and A. Bousquet, "Compensation for Occupational Injuries and Diseases: Its Effect Upon Prevention at the Workplace," *Ergonomics,* **38**(3), 475–486 (1995).
36. S. J. Bigos et al., "Back Injuries in Industry: A Retrospective study. III. Employee-Related Factors," *Spine,* **11**(3) (1986).
37. D. M. Spenger et al., "Back Injuries in Industry: A Retrospective study. I. Overview and Cost Analysis," *Spine,* **11**(3) (1986).
38. P. A. Nathan et al., "Obesity as a Risk Factor for Slowing of Sensory Conduction of the Median Nerve in Industry: A Cross-Sectional and Longitudinal Study Involving 429 Workers," *J. Occupat. Med.* **34**(4) (April 1992).
39. L. J. Fuortes, et al., "Epidemiology of Back Injury in University Hospital Nurses From Review of Workers' Compensation Records and a Case-Control Survey," *J. Occupat. Med.* **36**(9) (Sept. 1994).
40. P. Radecki, "The Familial Occurrence of Carpal Tunnel Syndrome," *Muscle and Nerve,* **17**, 325–330 (March 1994).
41. F. D. Ashbury, "Occupational Repetitive Strain Injuries and Gender in Ontario, 1986 to 1991," *J. Occupat. Environ. Med.,* **37**(4) (April 1995).
42. M. J. Lord, K. W. Ha, and K. S. Song, "Stress Fractures of the Ribs in Golfers," *Am. J. Sports Med.* **24**(1) 118–122 (1996)
43. A. K.-F., M. Kugler, et al., "Muscular Imbalance and Shoulder Pain in Volleyball Attackers," *Brit. J. Sports Med.* **30**: 256–259 (1996).
44. J. S. O. Manninen, and M. Kallinen, "Low Back Pain and Other Overuse Injuries in a Group of Japanese Triathletes," *Brit. J. Sports Med.* **30**: 134–139 (1996).
45. J. W. Frymoyer, et al., "Epidemiologic Studies of Low-Back Pain," *Spine* **5**(5) (Sept./Oct. 1980).
46. U.S. Department of Labor, *Lost-Worktime Injuries and Illnesses: Characteristics and Resulting Time Away From Work, 1996,* U.S. Department of Labor, April 23, 1998.
47. J. P. Leigh et al., "Occupational Injury and Illness in the United States," *Arch. Internal Med.* 157 (July 28, 1997).

48. M. C. T. G. M. DeKrom et al., "Risk Factors for Carpal Tunnel Syndrome," *Am. J. Epidemiology* **132**(6) (1990).
49. V. J. Derebery, "Etiologies and Prevalence of Occupational Upper Extremity Injuries," in M. L. Kasdan, ed. *Occupational Hand & Upper Extremity Injuries and Diseases*, Hanley & Belfus, Philadelphia, PA, 1998.
50. C. M. Rystrom and W. W. Eversmann, Jr., "Cumulative Trauma Intervention in Industry: A Model Program for the Upper Extremity," in M. L. Kasdan, ed. *Occupational Hand & Upper Extremity Injuries and Diseases*, Hanley & Belfus, Philadelphia, PA, 1998.
51. A. J. Rubens, W. A. Oleckno, and L. Papaeliou, "Establishing Guidelines for the Identification of Occupational Injuries: A Systematic Appraisal," *J. Occupat. Environ. Med.*, **37**(2) (Feb. 1995).
52. E. M. McNeely, "Who's Counting Anyway? The Problem with Occupational Safety and Health Statistics," *J. Occupat. Med.*, **33**(10), 1071–1075 (Oct. 1991).
53. L. P. Hanrahan and M. B. Moll, "Injury Surveillance," *AJPH*, **79** (Suppl) (Dec. 1989).
54. B. S. Webster and S. H. Snook, "The Cost of 1989 Workers' Compensation Low Back Pain Claims," *Spine* **19**(10), 1111–1116 (1994).
55. G. B. J. Andersson et al., "Epidemiology and Cost," in *Occupational Low Back Pain: Assessment, Treatment, and Prevention*, Mosby-Year Book, St. Louis, MO, 1991.
56. B. S. Webster and S. H. Snook, "The Cost of Compensable Upper Extremity Cumulative Trauma Disorders," *J. Occupat. Med.*, **36**(7) (July 1994).
57. G. B. J. Andersson, L. J. Fine, and B. A. Silverstein, "Musculoskeletal Disorders," in B. S. Levy and D. H. Wegman, eds., *Occupational Health: Recognizing and Preventing Work-Related Disease*, Little, Brown, and Co., Boston, 1995.
58. T. J. Armstrong et al., "A Conceptual Model for Work-Related Neck and Upper-Limb Musculoskeletal Disorders," *Scand. J. Work Environ. Health*, **19**, 73–84 (1993).
59. W. Herzog, and B. Loitz, "Definitions and Comments," in B. M. Nigg and W. Herzog, eds., *Biomechanics of the Musclo-Skeletal System*, John Wiley & Sons, West Sussex, England, 1994.
60. D. P. Anagnos, A. C. Berger, and J. M. Kleinert, "Vascular Injuries and Disease," in M. L. Kasdan, ed., *Occupational Hand & Upper Extremity Injuries and Diseases*, Hanley & Belfus, Philadelphia, PA, 1998.
61. M. H. Pope, J. W. Frymoyer, and T. R. Lehmann, "Structure and Function of the Lumbar Spine," in M. H. Pope, et al., ed., *Occupational Low Back Pain: Assessment, Treatment, and Prevention*, Mosby-Year Book, St. Louis, MO, 1991.
62. A. A. White and M. M. Panjabi, *Clinical Biomechanics of the Spine*, Lippincott Co., Philadelphia, PA, 1978
63. P.-M. Brisson, M. Nordin, and C. Zetterberg, "The Musculoskeletal System and Occupational Syndromes," in W. N. Rom, ed., *Environmental and Occupational Medicine*, Little, Brown and Co: Boston, MA, 1992.
64. Ayoub, *Industrial* Ergonomics, Chapt. 1, 1985.
65. U.S. Department of Health and Human Services, *DHHS (NIOSH) Publication No. 81-122, Work Practices Guide for Manual Lifting*, March, 1981.
66. S. Pleasant, *Bodyspace: Anthropometry, Ergonomics and* PA, 1996.
67. R. Drillis, and R. Contini, *Body Segment Parameters*, Vocational Rehabilitation Administration, 1966.
68. Accredited Standards Committee Z365, *Control of Cumulative Trauma Disorders Working Draft*, National Safety Council, Itasca, IL, 1997.

69. T. J. Armstrong and D. B. Chaffin, "Carpal Tunnel Syndrome and Selected Personal Attributes," *J. Occupat. Med.* **21**(7) (July 1979).
70. B. A. Silverstein, L. J. Fine, and T. J. Armstrong, "Occupational Factors and Carpal Tunnel Syndrome," *Am. J. Ind. Med.* **11**, 343–358 (1987).
71. M. Hagberg, H. Morgenstern, and M. Kelsh, "Impact of Occupations and Job Tasks on the Prevalence of Carpal Tunnel Syndrome," *Scand. J. Work Environ. Health* **18** 337–345 (1992).
72. B. A. Silverstein, L. J. Fine, and D. Stetson, "Hand-Wrist Disorders Among Investment Casting Plant Workers," *J. Hand Surgery* **12A** (5, Part 2) (Sept. 1987).
73. S. Kiken et al., *NIOSH Investigation: Perdue Farms, Inc.*, National Institute for Occupational Safety and Health, Feb. 1990.
74. B. A. Silverstein and R. E. Hughes, "Upper Extremity Musculoskeletal Disorders at a Pulp and Paper Mill." *Appl. Ergonomics* **27**(3), 189–194 (1996).
75. J. A. Wells, et al., "Musculoskeletal Disorders Among Letter Carriers: A Comparison of Weight Carrying, Walking, and Sedentary Occupations," *J. Occupat. Med.* **25**(11) (Nov. 1983).
76. V. H. Hildebrandt, "Back Pain in the Working Population: Prevalence Rates in Dutch Trades and Professions," *Ergonomics* **38**(6), 1283–1298 (1995).
77. G. B. J. Andersson, "Epidemiologic Aspects on Low-Back Pain in Industry," *Spine* **6**(1), (Jan./Feb. 1981).
78. M. H. Pope, "Risk Indicators in Low Back Pain." *Ann. Med.* **21**, 387–392 (1989).
79. H., Riihimäki, S. Tola, T. Videman, and K. Hänninen, "Low-Back Pain and Occupation: A Cross-Sectional Questionnaire Study of Men in Machine Operating, Dynamic Physical Work, and Sedentary Work," *Spine* **14**(2), (1989).
80. H. Svensson and G. B. J. Andersson, "The Relationship of Low-Back Pain, Work History, Work Environment, and Stress: A Retrospective Cross-Sectional Study of 38- to 64-Year-Old Women," *Spine* **14**(5) (1989).
81. W. S. Marras, et al., "Biomechanical Risk Factors for Occupationally Related Low Back Disorders," *Ergonomics* **38**(2), 377–410 (1995).
82. J. H. Dobyns et al., "Bowler's Thumb: Diagnosis and Treatment," *J. Bone Joint Surgery* **45**-A(4) (June 1972).
83. M. P. F. Finelli, "Mononeuropathy of the Deep Palmar Branch of the Ulnar Nerve," *Arch. Neurology* **32** (Aug. 1975).
84. J. Hoffman and P. L. Hoffman, "Staple Gun Carpal Tunnel Syndrome," *J. Occupat. Med.*, **27**(11) (Nov. 1995).
85. G. M. Hägg, J. Öster, and S. Byström, "Forearm Muscular Load and Wrist Angle Among Automobile Assembly Line Workers in Relation to Symptoms," *Appl. Ergonomics* **28**(1), 41–47 (1997).
86. H. Sakakibara et al., "Overhead Work and Shoulder-Neck Pain in Orchard Farmers Harvesting Pears and Apples." *Ergonomics* **38**(4), 700–706 (1995).
87. A. Bjelle, M. Hagberg, and G. Michaelsson, "Occupational and Individual Factors in Acute Shoulder-Neck Disorders Among Industrial Workers," *Brit. J. Ind. Med.* **38**, 356–363 (1981).
88. K. Harms-Ringdahl and J. Ekholm, "Intensity and Character of Pain and Muscular Activity Levels Elicited by Maintained Extreme Flexion Position of the Lower-Cervical-Upper-Thoracic Spine," *Scand. J. Rehab. Med.* **18**, 117–126 (1986).
89. D. B. Chaffin, "Localized Muscle Fatigue: Definition and Measurement," *J. Occupat. Med.* **15**, 346–354 (1973).

90. W. S. Marras and R. W. Schoenmarklin, "Wrist Motions In Industry," *Ergonomics* **36**(4), 341–351 (1993).
91. T. Nilsson, Burström, and M. Hagberg, " Risk Assessment of Vibration Exposure and White Fingers Among Platers," *Intern. Arch. Occupat. Environ. Health* **61**, 473–481 (1989).
92. L. J. Cannon, E. J. Bernacki, and S. D. Walter, "Personal and Occupational Factors Associated with Carpal Tunnel Syndrome," *J. Occupat. Med.*, **38**(4) (April, 1981).
93. G. Wieslander et al., "Carpal Tunnel Syndrome (CTS) and Exposure to Vibration, Repetitive Wrist Movements, and Heavy Manual Work: A Case-Referent Study," *Brit. J. Ind. Med.* **46**, 43–47 (1989).
94. U.S. Department of Health and Human Services, *Criteria for a Recommended Standard: Occupational Exposure to Hand-Arm Vibration* National Institute for Occupational Safety and Health, Cincinnati, OH, 1989.
95. T. J. Armstrong et al., "Ergonomics and the Effects of Vibration in Hand-Intensive Work," *Scand. J. Work Environ. Health* **13**, 286–289 (1987).
96. H. C. Boshuizen, P. M. Bomgers, and C. T. J. Hulshof, "Self-Reported Back Pain in Tractor Drivers Exposed to Whole-Body Vibration," *Intern. Arch. Occupat. Environ. Health* **62**, 109–115 (1990).
97. H. C. Boshuizen, P. M. Bomgers, and C. T. J. Hulshof, "Self-Reported Back Pain in Fork-Lift Truck and Freight-Container Tractor Drivers Exposed to Whole-Body Vibration," *Spine* **17**(1) (1992).
98. T. Brendstrup and F. Biering-Sorenson, "Effect of Fork-Lift Truck Driving on Low-Back Trouble," *Scand. J. Work Environ. Health* 1987. **13** 445–452 (1987).
99. K. Miyashita ct al., "Symptoms of Construction Workers Exposed to Whole Body Vibration and Local Vibration," *Health* (1992).
100. M. Bovenzi and A. Zadini, "Self-Reported Low Back Symptoms in Urban Bus Drivers Exposed to Whole-Body Vibration," *Spine* **17**(9) (1992).
101. P. M. Bongers, et al., "Back Disorders in Crane Operators Explosed to Whole-Body Vibration," *Intern. Arch. Occupat. Environ. Health* **60** 129–137 (1988).
102. J. L. Kelsey and R. J. Hardy, "Driving of Motor Vehicles as a Risk Factor for Acute Hernated Lumbar Intervertebral Disc," *Am. J. Epidemiology* **102**(1) (1975).
103. H. Dupuis and G. Zerlett, "Whole-Body Vibration and Disorders of the Spine," *Inter. Arch. Occupat. Environ. Health* **59**, 323–336 (1987).
104. W. M. Keyserling et al., "A Checklist for Evaluating Ergonomic Risk Factors Associated with Upper Extremity Cumulative Trauma Disorders," *Ergonomics*, **36**(7), 807–831 (1993).
105. D. Ranney, R. Wells, and A. Moore, "Upper Limb Musculoskeletal Disorders in Highly Repetitive Industries: Precise Anatomical Physical Findings," *Ergonomics* **38**(7), 1408–1423 (1995).
106. T. J. Armstrong et al., "Some Histological Changes in Carpal Tunnel Contents and Their Biomechanical Implications," *J. Occupat. Med.* **26**(3) (March 1984).
107. I. Kuorinka and P. Koskinen, "Occupational Rheumatic Diseases and Upper Limb Strain in Manual Jobs in a Light Mechanical Industry," *Scand. J. Work Environ. Health* **5 (suppl 3)**, 39–47 (1979).
108. S. H. Kim and M. K. Chung, "Effects of Posture, Weight, and Frequency on Trunk Muscular Activity and Fatigue During Repetitive Lifting Tasks," *Ergonomics* **38**(5), 853–863 (1995).

109. C. Brisson et al., "Effect of Duration of Employment in Piecework on Severe Disability Among Female Garment Workers," *Scand. J. Work and Environ. Health* **15**, 329–334 (1989).
110. K. Ohlsson, R. Attewekk, and S. Skerfving, "Self-Reported Symptoms in the Neck and Upper Limbs of Female Assembly Workers," *Scand. J. Work Environ. Health* **15**, 75–80 (1989).
111. H. T. Zwahlen and C. C. Adams, in S. S. Asfour, ed. *"A Quantitative Work-Rest Model for VDT Work, in Trends in Ergonomics/Human Factors,"* Elsevier Science, North-Holland, 1987.
112. M. Hagberg and G. Sundelin, "Discomfort and Load on the Upper Trapezius Muscle When Operating a Word Processor," *Ergonomics* **29**, 1637–1645 (1986).
113. M. Waersted and R. H. Westgaard, "Working Hours as a Risk Factor in the Development of Musculoskeletal Complaints," *Ergonomics 34*, **3** 265–276 (1991).
114. E. Grandjean, *Fitting the Task to the Man: A Textbook of Occupational Ergonomics*, 4th ed., Taylor & Francis, Philadelphia, PA, 1988.
115. M. S. Sanders and E. J. McCormick, Human Factors in Engineering and Design, 7th ed. McGraw-Hill, New York, 1993.
116. D. R. Clark, "Workstation Evaluation and Design," in A. Bhattacharya and J. D. McGlothlin, eds. *Occupational Ergonomics,* Marcel Dekker, New York, 1996.
117. Office of the Federal Register National Archives and Records Administration, *Code of Federal Regulations*, **29**, Parts 1900–1910.999 Government Printing Office (1996).
118. A. L. Cohen et al., *DHHS (NIOSH) Publication No. 97-117, Elements of Ergonomics Programs: A Primer Based on Workplace Evaluations of Musculoskeletal Disorders*, March 1997.
119. S. H. Rodgers, "Job Evaluation in Worker Fitness Determination," *Occupat. Med. State of the Art Reviews*, 1998, June. **3**(2), 219–239 (June 1998).
120. McAtamney, and E. N. Corlett, "RULA: A Survey Method for the Investigation of Work-Related Upper Limb Disorders," *Appl. Ergonomics*, **24**(2), 91–99 (1993).
121. U.S. Department of Health and Human Services, *DHHS (NIOSH) Publication No. 94-110, Applications Manual for the Revised NIOSH Lifting Equation*, January 1994.
122. T. R. Waters et al., "Revised NIOSH Equation for the Design and Evaluation of Manual Lifting Tasks," *Ergonomics* **36**(7), 749–776 (1993).
123. University of Michigan, *Center for Ergonomics Three Dimensional Static Strength Prediction Program*, University of Michigan College of Engineering, Ann Arbor, MI, 1998.
124. ErgoWeb, ErgoWeb, Inc. Home Page, ErgoWeb, Inc., Salt Lake City, UT, 1995–1998.
125. B. Bink, "The Physical Work Capacity in Relation to Working Time and Age," *Ergonomics*, **5**(1), 25–28 (1962).
126. F. Bonjer, "Actual Energy Expenditure in Relation to the Physical Work Capacity," *Ergonomics* **5**(1), 29–31 (1962).
127. T. E. Bernard, *Metabolic Heat Assessment*, University of South Florida College of Health: Tampa, FL, May 1991.
128. D. B. Chaffin, "The Prediction of Physical Fatigue During Manual Labor," *J. Methods-Time Measurement* **11**(5), 25–31 (1966).
129. S. H. Snook and V. M. Ciriello, "The Design of Manual Handling Tasks: Revised Tables of Maximum Acceptable Weights and Forces," *Ergonomics*, **34**(9), 1197–1213 (1991).
130. Bureau of National Affairs, "Occupational Safety and Health Act of 1970," in *Occupat. Safety Health Reporter* (1993).
131. "Landis to Start Ergo Program," *CTD News*, 2–3 (July 1998).

132. Occupational Safety and Health Review Committee, *Secretary of Labor v. Dayton Tire, Bridgestone/Firestone. Decision and Order*. www.oshrc.gov/html 1998/93-3327.html, 1998.
133. U.S. Department of Labor, *OSHA Publication 3123, Ergonomics Program Management Guidelines for Meatpacking Plants*, 1991.
134. American National Standards Institute, *ANSI Technical Report for Machine Tools: Ergonomic Guidelines for the Design, Installation, and Use of Machine Tools, ANSI/B11/TR.1-1993*, American National Standards Institute, New York, 1993.
135. American Conference of Governmental Industrial Hygienists, *Threshold Limit Values for Chemical Substances and Physical Agents*, American Conference of Governmental Industrial Hygienists Publication Dept., Cincinnati, OH, 1998.
136. Equal Employment Opportunity Commission, *Technical Assistance Manual on the Employment Provisions (Title I) of the Americans With Disabilities Act. EEOC-M-1A*, 1992.
137. U.S. Department of Health and Safety, *DHHS (NIOSH) Publication No. 94-122, Workplace Use of Back Belts*. 1994.
138. J. F. Kraus et al., "Reduction of Acute Low Back Injuries by Use of Back Supports," *Intern. J. Occupat. Environ. Health*, 1996. **2**(4), 264–273 (1996).
139. "Ergonomics Certification Program Now Under Development Will Enhance Professional Status of Providers in Practicing Prevention," *Work Injury Management*, December, 1996. **5**(10), 3–4 (Dec. 1996).

CHAPTER FIFTY-FIVE

Occupational Safety

Donald S. Bloswick, Ph.D., PE, CPE and Richard Sesek, Ph.D., CSP

1 INTRODUCTION

The industrial revolution has had a profound impact on production methods in the United States. Since the last half of the nineteenth century, production processes have changed from craft shops to mechanized processes. These changes have expanded the quantity and variety of products available to the average American but have also expanded the types of hazards which are present in the industrial workplace. This has resulted in an increased need for and development of industrial workplace safety programs.

The Bureau of Labor Statistics (1) indicates that in 1996 there were nearly 1.9 million injuries and illnesses requiring time away from work. The National Safety Council (2) estimates that in 1997 there were 5,100 work related fatalities (including 2,100 involving motor vehicles) and 3,800,000 disabling injuries.

Safety professionals must realize that the incorporation of safety into the production process is not only "good for the worker" or "good public relations" but cost effective. Accident costs include medical expenses, worker's compensation costs, production down time, product loss, decreased employee morale, and administrative cost related to accident investigation so a reduction in accidents can increase the profitability of the enterprise.

1.1 Definitions

Definitions are adapted from Hammer (3) and Firenze (4, 5).

Accident. An event which is unexpected, interrupts the work process, and may result in injury or damage.

Patty's Industrial Hygiene, Fifth Edition, Volume 4. Edited by Robert L. Harris.
ISBN 0-471-29749-6

Hazard. A condition which has the potential to cause injury, damage to equipment or facilities, property or material loss, or an adverse affect on mission accomplishment.

Danger. Relative exposure to a hazard. For example, a rotating piece of machinery is a significant hazard but may present little danger if properly guarded.

Damage. Measure of the severity of injury or magnitude of loss which results from an uncontrolled hazard.

Risk. A combination of the probability of loss and magnitude of potential loss.

Safety. A measure of the absence of exposure to hazards. Safety is sometimes thought of as the control of hazards to an acceptable level.

2 SAFETY MANAGEMENT

2.1 Regulatory Environment

2.1.1 History

The first safety guideline (which, in fact, includes a penalty clause) was written by Hammurabi in approximately 1750 B.C. It indicates that "if a builder constructs a house for a man and does not make it firm and the house collapses and causes the death of its owner, the builder shall be put to death" (3).

Many of the first modern industrial safety standards were established by the American Society of Mechanical Engineers (ASME) and were adopted as a voluntary code for the development and maintenance of boilers. The first general industry safety conference was held in Milwaukee in 1912 and the following year the National Council for Industrial Safety was organized in New York. The organization was soon enlarged to include issues other than boiler safety and the name was changed to the National Safety Council, from which the American National Standards Institute (ANSI) evolved. ANSI assists with the development of consensus standards by serving as a locus of coordination by industry, labor, and government entities which have a common interest in a particular product or safety area. ANSI standards are not, in themselves, regulatory. In fact, at the beginning of ANSI standards it is noted that:

> An American National Standard implies a consensus of those substantially concerned with its scope and provisions. A National Standard is intended as a guide to aid the manufacturer, the consumer, and the general public. The existence of an American National Standard does not in any respect preclude anyone, whether he has approved the standard or not, from manufacturing, marketing, purchasing, or using products, processes, or procedure not conforming to the standard.

Many ANSI standards, however, have been adopted by state and federal agencies as the basis for government regulations or have been included, by reference, into regulatory documents and are, therefore, regulatory.

2.1.2 Workers Compensation

Workers Compensation programs are developed and operated by individual states. The first workers compensation laws began in 1911 in Wisconsin and New Jersey and by 1915, thirty states had passed some type of workers compensation legislation. The first workers compensation laws were declared a violation of the 14th amendment in that they required an employer to pay damages without regard to fault. This was considered as taking property "without due process of law". The U.S. Supreme Court ruled (White vs. the New York Central Railroad, 1917), that such taking of property was within the State's police powers because of the extreme degree of public interest involved. After this decision, the remaining states quickly passed their own workers compensation laws and at present there are 54 workers compensation laws (50 states, District of Columbia, Guam, Puerto Rico, and the Virgin Islands).

2.1.3 OSHA

The Walsh Healey public contract act of 1936 required that contracts in excess of $10,000 entered into by any United States government entity prohibit the use of materials "manufactured in working conditions which are unsanitary or dangerous to the health and safety of the employees." The Construction Safety Act of 1969 required that all federal, federally financed, or federally assisted projects in excess of $2,000 comply with established safety and health standards. The Federal Metal and Non-Metallic Mine Safety Act of 1966 and the Federal Coal Mine Health and Safety Act of 1969 focused attention on occupational safety and health issues, and the Federal Mine Safety and Health Act promulgated in 1977 established a single mine safety and health agency for all mining operations. This was enforced by the Mine Safety and Health Administration (MSHA) which is located within the Department of Labor.

On October 29, 1970, Richard Nixon signed Public Law 91-596, The Williams-Steiger Occupational Safety and Health Act of 1970, which became effective on April 28, 1971 (6). The stated purpose of this act is to:

> To assure safe and healthful working conditions for working men and women; by authorizing enforcement of the standards developed under the Act; by assisting and encouraging the States in their efforts to assure safe and healthful working conditions; by providing for research, information, education, and training in the field of occupational safety and health; and for other purposes.

The Occupational Safety and Health Administration (OSHA) is located within the Department of Labor and was established to enforce the OSHAct. OSHA is primarily an enforcement, or compliance, agency and is responsible to assure that the provisions of the OSHAct are met by private and public entities. The National Institute for Occupational Safety and Health (NIOSH) was established within the Department of Health, Education, and Welfare (currently the Department of Health and Human Services) by the OSHAct. The primary functions of NIOSH are to perform safety and health research, develop and establish recommended standards, and facilitate the education of personnel to implement the provisions of the OSHAct.

There are three separate sets of OSHA standards: General Industry (**29** *CFR* 1910), Construction (**29** *CFR* 1926), and Maritime Employment (**29** *CFR* 1915–1919). Summaries of major portions of the OSHA standards have been prepared by OSHA and are available in digest form. OSHA publication 2201 is a summary of the OSHA General Industry standards (7) and OSHA publication 2202 is a summary of the OSHA Construction standards (8).

Even though the three sets of standards noted above are extensive, they do not address all hazards. Hazards not specifically included in these OSHA standards can be addressed by Section 5 (A) of the OSHAct or the "General Duty Clause" which states that:

Each employer:

1. Shall furnish to each of his employees employment and a place of employment which are free from recognized hazards that are causing or are likely to cause death or serious physical harm to his employees.
2. Shall comply with occupational safety and health standards promulgated under this Act.

Historically one measure of OSHA's performance has been the number of citations issued. OHSA recently implemented a new initiative for achieving it's mission to "assure so far as possible every working man and woman in the nation safe and healthful working conditions" (9). These initiatives include a shift from enforcement to partnership with industries and an increased cooperation with business and labor in the promulgation of regulations. These initiatives also focus on the results of safety programs. One example of this effort is the Maine Top 200 Program which emphasizes partnership. Initially OSHA selected 200 companies in Maine which registered the highest workers compensation claims. While these firms accounted for approximately 1% of the state's employers, they accounted for approximately 45% of the workplace injuries, illnesses, and fatalities (9). Companies were given the choice between partnership with OSHA and the traditional enforcement process. Most companies chose partnerships which included comprehensive self-audits and quarterly reports to OSHA. In the eight years prior to the Maine Top 200 Program OSHA identified approximately 37,000 hazards at 1,316 work sites. In three years, under the Maine Top 200 Program, employers participating in the Maine Program identified over 180,000 workplace hazards and corrected 128,000 of them (10, 11).

Four governmental units have the primary responsibility to carry out the OSHAct:

1. *The Occupational Safety and Health Administration (OSHA)* is concerned with national, regional and administrative programs for developing, and ensuring compliance with, safety and health standards. It also trains OSHA personnel. U.S. Department of Labor, Department of Labor Building, 200 Constitution Avenue, NW., Washington, DC20210; (202) 219-8148. http://www.osha.gov
2. *The Occupational Safety and Health Review Commission (OSHRC)*, reviews citations and proposed penalties in enforcement actions contested by employers or employees. 1120 20th Street NW 9th floor, Washington DC, 20036; (202) 606-5100.
3. *The National Institute for Occupational Safety and Health (NIOSH)* is a research, training and education agency. U.S. Department of Health and Human Services,

4676 Columbia Parkway, Cincinnati, OH, 45226-1998; (800) 356-4674. *http://www.cdc.gov/niosh.*

4. *The Bureau of Labor Statistics (BLS)* conducts statistical surveys and establishes methods for acquiring injury and illness data. Division of Information Services, 2 Massachusetts Avenue NE Room 2860, Washington, DC, 20212; (202) 606-5886. *http://www.bls.gov*

2.1.4 Accident Statistics

OSHA requires that illnesses or injuries which involve lost time, restricted work, or medical care other than minor first aid be recorded in order to allow comparison between departments, facilities, or industries. A base of 200,000 person-hours is used to determine recordable incident rate. For example, if there were five recordable incidents in a year when there were 500,000 hours worked, the OSHA incident frequency rate would be:

$$\frac{(\text{Number of incidents}) \times (200{,}000)}{\text{Hours of employee exposure}}$$

or

$$\frac{(5) \times (200{,}000)}{500{,}000} = 2.0 \text{ incidents per 200,000 hours worked}$$

The same procedure may be used to determine the number of lost work day incidents. For example in the above example if three of the five incidents resulted in lost work days, the lost work day injury (or LWDI) rate would be:

$$\frac{(3) \times (200{,}000)}{500{,}000} = 1.2 \text{ LWD injury per 200,000 hours worked}$$

An injury severity rate can be calculated by using a measure which involves total days lost from the job or on restricted work activity. For example, if in the above example the three lost work day incidents resulted in a total of 30 days away from work or restricted work activity a measure of severity would be:

$$\frac{(30) \times (200{,}000)}{500{,}000} = 12 \text{ lost (or restricted) work days per 200,000 hours worked}$$

Injury and illness records must be kept by all employers subject to the OSHAct except those noted below and even these employers may be required to maintain records by specific state or local regulations or may be required by OSHA to maintain records for a specified time period.

1. Employers with a total of 10 or fewer full-time or part-time employees during the previous calendar year at all the employer's work sites;

2. Employers who conduct primary business in one of the Standard Industrial Classifications (SIC codes) specifically exempted by OSHA. OSHA Form 200 is the basic log or summary of occupational injuries and illnesses and contains information relating to the employer, employee name, employee work location, type of injury or illness, and the severity and outcome of the injury or illness. There are several publications which provide additional information on OSHA recordkeeping requirements (12). It should also be noted that OSHA has proposed a new, simplified recordkeeping system which is intended to simplify the forms and classification of injuries and illnesses (13). The reader is encouraged to monitor developments in this area. OHSA may also require that employers maintain records for specific programs such as lockout/tagout, hazard communication, confined space entry, bloodbourne pathogens, and so on. These recordkeeping requirements include issues such as training, logs, certification documents, and entry permits.

2.2 Accident Investigation

The two primary goals of accident investigation are to determine the cause(s) of the accident and to prevent the accident (or similar types of accidents) from happening again.

The accident investigation process involves the investigation of all factors relating to an accident in order to determine cause(s) of the accident and events leading up to the accident. The accident investigation process must consider personal, environmental, procedural, physical, and other factors. It is also important to determine if a violation of a safety standard, recommended practice, or policy was involved so that training and/or enforcement procedures can be modified/developed or an appropriate guideline developed.

Basic equipment required for accident investigations includes: personal protective equipment (PPE) necessary to enter the workplace and accident scene (i.e., safety glasses, respirators, safety shoes, hearing protection, etc.); photographic and/or sketching equipment; sample collection media (i.e., containers, bags, industrial hygiene sampling gear, etc.); measuring and data collection devices (i.e., stop watch, tape measure, compass, tape recorder, etc.); and any specific logs, forms, or checklists used by the company to ensure a thorough investigation.

The scene of an accident should first be "controlled" by eliminating the hazard that caused or contributed to the accident. While any hazards resulting from the accident should be minimized, the scene should be left as undisturbed as possible to facilitate investigation of the conditions that caused the accident. The investigation of "minor" or "near/miss accidents" can often provide information that can be used in the prevention of more serious accidents.

Accident investigation forms may be tailored to the needs of a specific organization or worksite. In general, however, forms should include the following information:

1. Information about the injured person (name, employee identification number, nature of injuries, etc.).
2. A narrative description of the accident including the activity of the injured person at the time of the accident and, if different, what was the person supposed to be doing.

3. List of standard operating procedures (SOPs) for the activity.
4. Training the employee had received.
5. An examination and description of tools, equipment, and personal protective equipment used.
6. Physical conditions and work environment existing at the time and place of the accident.
7. Past accident record of the employee and work area.
8. Corrective actions outlined to prevent reoccurrence and, if possible, responsible person and date correction is to be accomplished.

2.3 Training

It is frequently the case that personnel injured at work lack the information, knowledge, and skills required to protect themselves. OSHA (14) notes that various surveys by the Bureau of Labor Statistics have found that:

- 27% of 724 workers injured while working with scaffolds indicated that they had received no information on the use of the scaffold they were using,
- 71% of 868 workers suffering head injuries said that they had received no training regarding the use of hard hats.
- 61%of 554 workers hurt while maintaining equipment indicated that they had received no training on lockout procedures (procedures which control hazards during equipment maintenance).

OSHA requires specific training in many types of hazardous work. Safety training material is available from a number of commercial sources. In addition, information on safety and health training from government sources is available as noted below:

1. A catalog of "OSHA Publications and Audiovisual Materials" is available from the OSHA Publications Office, U.S. Department of Labor, 200 Constitution Avenue NW Room N3101, Washington, DC, 20210;(202) 523-9655; http://www.osha.gov/oshpubs
2. A six-part self-study program "Principles and Practices of Occupational Safety and Health" for first-line supervisors is available from the Government Printing Office, Superintendent of Documents, P.O. Box 371954, Pittsburgh, PA, 15250-7954; (202) 512-1800. http://www.access.gpo.gov/su_docs
3. A book "A Resource Guide to Worker Education Materials in Occupational Safety and Health" lists training materials and publications on safety and health which are available from some public and private organizations. It may be purchased from the Government Printing Office, Superintendent of Documents, P.O. Box 371954, Pittsburgh, A, 15250-7954; (202) 512-1800; http://www.access.gpo.gov/su_docs
4. The OSHA Training Institute provides training for industrial personnel in the areas of general industry and construction safety. Tuition is modest, but participants must

register in advance. Space may be limited as there is high demand for some classes. Information is available from The OSHA Training Institute, 1555 Times Drive, Des Plaines, IL, 60018-1548; (847) 297-4913. http://www.osha-slc.gov/Training

2.4 Safety Organizations

The National Safety Council (NSC) is the largest single safety organization in the world with over 18,000 corporate members representing more than 30 million employees. The NSC also has a network of over 70 local safety councils in the United States and Canada which are run by local boards of directors. The local safety councils sponsor safety seminars and courses and sell nationally sanctioned promotional items. Organized largely to support industrial safety, the National Safety Council is backed by a staff of 300 full time employees.

National Safety Council (NSC)
1121 Spring Lake Drive
Itasca, IL 60143-3201
(630) 285-1121
http://www.nsc.org

The American Society of Safety Engineers (ASSE) is the primary professional safety group in the United States with over 30,000 members. It has 148 chapters within 8 national regions. These chapters offer professional development and networking opportunities to individuals in the safety profession.

American Society of Safety Engineers (ASSE)
1800 E. Oakton St.
Des Plaines, IL 60018
(847) 699-2929
http://www.asse.org

2.5 License and Certification

There are several certificates or licenses that help one measure the experience and knowledge level of a safety person.

Every state has a registration board for professional engineers and many states have special PE *safety* registration. Registration requires proof of experience and passing rigorous written tests. The license must be renewed at set periods and the state monitors licensee performance.

National Society of Professional Engineers
1420 King St
Alexandria VA 22314-2794

(888) 285-6773
http://222.nspe.org

The best known certificate is the Certified Safety Professional (CSP). This reflects the equivalent of 15 years of full-time professional experience. For some experience and college degrees the requirement can be reduced, sometimes to a very few years. At least two thorough, monitored exams must be taken. Once designated a CSP, holders enter a continuance program. The program requires building credits through academic activities, further examinations, or professional pursuits. This ensures currency in the field.

Board of Certified Safety Professionals
208 Burwash Ave
Savoy IL 61874
(217) 359-9263
http://www.bcsp.com

The Board of Certified Safety Professionals and the American Board of Industrial Hygiene have joined to administer a certificate known as the Occupational Safety and Health Technologist. This illustrates the close relationship between safety and industrial hygiene practice at the technologist/technician level.

3 SAFETY ENGINEERING

3.1 Electrical Hazards

This material is adapted from Ref. 15. The reader is referred to Ref. 15 and Code of Federal Regulations, **29** *CFR* 1910, Subpart S: Electrical for further information.)

Major electrical concepts are current and voltage which are analogous to the volume and pressure of water flowing through a hose. The electrical power source is the "pump", the current (amperes) is the volume of "water" flowing, and the voltage (volts) is the "pressure". The resistance to the flow of electricity is measured in ohms and is a function of the type, cross sectional area, and temperature of the material through which the current is flowing.

3.1.1 Electrical Shocks

When the body is a part of the circuit through which electrical current is passing, electrical shock may result. The severity of this shock is generally a function of the amount of current flowing through the body, the duration of the current flow, and the path that the current flows through the body. Other factors which affect the severity are the frequency of the current, the phase of the heartbeat when the current flows through the body, and the general health of the individual. While there are no specific levels of current which will cause the same harm or sensation in all individuals, Table 55.1 indicates the general effect of a 60 cycle current passing from the hand to the foot for a duration of 1 second.

Table 55.1. Effects of Electrical Current in the Human Body[a]

Current	Reaction
1 milliampere	Perception level: Just a faint tingle.
5 milliamperes	Slight shock felt; not painful but disturbing. Average individual can let go. However, strong involuntary reactions to shocks in this range can lead to injuries.
6–25 milliamperes (women) 9–30 milliamperes (men)	Painful shock, muscular control is lost. This is called the freezing current or "let-go"[b] range.
50–150 milliamperes	Extreme pain, respiratory arrest, severe muscular contractions.[b] Individual cannot let go. Death is possible.
1–4.3 Amperes[c]	Ventricular fibrillation. (The rhythmic pumping action of the heart ceases.) Muscular contraction and nerve damage occur. Death is most likely.
10+ Amperes	Cardiac arrest, severe burns and probable death.

[a]From Ref. 15.
[b]If the extensor muscles are excited by the shock, the person may be thrown away from the circuit.
[c]Where shock durations involve longer exposure times (5 seconds or greater) and where only minimum threshold fibrillation currents are considered, theoretical values are often calculated to be as little as 1/10 the fibrillation values shown.

It should be noted that current above approximately 50 milliamperes is above the "let go" level and can cause respiratory arrest. This means that, if unaided, the individual may not be able to release the contact and (due to involuntary muscle contractions) may suffer severe injury or even death. An individual's exposure to electrical shock may also result in a secondary injury (sometimes called body reaction injury) due to involuntary muscle reaction and fall. In addition, injuries and property damage may also result from fires caused by electrical arcing or explosions.

3.1.2 Correcting Electrical Hazards

Electrical hazards can be minimized through the use of insulation, guarding, grounding, mechanical safeguards, and safe work practices. A conductor may be insulated by covering it with a material which has a very high resistance to electrical current flow. Materials often used as insulators are glass, mica, rubber, and plastic. OSHA requires that circuit conductors be insulated with a material suitable for the voltage and conditions (temperature, moisture, contaminants, etc.). Guarding is required for indoor electrical installations over 600 volts which are accessible to unqualified persons. These must be enclosed in a lock controlled area or metal case.

Guarding of live conductors of 50 volts or more may be done by:

1. Location in a room or enclosure accessible only to qualified personnel.
2. Installation of permanent, substantial screens or partitions to exclude unqualified personnel.

3. Location of the energized part on a balcony, gallery or elevated platform and configured to exclude unqualified personnel.
4. Elevation of at least 8 feet (2.4 meters) above the floor.

Grounding is a measure which provides a low resistance path to ground (often the earth) so that any current which "leaks" from the circuit will use this low resistance path and not the human body as a route to complete the circuit. An "equipment ground" provides a low resistance to ground for a specific tool or machine. For example, the case of an electrical handtool is often grounded so that if there is a short to the case, the ground wire will route the current to ground. A "service ground" or "system ground" consists of one wire grounded at the transformer and at the service entrance of the building.

Fuses and circuit breakers monitor the current flow in a circuit and open or break the circuit when this current flow exceeds a predetermined amount. Although these devices prevent or reduce direct damage to conductors and equipment, and reduce the potential for fires, they do little to protect operators from direct shock hazards. Ground fault circuit interrupters (GFCI) stop the electrical power in a circuit when there is a current loss (due to a short circuit for example) above some specified level. The GFCI senses when the current loss is small (0.005 amperes) and breaks the circuit within as little as 0.025 seconds (16).

Employee safe work practices to reduce electrical hazards include:

1. De-energize all electrical equipment before performing maintenance operations.
2. Use only electrical tools which are safe and properly maintained.
3. Follow applicable safety guidelines and use good judgement when working near energized lines.
4. Use only adequate, properly maintained personal protective equipment.
5. Keep a distance of at least 10 feet (3 meters) between overhead power lines and ladders, cranes, or other equipment.

An excellent resource in this area is *An Illustrated Guide for Electrical Safety* which is available through the American Society of Safety Engineers (17).

3.2 Fires and Explosions

This material is adapted from Refs. 3 and 18. The reader is referred to Refs. 3 and 18 and *Code of Federal Regulations*, **29** *CFR* 1910, Subpart L: Fire Protection for further information.

3.2.1 Fires

Fire initiation requires fuel and oxidizer (in proper position to support combustion), heat, and a chemical chain reaction. Generally fires pass through four stages. The *incipient* stage generates no visible smoke, flame, or significant heat but does generate combustion particles which can be detected by ionization fire detectors. During the *smoldering* stage the amount of combustion particles increases and photo electric fire detectors may be used to detect the smoke. When the point of ignition is reached, the *flame* stage begins and the resulting infrared energy may be detected by infrared fire detectors. The flame stage usually becomes the *heat* stage very quickly and this heat energy may be detected by thermal fire detectors. Fires are generally divided into one of the four classes based on the type of fuel involved:

Class A Solids such as coal, paper, and wood which produce char or glowing embers.

Class B Gasses and liquids which require vaporization for combustion.

Class C Class A or B fires which involve electrical equipment.

Class D Fires involving magnesium, aluminum, titanium, zirconium, or other easily oxidized metals.

The *flash* point of a liquid is the temperature above which it will give off sufficient vapor to momentarily ignite and burn but the burning stops as soon as the vapors are consumed. When the temperature rises above the *fire* point, the burning continues after ignition. Different organizations with an interest in fire safety specify whether liquids are flammable (flash point below a specified temperature) or combustible (flash point above that temperature). The temperatures which specify the flammable and combustible point for liquids are different for different organizations.

Once started, a fire can be stopped by: (*1*) removal or isolation of the fuel from the oxidizer (usually air), (*2*) increasing the volume of inert gas in the oxidizer, (*3*) quenching or cooling the heat of combustibles, or (*4*) inhibition of the combustion chain reaction. The type of extinguishing devise to be used is dependent on the class and size of the fire. The improper use of fire suppression equipment can result in employee injury or an increase in property loss so only those trained in the proper use of fire suppression equipment should be allowed to fight fires. Hand fire extinguishers are only effective during the initial stages of a fire and should be immediately accessible to trained personnel. Since the fire fighting capacity of an extinguisher drops to 40% in the hands of an untrained user (18), it is particularly important that all employees who are expected to operate fire extinguishers receive training in their proper use.

3.2.2 Explosions

An explosion is a sudden, violent release or expansion of a gas. It can be caused by a chemical reaction or a sudden release of compressed gas. Damage and injuries from the explosion may result from material fragments, body movements resulting from the shockwave, or the shockwave itself. Methods of preventing explosions and the damage resulting from explosions include:

1. Reducing the use of and storage requirements for, explosive or pressurized materials.
2. Isolating explosive materials and processes from people and valuable equipment through barriers or other protective features.
3. Use of pressure release valves or other such devices.

4. Use of suppressants which inhibit the chain reaction involved in chemical explosions.
5. Elimination and/or control of explosive dust concentrations.

3.3 Pressure

Pressurized liquids and gases are commonly utilized by industry. In some cases, they are a by-product of the manufacturing process and are not desirable. In either case, pressures must be controlled to prevent hazardous conditions. Some of the hazards associated with pressure include: pressure vessel rupture, propelled vessels and whipping hoses, projectile debris, high pressure cuts, air bubbles in the blood stream, and system malfunction and damage from high or low (vacuum) pressures.

Pressure vessels are subject to rupture when the fluid force within the container exceeds the container's strength. Both fired vessels, such as boilers, and unfired vessels, such as cylinders, can rupture. When failure is rapid, destruction of the container can produce fragmentation and even generate a shock wave. Such failures are violent and can result in serious injuries and property damage including fatalities and complete system or facility loss. Some sources of over-pressurization that could lead to rupture include heat from boilers and other intentionally generated sources, heat from the sun, heat from other processes or equipment, chemical reactions within the vessel, and pumps and compressors used to pressurize the vessel. Some safeguards that could minimize the likelihood of dangerous overpressurization include pressure relief valves, interlocked thermostats, regular preventive maintenance, and periodic system inspection.

Dynamic pressure hazards (propelling of the vessel itself) can occur when pressure is suddenly released, such as when a compressed gas cylinder valve is broken off. A compressed gas cylinder can generate 20–50 times its weight (200 lb) in thrust which can accelerate a 2500 psig cylinder to a velocity of 50 feet per second in 1/10 of a second (3). Because of the widespread use of compressed gas cylinders, safety engineers should be familiar with the significant hazards that they can present and employ appropriate precautionary measures. Personnel should be familiar with the hazards of both the materials within the cylinders and the cylinders themselves. Cylinders should be capped when not in use and secured to prevent tipping (19). Cylinders should be used only in well ventilated areas and stored away from incompatible materials and ignition and heat sources.

Pressurized gas and liquid lines could begin to whip if connections fail. Such accidents have resulted in serious injury and even fatalities. Lines capable of "whipping" should be restrained with weights, leashes, or clamps. Hoses should be as short as possible.

High pressure can cut like a knife. Employees should never use their hands or fingers to probe for a leak. A small amount of cloth (streamer) on a stick or a soapy solution can be used to detect such leaks. When maintenance must be performed on pressurized lines, the lines should be locked and tagged out as described in section 3.6 of this chapter.

Compressed air, particularly air used for cleaning, can project debris into a worker's skin or eyes. OSHA limits compressed air used for cleaning to 30 psi or less and requires that "chip guarding" be provided to prevent dirt and debris from bouncing back at the operator (20). If compressed air (or another gas) enters the body through cuts in the skin (or, when pressure is high enough, creates a cut in the skin), internal gas bubbles can cause health effects ranging from discomfort and swelling to death from stroke.

3.4 Hand and Power Tools

This material is adapted from Ref. 21. The reader is referred to Ref 21 and *Code of Federal Regulations*, **29** *CFR* 1910, Subpart P: Hand and Portable Powered Tools and Other Hand-Held Equipment for further information.)

The hazard from hand tools is often a result of misuse and improper maintenance. Saw blades and knives and other tools with sharp edges must be directed away from the user and other employees. Cutting tools such as knives and scissors must be kept sharp since dull tools require more force to use and are frequently more hazardous than sharp tools. When necessary mesh gloves, hand and arm guards, protective aprons, or other personal protective equipment should be worn when workers are using cutting tools. Spark resistant tools should be used in situations where sparks from traditional metal tools present an explosion hazard.

Energy sources for power tools include electric, pneumatic, liquid fuel, hydraulic, and powder activation. General power tool precautions include the following:

1. Never use the cord or hose to carry a tool.
2. Never yank the cord or hose to remove the connection from the receptacle.
3. Keep cords and hoses away from heat, oil, and sharp edges.
4. Disconnect tools when servicing and changing accessories.
5. Keep other workers or bystanders at a safe distance.
6. Secure the workpiece so two hands can be used to operate the tool.
7. Avoid accidental starting.
8. Maintain tools properly.
9. Maintain good footing and balance during tool use.
10. Do not wear loose clothing.
11. Properly tag damaged tools and repair or discard promptly.

3.5 Machine Guarding

This material is adapted from Ref. 22. The reader is referred to this document and *Code of Federal Regulations*, **29** *CFR* 1910, Subpart O: Machinery and Machine Guarding for further information.)

Guarding is required at the point of operation (punching, pressing, cutting, shaping, forming, etc.), around power transmission apparatus, and around other moving or hazardous parts.

Guards must prevent contact between the worker and dangerous moving parts and must be firmly attached to the machine and discourage tampering. In addition, guards must (*1*) provide protection from inadvertent insertion or dropping of foreign objects, (2) create no new hazards, (*3*) create minimum interference with job performance, and (*4*) allow safe maintenance.

Machine guarding methods may be grouped into five general classifications: (*1*) guards, (*2*) devices, (*3*) location and distance, (*4*) feeding and ejection mechanisms, and (*5*) miscellaneous aids.

Guards. Guards may be fixed, interlocked, adjustable or self-adjusting. A *fixed* guard generally encloses the entire point of operation and is a permanent part of the machine. This type of guard is often preferred because of its simplicity, however, care must be taken to allow safe access for inspection and maintenance. An *interlocked* guard is one which when opened or removed automatically shuts off the machine or prevents the machine from starting a cycle until the guard is replaced. *Adjustable* guards accommodate parts of different sizes or shapes. *Self adjusting* guards adjust automatically to the movement of the tool or stock which is being inserted.

Devices. Devices may be presence sensing, pullback, restraint, safety controls, or gates. *Presence sensing devices* monitor the operating area and interrupt the operating cycle when a foreign object is in the detection zone. Presence sensing devices must be "fail safe" so that a failure within the detection system prevents operation of the machine (23). *Pullback devices* attach to the operator's hands, wrists, or arms and withdraw the body member from the point of operation when the machine cycles. *Restraint devices* attach to the operator, usually to the wrists, and keep the operator's hands away from the point of operation altogether. *Safety controls* may be of several types. Safety trip controls (bars, tripwires, triprods) allow the operator to stop the machine quickly in an emergency situation. Two-hand controls require the concurrent use of both hands to cycle a machine and a two-hand trip requires the simultaneous use of two hands to start a machine cycle. A *gate* is a movable barrier which must be in place at the point of operation before the machine cycle can start.

Location/Distance. Guarding by location/distance locates the dangerous parts of the machine so that they are not accessible or do not present a hazard to the operator during normal operation. Dangerous parts may be located high enough to be out of any possible reach or workers may be protected by a wall or other barrier. Operator controls may be located at a safe distance from the machine if the operator is not required to tend the process.

Feeding/Ejection Mechanisms. Feeding and ejection mechanisms protect the operator by eliminating the need for the operator to place his or her hands in the point of operation, however, guards and devices may still be required to fully protect the operator.

Miscellaneous Aids. These include awareness barriers which remind workers of dangers or dangerous areas, protective shields, and tools which may be used to insert and remove stock from the point of operation.

A list to remind the reader of important machine guarding issues is shown in Figure 55.1.

3.6 Lockout/Tagout

All machines and equipment use or manipulate energy to perform work. That energy may be electrical, mechanical, hydraulic, pneumatic, chemical, or thermal. That energy may be

Machine Guarding Checklist

The following questions should help the interested reader determine the adequacy of safeguarding methods in the workplace.

Requirements for All Safeguards

	Yes	No
1. Do the safeguards provided meet the minimum OSHA requirements?	______	______
2. Do the safeguards prevent workers hands, arms, and other body parts from making contact with dangerous moving parts?	______	______
3. Are the safeguards firmly secured and not easily removable?	______	______
4. Do the safeguards ensure that no objects will fall into the moving parts?	______	______
5. Do the safeguards permit safe, comfortable, and relatively easy operation of the machine?	______	______
6. Can the machine be oiled without removing the safeguard?	______	______
7. Is there a system for shutting down the machinery before safeguards are removed?	______	______
8. Can the existing safeguards be improved?	______	______

Mechanical Hazards

The point of operation:

	Yes	No
1. Is there a point-of-operation safeguard provided for the machine?	______	______
2. Does it keep the operators hands, fingers, body out of the danger area?	______	______
3. Is there evidence that the safeguards have been tampered with or removed?	______	______
4. Could you suggest a more practical, effective safeguard?	______	______
5. Could changes be made on the machine to eliminate the point-of-operation hazard entirely?	______	______

Power transmission apparatus:

	Yes	No
1. Are there any unguarded gears, sprockets, pulleys, or flywheels on the apparatus?	______	______
2. Are there any exposed belts or chain drives?	______	______
3. Are there any exposed set screws, key ways, collars, etc.?	______	______
4. Are starting and stopping controls within easy reach of the operator?	______	______
5. If there is more than one operator, are separate controls provided?	______	______

Other moving parts:

	Yes	No
1. Are safeguards provided for all hazardous moving parts of the machine including auxiliary parts?	______	______

Nonmechanical Hazards

	Yes	No
1. Have appropriate measures been taken to safeguard workers against noise hazards?	______	______
1. Have special guards, enclosures, or personal protective equipment been provided, where necessary, to protect workers from exposure to harmful substances used in machine operation?	______	______

Figure 55.1. Important machine guarding issues. Reprinted from Ref. 22.

Electric Hazards

1. Is the machine installed in accordance with National Fire Protection Association and National Electric Code requirements?	______	______
2. Are there loose conduit fittings?	______	______
3. Is the machine properly grounded?	______	______
4. Is the power supply correctly fused and protected?	______	______
5. Do workers occasionally receive minor shocks while operating any of the machines?	______	______

Training

1. Do operators and maintenance workers have the necessary training in how to use safeguards and why?	______	______
2. Have operators and maintenance workers been trained in where the safeguards are located, how they provide protection, and what hazards they protect against?	______	______
3. Have operators and maintenance workers been trained in how and under what circumstances guards can be removed?	______	______
4. Have workers been trained in the procedures to follow if they notice guards that are damaged, missing, or inadequate?	______	______

Protective Equipment and Power Clothing

1. Is protective equipment required?	______	______
2. If protective equipment is required, is it appropriate for the job, in good condition, kept clean and sanitary, and stored carefully when not in use?	______	______
3. Is the operator dressed safety for the job (i.e., no loose-fitting clothing or jewelry)?	______	______

Machinery Maintenance and Repair

1. Have maintenance workers received up-to-date instruction on the machines they service?	______	______
2. Do maintenance workers lock out the machine from its power sources before beginning repairs?	______	______
3. Where several maintenance persons work on the same machine, are multiple lockout devices used?	______	______
4. Do maintenance persons use appropriate and safe equipment in their repair work?	______	______
5. Is the maintenance equipment itself properly guarded?	______	______
6. Are maintenance and serving workers trained in the requirements of 29 CFR 1910.147, lockout/tagout hazard, and do the procedures for lockout/tagout exist before they attempt their tasks?	______	______

Figure 55.1. Continued.

stored, even after the equipment has been shut off. For example, energy may be stored in springs, pressurized air or liquid, electrical capacitors, or suspended weights, to name a few. An employee can be seriously injured if this energy is accidentally discharged during servicing or maintenance.

OSHA Standard **29** *CFR*1910.147 "The control of hazardous energy (lockout/tagout)." covers the servicing and maintenance of machines and equipment in which the unexpected energization or start up of the machines or equipment or release of stored energy could cause injury to employees. Please refer to the actual standard for complete information (24).

Injuries that occur during servicing are often very serious since the victim is typically in a vulnerable position (i.e., head, arm, or complete body inside a piece of equipment).

Lockout/tagout means physically locking and tagging the controls to all power sources (i.e., circuit breakers, switches, valves, etc.) in the "OFF" (deenergized) position. The purpose of lockout/tagout is to prevent the accidental release of stored energy by locking all energy sources and controls in the "OFF" or deenergized position and discharging all potential energy sources.

This standard does not apply to normal production. However, servicing and maintenance which takes place during normal production operations are covered by this standard whenever: an employee is required to remove or bypass a guard or other safety device; or an employee is required to place any part of his or her body into a point of operation or other hazardous location.

When alternative means which provide effective protection are employed, minor tool changes and adjustments, or other minor servicing activities, which take place during normal production operations and are routine, repetitive, and integral to the use of the equipment for production, do not require lockout/tagout. Alternative means of protection may include combinations of interlocks, presence sensing devices, physical barricades that prevent machine actuation, and special devices for removing items from the point of operation (such as extended tongs).

Work on cord and plug connected electric equipment for which exposure is controlled by the unplugging of the equipment from the energy source is also exempted, provided the plug remains under the exclusive control of the employee performing the servicing.

Where lockout/tagout is required, employers must implement an "Energy Control Program" (lockout/tagout program) that includes: written energy control procedures, employee training, and periodic inspections of the lockout program and methods (24). The program must ensure that before employees perform any servicing or maintenance on a machine or equipment where the unexpected energizing, startup, or release of stored energy could occur, the machine or equipment will be isolated from all energy sources, rendered inoperative, and secured (locked out) to prevent accumulation of energy.

The OSHA Lockout/Tagout standard allows tags to be used in lieu of locks whenever "the employer can demonstrate that the tagout program will provide a level of safety equivalent to that obtained by using a lockout program." In reality, tags alone never provide a level of safety equivalent to the use of both locks and tags and tagging alone is not recommended.

OSHA requires written energy control procedures to contain the following (24):

- A specific statement of the intended use of the procedure.
- Specific procedural steps for shutting down, isolating, blocking and securing machines or equipment to control hazardous energy.
- Specific procedural steps for the placement, removal and transfer of lockout devices or tagout devices and the responsibility for them.
- Specific requirements for testing a machine or equipment to determine and verify the effectiveness of lockout devices, tagout devices, and other energy control measures.

The locks, tags, chains, and other hardware provided for isolating, equipment from energy sources must be unique (singularly identifiable) and may be used only for lockout and tagout (i.e., lockout locks should not be used to secure lockers or tool boxes). Further, lockout devices must be durable, standardized, substantial, and identifiable. Tagout devices must clearly warn against hazardous conditions if the machine or equipment is energized and must include a legend such as the following: "Do Not Start. Do Not Energize. Do Not Operate."

Employers must provide training to all employees that may be affected by lockout/tagout procedures to ensure that the purpose and function of the energy control program are understood, particularly the prohibition relating to attempts to restart or reenergize equipment which is locked out. Employees authorized to actually perform lockout operations must receive more involved training which includes: training in the recognition of applicable hazardous energy sources, the type and magnitude of the energy available in the workplace, and the methods and means necessary for energy isolation and control.

The established procedures for the application of energy control (lockout procedures) must contain the following elements:

- Preparation for shutdown, notification of affected employees.
- Machine or equipment shutdown.
- Machine or equipment isolation.
- Lockout/tagout device application.
- The relieving of stored energy.
- Verification of isolation.

If reaccumulation of stored energy to hazardous levels is possible, verification of isolation must be continued until the servicing or maintenance is completed.

Similarly, before lockout and tagout devices are removed and energy is restored to the machine or equipment, procedures must be followed and actions taken by the authorized employee(s) to ensure the following:

- Inspection of machine or equipment.
- Employee notification.
- Removal of lockout/tagout devices.

Each employee must place his or her locks on all lockout devices. When servicing is performed by a crew, department or other group, they may utilize a procedure which affords the employees a level of protection equivalent to that provided by the implementation of personal lockout devices on all energy sources. Each authorized employee may instead affix a personal lockout device to a group lockout device, such as a group lockbox or comparable mechanism, prior to beginning work. The group lockbox contains the keys to all of the locks used to secure the energy isolating devices and cannot be opened to retrieve these keys until all employees have removed their locks from the lockbox. Before beginning work, each employee should personally inspect each lockout device to verify that the equipment has been properly isolated. This technique can simplify lockout when there are many employees and many energy sources requiring lockout devices. For example, five employees locking out five energy sources would need 25 locks, while using a lockbox would require only 10 locks (five locks for the energy sources and one lock each for the lockbox). If maintenance activities span multiple work shifts, specific procedures must be utilized during the shift change to ensure the continuity of lockout protection.

3.7 Confined Space

A confined space is a space that is large enough and so configured that an employee can bodily enter and perform work, has limited or restricted means for entry or exit, and is not designed for continuous occupancy

Confined spaces include: storage tanks, bins, boilers, ventilation and exhaust ducts, pits, manholes, vats, reactor vessels, silos, pipelines, and septic tanks and sewers. Even open-topped spaces, such as dikes or embankments (capable of collecting and containing gasses) can be considered confined spaces.

A "nonpermit-required" confined space is a space that meets the above definition but, after evaluation, does not contain or have the potential to contain any hazard capable of causing death or serious physical harm. A "permit-required" confined space is a confined space that contains or has the potential to contain one or more of the following hazards:

- Atmospheric hazard.
- Physical hazard.
- Configuration hazard.
- Any other recognized serious safety or health hazard.

OSHA standard **29***CFR*1910.146 contains the specific requirements for the entry of confined spaces (25). In addition to the inherent dangers, confined space entry operations are more hazardous than similar operations performed in open areas because difficulty with communication and entry and egress may greatly impact preventive measures and rescue operations. This risk cannot be overstated, particularly given the fact that 36% of confined space entry fatalities are would-be rescuers (26).

Atmospheric hazards include: oxygen deficiency (<19.5%) or oxygen enrichment (>23.5%); the presence of flammable or combustible gases; the presence of toxic gases or materials; and biological hazards. In a study by NIOSH, nearly 80% of confined space incidents involved hazardous atmospheres (26). Often these hazards are not visible or otherwise detectable and victims are overcome before they even realize they are in danger. Likewise, would-be rescuers cannot see the dangerous atmosphere, only their collapsed co-workers.

Physical hazards that may exist in a confined space include: electrical shock hazards; grinding, crushing, and mixing mechanisms; contact with hazardous chemicals (corrosives); and environmental extremes (high temperatures, pressures, noises, etc.). Physical hazards such as these require that lockout/tagout procedures be followed as described in Section 3.6 of this Chapter.

Configuration hazards include: the presence of an engulfing or drowning medium (such as grain in a silo); sloping, inwardly converging, or tapering floors and walls that could trap or asphyxiate; and fall hazards. These configuration hazards also complicate rescue efforts.

Other potential hazards include rodents, snakes, spiders, poor visibility, noise, etc.

OSHA requires employers with permit-required confined spaces to establish a confined space program with the following basic elements (25):

- Written program and record keeping.
- Identification and evaluation of confined spaces.
- Written entry permit system.
- Air monitoring (initial and continuous).
- Personnel training.
- Personal protective equipment (PPE) selection and use.
- Emergency response preparation.

Confined space hazards may be caused by previously stored products or chemicals; unstable materials (engulfing/drowning hazard); or chemical reactions, including manufacturing processes, incompatible stored materials, the drying of paints, oxidation/reduction, cleaning with acids or solvents, the rusting of metals, or rotting, decomposition, or fermentation.

The operations performed within the space can also create hazardous conditions.

For example, welding, painting, scraping, sand blasting, and sludge removal can all create hazardous conditions.

Safe entry procedures must be developed for permit-required spaces before personnel are allowed to enter these spaces. It is important to anticipate potential emergencies and consider emergency response and rescue procedures in employee training. The number of entrants allowed into the space should be minimized. All lockout/tagout (isolation) provisions, as well as preparations for entry and emergency response, should be verified prior to entry. If emergency assistance or rescue is deemed necessary, entry conditions should be reevaluated prior to entry. A formal entry permit must be completed prior to entry. The entry permit is an authorization and approval in writing that specifies the location and type of work to be done, certifies that all existing hazards have been evaluated by a qualified person, and that necessary protective measures have been taken to ensure the safety of each worker.

Confined space entry training must emphasize preventing hazardous conditions during confined-space entries and preparing for emergency response, both self-help and emergency rescue. Training should emphasize that serious emergencies can develop quickly in confined spaces. Trained personnel, preplanned responses, and preplaced (standby) rescue equipment are important for immediate and effective action when emergencies occur. Training must be conducted annually for confined-space-entry and emergency rescue. Support courses such as First Aid and CPR should also be conducted to keep worker certifications current.

All permit-required confined spaces must be tested by a qualified person before entry to determine whether the confined space atmosphere is safe for entry. Tests should be made for oxygen level, flammability, and known or suspected toxic substances. The confined space should be continuously monitored to determine whether the atmosphere has changed while the work is being performed.

The atmosphere in a confined space should always be tested in the following order: oxygen level, then flammability (% of LEL), and then toxic gases.

The oxygen content of the air in a confined space is the first and most important constituent to measure before entry is made. The acceptable range of oxygen is between 19.5% and 23.5%. This content is measured before flammability is tested because rich mixtures of flammable gases or vapors may give erroneous measurement results. For example, a mixture of 90% methane and 10% air (2% oxygen) may test nonflammable because there is not enough oxygen to support the combustion process in flammability meters. This mixture will not support life and may soon become explosive if ventilation is provided to the space. Before entry, spaces must be ventilated until both oxygen content and flammability are acceptable. Tests for toxic contaminants must be specific for the target toxin. It is important to know the history of the confined space so proper tests can be performed. All possible contaminants that could be in the confined space should be identified.

3.8 Material Handling and Storage

Material handling and storage takes place on some level in every business. The mishandling of materials is considered the single largest cause of accidents and injuries in the workplace (27).

Material handling is estimated to account (at least in part) for 20 to 45% of all occupational accidents (28).

Handling materials manually is typically the slowest and least efficient method. More important, it is often the most hazardous method as well. Back injuries are among the most common material handling injuries, with the total annual cost estimated to be between \$50 and \$100 billion per year (29). The total cost goes up considerably when combined with other types of material handling injuries, such as overexertions, upper extremity cumulative trauma disorders, slips/trips/falls, and accidents involving machines. Mechanized methods typically result in fewer accidents and injuries, but they create a new set of hazards which often have more severe consequences.

Several areas of material handling are covered directly by the Occupational Safety and Health Act (OSH Act). For example, OSHA contains specific regulations regarding the use of industrial trucks (fork lifts), cranes, derricks, hoists, and other mechanical equipment.

The best way to reduce material handling injuries is to eliminate unnecessary handling. This will increase operational efficiency while decreasing employee exposure to manual handling risks. When unnecessary handling has been eliminated, the goal should be to mechanize as much of the necessary handling as practical and feasible with priority given to the those strenuous tasks that present the biggest ergonomic risk to workers.

3.8.1 Manual Handling

When manual handling and lifting of materials is necessary, ergonomic principles should be employed to ensure that the manual handling is performed as safely as possible. The consideration of ergonomics in the design of manual material handling tasks can result in reduced physical stress and injury costs. Please refer to Chapter 57 of this volume for more information on ergonomics. General lifting guidelines to reduce musculoskeletal stresses during manual material handling include:

- Keep the load as close to the body as possible.
- Use the most comfortable posture.
- Lift slowly and evenly (Do not jerk).
- Avoid twisting the back while lifting or exerting.
- Securely grip the load.
- Use lifting aids and get help when necessary.

General design considerations to reduce musculoskeletal stresses during manual material handling include:

- Position strenuous lifts near waist level.
- Machines should facilitate loading and unloading.
- Material handling aids that are accessable and usable.
- Optomizing containers (size, handles, etc.) for the materials to be handled.
- Walking/working surfaces clean and non-slip.
- Acceptable task frequencies.

Back belts are often issued in the hope that they will reduce back injury rates among employees who manually handle materials. One study has indicated a positive effect of back belt use (30), but most studies fail to consistently show any clear biomechanical advantage to using back belts (31). There may be times when a back belt could be advantageous, particularly when prescribed by a physician, but at this time there is not sufficient proof to warrant supplying them to all workers prophylactically.

3.8.2 Equipment

The size, weight, and shape of the material moved dictates what types of equipment should be used to move it safely. While most pieces of equipment have specific requirements for their safe operation and use, there are some basic rules that apply to each broad class or type of equipment. Nearly all materials handling equipment requires specialized training for safe operation. Some types of equipment require extended courses and certification. Some representative types of equipment, including industrial trucks, cranes, hoisting equipment, conveyors, and automated material handling systems are presented. These brief overviews are intended only to expose the reader to some of the critical safety considerations involved in materials handling equipment and some of the basic requirements for their safe use.

3.8.3 Manual Hand Trucks

One of the most common devices used to assist in manual material handling is the hand truck or dolly. Hand trucks can be purchased or fabricated for objects of various sizes and shapes, but should only be used for the purpose for which they were designed. For example, a hand truck designed for boxes may not be appropriate for lifting and moving drums or cylinders. Common two-wheeled hand trucks appear easy to handle but can present problems for the unskilled user. The guidelines presented below can be helpful for hand truck users (32):

- Keep the load's center of gravity as low as possible.
- Place the load such that the weight is carried by the axle and not the handles. The truck should "carry" the load with the operator only balancing and pushing.
- Stabilize the load to prevent shifting, slipping, or tipping. For bulky or awkward loads strap, chain, or otherwise secure the load to the truck.
- Load the truck to a height that allows a clear view when traveling.
- Avoid walking backwards with a hand truck.
- When moving up or down an incline, keep the truck down slope from you.
- Move the truck at a safe speed that can always be quickly and safely stopped.
- Use hand trucks only for the purpose for which they were designed.

3.8.4 Powered Industrial Trucks

The powered industrial truck or "fork lift" is one of the most common pieces of material handling equipment. Its power and versatility allow its use in a variety of work environments. The power and weight of these vehicles is often underestimated since they are highly maneuverable and relatively easy to control. Every year, forklifts are involved in approximately 75–100 fatal injuries and an estimated 13,000 workers' compensation claims (33).

Many industrial trucks are powered by batteries. The most common batteries are lead and nickel–iron containing corrosive chemical solutions (electrolyte). Hydrogen and oxygen gases are given off during the charging process. Under certain conditions, these gases can reach explosive concentrations (34). The charging process must be conducted only in designated areas that are equipped with facilities for flushing and neutralizing spilled electrolyte, proper fire protection, and adequate ventilation to prevent the build up of hazardous air contaminants. In addition, whenever there is the possibility that the eyes or body may be exposed to the corrosive electrolyte solutions, eye washes and showers must be provided for quick flushing of the eyes and drenching of the body (35).

OSHA permits only trained and authorized operators to operate industrial trucks. Training should include practical proficiency testing and should be tailored to the specific facility in which it will be used. Videos and training booklets are good supplements to a training program, but should not be used alone and cannot take the place of hands-on training.

Following is a list of basic industrial truck safety rules:

- Employees should never be permitted to stand under the elevated portion of the truck whether loaded or unloaded.
- Riders should not be permitted on industrial trucks unless the truck is designed for it, has a safe place provided for riding, and riding of trucks is authorized.
- Unattended trucks should have the load engaging means (forks or lifting attachment) fully lowered, the brakes set, and the power off. Wheels should be blocked if the truck is on an incline.
- High lift trucks must be provided with overhead guards for protection against falling objects.
- Trucks should not be used for lifting personnel unless they are specially fitted and equipped for doing so.
- Trucks should abide by all traffic regulations, such as plant speed limits and safe following distances (approximately three truck lengths).
- Other trucks should not be passed at intersections, blind spots, or other dangerous locations.
- Operators should slow and sound their horns where aisles cross and where vision is obstructed.
- Operators should travel with the load trailing whenever the load obstructs vision.
- Only stable, safe loads should be handled. The rated capacity should never be exceeded and must be adjusted downward for odd shapes and sizes of loads.

For more information on industrial trucks and their safe operation refer to Refs 36 and 37.

3.8.5 Powered Hand Trucks

A powered hand truck is a self-propelled truck that is controlled by a walking operator or standing rider. These are often referred to as "pallet jacks" or "mules." The OSHA standard for industrial trucks, **29** *CFR* 1910.178, specifically includes powered hand trucks (37) and most of the principles that apply to rider-operated powered industrial trucks also apply to powered hand trucks.

In addition, all powered hand trucks should be equipped so that when the handle is fully raised or lowered the brakes are applied (38). When released, the handle should automatically return to a position that activates the brakes. Hand guards can be installed around the controls to prevent the hands and/or controls from contacting obstacles in tight quarters.

3.8.6 Front-End Loaders

Just as the fork lift truck is the jack-of-all-trades in the manufacturing environment, so is the front-end loader to the construction and mining industries. These multipurpose vehicles are most commonly involved in accidents while backing up, but often the most serious accidents occur when unloaded machines are driven at high speed (39). These vehicles are typically much less stable when traveling empty and, at high speeds, have a tendency to bounce and weave. Operators must be aware of this limitation. Care must also be exercised when going up or down slopes or when traversing uneven ground, particularly with the load carried high where it can create a lever arm and topple the loader. As with fork trucks, the bucket should not be left unattended in the raised position nor should it be used as a work platform. The rules of operation and handling that apply to industrial trucks should also be applied to front-end loaders.

3.8.7 Loading Highway Trucks and Railroad Cars

Most facilities unload their raw materials from or load their finished goods onto highway trucks and railroad cars.

One of the most serious accidents that can occur while loading or unloading highway trucks or trailers results from movement or "creeping" of the tractor and trailer. If a trailer moves forward due to the weight of fork lift and load, the fork lift can fall between the loading dock and trailer. To prevent movement of the tractor and trailer, the truck should be left in gear, the tractor's brakes should be set, and wheel "chocks" or blocks should be used to secure the rear wheels of the trailer. "Dock locks" or other recognized positive means for preventing the trailer from moving can be used in addition to or in lieu of wheel chocks. If the trailer is loaded or unloaded without the tractor, jacks should be placed behind the rear wheels to prevent lifting of the front of the trailer during entry and, if necessary, near the trailer support for extra stability.

Similarly, wheel stops or other recognized protection should be in place to prevent railroad cars from shifting or moving during loading and unloading with industrial trucks. Positive protection and appropriate warning placards (typically a blue flag) should be in place to prevent the cars from being moved by a switch engine during loading or unloading (40).

3.8.8 Cranes and Lifting Devices

A crane is simply a machine for lifting, lowering, and horizontal movement of loads. Cranes are not designed for transporting loads great distances and are safe only when used within fairly narrow operating parameters. Cranes and lifting devices should not be used to slide or pull materials horizontally. Items should be lifted, then moved. Nonvertical lifting can lead to "swinging" of the load resulting in an accident. Also, nonvertically loaded cranes can become damaged and fail at significantly lower loads than their rated capacities. Because of the potential for injury, property damage, and product loss, only persons thoroughly familiar with crane functions and operations should be involved with their use. OSHA standards (41) contain some general requirements for cranes which include the following items:

- Rated loads (lifting capacities) must be clearly marked on each side of the crane and on each lifting device used with that crane.
- Only employees designated, qualified, and authorized may operate cranes.
- Cranes must be thoroughly tested and inspected, including a load test, prior to use and after alteration.

- A preventive maintenance program based on the manufacturers recommendations must be established and followed.
- Cranes must be inspected daily before use and, based on level of service and operating conditions, must be periodically inspected more thoroughly.

All crane operators must be thoroughly trained. Qualifications for operators vary with type of crane, but all limit certification to the specific type of equipment on which the operator is trained and has been examined (42).

All cranes and hoists must be inspected daily by operators prior to use. These daily inspections should include both visual and functional tests. Abnormalities should be promptly reported to qualified maintenance personnel. When necessary, hoists should be removed from service with controls locked and tagged to prevent use until repairs can be made.

3.8.9 Hoisting Apparatus

All slings and "below-the-hook" lifting devices must be inspected daily prior to use and more thoroughly inspected periodically by a qualified individual (43). Periodic inspections are a function of frequency of sling use, severity of service conditions, the nature of the lifts being made, and experience based on the service life of slings used in similar circumstances (43). All employees who use slings or other lifting devices must be trained in how to inspect these items for wear, damage, or other defects. The following factors must be considered while following the safe operating practices required by OSHA (43):

- Never use damaged or defective slings.
- Slings must not be shortened with knots, bolts, or other makeshift devices.
- Slings should never be loaded beyond their rated capacity.
- Slings must be padded or otherwise protected from loads with sharp edges.
- Suspended loads must be kept clear of all obstructions and never be lifted over or near employees.
- Shock loading is prohibited as is pulling the sling from under a load.

Ref. (44) provides more detailed guidance on the rating, inspection, and modification of lifting devices.

3.8.10 Conveyors

Hazards associated with conveyors include: pinch points between moving conveyor components or items moving on the conveyor and stationary portions of the conveyor, pinch points caused by items "bumping-up" against one another on the conveyor, objects falling from the conveyor, falling components of the conveyor itself, workers falling from conveyors they attempt to cross or ride, bumping into conveyor components, accidental starting during maintenance or trouble-shooting, and contact with power transmission components such as belts, chains, gears, or pulleys (45). Hazards associated with bulk material conveyors can include fires from friction, overheating, or overload of electric motors;

health hazards associated with irritating dusts; explosions from build up of combustible dusts; and increased potential for falling materials (46).

The most serious injuries typically occur when an individual becomes entrapped in a pinch point or in-running nip point. Usually employees are pulled in when their hands, clothing, or tools are caught in a pinch point (46). Employees working with or near conveyors should wear close-fitting clothing that cannot easily become caught in the moving components of the conveyor. Where there is the potential for objects falling from conveyors safety shoes should be required. Some bulk material handling environments may also necessitate the use of hard hats, goggles, respirators or other personal protective equipment.

For both safety and convenience, many conveyors can be started and stopped from multiple locations. All controls should therefore be clearly marked and protected by recessing them or otherwise guarding them from accidental operation by employees or conveyed materials. Whenever there are two or more stop buttons on a conveyor, the conveyor should be wired to require a manual reset before the conveyor can be restarted after activation of a stop button. During restart, workers should be instructed to stand clear.

All conveyors should be periodically inspected to ensure their continued safe operation with particular emphasis on safety features such as emergency stops, braking devices, and guard rails. Walking surfaces and passageways around, over, or under conveyors must comply with the ANSI/ASME B.20.1 Conveyor standard and the NFPA 101 Life Safety Code (47). Hinged sections, to allow employees to pass through a conveyor line, should be fitted with interlocks to stop the conveyor while the section is up, be made of light weight materials, counter balanced, or otherwise mechanically assisted to simplify access through the conveyor line, and be fitted with a catch mechanism to hold them in place while they are open (46).

Screw conveyors are a special type of bulk material conveyor that move material with a continuous spiral screw mounted within a stationary, typically U-shaped, trough. An unenclosed screw conveyor presents a serious entrapment hazard. When accidents occur, they are generally very serious and include amputations and death (48). Covers and gratings must be securely fastened prior to operation of the conveyor. Feed openings, where materials enter the conveyor, should be fitted gratings. When unguarded housing or feed openings are functionally necessary, the conveyor should be otherwise guarded such as by fencing the area to prevent access to the screw mechanism. Employees should never walk on covers, gratings, or guards.

Manual roller conveyors rely on employee muscle or the force of gravity to move items. As a result, workers sometimes fail to exercise necessary precaution when working around them. Employees can be injured if they attempt to climb up on roller conveyors to release stuck or jammed packages or try to cross them and should be prohibited from doing so. Steel or wood plates can be installed between rollers to prevent an employee's limbs from being caught between rollers. The principal hazards associated with roller conveyors are loads falling off the conveyor and loads that "run away" (46).

3.8.11 Automated Guided Vehicles and Automated Storage/Retrieval Systems

Automated Guided Vehicles (AGVs) and Automated Storage/Retrieval Systems (AS/RS's) are becoming increasingly common. An automated guided vehicle (AGV) is a material transport vehicle that travels over prearranged routes. Its movement is controlled by electromagnetic wires buried in the floor, optical guidance, infrared, inertial guidance (gyroscope), position-referencing beacons, or computer programming (49).

AGVs must be equipped with a means for stopping if a person or object is in the path. This is usually achieved with a lightweight, flexible bumper that shuts off power and applies the brakes when contacted (28). AGV bumpers should not need hardware or software logic or signal conditioning in order to operate. AGVs in automatic mode must stop immediately when they lose guidance (49) and most vehicles are programmed to require manual reset before resuming motion (50). AGVs must have clearly marked and unobstructed aisles for operation (28). Blinking or rotating lights and/or warning bells can alert workers to the presence of AGV's in their work area and turn signals can indicate which way an AGV will be turning.

An automated storage/retrieval system (AS/RS) is defined by the Materials Handling Institute as "A combination of equipment and controls which, handles, stores, and retrieves materials with precision, accuracy, and speed under a defined degree of automation (51). Typically AS/RS are comprised of a series of storage aisles with one or more storage/retrieval (S/R) machines (usually one per aisle) to deliver materials. Some AS/RSs are not directly controlled by operators and access to the storage areas must be interlocked or otherwise guarded to protect workers from the moving equipment. More information on AS/RS machines can be found in the ASME B30.13 Standard "Storage/Retrieval Machines and Associated Equipment" (36).

3.8.12 Storage

The storage of raw materials and finished products is as important to the safety and efficiency of an operation as the handling and transportation of those goods. The storage of materials may interfere with plant operations, impede emergency exit, or present a hazard itself. For example, sharp or protruding edges, flammable or combustible materials, unstable or shifting materials, and material blocking passage ways all present hazards. In addition, stored materials should not restrict access to machinery controls or safety equipment such as fire extinguishers, eye wash stations, fire alarms, or first aid kits.

When designating storage areas, consider: the materials being stored, their proximity to other materials and processes, the storage methods to be employed (racks, pallets, bins, tanks, etc.), and paths to exits. Good housekeeping is critical to safe storage. Rubbish, trash, and other waste should be disposed of at regular intervals to prevent fire and tripping hazards.

Adequate clearance must be provided from sprinklers to ensure proper function. Materials stacked too closely to sprinklers will block the flow of water and limit effectiveness. Typically, a minimum of at least 18″ of clearance is needed. Depending on the class, quantity, and height of rack storage, in-rack sprinklers may also be necessary (52). Exits and paths to exits must be maintained free of obstructions and other hazards. Paths to exits should be clearly marked and visible from all locations in the storage area. Signs indicating escape routes should not be blocked by stored materials.

Sufficient safe clearance must be provided for aisles and passageways in material storage areas. Where mechanical materials handling equipment is used, additional space is needed, including the separate designation of vehicle and pedestrian paths. Permanent aisles should be appropriately marked. Ideally, physical barriers such as chains or guard rails are used to separate pedestrian paths from vehicular traffic.

Employees should travel through doorways provided for pedestrian use, not bay or dock doors used by forklifts and other handling equipment. These doorways are especially dangerous since the visibility of both the pedestrian and the equipment operator may be impaired by changes in lighting or by flexible doors (such as overlapping plastic slats). The use of pedestrian doors should be enforced and vehicle operators should slow, stop, and sound their horns as they enter doorways, intersections, or other limited visibility areas.

Strict housekeeping must be enforced to ensure that storage areas remain free from accumulation of materials that constitute trip, fire, explosion, pest harborage, or other hazards (53). Poor housekeeping is often indicative of poor safety practices and should provide a warning that the workplace is not only operating inefficiently, but possibly in an unsafe manner as well. Good housekeeping helps make unsafe conditions more visible and obvious, providing an atmosphere more conducive to safe behavior.

3.8.13 Hazardous Materials

Hazardous materials such as flammable, explosive, corrosive, extremely high or low temperature, toxic, and carcinogenic materials all present hazards beyond those associated with the handling of the materials themselves. The consequences of a material handling accident are more severe, since they could result in a release and subsequent exposure of workers to the hazardous materials.

Special consideration must be given to material handling operations that involve hazardous materials including preventive measures that include safe handling, spill prevention, and the use of personal protective equipment and clothing; emergency showers and eye wash stations; and respiratory protection equipment. Preventive measures should also include substitution of less toxic or corrosive materials, isolation of the hazardous process by the use of enclosures, and provision of adequate exhaust ventilation.

Working with hazardous materials requires a familiarity with safe work practices and current regulations as well as the regulatory details governing the shipping and receiving of hazardous materials and wastes. The assistance of staff or consultants with special expertise may be necessary to achieve full compliance.

The environment in which the hazardous materials are stored, transported, and processed often must be controlled. Some locations where flammable materials are stored, handled, or used may require the use of explosion proof, intrinsically safe, or otherwise protected equipment depending on the classification of the that area (54). Classification is a function of the properties of the flammable vapors, liquids, gases, combustible dusts, or fibers that may be present and the likelihood that a flammable or combustible concentration or quantity is present (54). The equipment used to move the materials must also meet or exceed these hazard classification requirements.

3.9 Construction Safety

In addition to normal industrial safety hazards, workers in the construction industry must be protected from additional hazards more common to construction sites such as open excavations, falling from elevations, falling objects, temporary wiring, excessive dust and noise, and heavy construction machinery. Construction safety programs may be thought of as providing for worker safety during the processes of transportation, excavation, fabrication, erection, and demolition. OSHA standards relating to construction are contained in the Code of Federal Regulations (29 CFR 1926). Ref. 7 is a summary of the OSHA Construction standards.

Transportation. Traffic patterns in a construction site are often unclear and may vary from day-to-day. Efforts should be made to establish clear traffic flow patterns and communicate this to all affected personnel. Trucks and other transportation equipment must not collide with or contact power lines, other vehicles, structures, or excavation. Vehicles with obstructed views to the rear must be equipped with a reverse signal alarm audible above the surrounding noise level or an observer ("spotter") must signal when it is safe to back up (55). Transporting and handling materials on a construction site is often more dangerous than similar transport in manufacturing environments because there is less control of the workplace due to uneven terrain, weather, multiple contractors, equipment that must be transported and assembled, and deliveries by outside parties.

Excavation. This material is adapted from Ref. 56. The reader is referred to this document and *Code of Federal Regulations*, **29** *CFR* 1926, Subpart P: Excavations, for further information.

OSHA requires that all trenches over 5 feet (1.5 meters) deep, except those in solid rock, be sloped, shored, sheeted, braced or otherwise supported. Trenches less than 5 feet (1.5 meters) deep must be protected if hazardous ground movement is expected. Factors to be taken into account when determining the design of a support system are soil structure, depth of cut, water content of soil, weather and climate, superimposed loads, vibrations, and other operations in the vicinity. The approximate angle of repose for sloping on the sides of excavations ranges from vertical (90° from horizontal) for solid rock, shale or cemented sand and gravel to a 1.5:1 slope (34° from vertical) for well rounded loose sand (57). The costs and safeguards for excavation projects are dependent on the traffic, proximity of structures, type of soil, surface and ground water, water table, underground utilities, and weather.

Erection. This material is adapted from Ref. 8. The reader is referred to this document and *Code of Federal Regulations*, **29** *CFR* 1926, Subparts E, L, M, and R for further information.) Scaffolds, guardrails and toeboards, ladders, safety nets, and steel erection are important issues during the erection of structures at construction sites. Scaffolds must be able to accept at least four times the maximum intended load and must be erected on a sound footing able to accept the maximum intended load without settling (58). Employees working on scaffolds or platforms must be protected by the use of personal fall arrest systems or guardrail systems. Toeboards and handrails provide protection from falling objects. Toeboards must be of substantial construction with openings not to exceed 1 inch (2.5 cm) (58). Ladders must have uniformly spaced rungs with slip resistant steps and ladders should extend at least 36 in. (91.4 cm) above the landing. Portable metal ladders must not be used for electrical work (59). Safety nets must be provided when workplaces are higher than 25 feet (7.6 meters) above the floor if the use of scaffolds, temporary floors, or other safer procedure is not practical (60). During steel erection a temporary floor must be maintained within two stories or 30 feet (9.1 meters) (whichever is less) directly below where work is being performed. A 1/2 inch (1.3 cm) wire rope or equivalent must be installed at a height of approximately 42 inches (106.7 cm) around the perimeter of temporary floors. Except when structural integrity is maintained by the design of the building, a permanent floor must be installed so that there are no more than eight stories between the erection floor and the highest permanent floor (61).

Demolition. Demolition should be performed by specialists who are familiar with the relevant regulations and procedures required to protect themselves, other workers and the general public.

3.10 Hazardous Material

This material is adapted from Ref. 62. The reader is referred to Ref. 62 or *Code of Federal Regulations*, **29** *CFR* 1910, Subpart H for further information).

There are approximately 32 million workers potentially exposed to chemical hazards. These exposures may result in heart ailments, kidney and lung damage, sterility, cancer, burns, and rashes. In addition, these chemicals may cause fires or explosions.

Each hazardous material container in the workplace must be tagged, labeled, or marked with the identity of the material it contains and must include prominently displayed warnings in written, picture, or symbol forms that convey the hazards of the chemical.

Chemical manufacturers and importers must develop material safety data sheets (MSDS) which include the name of the hazardous chemical, its specific chemical identity, physical and chemical characteristics, known acute and chronic health effects, exposure limits, if the chemical is considered to be a carcinogen, precautionary measures, emergency and first-aid procedures, and the organization that prepared the MSDS. The MSDS for chemicals in a particular work area must be readily accessible to employees in that area and employers must establish a training and information program for all employees exposed to hazardous chemicals.

3.11 Noise

Excessive noise (unwanted sound) may result in (*1*) decreased hearing sensitivity, (*2*) immediate physical damage, (*3*) interference or masking of particular sounds, (*4*) annoyance, (*5*) distraction, and (*6*) contribution to other types of disorders (3).

The reduction of the adverse effects of noise in the workplace is best accomplished through early planning and equipment/process design which minimizes noise and reduces the potential for noise exposure. The reduction of noise at the source (through equipment modification) and the use of insulation can be very costly. The use of personal protective equipment (PPE) may be difficult and cause worker discomfort and is not always effective, particularly if not properly used.

OSHA requires that hearing protection be made available to employees who are exposed to an 8-hour time weighted average (TWA) of 85 dB(A) or above, and that hearing protection must be worn by employees who are exposed to an 8-hour TWA of 90 dB(A) or above. The hearing protection must attenuate employee exposure to an 8-hour TWA of 90 dB(A).

3.12 Personal Protective Equipment

This material is adapted from Ref. 63. The reader is referred to Ref. 63 and *Code of Federal Regulations*, **29** *CFR* 1910, Subpart I: Personal Protective Equipment for further information). OSHA requires that employers determine if hazards are present (or likely to be present) which necessitate the use of personal protective equipment (PPE) and select/enforce the use of PPE which will protect workers from those hazards (64). OSHA notes that employers must enforce the use of PPE when there is a "reasonable probability" that the use of such equipment would prevent injury. In addition, employers must maintain records which document that the workplace has been evaluated and that the necessary PPE has been selected. It is important that workers are trained in the proper use and care of any PPE required as a result of this process.

The selection criteria for eye and face protection includes the type of hazard present, the degree of exposure, comfort, snugness of fit, durability, and maintainability. Hard hats must absorb the shock associated with a blow to the head and resist penetration by falling or other foreign objects. Leg and foot protection may be required to protect against rolling or falling material, sharp objects, molten metal, or hot surfaces. Safety shoes must be sturdy, have an impact resistant toe, and provide adequate slip resistance when used on wet and slippery surfaces. A variety of other PPE is available to protect the arms, hands, and torso from cuts, heat, splashes, impacts, acids, and radiation. This equipment must be selected to fit the specific task. Hearing protection includes wax, cotton, foam, or glass fiber which are self-forming or devices which are molded or fitted to the individual. Non-disposable ear plugs should be properly maintained and disposable ear plugs should be discarded after one use. Respiratory protection is required when employees are exposed to hazardous dusts, fogs, fumes, mists, gasses, smokes, sprays, or vapors in excess of OSHA permissible exposure limits (PELs). A respirator program which outlines the requirements for selection, use, fitting, inspection, and user medical status must be in place where respirators are used to control exposures to the above hazards.

It must again be emphasized that employees must be trained in the proper use and maintenance of PPE. They must be made aware that the use of PPE does not eliminate the hazard that harmful exposure may result if the PPE fails or is improperly used.

3.13 Radiation

Radiation is energy transmitted through space. Ionizing radiation such as x-rays, gamma-rays, and cosmic rays convert atoms into ions by adding or removing electrons. In general, a substance that emits ionizing radiation is called radioactive. The exposure to ionizing radiation may result in adverse affects such as cancer, birth defects in future children of exposed parents, and cataracts. While there is not conclusive evidence of a cause–effect relationship between current levels of occupational radiation exposure and adverse health effects, it is advisable to assume that some adverse health effects may occur at occupational exposure levels. The Nuclear Regulatory Commission requires that the exposure of workers and the public to ionizing radiation be kept as low as reasonably achievable (ALARA) (65). The risk of adverse effect from ionizing radiation can be reduced through (*1*) the reduction of exposure time, (*2*) increase in distance between the worker and radiation source, and (*3*) appropriate shielding between the worker and the radiation source.

Nonionizing radiation includes ultraviolet, visible light, infrared, microwave, and lasers. The risks from these sources can also be reduced through the time, distance, and shielding measures noted above, as well as the use of appropriate glasses, skin creams, clothing, gloves, and face masks.

3.14 Slips and Falls

Slips and falls from elevation were discussed in the earlier presentation of construction safety issues. Falls on the same level tend to result from a slip between the shoe and the walking surface or a stumble or contact of the foot or leg with an unexpected obstruction. Slips can be prevented by the proper selection and maintenance of foot wear and work surfaces which optimize the slip resistance between the foot wear and the walking surface. Some standards suggest that an adequate slip resistance is defined by a coefficient of friction of 0.5, however, there is a lack of consensus as to how this slip resistance is to be measured. A higher slip resistance may be required when pushing or pulling loads or when walking on ramps. In addition, this slip resistance is a function of the shoe sole type as well as the floor surface so the specification of a slip resistance for a particular floor surface (when the type of shoe is unknown) is questionable. An extremely high slip resistance between the shoe and the floor may result in stumbles or stresses at the knee when the upper body pivots about the foot. Stumbles are best prevented by good housekeeping principles such as adequate lighting, proper marking of walkways, and load carrying techniques which do not adversely affect worker visibility or overload the worker.

3.15 Robot Safety

This material is adapted from Ref. 66. The reader is referred to Ref. 66 and Ref. 67.

Although workers may recognize the hazards associated with the working end of the robot arm, they may not recognize the dangers associated with robot maintenance or the movement of other parts of the robot. Robotic system safety includes robot design, worker training, and worker supervision.

Proper robot design should include:

1. Physical barriers with interlocked gates.
2. Motion sensors, light curtains, or floor sensors that stop the robot when a worker crosses the barrier.
3. Barriers between robotic equipment and freestanding objects to eliminate "pinch points."
4. Adequate clearance around all moving components.
5. Remote diagnostic instrumentation to facilitate troubleshooting away from the moving robot.
6. Adequate illumination of control and operational areas.
7. Marks on the floor and working surfaces to indicate the movement area of the robot (work envelope).

The training of operators and maintenance personnel should include:

1. Familiarity with all working aspects of the robot including range of motion, known hazards, programming, emergency stop methods, and safety barriers.
2. Importance of staying out of the reach of the robot during operation.
3. Necessity of operating at reduced speed and awareness of all pinch points during programming.

Supervisors of operators and maintenance personnel should:

1. Assure that no one is allowed within the operational area of the robot without first shutting down or reducing the speed of the robot.
2. Recognize that, over time, workers may become complacent or inattentive to the hazards inherent in robotic equipment.

3.16 Systems Safety

System safety analysis includes the analysis and reduction of risk through the acquisition, review, and analysis of specific information relevant to a particular system. This process is systematic and is intended to facilitate informed management decisions. Five often used system safety techniques are Preliminary Hazards Analysis, Job Safety Analysis, Failure Methods and Effects Criticality Analysis, Fault Tree Analysis and Management Oversight and Risk Tree Analysis.

Preliminary Hazard Analysis (PHA). Preliminary Hazard Analysis is a qualitative analysis of the basic hazards which exist in the system and is conducted during the conceptual or early developmental phases of a system's life. It must be emphasized that the hazards may not be well defined at this stage. The analyst attempts, therefore, to identify general hazard areas so they can be corrected or monitored during system development when they can be corrected. The PHA is intended to:

1. Identify known hazardous conditions and potential failures.
2. Determine the cause(s) of these conditions and potential failures.
3. Determine the potential effect of these conditions and potential failures on personnel, equipment, facilities, and operations.
4. Establish initial design and procedural requirements to eliminate or control these hazardous conditions and potential failures.

In some cases an additional step, the estimation of the probability of an accident due to the hazard, is performed between steps 3 and 4 above.

Job Safety Analysis (JSA). Job Safety Analysis is a written procedure designed to uncover hazards and recommend safe job procedures through systematic review of job methods. Smith (68) notes the following four basic steps in making a JSA:

1. Select the job, usually basing selection on potential hazards or high incidence rates.
2. Break the job down into a sequence of steps. Job steps are recorded in their normal order of occurrence.
3. Identify the potential hazards by observing the job, discussing the job with the operator, and checking accident records.
4. Recommend safe job procedures to eliminate or minimize the potential accident.

A basic JSA form should also include the name of the person performing the analysis, name of the operator and supervisor, or the name of reviewers or approvers of the analysis.

Failure Modes and Effects Criticality Analysis (FMECA). Failure Modes and Effects Criticality Analysis is used to identify how different failure types (modes) in system components will affect the safe and successful life of the system or prevent a system from accomplishing its intended mission. The FMECA addresses the following issues:

1. System component—What components make up the system?
2. Failure modes—How could a component fail?
3. System causes—What would cause the component failure or malfunction?
4. System effects—What would be the effect of such a failure on the system and how would this failure affect other components in the system?
5. Severity index—Consequences are placed into severity categories. These are generally four or five categories ranging from Negligible to Catastrophic (5).

For example:

   a. Negligible—Probably would not affect personal safety or health but is a violation of a safety or health standard (score = 1).
   b. Marginal—May cause minor injury/occupational illness, or minor property damage (score = 2).
   c. Critical—May cause severe injury/occupational illness, or major system damage (score = 3).
   d. Catastrophic—May cause multiple injuries, death, system loss (score = 4).

6. Probability Index—How likely is the event to occur? This is based on such factors as accident experience, test results from component manufacturers, comparison with similar equipment, or engineering data.
   a. Extremely remote (Unlikely to occur (score = 1)).
   b. Remote (Possible to occur in time (score = 2)).
   c. Reasonably probable (Probably will occur in time (score = 3)).
   d. Probable (Likely to occur immediately or within a short period of time (score = 4)).
7. Criticality Index—This measure combines the severity and probability noted above to produce a risk index.

Action or Modification. After the failure modes, causes and effects, severity, probability, and risk have been established, it is necessary to modify the system to prevent or control the failure mode to what is deemed an "acceptable" level of risk.

An example of a FMECA is shown in Table 55.2. Note that two different failure modes are suggested for the commutator with very different system effects and criticality indexes.

Fault Tree Analysis (FTA). Fault Tree Analysis observes or postulates the possible failure of a system and then identifies combination of component conditions that can cause the failure. It reasons backwards from the end event to identify all of the ways in which such an event could occur and, in doing so, identifies the contributory causes. FTA uses Boolean logic and algebra to represent and quantify the interactions between events. The primary Boolean operators are AND and OR gates. With an AND gate, the output of the gate (the event that is at the top of the symbol) occurs only if *all* of the conditions feeding into the gate, coexist. With an OR gate, the output event occurs if *any* of the input events occur. Figure 55.2 (3) illustrates and describes the most common fault tree symbols and Figure 55.3 illustrates a basic FTA.

When the probabilities of initial events or conditions are known, it is possible to determine the probabilities of succeeding event through the application of Boolean algebra. For an AND gate the probability of the output event is the intersection of the Boolean probabilities (the product of the probabilities of the input events). For example, for an AND gate with three input events:

$$\text{Probability (output)} = (\text{Prob Input 1}) \times (\text{Prob Input 2}) \times (\text{Prob Input 3})$$

For an OR gate, the probability of the output event is the probability that *any* of the input events will happen. When the events are mutually exclusive (can not happen together) this is the sum of the input events. When probabilities are small this can be estimated by the sum of the probability of the input events. For nonmutually exclusive events the probability of an output event for an OR gate is the sum of the probabilities of the input events minus the redundant "intersections" which indicate how the input events can happen simultaneously. This calculation is tedious. Instead one can calculate the probability that *any or all* of the input events will happen or 1- (probability that *none* of the input events) happen. For an OR gate with these input events this is:

$$\text{Probability (any or all)} = 1 - (\text{probability none}).$$

$$\text{Probability (any or all)} = 1 - [(\text{prob 1 does not happen}) \cdot (\text{prob 2 does not happen})(\text{prob 3 does not happen})]$$

$$\text{Probability (output)} = 1 - [1 - (\text{Prob Input 1})][1 - (\text{Prob Input 2})] \cdot [1 - (\text{Prob Input 3})]$$

As an example, assume in Figure 55.3 that the following probabilities exist:

Table 55.2. Failure Mode and Effect Analysis of an Electric Drill[a]

Component	Failure Mode	System Causes	System Effects	Severity Index	Probability Index	Criticality Index	Action or Modification
On-off switch	Falls open	Normal wear	Drill disabled	1	3	4	None
	Fails closed	Normal wear	Drill continues to run	1	3	4	None
Reduction gear train	Functional disability	Normal wear, overload, lack of lubrication	Drill disabled	2	1	3	Preventive maintenance, avoid overload
Commutator	Functional disability	Normal wear, broken spring	Drill disabled	2	2	4	None
	Fault to case	Carbon buildup across insulating block	Fatality, if operator is grounded at a remote extremity	4	4	8	Ground the case or go to double insulated construction
Windings	Burnout	Overload, plus fuse failure to open	Drill disabled	2	3	5	Check fuse, avoid overload
	Fault to case	Insulation failure	Fatality, if operator is grounded at a remote extremity	4	3	7	Ground the case or go to double insulated construction

[a]Adapted from Ref. 69 with permission.

Note: The analysis of this drill system includes a criticality index (in this case, the sum of the severity and probability indices); therefore, it can be called a Failure Mode, Effect, and Criticality Analysis.

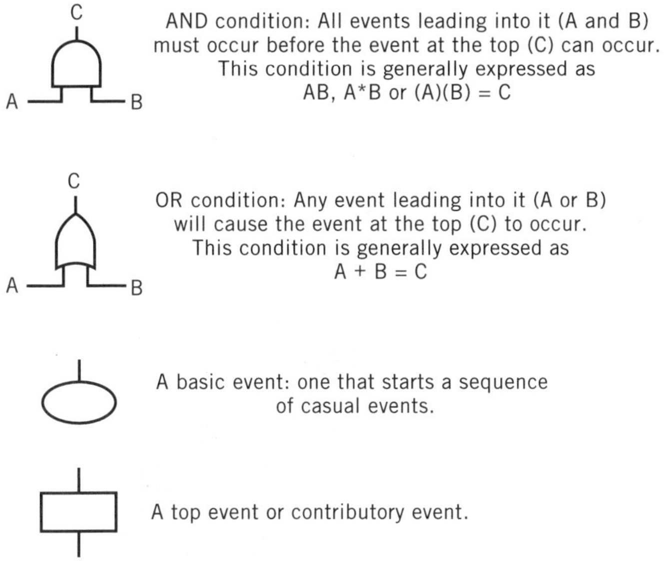
C
A
B
AND condition: All events leading into it (A and B) must occur before the event at the top (C) can occur. This condition is generally expressed as AB, A*B or (A)(B) = C
C
A
B
OR condition: Any event leading into it (A or B) will cause the event at the top (C) to occur. This condition is generally expressed as A + B = C
A basic event: one that starts a sequence of casual events.
A top event or contributory event.

Figure 55.2. Fault Tree symbols.

$$\text{Probability of power supply failing} = 0.00100$$
$$\text{Probability of switch open} = 0.00300$$
$$\text{Probability of blown fuse} = 0.00200$$
$$\text{Probability of light 1 out} = 0.03000$$
$$\text{Probability of light 2 out} = 0.04000$$

Since the power will be off if the power supply fails or the switch is open or the fuse is blown, the probability of the power being off is the "sum" of the probabilities of the power supply failing, the switch being open and the fuse being blown which can be estimated as $(0.00100) + (0.00300) + (0.00200) = 0.00600$. (The exact probability for this OR gate is $1 - (0.99900)(99700)(0.99800) = 0.00599$.) Both lights will be burned out if light 1 is out and light 2 is out. The probability of both lights being out is $(0.03000) \times (0.04000)$ or 0.00120. The room will be dark if the power is off or both lights are burned out which can be estimated as $(0.00600) + (0.00120)$ or 0.00720. (The exact probability for this OR gate is $1 - (1 - 0.00599)(1 - 0.00120)$ or $1 - (0.99401)(0.99880) = 0.0719$.)

The concepts of "cut sets" and "path sets" are useful in the analysis of fault trees. A cut set is any group of contributing elements which, if all occur, will cause the end event to occur. A path set is any group of contributing elements which, if none occur, will prevent the occurrence of the end event.

For the example in Figure 55.3 the end event will occur if:

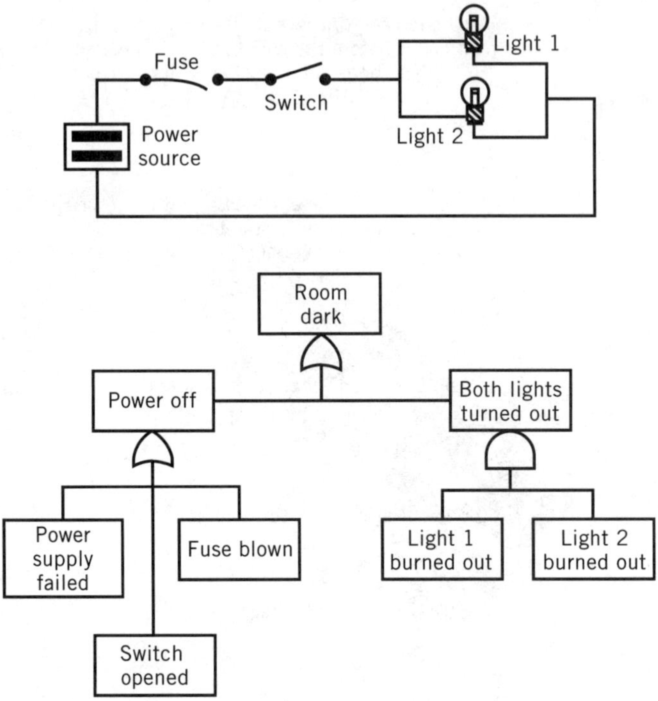

Figure 55.3. Fault Tree Analysis of light bulbs. Adapted from Ref. 70.

The power supply fails.
The switch is open.
The fuse is blown.
Light 1 *and* light 2 are burned out.

Each of the four sets above represents a cut set since if all of the events in any of the sets occur the end event (room dark) will occur.

For the example in Figure 55.5 the end event will not occur if none of the events in the following sets occur:

The power supply fails, the switch is open, the fuse is blown, light 1 is burned out.
The power supply fails, the switch is open, the fuse is blown, light 2 is burned out.

Each of the two sets above represents a path set since if none of the events in either of the sets occur the end event (room dark) can not occur.

Management Oversight and Risk Tree Analysis (MORT). Management Oversight and Risk Tree Analysis is defined as:

Table 55.3. Safety Related Publications

Publication	Publisher
	Trade Publications
Material Handling Engineering	Penton Publishing 1100 Superior Avenue Cleveland, OH 44114-2543 216/696-7000
Occupational Hazards	Penton Publishing 1100 Superior Avenue Cleveland, OH 44114-2543 216/696-7000
Occupational Health & Safety	Stevens Publishing 222 Rosewood Drive Danvers, MA 01923 508/750-8500
Professional Safety	ASSE 1800 E. Oakton St Des Plaines, IL 60018-2187 708/692-4121
Safety & Health	National Safety Council 1121 Spring Lake Drive Itasca, IL 60143-3201
	Journals
International Journal of Occupational Safety and Ergonomics	Central Institute for Labour Protection ul. Czerniakowska 16 00-701 Warszawa, Poland (48-22) 623-46-21 jose@ciop.waw.pl
Journals of Safety Management	National Safety Management Society 123 KyFields Weaverville, NC 28787 828/645-5229
Professional Safety	American Society of Safety Engineers 1800 E Oakton St Des Plaines, IL 60018-2187 847/699-2929 *vice@asse.org* www.asse.org
Journal of Injury Prevention	BMJ Publishing Group BMA House Tavistock Square London WC1H 9JR fax: +44 (0)171 383 6418/6299 tel: +44 (0)171 387 4499 email: bmj@bmj.com
Journal of Safety Research	Pergamon Press 660 White Plains Rd Tarrytown, NY 10591 212/989-5800

Table 55.3. (continued)

Organizations

The American Society of Safety Engineers 1800 E Oakton St Des Plaines, IL 60018 (847) 699-2929 www.asse.org	National Institute of Occupational Safety and Health (NIOSH) 1-800-35-NIOSH (4674). www.cdc.gov/niosh/homepage.html
Human Factors and Ergonomics Society P.O. Box 1369 Santa Monica, California 90406 (310) 394-1811 www.hfes.org	Occupational Safety & Health Administration (OSHA) U.S. Department of Labor 200 Constitution Avenue Washington, D.C. 20210 (202) 693-1999 www.osha.gov
National Fire Protection Association 1 Batterymarch Park, Quincy, MA 02269 (617) 770-3000 www.nfpa.org	American National Standards Institute 11 West 42nd Street New York, New York 10036 (212) 642-4900 www.ansi.org
National Safety Council 1121 Spring Lake Drive, Itasca, IL 60143 (630) 285-1121 www.nsc.org	Society for Risk Analysis 1313 Dolley Madison Blvd. Suite 402 McLean, VA 22101 703-790-1745 www.sra.org
System Safety Society P.O. Box 70 Unionville, VA 22567 (800) 747-5744 www.system-safety.org	

> A formalized, disciplined logic or decision tree to relate and integrate a wide variety of safety concepts systematically. As an accident analysis technique, it focuses on three main concerns: specific oversights and omissions, assumed risks, and general management system weaknesses (71).

MORT has three primary goals:

1. To reduce safety-related oversights, errors, and omissions.
2. To allow risk quantification and the referral of residual risk to proper organizational management levels for appropriate action.
3. To optimize the allocation of resources to the safety program and to organizational hazard control efforts.

It is essentially a series of fault trees with three basic subsets or branches that deal with:

1. Specific oversights and omissions at the worksite.
2. Management systems that establish policies and make them work.

3. Assumed risk that acknowledges that no activity is completely free of risk and that risk management functions must exist in any well managed organization.

MORT includes about 100 generic causes and thousands of criteria. The MORT diagram terminates in approximately 1,500 basic safety program elements that are required for a successfully functioning safety program. These are elements that prevent the undesirable consequences indicated at the top of the tree.

EG&G Idaho, Inc. (P.O. Box 1625, Idaho Falls, Idaho 83415) is one of the primary users and developers of MORT and offers training courses in MORT application.

3.17 Safety Through Design

The concept of Safety Through Design is relatively new. It focuses on the integration of safety considerations into the design of products or processes. Often systems safety analysis methods such as preliminary hazards analysis (PHA) and failure modes and effects analysis (FMEA) (discussed in section 3.15) are used. The consideration of safety issues early in the design process will reduce (*1*) injuries/illnesses, (*2*) damage to facilities and the environment, and (*3*) operating costs. In addition, the cost of product recall and retrofitting to address design shortcomings will be avoided (72).

4 SUMMARY

The high costs of injuries, facility damage, equipment downtime, and product loss due to accidents, make it critical for the manager and safety personnel to understand and control the conditions that can lead to accidents and illnesses.

To be effective, a safety program must have both management commitment and employee involvement. First-level supervisors must understand the importance of safety and be held accountable and responsible for enforcing employee compliance with safety standards and safe work practices. Management must take responsibility supporting safety personnel in the initiation and enforcement of a strong safety plan. Regular inspections of the work environment, equipment, and work practices as well as initial training and periodic retraining of employees will help ensure that work environment remains as safe as possible. Mechanical hazards should be designed out and safe operating and emergency response procedures should be considered during the design phase for equipment and facilities.

Some basic principles of occupational safety have been presented in this chapter. For further information relating to OSHA regulations the reader is directed to the OSHA General Industry Standards (**29** *CFR* 1910) and Construction Standards (**29** *CFR* 1926). The reader is also directed to the numerous standards and recommended practices developed by other organizations (often referenced in the OSHA standards) that apply to their particular applications. There is also a vast array of information available on the Internet. Internet addresses for many safety related organizations are noted in this chapter. For more information on this issue readers may use one of the many internet search engines or refer to Safety & Health on the Internet (73).

Safety related trade publications, journals, and organizations are listed in Table 55.3 Rich received B.S. degrees in General Engineering and Engineering Psychology and a M.S. in General Engineering from the University of Illinois. He also received a Masters in Public Health and Ph.D. in Mechanical Engineering (Ergonomics and Safety emphasis) at the University of Utah.

ACKNOWLEDGMENTS

The authors would like to acknowledge and thank Roanna Keough who provided invaluable assistance in preparing this chapter.

The preparation of this chapter was supported, in part, by NIOSH Training Grant No. T42/CCT810426.

BIBLIOGRAPHY

1. Bureau of Labor Statistics, *http://stats.bls.gov*, 1999.
2. National Safety Council, *Accident Facts*, 1998 ed., National Safety Council, Itasca, 1998.
3. W. Hammer, *Occupational Safety Management and Engineering*, 4th ed Prentice Hall, Englewood Cliffs 1989.
4. R. J. Firenze, *Evaluation and Control of the Occupational Environment*, U.S. Government Printing Office, Washington, DC, 1988.
5. R. J. Firenze, *The Impact of Safety on High-performance/High involvement Production Systems*, Creative Work Designs, Inc., 1991.
6. U.S. Department of Labor, *Public Law 91-596*. Occupational Safety and Health Act, U.S. Government, Dec. 29, 1970.
7. U.S. Department of Labor, *OSHA Publication 2202, General Industry-Occupational Safety and Health Standards Digest*, 1994.
8. U.S. Department of Labor, *OSHA Publication 2202, Construction Industry-Occupational Safety and Health Standard Digest*, 1994.
9. U.S. Department of Labor, *The New OSHA: Reinventing Worker Safety and Health*, 1997, OSHA World Wide Web Page [On-line], http://spider.osha.gov/oshinfo/reinvent/reinvent:html.
10. U.S. Department of Labor, *OSHA Regulations (Standards) Part 1904 Recording and Reporting Occupational Injuries and Illnesses*, 1997.
11. U.S. Department of Labor, *OSHA Regulations (Standards) Part 1910 Subpart I: Personal Protective Equipment, Section 134: Respiratory Protection*, 1997.
12. Bureau of Labor Statistics, *Recordkeeping Guidelines for Occupational Injuries and Illnesses*, U.S. Government Printing Office, Washington, DC, 1986.
13. Federal Register, *Occupational Injury and Illness Recording and Reporting Requirements, Federal Register* **61**, 4029–4067 (Feb. 2, 1996).
14. U.S. Department of Labor, *OSHA Fact Sheet 87-07, Improving Workplace Protection for New Workers*, 1987.
15. U.S. Department of Labor, *OSHA Publication 3075, Controlling Electrical Hazards*, 1986.

16. U.S. Department of Labor, *OSHA Publication 3007, Ground-Fault Protection on Construction Sites*, 1992.
17. American Society of Safety Engineers, *An Illustrated Guide to Electrical Safety*, Des Plaines, IL, 1993.
18. National Safety Council, "Fire Protection," in G. R. Krieger and J. F. Montgomery, ed. *Accident Prevention Manual for Business and Industry: Engineering and Technology*, National Safety Council, Itasca, IL, 1997.
19. U.S. Department of Labor, *OSHA Regulations (Standards) Part 1910 Subpart Q: Welding, Cutting, Brazing, Section 253: Oxygen-Fuel Gas Welding and Cutting*, 1996.
20. U.S. Department of Labor, *OSHA Regulations (Standards) Part 1926 Subpart P: Hand and Portable Powered Tools, Section 242: Hand and Portable Powered Tools and Equipment, General*, 1996.
21. U.S. Department of Labor, *OSHA Publication 3080, Hand and Power Tools*, 1986.
22. U.S. Department of Labor, *OSHA Publication 3067, Concepts and Techniques of Machine Safeguarding*, 1992.
23. U.S. Department of Labor, *OSHA Regulations (Standards) Part 1910 Subpart O: Machinery and Machine Guarding, Section 217: Mechanical Power Presses*, 1997.
24. U.S. Department of Labor, *OSHA Regulations (Standards) Part 1910 Subpart J: General Environmental Controls, Section 147: The Control of Hazardous Energy (lockout/tagout)*, 1996.
25. U.S. Department of Labor, *OSHA Regulations (Standards) Part 1910 Subpart J: General Environmental Controls, Section 146: Permit-Required Confined Spaces*, 1996.
26. U.S. Department of Health and Human Services, *DHHS (NIOSH) Publication No. 94-103. Worker Deaths in Confined Spaces: A Summary of Surveillance Findings and Investigative Case Reports*, Jan. 1994.
27. U.S. Department of Labor, *OSHA Publication 3072, Sling Safety*, revised, 1996.
28. National Safety Council, "Powered Industrial Trucks," in G. R. Krieger and J. F. Montgomery, Eds., *Accident Prevention Manual for Business and Industry: Engineering and Technology*, National Safety Council, Itasca, IL, 1997, pp. 506–528.
29. U.S. Department of Health and Human Services, *DHHS (NIOSH) Publication 96-115*. National Occupational Research Agenda, April 1996.
30. J. F. Krause, et al., "Reduction of Acute Low Back Injuries by Use of Back Supports," *Intern. J. Occupat. Environ. Med.* **2**, 264–273 (1996).
31. M. S. Perkins and D. S. Bloswick, "The Use of Back Belts to Increase Intraabdominal Pressure as a Means of Preventing Low Back Injuries: A Survey of the Literature," *Intern. J. Occupat. Environ. Health* **1**, 326–335 (1995).
32. National Safety Council, "Materials Handling and Storage," in G. R. Krieger and J. F. Montgomery, eds. *Accident Prevention Manual for Business and Industry: Engineering and Technology*, National Safety Council: Itasca, IL 1997, pp. 375–412.
33. A. Suruda, et al., A Study of Fatal Work Injuries Involving Forklifts, *J. Occupat. Environ. Med.*, in press.
34. National Fire Protection Association, *Fire Safety Standard for Powered Industrial Trucks Including Type Designations, Areas of Use, Conversions, Maintenance, and Operation*. (NFPA 505), National Fire Protection Association, Quincey, MA, 1996.
35. U.S. Department of Labor, *OSHA Regulations (Standards) Part 1910 Subpart K: Medical and First Aid, Section 151: Medical Services and First Aid*, 1996.

36. American Society of Mechanical Engineers, *ASME B20.1-1993, Safety Standard for Conveyors and Related Equipment*. American Society of Mechanical Engineers, New York, 1993.
37. U.S. Department of Labor, *OSHA Regulations (Standards) Part 1910 Subpart N: Materials Handling and Storage, Section 178: Powered Industrial Trucks*, 1996.
38. National Safety Council, *Powered Hand Trucks*, in (*DataSheet I-317 Rev. 91*) 1991, National Safety Council, Itasca, IL.
39. National Safety Council, *Front-End Loaders*, in (*DataSheet I-589 Rev. 90*) National Safety Council, Itasca, IL, 1990.
40. National Safety Council, *Dock Plates and Gangplanks*, in (*Data Sheet I-318-Reaf. 90*), National Safety Council, Itasca, IL, 1990.
41. U.S. Department of Labor, *OSHA Regulations (Standards) Part 1910 Subpart N: Materials Handling and Storage, Section 179: Overhead and Gantry Cranes*, 1996.
42. American Society of Mechanical Engineers, *ASME B30.2-1996, Overhead and Gantry Cranes: Top Running Bridge, Single or Multiple Girder, Top Running Trolley Hoist*, American Society of Mechanical Engineers, New York, 1996.
43. U.S. Department of Labor, *OSHA Regulations (Standards) Part 1910 Subpart N: Materials Handling and Storage, Section 184: Slings*, 1996.
44. American Society of Mechanical Engineers, *ASME B30.20-1993, Below-the-Hook Lifting Devices: Safety Standard for Cableways, Cranes, Derricks, Hoists, Hooks, Jacks, and Slings*, American Society of Mechanical Engineers, New York, 1994.
45. National Safety Council, *Belt Conveyors for Bulk Materials*, in (*Data Sheet I-569 Reaf. 90*), National Safety Council, Itasca, IL, 1990.
46. National Safety Council, *Roller Conveyors*, in (*Data Sheet I-528 Rev. 91*) National Safety Council, Itasca, IL, 1991.
47. National Fire Protection Association, *Life Safety Code (NFPA 101)*, 1997.
48. National Safety Council, "Hoisting and Conveying Equipment," in G. R. Krieger and J. F. Montgomery, eds., *Accident Prevention Manual for Business and Industry: Engineering and Technology*, National Safety Council: Itasca, IL, 1997, pp. 413–473.
49. National Safety Council, "Automated Lines, Systems," and Processes, in G. R. Krieger and J. F. Montgomery, eds. *Accident Prevention Manual for Business and Industry: Engineering and Technology*, National Safety Council, Itasca, IL, 1997, pp. 732–753.
50. M. P. Groover, "Automated Material Handling," in *Automation, Production, Systems, and Computer-Integrated Manufacturing*, Prentice-Hall, Englewood Cliffs, NJ, 1987, pp. 361–403.
51. M. P. Groover, "Automated Storage Systems," in *Automation, Production, Systems, and Computer-Integrated Manufacturing*, Prentice-Hall, Englewood Cliffs, NJ, p. 404–430.
52. National Fire Protection Association, *Standard for Rack Storage of Materials*, (NFPA 231C), National Fire Protection Association, Quincey, MA, 1995.
53. U.S. Department of Labor, *OSHA Regulations (Standards) Part 1910 Subpart N: Materials Handling and Storage, Section 176: Handling Material-General*, 1996.
54. National Fire Protection Association, *National Electric Code 1996 Edition (NFPA 70)*. National Fire Protection Association, Quincey MA, 1995.
55. U.S. Department of Labor, *OSHA Regulations (Standards) Part 1926 Subpart O: Motor Vehicles, Mechanized Equipment, and Marine Operations, Section 601: Motor Vehicles*, 1997.
56. U.S. Department of Labor, *OSHA Publication 2226, Excavating and Trenching Operations*, 1995.

57. U.S. Department of Labor, *OSHA Regulations (Standards) Part 1926 Subpart P: Excavations, Section 652: Requirements for Protective Systems*, 1997.
58. U.S. Department of Labor, *OSHA Regulations (Standards) Part 1926 Subpart L: Scaffolds, Section 451: General Requirements*, 1997.
59. U.S. Department of Labor, *OSHA Regulations (Standards) Part 1926 Subpart X: Stairways and Ladders, Section 1053: Ladders*, 1997.
60. U.S. Department of Labor, *OSHA Regulations (Standards) Part 1926 Subpart E: Personal Protective and Life Saving Equipment, Section 105: Safety Nets*, 1997.
61. U.S. Department of Labor, *OSHA Regulations (Standards) Part 1926 Subpart R: Steel Erection, Section 750: Flooring Requirements*, 1997.
62. U.S. Department of Labor, *OSHA Publication 3084, Chemical Hazard Communication*, 1988.
63. U.S. Department of Labor, *OSHA Fact Sheet 86-08, Protect Yourself With Personal Protective Equipment*, 1986.
64. U.S. Department of Labor, *OSHA Regulations (Standards) Part 1910 Subpart I: Personal Protective Equipment, Section 132: General Requirements*, 1997.
65. Nuclear Regulatory Commission, *Title 10 Energy, Part 20: Standards for Protection Against Radiation, Subpart B: Radiation Protection Programs*, Department of Energy, 1997.
66. National Institute for Occupational Safety and Health, *NIOSH Alert-Request for Assistance in Preventing the Injury of Workers by Robots*, 1984.
67. National Institute for Occupational Safety and Health, *Safe Maintenance Guide for Robotic Workstations*, 1988.
68. L. C. Smith, "The J Programs." *National Safety News* (Sept. 1980).
69. ASSE, *Professional Safety* (March 1981.)
70. Idaho National Engineering Lab, EG&G Safety Course Materials.
71. *EG&G, Publication 76-45/28. Glossary of Terms and Acronyms*, EG&G, Idaho Falls, ID, 1984.
72. D. S. Bloswick and B. S. Joseph, "Pro-Active Ergonomics and Designing for Error-Free Work," in W. Christensen, ed. *Safety Through Design*, National Safety Council: Itasca, IL, 1999.
73. R. B. Stuart, *Safety & Health on the Internet.* Government Institutes, Rockville, MD, 1997.

CHAPTER FIFTY-SIX

Explosion Hazards of Combustible Gases, Vapors, and Dusts

Martin Hertzberg, Ph.D., Consultant

1 INTRODUCTION

1.1 Definition of An Explosion

An explosion is a gas-dynamic process by which a marked increase in system pressure generates destructive forces. The cause of the pressure rise is a rapid, exothermic chemical reaction, which propagates as a combustion wave, and which markedly increases the gas temperature behind the wave. Those destructive forces can damage or destroy structures, and endanger personnel nearby. Such chemical explosions are a major hazard in homes, commerce, and industry, and it is essential to understand the factors that cause them in order to prevent their occurrence. The rapid chemical reaction that generates the explosion is usually the oxidation of a fuel in air, which leads to such a rapid energy release from the exothermic reaction that the system temperature increases markedly. Whether the reacting fuel is a gas, a vaporizing droplet, or a volatilizing dust, the combustion products are normally gases, and the explosion process is most simply understood in terms of the ideal gas law:

$$pV = nRT = (m/M)RT \tag{1}$$

where the system volume is V, the system pressure is p, and T is the absolute temperature. The number of moles of gas is n, and R is the universal gas constant. The number of moles

Patty's Industrial Hygiene, Fifth Edition, Volume 4. Edited by Robert L. Harris.
ISBN 0-471-29749-6

of gas, n, is equal to the mass of gas, m, divided by the average molecular weight, M. For most accidental explosions, it is the oxygen of the air that reacts with the fuel that is dispersed in gaseous, droplet, or dust form. Since air is mostly nitrogen, there is usually little change in the molecular weight as the combustion reaction proceeds, and the number of moles in the system, n, remains essentially constant for most accidental explosions. The reaction occurs so rapidly that the internal chemical energy released as heat is retained by the nitrogen of the air and the combustion products. As a result, the system temperature increases from its initial temperature, T, to a much higher, burned gas temperature, $T(b)$. If the system volume is fixed at V, and R is a constant, then according to equation 1, the system pressure will increase from its initial value, p, to the explosion pressure, $p(\max)$, in direct proportion to the ratio of the temperature increase:

$$p(\max)/p = T(b)/T \tag{2}$$

Examples of such explosions in a 20-L volume are shown in Figure 56.1. In A, coal dust at a concentration of 500 g/m^3, was initially dispersed uniformly in air at atmospheric

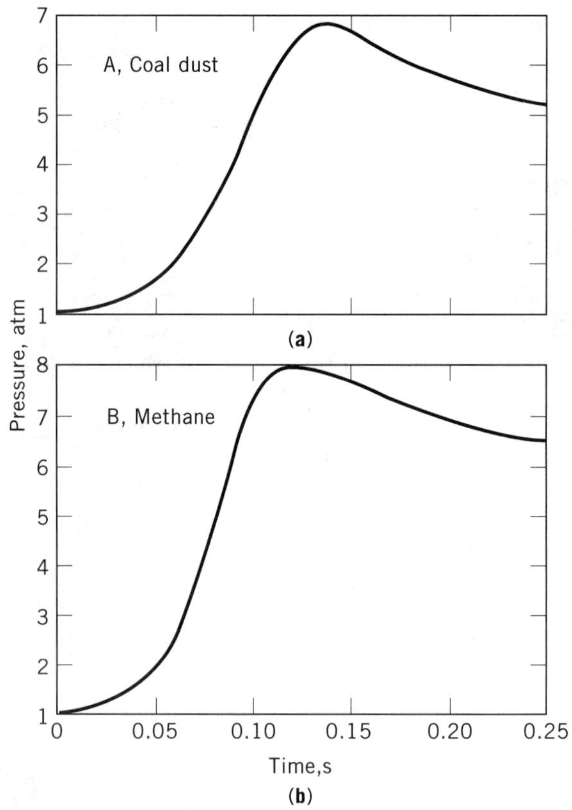

Figure 56.1. Examples of the pressure-time traces from explosions in a 20-L chamber. (**a**) A coal dust explosion in air; and (**b**) a methane gas explosion in air (from Fig. 1 of Ref. 1).

pressure. It was ignited near the center of the volume, and the exothermic chemical reaction occurs in the form of a spherical combustion wave (2), or "fireball". The hot, burned gases at the fireball surface ignite the unburned mixture outside of the fireball layer by layer, and the fireball thus grows in size at a characteristic flame speed, S_b. When it reaches the wall, the fireball has consumed the entire flammable mixture. In A, the combustion reaction is completed in about 0.14 s as the reaction wave reaches the wall of the 20-L volume. The burned gas fireball temperature at that instant is about $T(b) = 2100$ K (3). Since the initial temperature was ambient, $T = 300$ K, the ratio $T(b)/T$ is about 7. As seen in A, the measured maximum explosion pressure is about 7 atm, as is to be expected from equation 2.

An explosion in a near stoichiometric gaseous mixture of methane in air is shown in B for comparison. The pressure time curve is similar to that for A, and the maximum explosion pressure is similarly predictable. An explosion from a predispersed, heavy hydrocarbon vapor spray would also behave similarly in the 20-L chamber.

Because the chamber is made of thick steel, it is designed to withstand the high pressure, and the explosion force is contained within the chamber volume. Eventually, the burned gases cool down to room temperature by conduction and convection heat losses to the chamber wall, and the system pressure is returned to near atmospheric. If, however, the chamber had been made of typical glassware, or of thinner metal, or of brick, or concrete, it would have ruptured during the explosion. Such structural failure would be a threat to nearby personnel. Now, if the system volume had not been fixed at V by a confining structure, then the pressure forces generated by the combustion reaction wave would have been determined by gas–dynamic motions—by the velocity of the expanding flow outside the fireball that tends to relieve the pressure developing within the fireball, relative to the velocity of the combustion wave that generates the fireball. Even if the volume of the system is not fixed, and the burned gases are not confined but are free to expand toward a vent or the open end of a tube, the reaction rate during the explosion can be so rapid that gas dynamic motions may be too slow to relieve the developing pressure, and structural failure may still result. Explosions with destructive overpressures can thus occur even if the initial flammable volume is completely unconfined. As indicated in our definition, an explosion is a gas-dynamic phenomenon, and as such, the degree of confinement is not entirely determined by geometric boundary constraints such as confining walls. As combustion wave velocities approach (or exceed) the velocity of sound in the unburned mixture ahead of the wave, expanding gas motions become too slow to provide for pressure relief, and the system becomes confined by virtue of its own dynamic state quite apart from the initial geometric structure of its boundaries. The explosion then transits to an even more destructive process; a *detonation* (4–7).

1.2 Exclusions

In a previous edition of this volume (8) the title used for this section was "Fire and Explosion Hazards of Combustible Gases, Vapors, and Dusts," however, in reviewing the material presented therein, this author has concluded that the fire problem was not addressed in sufficient depth to justify the use of the term "fire" in the title. Other reviews of this subject have similarly used the term "fire" even though the material covered involves mainly explosions. The term "fire" is generally used to describe a much slower combustion process in which the overall rate of fuel oxidation is severely limited by its rate of mixing with the surrounding air. The speed of that fuel–air mixing process depends on the scale of the fire, and for large scale fires, it is controlled by turbulent mixing induced by natural convection. Although these earlier reviews did consider some aspects of the fire problem: the flash point ignition temperature of hydrocarbon liquids, the layer ignition temperatures for smoldering dusts, the minimum irradiance for the ignition of solids, there was no genuine synthesis of all of the factors required to evaluate overall fire hazards. Critical factors in the fire problem such as fuel structure, combustion loading density, combustible configuration, flame spread rate in the vertical and horizontal direction, ceiling height, moisture content of the fuel, and other factors that influence fire growth, were not seriously considered. Accordingly, the fire problem is here considered to be beyond the scope of this article, and the new title reflects this article's major focus, namely explosions and explosion prevention. Accordingly, one is not herein considering the mixing limited combustion process that characterizes fires, but instead deal only with the rapid combustion of the premixed systems: namely, explosions.

Nevertheless, in real accident scenarios, the two processes can be interrelated: large scale fires in confined system can lead to explosions, and explosions can generate fires in their aftermath; however the two combustion processes are markedly different phenomena involving significantly different hazard magnitudes and prevention countermeasures. Thus, if escape routes are available, personnel can run or walk to safety from a developing fire; however they cannot run fast enough to escape an explosion. Typical sprinkler systems can be effective in extinguishing developing fires, but they are generally useless against developing explosions.

Another combustion phenomenon that is excluded from our consideration involves the safety hazards associate with the use of condensed phase explosives. High explosives contain fuel and oxidants that are premixed either on the molecular scale (TNT) or on a fine particle scale (ammonium nitrate/fuel oil mixtures). They are generally designed to detonate in the condensed phases in order to generate overpressures for use in mining, construction, and weapons. Safety considerations and effectiveness in their use involves factors such as their ignition sensitivity, detonability, and energy content. Such issues involving explosive safety are also excluded from consideration in this article.

1.3 Flame Propagation and Rate of Pressure Rise: Experimental Systems

A variety of experimental systems for measuring the explosion behavior of combustible gases, vapors, and dusts have evolved over the years. One of the earliest was the U. S. Bureau of Mines standard apparatus for limit-of-flammability determinations (9). It consisted of a 5-cm diameter tube some 1.5-meters in length that was filled with homogeneous gas mixtures to be tested. Spark ignition was used, and visual observations were made to determine the presence or absence of flame propagation in regions beyond the spark. The spark was located at the bottom of the tube to test for upward flame propagation, and at the top of the tube to test for downward flame propagation. The tube could also be oriented horizontally, and the system could be immersed in a heated oven to study liquid fuels that required elevated temperature for vaporization. For dusts, the earliest apparatus was the

U. S. Bureau of Mines 1.2-L Hartmann apparatus (10) in which the dust was dispersed by a small pulse of air as it was being spark-ignited. The results of these measurements for many gases, vapors, and dusts of interest have been published extensively (8, 11–14) and were summarized in the earlier editions of this volume.

Those tabulations will not be repeated in this article, instead those data will be combined with more recent studies to present a general overview of the major factors that determine the flammability or explosibility of fuels. Those early test systems had several limitations: the test volumes were generally rather small, the ignition sources were often too weak (especially for dusts), the visual criteria used to judge flame propagation were somewhat arbitrary and not sufficiently objective, and for dusts, the dispersion in air was generally inadequate to ensure a uniform distribution of fuel within the test volume. These limitations have been extensively discussed in the literature (15–17), and it is beyond the scope of this article to consider those issues in detail. Suffice to say that the data to be presented here will emphasize more recent measurements obtained in large test volumes that use pressure rise rather than visual criteria for flame propagation. In addition, for the newer measurements, adequate ignition sources were used and strong dispersion methods with adequate monitoring ensured that dust concentrations were uniform within the test volumes.

There are sufficient discrepancies between the more recent data and some of the earlier tabulations (especially for dusts) to justify considerable caution in using those earlier tabulations without back-up testing. Accordingly, it is essential to have samples of gases, vapors, and dusts that are of interest tested independently in modern laboratories using the more recently developed test apparatus, rather than to depend on earlier tabulations.

As a concrete example of the more recent test methods, consider the data for methane–air mixtures summarized in Figure 56.2 (18). Pressure versus time curves are shown for mixtures tested in a 120-L chamber, ignited centrally with a strong electric spark. Figure

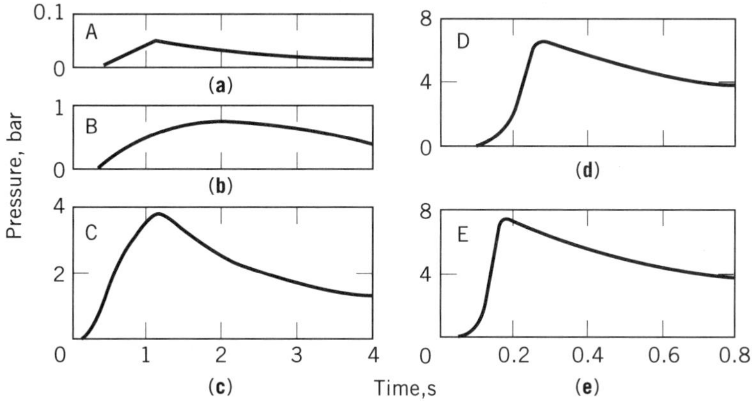

Figure 56.2. Pressure-time traces for mixtures of methane in air tested in a 120-L chamber, ignited centrally with a strong spark: (**a**) 5% methane; (**b**) 5.5% methane; (**c**) 6% methane; (**d**) 8% methane; (**e**) 10% methane (from Fig. 3 of Ref. 18).

56.2(**a**) shows the data for 5% methane in air. The pressure–time trace shows only a small pressure rise of only 0.06 bar above the initial pressure, and it is typical of mixtures at their limits of flammability. There is some burning near the spark, and the flame propagates only upward as the fireball rises by buoyancy, consuming only a small fraction of the test volume before it reaches the top of the chamber and propagation ceases. The data in Figure 56.2(**b**) is for a 5.5% methane–air mixture, and it shows a somewhat more significant pressure rise of 0.7 bar associated with upward and horizontal propagation. It too consumes only a small fraction of the test volume because of buoyancy limitations. The data in Figure 56.2(**c**) is for a 6% methane–air mixture which displays a very significant pressure rise of 4 bar, associated with flame propagation in all directions from the ignition source, which results in burning of almost all of the test volume. The data in Figure 56.2(**d**) for 8% methane–air is for near spherical propagation and it generates an explosion overpressure of 6 1/2 bar. The data in Figure 56.2(**e**) is for 10% methane–air, a near stoichiometric mixture, whose flame speed is so rapid that it consumes the entire test mixture in only 0.2 s. The measured peak explosion overpressure of 7 1/2 bar for the stoichiometric mixture is close to the calculated, adiabatic, equilibrium value for constant volume combustion, which is 7.9 bar.

Classical combustion theory (2) allows one to predict the pressure evolution from an explosion in a constant volume system. For the ideal case of spherical, adiabatic flame propagation, the pressure evolution, $p(t)$, is related to the volume, $V(t)$, occupied by the fireball during the propagation time, t:

$$[p(t) - p(o)]/[p(\max) - p(o)] = kV(t)/V \tag{3}$$

where $p(\text{o})$ is the initial pressure, $p(\text{max})$ is the maximum pressure (attained as the fireball reaches the chamber wall), V is the chamber volume, and k is a correction factor related to the difference in compressibility between the burned gases and the unburned gases. For spherical propagation from a point source during the propagation time, t:

$$V(t)/V = [r(t)/r]^3 = [S_b t/r]^3 \tag{4}$$

where $r(t)$ is the fireball radius, r is the chamber radius, and S_b is the flame speed, given by:

$$S_b = dr(t)/dt = [\rho_u/\rho_b]S_u \tag{5}$$

where ρ_u/ρ_b is the density ratio of unburned to burned gases (at constant pressure), S_b is the flame speed relative to the laboratory observer, and S_u is the burning velocity (the rate of flame propagation relative to the unburned gas motion). For spherical propagation in a spherical chamber, $p(\text{max})$ is reached at the instant the fireball contacts the wall, and at that instant, only burned gas is present in the system, and $k = 1$. Differentiating equation 3 with respect to time and substituting equations 4 and 5 into the result, gives:

$$dp(t)/dt = 3[p(\max) - p(o)][\rho_u/\rho_b][r(t)^2/r^3]S_u \tag{6}$$

Equation 6 shows that the maximum rate of pressure rise should occur at the instant the fireball contacts the chamber wall. At that instant, $r(t) = r = [3V/4\pi]^{1/3}$, and setting $\rho_u/\rho_b = T_b/T = p(\max)/p(o)$, gives:

$$K_G \text{ (or } K_{St}) = [dp(t)/dt]_{\max} V^{1/3} = 4.84[p(\max)/p(o) - 1]p(\max)S_u \tag{7}$$

Equation 7 is the so-called "cubic law" and the *K*s are the size-normalized rates of pressure rise. The subscript *G* refers to gaseous mixtures, and the subscript *St* refers to dust mixtures, and is taken from the German word for dust: "Staub". The *K*-values are used in the practical design of rapid-acting, venting systems that limit system overpressures during explosions so that damage to equipment is minimized in the event of accidental explosions (19). Because the *K*-values are size-normalized, they can be measured in systems of relatively small volumes, but reliably applied to much larger systems used in industry. They can be used reliably only if the propagation is near spherical at a constant flame speed but cannot be used reliably if the propagation becomes turbulent and the flame speed accelerates rapidly.

A graphical summary of all the recent data obtained for mixtures of methane in air tested in both the 120-L chamber and in a smaller 20-L chamber, are shown in Figure 56.3. The top figure shows the measured K_G-values, the lower figure shows the measured pressure ratio $p(\max)/p(o)$. Both are plotted as a function of the methane concentration. Also shown in the lower figure (as the dashed line) is the calculated adiabatic value for the explosion pressure ratio. These data are typical of those for all flammable gases. For methane, there are sharp discontinuities at the lower flammability limit near 5% and at the upper flammability limit near 16% methane. Flame propagation is not possible beyond these limits even though calculated, adiabatic explosion pressures are still quite high at lean limit concentrations and below, and at rich limit concentrations and above. Each gaseous fuel has its own characteristic lean and rich limits.

2 LIMITS OF FLAMMABILITY AND APPLICATION TO PRACTICE

As shown above, limits of flammability are the extremes of concentration between which fuel–air mixtures can sustain combustion-driven explosions when subjected to an effective ignition source. Beyond these composition limits, flame propagation is not possible. For example, while methane–air mixtures can be readily oxidized on a heated or catalytic surface at concentrations well below the 5% lean limit, such combustion is confined to the surface and cannot propagate into the surrounding mixture to generate significant explosion pressures. Only mixtures in the 5–16% range can propagate beyond the heated surface. The lower limit is referred to as the lean limit of flammability, and the upper limit is referred to as the rich limit of flammability. In general, those limits for a particular pre-mixed system of fuel and oxidant are affected by the initial temperature and pressure of the mixture. They are also affected by the direction of flame propagation, and other special factors such as the selective diffusion of reactants into the flame front. For dusts, there are

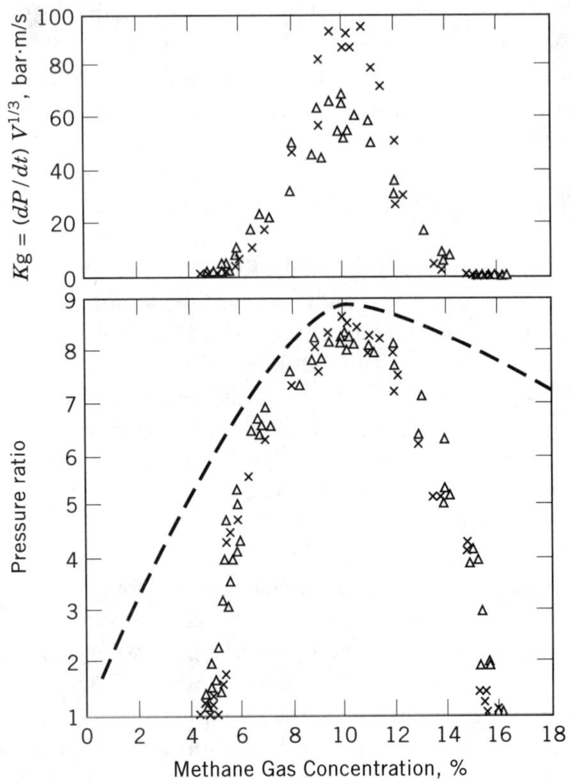

Figure 56.3. A summary of the flammability data for methane–air mixtures tested in a 20-L and a 120-L chamber, plotted as a function of methane concentration. Dashed line in the lower figure is the calculated, adiabatic explosion pressure ratio (from Fig. 5 of Ref. 18). x = 120 L; Δ = 20 L.

additional factors involved: the particle size of the dust, its combustible volatile content, and its rate of devolatization within the flame front. In addition, dusts differ markedly from gases in their rich limit behavior: dusts generally do not display a well-defined rich limit of flammability.

In considering the question of preventing accidental explosions, it is impractical to differentiate between mixtures on the basis of the amount of violence produced in a laboratory test, and accordingly one has to be very cautious in applying the *K*-values considered earlier to real systems. Much more than mixture composition is involved in determining the destructive effect of an explosion. Consider, for example, a mixture just within the flammability limit (say 6% methane) that is confined to a long tube, open at one end and ignited at the open end. In such a system flame will propagate quietly and slowly through the tube generating trivial overpressures as the burned gases vent through the open end of the tube. If the identical mixture is ignited at the closed end of the same tube, flame propagation is accelerated by the turbulence generated in the unburned mixture flowing outward through the open end. The flame accelerates rapidly and it soon becomes impossible for the venting flow to balance the speed of flame propagation. Explosion pressures of 3–5 atmospheres can then be attained in that same mixture, and if the tube is long enough and wide enough, even higher pressure detonations can result even though the mixture composition was barely within the limit of flammability. Clearly the configuration of the flammable volume relative to the ignition point plays a critical role, and much more than mixture composition is involved in determining the destructive effect of an explosion. For ignition at the closed end, pressures are high enough to cause the failure of the tube, as evidenced in Table 56.1, which lists the peak overpressures for the failure of typical structural elements.

Thus no comfort can be taken in the expectation that only a limit composition can be encountered in a specific operation. Any unwanted flame propagation is inherently disastrous. Consequently, good safety practice prohibits closely skirting the flammability limit boundaries. In mines, for example, the precautionary measure of shutting down machinery that can cause ignition is mandated (21) at methane concentrations that are within 20% of the lean limit concentration (20% of 5% methane = 1% methane). At methane concentrations of 1.5%, evacuation of the mine is required. Such margins are well-advised for a variety of reasons. The flammability limits measured in standard tests are for completely mixed fuel–air compositions. In real situations in homes, commerce, industry, and nature, the fuels are generally incompletely mixed, and local flammable mixtures can exist in a part of the volume under consideration even though a completely mixed system would be nonflammable. For example, the release of 1 liter of pure methane gas in a 100-liter container of air would give only a 1% methane mixture which would be nonflammable if the mixing were complete throughout the entire volume; however, methane is lighter than air, and if the gas is released near the top of the container, and mixes with only the upper 10 L of the available air, it would generate a stoichiometric mixture at the top of the container. If ignited, the combustion of that upper layer could generate an overpressure that would be disastrous even though most of the volume below is nonflammable. A layering hazard from such partial mixing is possible not only for lighter-than-air fuels that form roof-hugging layers, but also from flammable gases that are heavier than air such as propane, butane, and gasoline. Such heavier fuels can generate disasters by way of ground-hugging layers that travel considerable distances from their point of release, mixing with small volumes of air near the ground before they find ignition sources. Flammable dust

Table 56.1. Peak Overpressures for the Failure of Structural Materials[a]

Structural Material	Usual Failure Mode	Peak Overpressure (bar)
Glass windows	Shattering with frame failure	0.03–0.07
Corrugated siding (steel or aluminum)	Connection failure and buckling	0.07–0.14
Wood siding (standard house panels)	Connection failure and buckling	0.07–0.14
Concrete walls (not reinforced)	Shattering	0.14–0.20
Brick walls (8–10 in. not reinforced)	Shearing and flexure failure	0.48–0.54

[a]From Ref. 20.

layers collecting on surfaces are even more dramatic examples of surface-hugging fuel layers than can readily generate explosive mixtures if they are partially dispersed within their system volumes.

Nor should much comfort be taken in operating systems at fuel concentrations above the rich limit of flammability. Such systems become explosive when they are diluted with air. Thus, the ullage space above the liquid fuel level in an automobile gasoline tank generally contains a high enough hydrocarbon vapor concentration to normally be above the rich limit of flammability. But an open tank, or one that has just been ruptured by collision, causes a dilution of the rich mixture to a fuel concentration that is below the rich limit and therefore flammable. The dilution with air can generate a large flammable volume, leading to disastrous consequences upon ignition.

2.1 Homogeneous Gas Mixtures

Over a century ago, the early investigators of the flame propagation process (22, 23) noted that the lean limit mixture heats of combustion for a variety of saturated hydrocarbon fuels were approximately constant at about 11–12 kcal per mole of limit mixture. If L is the lean limit concentration and ΔH is the heat of combustion per mole of fuel, then their product $L\Delta H$ is the lean limit mixture heat of combustion. If the heat of combustion is expressed in kcal per mole of fuel, and if L is expressed in volume ratio of fuel to total mixture volume (or its equivalent, mole ratio of fuel to total number of moles in the mixture), then the product $L\Delta H$ is the lean limit mixture heat of combustion in kcal per mole of fuel–air mixture. It is also referred to as the limit calorific value. Since those mixtures contain mostly air whose heat capacity is constant, a constant limit calorific value corresponds to a constant limit flame temperature of about $T(L) = 1500$ K.

There has been a temptation over the years to use those values of 11–12 kcal/mole, and 1500 K as kinds of "magic numbers" to predict whether some untested mixtures can involve explosion hazards. For methane, the limit calorific value for upward propagation at room temperature and pressure is 9.6 kcal/mole of mixture. For downward propagation, the value for methane is 11.0 kcal/mole of mixture. With increasing carbon number, the values for the heavier saturated hydrocarbons from ethane to pentane show a small increase in limit calorific value, and the values for upward and downward propagation begin to converge. Above hexane, the values level off to about 11.6 kcal/mole of mixture, remaining constant at that value for all the heavier saturated hydrocarbons (24).

One is thus tempted to argue that pure thermodynamics would do a good job in hazard prediction; however, there is a hint that reaction kinetic effects cannot be ignored from the limit calorific values for the branched-chain isomers of pentane and hexane. These more reactive isomers can sustain flame propagation at significantly lower limit calorific values than their straight chain counterparts.

The chemical kinetic effects become even more dramatic if one considers a variety of other fuels: amines, alkenes, alkynes, and hydrogen. Ammonia, the zero carbon number member of the primary amines, has a limit calorific value of 12.4 kcal/mole of mixture, and requires a limit flame temperature of 1800–1900 K to sustain flame propagation. It is clearly less reactive than the hydrocarbons. As the carbon number of the primary amine increases, and the effect of the amine group become smaller relative to the hydrocarbon portion of the compound, their limit calorific values converge to the hydrocarbon value of 11.6.

The unsaturated alkenes from ethylene to butene are all more reactive than their saturated counterparts. Ethylene's limit calorific value for downward propagation is 9.4 compared to ethane's value of 11.1 (the units of kcal/mole of mixture are hereafter to be omitted). The value for propene is comparably lower than the propane value, and similarly for butene relative to butane.

The alkynes are even more reactive; acetylene's limit calorific value for downward propagation is about 8.2 compared to ethane's 11.1. For all the double- or triple-bonded unsaturated hydrocarbons that contain only one unsaturated group, the limit calorific values converge to the saturated hydrocarbon value of 11.6 as the carbon number increases. Similarly for the amines, alcohols, and other functional groups. Clearly, the higher the molecular weight of the compound, the less significant the functional group relative to the paraffin chain, and the more closely the limit calorific value approaches 11.6. For the saturated hydrocarbons themselves, the reactivity factors vary somewhat for the lower carbon number compounds whose H/C ratios vary significantly from 4 for methane to 3 for ethane, to 2⅔ for propane, to 2½ for butane. Above hexane, where the H/C ratio becomes fairly constant, near 2.0, the reactivity factors remain constant and the limit calorific value is also constant at 11.6. The generalization even applies to the saturated hydrocarbon *solid*, polyethylene, whose carbon number is nearly infinite. The constant reactivity for the saturated alkanes means that the chemical kinetics of their oxidation processes within their flame fronts must all be quite similar. A study of their burning velocity behavior has, in fact, shown that the higher molecular weight alkanes all rapidly degrade to lower molecular weight intermediates as they enter the flame front. The rate-limiting reactions involve the oxidation of those lower molecular weight intermediates, and hence their reactivities are independent of the molecular weight of the starting alkane (25).

The most dramatic effect of chemical kinetics is for hydrogen itself: its limit calorific value is only about 2.5 kcal/mole of mixture for upward flame propagation, and about 5.0 for downward flame propagation in quiescent mixtures. Hydrogen also displays a most dramatic difference in lean limit compositions depending upon the direction of propagation. That effect is caused by the marked tendency of the lighter hydrogen molecules to diffuse more rapidly into the flame front than the heavier oxygen molecules. As a result of that enrichment in hydrogen, the flame behaves as though it is much richer in hydrogen than one would expect from its initial concentration in the unburned mixture. This phenomenon of selective diffusional demixing also plays an important role in the rich limit behavior of the heavy hydrocarbons and tends to suppress the difference in rich limits for their upward versus downward propagation. Such complexities (26) are beyond the scope of this article and will not be considered in detail here. Low limit calorific values comparable to the hydrogen values of 2.5 to 5.0 also characterize very reactive fuels such as hydrazine and diborane. Such reactive fuels can propagate flame at limit flame temperatures of only 800 K or even lower.

The above generalizations for limit calorific values apply to the normal, run-of-the-mill fuels: those that contain abundant amounts of hydrogen atoms in their structure. Such fuels generate abundant amounts of H-atoms, OH and HO_2 radicals, which play key roles as chain carriers in the flame propagation process. There are some fuels, however, that are nonhydrogen bearing, such as carbon monoxide, cyanogen, and carbon disulfide. The limit behavior of such fuels show an anomalous sensitivity to hydrogen or hydrogen-bearing impurities such as moisture. They are far less reactive in their pure state and their flame propagation rate is markedly catalyzed by providing a source of H-atom chain carriers. For such hydrogen-starved fuels in their ultrapure state, kinetic factors seem to dominate over the thermodynamic ones.

Because such kinetic effects are difficult to predict in advance, particularly as they might relate to mixtures of fuels with varying kinetic factors, the earlier caveat regarding the uncritical use of previously published tabulations needs to be reemphasized. *Potentially hazardous mixtures should be tested directly and independently, in modern test laboratories, in large test volumes, with strong ignition sources, and under conditions that accurately simulate the conditions of their use in industry.*

Although, as shown above, limit calorific values vary markedly for the different fuels, research by this author has shown that there is one parameter that is essentially invariant at the limit composition for different fuel types. That parameter is the ideal burning velocity of the mixture at the limit composition. At the limit composition for upward flame propagation, that limit burning velocity is about 3 cm/s, independent of fuel type. The mechanism for flame quenching at that limit involves the buoyancy of the burned gas fireball, whose upward velocity induces natural convective flow gradients that stretch the flame front, forcing it to "swallow" more cold, unburned gas than it can ignite. For horizontal flame propagation, the limit burning velocity is about 6 cm/s. Such a detailed theoretical analysis of the causes of the existence of flammability limits, is clearly beyond the scope of this article. Readers with an interest in the question are referred to the author's earlier publications (24, 26–29).

2.2 Dust Dispersions in Air

Solid dust fuels, such as the coal dust–air mixture shown in Figure 56.1(**a**) can also sustain highly exothermic chemical reactions with air, and are thus capable of generating explosions similar to those for homogeneous fuel gases in air. Examples of such dust explosions for a variety of fuel dusts are shown in Figure 56.4 (30). Shown are the pressure–time traces for eight different dusts dispersed in air, each at a concentration of 600 grams per cubic meter, and ignited in a 20-L chamber. The traces are for two carbonaceous dusts: coal dust and polyethylene, and for six metallic dusts: aluminum, magnesium, titanium, iron, zinc, and tantalum. The relative reactivities of the various dusts can be estimated either from their peak explosion pressures, or from their maximum rates of pressure rise. The aluminum dust shows the highest reactivity, but that does not necessarily mean that aluminum is intrinsically the most reactive. Its particle size was much finer than for any of the other dusts shown.

For a dust, there is an additional variable to consider: its reactivity in the flame front depends not only on the intrinsic reactivity of it fuel component, but also on its specific particle size, or size distribution.

As the concentration of any of the dusts shown in Figure 56.4 is reduced from its 600 grams per cubic meter value to lower values, the exothermicities of the reactions eventually

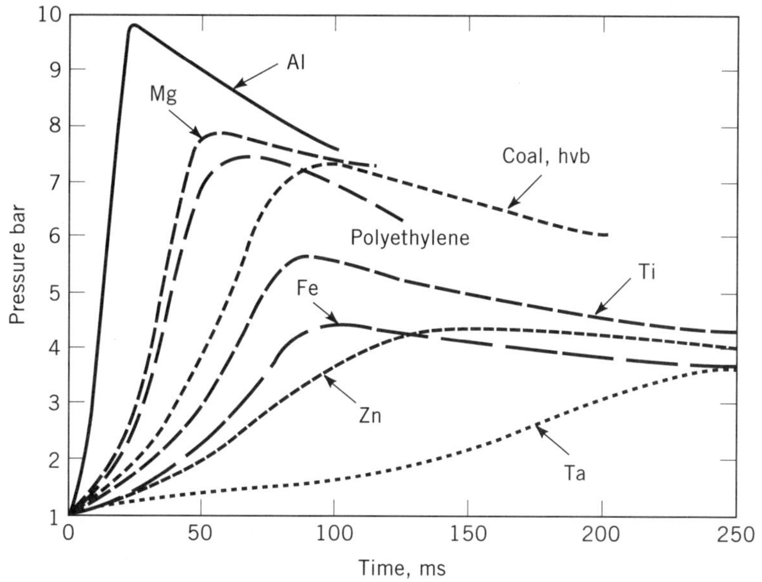

Figure 56.4. Examples of dust explosion data for a variety of fuel dusts: pressure-time traces for carbonaceous and metallic dusts.

diminish as the fuel concentrations decrease. At some limiting dust concentration, the mixture is no longer capable of sustaining flame propagation and explosions are not possible below that limiting concentration. For dusts this lean limit concentration is also referred to as the minimum explosive concentration (MEC). Just as each homogeneous gaseous fuel–air mixture has its characteristic lean limit concentration, so does each fuel dust mixture in air have its own characteristic MEC. However, dusts differ from gases in that its MEC or lean limit depends on an additional variable: the particle size.

Dust dispersions differ from gaseous mixtures in still another way, as illustrated in Figure 56.5. The data for methane–air explosions are compared with the data for polyethylene dust and for high volatile bituminous (hvb) coal dust explosions. The data are maximum explosion pressures as a function of fuel concentration. The methane data are replotted in terms of mass concentration of methane per unit volume of air so that they can be compared directly with the dust data. The methane plot, as considered earlier, shows its well-defined lean limit of flammability at 35 grams per cubic meter (5% methane by volume), and its rich limit of flammability at 130 grams per cubic meter (16% methane by volume). Methane's maximum explosion pressure occurs at a stoichiometric concentration. While the dusts display similar lean limit behavior at their characteristic MECs, they do not show the "normal" rich limit behavior. At high dust concentrations, where the mixtures are nominally fuel rich, the explosion pressures do not decline as they do for methane. Instead, they remain at the high levels that characterize the near-stoichiometric mixtures. These data for polyethylene and coal dust are typical of the behavior of all

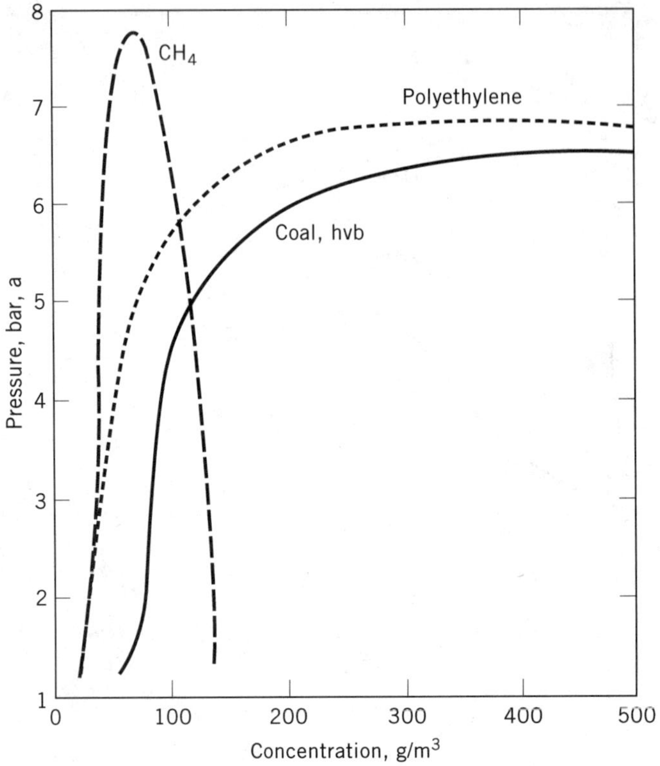

Figure 56.5. Explosion pressure data for polyethylene dust and high volatile bituminous (hvb) coal dust in air compared with the data for methane gas in air (from Fig. 7 of Ref. 30).

explosive dusts, and they are a reflection of the phase heterogeneity of the reacting system. It is the same effect that causes the MEC of a dust to be dependent on particle size.

For dusts, the inherent phase heterogeneity of the system, introduces two additional rate processes that must be completed before the "normal" combustion reactions can proceed: the first is that the solid phase fuel particle must first devolatilize, and the second is that the volatiles must mix with the air surrounding each particle. These processes must be completed within the flame front before flame can propagate through the dust–air mixture. These additional rate process requirements are the cause of both effects: the dependence of the MEC on particle size, and the absence of rich limit behavior. In the former case, the additional requirement limits the flame propagation rate as particle sizes increase. In the latter case, the additional requirement prevents the accumulation of excess fuel that would normally lead to a rich limit.

Devolatilization is the pyrolysis or thermal decomposition of the dust particle resulting in the generation of combustible volatiles. Only after devolatilization, and mixing of those volatiles with air, can the normal, gas-phase combustion reactions proceed. As dust concentrations increase, the devolatilization rate becomes the limiting rate; that is, it becomes too slow relative to the rate of the combustion reactions in the gas phase to allow each particle to devolatilize completely. The flame front then propagates rapidly through a near stoichiometric concentration of volatiles as soon as that concentration is generated and before it can be diluted with excess fuel. As a result, the flame front "rides the crest" of a near stoichiometric concentration, leaving the partially devolatilized dust in its wake. A near stoichiometric concentration is thus reacting within the flame front regardless of the initial dust loading; hence the absence of a "normal" rich limit. Eventually, at very heavy dust loadings of several thousands of grams per cubic meter, the large mass of excess unreacting dust will provide a sufficient heat sink for the flame to be quenched. Such quenching is not a normal rich limit, and it occurs at such a high dust loading that for practical purposes dusts can be considered to have no rich limit of explosibility (31).

It should also be noted that the data for polyethylene show that its lean limit (MEC) occurs at virtually the same mass concentration as for methane (and the other normal paraffin hydrocarbons). That is no coincidence because polyethylene volatilizes rapidly and completely during combustion at the limit burning velocity (but not at a near-stoichiometric burning velocity). Polyethylene thus behaves similar to the saturated alkanes at their lean limit concentrations, with the same combustion chemistry, the same limit calorific value, and the same lean limit flame temperature of 1500 K.

For the coal dust data shown in Figure 56.5, the MEC is considerably higher, about 80 grams per cubic meter because it is not completely volatilizable. There is always a char residue from coal dust, and a small fraction of its volatiles are noncombustible. The hvb coal has a volatility of 37% (actually somewhat higher at the heating rates within a flame front), so that a higher dust loading is required in order for it to generate the minimum combustible volatile content of hydrocarbon fuel. The lower volatility coals display proportionately higher MECs consistent with their lower volatilities.

From this discussion, it is clear that the explosion behavior of carbonaceous dusts is quite consistent with the well established mechanism of flame propagation that applies to homogeneous gas flames provided that additional processes are added: the heating and devolatilization of the solid dust particles, and the mixing of those volatiles with air. Those additional processes most clearly distinguish dust explosions from homogeneous gas explosions, and they can become rate-controlling at high burning velocities, for rich dust concentrations, and for coarse particle sizes. Most of the available data for metallic dusts indicate similar behavior, but over a different range of flame temperatures and devolatilization rates (32).

2.3 Liquid Droplet Sprays, Flash Points

Sprays of evaporating liquid droplet fuels dispersed in air contain both a phase heterogeneous mixture of liquid droplets in air, and a gaseous mixture of fuel vapor in the air space between the droplets. Data on the flammability limits of such systems are rare because it is very difficult to independently control the important variables: the mass concentration of dispersed fuel droplets, their particle size distribution, and the vapor phase concentration of the evaporating gaseous fuel. Since they are hybrid mixtures of a condensed phase, liquid droplet suspended in air, and a gaseous fuel–air mixture, their flammability behavior should be similar to that of hybrid mixtures of fuel dusts and gases which are discussed in the next section (2.4). For liquid fuel sprays, however, the droplets will generally volatilize more completely in the flame front than will solid dusts, and their rate of devolatilization will be more rapid. Accordingly, fuel droplet sprays will generally be more flammable and easier to ignite than hybrid dust–gas mixtures dispersed in air.

For a nondispersed liquid fuel in a liquid pool form, there is a traditional measurement that has been used to estimate its "flammability" hazard; namely, the "flash point." This has been used with a standardized test method to compare the flammability hazard of various liquid fuels to one another (33, 34). Flash points are relatively easy to measure and have been compiled for a wide variety of pure organic compounds, fuel mixtures, lubricants, and other mixtures of commercial interest. The main problem with the classification is its failure to distinguish between the fire hazard and the explosion hazard associated with the fuel. Generally, fuels with low flash point temperatures below room temperature, represent explosion hazards, whereas fuels with flash point temperatures above room temperature are mainly fire hazards. But "room temperature" can change markedly under different ambient conditions and the borderline depends upon the particular industrial use.

The most reliable form of the measurement of the flash point involves the "closed-cup" method. A given volume of the liquid sample is heated and a given volume of air is isolated above it. The heating is at a prescribed rate so that the temperature of the liquid reservoir increases slowly. As its temperature increases, the liquid vaporizes generating a fuel vapor–air mixture in the space above the liquid surface whose concentration is determined by the equilibrium vapor pressure of the liquid. In the closed cup method, a small opening in the air volume above the liquid is periodically uncovered and exposed to a "test" ignition flame or a spark. At some point the test flame will ignite the fuel–air mixture above the liquid surface, and flame will propagate freely across the surface of the liquid as a "flash". The liquid reservoir temperature at which that happens is called the "flash point". The propagation of the flame in the gas phase may or may not provide sufficient preheating to generate a pool flame in its wake. If not, the heating, continues until the liquid pool surface sustains a flame, and that temperature is referred to as the "fire point" (34). Clearly flame can propagate in the fuel air space above the liquid only when enough fuel has been vaporized to generate a fuel vapor concentration in the air space that is at least equal to the fuel's lean limit of flammability. Thus flash points and lean limits of flammability are related to each other through the liquid's equilibrium vapor pressure. At the flash point, the partial equilibrium vapor pressure of the fuel in air (in percent by volume) is equal to its lean limit concentration in volume percent of fuel in air.

Clearly, the higher the flash point of a fuel, the lower its flammability hazard. The important safety parameter in the event of a liquid spill, is the liquid's flash point temperature relative to the highest ambient temperature that may be encountered in the storage, transportation, or use of the fuel. It must however be clearly recognized that some uses present special circumstances, for example, if the liquid fuel is pressurized. Such pressurized fuels can generate fuel sprays if their containers are ruptured, and such sprays dispersed in air are highly explosive regardless of the liquid's flash point. Thus, for example, diesel fuels with high flash points would normally represent only a fire hazard in a normal spill of the liquid; however, that same diesel fuel in a pressurized system that ruptures and disperses the fuel in droplet form, is also an explosion hazard. That is, in fact, how the fuel is dispersed within the diesel engine, to autoignite in internal explosions that generate the engine's motive power.

2.4 Mixtures of Fuels

2.4.1 Homogeneous Fuel Gas Mixture in Air

For mixtures of various fuel gases in air, the limits of flammability for the fuel mixture can be calculated from the limits of flammability for each individual fuel component of the mixture by using the simple additive formula proposed by LeChatelier (22). Consider a binary mixture of two fuel gases in air. If the lean limit concentrations (in volume percent) for each fuel measured separately are $L(1)$ and $L(2)$ respectively, and if their concentrations (in volume percent) in the air mixture of the two fuels are $v(1)$ and $v(2)$ respectively, then the condition for the binary fuel mixture to be flammable is:

$$v(1)/L(1) + v(2)/L(2) > \text{or} = 1 \tag{8}$$

The equality condition delineates all possible lean limit concentrations for the binary fuel mixture in air. This LeChatelier law (not to be confused with the other LeChatelier law for chemical equilibrium) indicates that each fuel contributes linearly to the lean limit of the mixture in proportion to its concentration relative to its lean limit concentration. The ratio v/L is a dimensionless measure of its concentration, normalized to its individual lean limit. Each fuel thus contributes its fractional part, v/L, to the overall flammability of the binary fuel mixture. Thus, for example, a mixture of methane and ethane in air that contains one-half the amount of methane required to form a lean limit mixture of methane [that is, $v(\text{methane})/L(\text{methane}) = 0.5$] and also contains one-half the amount of ethane required to form a lean limit mixture of ethane [that is, $v(\text{ethane})/L(\text{ethane}) = 0.5$], would then be at the lean limit of flammability for the mixture of the two fuels. That mixture would consist of 2.5% methane and 1.55% ethane, and the remainder would be air. A mixture containing one-quarter of the amount methane needed to form its lean limit in air, and three-quarters of the amount of ethane needed to form its lean limit in air, would also be at the lean limit of flammability for the mixture of two fuels. That mixture would consist of 1.25% methane and 2.33% ethane.

All possible concentrations of the binary mixture of fuels 1 and 2 can be represented graphically as points in a two-dimensional space with $v(1)$ as one coordinate, and $v(2)$ as the other coordinate. The lean limit concentration for one of the pure fuels is the point $v(1) = L(1)$ on one axis, and the lean limit concentration for the other pure fuel is the point $v(2) = L(2)$ on the other axis. Equation 9 represents the straight line between those two points or intercepts, and it delineates all possible lean limit mixtures of the two fuels in air.

$$v(1)/L(1) + v(2)/L(2) = 1 \tag{9}$$

The LeChatelier formula may be generalized to apply to mixtures containing any number of fuels, so that all possible lean limit mixtures for the multicomponent fuel mixture in air is represented by the condition:

$$v(1)/L(1) + v(2)/L(2) + v(3)/L(3) + \ldots = 1 \qquad (10)$$

where the summation is taken over all the fuels in the mixture.

A more useful form of LeChatelier's law (9) expresses the lean limit concentration for the mixture of fuel gases, $L(x)$, (considered as a single fuel component) in terms of the percentages of each pure fuel component in the mixture.

$$L(x) = \frac{100}{p(1)/L(1) + p(2)/L(2) + P(3)/L(3) + \ldots} \qquad (11)$$

where $p(1)$, $p(2)$, $p(3)$, . . . are the percentages of each fuel in the fuel mixture (excluding air), so that $p(1) + p(2) + p(3) + \ldots = 100$, and $L(x)$ is the lean limit concentration of the total fuel mixture. $L(x)$ is its volume percent in air considering the fuel mixture as a single component.

Thus, for example, consider a natural gas mixture containing the following components:

	Percent in natural gas mixture	Lean limit of pure component (% in air)
methane	80	5
ethane	15	3.1
propane	4	2.1
butane	1	1.8

The mixture's lean limit concentration (percent natural gas in air) would be:

$$L(x) = \frac{100}{80/5.0 + 15/3.1 + 4/2.1 + 1/1.8} = 4.3\%$$

The accuracy of LeChatelier's formula has been tested for many fuel mixtures, and while it is reasonably accurate for run-of-the-mill fuels such as those considered in the above example, there are many exceptions. The formula is derivable using the well-established concept of a constant limit flame temperature for a given class of fuels. For the hydrocarbon mixtures considered in the example, all the fuels in the mixture have approximately the same limit flame temperature of 1400–1500 K, and their combustion reactions all involve similar chemical kinetic reaction rates and diffusion processes; hence, LeChatelier's law is obeyed. However, hydrogen has a limit flame temperature of only 800 K, and one must be very cautious in applying the law to mixtures of hydrogen with other fuels that have much higher limit flame temperatures. Our previous caveat with respect to the use of early literature tabulations applies doubly if those uncertainties are further compounded by extending the data to fuel mixtures. *By no means should the lean limits estimated from LeChatelier's law be used as a substitute for actual measurements for the fuel mixtures of interest.* Direct flammability limit measurements in modern laboratories using the most recent test systems are to be preferred whether one is dealing with a pure fuel or a mixture of fuels.

2.4.2 Hybrid Gas/Dust Mixtures and Mixtures of Dusts.

The most extensive measurements for hybrid gas/dust mixtures involve the lean limit data for methane addition to coal dust, which has been obtained in both laboratory systems and in full scale mine experiments (1, 35). The data show that the lean limit concentrations for the hybrid mixtures involve the same linear weighting factors as for gaseous mixtures, and that LeChatelier's law applies to mixtures of methane and coal dust in air. LeChatelier's law was also valid for mixtures of methane with oil shale dust (36). The law was also shown to be valid for a mixture of gilsonite dust and coal dust (1). In all of these cases, the pure fuels all have about the same limit flame temperature and involve similar hydrocarbon fuels and gas-phase combustion reaction mechanisms; hence they obey LeChatelier's law.

It should be emphasized, however, that more is involved than the flammability limit behavior in estimating the hazard involved in adding methane to the coal dust accumulations in a mine. The addition of a fuel gas to a dust–air mixture will also increase the explosion hazard because the gaseous fuel is much more easily ignited than the dust. This question of ignitability will be considered in more detail in subsequent sections. Suffice to say at this point that the spark ignition energy for the gaseous fuel is so much lower than that for the dust that the addition of a small amount of fuel gas may have a much more profound effect on the ignitability of the mixture than would be predicted by a simple linear relationship. There are complex synergistic effects among the various types of ignition sources: thermal, electrical, and chemical. Since the gaseous fuel is much more easily ignited by any of those sources, the gas sensitizes the mixture to ignition to a greater extent than it influences the flammability limit, and the overall hazard is magnified even more by the presence of the gas.

By contrast to the methane addition data, the data for hydrogen addition to cornstarch (37), show significant departures from the linear relationship in LeChatelier's law. The limit flame temperatures for methane and coal dust are of comparable magnitude because their combustion reactions involve similar chemical kinetic reaction mechanisms and diffusion processes. Accordingly, there are no significant changes in flame temperature or reaction rates in going from pure methane to pure coal dust, and the linear relationship in LeChatelier's law is maintained. By contrast, the limit flame temperature for hydrogen is much lower than the limit flame temperature for the cornstarch dust. There are, therefore, rather drastic changes in flame temperature and reaction rates in going from pure hydrogen to pure cornstarch dust. Although the hydrogen fuel can maintain its combustion rate at the lower temperatures, the devolatilization rate of the cornstarch dust, or the reaction rate of its volatiles, or both, are too slow at those temperatures for the linear relationship to be maintained. The required temperature of the limit mixture must therefore be higher than the linear weighted average, and more fuel dust is required to render the system flammable, much more than is predicted on the basis of LeChatelier's law.

2.5 Other Factors Influencing the Limits of Flammability

2.5.1 Initial Temperature

Generally both gases and dusts become more flammable at elevated initial temperatures. Preheating the mixture supplies additional enthalpy to the system, and hence less combustion energy is required for flame propagation. This allows flame to propagate at leaner fuel–air mixtures that supply less combustion energy. At the standard ambient temperature of 25°C, where the lean limit concentration is $L(25)$, the minimum enthalpy that must be supplied by the combustion reaction to sustain flame propagation is $L(25)\Delta H$, where ΔH is the heat of combustion per mole of fuel. That quantity was referred to earlier as the lean limit mixture heat of combustion, or the limit calorific value. At some elevated temperature, T, the total enthalpy available to the system is the sum of the combustion enthalpy and the sensible enthalpy, which is: $L(T)\Delta H + C(T - 25)$, where C is the heat capacity of the unburned mixture, and $L(T)$ is the lean limit concentration at that higher temperature. Equating that higher total enthalpy to the minimum required for flame propagation gives:

$$L(T)\Delta H + C(T - 25) = L(25)\Delta H \quad (12)$$

Solving for the ratio of the lean limit at T to the lean limit at 25°C gives:

$$L(T)/L(25) = 1 - C(T - 25)/L(25)\Delta H = 1 - 0.00072(T - 25) \quad (13)$$

where our earlier value of 11.6 kcal/mole of mixture was used for the limit calorific value for typical hydrocarbons in order to obtain the constant of 0.00072 C^{-1}. Equation 13 is the Burgess-Wheeler law, which applies reasonably well to typical hydrocarbon fuel vapors in air (38, 39). Essentially the same derivation applies to other classes of fuels provided that the constant in the equation is adjusted to reflect their own limit calorific values. For hydrogen–air mixtures, for example, the limit calorific value is much lower and the predicted constant in equation 13 is therefore considerably higher than for the hydrocarbons. For hydrogen that constant is 0.0011 C^{-1}, and is in fair agreement with the data (9).

For dusts, the same derivation applies, and if the Burgess-Wheeler law is expressed in terms of the mass concentration of the fuel dust, it becomes:

$$C(T)/C(25) = [1 - \beta(T - 25)][298/(T + 273)] \quad (14)$$

where $C(T)$ is the dust's lean limit concentration at the elevated temperature, and $C(25)$ is the lean limit concentration at 25°C. The value of the constant β should again depend on the type of the dust fuel and its limit calorific value. For carbonaceous dusts, its value is 0.0006 to 0.0007 C^{-1}, comparable to its value for hydrocarbon gases. Equation 14 has been shown to be reasonably valid for Lycopodium dust (40), for sulfur and benzoic acid dusts (41), and for a variety of coal dusts and agricultural dusts (42).

Aside from the question of flammability limits, the effect of initial temperature on ignitability is also an important consideration in assessing explosion hazards. The available data shows that the effect of elevated initial temperature on ignitability is much more marked than on the limits of flammability. For example, at a coal dust concentration of 400 g/m^3, an increase in initial temperature from 75 to 200°C results in an order of magnitude reduction in the spark ignition energy while the lean limit decreases by only a factor of two (40). There is clearly a strong synergy between the thermal energy content and the spark ignition energy which magnifies the ignition hazard at elevated temperatures.

And finally, as indicated earlier, for a liquid fuel in its reservoir, or in a spill, the ambient temperature has a profound effect on the magnitude of the hazard. A fuel in liquid form is normally only a fire hazard if the ambient temperature is well below its flash point temperature. However, as the ambient temperature rises to approach or exceed the flash point, the hazard magnifies markedly from only a fire hazard to both a fire and explosion hazard.

2.5.2 *Initial Pressure*

Another important factor that can influence the lean limit concentration of a fuel and its overall explosibility behavior, is the operating pressure of the system. For hydrocarbons, the lean limits, *expressed in volume percent*, vary very little as pressures are increased above their ambient value. For methane in air, for example, the lean limit is constant at 5% methane for pressures up to 100 bar (9, 39). If the methane concentration is expressed in mass concentration per unit volume, the lean limit varies linearly with initial pressure. The same linear dependence is observed for dusts such as polyethylene whose limits are usually measured in terms of mass concentration. In fact, the lean limit versus pressure curves for methane and polyethylene are virtually congruent in terms of their mass concentration (1) reflecting their similar kinetic mechanisms of combustion. The same linear dependence is observed for various ranks of coal dust (42).

2.5.3 *Inerting and Minimum Oxygen Concentration*

One method of preventing a gas or dust explosion involves the addition of an inert gas to the air space of interest so that the oxygen concentration is reduced to such a low level that flame can no longer propagate in the region regardless of the fuel concentration. The combustion reactions that normally generate explosions involve both fuel and the oxygen of the air as the reactants. Just as there is a minimum fuel concentration required for flame propagation when sufficient air is present (the lean limit), so too is there a minimum oxygen concentration required for flame propagation when there is sufficient fuel present. That limiting oxygen level is referred to as the minimum oxygen concentration, and its value depends on the reactivity of the fuel and on the nature of the inerting gas added to the air in order to reduce is oxygen content.

For nitrogen added as the inert diluent to air, the minimum oxygen concentration for typical hydrocarbon fuels is in the range of 10 to 12% oxygen by volume (9, 34). Similar minimum oxygen values are measured for carbonaceous dusts (30). Carbon dioxide, with its higher heat capacity, is somewhat more effective than nitrogen, and if it is used as the inerting additive, the minimum oxygen concentration is in the range of 13 to 15% for typical hydrocarbons.

For the more reactive fuel, hydrogen, much lower oxygen levels are needed: the minimum is 5% for nitrogen inerting and 6% for carbon dioxide inerting.

If inerting is used, a reasonable safety margin is recommended, and the safe inerting level is suggested to be 8% oxygen by volume for carbonaceous or organic dusts at ambient temperature and pressure when they are inerted with nitrogen (43). For the more reactive metal dusts, however, much lower levels are required.

In addition to carbon dioxide being a more effective inertant than nitrogen, water vapor also has a higher heat capacity and is more effective against carbonaceous fuels. However, neither of those inertants can be used for the more reactive metallic dusts since the metals have a stronger affinity for the bound oxygen atoms in those molecules than carbon or hydrogen. Those metals react exothermically with carbon dioxide and water vapor and hence those molecules are no longer inertants. The most notorious case is zirconium metal, which is still used as the primary heat-transfer interface in water-cooled nuclear reactors even though its reaction with water vapor is highly exothermic, as evidenced by the Three Mile Island Accident. There are even some metals that react exothermically with nitrogen to form nitrides, and for those metals, only the rare gases such as argon or helium are effective inertants.

For hydrocarbons, water vapor is effective, and for a stoichiometric mixture of methane–air, the addition of 27% by volume to the mixture is sufficient to inert the gaseous hydrocarbon (39). Of course, one cannot maintain such a high water vapor content in a system at room temperature because the saturation vapor pressure of water is at most 3–4% by volume at room temperature. If steam were added at room temperature in order to inert such a system, it would be ineffective because the water vapor would simply condense on the walls. However, if the system were continuously maintained at an elevated temperature, as for example a coal pulverizer which would normally operate at about 75°C, then saturation of the system would give a water vapor content that exceeds the inerting level for either coal dust or methane. Water vapor has been used effectively in such systems (42, 44, 45).

2.5.4 Inert Dust Addition

In addition to diluting the oxidant with inert gases, a dust fuel can be inerted directly by mixing it with a sufficient quantity of an inert dust. The inertant whose use is required by law in coal mines (21) of the United States and most other nations, is rock dust, which is mainly the calcium carbonate or limestone. The rock dust is pulverized to a size that is typically as fine or finer than the coal dust that accumulates in mines. Complete inerting requires a rock dust concentration of 75–80% by weight in the mixture of coal dust and rock dust (35). However, rock dust is virtually useless in preventing methane gas explosions (46, 47). As a result, the presence of a relatively small concentration of methane in a mine is profoundly deleterious to the inerting effectiveness of the rock dust. Rock dust is nevertheless beneficial and practical for mine use because it is insoluble in water, its cost is low, and it is readily available. Nevertheless, it is considerably less effective in preventing coal dust or methane explosions than other extinguishant dusts such as ammonium dihydrogen phosphate. This subject will be considered in more detail in a subsequent section dealing with the rapid release of extinguishants to suppress developing explosions.

2.5.5 Particle Size

A major factor in determining the explosion hazard of a dust is a variable that is often "hidden". Dusts differ markedly from gases in that they have an additional degree of freedom in their composition variable that does not exist for gases; namely, their particle diameter. A knowledge of the particle diameter is essential for an adequate description of the fuel's distribution in space on the essential microscopic scale. Typical data (48) obtained with narrowly sized, "monodispersed" dusts, show that the lean limits are insensitive to particle diameter below some characteristic diameter. For particle diameters above the characteristic diameter, the lean limit concentrations increase rapidly with increasing diameter until a critical size is reached above which the dust is nonexplosive for any concentration. For polyethylene, the characteristic diameter is 80 micrometers and the critical diameter is 120 microns. For the hvb coal from the Pittsburgh seam, the characteristic diameter is about 40 micrometers and its critical diameter is 95 micrometers. For a low volatile bituminous coal from the Pocahontas seam, the characteristic and critical diameters are even lower: 10 micrometers for the former, and 50 micrometers for the latter. For iron dust, the characteristic diameter is 20 micrometers and the critical diameter is near 60 micrometers (30).

Both the characteristic diameter and the critical diameter (the coarse size limit of flammability) were observed to increase with increasing oxygen content of the air and to decrease with decreasing oxygen content.

All the data quoted above were for idealized tests in which very narrow size distributions were used. In full scale experiments, and in industrial situations of accidental explosions, the dusts involved usually have very broad particle size distributions. Although tests conducted with broad size distributions show the same general trends as those reported above for the monodispersed dusts, the strong size dependences observed for those monodispersed dusts are inevitably blurred and smoothed out for broad size distributions.

The existence of a characteristic diameter for dusts is simply a reflection of the size at which the particle diameter is so large that the devolatilization rate process becomes rate limiting at the low burning velocities that characterize near limit concentrations. Below the characteristic diameter, the devolatilization rate is not limiting, the limit concentration is insensitive to particle diameter, and the combustion process is effectively "homogeneous" with a behavior that is similar to that of a premixed gas. Above the characteristic diameter, the combustion process becomes devolatilization rate limited, and only the surface regions of the particle, or its sharpest corners, can contribute volatiles to the flame front in the time available for flame front passage through the dust–air mixture. A larger mass loading of dust is then required to generate a lean limit concentration of combustible volatiles since only a portion of each dust particle contributes to the flame. Eventually, at the critical diameter, the dust is so coarse that its devolatilization rate is too slow to generate a lean limit concentration of combustible volatiles in the time available, and the coarse dust mixture is nonexplosive.

In a subsequent section dealing with the question of ignition, autoignition temperatures of dusts and gases will be discussed. Suffice to say here that minimum autoignition temperatures (AITs) of dusts show the same type of behavior with respect to particle size. There is a characteristic diameter below which the minimum AIT is insensitive to particle diameter and above which the minimum AIT increases with increasing diameter (49). The preheating of the dust particles to elevated initial temperature which is required in order to measure their minimum AITs, eases the devolatilization rate limitations, and accordingly, the characteristic and critical diameters for the AIT's are larger than their respective values for the lean limits of the same dusts.

3 PREVENTION AND CONTROL

3.1 Explosion Probabilities

Having defined and discussed the conditions under which potentially explosive concentrations of gases, dusts, and their hybrid mixtures can exist, it is necessary to realize that the possible existence of a flammable concentration that exceeds the lean limit of flammability, is only one of three conditions that must be satisfied before an explosion can occur in any given region of space. The three conditions that must be met for an explosion to occur are

1. The fuel gas or dust (or droplet spray) contained within a volume must be dispersed and mixed with the air.
2. The concentration of the dispersed fuel must be above the lean limit of flammability and below the rich limit of flammability.
3. An ignition source must be present of sufficient power density and total energy content to initiate the combustion wave whose propagation generates the explosion.

These three requirements naturally lead to a method of quantifying the magnitude of the explosion hazard in any region of space. If it is assumed that satisfying these three conditions are mutually exclusive events (which is not always the case, as will be discussed later), then the probability of an explosion occurring can be quantified in terms of the product of the probabilities of each of those three conditions being satisfied, namely:

$$Pr(expl) = Pr(d) \times Pr(f) \times Pr(i) \tag{15}$$

where $Pr(d)$ is the probability of the fuel gas or dust being dispersed adequately and mixed with air, $Pr(f)$ is the probability of that mixture being in the flammable concentration range, and $Pr(i)$ is the probability of having an adequate ignition source present in the system.

It is only the second of those probabilities, $Pr(f)$ that is determined by the lean limit concentration. That probability of the existence of a flammable volume, $Pr(f)$, is simply quantified for gases or dusts as the fraction of time that the system exists in a composition domain that is between its lean and rich limits of flammability. For fuels in air, the wider the domain of flammability, the more hazardous the substance. For dusts, which essentially have no rich limit, only the lean limit (minimum explosive concentration) is determining, and the lower that lean limit concentration, the more hazardous the dust.

Equation 15 can serve as a basis for quantifying the explosion hazard problem, and in this article, it will also serve as the basis for considering the two additional factors that are to be considered, namely, fuel dispersion and ignition.

3.2 Dispersion of the Fuel: *Pr*(*d*)

For a given dust loading on the interior surfaces of a mine, factory, or other facility, and in the presence of a given aerodynamic disturbance, the ease with which a dust can be dispersed into the air is a function of several factors: the individual density of the dust particles, their shape and diameter, their cohesive properties with respect to each other, and their adhesive properties with respect the surfaces on which they are accumulated. External factors that play a role in the dispersion process are: the structure and intensity of the aerodynamic disturbance, the location of the dust (whether it is on the ceiling, floor, walls, or shelves), the geometric structure of those surfaces, and other factors related to the mine or factory. The major factor that magnifies the dust explosion hazard and intensifies its destructive effect is the fact that the dispersion probability is not independent of the overall explosion probability. In most dust explosion scenarios, it is the existence of a dust explosion in one region of the plant or mine that provides the aerodynamic disturbance that disperses the dust in all other regions of the mine or plant, leading to its total destruction. This coupling or dependence of $Pr(d)$ in one region with $Pr(expl)$ in another region is the usual situation in typical coal dust explosions in mines or power plants, and for typical agricultural dust explosions in grain elevators (50).

The details of dust dispersion dynamics will not be considered in detail here. However the problem becomes moot for systems in which the dust is dispersed by design, as in a coal pulverizer, or in a pneumatic transport line whose function is to transport dust in an airstream. In the pulverized coal-firing system of typical electric utility power plant, or in a cement kiln using air as the transport medium, $Pr(d) = 1$. Under normal operating conditions in such systems, and for typical coal to air feed rates, a flammable dust air mixture is present continuously, so that $Pr(f) = 1$. In those systems, according to equation 15; the explosion probability is then determined exclusively by the ignition probability, $Pr(i)$. The question of ignition will be considered in a subsequent section.

Equation 15 is equally valid for a fuel gas explosion as for a dust explosion; however, the dispersion and mixing problem for gases differs markedly from the dispersion and mixing process for dusts. The question of the tendency of lighter-than-air fuel gases to form roof layers, and the tendency of heavier-than-air fuel gases to form ground hugging layers, has already been mentioned. Such layers will eventually mix under natural convective flows, requiring only relatively mild aerodynamic disturbances for mixing to occur readily. But the mass density of the solids from which the dusts are generated are typically a factor of one thousand greater than the density of the air into which they are dispersed. Accordingly, the everpresent gravitational force tends to segregate the dust from the air at a rate that is characterized by the settling velocity of the dust particles, or their agglomerates. Intense airflows are usually required to redisperse the dust against gravity and to maintain the dust–air mixture in a uniformly dispersed state. By contrast, for gaseous fuels, their molecular sizes and densities do not differ that much from the density of air, so that airflows associated with the fuel's initial velocity of injection into the air, or even the everpresent natural convective eddies, are sufficient to mix the gas rather rapidly into the surrounding air. Furthermore, for the gas, once the flammable mixture is generated, the mixing is intimate and on the molecular scale, and the presence of some external force that is capable of resegregating the mixture is improbable. Thus a flammable gas–air mixture in a given enclosure will remain explosive indefinitely. All that is then needed to generate and explosion is an ignition source. For a dust–air mixture, on the other hand, if the dispersing flow is stopped, the dust will settle out. The dust thus resegregates on surfaces within the system. In that resegregated system, the presence of an ignition source, by itself, will not generate an explosion unless the dust is redispersed into the ignition source. In view of that contrasting behavior between dusts and gases, it can be argued that dust fuels are intrinsically less hazardous than gaseous fuels. Furthermore, as will be shown in the section dealing with ignition, dusts are less hazardous because they require much stronger ignition sources to initiate a combustion wave. These more difficult dispersion and ignition requirements for fuel dusts generally provide the dust fuels with an inherently larger margin of safety than exists for gaseous fuels.

However, it must also be realized that the ease of dispersion of a gas relative to that of a dust, is not just a disadvantage but is simultaneously an advantage from a safety point of view. The same rapid and irreversible mixing process that facilitates the generation of a flammable volume from a gas fuel leak into the air, also facilitates the dilution of the fuel to concentrations well below the lean limit of flammability, *provided that the ventilating flow is of sufficient magnitude relative to the leak.*

Consider a mine or factory through which air can flow at some fixed volumetric rate that is maintained either by the forced convection of a fan, or even by natural convection through its various vents to the ambient atmosphere. The ventilating air currents ensure the rapid mixing of fuel gas leaks with air. If the volumetric flow of ventilating air is much larger than the volumetric flow of the fuel leak, there will be adequate dilution of the gaseous fuel to concentrations well below its lean limit concentration. In that case, $Pr(f) = 0$, and $Pr(expl) = 0$ regardless of the presence or absence of an ignition source. Once the fuel gas is diluted adequately, there is no opportunity for its reconcentration into a dangerous accumulation. There can be no subsequent demixing of the fuel from the air stream, and the ventilating flow will "dilute, carry away, and render harmless" (21) whatever fuel leaks have occurred. Such dilution by adequate ventilation is clearly an effective means of preventing gas explosions from a fuel source of small to moderate intensity.

For dusts, however, the normal ventilation velocities in most regions of a mine or other facility are generally much too low to transport the dust out of the system. For most dust sizes, gravitational settling velocities are too high relative to ventilation velocities for the flow to remove the dust. Instead the dusts settles out on surfaces and accumulates in time until a loading density is reached that readily exceeds an explosive loading even if the dust generation rate is rather low.

In summary then, for a gaseous fuel source or leak into a system containing air, adequate ventilation can ensure that $Pr(f)$ remains at zero; however for a dust source, the normal ventilation is generally not effective, and after some time interval a sufficiently high dust loading accumulates, which, if dispersed, will generate a flammable dust concentration to give $Pr(f) = 1$. For the dust, some countermeasure other than normal ventilation must be employed to remove the dust accumulation or to neutralize its presence.

3.2 Probability of the Existence of a Flammable Volume, *Pr*(*f*)

The probability of the existence of a flammable volume during any specified time interval is simply quantified as the fraction of time that the system exists in a composition domain that is between the lean and rich limits for gases, or simply above the lean limit concentration for dusts. Although this definition of $Pr(f)$ appears simple, there is a hidden ambiguity for dusts that requires further clarification. Is the dust concentration referred to above the actual *dispersed* dust concentration, or is it the potential dust concentration if all the dust accumulated on surfaces were dispersed? The dispersed dust concentration is readily expressed in normal volumetric concentration units of grams per cubic meter of system volume. If the dust is initially present on surfaces within the system volume, then its distribution is more accurately described by the surface loading density, which is the mass accumulation per unit surface area. The volumetric concentration that would be achieved if that surface loading of dust were distributed throughout the system volume can be calculated, but that concentration is achieved only if there is perfect dispersion. If the dispersion is imperfect, higher concentrations are present in some regions and lower concentrations in other regions. In many actual situations, that uncertainty is always present, and it can play an important role in the hazard evaluation.

In modern laboratory experiments designed to measure the flammability limits of dusts, such uncertainties must be avoided. Great care is taken to effectively disperse a known mass of dust into a known volume so that its concentration is uniform throughout the test volume (51, 52). In that case, $Pr(d) = 1$. Furthermore, in such experiments, and effective ignition source must be present so that $Pr(i) = 1$. In such laboratory experiments then, according to equation 15, $Pr(expl) = Pr(f)$, and the occurrence or nonoccurrence of an explosion is determined uniquely by the existence or nonexistence of a flammable dust concentration. In the older test systems, such precautions were rarely taken. Dispersion effectiveness was generally not monitored and the spark ignition sources used were generally not sufficiently energetic to give reliable data for dusts. Accordingly, one must be very cautious in using the older tabulations of data. Now there are some industrial systems whose dispersion characteristics approximate those in the modern laboratory test systems: the pneumatic dust feed lines between a coal pulverizer and the burner in a coal-fired boiler, the pulverized coal feed lines in a cement kiln, or a dust drier in which the dust and air are uniformly mixed. In those systems, $Pr(d) = 1$, and the dust concentration is given by the ratio of the coal dust mass feed rate to the volumetric flow rate of the air.

In other industrial systems, the dust accumulation is incidental and it is present mainly on surfaces within the plant enclosure. When attempts are made to evaluate the hazard associated with such dust accumulations, there is an inevitable ambiguity or uncertainty about the volume through which it will be dispersed. Will an aerodynamic disturbance disperse the dust throughout the *entire* volume of the enclosure, or will it disperse the dust through some smaller fraction of that volume? The properties of the dispersion system can now play a role in determining whether a lean limit concentration is obtained within any given region of space, and the two probabilities, $Pr(d)$ and $Pr(f)$ are no longer mutually independent of one another.

As an example, consider a system whose air volume is one cubic meter and which contains 60 grams of pulverized Pittsburgh seam coal dust lying on the floor of the chamber. The lean limit concentration for that coal dust is 90 grams per cubic meter. What then is $Pr(f)$ for that system? If the dust is uniformly dispersed throughout the volume by a strong aerodynamic disturbance, the concentration would be 60 gram per cubic meter, which is below the lean limit. Thus for the period of time that the strong aerodynamic disturbance is maintained, $Pr(f) = 0$. If, however, the aerodynamic disturbance were weaker so that dispersion of the dust were limited to the lower half of the chamber (but still strong enough to lift all of the dust), then the concentration would be 120 grams per cubic meter, which is above the lean limit. For the weaker disturbance in the lower half of the chamber, for as long as the disturbance was maintained, $Pr(f) = 1$. An adequate ignition source in the lower half of the chamber could then initiate a dust explosion in the system. Since in that case, the flammable volume occupies only half the chamber volume, the explosion pressure would be about half of the maximum explosion pressure, which is still high enough to destroy a typical plant enclosure. For the perfectly dispersed case, such an explosion would have been impossible.

On the other hand, if the dust loading had been doubled to 120 grams, explosions would have been possible for both the weak and the strong disturbances in the one cubic meter volume.

Based upon full-scale mine experiments in the U.S. Bureau of Mines facilities in Bruceton and Lake Lynn, there seems to be about a factor of two difference between the lean limit concentration of a predispersed dust cloud and the concentration equivalent of the minimum surface loading required to propagate an explosion. For hvb coal from the Pittsburgh seam, the lean limit concentration for the uniformly predispersed system is 90 grams per cubic meter, whereas the minimum surface loading required to propagate an explosion corresponds to about 35 to 60 grams of dust per cubic meter of dusted volume. The exact value depends on how the dust is loaded: loading on shelves near the roof of the mine giving the lower value, and floor loading giving the higher value. Thus for practical evaluation one should consider the possibility that dispersion can occupy only half of the available volume containing the surface loading of dust. Reasonable judgements are required to estimate the expected dispersion volume, and such estimates should take into consideration the detailed geometry and structure of the system and the magnitude of the aerodynamic disturbance that might be expected to disperse the dust.

3.3 Probability of Ignition, *P*(*i*)

Ignition sources can be characterized according to the type of energy they introduce into the system: chemical, electrical, or purely thermal. In general, an ignition source can have a variety of geometric sizes and shapes for the manner in which energy is delivered into the system, as well as a variety of time dependencies for its energy delivery rate or power density. The effectiveness of an ignition source is defined by whether or not it ignites the flammable volume, and that effectiveness is generally a function of all of those factors: the magnitude of the ignition energy, the size and shape of the ignition source, and the time dependence of its power density. As with the other probabilities, $Pr(i)$ for any given time interval, may be simply defined as the fraction of time an effective ignition source is present within the system volume.

Examples of some of the data obtained for the effectiveness of chemical pyrotechnic ignitors in igniting several dusts, are shown in Figure 56.6. The source energy is plotted as a function of the dust concentration, and the curves delineate the boundary between ignition and nonignition. Ignition occurred above and to the right of the curves, at higher energies and higher concentrations. There was no ignition below and to the left, at lower energies and lower concentrations. In these experiments, the source energy by itself is an adequate description of the source intensity so long as the energy is delivered rapidly enough in a sufficiently concentrated volume. For a source which delivers its energy more

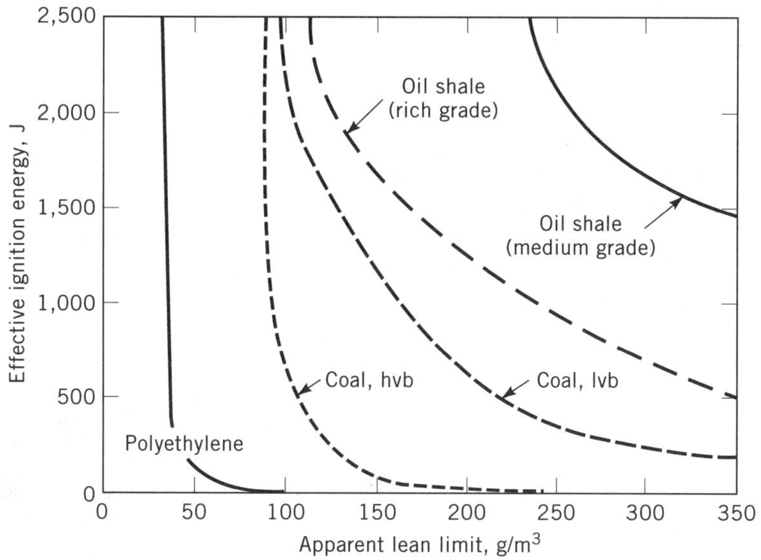

Figure 56.6. Effectiveness of chemical pyrotechnic ignitors for various dusts. Apparent lean limits as a function of ignition energy (from Fig. 9 of Ref. 1).

slowly, or which is too extended spatially, it is the power density rather than the source energy that determines its effectiveness. It is clearly beyond the scope of this article to consider all possible geometries and time dependencies for the energy delivered by ignition sources.

It should be noted that the curves shown in Figure 56.6 approach their energy independent, asymptotic lean limit of flammability only at high energies. The lower volatility dusts require higher ignition energies to attain their asymptotic limits. The data reemphasize the necessity of using modern test equipment with such strong, pyrotechnic ignition sources in order to obtain accurate values for the lean limits of dusts. Earlier test systems that used electrical spark sources often do not give reliable data because those spark energies are too weak for dusts.

By contrast, for flammable gases, spark sources are usually sufficient to obtain accurate limit data. That generalization does not apply for all possible gas mixtures, and a careful investigator will carry all lean limit measurements to their asymptotic limits at high energies and large chamber volumes. For hydrocarbon gases in air, graphs of ignition energy as a function of fuel concentration yield asymptotic lean (or rich) limits at spark ignition energies in the range of only tens of *milli*joules (2). Thus, the energy requirements for the ignition of hydrocarbon fuels at their lean limit concentrations are *five orders of magnitude lower* than the energy requirements for the ignition of carbonaceous dusts at their lean limit concentrations. *That difference further emphasizes the greater hazard of fuel gas mixture and the effect of fuel gas additions in magnifying the dust explosion hazard.* As concentrations of either dusts or gases increase above their lean limit values, the required ignition energies decrease until at near stoichiometric concentrations for gases, and somewhat higher concentrations for dusts, ignition energies reach their *minimum* values. Those minimum ignition energies are in the range of 0.1 joule for the carbonaceous dusts (40), and in the range of 0.1 *miili*joule for hydrocarbon gases (2), a *three order of magnitude* difference between dusts and gases. Thus the marked difference in ignitability between gases and dusts remains even if they are measured at their optimum concentrations for flame propagation. Ignition energy values for dusts show a wider range of variability related mainly to their devolatilization rate requirements (40). For gases, the spark energy need only initiate the exothermic gas phase combustion reactions in some minimum volume that is large enough to sustain flame propagation into the surrounding mixture. Since those reactions are quite similar for the hydrocarbon fuels, there is little difference in their minimum ignition energies. For dusts, the spark energy must first devolatilize enough dust to generate a gas phase, stoichiometric mixture of vapors, and it is only after devolatilization that the gas phase combustion reactions can be initiated. Devolatilization requirements can thus differ markedly for the different dusts depending on the details of their chemical structure which determines the particle temperature that must be reached before they can devolatilize.

In the experiments cited above, the source energy by itself was an adequate description of the ignition source intensity since the energy was delivered rapidly enough and was sufficiently concentrated spatially. At the other end of the spectrum of possible ignition sources is one that is extended in both space and time. The simplest of such an extended source is a purely thermal source that is geometrically uniform in its spatial extension and steady state in time. Such a source is isothermal in time and space, and its ignition behavior can be uniquely characterized by its temperature. The minimum temperature at which the most ignitable concentration of the fuel-air mixture ignites when its temperature is raised under those conditions is referred to as the minimum autoignition temperature (minimum AIT).

The minimum AITs for a variety of fuel gases, dusts, and vapors are summarized in Table 56.2 and Table 56.3.

There are two important factors that have a considerable effect on the minimum AIT that do not appear in Table 56.2 and Table 56.3. They are the concentration at which the minimum occurs and the particle size variable in the case of dusts. For the paraffin hydrocarbons, the lower carbon number members of the homologous series have much higher ignition temperature than the heavier hydrocarbons. Methane's value of 600°C is anamolously high, and the minimum AIT drops markedly with increasing carbon number until for the heaviest paraffin hydrocarbons an asymptotic value of 200°C is reached. Methane's high value reflects the high strength of the C—H bond that must be broken in order to initiate the chain reactions for combustion. With increasing carbon number, it is the weaker C—C bond that is broken to initiate the combustion reactions, and there are many more C—C bonds available for the higher carbon number molecules, thus increasing the probability that the overall vibrational energy stored in the system can be concentrated into one bond to cause its rupture. For methane, the minimum AIT is observed for the leaner mixtures, whereas for the heavier hydrocarbons the minimum is observed for rich mixtures and is associated with the "cool flame" phenomenon (2). Despite hydrogen's low limit calorific value, its minimum AIT is relatively high, 500°C. The high value reflects

Table 56.2. Minimum Autoignition Temperatures for Gases in Air[a]

Class of Fuel	Fuel	Minimum AIT, °C
n-Paraffin hydrocarbons	Methane	601
	Ethane	515
	Propane	450
	n-Butane	370
	n-Hexane	225
	n-Octane	220
	n-Decane	210
	n-Tetradecane	200
Alkenes	Ethylene	490
	1-Butene	385
	1-Pentene	275
Alkynes	Acetylene	305
	Propyne	340
Aromatic hydrocarbons	Benzene	560
	Toluene	480
	Ethylbenzene	430
Alcohols	Methanol	385
	Ethanol	365
	n-Pentanol	300
Ethers	Dimethyl ether	350
	Methyl ethyl ether	190
	Diethyl ether	160
Aldehydes	Formaldehyde	430
	Acetaldehyde	175
Ketones	Acetone	465
	Diethyl ketone	450
Amines	Methyl amine	430
	Ethyl amine	385
	n-Butyl amine	312
Halogenated hydrocarbons	Methyl chloride	632
	n-Propyl chloride	520
Sulfur compounds	Hydrogen sulfide	260
	Dimethyl sulfide	205
	Carbon disulfide	100
Miscellaneous	Ammonia	650
	Hydrogen	500
	Hydrazine	270
	Methyl hydrazine	185
	Diborane	40

[a]Refs. 34 and 49.

Table 56.3. Minimum Autoignition Temperatures for Dusts, and Vapors in Air[a]

Class of Dust	Dust (volatility, %)	Particle Diameter, D_s(μm)	Minimum AIT, °C
Coals	Anthracite (4–5%)	6–12	675–780
	Bituminous		
	Pocahontas (16%)	16	635
	Sewell (29%)	29	560
	Pittsburgh (36%)	28	540
	Subbituminous		
	Wyoming (38%)	31	535
	Western (35%, dried)	25	450
Gilsonite	Gilsonite (85%)	20	480
Mineral dusts	Green River Oil Shale (22–25%)	23	475
	Green River Oil Shale (9–19%)	25	500
	Sulfide Ore	27	550
Agricultural dusts	Lycopodium (85%)	27	435
	Cornstarch (87%)	18	400
Miscellaneous chemicals and plastics	Anthraquinone (99%)	28	740
	Benzoic acid (100%)	—	575
	Paraformaldehyde (100)	—	475
	Polyethylene (100%)	26	400
	Sulfur	50	290
	Decane (liquid spray)	—	275

[a]Refs. 34 and 49.

the difficulty in breaking the strong O=O or H—H bonds by thermal collisions. In a flame that has been ignited by a spark, there is no such problem: H-atoms are already present in the initiating flame kernel, and they diffuse rapidly into the unburned mixture to initiate the combustion chain reactions. For purely thermal initiation, H- or O- atoms must be generated by collision *ab initio*, so that the AIT value for hydrogen is high.

Ethers have low AITs reflecting the relative ease with which their C—O—C bond is broken. Ammonia's AIT behavior is similar to methane: its relatively high AIT reflecting the initial strength of the N—H bond. The organic sulfides have low AITs, a reflection of the ease with which the weaker C—S bond can be broken. Diborane's extraordinarily low AIT of 40°C, is so close to room temperature that the compound should be considered spontaneously flammable or pyrophoric when mixed with air. Its extraordinarily low minimum AIT most probably reflects the weakness of the unusual, three-centered, B—H—B bridge bonds in the center of the molecule.

For the dusts, the minimum AIT values occur at very high dust concentrations as shown in Figure 56.7. The data for lycopodium powder shows that the minimum AIT occurs at dust concentrations above 1000 grams per cubic meter, much higher than the nominally stoichiometric concentration for the dust. That behavior is characteristic of all dusts, and is consistent with the previous observation that flammable dusts do not display rich limit behavior for the reasons explained earlier. Figure 56.7 also relates the minimum AIT

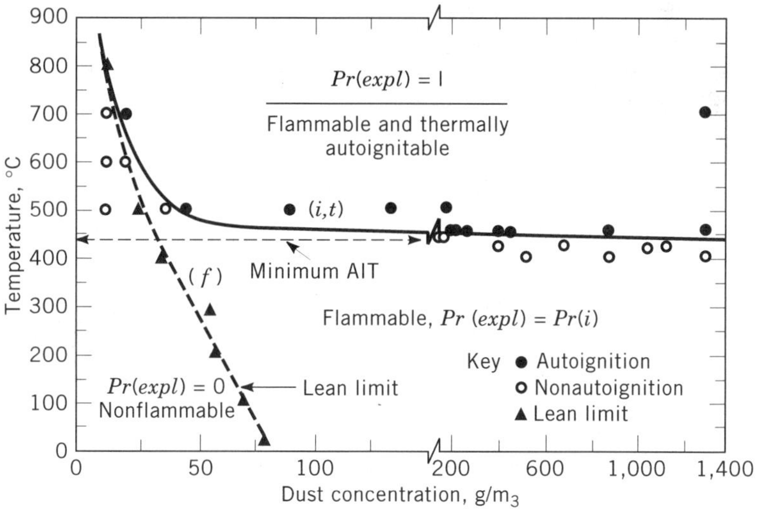

Figure 56.7. Domains of flammability and thermal autoignitability for lycopodium dust disperse in air (from Fig. 1 of Ref. 49). ●, Autoignition; ○, nonautoignition; ▲ lean limits.

measurements, and their concentration dependence to the lean limit concentration and its temperature dependence. For all the data shown in Figure 56.7, the dust is uniformly dispersed initially so that *Pr*(*d*) = 1. The curve labeled (*i*, *t*) is the thermal autoignition boundary above which the explosion probability is unity because all mixtures above and to the right of that boundary are flammable mixtures with *Pr*(*f*) = 1, and they will autoignite spontaneously once they are dispersed, so that *Pr*(*i*) = 1. Thus above and to the right of the (*i*, *t*) contour, *Pr*(*exp*) = 1. The lower dashed curve, labeled (*f*) is the lean flammability limit curve in the same concentration vs temperature space. It shows that the lean limit concentration for lycopodium of about 80 grams per cubic meter at room temperature decreases to lower concentrations as the temperature is raised (in accordance with the Burgess-Wheeler law). All states of the system below and to the left of the (*f*) contour are nonexplosive because *Pr*(*f*) = 0. For states between the (*f*)-contour and the (*i*, *t*)-contour, flammable mixtures are present, so that *Pr*(*f*) = 1, and the explosion probability becomes equal to the ignition probability associated with the presence of some other source of energy in the system.

For the dusts, the minimum AITs become particle size dependent above some characteristic diameter. Below the characteristic diameter, the minimum AITs are size invariant, as was observed for their lean limits of flammability. As was observed for the lean limits, the higher volatility dusts tend to have larger characteristic diameters, whereas the lower volatility dusts have smaller characteristic diameters (49). Such behavior is again a manifestation of the devolatilization rate being the rate controlling process for thermal autoignition as it was for flame propagation at the lean limit. Finer particles devolatilize more rapidly than coarser particles, and dusts with a higher total volatility tend to devolatilize more rapidly and at a lower temperature than dusts with a lower total volatility. Thus for the higher volatility dusts, the characteristic diameter is greater than 20–30 micrometers, and the AITs listed in the Table 56.3 for particle sizes in that range, or below, are truly representative of their lowest possible AITs. On the other hand, for the lower volatility dusts, the characteristic diameters are much smaller than 20–30 micrometers, and their AITs for their finer dust sizes may be considerably lower than the values listed in Table 56.3.

A final caveat needs to be considered in using the temperature values reported in Tables 56.2 and 56.3. The values reported are for relatively small volumes, typically in the range of one liter or so. It remains to be seen whether still larger test volumes would further reduce the measured minimum AITs. Considerable caution should therefore be exercised in attempting to extrapolate the absolute values of these minimum AITs to the much larger operating volumes in plants or factories. In the much larger heated volumes of pulverizers, driers, and reactors, the minimum AIT's may be considerably lower than those reported here.

Having presented the data on thermal autoignition, the next logical question to be considered is how should one use the data to estimate, *Pr*(*i*), the ignition probability in any system of interest. In an earlier consideration of this problem (17), the issue was discussed in the context of a critique of the so-called "explosibility index" for dusts which was proposed some 40 years ago (11), and which was used in an earlier version of this article. It is beyond the scope of our considerations here to go into detail regarding the deficiencies of that "explosibility index," sufficc to say that it is the consensus view of virtually all current researchers in the field that it should no longer be used.

In that earlier consideration of the problem of the the autoignition probability, it was proposed that the critical factor in determining *Pr*(*i*, thermal) was the difference between the autoignition temperature and the *operating* temperature of the system whose hazard is being evaluated. If *T*(min) is the minimum autoignition temperature, and *T*(o) is the operating temperature, it was suggested that *Pr*(*i*, thermal) should depend on the difference: [*T*(min) − *T*(o)], and that the functional dependence should be an exponential one, analagous to the familiar Boltzmann distribution function. In the real world of operating systems, a precise knowledge of the fraction of time an operating system spends at each temperature is lacking, and usually the important criterion is the maximum or extreme operational temperature under the transient conditions of startup and shutdown, or emergency trips of the system, or with variations in the material feed, or with seasonal variations in ambient temperature and humidity. The problem of determining the fraction of time that the system's temperature will exceed the critical value for ignition is mathematically quite analagous to the well known statistical mechanical problem of determining the fraction of atomic or molecular particles whose energies exceed some critical energy level. Naturally the "statistics" that determine the temperature fluctuations in a given operating system are less predictable than those involved in the random motion of molecules exchanging kinetic energy by collision. Clearly, the proportionality constant in the exponential function that would describe the operational system would not be the Boltzmann constant, nor can the constant for the operating system be predicted *a prior*, nor would the constant for one system be the same as that for some other system, nevertheless an exponential function of $[T(\text{min}) - T(\text{o})]$ for $Pr(i$, thermal) would appear to be reasonable.

One final point relates to the copresence of other types of energy sources in the system: electrical or thermal, which would magnify the ignition probability considerably. As indicated earlier there is a strong synergism in the copresence of ignition sources. An elevated temperature markedly reduces the spark ignition energy of dusts. A similar synergism probably exists between chemical ignition sources and the operating temperature. For a coal pulverizer, for example, such chemical sources might involve the oxidation of metal particles in the feed, the "thermite" reaction between tramp aluminum and hot steel particles generated by frictional wear, or even unreacted explosives residues left in the coal during its conventional mining.

3.4 Countermeasures for Preventing Explosions or Minimizing the Hazards from Explosions

In an earlier edition of this article (8), the following countermeasures were discussed for preventing explosions or minimizing the hazard resulting from their occurrence:

1. Reducing the oxygen content of the atmosphere.
2. Operating outside of the range of flammability.
3. Use of less flammable substances or chemical flame inhibitors.
4. Adequate ventilation.
5. Use of combustible gas indicators.
6. Elimination of ignition sources.
7. Segregation of hazardous operations.
8. Use of flame arrestors.
9. Venting explosions.

Countermeasures (1) and (2) have already been considered in the discussion of the concentration limits of flammability (Sections 2, through 2.5.2), and the discussion of minimum oxygen concentrations (Section 2.5.3). Reducing the oxygen content with an inerting gas is realistic only in fixed volume reactor, and in regions that are not frequented by personnel. Its use in occupied spaces is not possible, and in many intermediate situations such as the ullage spaces of the holds of oil tankers, the use of inerting has led to accidental deaths from anoxia when workers entered such spaces in order to clean them without realizing that they were still inerted. Countermeasure (3) for dusts was considered in Section 2.5.4 in the context of diluting flammable dust accumulations with inerting dusts. For gases, the earlier edition (8) considered the use of halogenated hydrocarbons; however, current research suggests that their effectiveness against explosions is rather limited in comparison to their effectiveness against fires (46). Countermeasure (4), adequate ventilation, has been discussed earlier in this article, and the use of combustible gas indicators, countermeasure (5), is the obvious method of monitoring the effectiveness of ventilation in diluting flammable gas leaks to concentrations below their lean limits of flammability.

All of the above countermeasures, (1)–(5), reduce the explosion probability by maintaining a low to zero value for $Pr(f)$.

Countermeasure (6), the elimination of ignition sources, attempts to eliminate explosions by maintaining a low to zero value for $Pr(i)$. Some representative ignition sources listed in the earlier edition of this article were:

a. Electric sparks or arcs generated by short circuits, lightning, or static electricity.
b. Open pilot flames or burner flames, matches or cigarette lighters, space heater flames, stoves, hot water heaters, burning materials, incinerators, etc.
c. Hot surfaces or burning metal "sparks" generated by friction, heated wires or fragments, hot vessels or pipes, glowing metals, hot cinders, overheated bearings, and shattering light bulbs that expose their hot filaments.
d. Hot gas jets, or hot gases generated by adiabatic compression or shock compression.
f. Lasers.
g. Pyrophoricity or catalytic surfaces.
h. Self heating or spontaneous combustion.

The general problem of ignition has been considered in the previous section. The above sources of ignition can be controlled by proper safety practices such as using flameproof and vaporproof electrical equipment, and building industrial plants so that boilers, water heaters, incinerators, and other equipment with open flames and incandescent materials are isolated at safe distances from operations involving flammable materials. Such isolation is countermeasure (7) above. Static electricity is more of a hazard in dry atmospheres, and its hazard can be minimized by maintaining a high relative humidity. Frictional heating can be caused by belt slippage and inadequately lubricated bearings. Static electricity can also be generated by pneumatic dust transport at high flow velocity, by impact or cleavage, or by induction. Proper grounding of machinery and piping is advisable in cases where flammable gases can be ignited by static electricity. The use of conducting floors and shoes is even recommended for the grounding of personnel in some circumstances.

As a countermeasure, however, minimizing $Pr(i)$ is generally more difficult and less cost effective than minimizing $Pr(f)$. Ideally, both should be minimized, however, from a practical point of view, operating outside of the range of flammability offers a greater margin of safety than attempting to trace down and eliminate every possible ignition source in a system. In practice one often finds extraordinary efforts expended in minimizing $Pr(i)$, when considerably less effort would be required to minimize $Pr(f)$. Those extreme efforts are often inspired by "conventional wisdom" which attributes the "cause" an explosion to the event which preceded it most closely in time. Since all explosions are preceded by ignitions, conventional wisdom tends to consider the ignition event as the "cause" of the explosion. In many accidents involving fuels that can be readily ignited, the offending ignition source may be impossible to find. Conventional wisdom overemphasizes the importance of the ignition source, and if this is done to the exclusion of the other factors involved, it often leads to the unproductive approach of trying to exclude all possible ignition sources, when, in fact, that may be the least productive approach.

Consider, for example, the air or ullage space above the fuel level in the wing tanks of an operational aircraft. The wing tanks normally contain jet fuel at a level that varies markedly between refueling and landing. The fuel vapor concentration in that air space can vary considerably during the operation of the aircraft. The flash point of such jet fuels vary from as low as $-18°C$ for "Jet A" to as high as 52°C for "Jet B" (34). The ambient temperatures to which the aircraft is exposed can vary from as low as $-40°C$ in the arctic regions or at high altitudes, to as high as 48°C on a tropical runway in the afternoon. If that space above the fuel level is not inerted, and depending on how the pressure in that wing tank is controlled, that space can contain a range of concentrations: from below the lean limit, to above the rich limit, to flammable concentrations between those limits. Assume now for the sake of argument, that during the operating life of a given aircraft, the fuel tank contains a flammable concentration for 1% of the time. In that case, $Pr(f) = 0.01$. Considerable care is normally taken in those fuel tanks to eliminate all possible ignition sources, but for the sake of argument, assume that some improbable defect in the electrical wiring leading to the explosion proof fuel pump generates a short circuit spark in the wing tank's air space, and that the spark lasts for only 10 milliseconds before a circuit breaker is tripped to inactivate the fuel pump. If the operating life of the aircraft is taken to be 10 years, one can calculate the ignition probability as $Pr(i) = 0.01$ sec/10 yrs $= 3.2 \times 10^{-11}$. If that spark occurs during the 1% of the time that the concentration is within the flammable range, it will initiate an explosion that would destroy the aircraft.

Further assume that the accident investigators recover the "black box" after the accident, and that the records of all the operating systems clearly show the trip of the fuel pump circuit breaker as the only anomalous event immediately preceding the explosion. Is there any doubt that the accident report will, in accordance with conventional wisdom, attribute the "cause" of the explosion to the short circuit that ignited the flammable mixture above the fuel level? Yet, if one stands back and looks at the situation in its true perspective, one realizes that conventional wisdom is incorrect in emphasizing the *least* frequent of the events as the cause of the explosion. The least frequent ignition event does indeed trigger the explosion, and because it immediately precedes the dramatic explosion event, there is the inevitable subjective temptation to isolate it as *the* cause. *But if one is forced to choose a single cause for the explosion, is it not more reasonable to attribute the cause to the offending condition or event with the highest probability rather than to the one with the lowest*? For after all, in the case considered, $Pr(f)$ was over nine orders of magnitude larger than $Pr(i)$! It is therefore more logical to consider the existence of the flammable volume as a far more significant cause than the explosion event. Conventional wisdom overemphasizes the final event at the expense of the entire chain of precipitating events that caused the explosion.

In the final section of this article, a variety of other "typical" explosion scenarios will be considered in the context of some actual evaluations of overall explosion probabilities.

Countermeasure 7, the isolation of hazardous operations, has already been mentioned earlier, and involves the spacing of equipment and operations at minimal safe distances from one another so that a fire or explosion in one part of the plant does not result in a fire or explosion in an adjacent part of the plant. This may involve the use of suitable dikes so that the accidental spillage of flammable liquids is contained. It also involves the location of storage tanks at minimal distances from one another so that a fire in one tank does not spread to adjacent tanks. One of the worst examples of a structure that fails to isolate hazardous operations is the "traditional" design of grain storage facilities: huge concrete enclosures containing storage silos, conveyor belts, bucket elevators, and the offices of operating personnel immediately above the elevator. Typically, corridors lead from one space to the other within a single concrete enclosure that contains them all. Grain dust explosions in such facilities inevitably result in complete disasters as dust flames initiated in one region accelerate from their beginning in one corridor into adjacent corridors and spaces that contain grain dust that is predispersed by the aerodynamics disturbance from the initiating region. The bucket elevator is often the initiator of the explosion, and the location of office spaces and personnel above the elevator, in effect, places them at the mouth of a potential cannon.

Modern designs would isolate each function in separate structures that are lightly constructed with sheet metal siding and roofs that vent readily. The conveying structures leading from one region to the other should be essentially open except for a light covering for weather protection.

The use of flame arrestors, countermeasure 8, involves the use of an assembly of narrow passages or apertures through which gases can flow, yet small enough to quench flames by thermal wall losses (28). Such flame arrestors are generally made of wire gauze, crimped metal, sintered metal, perforated metal plates, or pebble beds. They can be used in conjunction with pressure release vents and are designed to prevent explosions from passing from one vessel to another. Their effectiveness depends on the overpressure driving the explosion from one enclosure to another. They can be effective if the initiating enclosure is vented adequately so that the overpressure driving the flame through the flame arrestor is minimal; however if the driving pressure is too large, their effectiveness is lost as burned gas flows from the fireball in the initiating enclosure overwhelm the ability of the flame arrestor to cool them. The hot jet of gases from the first enclosure can then initiate an explosion in the adjacent enclosure.

The flame arrestors described above are "passive" devices. More recent developments in prevention methods involve the use of "active" devices such as rapid acting valves, either of the gate or butterfly type, that are triggered by an optical or overpressure sensor that detects the developing explosion in an enclosure (43). The rapid sealing of the enclosure in which the explosion develops protects adjacent enclosures. In another design, protection is afforded by a float type valve, which is activated to closure by the kinetic energy of the pressure wave which precedes the flame front.

The last countermeasure listed, countermeasure 9, is explosion venting. This can be done "passively" by designing structures so that their enclosures are made of light paneling that readily vent to the surrounding at a low overpressure. For chemical reactors that require pressurization or for other unit processes that require pressure vessels, passive methods are not possible, and active venting is required. Explosion venting involves countermeasures that allow initially closed vessels to open in a safe direction during a developing explosion and thereby prevent the buildup of pressure that would rupture the vessel. Guidelines for the required venting area depend on the volume of the vessel to be protected, the size normalized rate of pressure rise for the gas or dust flame (its K_G or K_{St} value, as discussed earlier), and the design pressure of the vessel which is the maximum pressure the vessel is designed to withstand. Guidelines for such venting systems have been developed in nomograms published by VDI (19), and details concerning their use have been extensively discussed elsewhere (43). The venting system must act rapidly, and is typically triggered by a pressure sensor which detects the overpressure from the developing explosion. An alternative to venting involves the use of such sensors to trigger a pressurized extinguishant which is rapidly released into the vessel in order to extinguish the developing flame front. The extinguishant is typically a free-flowing, fine powder such as "ABC" (ammonium dihydrogen phospate).

4 EXPLOSION PROBABILITY EVALUATIONS IN TYPICAL SCENARIOS

The net explosion probability given by equation 15 was previously expressed in terms of the product of three separate probabilities: the dispersion probility, $Pr(d)$; the presence of a flammable volume probability, $Pr(f)$; and the ignition probability, $Pr(i)$. It was implicitly assumed that the three events or conditions determining those probabilities were mutually independent of one another. In that case, the events or conditions that determine each probability are temporally randomly distributed with respect to each other. Each probability was defined as the fraction of time that the system existed in a state in which those separate events or conditions were present. For dusts, $Pr(d)$ was the fraction of time the dust was dispersed; $Pr(f)$ was the fraction of time a flammable concentration was present, and $Pr(i)$ was the fraction of time an effective ignition source was present. The requirement of mutual independence does not mean that the events or conditions that determine any one probability must itself be randomly distributed in time, but only that they have a random relationship to the events or conditions that determine the other two probabilities.

4.1 Examples A and B: Baghouse Dust Collector

Consider a filter baghouse dust collector that is periodically cleaned every hour by a reverse pressure pulse, which generates a flammable dust cloud that is airborne for three minutes before the dust settles by gravity to the bottom of the baghouse, where it is removed by a rotary feeder. In that system, $Pr(d) = 3$ min./60 min. $= 0.05$. The cleaning pulse, by itself, is quite periodic and not at all randomly distributed in time. But that, by itself, says nothing about its randomness with respect to the other two probabilities. If, for example, the events that cause the presence of an ignition source in the baghouse volume are independent of that pulse cleaning cycle, and randomly distributed with respect to that cycle, then $Pr(i)$ and $Pr(d)$ are mutually independent of one another. Such would be the case, for example, if the only ignition source possible were hot embers from an upstream grinder. For example A, assume that such a hot ember ignition source was present once a month, on the average, and that the ember cooled rapidly so that it was hot enough to ignite the dispersed dust for a time interval of only 10 seconds after entering the baghouse. With those assumptions, one can calculate the ignition probability as $Pr(i) = 10$ sec/1 month $= 3.9 \times 10^{-6}$. It is assumed that the total loading of flammable dust in the baghouse is always sufficient to exceed its lean limit of flammability, so that $Pr(f) = 1$. One can now calculate the explosion probability for example A as:

$$Pr(expl) = (1) \times (0.05) \times 3.9 \times 10^{-6} = 2 \times 10^{-7}$$

But what precisely does this number mean? From the viewpoint of predicting explosion hazards, it is not the dimensionless probability that is needed but rather the predicted frequency of occurrence. For an explosion frequency prediction, only two of the conditions should be expressed as probabilities, and the third should be expressed in terms of its frequency of occurrence. In example A, the presence of an ignition source was the least frequent event of the three events or conditions, and it is therefore chosen as the third factor to be expressed in terms of its frequency of occurrence, which was once per month (1/month). Hence for example A:

$$\begin{aligned} f(expl) &= Pr(d) \times Pr(f) \times f(i) = (0.05) \times (1) \times (1/\text{month}) \\ &= 0.05/\text{month} = 0.6 \text{ per year} \end{aligned}$$

That explosion frequency corresponds to a predicted period of one explosion every 1.7 years, on the average. It should be noted that if the dispersion frequency is chosen as the third factor, then the result would be $f(\text{expl}) = Pr(d) \times Pr(i) \times f(d) = (1) \times (3.9 \times 10^{-6}) \times (1/\text{hour}) = 0.34$ per year, which corresponds to one explosion every three years. As will be seen the choice of which event is to be chosen as the frequency factor is not arbitrary.

This example was presented for illustrative purposes only. It should not be considered as a real evaluation. Clearly, such explosion frequencies are too high to be considered acceptable. In the $f(d)$ calculation, a reduction in the duration of the airborne dust cloud's presence from 3 min. to a shorter time would reduce that explosion frequency proportionately.

For example B, assume that the dust in the same baghouse was easily ignited by an electric spark discharge, and that such a spark was initiated by static electricity every time the reverse air pulse was turned on. The two probabilities of dispersion and ignition would no longer be independent of one another. If the spark lasted for only 10 microseconds, then, at a cleaning rate of once per hour, one could calculate $Pr(i) = 10^{-5}$ sec/3600 sec $= 2.8 \times 10^{-9}$, but that value would be irrelevant to the problem since the presence of the ignition source correlates perfectly with the dispersion of the flammable dust. Since the ignition source is always present at each dispersion pulse, the explosion frequency is $f(expl) = Pr(f) \times f(d) = 1 \times (1/\text{hour}) = 1/\text{hour}$, which gives an explosion each time that the pulse cleaner is turned on. Naturally, such a bag house could never function successfully.

The above examples were designed to illustrate some of the complexities involved in obtaining realistic hazard evaluations. A cautious approach needs to be taken: the complicated interrelationships between the various processes and events involved need to be fully understood, and any evaluation should clearly emphasize the limitations and uncertainties involved.

4.2 Examples C and D, a Pneumatic Dust Transport System

Consider a simpler case, example C, a relatively dilute dust-bearing system in which a predispersed dust concentration above the lean limit of flammability is present rather infrequently, let us say for 1% of the time. Then $Pr(f) = 0.01$. If the dispersion is continuous, then $Pr(d) = 1$. Consider the presence of an effective ignition source whose duration is very short compared to the period of those concentration fluctuations. If that ignition source is present at a frequency of once per year, then the expected explosion frequency for the system is: $f(expl) = (0.01) \times (1) \times (1/\text{year}) = 0.01$ per year, which corresponds to one explosion every 100 years, on the average. In this instance, the duration of the effective ignition source is irrelevant. If it lasts for only ten milliseconds at a frequency of once per year, the calculation of $Pr(i) = 10$ milliseconds/1 year $= 3 \times 10^{-13}$ is irrelevant for the problem, and only the frequency of occurrence of the ignition event is significant. In our previous examples (A and B), there appeared to be some arbitrariness in the choice of whether the ignition event or the dispersion event was selected as the discrete event whose frequency should be chosen to determine $f(expl)$. This example makes clear that the *least* frequent event, or the event with the lowest *Pr* should be chosen as the discrete one whose frequency is to be used to determine $f(expl)$. The other events or conditions with the higher probabilities are expressed in terms of their *Pr*s.

Consider case D, for example, the same pneumatic transport system that was considered in case C, but in which $Pr(i)$ is larger than $Pr(f)$. If an ignition source is present 1% of the time, then $Pr(i) = 0.01$. If a lean limit concentration were randomly exceeded only once every 6 months for a duration of say 10 minutes on the average, then $Pr(f) = 10$ min./6 months $= 4 \times 10^{-5}$. Since the probability of the existence of a flammable volume is now the lowest of the probabilities, it is its frequency of occurrence that counts, and the expected explosion frequency is $f(expl) = Pr(d)Pr(i)f(f) = (1)(0.01)(1/0.5 \text{ years}) = 0.02$ per year, which corresponds to one explosion every 50 years. Since the existence of the flammable volume is the rarer event, it was chosen to be expressed in terms of its frequency of occurrence, quite independent of its duration (so long as the duration is such that $Pr(f)$ does not exceed either $Pr(i)$ or $Pr(d)$. It is again an implicit assumption that the events in question are randomly related to one another, as discussed earlier.

4.3 Example E: A Grain Storage Facility

Consider another specific example: a dusty grain storage facility containing a nominal dust loading on the floor, walls, and roof of a corridor that contains a conveyor belt run. Assume that housekeeping is poor and that grain dust accumulates to give a loading of dust that would easily generate a flammable concentration if it were dispersed. In that case $Pr(f) = 1$. Consider that for one reason or another, an unsheilded space heater is located within the conveyor run passageway, and that it is operated mainly during the winter months. Assume that the surface temperature of its heating elements are well above the cloud ignition temperature of the grain dust, and that the heater is operating for about 5% of the time. The heater is not in direct contact with the dust accumulations on the floor, walls, or roof of the passageway, but is several feet above the floor and dispaced about a foot from the wall. For that ignition source, $Pr(i) = 0.05$. In this case, it is the dispersion of the dust into the air space surrounding the heater that is the least probable event, and hence it is treated as the discrete event that is "rate limiting". If some aerodynamic disturbance capable of dispersing a flammable dust cloud concentration into the heater is present, on the average, only about once in nine months, and if such disturbances are randomly distributed temporally relative to the operation of the space heater, then:

$$f(expl) = Pr(f)Pr(i)f(d) = (1) \times (0.05) \times 1/0.75 \text{ years} = 0.067 \text{ per year}$$

and on the average, an explosion would occur once every 15 years. The offending aerodynamic disturbance could be a ruptured compressed air line, a large object falling on the dusty floor, a slammed door, a sweeping broom, or a wind gust blowing through an open window. In such instances, conventional wisdom attributes the "cause" of the explosion to the event with the lowest probability. If the conveyor belt were continuously monitored, nothing would be seen to be happening until the aerodynamic disturbance "triggers" the explosion. Conventional wisdom would thus attribute the cause of the explosion to the event preceding it most directly in time. Even though an effective ignition source was present 5% of the time, and an explosive dust loading was present 100% of the time, nothing would be seen to be happening until the dispersion event occurs. The instant the dust is dispersed, the explosion occurs, and hence conventional wisdom considers the dispersion event as the cause.

As with our earlier example of the aircraft wing tank explosion, conventional wisdom thus focuses on the "straw that breaks the camel's back," ignoring the enormous loads that have been previously piled upon the poor camel.

4.4 Correlation of Events vs Randomness

As indicated earlier, in cases where the individual events or conditions that determine each of the three probabilities (or the one frequency) are not randomly distributed with respect to one another, they cannot be considered to be mutually independent, and equation 15, or its frequency counterpart, cannot be used directly. If there are time correlations between the conditions or events involved, there is a magnification of the hazard if the events or conditions tend to occur in phase with one another. On the other hand, there is a diminution of the hazard if the events or conditions tend to occur out of phase with one another. For example E, the conveyor belt run in the grain storage facility, the space heater ignition source is more likely to be operating during the winter months when the humidity is low and the grain is drier. A drier grain dust has a lower lean limit, a lower thermal ignition temperature, and a greater dispersibility in the presence of a given aerodynamic disturbance. A drier grain will have a higher rate of dust generation per unit mass of grain transported along the belt. These in-phase conditions increase the hazard, and give an explosion frequency that can be much higher than that estimated on the basis of randomness.

The use of proper safety practices will, in fact attempt to reduce explosion hazards by the purposeful suppression of randomness. In proper safety practice, the hazardous events or conditions are controlled so that they occur out of phase with one another. Consider, for example, a coal pulverizer and its associated pneumatic dust transport system that feeds the burner system of a coal-fired boiler or cement kiln. For typical operating conditions, the coal and air flow ratios generate dust concentrations that are well above the lean concentration limits, and since the dust is predispersed in the flow, $Pr(f) = 1$ and $Pr(d) = 1$ within the pulverizer and pneumatic feed lines to the burners.

Now a welding torch is a very effective ignition source, and if such an ignition source were present within the system at a frequency $f(i)$, the explosion frequency would be equal to the frequency of its presence if its presence were a random event. Welding torches are used rather frequently to repair, maintain, or modify those systems, but proper safety procedures require that the system be shut down and the coal dust removed before any welding operations can begin. Safe maintenance procedures require that the ignition source's presence be ordered so that it is out of phase with the other two conditions, so that $Pr(d) = 0$ and $Pr(f) = 0$ when $Pr(i) = 1$. Operating under those safety rules, the events in the system are not randomly distributed with respect to one another, and $f(expl)$ will remain zero regardless of the frequency of welding. The system is safe so long as the "lockout" procedures are followed; that is, so long as there is never an overlap in the time intervals that $Pr(d)Pr(f)$ and $Pr(i)$ are simultaneously unity. Abundant cases exist of explosions that were initiated by welding on the *outside* of vessels that contained flammable fuel air mixtures in which the proper safeguards were not exercised.

In the above example, the lack of randomness is purposeful and it is arranged by the system operator in order to minimize the hazard. Unfortunately, there are many instances where the events themselves are fortuitously ordered in a "malevolent" manner, such that they exhibit an in phase time correlation or simultaneity which magnifies the explosion hazard. An example of such malevolent simultaneity is the "gas ignition" problem near the face of an operating coal mine. Consider a coal mine which contains a coal dust loading on its floor, ribs, and roof that would, if dispersed, generate an explosive concentration. For that mine $Pr(f) = 1$. Although such a situation is now forbidden by law which requires that such a dust loading be inerted with rock dust (21), such noninerted dust loadings were quite common in coal mines in the early 1900s. If it is a relatively deep mine, significant methane generation is inevitable, and it leads to a significant "face ignition" frequency. The term "face ignition" refers to small scale "flashes" or "puffs" caused by the burning of methane in limited volumes near the mining machine. As the mining machine cuts into the coal seam, the methane pressure gradient in the coal seam steepens, generating a high methane flow, and a flammable volume develops near the face. It can be ignited by the frictional heating of hard rock inclusions being struck by the steel bits of the mining machine. Some fraction of those "face ignitions" may create a large enough aerodynamic disturbance to disperse the coal dust accumulations nearby. If the average frequency of such a strong ignition is $f(i)$, and if the ignition also disperses the coal dust to give $Pr(d) = 1$, the explosion frequency is $f(expl) = (1) \times (1) \times f(i) = f(i)$. If the dispersion conditions were randomly distributed relative to the ignition, the explosion probability would be much lower. For example, if the duration of the "face ignitions" were about 2 sec, and there were 5 such strong ignitions per year, one can calculate that $Pr(d) = 5 \times 2$ sec/1 year $= 3 \times 10^{-7}$. If that same duration of dispersion were truly randomized relative to the ignition process, then the explosion frequency would be $f(expl) = 1 \times 3 \times 10^{-7} \times f(i)$, which would be over one million times smaller than the value obtained earlier. The earlier value is the correct one because the ignition and dispersion events occur in phase and are not randomly distributed temporally. The same face ignition event that is capable of igniting a dispersed dust cloud also provides the aerodynamic disturbance that is sufficient to disperse the dust accumulations. The hazard is thus markedly increased by the fact that dispersion and ignition occur at the same instant of time. By contrast, some other ignition source that does not simultaneously generate an aerodynamic disturbance is far less hazardous. A welder's torch or a trolley wire arc, either of which would be capable of igniting a coal dust cloud, are not nearly as dangerous so long as they do not disperse the dust simultaneously, and occur randomly in time relative to any dispersion processes that may be present.

In the early 1900s, there was a prevalent misapprehension that coal dust, by itself, was not an explosive fuel. Because most mine explosions were associated with "gassy" mines, it was argued that the earlier disasters were caused by methane gas explosions. So prevalent was that mistaken belief that the loose coal dust accumulations in mines were routinely used to pack or "stem" explosives in their boreholes. The explosives used in those early days were readily capable of igniting coal dust. The result of such practices were extraordinarily high disaster rates even in relatively shallow mines that contained little methane. Such hazardous blasting practices created a situation in which $Pr(f)$, $Pr(d)$, and $Pr(i)$ were intimately coupled in phase so that they reached their peaks simultaneously in time.

It should be noted that for dust explosions where dispersion of the dust can be a limiting factor in the overall explosion probability, there is an inevitable coupling between the existence of an explosion in one region of space and the dispersion of dust in adjacent regions. The burned gases near the ignition source in one region of space expand outward rapidly and push the unburned, surrounding gas outward. An aerodynamic disturbance in adjacent regions is thus inevitable, and that disturbance disperses the dust in adjacent regions. In most cases, that flow also becomes turbulent, which markedly increases the flame propagation rate, and as the flame front accelerates, the aerodynamic disturbance ahead of it becomes more intense, accelerating it further and dispersing more dust. The process is self accelerating and can lead to a detonation.

Many dust explosions that occur in mines, grain storage facilities, factories, and power plants, are actually such secondary explosions initiated by primary explosions of gas or dust in other regions of the facility. The primary explosion generates flows that vent into the dusty spaces of adjacent regions predispersing the dust in those regions and allowing the primary explosion to propagate further and further until the entire facility is destroyed. Again the hazard is magnified because ignition and dispersion are not randomly distributed in time, but are intimately coupled in phase with one another. For such secondary explosions, the turbulence generated by the flows further reenforces the dispersion and flame acceleration processes, magnifying the hazard. As indicated previously, it is the complex dynamics of such processes that ultimately determine the real severity of a dust explosion, and such a process is not accurately simulated in smaller, laboratory-scale tests that are usually limited to spherical flame propagation.

BIBLIOGRAPHY

1. M. Hertzberg and K. Cashdollar, "Introduction to Dust Explosions," in M. Hertzberg and K. Cashdollar, eds., *Industrial Dust Explosions*, ASTM 958, American Society for Testing and Materials, Philadelphia, 1987, pp. 5–32.

2. B. Lewis and G. von Elbe, *Combustion, Flames, and Explosions of Gases*, Academic Press, New York, 1961.
3. K. L. Cashdollar and M. Hertzberg,*Combustion and Flame* **51**, 23–25 (1983).
4. M. Berthelot and P. Vieille, *Comptes Rendus de l'Academie des Sciences* **93**, 18–22 (1881).
5. E. Mallard and H. LeChatelier, *Comptes Rendus de l'Academie des Sciences* **93**, 145–148 (1881).
6. G. I. Taylor, *Proc. Roy. Soc.* A200, 235–247 (1950).
7. I. B. Zeldovich and A. S. Kompaneets, *Theory of Detonations*, Academic Press, New York, 1960.
8. J. Grumer, "Fire and Explosion Hazards of Combustible Gases, Vapors, and Dusts" in G. D. Clayton and F. E. Clayton, eds., Vol I B, *Patty's Industrial Hygiene and Toxicology*, 4th Ed., John Wiley & Sons, New York, 1991, pp 883–947.
9. H. F. Coward and G. W. Jones, "Limits of Flammability of Gases and Vapors," *U S Bureau of Mines Bulletin No. 503*, 1952, 155 pp.
10. H. G. Dorsett, Jr., M. Jacobson, J. Nagy, and A. P. Williams, "Laboratory Equipment and Test Procedures for Evaluating Explosibility of Dusts", *U.S. Bureau of Mines Report of Investigations No. 5624*, 1960, 21pp.
11. M. Jacobson, J. Nagy, A. R. Cooper and F. J. Ball, "Explosibility of Agricultural Dusts", *U.S. Bureau of Mines Report of Investigations No. 5753*, 1961, 23pp.
12. J. Nagy,H. G. Dorsett Jr., and A. R. Cooper, "Explosibility of Carbonaceous Dusts", *U.S. Bureau of Mines Report of Investigations No. 6597*, 1965, 30pp.
13. J. Nagy, A. R. Cooper, and H. G. Dorsett Jr., "Explosibility of Miscellaneous Dusts", *U.S. Bureau of Mines Report of Investigations No. 7208*, 1968, 31pp.
14. National Fire Protection Association (NFPA), *Fire Protection Handbook* 14th ed., Sec. 3 Chapt. 8, NFPA, Boston, 1976, pp 3–106 to 3–118.
15. L. A. Eggleston and A. J. Prior, *Fire Technology* **3**, 77 (1967).
16. W. Bartknecht, *Staub-Reinhalt. Luft* **31**, 112 (1971).
17. M. Hertzberg, "A Critique of the Dust Explosibility Index: An Alternative for Estimating Explosion Probabilities," *U.S. Bureau of Mines Report of Investigations No. 9095*, 1991, 24 pp.
18. K. Cashdollar, I. A. Zlochower, G. M. Green, R. A. Thomas, and M. Hertzberg, "Flammability of Methane, Propane, and Hydrogen Gas," *Proc. of the Int. Symp. on Hazards, Prevention, and Mitigation of Industrial Explosions*, Schaumburg, IL, Sept. 21–25, 1998 (to be published).
19. Verein Deutscher Ingenieure (VDI—Association of German Engineers), "Pressure Relief of Dust Explosions, VDI 3673, in *Handbuch Reihaltung der Luft*, Vol. 6, 1983
20. S. Glasstone, "The Effects of Nuclear Weapons," U.S. Atomic Energy Commission, Apr. 1962, 730pp.
21. U.S. Congress, "Federal Coal Mine Health and Safety Act of 1969," Public Law 91-173, December 30, 1969, and its modification, "Federal Coal Mine Health and Safety Act of 1977," Public Law 95-164, November 9, 1977, Code of Federal Regulations 30, Chapter 1, Part 75
22. H. LeChatelier and O. Boudouard, *Bull. Soc. Chim., Paris* **19**, 483 (1898).
23. H. LeChatelier and O. Boudouard, *Comptes Rendus de l'Academie des Sciences* **126**, 1510 (1898).
24. M. Hertzberg, "The Flammability Limits of Gases, Vapors, and Dusts: Theory and Experiment" in J. H. S. Lee and C. M. Guirao, eds., *Fuel-Air Explosions*, University of Waterloo Press, 1982, pp. 3–48.

25. J. Warnatz, *Eighteenth Symposium (International) on Combustion*, The Combustion Institute, Pittsburgh, 1981, p. 369.
26. M. Hertzberg, "Selective Diffusional Demixing: Occurrence and Size of Cellular Flames," *Prog. Energy Combus. Sci.* **15**, 203–239 (1989).
27. M. Hertzberg, "The Theory of Flammability Limits, Natural Convection," *U.S. Bureau of Mines Report of Investigations No. 8127*, 1976, 15 pp.
28. M. Hertzberg, "The Theory of Flammability Limits. Conductive-Convective Wall Losses and Thermal Quenching," *U.S. Bureau of Mines Report of Investigations No. 8469*, 1980, 25 pp.
29. M. Hertzberg, "The Theory of Flammability Limits. Flow Gradient Effects and Flame Stretch," *U.S. Bureau of Mines Report of Investigations No. 8865*, 1984, 36 pp.
30. K. L. Cashdollar, "Dust Explosion Overview," *Proc. of the Int. Symp. on Hazards, Prevention, and Mitigation of Industrial Dust Explosions*, Schaumburg, IL, Sept 21–25, 1998 (to be published).
31. P. Wolanski, "Dust Explosion Research in Poland," *Powder Technology* **71**, 197–206 (1992).
32. M. Hertzberg, I. A. Zlochower, and K. L. Cashdollar, *Twenty-Fourth Symposium (International) on Combustion*, The Combustion Institute, Pittsburgh, 1992, pp. 1827–1835.
33. National Fire Protection Association (NFPA), *Flash Point Index of Trade Name Liquids*, NFPA No. 325A, 1972, 258 pp.
34. J. M. Kuchta, "Investigation of Fire and Explosion Accidents in the Chemical, Mining, and Fuel-Related Industries—A Manual," *U.S. Bureau of Mines Bulletin No. 680*, 1985, 84 pp.
35. K. L. Cashdollar, M. J. Sapko, E. S. Weiss, and M. Hertzberg, "Laboratory and Mine Dust Explosion Research at the Bureau of Mines," in M. Hertzberg and K. Cashdollar, eds., *Industrial Dust Explosions*, ASTM 958, American Society for Testing and Materials, Philadelphia, 1987, pp. 107–123.
36. J. K. Richmond, M. J. Sapko, and L. F. Miller, "Fire and Explosion Properties of Oil Shale," *U S Bureau of Mines Report of Investigations No. 8726*, 1982.
37. M. Gaug, R. Knystautus et. al., *Progress in Astronautics and Aeronautics*, **105**(II), 155–168 (1986).
38. M. J. Burgess and R. V. Wheeler, *J. Chem. Soc.* **99**, 2013–2030 (1911).
39. M. G. Zabetakis, "Flammability Characteristics of Combustible Gases and Vapors," *U.S. Bureau of Mines Bulletin 627*, 1965, pp. 21–24.
40. M. Hertzberg, R. S. Conti, and K. Cashdollar, "Electrical Ignition Energies and Thermal Autoignition Temperatures for Evaluating the Explosion Hazard of Dusts," *U.S. Bureau of Mines Report of Investigations No. 8988*, 1985, 41 pp.
41. A. Y. Korol'chenko, A. V. Perov, and Y. N. Shebeko, *Combustion, Explosions, and Shock Waves* **18**, 112–113 (1982).
42. W. Wiemann, "Influence of Temperature and Pressure on the Explosion Characteristics of Dust/Air and Dust/Air/Inert Gas Mixtures," in M. Hertzberg and K. Cashdollar, eds., *Industrial Dust Explosions, ASTM 958*, American Society for Testing and Materials, Philadelphia, 1987, pp. 33–44.
43. W. Bartknecht, "Prevention and Design Measures for Protection Against the Danger of Dust Explosions," in M. Hertzberg and K. Cashdollar, eds., *Industrial Dust Explosions, ASTM 958*, American Society for Testing and Materials, Philadelphia, 1987, pp. 158–190.
44. R. C. Carini and K. R. Hules, "Coal Pulverizer Explosions," in M. Hertzberg and K. Cashdollar, eds., *Industrial Dust Explosions, ASTM 958*, American Society for Testing and Materials, Philadelphia, 1987, pp. 202–216.

45. P. M. DeGariele, H. Causilla, and J. P. Henschel, *Fire Technology* **16**, 212–226 (1980).
46. M. Hertzberg, K. Cashdollar, I. Zlochower, and D. Ng, *Twentieth Symposium (International) on Combustion*, The Combustion Institute, Pittsburgh, 1985, pp. 1691–1700.
47. M. Hertzberg, I. Zlochower, and K. Cashdollar, *Twenty First Symposium (International) on Combustion*, The Combustion Institute, Pittsburgh, 1987, pp. 325–333.
48. M. Hertzberg, K. Cashdollar, D. Ng, and R. Conti, *Nineteenth Symposium (International) on Combustion*, The Combustion Institute, Pittsburgh, 1982, pp. 1169–1180.
49. R. Conti and M. Hertzberg, "Thermal Autoignition Temperatures from the 1.2-L Furnace and Their Use in Evaluating the Explosion Potential of Dusts," in M. Hertzberg and K. Cashdollar, eds., *Industrial Dust Explosions, ASTM 958*, American Society for Testing and Materials, Philadelphia, 1987, pp. 45–59.
50. C. W. Kauffman, "Recent Dust Explosion Experiences in The U S Grain Industry," in M. Hertzberg and K. Cashdollar, eds., *Industrial Dust Explosions, ASTM 958*, American Society for Testing and Materials, Philadelphia, 1987, pp. 243–264.
51. M. Hertzberg, K. Cashdollar, and J. Opferman, "The Flammability Limits of Coal Dust-Air Mixtures: Lean Limits, Flame Temperatures, Ignition Energies, and Particle Size Effects," *U.S. Bureau of Mines Report of Investigations No. 8360*, 1979, 70 pp.
52. K. Cashdollar and M. Hertzberg, *Rev. Sci. Instrum.* **56**, 596–602 (1985).

CHAPTER FIFTY-SEVEN

Environmental Control in the Workplace: Water, Food, Insects, and Rodents

Maurice A. Shapiro and Gerald M. Barron, MPH

1 INTRODUCTION

In the second and third editions of this book, the chapter headed "Industrial Sanitation" was introduced as follows: "Industrial sanitation is essentially a specialized application of community environmental health services. Within the purview of industrial sanitation are the principles involved in controlling the spread of infection or other insults to the health of the employee not inherent in the manufacturing process *per se*." The fourth edition chapter heading is changed in the current revision since a new chapter, Chapter 61, HAZARDOUS WASTE has been added. This addition is emblematic of the subject matter and practice of environmental control in the work place at the threshold of the twenty-first century. Nevertheless, the definition utilized in the second and third editions is still valid, and because the objective of industrial hygiene is to safeguard the health of working people, environmental control in the workplace should be an intrinsic function of occupational safety and health.

Environmental control consists of maintaining and improving the general environment of the workplace. Exposures to pathogenic organisms and toxic substances in the environment, e.g., in drinking water, can and do lead to illness among employees. Moreover, an industrial establishment which provides its workers with a food service lacking sanitary food handling facilities and practices that afford protection, invites the possible disaster of

Patty's Industrial Hygiene, Fifth Edition, Volume 4. Edited by Robert L. Harris.
ISBN 0-471-29749-6

widespread foodborne infection by harmful organisms and their toxins, or by other poisonous materials.

Since the publication of the fourth edition of *Patty's Industrial Hygiene and Toxicology* new legislation in the United States affecting environmental health regulation has been enacted. However, unchanged, is the Williams-Steger Occupational Safety and Health Act of 1970 which includes sections relating to "General Environmental Controls," "Sanitation", "Temporary Labor Camps", and "Non-Water Carriage Disposal Systems." A decade after the enactment of the Safe Drinking Water Amendments (SDWA) of 1986 (PL 99-339) another set of amendments to the Safe Drinking Water Act became law on August 6, 1996. The 1986 amendments to the original 1974 SDWA were enacted to respond to the concerns about unregulated contaminant pollution of ground water sources by discharges of wastes containing solvents and pesticides. In addition to this, most recently there has been the growing problem of drinking water source contamination by pathogenic parasites such as *Giardia lambia* and *Cryptosporidium parvum*. The Centers for Disease Control and Prevention (CDC) stated that during 1976–1982 Cryptosporidiosis was reported rarely and occurred predominantly in immunocompromised people (1). The number of reported cases began to increase in 1982. It is thought that the reason for this increase was the start of the Acquired Immunodeficiency Syndrome (AIDS) epidemic. In 1993 an outbreak of Cryptosporidiosis in Milwaukee, Wisconsin, which affected some 403,000 persons and resulted in a 150 deaths, was found to be associated with the community water supply. The municipal water system was operating within the limits of state and federal regulations. This knowledge spurred a nationwide evaluation of water source microbiological contamination. In addition it highlighted the need for improved surveillance systems and epidemiological study designs. The finding that *Cryptosporidium oocysts* are present in 65–97% of surface water, i.e., rivers, streams and lakes in the United States, and that the parasite is highly resistant to the chemical disinfection used in drinking water treatment has placed the burden on other unit processes in the treatment system. These, such as coagulation, sedimentation, and filtration result in the physical removal of the pathogen. However, many cities do not use filtration as a component of their water treatment system. At the same time no existing water treatment methods can guarantee a *Cryptosporidium* free finished water.

The 1986 SDWA revisions included requirements that mandated standards for 83 contaminants be established and regulated by the Environmental Protection Agency (EPA). An additional requirement was that every three years thereafter standards for an additional 25 contaminants be established. The 1996 SDWA cancelled these requirements and in its place the EPA is now authorized to set new standards based upon "risk" and "cost benefit" analysis.

The grandfather of environmental legislation, the Clean Water Act (amended by P.L. 95-217, PL 97-117 and PL 100-4, originally the Federal Water Pollution Control Act (P.L. 92-500) enacted in 1948) is the principal law governing water quality in the nation's streams, rivers, lakes and estuaries. The act has two major sections. First, it contains regulatory provisions that impose progressively more stringent requirements on municipalities and industries to reduce their liquid wastes and move them towards the goal of zero discharge of pollutants. Secondly the Act provided financial assistance to publicly owned wastewater treatment facilities. Because industrial establishments are significant wastewater contributors to publicly owned sewage treatment systems, the Clean Water Act includes provisions that require industrial wastewater sources to pretreat wastes that could disrupt the biological processes utilized to treat municipal wastewater or would otherwise interfere with the process. The National Pollution Discharge and Elimination System (NPDES), as codified in **40** *CFR* Parts 122 and 123, is a rule making which requires that permit renewal applications include listed effluent data. The 1987 amendments to the Clean Water Act (PL 100-4) added a section 319 which requires states to develop and implement control of "non-point" pollution sources including construction and mining sites, and to focus on urban areas where stormwater flows are pollution sources. Experience has taught that lack of concern about the consequences of pollutant and contaminant discharges to land, water or air has resulted in great harm to the environment and public health. The United States Congress recognized this fact and enacted the Pollution Prevention Act of 1990. The Act established pollution prevention as a "national objective" and the most important component of the environmental management system. This means that U.S. national policy declares that the creation of potential pollutants should be prevented or reduced to a minimum during the production cycle whenever feasible. Moreover, the prevention of pollution has been demonstrated to reduce both operating costs and potential liabilities, in addition to preserving the environment.

The Toxic Substance Control Act (TSCA) (PL 94-469) was enacted into law in 1976. Its aim was to address risks from hazardous chemicals. Continuing environmental contamination by toxic substances highlighted the fact that prior to the enactment of TSCA damaging chemical releases were handled on an *ad hoc* basis under the aegis of the Clean Air Act; the Consumer Product Safety Act, or Occupational Safety and Health Act. TSCA legislation, originally proposed in 1971, was enacted in 1976, provides the EPA with authority to:

- Induce testing of existing chemicals—those currently in widespread commercial production or use.
- Prevent future chemical risks through premarket screening and regulatory tracking of new chemical products.
- Control of unreasonable risks of chemicals already known or as they are discovered.
- Gather and disseminate information about chemical production, use and possible adverse effects on human health and the environment.

In 1965 President Lyndon B. Johnson called for "better solutions to the disposal of wastes". The President recommended that Congress pass legislation that would "(*1*) assist the States in developing comprehensive programs for some forms of solid waste disposal: and (*2*) provide for research and demonstration projects leading to more effective methods of disposing or of salvaging solid waste".

Later in 1965 the Congress passed Public Law 89-272 amending the Clean Air Act (CAA). Title I amended the CAA and established a new title, which monitors the pollution caused by a motor vehicle from the time it is produced to its final disposal. Title II created the Solid Waste Disposal Act (SWDA) as a separate environmental statute. The Resource Conservation and Recovery Act (RCRA) enacted in 1976 (PL 94-580), supplanted The

Solid Waste Disposal Act, it established the federal program regulating solid and hazardous waste. Subtitle C of the Act regulates hazardous waste and establishes "cradle to grave" management procedures for these wastes from generation, through transport, to treatment, and/or disposal stages. Solid waste, including solid waste generated by industry, is regulated under Subtitle D of RCRA. The regulation establishes criteria with minimum technical requirements for acceptable operation of municipal solid waste facilities. Currently municipal solid waste management practices are (a) landfilling, (b) incineration, and (c) recycling/reuse. In most communities a combination of at least two disposal methods, and sometimes all three are used. As it is demonstrated that recycling/reuse is technically feasible and economically viable, a greater percentage of the solid waste stream is being recycled/reused. It is estimated that in 1976 there were 20,000 landfills operating in the United States. By 1998 there were 3,563. The issues facing legislators, communities and industry are the declining availability of environmentally viable landfilling disposal sites and who, under what auspices, should develop new solutions.

By 1998 it had become evident that communities are being forced to transport the municipal solid waste and sewage sludge they generate to distant disposal sites. Larger cities such as New York City face major decisions. The "Fresh Kills" landfill on Staten Island is slated for closure in the year 2000. The manner in which New York City solves its solid waste disposal problem is worthy of attention and review. It should provide evidence of new disposal methodology, the politics of solid wastes. It may also serve to demonstrate a greater need for conservation.

A special solid waste case is that designated "hazardous waste." Title C of RCRA details four classes to define a hazardous waste. These are

1. Irritability.
2. Reactivity.
3. Corrosiveness.
4. Toxicity.

Managing these wastes and the need for dealing with hazardous waste emergencies is being dealt with in a new chapter of this volume entitled HAZARDOUS WASTE, Chapter 61.

As it relates to emerging environmental health problems human history is replete with evidence of an ecological struggle with the microbial environment. Its pathogenic components have been major causes of sickness and death. Infectious diseases have seriously limited well being and life span.

In the late 19th century, when scientific knowledge allowed, there began a process of intervention in the ecological balance of humanity and microbes, by intent and intelligence it was changed in favor of human beings. It was discovered that microbes caused infectious diseases. One learned how to interpose barriers in the transmission process, to keep them out of food, water, and the environment. Then it was learned how to prevent infectious disease by immunization and by destroying invading pathogens in our bodies when they caused disease. Early in the 20th century the advent of public health and environmental health developed measures against environmental transmission of infectious agents. These included water treatment and disinfection, wastewater collection, treatment and disinfection, food sanitation in the processing and preparation stages and milk pasteurization. In the last 50 years antimicrobial agents were developed and designed to kill microbes and cure infectious diseases. By applying the growing understanding of microbiology and immunology we developed very effective vaccines against major bacterial and viral diseases. This has been the foundation of preventing poliomyelitis, measles and hepatitis B. Knowledge and a worldwide campaign resulted in the eradication of small pox as declared by the World Health Organization (WHO) on May 8, 1980. So far (1999) the only disease to be so designated. This was the culmination of a century of achievements against infectious diseases beginning with the discovery of the cause of tuberculosis by Koch in 1882.

Recent experience with the assurance of safe drinking water supplies has emphasized that emerging environmental health problems recapitulate past experience. Human history is replete with evidence of an ecological struggle with the microbial environment. By the 1990s it was learned that even the best measures available to intervene in the ecology of pathogens, with the exception of one disease small pox, are imperfect.

In the current scientific panorama of infectious disease there appears a concept such as "emerging infectious diseases". The word "emerging" means "newly arisen". In that generic sense, infectious diseases are characterized by Monto Ho (2) as shown in Table 57.1.

There are some infectious diseases that are new to humanity. Acquired Immunodeficiency Syndrome (AIDS) (Table 57.1, Category 1) is such a disease. It is a new, previously unknown disease. AIDS, a fatal disease was first discovered in 1981 and in 1984 was demonstrated to be caused by a completely new virus, HIV. Analysis of the Human Immunodeficiency Virus (HIV) gene-genome suggests a simian source for the virus. This new agent has caused a world pandemic, which has killed millions in the past 18 years. Newly developed mutants of old microbes pose a threat to humanity because of its lack of herd immunity to them (Table 57.1, Category 2). New mutants of Influenza A are important examples of this danger. These are a constant threat. The latest known mutant is a HN strain identified in Hong Cong, presumably a recombinant strain of a chicken virus, which caused the preventive destruction of large numbers of fowl. There is a new epidemic strain of cholera named 0139, which is supplanting the old 01 strain, that was endemic on the Indian subcontinent for many years. This new cholera strain has now spread to Latin America.

Table 57.1. Emerging Infectious Diseases

1. Agents new to humanity-recently identified: HIV, AIDS
2. New mutants of well known agents: Periodic new emergent variants of Influenza A virus, as HN, and vibrio cholera 0139.
3. Newly identified agents, which are probably old to humanity: Legionella, Lassa, Ebola, Hemorrhagic fevers, Hantavirus, "Four Corners Disease", Cryptosporidiosis, Hepatitis C, HTLV-1 leukemia, bovine Creutsfeld-Jacob agent, and *E. coli* 0157:H7
4. Old disease agents with a "new" face: Dengue Fever in Taiwan, Multiresistant Tuberculosis, antibiotic resistant bacterial infections, hospital acquired infections (nosocomial infections).
5. Old diseases never fully controlled and now re-emerging. Salmonellosis in Great Britain, Diphtheria in Russia, Japan Encephalitis, diarrheal diseases and Tuberculosis in Taiwan.

Another category of diseases is caused by newly discovered agents, which probably are not new (Table 57.1, Category 3). Their discovery was facilitated by improved detection methods and because some of these diseases have become more prevalent. These newly discovered agents of filovirus and arenavirus cause highly transmissible, fatal hemorrhagic fevers. These diseases are not only a threat in Africa and Latin America where they are endemic, but in the age of rapid transportation to the entire world. Of these the Marburg virus and the Ebola virus are filoviruses and Lassa virus is an arenavirus. In 1980 HTLV-I was found to be endemic in Japan and the Caribbean, and to be the cause of T-cell leukemia/lymphoma. In 1977 Hantavirus was discovered to be the agent of Korean hemorrhagic fever. This virus group was found to be endemic on the entire Eurasian continent, where it has caused disease syndromes with neuropathy. In the United States, Hantavirus appeared in 1993 in a new guise, a form of highly fatal pulmonary disease. The Hantavirus, so called because one strain of the virus infected about 3,000 U.S. soldiers along the Hanataan River during the Korean War, has probably killed many in the American West for centuries.

Other newly discovered diseases and their infective agents include Legionnaires' disease, which is discussed later in this chapter, was discovered in the United States in 1977; Hepatitis C; Cryptosporidiosis; and the hemorrhagic fevers caused by a group of viruses. In the past few years bovine spongiform encephalitis, or Mad Cow disease caused by a protein-prion, appeared. In humans it may manifest itself as a Creutzfeldt-Jakob like disease. Bovine spongiform encephalitis is not only a very serious and difficult problem, in terms of human illness and mortality, but it also is a huge economic burden. In Great Britain, where the bovine epidemic was most pronounced, apart from decrees about the composition of animal feed, hundreds of thousands of cattle were destroyed. In 1982 a serotype of the common *E. coli*, type 0157:H7, was discovered to cause hemorrhagic uremic syndrome. The source of the organism is often difficult to detect. The protozoan *Cryptosporidium* is ubiquitous in the environment. Its spores are derived from animals and easily enter bodies of water. When the disease, Cryptosporidiosis appeared it was thought to be an opportunistic disease of individuals with HIV/AIDS. Since then it has been demonstrated to cause gastroenteritis among immunocompetent persons as well.

2 PROVISION OF A SAFE, POTABLE, AND ADEQUATE WATER SUPPLY

2.1 Source and Regulatory Control: The Safe Drinking Water Act

In the case of large industries, the facilities for providing a safe and adequate water supply rival the size and complexity of many a community system. Although the advent of the Safe Drinking Water Act of 1974 (PL 93-523) (SDWA) brought about major changes in water supply regulation, even more extensive and demanding requirements were enacted in the Safe Drinking Water Act Amendments of 1986 (PL 99-339) and 1996 (PL 104-182).

New events and conditions such as outbreaks of waterborne giardiasis and cryptosporidosis, previously relatively unknown diseases, brought pressure to assure that they be excluded from public water supplies. *Giardia lamblia* is more resistant than other pathogens to the most common disinfection process, chlorination, and therefore, if allowed to reach the final disinfection point, sufficient numbers of this pathogen to cause disease may survive in the distribution or plumbing systems. Although disinfection alone is not sufficient, when combined with coagulation, sedimentation, and filtration they are effective in removing and inactivating the organism. Besides such large communities as Seattle, Washington and New York, which employ extensive water quality surveillance, there are many small filtered and unfiltered public water supplies in the United States that pose a danger of explosive giardiasis and cryptosporidiosis outbreaks. As mentioned previously *Cryptosporidium parvum*, was responsible for a massive waterborne disease outbreak in Milwaukcc, Wisconsin, in 1993. An estimated 403,000 persons became ill with about 150 deaths occurring in immunocompromised individuals. Unlike giardiasis, as of this writing, there is no effective medication for cryptosporidiosis.

Giardia cysts and *Cryptosporidium* oocysts are found in surface water sources, groundwater sources under the direct influence of surface water such as infiltration galleries and springs, and improperly constructed and protected groundwater sources. Sources of the parasite are runoff from animal feedlots, bypasses at wastewater treatment plants, and combined sewer overflows. Therefore new filtration criteria for surface water supplies leading to regulations have been mandated.

Other developments, such as the growing awareness of the widespread contamination of drinking water sources such as shallow groundwater aquifers, by a variety of improperly disposed solvents, pesticides, and industrial chemicals, led to enactment of the extensive amendment embodied in the Safe Drinking Water Act of 1996 (PL 104-182). In addition, radon, a radionuclide naturally occurring in soil and groundwater, has been found in a number of localities. The public health significance of this contaminant, owing to the lung cancer that can result from long-term exposure to the gas, may be as great as any other contaminant.

To reduce the threat of waterborne disease outbreaks, regulations require:

- Priority should be given to selection of the purest source.
- Evaluation of groundwater sources for the direct influence of surface water
- A multiple barrier approach for surface water sources of coagulation, sedimentation, filtration, and disinfection, with each unit treatment process optimized.
- Specific criteria have to be met if the water supply is to avoid treatment by filtration. This requires source water protection and monitoring.
- Monitoring for turbidity performance.
- Monitoring for coliform bacteria in the distribution system as a indicator of disinfection efficiency and for contamination within the distribution system.

Lead in drinking water is a major concern due to the health effects of lead, although other sources of lead in the environment include lead paint and lead fumes which could be present in an industrial setting. To address the potential for lead in water, all community and nontransient noncommunity water systems started programs of tap sampling for lead and copper in 1992 or 1993, depending on the number of persons served by the water system. Samples were from taps selected to be at risk, and were taken after a 6–12 hour standing period for a worst case sample. Action levels are set for lead and copper (15 ppb and 1.3 ppm, respectively) with the 90th percentile result for each contaminant required to meet these action levels. Water systems that failed to meet the action levels were required to modify treatment to make the water less corrosive, and conduct public education in the case of a failure to meet the lead action level. Industries and institutions should also evaluate the drinking fountains for lead content. In the January 18, 1990 *Federal Register* a listing of "Drinking Water Coolers That Are Not Lead-Free" was published. Replacement of existing lead water supply lines, lead solder and brass components is not required by the SDWA unless the water system does not meet the action levels in follow-up rounds of monitoring. The Lead Contamination Control Act of 1988, now incorporated into the SDWA, requires use of lead free pipe, solder and flux in public water systems and in plumbing providing water for human consumption in residences and buildings. Flux and solder may not have more that 0.2% lead and pipe and fittings not more than 8% lead. Effective June 19, 1988, each state is required to enforce this provision through state and local plumbing codes. EPA adopted the American National Standard Institute/National Sanitation Foundation (ANSI/NSF) Standard 61 for faucets, drinking fountains and other drinking water system components, effective August 6, 1998. This standard establishes maximum lead leaching levels from new plumbing fittings and fixtures that are intended by the manufacturer to dispense water for human consumption.

Additional monitoring of groundwater sources has detected contamination of shallow groundwater aquifers with volatile organic chemicals (VOCs), synthetic organic chemicals (SOCs), and nitrates. Improper disposal of spent solvents, extensive pesticides use and leaking underground storage tanks located within the area of influence of a well can lead to contamination of a water sources. Treatment options for water obtained from contaminated sources includes: air stripping of volatile organic chemicals, adsorption on to granular activated carbon, or abandoning the well.

A public water system is defined as one providing water for human consumption through constructed conveyances to at least 15 service connections or an average of 25 individuals daily at least 60 days a year. Originally the water system to be regulated was restricted to "piped water systems." Effective August 6, 1998 the definition was broadened to include canals and other constructed water conveyance systems. Some industries that, in the past, were exempt from SDWA regulation may now qualify as public water systems. USEPA interprets "human consumption" to include drinking, bathing, showering, cooking, dishwashing and maintaining oral hygiene. Therefore, a facility can not avoid regulation under the SDWA by providing bottled water dispensers for drinking when the piped water source in the facility is used for other human consumption, such as showering.

Public water systems are classified as community (serving at least 15 service connection used by year-round residents or at least 25 year-round residents), nontransient noncommunity (not a community and regularly serving at least 25 of the same persons over six months of the year, for example a school or a factory) or noncommunity (not a community, for example a restaurant). Some industrial and institutional facilities that consist of several buildings with water pipelines, valves, fire hydrants and/or storage tanks do not fall under the scope of the SDWA. The regulations do not apply to a system

1. Which consists only of storage and distribution facilities (and does not have any collection and treatment facilities).

2. Which obtains all of its water from, but is not owned or operated by, a public water system to which such regulations apply.
3. Which does not sell water to any person;
4. Which is not a carrier which conveys passengers in interstate commerce.

An industrial or institutional facility may be compelled to obtain its supply of water from multiple sources within its boundary, depending on growth needs, source capacity and economic viability. When there are two sources of drinking water, i.e., one public and the other the industrial establishment there shall be no direct connection between them. This is especially important if one is a potable water supply and the other a separate fire-fighting system. Water may be purchased from a municipal supply or obtained from an on-site source and treated. Compliance with plumbing codes and installation of backflow protection devices where needed should minimize hazards of backflow. Community water systems may require high-risk customers, e.g., metal plating industries, to install backflow protection devices at the meter to contain any contamination within the facility.

To protect the health of the public, the primary contaminants each have a maximum contaminant level goal (MCLG) and maximum contaminant level (MCL) set by the regulations. The MCLG is set at the level at which no known or anticipated adverse effects on the health of person occur and which allows an adequate margin of safety. The MCL is set as close to the MCLG as is feasible, utilizing best available treatment technology and considering costs, and is the enforceable limit. Secondary MCLs are set for contaminants with no health effects, but which affect the esthetic quality of the water. Enforcement of the secondary MCLs varies from state to state.

New contaminants are proposed for regulation by listing on the Drinking Water Priority List. Available data on health effects, occurrence in water supplies and cost effectiveness of treatment technologies are considered in the development of a regulation to control a contaminant. As knowledge of contaminant health effects improves their occurrence in water supplies and cost effectiveness of treatment technologies determined new MCLGs and MCLs. Updated standards are available on the Internet at: http://www.epa.gov/ogwdw/wot/appa.html

All public water systems must monitor their supply for the regulated contaminants. Monitoring frequency and number of samples collected vary according to the type of Public Water Supply (PWS), the system population served and the results of previous monitoring. On July 30, 1992 the number of regulated inorganic and volatile organic compounds was increased and a standardized monitoring framework was established. Each 9-year compliance cycle had three 3-year compliance periods, beginning January 1, 1993. This eliminated some confusion, as the monitoring frequency of most contaminants is monthly, quarterly, annually or triennially. A Non-Transient Non-Community (NTNC) water system is required to perform the same monitoring as a community water system of the same size.

If a public water system does not comply with the regulations by failing to complete required monitoring, or exceeds an MCL, or does not utilize a treatment technique set by the regulations, or does not comply within a variance schedule, it must publicly notify its customers or users. For a Non-Transient Non-Community (NTNC), the public notice containing required health effects language in the case of an MCL exceedance, can typically be posted in conspicuous places within the facility or hand delivered to system users. Posting must continue for as long as the violation or failure exists. If a PWS fails to undertake public notification, the state regulatory agency may conduct the notification.

Congress amended the SDWA on August 6, 1996, directing the EPA to promulgate new regulations over the following years. Regulations and provisions included in this legislation of interest to the Non-Transient Non-Community (NTNC) water system include:

- New guidelines requiring states to develop operator certification programs, which mandate that NTNC water systems are operated by persons certified as a waterworks operator.
- Funding for states to develop training and technical assistance programs for all public water systems.
- Contaminants yet to be assessed and for which regulations are to be promulgated are arsenic, sulfate, and radon.
- A provision that States must develop source water protection programs, with the approval of EPA, including watershed surveillance and protection for surface water sources, and wellhead protection programs for groundwater sources.

A public water supply serving an industry is required to prepare and provide Consumer Confidence Reports (CCRs) for one calendar year in the following one, e.g., for 1998 in 1999. An industrial establishment supplied by a community water supply system must be informed about contaminants detected, monitoring violations and be provided with general information about the source and treatment of the water supplied.

All current drinking water regulations that have been promulgated as of July 1 in any year may be found in the *Code of Federal Regulations* (*CFR*), Volume 40, Parts 141, 142 and 143. Regulations published between *CFR* editions may be found in the *Federal Register*. It is important to contact the individual state agency for the regulations and guidance documents effective in that state. State regulations may be stricter than the Federal regulations, but may not be less strict. The EPA has transferred primary enforcement responsibility (primacy) to all states except Wyoming.

The Internet is an excellent source of up-to-date regulatory information and technical assistance. The EPA (http://www.epa.gov) web site includes regulation updates, listing of current regulated contaminants and their maximum contaminant levels, health effects fact sheets and environmental databases. State regulatory agency web-sites can be searched for specific state guidelines and assistance. The American Waterworks Association (http://www.awwa.org) site includes regulatory updates and an on-line forum for technical assistance. The EPA sponsors the Safe Drinking Water Hotline, 1-800-426-4791 (9:30 am–5:30 pm EST Monday through Friday) to answer questions and provide drinking water information.

2.2 Drinking Water Supply

A factory or plant with its own source of drinking water and with 25 or more employees is subject to the regulations promulgated under the Safe Drinking Water Act (SDWA) and classified as a Non-Transient Non-Community (NTNC) public water system. The current National Primary Drinking Water Standards (NPDWS) set maximum contaminant levels (MCLs) for 12 inorganic chemicals, 21 volatile organic chemicals, 32 synthetic organic chemicals, turbidity, total trihalomethanes, radiological contaminants and microbiological contaminants. NTNC water systems are required to monitor their supply for regulated contaminants. However they are exempt from monitoring for radiological contaminants and disinfection byproducts. If there is a possibility of in-plant radiological contamination, this monitoring should be conducted voluntarily. As new regulations are promulgated, additional monitoring requirements will be added to the responsibilities of the operator of a NTNC public water system.

In addition to assuring safety, industrial plants should provide water for drinking and cooking that is acceptable to its employees. Drinking water should be supplied at a temperature within a range of 40 to 80°F (the optimal range is 45 to 50°F). When cooling is needed, mechanical refrigeration or ice can be used. The ice must be produced from water meeting drinking water standards and maintained to prevent postproduction contamination. In nonfood purveying areas, ice that is used to cool drinking water should not be allowed to come in contact with the water. In cafeterias and other food serving areas, the advent of ice-making machines makes it feasible to use ice to cool water and beverages.

Water supplied in drinking fountains and food preparation centers of the plant must be safe, clean, potable, and cool. A sanitary drinking fountain of approved design is the most efficient method of providing drinking water for employees. The American National Standard Minimum Requirements for Sanitation in Places of Employment (3) state that: "Sanitary drinking fountains shall be of a type and construction approved by the health authorities having jurisdiction" or meet local plumbing code requirements. "New installations shall be constructed in accordance with the requirements of the health authorities having jurisdiction. If there are no such requirements, installation should be in accordance with American National Standards Institute Standard for Drinking Fountains, ANSI/ARI 1010-82 and ANSI/NSF 61, or the latest revision approved by the American National Standards Institute" (4). To keep possible disease transmission to a minimum, individual disposable drinking cups may be supplied. Whenever it is not feasible to have a drinking fountain connected to the supply, an approved drinking water container with an approved fountain and individual disposable cups should be furnished. In general, location of fountains may be determined by an overall standard of one drinking fountain for each 50 employees. However, the distance the employee must travel to the nearest source may be a controlling factor in locating drinking water sources (The ANSI/ARI Standard 1010-82 requires that this distance be no more than 200 ft). Similarly, wherever the employees are subjected to above normal heat stress, this fact should be the controlling criterion for the location of a drinking water source.

The 1972 NIOSH criteria document on heat stress in the working environment (5) recommend a series of work practices to ensure that the employee body core temperature does not exceed 38°C (100.4°F). Among the seven work practices specified, the fifth is titled "Enhancing Tolerance to Heat."

> V. Regular breaks, consisting of a minimum of one every hour, shall be prescribed for employees to get water and replacement salt. The employer shall provide a minimum of eight

quarts of cool potable 0.1%-salted drinking water or a minimum of eight quarts of cool potable water and salt tablets per shift. The water supply shall be located as near as possible to the position where the employee is regularly engaged in work, but never more than 200 feet, except where a variance had been granted.

In 1992 the National Institute for Occupational Safety and Health (NIOSH) developed revised criteria for a recommended standard for occupational exposure to hot environments (6). Two of the recommendations update prior recommendations for salinization of the drinking water supply in hot weather or hot environments. The 1992 revised criteria are

c. to ensure that water lost in the sweat and urine is replaced (at least hourly) during the workday, an adequate water supply and intake are essential for heat tolerance and prevention of heat induced illnesses.

d. Electrolyte balance in the body fluids must be maintained to prevent some of the heat-induced illnesses. For heat-unacclimatized workers who may be on a restricted salt diet, additional salting of the food, with a physician's concurrence, during the first two days of heat exposure may be required to replace the salt lost in the sweat. The acclimatized worker loses relatively little salt in the sweat; therefore, salt supplementation of the normal U.S. diet is usually not required.

2.3 Water Uses in Industry

In addition to drinking water supply, industrial plants require water for cooling, processing, and cleaning. Different quality demands are associated with each use. A pharmaceutical manufacturing plant, for example, may require deionized pyrogen-free water, whereas a steel mill uses very large amounts of cooling water with far different quality demands.

The U.S. Geological Survey's report, "Estimating Use of Water in the United States in 1995" (7) records that nearly 341 billion gallons of fresh water and 60.8 billion gallons of saline water were withdrawn per day in 1995 in the United States. Mining activities withdrew 2.6 billion gallons of fresh water and 1.2 billion gallons of saline water per day, while thermoelectric power generation withdrew 132 billion gallons per day of fresh water and 58 billion gallons of saline water. Other industrial withdrawals were 20.7 billion gallons per day of fresh water and 1.7 billion gallons per day of saline water. Thus, the total withdrawal of fresh and saline water by industry, mining, and thermoelectric power facilities was estimated to be 216 billion gallons per day. The gross need is higher and it is met through a combination of once through and recycling of industrial effluents.

Ranking of industrial water uses by category is as follows:

1. Cooling.
2. Steam generation.
3. Solvent.
4. Washing.

5. Conveying medium.
6. Air scrubbing.

Cooling is the largest volume use, and thermoelectric plants are the major users of cooling water. Because cooling water typically is separated from process water (i.e., water utilized in the manufacturing process), it usually does not come in contact with the product. Except for the addition of heat, the other quality characteristics of cooling water may not be significantly changed. Additives designed to reduce fouling due to bacterial and algal growths, when recycling is practiced, do change its characteristics.

Water used for steam generation becomes contaminated with chemical additives or intermediate products that must be removed before reuse or discharge. In the petroleum refining industry, "sour water strippers" are used to remove sulfur and other polluting substances from condensed steam that has been employed to heat crude oil during the distillation process.

Water, the universal solvent, is extensively used in industry to dissolve compounds. To effect reuse, the renovation of such water demands careful evaluation of its contaminants. Washwater picks up and entrains a wide variety of contaminants that, if the water is to be reused, must be removed or at least greatly reduced in concentration. In addition, washing and degreasing compounds and solvents may be introduced into the washwater. Water is used extensively as a conveying medium and in the process is contaminated with solids removed from the material conveyed, e.g., soil from agricultural products and any other biological, organic, and inorganic matter that may be mechanically removed or dissolved.

With the advent of more stringent requirements to control the emission of air pollutants, air-scrubbing devices that use water to entrap and entrain pollutants are being used in increasing numbers. These devices use large quantities of water and are a good example of the "cross-media" problem engendered. The material scrubbed out of the airstream is essentially a concentrated pollutant that must be dewatered and disposed of on land or otherwise utilized as a solid waste.

To be able to establish the most efficient wastewater treatment and renovation processes, industrial establishments, at all times, must be able to determine the following:

1. Where wastewater streams originated.
2. The contaminants present, their concentrations, and their variability.
3. The diurnal, weekly, and so on, variability of the wastewater volume.

McClure (8) reports that in the case of steel mill-cooling water containing a variety of suspended solids and iron, removal of the offending material is accomplished by precipitation and subsequent cyanide destruction. This is followed by alkaline chlorination and phenol reduction by chlorination of the clarifier blowdown to enable utilization of a closed-cycle cooling system. Thus, only the blowdown from the clarifier or thickener has to be treated, thereby minimizing makeup water needs.

In general, the distribution by category of water use in manufacturing industries in the United States is as follows:

Use	Percentage
Process water	28.3
Boiler feed water; sanitary	4.8
Heat exchange	3.2
Air conditioning	
Cooling	
Steam, thermoelectric	12.1
Other, condensing	51.6
Total	100.0

Some of this water is not returnable because of evaporation, incorporation into products, leaks, and other losses.

2.4 Conservation and Reuse of Water in Industry

Another factor controlling the use of water in industry has been the growing necessity to conserve water resources, primarily but not exclusively in arid and semiarid regions. An example of growing reuse practice is what is happening in Florida, a state with an average annual rainfall of 54 inches. A new impetus of major importance was the advent of the Water Pollution Control Act Amendments of 1972 (PL 92-500). A most important element of the law, and the program of water pollution control it has spawned, is the system of effluent limitations and required permits under the National Pollutant Discharge Elimination System (NPDES) applicable to discharges of industrial and other wastes into the navigable waters of the United States. The EPA has been empowered to issue wastewater discharge permits to individual industrial establishments, power-generating plants, refineries, municipal wastewater treatment plants, and similar facilities that are based on national effluent limitation guidelines. These guidelines designate the quantity and chemical, physical, and biological characteristics of effluent that industry may discharge into receiving bodies of water. Of crucial importance are the specific goals and objectives of the Amendments and the schedule of reaching water quality levels in the nation's waters. Levels, which will provide for the protection and propagation of fish, shellfish, and wildlife, and for recreation in and on the water. The Water Pollution Control Act Amendments of 1972 mandated the elimination of the discharge of pollutants into navigable waters by 1985.

A major review of the act and its goals was authorized by Congress, which established the National Commission on Water Quality and charged it to ". . . make a full and complete investigation and study of all the technological aspects of achieving, and all aspects of the total economic, social and environmental effects of achieving or not achieving, the effluent limitations and goals set forth for 1983 in section 301(b)(2) of this Act" (9).

Studies conducted by the National Commission on Water Quality have indicated that inplant changes, such as process modification and better internal control, can be joined to play an important role in meeting abatement goals. Flow reduction measures by means of increased water recycling and reuse are common to all the postulated strategies, as are better housekeeping procedures.

Since Congress mandated waste minimization as a policy in the 1984 Hazardous and Solid Waste Amendments to RCRA, the impetus to institutionalize such procedures and practices has increased. In addition, waste minimization can contribute to the reduction of a generator's liabilities under the provisions of the Comprehensive Response, Compensation and Liabilities Act (CERCLA or Superfund). The EPA working definition of waste minimization is that it consists of "source reduction and recycling." To assist in developing good hazardous waste management, EPA has published its Waste Minimization Opportunity Assessment Manual (10).

An example of reuse technology reported by Renn (11) suggests that a relatively low-level technology storage system for secondary treated wastewater is feasible, particularly when water is in short supply. In this case the recycled water was utilized for air conditioning heat exchange, cooling towers, and a variety of machine tool cutting operations. Successful operation required modification of the industrial processes to accept water containing varying concentrations of organic matter, color, suspended matter, and particulates generated during storage or in transmission.

Storage and reuse of treated wastewater during drought periods is a practical method of extending the available water supply for some industries and communities. It also is a device for achieving zero effluent discharge during critical, low stream flow periods. Ultimately, however, the storage system must be discharged (12).

Conservation of water in industry refers to reduction in "net-intake" water requirements. This should be differentiated from "gross water applied" or "gross water use," terms that are directly related to production. Significant reduction of "gross water use" is desirable and may be achieved by new technology such as substitution of other fluids for heat transfer or process water. Conservation not only entails technical measures, such as leak detection and elimination, avoidance of spills, and reduction of evaporation losses; it also demands awareness and alertness on the part of individuals and groups of employees. In general, the techniques utilized in water conservation practice may be classified as follows: (a) using less water by avoiding waste; (b) recycling by using water over again in the same process; (c) multiple or successive use—using water from one process for one or more additional processes in the same establishment, or where possible using nonpotable waters; and (d) nonevaporative cooling techniques by using special methods to reduce the amount of water required for cooling.

Use of reclaimed water from sewage and industrial wastes has been practiced for many years in many parts of the world. In industry, its use is not yet as widespread or as extensive as in agriculture. Reclaimed waters have in the past been most suited for cooling purposes, because in this instance the quantity and temperature of the water are of greater importance than its quality. Industry can employ reclaimed water for plant-processing water, boiler-feed waters, certain sanitary uses, fire protection, air conditioning, and other miscellaneous cooling.

In addition to the general public health hazards involved in the utilization of a reclaimed effluent from a wastewater treatment plant, there are quality considerations that arise from the concentration effects of the treatment process. The concentration of total salts may be increased measurably, and the hardness of the water thus reclaimed can make its utilization impractical or costly.

However, recent developments in wastewater treatment and so-called advanced wastewater treatment argue well for the possibility of using renovated water in the growing of food crops, recharge of underground aquifers, and direct reuse in a wider variety of manufacturing plants. A study (13) reports that in 1975 reuse of treated municipal wastewater in industry was continuous in 358 locations in the United States. Approximately 95% located in the semiarid southwest. Arnold et al. (14) in their report on the reclamation of water from sewage and industrial wastes in Los Angeles County, drew the following, still valid conclusions:

(a) There are a number of important factors limiting the direct use of water reclaimed from wastewater in industry. In addition to public health hazards and total salt concentration, considerations of the hardness of the water may be of material importance.

(b) In general, the direct reuse in industry of acceptable water reclaimed from wastewater is tolerated and encouraged in manufacturing processes that do not involve contact between the reclaimed water and human beings or foods to be consumed by humans.

(c) The most obvious direct reuse for reclaimed sewage water is for cooling and condensing operations.

With the advent of greatly increased effluent quality requirements, recycling of industrial wastewater after treatment to meet future effluent guidelines should become more prevalent. Kollar and Brewer (15) reviewing the "20 best of file" plants with high recycling rates, concluded that treated wastewater is being recycled at a high rate by many of the major water-using industries.

The concept that highest quality water be reserved for drinking, culinary, and washing uses, with a lower-quality water derived from either nonpotable source or reclaimed waste water used to meet other needs, is not new. Sextus Frontinus, the water commissioner of Rome in the Aqueducts of Rome (16), describes the supply of lower-grade water delivered to the fountains and public baths of the city. Reclaimed water has long been used as a second supply to flush toilets in water-deficient areas such as the facilities at the north rim of the Grand Canyon. An early example of reclaimed water use in industry was the Bethlehem Steel Company plant at Sparrows Point, Maryland use of further-treated, City of Baltimore Back River treatment plant effluent for cooling and other nonpotable uses. Until relatively recently such community and industrial use was highly restricted.

Bogly (17) has distinguished among the various types of reuse, a major source for the lower quality portion of a dual water supply:

- *Indirect potable reuse*. The abstraction of water for drinking and other purposes from a surface or underground source into which treated or untreated wastewaters have been discharged.
- *Direct potable reuse*. The piping of treated wastewater into a water supply system that provides water for drinking.
- *Indirect nonpotable reuse*. The abstraction of water for one or more nonpotable purposes from a surface or underground source into which treated or untreated wastewaters have been discharged.

- *Direct nonpotable reuse*. The piping of treated wastewater directly into a water supply system that provides water for one or more nonpotable purposes.

Okun (18) further states that direct nonpotable reuse in a separate system is being adopted throughout the world. In the United States the practice is widespread in the southeast, southwest, and west. In contrast with a total of 358 reuse projects in the United States in 1975, in California alone by 1985 there were 380 projects in operation providing water for agricultural irrigation, cooling, industry, toilet flushing, construction, and other purposes not requiring drinking water quality.

In a more recent article on the advantages of developing dual water supplies, one for potable and the second for reclaimed water, Okun (19) describes current industrial uses of reclaimed water: "industrial uses include makeup water for evaporative cooling towers, boiler feedwater, process water, and irrigation of plant grounds. Evaporative cooling is the most widely used application in industry. Industrial process applications that have found reclaimed water suitable for their purposes include pulp and paper, chemicals, steel manufacturing, textiles, petroleum and coal products, and industrial production in which the product poses no health risk to customers. When an industrial area can use a substantial amount of reclaimed water a dedicated line may be provided, with the reclaimed water quality suited to the needs of the customers. More commonly industries are scattered throughout a community, and the reclaimed water is distributed to all residential, commercial, and industrial customers. Industries that need special high quality water, such as may be required for special papers, instruments, and the like, would be obliged to provide this themselves as is generally the case when potable water supplies are used."

As the practice inevitably grows, both public and private agencies must engage in study and research to provide a better foundation for increasing safety of such reuse. Among the problems to be addressed are quality standards for various nonpotable uses, treatment required to achieve these standards, monitoring requirements, plumbing codes for such supplies, distinguishing between the different supplies by the use of universal color codes, and managing the dual systems.

An American Water Works Association committee established to evaluate the potential for dual distribution water systems defined a dual distribution system, which although developed primarily with community conditions in mind, is similarly applicable for industrial establishments. The definitions are

> *Potable water*. Water of excellent quality intended for drinking, cooking, and cleaning uses. This grade of water would conform to the quality requirements of state and federal regulatory agencies.
>
> *Nonpotable water*. Water acceptable for uses other than potable. This water would be safe for occasional inadvertent human consumption and would generally conform to the water quality requirements of state and federal regulatory agencies for nonpotable water.

2.5 Cross-Connections and Other Means of Contamination

The great variety of water supplies and uses within a plant, as well as the growing need for reuse and recycling of industrial water, present actual and potential hazards to the cleanliness, safety, potability, and coolness of the drinking and culinary water supplies. Such danger occurs when there are any connections between a water supply of known potability (primary source) and a supply of unknown or lesser quality, and when there are plumbing defects that may allow wastewater and toxic materials to enter the drinking water distribution system. There are many possibilities for contaminating a potable supply by a nonpotable supply when unauthorized or unsafe connections are made. Pressure variations in distribution systems are not uncommon, and when reduced pressure conditions occur in a potable system connected with another of unknown potability, having an even momentary higher pressure, the flow will be from the system of unknown potability into the potable system.

The Centers for Disease Control and Prevention's (CDC) Waterborne Disease Outbreaks in the United States surveillance report for 1995–1996 indicates that for the period 1995–1996, 13 states reported a total of 22 outbreaks associated with drinking water. Four (18%) outbreaks occurred because of back flow or back siphonage through a cross-connection. For three of the four outbreaks caused by contamination from a cross-connection, an improperly installed vacuum breaker or a faulty back flow prevention device was identified; no protection against back siphonage was found for the fourth outbreak (20). This information is in agreement with the general observation that cross-connections between safe and raw or unsafe water systems can and do occur frequently.

Contamination of the potable water supply by means of cross-connections or defective plumbing fixtures by back flow may take place by the occurrence of a vacuum in the supply lines, causing back siphonage of contaminated material; or by the development in the fixture, appliance, or piping system to which the water supply is connected of pressure that is greater than that in the supply system itself; and through the activities or actions of vermin, birds, or small animals in parts of the supply system not under pressure, or by dust reaching water-holding devices.

Back flow can occur under two conditions: (a) back pressure and (b) back siphonage. Back pressure in a supply system can take place whenever water temperature is elevated, or pressure or pumps are used in the system. Under such conditions it is possible that the pressure at the discharge point is higher than the supply pressure. Back siphonage occurs when atmospheric pressure exceeds supply pressure. The classic example of back siphonage is when the high-volume pumping of fire-fighting equipment drastically reduces the supply pressure.

In general, cross-connection control requires, first, a thorough knowledge of the water supply systems in the plant and, second, the other liquid purveying and removal systems. Because changes to these systems are usually quite frequent, a record of such changes must be kept current and readily retrievable. Other requirements include the use of back flow prevention devices, hazardous activity isolation, air-gap separation and installation of atmospheric and pressure-type vacuum breakers, and double-check valve assemblies. Certain industries have special cross-connection conditions such as water-using equipment that deserve special attention.

The following is an example of the damage that can result when back siphonage occurs. An acute nickel toxicity episode among electroplating workers who ingested a solution of nickel sulfate and nickel chloride was described by Sunderman et al. (21). The description of the exposure illustrates the perils of cross-connections very well.

The episode began on Saturday afternoon, June 13, 1987 when the main water supply to a metal-plating factory was temporarily shut down for repairs. The plant, which employed 338 workers, operated several automated electroplating lines used primarily to refinish automobile bumpers. Except for a few maintenance and repair personnel, the plant was unstaffed over the weekend until the evening shift reported for work at 1:00 P.M. on Sunday the 14th of June. After 3:00 P.M. several workers on a nickel-plating production line began to feel ill with symptoms that some attributed to bitter-tasting water from a drinking fountain. The supervisor on the night shift tested water samples from seven drinking fountains and found that the sample from the suspect fountain was green and contained nickel (approximately 2 g/L) while the samples from the other fountains were colorless and uncontaminated with nickel. The water supply to the contaminated fountain, which was adjacent to a nickel-plating tank, was flushed and the fountain was disconnected at 3:00 A.M. on Monday, the 15th of June. Since the ambient temperature was warm, most workers on the evening and night shifts sweated profusely; the workers who developed symptoms evidently had ingested 0.5–1.5 L of water from the contaminated fountain. The green tint of the water was inapparent under the factory lighting when viewed against the metallic gray background of the stainless-steel bowl.

The nickel contamination of the drinking water was caused by back siphonage from a water recirculation system that cooled filtration pumps for the nickel-plating tank. Leakage of pump gaskets had allowed the nickel-plating solution (nickel sulfate, $NiSO_4.6H_2O$, 310 g/L; nickel chloride, $NiCl_2.6H_2O$, 127 g/L; boric acid, H_3BO_4, 36 g/L) to seep into the water recirculation system. The recirculation system was connected via an open valve to a freshwater line. When the water main was shut down, nickel-contaminated water was sucked from the recirculation system into the freshwater line, from which the drinking fountain was directly tapped.

Periods when reduced pressure or partial vacuum situations may occur within a potable water system result from the following circumstances or by a combination of a number of them:

1. Interruption of the supply for maintenance of the municipal or services supply main.
2. Excessive demand placed on the supply mains during fire or other emergencies.
3. Complete failure of the supply due to breaks in the mains, earthquake damage, interruption of pumping by either mechanical failure or power shutoff, or deficient water supply.
4. Freezing of mains or service lines in extremely cold weather.
5. Excessive friction losses due to inadequate size of mains or service lines, or due to reduction of the effective diameter of pipes caused by deposits and encrustation.
6. Occurrence of negative water hammer pressure waves.
7. Operations of long pump suction lines.
8. Condensation of steam within boilers, hot water systems, or units such as hospital sterilizers.

2.6 Elimination of Cross-Connections

The safest method of making two water supplies available to a building is by the use of an unobstructed vertical fall through the free atmosphere between the lowest opening from any pipe supplying water to a reservoir and the highest possible level the water may reach in the reservoir. The minimum air gap thus provided should be twice the diameter of the effective opening and in no case less than 4 in. In addition, a float valve arrangement should be installed to cut off the supply when the water reaches the free overflow level.

The overflow channel or pipe should allow free discharge to the atmosphere with no enclosed connection to the sewer. This is the recommended method of the Building Officials and Code Administrators International (the National BOCA Code) for safe water supply to tanks or cooling tower basins.

To prevent back flow from fixtures in which the outlet end may at times be submerged, such as a hose and spray, direct flushing valves, and other devices in which the surface of the water is exposed to atmospheric pressure, a vacuum breaker should be installed. There are two general types of vacuum breaker: moving part and nonmoving part. The type selected should meet test requirements of the American National Standard for Back flow Preventers in Plumbing Systems ANSI 40.6-1943 and the American Water Works Association (AWWA) Manual M-14, "Recommended Practice for Backflow Prevention and Cross Connection Control" (22), and the AWWA Standards for "Backflow Prevention Devices" (23).

All fixtures supplied by a faucet should have an air gap, which is measured vertically from the end of the faucet spout or supply pipe to the flood level rim of the fixture or vessel. The minimum air gap provided should be twice the diameter of the effective opening, but no less than 1 inch. When affected by a near wall, the minimum air gap should be 1½ inches. Other conditions should meet the requirements of the American National Standard for "Air Gaps in Plumbing Systems," ASA40.4-1942 (revised 1973).

Connections to condensers, cooling jackets, expansion tanks, overflow pans, and other devices that waste clear water only should be discharged through a waste pipe connected to the drainage system with an air gap.

2.7 *Legionella pneumophila* and Legionnaires' Disease

The fact that the causative agent of Legionnaires' disease is a common inhabitant of aquatic environments became known after *Legionella pneumophila* was identified. The organism is detected in high numbers in the plumbing systems of hospitals, dwellings, manufacturing plants, and service facilities. Two major sources of amplification of the organism are hot water heaters and cooling towers. Although Legionella has not always been associated with disease, the consensus is that Legionnaires' disease is, in the majority of cases, water related (24).

2.7.1 Legionella

The bacterium Legionella can cause two distinct diseases: Legionnaires' disease and Pontiac fever. Legionnaires' disease is characterized by a pneumonia resulting in a 15–30% fatality rate (25). In contrast, Pontiac fever is a nonpneumonic, nonfatal, febrile disease with a very high attack rate. *Legionella pneumophila* is a Gram-negative, rod-shaped, fastidious, ubiquitous organism and can be recovered from various environmental sources, including soil, mud, showerheads, nebulizers, dehumidifiers, humidifiers, potable water, streams, lakes, evaporative condensers, and cooling towers. It has been recovered from industrial, commercial, hospital, and hotel plumbing systems and hot-water recirculating systems. Metabolic products of protozoa, algae, and various bacterial species support growth and multiplication of this fastidious organism in the environment.

In 1976 there was an outbreak of acute respiratory illness among members of the Pennsylvania American Legion, who had attended a state convention in Philadelphia. Approximately 7% of the attendees became ill and there was a 15% mortality rate among these cases. Laboratory investigation did not reveal the cause of the disease until six months after the beginning of the outbreak. In 1977, workers from the Centers for Disease Control isolated and characterized a bacterium that was later named *Legionella pneumophila*. The thoughtful preservation of sera and bacterial specimens from earlier explosive outbreaks of pneumonia disclosed that *Legionella pneumonia* was not a new syndrome, but had occurred uncharacterized for some time.

As reported by States et al. (26) although both legionellosis syndromes appear to be caused by the same group of bacteria, the disease occurs in two distinct clinicoepidemiologic forms. The first is *Legionella* pneumonia (Legionnaires' disease), a severe, acute pneumonia with multisystem involvement and an incubation period ranging from 2 to 10 days. Attack rates, which range between 0.1 and 4.0% of those exposed, are highest in immunocompromised individuals. Other risk factors include being male, being over 50 years old, being a smoker, heavy alcohol consumption, and underlying chronic disease. The other form of Legionellosis, Pontiac fever, is a nonpneumonic, self-limited, nonfatal influenza-like syndrome of fever, myalgia, and headache. The incubation time is 5 to 66 hr. Pontiac fever is much less dependent on host factors, with an attack rate approaching 100%.

Although Legionellosis is a reportable disease, it is difficult to ascertain its actual incidence. However, for the first 10 months of 1998, the Centers for Disease Control reported a total of 919 cases in the United States.

2.7.2 Pontiac Fever

The first known outbreak of Pontiac fever occurred in July of 1968 in Pontiac, Michigan, and was described by Glick et al. in 1978 (27). The early investigations of this outbreak failed to recognize an etiologic agent. Since these early investigations, a bacterium was isolated from guinea pigs exposed in the Pontiac Health Department building during the 1968 outbreak.

In July of 1973, ten previously healthy males spent nine hours cleaning a steam boiler condenser located on the James River in Virginia. All subsequently became sick for two days. No patient had pneumonia or objective evidence of other organ system involvement. The James River outbreak appears clinically and epidemiologically to resemble the Pontiac fever syndrome of Legionellosis. The high attack rate, short incubation period, and absence of pneumonia are typical. The demonstration that five of the patients showed seroconversion and that three others had convalescent phase titers of 64 confirms the association with *Legionella*.

Another outbreak occurred among employees of an engine assembly plant in Windsor, Ontario, Canada. The outbreak occurred from August 15 to August 21, 1981. In the study of the outbreak, a case was defined as a worker who experienced and reported at least three of four symptoms: fever, chills, headache, and myalgia. A total of 695 employees (85%) completed a questionnaire. Of these, 317 (46%) met the case definition, 270 (39%) were not ill, and 108 (16%) were ill, but did not meet the case definition. The illness had a mean maximum incubation period of 46 hr and was characterized by fever ranging from 99.5 to 104°F, severe myalgia, headache, and extreme fatigue. The illness was short (median duration of three days), but was severe enough to cause nearly 30% of the workers to miss work (median days of sick leave, two). There were no fatalities, and only four of the workers reported similar illness in family members within 72 hr after the onset of the worker's illness.

A *Legionella*-like organism, designated WO44C, was isolated from a sample of coolant obtained on August 19 from system 17 in the piston department. No other *Legionellae* or *Legionella*-like organisms were isolated from any of the other environmental samples. Like other *Legionella*, the organism did not grow on blood agar, required cysteine for growth, and produced catalase, but did not produce urease, reduce nitrate to nitrite, or produce acid from carbohydrates. Unlike other previously described legionellae, it did not produce gelatinase, and WO44C and *L. pneumophila* are the only two species that hydrolyze hippurate (28). Pontiac fever is a relatively mild influenza-like, short-term disease with no fatalities. Because of these characteristics, it is believed that many outbreaks of Pontiac fever have not been recorded.

2.7.3 Legionella pneumophila

Legionella spp. are widely distributed in nature and commonly found in many types of water systems; however, for reasons that are yet unexplained, infection in association with these systems appears to be the exception rather than the rule. Legionnaires' disease is pneumonia caused by a bacterium in the family *Legionellaceae*. The most common species involved in human infection is *Legionella pneumophila*, although 20 other species (e.g., *Legionella micdadei*) have also caused pneumonia. There are many serogroups of *L. pneumophila*, although serogroups 1,4,6 are the most pathogenic. The clinical manifestation of Legionnaires' disease is that of bacterial pneumonia. There can be a diverse clinical presentation ranging from mild cough and slight fever to stupor with widespread pulmonary infiltrates and multisystem failure. In the early stages of illness, symptoms are nonspecific including fever (often exceeding 104°F), malaise, myalgia, and headache. Cough is initially mild and slightly productive. Chest pain, occasionally pleuritic, may be present. Gastrointestinal symptoms (nausea, vomiting, and abdominal pain) are often prominent; diarrhea is seen in 25–50% of cases. The chest x-ray shows pneumonia. Laboratory findings may include abnormal liver function tests, hypophosphatemia, hematuria, and thrombocytopenia. Hyponatremia (sodium less than 130 mEq/L) is very common. Sputum gram stain shows many neutrophils but few, if any, bacteria. Failure to respond to beta-lactam antibiotics (penicillins or cephalosporins) should raise the possibility of Legionnaires' Disease. Specialized laboratory tests are the key feature for diagnosing this infection since the clinical presentation is nonspecific (29).

It is not possible to treat routinely all potential sources of Legionella in terms of practicability and cost, but if good engineering practices are followed in the operation and maintenance of water-cooled air conditioning, ventilation systems and hot water tanks the possibility of *Legionella* sp. infections can be reduced. Cleaning is indicated if organic growth is evident. Particular attention should be paid to water storage tanks, which should normally be kept covered. Cooling water systems should be treated to prevent a buildup of slime, algae, and scale. Such treatment thus can not only minimize the loss of efficiency by the system, but also inhibit rust, reducing the nutritional sources of the growth of *Legionella* sp.

Legionella may also colonize hot water heaters, pipes, water softeners and filters, and outlets including taps, showerheads, and other appliances. Proliferation can occur whenever conditions are ideal for growth. Factors that promote Legionella colonization in potable water systems include: temperature (<122°F or <50°C); scale and sediment accumulation; stagnation; and presence of commensal microflora (free-living protozoa and other bacteria). Although there is no current consensus as to which control measure industries or other establishments should take the following should be explored (29).

Copper–Silver Ionization. Positively charged metal ions kill bacteria by bonding with negatively charged sites on the microorganism denaturing cellular protein and distorting cellular permeability. An ionization unit is installed at the hot water recirculating line. Copper and silver ions are released into the plumbing system by hot water passing within a flow-through ionization chamber when electrical current is applied to copper/silver electrodes. The typical levels used are 0.2–0.8 ppm copper and 0.02–0.08 ppm silver. Copper–silver ionization systems have proven to be quite effective and their use is becoming widespread. While this approach usually does not completely eradicate legionellae from the plumbing system, it keeps their numbers at acceptably low levels. Experience has shown that it is important to maintain the units and periodically monitor for metal concentrations and legionellae occurrence. Ion levels can be best determined by atomic absorption spectroscopy: low levels will be ineffective while high levels may cause discoloration of water. Hot water samples should be collected from a distal site and the hot water recirculating line. It is important to note that ionization systems control legionellae in showerheads, taps, and other parts of the plumbing that are regularly in use.

Thermal Eradication. This is the "*Heat-and-Flush*" method in which hot water tank temperatures are elevated to greater than 158°F (70°C) followed by a flushing of all faucets and showerheads with hot water to kill *Legionella* colonizing these distal sites. All outlets are flushed for a period of 20–30 minutes, depending upon the temperature of the water when it reaches the outlets. At 140°F (60°C) each outlet must be flushed for at least 30 minutes. For the technique to be effective the water at distal outlets should reach at least 140°F (60°C). This procedure may need to be repeated periodically based on the results of routine culturing of the plumbing. Recolonization can be minimized by maintaining the tank temperature at 140°F (60°C). An advantage of this method is that no special equipment is required so the procedure can be initiated expeditiously—an advantage in outbreak situations. However, the technique is time consuming and labor intensive, and therefore costly. Also scalding is a potential hazard.

Chlorination. Chlorination was one of the first modalities used for *legionellae* disinfection. With long-term experience, some logistic disadvantages have emerged so that this approach should be reserved as an alternative option, should other methods prove unsatisfactory. The possibility that it can be used in concert with copper–silver ionization remains to be explored since chlorine interacts synergistically with copper and silver ions against *Legionella.* The disadvantage of chlorination include high expense, the necessity for unusually stringent monitoring of *Legionella* and chlorine levels, and corrosion with development of pinhole leaks in the piping.

Instantaneous Steam Heating Systems. These systems operate by "flash heating" water to a temperature >190°F(88°C) and then blending the hot water with a proportionate volume of cold water to achieve the desired temperature. Instantaneous heating systems are most effective when installed in the original heating system of a new building that has not previously been colonized with *Legionella.* The system eliminates the need for hot water storage tanks, which are well documented breeding sites for *Legionella.* This system cannot be used as the sole disinfection modality when *Legionella* has already colonized.

Ultraviolet Irradiation. A UV sterilizer is a flow-through device installed on a water line to kill *Legionella* as water flows through the unit. UV light is most effective if disinfection can be localized to a specific area within the building. Since UV lamps are housed in quartz sleeves, the sleeves need to be kept clean to allow adequate light penetration. If the water is hard or turbid, either a manual wipe system or an automatic motorized wiper system will be helpful. A prefiltration unit can also assist in preventing scale buildup. Additionally, a sensor should be utilized to ensure adequate intensity of UV radiation. Advantages of UV systems are that UV is an effective bactericide, which produces no disinfectant byproducts and does not damage the plumbing system. Disadvantages include the fact that UV light, unlike chlorine or copper–silver ionization, provides the water with no residual disinfectant and, therefore, will not eliminate legionellae colonizing portions of the plumbing system downstream of the unit. Thus, it cannot bc uscd as a solc modality for disinfection for an entire building. Additionally, high water temperatures and scale buildups reduce disinfection efficacy and increase the maintenance requirement.

Fraser (30), summarizing the Second International Symposium on *Legionella*, postulated that the following six links in the chain of causation that lead from environmental sites to infection in humans are necessary: (*1*) there exists an environmental reservoir where *legionellae* live; (*2*) there are one or more amplifying factors that allow legionellae to grow from low concentration to high ones; (*3*) that there is some mechanism of dissemination of legionellae from the reservoir so as to expose people; (*4*) the strain of *Legionella* that is disseminated is virulent for humans; (*5*) the organism is inoculated at the appropriate site on the human host; and (*6*) the host is susceptible to *Legionella* infection.

The ubiquity of *L. pneumophila* makes the use of epidemiological techniques all the more important in defining environmental sources for human infection. Air conditioning cooling towers or evaporative condensers have been clearly implicated as the source of outbreaks in the United States (Pontiac, Michigan; Burlington, Vermont; Memphis, Tennessee; Eau Claire, Wisconsin and Pittsburgh, Pennsylvania). The risk of illness is directly related to the degree of exposure to the droplet-laden drift from the cooling tower or evaporative condenser. In one outbreak infection stopped after exposure ceased. In several other outbreaks, a cooling tower or evaporative condenser water was found to be contaminated with *L. pneumophila* and may plausibly have been the source of infection, but the epidemiological proof was uncertain. It is recommended that in new buildings, cooling towers should not be located in such a position that the "drift" would readily enter a building through a ventilation system or by any other route.

In order to control *Legionella* in plumbing systems one or more of the following design, operation, and maintenance measures should be evaluated and implemented after considering the results of environmental monitoring and case epidemiology:

1. *Faucet Aerators.* The use of aerators on faucets should be avoided since Legionella grow well in the sediment that accumulates there.
2. *Drains.* Hot water storage vessels and hot water service boilers should be fitted with a drain valve located in an accessible position at the lowest point so that accumulated sludge may be removed. Horizontal vessels should be slanted to enable the vessel to be emptied completely. The drain in heating vessels should be located preferably at the heating coil end.
3. *Distribution System Design.* The distribution system should be designed to minimize the length of and dead-legs (i.e., laterals that are capped-off or infrequently used). Legionella can grow in these stagnant sections of the distribution system and subsequently contaminate the rest of the system.
4. *Storage and Distribution Temperatures.* The temperature of water as it leaves the hot water heater or hot water storage vessel should be 140°F(60°C ±2.5°C). Ideally, this temperature should be maintained for at least five minutes prior to discharge into the hot-water distribution system. At this temperature, *Legionella* survive for only approximately two minutes. Therefore, it is unlikely that the organism will be distributed into the hot water distribution system in sufficient numbers to be harmful. Temperatures greater than 140°F(60°C) are undesirable because the risk of accidental scalding increases considerably above 140°F(60°C), and dramatically above 149°F(65°C). The minimum temperature in the water returning to the hot water heater should be 122°F(50°C). Hot water at sinks and baths should reach a steady static temperature of between 122°F(50°C) and 140°F(60°C) within 1 minute at full flow. Cold water at sinks and baths should ideally reach less than 60°F(20°C) within two minutes. *Legionella* growth increases at temperatures >60°F(20°C).
5. *Stagnation in Hot Water Heaters.* Stagnation can occur in a hot water heater if the cold feed and/or return water connections are incorrectly sited. These can be modified by relocating the tapping or by sparge pipes. A degree of stagnation also occurs in concave-base vertical calorifiers. This can only be eliminated by replacement or by addition of pumps.
6. *Routine Cleaning and Maintenance.* All cold water storage and feed tanks should be regularly examined, cleaned, and disinfected annually. Hot-water heaters should be drained quarterly to minimize the accumulation of sludge. This frequency can be increased or decreased based on the amount of debris detected during inspection.
7. *Interruption of Services.* If a hot water tank or substantial part of the hot-water system is out of use for a week or longer for maintenance or other purposes, water should be raised to the operating temperature of 140°F(60°C) throughout for at least one day before being brought back into use (29).

As concluded by States et al. (26), research has demonstrated that *Legionella* is a common inhabitant of aquatic environments and is capable of surviving and multiplying in water treated to meet drinking water standards. *Legionella* can proliferate within certain systems and bacterial amplifiers. Following amplification in cooling towers, showerheads etc., it is currently believed the organism is able to infect individuals, particularly those at high risk due to immunosuppression, underlying illness, advanced age, heavy smoking, or alcoholism.

3 ASSURANCE OF A SAFE FOOD SUPPLY

The objective of food sanitation is the prevention of foodborne disease. Technological and economic considerations of purveying food play an important role in the application of effective control measures. However, the individual food worker has a crucial role in the safety of food that begins with its production, continues during delivery to the facility, storage, processing, preparation, hot/cold holding, cooling, reheating and ends when the food is served or otherwise made available to the employees. In addition, a secondary objective of a food sanitation programs is to prevent disease without impairing the nutritional value of the food.

For more detail on the requirements for the operation of a commercial kitchen see the U.S. Public Health Service/FDA 1999 Food Code (31).

3.1 Foodborne Disease

Diseases caused by ingestion of contaminated food may be divided into four major classes: (a) infections caused by the ingestion of living pathogenic organisms (bacterial or viral) contained in the food, (b) intoxication resulting from ingestion of food in which preformed microbial toxins have developed, (c) animal parasitism, and (d) poisoning by chemically contaminated food. In addition, poisonous plants and animals are implicated.

Foodborne disease reporting in the United States began more than 60 years ago, and in 1966 a system of foodborne disease surveillance was established by the Centers for Disease Control (CDC), which is now known as the Centers for Disease Control and Prevention. Reports of enteric disease outbreaks attributed to microbial or chemical contamination of food and water received by the CDC are incorporated into annual summaries. The Surveillance Report for Foodborne Disease Outbreaks in the United States between 1988–1992 (32) lists the annual numbers of reported outbreaks to range from 407–532. For the reporting period there were a total of 2,423 outbreaks which caused a reported 77,373 persons to become ill. Prior to 1992 an outbreak was defined as an incident in which (*1*) two or more persons experience a similar illness, usually gastrointestinal, after ingestion of a common food; and (*2*) epidemiological analysis implicates the food as a source of the illness. An exception to this definition, for example, was that one case of botulism or chemical poisoning constitutes an outbreak. In 1992 the exception to the definition was removed.

Estimates of the total annual number of foodborne disease cases in the United States are quite varied. However in 1994 the Council for Agriculture Science and Technology estimated 6.5 million to 33 million people become ill from microorganisms in food, resulting in as many as 9,000 deaths every year. While the exact cost of foodborne illnesses is unknown, it is estimated that its annual cost in terms of pain and suffering, reduced productivity, and medical costs is between $10 and $83 billion.

In the 5-year period 1988–1992, the responsible pathogen was identified in 41% of foodborne disease outbreaks reported to CDC. As reported in the Annual Summary for 1988–1992, many pathogens are not identified because of late or incomplete laboratory investigation. In other cases the responsible pathogen may have escaped detection despite a thorough laboratory investigation. Sometimes the responsible pathogen cannot yet be identified by available laboratory techniques. There is a need for further clinical, epidemiologic, and laboratory investigations to permit the identification of these pathogens or toxic agents and the institution of suitable measures to control diseases caused by them. Pathogens suspected of being, but not yet determined to be, etiologic agents in foodborne disease include group *D Streptococcus*, *Citrobacter*, *Enterobacter*, *Klebisiella*, *Pseudomonas*, and the presumed viral agents of acute infectious nonbacterial gastroenteritis.

3.1.1 Microbial Infections and Chemical Intoxication

The majority of reported and investigated foodborne disease outbreaks in the United States are caused by bacterial contamination. In 1988–1992, 2,423 outbreaks of foodborne disease were reported to CDC and the etiologic agent was confirmed, as stated above in only 41% of the reported outbreaks. Bacterial pathogens accounted for 796 (79%) of the total of 1001 investigated outbreaks where an agent was identified. *Salmonella* (69%) was the most frequently isolated organism (32). In 1982 two outbreaks of a previously unrecognized foodborne disease organism, *Escherichia coli*, serotype 0157:H7, which causes a severe diarrhea characterized by grossly bloody stools, were reported (33). Other etiologic agents are *Clostridium botulinum*, *Bacillus cerus*, *Campylobacter*, *Shigella*, *Staphylococcus aures*, *Clostridium perfringens*, *Listeria monocytogenes*, *Vibrio cholerae*, *Vibrio parahemolyticus*, and *Vibrio vulnaficus*.

In the period 1988–1992 chemicals, including heavy metals and toxins, accounted for 14% (32) of the outbreaks. Parasitic infection was involved in 2% and viruses in 4%. The relative incidence of foodborne disease of various etiologies is still unknown. Disorders characterized by short incubation periods, such as *Staphylococcus* infection, are more likely to be recognized as due to common source foodborne disease outbreaks than those with longer incubation periods. Viral disease, such as hepatitis A, with its typical incubation period of several weeks, is most likely to escape detection. Other detection problems are due to difficulties in the transportation and culturing of anaerobic specimens. Also pathogens that cause mild illness will be under-represented, while those causing serious illness are more likely to be identified and reported. Similarly, outbreaks associated with restaurants, workplace feeding sites or commercial products are more likely to be reported.

The *Enterobacteriaceae* of which *E. coli* is a member are facultative Gram-negative rods that ferment glucose (34, 35). *Enterobacteriaceae* are widely distributed in the environment, on plants, in soil, and in the intestines of humans and animals. Organisms in this family are associated with many types of human infections, including abscesses, pneumonia, meningitis, septicemia, and intestinal, urinary, and wound infections. These bacteria account for 60 to 70% of bacterial enteritis cases. As far as intestinal infections are concerned, although many *Enterobacteriaceae* have been implicated as a cause of diarrhea, only members of the genera *Escherichia*, *Salmonella*, *Shigella*, and *Yersinia* are clearly established as enteric pathogens. The nonpathogenic strain of *E. coli* inhabits the lower bowel of humans and animals, where it is often present in concentrations of 10 million or more viable organisms per gram of fecal material. For many decades it has been used as an indicator organism for fecal contamination of drinking water supplies as well as in food.

However, *E. coli* is also associated with at least four types of human enteric disease, (*1*) enteropathogenic, (*2*) enterotoxigenic, (*3*) enteroinvasive, and (*4*) hemorrhagic. Of these, for example, hemorrhagic colitis was recognized as being acquired by consumption of undercooked hamburger meat.

When contamination is heavy or when the particular bacteria are highly virulent, infection may follow ingestion of the food even though it has been stored under optimal conditions. More frequently, however, the bacteria originally contaminating the food multiply to dangerous numbers during prolonged storage at a temperature between 41°F (5°C) and 140°F (60°C), the so-called incubation or danger zone. Foods most commonly found to be vehicles of the bacterial infectious agents are milk and cream, ice cream, seafoods, meats, poultry, eggs, salads, custards, cream-filled pies, and eclairs and other filled pastries. Unfortunately, food contaminated by disease agents does not necessarily decompose or alter in taste, odor, or appearance.

Bacterial contamination of foods may occur in different ways. Infected persons may transmit bacteria by droplets (sneeze or cough) from the respiratory tract, by discharges from skin infections of the hand, and by feces contaminated hands, nasopharyngeal secretions, and discharges from open or draining sores elsewhere on the body (i.e., by failure to keep these adequately covered). In some instances, the infected person is without symptoms or signs of illness; therefore the infection is difficult to detect and control by, for example, assignment to other work. Another source of bacterial contamination of foods is the multiplication of bacteria in improperly cleaned utensils, which are then transferred to food subsequently prepared, stored, or served in them.

Mice, rats, and roaches may contaminate foods and utensils by bacteria carried mechanically on their feet and bodies from an infected to a noninfected area, or they may contaminate food with their urine and feces, which contain pathogenic organisms such as, in the case of rats, *Salmonella* and *Leptospira interrogans* (cause of Weil's disease). The source of infection in meat and poultry may be infection before killing or contamination may occur subsequent to killing, as with other foods.

The clinical symptoms of acute gastric enteritis due to bacterial infection usually have their onset 6 to 24 hours after ingestion of the infected food. Cramping, abdominal pain, diarrhea, nausea, and vomiting are the chief manifestations; often headache, and general malaise accompany these symptoms.

3.1.2 Bacterial Food Intoxication

Bacteria that, although usually harmless to humans, produce toxic substances when they grow in food cause bacterial food intoxication. When ingested in sufficient amount these toxins give rise to serious or fatal disease, even though after ingestion there is no further multiplication of the organisms in the body. Bacteria of the *Staphylococcus* group, often found in skin infections (pimples, boils, carbuncles) and respiratory tract infections, may form a virulent enterotoxin when allowed to grow in food. The symptoms of staphylococcal intoxication are nausea, vomiting, cramping, and abdominal pain, and diarrhea. In severe cases, blood and mucus may be found in the stool and vomit. The chief characteristic distinguishing food poisoning from toxins and enteric disease from that due to pathogenic bacteria carried in food is the time of onset of symptoms. In staphylococcal intoxication, this is often less than three hours after the ingestion of the contaminated food and rarely longer than six hours. On the other hand, in bacterial infections a delay of 6 to 24 hours between ingestion of food and onset of symptoms is more likely. The rapidity of onset and the uniform distribution of food poisoning symptoms in a group of people make it possible to determine the meal and particular foods in the meal responsible for poisoning.

3.1.3 Animal Parasitism

Some of the diseases caused by ingestion of food contaminated with animal parasites are amoebiasis, trichinosis, and tapeworm infestation. In the instance of infection with *Entamoeba histolytica* (causative organism of amoebic dysentery), the source and transmission of infection is similar to that of other enteric pathogens such as *Shigella* and *Salmonella*. The problem is somewhat different for *Trichinella spiralis* and the tapeworms *Taenia saginata* (beef tapeworm) and *Taenia solium* (pork tapeworm), for in these cases the infectious organisms are present in cattle or swine before the time of killing. Prevention of disease caused by these parasites, as well as their control, depends on inspection and selection of food at the time of purchase as well as proper preparation, as the heat of prolonged cooking will destroy the organisms.

3.1.4 Toxic Chemicals

Natural occurring toxins in food and accidental contamination of foods by toxic chemicals is an additional hazard. Acute poisonings are characterized by sudden onset, from a few minutes to two hours after ingestion of the chemical. The more common poisons reported as foodborne are scombroid toxin (histamine) found in certain types of fish such as tuna and mahi-mahi, which have been mishandled. Ciguatoxin poisoning is an example of an intoxication caused by eating contaminated tropical reef fish. The toxin is found in algae, which live among certain coral reefs. When small reef fish eat the toxic algae, it is stored in their flesh, skin, and organs. When bigger fish such as mackerel, mahi-mahi, bonito jackfish, and snapper eat the small reef fish, the toxin accumulates in the flesh and skin of the consuming fish. The toxin does not affect the contaminated fish. The toxin is heat stable and therefore not destroyed by cooking. Currently there is no commercially known method to determine if ciguatoxin is present in a particular fish. Other types of toxin are paralytic or neurotoxic shellfish poisons associated with certain species of dinoflagellates in the water where the implicated mollusks are gathered, and mushroom toxins. Common toxic chemicals, which have caused foodborne illnesses are cadmium, sodium flouride, and arsenic. Other substances frequently mentioned are antimony, zinc, lead, and copper.

Compounds such as aliphatic and aromatic amines used as boiler feedwater conditioners may present an additional hazard in the preparation of foods. Wherever steam may come in direct contact with food, the toxicity, type, and quantity of boiler feedwater conditioners should be ascertained. Complete physical separation of the food and steam from the boiler is recommended.

The OSHA standard in Section 1910.141 (Sanitation) states:

(g) Consumption of food and beverage on premises:

1. *Application.* This paragraph shall apply only where employees are permitted to consume food or beverages or both, on the premises.

2. *Eating and Drinking Areas.* No employee shall be allowed to consume food or beverages in a toilet room nor in any area exposed to a toxic material.

3. *Waste Disposal Containers.* Receptacles constructed of smooth, corrosion-resistant, easily cleanable or disposable materials, shall be provided and used for the disposal of waste food. The number, size, and location of such receptacles shall encourage their use and not result in overfilling. They shall be emptied not less frequently than once each working day, unless unused, and shall be maintained in a clean and sanitary condition. Receptacles shall be provided with a solid tight-fitting cover unless sanitary conditions can be maintained without use of a cover.

4. *Sanitary Storage.* No food or beverages shall be stored in toilet rooms or in an area exposed to a toxic material.

5. *Food Handling.* All employee food service facilities and operations shall be carried out in accordance with sound hygienic principles. In all places of employment where all or part of the food service is provided, the food dispensed shall be wholesome, free from spoilage, and shall be processed, prepared, handled, and stored in such a manner as to be protected against contamination.

The revised American National Standards Institute Minimum Requirements for Sanitation in Places of Employment (3) states in Item 10.1.1 that:

In all places of employment where employees are permitted to lunch on the premises, an adequate space suitable for the purpose shall be provided for the maximum number of employees who may use such space at one time. Such space shall be separate from any location where there is exposure to toxic materials.

Furthermore, Item 10.1.3 states that "No employee shall be permitted to store or eat any part of his/her lunch or eat other food at any time where there is present any toxic material or other substance that may be injurious to health." Item 10.1.4 is as follows:

In every establishment where there is exposure to injurious dusts or other toxic materials, a separate lunchroom shall be maintained unless it is convenient for the employees to lunch away from the premises. The following number of square feet per person, based on the maximum number of persons using the room at one time, shall be required:

Persons	Area, square-feet/person
25 and less	8
26–74	7
75–149	6
150–499	5
500 and more	4

3.2 Food Handling

Food workers educated in the proper methods of handling foods, and in good personal hygiene are the cornerstones of a safe food supply. In food sanitation, the health of the food handler is of primary importance, and every effort should be made to exclude from food handling work any sick or injured employee who has a discharging wound or lesion. This statement does not constitute recommendation of a system of frequent medical and laboratory examinations of food handlers. On the contrary, these procedures have not been found to be cost effective. Education of the food handler to report promptly any illness is a more effective means of reducing foodborne disease. In many localities health departments are prepared to provide preemployment or in-service food handler training programs, and a variety of training manuals and visual aid materials are available.

In the handling of foods, particularly those previously mentioned as being sources of infection or intoxication, care must be taken to keep the foods at a temperature out of the incubation zone; that is, they must be refrigerated at a temperature below 41°F(5°C) or kept hot at 140°F(60°C) or above. The exception to this rule is that during preparation, serving, or transportation, food may be kept at intermediate temperature for a period not exceeding two hours. Indicating thermometers of proven accuracy should be installed in all refrigerators. Common deficiencies found in refrigerated food storage are the overloading of the refrigerator, which prevents the free circulation of cold air, and the storage of foods in containers so deep that the total mass of the food cannot be cooled to 70°F(21°C) in two hours and 41°F(5°C) or less within four additional hours. Further provisions for continual vigilance against possible chemical contamination must be exercised by prohibiting the use of cadmium-plated utensils, prohibiting the use of galvanized utensils for cooking, and banning the use in food preparation and serving areas of roach powders containing sodium floride, cyanide metal polish, and other hazardous substances (36).

In view of the widespread custom, especially in the United States, of providing cubed and crushed ice in drinking water and beverages, it is of interest to note two reports on the sanitary quality of such ice (37). Both studies report that such ice often fails to meet sanitary standards and is contaminated with *Escherichia coli*, *Clostridia*, *Micrococci*, and *Streptococci*. All ice should be made from potable water, and stored in a container that is easy to clean, which is covered to protect the ice from contamination. The container should be routinely cleaned and disinfected similarly to other food contact surfaces.

3.3 Kitchen and Kitchen Equipment Design

The ordinance and code prepared by the Public Health Service (31) and adopted by the American National Standards Institute as a minimum standard, whenever local regulations do not apply to the industrial food handling establishment, provides for certain materials of construction and gives minimum lighting and ventilation requirements. The design of kitchens represents a problem in materials handling, batch preparation, and small quantity distribution. The designer must provide for the greatest ease in performing a large variety of tasks, usually in crowded quarters.

It behooves the kitchen designer to allow sufficient space in this important working area to permit the sanitary performance of the tasks of preparing food, washing of dishes and utensils and handwashing facilities. Separation of activities i.e., food preparation, and dishwashing, is preferable to combining them in one room. The location of kitchen equipment in relation to ease of cleaning is of great importance. Too often stoves, steam kettles, and mixers are placed close to walls, allowing dirt to accumulate and providing excellent breeding and hiding places for roaches.

Tables and stands should have shelves no less than 6 in. off the floor. Fixed kitchen equipment should be sealed with impervious material when attached to the wall or, alternatively, kept far enough away from the wall to allow for easy cleaning. All work surfaces should be made of impervious, noncorrosive, and easily cleaned material. Cutting, peeling, and other mechanical equipment should be designed to eliminate dirt-accumulating crevices and should be readily disassembled for cleaning. A voluntary organization, the National Sanitation Foundation, representing the public health profession, business, and industry, publishes a series of standards for food service equipment, and spray-type dishwashing machines; others are in preparation (38). These standards are of value in identifying equipment that has been tested to meet exacting food sanitation requirements.

3.4 Food Vending Machines

Vending machines are available that serve hot beverages, cold carbonated and noncarbonated beverages, sandwiches, and pastries. A recommended ordinance and code has been published by the Department of Health and Human Services (31). The following considerations may serve as a guide for the selection of automatic food vending machines.

All surfaces that come in contact with the food should be constructed of smooth, noncorrosive, nontoxic materials. All flexible piping should be nonabsorbent, nontoxic, and easily cleanable; treatment with the more common disinfecting agents should not harm it.

Vending machine purveying perishable food should be refrigerated. Their design should include an automatic temperature-controlled shutoff that will prevent the machine from operating whenever the temperature in the storage compartment rises above 41°F (5°C). In general, the design of the machine should allow easy accessibility for cleaning and disinfecting all surfaces that come in direct contact with the beverage or food.

Whenever disposable cups are furnished they should be stored in protective devices, and it should be possible to reload these without touching the lips or interior surface of the cups.

The drip receptacles in beverage vending machines should be provided with a float switch that would prevent the machine from operating when the liquid in the receptacle reaches a certain level. The vending area of the machine should be protected from dust, dirt, vermin, and possible mishandling by patrons by means of self-closing gates or pans.

The cleaning of tanks, containers, and other demountable equipment should be done at a central location. Whenever cleaning is accomplished at the vending machine, a portable three-compartment sink should be provided, to permit proper washing, rinsing, and disinfecting of all demountable equipment. At such points water and sewer facilities should be provided.

The food and drink for the vending machines should be stored and handled as they would have to be to meet the requirements of food handling establishments. A good rule is to offer for sale only packaged foods that were prepared at a central commissary. All perishable food should be coded to show place of origin and date of preparation. Perishable food that remains unsold, even though it has been properly refrigerated, should be disposed of at frequent intervals.

No vending machines should be located in areas where there is exposure to toxic materials.

4 CONTROL AND ELIMINATION OF INSECTS AND RODENTS

4.1 Insect Control

In general, insect control in an industrial setting requires that a high level of sanitation be practiced in and surrounding the plant. To control insect infestation, the primary prevention strategy is to eliminate breeding and harborage locations and to deprive the organisms of food. For example, depriving the *Aedes aegypti* mosquito, the vector of yellow fever, of accumulated standing water assures its control. However, any small puddle, or rainwater accumulation in discarded cans, tires, or other containers, provides near-ideal breeding locations for this mosquito.

Whenever unrecognized infestation occurs it is possible to obtain identification assistance and advice on control measures from the State or County Extension Service.

The insects of public health significance are flies, fleas, lice, mosquitoes, mites, ticks, and roaches. To be able to prevent or control infestation effectively one must possess an understanding of insect biology, morphology, breeding habits, and disease spreading mechanisms. Based upon such knowledge it is possible to mount the most effective chemical, physical, or biological control.

Except for the food and other industries dealing with products that may serve as nourishment for insects, the problem of controlling these pests depends largely on good housekeeping. The more general problem of control of arthropods and arthropodborne disease cannot be accomplished on a restricted scale as in a single industrial establishment. To be effective, it must be a community-wide or regional undertaking of considerable magnitude. It is axiomatic that an industry located in an infested area should cooperate with the governmental agency responsible for effectuating control.

The number and prevalence of flies and roaches are good general indexes of the "sanitation" practiced in the establishment. Of direct interest is the proof obtained of the relation between fly prevalence and the infectious diseases, and deaths caused by the bacillary dysentery organism (39). Field studies showed that bacillary dysentery was materially reduced in a community when effective fly control measures were undertaken. The most direct method of dealing with an insect infestation problem is to eliminate breeding places. Wherever garbage, rubbish, or other organic matter is allowed to accumulate and become exposed to insects, it produces ideal conditions for their development. In food handling areas, nourishment is available, and cracks, crevices, voids, and other unused spaces harbor the pests.

The most common insect invaders are the roaches—German cockroach, *Blatella germanica*; American cockroach, *Periplaneta americana*; Australian cockroach, *Periplaneta australasiae*; Oriental cockroach, *Blatta orientalis*; brown-banded cockroach, *Supella su-*

pellectilium—and flies, of which 90 to 95% are houseflies, *Musca domestica*, and blowflies, *Phaenicia* (39).

The use of insecticides, which should be undertaken only after the environmental conditions favoring breeding have been eliminated, poses health problems to the applicator and others. The problems and their prevention are extensively discussed in the literature. New insect control technology is being rapidly developed; for example, sex attractants and juvenile hormones are proposed as insect control materials. However, they too may have adverse health effects on mammalian species, including humans. On the other hand, biological insect control may have beneficial public health and ecological consequences because it could prevent the positive feedback aspects inherent in the use of chemical insecticides.

Other biological control methods include the utilization of specific parasites, insect predators, and pathogens. The release of radiation-sterilized males, as practiced in the attempt to control Mediterranean fruit fly infestation, is an advanced biological control technique. Other means considered are use of genetic engineering and biologically produced compounds. Plant insecticides have been an extremely important group that includes derris and its chief constituent rotenone, and pyrethrum, which contains pyrethrins. The latter are nontoxic to warm-blooded animals and have a very rapid "knock-down" effect on a large variety of insects. Because they are highly volatile and are broken down by light, their effect is transitory. They may be stabilized and their effect intensified by the addition of piperonyl butoxide and other compounds.

4.2 Rodent Control

Apart from the health hazards rodents produce as vectors of disease (Bubonic Plague, endemic or murine typhus fever, Salmonellosis, and Weil's disease), their infestation of an industrial plant can be a tremendous economic drain. As a general principle, in the community the destruction of food, crops, merchandise, and property by rats is serious enough to justify suppressive measures, even if rats are not responsible for human disease. Due to gnawing the insulation from electric wires many fires of unknown origin must be attributed to rats. Because of their destructiveness and their pollution of food, rats have gained the unenvied fame of being the worst mammalian pest in civilization.

Sewer systems in cities and large industrial facilities provide a major habitat for rodents. In particular this is the case in older sewer systems in need of repair. With the low state of repair of the nation's infrastructure, including its sewer system, the sewer rodent infestation problem is of consequence. Sewers provide rodents with food, shelter, water, and an equable temperature year round. Sewers also provide easy egress and ingress. This is especially true in the case of combined sewers, which convey sanitary wastes and storm water in a single pipe. A separate sewer system, a requirement in all new construction, is less favorable to rodent infestation. Sewer rodents living in sewers are exposed to a variety of infectious diseases and thus afford, once in contact with people, an increased opportunity for spreading disease.

When old and deteriorated structures are to be demolished it is necessary first to conduct a rodent infestation survey and then to institute a program to rid it of rodents before they migrate.

In the industrial environment, other than the destruction of food, rodents pose a health problem with the possible transmission of Weil's disease or Leptospirosis, a relatively rare disease caused by a spirochete bacterium. The organism is harbored in the rodent and excreted in the urine. The individual becomes infected by direct or indirect contact with infected rodent urine. The organism enters the body through the mucous membranes, cuts, or skin abrasions. There are about 100 cases of Weil's disease per year in the United States, with a few deaths.

It is estimated that there is one rat for every two people in the United States. That would be are approximately 125,000,000 rats (40). As pointed out by Davis, the regulatory factors in any problem of species management can be classed in three main groups: (a) environment, which includes availability of food and protective shelter, (b) predation, which includes animals that feed on the species, trapping, poisoning, and disease; and (c) competition, the fight for a limited supply of environmental necessities.

Control of the rodent population should be based primarily on a change of the environmental conditions so that the rodents are deprived of food and shelter. When this is accomplished, the rodent population is reduced by intraspecies competition. Because the average reproductive rate of Norway rats (they reproduce during all seasons of the year, with a maximum in spring and fall occurring in many areas) is 20 rats per year, it is evident that killing procedures will have little effect on reducing the population and maintaining it at low level if environmental conditions are favorable to rats (41).

4.3 Ratproofing (Building out the Rat)

"Ratproofing" is a term applied to procedures for controlling the portals of entry of rats into a structure. Being extremely cautious, nocturnal mammals, rats prefer to enter a building through small openings. Only when they are subjected to peril do they venture away from normal pathways. The points of ingress and egress of rats that must be guarded against in particular are any openings around pipes, unused stacks and flues, ventilators, and hatches.

All new buildings should be designed and constructed to be ratproof. Foundations must be continuous, extending not less than 18 in. below the ground level, and they should always be flush with the under surface of the floor above. Floor joists should be embedded in the wall and the space between the joists filled in and completely closed to the floor level. All construction materials should be as ratproof as possible (41). In industry, the most frequent shelters for rats are areas surrounding the cafeteria, lunchrooms, and other eating areas, although no place that affords attractive food and shelter is immune to rat infestation. The office in which employees are permitted to eat their lunch is often rat infested, the scraps left in the wastebaskets providing the rats with a good food source. All other organic wastes, such as those found in sewers where they harbor, are a good additional source of food for rats.

4.3.1 Killing Rats

Trapping by means of snap and cage traps is a method of killing rats that requires great skill and ingenuity, for rats are rather wary mammals. Killing by use of rodenticides has advanced greatly in the past two decades. There are many old rodenticides, some which were still in use as late as the 1970s. These included Red Squill, Antu (which was highly toxic to rats, dogs and pigs, but less so to mice and cats), arsenic, barium carbonate, phosphorous paste, sodium fluoroacetate (1080), strychnine, and many fumigant gases. Only zinc phosphide is still registered for use in the control of rodents outdoors in orchards and nursery stock. The rest are either no longer registered, or rats developed bait shyness or chemical resistance to the toxicant long ago, or are rarely used because of legal restrictions. The same is true for some of the early anticoagulants such as coumafuryl, pindone and isovalery indandione (Pival). They have all been replaced by the more effective and less hazardous rodenticides of today.

Anticoagulant rodenticides are the most widely used materials today and comprise about 90% of the baits used by pest control operators and the general public. They cause internal hemorrhage by destroying the blood's ability to form clots. Once blood cells begin leaking out of porous capillaries, the animal dies of shock in a few days.

Modern rodenticides are some of the most powerful toxicants still available to the public, yet other than intentional exposures and suicides, they have a good safety record. This is due to the very low dosage used in baits (0.025%), the symptoms of poisoning are widely recognized, the antidote (vitamin K) is readily available, and some new additives (Bitrex) make them virtually unpalatable to anything, but rodents. Baits come in many formulations to fit the particular environment in which they are used, such as cereals and pellets for general use, wax blocks for outdoors and sewers, and liquids, gels, and dusty tracking powders for dry indoor sites and rodent burrows.

Multiple dose anticoagulants, such as warfarin, chlorophacinone and diphacinone are slow acting and so named because rats must ingest repeated doses over several days in succession. Once a lethal dose is taken, death follows in 3 to 10 days. Mice, known for nibbling and storing food in a cache, may take even longer to die. Bait shyness does not occur with these products, but enough bait must be available continuously for an extended baiting period.

Second generation "single dose" anticoagulants (brodifacoum, bromadiolone and difethiolone) are more toxic than warfarin-type products and are more hazardous to nontarget animals, but due to their effectiveness, save time and labor costs. Their advanced chemistry can cause death from a single feeding, but the mode of action is similar in that it still takes 3 to 7 days or more to destroy the animal's blood clotting factors. They are particularly useful in places where rodents have many alternate choices.

There are newer rodenticides with much different modes of action than the anticoagulants. The most common in use today are zinc phosphide, still used outdoors in crops and orchards, bromethalin, which inhibits adenosine triphosphate (ATP), the chief energy-carrying chemical in the body, and cholecalciferol, a form of vitamin D that withdraws calcium from the bones and dumps it into the circulatory system, causing hypercalcemia and kidney failure. They are most effective against anticoagulant-resistant rodents.

There are many new methods and materials available to improve the safety and effectiveness of rat control, but the list is too great for this discussion. If rodent control is undertaken by the industry itself, those conducting the service should be well trained and familiar with the building and its production. In general, employees of the industry do not need a special state license or certification to provide pest control within the limits of their employer's property. However, if the work covers areas open to the public (parks, athletic fields, etc.) or requires the use of restricted-use pesticides, then a state pesticide certification is required. Questions about these limitations can be directed to the state agency (Department of Agriculture, etc.) that is responsible for pesticide certification. Hiring a licensed pest control operator is another option.

Finally control of rats and other rodents requires (a) environmental control to eliminate sources of food and harborage, (b) effective ratproofing, and (c) efficient rat killing programs. Controlling rat populations, not individual rats, is the key to a successful rodent control program.

5 GENERAL ENVIRONMENTAL CONTROL AND PROVISION OF ADEQUATE SANITARY FACILITIES AND OTHER PERSONAL SERVICES

The Occupational Safety and Health Standards promulgated under the Williams-Steiger Act of 1970 (84 Stat. 1593) were published in the Federal Register of June 27, 1974 (42). Subpart J (General Environmental Controls) contains Section 191D.141 (Sanitation), Section 1910.142 (Temporary Labor Camps), and Section 1910.143 (Non-Water-Carriage Disposal Systems).

To be applicable in a wide variety of circumstances, the regulations in these sections are, for the most part, very general. For example, insect and rodent control, which is described as "vermin control," requires that "enclosed workplaces shall be constructed, equipped and maintained, so far as reasonably practicable, as to prevent the entrance and harborage of rodents, insects and other vermin. Furthermore, a continuing and effective extermination program shall be instituted where their presence is detected." Such a statement provides little that is helpful to the plant manager and less to OSHA personnel inspecting the premises. On the other hand, the requirements for toilet and washing facilities (e.g., lavatories and showers) are enumerated and relatively detailed.

The personal hygiene practices of employees carry many health implications without reference to the individuals' particular employment. For instance, the disease-transmitting potential from failure to clean the hands after defecating, or the skin diseases resulting from poor body cleansing, may not be related to the person's job. There are, however, health problems relating to personal hygiene practices that occur in industry exclusively. Such problems are related to ingestion of chemical toxins or disease-producing organisms and to local skin, conjunctiva, and mucous membrane inflammation, due to sensitivity or direct irritative action of industrial chemicals.

Schwartz et al. (43) have stated that uncleanliness "is probably the most important predisposing cause of occupational dermatitis." Lack of cleanliness in the working environment exposes the worker to large doses of external irritants. Personal uncleanliness not only does the same, but also permits external irritants to remain in prolonged contact with the skin. Workers wearing or carrying to their homes their dirty work clothes may even cause dermatitis in other members of the family who come in contact with soiled clothes, or among unsuspecting workers who clean them. Safeguards such as protective creams help, but personal hygiene is a necessity.

Where the employee is exposed to toxic materials or is a food handler, the need for the optimum in clean, well-lighted, and well-ventilated washing and locker facilities becomes imperative. From the point of view of protecting the health of the individual employee and minimizing the possibility of transmitting infections to others, toilet and washroom facilities should be easily cleanable, adequate in size and number, and accessible, or the personal hygiene habits of employees in industry will suffer. The American National Standards Institute Minimum Requirements for Sanitation (3) considers that "ready accessibility" has not been provided when an employee has to travel more than one floor-to-floor flight of stairs to or from a toilet facility. Some minimum standard on the basis of distance must be set. However, the advent of complex automatic and remote controls and unit operations may bring about modification of some distance standards. In Section 1910.141 (Sanitation) of the OSHA June 27, 1974, Standards item (c) Toilet Facilities (ii) has taken this into account and exempts establishments in which mobile crews work in normally unattended work locations "so long as employees working in these locations have transportation immediately available to nearby toilet facilities which meet the usual requirements."

The OSHA standards state that whenever employees are required by a particular standard to wear protective clothing because of the possibility of contamination with toxic materials, change rooms equipped with storage facilities for the protective clothing shall be provided. Although recommended a number of years ago, the following arrangement, which differs from the current OSHA requirement of locating work and street clothing lockers side by side, is good practice.

A good arrangement is a room or building divided into two sections: a street clothes section and a work clothes section, with bathing and toilet facilities between. The street clothes section has an outside entrance and lockers for street clothes, toilets and wash basins or wash fountains, and changing facilities for supervisors. The work clothes section has room for work clothes, toilets and wash basins or wash fountains, showers, and laundry. Three connections between the street clothes section and the work clothes section are (a) through the supervisors' change room, (b) through the main shower room, and (c) a hall with "one-way traffic doors" from street to work clothes sections.

In the mineral, petroleum, and allied industries, the change house may be located at the entrance gate or near the parking lot to make it readily accessible. It may be a separate building connected to the working area by a covered walkway, and it should be constructed of fireproof or fire-resistant materials and be properly heated, lighted, and ventilated to meet the requirements of Section 4 of the American Standard Minimum Requirements for Sanitation (3). The interior arrangement of the equipment and facilities in change houses and in locker and toilet rooms is not standardized, but is a matter of proper design to meet the space, cost, and other requirements of the individual plant.

The minimum facilities required for places of employment are set forth in the various state regulations of Departments of Labor and Divisions of Industrial Hygiene. In addition, various industrial associations have established, through industry-wide sanitation committees, practice codes related to the particular industry. An example is the Code of Recommended Practices for Industrial Housekeeping and Sanitation of the American Foundrymen's Association (44). The Association of Food Industry Sanitarians, in cooperation with the National Canners Association, has published a comprehensive manual, Sanitation for the Food-Preservation Industries (45). For all places of employment, minimum requirements for housekeeping, light and ventilation, water supply, toilet facilities, washing facilities, change rooms and retiring rooms for women, and food handling requirements are set forth in the American Standard Minimum Requirements for Sanitation (3).

The OSHA requirements (OSHA Standards June 27, 1974, Section 1910.141, Sanitation) are somewhat less stringent for establishments with fewer than 150 employees, nor does the OSHA regulation require the provision of retiring rooms. In general, minimum standards for toilet facilities can be delineated quite specifically with respect to their construction and provision for ease of cleaning. Critics point out that in some instances, at least, the standard lavatories installed in washrooms fail to meet the need for ease of washing. However the wash fountain with its foot pedal supply control or automatic (infrared) water flow control fixture, basin large enough to permit easy bathing of arms, face, and upper torso, free-flowing stream of thermostatically tempered water, and central dispenser of cleansing material offers a practical solution to the problem of providing good washing facilities. In addition, the wash fountain requires less floor space, initial installation is less complicated and costly, and it is maintained with greater ease. Experience with wash fountains in washrooms for female employees indicates that they are acceptable there.

6 MAINTENANCE OF GENERAL CLEANLINESS OF THE INDUSTRIAL ESTABLISHMENT

Housekeeping and plant cleanliness in their fullest sense imply not only the absence of clutter and debris in the working area and passageways, but also the maintenance of painted surfaces in a clean condition, the frequent removal of dust from lighting fixtures, and in general making the plant "a better place to work." The industrial plant that processes hazardous or potentially hazardous materials that cannot be prevented from escaping into the general atmosphere has a double responsibility to provide the utmost in housekeeping and plant cleanliness.

It is no longer sufficient to delegate responsibility for cleanliness of the plant and surroundings to a foreman or supervisor whose primary duties, interests, and training lie elsewhere. The complexity and cost of housekeeping machinery and supplies, the technical knowledge required for their proper and effective use, and the extensiveness of the work load make the establishment of a housekeeping department in large plants a virtual necessity. Working standards, schedules, and quality controls are also required in this routine plant operation. Additional evidence of the economic importance of this operation is the fact that fire insurance premiums are in part determined by the status of plant cleanliness.

Section 1920.22 (General Requirements) of the OSHA regulations sets very general standards for housekeeping. For a continuing process, too often relegated to a low level of managerial supervision, it is possible to establish quantitative measures of housekeeping effectiveness by such means as dust counts, bacteria counts, color intensity, and light transmission measurements.

The food industry presents special problems of frequent or even continuous cleaning operations as well as at the end of each shift. Thus this industry, with its combination of wet- and dry-cleaning requirements, represents an extreme example for other industries. Perhaps the ultimate in "housekeeping" requirements are those for the "clean rooms" of the electronics and pharmaceutical industries, where continuous sanitation of the total environment is absolutely essential.

Housekeeping must be scheduled, and it must be the assigned responsibility of a trained group under proper supervision. A staggered work arrangement, for example, allows the use of a shift permitting the housekeeping crew to start work one to two hours after the production personnel. This permits some general cleaning during the morning before very much material has accumulated. It also permits scheduling cleaning during the regular lunch hour shutdown and completing the schedule after the day's operations have ended.

Housekeeping, a continuing indirect cost in industry, deserves uninterrupted scrutiny. The activity must be evaluated periodically with respect to effectiveness and cost.

7 THE ENVIRONMENTAL ENGINEER IN INDUSTRY

The environmental factors described in this chapter are inherent in varying degrees in every industrial establishment. The plants whose products are intended for human food and pharmaceutical manufacturing may be placed in a special category, for they have a paramount responsibility to protect their product from contaminants that endanger the health of the consumer. Because of this responsibility, the roles of the environmental engineer and sanitarian in such industries should be obvious. There is great need to conserve the nation's water resources through effective minimization and treatment of industrial wastes, to provide production and management personnel with safe food and potable water at all times, and to maintain a sanitary working environment to protect the employee from the diseases that are not directly of occupational origin. Legislation emphasizes the need for such specialized professionals in industry. Programs such as the Environmental Management System (EMS) of the Kentucky Pollution Prevention Center provide such services as on-site gap analysis, workshops and training for management and employees. Such programs allow better management of a company's environmental deficiencies. It provides responsible executives with a better understanding of environmental planning and control (46).

The standards and regulations promulgated to meet environmental goals and objectives are in a state of flux. Timetables are under administrative and legislative review, and the technology to achieve them is evolving. In addition, it is the intent of Congress to have increasing public participation in the standard setting process. Most generally, the environmental engineer is employed by industry to apply his or her knowledge to the solution of problems of water supply and their corollary, liquid waste disposal, solid and hazardous waste management, and air pollution abatement. However, engineering and public health skills are additionally applicable in such areas as food sanitation, insect and rodent control, and plant maintenance. Although the environmental engineer may or may not be directly associated with the occupational health section of the industry, the roles filled by the engineer and by the industrial hygienist are directed toward the same end. The parallel may be further drawn by pointing out that just as the industrial hygienist aims to recognize hazardous working conditions in the design state, to be able to eliminate them completely or to make the application of control measures more effective, convenient, and economical, the environmental and environmental health engineer in industry applies the same principle to the control of environmental factors other than those related to the occupation *per se*.

BIBLIOGRAPHY

1. Centers for Disease Control, "Surveillance for Waterborne Disease Outbreaks-United States, 1976–1982, Atlanta, GA.
2. M. Ho, "Current Problems of Infectious Diseases in Taiwan," *J. Infectious Disease, Microbiology, Immunology* (March 1998).
3. American National Standards Institute, *Sanitation in Places of Employment-Minimum Requirements*, ANSI-Z4.1, 1986.
4. American National Standards Institute, *Drinking Water Fountains and Self Contained Mechanically Refrigerated Drinking Water Coolers*, ANSI/ARI 1010-82, and ANSI/NSF 61.
5. U.S. Department of Health and Human Services-Public Health Service-Centers for Disease Control-National Institute for Occupational Safety and Health, *Occupational Exposure to Hot Environments Criteria 1972.*
6. U.S. Department of Health and Human Services-Public Health Service-Centers for Disease Control-National Institute for Occupational Safety and Health, *Occupational Exposure to Hot Environments-Revised Criteria 1986.*
7. U.S. Geological Survey, *Estimated Use of Water in the United States in 1995*, USGS Circular 1200.
8. A. F. McClure, *J. Am. Water Works Assoc.* **66**, 240 (1974).
9. National Commission on Water Quality, Report, Washington, DC, Nov. 1975.
10. U.S. Environmental Protection Agency, *Waste Minimization Opportunity Assessment Manual*, EPA/625/7-88/003, Hazardous Waste Engineering Laboratory, Cincinnati, OH, 1988.
11. C. E. Renn, *Management of Recycled Waste-Process Water Ponds*, Project WPD117 Environmental Technology Series, U.S. Environmental Protection Agency R2-73-223, 1973.
12. W. C. Ackerman, *J. Am. Water Works Assoc.* **12**, 691 (1975).
13. C. J. Schmidt et al., *J. Water Pollut. Control Fed.* **47**, 2229 (1975).
14. C. E. Arnold, H. E. Hedger, and A. M. Rawn, "Report Upon the Reclamation of Water from Sewage and Industrial Wastes in Los Angeles County, California," prepared for the County of Los Angeles, 1949.
15. K. L. Kollar and R. Brewer, *J. Am. Water Works Assoc.* **67**, 686 (1975).
16. Sextus J. Frontinus, *The Aqueducts of Rome* (translated by Bennett), William Heineman, London, 1925.
17. Bogly W. J., Jr., *Experiments with Dual Water Distribution Systems*, Proceedings AWWA Seminar, AWWA Publication 20189, June 23, 1985, p. 5.
18. D. A. Okun, *Overview of Dual Water Systems*, Proceedings AWWA Seminar, AWWA Publication 20189, June 23, 1985, p. 1.
19. D. A. Okun, "Distributing Reclaimed Water Through Dual Systems" *J. Am. Water Works Assoc.* **89**, 52 (1997).
20. Centers for Disease Control and Prevention (CDC), *Surveillance for Waterborne Disease Outbreaks, United States, 1995–1996*, Dec. 11, 1998/Vol. 47/No. SS-5

21. F. W. Sunderman Jr., B. Dingle, S. M. Hopfer, and T. Swift. "Acute Nickel Toxicity in Electroplating Workers Who Accidentally Ingested a Solution of Nickel Sulfate and Nickel Chloride," *Am. J. Ind. Med.* **14**, 257 (1988).
22. American Water Works Association, Manual M-14, AWWA No. 30014, 1966, 22 pp.
23. American Water Works Association, *Standards for Backflow Prevention Devices-Reduced Pressure Principle and Double Check Valve Types*, Standard c506-78/(R83), AWWA No. 43506, 1983, 20 pp.
24. G. Keleti and M. A. Shapiro, "Legionella and The Environment," *Environ. Controls* **17**(2), 133 (1987).
25. J. E. McDade, C. C. Shepared, D. W. Fraser, T. F. Tsai, M. A. Redus, W. R. Dowdie, and the Laboratory Investigative Team, Legionnaires Disease, "Isolation of a Bacterium and Demonstration of its Role in Other Respiratory Disease," *New Engl. J. Med.* **297**, 1197 (1977).
26. J. J. States, R. M. Wadowsky, J. M. Kuchta, R. S. Wolford, L. F. Conley, and R. B. Yee, "Pathogenic Organisms in Drinking Water: Legionella," in G. A. McFeters, ed., *Advances in Drinking Water Microbiology Research*, Science Tech Publishing, Madison, WI, 1989.
27. T. H. Glick, M. B. Gregg, G. W. Mallison, W. W. Rhodes, and B. Kassanoff, "Pontiac Fever: An Epidemic of Unknown Etiology in a Health Department. Clinical and Epidemiologic Aspects," *Am. J. Epidemiol.* **107**, 149 (1978).
28. L. A. Herwaldt, G. W. Gorman, A. W. Hightower, B. B. Brake, H. Wilkinson, A. L. Reingold, P. A. Boxer, T. McGrath, D. J. Brenner, C. W. Moss, and V. V. Broome. "Pontiac Fever in an Automobile Assembly Plant, in Legionella," in C. Thornberry, A. Balows, J. C. Feeley, and W. Jakubowski, eds., *Proc. 2nd International Symposium*, ASM, Washington, DC, 1984, p. 246.
29. Allegheny County Health Department (Pennsylvania), "Approaches to Prevention and Control of Legionella Infection in Allegheny County Health Care Facilities", Jan. 1997.
30. D. W. Fraser, "Sources of Legionellosis" in C. Thornberry, A. Balows, J. C. Feely, and W. Jakubowski, eds. *Proc. 2nd International Symposium*, ASM, Washington, DC, 1984, p. 277.
31. U.S. Department of Health and Human Services, Public Health Service, Food and Drug Administration, *Food Code-1999.*
32. Centers for Disease Control (CDC), *Surveillance for Foodborne-Disease Outbreaks-United States, 1988–1992*, Vol. 45/No.SS-5 October 25, 1996.
33. Centers for Disease Control (CDC), *Annual Summary of Foodborne and Waterborne Disease Outbreaks*, 1982, p. 198.
34. M. T. Kelly, D. J. Brenner, and J. J. Farmer III, "Enterobacteriaceae," in E. H. Leunette, A. Balows, W. J. Housler, Jr., and H. J. Shadony, eds., *Manual of Electrical Microbiology, IV* ASM, Washington, DC, 1985, pp. 263–277.
35. C. F. Clancy, in W. M. O'Leary, ed., *Enterobacteriaceae Practical Handbook of Microbiology*, CRC Press, Boca Raton, FL, 1989, pp. 71–80.
36. U.S. Department of Health and Human Services, U.S. Public Health Services, *Sanitary Food Service Instructor's Guide*, FDA 78-2081, 1976 Revision.
37. E. W. Moore, E. W. Brown, and E. M. Hall, "Sanitation of Crushed Ice for Iced Drinks," *J. Am. Public Health Assoc.* **43**, 1265 (1953); V. D. Foltz, "Sanitary Quality of Crushed and Cubed Ice as Dispensed to the Consumer," *Public Health Rep.* **68**, 949 (1953).
38. National Sanitation Foundation, "Soda Fountain and Luncheonette Equipment" Standard No. 1; "Food Service Equipment" Standard No. 2; "Spray-Type Dishwashing Machines," A Sanitation Ordinance and Code, Publication No. 546, 1965.

39. D. E. Lindsay, W. H. Stewart, and J. Walt, "Effect of Fly Control of Diarrheal Disease in an Area of Moderate Morbidity," *Public Health Rep.* **68**, 361 (1953).
40. D. E. Davis, "Control of Rates and Other Rodents," in K. F. Maxcy, ed., *Preventive Medicine and Hygiene*, 8th ed., Appelton-Century-Crofts, New York, 1956, Chapt. 8.
41. Centers for Disease Control, *Rat-Borne Disease: Prevention and Control*, CDC, Atlanta, GA, 1969.
42. Occupational Safety and Health Standards, U.S. Department of Labor, Occupational Safety and Health Administration, *Fed. Reg.* **39**(125), 23502–23828 (June 24, 1974).
43. L. Schwartz, L. Tulipaw, and D. J. Birmingham, *Occupational Diseases of the Skin*, 3rd ed., Lea & Febiger, Philadelphia, 1957.
44. F. E. Cash, "Suggested Standards for Change Houses," paper presented at the annual meeting of the American Public Health Association, October 1951.
45. American Foundrymen's Association, "Code of Recommended Practices for Industrial Housekeeping and Sanitation," AFA, Chicago, IL, 1944.
46. Kentucky Pollution Prevention Center, *The Bottom Line* **11**(2) (Fall 1998).

CHAPTER FIFTY-EIGHT

Air Pollution

David J. McKee, Ph.D. and John D. Bachman

1 INTRODUCTION

Air pollution has been affecting human health since the dawn of mankind. Whether caused by human activity, such as the burning of carbonaceous fuels, or natural events, such as wildfires and volcanic eruptions, it has been recognized for millenia that air pollution can adversely affect human health and even result in the death of some individuals. Some of the best known air pollution disasters occurred in Europe and the U.S. (e.g., Meuse Valley, Belgium in 1930; Donora, Pennsylvania in 1948; London in 1952) during this century and were documented better than any previous air pollution events. These events and growing public support in the 1960s provided the impetus for passing the Clean Air Act of 1970, legislation designed specifically to improve air quality in the U.S. This chapter documents results of implementing the Clean Air Act and health effects of many of the air pollutants regulated by the Act.

Section 2 of this chapter provides a summary of sources, health/environmental effects, and trends in air quality of the six principal ambient air pollutants (commonly known as air quality criteria pollutants) and discusses other air pollution problems, including global warming, stratospheric ozone depletion, and visibility impairment. A more detailed discussion of the health effects of the six principal air pollutants follows.

2 SUMMARY OF AIR QUALITY AND EMISSIONS TRENDS

This section is extracted from Ref. 7. See Appendix A for further information, including websites, contacts, and a list of acronyms used. The *Trends Report* documents air quality

Patty's Industrial Hygiene, Fifth Edition, Volume 4. Edited by Robert L. Harris.
ISBN 0-471-29749-6

trends over the past decade and provides air pollution information from cities throughout the U.S.

EPA tracks two kinds of trends: air concentrations based on actual measurements of pollutant concentrations in the ambient (outside) air at selected monitoring sites throughout the country, and emissions based on engineering estimates of the total tonnage of these pollutants released into the air annually. In addition, starting in 1994, under the Acid Rain Program, EPA began tracking emissions of SO_2 and NO_x based on data from continuous emission monitors for the electric utility industry.

Generally there are similarities between air quality trends and emission trends for any given pollutant. However, in some cases, there are notable differences between the percent of change in ambient concentrations and the percent of change in emissions. These differences can mainly be attributed to the location of air quality monitors. Because most monitors are positioned in or near urban areas, trends in air quality tend to more closely track changes in urban emissions rather than changes in total national emissions.

Each year, EPA gathers and analyzes air quality concentration data from thousands of monitoring stations around the country. Monitoring stations are operated by state, Tribal, and local government agencies as well as some Federal agencies, including EPA. Trends are derived by averaging direct measurements from these monitoring sites on a yearly basis. During the last 10 years (1988 through 1997), air quality has continued to improve.

The most notable improvements are a 67% decrease in Pb concentrations, a 38 percent decrease in CO concentrations, and a 39% decrease in SO_2 concentrations. Improvements in measured concentrations are also noted for the other principal pollutants, including NO_2 ozone, and PM-10 during this same time frame.

EPA estimates nationwide emissions trends based on actual monitored readings or engineering calculations of the amounts and types of pollutants emitted by automobiles, factories, and other sources. Emission trends are based on many factors, including the level of industrial activity, technology developments, fuel consumption, vehicle miles traveled, and other activities that cause air pollution.

Emissions trends also reflect changes in air pollution regulations and installation of emissions controls. For almost three decades, emissions have shown improvement (decreased emissions) for all six principal air pollutants. Between 1970 and 1997, the U.S. population increased 31%, vehicle miles traveled increased 127%, and gross domestic product increased 114%. At the same time, total emissions of the six principal air pollutants decreased 31%.

The dramatic improvements in emissions and air quality occurred simultaneously with significant increases in economic growth and population. The improvements are a result of effective implementation of clean air laws and regulations, as well as improvements in the efficiency of industrial technologies. Despite great progress in air quality improvement, in 1997 there were still approximately 107 million people nationwide who lived in counties with monitored air quality levels above the primary national air quality standards, suggesting the need for the continuing efforts now being implemented to improve air quality nationwide.

2.1 Six Principal Pollutants

2.1.1 Carbon Monoxide

2.1.1.1 Nature and Sources of the Pollutant. Carbon monoxide (CO) is a colorless, odorless and at high levels, a poisonous gas, formed when carbon in fuel is not burned completely. It is a component of motor vehicle exhaust, which contributes about 60% of all CO emissions nationwide. High concentrations of CO generally occur in areas with heavy traffic congestion. In cities, as much as 95% of all CO emissions may come from automobile exhaust. Other sources of CO emissions include industrial processes, nontransportation fuel combustion, and natural sources such as wildfires. Peak CO concentrations typically occur during the colder months of the year when CO automotive emissions are greater and nighttime inversion conditions (where air pollutants are trapped near the ground beneath a layer of warm air) are more frequent.

2.1.1.2 Health Effects. Carbon monoxide enters the bloodstream through the lungs and reduces oxygen delivery to the body's organs and tissues. The health threat from lower levels of CO is most serious for those who suffer from cardiovascular disease, such as angina pectoris. At much higher levels of exposure, CO can be poisonous and even healthy individuals may be affected. Visual impairment, reduced work capacity, reduced manual dexterity, poor learning ability, and difficulty in performing complex tasks are all associated with exposure to elevated CO levels.

2.1.1.3 Trends in CO Levels. Long-term improvements in CO continued between 1988 and 1997. Ambient CO concentrations decreased 38%, and the estimated number of exceedances of the national standard decreased 95%. While CO emissions from highway vehicles alone have decreased 29%, total CO emissions decreased only 25% overall. Long-term air quality improvement in CO occurred despite a 25% increase in vehicle miles traveled in the United States during this 10-year period. Between 1996 and 1997, ambient CO concentrations decreased 7%, while CO emissions decreased 3%. Transportation sources (including highway and off-highway vehicles) now account for 77% of national total CO emissions.

2.1.2 Lead

2.1.2.1 Nature and Sources of the Pollutant. In the past, automotive sources were the major contributor of lead (Pb) emissions to the atmosphere. As a result of EPA's regulatory efforts to reduce the content of Pb in gasoline, the contribution from the transportation sector has declined over the past decade. Today, metals processing is the major source of Pb emissions to the atmosphere. The highest air concentrations of Pb are found in the vicinity of nonferrous and ferrous smelters, and battery manufacturers.

2.1.2.2 Health and Environmental Effects. Exposure to Pb occurs mainly through inhalation of air and ingestion of Pb in food, water, soil, or dust. It accumulates in the blood, bones, and soft tissues. Lead can adversely affect the kidneys, liver, nervous system, and other organs. Excessive exposure to Pb may cause neurological impairments, such as seizures, mental retardation, and behavioral disorders. Even at low doses, Pb exposure is associated with damage to the nervous systems of fetuses and young children, resulting in learning deficits and lowered IQ. Recent studies also show that Pb may be a factor in high blood pressure and subsequent heart disease. Lead can also be deposited on the leaves of plants, presenting a hazard to grazing animals.

2.1.2.3 Trends in Pb Levels. Between 1988 and 1997, ambient Pb concentrations decreased 67%, and total Pb emissions decreased 44%. Since 1988, Pb emissions from highway vehicles have decreased 99% due to the phase-out of leaded gasoline. The large reduction in Pb emissions from transportation sources has changed the nature of the pollution problem in the United States. While there are still violations of the Pb air quality standard, they tend to occur near large industrial sources such as lead smelters. Between 1996 and 1997, Pb concentrations and emissions remained unchanged.

2.1.3 Nitrogen Dioxide

2.1.3.1 Nature and Sources of the Pollutant. Nitrogen dioxide (NO_2) is a reddish brown, highly reactive gas that is formed in the ambient air through the oxidation of nitric oxide (NO). Nitrogen oxides (NO_x), the term used to describe the sum of NO, NO_2 and other oxides of nitrogen, play a major role in the formation of ozone. The major sources of man-made NO_x emissions are high-temperature combustion processes, such as those occurring in automobiles and power plants. Home heaters and gas stoves also produce substantial amounts of NO_2 in indoor settings.

2.1.3.2 Health and Environmental Effects. Short-term exposures (e.g., less than 3 hours) to current nitrogen NO_2 concentrations may lead to changes in airway responsiveness and lung function in individuals with pre-existing respiratory illnesses and increases in respiratory illnesses in children (5–12 years old). Long-term exposures to NO_2 may lead to increased susceptibility to respiratory infection and may cause alterations in the lung. Atmospheric transformation of NO_x can lead to the formation of ozone and nitrogen-bearing particles (most notably in some western urban areas) which are both associated with adverse health effects. Nitrogen oxides also contribute to the formation of acid rain. Nitrogen oxides contribute to a wide range of environmental effects, including potential changes in the composition and competition of some species of vegetation in wetland and terrestrial systems, visibility impairment, acidification of freshwater bodies, eutrophication (i.e., explosive algae growth leading to a depletion of oxygen in the water) of estuarine and coastal waters (e.g., Chesapeake Bay), and increases in levels of toxins harmful to fish and other aquatic life.

2.1.3.3 Trends in NO_2 Levels. Over the past ten years, ambient NO_2 concentrations decreased 14%. Between 1996 and 1997, national average annual mean NO_2 concentrations remain unchanged. In the last 10 years, NO_x emissions levels have remained relatively constant. Between 1988 and 1997, NO_x emissions declined 1%, while they increased slightly (by 1%) between 1996 and 1997. However, over the longer term since 1970, total NO_x emissions have increased 11% and NO_x emissions from coal-fired power plants have increased 44%.

2.1.4 Ground-Level (Tropospheric) Ozone

2.4.4.1 Nature and Sources of the Pollutant. Ground-level ozone (O_3) (the primary constituent of smog) continues to be a pervasive pollution problem throughout many areas of the United States. Ozone is not emitted directly into the air but is formed by the reaction of volatile organic compounds (VOCs) and NO_x in the presence of heat and sunlight. Ground-level O_3 forms readily in the atmosphere, usually during hot summer weather. Volatile organic compounds are emitted from a variety of sources, including motor vehicles, chemical plants, refineries, factories, consumer and commercial products, and other industrial sources. Nitrogen oxides are emitted from motor vehicles, power plants, and other sources of combustion. Changing weather patterns contribute to yearly differences in O_3 concentrations from city to city. Ozone and the precursor pollutants that cause O_3 also can be transported into an area from pollution sources found hundreds of miles upwind.

2.1.4.2 Health and Environmental Effects. Short-term (1–3 hours) and prolonged (6–8 hours) exposures to ambient O_3 have been linked to a number of health effects of concern. For example, increased hospital admissions and emergency room visits for respiratory causes have been associated with ambient O_3 exposures. Repeated exposures to O_3 can make people more susceptible to respiratory infection, result in lung inflammation, and aggravate pre-existing respiratory diseases such as asthma. Other health effects attributed to O_3 exposures include significant decreases in lung function and increased respiratory symptoms such as chest pain and cough. These effects generally occur while individuals are engaged in moderate or heavy exertion. Children active outdoors during the summer when O_3 levels are at their highest are most at risk of experiencing such effects. Other at-risk groups include adults who are active outdoors (e.g., outdoor workers), and individuals with pre-existing respiratory disease such as asthma and chronic obstructive lung disease. In addition, longer-term exposures to moderate levels of O_3 present the possibility of irreversible changes in the lungs which could lead to premature aging of the lungs and/or chronic respiratory illnesses.

Ozone also affects vegetation and ecosystems, leading to reductions in agricultural and commercial forest yields, reduced growth and survivability of tree seedlings, and increased plant susceptibility to disease, pests, and other environmental stresses (e.g., harsh weather). In long-lived species, these effects may become evident only after several years or even decades, thus having the potential for long-term effects on forest ecosystems. Ground-level O_3 damage to the foliage of trees and other plants also can decrease the aesthetic value of ornamental species as well as the natural beauty of our national parks and recreation areas.

2.1.4.3 Revised Ozone Standards. In 1997, EPA revised the national ambient air quality standards for O_3 by replacing the 1-hour O_3 standard of 0.12 parts per million (ppm) with a new 8-hour 0.08 ppm standard. The revision to the O_3 standard was set such that the 1-hour standard will no longer apply once an area has air quality data meeting the 1-hour standard. Although areas that do not meet the new 8-hour standard will not be designated "nonattainment" until the year 2000, EPA is beginning to track trends in 8-hour levels of O_3.

2.1.4.4 Trends in Ozone Levels. Ambient O_3 trends are influenced by year-to-year changes in meteorological conditions, population growth, loadings of VOCs and NO_x in the atmosphere, and by changes in emissions from ongoing control measures. Between 1988 and 1997, ambient O_3 concentrations decreased 19%, based on the pre-existing standard. Between 1996 and 1997, ambient O_3 concentrations did not change based on the pre-existing standard. Nationally, 8-hour levels of O_3 have decreased 16 percent over the past 10 years. Between 1996 and 1997, O_3 concentrations decreased 1% based on the revised standard.

In order to address O_3 pollution, EPA has traditionally focused its control strategies on reducing emissions of VOCs in nonattainment areas. However, EPA and the states have recognized a need for an aggressive program to reduce regional emissions of NO_x. In 1998, EPA issued a rule that will significantly reduce regional emissions of NO_x in 22 states and the District of Columbia, and, in turn, reduce the regional transport of O_3.

National trends in emissions of NO_x and VOCs underscore the importance of this new approach. Volatile organic compound emissions decreased 20% between 1988 and 1997, while NO_x emissions decreased only 1%. VOCs emissions from highway vehicles have declined 38% since 1988, while highway vehicle NO_x emissions have declined 8% since their peak level in 1994. Further, between 1970 and 1997 emissions of VOCs have decreased 38% whereas emissions of NO_x have increased 11% and NO_x emissions from coal-fired power plants have increased 44%.

2.1.5 Particulate Matter

2.1.5.1 Nature and Sources of the Pollutant. Particulate matter (PM) is the general term used for a mixture of solid particles and liquid droplets found in the air. Some particles are large or dark enough to be seen as soot or smoke. Others are so small they can be detected only with an electron microscope. These particles, which come in a wide range of sizes ("fine" particles are less than 2.5 micrometers in diameter and coarser-size particles are larger than 2.5 micrometers), originate from many different stationary and mobile sources as well as from natural sources. Fine particles (PM-2.5) result from fuel combustion from motor vehicles, power generation, and industrial facilities, as well as from residential fireplaces and wood stoves. Coarse particles (PM-10) are generally emitted from sources, such as vehicles traveling on unpaved roads, materials handling, and crushing and grinding operations, as well as windblown dust. Some particles are emitted directly from their sources, such as smokestacks and cars. In other cases, gases such as sulfur oxide and SO_2, NO_x and VOC interact with other compounds in the air to form fine particles. Their chemical and physical compositions vary depending on location, time of year, and weather.

2.1.5.2 Health and Environmental Effects. Inhalable PM includes both fine and coarse particles. These particles can accumulate in the respiratory system and are associated with numerous health effects. Exposure to coarse particles is primarily associated with the aggravation of respiratory conditions, such as asthma. Fine particles are most closely associated with such health effects as increased hospital admissions and emergency room visits for heart and lung disease, increased respiratory symptoms and disease, decreased lung function, and even premature death. Sensitive groups that appear to be at greatest risk to such effects include the elderly, individuals with cardiopulmonary disease, such as asthma, and children. In addition to health problems, PM is the major cause of reduced visibility in many parts of the United States. Airborne particles also can cause damage to paints and building materials.

2.1.5.3 Revised Particulate Matter Standards. In 1997, EPA added two new PM-2.5 standards, set at 15 micrograms per cubic meter (μg/m3) and 65 μg/m3, respectively, for the annual and 24-hour standards. In addition, the form of the 24-hour standard for PM-10 was changed. EPA is beginning to collect data on PM-2.5 concentrations. Beginning in 2002, based on three years of monitor data, EPA will designate areas as nonattainment that do not meet the new PM-2.5 standards.

2.1.5.4 Emissions Trends. Between 1988 and 1997, average PM-10 concentrations decreased 26 percent. Short-term trends between 1996 and 1997 showed a decrease of 1% monitored PM-10 concentration levels.

Emissions of PM-10 based on estimates of anthropogenic emissions including fuel combustion sources, industrial processes, and transportation sources, account for only 6% of the total PM-10 emissions nationwide. Between 1988 and 1997, PM-10 emissions for these sources decreased 12%. Emissions of PM-10 between 1996 and 1997 decreased 1%. These emissions estimates do not include emissions from natural and miscellaneous sources, such as fugitive dust (unpaved and paved roads), agricultural and forestry activities, wind erosion, wildfires, and managed burning. These emissions estimates also do not account for PM that is secondarily formed in the atmosphere from gaseous pollutants (i.e., SO_2 and NO_x).

2.1.6 Sulfur Dioxide

2.1.6.1 Nature and Sources of the Pollutant. Sulfur dioxide (SO_2) belongs to the family of sulfur oxide gases. These gases are formed when fuel containing sulfur (mainly, coal and oil) is burned and during metal smelting and other industrial processes. Most SO_2 monitoring stations are located in urban areas. The highest monitored concentrations of SO_2 are recorded in the vicinity of large industrial facilities.

2.1.6.2 Health and Environmental Effects. High concentrations of SO_2 can result in temporary breathing impairment for asthmatic children and adults who are active outdoors. Short-term exposures of asthmatic individuals to elevated SO_2 levels while at moderate exertion may result in reduced lung function that may be accompanied by such symptoms as wheezing, chest tightness, or shortness of breath. Other effects that have been associated with longer-term exposures to high concentrations of SO_2, in conjunction with high levels of PM, include respiratory illness, alterations in the lungs' defenses, and aggravation of existing cardiovascular disease. The subgroups of the population that may be affected under these conditions include individuals with cardiovascular disease or chronic lung disease, as well as children and the elderly.

Together, SO_2 and NO_x are the major precursors to acidic deposition (acid rain), which is associated with the acidification of soils, lakes, and streams, accelerated corrosion of buildings and monuments, and reduced visibility. Sulfur dioxide also is a major precursor to PM-2.5, which is a significant health concern as well as a main pollutant that impairs visibility.

2.1.6.3 Trends in SO_2 Levels. Between 1988 and 1997, national SO_2 concentrations decreased 39% and SO_2 emissions decreased 12%. Between 1996 and 1997, national SO_2 concentrations decreased 4% and SO_2 emissions increased 3%. Sulfur dioxide emissions from electric utilities decreased 12% between 1994 and 1997. These recent reductions are due, in large part, to controls implemented under EPAs Acid Rain Program. The 3% increase that occurred between 1996 and 1997 is primarily due to increased demand for electricity.

2.2 Global Warming and Climate Change

2.2.1 Nature and Sources

The Earth's climate is fueled by the Sun. Most of the Sun's energy, called solar radiation, is absorbed by the Earth, but some is reflected back into space. A natural layer of atmospheric gases absorbs a portion of this reflected solar radiation, eventually releasing some of it into space, but forcing much of it back to Earth. There it warms the Earth's surface creating what is known as the natural "greenhouse effect," as illustrated in Figure 58.1. Without the natural greenhouse effect, the Earth's average temperature would be much colder, and the planet would be covered with ice.

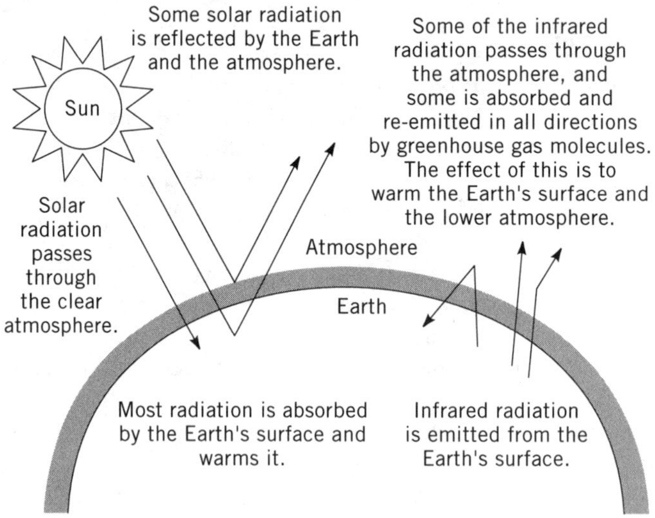

Figure 58.1. The Greenhouse Effect. From U.S. Department of State, 1992.

Recent scientific evidence shows that the greenhouse effect is being increased by release of certain gases to the atmosphere that cause the Earth's temperature to rise. This is called "global warming." Carbon dioxide (CO_2) accounts for about 85% of greenhouse gases released in the United States. Carbon dioxide emissions are largely due to the combustion of fossil fuels in electric power generation. Methane (CH_4) emissions, which result from agricultural activities, landfills, and other sources, are the second largest contributor to greenhouse gases in the United States.

Industrial processes such as foam production, refrigeration, dry cleaning, chemical manufacturing, and semiconductor manufacturing produce other greenhouse gas emissions, such as hydrofluorocarbons (HFCs). Smelting of aluminum produces another greenhouse gas called perfluorinated compounds (PFCs). Emissions of NO_x and VOC from automobile exhaust and industrial processes contribute to the formation of ground-level O_3 or smog, also a greenhouse gas.

2.2.2 Health and Environmental Effects

Greenhouse gas emissions could cause a 1.8 to 6.3°F rise in temperature during the next century, if atmospheric levels are not reduced. Although this change may appear small, it could produce extreme weather events, such as droughts and floods; threaten coastal resources and wetlands by raising sea level; and increase the risk of certain diseases by producing new breeding sites for pests and pathogens. Agricultural regions and woodlands are also susceptible to changes in climate that could result in increased insect populations and plant disease. This degradation of natural ecosystems could lead to reduced biological diversity.

2.2.3 International Developments

In 1988, the Intergovernmental Panel on Climate Change (IPCC) was formed to assess the available scientific and economic information on climate change and formulate response strategies. In 1995, the IPCC published a report representing the work of more than 2,000 of the world's leading scientists. The IPCC concluded that humans are changing the Earth's climate, and that "climate change is likely to have wide-ranging and mostly adverse impacts on human health, with significant loss of life."

In 1992, 150 countries signed the "Framework Convention on Climate Change" (FCCC), which has the objective of stabilizing the concentration of greenhouse gases in the atmosphere at levels that would prevent dangerous interference with the climate system. Under the FCCC, industrialized countries agreed to aim to reduce greenhouse gas emissions to 1990 levels by the year 2000. It now appears that most industrialized countries, including the United States, will not meet this target. In light of the 1995 scientific findings of the IPCC and the continued rise in greenhouse gas emissions, parties to the FCCC formulated the "Kyoto Protocol" at a 1997 conference held in Kyoto, Japan. The Kyoto Protocol includes greenhouse gas emission targets for industrialized countries for the period of 2008–2012. The average reduction target for all industrialized countries for this period is 5% below 1990 emission levels. The reduction target varies across countries to account for differing circumstances, with the United States' target being a 7% reduction below 1990 levels. The Kyoto Protocol also provides for market-based measures, such as international emissions trading, to help countries meet their commitments at the lowest possible cost. (The U.S. Administration will seek the Senate's consent for ratification of the Kyoto Protocol after working for further progress on the details of the market mechanisms and on the involvement of key developing countries.)

U.S. Programs to Mitigate Climate Change: The United States adopted a Climate Change Action Plan (CCAP) in 1993 to reduce greenhouse gas emissions. Thousands of companies and nonprofit organizations are working together to effectively reduce their emissions. The Plan involves more than 40 programs implemented by EPA, the Department of Energy, the Department of Agriculture, and other government agencies. In 1997, these voluntary programs reduced greenhouse gas emissions by more than 15 million tons of carbon, while partners saved over $1 billion from energy bill savings.

2.3 Stratospheric Ozone

2.3.1 Nature and Sources of the Problem

The stratosphere, located about 6 to 30 miles above the Earth, contains a layer of O_3 gas that protects living organisms from harmful ultraviolet radiation (UV-b) from the sun. Over the past two decades, however, this protective shield has been damaged. Each year, an "ozone hole" forms over the Antarctic, and O_3 levels fall to 70% below normal. Even over the United States, O_3 levels are about 5% below normal in the summer and 10% below normal in the winter. The trend line shows a 3.4% decrease per decade in average total O_3 over Northern Hemisphere mid-latitudes since 1979.

As the O_3 layer thins, more UV-b radiation reaches the Earth. In 1996, scientists demonstrated for the first time that UV-b levels over most populated areas have increased. Scientists have linked several substances associated with human activities to O_3 depletion, including the use of chlorofluorocarbons (CFCs), halons, carbon tetrachloride, and methyl chloroform. These chemicals are emitted from home air conditioners, foam cushions, and many other products. Strong winds carry them through the lower part of the atmosphere, called the troposphere, and into the stratosphere. There, strong solar radiation releases chlorine and bromine atoms that attack protective ozone molecules. Scientists estimate that one chlorine atom can destroy 100,000 O_3 molecules.

2.3.2 Health and Environmental Effects

Some UV-b radiation reaches the Earth's surface even with normal O_3 levels. However, because the O_3 layer normally absorbs most UV-b radiation from the sun, O_3 depletion is expected to lead to increases in harmful effects associated with UV-b radiation. In humans, UV-b radiation is linked to skin cancer, including melanoma, the form of skin cancer with the highest fatality rate. It also causes cataracts and suppression of the immune system.

The effects of UV-b radiation on plant and aquatic ecosystems are not well understood. However, the growth of certain food plants can be slowed by excessive UV-b radiation. In addition, some scientists suggest that marine phytoplankton, which are the base of the ocean food chain, are already under stress from UV-b radiation. This stress could have adverse consequences for human food supplies from the oceans. Because they absorb CO_2

from the atmosphere, significant harm to phytoplankton populations could increase global warming.

2.3.3 Programs to Restore the Stratospheric Ozone Layer

In 1987, 27 countries signed the Montreal Protocol, a landmark treaty that recognized the international nature of O_3 depletion and committed the world to limiting the production of O_3-depleting substances. Today, over 160 nations have signed the Protocol, which has been strengthened four times and now calls for the elimination of those chemicals that deplete O_3.

The 1990 Clean Air Act Amendments established a U.S. regulatory program to protect the stratospheric O_3 layer. In January 1996, U.S. production of many O_3-depleting substances virtually ended, including CFCs, carbon tetrachloride, and methyl chloroform. Production of halons ended in January 1994. Many new products that either do not affect or are less damaging to the O_3 layer are now gaining popularity. For example, computer-makers are using O_3-safe solvents to clean circuit boards, and automobile manufacturers are using HFC-134a, an O_3-safe refrigerant, in new motor vehicle air conditioners. In some sectors, the transition away from O_3-depleting substances has already been completed. EPA is also emphasizing new efforts like the UV Index, a daily forecast of the strength of UV radiation people may be exposed to outdoors, to educate the public about the health risks of overexposure to UV radiation and the steps they can take to reduce those risks.

2.3.4 Trends in Stratospheric Ozone Depletion

Scientific evidence shows that the approach taken under the Montreal Protocol has been effective. In 1996, measurements showed that the tropospheric concentrations of methyl chloroform had started to fall, indicating that emissions had been greatly reduced. Tropospheric concentrations of other O_3-depleting substances, like CFCs, are also beginning to decrease. It takes several years for these substances to reach the stratosphere and release chlorine and bromine. For this reason, stratospheric chlorine levels are expected to continue to rise, peak by the year 2000, and then slowly decline. Because of the stability of most O_3-depleting substances, chlorine will be released into the stratosphere for many years, and the O_3 layer will not fully recover until well into the next century.

In 1996, scientists developed a new technique allowing them to draw conclusions about UV-b radiation at ground level. According to satellite-based trend analyses, major populated areas have experienced increasing UV-b levels over the past 15 years. As shown in Figure 58.2, at latitudes that cover the United States, UV-b levels are 4 to 5% higher than they were 10 years ago.

2.4 Toxic Air Pollutants

2.4.1 Nature and Sources

Toxic air pollutants are those pollutants that cause or may cause cancer or other serious health effects, such as reproductive effects or birth defects, or adverse environmental and ecological effects. The Clean Air Act requires EPA to address 188 toxic air pollutants.

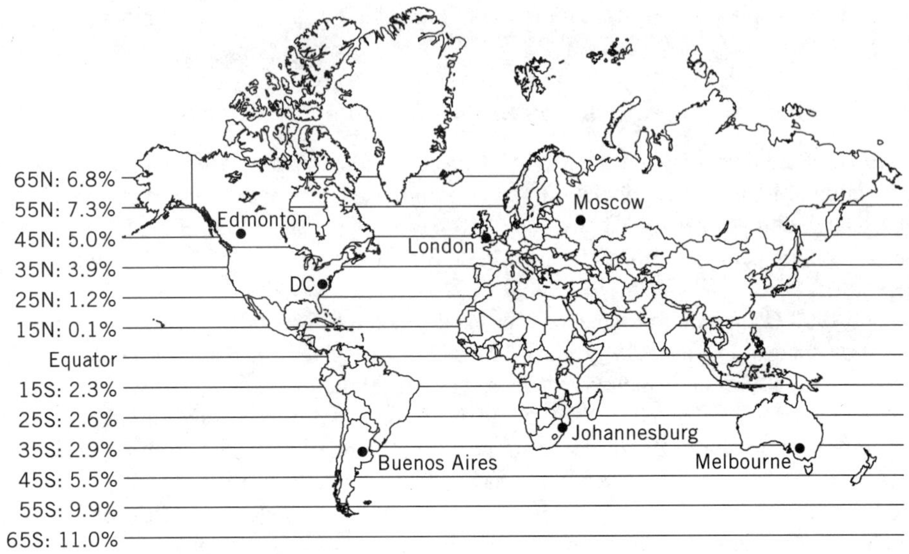

Figure 58.2. UV-b radiation increases by latitude.

Examples of toxic air pollutants include benzene, found in gasoline; perchloroethylene, emitted from some dry-cleaning facilities; and methylene chloride, used as a solvent and paint stripper by a number of industries. Some air toxics are released from natural sources such as volcanic eruptions and forest fires. Most, however, originate from human-made sources, including both mobile sources (e.g., cars, trucks, buses) and stationary sources (e.g., factories, refineries, power plants).

2.4.2 Health and Environmental Effects

People exposed to toxic air pollutants at sufficient concentrations and for sufficient durations have an increased chance of getting cancer or experiencing other serious health effects. These health effects can include damage to the immune system, as well as neurological, reproductive (i.e., reduced fertility), developmental, respiratory and other health problems. Some toxic air pollutants pose particular hazards to people at a certain stage in life, such as young children or the elderly. Some health problems occur very soon after a harmful exposure. Health effects associated with long-term exposures to toxic air pollutants, however, may develop slowly over time or not appear until many months or years after the initial exposure.

Toxic pollutants in the air or deposited on soils or surface waters can have a number of environmental impacts. Like humans, animals experience health problems if exposed to sufficient concentrations of air toxics over time. Persistent toxic air pollutants that can accumulate in plants and animals are of particular concern in aquatic ecosystems because the pollutants accumulate in tissues of animals, magnifying up the food chain to levels many times higher than in the water. Toxic pollutants that disrupt the endocrine system also pose a threat. In some wildlife, for example, exposures to pollutants such as DDT, dioxins, and mercury have been associated with decreased fertility, decreased hatching success, damaged reproductive organs, and altered immune systems.

2.4.3 Program Structure

Control of toxic air pollutants differs from the control of the six principal pollutants for which EPA has established national air quality standards. For the six principal pollutants, the Clean Air Act requires states to develop plans to meet the national air quality standards by specific deadlines. In contrast, for toxic air pollutants, the Act requires EPA to have a two-phased program. The first phase consists of identifying the sources of toxic pollutants and developing technology-based standards to significantly reduce their emissions. The second phase consists of strategies and programs for evaluating the remaining risks and ensuring that the overall program has achieved substantial reduction in risks to public health and the environment. The objective is to ensure that on a national basis sources of toxic air pollution are well controlled and that risks to public health and the environment are substantially reduced.

In addition, the toxic air pollutant program is important in reducing highly localized emissions near industrial sources and in controlling pollutants that are toxic even when emitted in small amounts. Companies handling extremely hazardous chemicals are required by EPA to develop plans to prevent accidental releases and to contain any releases in the event they should occur.

2.4.4 Trends in Toxic Air Pollutants

EPA is using the National Toxics Inventory (NTI) to track nationwide emissions trends for toxic air pollutants listed in the Clean Air Act. The NTI includes emissions information for 188 hazardous air pollutants from more than 900 stationary sources based on a 1993 survey (1). There are approximately 8.1 million tons of air toxics released to the air each year according to NTI. As illustrated in Figure 58.3, NTI includes emissions from large industrial or "point" sources, smaller stationary sources called "area" sources, and mobile sources. The NTI estimates of the large point source and mobile source contributions to the national emissions of toxic air pollutants are approximately 61 and 21%, respectively.

Currently, EPA has issued 27 air toxics emissions standards under the first (technology-based) phase of the regulations program. These standards affect 52 categories of major industrial sources, such as chemical plants, oil refineries, aerospace manufacturers, and steel mills, as well as eight categories of smaller sources, such as dry cleaners, commercial sterilizers, secondary lead smelters, and chromium electroplating facilities. EPA has also issued two standards to control emissions from solid waste combustion facilities (1). Together these standards reduce emissions of over 100 different air toxics. When fully implemented, these standards will reduce air toxics emissions by about 1 million tons per year—almost ten times the reductions achieved prior to 1990. In addition, controls for toxic air pollutants will also reduce VOC and PM emissions by more than 2.5 million tons per year over the same time period.

EPA is now moving into the second (risk-based) phase of the regulatory program. EPA has recently published a draft integrated strategy that will address cumulative risks from

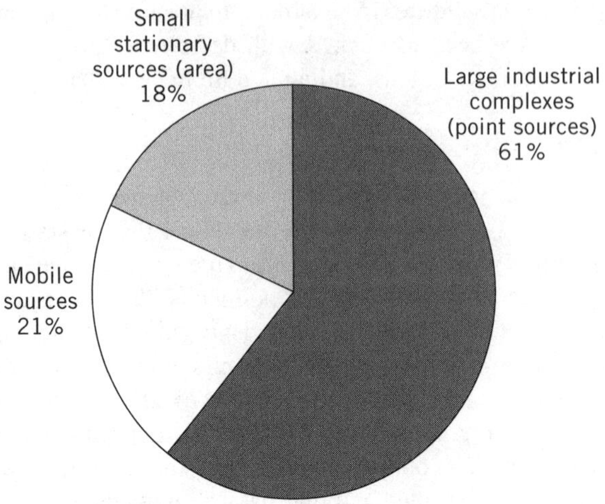

Figure 58.3. National toxic air pollutant emissions by source, 1993.

multiple sources (both stationary and mobile) of toxic pollutants and from combined exposures of these pollutants in urban areas.

EPA collects data through its Photochemical Assessment Monitoring Stations (PAMS) program on concentrations of O_3 and its precursors in 22 areas across the nation with the most significant ozone problems. The PAMS program requires routine measurement of ten pollutants that contribute to O_3 formation and which also are toxic air pollutants: acetaldehyde, benzene, ethyl benzene, formaldehyde, hexane, styrene, toluene, *m*/*p*-xylene, *o*-xylene, and 2,2,4-trimethylpentene. Preliminary analysis of the monitoring data indicates that concentrations of some of these toxic VOC in the areas monitored are declining. Monitoring networks now being set up will provide more toxics data in the future.

2.5 Visibility

2.5.1 Nature and Sources of the Problem

Visibility impairment occurs as a result of the scattering and absorption of light by air pollution, including particles and gases. In addition to limiting the distance that can be seen, the scattering and absorption of light caused by air pollution can also degrade the color, clarity, and contrast of scenes. The same particles that are linked to serious health effects can also significantly affect our ability to see.

Both primary emissions and secondary formation of particles contribute to visibility impairment. "Primary" particles, such as dust from roads or elemental carbon (soot) from wood combustion, are emitted directly into the atmosphere. "Secondary" particles are formed in the atmosphere from primary gaseous emissions. Examples include sulfate, formed from SO_2, and nitrates, formed from NO_x. In the Eastern United States, reduced visibility is mainly attributable to secondarily-formed particles. While these secondarily-formed particles still account for a significant amount in the West, primary emissions from sources like wood smoke contribute a larger percentage of the total particulate loading than in the East.

Humidity can significantly increase the effect of pollution on visibility. Some particles, such as sulfates, accumulate water and grow in size, becoming more efficient at scattering light and causing visibility impairment. Annual average relative humidity levels are 70–80% in the East as compared to 50–60% in the West. Poor summer visibility in the eastern United States is primarily the result of high sulfate concentrations combined with high humidity levels.

2.5.2 Long-term Trends

Visibility impairment has been analyzed using visual range data collected since 1960 at 280 monitoring stations located at airports across the country. At these stations, measurements of visual range (the maximum distance at which an observer can discern the outline of an object) were recorded. The maps below show the amount of haze during the summer months of 1970, 1980, and 1990. The dark blue color represents less haze and red represents more haze. Overall, maps show that visibility impairment in the Eastern United States increased greatly between 1970 and 1980 and decreased slightly between 1980 and 1990. This follows the overall trend in emissions of SO_x during these periods.

2.5.3 Visibility Monitoring Network

In 1987, a visibility monitoring network was established as a cooperative effort between the EPA, states, National Park Service, U.S. Forest Service, Bureau of Land Management, and U.S. Fish and Wildlife Service. The network is designed to track progress toward the Clean Air Act's national goal of remedying the existing and preventing future visibility impairment in the 156 national parks and wilderness areas. Each of these sites contains over 5,000 acres. The network is the largest in the country devoted to fully characterizing visibility. It also provides information for determining the types of pollutants and sources primarily responsible for reduced visibility.

Visibility impairment is generally worse in the rural East compared to most of the West. This is primarily due to higher concentrations of man-made pollution, slightly higher background levels of fine particles, and higher relative humidity levels in the East. The chart below shows the relative levels of pollutants that contribute to visibility impairment in the Eastern and Western parts of the United States.

Pollutant, %	West	East
Sulfates	25–65	>60
Organic carbon	15–35	10–15
Nitrates	5–45	10–15
Elemental carbon (soot)	15–25	10–15
Crustal material (soil dust)	10–20	10–15

2.5.4 Programs to Improve Visibility

EPA proposed a new regional haze program in 1997 to address visibility impairment in national parks and wilderness areas caused by numerous sources located over broad regions. When finalized, the program will lay out a framework within which states develop implementation plans to achieve "reasonable progress" toward the national visibility goal of remedying any existing and preventing any future human-caused impairment. These plans will include emission management strategies to improve visibility over time in national parks, particularly for the worst visibility days. States will be required to periodically track progress and revise any strategies as necessary. Because fine particles are frequently transported hundreds of miles, pollution that occurs in one state may contribute to the visibility impairment in another state. Thus, to effectively address the regional haze problem, states are encouraged to coordinate with each other in developing strategies to improve visibility and to comply with the PM-2.5 and ozone NAAQS.

Other air quality programs are expected to lead to emission reductions that will improve visibility in certain regions of the country. The Acid Rain Program is designed to achieve significant reductions in SO_x emissions, which is expected to reduce sulfate haze, particularly in the Eastern United States. Additional control programs on sources of NO_x to reduce formation of O_3 can also improve regional visibility conditions. In addition, programs, such as the national ambient air quality standards, controls on diesel-powered mobile sources, and programs to improve wood stove efficiency can benefit areas adversely impacted by visibility impairment.

3 HEALTH EFFECTS OF THE SIX PRINCIPAL AMBIENT AIR POLLUTANTS

This section is a reproduction of the chapter "Critical Health Issues of Criteria Air Pollutants" (from Ref. 2) and is reproduced with copyright permission of Taylor & Francis.

This section summarizes key health information for the six principal outdoor air pollutants that can cause adverse health effects at current or historical ambient levels in the United States. Of the thousands of air pollutants, very few meet this definition. The Clean Air Act (CAA) and Amendments (see Appendix B) of the United States requires that the U.S. Environmental Protection Agency (EPA) identify such pollutants (called *criteria pollutants*) and set standards (National Ambient Air Quality Standards [NAAQS]) to protect sensitive subpopulations from the adverse effects of these compounds. The criteria pollutants are ozone, nitrogen dioxide (NO_2), sulfur dioxide (SO_2), particulate matter (PM), carbon monoxide (CO), and lead. The World Health Organization (WHO) has developed air quality guidelines (AQG) for Europe in reference to the health effects of these pollutants (3).

Exposures to these pollutants are widespread because of the wide diversity of their sources (Table 58.1). Table 58.2 shows their major classes of effects, susceptible subpopulations, the U.S. NAAQS, and the WHO AQG. The NAAQS are set according to a complex process involving substantial review of the literature and several reviews of the resultant interpretations by committees of experts, environmental groups, industry groups, and the interested public. The CAA requires that the NAAQS be set to protect susceptible

Table 58.1. Sources of Criteria Air Pollutants

Pollutant	Predominant Sources
Ozone	Photochemical oxidation of nitrogen oxides (primarily from combustion) and volatile organic compounds (from stationary and mobile sources and natural sources)
Nitrogen dioxide	Photochemical oxidation of nitric oxide (primarily from combustion of fossil fuels) and direct emissions (primarily from combustion of natural gas)
Sulfur dioxide	Primarily combustion of sulfur-containing fossil fuels; also smelters, refineries, and others
Particulate matter	Combustion of fuels and natural sources
Carbon monoxide	Combustion of fuels, especially by mobile sources
Lead	Leaded gasoline (prior to phase-out in gasoline); currently no significant air sources

subpopulations with an adequate margin of safety, without regard to the economic impact, and it requires they be reevaluated every five years. The WHO AQG are set by panels of experts with subsequent peer review. The AQG are not standards, *per se*. Rather, they provide guidance to many nations and the European Community as they seek to develop their own standards for protection of the public health.

The complete health effects databases for these pollutants are far too extensive for recounting here. Rather, this summary focuses on the primary effects of concern and provides references to more complete reviews.

3.1 Ozone

Animal toxicology, controlled human clinical, field, and epidemiological studies with O_3 have shown effects on lung function, inflammation, host defenses, biochemistry, and structure, as well as several indices of morbidity. These and other effects recently have been reviewed comprehensively (5).

Pulmonary host defenses are very sensitive to acute O_3 exposure (5). Gardner (6) summarizes the observations in animals, which include effects of 3-hr exposures to levels of O_3 as low as 0.08 ppm that increase susceptibility to bacterial lung infections in mice. Human alveolar macrophage functions also are impaired in humans exposed to levels as low as 0.08 ppm (6.6 hr, moderate exercise) (5).

Ozone exposure produces many functional responses in humans, including decrements in lung volumes and dynamic spirometry, increased airway resistance, increased specific and nonspecific airway reactivity, and a variety of respiratory symptoms. In healthy people, O_3 causes spirometric lung function effects at concentrations as low as 0.08 ppm for prolonged (6 to 8 hr) exposures involving several hours of moderate exercise or as low as 0.12 ppm for 1 hr of continuous heavy exercise. There is a strong consistent dose–response relationship (7), where exposure dose is calculated as a function of O_3 concentration, duration, and ventilation (V). The level of exertion during the exposure is the prime de-

Table 58.2. Summary of Effects, Populations of Concern, and Standards

Pollutant	Key Effects of Concern	Subpopulation of Concern	NAAQS[a]	AQG
Ozone	Decreased pulmonary function, lung inflammation, increased respiratory hospital admissions	Children, people with preexisting lung disease, outdoor-exercising healthy people	1-hr avg. 0.12 ppm (235 μg/m^3) 8-hr avg. 0.08 ppm (157 μg/m^3)	8-hr avg. 0.06 ppm (120 μg/m^3)
Nitrogen dioxide	Respiratory illness, decreased pulmonary function	Children, people with preexisting lung disease	Annual arithmetic mean 0.053 ppm (100 μg/m^3)	1-hr avg. 0.1 ppm (200 μg/m^3) annual 0.02 ppm (40 μg/m^3)
Sulfur dioxide	Respiratory morbidity, mortality, decreased pulmonary function	Children, people with preexisting lung disease (especially asthma)	Annual arithmetic mean 0.03 ppm (80 μg/m^3), 24-hr avg. 0.14 ppm (365 μg/m^3)	10-min. avg. 0.175 ppm (500 μg/m^3), 24-hr avg. 0.05 ppm (125 μg/m^3), annual 0.02 ppm (50 μg/m^3)
Particulate matter	Mortality and morbidity	People with preexisting heart and lung disease, children	PM_{10}, annual arithmetic mean 50 μg/m^3, 24-hr avg. 150 μg/m^3; $PM_{2.5}$, annual arithmetic mean 15 μg/m^3, 24-hr avg. 65 μg/m^3	Exposure response used[b]
Carbon monoxide	Shortening of time to onset of angina and other heart effects	People with coronary artery disease	8-hr avg. 9 ppm (10 mg/m^3), 1-hr avg. 35 ppm (40 mg/m^3)	8-hr avg. 10 ppm (10 mg/m^3), 1-hr avg. 25 ppm (30 mg/m^3), 30-min. avg. 50 ppm (60 mg/m^3), 15 min. avg. 90 ppm (100 mg/m^3)
Lead	Developmental neurotoxicity	Children	Quarterly avg. 1.5 μg/m^3	Annual 0.5 μg/m^3

[a]NAAQS are specified by concentration, averaging time, and "form". The latter typically is expressed in a fashion to enable determination of compliance and is too complex for this overview (details are in U.S. EPA (4)).
[b]The WHO described the exposure response to PM, allowing users to determine what specific guidelines would be best for their countries (3).

terminant of V. The dynamics of the spirometric response have been clarified by a series of prolonged exposure studies (8–10). Recovery from an O_3-induced spirometric response proceeds rapidly over the first few hours after exposure, but it may require in excess of 24 hr for complete functional recovery to occur if the initial response is large (e.g., ≥30% decrease in forced expiratory volume in 1 second) (11).

Symptoms and lung function changes are well correlated within individuals, but symptoms are not useful predictors of lung function responses because of the wide range of symptom-response slopes among individuals. Spirometry measures, airway resistance, and airway reactivity responses to O_3 exposure are not well correlated (12, 13). It is likely that these responses each evolve from independent response mechanisms (5). For example, spirometric responses appear to result primarily from a restriction of maximum inhaled volume (reduced forced vital capacity or total lung capacity). Ozone exposure causes shedding of airway epithelial cells (14, 15), and release of substance P, presumably from bronchial C fibers (16). Stimulation of bronchial C fibers appears to be an important pathway for some O_3-induced responses (17).

Increasing attention has been paid to the differing responses of sensitive individuals, especially those with respiratory disease. It has been known for some time that O_3 causes an increase in airway responsiveness to bronchoconstrictors such as methacholine or histamine (18). Evidence that O_3 also causes increased specific airway reactivity in allergic asthmatics has emerged more recently (19–21). The original observation by Molfino et al. (19) of increased response to ragweed/grass antigen after a resting exposure for 1 hr to 0.12 ppm O_3 has not been replicated (22) under these exposure conditions. However, with a higher O_3 exposure dose (3 hr at 0.25 ppm with intermittent exercise), Jörres et al. (20) found a significant increase in response to inhaled allergen in asthmatics within 2 to 3 hr after exposure. Kehrl et al. (21) found increased specific airway reactivity of asthmatics to dust mite antigen 18 hr after exposure to O_3 (0.16 ppm for 8 hr with exercise). These observations of increased airway reactivity to common antigens following O_3 exposure suggest a plausible mechanism for the findings in epidemiological studies of increased asthma attacks, emergency room visits for asthma (23), and hospital admissions for respiratory disease (24) that are associated with increased ambient O_3 levels. It is estimated that O_3 may account for as much as 15 to 20% of summertime asthma hospital admissions (25).

Asthmatics appear to have more pronounced inflammatory responses following exposure to O_3 in the range of 0.16 to 0.20 ppm for 4.0 to 7.6 hr (26–28). In bronchoalveolar lavage (BAL) conducted 18 hr after these ozone exposures, more BAL neutrophils were found in asthmatics compared with similarly exposed nonasthmatics. Typically after ozone exposure, nonasthmatics had 7 to 9% of BAL cells as neutrophils, whereas asthmatics ranged from 8 to 16% neutrophils. In one of these studies (29), asthmatics tested specifically for allergy to house dust mites had both eosinophillic and neutrophillic inflammation.

It has been determined that the spirometric and symptom responses to O_3 are attenuated with 3 to 5 days of repeated exposure; that this attenuation persists for up to a week; and that the rate of response of attenuation is a function of individual sensitivity, the response endpoint, and O_3 exposure conditions (5). It recently has been reported that a response attenuation also occurs in asthmatics (29). In a study utilizing a time course of exposure similar to the human "adaptation" studies, Tepper et al. (30) found that functional lung responses disappeared after four to five days of repeated exposure in rats, but that inflammatory responses were still evident. This prompted the examination of the inflammatory response in humans exposed repeatedly to O_3. Devlin et al. (15) found that neutrophillic inflammation was much less after 5 days of O_3 exposure (2 hr/day at 0.40 ppm with exercise) than after a single exposure. Although lung function responses were minimal, symptoms were absent, and inflammatory cells were much reduced, in BAL fluid collected 1 hr postexposure, the indicators of cell damage (i.e., lactate dehydrogenase activity and protein) were not reduced compared with the response seen after a single O_3 exposure. These observations of reduced neutrophil infiltration have been substantiated by Christian et al. (31), using a different exposure protocol. They performed BAL some 20 hr after one and four consecutive days of exposure and observed that indicators of epithelial cell injury were similar at both time points. These studies indicate that, even in the absence of spirometry changes, symptoms, and now inflammatory cells, it cannot be assumed that the lung is protected from oxidant damage, despite the "adaptation" after repeated exposure. These findings are substantiated by the earlier research in animals that raised the issue of tolerance (a form of adaptation) to O_3. Using a unilateral lung exposure technique, Gardner et al. (32) found that an initial acute exposure of rabbits to O_3 produced a tolerance against the development of pulmonary edema after repeated exposures. However, there was no protection against the cytotoxic effects of O_3 (i.e., effects on numbers, viability and enzymatic activity of alveolar macrophages). Inflammation, as measured by the presence of neutrophils, was increased twofold in the "tolerant" lung.

Numerous toxicology studies in several animal species, including nonhuman primates, have shown that long-term exposure to O_3 results in several characteristic changes in the centriacinar region (where the conducting airways and gas exchange region join) (5). Monkeys and rats exposed to levels as low as 0.15 ppm (8 hr/day, 6 to 90 days) experience distal airway remodeling, in which Type 1 cells are replaced by Type 2 cells, bronchiolization (airway cells extend into the alveoli) occurs, and the interstitium becomes thicker (because of the accumulation of collagen fibers and amorphous material). In both monkey and rat studies, the increased collagen persisted even after exposure ceased. Tyler et al. (33) observed that monkeys receiving a "seasonal" exposure (1 mo of ozone, 0.25 ppm, 8 hr/day, 1 mo of clean air, cycled over 18 mo) had increased lung collagen content, increased chest wall compliance, and increased inspiratory capacity, compared to monkeys receiving 18 mo of O_3 exposure. Both groups had respiratory bronchiolitis and other changes in common. This and a few other studies of similar design show that the exposure pattern to O_3 can have a profound influence on the health outcome (5).

Because of the wide array of effects demonstrated in animals, there always has been an interest in animal-to-human extrapolation. The qualitative extrapolation generally is accepted, but the quantitative extrapolation is still a major topic of investigation. Because the structural changes in animals exposed for months to years can be interpreted as being consistent with incipient peribronchiolar fibrogenesis within the lung interstitium, a quantitative extrapolation of the related structural changes was performed (5). The process entailed using an extrapolation model with numerous assumptions to calculate the delivered dose of O_3 to the proximal alveolar region of a hypothetical child and outdoor worker exposed to ozone over an O_3 season (April through October 1991) in New York City. These calculations then were compared to the calculated O_3 doses that caused increased total interstitial and acellular thickness in the proximal alveolar regions in two rat and two monkey studies. It was concluded that the animal data provide reasons to be concerned about the potential for chronic effects in humans that may be irreversible. Epidemiological studies have not added significantly to the understanding of the chronic effects of O_3. However, Künzli et al. (34) recently reported the results of a pilot study that indicated that, in freshmen from Southern California entering the University of California (Berkeley), a high lifetime exposure to O_3 (presumed from their living in a high-O_3 region) was associated with poorer performance on spirometric tests of lung function. Although the possibility that chronic ambient O_3 exposure results in chronic effects on the lung in humans has not been demonstrated clearly, the above study provides a provocative suggestion of such an effect in Southern California children.

3.2 Nitrogen Dioxide

The scientific literature concerning NO_x has been reviewed comprehensively (35, 36). Key effects of concern are respiratory illness in children, lung function effects in asthmatics, and the potential for alterations in lung host defenses and structure.

Because homes with unvented gas combustion devices (principally gas stoves) have elevated levels of NO_2, they have been the subject of many epidemiological studies (35, 36). A meta-analysis by Hasselblad et al. (37) of nine studies of homes with and without gas stoves showed that there was an estimated increased risk of approximately 20% for respiratory symptoms and disease in 5- to 12-year-old children for each increase of 0.015 ppm NO_2 (combined odds ratio of about 1.2, with confidence intervals ranging from about 1.1 to 1.3). The study by Neas et al. (38), which was included in the meta-analysis mentioned above, was one of the most definitive. Two 2-week measurements of NO_2 in the home of each child (1,286 children, 7 to 11 years of age) were made, one in the heating season and another in the cooling season. There was a linear concentration–response relationship between NO_2 and lower respiratory symptoms in the children. The term *lower respiratory symptoms* indicates the presence of any one or more of several symptoms, including shortness of breath, persistent wheeze, chronic cough, chronic phlegm, or bronchitis. Of the concentrations composing the analysis, only the range of 0.02 to 0.0782 ppm (0.031 ppm mean) was significantly associated with symptoms (odds ratio 1.65, 95% confidence interval 1.03 to 2.63). There are many uncertainties in all the analyses, including exposure measurement errors, which, among other things, make it difficult to ascertain the exposure pattern associated with the effects. Even so, the general interpretation is that the changes are associated with long-term exposures, rather than with an acute exposure. A meta-analysis of seven studies of children less than two years old living in homes with and without gas stoves found no statistically significant increase in respiratory disease (35).

Although there are many "outdoor" epidemiological studies of NO_2, they are difficult to interpret (35, 36). Those few community studies with quantitative estimates suggest the potential for associations between respiratory endpoints in children and NO_2 exposure, but effects likely are confounded by the presence of other outdoor air pollutants. The time series analysis of air pollution effects from the European multicity epidemiology project found mixed results with NO_2, with no clear indication of significant effects of NO_2 (39).

Even at the higher concentrations that have been observed infrequently over short averaging times in ambient air (around 0.50 ppm), NO_2 appears to have minimal acute effects on lung mechanics, symptoms, and response to bronchoconstrictors in healthy young adults (35). For example, concentrations of 1.5 ppm or higher are required to cause functional effects in such populations (36). However, in asthmatics exposed for less than a few hours to low concentrations of NO_2 (as low as 0.3 ppm), acute responses, such as small decreases in spirometry, increased respiratory symptoms, and increased airway resistance, have been seen (35, 36, 40). Nevertheless, many studies (including some at concentrations as high as 4 ppm) have reported no changes in these measures (35, 36, 41–43). A meta-analysis found no dose response (35).

The most consistently observed finding with regard to NO_2 exposure in asthmatics is an increase in airway reactivity to bronchoconstrictors (35, 36). However, a concentration–response relationship was not observed at the lower end of the concentrations used (about 0.2 to 0.6 ppm) (35). This response appears to be most pronounced several hours after exposure (4 to 24 hr) to NO_2 ranging from 0.26 to 0.40 ppm (44–46). A number of studies have investigated responses of asthmatics to allergen challenge following a period of NO_2 exposure. Collectively, these studies suggest that NO_2 causes an increased early-phase response to inhaled antigen. With prolonged exposure to higher concentrations (6 h at 0.4 ppm), responses can be increased significantly immediately after exposure (42, 46). However, after brief exposures (30 min), the response appears to peak after some 4 to 6 hr (41, 44, 45, 47) and may persist for at least 24 hr (46). In addition to the early phase response, NO_2 exposure also exacerbates the late phase response to inhaled antigen (41, 45, 47). Nitrogen dioxide exposure after immunization with dust mite antigen caused an increase of antigen-specific serum IgE, as well as increased inflammatory cells in the lungs of animals (48).

An inflammatory response was seen in healthy subjects about 6 hr after a 4-hr exposure to NO_2 (2 ppm) (49). At lower NO_2 concentrations, no inflammatory response was seen in sputum samples 2 hr after the exposure in healthy, asthmatic, or chronic obstructive pulmonary disease patients (40, 43).

Research with laboratory animals is consistent with, although not directly related to, the human clinical and epidemiologic studies. Animal studies in several species have shown that the portion of the lung that is most susceptible to damage from NO_2 is the centriacinar region (35, 36). Epithelial cells (both Type 1 and ciliated bronchiolar epithelium) most often are damaged and subsequently replaced by more resistant cells (e.g., Type 2 epithelium, nonciliated cells). Longer exposures typically cause effects at lower concentrations. Prolonged exposures to very high concentrations of NO_2 can lead to emphysema-like changes in several animal species. Nitrogen dioxide increases mortality from pulmonary bacterial infections in mice and other animal species (35, 50). Extensive studies in mice have demonstrated that the concentration of NO_2 appears to be a more important factor than does duration of exposure in reducing lung host defenses. The few human clinical studies of host defenses are not clear, but they do indicate that host defense effects are possible (35).

3.3 Sulfur Dioxide

Sulfur dioxide is of concern because of its interactions with PM, its associations with morbidity and mortality, and the ability of short exposures (e.g., $<$10 min) to affect asthmatics (51, 52). Association of SO_2 with mortality and various morbidity end points have been observed in epidemiologic studies (52). Increased mortality related to both respiratory and cardiovascular causes has been associated with SO_2 (53, 54). Sulfur dioxide levels also are associated with hospital admissions and emergency room visits for respiratory causes, including asthma (55). Nevertheless, there also have been studies that show little or no association between SO_2 and mortality (56). Also, high levels of SO_2 are seldom present without high levels of PM. In U.S. studies, at least, associations of mortality and morbidity with short-term PM levels generally have been stronger than associations with short-term SO_2 levels. The largest body of dose–response data for humans exposed to SO_2 is for asthmatics. Exposures were for brief periods, typically while exercising. It has been estimated that people with asthma are approximately 10 times more sensitive to SO_2 than are typical healthy individuals (51, 52). The sensitivity of asthmatics does not necessarily relate to the severity of their disease (52). Although the prevalence of asthma in African Americans is higher than that for Americans of European descent, there does not appear to be a difference in their response to SO_2 (57).

In asthmatics exposed to SO_2 for brief periods (less than 10 min), the predominant response is bronchoconstriction. This physiologic change is associated with the symptoms of wheezing, chest tightness, and shortness of breath. These responses are typically self-limiting and may resolve spontaneously in 30 to 60 min. However, in many cases, a bronchodilator may be necessary to relieve symptoms. The direct smooth-muscle-relaxant effect of a beta-2 sympathomimetic agonist generally will reverse the bronchoconstriction caused by SO_2, and a long-acting beta-2 agonist can provide protection from SO_2-induced bronchoconstriction for at least 12 hr in asthmatics (58). Sulfur dioxide-induced bronchoconstriction can be inhibited (in a dose-dependant manner) by prophylactic use of cromolyn sodium or nedocromil sodium (59), suggesting that mediator release from mast cells may play a role in the response. Muscarinic receptor antagonists (atropine and ipratropium bromide), used in combination with cromolyn or nedocromil sodium, are more effective than either agent alone in reducing SO_2-induced bronchoconstriction. Responses to SO_2 also can be attenuated by morphine and indomethacin (60). Lazarus et al. (61) demonstrated that the specific leukotriene receptor antagonist zafirlukast inhibited SO_2-induced bronchoconstriction. This suggests that leukotrienes are involved in the response to SO_2.

Responses to SO_2 in asthmatics are exacerbated by exercise because of both the exercise, per se, which often causes an increase in airway resistance, and the increased delivery of SO_2 to the lower respiratory tract. Gong et al. (62) noted that SO_2 concentrations greater than 0.5 ppm for 10 min. had more impact than moderate exercise alone without SO_2. The pragmatic purpose of their study was to determine whether SO_2-induced responses had comparable or greater effects on asthmatics than did the impact of "everyday stresses."

Increased ventilation, and hence increased inspiratory flow, reduces the residence time of SO_2 in the upper airway, which results in reduced absorption of this highly soluble gas by airway surface fluids. Reduced absorption, in combination with increased overall ventilation, results in considerable increase in delivery to the intrathoracic airways. Responses to SO_2 also are exacerbated by-breathing air with a low absolute humidity (either dry or cold), presumably because of decreased airway surface fluid leading to decreased absorption of SO_2. Sulfur dioxide generally has not been found in controlled exposure studies to cause inflammation or to increase airway responsiveness to nonspecific bronchoconstricting agents. However, Rusznak et al. (46) demonstrated an increased response to inhaled allergen after exposure to a combination of NO_2 (0.4 ppm) and SO_2 (0.2 ppm), both immediately and 24 and 48 hr after exposure. Previous work had shown an increased allergen response immediately after exposure to an NO_2/SO_2 combination that was not significant for the individual pollutants (42). An increase in bronchial responsiveness was found for 24-hr mean increases in levels of ambient SO_2 (in combination with NO_2 and Black Smoke) in a panel of 38 asthmatics. The concentrations were at or below the WHO AQG for these pollutants (63).

3.4 Particulate Matter

Historically, PM has been recognized as posing serious public health threats, with acute exposures to high levels of ambient PM during widely publicized air pollution episodes of the 1930s (Meuse Valley, Belgium), 1940s (Donora, PA), and 1950s (London, UK) being associated with dramatic increases in mortality and morbidity among the general population. Such episodes, and others involving photochemical smog components discussed elsewhere in this chapter, stimulated environmental research on the physical and chemical properties of airborne particles, their distribution in the atmosphere, resulting human exposures, and associated human health effects. Many of the more salient findings and conclusions derived from U.S. assessments of this research (64) are summarized here, with emphasis on those arising from the most recent evaluations.

Atmospheric PM is comprised of a complex mixture of organic and inorganic components, typically distributed by mass and composition into two main modes, as shown in Figure 58.4. Fine-mode particles ranging in size up to 1 to 2.5 μm (aerodynamic diameter) are derived mainly from high temperature smelting or combustion processes, recondensed organic and metal vapors, and secondarily formed aerosols (e.g., sulfates, nitrates), resulting from gas to particle conversion. Larger, coarse-mode particles, ranging in size down to 1 to 2.5 μm, primarily contain crustal materials, road dust, and other substances derived from anthropogenic industrial activities that involve crushing or grinding operations, such as mining, cement making, and agricultural activities.

When the NAAQS for total suspended particles were first established in 1971, little explicit distinction was made between atmospheric particles of varying size or chemical composition with regard to attempting to protect against human health effects associated with exposures to ambient (outdoor) PM. In 1987, the original NAAQS were revised to protect against adverse health effects of inhalable airborne particles with an upper 50% cutpoint of 10 μm aerodynamic diameter (PM_{10}), which can be deposited in the lower (thoracic) respiratory tract. The most recent EPA PM criteria document and PM staff paper (64, 65) characterize key elements of the scientific bases for EPA's promulgation of decisions to retain, in modified form, the PM_{10} NAAQS and to add new $PM_{2.5}$ NAAQS (see Table 58.2) to protect against adverse health effects associated with exposures to fine-mode particles.

The scientific bases for the 1997 PM NAAQS decisions include important distinctions made between fine- and coarse-mode ambient air particles with regard to size, chemical composition, sources, and transport (Table 58.3). Besides differences in the major sources and formation mechanisms noted above for fine and coarse particles, other distinctions

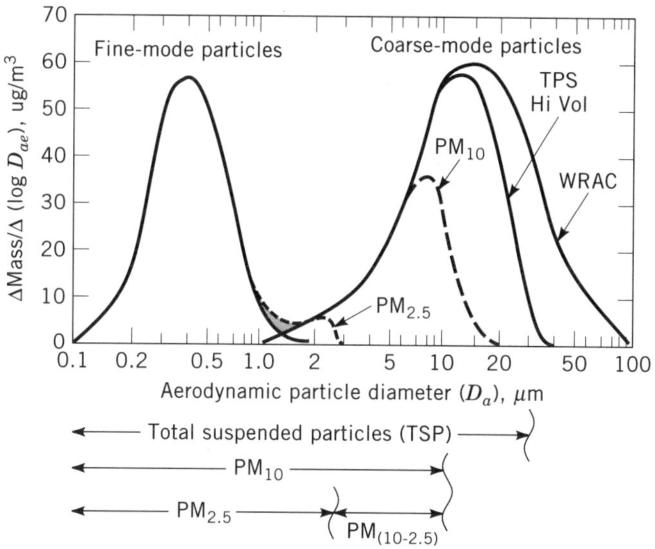

Figure 58.4. An idealized distribution of ambient particulate matter showing fine and coarse mode particles and the fractions collected by size-selective samplers. From Ref. 64.

should be emphasized, including the specific types of substances comprising particles in each mode, the usually greater solubility of fine particles compared with coarse particles, the notably longer atmospheric half-life of fine particles (days to weeks) compared to coarse particles (minutes to hours), and greater transport distances for fine particles (hundreds to thousands of kilometers) compared with coarse particles (less than one to tens of kilometers). The latter two characteristics result in coarse-mode particles being more highly concentrated around originating emission sources in contrast to more uniform patterns of distribution across urban areas (or larger multistate areas of the United States or multinational areas) and long distance (even transatlantic or transpacific) transport of fine particles. Conclusions regarding relative contributions of particles of ambient air origin (as measured at outdoor community monitoring stations) to total human exposures to PM (which also typically include indoor air exposures to particles generated from indoor sources and to particles of ambient origin that infiltrate indoor microenvironments) were discussed in Chapter 7 of U.S. EPA (64) and form another key element considered in setting the $PM_{2.5}$ NAAQS.

The assessment and interpretation of newly emerging findings on airborne particle health effects, especially as derived from newly published community epidemiologic studies, represent the key health effects information underlying derivation of the new $PM_{2.5}$ NAAQS. More than 80 such epidemiology studies, most of which found significant associations between human health effects and PM levels at or below the 1987 PM_{10} NAAQS (150 $\mu g/m^3$, 24 hr; 50 $\mu g/m^3$, annual), were assessed by U.S. EPA (64, 65). Health effects of concern shown by community epidemiology studies to be statistically associated with

Table 58.3. Comparison of Ambient Fine- and Coarse-Mode Particles[a]

	Fine	Coarse
Formed from	Gases	Large solids and droplets
Formed by	Chemical reaction Nucleation Condensation Coagulation Evaporation of fog and cloud droplets in which gases have dissolved and reacted	Mechanical disruption (crushing, grinding, and abrasion of surfaces, etc.) Evaporation of sprays Suspension of dusts
Composed of	Sulfate (SO_4^-) Nitrate (NO_3^-) Ammonium (NH_4^+) Hydrogen ion (H^+) Elemental carbon Organic compounds (e.g., PAHs, PNAs) Metals, (e.g., Pb, Cd, V, Ni, Cu, Zn, Mn, Fe) Particle-bound water	Resuspended dusts (soil dust and street dust) Coal and oil fly ash Oxides of crustal elements, (Si, Al, Ti, Fe) $CaCO_3$, NaCl, and sea salt Pollen, mold, and fungal spores Plant and animal fragments Tire wear debris
Solubility	Largely soluble, hygroscopic, and deliquescent	Largely insoluble and nonhygroscopic
Sources	Combustion of coal, oil, gasoline, diesel, and wood Atmospheric transformation products of NO_x, SO_2, and organic compounds including biogenic organic species (e.g., terpenes) High-temperature processes, smelters, steel mills, etc.	Resuspension of industrial dust and soil tracked onto roads and streets Suspension from disturbed soil (e.g., farming, mining, unpaved roads) Biological sources Construction and demolition, coal and oil combustion, and ocean spray
Atmospheric half-life	Days to weeks	Minutes to hours
Travel distance:	100s to 1,000s of km	<1 to 10s of km

[a]Adapted from Refs 64 and 66.

ambient PM exposures include increased mortality [especially for the elderly (over 65 years of age) and those with preexisting cardiopulmonary conditions] and morbidity (indexed by increased hospital admissions, respiratory symptom rates, and decrements in lung function).

Much controversy continues to surround assessment of the methodological soundness and interpretation of the community epidemiology studies. As noted in U.S. EPA (64), several viewpoints exist on how best to interpret the epidemiology data. One sees PM exposure indicators as surrogate measures of complex ambient air pollution mixtures, and reported PM-related effects represent those of the overall mixture. Another holds that reported PM-related effects are attributable to PM components (*per se*) of the air pollution mixture and reflect independent PM effects. Still another view is PM as both a surrogate indicator and a specific cause of health effects.

The consistency and coherence of the epidemiologic evidence, as discussed in U.S. EPA (64, 65), support the conclusion that exposure to ambient PM, acting alone or in combination with other pollutants, is likely a key contributor to the increased mortality and morbidity risks observed in the epidemiology studies. The EPA relied mainly on relative risks reported for increased mortality or morbidity, based on U.S. and Canadian PM studies, as providing the most directly pertinent quantitative risk estimates as inputs to PM NAAQS decisions. These included relative risk estimates for mortality or morbidity (indexed by increased hospital admissions, respiratory symptoms, etc.) associated with the 50 $\mu g/m^3$ increases in 24-hr PM_{10} concentrations shown in Table 58.4 or with the variable increases in fine particle indicators (e.g., 25-$\mu g/m^3$ increment in 24 hr $PM_{2.5}$ concentrations) shown in Table 58.5 and analogous relative risk estimates for health effects related to specified increments in annual mean or median levels of fine particle indicators (Table 58.6). The study results summarized in these tables were judged, overall, to warrant the conclusion that significant associations of increased mortality and morbidity risks likely are attributable to fine particles, as indexed by various particle indicators (e.g., $PM_{2.5}$, sulfates). However, it also was concluded that possible toxic effects of the coarse fraction of PM_{10} cannot be ruled out. The PM_{10} coarse fraction particles do reach sensitive areas of the lower respiratory tract, and health effects of concern (e.g., aggravation of asthma and increased respiratory illness), particularly in children, have been suggested by some epidemiology studies cited in U.S. EPA (64) but not substantiated in others (e.g., Refs. 100, 101).

The recently completed update of the WHO AQG for Europe (3) arrived at much the same conclusions regarding PM epidemiologic evidence as noted above. The WHO assessment included discussion of newly published results from the so-called APHEA project, a European multicity epidemiology project (39). The APHEA studies were noted by WHO (3) as showing significant associations between increased mortality or morbidity and various air pollution components (including PM) in several, but not all, of the European cities studied. The WHO (3) assessment includes several tables summarizing relative risk estimates for various types of health effects (e.g., increased mortality, respiratory hospital admissions, respiratory symptoms, medication usage) associated with increments (e.g., 10 $\mu g/m^3$) of ambient PM_{10} or $PM_{2.5}$. Also presented are estimated numbers of persons (in a city of 1 million) projected as likely to experience one or another of such health effects at 50 or 100 $\mu g/m^3$ of PM_{10} or $PM_{2.5}$.

U.S. EPA (64, 65) noted the very limited extent of available toxicologic findings by which to identify key PM constituents of urban ambient air mixes that may be causally related to the types of mortality and morbidity effects observed in the community epidemiologic studies or to delineate plausible biological mechanisms by which such effects could be induced at the relatively low ambient PM concentrations evaluated in the epidemiologic studies. Several types of mechanisms have been shown, as discussed in U.S. EPA (64) to underlie toxic effects observed with acute or chronic exposures to various PM species or mixtures (e.g., acute lung inflammation, particle accumulation and overload, impaired respiratory function, impaired pulmonary defense mechanisms), but generally at

Table 58.4. Effect Estimates per 50 μg/m^3 Increase in 24-hour PM_{10} Concentrations from U.S. and Canadian Studies

Study Location	Relative Risk (±CI) Only PM in Model	Relative Risk (±CI) Other Pollutants in Model	Reported PM_{10} μg/m^3 Mean (Min/Max)[a]
	Increased Total Acute Mortality		
Six Cities[b]			
Portage, WI	1.04 (0.98, 1.09)	—	18 (±11.7)
Boston, MA	1.06 (1.04, 1.09)	—	24 (±12.8)
Topeka, KS	0.98 (0.90, 1.05)	—	27 (±16.1)
St. Louis, MO	1.03 (1.00, 1.05)	—	31 (±16.2)
Kingston/Knoxville, TN	1.05 (1.00, 1.09)	—	32 (±14.5)
Steubenville, OH	1.05 (1.00, 1.08)	—	46 (±32.3)
St. Louis, MO[c]	1.08 (1.01, 1.12)	1.06 (0.98, 1.15)	28 (1/97)
Kingston, TN[c]	1.09 (0.94, 1.25)	1.09 (0.94, 1.26)	30 (4/67)
Chicago, IL[d]	1.04 (1.00, 1.08)	—	37 (4/365)
Chicago, IL[e]	1.03 (1.02, 1.04)	1.02 (1.01, 1.04)	38 (NR/128)
Utah Valley, UT[f]	1.08 (1.05, 1.11)	1.19 (0.96, 1.47)	47 (11/297)
Birmingham, AL[g]	1.05 (1.01, 1.10)	—	48 (21, 80)
Los Angeles, CA[h]	1.03 (1.00, 1.055)	1.02 (0.99, 1.036)	58 (15/177)
	Increased Hospital Admissions (for elderly, >65 years of age)		
Respiratory Disease			
Toronto, Canada[i]	1.23 (1.02, 1.43)[j]	1.12 (0.88, 1.36)[j]	30–39[k]
Tacoma, WA[l]	1.10 (1.03, 1.17)	1.11 (1.02, 1.20)	37 (14, 67)
New Haven, CT[l]	1.06 (1.00, 1.13)	1.07 (1.01, 1.14)	41 (19, 67)
Cleveland, OH[m]	1.06 (1.00, 1.11)	—	43 (19, 72)
Spokane, WA[n]	1.08 (1.04, 1.14)	—	46 (16, 83)
COPD			
Minneapolis, MN[o]	1.25 (1.10, 1.44)	—	36 (18, 58)
Birmingham, AL[p]	1.13 (1.04, 1.22)	—	45 (19, 77)
Spokane, WA[q]	1.17 (1.08, 1.27)	—	46 (16, 83)
Detroit, MI[q]	1.10 (1.02, 1.17)	—	48 (22, 82)
Pneumonia			
Minneapolis, MN[o]	1.08 (1.01, 1.15)	—	36 (18, 58)
Birmingham, AL[p]	1.09 (1.03, 1.15)	—	45 (19, 77)
Spokane, WA[n]	1.06 (0.98, 1.13)	—	46 (16, 83)
Detroit, MI[q]	—	1.06 (1.02, 1.10)	48 (22, 82)
Ischemic HD			
Detroit, MI[r]	1.02 (1.01, 1.03)	1.02 (1.00, 1.03)	48 (22, 82)

Table 58.4. (continued)

Study Location	Relative Risk (±CI) Only PM in Model	Relative Risk (±CI) Other Pollutants in Model	Reported PM_{10} μg/m^3 Mean (Min/Max)[a]
	Increased Respiratory Symptoms		
Lower Respiratory			
Six Cities[s]	2.03 (1.36, 3.04)	Similar relative risk	30 (13, 53)
Utah Valley, UT[t]	1.28 (1.06, 1.56)[u]	—	46 (11/195)
	1.01 (0.81, 1.27)[v]		
Utah Valley, UT[w]	1.27 (1.08, 1.49)	—	76 (7/251)
Cough			
Denver, CO[x]	1.09 (0.57, 2.10)	—	22 (0.5/73)
Six Cities[s]	1.51 (1.12, 2.05)	Similar relative risk	30 (13, 53)
Utah Valley, UT[x]	1.29 (1.12, 1.48)	—	76 (7/251)
Decrease in Lung Function			
Utah Valley, UT[t]	55 (24, 86)[y]	—	46 (11/195)
Utah Valley, UT[w]	30 (10, 50)[y]	—	76 (7/251)
Utah Valley, UT[z]	29 (7,51)[a]	—	55 (1, 181)

[a]Min/max 24-h PM_{10} in parentheses unless noted otherwise as standard deviation (±S.D.), 10_{th} and 90_{th} percentile (10, 90). NR = not reported.
[b]Ref. 67.
[c]Ref. 68/ozone.
[d]Ref. 69.
[e]Ref. 70/ozone.
[f]Ref. 71, 72/ozone.
[g]Ref. 73.
[h]Ref. 74/ozone, CO.
[i]Ref. 75/ozone.
[j]Relative risk refers to total population, not just those >65 years.
[k]Means of several cities.
[l]Ref. 76.
[m]Ref. 77.
[n]Ref. 78.
[o]Ref. 79.
[p]Ref. 80.
[q]Ref. 81.
[r]Ref. 82/ozone, CO, SO_2.
[s]Ref. 83.
[t]Ref. 84.
[u]Children.
[v]Asthmatic children and adults.
[w]Ref. 85.
[x]Ref. 86.
[y]PEFR (peak expiratory flow rate) decrease in mL/s.
[z]Ref. 87.
[a a]FEV_1 decrease.

Table 58.5. Effect Estimates per Variable Increments in 24-hour Concentrations of Fine Particle Indicators from U.S. and Canadian Studies

	Fine Particle Indicator	Relative Risk (±CI) per 25 g/m^3 PM Increase	Reported PM μg/m^3 Mean (Min/Max)[a]
		Acute Mortality	
Six cities[b]			
Portage, WI	$PM_{2.5}$	1.030 (0.993, 1.071)	11.2 (±7.8)
Topeka, KS	$PM_{2.5}$	1.020 (0.951, 1.092)	12.2 (±7.4)
Boston, MA	$PM_{2.5}$	1.056 (1.038, 1.0711)	15.7 (±9.2)
St. Louis, MO	$PM_{2.5}$	1.028 (1.010, 1.043)	18.7 (±10.5)
Kingston/Knoxville, TN	$PM_{2.5}$	1.035 (1.005, 1.066)	20.8 (±9.6)
Steubenville, OH	$PM_{2.5}$	1.025 (0.998, 1.053)	29.6 (±21.9)
		Increased Hospitalization	
Ontario, Canada[c]	SO	1.03 (1.02, 1.04)	R = 3.1–8.2
Ontario, Canada[d]	SO	1.03 (1.02, 1.04)	R = 2.0–7.7
	Ozone	1.03 (1.02, 1.05)	
NYC/Buffalo, NY[e]	SO	1.05 (1.01, 1.10)	NR
Toronto[e]	H^+ (Nmol/m^3)	1.16 (1.03, 1.30)[f]	28.8 (NR/391)
	SO	1.12 (1.00, 1.24)	7.6 (NR, 48.7)
	$PM_{2.5}$	1.15 (1.02, 1.78)	18.6 (NR, 66.0)

		Increased Respiratory Symptoms	
Southern California[g]	SO	1.48 (1.14, 1.91)	R = 2–37
Six Cities[h]	$PM_{2.5}$	1.19 (1.01, 1.42)[i]	18.0 (7.2, 37)[j]
(Cough)	$PM_{2.5}$ sulfur	1.23 (0.95, 1.59)[i]	2.5 (3.1, 61)[j]
	H^+	1.06 (0.87, 1.29)[i]	18.1 (0.8, 5.9)[j]
Six Cities[k]	$PM_{2.5}$	1.44 (1.15–1.82)[i]	18.0 (7.2, 37)[j]
(Lower respiratory symptoms)	$PM_{2.5}$ sulfur	1.82 (1.28–2.59)[i]	2.5 (0.8, 5.9)[j]
	H^+	1.05 (0.25–1.30)[i]	18.1 (3.1, 61)[j]
		Decreased Lung Function	
Uniontown, PA[k]	$PM_{2.5}$	PEFR 23.1 (0.3, 36.9) (per 25 g/m^3)	25/88 (NR/88)

[a]Min/Max 24-h PM indicator level shown in parentheses unless otherwise noted as (± S.D.), 10_{th} and 90_{th} percentile (10, 90) or R = range of values from min-max, no mean value reported.
[b]Ref. 67.
[c]Ref. 24.
[d]Ref. 88/ozone.
[e]Refs. 26, 76.
[f]Change per 100 nmoles/m^3.
[g]Ref. 89.
[h]Ref. 83.
[i]Change per 20 g/m^3 for $PM_{2.5}$; per 5 g/m^3 for $PM_{2.5}$ sulfur; per 25 nmoles/m^3 for H^+.
[j]50th percentile value (10, 90 percentile).
[k]Ref. 90.

Table 58.6. Effect Estimates per Increments in Annual Mean Levels of Fine Particle Indicators from U.S. and Canadian Studies

Type of Health Effect and Location	Indicator	Change in Health Indicator per Increment in PM[a]	Range of City PM Means (g/m^3)
	Increased Total Chronic Mortality in Adults	Relative Risk (95% CI)	
Six Cities[b]	$PM_{15/10}$	1.42 (1.16–2.01)	18–47
	$PM_{2.5}$	1.31 (1.11–1.68)	11–30
	SO	1.46 (1.16–2.16)	5–13
ACS Study[c]	$PM_{2.5}$	1.17 (1.09–1.26)	9–34
(151 U.S. Standard Metropolitan Statistical Areas)	SO	1.10 (1.06–1.16)	4–24
	Increased Bronchitis in Children	Odds Ratio (95% CI)	
Six Cities[d]	$PM_{15/10}$	3.26 (1.13, 10.28)	20–59
Six Cities[e]	TSP	2.80 (1.17, 7.03)	39–114
24 Cities[f]	H^+	2.65 (1.22, 5.74)	6.2–41.0
24 Cities[f]	SO	3.02 (1.28, 7.03)	18.1–67.3
24 Cities[f]	$PM_{2.1}$	1.97 (0.85, 4.51)	9.1–17.3
24 Cities[f]	PM_{10}	3.29 (0.81, 13.62)	22.0–28.6
Southern California[g]	SO	1.39 (0.99, 1.92)	—
	Decreased Lung Function in Children		
Six Cities[d,g]	$PM_{15/10}$	NS changes[h]	20–59
Six Cities[e]	TSP	NS changes[h]	39–114
24 Cities[i,j]	H^+ (52 nmoles/m^3)	3.45% (4.87, 2.01) FVC	—
24 Cities[i]	$PM_{2.1}$ (15 g/m^3)	3.21% (4.98, 1.41) FVC	—
24 Cities[i]	SO (7 g/m^3)	3.06% (4.50, 1.60) FVC	—
24 Cities[i]	PM_{10} (17 g/m^3)	2.42% (4.30, 0.51) FVC	—

[a]Estimates calculated annual-average PM increments assume a 100-g/m^3 increase for TSP; a 50-g/m^3 increase for PM_{10} and PM_{15}; a 25-g/m^3 increase for $PM_{2.5}$; a 15-g/m^3 for SO, except where noted otherwise; and a 100-nmole/m increase for H^+.
[b]Ref. 91.
[c]Ref. 92.
[d]Ref. 93.
[e]Ref. 94.
[f]Ref. 95.
[g]Refs. 96–98.
[h]NS = not significant.
[i]Ref. 99.
[j]Pollutant data same as in Ref. 95.

much higher PM concentrations (often $\geq$ 0.5 to 1.0 mg/m^3) than now seen in the United States. Currently, much research attention is being focused on several fine-particle constituents hypothesized as likely being important contributors to ambient PM effects, including acid aerosols (indexed by sulfates; hydrogen ions, etc.), transition metals (e.g., iron, manganese), and ultrafine particles (based mainly on the possibility that absolute number or larger total surface area of extremely fine particles, rather than particle mass alone, may be crucial in causing some PM-related toxic effects).

Numerous studies are underway with humans and laboratory animals that seek to elucidate the mechanisms of action of PM. For brevity, only a few examples are provided here. For example, a series of intratracheal instillation studies discussed by Costa and Dreher (102) implicate bioavailable transition medals in PM as mediating cardiopulmonary injury in healthy and compromised animal models (pulmonary vasculitis and hypertension). Osier and Oberdörster (103), in comparing responses of rats exposed by intratracheal inhalation to "fine" ($\approx$200 μm) and "ultrafine" ($\approx$21 μm) titanium oxide particles to exposures to similar doses by intratracheal instillation, found differences in pulmonary responses to an inhaled dose compared with instilled doses, possibly resulting from differences in dose rate, in regional deposition of particles in the respiratory tract, or altered clearance between the two methods. These latter findings argue for caution in trying to use instillation results to extrapolate quantitatively to humans. Other studies using exposures of compromised animal models to concentrated particles from ambient air PM samples (e.g., Ref. 104) hold much promise for helping to identify specific PM components that may be causally related to increased mortality and morbidity effects observed in epidemiologic studies. Lastly, other studies (e.g., Refs. 105 and 106) evaluating potential differences in respiratory tract regional deposition, clearance, or retention of particles of varying size and composition in men, women, children, and individuals with cardiopulmonary diseases can be expected to aid in the identification of susceptible human population groups and the factors that put them at increased risk for experiencing adverse health effects caused by exposures to ambient PM.

3.5 Carbon Monoxide

The health effects of CO have been reviewed extensively (107–111). The public health significance of CO in the air results largely from CO being readily absorbed from the lungs into the bloodstream, there forming a slowly reversible complex with hemoglobin (Hb) known as carboxyhemoglobin (COHb). The presence of significant COHb saturation in the blood causes hypoxia (i.e., reduced availability of oxygen to body tissues). The blood COHb level, therefore, represents a useful physiological marker to predict the potential health effects of CO exposure. The amount of COHb formed is dependent on the CO concentration and duration of CO exposure, exercise (which increases the amount of air inhaled and exhaled per unit time and the pulmonary diffusing capacity for CO), ambient pressure, health status, and the specific metabolism of the exposed individual. The formation of COHb is a reversible process, but, because of tight binding of CO to Hb, the elimination half-time is quite long, varying from 2 to 6.5 hr depending on the initial COHb levels. This may lead to accumulation of COHb, and even relatively low inhaled concentrations of CO could produce substantial blood levels of COHb. Because COHb measurements are not readily available in the exposed population, mathematical models, based on the Coburn-Forster-Kane equation, have been developed to predict COHb levels from known CO exposures under a variety of circumstances (107, 108).

Evaluation of human CO exposure situations indicates that occupational exposures in some workplaces and exposures in homes with faulty or unvented combustion appliances can exceed 100 ppm CO, often leading to COHb levels of 4 to 5% with 1-hr exposure and 10% or more with continued exposure for ≥8 hr (see Table 58.7). Such high exposure levels are encountered much less frequently by the general public exposed under ambient conditions. More frequently, exposures to less than 25 to 50 ppm CO for an extended period of time occur among the general population, and, at the low exercise levels usually engaged in under such circumstances, resulting COHb levels typically remain below 2 to 3% among nonsmokers. Those levels can be compared to the physiological baseline for nonsmokers, which is estimated to be in the range of 0.3 to 0.7% COHb. Baseline COHb saturations in smokers, however, average 4% with a usual range of 3 to 8% for one- to two-packs-per-day smokers, reflecting absorption of CO from inhaled smoke. Levels as high as 15% COHb have been reported for chain smokers. As a result of higher baseline COHb levels, smokers actually may be exhaling more CO into the air than they are inhaling from the ambient environment.

The key human health effects most clearly demonstrated to be associated with exposure to CO are summarized in Table 58.8. The lethality of CO that results from exposure to very high concentrations is well known. The effects of exposure to low concentrations, such as the levels found in ambient air, are far more subtle and considerably less threatening than those occurring in frank poisoning from high CO levels. Because the COHb level of the blood is the best indicator of potential health risk, symptoms of exposures to excessive ambient air levels of CO are described here in terms of associated COHb levels.

Maximal exercise duration and performance in healthy individuals have been shown to be reduced at COHb levels of ≥2.3% and ≥4.3%, respectively. The decrements in performance at these levels are small and likely to affect only competing athletes rather than people engaged in everyday activities. In fact, no effects were observed during submaximal exercise in healthy individuals at COHb levels as high as 15 to 20%.

Adverse effects have been observed in individuals with coronary artery disease (CAD) at 3 to 6% COHb. At these levels, individuals with reproducible exercise-induced angina

Table 58.7. Predicted Carbon Monoxide Exposures in the Population

Exposure Conditions	Predicted COHb Response[a,b] 1 hr Light Exercise	8 hr Light Exercise
Nonsmoking adults exposed to 25- to 50-ppm peak CO levels	2 to 3%	4 to 7%
Workplace or home with faulty combustion appliances at ≈100 ppm	4 to 5%	12 to 13%

[a]From the Coburn-Forster-Kane equation (107, 108, 112).
[b]Light exercise at 20 L/min.

Table 58.8. Key Health Effects of Exposure to Carbon Monoxide[a]

Target Organ	Health Effect(s)[b]	Tested Populations[c]	References
Lungs	Reduced maximal exercise duration with 1-hr peak CO exposures resulting in ≥2.3% COHb	Healthy individuals	113–115
Heart	Reduced time to ST segment change of the electrocardiogram (earlier onset of myocardial ischemia), with peak CO exposures resulting in ≥2.4% COHb	Individuals with coronary artery disease	116
Heart	Reduced exercise duration because of increased chest pain (angina), with peak CO exposures resulting in ≥3% COHb	Individuals with coronary artery disease	115–120
Heart	Increased number and complexity of arrhythmias (abnormal heart rhythm), with peak CO exposures resulting in ≥6% COHb (CO-Oximeter)	Individuals with coronary artery disease and high baseline ectopy (chronic arrhythmia)	121
Heart	Increased hospital admissions associated with ambient pollutant exposures	Individuals >65 years old with cardiovascular disease	82, 122–124
Brain	Central nervous system effects, such as decrements in hand-eye coordination (driving or tracking) and in attention or vigilance (detection of infrequent events), with 1-hr peak CO exposures (≈5 to 20% COHb)	Healthy individuals	125–129
	Neurological symptoms can occur, ranging from headache, dizziness, weakness, nausea, confusion, disorientation, and visual disturbances to unconsciousness and death, with continued exposure to high levels in the workplace or in homes with faulty combustion appliances (≥20% COHb)	Healthy individuals	130

[a]Modified from Refs. 4, 107, 108.

[b]U.S. EPA has set significant harm levels of 50 ppm (8-hr average), 75 ppm (4-hr average), and 125 ppm (1-hr average). Exposure under these conditions could result in COHb levels of 5 to 10% and cause significant health effects in sensitive individuals. Measured blood COHb level after CO exposure.

[c]Fetuses, infants, pregnant women, elderly people, and people with anemia or with a history of cardiac or respiratory disease may be particularly sensitive to CO.

(chest pain) are likely to experience a reduced capacity to exercise because of decreased time to onset of angina. The indicators of myocardial ischemia during exercise, which is detectable by electrocardiographic changes (ST depression) and associated angina, were statistically significant in one study at ≥2.4% COHb and showed a dose–response relationship with increasing COHb. An increase in the number and complexity of exercise-related arrhythmias also has been observed at ≥6% COHb (121) in some people with CAD and high levels of baseline ectopy (chronic arrhythmia), but not at lower levels in three other studies (4, 108). The results from the above controlled studies, combined with epidemiologic evidence of CO-related morbidity (122–124, 131) and mortality (e.g., Refs. 132, 133), and the morbidity and mortality studies of workers who are exposed routinely to combustion products (e.g., Refs. 134–136) suggest that CO exposure may provide an increased risk of hospitalization or death in patients with more severe heart disease.

Central nervous system effects, including reductions in hand-eye coordination (driving or tracking) and in attention or vigilance, have been reported at peak COHb levels of 5% and higher, but later work indicates that significant behavioral impairments in healthy individuals should not be expected until COHb levels exceed 20% (108, 137, 138). It must be emphasized, however, that even a 5% COHb level is associated with 1-h CO concentrations of 100 ppm or higher. Thus, at typical ambient air levels of CO, no observable central nervous system effects would be expected to occur in the healthy population.

Persons with a history of cardiovascular disease are believed to be particularly sensitive to ambient levels of CO. Other people who may be sensitive, because of the effects of preferential binding of CO to Hb, include fetuses; infants; pregnant women; the elderly; and patients with congestive heart failure, peripheral vascular or cerebrovascular disease, hematological diseases such as anemia, and obstructive lung diseases. Some people also may be at increased risk from exposure to CO because of medicinal or recreational drug use that affects the brain or cerebrovasculature, exposure to other pollutants (e.g., solvents) that increase endogenous production of CO, or initial exposure to high altitude and CO. Unfortunately, little empirical evidence is available by which to specify health effects associated with ambient or near-ambient CO exposures in these probable risk groups.

The current NAAQS for CO are intended to keep COHb levels below 2.1% to protect the most sensitive members of the general population (i.e., individuals with CAD). Individuals in motor vehicles are at the greatest risk from ambient exposure, followed by pedestrians, bicyclists, and joggers in the proximity of roadways and the rest of the general urban population exposed to vehicle exhaust. Several hours of exposure to peak ambient CO concentrations found occasionally at downtown urban sites during periods of heavy traffic would be required to produce COHb levels of concern in the most sensitive nonsmokers. Carbon monoxide levels occurring outside the downtown urban locations are expected to be lower and are probably representative of levels found in residential areas where most people live. Significant health effects from ambient CO exposure are not likely under these latter exposure conditions.

3.6 Lead

The health effects of lead have been the subject of extensive scientific investigation and intensive public health debate. Several health risk assessments of lead have been prepared in recent years as new health effects information became available and pointed to adverse health effects related to progressively lower levels of lead exposure. Perhaps the most comprehensive of these assessments was prepared by the U.S. EPA (139), with a subsequent update focusing on the hypertensive and developmental neurotoxic effects of lead (140). Other agencies also have examined the health effects of lead (e.g., Ref. 141), in some cases focusing especially on the effects of lead on children (e.g., Refs. 142, 143). As the database on lead health effects has grown in recent years, so has the consensus of scientific and public health opinion on the existence of low-level lead effects. Consequently, efforts largely have shifted from further documentation of the effects of lead to evaluating and reducing lead exposure (144).

Although the database on lead health effects is massive and pertains to essentially every body system, health risk assessments of lead have tended to center on developmental neurotoxicity in children because of evidence indicating that such effects have among the lowest if not the lowest exposure-effect levels of any health effect of lead and because of the *prima facie* adverse nature of perturbations in the neurobehavioral development of children. Lead long has been recognized as toxic to the developing fetus and as a particularly potent neurotoxicant for children. However, establishing the lowest levels of exposure at which lead adversely affects children has occupied several decades of research. Numerous epidemiological studies (reviewed by U.S. EPA (139), and Grant and Davis, (145) were devoted to examining the relationship between measures of lead exposure (usually represented by blood lead level) and neurobehavioral development (especially but not exclusively quantified as intelligence quotient). Collectively, these studies suggested that lead could affect neurobehavioral performance at low levels of exposure [i.e., below an average blood lead level of 30 μg/dL and possibly down to 15 to 20 μg/dL (146)]. However, the studies varied considerably in design and quality, and it was difficult to draw definitive conclusions from the results of many of these studies.

It was not until the advent of a group of independent prospective epidemiological studies that a firmer basis existed by which to identify a blood lead level of concern for adverse effects on neurobehavioral development in children (146). These studies offered significant advantages over cross-sectional epidemiological studies in general. A particular strength of the prospective studies was a more precise characterization of the history of lead exposure during early development, owing to repeated blood lead measurements both postnatally and, in most studies, prenatally via the mother's blood or at least the placental cord blood. Another advantage of the prospective studies as a group was that, although independent in design and conduct, they tended to use the same evaluation instruments, such as the Bayley Scales of Infant Development, which permitted more direct comparisons of the results among studies. Yet another strength of the prospective lead studies, albeit not necessarily unique to prospective studies, was the effort to take into consideration numerous covariates and confounders that had plagued earlier cross-sectional studies.

Key findings from prospective studies of lead-exposed children are summarized in Table 58.9. Although many of these results did not achieve statistical significance and a few are not even in the expected direction, the overall pattern of findings points to an effect of lead on neurobehavioral development in children at blood lead levels around 10 to 15 μg/dL. In addition, these prospective study findings are supported not only by evidence from several of the cross-sectional studies of lead-exposed children noted above, but by a large

Table 58.9. Summary of Selected Cognitive Effects from Prospective Studies of Early Developmental Lead Exposure[a]

	Blood Lead Measure					
Study[b]	Mean (range/± S.D.; µg/dL)	Time/Source	N	Endpoint	Effect	P Value
Boston	7.4 (±5.5)	Delivery/cord	201	6-mo MDI	−4.3 pts. at PbB ≥ 10 µg/dL	0.095
Bellinger et al. (149)	7.4 (+5.5)	Delivery/cord	199	12-mo MDI	−5.8 pts. at PbB ≥ 10 µg/dL	0.020
	7.4 (±5.5)	Delivery/cord	187	18-mo MDI	−6.7 pts. at PbB ≥ 10 µg/dL	0.049
	7.4 (±5.5)	Delivery/cord	182	24-mo MDI	−7.8 pts. at PbB ≥ 10 µg/dL	0.006
	6.8 (±6.3)	Delivery/cord	170	57-mo GCI	−6.9 pts. at PbB ≥ 10 µg/dL	0.040
	6.5 (±4.9)	24-mo/child	148	10-year WISC-R	−5.8 pts per 10 µg/dL PbB	0.007
		24-mo/child	148	(FS)	−8.9 pts per 10 µg/dL PbB	0.0003
				10-year K-TEA		
Cincinnati	6.3 (±4.5)	Delivery/cord	96	3-mo MDI	−6.0 pts per 10 µg/dL PbB	0.02
Dietrich et al. (150)	8.0 (±3.7)	Prenatal/mother	266	3-mo MDI	−3.4 pts per 10 µg/dL PbB	0.05
	8.0 (±3.7)	Prenatal/mother	249	6-mo MDI	−8.4 pts per 10 µg/dL PbB (males)	≤0.01
	4.6 (±2.8)	10-day/child	283	6-mo MDI	−7.3 pts per 10 µg/dL PbB (lower SES)	≤0.03
	4.6 (±2.8)	10-day/child	257	12-mo MDI	−6.2 pts per 10 µg/dL PbB	0.04
		Prenatal/mother	237	24-mo MDI	+5.1 pts per 10 µg/dL PbB	0.022
		10-day/child	270	2.4-mo MDI	−0.2 pts per 10 µg/dL PbB	0.948
		10-day/child	247	4-year K-ABC	−6.3 pts per 10 µg/dL PbB	≤0.01
		10-day/child	258	(MPC)	−5.0 pts per 10 µg/dL PbB	≤0.05
	14.1 (7.3)	4-year/child	259	4-year K-ABC (SIM)	−2.0 pts per 10 µg/dL PbB	≤0.05
	<12 (<12)	6-year/child	231	5-year K-ABC (SIM)	−3.3 pts per 10 µg/dL PbB	≤0.05
	<2 (<12)	6-year/child	231	6-year WISC-R (FS)	−5.2 pts per 10 µg/dL PbB	≤0.01
				6-year WISC-R (P)		

Cleveland	5.8 (2.6–14.7)	Delivery/cord	132	>1-day Soft Signs	− ? at PhB < 15 μg/dL	0.008
Ernhart et al. (151)				12-mo MDI	Sig. assoc. with Soft Signs	?
	6.5 (2.7–11.8)	Delivery/mother	119	6-mo KID	− ? at PbB < 12 μg/dL	0.002
	6.5 (2.7–11.8)	Delivery/mother	127	6-mo MDI	− ? at PbB < 12 μg/dL	<0.05
	6.5 (2.7–11.8)	Delivery/mother	145	12-mo MDI	− ? at PbB < 12 μg/dL	N.S.
	6.5 (2.7–11.8)	Delivery/mother	142	24-mo MDI	+ ? at PbB < 12 μg/dL	N.S.
	6.5 (2.7–11.8)	Delivery/mother	138	36-mo S-B IQ	+ ? at PbB < 12 μg/dL	N.S.
	6.5 (2.7–11.8)	Delivery/mother	134	58-mo WPPSI	− ? at PbB < 12 μg/dL	N.S.
Port Pirie	14.4 g.m. (?)	6-mo/child	575	24-mo MDI	− 1.6 pts. per 10 μg/dL PbB	0.07
Baghurst et al. (152)	21.2 g.m. (5–57)	24-mo/child	534	48-mo GCI	− 3.1 pts. per 10 μg/dL PbB	0.04
	17.6 g.m. (?)	0–4 year avg./child	463	48-mo GCI	− 3.8 pts. per 10 μg/dL PbB	0.04
	17.4 g.m. (?)	0–3 year avg./child	≤494	7.5-year WISC-R	− 1.2 pts. per 10 μg/dL PbB	0.04
Sydney	9.1 g.m. (3–28)	Delivery/mother	235	6-mo MDI	?	>0.25
Cooney et al. (153)	8.1 g.m. (1–36)	Delivery/cord}	235	12-mo MDI	?	>0.30
				24-mo MDI	?	>0.70
	≈12.5 g.m. (?)	Prior and current/child		36-mo GCI	?	>0.70
			207	48-mo GCI	?	0.14
Yugoslavia	14.4 (± 10.4)	Delivery/cord	348	24-mo MDI	− 0.8 pts. per 10 μg/dL PbB	0.12
Wasserman et al. (154)	15.1 (± 11.8)	6-mo/child	291	24-mo MDI	− 0.6 pts. per 10 μg/dL PbB	0.34
	18.3 (± 13.8)	12-mo/child	268	24-mo MDI	− 0.4 pts. per 10 μg/dL PbB	0.17
	22.0 (± 16.7)	18-mo/child	273	24-mo MDI	− 0.4 pts. per 10 μg/dL PbB	0.16
	24.3 (± 17.0)	24-mo/child	312	24-mo MDI	− 1.3 pts. per 10 μg/dL PbB	0.03

[a]Table from Ref. 148. Abbreviations: ?, not reported or was inferred from other information; GCI, McCarthy General Cognitive Index; g.m., geometric mean; K-ABC, Kaufman Assessment Battery for Children (MPC = Mental Processing Composite; SIM = Simultaneous Processing); KID, Kent Infant Development Scale; K-TEA, Kaufman Test of Educational Achievement; MDI, Bayley Mental Development Index; NS not significant at $p < 0.05$, two-tailed; PbB, blood lead level; S-BIQ, Stanford-Binet Intelligence Quotient; Soft Signs, Graham-Rosenblith Neurological Soft Signs Scale; WISC-R, Wechsler Intelligence Scale for Children, Revised (FS = Full Scale; P = Performance); WPPSI, Wechsler Preschool and Primary Scales of Intelligence.

[b]Representative references have been selected for these studies; additional reports of the results from these longitudinal studies may be found in the papers cited above and in the review by Davis and Elias (148).

body of experimental animal toxicity studies that also clearly demonstrate the neurotoxic effects of lead, even at quite low levels of exposure (155). Taken together, these various studies supported a conclusion that 10 to 15 μg/dL constitutes a blood lead level of concern for neurotoxicity in developing children (139, 140, 141–143, 146). It is important to note that 10 μg/dL does not necessarily represent a threshold for the biological or toxicological effects of lead, rather, it "reflects scientific judgment for purposes of protecting public health" (147). Thus, 10 μg/dL serves as a practical point of reference. It is also important to point out that the effects of lead on neurobehavioral development, although generally difficult to detect clinically in any given individual, are nonetheless significant in terms of population distributions and public health impacts. To illustrate, a shift of 4 points in the mean of a normal distribution of intelligence quotient scores would yield a 50% increase in the number of scores more than one standard deviation below the mean (see Ref. 148).

APPENDIX A. CLEAN AIR ACT AMENDMENTS OF 1990.

The information contained in Section II of Chapter 29 is available on the Internet at: http://www.epa.gov/oar/aqtrnd97/brochure

For Further Information:
National Air Pollutant Emission Trends, 1900–1997
Call: (919) 541-5285
Internet: http://www.epa.gov/oar/oaqps/efig
National Air Quality and Emissions Trends Report, 1997 (EPA-454/R-98-016)
Call: (919) 541-5558
Internet: http://www.epa.gov/oar/aqtrnd97
Acid Rain Hotline: (202) 564-9620
Energy Star (Climate Change) Hotline: (888) STAR-YES
Mobile Sources National Vehicles and Fuel Emissions Lab: (734) 214-4200
Stratospheric Ozone Hotline: (800) 296-1996

The following is a list of acronyms used in the National Air Quality and Emissions Trends Report.

AIRS = Aerometric Information Retrieval System
AIRMoN = Atmospheric Integrated Assessment Monitoring Network
AQRV = Air-Quality Related Values
BEIS = Biogenic Emissions Inventory System
CAA = Clean Air Act
CAAA = Clean Air Act Amendments
CARB = California Air Resources Board
CASAC = Clean Air Scientific Advisory Committee
CASTNet = Clean Air Status and Trends Network
CEMs = Continuous Emissions Monitors
CEP = Cumulative Exposure Project

CFR = *Code of Federal Regulations*
CO = Carbon monoxide
CMSA = Consolidated Metropolitan Statistical Area
DRI = Desert Research Institute
DST = Daylight Savings Time
EPA = Environmental Protection Agency
FRM = Federal Reference Method
GDP = Gross Domestic Product
HAPs = Hazardous Air Pollutants
IADN = Integrated Atmospheric Deposition Network
IMPROVE = Interagency Monitoring of PROtected Environments
LMOS = Lake Michigan Ozone Study
MACT = Maximum Achievable Control Technology
MARAMA = Mid-Atlantic Regional Air Management Association
MDN = Mercury Deposition Network
MSA = Metropolitan Statistical Area
NAAQS = National Ambient Air Quality Standards
NADP = National Atmospheric Deposition Program
NADP/NTN = National Atmospheric Deposition Program/National Trends Network
NAMS = National Air Monitoring Stations
NAPAP = National Acid Precipitation Assessment Program
NARSTO = North American Research Strategy for Tropospheric Ozone
NESCAUM = Northeast States for Coordinated Air Use Management
NET = National Emissions Trends Inventory
NMOC = Non-Methane organic compound
NO_2 = Nitrogen dioxide
NO_x = Nitrogen oxides
NOAA = National Oceanic and Atmospheric Administration
NPS = National Park Service
NTI = National Toxics Inventory
O_3 = Ozone
OTAG = The Ozone Transport Assessment Group
OTC = Ozone Transport Commission
PAHs = Polyaromatic hydrocarbons
PAMS = Photochemical Assessment Monitoring Stations
Pb = Lead
PCBs = Polychlorinated biphenyls
PM10 = Particulate Matter of 10 micrometers in diameter or less
PM2.5 = Particulate Matter of 2.5 micrometers in diameter or less
POM = Polycyclic organic matter
ppm = parts per million
PSI = Pollutant Standards Index
RFG = Reformulated gasoline
RVP = Reid Vapor Pressure
SAMI = Southern Appalachian Mountain Institute

SIP = State Implementation Plan
SLAMS = State and Local Air Monitoring Stations
SNMOC = Speciated Non-Methane Organic Compound
SO_2 = Sulfur dioxide
SO_x = Sulfur oxides
SOS = Southern Oxidant Study
STP = Standard Temperature and Pressure
TNMOC = Total Non-Methane Organic Compound
TRI = Toxic Release Inventory
TSP = Total Suspended Particulate
UATMP = Urban Air Toxics Monitoring Program
VMT = Vehicle Miles Traveled
VOCs = Volatile Organic Compounds
μg/m3 = Micrograms per cubic meter

APPENDIX B. 1990 CLEAN AIR ACT AMENDMENTS SUMMARIZED

The following material summarizes major provisions of the 1990 Amendments to the Clean Air Act. It is based on information provided on the EPA web site for the Office of Air and Radiation; more detail can be found on that web site at http://www.epa.gov/ttn/oarpg/amend.html.

Title I—Nonattainment

- Divides cities into six categories for ozone (3 years to attain—marginal; 6 years—moderate; 9 years—serious, 15 to 17 years—severe; 20 years—extreme) and two categories for carbon monoxide.
- % REDUCTION: Applies to ozone only. Moderate areas and above must achieve 15% volatile organic compounds (VOC) reduction within 6 years of enactment. For serious and above, average of 3% VOC per year thereafter until attainment. Annual VOC and nitrogen oxides (NO_x) reductions as needed to attain. The 15% and 3% is from an adjusted baseline and all reductions except those from existing federal motor vehicle control program (FMVCP), gasoline volatility, reasonably available control technology (RACT) and inspection/maintenance (I/M) fixups are creditable. Possible exemption from % reduction based on technological feasibility, if state implementation plan (SIP) adopts measures similar to those in next higher category and if all feasible measures are adopted in the first 6 years. NO_x substitution possible after 6 years.
- PRESCRIBED MEASURES: Major NO_x sources meet same requirements as major VOC sources unless EPA finds no benefit. All ozone nonattainment areas correct existing RACT rules and I/M programs. Moderate areas add basic I/M, Stage II and RACT on new and existing control technology guidelines (CTG) and 100 ton non-CTG sources, and make an attainment demonstration. Serious areas add enhanced

I/M, RACT on 50 ton non-CTG sources, a fleet vehicle program in areas of 250,000 and up, TCMs needed to offset vehicle growth, special rules for source modifications, and photochemical modeling attainment demonstration. Severe areas add RACT for 25 ton VOC non-CTG sources and provisions requiring adoption of TCMs, if necessary to meet progress requirements and employer trip reduction provisions. Extreme areas add RACT on 10 ton sources, eliminate feasibility exemption from 15% and 3%, add NO_x reductions from clean fuels or advanced technology, have peak hour traffic controls; can get SIP approved based on anticipated new technology.

- FEDERAL MEASURES: EPA issues 11 new CTGs plus CTGs for aerospace coatings, shipbuilding and repair, marine vessels rule and consumer products rules. Requires an ACT for 25 ton NO_x and VOC sources.
- SANCTIONS: Grace period of 18 months to cure planning failure. Then must apply 1 of 2 sanctions (modified highway ban or 2:1 offset). Air grants are available. Existing construction bans remain, but no new ones.
- FEDERAL IMPLEMENTATION PLANS (FIPs): Within 2 years of state failure to develop an adequate SIP, mandatory attainment FIPs required.
- TRANSPORT: Sets up 11 state NE transport commission. Requires transport states to adopt RACT for existing and new CTGs, RACT on major (50-ton) non-CTG sources, enhanced I/M in MSAs above 100,000 and Stage II or equivalent. No opt-out of VOC measures. Major NOx sources meet same requirements as major VOC sources unless EPA finds no benefit.
- CO AND PM-10: Wintertime oxygenated fuels in all CO areas > 9.4 ppm. Areas > 12.7 ppm add VMT forecast, enhanced I/M and demonstrate attainment. Serious CO areas add TCMs as in severe ozone areas. PM-10 areas initially designated nonattainment must attain by 12/94 (possible extension to 2001). Moderate areas adopt RACM; serious areas add BACM. Serious CO and PM-10 areas adopt measures to achieve 5% reduction per year effective upon failure to attain.

Title II—Mobile Sources

- TAILPIPE STANDARDS: Cars and light trucks: Tier I is 0.25 non-methane hydrocarbons (NMHC), 3.4 CO and 0.4 NOx. Possible Tier II is 0.125 NMHC, 1.7 CO and 0.2 NOx. Tier I phased in 1994–1996. Effectiveness of Tier II in 2004 depends on EPA study of need, feasibility, and cost-effectiveness. Useful life extended to 100,000 miles for most emission standards.
- COLD TEMPERATURE CO: Phase-in beginning in 1994 of 10 gpm at 20 degrees F for cars. A 3.4 gpm standard takes effect in 2002 if 6 or more cities are in CO nonattainment in mid-1997.
- CLEAN FUELS: In 1998 all centrally fueled fleets in 26 areas must buy 30% of the new vehicles that meet standards of 0.075 gpm VOC and 0.2 NOx; no toxic standards. If such vehicles are not being offered for sale in California the program is delayed possibly until 2001. Purchase requirements increase to 70% in 3rd year. In 1996, 150,000 clean fuel cars are required to be sold in California; increasing to 300,000

per year by 1999. These cars must meet a standard of 0.125 gpm VOC. Phase 2 begins in 2001 with cars meeting fleet-type standards. Other cities can opt-in to program.

- REFORMULATED GASOLINE: Beginning in 1995 reformulated gasoline is required in the 9 worst ozone areas; minimum oxygen content (2.0%), benzene (1.0%), aromatics (25%), VOCs and toxics reductions (15%, up to 20–25% in 2000). Cities can opt-in.
- OXYFUELS: Beginning in 1992, gas in 41 CO areas must have 2.7% oxygen level in winter months.
- URBAN BUSES: Delays diesel particulate standard from 1991 to 1993. Beginning in 1994 all buses must meet a PM standard of 0.05 g/hphr (if not feasible EPA will set at 0.07). Based on performance EPA may implement a low polluting bus program in larger cities.
- REFUELING: After consultation with DOT on safety issues, EPA required to promulgate onboard controls. Stage II requirements vary by classification.
- VOLATILITY: 9 psi in most of the country beginning 1992; EPA can set lower levels in warmer areas, but cannot require any standard below 9 psi in attainment areas.
- DESULFURIZATION: Diesel fuel highway use limited to 0.05% sulfur by weight.
- AIR TOXICS: Based on a study of mobile source-related toxics, EPA will regulate, at a minimum, emissions of benzene and formaldehyde.
- NON-ROAD ENGINES: Based on a study, EPA may regulate any category of non-road engines that contribute to urban air pollution. At a minimum, EPA must control locomotive emissions.
- LEAD IN GASOLINE: As of January 1, 1996, lead banned from use in motor vehicle fuel.

Title III—Air Toxics

- The Environmental Protection Agency (EPA) has developed the General Provisions rule to establish a consistent set of requirements and increase the clarity and certainty for industries and other sources that the Agency will regulate under the air toxic provisions of the Clean Air Act over the next several years.
- The Clean Air Act Amendments of 1990 require EPA to issue standards over a 10 year period regulating emissions of 189 toxic air pollutants from various industries and other sources.
- In the course of planning for and developing these standards to meet the requirements of the 1990 Amendments, EPA determined that it would issue a "General Provisions" rule to address general information and requirements that would apply to all of these air toxics rules. [For example, many of the requirements in the General Provisions rule will apply to those sources affected by the chemical manufacturing ("HON") final rule also being issued today.]
- The General Provisions rule creates the technical and administrative framework for implementing the air toxics requirements of the Act. It also serves as the primary

vehicle for informing owners and operators of their basic compliance responsibilities and of EPA's administrative and enforcement responsibilities under the Act. The rule also eliminates the need for EPA to repeat general information and requirements in future air toxics rules.

- EPA issued the proposed General Provision rule on August 11, 1993. The Agency received public comment, made adjustments to the rule based on those comments and issued the rule in final form.

What does the General Provisions Rule Require?

- The General Provisions rule will apply to all sources that will be regulated under the air toxics standards EPA issues under the Clean Air Act. However, when appropriate, EPA can override or modify the General Provisions requirements in individual air toxics rules it issues in the future.
- The General Provisions include generic information, such as definitions of terms and sections that spell out EPA's administrative responsibilities. The rule also specifies compliance dates and outlines the compliance responsibilities of owners or operators who are subject to an air toxic emission standard or other requirement.
- Once a source becomes subject to a certain EPA-issued air toxics rule, the General Provisions rule will require applications for approval of new construction and reconstruction, as well as performance testing, monitoring, record keeping, and reporting.
- The rule also provides flexibility to industry and lessens the compliance burden by allowing certain qualified sources to reduce the frequency of testing and reporting. The rule also encourages technological innovation by allowing alternative means of compliance.
- The General Provisions rule also cross-references and helps coordinate the various air toxics, operating permit and enforcement requirements found in different sections of the Clean Air Act Amendments of 1990.

Title IV—Acid Rain

- SO_2 REDUCTION: A 10 million ton reduction from 1980 levels, primarily from utility sources. Caps annual utility SO_2 emissions at approximately 8.9 million tons by 2000.
- ALLOWANCES: SO_2 reductions are met through an innovative market-based system. Affected sources are allocated allowances based on required emission reductions and past energy use. An allowance is worth one ton of SO_2 and it is fully marketable. Sources must hold allowances equal to their level of emissions or face a \$2000/excess ton penalty and a requirement to offset excess tons in future years. EPA will also hold special sales and auctions of allowances.
- PHASE I: SO_2 emission reductions are achieved in two phases. Phase I allowances are allocated to large units of 100 MW or greater that emit more than 2.5 lb/mmbtu in an amount equal to 2.5 lb/mmBtu $\times$ their 1985–1987 energy usage (baseline).

Phase I must be met by 1995 but units that install certain control technologies may postpone compliance until 1997, and may be eligible for bonus allowances. Units in Illinois, Indiana or Ohio are allotted a pro rata share of an additional 200,000 allowances annually during Phase I.

- PHASE II: Phase II begins in 2000. All utility units greater than 25 MW that emit at a rate above 1.2 lbs/mmBtu will be allocated allowances at that rate × their baseline fuel consumption. Cleaner plants generally will be provided with 20% more allowances than would have been received based on their baseline consumption. 50,000 bonus allowances are allocated to plants in 10 midwestern states that make reductions in Phase I.
- NO_x: Utility NO_x reductions will help to achieve a 2 million ton reduction from 1980 levels. Reductions will be accomplished through required EPA performance standards for certain existing boilers in Phase I, and others in Phase II. EPA will develop a revised NO_x NSPS for utility boilers.
- REPOWERING: Units repowering with qualifying Clean Coal Technologies receive a 4 year extension for Phase II compliance. Such units may be exempt from New Source Review requirements and New Source Performance Standards.
- ENERGY CONSERVATION AND RENEWABLE ENERGY: These projects may be allocated a portion of up to 300,000 incentive allowances.
- CLEAN COAL TECHNOLOGIES (CCT): Certain CCT demonstration projects may be exempt from NSPS, NSR, and Title I nonattainment requirements.
- MONITORING: Requires continuous emission monitors or an equivalent for SO_2 and NO_x and also requires opacity and flow monitors.

Regulation to Accelerate the Phaseout of Ozone-Depleting Substances

Section 606 of the Clean Air Act Amendments of 1990

Background. In July 1992, EPA issued its final rule implementing Section 604 of the Clean Air Act Amendments of 1990. That section limits the production and consumption of a set of chemicals known to deplete the stratospheric ozone layer. EPA controls production and consumption by issuing allowances or permits that are expended in the production and importation of these chemicals. These allowances can be traded.

The July, 1992 rule required producers of class I substances (chlorofluorocarbons, halons, carbon tetrachloride, and methyl chloroform) to reduce gradually their production of these chemicals and to phase them out completely as of January 1, 2000. In addition to these production limits, the rule required a similar reduction in consumption, defined as production plus imports minus exports.

On February 11, 1992, the United States, responding to recent scientific findings, announced that the production of chlorofluorocarbons (CFCs), halons, carbon tetrachloride, and methyl chloroform would be accelerated and that these substances would be phased out by December, 31, 1995. It was also stated that the U.S. will consider recent evidence suggesting the possible need to phase out methyl bromide. At the same time, the Agency received petitions from environmental and industry groups to accelerate the phaseout of these chemicals.

In addition, the fourth meeting of the Parties to the Montreal Protocol took place in Copenhagen in November, 1992. At this meeting the Parties made a number of decisions which are reflected in this final regulation. This regulation implements the United States' obligation to the recent agreements by the Parties in Copenhagen to the Protocol, and implements the accelerated phaseout of ozone- depleting substances, while responding to the petitions received by the Agency from environmental and industry groups.

The Parties to the Protocol agreed to phase out hydrobromofluorocarbons by the end of 1995 and the production and consumption of halons by the end of 1993. In accordance with these agreements, this regulation schedules the phaseout of these chemicals by these dates.

The recent United Nations Environmental Program (UNEP) Scientific Assessment identified methyl bromide, widely used as a soil fumigant, as a significant ozone-depleting compound. The parties set an ozone-depleting potential of 0.7 for this chemical. This regulation freezes the production and consumption of this chemical at 1991 levels through the year 2000. At that point, the Agency is obligated under the Clean Air Act to phase out this chemical by the year 2001.

Specifically, EPA is reducing production and consumption levels and phasing out the production and consumption of the major ozone-depleting chemicals. In addition, methyl bromide and the HBFCs are added to the list of class I substances and scheduled for phaseout. The phaseout schedule, in terms of percentage of baseline production allowed, is presented as follows:

Class I Substances

Date (Jan. 1)	CFCS (%)	Halons (%)	Carbon Tetrachloride (%)	Methyl Chloroform (%)	Methyl Bromide (%)	HBFCS (%)
1994	25	0	50	50	100	100
1995	25	0	15	30	100	100
1996	0	0	0	0	100	0
1997	0	0	0	0	100	0
1998	0	0	0	0	100	0
1999	0	0	0	0	100	0
2000	0	0	0	0	100	0
2001	0	0	0	0	0	0

The Parties also agreed in Copenhagen to phase out the production and consumption of hydrochlorofluorocarbons (HCFCs) by the year 2030. Production and consumption would be limited to a cap equal to 3.1 percent of CFCs consumed in 1989, weighted by ozone-depleting potential, plus the consumption of HCFCs in the same year, also weighted by ozone-depleting potential. Parties would reduce this cap by specified reductions, culminating in a complete phaseout by 2030. Although less harmful than CFCs, these chemicals do deplete the ozone and, if left unchecked, would contribute to this problem.

The Agency intends to meet these limits by accelerating the phaseout of HCFC-141b, HCFC-142b and HCFC-22. These are the most damaging of the HCFCs. By eliminating these chemicals by the specified dates, the Agency believes that it will meet the requirements of the cap approved by the Parties. The controls on HCFC-141b, HCFC-142b and HCFC-22 are specified below.

Class II Substances

Affected Compounds Restriction

January 1, 2003
HCFC-141b
Ban on production and consumption, except for specified exemptions.

January 1, 2010
HCFC-142b, HCFC-22
Production and consumption frozen at baseline levels.
HCFC-142b, HCFC-22
Ban on the production and consumption of virgin chemical unless used as feedstock or refrigerant in appliances manufactured prior to January 1, 2010.

January 1, 2015
All other HCFCs
Production and consumption frozen at baseline level.
Ban on the production and consumption of virgin chemical unless used as feedstock or refrigerant in appliances manufactured prior to January 1, 2020.

January 1, 2020
HCFC-142b, HCFC-22
Ban on production and consumption, except for specified exemptions.

January 1, 2030
All other HCFCs
Ban on production and consumption, except for specified exemptions. In addition to the above phaseout schedule changes, the final regulation implements a number of additional amendments to the existing phaseout regulations as follows:

The final rule now permits an exemption from the allowance requirements for the production of ozone-depleting chemicals if such production is inadvertent or coincidental during a manufacturing process. Also, these inadvertent or coincidentally produced chemicals are not considered controlled substances or products if they are present in trace quantities as a result of the use of these chemicals as a process agent.

The regulation permits production of controlled substances for transformation or destruction outside of the production and consumption allowance requirements, if the destruction is achieved by one of the processes approved by the Parties to the Montreal Protocol. The following processes have been approved: 1) liquid injection incineration, 2) reactor cracking, 3) gaseous/fume oxidation, 4) rotary kiln incineration, and 5) cement kilns

The transshipment of bulk controlled chemicals from one foreign country to another, through the United States, will not count as consumption under these regulations by the United States.

The import and export of recycled or used controlled substances will no longer be considered consumption by the United States.

The regulation implements a new definition of importer to ensure that the owner, not necessarily the "importer of record", is responsible for the import.

The regulation simplifies and reduces the reporting and recordkeeping requirements for companies dealing in controlled substances, and exempts controlled substances used for feedstock purposes from the requirements to expend allowances when producing or importing these substances for feedstock uses.

Finally, this regulation includes various trade provisions required by the Montreal Protocol to encourage countries to join the Protocol by ??0000??.

BIBLIOGRAPHY

1. *National Air Quality and Emissions Trends Report*, EPA 454/R-98-016, 1997
2. J. A. Graham, L. Folensbee, J. M. Davis, J. Raub, and L. Grant, "Critical Heated Issues of Criteria Air Pollutants," in *Toxicology of the Lung*, Taylor & Francis
3. World Health Organization 1999. *WHO Air Quality Guidelines for Europe*, 2nd ed. WHO Regional Office for Europe, Copenhagen, Denmark: 1999, in press
4. U.S. Environmental Protection Agency. *National Air Quality and Emissions Trends Report, 1996*, Office of Air Quality Planning and Standards, Research Triangle Park, NC, EPA-454/R-97-013, 1998.
5. U.S. Environmental Protection Agency *Air Quality Criteria for Ozone and Related Photochemical Oxidants.* Office of Research and Development, Research Triangle Park, NC, EPA/600/AP-93/004aF-cF 3v, 1996.
6. D. E. Gardner. "Use of Experimental Airborne Infections for Monitoring Altered Host Defenses," *Environmental Health Perspectives* **43**, 99–107 (1982).
7. W. F. McDonnell, P. W. Stewart, S. Andreoni, E. Seal, Jr., H. R. Kehrl, D. H. Horstman, L. J. Folinsbee, and M. V. Smith. "Prediction of Ozone-induced FEV_1 Changes: Effects of Concentration, Duration, and Ventilation," *Am. J. Respiratory Critical Care Med.* **156**, 715–722 (1997).
8. L. J. Folinsbee, W. F. McDonnell, and D. H. Horstman. "Pulmonary Function and Symptom Responses after 6.6-hour Exposure to 0.12 ppm Ozone with Moderate Exercise," *JAPCA*, **38**, 28–35 (1988).
9. D. H. Horstman, L. J. Folinsbee, P. J. Ives, S. Abdul-Salaam, and W. F. McDonnell. "Ozone Concentration and Pulmonary Response Relationships for 6.6-hour Exposures with Five Hours of Moderate Exercise to 0.08, 0.10, and 0.12 ppm," *Am. Rev. Respiratory Disease* **142**, 1158–1163 (1990).

10. M. J. Hazucha, L. J. Folinsbee, and E. Seal, Jr. "Effects of Steady-State and Variable Ozone Concentration Profiles on Pulmonary Function," *Am. Rev. Respiratory Disease* **146**, 487–1493 (1992).

11. L. J. Folinsbee and M. J. Hazucha. "Persistence of Ozone-induced Changes in Lung Function and Airway Responsiveness," in T. Schneider, S. D. Lee, G. J. R. Wolters, and L. D. Grant, eds., *Atmospheric Ozone Research and its Policy Implications: Proceedings of the 3rd US-Dutch International Symposium*, Elsevier Science Publishers, (*Studies in Environmental Science 35*), Amsterdam 1989, pp. 483–492.

12. R. M. Aris, I. Tager, D. Christian, T. Kelly, and J. R. Balmes. "Methacholine Responsiveness is not Associated with O_3-induced Decreases in FEV_1," *Chest* **107**, 621–628 (1995).

13. J. R. Balmes, L. L. Chen, C. Scannell, I. Tager, D. Christian, P. Q. Hearne, T. Kelly, and R. M. Aris. "Ozone-induced Decrements in FEV_1 and FVC do not Correlate with Measures of Inflammation," *Am. J. Respiratory Critical Care Med.* **153**, 904–909 (1996).

14. A. Torres, M. J. Utell, P. E. Morow, K. Z. Voter, J. C. Whitin, C. Cox, R. J. Looney, D. M. Speers, Y. Tsai, and M. W. Frampton. "Airway Inflammation in Smokers and Nonsmokers with Varying Responsiveness to Ozone," *Am. J. Respiratory Critical Care Med.* **156**, 728–736 (1995).

15. R. B. Devlin, L. J. Folinsbee, F. Biscardi, G. Hatch, S. Becker, M. C. Madden, M. Robbins, and H. S. Koren. "Inflammation and Cell Damage Induced by Repeated Exposure of Humans to Ozone," *Inhalation Toxicology* **9**, 211–235 (1997).

16. M. T. Krishna, A. Blomberg, G. L. Biscione, F. Kelly, T. Sandström, A. Frew, and S. Holgate. "Short-Term Ozone Exposure Upregulates P-Selectin in Normal Human Airways," *Am. J. Respiratory Critical Care Medicine* **155**, 1798–1803 (1997).

17. E. S. Schelegle, M. L. Carl, H. M. Coleridge, J. C. G. Coleridge , and J. F. Green, "Contribution of Vagal Afferents to Respiratory Reflexes Evoked by Acute Inhalation of Ozone in Dogs," *J. Appl. Physiology* **74**, 2338–2344 (1993).

18. M. J. Holtzman, J. H. Cunningham, J. R. Sheller, G. B. Irsigler, J. A. Nadel, H. A. Boushey. "Effect of Ozone on Bronchial Reactivity in Atopic and Nonatopic Subjects," *Am. Rev. Respiratory Disease* **120**, 1059–1067 (1979).

19. N. A. Molfino, S. C. Wright, I. Katz, S. Tarlo, F. Silverman, P. A. McClean, J. P. Szalai, M. Raizenne, A. S. Slutsky, and N. Zamel. "Effect of Low Concentrations of Ozone on Inhaled Allergen Responses in Asthmatic Subjects," *Lancet* **3388**(761), 199–203 (1991).

20. R. Jörres, D. Nowak, H. Magnussen, P. Speckin, and S. Koschyk. "The Effect of Ozone Exposure on Allergen Responsiveness in Subjects with Asthma or Rhinitis," *Am. J. Respiratory Critical Care Medicine* **153**, 56–64 (1996).

21. H. R. Kehrl, D. B. Peden, B. A. Ball, L. J. Folinsbee, and D. H. Horstman. "Increase Specific Airway Reactivity of Mild Allergic Asthmatics Following 7.6 hr Exposures to 0.16 ppm Ozone," *Am. J. Respiratory Critical Care Medicine* [submitted], 1998.

22. B. A. Ball, L. J. Folinsbee, D. B. Peden, and H. R. Kehrl. "Allergen Bronchoprovocation of Patients with Mild Allergic Asthma after Ozone Exposure," *J. Allergy Clinical Immunology* **98**, 563–572 (1996).

23. M. C. White, R. A. Etzel, W. D. Wilcox, and C. Lloyd. "Exacerbations of Childhood Asthma and Ozone Pollution in Atlanta," *Environ. Res.* **65**, 56–68 (1994).

24. R. T. Burnett, R. E. Dales, M. E. Raizenne, D. Krewski, P. W. Summers, G. R. Roberts, M. Raad-Young, T. Dann, and J. Brook. "Effects of Low Ambient Levels of Ozone and Sulfates on the Frequency of Respiratory Admissions to Ontario Hospitals," *Environ. Res.* **65**, 172–194 (1994).

25. G. D. Thurston, K. Ito, P. L. Kinney, and M. Lippmann. "A Multi-Year Study of Air Pollution and Respiratory Hospital Admissions in Three New York State Metropolitan Areas: Results for 1988 and 1989 Summers," *J. Exposure Analysis Environ. Epidemiology* **2**, 429–450 (1992).

26. M. A. Basha, K. B. Gross, C. J. Gwizdala, A. H. Haidar, and J. Popovich, Jr. "Bronchoalveolar Lavage Neutrophilia in Asthmatic and Healthy Volunteers after Controlled Exposure to Ozone and Filtered Purified Air," *Chest* **106**, 1757–1765 (1984).

27. C. Scannell, L. Chen, R. M. Aris, I. Tager, D. Christian, R. Ferrando, B. Welch, T. Kelly, and J. R. Balmes. "Greater Ozone-Induced Inflammatory Responses in Subjects with Asthma," *Am. J. Respiratory Critical Care Med.* **154**, 24–29 (1996).

28. D. B. Peden, B. Boehlecke, D. Horstman, and R. Devlin. "Prolonged Acute Exposure to 0.16 ppm Ozone Induces Eosinophilic Airway Inflammation in Asthmatic Subjects with Allergies," *J. Allergy Clinical Immunol.* **100**, 802–808 (1997).

29. H. Gong, Jr., M. S. McManus, and W. S. Linn. "Attenuated Response to Repeated Daily Ozone Exposures in Asthmatic Subjects," *Archives of Environmental Health* **52**, 34–41 (1997).

30. J. S. Tepper, D. L. Costa, J. R. Lehmann, M. F. Weber, and G. E. Hatch. "Unattenuated Structural and Biochemical Alterations in the Rat Lung during Functional Adaptation to Ozone," *Am. Rev. Respiratory Disease* **140**, 493–501 (1989).

31. D. L. Christian, L. L. Chen, C. H. Scannell, R. E. Ferrando, B. S. Welch, and J. R. Balmes. "Ozone-induced inflammation is attenuated with multiday exposure," *Am. J. Respiratory Critical Care Med.* **158**, 532–537 (1998).

32. D. E. Gardner, T. R. Lewis, S. M. Alpert, D. J. Hurst, and D. L. Coffin. "The Role of Tolerance in Pulmonary Defense Mechanisms," *Archives of Environmental Health* **25**, 432–438 (1972).

33. W. S. Tyler, N. K. Tyler, J. A. Last, M. J. Gillespie, and T. J. Barstow. "Comparison of Daily and Seasonal Exposures of Young Monkeys to Ozone," *Toxicology* **50**, 131–144 (1988).

34. N. Künzli, F. Lurmann, M. Segal, L. Ngo, J. Balmes, and I. B. Tager. "Association between Lifetime Ambient Ozone Exposure and Pulmonary Function in College Freshmen—Results of a Pilot Study," *Environmental Research* **72**, 8–23 (1997).

35. U.S. Environmental Protection Agency, *Air Quality Criteria for Oxides of Nitrogen*, Office of Health and Environmental Assessment, Environmental Criteria and Assessment Office, Research Triangle Park, NC, EPA/600/8-91/049aF-cF 3v, 1993.

36. World Health Organization. "Nitrogen Oxides," 2nd ed., *Environmental Health Criteria 188*, World Health Organization, Geneva, Switzerland.

37. V. Hasselblad, D. M. Eddy, and D. J. Kotchmar. "Synthesis of Environmental Evidence: Nitrogen Dioxide Epidemiology Studies," *J. Air Waste Manag. Assoc.* **42**, 662–671 (1992).

38. L. M. Neas, D. W. Dockery, J. H. Ware, J. D. Spengler, F. E. Speizer, and B. G. Ferris, Jr. "Association of Indoor Nitrogen Dioxide with Respiratory Symptoms and Pulmonary Function in Children," *Am. J. Epidemiology* **134**, 204–219 (1991).

39. K. Katsouyanni, J. Schwartz, C. Spix, G. Touloumi, D. Zmirou, A. Zanobetti, B. Wojtyniak, J. M. Vonk, A. Tobias, A. Pönkä, S. Medina, L. Bachárová, and H. R. Andersen. "Short Term Effects of Air Pollution on Health: A European Approach using Epidemiology Time Series Data: the APHEA Protocol," *J. Epidemiology Community Health* **50**(Suppl. 1), S12–S18 (1996).

40. R. Jörres, D. Nowak, F. Grimminger, W. Seeger, M. Oldigs, and H. Magnussen. "The Effect of 1 ppm Nitrogen Dioxide on Bronchoalveolar Lavage Cells and Inflammatory Mediators in Normal and Asthmatic Subjects," *Eur. Respiratory J.* **8**, 416–424 (1995).

41. W. S. Tunnicliffe, P. S. Burge, and J. G. Ayres. "Effect of Domestic Concentrations of Nitrogen Dioxide on Airway Responses to Inhaled Allergen in Asthmatic Patients," *Lancet* **344**, 1733–1736 (1994).

42. J. L. Devalia, C. Rusznak, M. J. Herdman, C. J. Trigg, H. Tarraf, and R. J. Davies. "Effect of Nitrogen Dioxide and Sulphur Dioxide on Airway Response of Mild Asthmatic Patients to Allergen Inhalation," *Lancet* **344**, 1668–1671 (1994).

43. B. Vagaggini, P. L. Paggiaro, D. Giannini, A. D. Franco, S. Cianchetti, S. Carnevali, M. Taccola, E. Bacci, L. Bancalari, F. L. Dente, and C. Giuntini. "Effect of Short-Term NO_2 Exposure on Induced Sputum in Normal, Asthmatic and COPD Subjects," *Eur. Respiratory J.* **9**, 1852–1857 (1996).

44. V. Strand, P. Salomonsson, J. Lundahl, and G. Bylin, "Immediate and Delayed Effects of Nitrogen Dioxide Exposure at an Ambient Level on Bronchial Responsiveness to Histamine in Subjects with Asthma," *Eur. Respiratory J.* **9**, 733–740 (1996).

45. V. Strand, M. Svartengren, S. Rak, C. Barck, and G. Bylin. "Repeated Exposure to an Ambient Level of NO_2 Enhances Asthmatic Response to Nonsymptomatic Allergen Dose," *Eur. Respiratory J.* **12**, 6–12 (1998).

46. C. Rusznak, J. L. Devalia, and R. J. Davies, "Airway Response of Asthmatic Subjects to Inhaled Allergen after Exposure to Pollutants," *Thorax* **51**, 1105–1108 (1996).

47. V. Strand, S. Rak, M. Svartengren, and G. Bylin, "Nitrogen Dioxide Exposure Enhances Asthmatic Reaction to Inhaled Allergen in Subjects with Asthma," *Am. J. Respiratory Critical Care Med.* **155**, 881–887 (1997).

48. M. I. Gilmour. "Interaction of Air Pollutants and Pulmonary Allergic Responses in Experimental Animals," *Toxicology* **105**, 335–342 (1995).

49. A. Blomberg, M. T. Krishna, V. Bocchino, G. L. Biscione, J. K. Shute, F. J. Kelly, A. J. Frew, S. T. Holgate, and T. Sandström. "The Inflammatory Effects of 2 ppm NO_2 on the Airways of Healthy Subjects," *Am. J. Respiratory Critical Care Med.* **156**, 418–424 (1997).

50. D. E. Gardner. "Influence of Exposure Patterns of Nitrogen Dioxide on Susceptibility to Infectious Respiratory Disease," in S. D. Lee, ed., *Nitrogen Oxides and their effects on Health*, Ann Arbor Science Publishers, Inc. Ann Arbor, MI, 1980, pp. 267–288.

51. U.S. Environmental Protection Agency, *Second Addendum to Air Quality Criteria for Particulate Matter and Sulfur Oxides (1982): Assessment of Newly Available Health Effects Information.* Office of Health and Environmental Assessment, Environmental Criteria and Assessment Office, Research Triangle Park, NC, EPA-600/8-86-020F, 1986.

52. U.S. Environmental Protection Agency *Supplement to the Second Addendum (1986) to Air Quality Criteria for Particulate Matter and Sulfur Oxides (1982): Assessment of New Findings on Sulfur Dioxide Acute Exposure Health Effects in Asthmatic Individuals.* Office of Health and Environmental Assessment, Environmental Criteria and Assessment Office, Research Triangle Park, NC, EPA-600/FP-93/002, 1994.

53. D. Zmirou, T. Barumandzadeh, F. Balducci, P. Ritter, G. Laham, and J.-P. Ghilardi, "Short Term Effects of Air Pollution on Mortality in the City of Lyon, France, 1985–1990, " *J. Epidemiology and Community Health* **50**(Suppl. 1), S30–S35 (1996).

54. J. Sunyer, J. Castellsagué, M. Sáez, A. Tobias, and J. M. Antó, "Air Pollution and Mortality in Barcelona," *J. Epidemiol. Community Health* **50** (Suppl. 1), S76–S80 (1996).

55. W. Dab, S. Medina, P. Quénel, Y. Le Moullec, A. Le Tertre, B. Thelot, C. Monteil, P. Lameloise, P. Pirard, I. Momas, R. Ferry, and B. Festy. "Short Term Respiratory Health Effects of Ambient Air Pollution: Results of the APHEA Project in Paris," *J. Epidemiol. Community Health* **50**(Suppl. 1), S42–S46 (1996).

56. L. Bachárová, K. Fandáková, J. Bratinka, M. Budinská, J. Bachár, and M. Gudába. "The Association between Air Pollution and the Daily Number of Deaths: Findings from the Slovak Republic Contribution to the APHEA Project," *J. Epidemiol. Community Health* **50**(Suppl. 1), S19–S21 (1996).
57. S. K. Heath, J. Q. Koenig, M. S. Morgan, H. Checkoway, Q. S. Hanley, and V. Rebolledo. "Effects of Sulfur Dioxide Exposure on African-American and Caucasian Asthmatics," *Environmental Research* **66**, 1–11 (1994).
58. H. Gong, Jr., W. S. Linn, D. A. Shamoo, K. R. Anderson, C. A. Nugent, K. W. Clark, A. E. Lin. "Effect of Inhaled Salmeterol on Sulfur Dioxide-induced Bronchoconstriction in Asthmatic Subjects," *Chest* **110**, 1229–1235 (1996).
59. B. Bigby and H. Boushey. "Effects of Nedocromil Sodium on the Bronchomotor Response to Sulfur Dioxide in Asthmatic Patients," *J. Allergy Clinical Immunol* **92**, 195–197 (1993).
60. P. I. Field, R. Simmul, S. C. Bell, D. H. Allen, and N. Berend. "Evidence for Opioid Modulation and Generation of Prostaglandins in Sulphur Dioxide $(SO)_2$-induced Bronchoconstriction," *Thorax* **51**, 159–163 (1996).
61. S. C. Lazarus, H. H. Wong, M. J. Watts, H. A. Boushey, B. J. Lavins, and M. C. Minkwitz. "The Leukotriene Receptor Antagonist Zafirlukast Inhibits Sulfur Dioxide-induced Bronchoconstriction in Patients with Asthma," *Am. J. Respiratory Critical Care Med.* **156**, 1725–1730 (1997).
62. H. Gong, Jr., P. A. Lachenbruch, P. Harber, and W. S. Linn. "Comparative Short-term Health Responses to Sulfur Dioxide Exposure and Other Common Stresses in a Panel of Asthmatics," *Toxicology and Industrial Health* **11**, 467–487 (1995).
63. S. C. O. Taggart, A. Custovic, H. C. Francis, E. B., Faragher, C. J. Yates, B. G. Higgins, and A. Woodcock, "Asthmatic Bronchial Hyperresponsiveness Varies with Ambient Levels of Summertime Air Pollution," *Eur. Respiratory J.* **9**, 1146–1154 (1996).
64. U.S. Environmental Protection Agency. *Air Quality Criteria for Particulate Matter*. National Center for Environmental Assessment-RTP Office, Research Triangle Park, NC, EPA/600/P-95/001aF-cF.3v, 1996.
65. U.S. Environmental Protection Agency, *Review of the National Ambient Air Quality Standards for Particulate Matter: Policy Assessment of Scientific and Technical Information. OAQPS staff paper*. Office of Air Quality Planning and Standards, Research Triangle Park, NC, EPA/452/R-96-013, 1996.
66. W. E. Wilson, and H. H. Suh, "Fine Particles and Coarse Particles: Concentration Relationships Relevant to Epidemiologic Studies," *J. Air Waste Management Assoc.* **47**, 1238–1249 (1997).
67. J. Schwartz, D. W. Dockery, and L. M. Neas, "Is Daily Mortality Associated Specifically with Fine Particles?" *J. Air Waste Manag Assoc* **46**, 927–939 (1996).
68. D. W. Dockery, J. Schwartz, and J. D. Spengler. "Air Pollution and Daily Mortality: Associations with Particulates and Acid Aerosols," *Environmental Research* **59**, 362–373 (1992).
69. P. Styer, N. McMillan, F. Gao, J. Davis, and J. Sacks, "Effect of Outdoor Airborne Particulate Matter on Daily Death Counts," *Environ Health Perspectives* **103**, 490–497 (1995).
70. K. Ito and G. D. Thurston. "Daily PM_{10}/Mortality Associations: An Investigation of At-Risk-Subpopulations," *J. Exposure Analysis and Environmental Epidemiology* **6**, 79–95 (1996).
71. C. A. Pope, III, J. Schwartz, and M. R. Ransom, Daily Mortality and PM_{10} Pollution in Utah Valley. *Archives of Environ. Health* **47**, 211–217 (1992).
72. C. A. Pope, III. "Particulate Pollution and Mortality in Utah Valley," prepared for *Critical Evaluation Workshop on Particulate Matter—Mortality Epidemiology Studies, Raleigh, NC*, Brigham Young University, Provo, UT, 1994.

73. J. Schwartz. "Air Pollution and Daily Mortality in Birmingham, Alabama", *Am. J. Epidemiol.* **137**, 1136–1147 (1993).
74. P. L. Kinney, K. Ito, and G. D. Thurston. "A Sensitivity Analysis of Mortality/PM_{10} Associations in Los Angeles," *Inhalation Toxicology* **7**, 59–69 (1995).
75. G. D. Thurston, K. Ito, C. G. Hayes, D. V. Bates, and M. Lippmann, "Respiratory Hospital Admissions and Summertime Haze Air Pollution in Toronto, Ontario: Consideration of the Role of Acid Aerosols," *Environ. Res.* **65**, 271–290 (1994).
76. J. Schwartz, "Short-Term Fluctuations in Air Pollution and Hospital Admissions of the Elderly for Respiratory Disease," *Thorax* **50**, 531–538 (1995).
77. J. Schwartz, C. Spix, G. Touloumi, L. Bachárová, T. Barumamdzadeh, A. le Tertre, T. Piekarksi, A. Ponce de Leon, A. Pönkä, G. Rossi, M. Saez, and J. P. Schouten, "Methodological Issues in Studies of Air Pollution and Daily Counts of Deaths or Hospital Admissions," *J. Epidemiol. Community Health* **50**(Suppl. 1), S3–S11 (1996).
78. J. Schwartz, "Air Pollution and Hospital Admissions for Respiratory Disease," *Epidemiology* **7**, 20–28 (1996).
79. J. Schwartz, "PM_{10}, Ozone, and Hospital Admissions for the Elderly in Minneapolis, MN," *Archives of Environ. Health* **49**, 366–374 (1994).
80. J. Schwartz, "Air Pollution and Hospital Admissions for the Elderly in Birmingham, Alabama," *Am. J. Epidemiol.* **139**, 589–598 (1994).
81. J. Schwartz, "Air Pollution and Hospital Admissions for the Elderly in Detroit, Michigan," *Am. J. Respiratory Critical Care Med.* **150**, 648–655 (1994).
82. J. Schwartz, and R. Morris, "Air Pollution and Hospital Admissions for Cardiovascular Disease in Detroit, Michigan," *Am. J. Epidemiol.* **142**, 23–35 (1995).
83. J. Schwartz, D. W. Dockery, L. M. Neas, D. Wypij, J. H. Ware, J. D. Spengler, P. Koutrakis, F. E. Speizer, and B. G. Ferris, Jr. "Acute Effects of Summer Air Pollution on Respiratory Symptom Reporting in Children," *Am. J. Respiratory Critical Care Med.* **150**, 1234–1242 (1994).
84. C. A. Pope, III, D. W. Dockery, J. D. Spengler, and M. E. Raizenne, "Respiratory Health and PM_{10} Pollution: A Daily Time Series Analysis," *Am. Rev Respiratory Disease* **144**, 668–674 (1991).
85. C. A. Pope, III, and D. W. Dockery, "Acute Health Effects of PM_{10} Pollution on Symptomatic and Asymptomatic Children,". *Am. Rev. Respiratory Disease* **145**, 1123–1128 (1992).
86. B. D. Ostro, M. J. Lipsett, M. B. Wiener, and J. C. Selner, "Asthmatic Responses to Airborne Acid Aerosols," *Am. J. Public Health* **81**, 694–702 (1991).
87. C. A. Pope, III, and R. E. Kanner, "Acute Effects of PM_{10} Pollution on Pulmonary Function of Smokers with Mild to Moderate Chronic Obstructive Pulmonary Disease," *Am. Rev. Respiratory Disease* **147**, 1336–1340 (1993).
88. R. T. Burnett, R. Dales, D. Krewski, R. Vincent, T. Dann, and J. R. Brook. "Associations between Ambient Particulate Sulfate and Admissions to Ontario Hospitals for Cardiac and Respiratory Diseases," *Am. J. Epidemiol.* **142**, 15–22 (1995).
89. B. D. Ostro, M. J. Lipsett, J. K. Mann, A. Krupnick, and W. Harrington, "Air Pollution and Respiratory Morbidity among Adults in Southern California," *Am. J. Epidemiology* **137**, 691–700 (1993).
90. L. M. Neas, D. W. Dockery, P. Koutrakis, D. J. Tollerud, and F. E. Speizer, "The Association of Ambient Air Pollution with Twice Daily Peak Expiratory Flow Rate Measurements in Children," *Am. J. Epidemiology* **141**, 111–122.

91. D. W. Dockery, C. A. Pope, III, X. Xu, J. D. Spengler, J. H. Ware, M. E. Fay, B. G. Ferris, Jr., and F. E. Speizer. "An Association between Air Pollution and Mortality in Six U.S. Cities," *New Engl. J. Med.* **329**, 1753–1759.
92. C. A. Pope, III, M. J. Thun, M. M. Namboodiri, D. W. Dockery, J. S. Evans, F. E. Speizer, and C. W. Heath, Jr. "Particulate Air Pollution as a Predictor of Mortality in a Prospective study of U.S. Adults," *Am. J. Respiratory Critical Care Med.* **151**, 669–674 (1995).
93. D. W. Dockery, F. E. Speizer, D. O. Stram, J. H. Ware, J. D. Spengler, and B. G. Ferris, Jr. "Effects of Inhalable Particles on Respiratory Health of Children," *American Review of Respiratory Disease* **139**, 587–594 (1989).
94. J. H. Ware, B. G. Ferris, Jr., D. W. Dockery, J. D. Spengler, D. O. Stram, and F. E. Speizer, "Effects of Ambient Sulfur Oxides and Suspended Particles on Respiratory Health of Preadolescent Children," *Am. Rev. Respiratory Dis.* **133**, 834–842.
95. D. W. Dockery, J. Cunningham, A. I. Damokosh, L. M. Neas, J. D. Spengler, P. Koutrakis, J. H. Ware, M. Raizenne, and F. E. Speizer. "Health effects of acid aerosols on North American children: respiratory symptoms," *Environmental Health Perspectives* **104**, 500–505.
96. D. E. Abbey, M. D. Lebowitz, P. K. Mills, F. F. Petersen, W. L. Beeson, and R. J. Burchette. "Long-term Ambient Concentrations of Particulates and Oxidants and Development of Chronic Disease in a Cohort of Nonsmoking California Residents," *Inhalation Toxicology* **7**, 19–34 (1995).
97. D. E. Abbey, B. E. Ostro, F. Petersen, and R. J. Burchette. "Chronic Respiratory Symptoms Associated with Estimated Long-term Ambient Concentrations of Fine Particulates Less than 2.5 microns in Aerodynamic Diameter (PM2.5) and Other Air Pollutants," *J. Exposure Analy. Environ. Epidemiology* **5**, 137–159 (1995).
98. D. E. Abbey, B. E. Ostro, G. Fraser, T. Vancuren, and R. J. Burchette. "Estimating Fine Particulates Less that 2.5 microns in Aerodynamic Diameter (PM2.5) from Airport Visibility Data in California," *J. Exposure Analy. Environ. Epidemiology* **5**, 161–180 (1995).
99. M. Raizenne, L. M. Neas, A. I. Damokosh, D. W. Dockery, J. D. Spengler, P. Koutrakis, J. H . Ware, and F. E. Speizer, "Health Effects of Acid Aerosols on North American Children: Pulmonary Function," *Environ. Health Perspectives* **104**, 506–514 (1996).
100. W. Roemer, *Pollution Effects on Asthmatic Children in Europe: The PEACE Study* [dissertation]. Landbouwuniversiteit [Agricultural University] Wageningen, Wageningen, The Netherlands, 1998.
101. W. Roemer, G. Hoek, B. Brunekreef, J. Haluszka, A. Kalandidi, and J. Pekkanen, "Daily Variations in Air Pollution and Respiratory Health in a Multicentre Study: The PEACE Project," *Europ. Respiratory J.*, 1998.
102. D. L. Costa and K. L. Dreher. "Bioavailable Transition Metals in Particulate Matter Mediate Cardiopulmonary Injury in Healthy and Compromised Animal Models," *Environ. Health Perspectives Suppl.* **105**(5), 1053–1060 (1997).
103. M. Osier, and G. Oberdörster, "Intratracheal Inhalation vs Intratracheal Instillation: Differences in Particle Effects," *Fundamental Appl. Toxicol.* **40**, 220–227 (1997).
104. J. J. Godleski, C. Sioutas, M. Katler, and P. Koutrakis. "Death from Inhalation of Concentrated Ambient Air Particles in Animal Models of Pulmonary Disease," *Am. J. Respiratory Critical Care Med.* **153**, A15 (1996).
105. W. D. Bennett and K. L. Zeman. "Deposition of Fine Particles in Children Spontaneously Breathing at Rest," *Inhalation Toxicol.* **10**, 831–842 (1998).
106. C. S. Kim and S. C. Hu. "Regional Deposition of Inhaled Particles in Human Lungs: Comparison between Men and Women," *J. Appl. Physiology* **84**, 1834–1844 (1998).

107. U.S. Environmental Protection Agency, *Air Quality Criteria for Carbon Monoxide*, Office of Health and Environmental Assessment, Environmental Criteria and Assessment Office, Research Triangle Park, NC, EPA/600/8-90/045F, 1991.

108. U.S. Environmental Protection Agency, *Air Quality Criteria for Carbon Monoxide [External Review Draft]*. National Center for Environmental Assessment, Research Triangle Park, NC, EPA/600/R-98/148, 1998.

109. M. T. Kleinman. "Health Effects of Carbon Monoxide," in M. Lippmann, ed., *Environmental Toxicants: Human Exposures and Their Health Effects*, Van Nostrand Reinhold, New York, 1992, pp. 98–111.

110. R. Bascom, P. A. Bromberg, D. L. Costa, R. Devlin, D. W. Dockery, M. W. Frampton, W. Lambert, J. M. Samet, F. E. Speizer, and M. Utell. "Health Effects of Outdoor Air Pollution [part 2]," *Am. J. Respiratory Critical Care Med.* **153**, 477–498 (1996).

111. D. G. Penney, ed. *Carbon Monoxide*, CRC Press, Boca Raton, FL, 1996.

112. R. F. Coburn, R. E. Forster, and P. B. Kane. "Considerations of the Physiological Variables that Determine the Blood Carboxyhemoglobin Concentration in Man," *J. Clinical Investigation* **44**, 1899–1910 (1965).

113. B. L. Drinkwater, P. B. Raven, S. M. Horvath, J. A. Gliner, R. O. Ruhling, N. W. Bolduan, and S. Taguchi. "Air pollution, exercise, and heat stress," *Archives of Environmental Health* **28**, 177–181.

114. P. B. Raven, B. L. Drinkwater, R. O. Ruhling, N. Bolduan, S. Taguchi, J. Gliner, and S. M. Horvath, "Effect of Carbon Monoxide and Peroxyacetyl Nitrate on Man's Maximal Aerobic Capacity," *J. Appl. Physiology* **36**, 288–293 (1974).

115. S. M. Horvath, P. B. Raven, T. E. Dahms, and D. J. Gray. "Maximal Aerobic Capacity at Different Levels of Carboxyhemoglobin," *J. Appl. Physiology* **38**, 300–303 (1975).

116. E. N. Allred, E. R. Bleecker, B. R. Chaitman, T. E. Dahms, S. O. Gottlieb, J. D. Hackney, M. Pagano, R. H. Selvester, S. M. Walden, and J. Warren. "Short-Term Effects of Carbon Monoxide Exposure on the Exercise Performance of Subjects with Coronary Artery Disease," *New Engl. J. Med.* **321**, 1426–1432 (1989); "Acute Effects of Carbon Monoxide Exposure on Individuals with Coronary Artery Disease," *Research Report No. 25*. Health Effects Institute, Cambridge, MA, (1989); "Effects of Carbon Monoxide on Myocardial Ischemia," *Environ. Health Perspectives* **91**, 89–132 (1991).

117. E. W. Anderson, R. J. Andelman, J. M. Strauch, N. J. Fortuin, and J. H. Knelson. "Effect of Low-level Carbon Monoxide Exposure on Onset and Duration of Angina Pectoris: A Study in Ten Patients with Ischemic Heart Disease," *Ann. Internal Med.* **79**, 46–50 (1973).

118. K. F. Adams, G. Koch, B. Chatterjee, G. M. Goldstein, J. J. O'Neil, P. A. Bromberg, D. S. Sheps, S. McAllister, C. J. Price, and J. Bissette. "Acute Elevation of Blood Carboxyhemoglobin to 6% Impairs Exercise Performance and Aggravates Symptoms in Patients with Ischemic Heart Disease," *J. Am. College Cardiology* **12**, 900–909 (1988).

119. D. S. Sheps, K. F. Adams, Jr., P. A. Bromberg, G. M. Goldstein, J. J. O'Neil, D. Horstman, and G. Koch, "Lack of Effect of Low Levels of Carboxyhemoglobin on Cardiovascular Function in Patients with Ischemic Heart Disease," *Arch. Environ. Health* **42**, 108–116 (1987).

120. M. T. Kleinman, D. M. Davidson, R. B. Vandagriff, V. J. Caiozzo, and J. L. Whittenberger. "Effects of Short-Term Exposure to Carbon Monoxide in Subjects with Coronary Artery Disease," *Archives of Environmental Health* **44**, 361–369 (1989): M. T. Kleinman, D. A. Leaf, E. Kelly, V. Caiozzo, K. Osann, and T. O'Neill. "Urban Angina in the Mountains: Effects of Carbon Monoxide and Mild Hypoxemia on Subjects with Chronic Stable Angina," *Archives of Environmental Health* **50**(Nov./Dec., 1998).

121. D. S. Sheps, M. C. Herbst, A. L. Hinderliter, K. F. Adams, L. G. Ekelund, J. J. O'Neil, G. M. Goldstein, P. A. Bromberg, J. L. Dalton, M. N. Ballenger, S. M. Davis, and G. G. Koch, "Production of Arrhythmias by Elevated Carboxyhemoglobin in Patients with Coronary Artery Disease," *Ann. Internal Med.* **113**, 343–351 (1990).
122. R. D. Morris, E. N. Naumova, and R. L. Munasinghe. "Ambient Air Pollution and Hospitalization for Congestive Heart Failure among Elderly People in Seven Large US Cities," *Am. J. Public Health* **85**, 1361–1365 (1995).
123. J. Schwartz, "Air Pollution and Hospital Admissions for Cardiovascular Disease in Tucson," *Epidemiology* **8**, 371–377 (1997).
124. R. T. Burnett, R. E. Dales, J. R. Brook, M. E. Raizenne, and D. Krewski. "Association between Ambient Carbon Monoxide Levels and Hospitalizations for Congestive Heart Failure in the Elderly in 10 Canadian Cities," *Epidemiol.* **8**, 162–167 (1997).
125. S. M. Horvath, T. E. Dahms, and J. F. O'Hanlon. "Carbon Monoxide and Human Vigilance: A Deleterious Effect of Present Urban Concentrations," *Archives of Environmental Health* **23**, 343–347 (1951).
126. G. G. Fodor and G. Winneke. "Effect of Low CO Concentrations on Resistance to Monotony and on Psychomotor Capacity," *Staub-Reinhaltung der Luft* **32**, 46–54 (1972).
127. V. R. Putz, B. L. Johnson, and J. V. Setzer, *Effects of CO on Vigilance Performance: Effects of Low Level Carbon Monoxide on Divided Attention, Pitch Discrimination, and the Auditory Evoked Potential*, U.S. Department of Health, Education, and Welfare, National Institute for Occupational Safety and Health, Cincinnati, OH, Report No. NIOSH-77-124, 1976.
128. V. R. Putz, B. L. Johnson, and J. V. Setzer, "A Comparative Study of the Effects of Carbon Monoxide and Methylene Chloride on Human Performance," *J. Environ. Pathology Toxicol.* **2**, 97–112 (1979).
129. V. A. Benignus, K. E. Muller, C. N. Barton, and J. D. Prah. "Effect of Low Level Carbon Monoxide on Compensatory Tracking and Event Monitoring," *Neurotoxicology and Teratology* **9**, 227–234 (1987).
130. M. J. Ellenhorn and D. G. Barceloux, eds. Carbon monoxide. In *Medical Toxicology: Diagnosis and Treatment of Human Poisoning*, pp. 820–828. New York, NY: Elsevier.
131. J. Schwartz and R. Morris, "Air Pollution and Hospital Admissions for Cardiovascular Disease in Detroit, Michigan," *Am. J. Epidemiol.* **142**, 23–35 (1995).
132. R. T. Burnett, S. Cakmak, and J. R. Brook. "The Effect of the Urban Ambient Air Pollution Mix on Daily Mortality Rates in 11 Canadian Cities," *Canadian J. Public Health* **89**, 152–156.
133. R. T. Burnett, S. Cakmak, M. E. Raizenne, D. Stieb, R. Vincent, D. Krewski, J. R. Brook, O. Philips, and H. Ozkaynak. "The Association between Ambient Carbon Monoxide Levels and Daily Mortality in Toronto, Canada," *J. Air Waste Manage. Assoc.* **48**, 689–700.
134. F. B. Stern, R. A. Lemen, and R. A. Curtis, "Exposure of Motor Vehicle Examiners to Carbon Monoxide: A Historical Prospective Mortality Study," *Archives Environ Health* **36**, 59–66 (1981).
135. F. B. Stern, W. E. Halperin, R. W. Hornung, V. L. Ringenburg, and C. S. McCammon, "Heart Disease Mortality among Bridge and Tunnel Officers Exposed to Carbon Monoxide," *Am. J. Epidemiol.* **128**, 1276–1288 (1988).
136. J. Ström, L. Alfredsson, and T. Malmfors, "Carbon Monoxide: Causation and Aggravation of Cardiovascular Diseases—A Review of the Epidemiologic and Toxicologic Literature," *Indoor Environ.* **4**, 322–333 (1995).

137. U.S. Environmental Protection Agency. *National Air Quality and Emissions Trends Report, 1996*, Office of Air Quality Planning and Standards, Research Triangle Park, NC, EPA-454/R-97-013, 1998.

138. V. A. Benignus, 1994. "Behavioral Effects of Carbon Monoxide: Meta Analyses and Extrapolations," *J. Appl. Physiol.* **76**, 1310–1316 (1994).

139. U.S. Environmental Protection Agency. *Air Quality Criteria for Lead.* Office of Health and Environmental Assessment, Environmental Criteria and Assessment Office, Atlanta, GA, EPA-600/8-83/028aF-dF 4v, 1986.

140. U.S. Environmental Protection Agency, *Air Quality Criteria for Lead: Supplement to the 1986 Addendum*, Office of Health and Environmental Assessment, Environmental Criteria and Assessment Office, Research Triangle Park, NC, EPA/600/8-89/049F, 1990.

141. World Health Organization. "Inorganic Lead," *Environmental Health Criteria 165.* World Health Organization, Geneva, Switzerland, 1995.

142. Agency for Toxic Substances and Disease Registry (ATSDR), *The Nature and Extent of Lead Poisoning in Children in the United States: A Report to Congress.* U.S. Department of Health and Human Services, Public Health Service, Atlanta, GA, 1988, available from: NTIS, Springfield, VA, PB89-100184/XAB.

143. U.S. Centers for Disease Control, *Preventing Lead Poisoning in Young Children: A Statement by the Centers for Disease Control—October 1991.* U.S. Department of Health & Human Services, Public Health Service, Atlanta, GA, 1991.

144. J. M. Davis, R. W. Elias, and L. D. Grant. "Efforts to Reduce Lead Exposure in the United States," in M. Yasui, M. J. Strong, K. Ota, and M. A. Verity, eds., *Mineral and Metal Neurotoxicology*, CRC Press, Inc., Boca Raton, FL, 1996, pp. 285–293.

145. L. D. Grant and J. M. Davis. "Effect of Low-Level Lead Exposure on Paediatric Neurobehavioural Development: Current Findings and Future Directions," in M. Smith, L. D. Grant, and A. Sors, eds., *Lead Exposure and Child Development: An International Assessment*, Kluwer Academic Publishers, Boston, MA, 1989, pp. 49–115.

146. J. M. Davis and D. J. Svendsgaard. "Lead and Child Development," *Nature (London)* **329**, 297–300 (1987).

147. J. M. Davis. "Risk Assessment of the Developmental Neurotoxicity of Lead," *Neurotoxicology* **11**, 285–292 (1990).

148. J. M. Davis and R. W. Elias, "Risk Assessment of Metals," in L. W. Chang, L. Magos, and T. Suzuki, eds., *Toxicology of Metals*, CRC Lewis Publishers, Boca Raton, FL, 1996, pp. 55–67.

149. D. C. Bellinger, K. M. Stiles, and H. L. Needleman. "Low-level Lead Exposure, Intelligence and Academic Achievement: A Long-Term Follow-up Study," *Pediatrics* **90**, 855–861 (1992).

150. K. N. Dietrich, O. G. Berger, P. A. Succop, P. B. Hammond, and R. L. Bornschein. "The Developmental Consequences of Low to Moderate Prenatal and Postnatal Lead Exposure: Intellectual Attainment in the Cincinnati Lead Study Cohort Following School Entry," *Neurotoxicology and Teratology* **15**, 37–44 (1993).

151. C. B. Ernhart, M. Morrow-Tlucak, M. R. Marler, and A. W. Wolf. Low level lead exposure in the prenatal and early preschool periods: early preschool development. *Neurotoxicology and Teratology* **9**, 259–270.

152. P. A. Baghurst, A. J. McMichael, N. R. Wigg, G. V. Vimpani, E. F. Robertson, R. J. Roberts, and S.-L. Tong. "Environmental Exposure to Lead and Children's Intelligence at the Age of Seven Years; the Port Pirie cohort study," *New England Journal of Medicine* **327**, 1279–1284.

153. G. H. Cooney, A. Bell, W. McBride, and C. Carter. "Low-Level Exposures to Lead: the Sydney Lead Study," *Develop. Med. Child Neurology* **31**, 640–649 (1989).

154. G. Wasserman, J. H. Graziano, R. Factor-Litvak, D. Popovac, N. Morina, A. Musabegovic, N. Vrenezi, S. Capuni-Paracka, V. Lekic, E. Preteni-Redjepi, S. Hadzialjevic, V. Slavkovich, J. Kline, P. Shrout, and Z. Stein, "Independent Effects of Lead Exposure and Iron Deficiency Anemia on Developmental Outcome at age 2 Years," *J. Pediatrics* **121**, 695–703 (1992).

155. J. M. Davis, D. A. Otto, D. E. Weil, and L. D. Grant. "The Comparative Developmental Neurotoxicity of Lead in Humans and Animals," *Neurotoxicology and Teratology* **12**, 215–229 (1990).

CHAPTER FIFTY-NINE

Air Pollution Controls

Mark A. Golembiewski, CIH and Fredrick I. Cooper

This chapter focuses on the requirement for air pollution control, principles of air cleaning, selection of suitable control methods, and economic and energy resources needed to install and operate emission control systems of various types.

1 RELATION OF ATMOSPHERIC EMISSIONS TO WORKPLACE AIR QUALITY

Although the association between air pollution controls and industrial hygiene should be obvious to a practicing industrial hygienist, it is worthwhile to review the basis for this important concept. In this day of specialization, some tend to think of air pollution engineering as a field of endeavor somewhat removed from industrial hygiene. This unfortunate compartmentalization, found all too often in governmental and industrial organizations, can detract significantly from a full understanding of air pollution and industrial hygiene, fostering less than a total approach to effective problem solving in either air cleaning or occupational health. Only when an industrial hygienist applies expertise to achieve the optimal overall control strategy will due regard be given to controlling occupational exposures and maintaining acceptable ambient air quality.

The most basic concept common to air pollution control and industrial hygiene is the "source-process," which represents a common denominator between workroom air quality and atmospheric emissions. With this concept as a basis, the interrelating effect of the source-process falls into two general categories: operations with and without direct atmospheric exhaust.

Patty's Industrial Hygiene, Fifth Edition, Volume 4. Edited by Robert L. Harris.
ISBN 0-471-29749-6

1.1 Operations Exhausted Directly to Atmosphere

When an industrial operation or process is exhausted directly to the atmosphere, as with a hooded metallurgical furnace, workroom air quality is affected directly by the design and performance of the exhaust system. An improperly designed hood or a hood evacuated with a less-than-sufficient volumetric rate of air will allow contamination of the occupational environment and affect workers in the vicinity of the furnace. This simple, yet clear, illustration is but one form of the close relation between atmospheric emissions and occupational exposure.

Another example of this close relationship is the use of coke-side sheds adjacent to coke-oven batteries. These shed structures, when properly designed and evacuated, can provide effective capture of virtually all particulate and gaseous emissions emanating from the coke side of coke-oven batteries: door leaks, pushing emissions, and emissions from the coke car in transit to the quenching station. On the other hand, the same structures present a semiconfined space for containing coke-side emissions, and under conditions of inadequate exhaust, they can restrict the dilution and dispersion of emissions from the coke ovens, thereby jeopardizing the quality of the work environment for persons within the shedded structure. In this instance, the shed serves as the first of four components of the typical "local" exhaust system: hood, ductwork, air cleaner, and air-moving device (1).

1.2 Operations Exhausted Indirectly to Atmosphere

In some situations, the first step in the eventual, ultimate outdoor emission of materials generated from a source-process is the dispersion of the contaminant throughout an enclosed workplace, followed by significant release to the atmosphere throughout the general ventilation system, natural or mechanical. This delayed, decentralized mechanism for atmospheric release has been a reality in industry for decades, but it draws increasing attention as our regulatory standards impinge on the generation and emission of toxic materials, including those deemed "hazardous air pollutants" by the Administrator of the Environmental Protection Agency (EPA) (2). Materials such as beryllium, asbestos, and vinyl chloride are officially labeled "air pollutants" without regard to the location or mechanism of atmospheric release-dispersion and subsequent release through general ventilation, including natural exfiltration through doors, windows, and other openings in a building structure.

1.3 Recirculation of Exhaust Air

The recirculation of exhaust air back into the workroom provides a clear example of the close relationship between atmospheric emissions and workplace air quality. Industrial hygienists have long been concerned about the undesirable recirculation of exhaust air, particularly in the case of atmospheric emissions that are captured in the ambient airflow wake of a building. This phenomenon can occur when an exhaust discharge point is close to the building roof line and the contaminated air is drawn back into the building by virtue of natural or mechanical ventilation.

There has been a long-standing tradition in industrial hygiene engineering to disallow the intentional recirculation of any exhaust air, even after cleaning, if the air contains toxic materials. As the costs of heating and cooling make-up air have risen significantly, however, this maxim has been modified. In some cases, recirculation may be the only really feasible method of providing suitable general ventilation. In other cases, the economic and energy conservation considerations may be important enough to warrant the additional capital and operating expenses required to safely provide the desired air recirculation.

The feasibility of recirculation and the final design of a suitable system is now understood to depend on a number of industrial hygiene and engineering factors. The following criteria should be used to establish the feasibility of recirculation and to assure adequate worker protection (3):

1. The chemical, physical and toxicological characteristics of the contaminants in the airstream to be recirculated must be identified and evaluated. Exhaust air containing contaminants whose toxicity is unknown, or for which there is no established safe exposure level, should not be recirculated.
2. All federal, state, and local regulations regarding recirculation must be reviewed to determine if the recirculation system under review may be restricted or prohibited.
3. The availability of a suitable air cleaning device must be determined. The air cleaner must be capable of yielding a contaminant concentration in the effluent airstream that is sufficiently low to achieve acceptable workplace concentrations.
4. The effect of a malfunction of the recirculation system must be addressed. Recirculation should not be attempted if a malfunction could result in exposure levels that would cause serious worker health problems. Substances that can cause permanent damage or significant physiological harm from a short overexposure should not be recirculated.
5. The recirculation system must incorporate a monitoring system that provides an accurate warning signal if a malfunction should occur, and that is capable of initiating corrective action or process shutdown before harmful contaminant concentrations can build up in the workplace. Monitoring can be accomplished by a number of methods, depending on the type and hazardous nature of the substance. Examples include area monitoring for nuisance-type substances and in-line monitors or filter pressure drop indicators for more hazardous materials.

The permissible concentration of a contaminant in recirculated air may be calculated by the following equation if the system is operating under steady-state conditions (3):

$$C_r = 0.5(TLV - C_0)\frac{Q_t}{Q_r}\frac{1}{K} \tag{1}$$

where C_r = concentration of the contaminant in the recirculated air stream from the air cleaning system before any mixing

TLV = threshold limit value of the contaminant (4)

C_0 = concentration of the contaminant in the worker's breathing zone without recirculation

Q_t = total ventilation flow rate through the affected work space

Q_r = recirculated air flow rate
K = "effectiveness of mixing" factor, usually between 3 ("good mixing") and 10 ("poor mixing")

Figure 59.1 gives an example of a suitable system for the recirculation of air from a particulate air cleaning device.

2 EMISSION CONTROL REQUIREMENTS

Requirements for emission controls generally find their basis in two broad types of overlapping expectations, regulatory standards and engineering performance specifications. Most air pollution regulations in the United States relate to, or at a minimum conform to, federal regulations based on the Clean Air Act as amended (2). Engineering specifications for air pollution controls have developed into increasingly precise stipulations that often reference the regulatory constraints.

2.1 Federal Clean Air Act

The Clean Air Act (CAA) of 1970 provided the EPA with broad powers to adopt and enforce air pollution emission regulations. The agency subsequently promulgated primary

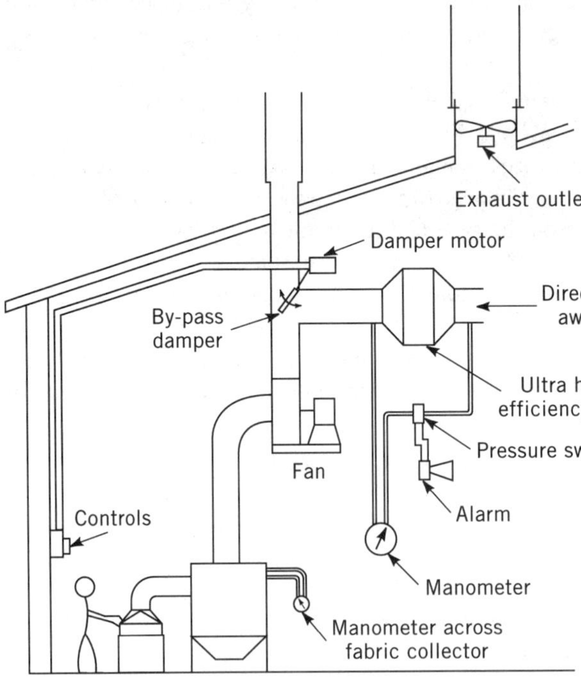

Figure 59.1. Example of a recirculation system from a particulate control device.

(designed to protect public health) and secondary (designed to protect public welfare) National Ambient Air Quality Standards (NAAQS). These standards have set maximum ambient concentrations for sulfur dioxide, particulate matter less than 2.5 microns (PM-2.5), carbon monoxide, nitrogen dioxide, ozone, and lead. The CAA authorizes the EPA Administrator to designate air quality-control regions (AQCRs) and states are required to list the status of NAAQS attainment for each criteria pollutant in each AQCR.

The principal method for attaining and maintaining the NAAQS is the state implementation plan (SIP), which is prepared by each state and subject to final approval by EPA (5). These plans are directed at source control and establish timetables for attaining the NAAQS. They are intended to provide the framework for determining whether individual facilities operate in a manner consistent with the achievement of air quality goals. States may adopt requirements that are more stringent than the federal regulations.

Amendments to the CAA in 1977 and 1990 have also produced requirements that address attainment areas and the prevention of significant deterioration of air quality (PSD) within those areas, as well as nonattainment areas (2). The PSD regulations apply to areas where ambient air quality is superior to the NAAQS and contain detailed permit provisions governing construction and modification of major stationary sources. The PSD provisions establish three classes of clean air areas and maximum allowable increases in pollution levels over base-line measurements. Accordingly, permission for the construction or expansion of industrial plants with major emissions is contingent on increasing the ambient pollutant levels by no more than the levels shown in Table 59.1. These facilities must use the best available control technology (BACT). The PSD provisions also establish requirements to protect visibility in PSD areas.

An area where the air quality is measured or estimated to be worse than the NAAQS is designated as a "nonattainment area." States must implement a program to ensure at-

Table 59.1. Significant Deterioration Limits

Pollutant	Deterioration Limitations (μg/m^3)		
	Area Class I	Area Class II	Area Class III
	Particulate Matter (PM-10)		
Annual	4	17	34
24-hour maximum	8	30	60
	Sulfur Dioxide		
Annual geometric mean	2	20	40
24-hour maximum	5	91	182
3-hour maximum	25	512	700
	Nitrogen Oxides		
Annual	2.5	25	50

[a]Ref. 5.

tainment of the NAAQS in these areas, including a permit program similar to that established under the PSD regulations. New or modified emission sources in a nonattainment area, however, are subject to an emission offset policy. That is, emissions from the existing sources in the area must be reduced by an amount more than adequate to offset the new plant's emissions. In addition, the new source must use lowest achievable emission rate (LAER) technology.

The most direct and explicit requirements for emission controls set in motion by the CAA are those for a list of major sources and for highly toxic substances.

New Source Performance Standards (NSPS) are specific limitations for selected stationary source categories that might contribute significantly to pollution levels. This provision of the law was intended to ensure that new stationary sources be designed, built, equipped, and maintained to minimize emissions, regardless of whether the sources are located in an attainment or a nonattainment area. The NSPS are designed to reflect the degree of emission reduction that EPA determines to be achievable for each source category by applying the "best technological system of continuous emission reduction" that has been adequately demonstrated for the source.

National Emission Standards for Hazardous Air Pollutants (NESHAPs) have been established under Section 112 of the CAA. These regulations apply to specific substances for which no ambient air quality standards have been set, and that cause or contribute to air pollution that may result in an increase in mortality or cause serious irreversible or incapacitating illness. The limited list of such hazardous substances now includes asbestos, beryllium, mercury, vinyl chloride, benzene, radionuclides, inorganic arsenic, coke oven emissions, radionuclides, radon, and fugitive emissions from equipment leaks.

The Clean Air Act amendments enacted in November 1990 touch on virtually every aspect of air pollution law. They regulate mobile and stationary sources, large and small businesses, routine and toxic emissions, and consumer products.

The essence of the bill's air toxics section is a list of 189 Hazardous Air Pollutants (HAPs) that EPA must regulate. The Agency's task is to determine "maximum achievable control technology" (MACT) for these substances. EPA may add or delete chemicals from the list, and private parties may petition for additions and deletions. The Agency has 10 years to issue MACT standards for all sources of the 189 chemicals. Sources that emit more than 10 tons per year of any substance or 25 tons per year of any combination of listed chemicals are covered by this title. EPA must establish categories for these sources for the purpose of promulgating standards.

The MACT standards will differ for new and existing sources. MACT for new sources refers to the most stringent level currently achieved with a similar source. MACT for existing sources would be the average control in place for the best 12% of similar sources. For any category with fewer than thirty sources, MACT would be the average of the five best-performing sources. In addition, any source making a 90% reduction of a listed chemical will have a six-year extension in the deadline for achieving the MACT standard.

Six years from enactment, EPA must complete a study and report to Congress on the level of risk remaining (residual risk) after the MACT reductions are in place. If Congress does not act on these recommendations, EPA must develop health-based emission standards that would limit the cancer risk to exposed individuals to no more than one case in one million.

2.2 Typical Regulatory Requirements

Within today's multiplicity of regulatory constraints for abating air pollution from stationary sources, three basic requirements emerge to affect the operator or owner of emission sources. These are quantitative emission standards, subjective prohibitions, and installation and operating permits (air use approval).

2.2.1 Quantitative Emission Standards

Local, state, and federal regulations provide complex spectrum of emission standards, ranging from general specifications of engineering standards to very precise emission limitations in specific concentration or mass-effluent units. Increasingly, these emission limitations are related to specific sampling and analytical methods. Typical of this class of regulations are those referenced to process-weight, effluent grain loading, and efflux weight limitations.

2.2.2 Subjective Prohibitions

2.2.2.1 Plume Visibility. Regulations covering plume appearance are usually expressed in terms of an allowable equivalent opacity or Ringelmann number, a subjective measure of visual obscuration. The original use of plume visibility as a regulatory tool covered the evaluation of the density of black smoke. Today most visual plumes are neither black nor soot-containing; their visibility is due to fly ash, dust, fumes, and condensed water vapor, and they often are white or light colored.

Although plume visibility is widely used by enforcement agencies as a convenient index for monitoring emission sources, certain inconsistencies in the equivalent opacity concept are obvious to the objective mind. These include the dependence of plume opacity on particle size, shape, and refractivity, the typical lack of correlation between opacity and the amount of particulate matter in a plume, and the dependence of the opacity on background and relative position of the observer with respect to the sun.

To decrease the subjective element in plume appearance reading, the EPA and other agencies have developed "smoke schools," where observers are certified on their ability to read "calibrated" dark and light plumes accurately and consistently.

2.2.2.2 Nuisance and Trespass. Typical of the nuisance and trespass class of restriction are the portions of air pollution regulations prohibiting the emission of an air contaminant that causes or "will cause" detriment to the safety, health, welfare, or comfort of any person, or that causes or "will cause" damage to property or business. Such sections of regulations are typically known as "nuisance clauses," and they supplement the quantitative or objective restrictions, as well as providing for administrative and legal relief to receptors when emissions jeopardize aesthetic values, comfort, or well-being.

2.2.3 Installation and Operating Permits

Title V of the 1990 Clean Air Act Amendments establishes a federally enforceable permit system in which major facilities need detailed permits that contain all applicable emission control requirements.

States must also meet minimum criteria for permit enforcement programs. EPA has authority to object to a state-issued permit if it conflicts with Clean Air Act requirements. The permitting authority has 90 days to revise the permit; otherwise, EPA must issue a final ruling on the permit.

In addition to the Title V permitting requirements for major facilities, major and minor facilities are required to obtain permits from the state and local air pollution control agency. Agencies maintain a two-tiered permit system: An "installation permit" (or a "permit to construct") is a sanction granted by the regulatory agency before construction begins and after compliance of the proposed source with all relevant regulations is assured. An "operating permit" (or a "permit to operate") is the second permit issued after the source is constructed and begins operation.

2.2.3.1 Predictive Air Quality Modeling. Facility applications for construction permits often employ air quality dispersion modeling to predict the impact of the proposed emission source (and emission control systems) in terms of expected downwind concentrations of the contaminant. The calculated ambient air contributions of the proposed source may be compared with relevant standards, such as the national ambient air quality standards; PSD limits for sulfur dioxide, nitrogen dioxide, and particulate matter; or threshold ambient concentrations established for toxic air contaminants. Modeling may be mandated by the EPA for the various state agencies as they administer the EPA-approved implementation plans designed to achieve the ambient concentrations specified in the national air quality standards and to restrict significant increases in the atmospheric burden in regions of superior air quality.

2.2.3.2 Certification. After installation of the source or control system is complete and the system begins to operate, a second permit must be obtained in many jurisdictions, attesting that the source operates in a manner satisfactory to the control agency. An operating permit is granted after inspection and observation of the operating facility by a representative from the air pollution control agency. In many instances, compliance source testing is required to document the level of effluent with reference to the applicable emission limitations.

2.3 Engineering Performance Specifications

Today's requirements for an emission control system usually include engineering performance specifications. It is technically sound, in general, that the performance specifications of air or gas cleaning equipment for a given application include any or all of the following factors:

1. Range of air-cleaning capacity in terms of exhaust air volumetric flow rate
2. The exit concentration of material in the system exhaust gases
3. Mass emission rate of discharged materials
4. Efficiency of control system based on mass flow rate
5. Acceptance test methods

The value of specifying these parameters among such traditional specifications as materials of construction, fabrication, and inspection is obvious, yet these factors are important enough to warrant review.

2.3.1 Range of Air-Cleaning Capacity

An emission control system must be designed to operate at sufficient capacity to exhaust all the gases that come from the contaminant-generating source. The application of even the latest technology to a given source is rather meaningless if the control unit is undersized and unable to capture and exhaust all the contaminants. Examples of inadequate system design are commonplace in industry, especially in primary and secondary metallurgical industries.

2.3.2 Exit Concentration of Material

If the object of an air-cleaning system is the recovery of materials, it is important to know the percentage or amount of material collected by the equipment. Usually, however, even though the recovery of material is an important consideration, the percentage of material collected is only incidental. If the object is to produce air of better quality, either for supply to occupied spaces or for discharge outdoors, performance should be specified in terms of the concentration of contaminant in the effluent airstream.

2.3.3 Emission Rate of Discharged Material

The emission rate of material (e.g., mass per unit time) is defined by the concentration of material in the exit gas stream and the volumetric flow rate of the gas stream. Specification of only one or the other of these two factors is insufficient to characterize the expected performance of an air pollution control system. In many applications, poor maintenance or other factors may result in an exit flow rate from a control system that is greater than the inlet flow rate; therefore, specification of the concentration alone could be misleading because of a dilution effect caused by the increased outlet flow rate.

Furthermore, the real impact of air pollution must be measured in terms of the emission rate to the atmosphere. Only this parameter provides a meaningful index of mass contribution to the atmospheric burden.

Finally, it is important to specify the emission rate of an air pollution control system either directly or indirectly (by flow rate and by concentration) because emission regulations are invariably expressed in terms of mass emission rate or emission factors that depend on both the mass emission rate and the process rate. When combined, mass emission rate and process rate yield such dimensional emission terms as "pounds per ton of product produced", or "pounds per ton of feed material".

2.3.4 Collector Efficiency

The term "efficiency" in air pollution control does not refer to the proportion of energy used effectively in the cleaning process. Rather, "efficiency" is used almost exclusively to indicate the proportion of material removed from an airstream, without regard to the amount of power required. Energy consumption, an important consideration in air pollution control, is treated separately in Section 3.

If an air contaminant were completely removed from an airstream by an air-cleaning device, the collection efficiency would be rated as 100%. The general definition of "collection efficiency" with respect to the degree of separation affected by an air pollution control system is the "emission rate entering the collector minus the emission rate leaving the collector divided by the emission rate entering the collector." For example, if the mass flow rate of material in a contaminated airstream entering a collector is 100 kg/h and the emission rate is 5 kg/h, the "efficiency" of the control device is

$$100\left(\frac{100 - 5}{100}\right), \quad \text{or } 95\%$$

2.3.5 *Acceptance Test Methods*

Measurement of the control efficiency of a collector is accomplished with source sampling equipment operating as a miniature-scale separation and collection system with very high inherent effectiveness. The measured emissions, however, depend very much on the method of sampling. Therefore, the performance of an air- or gas-cleaning system must be specified with reference to the sampling method to be used to determine the nature, amount, and mass flux of contaminants in the inlet and outlet of the system. Accordingly, the specific method of testing should be named in a statement of expected performance. If the specified method is not a standard method, such as an ASTM or EPA method, the sampling device or method should be described in enough detail to permit replication.

3 ECONOMIC IMPACT OF AIR POLLUTION CONTROLS

Air pollution represents a problem with inherent conflict because the goals of public policy are usually "noneconomic," yet have important economic aspects. Federal, state, and local governments have major programs aimed at the prevention and abatement of atmospheric pollution, but many economic issues play an increasingly important role in the promulgation of these regulations. The standard setting process often involves a balancing of costs and environmental impacts, of costs and dollar valuation of benefits, or of environmental impacts and the economic consequences of control costs. Cost analysis is also used to select the most suitable regulatory option from among alternatives that impose the same level of stringency. Once the environmental goal is established, a cost-effectiveness study is conducted to determine the method of achieving the goal with the least economic impact.

A macroeconomic analysis of air pollution control issues, although a very interesting and relevant topic, is not the subject of this section. Rather, the goal here is to present some macroeconomic considerations in air pollution control in terms of the cost-benefit relationships, the resources necessary to apply certain types of control, and the minimum requirements of an economic feasibility analysis needed to specify the most "economic" control methods for a given stationary source of air pollution from among suitable technical alternatives.

Many air pollution control problems can be solved by more than one alternative measure, but to identify the optimal method for controlling emissions, each technically adequate solution must be evaluated carefully before implementation. Sometimes basic measures, such as the substitution of fuels or raw materials and the modification or replacement of processes, provide the most cost-effective solution. These fundamental avenues of "front end" modification should never be overlooked (see Section 4).

3.1 Cost-Effectiveness Relationships

A cost-effectiveness "variable" measures all costs associated with a given project as a function of achievable reduction in pollutant emissions. For example, when estimating the total cost of an emission control system one must consider raw materials and fuel used in the control process, needed alterations in process equipment, control hardware, auxiliary equipment, disposal or reuse of collected materials, control equipment maintenance, and similar factors.

Figure 59.2 plots a typical cost-effectiveness relationship in stationary source emission control. The cost of control is represented on the vertical axis, and the quantity of material discharged is represented on the horizontal axis. Point P indicates the uncontrolled state at which there are no control costs. As control efficiency improves, the quantity of emissions is reduced and the cost of control increases. In most cases, the marginal cost of control is smaller at lower levels of efficiency (higher emissions) near point P of the curve. This cost-effectiveness curve demonstrates an important reality: as the degree of control increases, greater increments of costs are usually required for corresponding increments in emission abatement.

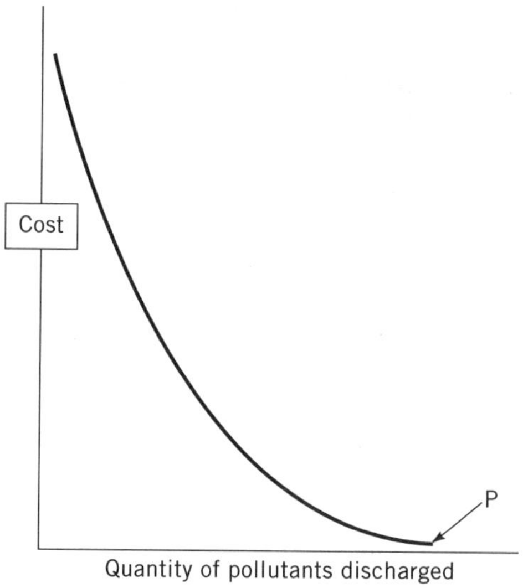

Figure 59.2. Cost-effective relationship in stationary source emission control (6).

Obviously, cost-effectiveness information is needed for sound decision making in emission control. Although several technically feasible measures may be available for controlling an emission source, the least-cost solution for a source can be calculated at various levels of control in most cases. After each alternative is evaluated, and future process expansions and the likelihood of more rigid control restrictions are considered, sufficient information should be available on which to base an intelligent control decision.

Other economic considerations that may impact the selection of air pollution controls include potential liability issues, reduction or elimination of regulatory concerns, public relations (company image), and employee morale. These factors are not typically included in the engineer's economic analysis because they are difficult to quantify. However, they are beginning to come into play more often in the selection process as management weighs all potential impacts of their control decisions.

3.2 Control Cost Elements

The cost of installing and operating an air pollution control system depends on many direct and indirect cost factors. An accurate analysis of control costs for a specific emission source must include an evaluation of all relevant factors. The definable control costs are those directly associated with the installation and operation of control systems. These cost elements have been compiled into a logical organization for accounting purposes and are presented in Table 59.2. The control cost elements that receive attention in this discussion are capital investment, maintenance and operating costs (including energy consumption), and conservation of material and energy in the design and operation of air pollution control systems.

3.3 Capital Investment

The total capital investment required for a given control system includes all costs of purchasing the necessary physical equipment, the costs of labor and materials for installing that equipment, costs for site preparation and for new buildings or modification of existing facilities, and certain other costs that are termed indirect installation costs.

The *installed cost* quoted by a manufacturer of air pollution control equipment should result from an analysis of the specific emission source. This cost usually includes three of the capital investment items listed in Table 59.2, i.e., instrumentation costs, auxiliary equipment costs, and direct costs for field installation. Basic control hardware usually includes built-in instrumentation and pumps.

Purchase costs are the amounts charged by a manufacturer for equipment constructed of standard materials. Purchase costs depend to a large degree on the size and collection efficiency of the control device. In addition, equipment fabricated with special materials for extremely high temperatures or corrosive applications generally costs much more than equipment constructed from standard materials.

3.4 Maintenance and Operating Costs

The costs of operating and maintaining air pollution controls depend not only on the inherent characteristics of the different control methods, but also on a wide range of quality

Table 59.2. Control Cost Elements[a]

Capital investment
Purchased equipment cost
Primary control device
Auxiliary equipment
Instrumentation
Freight
Direct installation cost
Foundation and supports
Handling and erection
Electrical, piping, insulation, and painting
Indirect installation cost
Engineering
Construction and field expenses
Startup
Performance testing
Contingencies
Land
Site preparation
Buildings or structure modifications
Maintenance and operation
Utilities
Supplies and materials (includes both raw and maintenance materials)
Replacement parts
Labor
Treatment and disposal of collected material
Capital charges
Taxes
Insurance
Interest

[a]Ref. 6

and suitability of control equipment, the user's understanding of its operation, and his/her commitment to maintaining it in reliable operation.

Maintenance costs represent the expenditures required to sustain the operation of a control device at its designed efficiency, using a scheduled maintenance program and necessary replacement of any defective component parts. Simple control devices of low efficiency have low maintenance costs, and complex, highly efficient control systems generally have high maintenance costs.

Annual operating costs refer to the yearly expense of operating an emission control system at its designed collection efficiency. These costs depend directly on the following factors:

1. Volumetric rate of gas cleaned.
2. Pressure drop across the control system.

3. Annual operating time.
4. Consumption rate and cost of electricity.
5. Mechanical efficiency of the air-moving device and, where applicable, two additional items:
6. Scrubbing liquor consumption and replacement costs.
7. Consumption rate and cost of fuel.

The energy requirement for air pollution control equipment is an important and significant component of operating costs. The role of energy conservation in industry is now generally well established and is particularly applicable to non-revenue-generating production equipment such as pollution control devices. Traditionally, the energy requirements for emission control equipment have been included in an overall consideration and expression of operating costs, but the continuing emphasis on reducing energy costs leads logically to the isolation of this cost element for at least two purposes: (*1*) to relate operating costs to energy costs, and (*2*) to locate potential areas for energy conservation.

Because air pollution control equipment ranges from natural-draft stacks to combinations of collectors (cyclones, filters, electrostatic precipitators, scrubbers, and stacks with induced or forced-draft fan units), the energy consumption requirements of the various types of pollution control equipment must be examined separately. Table 59.3 gives the most commonly used air pollution control equipment and the associated power requirements. Power requirements are listed in kilowatts, because most pollution control equipment operates on electricity. Pressure drop is expressed in inches of water. The heating requirements for afterburners and regeneration of adsorption beds are given in British thermal units per hour. The typical requirements in Table 59.3 can be added directly to obtain an estimate of total energy, pressure drop, and heating requirements for a complex chain of combined equipment. The total pressure drop can then be used to estimate the total fan power requirement.

As an example, consider a 120-MW coal-burning power plant (7). The boiler exhaust emission controls could include an electrostatic precipitator for removing fly ash particulates, an alkali scrubber system for removing sulfur dioxide, and a fan and a tall stack to help achieve acceptable air quality by dispersion. The equipment power requirements for this system could then be estimated as follows:

Item	Electrical Energy (kW/1000 scfm)
Electrostatic precipitator	0.3
Limestone scrubber	12.1
Tall stack	0
Fan	0.1
Total exhaust system power	*12.5*

For a 120-MW power plant with a stack flow rate of 300,000 standard cubic feet per minute (scfm), the electrical power required would be $300 \times 12.5 = 3750$ kW.

3.5 Material Conservation

Except for precious metal applications, most air pollution control systems historically have collected material of little economic worth. Nevertheless, as resource recovery becomes more feasible because of current and expected shortages of many material resources, the economic worth of material collected by emission controls may gain in significance in the economic analysis of air pollution controls.

The alternatives for handling collected particulate remain as follows:

1. Recycle material to the process.
2. Sell material as collected.
3. Convert material to salable products.
4. Discard material in the most economical (and environmentally acceptable) manner.

In some process operations, collected material is sufficiently valuable to warrant its return to the process, and the value of the recovered material can partially or wholly pay for the collection equipment. In many applications, however, the cost of the high-efficiency control systems necessary to achieve desired ambient air quality would be greater than the revenue returned for recovery of the material collected.

The cement industry provides an example of collected material being routinely returned to the process. Not only does recovered dust, in such situations, have value as a raw material, its recovery also reduces disposal costs, as well as costs related to the preparation of raw materials used in the process.

Although material collected by air pollution control equipment may be unsuitable for return to a process within that plant, it may be suitable for another manufacturing activity. Hence, it may be treated and sold to another firm that can use the material. Untreated, pulverized fly ash, for example, cannot be reused in a furnace, but it can be sold as a raw material to a cement manufacturer. It can also be used as a soil conditioner, as an asphalt filler, or as landfill material. When treated, pulverized fly ash can yield an even more valuable product. Some utilities, for example, sinter pulverized fly ash to produce a lightweight aggregate that can be used to manufacture bricks and lightweight building blocks.

3.6 Energy Recovery

Although there has been much comment about the need for industrial energy conservation, the fact remains that most exhaust stacks in industry are hot, suggesting that much energy is being wasted.

The most economical approach to energy conservation and environmental control is a combined approach whenever possible. Real savings can be gained by combining heat recovery with air pollution control. Whereas in the past, low fuel costs made it uneconomical to install heat recovery equipment in many industrial applications, such as on stacks of direct-fired furnaces, the increased emphasis on energy conservation and the global reduction of "greenhouse" gases today has turned the spotlight on fuel economy.

A major advantage in the combined approach to energy and air pollution controls is the drop in temperature, caused by the recovery of sensible heat energy from an airstream,

Table 59.3. Power Requirements for Air Pollution Equipment[a]

Item	Pressure (Drop) (in. H_2O)	Power[b] Required (kW per 1000 scfm), Typical Range	Remarks
Exhaust stacks	0	0	[c]
Filters and separators			
Baghouses, cloth filters	1–30	0.1–1.0	Power function of rapping rate, particle size, cloth and dust-layer pressure drop
Impingement and gravitational separators	0.1–1.5	None	
Cyclones			
Single	0.1–2	None	$\Delta P = 1$ to 20 inlet velocity heads
High efficiency	2–10	None	
Dry centrifugal	2–4	None	Combined separator-fan effect
Packed bed	1–10	None	
Electrostatic precipitators			
High voltage	0.1–1	0.2–0.6	
Low voltage	0.1–1	0.01–0.04	
Scrubbers		1–12	Power requirements for fan and pump
Centrifugal, mechanical venturi	2–8	None	
High pressure drop	10–60	None	3 gal/min per 1000 cfm typical liquor rate
Low pressure drop	0.5–100	None	3 gal/min per 1000 cfm typical liquor rate

Spray tower	1–2	None	
Impingement and entrainment	4–20	None	
Packed-bed absorption	0.5 per foot of thickness	None	
Plate absorption plus liquor circulation pump[d]	1–3 per plate 1–60 ft	None (see manufacturer's pump curve)	1–20 gal/min liquor, some sprays to 600 psi
Cyclone	0.1–10	None	
Adsorption beds	1–30	100–200 Btu/lb adsorbent to heat to 600°F without heat exchange	Regeneration rate dependent on adsorbent and pollutant concentration; fan power may also be required for hot gas regeneration 1500°F, with 80% heat exchanger (2×10^5 Btu/h); 1500°F, no regeneration, heat exchanger (14×10^5 Btu/h)
Acoustic fume afterburners	0.1–2 (0.05–1 per in. of bed)	2 to 14×10^5 Btu/h	
Catalytic oxidizers	0.5–2 (0.05–1 per in. of bed)	0.1 to 1.0×10^6 Btu/h	300–900°F, 80% heat exchanger to 900°F, without heat exchanger depends on inlet temperature

[a]Ref. 7
[b]"None" indicates that no direct power is required; the only power requirements is indirect from fans (due to pressure drop in equipment).
[c]The rise of hot gas in the stack creates a pressure head of about 0.5 in. H_2O, but this is counteracted by friction losses in the stack plus the desired velocity head leaving the stack for dispersions.
[d]Add scrubber pump power to scrubber power requirements.

that can lead to a dramatic reduction in the volumetric exhaust rate and the requisite cost and size of air pollution control equipment. Thus, it is possible to save not only on capital expenses, but also on operating costs, as well as in the consumption of increasingly expensive fuel. Recovered heat can be used to reduce fuel requirements by preheating primary combustion air, preheating air going to another process, or preheating other fuel; or it can be used in heating buildings.

3.7 Economic Analysis Models

Useful models for estimating operation and maintenance costs are presented for five general types of control equipment: *1*. Gravitational and inertial collectors; *2*. Fabric filters; *3*. Wet scrubbers; *4*. Electrostatic precipitators; and *5*. Thermal oxidizers.

These models are intended to provide a means of generating preliminary cost estimates for the purpose of comparing control alternatives. Other effective, and more detailed, cost estimating methodologies have been developed, such as that described by Vatavuk (8). Using a five-step procedure, the estimator arrives at the cost for a particular system by (1) obtaining the facility parameters and regulatory options for the facility under study, (2) roughing out the control system design, (3) sizing the control system components, (4) estimating the costs of these individual components, and (5) estimating the costs (capital and annual) of the entire system. Vatavuk also provides useful discussions of the various parameters and considerations that are inherent in each step of the process. Such elaborations are outside the scope of this discussion.

3.7.1 Gravitational and Inertial Collectors

In general, the only significant cost for operating mechanical collectors is the electric power cost, which varies with the unit size and the pressure drop. Because pressure deep in gravitational collectors is low, operational costs associated with these units are considered to be insignificant. Maintenance cost includes the costs of servicing the fan motor, replacing any lining worn by abrasion, and, for multiclone collectors, flushing the clogged small diameter tubes.

The theoretical annual cost G of operation and maintenance for centrifugal collectors can be expressed as follows (6):

$$G = S\left(\frac{0.7457PHK}{6356E} + M\right) \tag{2}$$

where S = design capacity of the collectors (actual cubic feet per minute, acfm)
P = pressure drop (in. water)
E = fan efficiency, assumed to be 60% (expressed as 0.60)
0.7457 = a constant (1 hp = 0.7457 kW)
H = annual operating time (assumed 8760 h)
K = power cost (\$/kW·h)
M = maintenance cost (\$/acfm)

For computational purposes, the cost formula can be simplified as follows (6):

$$G = S(195.5 \times 10^{-6}PHK + M)$$

3.7.2 Fabric Filters

Operating costs for fabric filters include power costs for operating the fan and the bag cleaning device. These costs vary directly with the sizes of the equipment and the pressure drop. Maintenance costs include costs for servicing the fan and shaking mechanism, emptying the hoppers, and replacing the worn bags.

The theoretical annual cost G for operation and maintenance of fabric filters is as follows (6):

$$G = S\left(\frac{0.7457PHK}{6356E} + M\right) \tag{3a}$$

with the units as defined as in Equation 2.

For computational purposes, the cost formula can be simplified as follows (6):

$$G = S(195.5 \times 10^{-6}PHK + M)$$

3.7.3 Wet Collectors

The operating costs for a wet collector include power and scrubbing liquor costs. Power costs vary with equipment size, liquor circulation rate, and pressure drop. Liquor consumption varies with equipment size and stack gas temperature. Maintenance includes servicing the fan or compressor motor, servicing the pump, replacing worn linings, cleaning piping, and any necessary chemical treatment of the liquor in the circulation system.

The theoretical annual cost G of operation and maintenance for wet collectors can be expressed as follows (6):

$$G = S\left[0.7457HK\left(\frac{P}{6356E} + \frac{Qg}{1722F} + \frac{Qh}{3960F}\right) + WHL + M\right] \tag{4}$$

where S = design capacity of the wet collector (acfm)
0.7457 = a constant (1 hp = 0.7457 kW)
H = annual operating time (assumed 8760 h)
K = power costs (dollars per kW-h)
P = pressure drop across fan (in. water)
Q = liquor circulation (gal/acfm)
g = liquor pressure at the collector (psig)
h = physical height liquor is pumped in circulation system (ft)
W = makeup liquor consumption (gal/acfm)
L = liquor cost ($/gal)
M = maintenance cost ($/acfm)

E = fan efficiency, assumed to be 60% (expressed as 0.60)
F = pump efficiency, assumed to be 50% (expressed as 0.50)

This equation can be simplified according to Semrau's total "contacting power" concept (9). Semrau shows that efficiency is proportional to the total energy input to meet fan and nozzle power requirements. The scrubbing (contact) power factors in Table 59.4 were calculated from typical performance data listed in manufacturers' brochures. These factors are in general agreement with data reported by Semrau. Using Semrau's concept, the equation for operating cost can be simplified as follows (6):

$$G = S\left[0.7457HK\left(Z + \frac{Qh}{1980}\right) + WHL + M\right] \tag{5}$$

where Z = contact power, that is, total power input required for collection efficiency (hp/acfm; see Table 59.4). It is a combination of:

1. Fan horsepower per acfm $= \dfrac{P}{6356E}$ (6)

2. Pump horsepower per acfm

$$= \frac{Qg}{1722F} \text{ the power to atomize water through a nozzle} \tag{7}$$

The pump horsepower, $Qh/1980$, required to provide pressure head is not included in the contact power requirements.

3.7.4 *Electrostatic Precipitators*

The only operating cost considered in running electrostatic precipitators is the power cost for ionizing the gas and operating the fan. Depending on the length and design of the upstream and downstream ductwork, the fan energy requirements may be relatively small. Because the pressure drop across the equipment is usually less than 0.5 in. water, the cost of operating the fan is assumed to be negligible if minimal ductwork is used. The power cost varies with the efficiency and the size of the equipment.

Table 59.4. Contact Power Requirements for Wet Scrubbers[a]

Scrubbler Efficiency	"Scrubbing" (Contact) Power (hp/acfm)
Low	0.0013
Medium	0.0035
High	0.015

[a]Ref. 6.

Maintenance usually requires the services of an engineer or highly trained operator, in addition to regular maintenance personnel. Maintenance includes servicing fans and replacing damaged wires and rectifiers.

The theoretical annual cost G for operation and maintenance of electrostatic precipitators is as follows (6):

$$G = S(JHK + M) \tag{8}$$

where S = design capacity of the electrostatic precipitator (acfm)
J = power requirements (kW/acfm)
H = annual operating time (assumed 8760 h)
K = power cost ($/kW-h)
M = maintenance cost ($/acfm)

3.7.5 Afterburners

The major operating cost item for afterburners is fuel. Fuel requirements are a direct function of the gas volume, the enthalpy of the gas, and the difference between inlet and outlet gas temperatures. For most applications, the inlet gas temperature at the source ranges from 50 to 400°F. Outlet temperatures may vary from 1200 to 1500°F for direct flame afterburners and from 730 to 1200°F for catalytic afterburners. The use of heat exchangers may bring about a 50% reduction in the temperature difference.

The equation for calculating the operation and maintenance costs G is as follows (6):

$$G = S\left(\frac{0.7457PHK}{6356E} + HF + M\right) \tag{9}$$

where S = design capacity of the afterburner (acfm)
P = pressure drop (in. water)
E = fan efficiency, assumed to be 60% (expressed as 0.60)
0.7457 = a constant (1 hp = 0.7457 kW)
H = annual operating time (assumed 8760 h)
K = power cost ($/kW-h)
F = fuel cost ($/acfm-h)
M = maintenance cost ($/acfm)

For computational purposes, the cost formula is simplified as follows (6):

$$G = S(195.5 \times 10^{-6}PHK + HF + M)$$

4 PROCESS AND SYSTEM CONTROL

Process and system control implies a careful review of a production unit operation within the context of air pollution control to examine whether the manufacturing process is optimal, considering the emission rate or necessary control and treatment of the process effluents.

A fundamental notion in air pollution control asserts, "The problem can be solved best if it is solved at the source of emissions." This simple rule, if applied consistently, could greatly reduce the number and complexity of emission problems in industry. Methods for total or partial "control at the source" include elimination of emissions, minimization of emissions, and concentration of contaminants before discharge.

4.1 Elimination of Emissions

An excellent example of the potentially beneficial effect of material substitution is conversion from a "dirty" fuel to a "clean" fuel, such as a switch from coal to gas in utility boilers. In the late 1960s and early 1970s, many state and local pollution control agencies enacted regulations to curb sulfur dioxide emissions from coal- and oil-burning power plants. The effect of such regulations was a trend toward the use of low sulfur oil and coal and, where possible, to the use of the cleanest fossil fuel, natural gas. Because of the shortages and higher costs of natural gas and low sulfur coal and oil, however, the achievement and maintenance of air quality goals cannot be accomplished solely by switching to naturally available low sulfur fuels. It is apparent that coal with medium to high sulfur content will have to be burned in increasing quantities by power plants to meet growth requirements and to safeguard an adequate supply of clean fuels for residential, commercial, and industrial use. However, the identification of acid rain as a serious air pollution problem in certain regions of North America has resulted in regulatory mandates for stricter limitations on sulfur dioxide and nitrogen dioxide emissions from power plants.

4.1.1 Process Changes

Process changes can be as effective as material substitution in eliminating air polluting emissions. It has often been possible and even profitable in the chemical and petroleum industries to control the loss of volatile organic materials to the atmosphere by condensation, leading to the "reuse" of otherwise fugitive vapors. The compressors, absorbers, and condenser units on petroleum product process vessels provide a good example of such process modification.

In the case of burning fossil fuels, process change can also be construed to encompass the addition of plants auxiliary to the main process that generate substitute fuels. One familiar example is the recycling of distilled volatiles in coke production. When high sulfur coals are distilled destructively, they produce gas streams that when treated, are useful energy sources, producing low sulfur oxides emissions on burning. Recycling a portion of this fuel to the coke ovens supplies energy for further fuel gas production from the next charge of coal in the ovens while also providing product coke.

The addition of a step to reduce atmospheric emissions has been commonplace in brass foundry practice when indirect fired furnaces are employed. A fluxing material applied to the surface of the molten brass serves as an evaporation barrier and reduces the evolution of brass fumes; this additional step was developed strictly as an air pollution control measure.

4.1.2 *Equipment Substitution*

An example of the benefit to be derived from equipment substitution was a trend in the polyvinyl chloride (PVC) industry in the 1970s to change to large polymerization reactors, which not only boosted productivity but helped to stem fugitive emissions of vinyl chloride monomer. A large PVC reaction vessel has only half the possible fugitive emissions leak points of two smaller units of the same total capacity (10). As late as 1972, the PVC industry relied on reactors of capacity smaller than 7500 gallons for 85% of its output. By the end of the decade, 20,000-gal vessels and even some massive 55,000-gal reactors were employed in the PVC industry to control vinyl chloride emissions.

4.2 Reduction of Emissions

One way to reduce air pollution control costs is to minimize, if not eliminate, release of the contaminant. Such action is cost effective both for fugitive emissions and process emissions.

4.2.1 *Fugitive Emissions*

Leaking conveyor systems can be modified to eliminate spilling and dusting. Airtight enclosures can be built around conveyors, or cover housings can be equipped with flanges and soft gaskets. Shaft bearings can be redesigned and relocated inside the conveyor housing to prevent dust leakage around the shaft.

Tanks and bins should have airtight covers and joints sealed with cemented strips of plastic or rubber sheeting. Each bin or tank normally should have only one vent where temperature difference can create a natural draft from the vent; other openings should be kept closed.

Where possible, interconnecting a series of vents from several tanks should be considered. While one tank is being filled, another can be emptied, thus reducing the need for exhausting air to the atmosphere. It may also be possible to have one small dust collector serve a number of units.

4.2.2 *Process Emissions*

When contaminants are a by-product of a production operation, generation can often be minimized by changes in operating conditions. Lower combustion temperatures will reduce nitrogen oxides formation. Levels of fluorine-containing compounds, released when some ores are heated, frequently can be decreased if water vapor is eliminated from the atmosphere.

Reducing excess air in coal- or oil-fired boiler systems can result in a substantial reduction in the conversion of sulfur dioxide to sulfur trioxide, thus reducing sulfuric acid emissions.

4.3 Concentrating Pollutants at the Source

One major phase of process and system control involves centralizing and decreasing the number of emission points and reducing the effluent volume of the exhaust gas to be treated.

4.3.1 Reducing the Number of Emission Points

Many industrial operations, such as machining in metalwork industries, are characterized by the replication of identical or similar operations. Examples include grinding, sawing, and cutting. Control of worker exposures and effective, economical emission control can often be best achieved by clustering the operations together and exhausting them locally and simultaneously to a common collection system, rather than exhausting each operation to an individual collector. The concept of centralized collection and abatement offers the advantage of economy of scale for collection systems; this tends to decrease control costs per unit weight of material collected. This advantage is due in part to the relatively high concentration of pollutants at a combined source compared to the more dilute concentrations, which would be experienced from collecting and exhausting materials from each similar operation independently.

4.3.2 Volume Reduction by Cooling

Many dust collectors such as bag filters and precipitators are sized more on a volumetric flow basis than on a mass flow basis. If the effluent is a hot, dusty gas, cooling can reduce the volume appreciably. Cooling by the addition of cold air is the poorest method from the standpoint of reducing cost. Radiation panels, fin surfaces, waste heat boilers, forced convection, heat interchange, and direct spray cooling with water are all possible means of effecting volume reduction.

4.3.3 Optimal Local Exhaust Ventilation

Well-designed and properly installed local exhaust systems are an important aspect of control measures to minimize exhaust flow rate and maximize concentrations in exhaust gases to be cleared before discharge to the atmosphere. By careful design, local exhaust hoods can be applied to assure complete capture of all contaminants. Close-fitting hoods not only ensure better control of materials that may escape into the work environment, they also provide minimal exhaust ventilation of processors that produce contaminants.

5 TECHNICAL CRITERIA FOR SELECTING AIR-CLEANING METHODS

The cost effectiveness of an air pollution control system is closely related to its degree of control, its capacity, and the type of control system. Often more than one type of control could be used to solve a specific emission problem. Nevertheless, there is generally only one type that is optimal from both technical and economic standpoints.

Some of the most basic technical factors affecting selection of equipment are as follows:

1. System volumetric flow rate
2. Concentration (loading) of contaminant
3. Size distribution of particulate contaminants
4. Degree of cleaning required

5. Conditions of air or gas stream with reference to temperature, moisture content, and chemical composition
6. Characteristics of the contaminant, such as corrosiveness, solubility, reactivity, adhesion or packing tendencies, specific gravity, surface, and shape

This section focuses on these and other technical criteria that are helpful in selecting emission controls.

5.1 Performance Objectives

The prime factor in the selection of control equipment is the maximum amount or rate of contaminant to be discharged to the atmosphere. Knowledge of this amount, together with knowledge of the amount of contaminant entering a proposed collection system, defines the required collection efficiency, this begins the technical selection of air-cleaning methods.

If the material to be collected is particulate, it is essential to understand that collectors have different efficiencies for different-sized particles. Therefore the particle size distribution of the emission must be known before the collector efficiency required for each particle size range can be determined. Collector efficiency also varies with gas flow rate and with properties of the carrier gas, which may fluctuate with flow rate or time. Such variations must be considered carefully in determining collector efficiency.

When the material to be collected is a gas or vapor, it is necessary to know to what extent the material is soluble in the scrubbing liquid or retainable on the absorbing surfaces. The determinations must be made for the concentrations expected in the collector inlet and outlet streams and for the conditions of temperature, pressure, and flow rate expected.

5.2 Contaminant Properties

Contaminant properties, as distinguished from carrier gas properties, comprise both chemical and physical characteristics of the material to be removed from an exhaust airstream.

5.2.1 Loading

Contaminant loading from many processes varies over a wide range for an operating cycle. Variations of an order of magnitude in concentration are not uncommon. Well-known examples of such variation include the basic oxygen furnace in steel making and soot blowing in a steam boiler. Contaminant loading may also vary with the carrier gas flow rate. Particularly in the case of gases, concentration is all-important in predicting removal efficiency and specifying system design parameters.

5.2.2 Composition

The composition of a contaminant affects both physical and chemical properties. Chemical properties, in turn, affect physical properties. For example, if collected material is to be used in a process or shipped in a dry state, a dry collector is indicated. If the collected material has a very high intrinsic value, a very efficient collector is called for.

Because chemical and physical properties vary with composition, a collector must be able to cope with both expected and unexpected composition changes. For example, in the secondary aluminum industry a collector must be able to deal with the evolution of aluminum chloride during chlorine "demagging." The aluminum chloride levels vary widely throughout the cycle; peak levels last for only a few minutes but continue to develop in decreasing amounts throughout a full cycle of 16 hours or more.

Further examples include absorption, in which solubility may be important to the ease with which the absorbent may be regenerated, and scrubbing to remove particulate matter, in which wettability of the dust by the scrubbing liquor aids the collector mechanisms and the basic separating forces that determine scrubber performance.

5.2.3 *Combustibility*

Generally, it is not desirable to use a collection system that permits accumulation of "pockets" of contaminant when the contaminant collected is explosive or combustible. Systems handling such materials must be protected against accumulation of static charges. Electrostatic precipitators are not suitable for such contaminants because of their tendency to spark. Wet collection by scrubbing or absorption methods is especially appropriate. Some dusts, however, such as magnesium, are pyrophoric in the presence of small amounts of water. In combustion (with or without a catalyst), explosibility must always be considered.

5.2.4 *Reactivity*

Certain obvious precautions must be taken in the selection of equipment for the collection of reactive contaminants. In filtration, selection of the filtering media can present a special problem. In absorption, because certain applications require that the absorbed contaminant react with the absorbent, the degree of reactivity is important. Where scrubbers are used to remove corrosive gases or particulates, the potential corrosiveness must be balanced against the potential savings when using corrosion-resistant construction. The decision is thus whether to use more expensive materials or to incur higher maintenance costs.

5.2.5 *Electrical and Sonic Properties*

The electrical properties of the contaminant can influence the performance of several types of collector. Electrical properties are considered to be a factor influencing the buildup of solids in inertial collectors. In electrostatic precipitators, such electrical properties of the contaminant as the resistivity are of paramount importance in determining collection efficiency and precipitator size, and they influence the ease with which collected particulate is removed by periodic cleaning of the collection surfaces. In fabric filtration, electrostatic phenomena may have direct and observable effects on the process of cake formation and the subsequent ease of cake removal. In spray towers, where liquid droplets are formed and contact between these droplets and contaminant particles is required for particle collection, the electrical charge on both particles and droplets is an important parameter in determining collection efficiency. The process is most efficient when the charges on the droplet attract rather than repel those on the particle. Sonic properties are significant where sonic agglomeration is employed.

5.2.6 Toxicity

The degree of contaminant toxicity influences collector efficiency requirements and may necessitate the use of equipment that will provide ultrahigh efficiency. Toxicity also affects the means for removal of collected contaminant from the collector and its disposal, as well as the means of servicing and maintaining the collector. However, toxicity of the contaminant does not influence the removal mechanisms of any collection technique.

5.2.7 Particle Size, Shape, and Density

Size, shape, and density are three properties of particulate matter that determine the magnitude of forces resisting movement of a particle through a gas stream. These are the major factors determining the effectiveness of removal in inertial collectors, gravity collectors, venturi scrubbers, and electrostatic precipitators. In these collections, resisting forces are balanced against some removal force (e.g., centrifugal force in cyclones) that is applied in the control device, and the magnitude of the net force tending to remove the particle determines the effectiveness of the equipment.

Size, shape, and density of a particle can be related to terminal settling velocity, which is a useful parameter in the selection of equipment for particulate control. Settling velocity is derived from Stokes' law, which equates the velocity at which a particle will fall at constant speed (because of a balance of the frictional drag force and the downward force of gravity) to the properties of the particle and the viscosity of the gas stream through which it is settling.

Terminal settling velocities can be determined by any of a number of standard techniques and used in the quantitative evaluation of the difficulties to be anticipated in designing particulate removal equipment.

The aerodynamic diameter of a particle, which is directly related to the terminal settling velocity, is another commonly used parameter to address size, shape, and density properties. It is defined as the diameter of a sphere with unit density (i.e., equal to the density of water) that will settle in still air at the same rate as the particle in question.

Because the aerodynamic diameter is related to the ease with which individual particles are removed from a gas stream, it is apparent that size distribution largely determines the overall efficiency of a particular piece of control equipment. Generally, the smaller the aerodynamic particle size to be removed, the greater the expenditure required for power or equipment or both. To increase the efficiencies obtainable with scrubbers, it is necessary to expend additional power either to produce high gas stream velocities, as in the venturi scrubber, or to produce finely divided spray water.

Cyclones call for the use of a larger number of small diameter units for higher efficiency in a given situation. Both the power cost and equipment cost are then increased. Achieving higher efficiencies for electrostatic precipitators necessitates the use of a number of units or fields in series because there is an approximately inverse logarithmic relationship between outlet concentration and the size of collection equipment. A precipitator giving 90% efficiency must be approximately doubled in size to give 99% efficiency and approximately tripled to give 99.9% efficiency.

5.2.8 Hygroscopicity

Although not specifically related to any removal mechanism, hygroscopicity may be a measure of how readily particulates will cake or tend to accumulate in equipment if moisture is present. If such accumulation occurs on a fabric filter, it may completely blind it and prevent gas flow.

5.2.9 Agglomerating Characteristics

Collectors are sometimes used in series, with the first collector acting as an agglomerator and the second collecting the particles agglomerated in the first one. In carbon black collection, for example, extremely fine particles are first agglomerated; then they can be collected practicably.

5.2.10 Flow Properties

Flow properties of the material are mainly related to the ease with which the collected dust may be discharged from the collector. Extreme stickiness may eliminate the possibility of using equipment such as fabric filters. Hopper size and shape depend in part on the packing characteristics or bulk density of the collected material and its flow properties. Hygroscopic materials, or collected dust tending to cake, exhibit flow properties that change with time as the dust remains in the hopper. Hopper heaters or rappers may be required.

5.3 Carrier Gas Properties

Carrier gas properties are important insofar as they affect the selection of control equipment, especially with reference to composition, reactivity, conditions of temperature, pressure, moisture, solubility, condensability, combustibility, and toxicity.

5.3.1 Composition

Gas composition affects physical and chemical properties, which are important to the extent that there may be chemical reactions between the contaminants and the collector, either in its structure or in its contents. One common reaction between gaseous components and equipment is the corrosion of metallic parts of collectors when gases contain sulfur oxides and water vapor.

Composition, concentration, and chemical reaction properties of the inlet stream determine the collection efficiency in packed-tower scrubbers removing gaseous or vapor phase contaminants.

5.3.2 Temperature

The temperature of the carrier gas principally influences the volume of the carrier gas and the materials of construction of the collector. The volume of the carrier gas influences the size and cost of the collector and the concentration of the contaminant per unit volume; concentration in turn is the driving force for removal. In addition, viscosity, density, and other gas properties are temperature dependent. Temperature also affects the vapor–liquid equilibria in gaseous contaminant scrubbing such that scrubbing efficiency for partially soluble gases decreases with increasing temperature.

Adsorption processes are generally exothermic and are impracticable at higher temperature, the absorbability being inversely proportional to the temperature (when the reaction is primarily physical and is not influenced by an accompanying chemical reaction). Similarly, in absorption (where gas solubility depends on the temperature of the solvent), temperature effects may have significance if the concentration of the soluble material is such that an appreciable temperature rise results. When combustion is used as a means for contaminant removal, the gas temperature affects the heat balance, which is the vital factor in the economics of the process. In electrostatic precipitation, both dust resistivity and the dielectric strength of the gas are temperature dependent.

Wet processes cannot be used at a temperature that would cause the liquid to freeze, boil, or evaporate too rapidly. Filter media can be used only in the temperature range within which they are stable. The filter structure must retain structural integrity at the operating temperature.

Finally, cool gases flowing from a stack downstream of control equipment disperse in the atmosphere less effectively than do hot gases. Consequently, benefits derived from partial cleaning accompanied by cooling may be offset if the cooler exhaust gas cannot be well dispersed. This is a factor of importance in wet cleaning processes for hot gases, where the advantage gained by cleaning is sometimes offset by downwash from the stack near the plant because the exhaust gas is cooled. In the case of wet collection devices, the effluent gases may present a visually objectionable steam plume or even "rain out" in the stack vicinity. Raising the discharge gas temperature may eliminate or reduce these problems.

5.3.3 Pressure

In general, carrier gas pressure much higher or lower than atmospheric pressure requires that the control equipment be designed as a pressure vessel. Some types of equipment are much more amenable to being designed into pressure vessels than others. For example, catalytic converters are incorporated in pressure processes for the production of nitric acid and provide an economical process for the reduction of nitrogen oxides before release to the atmosphere.

Pressure of the carrier gas is not of prime importance in particulate collection except for its influence on gas density, viscosity, and electrical properties. It may, however, have importance in certain special situations, for example, when the choice is between high-efficiency scrubbers and other devices for collection of particulate matter. The available source pressure can be used to overcome the high pressure drop across the scrubber, reducing the high power requirement that often limits the utilization of scrubbers. In adsorption, high pressure favors removal and may be required in some situations.

5.3.4 Viscosity

Viscosity is important to collection techniques in two respects. First, it is important to the removal mechanisms in many situations (inertial collection, gravity collection, and electrostatic precipitation). Particulate removal technique involves migration of the particles through the gas stream under the influence of some removal force. Ease of migration decreases with increasing viscosity of the gas stream. Second, viscosity influences the pressure drop across the collector, thereby becoming a major parameter in power requirement computation.

5.3.5 Density

Density of the carrier gas, for the most part, has no significant effect in most real gas cleaning processes, although the difference between particle density and gas density appears as a factor in the theoretical analysis of all gravitational and centrifugal collection devices. Particle density is so much greater than gas density that the usual changes in gas density have negligible effects. Carrier gas density does influence fan power requirements and therefore is important to fan selection and operating cost. Furthermore, special precautions must be taken in "cold start-up" of a fan designed to operate at high temperature, to ensure that the motor capacity is not exceeded.

5.3.6 Humidity

Humidity of the carrier gas stream may affect the selection and performance of control equipment in any of several basically different ways. High humidity may contribute to accumulations of solids and lead to the caking and blocking of inertial collectors as well as the caking of filter media. It can also result in cold spot condensation and aggravation of corrosion problems. In addition, the water vapor may act on the basic mechanism of removal in electrostatic precipitation and greatly influence resistivity. In catalytic combustion it may be an important consideration in the heat balance that must be maintained. In adsorption it may tend to limit the capacity of the bed of water preferentially or concurrently adsorbed with the contaminant. Even in filtration it may influence agglomeration and produce subtle effects.

The above-mentioned considerations are the main limitations on the utilization of evaporative cooling in spite of its obvious power advantage. When humidity is a serious problem for one of the foregoing reasons, scrubbers, or adsorption towers may be particularly appropriate devices. Humidity also affects the appearance of the stack exhaust gas discharged from wet collector devices. Because steam plumes are visually objectionable, sometimes it is necessary to heat exhaust gases to raise their dew point before discharge. This can add considerable operating expense to the total system.

5.3.7 Combustibility

The handling of a carrier gas that is flammable or explosive requires certain precautions. The most important is making sure that the carrier gas is either above the upper explosive limit or below the lower explosive limit for any air admixture that may exist or occur. The use of water scrubbing or adsorption may be an effective means of minimizing the hazards in some instances. Electrostatic precipitators are often impractical, for they tend to spark and may ignite the gas.

5.3.8 Reactivity

A reactive carrier gas presents special problems. In filtration, for example, the presence of gaseous fluorides may eliminate the possibility of high-temperature filtration using glass fiber fabrics. In adsorption, carrier gas must not react preferentially with the adsorbents. For example, silica gel is not appropriate for adsorption of contaminants when water vapor is a component of the carrier gas stream. Also, the magnitude of this problem may be greater when a high-temperature process is involved. On the one hand, devices that use water may be eliminated from consideration if the carrier gas reacts with water. On the other hand, scrubbers may be especially appropriate in that they tend to be relatively small and require small amounts of construction material, permitting the use of corrosion-resistant components, with lower relative increase in cost.

5.3.9 Toxicity

When the carrier gas is toxic or is an irritant, special precautions are needed in the construction of the collector, the ductwork, and the means of discharge to the atmosphere. The entire system, including the stack, should be under negative pressure and the stack must be of tight construction. Because the collector is often under "suction," special means such as "airlocks" must be provided for removing the contaminant from the hoppers if collection is by a dry technique. Special precautions may also be required for service and maintenance operations on the equipment.

5.3.10 Electrical and Sonic Properties

Electrical properties are important to electrostatic precipitation because the rate or ease of ionization will influence removal mechanisms.

Generally speaking, intensity of Brownian motion and gas viscosity increase with gas temperature. These factors are important gas stream characteristics that relate to the "sonic properties" of the stream. Increases in either property tend to increase the effectiveness with which sonic energy can be used to produce particle agglomeration.

5.4 Flow Characteristics of Carrier Gas

5.4.1 Volumetric Flow Rate

The rate of evolution from the process, the temperature of the effluent, and the degree and the means by which it is cooled, if cooling is used, fix the rate at which carrier gases must be treated, and, therefore, the size of removal equipment and the rate at which gas passes through it. For economic reasons, it is desirable to minimize the size of the equipment. Optimizing the size and velocity relationship involves consideration of two effects: (*1*) reduction in size results in increased power requirements for handling a given amount of gas because of increased pressure loss within the control device, and (*2*) velocity exerts an effect on the removal mechanisms. For example, higher velocities favor removal in inertial equipment up to the point of turbulence, but beyond this, increased velocity results in decreased efficiency. In gravity settling chambers, flow velocity determines the smallest size that will be removed. In venturi scrubbers, efficiency is proportional to velocity through the system. In absorption, velocity affects film resistance to mass transfer. In filtration, the resistance of the medium often varies with velocity because of changes in dust cake permeability with flow. In adsorption, velocity across the bed should not exceed the maximum that permits effective removal. Optimal velocities have not generally been established with certainty for any of the control processes because they are highly influenced by the properties of the contaminant and carrier gas as well as by the design of the equipment.

5.4.2 Variations in Flow Rate

Rate variations result in velocity changes, thereby influencing equipment efficiency and pressure drop. Various control techniques differ in their abilities to adjust to flow changes. When rate variations are inescapable, it is necessary to (*1*) design for extreme conditions, (*2*) employ devices that will correct for flow changes, or (*3*) use a collector that is inherently positive in its operation. Filtration is most easily adapted to extreme rate variations because it presents a positive barrier for particulate removal. This process is subject to pressure drop variations, however, and generally the air-moving equipment will not deliver at a constant rate when pressure drop increases. In most other control techniques, variations in flow produce a change in the effectiveness of removal.

One means of coping with rate variation is the use of two collectors in series, one that improves performance with increasing flow (e.g., multicyclone) and one whose performance decreases with increasing flow (e.g., electrostatic precipitator). Some venturi scrubbers are equipped with automatically controlled, variable size throats. Changes in gas flow rate are automatically sensed, and the throat's cross-sectional area is changed correspondingly to maintain a relatively constant pressure drop and efficiency over a relatively wide range of conditions.

6 CLASSIFICATION OF AIR POLLUTION CONTROLS

In a broad sense, air pollution controls for stationary source emissions include process and system control, air-cleaning methods, and the use of tall stacks. Process and system control were covered in Section 4. This section focuses on the other types of air pollution control and offers an overview of the role of each method. More detailed information on air-cleaning methods follows in Sections 7 through 12.

6.1 Stack Height

In the context of air pollution control, a tall stack has the simple function of discharging exhaust gases at an elevation high enough to reduce the maximum concentration of contaminants experienced at ground level to within acceptable limits. However, it is unwise and often illegal to rely only on tall stacks for solving air pollution problems. Furthermore, the use of a tall stack may serve to satisfy ground-level concentration requirements, but does nothing to reduce the quantity of pollutants being emitted into the atmosphere and may contribute to air quality problems at distances far downwind from the emission source.

Through the Clean Air Act, the EPA regulates permissible stack heights for new emission sources (2). The agency has defined a "good engineering practice (GEP) stack height" as the greater of either 65 meters or the height computed from a formula that takes into account the dimensions of nearby structures. This provision is intended to ensure that emissions from a stack do not result in excessive concentrations of pollutants as a result of atmospheric downwash, wakes, or eddy current effects.

Ground-level concentrations depend on the strength of the emission source, the physical and chemical nature of materials discharged, atmospheric conditions, stack and exhaust gas parameters, topography, and the aerodynamic characteristics of the physical surroundings. The state of the art in atmospheric dispersion modeling continues to advance, and air quality modeling is embodied in many state and local air pollution regulations and is sanctioned as well by the EPA (see Section 2.2).

The computational details of atmospheric dispersion modeling are outside the scope of this discussion. Several excellent publications (11–13) illustrate the rationale and methods for estimating ground level concentrations downwind from stationary sources.

6.2 Gravitational and Inertial Separation

Gravitational and inertial separation are the simpler forms of particulate collectors, but several different types and configurations are included. Each type is built to incorporate a hopper into which the collected material will eventually settle. The performance of this type of collector depends either on velocity reduction to permit settling or on the application of centrifugal force that increases the effective mass of particles. The common types of mechanical collectors include settling chambers, cyclones, and multiple cyclones.

6.2.1 Settling Chambers

A settling chamber may be nothing more than a long, straight, bottomless duct over a hopper, or it may consist of a group of horizontal passages formed by shelves in a chamber. A baffle chamber consists simply of a short chamber with horizontal entry and exit, using single or multiple vertical baffles. The inertia of the entrained particles causes them to strike the plates, enabling the dust to fall into the collection hopper. Settling chambers require a large space and are effective only on particles with diameters greater than 50 to 100 μm. However, their resistance to airflow, as measured by the pressure drop, is low. Collectors of this class are frequently used as precleaners for coarse particles, preceding other, more efficient types of collector.

6.2.2 Cyclones

In a cyclone, a vortex created within the collector propels particles to locations from which they may be removed from the collector. The devices can be operated either wet or dry. Cyclones may either deposit the collected particulate matter in a hopper or concentrate it into a stream of carrier gas that flows to another separator, usually of a different, more efficient type, for ultimate collection. As long as the interior of a cyclone remains clean, pressure drop does not increase with time. Up to a certain limit, both collection efficiency and pressure drop (usually less than 2 in. water) increase with flow rate through a cyclone. This type of collector is best applied in the removal of coarse dusts.

Cyclones are frequently used in parallel but seldom in series. When they are used in series it is to accomplish a special objective, such as to provide a backup in case the dust discharge of the primary cyclone fails to function. Small-diameter cyclones are more effective than larger ones, since centrifugal force for a given tangential velocity varies inversely with the radius. Accordingly, multiple cyclones (i.e., banks of small cyclones arranged in parallel) are used commonly to maximize particulate control efficiency within the inertial collector concept. Whereas single cyclones can be several feet in diameter, multiple cyclones are often less than 12 inches in diameter; pressure drop across this configuration is typically 3–8 in. water, gauge.

6.2.3 Other Devices

Other types of inertial separators include impingement collectors, consisting of a series of nozzles, orifices, or slots, each followed by a baffle or plate surface, and dynamic collectors, consisting of a power-driven centrifugal fan with the provisions and housing for skimming off a layer of gas in which dust has been concentrated.

6.3 Filtration

Filters are devices used for removal of particulate matter from gas streams by retention of particles in and around a porous structure through which the gas flows. The porous structure is most commonly a woven or felted fabric, but pierced, woven, or sintered metal can be used, as well as beds of a large variety of substances (fibers, metal turnings, coke, slag roll, sand, etc.).

Unless operated wet to keep the interstices clean, filters generally improve in retention efficiency as the interstices in the porous structure begin to be filled by collected material. These collected particles form a pore structure of their own, supported by the filter, and because of the increased surface area, they have the ability to intercept and retain other particles. This increase in retention efficiency is accompanied by an increase in pressure drop through the filter. To prevent decrease in gas flow, therefore, the filter must be cleaned periodically, or else replaced after a certain length of time.

For controlling air pollution, the fabric filter collector, commonly known as a "baghouse," together with its fan or air mover, can be likened to a giant vacuum cleaner. Baghouses utilize fabrics, woven or felted, from natural or synthetic fibers. In its most usual configuration, a fabric filter consists of a series of cylindrical bags or tubes, vertically mounted, and dust is deposited on either the inner or outer surfaces of the fabric. Fabric filters are generally useful where:

1. Very high particulate control efficiencies are desired.
2. Temperatures are below 550°F.
3. Gas temperatures are above the dew point.
4. Valuable material is to be collected dry.

Because of its high collection efficiency, simplicity of operation, and relatively low operating cost when compared to other basic collection systems, the baghouse is the leading choice for most particulate control applications.

6.4 Electrostatic Precipitation

Electrostatic precipitators are devices in which one or more high-intensity electrical fields cause particles to acquire an electrical charge and migrate to a collecting surface. The collecting surface may be either dry or wet. Because the collecting force is applied only to the particles and not to the gas, the pressure drop of the gas is only that due to flow through a duct having the configuration of the collector. Hence pressure drop is very low and does not tend to increase with time. Generally, collection efficiency increases with the length of passage through an electrostatic precipitator. Therefore, replicate precipitator sections are employed in series to obtain higher collection efficiency.

Electrostatic precipitators may incorporate pipes or flat plates as grounded electrodes onto which electrically charged dust particles are deposited for subsequent removal. In any electrostatic precipitator, three functional units exist by virtue of three specific operations tending to occur simultaneously.

1. Charging of particulate.
2. Collection of particulate.
3. Removal and transport of collected particulate.

Electrostatic precipitators find their greatest use in situations where:

1. Gas volumes are relatively large.
2. High efficiencies are required on fine particulate.
3. Valuable materials are to be collected dry.
4. Relatively high-temperature gas streams must be cleaned.

Although the energy required to separate particulates from the gas stream is considerably less than for other types of gas cleaning devices, reliability, operability and maintenance difficulties typically encountered with the electrostatic precipitator (such as electrode failures and ineffective collection plate cleaning) result in lowered collection efficiencies and higher maintenance costs.

6.5 Liquid Scrubbing

The prime means of collection in wet scrubbers is a liquid introduced into the collector for contact with the contaminant aerosol. Scrubbers are used to remove vapor phase contaminants and to separate particulate from the carrier gas. The scrubber liquid may simply wet the contaminant to facilitate removal or it may dissolve or react chemically with the contaminant collected.

Methods of effecting contact between scrubbing liquid and carrier gas include the following:

1. Spraying the liquid into open chambers containing various forms of baffles, grilles, or packing.
2. Flowing the liquid into these structures over weirs.
3. Bubbling the gas through tanks or troughs of liquids.
4. Forcing the gas to pass through sheets or films of liquid.
5. Using gas flow to create droplets from liquid introduced at a location of high gas velocity.

Scrubber liquid frequently can be recirculated after the collected contaminant is partially or completely removed. In other cases, all or part of the liquid must be discarded. In general, as long as the interior elements of the scrubber remain clean, the pressure drop does not increase with time. Collection efficiency tends to increase with increasing gas flow rate, provided the liquid feed keeps pace with gas flow and carryout or entrainment of liquid with the effluent gases is prevented effectively.

Scrubbers are generally of either the low-energy or high-energy type. Examples of low-energy scrubbers are spray chambers, centrifugal units, impingement scrubbers, most packed beds, and the submerged nozzle type of scrubber. High-energy scrubbers, especially the venturi scrubber, find wide application in situations calling for the removal of fine particulate matter at high efficiency levels. In the most common design, a venturi scrubber consists of a venturi-shaped air passage with radially directed water jets located just before or at the high-velocity throat. The water is broken into fine droplets by the action of the high velocity of the gas stream. The particulate matter is deposited on the water droplet by impaction, diffusion, and condensation. Coarse droplets of water, together with the entrained dust, are readily separated by a comparatively simple demister section on the venturi discharge.

Wet scrubbers can handle high temperatures and are often used simultaneously as particulate collectors and heat transfer devices. They generally are applicable and should be considered in a feasibility analysis if any one of the following situation descriptions applies:

1. The gaseous vapor phase contaminant is soluble in water (particulate need not be soluble).
2. The exhaust gas stream contains both gaseous and particulate contaminants.
3. Exhaust gases are combustible.
4. Cooling is desirable and an increase in moisture content is acceptable.

Since liquid scrubbers can remove both particulate matter and gases from the effluent gas, they provide many options for the control of industrial emissions. Although they are relatively simple devices, operating costs may be high if a high-energy type scrubber is required. Additionally, adequate precautions must be taken to meet wastewater discharge regulations before scrubber waste liquid is disposed.

6.6 Gas–Solid Adsorption

Adsorbers are devices in which contaminant gases or vapors are retained on the surface of a porous medium through which the carrier gas flows. Adsorption occurs primarily on the internal surfaces of the highly porous particles. The medium most commonly used for adsorption is activated carbon because of its nonpolar surface. Other adsorbent materials used industrially include silica gel, activated alumina (aluminum oxide), and zeolites (molecular sieves). The design of an adsorber parallels that of a filtration device for particulate matter in that the effluent gas flows through a porous bed. In the case of an adsorber, however, the adsorption bed should be preceded by a filter, to protect it from plugging due to particulate matter.

In true adsorption, there is no irreversible chemical reaction between the adsorbent and the adsorbed gas or vapor. The contaminant gas or vapor is merely being temporarily stored on the surface of the adsorbent. Therefore, the adsorbed contaminant can be driven off the adsorbent by heat, vacuum, steam, or other means and may be recovered in a highly concentrated state. In some adsorbers, the adsorbent is regenerated in this manner for reuse. In other applications, the spent adsorbent is discarded and replaced with fresh adsorbent. Such nonregenerable systems are normally used to control exhaust streams with low pollutant concentrations (less than 1.0 ppm) that are either highly odorous or toxic (14).

Pressure drop through an adsorber that does not handle gas contaminated by particulate matter does not increase with time but does increase with gas flow rate. The relationship between adsorption efficiency and gas flow rate depends entirely on adsorber design and on the characteristics of the material being collected.

Carbon adsorption is generally carried out in large horizontal fixed beds with depths of 3–25 feet. Such units can handle from 2,000 to about 50,000 cfm and are often equipped with blowers, condensers, separators, and controls. Typically, an installation includes at least two carbon beds; one is onstream while the other is being regenerated.

Although molecular sieves (zeolites) have reached commercial stage application in such processes as petroleum refining, activated carbon remains the most important dry adsorbent in gaseous emission control. Since molecular sieves have a crystalline structure and their pores are uniform in diameter, they can be used to capture or separate gases on the basis of molecular size and shape (14). Molecular sieves, like activated carbon, usually require fixed-bed units with sequence valves for switching the beds from adsorption to regeneration. The key element in these systems is molecular sieve-adsorbent blends: synthetic crystalline metallic-alumina silicates that are highly porous adsorbents for some liquids and gases.

Absorption using activated carbon appears to have an advantage over other collection methods (14):

1. When large amounts of air have to be utilized to capture the contaminant.
2. When concentrations of contaminants are exceptionally low so as to make other methods impractical, but the contaminant must still be removed.
3. In higher concentrations where the absorbed material has a recovery value.
4. Where, in smaller applications, space and economy dictate the use of activated charcoal filters over bulkier systems.

Activated carbon is used extensively in air purification systems and in solvent recovery systems that reclaim valuable solvents for reuse. Most of these regenerative solvent recovery systems pay for themselves in a short period of time.

6.7 Combustion

Many processes produce gas streams that bear organic materials having little recovery value or containing toxic or odorous materials that can be oxidized to less harmful combustion products. In such cases, thermal oxidation may be the optimal control route, especially if the gas streams are combustible. There are three methods of combustion in common use today: thermal oxidation, direct flame incineration, and catalytic oxidation.

Thermal oxidizers, or afterburners, can be used when the contaminant is combustible. The contaminated airstream is introduced to an open flame or heating device, then goes to a residence chamber where combustibles are oxidized to carbon dioxide and water vapor. Most combustible contaminants can be oxidized at temperatures between 100 and 1500°F. The residence chamber must provide sufficient dwell time and turbulence to allow complete oxidation.

Direct combustors differ from thermal oxidizers by introducing the contaminated gases and auxiliary air into the burner itself as fuel. Auxiliary fuel, usually natural gas or oil, is generally necessary for ignition and may or may not be required to sustain burning.

Catalytic oxidizers may be used when the contaminant is combustible. The contaminated gas stream is preheated, then passed through a catalyst bed that promotes oxidation of the combustibles to carbon dioxide and water vapor. Metals of the platinum family are commonly used catalysts, and they promote oxidation at temperatures between 700 and 900°F.

To use either thermal or catalytic oxidation, the combustible contaminant concentration must be below the lower explosive limit. Equipment specifically designed for control of gaseous or vapor contaminants should be applied with caution when the airstream also contains solid particles. Solid particulate matter can plug catalysts and, if noncombustible, it will not be converted in thermal oxidizers and direct combustors.

Thermal and direct combustors usually have lower capital cost requirements but higher operating costs because an auxiliary fuel is often required for burning. This cost is offset when heat recovery is employed. The catalytic approach has high capital cost because of the relatively expensive catalyst, but it needs less fuel. Either method provides a clean, odorless effluent if the exit gas temperature is sufficiently high.

6.8 Combination Systems

Within the wide spectrum of stationary emission sources, there are several applications for coupling complementary control devices into an effective combination system. Selection of the best combination and sequence of control methods depends to some degree on past experience, but excellent results can be achieved by careful analysis of the contaminated air or gas to be treated.

A good example of a combination system is the inertial collector-electrostatic precipitator system, applicable when a wide range of particle sizes occurs in the gas, or where dust loadings are high. Examples are blast furnaces and coal-fired boilers. Another innovative combination uses the carbon absorber and incinerator to minimize operating costs. In this control system, organic vapors are absorbed for several hours on activated carbon; during this time, the incinerator is turned off. When the carbon bed is regenerated, the desorbent vapors, including the steam used to heat the bed, can be sent directly to the incinerator for destruction. Not only is auxiliary fuel saved by this intermittent operation, but the desorbent may be self-sustaining during incineration.

Furthermore, incinerator size is reduced dramatically compared with direct-fired equipment. Capital costs may sometimes be offsetting, however.

Two principal arrangements of multiple-unit air-cleaning systems are recognized: parallel combination and serial combination. Parallel combination is simply a means of providing a wide range of flexibility in adjusting the capacity of a system to the airflow rate. It permits small, highly effective elements, such as centrifugal tubes (multiple cyclones), to be assembled in parallel arrangement for handling equal shares of the total gas flow. Serial combination presents the problem of determining the best sequence, if different types of cleaning units are to be used, or the best combination of preparation and separation units to meet the range of contaminants that must be removed from the airstream.

6.8.1 Two-Stage Cleaning Systems

Two-stage systems comprise a preparation stage and a separation stage. Examples of this arrangement include the following:

1. Two-stage electrostatic precipitators, with particle charging preceding precipitation.
2. Venturi scrubbers, with simultaneous agitation and liquid injection preceding cyclone separation.
3. Sonic agglomeration systems, with simultaneous agglomeration and gravitational precipitation followed by inertial or centrifugal separation.

6.8.2 Sequence of Separating Stages

The selectivity of particulate separators, or their fractionating tendencies, makes it advisable to combine stages of cleaning that are complementary. The sequence of treatment is generally, but not invariably, from the types of separator least able to remove submicrometer particles to those at the final stages most likely to succeed at this removal.

The agglomerating action of certain types of separator results in large aggregates of flocs blowing intermittently from dry collecting surfaces into the airstream. When this condition occurs, because of excessive air velocities or poor retentivity at the contact surfaces, it is good practice to interpose coarse particle separators to reduce the load on subsequent stages of air cleaning.

Some of the variation in airborne matter is gradually suppressed or damped as the air progresses through a multistage system of cleaning. The process fluctuations may not be eliminated, but they frequently are modified so greatly in the initial stages of air treatment that the final, and most critical, stages are protected against disturbing variations in the character of materials traveling with the air.

6.8.3 Concentration and Subdivision of Airstreams

As the air passes through a cleaning system, it may be handled as a single stream from inlet to outlet. There are numerous installations, however, in which the airstream is subdivided one or more times to specialize the task of cleaning and to make each stage more effective. A common example is conveyance of the small-volume, high-concentration effluent from the apex of the periphery of a cyclone to a secondary separator generating greater centrifugal force.

7 GRAVITATIONAL AND INERTIAL SEPARATION

The simplest type of particulate collector is commonly known as a "mechanical collector." Mechanical collectors utilize either gravity or inertia to remove relatively large particles from suspension in a moving gas stream. In these collectors, the gas stream is made to flow in a path that either enhances the gravitational separation or changes direction such that particles cannot easily follow because of their inertia. Most mechanical collectors operate in a dry condition, although water is sometimes used in conjunction with a mechanical collector to aid in continuous cleaning of the control device.

7.1 Range of Performance

Gravitational and inertial separation devices range in capture efficiency from less than 50 to about 90%, depending on particle size and type of collector. Generally, as the velocity through the collector increases, control efficiency increases, except for gravitational settling chambers where the opposite effect occurs.

Pressure drop across an inertial collector of this type is one indicator of relative efficiency. Although a direct relationship between drop and efficiency cannot always be calculated easily, there is usually a significant correlation between the two; control efficiency usually increases with pressure drop. One important aspect of this phenomenon is that pressure drop increases require greater amounts of energy to operate the collection system.

Table 59.5 compares characteristics of various kinds of gravitational and inertial separators.

7.2 Settling Chambers

Settling chambers represent the oldest form of air pollution control device. Because of their simplicity of design and relatively low operating costs, they are most commonly used today as precleaners for more efficient particulate control methods that consume higher amounts of energy.

Settling chambers use the force of gravity to separate dust and mist from the gas stream by slowing down the gas stream so that particles will settle out into a hopper (Fig. 59.3) or onto shelves from which they can be removed (Fig. 59.4).

From a practical standpoint, settling chambers are not employed for the removal of particles less than 50 μm in diameter because of the excessive size of the equipment

Table 59.5. Characteristics of Mechanical Collection Equipment[a]

Collector Type	Space Require-ments	Volume Range	Efficiency by Weight	Pressure Loss[b] (in. H_2O)	Temperature Limitations	Power[c] (hp per 1000 cfm gas)	Application Areas
Settling chamber	Large	Space available only limitation	Good above 500 μm	0.2–0.5	700–1000°F, limited only by materials of construction	0.04–0.12	Precollector for fly ash metallurgical dust, can be used for any large dust particles above 50 μm
Conventional cyclone	Large	Normal range up to 50,000 cfm	Approximately 50% on 20 μm	1–3	700–1000°F, limited only by materials of construction	0.24–0.73	Woodworking, paper, buffing, fibers, etc.; well suited for dry dust particles 20 μm and above
High-efficiency cyclone	Medium	Normal range up to 12,000 cfm	Approximately 80% to 10 μm	3–5	700–1000°F, limited only by materials of construction	0.73–1.2	Woodworking, material conveying, product recovery, etc.; well suited for dry dust particles 10 μm and above
Multitube cyclone	Small	Normal range up to 100,000 cfm	90% on 7.5 μm	4.5	700–1000°F	1.1	Precollector for electrostatic precipitator on fly ash, product recovery, etc.; well suited for dry dust particles 5 μm and above
Dynamic precipitator	Small	17,000 cfm	80% on 15 μm	No loss (true fan)	700°F	Power consumption will depend on selection point, mechanical efficiency in usual selection range from 40 to 50%	Woodworking, nonproduction buffing, metal working, etc.; well suited for dry dust particles 10 μm and above
Impingement separator	Small	Space available only limitation	90% on 10 μm	1–5	700°F	0.24–1.2	Certain types used for collecting coarse particles in boiler fly ash and cement clinker cooler; recent designs used for cleaning atmospheric air to diesel engines and gas turbines

[a]Ref. 15.
[b]Pressure drop is based on standard conditions.
[c]Power consumption figured from: horsepower = cfm × total pressure/6356 × ME; mechanical efficiency (ME) assumed to be 65%.

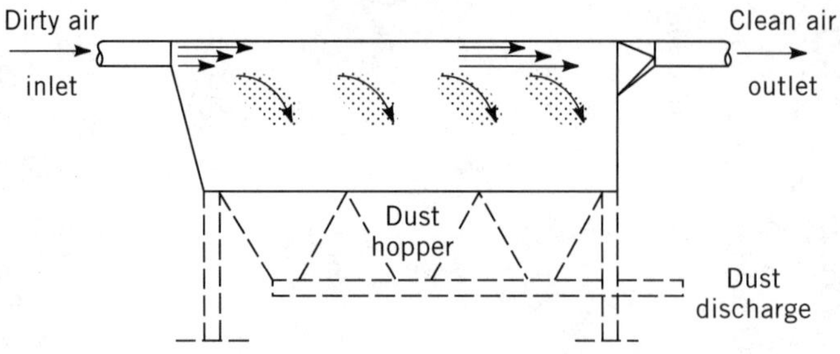

Figure 59.3. Gravity dust-settling chamber (15).

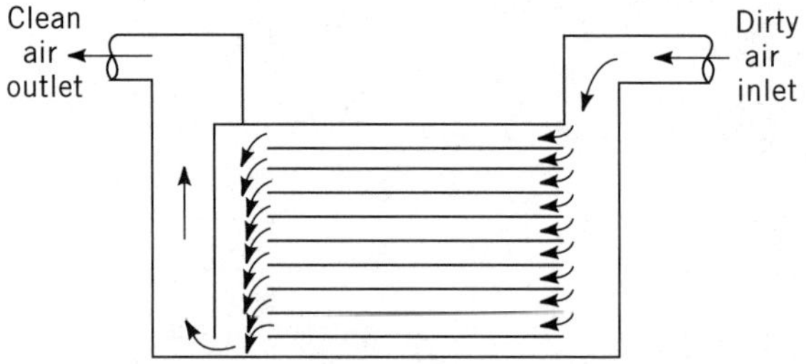

Figure 59.4. Multiple-tray dust collector (6).

required for collection below this size limit. To prevent re-entrainment of the separated particles, the velocity of the gas stream entering the settling chamber should not exceed 600 fpm.

The efficiency of a settling chamber can be calculated by the following equation (16):

$$E = \frac{100\ U_t WL}{Q} \tag{10}$$

where E = efficiency, weight percent of particles of settling velocity U_t (dimensionless)
U_t = terminal settling velocity of dust (ft/scc)
L = chamber length (ft)
W = chamber width (ft)
Q = gas flow rate (ft^3/sec)

This efficiency model assumes good distribution at the inlet and outlet of the chamber. Combining this equation with Stokes' law, the minimum particle size that can be completely separated from the gas stream can be calculated by the following equation (17):

$$d_p = \left[\frac{18\ \mu HV}{gL(\rho_p - \rho_g)}\right]^{1/2} \tag{11}$$

where d_p = minimum size particle collected completely (ft)
μ = gas viscosity (lb/ft-sec)
H = chamber height (ft)
V = gas velocity (ft/sec)
g = gravitational constant (32.2 ft/sec^2)
ρ_p = particle density (lb/ft^3)
ρ_g = gas density (lb/ft^3)

If there are horizontal plates or trays in the chamber, the effective settling distance is reduced, and the efficiency and minimum particle size that can be completely separated are given by Equations 12 and 13, respectively (17):

$$E = \frac{NU_tWL}{Q} \tag{12}$$

$$d_p = \left[\frac{18\ \mu Q}{gNWL(\rho_p - \rho_g)}\right]^{1/2} \tag{13}$$

where N is the number of plates and other units are as in Equations 10 and 11.

The chief advantages of settling chambers are their low operating cost, relatively low initial cost, simple construction, low maintenance costs, and low pressure drop in operation.

Their main disadvantages are large space requirements and low collection efficiency for smaller particles. When using shelves to increase efficiency, this characteristic also causes a problem with cleaning the settled particulate off the shelves; this is often done by rinsing the plates with water.

7.3 Inertial Separators

Inertial separators represent one of the simplest pollution control devices. They take up less room than settling chambers and are more efficient. Inertial separators are relatively inexpensive to acquire and operate, but they are not very efficient, and like settling chambers, they are used basically for precleaning of gas streams. Inertial separators include all dry-type collectors that utilize the relatively great inertia of the particles to effect particulate-gas separation. Two basic types of inertial separation equipment are simple impaction separators, which employ incremental changes of the carrier gas stream direction to exert the greater inertial effects of the particles, and cyclonic separators, which produce continuous centrifugal force as a means of exerting the greater inertial effects. The cyclones are discussed in a subsequent section.

Impingement or impaction separation occurs when the gas stream suddenly changes its direction because of the presence of an obstructing body. The impingement efficiency, defined as the fraction of particles in the gas volume swept by the obstructing body that will impinge on that body, is a direct function of the separation number. The separation number is given as follows:

$$N_s = \frac{D_p^2 V \rho_p}{18 \mu D_b} \tag{14}$$

where N_s = separation number (dimensionless)
D_p = particle diameter (ft)
μ = gas viscosity (lb/ft-sec)
ρ_p = particle density (lb/ft^3)
V = relative velocity, gas to target velocity (ft/sec)
D_b = equivalent diameter of obstructing body (ft)

Impingement separators vary with the configuration of the obstructing bodies and include baffle type, orifice impaction type, high-velocity gas reversal type, and louver type.

The baffled chamber (Fig. 59.5) is the simplest of the impingement separators. It uses one or two plates as impingement sites to stop larger particles and cause them to fall into a dust-collecting bin. This equipment can remove particles larger than 20 μm in diameter with pressure drops varying from 0.5 to 1.5 in. water.

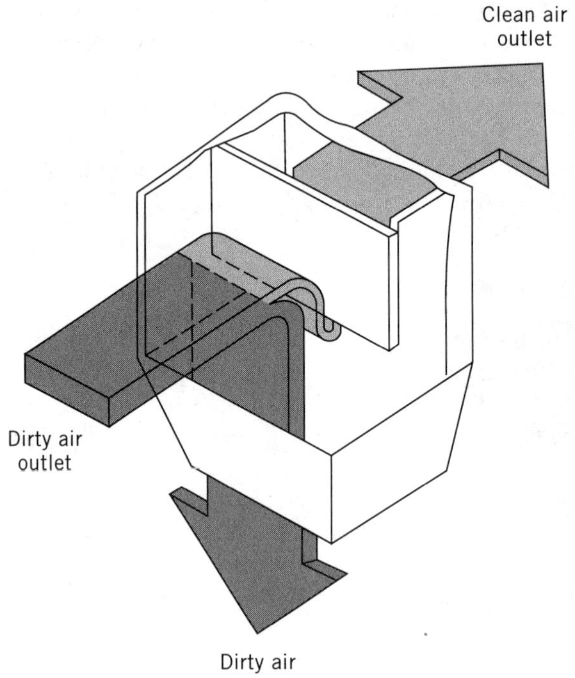

Figure 59.5. Baffled chamber.

The orifice impaction collector (Fig. 59.6) may consist of many nozzles, slots, or orifices followed by a plate or baffle for an impingement surface from which the particles can fall into the dust bin. This device can remove particles greater than 2 μm in diameter. It is more efficient than the simple baffle type but more expensive. Normal velocities through the orifices are about 50 to 100 ft/sec and the pressure drop is approximately 2.5 orifice velocity heads.

The louvered impingement separator uses many impingement surfaces on an angle, to rebound the particles into a secondary airstream and allow the cleaner air to pass through the louver. The particulate is thus concentrated into a secondary airstream whose flow rate is 5 to 10% that of the primary airstream. This concentrated airstream is usually passed through another more efficient cleaner for discharge into the atmosphere. The efficiency of this type of collector is basically a function of the louver spacing, closer spacings producing higher efficiencies. The two principal arrangements of louvers are the flat louver impingement separator (Fig. 59.7) and the conical louver impingement separator (Fig. 59.8).

7.4 Dynamic Separators

Dynamic separators, sometimes termed rotary centrifugal separators, are among the more recently developed gravitational and inertial devices used for collecting particulate. Dy-

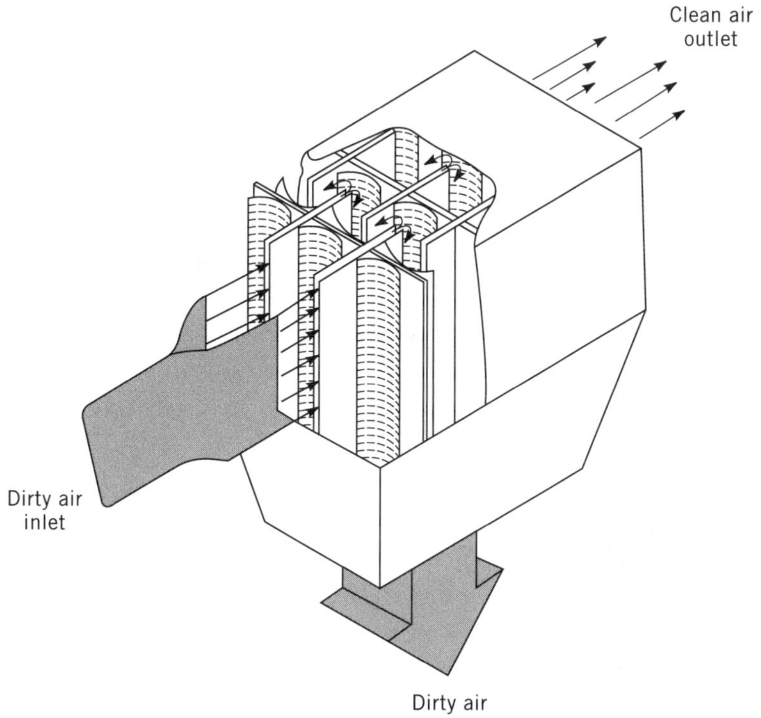

Figure 59.6. Orifice impaction collector.

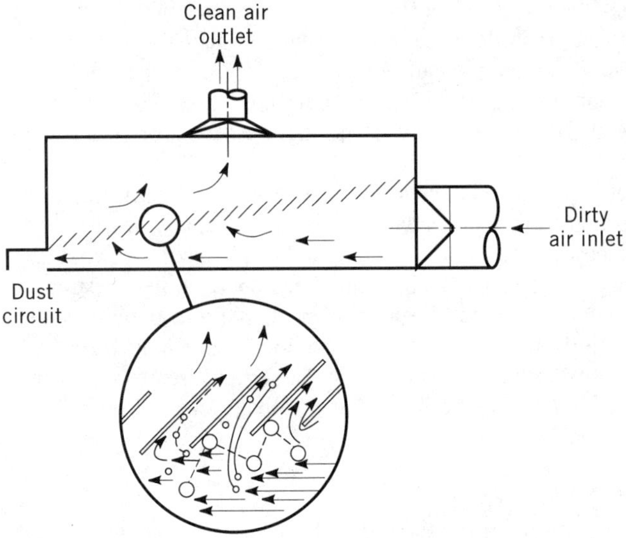

Figure 59.7. Flat louver impingement separator (18).

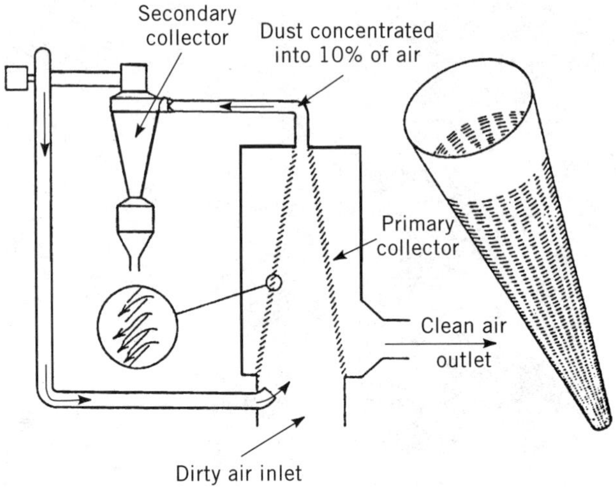

Figure 59.8. Conical louver impingement separator (18).

namic precipitators use centrifugal force to separate particulate from an airstream. The precipitator concentrates the dust around the impellers, and it is dropped into a hopper. These devices are about 80% efficient on particles larger than 15 μm in diameter.

The major advantage of the dynamic separator is its small size, which is most helpful when a facility needs many independently operated precipitators but has only limited space.

Because this separator is a true fan, there is no pressure drop across the device. Its major limitation is its tendency toward plugging and imbalance, as well as wearing of the blades.

A dynamic precipitator (see Fig. 59.9) separates particles by first drawing the gas stream into it; then as the particulate impacts the impeller, centrifugal force throws the heavier particles to the periphery of the housing. The lighter particles are impacted on the blades and glide along the blade surface to the outside edge where they are also thrown to the periphery of the housing. The particles are then discharged out of the annular slot to the dust collection bin. The cleaned air is discharged into a scroll-shaped discharger.

7.5 Cyclones

Cyclones have long been regarded as one of the simplest and most economical mechanical collectors. They can be used as precleaners or as final collectors. The primary elements of a cyclone are a gas inlet that produces the vortex, an axial outlet for cleaned gas, and a dust discharge opening. Several different types of cyclones (Figs. 59.10 through 59.13) are as follows:

1. Tangential inlet with axial dust discharge (most common).
2. Tangential inlet with peripheral dust discharge.

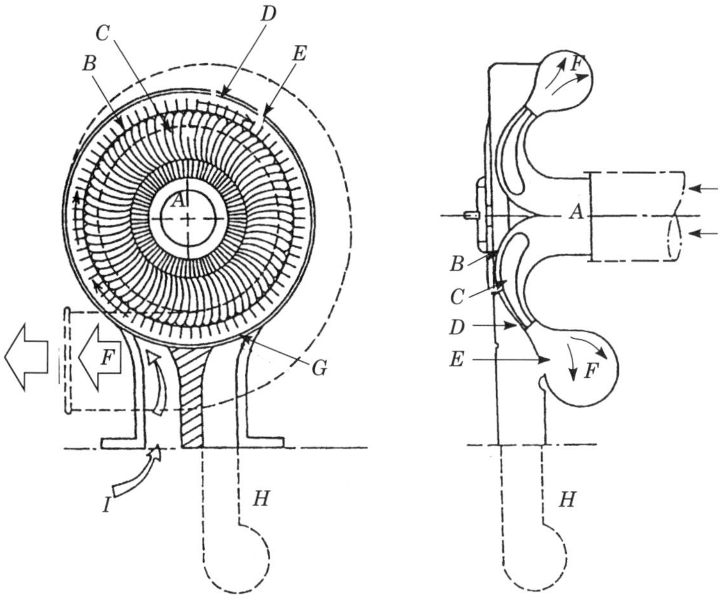

Figure 59.9. Dynamic separator: *A*, center inlet; *B*, interception disk; *C*, impeller blades; *D*, opening for dust escape; *E*, chamber; *F*, scroll-shaped discharge chamber; *G*, point of deflection into hopper *H*; *I*, secondary air return (18).

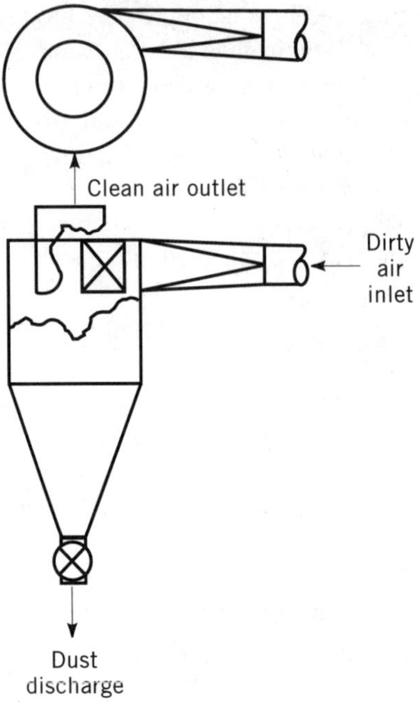

Figure 59.10. Cyclone with tangential inlet and axial discharge (17).

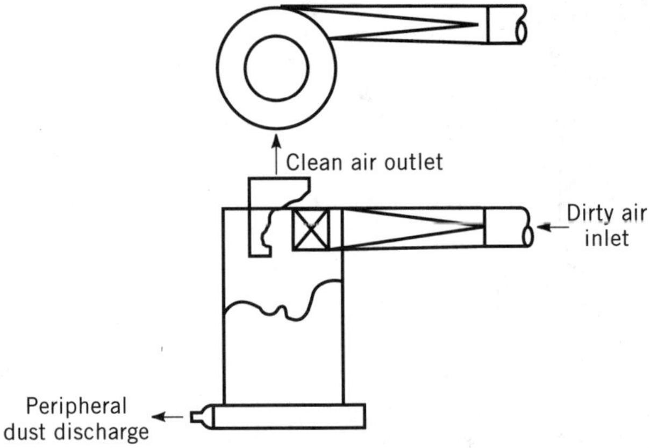

Figure 59.11. Cyclone with tangential inlet and peripheral discharge (17).

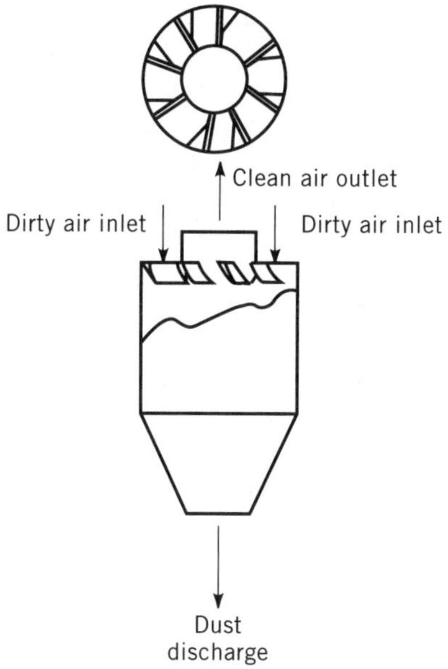

Figure 59.12. Cyclone with axial inlet and axial discharge (17).

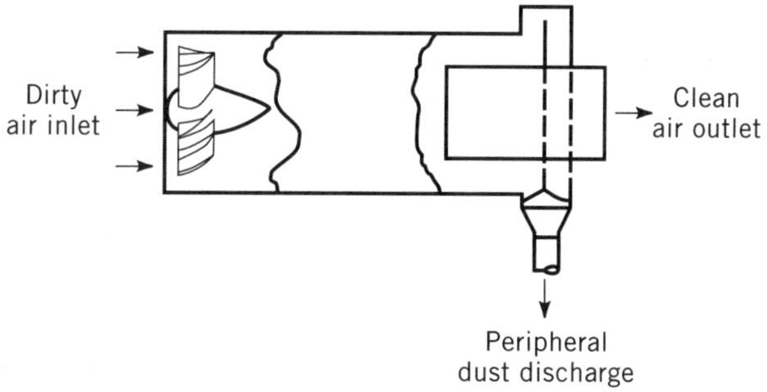

Figure 59.13. Cyclone with axial inlet and peripheral discharge (17).

3. Axial inlet through swirl vanes with axial dust discharge.
4. Axial inlet through swirl vanes with peripheral dust discharge.

Centrifugal force is the precipitating force for particulate and droplets. The air to be cleaned usually enters at the top of a cyclone, either by tangential flow or by curved vanes in an axial inlet. As the air turns, it flows down the cyclone; near the bottom, it reverses direction and flows up the center of the cyclone through the axial exhaust port. Because air is always spinning in the same direction, suspended particles are being forced to the outside wall, where they slide down to the discharge chute (Fig. 59.14). Standard cyclone design dimensions are given in Table 59.6.

Efficiency of a cyclone is defined as the fractional weight of particles-collected. The major parameter in the prediction of collection efficiency is particle size. The particle size that can be removed from the inlet gas stream at an efficiency of 50% in a cyclone is defined as the particle cut size d and is represented in the following relationship (20):

$$d_{pc} = \left[\frac{9\ \mu W}{2NV(\rho_p - \rho_g)\pi}\right]^{1/2} \qquad (15)$$

where d_{pc} = diameter cut size particle collected at 50% efficiency (ft)
μ = gas viscosity (lb mass/sec-ft = centipoise $\times$ 0.672 $\times$ 10^{-3})
W = cyclone inset width (ft)
N = effective number of turns within cyclone
V = inlet gas velocity (ft/sec)
ρ_p = true particle density (lb/ft^3)

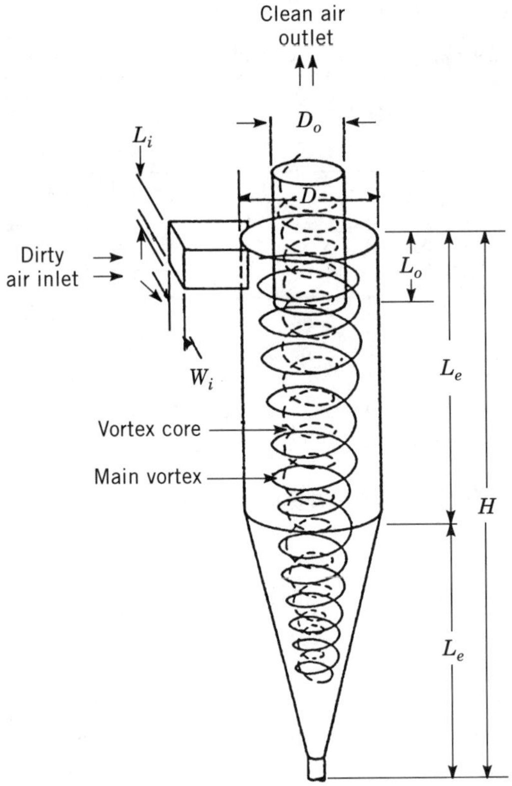

Figure 59.14. Typical cyclone; for abbreviations and standard dimensions, see Table 59.6 (19).

Table 59.6. Standard Basic Cyclone Design Dimensions

Parameter	Conventional Cyclone	High-Throughput Cyclone	High-Efficiency Cyclone
Cyclone diameter, D	D	D	D
Cyclone length, L_c	$2D$	$1.5D$	$1.5D$
Cone length, L_c	$2D$	$2.5D$	$2.5D$
Total height, H	$4D$	$4D$	$4D$
Outlet length, L_o	$0.675D$	$0.875D$	$0.5D$
Inlet height, L_i	$0.5D$	$0.75D$	$0.5D$
Inlet width, W_i	$0.25D$	$0.375D$	$0.2D$
Outlet diameter, D_o	$0.5D$	$0.75D$	$0.5D$

ρ_g = gas density (lb/ft^3)
π = constant, 3.1416

This equation, together with Figure 59.15, permits the accurate prediction of the collection efficiency of a cyclone when the particle size distribution is known. The design factor having the greatest effect on collection efficiency is cyclone diameter. For a given pressure drop, the smaller the diameter of the unit, the higher the collection efficiency obtained. The efficiency of a cyclone also increases with increased gas inlet velocity, cyclone body length, and ratio of body diameter to gas outlet diameter. Conversely, efficiency decreases with increased gas temperature, gas outlet diameter, and inlet area.

The pressure drop across cyclones is another factor affecting collection efficiency. It varies between approximately 2 and 8 in. water; pressure drop increases with the square

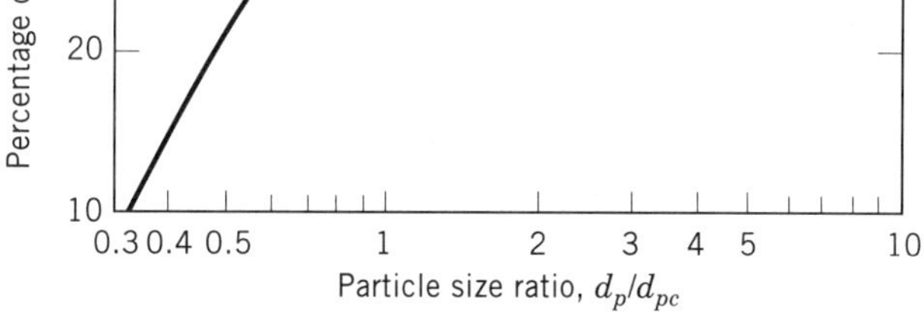

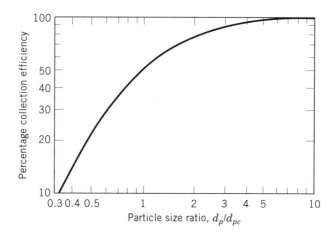

Figure 59.15. Cyclone collection efficiency as a function of particle size ratio (20).

of the inlet velocity. Collection efficiency also increases with the square of the velocity, but not as rapidly as does the pressure drop.

The inlet velocity can be increased only to the point (70 ft/sec) at which the turbulence is not causing re-entrainment of the particles being separated. After this point, the efficiency decreases with increased inlet velocity. Efficiency also increases as the dust loading increases, as long as the dust does not plug the cyclone or cause it to be eroded severely. In addition, the size of the particles affects the efficiency: the larger the particle, the higher the collection efficiency.

One of the greatest advantages of cyclones is that they can be made of many materials and thus are able to handle almost any type of contaminant. Other advantages of cyclones include their ability to handle high dust loads, relatively small size, relatively low initial and maintenance costs, and, excluding mechanical centrifugals, temperature and process limitations imposed only by materials and construction. Because cyclones have larger pressure drops than settling chambers or inertial separators, their cost of operation is higher, but still lower than that of most other cleaners. A disadvantage of cyclones is that efficiency drops off with decreased dust loading. Plugging can also occur, but if the particle characteristics are taken into consideration, this problem can be avoided. Low collection efficiencies for particles below 10 μm in diameter is another disadvantage.

7.6 Wet Cyclones

A wet cyclone is nothing more than a dry cyclone with an inlet for water or some other fluid to be impacted with the incoming particles, rinsing the particles from the cyclone. It costs more to operate than a dry cyclone because of the increased cost for the fluid introduced into the cyclone and for requisite cleaning of the discharge fluid to remove suspended or dissolved particulate.

The forces acting to remove the particles are the same as for dry cyclones. The smaller particles are made larger by impaction with fluid droplets, to increase the efficiency for smaller particles. Because the cyclone walls bear a liquid film, the chance for particulate re-entrainment is reduced; thus efficiencies are greater.

Wet cyclones offer the ability to take the liquid from any of several points other than the axial part of the vortex cone, which means that the vortex itself is not distributed. A wet cyclone usually has fewer erosion problems than a dry cyclone, but if corrosive dust or gas is present, water may activate a serious corrosion problem. Disposal or cleaning of the contaminated water or liquid is another problem to be solved in wet operations.

Wet cyclones entail some design problems that are not present with dry cyclones. First, the liquid tends to creep along the walls of the cyclone to the outlet, where droplets are sheared off into the outlet stream. This can be corrected by installing a skirt on the exit (Fig. 59.16) so that the droplets go into the vortex rather than the exit gas stream. If the velocity is too high in the cyclone, very small droplets cannot be recollected. The recommended inlet velocity is not to exceed 150 ft/sec at atmospheric pressure and for injected water; this estimate varies with different gases and liquids but is a general rule.

7.7 Multiple Cyclones

Multiple cyclones are simply many cyclones put in parallel or series. Cyclones in series are not used very often. Cyclones in parallel constitute the most common configuration

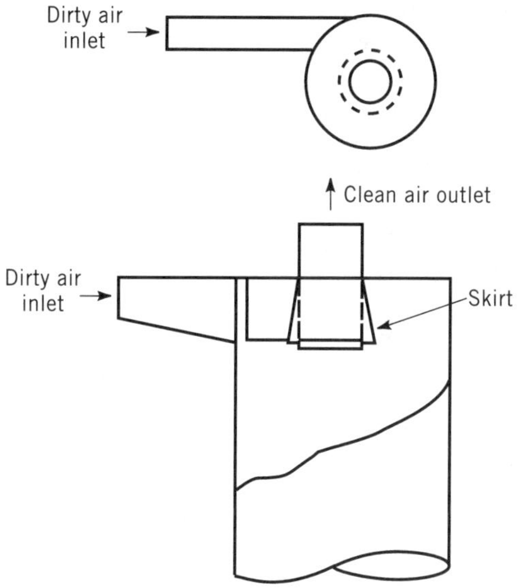

Figure 59.16. Gas outlet skirt for wet cyclone (17).

for multiple cyclones. Space requirements are moderate for multiple cyclones, but these devices are more expensive than regular single cyclones to purchase and design. Operational costs for multiple cyclones are about the same as those for single cyclones. The devices are usually used as a final control but also serve as precleaners in some applications.

Multiple cyclones make use of the same forces to separate the particles from the airstream as the single cyclone. The multiple cyclones usually consist of smaller cyclones that have more "separation force" for a given overall energy input. When many small (3- to 12-in. diameter) cyclones are used in parallel, they are more efficient than one larger cyclone of the same capacity. Their pressure drops are usually between 3 and 6 in. water.

Multiple cyclones offer the most significant advantage of ability to handle large airflows with heavy dust loadings and still exhibit good efficiency. When properly designed inlet ductwork provides even flow to all the individual cyclones, the multiple cyclone performs at peak design efficiency. This also ensures that there will not be any backflow through some of the cyclones, causing particles to flow from the dust bin to the exit plenum. Many multiple cyclones have several gas inlet and exit plenums and several dust exit valves; this allows any part of the multiple cyclone to be shut off from another part in the event that a cyclone plugs or needs repairs.

8 FILTRATION

Filtration represents the oldest, and inherently the most reliable, of the many methods by which particulate matter (dusts, mists, and fumes) can be removed from gases. Filters are especially desirable for extracting particulate matter from gases produced by industrial operations because they generally offer very high collection efficiencies with only moderate power consumption. Initial investment costs and maintenance expenditures can range from relatively low to comparatively high, depending on the size and density of the particulate matter being collected, as well as the quantity and temperature of the dusty gas to be cleaned.

Filters are most readily classified according to filtering media. For particulate emissions control, the different medial can be broadly categorized as (*1*) woven fibrous mats and aggregate beds, (*2*) paper filters, and (*3*) fabric filters.

8.1 Range of Performance

Filtration is highly effective for small particulates, even down to less than 0.05 μm in diameter, if the flow through the filter is kept low enough. Filtration can be used on almost any process emitting small particles of dust. Limiting factors, however, are the characteristics of the gas stream. If the unfiltered stream is too hot, it may be necessary to precool the gas stream. If the gas stream is too heavily loaded, a precleaner may be necessary to rid the stream of larger particles. The cost of auxiliary equipment required for such pretreatment may equal or even exceed that of the filter itself. There are also several causes for filter failure, including blinding, caking, burning, abrasion, chemical attack, and aging.

Designs of filters and filter enclosures are many. Filters can be configured as mats, panels, tubes, or envelopes. When tubular fabric filters are enclosed, the structure and its contents are often called a "baghouse."

8.2 Fibrous Mats and Aggregate Beds

Fibrous mats and aggregate-bed filters are characterized by high porosity; both are composed largely (97–99%) of void spaces. Such void space is usually much larger than the particles being collected; thus the mechanisms of sieving and straining of particles are of no significance in these filters. The predominant forces functioning in the cleaning of gas streams using large void filters include impaction, impingement, and surface attraction. Impaction and impingement are effective when the particles are made to change direction quickly and impact or impinge on the surface of the filter medium. Surface-attractive forces are mostly electrostatic, and before they contribute significantly to total collection, the dust particles must come within several particle diameters.

Efficiencies can be extremely high even with very low dust loadings in the input airstream. Early applications using sand as the filtering medium reported collection efficiencies of 99.7% for submicrometer particles (21). Somewhat more recently, glass fiber beds have been used and efficiencies of 99.99 percent have been reported for dust loadings of 0.00002–0.00004 grain/ft^3 (22).

Advantages of aggregate-bed filters include a longer life without frequent cleaning, high dust storage capacity with a modest increase in airflow resistance, and application to high-temperature emissions. One disadvantage is difficulty in cleaning; many fibrous mats are simply discarded. Large space requirements also pose a problem when these filters are used.

These filters range from very inexpensive mats that are changed and discarded to very expensive beds that are almost never replaced. Deep beds and fibrous mats are relatively inexpensive to operate because of low pressure losses (between 0.1 and 1 in. water) in most operations.

Designs for deep-bed filters are numerous. Almost any material can be chosen to make up the bed. Sand has been used most often, but glass fiber beds have been used more recently. Sulfuric and phosphoric acid plants use what are known as "coke boxes" for collection of acids. A lead or ceramic-lined box is filled with several feet of ¼- to ½-in. diameter coke to collect the mechanically produced mist at an efficiency of 80 to 90%. Condensed mist is generally too small to be collected by the coke box.

The mats are of many materials. Glass fibers, stainless steel, brass, and aluminum are all used in mats for different types of airstreams. Fibrous mats must be cleaned when their pressure loss becomes too high or the flow rate becomes too low. Cleaning can be accomplished by removing a portion of the filter bed continually and replacing it with new or cleaned portions. Some cleaning is accomplished by reversing the airflow and vibrating the bed or using shock waves. Some mats are self-cleaning and some are of a continuous-cleaning design.

Designs for self-cleaning beds are numerous. For example, an automatic viscous filter (Fig. 59.17) uses an oil film to ensure that the particles are not re-entrained in the airstream. The airstream must pass through the filter twice. The plates on the conveyer belt open and

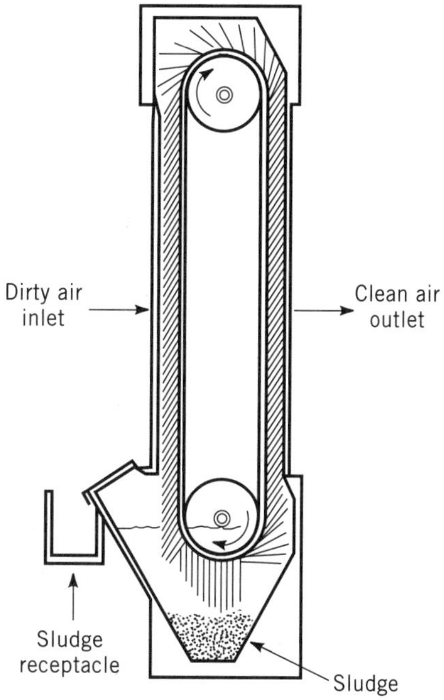

Figure 59.17. Self-cleaning automatic viscous filter (22).

close at the top and bottom of the filter. At the bottom, the plates are plunged into an oil reservoir and cleaned by agitation. Upon emerging from the oil bath, they are cleaned further and oiled.

Another type of self-cleaning mat is the water spray or wet filter (Fig. 59.18). The mats are sprayed continuously while the airstream flows through the filter. The particles are dislodged by the water and flow with it to the sump.

8.3 Paper Filters

Paper filters are of relatively recent design for air pollution control and find service chiefly where ultrahigh efficiencies are needed. These filters have come into wide use where very clean air is essential, as in "white rooms" of hospitals, data processing centers, "clean rooms" in the aerospace industry, food processing plants, biotechnology R&D laboratories, and semiconductor manufacturing.

Paper filters can be made of minerals, asbestos, or glass microfibers, with or without binders that add strength, formability, or water resistance. Glass microfilters are the most popular because of their fire resistance and availability. The frames that contain a set of filters known as "packs" or "plugs" can be made of steel, aluminum, hardboard, or plywood, depending on the application. These frames are normally fitted with gaskets on both sides to ensure a good seal; the seals are made of appropriate materials for the application (e.g., neoprene, mineral wool, asbestos, or rubber).

Paper filters may be flat, cylindrical, or any shape that is practical for the application and space available. Fluting of the paper medium increases the surface area of a filter compared to that of a flat fiber; this can save appreciably in equipment size used to house the filtering elements.

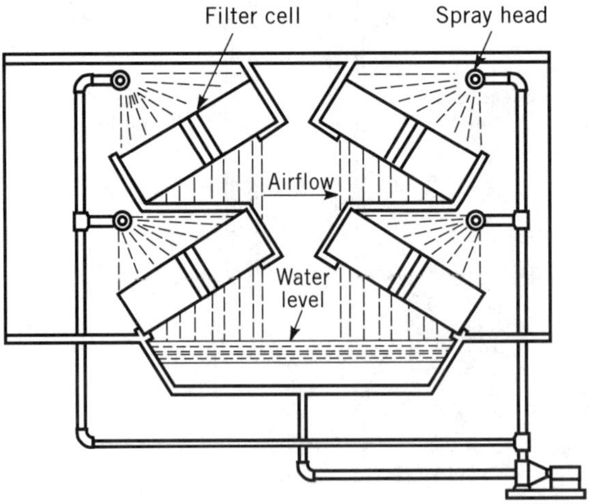

Figure 59.18. Self-cleaning wet filter (22).

Paper filter systems require moderate initial capital investment, and operating costs are relatively low. Variations in design, however, can make capital investment costs very high, depending on the application.

Paper filters use all mechanisms of particle capture and retention. The most significant is diffusion. Paper filters are often used as final filters to remove very small particles or dusts in very low concentrations. Any larger particles still present in the gas stream may be captured, but the life of a filter is shortened measurably if too many large particles are captured on a continuous paper filter. Many times high-efficiency, inexpensive precleaners are used before the paper filter to make the paper medium last longer. Because the filtration is mostly a surface action phenomenon, the dust storage capacity is a function of the surface area.

The efficiency of paper filter media is very high, 99.97% by weight or greater for the best commercial paper filters. The dioctyl phthalate (DOP) method for calculating the efficiency of a filter is employed for paper filters. The test method is the U.S. Army Chemical Corps. DOP Smoke Penetration and Air Resistance Test No. MIL-STD-282, Method 102.9.1 (23).

To ensure good collection efficiencies throughout their lifetime, paper filters must have larger pressure losses when new compared with other filter types. This means that the ratio of pressure loss in a new filter to a spent filter is larger. Flow velocity through the paper is usually around 5 ft/min.

Paper filters are advantageous in that they have a long life expectancy, usually 1 to 2 years, and are relatively inexpensive to replace. Paper filters cannot be cleaned and reused. Other disadvantages include the necessity to provide low flow rates and low dust loadings to the paper filtering element.

8.4 Fabric Filters

One of the most positive methods for removing solid particulate contaminants from gas streams is filtration through fabric media. A fabric filter is capable of providing a high collection efficiency for particles as small as 0.5 μm and will remove a substantial quantity of particles as small as 0.01 μm.

Fabric filters are usually tubular (bags) or flat and made of woven or felted synthetic fabric. The dirty gas stream passes through the fabric and the particles are collected on the upstream side by the filtration of the fabric. The dust retained on the fabric is periodically shaken off and falls into a collecting hopper.

The structure in which the bags hang is known as a "baghouse." The number of bags in a baghouse may vary from one to several thousand and the baghouse may have one compartment or many, making it possible to clean one while others are still in service. Bags typically have an average life of 18 to 36 months.

Removal of particulates from the gas stream is not a simple filtration or sieving process, because the pores of the fabric employed in fabric filters are normally many times the size of the particles collected, sometimes 100 times larger or more. The collection of particles takes place through interception and impaction of the particles on the fabric fibers, and through Brownian diffusion, electrostatic attraction, and gravitational settling within the fabric pores. Once a mat or a cake of dust is accumulated on the fabric, further collection is accomplished by the mechanism of mat or cake sieving, as well as by the foregoing mechanisms, thus improving the overall collection efficiency of the filter. Periodically the accumulated dust is removed, but some residual dust remains and serves as an aid to further filtering.

Direct interception occurs whenever the gas streamline, along with a particle, approaches a filter element. Inertial impaction occurs when a particle, unable to follow the streamline curving around an obstacle, comes closer to the filter element than it would have come if it had approached along the streamline. Small particles, usually less than 0.2 μm, do not follow the streamline because collision with gas molecules occurs, resulting in a random Brownian motion that increases the chance of contact between the particles and the filter element. Electrostatic attraction results from electrostatic forces drawing particles and filter elements together whenever either or both possess a static charge.

The major particulate removal mechanisms as they apply to a single fiber in a fabric filter are shown in the following tabulation:

Primary Collection Mechanism	Diameter of Particle (μm)
Direct interception	>1
Impingement	>1
Diffusion	<0.01–0.2
Electrostatic attraction	<0.01
Gravity	>1

Fabric filter systems (i.e., baghouses) are characterized and identified according to the method used to remove collected dust from the filters. This is accomplished by a variety of cleaning methods, including shaking (Fig. 59.19), reverse airflow (Fig. 59.20), reverse jet (Fig. 59.21), and reverse pulse (Fig. 59.22). Mechanically shaking the bags is the oldest cleaning technique and is used in all size baghouses, from very small collectors to the largest structural design units. Reverse-air cleaning is probably the gentlest cleaning method, using a separate fan to backflush the bags in one off-line compartment at a time with low-pressure air. Woven filter media are generally used with this type of system. Reverse pulse or "pulse jet" cleaning employs high-pressure compressed air to create a shock wave that travels down the bags, vigorously knocking away the dust cake. This method is normally used with felted media.

The reverse jet (or "blow ring") method of cleaning is today found only in limited applications. An ring with air jets surrounds each filter bag and travels up and down the length of the bag on a carriage to remove the dust cake. Other types of infrequently used cleaning methods include high-frequency agitation, sonic and ultrasonic cleaning, and the plenum pulse method.

The fundamental criterion in applying any baghouse to any application is the air-to-cloth ratio, defined as:

$$a/c = \frac{Q}{A} \tag{16}$$

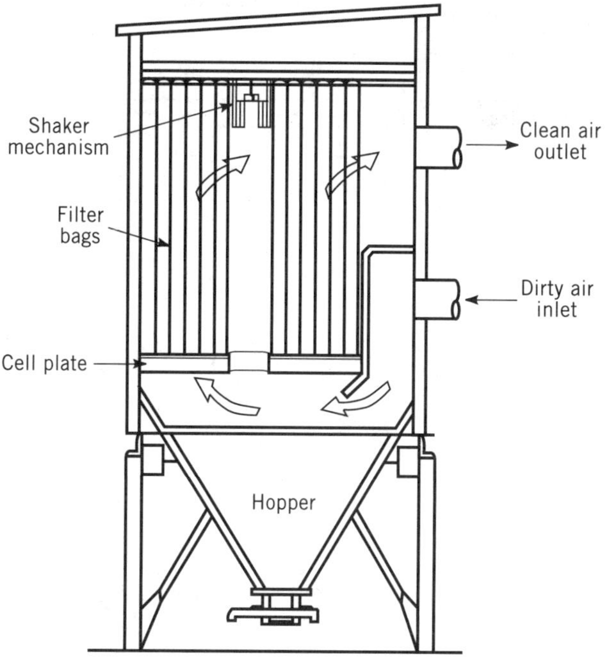

Figure 59.19. Shaker-type baghouse (15).

where a/c = air-to-cloth ratio (ft/min)
Q = volumetric gas flow rate (acfm)
A = net cloth area (ft^2)

Air-to-cloth ratio is equal to the superficial face velocity of the air as it passes the cloth. Shaker and reverse airflow baghouses usually operate at an air-to-cloth ratio of from 1 to 3, whereas a reverse pulse baghouse operates at about three to six times this range.

A second important factor in baghouse application is the type of filter material. Woven cloth is used in shaker and reverse airflow baghouses in which the dirty airstream flows from the inside of the bags to the outside. Newly installed woven fabric bags tend to "bleed" dust until a dust cake begins to build up. Baghouses containing woven filters are typically operated at low air-to-cloth ratios and with long filtration periods between cleaning cycles. Nonwoven or felted cloth is selected for reverse pulse baghouses, which offer high cleaning energy and high gas flow. Nonwoven cloths are less prone to "bleeding" and are effective collectors of heavier dusts, such as grain, dry sand, and limestone. They also tend to provide longer bag life but greater resistance to flow than woven fabrics.

Filtration fabrics are made from a wide variety of natural and synthetic fibers, ranging from cotton and wool to Nomex® and Teflon®. The properties of each type of fiber must be considered when selecting the proper fabric for a given cleaning application. These characteristics include tensile strength, abrasion resistance, chemical resistance, and op-

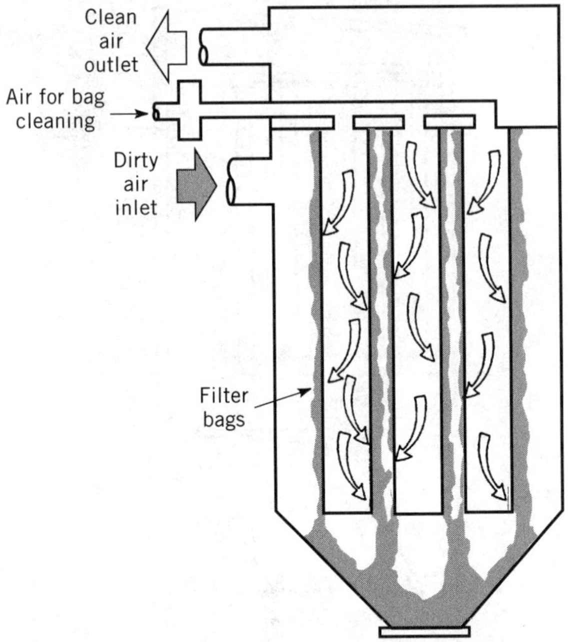

Figure 59.20. Reverse airflow baghouse.

erating temperature range. Bag manufacturers and baghouse vendors can provide media selection charts to assist in the decision making process.

The most recent advance in filtration media technology is the use of an expanded PTFE (polytetrafluoroethylene) membrane bonded to various substrates, essentially an artificial primary dust cake installed on the fabric surface (24). The stated goal of this new "surface filtration" technology is to permit air to pass through the filter media while trapping "all" particulate on the surface. The result has been improved fine particulate emission control, reduced operating pressure drop, higher air flows permitting higher air-to-cloth ratio designs, and improved service life. Since the surface is both hydrophobic and smooth, dust cake release characteristics of the PTFE membrane is superior to conventional filter media.

One of the important performance characteristics of filter fabrics is permeability, which is expressed as the air volume, in actual cubic feet per minute, passing through a square foot of clean cloth with a pressure differential of 0.5 in. water. The overall range of permeability is from 10 to 110 acfm/ft^2, but it usually ranges from 10 to 30 acfm/ft^2 for all types of baghouses.

Another important operating characteristic is the filter drag, given as

$$S = \frac{\Delta PA}{Q} \tag{17}$$

where S = filter drag (in. water/fpm)
ΔP = pressure drop across filter (usually 2 to 6 in. water)

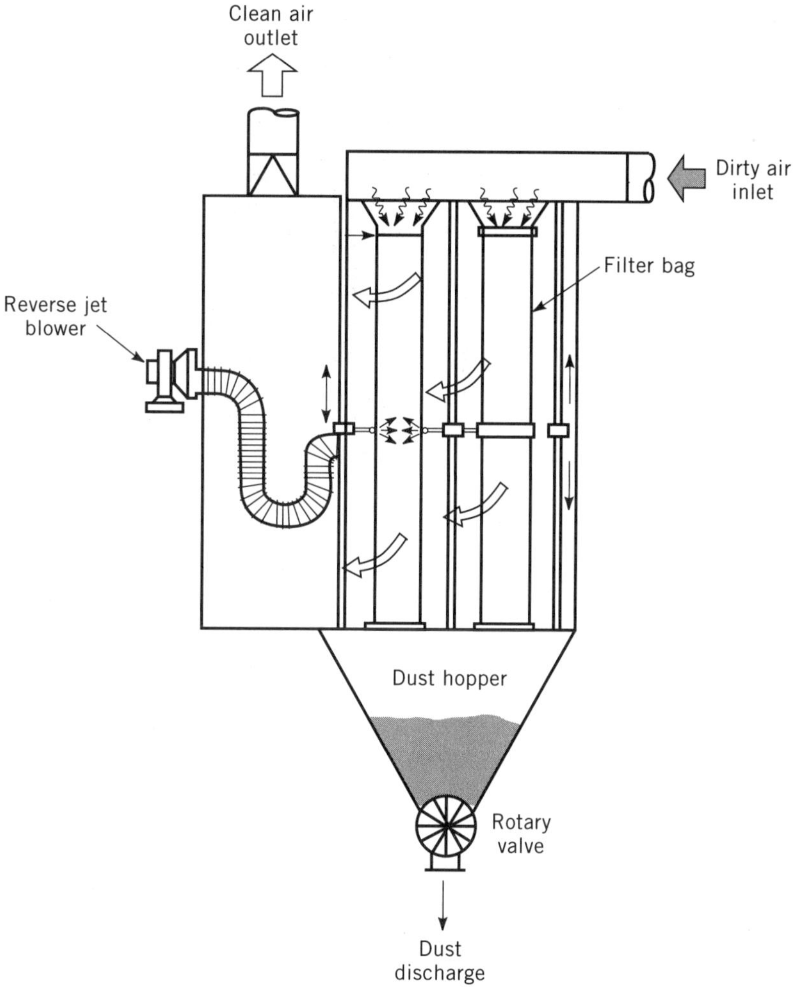

Figure 59.21. Reverse jet baghouse (22).

A = net cloth area (ft^2)
Q = volumetric gas flow rate (acfm)

Figure 59.23 shows the effect of filter drag on outlet concentration (25).

Baghouses are among the most maintenance-intensive of all air pollution control devices. Routine inspections and a preventive maintenance program are prerequisites for continued effective baghouse operation. Bags have a finite life expectancy and can fail from many different causes. Since locating and replacing failed bags is a labor intensive operation, an effective preventive maintenance program can reduce such costs and guard against unexpected emission exceedences.

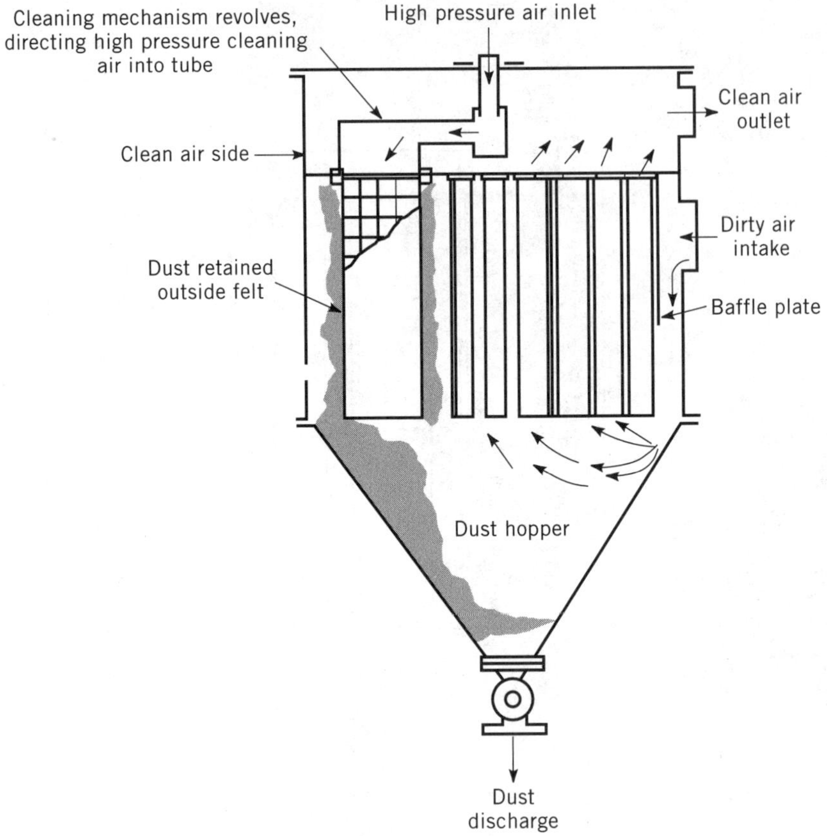

Figure 59.22. Reverse pulse baghouse (15).

9 LIQUID SCRUBBING

"Liquid scrubbing" denotes a process whereby soluble gases or particulate contaminants are removed from a carrier gas stream by contacting the contaminated gas stream with a suitable liquid to decrease the concentration of the contaminant. Scrubber geometry, contacting media, and the scrubbing liquor are design variables that have been the subject of years of investigation to optimize scrubber performance in a variety of applications.

Although the liquid used in liquid scrubbing is generally recirculated, the need to discharge some portion of the scrubbing liquid can create water pollution problems complex enough to render liquid scrubbing infeasible. Nonetheless, the application of liquid scrubbers to air pollution abatement strategies finds optimal use when soluble gaseous contaminants or fine particulate must be removed to high efficiency levels, when the gases involved are combustible, when cooling is desired, and when increased moisture content can be tolerated. When one or more of these prerequisites is met, liquid scrubbing may very

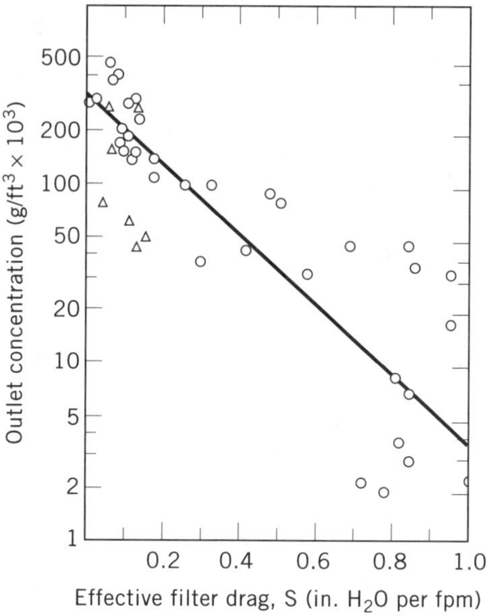

Figure 59.23. Baghouse outlet concentration as a function of filter drag (25).

well provide the only applicable air pollution abatement strategy within the contexts of economic and technical feasibility.

9.1 Range of Performance

The type of scrubber used in a particular application depends mostly on the characteristics of the contaminants being scrubbed and the degree of control required. Almost any device in which good contact is promoted between the scrubbing liquor and a contaminated gas stream will absorb, to some degree, both gaseous and particulate contaminants. The questions of what type of scrubber to specify in an individual case hinges on the efficiency desired, the properties of the contaminant, and the merging of these two factors within the context of minimum operating costs and energy consumption. For example, although a high-energy venturi scrubber using water as the scrubbing liquor will efficiently absorb some water-soluble gases, the energy requirement of the venturi scrubber can be an order of magnitude greater than that of a packed tower to promote the same degree of mass transfer.

The design parameters and established correlations useful in specifying and sizing liquid scrubbers can be subdivided into two general classes: (*1*) mass transfer dynamics for gases dissolving and/or reacting with the scrubbing liquor, and (*2*) momentum transfer dynamics for particulate matter colliding with, and being entrained in, the liquid scrubbing medium. In the first class, either gas-liquid equilibrium data and mass transfer coefficients must be established by experience with the actual species of chemical constituents being encountered, or existing data for chemically similar scrubbing system applications may be extrapolated, in the hope that useful design information can be extracted and the new system design parameters thereby established. In the second class, semitheoretical considerations provide useful design criteria in specifying scrubbers that utilize inertial impaction or Brownian motion to extract particulate matter. Even in this case, however, empirical design data supply the soundest base on which to specify the type and size of liquid scrubber needed to ensure efficient collection of particulate matter at minimum feasible capital and operating cost.

Packed towers using water as a scrubbing medium have found wide applications in the chemical industry to absorb water-soluble gases and vapors, both for pollution abatement and for product recovery. In the power industry, scrubbers are frequently used in the absorption of sulfur dioxide, and they function efficiently as long as the partial pressure of the sulfur oxides in the gas stream is greater than the vapor pressure of the sulfur oxides existing at the surface of the scrubbing liquor. Such a concentration difference provides a driving force to promote mass transfer of sulfur oxides from the gas into the liquid stream. Because the solubility of most water-soluble gases is limited by equilibrium relationships, the amount of scrubbing liquor specified per unit volume of contaminated gas is determined using mass balance calculations and equilibrium relationships and giving due consideration to any chemicals added to the scrubbing liquor to increase absorption efficiency.

A wide variety of scrubbers, operating in pressure drop ranges of 0.5 to more than 80 in. water have been used successfully to absorb particulate matter ranging in size from tens of micrometers down to material of submicrometer size. In any particulate scrubber, the design must provide as large a constant area as possible for scrubbing liquor and particulate matter to interface. In a venturi scrubber, for example, the submicrometer dust impacts on the surfaces of dispersed liquid droplets and penetrates into and becomes part of the liquid droplets, which are subsequently agglomerated, coalesced, and removed from the gas stream.

Although contact area is a prerequisite to particulate absorption in liquid scrubbing, the nature of the dust and its physical properties in relation to the scrubbing liquor determine whether the dust can be wetted by the scrubbing liquid or whether it can even penetrate the surface of the scrubbing liquor. For example, in the scrubbing of iron oxide dust using water, the relatively high surface tension of water does not permit iron oxide to be “wetted” well enough to assure permanent retention in the scrubbing water. Addition of “surfactants” to the scrubbing water decreases the surface tension, thereby permitting the particles to penetrate the “skin” of the droplets at the gas-liquid interface.

9.2 Spray Chambers

Spray-type scrubbers are useful in collecting particulate matter and gases in liquid droplets that are dispersed in a chamber using spray nozzles to atomize the liquid. The geometry of the spray chamber can vary from a simple straight cylinder to complex configurations designed to provide maximum contact area and contact time between droplets and gases passing through the atomized sprayer.

In the simplest type of spray chamber, a vertical tower is the container in which droplets and gas meet. In vertical spray chambers, the droplet velocity eventually becomes the terminal settling velocity for a given droplet. The droplet velocity relative to the particulate collision velocity determines efficiency. Characteristically, spray chambers exhibit low pressure drop, utilize high relative rates of scrubbing liquid per unit volume of gas, and are generally very low in cost because of their simple design. Generally, spray chambers exhibit high collection efficiency for particles that are larger than 10 μm in diameter.

Usually the gases are introduced to the lower section of a spray tower, and the sprays, positioned at the top of the tower, inject scrubbing liquid at some velocity determined by the nozzle orifice (Fig. 59.24). The residence time and the average relative velocity difference between the upward traveling gas and the downward traveling droplets determine the amount of contaminant removal that can be expected for a given tower size and ratio of scrubbing liquid to gas flow rate.

Some spray towers introduce the gases at the base of the tower in tangential fashion, thereby imparting a spiraling motion to the gases. Not only is longer contact time obtained in this manner, but spray droplets are driven to the walls of the spray tower, where they impinge on and drain down the walls as a liquid film.

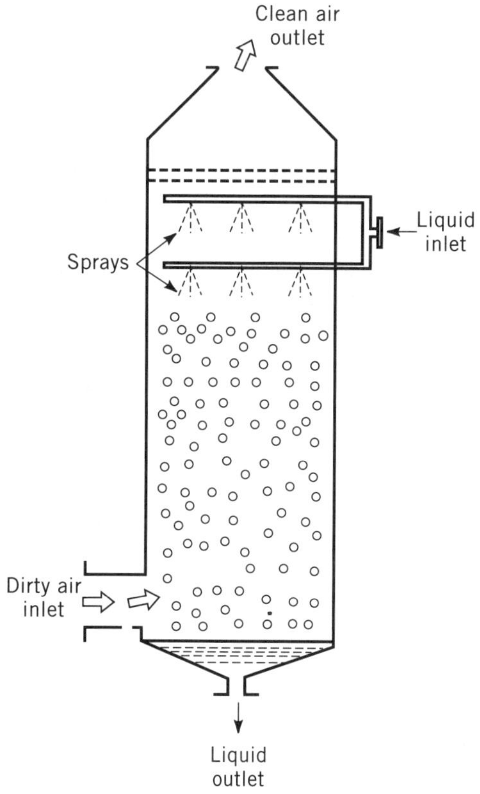

Figure 59.24. Spray tower (26).

In spray chambers, collection efficiency of particulate matter increases as droplet size decreases. The use of the high-pressure atomizing nozzles to produce fine droplets, enhances collection efficiency near the nozzle. These small droplets quickly decelerate, however, because of aerodynamic drag. The relative velocity difference between the spray droplets and the particulate traveling with the gas falls off so rapidly that the collection efficiency is almost zero in the rest of the tower a short distance from the nozzle orifices. An optimal droplet size is 500 to 1000 μm. These droplets in spray chambers scrub relatively efficiently over long enough path lengths throughout the spray tower to provide an optimal collection efficiency.

Because of their relatively simple design, spray towers also serve a useful function as precoolers or as quench chambers to drop the temperature of hot gases relatively quickly as they are emitted from a wide variety of processes. Very often these hot gases contain a relatively high solids content and the spray chamber is most likely to be the first piece of gas conditioning equipment the effluent stream encounters before further cleaning and final discharge to the atmosphere. The spray chamber can handle recycled scrubbing liquid with relatively high suspended solids concentrations because of its simple design and because the clearance in the spray nozzle orifice is relatively large. Therefore, the spray chamber is ideally suited both as an initial solid particulate remover and as a gas conditioner useful in preparing the gas for subsequent downstream processing equipment, which is probably more complex in design and more efficient in particulate extraction.

Specification of pertinent design parameters for a particular application depends to a large extent on the type of contaminant being scrubbed in the spray chamber. A rule of thumb, however, would indicate that in simple spray towers from 5 to 20 gal of scrubbing liquid per 1000 ft^3 of contaminated gas provides optimal liquid gas contact.

In applications featuring the scrubbing of gases, equilibrium considerations come into play and computation of the number of mass transfer units is necessary. These calculations are theoretical and contain proportionality constants that depend on the physical and chemical properties of the contaminants being scrubbed; therefore empirical data are required to provide reasonably accurate specifications. The calculations are similar to those made in sizing packed-tower scrubbers.

9.3 Packed Towers

Scrubbers containing packing material, such as Pall rings, Berl or Interlox saddles, or Raschig rings, are termed packed-bed scrubbers or packed towers. The packing provides high surface area of liquid films ideal for mass transfer. In general, packed-bed scrubbers can be classified as countercurrent, co-current, and cross-flow. In the countercurrent packed-bed scrubber (Fig. 59.25) scrubbing liquid, injected at the top of the packing, trickles down through the packing material as gases pass upward. Countercurrent scrubbers generally cannot be operated at liquid or gas flow rates as high as those used in co-current packed-bed scrubbers, where the gas and liquid phases flow in the same direction. Co-current scrubbers, in general, provide lower mass transfer per cubic foot of packing than countercurrent units.

Packed-bed scrubbers are normally used to remove gaseous contaminants from air streams. In general, when high loading of particulate is a possibility in such applications,

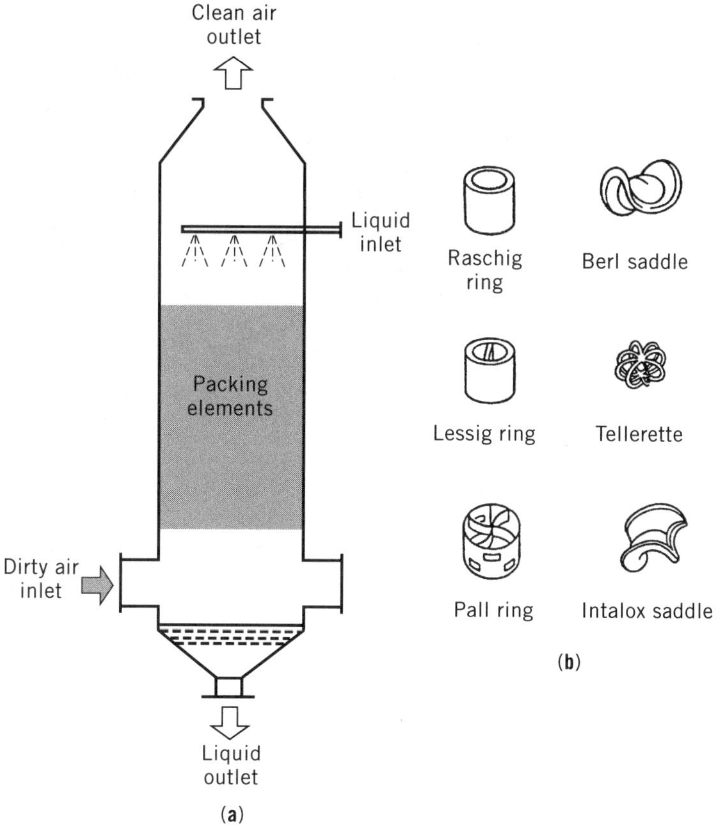

Figure 59.25. (**a**) Countercurrent packed-bed scrubber and (**b**) packings (26).

the packing media may be fouled by deposited particulate matter, resulting in eventual blockage of the unit. Because gaseous contaminant removal is limited by equilibrium considerations, the efficiency of the unit, as well as the absolute concentration of contaminant in the gas stream leaving the packed-bed scrubber, is affected by whether a co-current or countercurrent scrubber is specified.

The cross-flow scrubber is the most capable of the three types of packed-bed scrubber of coping with deposited solids on the tower packing. In these units, scrubbing liquid trickles through the packing from the top of the packing bed, while the dirty gas moves horizontally through the bed. This arrangement provides good washing of the media and simultaneously is a very stable configuration, difficult to flood.

Packed-bed scrubbers have been used historically in chemical operations in a wide variety of applications. By adding various chemical substances to the scrubbing liquor, pollutant gases can be removed to very low outlet concentrations if sufficient fan power is provided to overcome the bed pressure drop and if the absolute volume of packing and tower height are sufficiently large. Typically, empirical design data are necessary to determine mass transfer coefficients useful in predicting efficiencies for particular combinations of contaminant and scrubbing liquid. Pressure drops through packed-bed towers can be quite low. Although the pressure drop depends on the flow rate through the tower and the liquid-gas ratio, as well as the type of packing, typically 0.5 in. water per foot of packing is the pressure loss in packed-bed towers.

Packed beds offer the added advantage of being intrinsic mist eliminators. Because flow rate is characteristically low, entrainment of liquid into the gas leaving the tower is much less than in scrubbers of other types where the liquid phase is dispersed in fine droplets. Furthermore, because tower height is somewhat greater than packing height, scrubbing efficiency of the packed tower can be increased by simply adding more packing, at minimal cost to the overall system. Thus packed-bed scrubbers are relatively flexible to process changes that may alter the concentration or nature of gaseous contaminants in their inlet streams.

Section 9.6 discusses the pertinent design parameters and applicable equations necessary to size a packed-bed scrubber for a variety of conditions. In general, pertinent parameters include the concentration of the contaminant in the gas stream admitted to the scrubber, the required efficiency of the unit, the ratio of scrubbing liquor flow rate to inlet gas flow rate, an estimate of the mass transfer area per cubic foot of packing in the packed bed, and equilibrium data specifying the concentration of contaminant in the scrubbing liquor.

9.4 Orifice Scrubbers

Orifice scrubbers use the carrier gas velocity to promote dispersal of the scrubbing liquid and turbulent contact of the contaminated gas stream with the dispersed liquid (Fig. 59.26). Both the efficiency and the pressure drop depend on the gas flow rate through the scrubber and the ratio of the scrubbing liquor injected per unit volume of gas. Orifice scrubbers make use of internal geometric designs that attempt to supply scrubbing liquid uniformly across the cross section of gas flowing through the unit. The kinetic energy of the gas supplies the energy needed to disperse the liquid phase. In the turbulent contacting zone, usually a short distance downstream of the orifice, the intensely turbulent motions provide violent contact between particulate matter and the larger scrubbing liquid droplets, which possess high inertia.

Downstream of the contact zone, mist eliminators agglomerate and coalesce the droplets of scrubbing liquor containing the absorbed contaminant. Depending on the energy supplied to the unit, scrubbing efficiencies can range from 85% to more than 99.5%, depending on the nature and physical properties of the contaminant being removed. The optimum use of input energy to effect a given degree of contaminant removal is the subject of many innovative designs applied to a wide variety of industrial process emissions.

There are numerous geometries and internal designs for orifice scrubbers. This general classification includes all scrubbers that depend on contacting by passing the gas and liquid phases through some type of opening. In the more sophisticated designs, the opening may be varied to accommodate changes in requisite removal efficiency or gas flow rate. The basic principle of an orifice scrubber having a fixed orifice size dictates that scrubbing

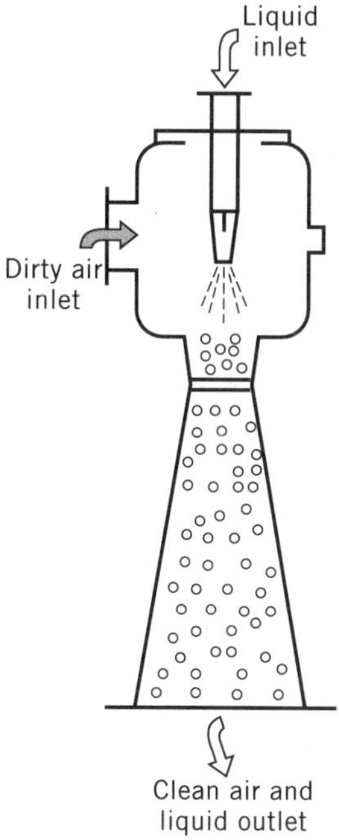

Figure 59.26. Ejector-type orifice scrubber (26).

efficiency will fall off markedly if gas velocity through the orifice is decreased. By providing pressure drop sensors to the variable pressure drop orifice type scrubber, automatic operation can compensate for wide variations in gas flow rate, thus ensuring relatively constant collection efficiency.

Another special type of orifice scrubber is the venturi scrubber (Fig. 59.27). Relatively high velocities between the gas and liquid droplets promote high collection efficiency in this unit. Such high velocities require a large energy input, raising operating costs to high levels, especially in applications processing large gas volumes. Venturi scrubbers typically use gas velocities of 200 to 400 ft/sec. At the venturi throat, the intensely turbulent action and shearing forces produce extremely fine water droplets that collect particulate matter

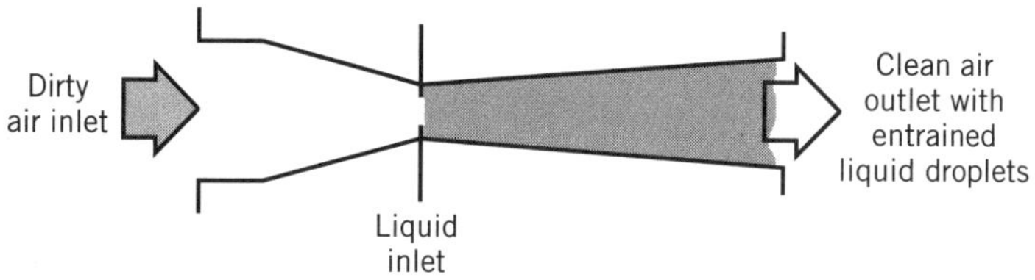

Figure 59.27. Venturi scrubber (26).

very efficiently. At the instant of formation, these droplets move relatively slowly, and the fast-moving dust particles collide with nearly 100% efficiency with any droplets they may encounter. In the diverging section of venturi scrubbers, the droplets are slowed in preparation for removal in downstream mist eliminators.

Ejector venturis use spray nozzles and high-pressure liquid to collect both particulate and gaseous contaminants and also to move the gas through the unit. These scrubbers have no fan to provide requisite gas flow. Thus fan power costs are eliminated, but relatively high energy consumption is still necessary because of the relatively high scrubber liquid pumping costs. Modifications to this type of device have featured sonic nozzles in the ejector venturi to give high collection efficiency while conserving somewhat on high liquid pumping costs.

The collection efficiency of the venturi scrubber increases with increased energy input. Usually a venturi scrubber operating in a pressure drop range of 30 to 40 in. water is capable of an almost total collection of particles ranging from 0.2 to 1.0 μm.

9.5 Mist Eliminators

Depending on the type of the scrubber used, the number and size of water droplets entrained in the gas stream emerging from the scrubber may present objectionable stack gas discharge. If adequate mist elimination is not practiced, "raining" of contaminated scrubbing liquid droplets near the discharge stack can mean noncompliance with both mass emission and opacity regulations.

Normally, the droplets emerging from a scrubber with inadequate mist elimination contain dissolved gases and/or captured particulate matter. Depending on the size of these droplets, fallout may occur a considerable distance beyond the point of discharge. In some cases, depending on ambient conditions and on the degree of saturation of the plume emerging from the scrubber, the droplets either grow because of further condensation or evaporate, thus liberating the captured contaminant.

Generally, droplets formed from tearing of liquid sheets of water (e.g., in orifice-type scrubbers) are 200 μm in diameter and larger. In extremely high-energy applications, droplets as small as 40 μm may be produced. In low-energy systems using coarse spray nozzles, the droplets may be as large as 1000 to 1500 μm. Clearly, the type of device that must be used to ensure adequate mist elimination is partly a function of the size of the droplets.

Mist eliminators are devices that are designed to use physical separation processes, relying on gravitational and inertial forces to "knock out" the droplets from the emerging gas stream. One such device appears in Fig. 59.28. Devices used include packed beds, cyclone collectors, gravitational settling chambers, and chambers containing baffles that make a tortuous "zigzag" path for the gases to follow, while more inert water droplets impinge on the solid surfaces. No matter what type of device is used, once the scrubbing liquid droplets are captured, if the mist eliminator is operated at proper velocities, the droplets coalesce and drain down to some quiescent zone in the mist eliminator for subsequent treatment.

One common problem with some mist eliminator designs is re-entrainment. If the liquid load is too great for the mist eliminator, the onset of re-entrainment begins at lower than normal gas velocities. That is, for a given gas velocity, as the amount of liquid removed

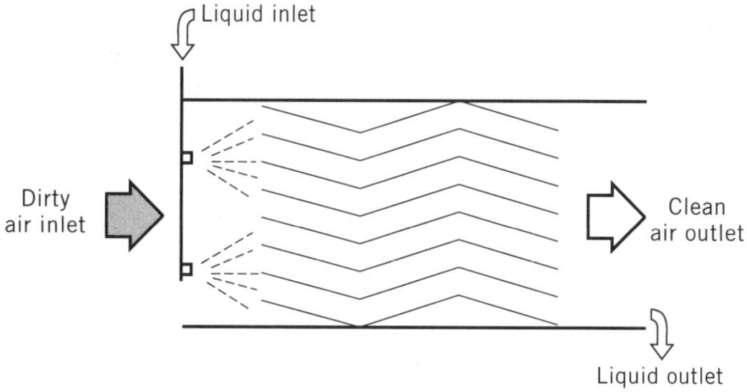

Figure 59.28. Baffle-type mist eliminator (26).

from the mist eliminator increases, the efficiency remains relatively constant up until the capacity of the mist eliminator is exceeded; then efficiently drops markedly.

Cyclone-type mist eliminators function basically on the same principles delineated in Section 7.5 and Section 7.6. Some departure from theory occurs, depending on re-entrainment or shearing effects at the cyclone walls, causing the liquid film to be sheared and reconverted to atomized droplets if cyclone operation is improper.

The fiber-type mist eliminator is a relatively new device. The fibrous bed provides a tortuous path for the carrier gas laden with entrained droplets. In one type of application using fibrous mist eliminators, an annular cylinder is used within a concentric shell. The entering wet gases pass readily through the annulus, and the liquid drains down the inner walls of the annular cylinder. Gases pass up through the center and exit through the top of the mist eliminator. The water is collected at the bottom of the outer shell and is drained off.

Specification of any mist eliminator must involve consideration of the pressure drop that the mist eliminator will add to the total control system. Frequently, manufacturers' specifications are the best source of such empirical data. When the emerging gas stream is saturated and still hot, liquid entrainment may have a problem if the gases are discharged to a relatively cool ambient environment. In such cases, it may be necessary to heat the exhaust gases to minimize production of steam plumes or condensation before the gases have a chance to mix and disperse in the atmosphere.

9.6 Design Parameters

The preceding sections discussed some design parameters for the particular types of scrubber delineated. Because of the largely empirical correlations, rules of thumb are frequently necessary to specify scrubbers, either for unique applications, or when the state of the art and data correlation for the particular unit are not yet sufficiently refined to permit accurate prediction of collection efficiencies.

There do exist, however, some concepts fundamental to nearly any type of wet scrubber designed to remove the general gaseous and/or particulate contaminant from the carrier gas stream. In the case of gaseous contaminant removal, the concepts of "transfer units" and "theoretically equivalent height of packing" form the basis of these correlations. In the case of particulate removal, inertial impaction parameters form the basis of predicting the size of droplets required to collect a dust of known size distribution.

The efficiency of a venturi scrubber in removing particulate emissions is a direct function of its pressure drop. The exact efficiency attained is also a function of downstream mist elimination and other factors. The equation below allows the prediction of the pressure drop:

$$P = v^2 \rho_g A^{0.133} L^{0.78} / 1{,}270 \tag{18}$$

where P = pressure drop (in. of water)
v = throat velocity (ft/sec)
ρ_g = gas density (lb/ft^3)
A = throat cross section (sq ft)
L = liquid:gas ratio (gallons per 1,000 cu ft)

9.6.1 *Particulate Scrubbing*

Particulate removal by wet scrubbing is accomplished by liquid droplet impaction and by mechanisms that depend on inertial forces. Statistical probability and impaction correlations are the basis of mathematical modeling in spray collector design. The impaction parameter is the basis of nearly all collection mechanisms that involve inertial impaction of particles on droplets. It also takes into account the departure of solid particles from gas streamlines in passing around, or colliding with, liquid droplets. A general rule of thumb, according to impaction correlations, is the higher the impaction parameter, the higher the collection efficiency.

9.6.2 *Gaseous Scrubbing*

In gaseous contaminant scrubbing, the design equations are based not on particle dynamics but on mass transfer characteristics. In its most general form, the design equation describing the height of the packed column required to perform a given gaseous separation is complex and involves mass transfer coefficients, vapor phase concentrations of the gaseous contaminant along the length of the column, and the tower cross-sectional area. The required height is essentially a function of the height of a "transfer unit" multiplied by the "number of transfer units" required. The relationships involved can be simplified so that for a given scrubbing efficiency, inlet gas flow rate, and entrained acid content, the volume of packing material required is completely determined if the scrubbing liquor flow rate is known and the mass transfer coefficient is defined.

10 ELECTROSTATIC PRECIPITATION

Electrostatic precipitation is a process by which particulate matter is separated from a carrier gas stream using electrostatic forces and deposited on solid surfaces for subsequent removal. Electrostatic precipitators find wide application in many segments of pollution abatement systems. One of the largest users of electrostatic precipitation is the electric power industry. The need to remove particulate matter from large volumes of combustion-exhaust gases produced by coal-fired boilers calls for a collector that can extract particulate matter efficiently with small consumption of energy. This all-important advantage renders electrostatic precipitation an extremely attractive control alternative from the viewpoint of low operating cost. Unfortunately the relatively large initial capital investment cost needed to construct and install these massive and complex steel structures may render the application somewhat less attractive from the point of view of annual cost.

10.1 Range of Performance

The high-voltage electrostatic precipitator has been applied at more installations emitting large gas volumes than any other class of high-efficiency particulate collecting device (6). Although the device can efficiently extract even submicrometer material, the precipitator commonly treats dusts whose particle size distributions indicate a maximum diameter of approximately 20 μm. In those cases, more coarse dusts are also included in the contaminated gas stream and the precipitator is often preceded by mechanical collectors.

Precipitators applied to the electric power industry have been designed to handle gas volumes ranging from 50,000 to 2×10^6 cfm (6), Both "cold-side" and "hot-side" electrostatic precipitators have been installed that efficiently remove particulate matter from gas streams ranging from ambient temperatures to 1650°F, respectively (27). Depending on the location of the precipitator in the control system (i.e., whether gas flows through the precipitator by way of an induced or forced-draft fan), the unit may function under negative or positive pressures ranging from several inches of water, gauge, to 150 psig (6). Characteristically, the pressure drop through electrostatic precipitators, mostly open chambers with nonrestrictive gas passages, is on the order of 0.5 to 2 in. water, gauge.

Typically, one stage of an electrostatic precipitator provides from 70 to 90% collection efficiency. By providing sequential stages, the overall precipitator efficiency can be increased to very high values. For example, in typical installations abating fly ash of moderate resistivity, one electrostatic precipitator stage is capable of producing nominally 80% collection efficiency. The remaining 20% of the entering particulate matter can be removed to an efficiency of 80%, approximately, in a second stage. The removal, then, of 80% plus 16% in the second stage (80% of the dust penetrating the first stage) provides an overall collection efficiency of 96% for the two-stage precipitator. Additional stages can be installed. Using this estimate of collection efficiency, a six-stage precipitator would provide a precipitator collection efficiency in excess of 99%.

10.2 Mechanisms of Precipitation

The electrostatic precipitator relies on three basic mechanisms to extract particulate matter from entering carrier gases. First, the entrained particulate matter must be electrically charged in the initial stages of the unit. Second, the charged particulate matter is propelled by a voltage gradient (induced between high voltage electrodes and the grounded collecting surfaces), causing the particulate matter to collect on the grounded surface. Third, the particulate matter must be removed from the collection surfaces to some external container such as a hopper or collecting bin.

The high-voltage potential across the discharge and collecting electrodes causes a corona discharge to be established in the region of the discharge electrode, and this forms a powerful ionizing field. When the particles in the gas stream pass through the field, they become charged and begin to migrate toward the collecting surfaces under the influence of the potential gradient existing between the electrodes.

Once the dust is deposited on the grounded surface, easy removal from the collecting surface depends on many factors. Dust resistivity is one prime factor dictating the collection efficiency and degree of re-entrainment that can be expected under a given set of design conditions. If dust resistivity is too low, particles once deposited on the grounded surface may lose their charge and reenter the gas stream from the electrode collecting surface. If resistivity is too high, particles once deposited can cause "corona quenching" or "back corona." In this condition, the voltage gradient between dust deposited on the plates and the high-voltage electrode begins to diminish because of the large voltage drop in the dust layer accumulated on the plate. This decreases collection efficiency, because the voltage gradient is all-important in determining the rate or migration of particulate matter toward the collecting surface.

In many electrostatic precipitators the dust is "rapped" from the collection surface by imparting a sharp blow to some area of the surface. The dust then cascades in a "sheeting action" down the collecting surface into a hopper. The extent to which this dust is removed in large pieces determines the degree of re-entrainment upon rapping. Plate design, rapper design, and use of modular sections of precipitators are among the widely varying approaches employed by precipitator manufacturers.

10.3 Plate-Type Precipitators

Two basic types of electrostatic precipitator have been used: plate type and pipe type. In plate-type devices the collecting surface consists of a series of parallel, vertical grounded steel plates between which the gas flows (Fig. 59.29). Alterations to the geometry of the plates are implemented by various manufacturers in an attempt to provide better collection and more efficient removal of collected dust upon rapping. Some plate-type precipitators are equipped with shielding-type structures installed vertically at incremental distances along the plate surface. These structures represent an attempt to minimize re-entrainment as the dust sheet falls behind the structure. The design of these protuberances affects the geometry and intensity of the electrostatic field set up between the high-voltage electrodes and the plates. Thus the spacing, size, and shape have been the subject of years of developmental work and research sponsored by precipitator manufacturers.

Most commonly, the plate-type precipitator finds application in the collection of solid particulate matter or dust. Plate-type precipitators consume approximately 200 W per 1000 cfm. This extremely low power requirement is affected also, however, by the resistivity of the dust and the geometry of the precipitator.

Sometimes the dust is removed from plate-type precipitators by washing. Strategically placed nozzles outside the electrostatic field are timed for actuation at various intervals,

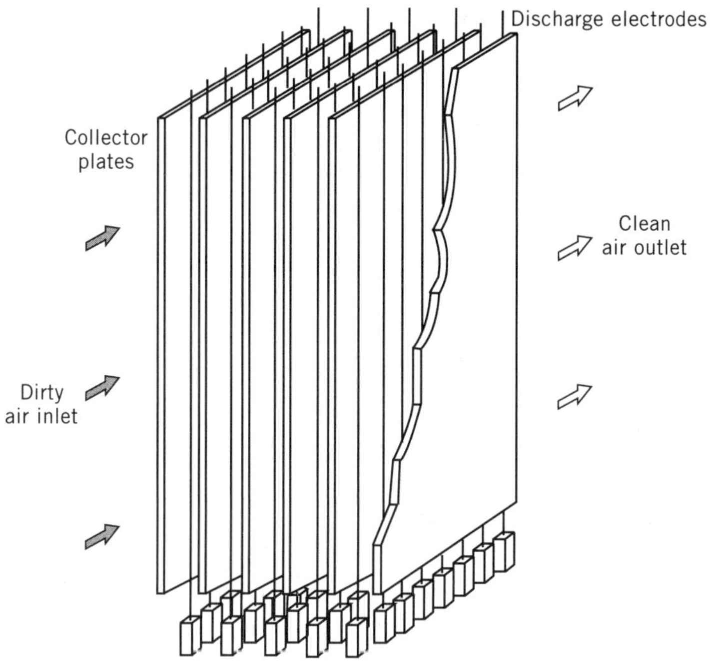

Figure 59.29. Plate-type electrostatic precipitator.

supplying irrigation to flush away dust such as sticky or oily particulate matter that is especially difficult to remove.

One disadvantage of plate-type precipitators with smooth plates (i.e., not equipped with a certain type of protuberance) is that they permit the gradual creepage of deposited particulate toward the precipitator outlet. Collected dust may be gradually moved by aerodynamic drag from the traveling gas stream, or particles may "jump" along the plate, depending on dust resistivity. In general, capital investment costs of the plate-type precipitator may be lower because of the simplicity of plate design as compared with pipe-type precipitators. The exact design, however, dictates actual assembly and construction costs.

10.4 Pipe-Type Precipitators

Pipe-type precipitators generally feature high-voltage electrodes positioned at the center line of a grounded pipe (Fig. 59.30). The gas flows through a bank of these pipes parallel to the high-voltage wire. Dust deposited on the inner walls of the grounded pipe is removed in much the same manner as in the plate-type precipitator, that is, by rapping.

Some studies indicate that the pipe-type precipitators should have superior electrical characteristics (16). The pipe-type precipitator is prone, however, to creepage of collected dust toward the clean gas exit. In general, the design parameters for the specification of pipe-type precipitators may vary in the actual numerical values used to specify the number

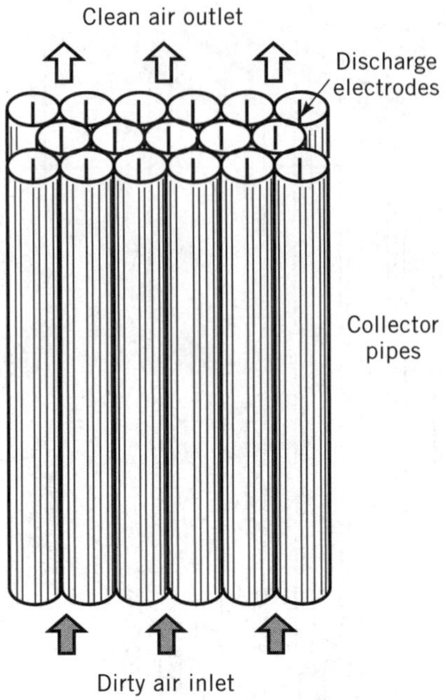

Figure 59.30. Pipe-type electrostatic precipitator.

of pipes. The theoretical models on which sizing correlations are based, however, still include consideration for dust resistivity, total plate collection area, and gas velocity through the pipe.

10.5 Design Parameters

Precipitator size is normally based on a predetermined estimation of the resistivity of the dust, knowledge of the volumetric flow rate of gas to be processed by the precipitator, and an estimate of the dust loading. Depending on dust resistivity, the cross-sectional area of the precipitator may be sized for the known gas flow rate to provided a line velocity through the precipitator parallel to the collection surface of 2 to 6 ft/sec. The number of collecting plates (i.e., plates that form gas passages or channels through which the dirty gas flows parallel to the plates) is fixed by this calculated cross-sectional area and by the plate spacing. In conventional precipitators, the plate spacing is predetermined by the maximum voltage to which the high-voltage electrode, located in the center of the channels, is subjected. Typically, 45,000 V applied to these center-line high-voltage electrodes, located typically 4 1/2 inches from the grounded electrode surface, produces a 10 kV/inch voltage gradient. This rule-of-thumb design parameter then forms the basis for the determination for the total number of gas passages in the precipitator and all necessary hardware needed to support and power the high-voltage electrodes.

Manufacturers claim higher efficiency for a given area of collecting surface depending on the design of the collecting electrodes, the design of the high-voltage electrodes, the wave form of the high voltage imparted to the electrodes, and the mechanism of particulate removal from the collecting surfaces. The collection efficiency of an electrostatic precipitator, however, is estimated using the Deutsch equation (28):

$$E = 100\left(1 - \exp\frac{-AW}{Q}\right) \tag{19}$$

where E = collection efficiency (%)
A = area of collecting electrodes (ft^2)
Q = gas flow rate (acfm)
W = particle migration velocity (ft/min)

Generally, the migration velocity is an empirically determined sizing factor for a given application, grain loading, and dust resistivity. This experience factor is often a closely guarded number among precipitator designers and manufacturers.

The design parameter of dust resistivity is a prime component in determining the migration velocity W. Dust resistivity measurements are frequently made on site. Resistivities vary from 109 to 1015 ohm-centimeters and are a function of the nature of the dust and the temperature. Some precipitator designers analyze collected dust in the laboratory for dust resistivity, yet dust resistivity measurements directly from the gas stream are indispensable in providing accurate design information. For example, it is not uncommon to find one to three orders of magnitude difference in measured dust resistivity when comparing field measurements to laboratory measurements of collected dust.

Rules of thumb indicate that the migration velocity W varies from 2 to 14 cm/sec, depending on the aforementioned design parameters. Typical values for coal-fired boilers range from 6 to 10 cm/sec. Clearly, from the foregoing equation, the required collecting plate area for a given gas volume and collection efficiency is inversely proportional to the migration velocity. Therefore, choosing a conservative migration velocity of 6, as opposed to 10 cm/sec, requires approximately 1.7 times the plate area.

11 GAS–SOLID ADSORPTION

Adsorption denotes a diffusion process whereby certain gases are retained selectively on the surface, or in the pores of interstices, of specially prepared solids. The principle has been used successfully for many years in the separation of mixtures and in product improvement through removal of odorous contaminants or through decolorizing.

In many applications, the primary purpose of adsorption is to render an operation economical by recovery of valuable materials for reuse or resale. A good illustration of this is the long-standing practice in several industries of recovering organic solvent vapors by adsorption. In such applications, the adsorbed materials may be "desorbed" by a variety of methods and concentrated and purified for reuse. The adsorbing solid is regenerated and can be resumed for many such cycles.

11.1 Range of Performance

In atmospheric pollution control as practiced today, the principle of adsorption is employed primarily to prevent odorous and offensive organic vapors from escaping into populated areas. A typical example is the removal of vapors generated by printing plants. Occasionally, however, adsorption is used to remove trace concentrations of highly toxic gaseous materials, such as radioactive iodine vapors.

Fixed-bed adsorbers are used when trace concentrations are encountered, whereas for heavier concentrations, regenerative adsorbers are needed. Product recovery from regenerative adsorbers (e.g., solvent recovery in the dry-cleaning field) may pay for the emission control system and its operation, or at least offset part of the annual cost.

Recently, carbon bed adsorption techniques have been applied to the goal of lessening the impact of energy use on processes that rely on incineration techniques to abate combustible process emissions. In earlier applications of incinerators, airstreams laden with vapors containing organic material have been directly incinerated; some of the incinerators use heat recovery to minimize the amount of auxiliary fuel required to support combustion and provide efficient abatement. In practice, either the organic vapor concentration in the airstream's incinerator was low due to process dynamics, or the concentration of organic vapors was necessarily limited to comply with safety code requirements (e.g., maintaining the concentration of combustible vapors at 25% of the lower explosive limit).

In recent applications, carbon bed adsorbers have been used to adsorb organic vapors over prolonged periods (e.g., 2 to 8 h). Upon saturation, the bed can be desorbed using saturated or superheated steam. The effluent from the bed, consisting mostly of water vapor and concentrated organic compounds, can be sent to an intermittently fired incinerator, which is designed to operate only during bed desorption. In most cases, the concentration of organic material has been raised sufficiently so that by mixing auxiliary air at the point of combustion, the combustion process sustains itself with little or no auxiliary fuel. Furthermore, because the desorbed stream consists of steam and organic compounds, the danger of explosion is eliminated in most cases (because the amount of air in the desorbed stream is very small, the concentration of organics is therefore above the upper explosive limit).

The nuisance aspect of industrial odorous emissions is becoming a more significant factor in the network of developing emission regulations. Characteristically, odors can be produced by organic compounds at very low concentrations and emission rates, which means that the odorous emission is not limited by organic compound emission regulations. Such low emissions are ideally suited to adsorption processes, especially those utilizing activated carbon. Furthermore, because the absolute amount of odorous substances is usually so low, the adsorbent can be used to extended periods and discarded with no significant impact on system operating cost. Thus, adsorption systems have found wide use in rendering plants and industrial processes emitting high molecular weight hydrocarbons.

In general, the range of performance of gas-solid adsorption systems is related to many factors controlling the effectiveness of physical adsorption in dynamic systems. These factors include surface area of the adsorbent, the affinity of the adsorbent for the adsorbate, the density and vapor pressure of the adsorbate, the concentration of the adsorbate, the system temperature and pressure, the dwell time, the bed packing geometry, the thermodynamics of adsorption of solvent mixtures, and the mechanism of removal of heat of adsorption (15). All these factors are interrelated, requiring careful analysis and specification before installing or even considering a carbon adsorption system as a viable pollution abatement system in a particular application.

Carbon adsorption has been used in many areas of the food processing industry: the handling and blending of spices, canning, cooking, frying, and fermentation processes, for example. In the manufacture and use of chemicals, process discharges have been abated by carbon adsorption to minimize objectionable emissions in the manufacture of odorous substances such as pesticides, glue, fertilizers, and pharmaceutical products, in paint and varnish production, and in the release of odorous vapors emitted to the atmosphere from tanks during filing operations (29). Carbon adsorption units are also commonly used to control atmospheric emissions from soil vapor extraction systems that remove organic vapors from areas of subsurface solvent contamination. Aside from activated carbon, gas-solid absorbers use other materials for reversible or irreversible removal of gaseous constituents. Among these are activated alumina, silica gel, and molecular sieves (30). Activated carbon, however, constitutes one of the most popular types of solid adsorbent, mostly because of its low cost and its ability to be regenerated.

11.2 Physical Adsorption

Physical adsorption denotes a type of adsorption process wherein multiple layers of adsorbed molecules accumulate on the surface of the adsorbing solid. In this process, the number of layers that can be made to accumulate on the activated carbon surface is related to the concentration of the adsorbate in the gas stream (31).

Physical adsorption processes generally entail reversible processes. In the case of adsorption of organic solvents on the surface and interstices of activated carbon, for example, the rate at which adsorbate molecules are "driven" from the relatively homogeneous gas phase into the pores of the activated carbon depends to some extent on the concentration of the gas. At the surface of the adsorbing solid, molecules are attracted and deposited, providing a concentration gradient useful in causing migration of solvent molecules into the activated carbon pores. From the standpoint of residence time and the ease of saturation of activated carbon, therefore, solvent concentration in the gas phase partly determines the amount of activated carbon required and the optimal placement or geometry of the carbon relative to the flowing gas stream. The more molecular layers that can be accumulated on the activated carbon interstitial area, the less carbon is required per pound of solvent to be adsorbed. Conversely, for a given residence time of contaminated gas in an activated carbon bed, the higher the concentration, the higher will be the extraction efficiency. These considerations apply in physical adsorption processes, whether the adsorbing solid is activated carbon, a siliceous adsorbent, or a metallic oxide adsorbent.

In the case of activated carbon adsorption, the micropore structure of the carbon surface provides an extremely large surface area for a given weight of carbon; typically, several hundred thousand square feet of available surface is distributed throughout the interstices of one pound of activated carbon (29). Generally, the adsorptive capacity of a particular activated carbon is measured by its ability to retain a given chemical entity at various concentrations of that chemical species. These empirical data are normally obtained by laboratory investigations performed by the suppliers of activated carbon. The data are useful for specifying the equilibrium concentrations theoretically obtainable at a given set of conditions, assuming that sufficient residence time is available in the bed (i.e., that sufficient carbon is present for a specified gas flow rate and velocity). From adsorption "isotherms" (supplied by activated carbon manufacturers), the minimum amount of carbon required can be determined for a given adsorption cycle time. The amount of solvent adsorbed on the carbon at this "saturation point" can be theoretically achieved, but only at the expense of low collection efficiency near the end of the adsorption cycle. Adsorption isotherm data do not necessarily specify adsorption efficiency, for efficiency may vary with the time since last regeneration or with the composition of mixtures being absorbed, as well as with bed geometry and solvent-laden gas flow rate through the bed.

As a rule, the particular geometric design and arrangement of activated carbon in an activated carbon bed adsorber is best specified by well-informed engineers who specialize in adsorption equipment. Even then, pilot-plant work may be required before a full-scale carbon bed absorber can be specified or designed. Generally, bed geometrics can be arranged in cylindrical canisters or in a variety of other shapes to provide optimal contact time for a given concentration of material and a given gas flow rate.

11.3 Polar Adsorption

Polar adsorption denotes a gas–solid adsorption process whereby molecules of a given adsorbate are attracted to and deposited on the surface of an adsorbing solid by virtue of the polarity of the adsorbate. Some siliceous adsorbents depend on the polarity of gas molecules for selective removal from a gas stream. For example, synthetic zeolites are adsorbents that can be produced with specific uniform pore diameters, giving them the ability to segregate molecules in the liquid or gas stage on the basis of the shape of the adsorbent molecule. Adsorbents of this type will not even adsorb organic molecules of the same size as their pores from a moist airstream, for the water molecules are differentially adsorbed because of the chemical structure of the adsorbing solid (29).

11.4 Chemical Adsorption

Chemical adsorption takes place when a chemical reaction occurs between the adsorbed molecule and the solid adsorbent, resulting in the formation of a chemical compound (31). Chemical adsorption may be contrasted with physical adsorption, where the forces holding the adsorbate to the solid surface are weak. In chemical adsorption, usually only one molecular layer is formed, and the chemical reaction is irreversible except in cases when the energy of reaction can be applied to the solid adsorbent to reverse the chemical reaction.

Many of the same factors useful in sizing the specifying physical adsorption equipment apply in chemical adsorption. As in applications featuring the removal of low-concentration, high molecular weight, odorous substances, the contaminated adsorption bed is usually discarded or shipped for external regeneration after its efficiency has degraded. The cost of operating chemical adsorption equipment may be prohibitive in applications involving either high flow rate or a concentration of adsorbate in the incoming gas stream that causes the bed to deteriorate quickly. Alternatively, chemical adsorption systems can be applied to streams where the contaminant is present in high concentrations but is emitted only intermittently.

11.5 Design Parameters

One of the first steps in specifying the quantity of activated carbon required for a given cycle time in an activated carbon absorber is to determine the chemical species and concentrations in the incoming gas stream. Adsorption isotherms are available from activated carbon suppliers either for individual chemical constituents being adsorbed, or for chemical species that are similar enough in chemical structure or physical properties to enable reasonable approximations regarding adsorbability and retention. Such empirical data, established by the suppliers, apply for the single chemical species being adsorbed on a given type of activated carbon. When mixtures of organic species are being processed, it can be expected that the higher molecular weight substances are preferentially adsorbed, and lower molecular weight species are adsorbed initially to a lesser efficiency than would be predicted if the lower molecular weight species were adsorbed alone. As the adsorption cycle proceeds, the effluent can be expected to be rich in low molecular weight compounds from the mixture, and high molecular weight compounds will be adsorbed at the highest efficiency.

Once the minimum bed size based on a given cycle time and equilibrium retention relation is calculated, the bed geometry and actual amount of carbon must be determined by consideration of the retention time in the absorber for a given adsorption efficiency, sufficient capacity to ensure an economical service life, low resistance to airflow to minimize overall system pressure drop, uniformity of airflow distribution over the bed to ensure full utilization, pretreatment of the air to remove particulate matter that would gradually impair and poison the bed, and provision for some manual or automatic mode for bed regeneration (29).

The geometry and depth of the absorbing bed depend primarily on the superficial gas velocity through the bed, the concentration of contaminants in the incoming airstream, and the cycle time. As adsorption continues, the portion of the activated carbon bed first encountered by the incoming gas becomes rich and eventually fully saturated in adsorbate. Deeper areas of the bed are progressively less and less saturated with adsorbate until some small but finite level of penetration of adsorbate can be observed at the bed exit. As the adsorption cycle continues, this "adsorption wave" proceeds through the bed until the concentration of the exit gas reaches a finite value in excess of allowable emissions. At this point "breakthrough" is said to have occurred, and it is necessary to regenerate the bed. By arranging the given amount of activated carbon in an optimal geometry, breakthrough can be postponed to the point at which minimum utilities are required for bed desorption and the maximum practical level of adsorbate is present in the desorption stream.

It is not possible to set down rules of design encompassing all applications of carbon bed absorption to pollution abatement processes. Because of the wide variation in possible solvent mixtures and concentrations, the bed geometry is usually specified from previous experience or extrapolated from performance data obtained in a similar type of installation.

The design of an adsorber system is a function of the contact time between the adsorbent and the vapor, and takes into account the kinetics and thermodynamics of both the vapor and the adsorbent. Adsorption curves for various vapors are usually provided by the supplier of the adsorbent material. The following equation, however, may be useful in estimating the maximum potential time until adsorption saturation is reached (32):

$$t = 2.9 \times 10^5 \, m/Qy_iM \tag{20}$$

where t = potential time to reach saturation (min)
m = mass of adsorbent (grams)
Q = gas flow rate (m^3/h)
M = average molecular weight of vapors
y_i = inlet vapor concentration (ppm)

12 COMBUSTION

Combustion, sometimes termed "incineration," oxidizes combustible matter entrained in exhaust gases, thus converting undesirable organic matter to less objectionable products, principally carbon dioxide and water.

In many instances, combustors are the most practical means for bringing equipment into compliance with air pollution control regulations. These devices have been employed to reduce or eliminate organic vapors, smoke, odors, and particulate matter in cases when such emissions might exceed the limits set by law.

Equipment in operation today essentially falls into two classifications, direct flame combustion and catalytic combustion. Direct flame incineration has been more widely used and has, in general, been more successfully applied. Increasingly, catalytic incineration units are being used to control organic vapors. Catalytic burners differ from direct flame burners in that they allow organic vapors to be oxidized at temperatures considerably below their autoignition point and without direct flame contact upon passage through certain catalysts.

With the increased cost of energy, efficient heat recovery systems must be employed with either catalytic or direct flame burners to make them economically sound. The arrangement of these devices is limited only by the type of operation and by other plant facilities. Effluent gases from these units often are used in dryoff ovens, water heaters, space heaters, and many other forms of heat exchange.

12.1 Range of Performance

Prior to the energy crisis of the early 1970s, any organic emissions not economically recoverable using the control techniques outlined in the preceding sections were often subjected to direct incineration. With the advent of higher energy costs and the unavailability of auxiliary fuel to support the combustion, the range of performance of combustion equipment has become the subject of innovative designs.

The myriad of combustible organic compounds emitted to the atmosphere from numerous manufacturing processes calls for the application of carefully designed combustion equipment. The design must be based on the particular contaminants being combusted, the concentration of these contaminants in the carrier gas stream, and the flow rate of the gas stream, which affects the physical size of the combustion equipment.

The specifically designed and engineered systems used to control the process emissions are as diverse as the industrial operations emitting combustible compounds. Typical examples of processes requiring combustion as a means for the control of hydrocarbon emissions are industrial drying processes, baking of paints, application of enamels and printing ink, application of coatings and impregnates to paper, manufacturing of fabric and plastic, and manufacturing of paints, varnishes, and organic chemicals, as well as synthetic fibers and natural rubber. In all these processes, flares, furnaces, and catalytic combustors have found suitable application, and the materials emitted may be oxidized to odorless, colorless, and innocuous carbon dioxide and water vapor (33).

The incineration of gases and vapors has found wide application in the control of both gaseous and liquid wastes. In the area of gaseous emissions, use of incineration to minimize odorous pollutants is often the only available control alternative. Because most odorous compounds are characteristically low in concentration, the choice of carbon bed adsorption, for example, may prove to be economically or even technologically infeasible because of low control efficiencies for certain compounds.

The opacity of plumes emitted from various processes has historically been reduced by passing the otherwise highly visible emission through a combustion apparatus. Example applications include coffee roasters, smokehouses, and some enamel baking ovens.

In some applications, the organic gases and vapors being emitted from a process are classified as "reactive hydrocarbons." These compounds, when discharged to the atmosphere, are known to contribute to the formation of smog and other irritating compounds. Afterburners can be used to convert most of these reactive hydrocarbons into carbon dioxide and water vapor.

Combustion equipment has been used extensively in refineries and chemical plants as a method of disposal for unusable waste and as a method of reducing explosion hazards. In these applications, ultimate destruction using a combustion-based process must be carefully designed to avoid creating a hazard to nearby, highly flammable storage tanks or to the process itself.

12.2 Flares

A flare is a method for the efficient oxidation of combustible gases when these are present in a stream that is within or about the limits of flammability (32). Flares usually find their widest application in petrochemical plants, especially in the disposal of waste gases that are often mixed with other inert gases such as nitrogen or carbon dioxide. Additionally, many of the chemical processing plants that produce, use, or otherwise discharge highly dangerous or toxic gases are often equipped with flares designed to be activated under emergency conditions, in the event that such toxic gases should require immediate discharge.

Generally, flares are designed after thorough analysis of the gas stream being incinerated. Flare heights are established by taking into account the heat and light emitted from the flare, and they are supposed to ensure sufficient mixing to prevent unburned organic compounds that may penetrate the flare from presenting a hazard or a nuisance to the surrounding area.

When flares are applied to installations where the concentration of combustible gases is not sufficient to ensure or maintain the persistence of the flare, auxiliary fuel must often be added to provide for efficient combustion and sustained burning.

In some applications, where the materials being combusted are difficult to burn, auxiliary air must be introduced at the point of combustion at the top of the flare, using steam jets or other satisfactory means to ensure good fuel-air mixing.

One design parameter in the sizing and specification of flares is the ratio of hydrogen to carbon in the materials being combusted. As an example, in low molecular weight, aliphatic hydrocarbons (e.g., methane), the high ratio of hydrogen to carbon guarantees burning with very little, if any, soot or smoke. In contrast, double-bonded or triple-bonded hydrocarbon molecules (e.g., acetylene) burn with a very sooty flame because the ratio of hydrogen to carbon is low. Normally, the discharge from a flare should be limited by design to carbon dioxide and water vapor and should certainly exclude the emissions of visually objcctionablc products of combustion such as carbonaceous soot.

When condensable mists are present in the gases to be flared, inertial separators should be provided at the base of the flare to permit automatic separation and automatic drainage of these liquefiable compounds, which may otherwise flow back through the flare and down its walls, causing flashback and/or an explosion.

12.3 Direct Flame Combustion

Direct combustors represent a broad classification of combustion equipment in which auxiliary fuel is added and heat recovery may or may not be practiced to ensure efficient combustion and minimization of operating costs. The designs of direct combustors are as varied as the manufacturer's experience. Generally, direct combustors rely on three factors to provide efficient conversion of the organic compounds to carbon dioxide and water:

1. Sufficient time in the incinerator to allow complete conversion to carbon dioxide and water vapor, all other conditions being specified.
2. Sufficient fuel to ensure combustion temperatures high enough to permit complete combustion within the allowable residence time in the incinerator.
3. Sufficient turbulence in the incinerator to provide for good mixing and contacting of the combustible materials with the active flame front.

Direct combustors equipped with heat-exchange units are often used to preheat the incoming gases to a temperature high enough to ensure more complete and efficient combustion and to save on total energy consumption needed to preheat the incoming gases to the combustion temperature. This heat exchange can be derived in a variety of ways, but it essentially causes the removal from the incinerator discharge of waste heat that would otherwise pass into the atmosphere, and recycles this heat to the incoming gases by way of some physical medium. Heat exchange efficiencies range from 20% to more than 90% depending on the design of the heat exchanger, the pressure drop through the device, and the materials of construction; these variables in design also affect, of course, the initial capital investment cost and the operating costs. Direct combustors are also useful in the removal of combustible particulate matter, particularly that which is odorous. When such particular matter is of submicrometer size, this method of control is often the only available or most technologically and economically feasible technique.

Incinerators are also quite widely used because they occupy relatively little space, have simple construction, and generate low maintenance requirements (6).

Multiple-chamber direct combustors provide a useful means for increasing combustion efficiency and overall conversion to carbon dioxide and water for a given quantity of fuel input. The multiple chambers provide additional residence time and therefore additional combustion efficiency, which may be required to convert difficult-to-burn organic compounds to carbon dioxide and water.

12.4 Catalytic Combustion

In catalytic combustors, reactions occur that convert compounds into carbon dioxide and water on the surface of a "catalyst," usually composed of platinum or palladium, with the end result that auxiliary fuel costs are minimized.

Catalytic combustion has come into wide use in installations where concentrations of hydrocarbons are relatively low and large amounts of auxiliary fuel would be required to sustain combustion. The systems do not find wide applicability in processes involving an inlet gas stream that can contain materials capable of "poisoning" the catalyst, thus rendering it ineffective. Additionally, systems containing high amounts of particulate matter often cannot be incinerated using catalytic systems, or else they must be equipped with high-efficiency precleaners to prevent the particulate matter from coating the surface of the catalyst, rendering it ineffective.

Basically, a catalytic system includes a preheat burner, exhaust fan, and catalyst elements, as well as control and safety equipment (28). The preheat burner usually raises the incoming gas stream to a temperature of 700 to 900°F. As the heated gases pass through the catalyst bed, the heat of reaction raises the temperature further to levels comparable to those found in direct combustors. The obvious advantage of minimal fuel input renders catalytic incineration an attractive alternative where applicable.

Typically, for a 10,000-scfm system, capital investment costs for catalytic incinerators may be approximately the same as direct flame incinerators; however, annual fuel costs may be reduced to about 2% of flame-type incinerators depending on hydrocarbon concentration and incoming gas temperature (28).

Catalytic incinerators may also be equipped with heat recovery devices, which may be used to recapture otherwise wasted heat and inject it into the incoming gas stream, thereby minimizing the already low preheat fuel requirements.

Even when well-controlled with precleaners, catalytic incinerators require periodic washing and maintenance to remove particulate matter that accumulates after long operating periods. Additionally, catalysts often require periodic reactivation because materials such as phosphorus, silica, and lead, even when present in trace amounts, shorten the active life of the catalysts (34).

12.5 Design Parameters

Depending on the type of incinerator selected, the sizing and specification of the unit rely primarily on thermodynamic calculations of auxiliary fuel requirements and system residence times. In general, the calculations are based on the simple concept that all incoming materials must be preheated to the combustion temperature, generally in excess of 1200°F and usually 1500°F, so that conversion to carbon dioxide and water vapor occurs if sufficient residence time is provided for the heated constituents to interact with the provided oxygen.

When concentrations of hydrocarbons in the gas stream are low, the specific heat of the gas stream may be regarded as equal to that of air. In general, the rate of heat input, Q, required to raise the temperature of the incoming gas from inlet conditions to 1200°F is expressed simply by

$$Q = S_s \rho_g C_p (1200 - T) \tag{21}$$

where Q = required rate of heat input (Btu/min)
S_s = gas flow rate at standard conditions (cfm)
ρ_g = gas density at standard conditions (lb/ft^3)
C_p = average specific heat for air over temperature range (Btu/lb/°F)
T = incoming gas temperature (°F)

To provide the requisite heat input, the total amount of auxiliary fuel may be determined by considering the heating value of the fuel being used as well as the heating value of any organic compounds present in the incoming stream. Assuming that the incoming stream has an average concentration C_a of hydrocarbons (lb/ft^3), and that this stream of hydrocarbons provides Q_{hc}, the quantity of heat per pound of hydrocarbons, the available heat per minute from the organic content of the stream Q_{oc} is calculated by the following equation:

$$Q_{oc} = S_s C_a Q_{hc} \tag{22}$$

Now, subtracting this available energy from the preheating requirements, the total amount of heat per minute required Q_{req} may be computed if the heating value of the available fuel, Q, is known:

$$Q_{req} = Q - Q_{oc} = F_a Q_f \tag{23}$$

where F_a = auxiliary fuel rate required (lb/min)
Q_f = heat content of auxiliary fuel (Btu/lb)

If heat recovery equipment is available, the quantity of available heat may be subtracted directly from the heat computed in Equation 23 and the fuel requirements thereby proportionally reduced. In practice, even though the heating requirements are small or even negative according to these simplistic energy balances, it is likely that auxiliary fuel will still be required because no account has been made in this analysis for heat losses in the total system.

Thermodynamic calculations are therefore useful in arriving at fuel requirements; however, the specification of total residence time or degree of turbulence involves an experience factor available from the supplier of the combustion equipment. As a rule, though, a residence time of 0.5 to 1 sec in the primary incinerator with at least 0.2 sec in its secondary chambers for direct flame combustion methods is considered to be adequate reaction time to convert oxidizable substances to carbon dioxide and water vapor, presuming sufficient or excess oxygen. The residence time required for efficient combustion in catalytic incinerators is reduced by at least one-fifth to one-tenth of the time required for direct flame combustion (35). This residence time must be computed knowing the cross-sectional area of the incinerator and the volume of gases in actual cubic feet per minute as determined from the combustion temperature.

To provide adequate turbulence, the internal design of the incinerator often features baffles and/or relatively tortuous turns. This allows good mixing while minimizing system pressure drop. The gas velocity required for good mixing is considered to be about 2100 ft/min (35).

13 SELECTING AND APPLYING CONTROL METHODS

The preceding sections of this chapter have presented and summarized the regulatory, economic, and technical considerations involved in the selection of the optimal method of emission control for a specific stationary source. The application of these basic factors in an orderly sequence of decisions may require not only a carefully conceived and applied methodology, but a perspective broader than that of this chapter.

The selection of the "best" control method for a specific emission source can be a very simple and obvious choice, or it can entail a complex decision-making procedure, especially important when the expected cost of control—usually related to the magnitude and strength of the emission—is large relative to the financial resources of the owner or operator of the source.

Because of technical uncertainties and ever-present alternatives, it would be very desirable to have an inclusive, systematic approach to the selection of a control system for a given application. Such a strategy should combine the type of information and methods presented in this chapter with an even more basic understanding of certain very crucial steps in a whole sequence of decisions necessary to ensure success of any emission cleanup effort. The ultimate performance of an emission control system rests very heavily on the avoidance of oversights during one of the planning steps, as well as the proper use and understanding of the basic fundamentals for selecting optimal control methods. Sound and accurate emissions data are essential, and skillful interpretation of all available data can spell the difference between a control system that functions efficiently and one that performs only marginally.

The overall strategy that should be incorporated when addressing the need to control a stationary emission entails defining the emission limit, identifying all related emission sources, investigating process modifications, defining the technical aspects of control problems, and selecting the optimal control system (28).

13.1 Defining the Emission Limit

The emission limit that applies for any given pollution control problem is the most basic information and, indeed, a building block with which to develop and specify the optimal control system. It is important to understand that the permissible emission often is not easily identified as a specifically designed regulatory emission limit. Defining the permissible emission limit is an effort that will likely be performed in two stages. First, the specifically defined regulatory limits (considering state and local, as well as federal regulations) should be defined. Such specifically defined limits may include such limits as particulate limitations by the process weight rule or specific emission performance specified in the federal New Source Performance Standards. In addition to these specific limits, the engineer must identify other, nonspecific limits that may be applicable to the proposed source. These may include such regulatory limits as BACT, MACT, or LAER requirements; opacity requirements; nuisance requirements; odor limitations; federal CAA residual risk requirements; state or local air toxics risk management requirements; and PSD requirements. At this first stage, anticipated or proposed regulations that are not yet promulgated should be identified. Finally, at the first stage of defining the permissible emission limit, the applicable regulation(s) that are likely to be the most stringent should be identified and allowable emission rates of all applicable contaminants should be estimated.

The second stage of the this effort may include detailed dispersion modeling analyses, literature reviews, and detailed regulatory calculations and demonstrations. Often these regulatory demonstrations are iterative efforts that are performed in concert with the engineering design.

13.2 Identifying all Related Emission Sources

Many times when emission controls are specified, it is later found that all emission sources have not been included in the "process envelope" of the control system because of insufficient care in the planning stage.

Normally, it is not enough to attach a well-designed control device to the main exhaust vent emitting the pollutant. With the very low emission levels required for the more toxic materials, emissions from sources other than the main process vent can, in total, overshadow even those from the main exhaust. It is good practice, particularly when dealing with toxic pollutants, to study the entire process and identify all points of emission and all possible solutions to the control problem. Some frequently overlooked emission points that have been found to contribute heavily to process emissions are as follows (28):

1. Accidental releases
   a. Spills
   b. Relief valve operation
2. Uncollected emissions
   a. Tank breathing
   b. Packing gland or rotary seal leakage
   c. Vacuum pump discharges
   d. Sampling station emissions
   e. Flange leaks
   f. Manufacturing area ventilation systems
3. Reemission of collected materials
   a. Vaporization from water wastes in ditches or canals
   b. Vaporization from aeration basins
   c. Reentrainment or vaporization from landfills
   d. Losses during transfer operations

13.3 Process Modification

This discussion of various air-cleaning methods began, as it always must, with a suggestion that process and system control represents a sound first step in making decisions in the selection and specification of emission controls. If the problem can be solved at its source, this approach offers several advantages, including a typical economic advantage, over the installation of some added or incremental type of control system.

Process modification is usually a most economical way to reduce emissions because little or no control equipment is needed. In addition, there can be improvements in operating efficiency that will reduce material or energy losses. The cost of a terminal control system, if one is ever required, can also be reduced because it is required to handle less material. It is important to retain the option of process modification from the very beginning to avoid costly repetition of an engineering analysis once the add-on controls have been specified and the cost has been estimated to be exorbitantly expensive. Some of the techniques (Section 4) that can be helpful include the following (28):

1. Substitution with a less toxic or less volatile solvent
2. Replacement of a raw material with a purer grade, to reduce the amount of inerts vented from the process or the formation of undesirable impurities and by-products
3. Changing the process operating conditions to reduce the amount of undesirable by-products formed
4. Recycling process streams to recover waste products, conserve materials, or diminish the formation of an undesirable by-product by the law of mass action
5. Enclosing certain process steps to reduce contact of volatile materials with air

13.4 Defining the Technical Aspects of the Control Problem

An important final step before selecting a control method is to define the properties of the exhaust gas stream; the basic data needed were reviewed in Section 5. Such data can come from several sources, however, and the information obtained must be viewed with objectivity if not skepticism. Laboratory data, for example, are never quite the same as results reported under actual plant conditions. Empirical prediction of emission factors or rates cannot replace actual source testing of stack gases. Particle size measurement is more reliable than literature data citing "typical" size distributions from various processes. In addition, the method of sampling where source testing is employed is a most important consideration that can yield differences both in reported results and the interpretation of those results.

Some common pitfalls to avoid in the technical definition of a control problem include failure to recognize the presence of another physical phase (e.g., condensed droplets with the gas or vapor), failure to recognize the presence of fine particulates, and failure to recognize variations in the characteristics of the emission (28).

If particulates, especially fines, are present when only a gaseous type of effluent is expected (or vice-versa) serious control efficiency problems can follow, because most devices designed for removing gaseous pollutants are far less efficient for particulates, and most particulate control devices are inefficient at removing gases. Only a few devices do both jobs well. Some packed-tower scrubbers or carbon adsorption units can be rendered inefficient by particulate matter, and baghouse filters can be corroded by even trace amounts of corrosive gases or vapors.

The problem is particularly complex when the particulate is an aerosol resulting from the condensation of a relatively nonvolatile material. In this case, a sample at the only available sampling point may give one answer, whereas a sample further upstream might show more gaseous material and one downstream might show more particulate.

13.5 Selection of Controls with a Technical–Economic Decision Model

Clearly, the basic strategy for the selection of the optimal control method must be an integrated approach that includes, at a minimum, identification of suitable technical alternative control methods, and selection of the alternative that is most attractive economically from among the suitable alternatives available.

It is not too much to expect the development of a computer program utilizing all relevant technical and cost-related information needed to select not only the most technically appropriate control methods, but also the most attractive economic choice. The need to use an overall systems analysis approach to the problem is obvious; indeed, it is relatively simple to identify and interrelate the various major types of information needed to achieve a logical stepwise set of decisions that will lead to the choice of the "best" method of control for a given emission problem. One such approach is outlined schematically in Figure 59.31.

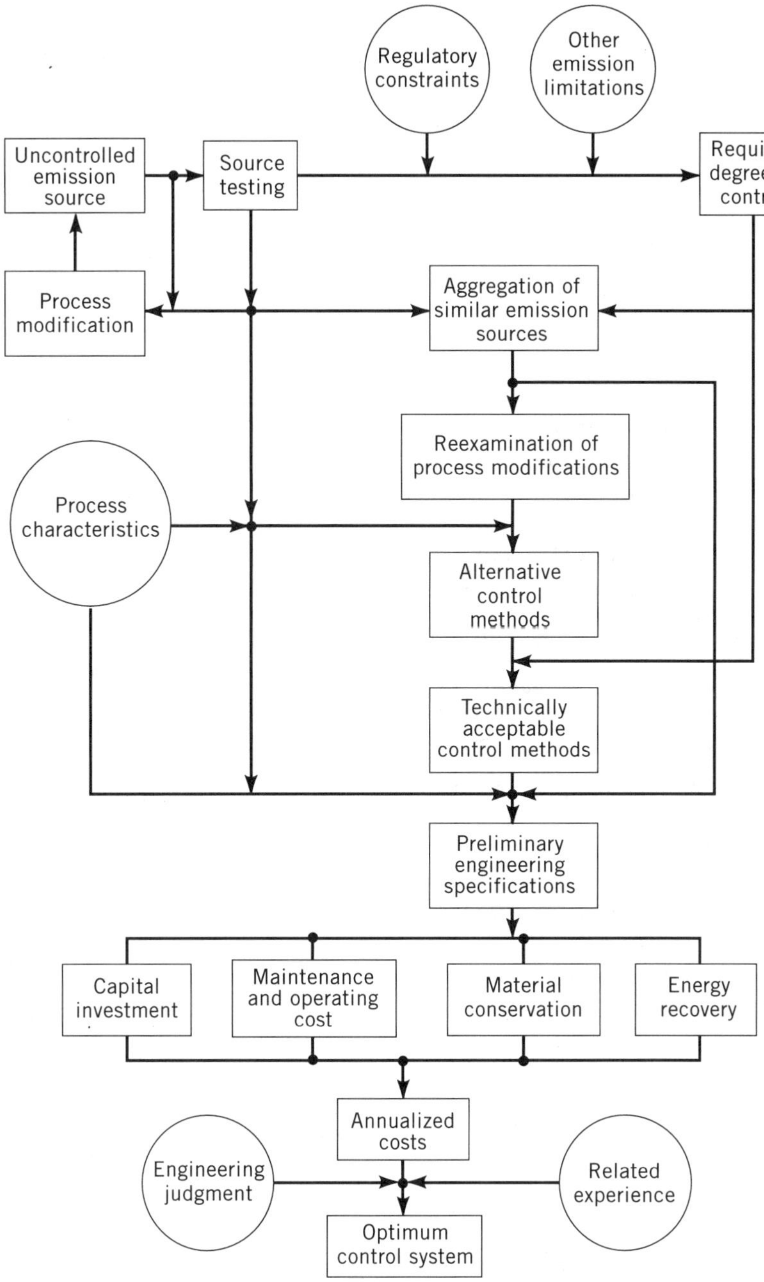

Figure 59.31. Technical-economic decision model for selecting emission controls; circled items denote required inputs of information data.

ACKNOWLEDGMENTS

The authors wish to thank John E. Mutchler, C.I.H., who was the original author of this chapter on air pollution controls in the third edition of *Patty's Industrial Hygiene and Toxicology* and served as co-author for the fourth edition. His original contributions remain the foundation of this chapter.

BIBLIOGRAPHY

1. J. E. Mutchler, "Principles of Ventilation," in *The Industrial Environment—Its Evaluation and Control*, U.S. Department of Health, Education, and Welfare, NIOSH, Cincinnati, OH, 1973.
2. Clean Air Act Amendments, Public Law 101 549, Titles I–IX, Nov. 15, 1990.
3. American Conference of Governmental Industrial Hygienists, *Industrial Ventilation—A Manual of Recommended Practice*, 23rd ed., ACGIH, Lansing, MI, 1998.
4. American Conference of Governmental Industrial Hygienists, *1998 TLVs® and BEIs®—Threshold Limit Values for Chemical Substances and Physical Agents, and Biological Exposure Indicies*, ACGIH, Cincinnati, OH, 1998.
5. Environmental Protection Agency Regulations on Approval and Promulgation of Implementation Plans, as amended through October 17, 1988, **40** *CFR* 52, *Environ. Rep.*, **S-820**, 125, 0201 (1988).
6. *Control Techniques for Particulate Air Pollutants*, Department of Health, Education and Welfare, National Air Pollution Control Administration, Washington, DC, 1969.
7. F. I. Honea, *Chem. Eng. Deskbook*, **81**, 55–60 (1974).
8. W. M. Vatavuk, "Cost Estimating Methodology," *OAQPS Cost Control Manual*, U.S. Environmental Protection Agency, Office of Air Quality Planning and Standards, Research Triangle Park, NC, 1990.
9. K. T. Semrau, *J. Air Pollut. Control Assoc.* **13**, 587–594 (1963).
10. "PVC Makers Move to Mop Up Monomer Emissions," *Chem. Eng.* **82**, 25–27 (1975).
11. D. B. Turner, *Workbook of Atmospheric Dispersion Estimates*, U.S. Environmental Protection Agency, Office of Air Programs, Publication No. AP-26 (revised), 1970.
12. U.S. Environmental Protection Agency, *Guideline on Air Quality Models (Revised)*, Office of Air Quality Planning and Standards, USEPA, Research Triangle Park, NC, 1986.
13. U.S. Environmental Protection Agency, *Guidelines for Air Quality Maintenance Planning and Analysis*, Vol. 10 (Revised) *Procedures for Evaluating Air Quality Impact of New Stationary Sources*, Office of Air and Waste Management, Office of Air Quality Planning and Standards, USEPA, Research Triangle Park, NC, 1977.
14. J. C. Mycock, J. D. McKenna, and L. Theodore, *Handbook of Air Pollution Control Engineering and Technology*, CRC Press, Boca Raton, FL, 1995.
15. American Industrial Hygiene Association, *Air Pollution Manual*, Part II, AIHA, Detroit, 1968.
16. A. H. Rose, D. G. Stephan, and R. L. Stenburg, "Control by Process Changes or Equipment," in *Air Pollution*, Columbia University Press, New York, 1961.
17. K. J. Caplan, "Source Control by Centrifugal Force and Gravity," in A. C. Stern, Ed., *Air Pollution*, Vol. 3, 2nd ed., Academic Press, New York, 1968.

18. C. J. Stairmand, *J. Inst. Fuel* **29**, 58–81 (1956).
19. H. E. Hesketh, *Understanding and Controlling Air Pollution*, 2nd ed., Ann Arbor Publishers, Ann Arbor, MI, 1974.
20. C. E. Lapple, *Chem. Eng.* **58**, 145–151 (1951).
21. U.S. Atomic Energy Commission, Hanford, WA, 1948.
22. K. Iinoya and C. Orr, Jr., "Source Control by Filtration," in A. C. Stern, Ed., *Air Pollution*, Vol. 3, 2nd ed., Academic Press, New York, 1968.
23. B. Goyer, R. Gruen, and V. K. Lamer, *J. Phys. Chem.* **58**, 137 (1954).
24. T. McKenna, "Expanded PTFE Membrane for Surface Filtration the Solution to Many Baghouse Problems," *in The User and Fabric Filtration VII, Proceedings of a Specialty Conference*, Air & Waste Management Association, Toronto, Ontario, 1994.
25. D. C. Drehmel, "Relationship between Fabric Structure and Filtration Performance in Dust Filtration," U.S. Environmental Protection Agency, Control Systems Laboratory, Research Triangle Park, NC, 1973.
26. S. Calvert, J. Goldshmid, D. Leith, and D. Mehta, *Wet Scrubber System Study*, Vol. 1, U.S. Environmental Protection Agency, Control Systems Division, Office of Air Programs, Research Triangle Park, NC, 1972.
27. A. B. Walker, *Pollut. Eng.* **2**, 20–22 (1970).
28. W. L. O'Connell, *Chem. Eng. Deskbook*, **83**, 97–106 (1976).
29. A. Turk, "Source Control by Gas-Solid Adsorption," in A. C. Stern, ed., *Air Pollution*, Vol. 3, 2nd ed., Academic Press, New York, 1968.
30. A. J. Buonicore and W. T. Davis, "Control of Gaseous Pollutants", in A. J. Buonicore, Ed., *Air Pollution Engineering Manual*, Van Nostrand Reinhold, New York, 1992.
31. P. N. Cheremisinoff and A. C. Moressi, *Pollut. Eng.* **6**, 66–68 (1974).
32. H. E. Hesketh, *Air Pollution Control: Traditional and Hazardous Pollutants*, Technomic Publishing Company, Inc., Lancaster, PA, 1991.
33. H. J. Paulus, "Nuisance Abatement by Combustion," in A. C. Stern, Ed., *Air Pollution*, Vol. 3, 2nd ed., Academic Press, New York, 1968.
34. G. L. Brewer, *Chem. Eng.* **75**, 160–165 (1968).
35. L. Thomaides, *Pollut. Eng.* **3**, 32–33, 1971.

CHAPTER SIXTY

Agricultural Hygiene

William Popendorf, Ph.D., CIH and Kelley J. Donham, DVM

1 AGRICULTURE

Agriculture was and in most places still is, a disciplined life style that has bred strong character and high morals in virtually every culture. But such attributes are probably more the necessities of successful agriculture than its natural benefit. Modern agriculture, as practiced by Western cultures, is not the bucolic, healthful working environment fantasized in media classics such as Laura Ingalls Wilder's *Farmer Boy* (1). Whatever it was, agriculture is changing or facing change everywhere. While some cultures (and indeed some crops within all cultures) have yet to make the transition from human to animal to mechanized power, or from chemical to genetically engineered tools, where these transitions have occurred, production per farmer has increased (2–4). Unfortunately, so too have health and safety stresses upon farmers. Twenty years ago agriculture was the third most hazardous occupation (5–7). Although conditions within general industry improved during the intervening years, agriculture has been exempt of mandated safety and health services (8, 9). Today agriculture is the most hazardous occupational group in the United States (4, 7, 10, 11). In fact, modern farm life has recently been described as "A Harvest of Harm." (4).

This chapter hopes to address the needs and interests of industrial hygienists, as opposed to other reviews of agricultural hazards (12–17). It first describes the dilemma of agriculture as both an industry and as a way of life. Selected health hazards that are either characteristic of or unique to agriculture are then reviewed. Table 60.1 gives a topical overview of hazardous processes, physical, biological and chemical agents, and diseases of concern in agriculture. The selected topics have been organized into acute and cumulative trauma, respiratory hazards, pesticides, veterinary chemicals, zoonoses, dermatoses, physical hazards, cancer, and mental stresses. The chapter concludes with a discussion of

Patty's Industrial Hygiene, Fifth Edition, Volume 4. Edited by Robert L. Harris.
ISBN 0-471-29749-6

Table 60.1 Overview of Safety and Physical Agents, Biological and Chemical Agents, and Diseases of Concern in Agricultural Hygiene.

Processes and Physical Hazards	Biological and Chemical Hazard	Diseases of Agricultural Interest
Commodity storage and transfer	Anhydrous ammonia	Arthritis
Electricity	Carbon monoxide	Carpal Tunnel syndrome
Ergonomics	Carbon dioxide	Dermatoses
Back injury (lifting)	Carcinogens (cancer agents)	Chemical induced dermatoses
Repetitive trauma	Confined space (e.g. silos)	Microbe induced-dermatoses
Farm machinery	Asphyxiation (see oxygen)	Insect induced dermatoses
Balers	Fumigation	Livestock induced dermatoses
Chain saws	Oxygen deficiency	Plant induced dermatoses
Combines	Dermal exposure and absorption	Immunologic diseases
Power take-off	Diesel exhaust	Allergic rhinitis
Roll-over protection [ROPS]	Dusts (inorganic aerosols)	Asthma
Safety guards	Hydrogen sulfide (manure gas)	Skin
Tractors	Biologic organisms	Lower back strain
Fire	Infectious microbes	Noninfectious diseases
Fuel storage (leaks and fires)	Noninfectious bioaerosols	Cancer
Illumination (lighting)	Parasites	Hypertension and heart disease
Lightning	Metal fumes	Organophosphate poisoning
Liquefied propane [LP gas]	Nitrogen dioxide (see silo filling)	Respiratory diseases
Livestock handling	Organic dusts	Asthma (also immunologic diseases)
Noise	Cotton dust	Bagassosis
Pesticide activities	Endotoxin	Bronchitis
Applicator exposure hazards	Grain dust	Byssinosis
Harvester reentry hazards	Mold spores (see farmer's lung)	Hypersensitivity pneumonitis
Nontarget/environmental hazards	Silo unloader's disease	Organic dust toxic syndrome (ODTS)
Psychologic stress	Sugar cane dust (see bagassosis)	Pneumoconiosis
Sanitation (field)	Wood dust	Silo filler's disease (from NO_2)
Silo (filling and uncapping)	Pesticides	Zoonotic diseases
Spray painting	Carbamates and thiocarbamates	
Thermal (heat and cold)	Chlorinated insecticides	
Transportation (on-road and off)	Organophosphates	
Ultraviolet (sun) exposure	Phenoxy-aliphatic acid herbicides	
Vibration	Triazine herbicides	
Welding	Waste handling (see hydrogen sulfide)	

how the traditional industrial hygiene components of anticipation, recognition, evaluation, and control might be applied to agriculture, potential policies to stimulate industrial hygiene and safety services on the farm and ranch, and approaches employed in other countries to deliver such services. While the problems discussed are world wide, this chapter will focus on U.S. agriculture.

1.1 Agriculture as an Industry

The fact that the number of farms and full-time farmers has been consistently decreasing since World War I does not belittle the fact that agriculture still represents the largest single occupational group in the United States, with over 10 million people, depending upon ones criteria for "agriculture worker." Consider the following (10, 18):

- There are about two million farms, each with a principal operator, although about half of these also work full or part-time jobs off the farm.
- There are about 2.7 million hired farm workers, of which about half are migrant or temporary farm workers.
- There are 6 million family members who also live and often work on these farms.
- In addition, there are an estimated 8 million workers in agribusiness who share many of the same occupational risks as production agriculture workers, e.g. grain inspectors (19), poultry processors (20), and veterinarians (21).

Knowledge of exposures and the applicability of traditional exposure limits to agriculture are complicated by the temporal variability of most agricultural processes and settings. Of course, outdoor work changes daily and even hourly with the weather. But even if the weather were constant, either the nature or/and the intensity of the work often changes as the animals or crop matures. Table 60.2 gives examples of the wide temporal diversity of manual labor inputs among various commodities. The mechanized crops that require minimum manual tending while growing and are also highly automated during harvest comprise the first and least labor-intensive category; such crops lend themselves to part-time farming. Bush and tree crops that are not labor intensive during their growing phases but require high manual labor during planting or/and harvest fall into the second category;

Table 60.2 Variation in Labor Intensity during Growing and Harvest Phases of Agricultural Production

Routine Intensity (e.g., While Growing)	Intensity Peaks (e.g., During Harvest)	
	Low	High
Low	(*1*) Mechanized crops, e.g. cotton, grains, processed tomatoes, nut crops	(*2*) range meats, bush and tree crops e.g., coffee, tobacco, stone fruit, citrus
High	(*3*) egg and milk production, enclosed livestock	(*4*) row fruits and vegetables, e.g. strawberries, lettuce, flowers

this category usually requires some group labor management. Livestock varies in the labor required. For instance, open range meat production with a couple of intense labor peaks also falls into the second category. In comparison, egg, milk, and enclosed swine and poultry production routinely require only low-levels of care per animal (category one), but full-time producers of these commodities comprise the third category of labor that is intensive throughout the production cycle with no particular harvest peak. Row crops that are labor intensive throughout most of their growing cycle yet also employ specialized harvest crews comprise the fourth category. Of course, production practices vary around the world or even within regions for various reasons, but these categories demonstrate the distinctively wide diversity in the intensity and temporal frequency of exposure to hazards that makes it difficult to generalize about agriculture as a whole.

1.2 Agriculture as a Way of Life

Although modern agriculture shares many of the hazards of Table 60.1 with general industry, agriculture is unlike general industry socially, economically, psychologically, and geographically. Its uniqueness has hindered the recognition of and response to its many health and safety hazards. Examples of its uniqueness include the following (6, 18, 22–24):

- The workplace and the residence are co-located. The "home place" is the emotional if not the productive heart of the typical family farm. The emotional commitment to farming leads the farmer to work "as long as it takes," and many of the hazards that affect the producer also affect the working family, including the children.
- Family farmers are self-employed. Management and labor are distinguishable. Family farmers do not have employee benefits such as sick leave, workers compensation (25); or even OSHA protections (9). Only in the case of highly labor-intensive crops (the fourth category in Table 60.2) have organized (union) activities contributed toward implementing significant worker benefit programs in agriculture.
- Although agriculture has few preselection barriers to job entry or retirement; it is the work and the agricultural economy that eventually forces farmers to be highly self-selective for the physically fit and psychologically motivated (22, 26).
- The intrinsically "risky" nature of agriculture as a business coupled with the inability of the farmer to set their sales price to reflect costs provides limited incentives to voluntarily purchase, install, or maintain preventive safety and health controls.
- Similar economic pressures cause most of the U.S. agricultural work force to be medically underinsured (often uninsured) and almost universally underserved in terms of its access to medical and preventive services.
- The psychological and cultural self-image of farming as a stoic, independent life style rather than a business further heightens the disinterest of farmers in preventive health services or distrust of external efforts to provide such services (18, 27).
- Agriculture is a geographically dispersed industry. Thus, most incidents of disease and injury occur in solitary, often isolated settings, not only aggravating epidemio-

logic "recognition" of hazards but also delaying the discovery of a victim, their access to immediate medical service, and the later provision of rehabilitation services.
- Despite their geographic isolation, farm families live in close-knit communities that share a common bond with the land and with each other.

Coupled with these characteristics is the rate of structural transition within American agriculture. A number of sources (3, 18, 28–30) describe the economic, political, institutional, and technological forces contributing to the consolidation and loss of roughly half of all individual and family farms every 20 years. These forces include inflation, foreign exchange rates, tax policies, governmental subsidy programs, economies of scale, farm organizations, vertical integration, the availability of human capital, computerization, and genetic engineering.

2 AGRICULTURAL HEALTH AND SAFETY ISSUES

2.1 General Health Status

The occupational health status of individual farmers is linked to their general health status, and that in turn is linked to their living environment and socioeconomic status. Although control programs should consider or address a farmer's socioeconomic status, the socioeconomic status of the U.S. farm population is extremely varied. On the low end (annual sales of <$100,000), are operators of small farms who often work full-time nonfarming jobs and of course migrant farm workers. In the middle (annual sales of $100,000–$250,000) are the medium-sized family farms of the greater Midwest and Northeast sections of the United States. At the highest end (>$250,000 annually) are the large corporate farms scattered across the Midwest, South, and Far West.

As a group, migrant workers have an increased incidence of infectious and parasitic diseases, and their living conditions are often marked by substandard housing and poor environmental sanitation (23, 31). Overall conditions have changed but not dramatically since 1962 when Irma West (32) wrote "Because of migrant status, seasonal work, language barriers, substandard education, marginal health, and poor hygiene, they [migrant workers] are the least able of any group to protect themselves against occupational hazards, particularly agricultural chemicals." Villarejo (33) pointed out that the requirement for seasonal workers in California has increased since 1975 even more than the 70% increase fresh vegetables production; this shift parallels the decreased use of farmer and unpaid family member labor as family farms are being displaced by large scale agricultural operations. Most of the OSHA regulations in Table 60.3 require employee training, but OSHA is only enforceable on large farms (with 11 or more employees). And training in agriculture is not always what it seems. For instance, a recent survey of Ohio migrant worker employers (34) found that 96% reportedly maintained a formal safety and health program, but only 17% included formal employee training as part of this program. Written rules and regulations were available from 44% of the employers, but less than 18% read safety rules to the largely Spanish speaking workers (34).

Table 60.3 OSHA Regulations Applicable to Agriculture

29 *CFR* 1910.1200	Hazard Communication (the same worker training and information as required in general industry)
29 *CFR* 1928.51	Roll-over Protective Structures (ROPS) for Tractors Used in Agricultural Operations (large employers must provide ROPS on tractors newer than 1976 for most uses)
29 *CFR* 1928.57	Guarding of Farm Field Equipment, and Cotton Gins (provides for safety protection guards, lock-out, and operator instructions)
29 *CFR* 1928.110	Farm Worker Protection Standard (covers general field sanitation, *viz.* toilets and drinking water)

Middle-class self-employed farmers actually have a more favorable outcome statistic for most major health conditions than the general population. They have lower overall rates of cancer (largely due to lower smoking prevalence), cardiovascular disease, and strokes (12, 16, 35–37). However, as pointed out by Knapp (22), it is not necessarily farming that makes a person healthy; rather, it's the farmer who must be a basically healthy person to farm.

2.2 Acute Trauma

Like in general industry, the rates of accidental injury and death in agriculture are better documented and comprise a more easily recognized problem than are occupational diseases. Agricultural health and safety as a whole suffers from the lack of a comprehensive or consistent reporting mechanism (8). Fatalities are better reported and accessed than are injuries, but in both cases it is often difficult to separate strictly work related cases from those due other activities on the farm and farming activities off the farm. Exclusion of accidental deaths from non-farming activities would tend to lower the reported rates; for instance, NSC estimates that about 50% of deaths to farmers are due to motor vehicles and 10% occur during off-farm public activities (7). On the other hand, since the NSC reported agricultural rates as per "farm resident," the principle operator is subject to a disproportionately high "work" portion of the reported risks. In fact, Stallones' study of death certificates in Colorado for the years 1980 through 1988 found that death rates for all accidental injuries, motor vehicle related injuries, suicide, and homicide were actually lower for males on farms compared to all Colorado males by rate ratios varying from 0.36 to 0.47; however, when motor vehicle deaths were excluded, the accidental death rates for males on farms were greater than those for all Colorado males, except for subjects older than 65 years (38). On the third hand, children are also victims of on-farm injuries and fatalities at elevated rates (38, 39).

Over forty years of death rates compiled by the National Safety Council (NSC) are plotted in Figure 60.1 (7). Examples of other sources of fatality data are given in Table 60.4. Nearly 50% of on-farm fatalities are from machinery, and almost all of these are due to tractors. Figure 60.1 shows how accidental death rates in U.S. agriculture have changed very slowly in comparison to falling rates since 1975 in construction and especially in

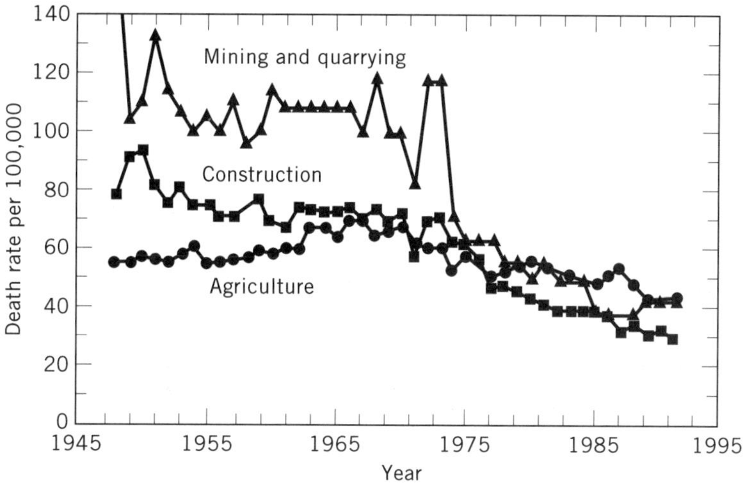

Figure 60.1. Fatality rates in mining, construction, and agriculture, 1948–1991 (7).

Table 60.4 Examples of Other Agricultural Fatality Studies

Author, Year	Sample *N*	Method and Selected Results
Richardson and May-Lambert, 1997[a]	2888 injuries and illnesses	Two BLS programs found the rate of fatal work injury in Texas agriculture in 1994 was 13 per 100,000 workers and that agriculture had the highest rate for skin disease and disorders of any industry
Richardson, et al., 1997[b]	Statewide medical examiners	An analysis of 228 fatal agricultural injuries in North Carolina found the rate rose from 20 per 100,000 worker years in 1977 to 56 per 100,000 worker years in 1991
Pratt et al., 1996[c]	8505 deaths in U.S.	National death certificate data showed occupational fatality rates from 1980 through 1989 were 7.5 in agriculture/forestry/fishing, 7.4 in mining, and 3.50 per 100,000 workers in construction.

[a]Ref. 41.
[b]Ref. 42.
[c]Ref. 43.

mining and quarrying. Although agricultural death rates may have risen since at least 1930 (40), Figure 60.1 makes it appear that it is the recent reduction in rates in other industrial sectors (covered by OSHA) that has allowed agriculture to become one of the most hazardous workplaces in America.

The continuing dominance of tractor related deaths despite the availability of an effective control is indicative of the nature of the agricultural health and safety problem. Tractor rollover protective structures (ROPS) are a well-known effective control that would virtually eliminate accidental deaths from rollovers. For instance, among 173 tractor accidents and 87 fatalities reported in Iowa over three years, no tractor overturn fatalities occurred involving tractors equipped with rollover protective structures (ROPS) (44). ROPS are not required to be installed on tractors in U.S. family farms, although since 1985 major manufacturers will no longer sell a new tractor over 20 horsepower without either ROPS or a cab. The cost of retrofitting ROPS onto existing tractors (ca. $300–1000 depending on size and installation costs) is often mistakenly compared to the low residual value of old tractors versus the value of losing the life of the principal farm operator (45). Because of the longevity of tractors, it took about ten years after Sweden required ROPS on all new tractors to cut rollover deaths in half, but it took their banning the **use** of tractors without ROPS to virtually eliminate tractor roll-over deaths (46). This last pro-active step has not occurred in the U.S.

The injury rate in agriculture appears high but less than in construction and manufacturing. The NSC estimated temporarily disabling injury rates in agriculture at just over 5% (7). These rates have also been relatively constant over the past 20 to 30 years. Other sources in Table 60.5 report higher overall injury and illness rates. The differences are probably due in part to the propensity for farmers to forego medical care unless or until more severe symptoms occur and in part to how the Bureau of Labor Statistics and workers compensation reporting requirements affect agriculture and general industry differently. Causes of injuries seem to differ slightly from fatalities, with animals and machines other than tractors often becoming more dominant.

2.3 Cumulative Musculoskeletal Trauma

Although there are limited data on the subject, historic indications that agricultural workers experience an elevated prevalence of chronic musculoskeletal problems from cumulative trauma have come from surveys of hospitalizations (51) and impairments (52, 53). A more recent NIOSH analysis of national survey data revealed 13.3% of farmers reported chronic diseases and impairments versus 7.8% for all other currently employed respondents and 1.4 times the age adjusted prevalence for arthritis (54).

In addition to general musculoskeletal injuries, certain groups of farmers appear to suffer from two specific conditions. Researchers in central Wisconsin have described degenerative osteoarthritis of the knee as a common condition among dairy farmers, for which they coined the term "milker's knee" (55, 56). French, Swedish, British, and Italian farmers have each been separately described as having an elevated prevalence of osteoarthritis of the hip (57–60). Low back pain and osteoarthritis of the hip are thought by some to be related to low-frequency vibration, perhaps coupled with twisted posture and sudden jolts transmitted through the tractor seat (58, 61–64). Furthermore, chronic vibration to the hands from tractor driving, and operation of chain saws and other hand tools may lead to Raynaud's phenomenon, ischemia to the hands following exposure to cold environments (65).

Table 60.5. Examples of Other Agricultural Injury Studies

Author, Year	Sample *N*	Method and Selected Results
Browning, et al. 1998[a]	70% of 1416 older farmers in Kentucky	A telephone survey found a crude rate of 9.0 injuries per 100 male farmers 55 years of age or older, and that a prior injury which limited ones ability to farm increased the risk of a new farm related injury.
Villarejo, 1995[b]	multi-year statewide reporting system	California's Dept. of Industrial Relations data indicated a rate of 5.9 disabling injuries per 100 full-time employees versus a rate of 15,500 injuries per 100,000 full-time employees filed with the state's Workers' Compensation Insurance Rating Bureau.
Gerberich, et al., 1994[c]	13,000 persons on 3,939 farms	A 1990 telephone and follow-up survey in five upper Midwest states found an incidence of 33.3 injuries per 100 farms, split about 18 to 16 between farming and nonfarming activities. Animals, nontractor machinery, and tractors were the major farm-related sources.
Demers and Rosenstock, 1991[d]	29,451 farm worker claims over 5 years to Washington state workers' compensation	Of 68% of claims submitted that received medical benefits, 37% were lost work time injuries, but only 27 resulted in permanent disability, 18 were fatalities, and 53 were related to OP poisoning and 54 to other pesticides. Ag workers' claim rate was almost 50% higher than nonagricultural workers' for sprains and strains, fractures, dislocations, concussions, amputations, dermatitis, systemic poisoning, tendinitis, respiratory diseases, and fatal injuries.
Hoskin and Miller, 1979[e]	4,176 cases from 24,703 farms in 21 states	Approximately 73% of farm injuries or illnesses were work related, and these resulted in the second largest number of days lost from work. The most frequent injuries occurred from animals while doing routine chores.

[a]Ref. 47.
[b]Ref. 33.
[c]Ref. 48.
[d]Ref. 49.
[e]Ref. 50.

Until recently there has been little research to evaluate and control ergonomic hazards in agriculture; therefore, few specific recommendations are available. However, there has been a good deal of research in the design of tractor seats in recent years, particularly in Germany and the United States (66). Certainly the newer designs of tractor seats that include back support and pneumatic or hydraulic suspension are much more "worker-friendly" than seats on older tractors. Milking parlors or strap-on spring-loaded milking stools, knee pads, and extended handles on milking machines may reduce the trauma to dairy farmers' knees (67, 68). One milking equipment manufacturer (Alfa Laval Agri) has developed a rail system to move equipment from one cow to another, reducing chronic stress to the back, shoulders and knees. Long-handled hoes would help spare the back of row-crop workers. Education to include back-hardening could help, if an effective dissemination system were in place.

2.4 Agricultural Respiratory Hazards

A wide range of respiratory hazards exists in agriculture (69–74). In fact, some think that respiratory hazards represent the greatest health hazard to farmers (72, 73). Table 60.6 groups these agents into gaseous versus particulate hazards, and the latter into predominantly inorganic versus organic constituents. None of these gaseous or inorganic agents are unique to agriculture; however, virtually all of the organic aerosols are associated with commodities and/or processes unique to agriculture. They all can create chronic health effects, and most can also present an acute hazard. The gases are as a whole, mainly an acute hazard. Some hazards occur only in highly enclosed spaces; others can occur in many indoor farm buildings; and a few occur in open spaces.

2.4.1 Gaseous Hazards

Anhydrous ammonia is the most heavily used fertilizer in production agriculture. It is stored and sold in liquid form under pressure by farm supply firms scattered in agricultural areas. It is typically transported to the field in small (ca. 1000 gal.) portable nurse tanks and injected into the soil while tilling with a tractor driven by either the farmer or an employee of the farm supply firm. Each step in the process has the potential for an acute hazard from an unexpected overturn or the failure of a quick-disconnect, hose, or relief valve (75). Because of this pattern, farmers as well as farm supply employees and local residents around the supply firm are at risk of exposure.

Anhydrous ammonia is hazardous because it is highly hygroscopic, highly caustic, and extremely cold (−28°F under pressure). When this material contacts the skin it desiccates, penetrates, and freezes tissue, with the severity depending on the extent of contamination. Anhydrous ammonia is particularly hazardous to the eyes, because almost any eye contact with this chemical results in permanent blindness (76). Ammonia is so irritating that an exposed person will physically avoid damaging vapor concentrations. However, concentrations that can result from a spill or if one is constrained from escaping can result in severe damage to the upper respiratory tract, resulting in bronchiectasis as a possible sequela (77).

Most of the occupational exposures occur during transfer of the chemical from the bulk tank to the nurse tank, and secondarily during field application. NIOSH conducted a Hazard and Operability (HAZOP) analysis to evaluate hazards associated with the storage, transfer, and application of anhydrous ammonia. Four categories were analyzed: training and procedures, equipment design, maintenance and inspection, and material compatibilities (75). Faulty couplings, bleeder valves, shutoff valves, worn hoses, and plugged applicator tips are common roots of an injury. Control should revolve around establishing a routine in-

Table 60.6. An Overview of Agricultural Respiratory Hazards[a]

Hazards	Sources or Examples
Gases and Vapors	
Ammonia (NH_3)	Animal urine, fertilizer
Hydrogen sulfide (H_2S)	Manure gas with other mercaptans
Nitrogen dioxide (NO_2 and dimers)	"Silo gas", arc welding, diesel exhaust
Oxygen deficiency (low O_2)	Confined space
Carbon dioxide (CO_2)	Animal respiration or plant decay
Carbon monoxide (CO)	Combustion from heaters or engines
Organic solvent vapors	Fuels, cleaning, painting
Airborne Pesticides	
Disinfectant derived gases	Formaldehyde, chlorine, cresylic acid
Fumigant pesticides	Phosphine, methyl bromide, chloropicrin
Inorganic (Dust) Aerosols	
Soil (dirt)	Soil contains silica, silicates, and microbes
Welding fumes	Especially welding galvanized steel
Organic Aerosols	
Livestock production buildings	Enclosed animal production buildings
Mycotoxins	Endotoxin, aflatoxin, fumonison, ochratoxin
Noninfectious microbes	Newcastle, Q fever, histoplasmosis
Pesticides	Organophosphates, carbamates, herbicides
Veterinary biologicals	Antibiotic feed additives and vaccines
Crop Specific Organic Aerosols	
Cotton dust	Mostly in processing and textiles
Grain dust	Both clean but particularly moldy grain
Mushrooms	Mold spores and endotoxin
Silo uncapping	Microbes causing FL; endotoxin, ODTS
Sugarcane byproducts	Microbes causing bagassosis
Stored woodchips	Microbes and endotoxin causing ODTS

[a]Adapted from Ref. 10.

spection and maintenance of bulk storage and nurse tanks at the farm supply firm. Hazard communication to employees and farmers is also very important to establish consistent wearing of eye protection and having clean water available to flush eyes and skin in case of contact.

Ammonia fertilizers have been used so extensively in modern agriculture, that groundwater supplies in intensively farmed areas have become contaminated with nitrates (78), increasing the potential for nitrate poisoning in infants (methemoglobinemia or the "blue baby syndrome") (79–82), and raising the question of the risk from carcinogenic nitrosamines that form in the drinking water secondary to nitrate contamination (83, 84, see also the section on cancer). As a result, there are efforts in many states to establish a policy to reduce the total use of ammonia fertilizers and to develop techniques that keep the chemical out of the groundwater (85).

Hydrogen sulfide (H_2S), other mercaptans, and organic acids of many sorts have been identified in the gases emanating from the anaerobic decay of liquid manure typically stored in a pit under most hog and some dairy or beef barns (86–88). These chemicals are part of the livestock odor from farms that some farmers refer to as "the smell of money." Under normal building conditions, H_2S is not at levels of great health concern (89, 90); but when the manure is agitated prior to pump-out to be returned to the fields as fertilizer, H_2S can upon rare occasions rapidly reach fatal levels (91–94). During agitation, the authors have measured levels of H_2S as high as 300 ppm in the work room and 1500 ppm in the pit. Manure gas deaths often involve multiple victims during futile rescue attempts (92, 94). Bundy et al. has shown some relief can be obtained with pit additives but has not tested them sufficiently to see if they can prevent the rare high emission conditions (95). And when they occur, the authors speculate that even the best ventilation would be inadequate to control exposures, especially at the hog breathing level.

The following history of silo gas is representative of the impact that fragmented information from a lack of systematic surveillance can have on an agricultural health hazard. Occupational hazards associated with silo gas were first reported in 1914 via case studies of four fatalities attributed to carbon dioxide (96). However, it was not until the 1950s that investigations revealed the presence and importance of nitrogen dioxide (97–99). The major portion of toxic NO_2 (and its dimer N_2O_4) is believed to be produced from organic nitrates, aggravated by the addition of heavy nitrate fertilizer and/or drought conditions (99). The process of NO_2 production takes several hours to begin and peaks in 2 to 5 days. Measurements of NO_2 concentrations have been sporadic; Peterson et al. (99) reported NO_2 as high as 150 ppm two days after filling. The hazard is primarily associated with vertical silos using corn as the silage (probably the most common but not exclusive combination). Recurrent reviews of the literature discuss the difficulty in diagnosing nonfatal cases of the disease owing to the multiple and usually delayed phases of its clinical manifestations (100, 101). In the 1980s, 14 definite cases and 6 probable cases with 4 fatalities identified from a review of New York hospital discharge records were used to estimate an incidence rate of 5.0 cases per 100,000 silo workers per year (101). The likelihood that only the very severe cases associated with silo filling sought treatment and were reported means this probably underestimates the true magnitude of the hazard. Visible gas was seen in seven cases. Few of the New York victims had followed standard safety procedures for working in and around silos. Of the 17 patients who survived long enough to be examined by a physician, dyspnea was the most common symptom, followed by weakness, cough, and headache. Fourteen of the 16 survivors recovered with no sequelae. Silo gas is less of a problem in the western U.S. where more farmers store silage in horizontal concrete bunkers or large plastic tubes called "ag bags" (from which the orange tell-tale sign of NO_2 can still be seen to seep out soon after filling, but can dissipate in the open air).

The other agricultural gas hazards like carbon dioxide, carbon monoxide, and oxygen depletion are not unique to agriculture and are adequately discussed elsewhere (see Patty's Toxicology volumes). Similarly, fumigants such as phosphine and a decreasing range of volatile chlorinated and/or brominated organics used in produce storage areas (which are often also confined spaces) are rarely unique to agriculture (102–105). These chemicals are listed here because they present primarily respiratory hazards, in comparison to the other pesticides to be listed in a later section that are a hazard to users via the dermal route of exposure. The disinfectants being used increasingly in indoor animal production facilities are rarely a respiratory hazard in themselves (106), but anecdotally, they can produce irritating gases if inappropriately mixed. For instance, mixing chlorine bleach with an active ingredient like o-phenyl phenol produces a noxious but unidentified gas.

2.4.2 Inorganic Aerosols

Inorganic soil-derived aerosols can be generated from both mechanized and manual farming operations as a function of activity and soil moisture (107–110). Data from the Midwest by Casterton (109) indicated dust levels within the plume of a variety of field implements ranged from 100 to 200 mg/m^3. Concentrations were of course, reduced by distance and wind to 10 to 20 mg/m^3 near the driver and by an enclosed cab to less than 2.5 mg/m^3, creating in effect a workplace protection factor of 5 to 10. More recent data from a variety of mechanized field tasks in California (110) showed more variable but somewhat lower total dust levels; median values within each task ranged from about 20 to 80 mg/m^3 total dust with cab protection factors of 10 to 50. The enclosed cabs in California seem to create their own micro-environment largely independent of outside activities; the total dust level within the California cabs was consistently about 1.5 mg/m^3. The respirable dust was between 0.2 and 0.4 mg/m^3 outside and again rather consistently about 0.1 mg/m^3 inside the cabs (110). Popendorf et al. (107) tracked the buildup of dust on foliage in the arid climate of central California for up to six months without rain. The actions of manual harvesters disturbing this foliage created median total aerosol concentrations of approximately 15, 20, and 30 mg/m^3 for peach, grape, and citrus harvesters, respectively. These aerosols ranged from 2 to 10% respirable by mass and were only partly correlated with foliar dust concentrations at the time of harvest.

Soil-derived dust can contain many respiratory hazards. The crystalline quartz content of dusts in central California ranged between 10 and 20% in the soil fines, 5 and 20% in the total aerosols, and 1 and 10% in the respirable aerosols, which was sufficient to exceed combined dust TLV 20 to 50% of the time for grape and citrus harvesters, respectively (107). Slightly higher quartz contents were recently reported from agricultural soils in North Carolina (111). There is some evidence that pneumoconiosis can also develop among farm workers from more common silicates (112). Asbestos can be a natural component of some soils (108). Coccidioidomycosis is associated with arthrospores in arid soils of the Southwest (113). Histoplasmosis and blastomycosis are additional soil-borne fungi which may result in chronic lung disease (114). Clearly, there is a significant potential for classical pneumoconiosis from long-term exposure to inorganic dust from both mechanized and manual agricultural operations in dry-to-arid climates.

2.4.3 Organic Aerosols

Organic dust is a more pervasive and increasingly recognized respiratory hazard in more humid climates (70–74, 115–119). Organic dust is a major hazard in many types of agricultural production including corn, wheat, and swine, poultry and dairy when raised indoors. There are numerous respiratory conditions known to be caused by exposure to organic dust including atopic asthma, non-allergic occupational asthma or asthma-like syndrome (118), bronchitis, mucous membrane irritation (MMI) (73), organic dust toxic syndrome (ODTS), and hypersensitivity pneumonitis [HP] or farmer's lung. Most of these diseases (such as atopic asthma, nonoccupational asthma, and bronchitis) are chronic, not specific to agriculture, but appear to be prevalent in agricultural populations. ODTS and HP, on the other hand, are acute responses largely specific to farm workers. And, a few respiratory diseases are caused by agricultural products but rarely occur among farm workers. For instance, byssinosis is a semi-acute syndrome that can develop after chronic exposures to cotton dust; chronic exposures are common in the textile industry but rarely happen on the farm. Its symptoms combine those of asthma, bronchitis, and ODTS and are most noticeable on Monday mornings following a return to work after a weekend (120, 121). Bagassosis is a well known HP caused by exposure to thermoactinomyces and other thermophillic organisms growing in stored sugarcane by-products, again usually occurring off the farm (122, 123).

Semi-enclosed animal production buildings are wide-spread in modern livestock economies and present chronic exposure hazards to a combination of organic dust and derived gases (71, 118). For instance, surveys of swine barns by Donham et al. (89) beginning in the mid-1970s found dust levels ranging from 3 to 15 mg/m^3 (124). Gas levels in these same buildings were more variable (90). Ammonia levels (from urea) exceeded its 25 ppm 8-TLV in about 50% of these swine buildings; levels of CO_2 (from animal respiration) exceeded its 5000 ppm 8-TLV in about 5% of the settings; while carbon monoxide (from unvented space heaters) approached but did not exceed its TLV, and H_2S (from anaerobic decay of manure) only approached about 10–20% of its TLV under these routine operating conditions (90). In similar studies in poultry barns, Mulhausen et al. (125) found total dust levels (about 18% respirable) ranged from 1 to 14 mg/m^3 and up to 28 mg/m^3 during "loadout" (when the birds are gathered and taken to market). Ammonia levels in poultry buildings were similar to those in swine and sometimes even exceeded its STEL of 35 ppm, while H_2S was undetected because of differences in manure and its handling. Median endotoxin levels in similar settings were 300 to 400 EU/m^3 but were the most variable of all with geometric standard deviations of 4–5 (126).

These largely organic aerosols are more biologically active than nuisance dusts and may act synergistically with irritating gases, producing symptoms similar to bronchitis. A significantly elevated proportion of the livestock workers in these buildings exhibit respiratory symptoms. The prevalence of the two most commonly reported respiratory symptoms in swine confinement workers, chronic cough (25%) and phlegm production (20%), were over twice the rate in a nonfarming comparison group, 10% and 7.4%, respectively (127). In addition, 10% of these swine workers reported chest tightness after two or more days away from work, a symptom identical to cotton processing workers. Similar studies found that pulmonary function of workers in turkey, broiler, egg, production and shackling operations (the first stage of poultry processing) was highly correlated to dust exposure in a dose–response relationship (20, 128). Such studies have lead some researchers to propose threshold levels for these workers of 2.5 mg/m^3 dust, 60 EU/m^3 endotoxin, and 12 ppm ammonia (129–130). Note that some authors report endotoxin as mass instead of "endotoxin units." S. A. Olenchock indicated that the conversion varies with the microbe of origin from roughly 1 to 8 EU/ng (131), while Reynolds and Milton found higher ratios of 15 to 100 EU/ng (132). Both groups of investigators recommend EU units for endotoxin.

The two acute effects from agricultural dusts have very similar symptoms but apparently different etiologies. ODTS is an acute influenza-like illness with headache, muscle aches and pains, fever, chills, and respiratory distress believed to be a direct effect of high doses of endotoxin (133, 134). The onset of these symptoms are delayed 4 to 6 hr following exposure to very high concentrations of organic dusts that often (but not always) have a high mold or other microbe content (127, 135). The individual often recovers from these exposures in 24 to 72 hr, with no known residual effects except possible bronchitis and some increased sensitivity to subsequent exposures. ODTS is a very common disease in the upper Midwest, where 35–40% of farmers report having experienced one or more incidents of acute ODTS (127).

The agricultural workers' version of HP is called Farmer's Lung (FL) or extrinsic allergic alveolitis. FL was first recognized back in the early 1930s (136). Although its acute clinical symptoms are similar to ODTS, FL patients demonstrate IgG antibodies to an antigen like *Faeni rectivirulga* (or *Micropolyspora faeni*) or Thermoactinomyces such as from moldy hay, silage, or grains (122, 123, 137). FL is generally more severe in the acute stage than ODTS and may lead to chronic lung interstitial fibrosis (138). Changing agricultural practices may have made FL much less common today (less than 1% prevalence in the U.S.), but it is still important to the long-term health of the farmer to make the differential diagnosis between FL and ODTS (122, 123, 137–139).

Based on case reports and symptoms surveys, one of this chapter's authors (Donham) hypothesizes that a chronic form of ODTS may occur. Chronic ODTS may explain the group of symptoms seen in many farmers that include fatigue, muscle aches and pains, and difficult breathing. It is felt these symptoms result from prolonged exposure to lower concentrations of microbial and total dust, with or without occasional high exposure peaks. Settings where acute and chronic organic dust exposures may originate are likely to differ.

Rare activities with the potential for acute symptoms such as ODTS and/or the hypersensitivity pneumonitis (e.g., Farmers Lung) include opening up silos (silo unloaders disease) and manually transferring moldy grain, wood chips, or compost. Expected aerosol levels in acute settings are shown as follows:

Total Aerosols	Endotoxin	Microbes
10–100 mg/m^3	>5000 EU/m^3	10^8–10^{10} CFU/m^3

Routine activities that might lead to chronic symptoms such as nonallergic asthma-like syndrome, bronchitis (cough with phlegm), and chronic inflammation of the mucus membranes include indoor swine or poultry production, work in dairy barns, mushroom production, or handling unspoiled grain. Expected aerosol levels in chronic settings are shown as follows:

Total Aerosols	Endotoxin	Microbes
1–10 mg/m^3	100–500 EU/m^3	10^4–10^7 CFU/m^3

Organic dust from agricultural operations is a complex mixture of biologically active materials (115, 117). It is usually quite difficult to identify which of the agents in the dust are responsible for the given condition(s). Some research has shown that grain mites or animal dander are important relative to atopic asthma (140–142). Endotoxin is the primary agent related to bronchitis, asthma-like syndrome, MMI, and ODTS (125, 144, 145). Glucans are emerging as agents of chronic inflammation like endotoxin and also like endotoxin are components of the cell wall of microbial organisms (74, 117, 145).

Control of agricultural dusts should follow the classic IH paradigm: rely first on reduction of the dust source, second on ventilation or other pathway control, and third on personal protection. Reduction of acute exposure to organic dusts may involve applying moisture to the top of the material to reduce its aerosolization when disturbed; to apply this principle to some farm operations (e.g., silo unloading) may require special techniques (146, 147). Good housekeeping to reduce dust accumulation can help. High ventilation of animal confinement or other farm buildings is often resisted by operators who prefer to conserve heat in cold winter climates (148, 149). And the application of negative exhaust pressures can pull hazardous agents into a building from covered manure pits or naturally exhausted heaters (148). The principles of respirator use in agriculture is similar to that in any other industry, except that there are few trained persons available either on an individual farm or even in the rural community to direct the selection, assure proper fit, and supervise the respiratory program (126, 150).

2.5 Pesticides

The term "agricultural chemicals" refers to fertilizers and pesticides. While fertilizers comprise the largest category of agricultural chemicals, pesticides are the traditional "whipping boy" of environmental and occupational health concerns for farmers, especially for farm workers. Although chemicals do indeed present hazards when misused, it should be apparent herein that pesticides and fertilizers represent only a narrow spectrum of the occupational risks within agriculture. The Federal Insecticide, Fungicide, and Rodenticide Act (FIFRA) refers to pesticides as "economic poisons" intended to prevent, destroy, repel, or mitigate "any insects, rodents, nematodes, fungi, or weeds or any other form of life declared to be pests, . . . and any substance or mixture of substances intended for use as a plant regulator, defoliant, or desiccant." Toxicologically, the major field-use agricultural pesticides can be broken down into six major chemical groups of organophosphate, carbamate and thiocarbamate, and chlorinated insecticides and phenoxy-aliphatics, triazines, and bipyridyl herbicides (70, 102, 105). Although reviews of their toxicities are readily available (e.g., 102–105), the industrial hygiene aspects of their use practices, levels of exposure, and the efficacy of exposure controls are less accessible.

Pesticides can present a hazard to applicators, to harvesters reentering a sprayed field, and to off-site rural residents via air, water, and even food contamination. Methods to assess exposure include direct methods via dermal patches (151–154), skin washes (151, 152, 154–156), dietary surveillance (157), and fluorescent tracers (154, 158). Indirect exposure assessment methods include biochemical responses such as change in cholinesterase activity (152, 159, 160), urinary excretion (152, 161–164), and DNA changes (160). And epidemiologic methods to assess response include morbidity (165–169) and mortality (165, 170–173) studies. Archetypical of the difficulty of investigating health effects among diversely exposed and dispersed populations was the discovery of testicular atrophy and sperm count depression among applicators of the nematocide dibromochloropropane (174, 175) following its initial discovery by and among pesticide formulators (176).

Differences in the above methods of assessment complicate comparisons among the multiple routes of exposure contributing to farmers' total doses. Dermal, inhalation, and ingestion are all possible routes of exposure; however when outdoors, the dermal route predominates over inhalation (>99% of the total dose) during application (177–181), harvest (152, 182, 183), and even in cases of local environmental contamination via spray drift (184–185). Indoor agricultural uses of pesticides such as grain fumigation (183, 186) and especially greenhouses (183, 187, 188) represents a specialized environment where airborne exposure can dominate. Ingestion of pesticides can occur through the contamination of food at work (189). Dermal exposure assessment has been a developing area of research. Some studies have found favorable comparisons between direct and indirect methods of assessment, while others have found differences or a lack of correlation. It is often forgotten that correlations should not be expected between measured chemicals that were prevented from actually reaching the skin by the collection media and any measured metabolic excretion of what is absorbed by the same subjects.

Not surprisingly, levels of exposure to pesticides during application vary by task. Broadly speaking, variations in exposures within the tasks shown in Table 60.7 are unrelated to the particular chemical being used but are a function of the pesticide formulation and concentration, the application process and equipment, clothing, personal techniques amenable to education, and uncontrolled conditions like weather and foliage (180). Taking advantage of this principle, the Pesticide Handlers Exposure Database (PHED) is a relatively new tool to predict dermal and inhalation exposure to mixers, loaders, applicators, and flaggers based on a database of previously measured values (181).

Studies of health effects from pesticides are predominantly mortality studies (see Section 2.10 on Cancer). Morbidity studies include acute poisoning reporting (105, 190, 191), a smaller number of investigations into subtle, chronic neurotoxic effects of pesticides (166, 168, 169), and even fewer on seasonal cholinesterase inhibitions (152, 192). For

Table 60.7. Ranges of Pesticide Dermal

Task	Range, mg/hr
Flaggers	0.03–300
Mixer–loaders	10–100
Applicators	2–10
Harvesters	0.5–30

instance, acute pesticide poisoning accounted for 10% of all hospital admittances of farmers and agricultural workers in Colorado, Iowa, and South Carolina during 1971–1973; this rate extrapolated to 9.1 per 100,000, for the 3-year study period. Organophosphates were responsible for 64% of these observed cases (189). Studies in both the United States and abroad have shown that only about 25% of the acute pesticide poisoning fatalities are of occupational origin. Of the remaining 75% in California, nearly 60% were children, frequently due to improperly stored insecticides (193); in the Third World, 75% were suicides (194). Elevated frequencies of suicides (even among U.S. farmers) (16, 195), indicates that rural life in general and farming in particular is stressful; in less developed countries, pesticides are merely an available, convenient, and perhaps economic vehicle for suicide.

Pesticide exposure controls for field applications include engineering/mechanical controls (110, 177; 196, 197) but in practice seem to stress personal protection from clothing, gloves, and respirators (156, 158, 180). In contrast to performance based programs implied by the TLV and regulated by OSHA via the PEL, pesticide usage as regulated by EPA is specified in the label instructions to users. All users of pesticides should be trained to read and follow the label instructions. Unfortunately, the historical focus of pesticide label instructions has been the respiratory route of exposure, despite the repeated finding by direct measurements that the dermal route tends to be at least 100 times larger than the respiratory route (152, 178–180). High rates of dermal depositions coupled with most insecticide's high rate of absorption through intact skin (by design), indicates that reducing dermal exposures is usually the most important component of control. Herbicides are not as dermally absorbable, but even for compounds like paraquat, the skin can be an important route of dosing if improper use practices are allowed (69). The impact of many but not all of these controls can be estimated from PHED (181). One of the unanswered questions is the impact of incomplete removal of pesticides by home laundering (198–201); at least one study (201) seems to be on the track of quantifying the fraction of the residual pesticide bound to clothing after machine washing that is transferable to the wearer, but the ability to predict a dose and an acceptable threshold remains elusive. Because of the acute danger when organophosphate (OP) pesticides are in use (and to a lesser degree carbamates), the existence of a cholinesterase monitoring program is important. Where surveillance is in place and protection breaks down, preplanned medical management is essential. Fortunately, guidelines for such monitoring and diagnosis of OP poisoning are well established (152, 158, 160, 202, 203).

The "reentry hazard" (going into a field after a pesticide application) represents a long unrecognized hazard to harvesters from field residues. It was initially implicitly assumed and only later shown that the rate of dermal exposures to harvesters is proportionate to the field residue, usually the foliar residue (152, 182, 183, 204). Thus in principle, exposure control for harvesters could be based on measured residues. However, applications must be planned to allow certain required field activities to be scheduled, and widespread field residue sampling has not been feasible. Therefore, the use of "reentry intervals" between application and harvest came be the protective tool. A poorly reported 25-year history of sporadic groups of harvesters having symptoms of organophosphate pesticide poisoning was often attributed to poor sanitation, water, or food poisoning (6, 189), thus delaying the recognition of a significant flaw in the reentry interval strategy (152). The classic study by Milby et al. (205) typifies the evolutionary impact of new analytic technologies upon the investigation and understanding of pesticide hazards. In this case using the then-new process of gas chromatography found for the first time that a more toxic "oxon" analogue of the applied thio-phosphate insecticide sometimes forms in field residues. Since that time, the high variability of oxon production in leaf residues (206) and its importance to harvester acute poisoning has been clarified and much better (although still perhaps incompletely) integrated into the field worker protection strategy (152, 183, 207).

A wide variety of disinfectants (also classified by EPA as a pesticide) are used in livestock operations, especially dairy farms and large hog buildings (106). They include chlorine, quaternary ammonia compounds, organic iodines, cresol-based compounds, and formaldehyde emitters, and often one of a variety of detergents. Certain individuals may develop contact dermatitis or an allergic contact dermatitis from these chemicals (208). Prevention of dermatoses can be based on selection and use of chemicals that are not known as irritants or sensitizers. Chemical resistant gloves should be worn as a rule during operations that require repeated contact with the chemicals, such as cleaning milking equipment. Protective hand creams are a better supplement to the use of gloves than they are an alternative (209). Hog farmers who use quaternary ammonium disinfectants have shown increased bronchial hyper-responsiveness (210).

2.6 Veterinary Biologicals, Antibiotics, and Pharmaceuticals

Biologicals are made from living products to enhance the immunity of an animal to a specific infectious disease or diseases. They may be live attenuated microbes, killed viruses (vaccines), killed bacteria (bacterins), or inactivated bacterial toxins (toxoids). All of the above products are intended to enhance the active immunity of the host. These products may also contain adjuvants which enhance the immunogenicity of the products. Another group of biologicals enhances the passive immunity of the host by injecting antibodies produced in another animal. These products may be crude blood sera from a hyperimmunized animal (antiserum), more refined globulin fractions of the sera, or genetically engineered products.

The main risk groups are those involved in livestock production and related veterinary care who administer these products to animals. Besides veterinarians and their assistants, farmers, ranchers, their family members, and employees all may be at risk (211). Operations involving swine, poultry, beef, dairy cattle, and sheep all may have an inherent risk for exposure. A government-regulated disease control program in effect for certain diseases (e.g., brucellosis, pseudorabies) requires that a veterinarian administer the biological. Otherwise the producer, as well as the veterinarian, may administer any of these biologicals.

The hazard is associated with either accidental inoculation, splashing the product into the eyes or mucous membrane, or contamination of the broken skin (211, 212). In one survey, almost 10% of veterinarians reported a needle stick injury per year (212). The result may be an infection (certain live products), inflammation, or an allergic reaction. Inflammation or allergic reactions may occur from inoculating either live or killed products, the adjuvant, or the foreign protein in the product (211). Inoculation may also introduce surface organisms beneath the skin where they can induce infection. And inoculation with a dirty needle has the extra risk of causing infections of environmental origin.

The primary products that have been associated with occupational illnesses include brucellosis strain 19, *Escherichia coli* bacterins, Jhone's disease bacterin, erysipelas vaccines, contagious ecthyma vaccine, and Newcastle disease vaccine. The most frequent reports of occupational illnesses associated with biologicals involve veterinarians using brucellosis strain 19, which is a live product containing an adjuvant (213). Veterinarians have become ill either by splashing the material in their eyes or by accidental needle sticks. The results may be infection, inflammation, and allergic reaction. The infection mimics the acute infection seen from acquisition of brucellosis directly from either cattle or swine (214). If the person had a previous exposure to brucellosis (many veterinarians practicing before the mid-60s had previous exposures), they may develop severe inflammatory and allergic reactions in addition to an infection (213). The reaction is characterized by severe localized swelling and pain extending from the site of the inoculation. The swelling and allergic reaction must be treated in addition to the infection in these cases. Disability may last for days to weeks in the worst cases.

Newcastle disease and contagious ecthyma (orf) vaccines are live products used in chickens and sheep, respectively. Newcastle vaccine is applied inside poultry buildings via a nebulizer. Workers who contaminate their eyes with this vaccine may acquire a moderate conjunctivitis with influenza-like systemic symptoms. Orf vaccine can cause the same pox-like lesions at the site of inoculation as a naturally acquired infection. Both of these diseases are self-limited and disability will only last for a few days, unless the orf lesions are numerous (215, 216).

Jhone's, *E. coli*, and most erysipelas biologicals are bacterins, and therefore injuries induced by these products are limited to the inflammatory response induced by the adjuvants.

Control of injuries associated with biologicals revolves around good animal handling techniques and facilities, because most of the accidental needle punctures are secondary to uncontrolled and untimely movements of stressed and improperly restrained animals. The proper construction of animal handling facilities has been reviewed by Grandin (217). The use of pneumatic syringes, lock-on needle hubs, and multiple-dose syringes will also help reduce injuries. Eye protection is indicated in many instances, and a full face respirator is necessary for aerosolized vaccines such as Newcastle.

Antibiotics are products derived (or synthesized) from living organisms, mainly mold species of the genus *Streptomyces*. Antibacterials are chemical compounds not from living organisms, but used in the same manner to treat infectious diseases therapeutically. They are also used widely at lower levels in livestock production to improve the rate of weight gain and feed efficiency in cattle, swine, and poultry. Livestock producers, veterinarians, and feed manufacturers and formulators are commonly exposed to these agents by direct contact with antibiotic-containing feeds, or via aerosol exposure within livestock buildings, within feed preparation areas on the farm, or in feed manufacturing plants. There are two main occupational hazards: allergic reactions and the development of antibiotic-resistant infections. There are many different products used as feed additives, but the main ones include penicillin, tetracycline, sulfamethazine, erythromycin, and virginiamycin. These same products plus many more are used therapeutically. Penicillin is the primary agent that may induce an allergic reaction manifest in the form of a skin reaction from direct contact, or possibly a systemic reaction from inhalation or inoculation.

A variety of these agents may induce development of resistant organisms in the gut flora of exposed individuals. The resulting health impact of this is not clear. However, there have been some cases of severe resistant salmonellosis traced to direct animal contact (218) and in people who were treated with antibiotics for a condition unrelated to salmonella. The latter case is a result of an overgrowth of the resistant organisms secondary to the antibiotic treatment.

Although the full importance of antibiotics as an agricultural health hazard is unknown, it is prudent to take some control measures. Feed formulation, grinding, mixing, and storing operations should be in closed systems. General dust control procedures should be utilized in both feed preparation areas and in animal feeding operations. Until dust control procedures are proven effective, dust masks should be worn in conjunction with other engineering and work practice procedures. In addition, an emphasis should be placed on removal of those antibiotics used in human health from feed additives and a rotation of the particular type of antibiotic used should be considered.

There are numerous pharmaceuticals used in livestock production and veterinary practice. These products are largely available without prescription and thus, workers at all levels may be using these, and like biologicals, accidental needle sticks or other exposures may occur. Two products in particular are of concern for pregnant women—Oxytocin and Protaglandins. Accidental inoculation with either of these products could cause abortion (213). These products are commonly used in swine, beef, and dairy production.

2.7 Zoonoses

Zoonoses are infectious diseases common to animals and man. At least 24 of the over 150 such diseases known worldwide are occupational hazards for agricultural workers in North America (114, 219–221). Some of these diseases may be contracted directly from animals, whereas many are contracted from the natural environment that is part of the farmer's workplace. A list of recognized agricultural zoonoses was prepared by Donham and Horvath elsewhere (12).

The agricultural worker's risk of acquiring a zoonotic infection varies with the type and species of animal and the geographic location (12, 220, 221). For example, dairy farmers in North America are at risk to acquire ringworm, milker's nodules, or leptospirosis. Beef cattle producers are more prone to acquire rabies, anthrax, or salmonella. Swine producers are at risk for contracting swine influenza, *streptococcus suis*, or *erysipelotrix rheusiopathiae* (erysipeloid). Besides livestock producers, those doing related service work (e.g., veterinarians) or animal processing are also at risk for certain zoonotic infections (220). Turkey processing workers are known to be at risk particularly for ornithosis, red-meat processing workers for brucellosis and leptospirosis, and hair and hide processors for anthrax (10, 114, 222).

Control of these infections in the production phase depends largely on an awareness of the specific hazards, good preventive veterinary care, hazard communication, and medical backup, especially in cases where serological monitoring of animals or people may be indicated. For livestock producers, close animal health monitoring and veterinary preventive practices are best. In processing, early identification of infected animals as they come into the plant and appropriate handling of them is important. In some cases sanitation and personal protection are important. The key is developing both an understanding of certain generic features of this group of diseases and an awareness of conditions and agricultural activities that increase infection risks within specific locations (as reviewed elsewhere, Refs. 114, 219, 223). Such an awareness is essential to enable the hygienist to anticipate, recognize, evaluate, and design a control program for zoonotic infections.

2.8 Skin Diseases

Diseases of the skin are very common in agriculture (208, 209, 224–228). Compared to other occupational groups, farmers have a proportionately higher prevalence of skin diseases (227), and in some regions skin diseases are the most common condition reported by agricultural workers (228). Common agricultural skin diseases, causative agents, and suggested methods of control are listed in Table 60.8

Irritant contact dermatitis is the most common type of agricultural dermatoses (209, 224–231). There is no particular subgroup of agricultural workers that is free from contacting a substance that may cause an inflammatory response to the skin. Irritant substances are ubiquitous and include ammonia fertilizers, several insecticides and fungicides, a few herbicides, soaps, petroleum products, and solvents (209, 226). Avoidance schemes must include work practices to eliminate or reduce exposure to the most irritative substances and/or the use of personal protection equipment.

Delayed allergic contact dermatitis is typified by poison ivy or poison oak reactions. These are exquisite sensitizers, and nearly 60% of the general population is capable of reacting to these allergens. Only a few herbicides and pesticides are sensitizers (208, 209). Several of these substances may produce a more immediate allergic response, but it is difficult to control exposure to sensitizers because just a small amount of the allergen may produce a reaction.

Sun-induced dermatoses include sunburn and skin cancers (173, 209, 232). Acute sunburn may be prevented by the use of sunscreens and protective clothing. More important is the cumulative effect of sun exposure, which may produce a variety of lesions about the face and arms. Skin thickening, wrinkling, and actinic keratoses are common in older farm workers; the latter is a precancerous lesion. Twenty-five percent of the preneoplastic lesions may develop into squamous cell carcinomas, the second most common skin cancer after basal cell carcinoma. Squamous cell carcinomas do not tend to be malignant unless they occur on the lip but usually require surgical removal. Basal cell carcinomas are more common but have a low tendency to become malignant. Melanomas have a high tendency to metastasize, but fortunately are the least common of these skin tumors. The risk for melanomas are related to the frequency of sunburns (high exposures), while the other sun induced skin lesions are due to the cumulative chronic (without burning) sun exposure.

Heat-induced dermatoses are not generally very serious, but they can be quite uncomfortable and recovery may take several days. The primary problem with heat is an inflammation of the eccrine sweat ducts, resulting in a pruritic eruption called prickly heat or *miliaria rubra* (233).

Infections of the skin are primarily a result of viruses and fungal agents of animal origin. Ringworm of cattle (*Trychophyton verrucosum*) is a common agricultural fungal skin infection (234). The pox viruses are the next major source of infection. The virus in sheep that produces contagious ecthyma (sore mouth) produces orf in humans, and the virus in cattle that produces pseudocowpox produces milker's nodules in humans (235).

Chiggers, grain mites, animal mites, bees, and wasps all can cause significant injury to the skin of agricultural workers (233, 236). The lesions vary from a mild self-limited skin rash from chiggers and mites to an anaphylactic reaction from stings of bees and wasps.

2.9 Physical Agents

It should come as no surprise that mechanization has had a major impact upon noise-induced hearing loss among farmers. Today, it is a novel experience (perhaps at a farm show) to watch a draft team pulling a plow and to hear the sod being broken and the soil turning over. Numerous surveys show that farmers today suffer a higher incidence of hearing loss compared to other occupational groups (e.g., Refs. 58, 237, 238). Even some nonmechanized farming practices can result in high noise exposure levels, as can be seen in Table 60.9. Sullivan et al. (241) conducted a year-long study of the noise environment of agricultural workers on six Nebraska farms and 67 farm workers. Thirty-eight percent of their machines produced sound levels in excess of 90 dB. To cope with farming's temporal variability, they time-weight averaged over monthly intervals and found 39% of farm workers exceeded 8-hour 90 dB OSHA limits for 15% of the months (241). Only slightly lower levels of exposure were reported more recently for New York dairy farmers (242).

Noise exposure is reduced in tractors with cabs; a study in Wisconsin found noise in 75% of tractors without cabs exceeded 90 dBA versus 18% with cabs (243). This study also found that partially opening the back cab windows increased the average noise level by 1.7 dB; operating the radio with the windows closed increased cab noise by an average of 3.1 dB; and completely opening the windows increased the average noise level by 4.5 dB. Other traditional methods used to prevent noise-induced hearing loss among general industrial workers (as described in CHAPTER 20) are broadly applicable to farmers. Obstacles to such interventions include the long-term capital investments characteristic of large mechanized pieces of equipment and the limited resources available to reach such a large, voluntary audience (244).

Heat, vibration, and ergonomic hazards are all prevalent in agriculture. Heat (and cold in many regions) is a seasonal stressor for outdoor workers generally (see also CHAPTER 23). Heat-induced illness is rarely reported for farmers (12). Among the few examples are elevated heatstroke reported by West (32), more than 2% of workers compensation claims for production agriculture were heat related reported by Jensen (245), and elevated PMRs from exposure to heat or cold reported by Une et al. (246). Given the exposure of farmers to extremes of both heat and cold, these few reports probably reflect the poor-to-no epidemiologic surveillance of this population, a large measure of self-selection within the work force, and perhaps a limited measure of self-control in their work hours.

Whole body vibration (WBV) is very common on tractors. A review of WBV on tractors found pathological radiological changes in the spine and that complaints of low-back pain were found to be associated with both total years and hours per year of tractor driving (61). A more recent investigation involving 577 of 732 male tractor drivers employed by two Dutch companies found the prevalence of self-reported back pain (most often in the

Table 60.8 Skin Conditions of Agricultural Workers: The Principal Source, Symptoms, and Prevention

Classification	Source or Agent (Examples)	Description of Condition	Control
Irritant contact dermatitis	Ammonia fertilizers Animal feed additives (ethoxquin, cobalt) Insecticides (inorganic sulfur, petroleum, coal tar derivatives) Plants (bulbs of tulips, hyacinths, onion, garlic, carrots, asparagus, celery, parsnips, lettuce) Herbicides (trichloroacetic acid, paraquat) Fumigants (ethylene oxide and methyl bromide)	Dermatitis mainly on hands, arms, and other points of contact	Assure proper dilution of chemicals; wear protective clothing; wash hands, arms, and other contact areas frequently
Allergic contact dermatitis	Herbicides (propachlor, thiram, maleic hydrazide, randox, barban, nitrofen, dazomet, lasso) Insecticides (pyrethrum, rotenone, malathion, phenothiazine, naled, ditalimfos, omite, dazomet, dinobuton) Antibiotics (penicillin, spiromycin, phenothiazine) Plants (poison ivy, poison oak, poison sumac, ragweed)	Acute inflammatory response with swelling, possibly reddish elevated eruptions, blisters, pruritus; usually on hands and arms	Same as above, plus: wash clothes that contact offending substances; any work practice change that will limit contact with offending substance
Photocontact dermatitis (including both photo irritant and photoallergic contact dermatitis)	Creosote Feed additives (phenothiazine) Plants containing furocoumarins (carrots, celery, parsley, parsnips, limes, lemons); ragweed, oleoresins	From skin exposure to agent followed by exposure to sunlight; dose dependent, furocoumarin causes blisters followed by hyperpigmentation in bizarre, streaked pattern	Wash hands and contact areas of skin; protective clothing (e.g., gloves and long-sleeved shirt)

Sun-induced dermatoses	Ultraviolet radiation	Includes sunburn; wrinkling of skin; actinic keratoses; squamous cell carcinoma; basal cell carcinoma	Protective clothing (wide-brimmed hat, long-sleeved shirt); sunscreen (e.g., paraaminobenzoic acid)
Infectious dermatoses	Cattle, swine, rodent animal ringworm (*Tichophyton verrucosum, Microsporum nanum, T. metagophytes*, respectively). Sheep pox virus (orf or contagious ecthyma). Cattle pseudocowpox virus (milker's nodules).	Ringworm: highly inflamed, scaley lesions on hands, arms, face, and head; Orf: lesions on hands and arms, develop as red papules, progress to an ulcerative lesion; Milker's nodules: multiple solitary, wartlike lesions on hands and arms.	Appropriate veterinary treatment and prevention, e.g., good sanitation of animal environment; wear protective clothing when handling infected animals.
Heat-induced dermatoses	Moist, hot environments	Miliaria rubra (prickly heat): an exanthematous eruption of the skin caused by inflammation of eccrine sweat glands mainly under the arms and around the belt line	Wear loose-fitting, well-ventilated clothing; ventilate the work environment; daily bathing with a good soap
Arthropod-induced dermatoses	Chiggers, animal mites, grain mites, Hymenoptera (bees, wasps, hornets, yellow jackets, fire ants).	Red maculas, papules, pruritic lesions, possibly vesicles; sensitivity may vary with repeated exposure; anaphylactic reaction possible	Wear light-colored, nonflowery clothing; avoid perfumes; use insect repellant (e.g., diethyltoluamide)

Table 60.9. Typical Noise Levels during Selected Farming Operation[a]

Activity	dBA
Chain saws	105–112
Vane-axial grain drying fan	100–110
Combine at full throttle	102–107
Corn grinder	94–103
Squealing sows	95–102
Bed chopper	94–102
Hay choppers and balers	95–100
Grain storage bin construction	60–98
Tractor at full throttle	
Next to tractor	102
On seat, no cab (75% > 90 dBA)	93
In cab	82–85
Harvestore® unloader/conveyer	85
Milking parlor	76–84

[a]Adapted from Refs. 238–240.

lower back) was approximately 10% higher in subjects who drove tractors exposed to WBV versus those not exposed, and the prevalence increased significantly with vibration dose (63). Thus, whole body vibration appears to interact with twisting of the spine and a prolonged sitting posture to increase the prevalence of lower back pain (60, 63, 64).

Vibration can be greatly reduced by properly designing the tractor seat (66, 247). Segmental vibration among farmers is most common from chain saws, although many hand tools also contribute. Exposure of farmers to such hand tools is usually limited to short periods (248), but the actual incidence of vibratory white fingers among farmers has not been reported. As discussed in Section 2.3, ergonomics is only beginning to have a major impact on agriculture (66–68), notably on the agricultural tractor cab, mechanized milking equipment, and the banning of the short hoe. Systematic study of additional hazards for disabled farmers returning to work is also in its infancy (249, 250).

2.10 Cancer

Compared to the general population, farmers have lower overall cancer rates (37, 173, 251–253). They also have lower rates for the most common cancers related to smoking, *viz.*, lung, esophageal, and mouth (251), consistent with the observation that only approximately 17% of farmers smoke compared to 34% of the general population (35, 173). In spite of this lower overall rate, positive associations often appear between farming or even rural life and several less common cancers including leukemia, non-Hodgkin's lymphoma, multiple myeloma, Hodgkin's disease, and cancers of the lip, skin, prostate, stomach, and brain (37, 173, 253–257).

According to the extensive 1991 review by Blair and Zahm, the statistical evidence for these associations is variable across studies and usually lacking a clear etiology (173).

Easily the strongest evidence is for lip cancer which, along with skin cancer, is quite clearly related to sun exposure (10, 50, 209). A small but innovative project tested the farmer's acceptability and sun protection characteristics of eleven different hats (258). Each hat had both positive and negative characteristics, and none was ideal. A baseball cap modified with a removable back flap was rated highest overall by the farmers. Many of these skin cancers are preventable through education adequate to break some strong cultural norms.

Determining the risk factors for other cancers has been a very difficult problem because of cancer's long latency periods, difficulties in obtaining accurate exposure-classification data, and probably the intrinsic variability in farming. Risk factors most extensively studied are for the reticuloendothelial cancers. A recent meta-analysis of leukemia found a pooled risk ratio of 1.09 for farming (256). Individual studies have found leukemia to be linked to exposure to dairy cattle, poultry, corn production, fertilizers, and animal pesticides (173). An early suspicion of a link to bovine leukemia virus via cattle has not been confirmed (259). Although farmers showed excess multiple myelomas in twelve of sixteen studies, they have been linked to pesticides in only two (173). The annual incidence of non-Hodgkin's lymphoma (NHL) in the United States rose from 5.9 per 100,000 people in 1950 to 9.3 in 1975, and 13.7 in 1989 (255). About half of the 21 cited studies showed excess NHL among farmers, and about half of these were significant but at a relative risk of less than two-fold (173). Non-Hodgkin's lymphoma was associated with exposure to the phenoxy acetic acid herbicide 2,4-D in North America but not in Europe or New Zealand (173). Hodgkin's disease seems typically to be slightly elevated but the least frequent of the reticuloendothelial-lymphatic neoplasms; risk factors have been linked to phenoxy acetic herbicides and to grain dust (173). No environmental factor was linked to prostrate cancer, but it is the most common cancer of those for which farmers are sometimes found to be at an elevated risk.

Clarifying the risk factors for agricultural cancers will require a great deal more research. Until further information is available, about the only thing the hygienist can tell farmers with certainty is that they can reduce their risk of skin cancer by wearing protective clothing, sun screen, and installing shade devices on their tractors and other pertinent equipment.

2.11 Mental Stress

Farmers die of suicide at a greater frequency (195) and suffer more frequent mental disability relative to other occupations (52). A 1985 study regarding social concerns within farm families suggests dysfunctional families, divorce, alcohol abuse, and children having problems are all more common within the farm community (260). The Iowa Farm Family Survey of 1988 indicated that farmers rated stress as one of their major concerns (85). Compounding inherent, endemic stressors in farming are episodic events such as the farm economic crisis of 1982–1987 and the drought of 1988–1989 (261), and increasing globalization, industrialization, and change in commodity support prices of the mid and late 1990s (18, 262). Mental stress not only should be considered an important occupational health issue for farm families but may also contribute to more frequent injuries (263).

Agriculture has always been plagued by economic uncertainty and the constant eroding of profit margins, requiring more to be produced with less labor to assure a livelihood.

Additionally, most types of farming include a series of work-cycle peaks (e.g., Table 60.2) that can be complicated by adverse weather conditions and machinery breakdowns (264). The stoic and independent nature of many farmers makes them reluctant to talk to anybody about these problems, let alone seek professional help (265). The organized support systems typically available in urban centers is not present in most rural communities, and the extended family and social makeup of the rural community is not the support structure that it once was (18, 27). To make ends meet economically in today's farm families, it is very common for one or both spouses to work full or part time off the farm (see Section 1.1). This increases family stress and creates a child-care problem. All too often children are found in the workplace, which is difficult to supervise and where too often they become accident victims (8, 38, 39).

Control of mental stresses is certainly difficult (27). A few innovative, largely proactive programs are being piloted in communities scattered around the country but there needs to be greater activity in this area (265, 266). One such program in Iowa is the "Sharing Help Awareness United Network (SHAUN)," which seeks out farm families in trouble such as having experienced an injured family member, and gets them together for discussion and mutual support. This promises to be a successful way to get help to the stoic independent farmer. The agricultural hygienist needs to be aware of "farm psychology" and mental health resources, and find ways to deal with this aspect of farm health (27).

2.12 Emerging Hazards

The ability to deliver effective prevention programs to the farm community should include the ability to anticipate developing occupational hazards. New genetically engineered crops, livestock, pesticides, and hormones will substantially increase productivity, forcing less efficient farmers out of business and concentrating agriculture even further into larger and fewer operations. Although consolidation will increase the need for hired labor, farm mechanization will eliminate many of the menial labor-intensive operations with which hired farm labor is primarily involved today. The trend for farm employees to become more technically skilled, and the need for temporary migrant labor to diminish, will probably continue (33). For example, the operation of indoor methods of livestock production requires rather specialized year-round labor, but the longer daily exposure to organic dust has created health hazards with which farmers have never before had to deal. The farm manager must become aware of these hazards and of the opportunities and responsibilities to control this working environment, the provision of preventive health and safety services will not only enhance productivity in these environments but may well be necessary for the farm's sustained profitability.

Whatever changes with technology, many risks in agriculture will continue to be of biological origin. For instance, Lyme disease, a tick-transmitted disease first recognized in the northeastern states, is now recognized in much of the upper Midwest (22, 219, 267, 268). This generalized illness from the organism *Borrelia burgdorferi* can result in prolonged arthritic disability. Although no studies of U.S. farmers have been reported, two British blood serology studies have shown farmers (particularly those with cattle) to be at high risk of exposure (269, 270). Its latency and dynamic environmental prevalence precludes an accurate assessment of its potentially large impact at this time. Hantavirus is a recurring problem in the arid regions of the southwestern states. Outbreaks of the resulting "hemorrhagic fever with renal syndrome" or "hantavirus pulmonary syndrome" have been reported among farmers elsewhere in the world (271), but none of the blood serology samples from fifty-seven randomly sampled farmers in New Mexico and Arizona tested positive for hantavirus antibodies (272). The risk of farmers to Creutzfeldt-Jakob Disease (CJD) from exposure to cattle sick with bovine spongiform encephalopathy (BSE) is best described as controversial. BSE has been a great concern to cattle production in the U.K. The authors of a third case report of a farmer who acquired CJD estimated the probability of three or more farmers in the U.K. acquiring CJD at between 0.002 and 0.09, leading them to conclude that farmers are at increased risk (273). However, another group of authors calculated the incidence of CJD in five European countries to be virtually the same (circa 0.75 cases per million person years), leading them to conclude that contact with BSE afflicted cattle is not a risk factor for farmers (274). The uncertainty with regard to exposure and the small numbers make it much too early to rely on either conclusion.

Aflatoxin is known to be an extremely toxic carcinogen in at least eight species of test animals but has long thought to represent only an oral risk to man. A recent follow-up of a small (60–70 persons) cohort exposed to roughly 5 pg/m^3 aflatoxin (that is 10^{-12} gram) on airborne organic dust in a peanut and linseed oil processing plant showed a 2.5 to 4.4 increased risk of cancer of all types for different exposed time periods (275). This diverse pattern of human toxic responses to aflatoxin is not inconsistent with the cancer findings previously described herein (69). The presence of aflatoxin in corn dust (276–278), its measured routine exposures of about 5 ng/m^3 within enclosed swine buildings and peak exposures of well over 1000 ng/m^3 while cleaning out grain bins (278), and the increased risk of *Aspergillus flavus* infestation during drought conditions (277) suggest that airborne agricultural exposures to aflatoxin should be a long-term concern. Another anticipated manifestation of climatic changes (including drought) is that continued ozone depletion will increase the ultraviolet light exposure to the farm population which will result in their greater risk for skin cancer.

Pesticides and nitrate fertilizers are known to contaminate rural water supplies (82–85), raising a hypothetical concern for increased toxic effects at some point in the future. This concern and others are likely to lead to decreasing use of pesticides and high-volume fertilizers, to be replaced by integrated pest management and genetically engineered tools, leading to decreased soil tillage and exposure to machinery. Whatever the potential health hazards of these new products, agricultural workers are likely to receive the highest exposures and exhibit the first adverse effects. Suffice it to say that not only should agricultural hygienists keep informed generally about new technologies and specific products that become available, but they are among the best qualified to anticipate their hazards and feasible controls.

3 INTERVENTION

3.1 Lessons from General Industry

The industrial hygiene paradigm of anticipation, recognition, evaluation, and control can, in principle, be applied to agriculture with the following translations:

- Anticipating health and safety hazards is the preventive application of a dose–response knowledge data base. Response data for hazards unique to agriculture must be generated by either mandatory or funded research surveillance systems. Transferring experience from other industries requires either a knowledge of dose (exposures generated by a given act), the ability to assess dose in real time, or an assumption of worst case. It is unfortunate that in widely diverse settings (characteristic of agriculture), the worst case is much worse than the average case. This "belt and suspenders" approach is characteristic of the specification standards implied by EPA's pesticide users labels. Overly restrictive controls for everyone necessary to protect against the worst case can be perceived as contrary to the intrinsically risk-taking philosophy of farming. Thus, a great deal of probability salesmanship would have to go into preventive programs based on anticipating the worst case, unless a solid understanding of exposure mechanisms is established.
- Recognizing the incidence of injury, disease, and fatalitics requires the systematic application of existing methods. Decades ago Knapp complained about the lack of good scientific epidemiologic studies in agriculture (22). Much of what was reported above is new, but data generated in agriculture (especially for health hazards) is fragmented in time and geography and reflects current and in some cases old technologies. Technologies evolve, and even if current risks were known, the dispersed and locally innovative nature of agriculture would bolster a natural bias toward the use of new unevaluated technologies (279).
- Methods exist to evaluate essentially all the agricultural hazards noted above (e.g., Refs, 70, 139, 131, 158, 240, 277). Although usually rewarding in the long run, agriculture is an inherently risky venture with a slow economic rate of return. The value of evaluation is dimmed by the psychological perception that risks are intrinsic to farming and the cost of voluntary prevention is not competitive in the short term. Even in general industry, a person with less wealth will be more willing to accept job risks; e.g., the average blue collar worker would accept perceived risky jobs for $900/yr more (in 1980 dollars) (279). The concept of evaluation within risk management can best be sold not only on the economic costs of retraining and even relocating, but also on psychological grounds imparted by the break with tradition and the loss of prestige and self-image of not being physically able to farm.
- Implementing controls requires resources. Among the resources readily accessible to the producer are time, land, water (usually), a wide variety of equipment, and the innovative skills to use them. Money is not on this list. The farmers' view of time and money have been inexorably bound up in the dichotomy of farming as a way of life versus a business. As one of Wilder's characters said about new technology in the 1800s, "All it saves is time, son. And what good is time with nothing to do" (1). Agricultural income is limited to those commodities producible by a given land and climate and marketable via the existing infrastructures. By and large, an individual farmer can increase income only by producing more (compatible with a strong work ethic typical of highly agricultural communities), producing better (a weakly marketable option such as "natural" produce), or producing cheaper (an option conducive to operating without "optional" protection features). The widespread use of exposure controls will require external policies and strategies.

3.2 Control Policies and Strategies

The provision of industrial hygiene (and expanded safety) services to agriculture could be initiated via some combination of governmental requirements, private economic incentives, and/or organized producer ("grass-roots") demands. The lack of U.S. governmental interest is dramatized by the 1986 comparison of federal expenditures in Table 60.10 and the long-standing fragmentation of responsibility for agricultural health and safety among multiple agencies (4, 280). National governmental interest subsequent to a major policy conference and report (85) may remain, but should disinterest return, one must look to private forces to initiate interest in or actually provide preventive occupational health and safety services to the agricultural industry. Generic options and approaches for any agency to implement such services are outlined below. The model most likely to succeed will depend on the local culture and broader political issues.

3.2.1 Research

Research on any or all of the traditional elements of anticipation, recognition, evaluation, or control can contribute to society's knowledge, but because of the technical and often interdisciplinary and specialized nature of health research and the equally segmented and organizationally flat nature of agriculture, such knowledge is either not available or not used by policy administrators and is often unusable by individual farmers who constitute agriculture. A large communication gap exists between agricultural, general industrial, environmental, and medical researchers who often publish in diverse literature. Thus collectively a great deal more is known by some than by any one or all.

3.2.2 Education Policies

Education is the least restrictive but most passive preventive measure. In the best of circumstances, education is the cornerstone of creating an interest to recognize, a desire to evaluate, and a commitment of resources to control health and safety hazards (282). Agricultural health and safety education has been federally supported, albeit weakly for over 50 years. Education's inability by itself to reduce injury and accidental death in pace with other industries as shown in Figure 60.1, may be attributed to a lack of resources for good, relevant research, the often dichotomous nature of dissemination between the technical literature and the farming literature, ineffective marketing of risk management to farmers, and the sheer magnitude of the target population (26, 283). The problem is not solely the fault of agriculture.

Table 60.10 Distribution of $210 M Total Federal Expenditures for Protective Labor Services in Fiscal Year 1986[a]

	$/Worker	$/Death	$/Disabling Injury
Mining	182.0	363,400	4540.00
General industry	4.34	39,770	231.00
Agriculture	0.30	606	5.71

[a]Ref. 281.

3.2.3 Certification

Certification of chemicals, equipment, implements, or structures via a voluntary standards process offers the next least restrictive option to prevent unsafe practices or environments. FIEI (the Farm Implement and Equipment Institute) is one organization that has acted to adopt consensus safety and health production standards for agriculture. In certain markets, labor-management agreements have become *de facto* certification requirements. Many insurers, but not all banks or lenders, actively enforce rudimentary on-farm certification requirements in their dealings with farmers. A variation on voluntary compliance is the delegation of Pesticide Certification Training (required by the U.S. EPA for the purchase of certain commercial pesticides) to many land grant universities. Although the threat of liability litigation can hinder expansion of voluntary certification standards, farm consolidation and incorporation can encourage them.

3.2.4 Cost Reduction

Deductions in the cost of doing business are a potential incentive to encourage safer production practices. One model for such incentives is a government tax subsidy (e.g., the 1970s energy conservation tax deduction in the United States and a 1980s workers compensation insurance safety equipment rebate program in Ontario). Another model is insurance discounts for farms meeting certain safety/health criteria (284). Related approaches could include reduced costs by financial lenders who might raise a pro-active farmer's credit rating or health care providers lowering costs for farm families who participate in preventive health and safety services. The potential benefits of this latter approach was tested in rural Iowa (285). The current policy of the major farm equipment manufacturers to offer retrofit safety options to their products at manufacturer's cost (without profit) is yet another example.

3.2.5 Taxation

Taxation can take various forms to create for producers either a financial disincentive to continue to use relatively unsafe practices or a financial incentive to choose to use relatively safe practices. The first form of this option is currently being used via governmental taxes to fund the Occupational Safety and Health Administration (OSHA) and preventive health and safety research (note that farmers too are paying for OSHA but are not receiving its benefits). Voluntary tax (sometimes referred to as a "mill tax") is an alternative program with control and benefits vested in the taxing organization such as a commodity group, the Farm Bureau, a rural coop business, or the integrated system of on-farm and clinical health services funded in Sweden via Lantbrukshälsan (286). An intermediate example can be found in the funding of the "Farmsafe" educational services through workers compensation fees in Ontario (287).

3.2.6 Regulation

Regulation in the United States implies either specification or performance standards, a system of inspectors, and (usually) financial penalties. Governmental control is characteristically political, bureaucratic, and restrictive. Although the passage of occupational health and safety legislation for general industry required broad political support, agriculture is the only sector to have purposefully precluded itself from most OSHA requirements (perhaps to its detriment). The U.S. EPA (as authorized by FIFRA legislation) has completely eliminated exposure to a small number of hazardous agents by cancellation of pesticide registrations, and has implemented specification standards via applicator certification and label use requirements (with only weak enforcement). The temporal and geographic diversity of agriculture (e.g., Table 60.2) creates a bureaucratic dilemma for any agency attempting to impose controls via a specification standard: the attempt to protect employees in one crop or region requires overly restrictive protection in other settings. The impact of OSHA's 1988 Field Sanitation standard (**29** *CFR* 1928.110) varies from dramatic to largely redundant, depending upon the setting. Although the OSHA Act states a preference toward performance standards, the level of exposure and compliance during an agricultural operation is difficult to assess or inspect because of its temporal transience. To the degree that agricultural regulations expand, perhaps a combination of specification standards assuring only a minimum level of protection with solid awareness education can be more palatable and therefore effective.

3.3 Model Programs

The development of preventive occupational health and safety services to U.S. agriculture suffers from a lack of both a clear governmental policy at the top and local leadership to express an interest at the bottom. Available services are fragmented, such as the various programs for migrant workers (which primarily stress acute medical services and occupational illness and injury education, but not hygiene); the Cooperative Extension Service (which is largely limited to one person per state to disseminate agricultural health and safety educational materials); and the Farm Bureau (which in part provides its members with services similar to the Extension Service). These activities are limited in scope to awareness-level information dissemination. They help but are clearly inadequate, considering the breadth and size of the industry's hazards.

In the late 1970s, Finland (62, 288) and Sweden (286) initiated model programs to deliver comprehensive occupational health services to their farm families. Sweden's Lantbrukshälsan clinics provide medical surveillance, medical treatment, preventive physiotherapy, education, and on-the-farm industrial hygiene and safety services. Via these voluntary but subsidized programs, the majority of farmers in these countries now have access to occupational services similar to those in general industry. These countries have been the example for Norway, Denmark, The Netherlands, and other countries who are establishing similar programs (289). France and Germany also have farm programs but they are not nearly so comprehensive; their programs are primarily through their insurance systems and concentrate on medical issues in France and equipment safety features in Germany. Australia is initiating a new program modeled after the Scandinavian approach. Ontario and Saskatchewan in Canada have well-developed programs based primarily on education but include some on-the-farm hygiene and safety services.

Suffice it to say that the small independent programs in the United States are quite behind all of these countries in providing services to farmers. The University of Iowa has had an active research and teaching activity at the Institute of Agricultural Medicine since

1955 (5). In 1987, two state-funded model projects were initiated to deliver comprehensive services through community hospitals with consultation, training of their medical staff, farm educational program development, and referral services provided by university based core staff. One of these programs (the Iowa Agricultural Health and Safety Network) has now been expanded to a total of twenty-two community sites (285). Evaluations are in progress to determine if this community mechanism is a feasible option to provide needed services. A further expansion of this program called the "Certified Safe Farm" is being initiated, whereby an operator will be eligible for insurance incentives if their farm safety inspection rating is sufficiently high to become "certified" (284).

The New York Center for Agricultural Medicine and Health at Cooperstown, New York was initiated in 1990, and is attempting to provide services out of their hospital, and network with several other regional hospitals (290). The Marshfield Medical Clinic at Marshfield, Wisconsin, has been active in treating farmers with occupational illnesses and doing research in agricultural lung illnesses (291). They have more recently expanded their activities in farmer education. These institutions (Iowa, New York, and Wisconsin) receive funding from NIOSH as part of their Agricultural Health Center programs, along with Kentucky, Florida, Texas, California, and Washington state). South Carolina has initiated a program that unites the land grant university extension safety specialists with the medical school to deliver health and safety education programs (292). This Agro-medicine Program expands the traditional extension approach, and programs have now expanded to several Atlantic and Midwestern states.

A greater emphasis must be placed on model programs offering comprehensive, interdisciplinary services rather than the piecemeal programs of the past. Policy strategies to implement services from research to education via consultation, certification, taxation, or regulation need coordination and strong leadership. Such calls have been made for decades (280). The fact that most new agribusiness and food employees are not agricultural school graduates (293) suggests that such leadership is likely to come from other backgrounds, perhaps even industrial hygiene.

4 CONCLUSION

The future of agricultural hygiene will continue to be affected by the economic and technologic forces that have promoted the progressive consolidation of farms into larger, more capital intensive operations (3, 18, 28, 29). New technologies will require more training, and from consolidation will evolve the stratification of agricultural producers into managers and hired employees. These economic and social forces should stimulate a growing interest in product safety, in occupational safety and health, and in more complete management services to both the traditional and the consolidated farm. Evidence for such interest is already seen internationally (289) and in the United States by increased funding at the national level, at the state levels, (e.g., in California, Iowa, Minnesota, and New York), and at local levels centered around rural community hospitals (285). A host of new and diverse professionals are becoming interested in the field, as noted by the recently formed National Coalition for Occupational Safety and Health (85). These forces and disciplines are already blending advantageously with the traditional extension services characteristic of U.S. land grant colleges and universities. It is hoped that this transition to more specialized and comprehensive services will include agricultural hygiene as a growing opportunity.

BIBLIOGRAPHY

1. L. I. Wilder, *Farmer Boy*, Harper and Row, New York, 1933/1971.
2. L. W. Knapp, *Arch. Environ. Hlth.* **13**(4), 501–506 (1966).
3. S. S. Batie and R. G. Healy, *Sci. Am.* **248**(2), 45–53 (1983).
4. T. A. Knudson, "A Harvest of Harm," an award-winning series of six feature articles in the *Des Moines Register*, Des Moines, IA, Sept. 16–30, 1984.
5. L. Lawhorne, *J. Iowa Med. Soc.* **66**(10), 409–418 (1976).
6. D. J. Murphy, *Prof. Safety* **26**(12), 11–15 (1981).
7. National Safety Council, *Accident Facts*, NSC, Chicago, IL, (1950–1998).
8. M. A. Purschwitz and W. E. Field, *Am. J. Ind. Med.* **18**(2), 179–192 (1990).
9. T. W. Kelsey, *Am. J. Publ. Hlth.* **84**(7), 1171–1177 (1994).
10. C. F. Mutel and K. J. Donham, *Medical Practices in Rural Communities*, Springer, New York, 1983, pp. 77–78.
11. K. J. Donham, *Am. J. Ind. Med.* **18**(2), 107–119 (1990).
12. K. J. Donham and E. Horvath, "Agricultural Occupational Medicine," in *Occupational Medicine*, 2nd ed., C. Zenz, 1988, pp. 933–957.
13. C. J. Chisholm, D. J. Bottoms, M. J. Dwyer, J. A. Lines, and R. T. Whyte, *Safety Sci.*, **15**(4–6), 225–248 (1992).
14. J. B. Sullivan, Jr., M. Gonzales, G. R. Krieger, and C. F. Runge, "Health-Related Hazards of Agriculture" in J. B. Sullivan, Jr., and G. R. Krieger, eds., *Hazardous Materials Toxicology, Clinical Principles of Environmental Health*, Williams and Wilkins, Baltimore, Maryland, 1992, pp. 642–666.
15. J. E. Zejda, H. H. McDuffie, and J. A. Dosman, *Western J. Med.* **158**(1), 56–63 (1993).
16. D. Pratt and J. May, "Agricultural Occupational Medicine" in C. Zenz, O. B. Dickerson, and E. P. Horvath, Jr., eds., *Occupational Medicine*, 3rd ed., Mosby-Year Book, Inc., St. Louis, MO, 1994, pp. 883–902.
17. J. Merchant, S. Reynolds, and C. Zwerling, "Work in Agriculture" in C. McDonald, ed., *Epidemiology of Work Related Diseases*, BMJ Publishing Group, London, 1995, pp. 267–292.
18. D. E. Albrecht and S. H. Murdock, *The Sociology of U.S. Agriculture*. Iowa State University Press, Ames, IA, 1990.
19. H. M. Deer, C. E. McJilton, and P. K. Harein, *Am. Ind. Hyg. Assoc. J.* **48**(6) 586–593 (1987).
20. B. Leistikow, W. Pettit, K. Donham, J. Merchant, and W. Popendorf, "Respiratory Risks in Poultry Farmers" in J. A. Dosman and D. W. Cockcroft, eds., *Principles of Health and Safety in Agriculture*, CRC Press, Inc., Boca Raton, FL, 1989, pp. 62–65.
21. C. S. Rhodes, "Health Concerns in Large Animal Veterinarians" in H. H. McDuffie, J. A. Dosman, K. M. Semchuk, S. A. Olenchock, and A. Senthilselvan, eds., *Agricultural Health and Safety: Workplace, Environment, Sustainability*, Lewis Publishers, Boca Raton, FL, 1995, pp. 339–342.

22. L. W. Knapp, *J. Occup. Med.* **7**(11), 545–553 (1965).
23. K. J. Donham and C. F. Mutel, *J. Family Pract.* **14**, 511–520 (1982).
24. M. B. Schenker, *J. Publ. Hlth. Policy* **17**(3), 275–305 (1996).
25. L. J. Fuortes, J. A. Merchant, S. Van Lier, L. F. Burmeister, and J. Muldoon, *Am. J. Ind. Med.* **18**(2), 211–222 (1990).
26. K. Thu, K. J. Donham, D. Yoder, and L. Ogilvie, *Am. J. Ind. Med.* **18**(4), 427–431 (1990).
27. J. L. Ellis and P. R. Gordon, *Occup. Med.: State Art Rev.* **6**(3), 493–502 (1991).
28. L. Tweeten, *Science* **219**, 1037–1041 (1983).
29. P. L. Martin and A. L. Olmstead, *Science* **227**, 601–606 (1985).
30. B. Kneen, *Farmageddon: Food and the Culture of Biotechnology*, New Society Publ., Stony, CT, 1999.
31. P. H. Poma, *Iowa Med. J.* **126**, 451–458 (1979).
32. I. West, *Arch. Environ. Hlth.* **9**, 92–98 (1964).
33. D. Villarejo, "Health Issues for Farm Employees in the United States," in H. H. McDuffie, J. A. Dosman, K. M. Semchuk, S. A. Olenchock, and A. Senthilselvan, eds., *Agricultural Health and Safety: Workplace, Environment, Sustainability*, Lewis Publishers, Boca Raton, FL, 1995, pp. 479–483.
34. L. K. Isaacs and T. L. Bean, "An Overview of the Ohio Migrant Farmworker Safety Needs Assessment," *J. Agric. Safety Hlth.* **1**(4), 261–272 (1995).
35. P. R. Pomrehn, R. B. Wallace, and L. F. Burmeister, *J. Am. Med. Assoc.* **248**, 1073–1076 (1982).
36. A. D. Stark, H.-G. Chang, E. F. Fitzgerald, K. Riccardi, and R. R. Stone, *Arch. Environ. Hlth.* **42**(4), 204–212 (1987).
37. A. Blair, S. H. Zahm, N. E. Pearce, E. F. Heineman, and J. F. Fraumeni, Jr., *Scand. J. Work, Envir. Hlth.* **18**(4), 209–215 (1992).
38. L. Stallones, "Fatal Injuries among Adult Males and Children on Colorado Farms, 1980–1988," in H. H. McDuffie, J. A. Dosman, K. M. Semchuk, S. A. Olenchock, and A. Senthilselvan, eds., *Agricultural Health and Safety: Workplace, Environment, Sustainability*, Supplement, University of Saskatchewan, Saskatoon, Saskatchewan, Canada, 1994, pp. 207–212.
39. F. P. Rivara, *Pediatrics*, **76**(4), 567–573 (1985).
40. S. P. Baker, B. O'Neill, and R. S. Karpf, *The Injury Fact Book*, Heath and Co., Lexington, MA, 1984.
41. S. Richardson, and S. May-Lambert, *J. Agromed.* **4**(3/4), 257–267 (1997).
42. D. Richardson, D. Loomis, S. H. Wolf, and E. Gregory, *Am. J. Ind. Med.* **31**(4), 452–458 (1997).
43. S. G. Pratt, S. M. Kisner, and J. C. Helmkamp, *J. Occup. Envir. Med.* **38**(1), 70–76 (1996).
44. C. J. Lehtola, K. J. Donham, and S. J. Marley, "Tractor Risk Abatement and Control: A Community-Based Intervention for Reducing Agricultural Tractor-Related Fatalities and Injuries." in H. H. McDuffie, J. A. Dosman, K. M. Semchuk, S. A. Olenchock, and A. Senthilselvan, eds., *Agricultural Health and Safety: Workplace, Environment, Sustainability*, Lewis Publishers, Boca Raton, FL, 1995, pp. 385–389.
45. J. R. Myers and K. A. Snyder, *J. Agric. Safety Hlth.* **1**(3), 185–197 (1995).
46. A. Thelin, *Am. J. Ind. Med.* **18**(4), 523–526 (1990).
47. S. R. Browning, H. Truszczynska, D. Reed, and R. H. McKnight, *Am. J. Ind. Med.* **33**(4), 341–353 (1998).

48. S. G. Gerberich, R. W. Gibson, P. D. Gunderson, L. R. French, F. Martin, J. A. True, J. Shutske, C. M. Renier, and W. P. Carr, "Regional Rural Injury Study (RRIS): A Population Based Effort." in H. H. McDuffie, J. A. Dosman, K. M. Semchuk, S. A. Olenchock, and A. Senthilselvan, eds., *Agricultural Health and Safety: Workplace, Environment, Sustainability*, Supplement, University of Saskatchewan, Saskatoon, Saskatchewan, Canada, 1994, pp. 195–200.
49. P. Demers and L. Rosenstock, *Am. J. Pub. Hlth.* **81**(12), 1656–1658 (1991).
50. A. F. Hoskin and T. A. Miller, *J. Safety Res.* **11**(1), 2–13 (1979).
51. J. A. Burkart, C. F. Egleston, and R. J. Voss, *The Rural Health Study: A Comparison of Hospital Experience Between Farmers and Non Farmers in a Rural Area of Minnesota*, DHEW Publ. No. (NIOSH) 78-184, 1978.
52. L. D. Haber, *J. Chronic Dis.* **24**, 482–483 (1971).
53. J. Kennedy and T. J. Fishback, *Occupational Characteristics of Disabled Workers: Social Security Disability Benefits Awards to Workers during 1969–1972*, U.S. DHHS Publ. No. (NIOSH) 80-145, U.S. Government Printing Office, Washington, DC, 1980.
54. R. M. Brackbill, L. L. Cameron, and V. Behrens, *Am. J. Epid.* **139**(11), 1055–1066 (1994).
55. J. Ekholm, R. Nisell, U. P. Arborelius, O. Svensson, and G. Nemeth, *Ergonomics*, **28**(4), 665–682 (1985).
56. C. L. Anderson, P. S. Treuhaft, W. E. Pierce, and E. P. Horvath, "Degenerative Knee Disease among Dairy Farmers," in J. A. Dosman and D. W. Cockcroft, eds., *Principles of Health and Safety in Agriculture*, CRC Press, Boca Raton, FL, 1989, pp. 367–379.
57. P. Louyot and R. Savin, *Rev. Rhum. Mal. Osteoartic.* **33**, 625–632 (1966).
58. A. Thelin, *Scand. J. Soc. Med.* **8**(Suppl. 22), 5–25 (1980).
59. P. Croft, D. Coggon, M. Cruddas, and C. Cooper, *Br. Med. J.* **304**(6837), 1269–1272 (1992).
60. G. Barbieri, S. Mattioli, S. Grillo, A. M. Geminiani, G. Mancini, and G. B. Raffi, "Spinal Diseases in an Italian Tractor Drivers Group" in H. H. McDuffie, J. A. Dosman, K. M. Semchuk, S. A. Olenchock, and A. Senthilselvan, eds., *Agricultural Health and Safety: Workplace, Environment, Sustainability*, Lewis Publishers, Boca Raton, FL, 1995, pp. 319–323.
61. C. Hulshof and B. V. vanZanten, *Int. Arch. Occup. Environ. Hlth.* **59**, 205–220 (1987).
62. K. Husman, V. Notkola, R. Virolainen, K. Tupi, J. Nuutinen, J. Penttinen, and J. Heikkonen, *Scand. J. Work. Env. Hlth.* **14** (Supp. 1), 118–120 (1988).
63. H. C. Boshuizen, P. M. Bongers, and C. T. J. Hulshof, *Intr. Arch. Occup. Envir. Hlth.* **62**(2), 109–115 (1990).
64. M. Bovenzi and A. Betta, *Appl. Ergonomics* **25**(4), 231–241 (1994).
65. G. B. Raffi, V. Lodi, G. Malenchini, M. Missere, M. Naldi, S. Tabanelli, F. Violante, G. Minak, Jr., V. D'Elia, and M. Montesi, *Arch. Ind. Hyg. Toxicol.* **47**(1), 19–23 (1996).
66. J. Matthews, *J. Soc. Occup. Med.* **33**(3), 126–136 (1983).
67. N. NevalaPuranen, *J. Occup. Rehab.* **6**(3), 191–200 (1996).
68. N. NevalaPuranen, *Appl. Ergonomics* **26**(6), 411–415 (1995).
69. W. Popendorf, K. J. Donham, D. N. Easton, and J. Silk, *Am. Ind. Hyg. Assoc. J.* **46**(3), 154–161 (1985).
70. W. Popendorf and S. J. Reynolds, "Industrial Hygiene Evaluations in Agriculture," in R. Langley et al., eds., *Health and Safety in Agriculture, Forestry, and Fisheries*, Government Institutes Inc., Rockville MD, 1997, pp. 439–468.
71. D. K. Olson and S. M. Bark, *AAOHN J.*, **44**(4), 198–204 (1996).
72. X. W. Li, *J. Environ. Sci. Hlth. Part A* **A32**(9–10), 2449–2469 (1997).

73. S. G. Von Essen and K. J. Donham, "Respiratory Diseases Related to Work in Agriculture." in R. Langley, W. Meggs, R. McLymore, and G. Roberson, eds., *Health and Safety in Agriculture, Forestry, and Fisheries*. Government Institution, Inc., Rockville, MD, 1997, pp. 353–384.
74. M. B. Schenker, D. Christiani, Y. Cormier, et al. *Am. J. Resp. Critical Care Med.* **158**(5), S1–S76 (1998).
75. A. B. Spencer and M. G. Gressel, *Am. Ind. Hyg. Assoc. J.* **54**(11), 671–677 (1993).
76. S. Helmers, F. H. Top, and L. W. Knapp, *J. Iowa Med. Soc.* **61**(5), 271–280 (1971).
77. I. Kass, N. Zamel, C. A. Dobry, and M. Holzer, *Chest*, **62**, 282–285 (1972).
78. J. A. Davies, Ed., *Ground Water Protection SW 8-86*, U.S. Environmental Protection Agency, 1980, pp. 1–19.
79. W. E. Donahoe, *Pediatrics* **3**, 308–311 (1949).
80. D. H. K. Lee, *Environ. Res.* **3**(5–6), 484–511 1970.
81. J. White, Jr., *J. Agric. Food Chem.* **23**, 886–891 (1975).
82. B. Gabel, R. Kozicki, U. Lahl, A. Podbielski, B. Stachel, and S. Struss, *Chemosphere* **11**(11), 1147–1154 (1982).
83. R. Zaldivar, *Experientia* **33**, 264–265 (1977).
84. R. C. Shank, *Toxicol. Appl. Pharmacology* **31**, 361–368 (1975).
85. J. A. Merchant, B. C. Kross, K. J. Donham, and D. S. Pratt, "Agriculture at Risk: A Report to the Nation," National Coalition for Agricultural Safety and Health, Iowa City, IA, 1989.
86. W. E. Burnett, *Environ. Sci. Technol.* **3**(8), 744–749 (1969).
87. J. A. Merkel, T. E. Hazen, and J. R. Miner, *Trans. ASAE* **12**, 310–315 (1969).
88. W. C. Banwart and J. M. Brenner, *J. Environ. Qual.* **4**(3), 363–366 (1975).
89. K. J. Donham, M. Rubino, T. D. Thedell, and J. Kammermeyer, *J. Occup. Med.* **19**(6), 383–387 (1977).
90. K. J. Donham and W. Popendorf, *Am. Ind. Hyg. Assoc. J.* **46**, 658–661 (1985).
91. D. L. Morese and M. A. Woodbury, *J. Am. Med. Assoc.* **245**(1), 63–64 (1981).
92. K. J. Donham, L. W. Knapp, R. Monson, and K. Gustafson, *J. Occup. Med.* **24**(2), 142–145 (1982).
93. S. R. Hagley and D. L. South, *Med. J. Australia* **2**, 459–460 (1983).
94. Anonymous, *MMWR* **38**(33), 583–586 (1989).
95. J. Zhu, D. S. Bundy, X. W. Li, and N. Rashid, *J. Environ. Sci. Hlth. Part A*: **A32**(3), 605–619 (1997).
96. E. R. Hayhurst and E. Scott, *J. Am. Med. Assoc.* **63**, 1570–1572 (1914).
97. R. R. Grayson, *Ann. Int. Med.* **45**, 393–408 (1956).
98. T. Lowry and L. M. Schuman, *J. Am. Med. Assoc.* **162**, 153–160 (1956).
99. W. H. Peterson, R. H. Burris, S. Rameshchandra, and H. N. Little, *Agric. Food Chem.* **6**, 121–126 (1958).
100. E. D. Horvath, G. A. do Pico, and R. A. Barbee et al., *J. Occup. Med.* **20**, 103–110 (1978).
101. F. L. Zwemer, Jr., D. S. Pratt, and J. J. May, *Am. Rev. Resp. Dis.* **146**(3), 650–653 (1992).
102. M. Moses, "Pesticides," in W. H. Rom, ed., *Environmental and Occupational Medicine*, Little, Brown, Boston, MA, 1983, pp. 547–571.
103. M. Moses, *AAOHN J.* **37**(3), 115–130 (1989).

104. J. A. Legaspi and C. Zenz, "Occupational Health Aspects of Pesticides. Clinical and Hygienic Principles" in C. Zenz, O. B. Dickerson, and E. P. Horvath, Jr., eds., *Occupational Medicine*, 3rd ed., Mosby-Year Book, Inc., St. Louis, MO, 1994, pp. 617–653.
105. C. S. Shaver and T. Tong, *Occup. Med. State Art Rev.* **6**(3) 391–413 (1991).
106. W. Popendorf and M. Selim, *Am. Ind. Hyg. Assoc. J.* **56**(11), 1111–1120 (1995).
107. W. Popendorf, A. Pryor, and H. R. Wenk, *Ann. Am. Conf. Gov. Ind. Hyg.* **2**, 101–115 (1982).
108. W. Popendorf and H. R. Wenk, "Chrysotile Asbestos in a Vehicular Recreation Area: A Case Study," in H. G. Wilshire and R. H. Webb, eds., *Environmental Effects of Off-Road Vehicles—Impacts and Management in Arid Regions*, Springer, New York, 1983, pp. 375–396.
109. R. H. Casterton, *Ann. Am. Conf. Gov. Ind. Hyg.* **2**, 121–127 (1982).
110. M. J. Nieuwenhuijsen and M. B. Schenker, *Am. Ind. Hyg. Assoc. J.* **59**(1), 9–13 (1998).
111. C. M. Stopford and W. Stopford, *Appl. Occup. Envir. Hyg.* **10**(3), 196–199 (1995).
112. R. P. Sherwin, M. L. Barman, and J. L. Abrahams, *Lab. Invest.* **40**(5), 576–582 (1979).
113. W. M. Johnson, *J. Occup. Med.* **23**(5), 367–374 (1981).
114. K. J. Donham, *Int. J. Zoonoses* **12**, 163–191 (1985).
115. R. Rylander, Y. Peterson, K. J. Donham, eds., *Am. J. Ind. Med.* **10**, 193–340 (1986).
116. R. Rylander and Y. Peterson, eds., *Am. J. Ind. Med.* **17**(1), 1–148 (1990).
117. Lacey and J. Dutkiewicz, *J. Aerosol Sci.* **25**(8), 1371–1404 (1994).
118. K. J. Donham, "Health Hazards of Pork Producers in Livestock Confinement Buildings: From Recognition to Control" in H. H. McDuffie, J. A. Dosman, K. M. Semchuk, S. A. Olenchock, and A. Senthilselvan, eds., *Agricultural Health and Safety: Workplace, Environment, Sustainability*, Lewis Publishers, Boca Raton, FL, 1995, pp. 43–48.
119. M. F. Carvalheiro, Y. Peterson, E. Rubenowitz, and R. Rylander, *Am. J. Ind. Med.* **27**(1), 65–74 (1995).
120. E. N. Schachter, "Byssinosis and Other Textile Dust-Related Lung Diseases," in L. Rosenstock and M. R. Cullen, eds., *Textbook of Clinical Occupational and Environmental Medicine*, Saunders Company, Philadelphia, 1994, pp. 209–224.
121. R. M. Niven and C. A. C. Pickering, *Thorax* **51**(6), 632–637 (1996).
122. L. A. Lindesmith, J. N. Fink, and E. P. Horvath, Jr., "Hypersensitivity Pneumonitis" in C. Zenz, ed., *Occupational Medicine: Principles and Practical Applications*, 2nd ed., Year Book Medical Publishers, Inc., Chicago, IL, 1988, pp. 226–234.
123. V. P. Kurup, *Immunology and Allergy Clinics of North America* **9**(2), 285–306 (1989).
124. K. J. Donham, L. J. Scallon, W. Popendorf, M. W. Treuhaft, and R. C. Roberts, *Am. Ind. Hyg. Assoc. J.* **47**(7), 404–410 (1986).
125. J. R. Mulhausen, C. E. McJilton, P. T. Redig, and K. A. Janni, *Am. Ind. Hyg. Assoc. J.* **48**(11), 894–899 (1987).
126. W. Popendorf, J. A. Merchant, S. Leonard, L. F. Burmeister, and S. A. Olenchock, *J. Appl. Occup. Envir. Hyg.* **10**(7), 595–605 (1995).
127. K. J. Donham, J. A. Merchant, D. Lassise, W. J. Popendorf, and L. F. Burmeister, *Am. J. Ind. Med.* **18**(3), 241–261 (1990).
128. K. J. Donham, *Sem. Resp. Med.* **14**(1), 49–59 (1993).
129. K. J. Donham, S. J. Reynolds, P. Whitten, J. A. Merchant, L. Burmeister, and W. J. Popendorf, *Am. J. Ind. Med.*, **27**(3), 405–418 (1995).
130. K. J. Donham, D. Cumro, S. J. Reynolds, P. Whitten, J. A. Merchant, and L. F. Burmeister, *J. Occup. Environ. Med.*, In press, 1999.

131. W. Popendorf, *Am. J. Ind. Med.* **10**(3), 251–259 (1986).
132. S. J. Reynolds and D. K. Milton, *Appl. Occup. Envir. Hyg.* **8**(9), 761–767 (1993).
133. S. Von Essen, R. A. Robbins, A. B. Thompson, and S. I. Rennard, *J. Toxicol. Clinical Toxicol.*, **2**(4), 389–420 (1990).
134. P. Malmberg and A. Rask-Andersen, *Sem. Resp. Med.* **14**(1), 38–48 (1993).
135. A. Rask-Anderson, *Br. J. Ind. Med.* **46**, 233–238 (1989).
136. J. M. Campbell, *Br. Med. J.* **2**, 1143–1144 (1932).
137. E. O. Terho, *Am. J. Ind. Med.* **10**, 329 (1986).
138. M. Arden-Jones, *Ann. Am. Conf. Gov. Ind. Hyg.* **2**, 172–182 (1982).
139. J. J. Marx, J. Guernsey, D. A. Emanuel, J. A. Merchant, D. P. Morgan, and M. Kryda, *Am. J. Ind. Med.* **18**(3), 263–268 (1990).
140. I. I. Lutsky, G. L. Baum, H. Teichtahl et al. *Eur. J. Respir. Dis.* **69**, 29–35 (1986).
141. A. D. Blainey, M. D. Topping, S. Ollier, and R. J. Davies, *Thorax* **43**, 697–702 (1988).
142. M. Van Hage-Hamsten, E. Ihre, O. Zetterstrom, and S. G. Johansson, *Allergy* **43**, 545–551 (1988).
143. K. J. Donham, *Am. J. Ind. Med.* **10**, 205–220 (1986).
144. S. A. Olenchock, J. J. May, D. S. Pratt, L. A. Piacitelli, and J. E. Parker, *Am. J. Ind. Med.* **18**(3), 279–284 (1990).
145. J. Milanowski, *Inhalation Toxicol.* **9**(4), 369–388 (1997).
146. R. D. Watson, *Am. J. Ind. Med.* **10**(3), 229–243 (1986).
147. D. S. Pratt, L. Stallones, D. Darrow, and J. J. May, *Am. J. Ind. Med.* **10**, 328–329 (1986).
148. J. J. R. Feddes and E. M. Barber, "Agricultural Engineering Solutions to Problems of Air Contaminants in Farm Silos and Animal Buildings," in H. H. McDuffie, J. A. Dosman, K. M. Semchuk, S. A. Olenchock, and A. Senthilselvan, eds., *Agricultural Health and Safety. Workplace, Environment, Sustainability*, CRC Press Inc., Boca Raton, FL, 1995, pp. 527–533.
149. X. W. Li, *J. Environ. Sci. Hlth. Part A* **A32**(9–10), 2449–2469 (1997).
150. W. Popendorf, *Nat. Hog Farmer* **35**(5), 38–48 (1990).
151. W. F. Durham and H. R. Wolfe, *Bull. World Health Org.* **26**, 75–91 (1962).
152. W. Popendorf and J. T. Leffingwell, *Residues Rev.* **82**, 125–201 (1982).
153. M. Hussain, K. Yoshida, M. Atiemo, and D. Johnston, *Arch. Environ. Contam. Toxicol.* **19**(2), 197–204 (1990).
154. G. Chester, in P. B. Curry, S. Iyengar, P. A. Maloney, and M. Maroni, eds., *Methods of Pesticide Exposure Assessment*, Plenum Press, New York, 1995, pp. 29–49.
155. R. R. Keenan and S. B. Cole, *Am. Ind. Hyg. Assoc. J.* **43**(7), 473–476 (1982).
156. J. E. Davies, V. H. Freed, H. F. Enos, R. C. Duncan, A. Barquet, C. Morgade, L. J. Peters, and J. X. Danauskas, *J. Occup. Med.* **24**(6), 464–468 (1982).
157. W. J. Hayes, "Monitoring Food and People for Pesticide Content," in *Scientific Aspects of Pest Control*, National Research Council, National Academy of Sciences, Washington, D.C., Publ. No. 1402, 1966, pp. 314–342.
158. R. A. Fenske and S. G. Bimbaum, *Am. Ind. Hyg. Assoc. J.* **58**(9), 636–645 (1997).
159. M. J. Coye, J. A. Lowe, and K. T. Maddy, *J. Occup. Med.* **28**(8), 619–627 (1986).
160. W. A. Anwar, *Environ. Hlth Perspectives*, **105**(Supp. 4), 801–806 (1997).
161. W. F. Durham, H. R. Wolfe, and J. W. Elliott, *Arch. Environ. Hlth* **24**, 381–387 (1972).

162. M. J. Coye, J. A. Lowe, and K. J. Maddy, *J. Occup. Med.* **28**(8), 628–636 (1986).

163. H. N. Nigg and J. H. Stamper, "Biological Monitoring for Pesticide Dose Determination. Historical Perspectives, Current Practices, and New Approaches." in R. G. M. Wang, C. A. Franklin, R. C. Honeycutt and J. C. Reinert, eds., *Biological Monitoring for Pesticide Exposure: Measurement, Estimation, and Risk Reduction*, ACS Symposium Series 382, American Chemical Society, Washington, DC, 1989, pp. 6–27.

164. W. J. Murray and C. A. Franklin, "Monitoring for Exposure to Anticholinesterase-Inhibiting Organophorsphorus and Carbamate Compounds by Urine Analysis," in B. Ballantyne and T. C. Marrs, eds., *Clinical and Experimental Toxicology of Organophosphates and Carbamates*, Butterworth-Heinemann, Ltd., Oxford, England, 1992, pp. 430–445.

165. D. P. Morgan, L. I. Lin, and H. H. Saikaly, *Arch. Environ. Contam. Toxicol.* **9**(3), 349–382 (1980).

166. J. L. De Bleecker, J. L. De Reuck, and J. L. Willems, *Clinical Neurology and Neurosurgery*, **94**(2), 93–103 (1992).

167. V. F. Garry, J. T. Kelly, J. M. Sprafka, S. Edwards, and J. Griffith, *Arch. Environ. Hlth.* **49**(5), 337–343 (1994).

168. D. J. Ecobichon, "Pesticide-Induced Chronic Toxicity: Fact or Myth?" in H. H. McDuffie, J. A. Dosman, K. M. Semchuk, S. A. Olenchock, and A. Senthilselvan, eds., *Agricultural Health and Safety: Workplace, Environment, Sustainability*, Lewis Publishers, Boca Raton, FL, 1995, pp. 119–126.

169. M. C. Keifer and R. K. Mahurin, *Occup. Med. State Art Rev.* **12**(2), 291–304 (1997).

170. H. H. Wang and B. MacMahon, *J. Occup. Med.* **21**(11), 741–744 (1979).

171. A. Blair, D. J. Grauman, J. H. Lubin, and J. F. Fraumeni, *J. Nat. Cancer Inst.* **71**(1), 31–37 (1983).

172. B. MacMahon, R. R. Monson, H. H. Wang, and T. Zheng, *J. Occup. Med.* **30**(5), 429–432 (1988).

173. A. Blair and S. H. Zahm, *Occup. Med. State Art Rev.* **6**(3), 335–354 (1991).

174. S. H. Sandifer, R. T. Wilkins, C. B. Loadholt, L. G. Lane, and J. C. Eldridge, *Bull. Envir. Contam. Toxicol.* **23**, 703–710 (1979).

175. R. I. Glass, R. N. Lyness, D. C. Mengle, K. E. Powell, and E. Kahn, *Am. J. Epidemiol.* **109**(3), 346–351 (1979).

176. D. Whorton, R. M. Krauss, S. Marshall, and T. H. Mibly, *Lancet* **7**, 1259–1261 (1977).

177. G. E. Carman, Y. Iwata, J. L. Pappas, J. R. O'Neal, and F. A. Gunther, *Arch. Environ. Contam. Toxicol.* **11**(6), 651–659 (1982).

178. H. N. Nigg and J. H. Stamper, *Arch. Environ. Contam. Toxicol.* **12**, 477–482, (1983).

179. J. M. Devine, G. B. Kinoshita, R. P. Peterson, and G. L. Picard, *Arch. Environ. Contam. Toxicol.* **15**, 113–119 (1986).

180. W. Popendorf, "Mechanisms of Clothing Exposure and Dermal Dosing during Spray Application," in S. Z. Mansdorf, R. Sager, and A. P. Nielsen, eds., *Performance of Protective Clothing: Second Symposium*, American Society for Testing and Materials, Philadelphia, 1988, pp. 611–624.

181. T. M. Leighton and A. P. Nielsen, *Appl. Occup. Environ. Hyg.* **10**(4) 270–273 (1995).

182. H. N. Nigg, J. H. Stamper, and R. M. Queen, *Am. Ind. Hyg. Assoc. J.* **45**(3), 182–186 (1984).

183. J. J. van Hemmen, Y. G. C. van Golstein Brouwers, and D. H. Brouwer, "Pesticide Exposure and Re-Entry in Agriculture." in P. B. Curry, S. Iyengar, P. A. Maloney, and M. Maroni, eds., *Methods of Pesticide Exposure Assessment*, Plenum Press, New York, 1995, pp. 9–19.

184. A. W. Taylor, *J. Air Pollut. Control Assoc.* **28**, 922–927 (1978).

185. W. M. Draper, R. D. Gibson, and J. C. Street, *Bull. Environ. Contam. Toxicol.* **26**, 537–543 (1981).

186. D. Zaebst, P. Morelli-Schroth, and L. Blade, "Summary of Recent Environmental Assessments of Exposure to Grain Fumigants at Export, Inland, and Country Elevators." in J. A. Dosman and D. W. Cockcroft, eds. *Principles of Health and Safety in Agriculture*, CRC Press, Inc., Boca Raton, FL, 1989, pp. 240–243.

187. J. H. Stamper, H. N. Nigg, W. D. Mahon, A. P. Nielsen, and M. D. Royer, *Chemosphere* **17**(5), 1007–1023 (1988).

188. J. Liesivuori, S. Liukkonen, and P. Pirhonen, *Scand. J. Work, Environ. Hlth* **14**(Suppl. 1), 35–36 (1988).

189. J. F. Armstrong, H. R. Wolfe, S. W. Comer, and D. C. Staiff, *Bull. Environ. Contam. Toxicol.* **10**(6), 321–327 (1973).

190. E. P. Savage, "Acute Pesticide Poisonings," in *Pesticide Residue Hazards to Farm Workers*, Proceedings of a Workshop Held February 1976, HEW Publ. No. (NIOSH) 76-191, 1976, pp. 63–65.

191. A. H. Hall and B. H. Rumack, "Incidence, Presentation and Therapeutic Attitudes to Anticholinesterase Poisoning in the USA," in B. Ballantyne and T. C. Marrs, eds., *Clinical and Experimental Toxicology of Organophosphates and Carbamates*, Butterworth-Heinemann, Ltd., Oxford, England, 1992, pp. 471–481.

192. W. Popendorf, *Am. J. Ind. Med.* **18**(3), 313–319 (1990).

193. K. T. Maddy and S. Edmiston, *Vet. Human Toxicol.* **30**(3), 246–254 (1988).

194. L. B. L. De-Alwis and M. S. L. Salgado, *Foren. Sci. Int.* **36**(1/2), 81–89 (1988).

195. S. Milham, *Occupational Mortality in Washington State, 1950–1971*, Vol. I–III, U.S. DHEW Publ, Nos. (NIOSH) 76-175 A,B,C, U.S. Government Printing Office, 1976.

196. E. F. Taschenberg, J. B. Bourke, D. F. Minnick, *Bull. Environ. Contam. Toxicol.* **13**(3), 263–268 (1975).

197. C. Lunchick, A. P. Nielsen, and J. C. Reinert, "Engineering Controls and Protective Clothing in the Reduction of Pesticide Exposure to Tractor Drivers." in S. Z. Mansdorf, R. Sager, and A. P. Nielsen, eds., *Performance of Protective Clothing: Second Symposium*, American Society for Testing and Materials, Philadelphia, 1988, pp. 605–610.

198. J. F. Stone and H. M. Stahr, *J. Environ. Hlth*, **51**(5), 273–276 (1989).

199. H. E. Braun, R. Frank, and G. M. Ritcey, *Bull. Envir. Contam. Toxic.* **44**(1), 92–99 (1990).

200. J. P. McBriarty and N. W. Henry, eds. *Performance of Protective Clothing: Fourth Volume*, American Society for Testing and Materials, Philadelphia, Pennsylvania, ASTM STP 1133, 1992.

201. Y. Yang and S. Li, *Arch. Envir. Contam. Toxicol.* **25**(2), 279–284 (1993).

202. J. E. Midtling, P. G. Barnett, M. J. Coye, A. R. Velasco, P. Romero, C. L. Clements, M. A. O'Malley, M. W. Tobin, T. G. Rose, and I. H. Monosson, *West. J. Med.* **142**(4), 514–518 (1985).

203. D. P. Morgan, *Recognition and Management of Pesticide Poisonings*, 4th ed., Health Effects Division, U.S. Environmental Protection Agency, U.S. Government Printing Office, Washington, DC, 1989.

204. R. C. Spear, W. Popendorf, W. F. Spencer, and T. H. Milby, *J. Occup. Med.* **19**(6), 411–414 (1977).

205. T. H. Milby, F. Ottoboni, H. W. Mitchell, *J. Am. Med. Assoc.* **189**(5), 351–356 (1964).

206. W. J. Popendorf and J. T. Leffingwell, *J. Agric. Food. Chem.* **26**(2), 437–441 (1978).
207. W. Popendorf, *Rev. Environ. Contam. Toxicol.* **128**, 71–117, 1992.
208. P. Lisi, *Clinics in Dermatology*, **10**(2), 175–184 (1992).
209. K. Abrams, D. J. Hogan, and H. I. Maibach, *Occup. Med. State of the Art Reviews* **6**(3), 463–492 (1991).
210. P. F. J. Vogelzang, W. J. van der Gulden, L. Preller, M. J. M. Tielen, C. P. van Schayck, and H. Folgering, *Intr. Arch. Occup. Envir. Hlth.* **70**(5), 327–333 (1997).
211. R. J. Geller, *Vet. Human Toxicol.* **32**(5), 479–480 (1990).
212. J. R. Wilkins, III, and M. E. Bowman, *Occup. Med.* **47**(8), 451–457 (1997).
213. W. W. Spink, *Brucella Abortus* **47**, 861–873 (1957).
214. I. Z. Trujillo, A. N. Zavala, J. G. Caceres, and C. Q. Miranda, *Infectious Disease Clinics of North America*, **8**(1), 225–241 (1994).
215. A. H. Keeney and M. C. Hunter, *Arch. Opthalmol.* **44**, 573–580 (1950).
216. U. W. Leavell, Jr., M. S. McNamara, R. Muelling, et al., *J. Am. Med. Assoc.* **204**, 109–116 (1968).
217. T. Grandin, *Vet. Clin. North Am. Food Anim. Pract.* **3**, 324–336 (1987).
218. R. W. Lyons, C. L. Samples, H. N. DeSilva et al., *J. Am. Med. Assoc.* **243**, 546–547 (1980).
219. P. N. Acha and B. Szyfres, *Zoonoses and Communicable Diseases Common to Man and Animals*, Scientific Publication No. 354, Pan American Health Organization, Washington DC, 1980.
220. D. Snashall, *Br. Med. J.* **313**(7056), 551–554 (1996).
221. K. J. Donham, "Infectious Diseases Common to Animals and Man of Occupational Significance to Agricultural Workers," in *Proceedings of Conference on Agricultural Health and Safety*, New York Society for Occupational and Environmental Health, Environmental Sciences Laboratory, New York, 1975, pp. 160–175.
222. K. Hedberg, K. White, J. Forfang, et al., *Am. J. Epidemiol.* **130**, 569–577 (1989).
223. P. R. Schnurrenberger and W. T. Hubbert, *An Outline of Zoonoses*, Iowa State University Press, Ames, IA, 1981.
224. D. Hogan and P. Lane, *Occup. Med. State Art Rev.* **1**, 285–300 (1986).
225. A. Cellini and A. Offidani, *Dermatology* **189**(2) 129–132 (1994).
226. M. A. O'Malley, *Occup. Med. State of the Art Reviews* **12**(2), 327–345 (1997).
227. C. L. Wand, *The Problem of Skin Diseases in Industry*, Office of Occupational Safety and Health Statistics, U.S. Department of Labor, U.S. Government Printing Office, Washington, DC, 1978.
228. W. B. Whiting, *J. Occup. Med.* **17**(3), 177–181 (1975).
229. J. C. TeLintum and J. P. Nater, *Dermatologica* **148**, 42–44 (1974).
230. D. Burrows, *Br. J. Dermatol.* **92**, 167–170 (1975).
231. R. D. Peachey, *Br. J. Dermatol.* **105**(Suppl. 21), 45–50 (1981).
232. J. C. Whitakar, W. R. Lee, and J. E. Downes, *Br. J. Ind. Med.* **36**, 43–51 (1979).
233. D. T. Harvey and D. J. Hogan, "Common Environmental Dermatoses." in S. M. Brooks, M. Gochfeld, J. Herzstein, R. J. Jackson, and M. B. Schenker, eds., *Environmental Medicine*, Mosby-Year Book, Inc., St. Louis, MO, 1995, pp. 263–281.
234. U. W. Leavell and J. A. Phillips, *Arch. Dermatol.* **111**, 1307–1311 (1975).
235. L. Chmel, J. Buchvald, and M. Valentova, *Int. J. Epidemiol.* **5**(3), 291–295 (1976).

236. W. L. Krinsky, *Int. J. Dermatol.* **22**(2), 75–91 (1983).
237. M. R. Reesal, L. Hagel, P. Pahwa, D. Domoney, H. McDuffie, and J. A. Dosman, "Hearing Loss in a Saskatchewan Farm Community" in H. H. McDuffie, J. A. Dosman, K. M. Semchuk, S. A. Olenchock, and A. Senthilselvan, eds., *Agricultural Health and Safety: Workplace, Environment, Sustainability, Supplement*, University of Saskatchewan, Saskatoon, Saskatchewan, Canada, 1994, pp. 201–206.
238. M. Feldman and C. D. E. Downing, *Can. Agric. Eng.* **14**(1), 2–5 (1972).
239. H. H. Jones and J. L. Oser, *Am. Ind. Hyg. Assoc. J.* **29**(2), 146–151 (1968).
240. D. Tharr, *Appl. Occup. Envir. Hyg.* **9**(8), 525–528 (1994).
241. N. W. Sullivan, R. D. Schneider, and K. Von-Bargen, *Prof. Safety* **26**(12), 16–21 (1981).
242. J. W. Dennis and J. J. May, "Occupational Noise Exposure in Dairy Farming" in H. H. McDuffie, J. A. Dosman, K. M. Semchuk, S. A. Olenchock, and A. Senthilselvan, eds., *Agricultural Health and Safety: Workplace, Environment, Sustainability*, Lewis Publishers, Boca Raton, FL, 1995, pp. 363–367.
243. J. J. Holt, S. K. Broste, and D. A. Hansen, *Laryngoscope* **103**(3), 258–262 (1993).
244. C. E. McJilton and R. A. Aherin, *Am. Ind. Hyg. Assoc. J.* **43**(6), 469–471 (1982).
245. R. C. Jensen, *Prof. Safety*, **28**(9), 19–24 (1983).
246. H. Une, S. H. Schuman, S. T. Caldwell, and N. H. Whitlock, *South. Med. J.* **80**(9), 1137–1140 (1987).
247. C. W. Suggs, L. F. Stikclcathcr, and C. F. Abrams, *Am Soc. Agric. Eng.* **13**(5), 608–611 (1970).
248. M. Takamatsu, T. Sakurai, and C. P. Chang, "Vibration Disease among Farmers Induced by Vibration Tools," *Proceedings of the VII International Congress of Rural Medicine*, Salt Lake City, Utah, Sept. 17–21, 1978, *Int. Assoc. Agric. Med.*, 188–190 (1978).
249. W. E. Field and R. L. Tormoehlen, *Appl. Ergonomics* **16**(3), 179–182 (1985).
250. P. B. Allen, W. E. Field, and M. J. Frick, *J. Agric. Safety Hlth.*, **1**(2), 71–81 (1995).
251. L. F. Burmeister, G. D. Everett, S. Van Lier, and P. Isacson, *Am. J. Epidemiol.* **118**, 72–77 (1983).
252. A. Blair, H. Malker, K. P. Cantor, L. Burmeister, and K. Wiklund, *Scand. J. Work Environ. Hlth.* **11**, 397–407 (1985).
253. M. Schenker and S. McCurdy, "Pesticides, Viruses, and Sunlight in the Etiology of Cancer Among Agricultural Workers," in C. E. Becker and M. J. Coye, eds., *Cancer Prevention: Strategies in the Workplace*, Hemisphere Publishing, Washington, DC, 1986, pp. 29–37.
254. K. P. Cantor, A. Blair, G. Everett, R. Gibson, L. F. Burmeister, L. M. Brown, L. Schuman, and F. R. Dick, *Cancer Res.* **52**(9), 2447–2455 (1992).
255. D. D. Weisenburger, *Ann. Oncology* **5**(Supp. 1), S19–S24 (1994).
256. J. E. Keller-Byrne, S. A. Khuder, and E. A. Schaub, *Envir. Res.* **71**(1), 1–10, 1995.
257. L. W. Figgs, M. Dosemeci, and A. Blair, *Am. J. Ind. Med.* **27**(6), 817–835 (1995).
258. B. Lee, B. Marlenga, and D. Miech, "Farmers' Caps and Hats Project" in H. H. McDuffie, J. A. Dosman, K. M. Semchuk, S. A. Olenchock, and A. Senthilselvan, eds.; *Agricultural Health and Safety. Workplace, Environment, Sustainability*, CRC Press Inc., Boca Raton, FL, 1995, pp. 535–539.
259. K. J. Donham, L. Burmeister, S. vanLier, and T. Greiner, *Am. J. Vet. Res.* **48**, 235–238 (1987).
260. E. Elam and R. Rauncy, *Rural Health Crisis* (A Project Report). Northwest Services, Inc., Mound City, MO, 1986.

261. R. R. Swisher, G. H. Elder, F. O. Lorenz, and R. D. Conger, *J. Hlth. Soc. Behavior.* **39**(1), 72–89 (1998).

262. D. J. Donham. "The Long Arm of the Farm: How an Occupation Structures Exposure and Vulnerability to Stressors Across Role Domains," *Sustainable Development in Intensive Livestock Production: Exposure Assessment and Health Outcomes*, abs. 4th International Symposium: Rural Health and Safety in a Changing World, held at The Center for Agricultural Medicine, University of Saskatchewan on October 18–22, 1998 in Saskatoon, Canada.

263. W. E. Field, *Effects of Stress on the Performance of Agricultural Equipment Operators*, SAE Tech. Paper Series, 800932, Warrendale, PA, 1980.

264. L. M. Haverstock, "Farm Stress: Research Considerations," in J. A. Dosman and D. W. Cockcroft, eds., *Principles of Health and Safety in Agriculture*, CRC Press, Boca Raton, FL, 1989, pp. 381–384.

265. C. N. Larson, S. Kuperman, and R. E. Smith, "Rural Psychiatry: A Definition of the Field," in J. A. Dosman and D. W. Cockcroft, eds., *Principles of Health and Safety in Agriculture*, CRC Press, Boca Raton, FL, 1989, pp. 385–388.

266. W. G. Hollister, "Innovations in Mental Health Service Delivery in Rural Areas," in J. A. Dosman and D. W. Cockcroft, eds., *Principles of Health and Safety in Agriculture*, CRC Press, Boca Raton, FL, 1989, pp. 399–401.

267. R. Reotutar, *J. Am. Vet. Med. Assoc.* **194**, 1387–1391 (1989).

268. B. S. Schwartz and M. D. Goldstein, *J. Occup. Med.* **31**, 735–742 (1989).

269. P. Morgan-Capner, S. J. Cutler, D. J. M. Wright, N. Hamlet, D. Nathwani, D. O. Ho-Yen, and E. Walker, *Lancet*, **1**(8641), 789–790 (1989).

270. A. G. Baird, J. C. M. Gillies, F. J. Bone, B. A. S. Dale, and N. T. Miscampbell, *Br. Med. J.* **299**(6703), 836–837 (1989).

271. T. F. Tsai, *Lab. Animal Sci.* **37**(4), 428–430 (1987).

272. P. S. Zeitz, J. M. Graber, R. A. Vorrhees, G. Kioski, L. A. Shands, T. G. Ksiazek, S. Jenison, and R. F. Khabbaz, *J. Occup. Envir. Med.* **39**(5), 463–467 (1997).

273. P. E. M. Smith, M. Zeidler, J. W. Ironside, P. Estibeiro, and T. H. Moss, *Lancet* **346**(8979), 898 (1995).

274. N. Delasnerie-Laupretre, S. Poser, M. Pocchiari, D. P. W. M. Wientjens, and R. Will, *Lancet* **346**(8979), 898 (1995).

275. R. B. Hayes, J. P. Van Nieuwenhuize, J. W. Raatgever, and F. J. W. Ten Kate, *Food Chem. Toxicol.* **22**(1), 39–43 (1984).

276. W. G. Sorensen, J. P. Simpson, M. J. Peach, T. D. Thedell, and S. A. Olenchock, *J. Toxicol Environ. Hlth* **7**, 669–672 (1981).

277. O. Shotwell and W. Burg, *Ann. Am. Conf. Gov. Ind. Hyg.* **2**, 69–86 (1982).

278. M. I. Selim, A. M. Juchems, and W. Popendorf, *Am. Ind. Hyg. Assoc. J.* **59**(4), 252–256 (1998).

279. W. K. Viscusi, *Risk By Choice: Regulating Health and Safety in the Workplace*, Harvard University Press, MA, 1983.

280. C. M. Berry, *Am. J. Public Hlth*, **55**(3), 424–428 (1965).

281. M. A. Purschwitz and W. E. Field, "Federal Funding for Farm Safety Relative to Other Safety Programs," in *Proc. of the 1987 National Institute for Farm Safety Summer Meeting*, National Institute for Farm Safety (NIFS), Columbia, MO, 1987.

282. L. J. Chapman, R. T. Schuler, T. L. Wilkinson, and C. A. Skjolaas, *Am. J. Ind. Med.* **28**(4), 565–577 (1995).

283. D. J. Murphy, N. E. Kiernan, and L. J. Chapman, *Am. J. Ind. Med.* **29**(4), 392–396 (1996).

284. S. VonEssen, K. Thu, K. J. Donham, "Insurance Incentives for Safe Farms" in K. Donham, R. Rautiainen, S. Scheuman, and J. Lay, eds. *Agricultural Health and Safety: Recent Advances.* Hawthorne Press, Binghamton, NY, 1997, pp. 125–127.

285. J. Gay, K. Donham, S. Leonard, *Am. J. Ind. Med* **18**(4), 385–389 (1990).

286. S. Höglund, *Am. J. Ind. Med.* **18**(4), 371–378, 1990.

287. Ontario Farm Safety Association, *Am. J. Ind. Med.* **18**(4), 409–411 (1990).

288. K. Husman, V. Notkola, R. Virolainen, J. Nuutinen, K. Tupi, J. Penttinen, and J. Heikkonen, *Am. J. Ind. Med.* **18**(4), 379–384 (1990).

289. S. Höglund, *Am. J. Ind. Med.* **18**(4), 365–370 (1990).

290. D. S. Pratt, *Am. J. Ind. Med.* **18**(4), 391–393 (1990).

291. D. A. Emanuel, D. L. Draves, and G. R. Nycz, *Am. J. Ind. Med.* **18**(2), 149–162 (1990).

292. S. H. Schuman, *Am. J. Ind. Med.* **18**(4): 405–408 (1990).

293. C. Tevis, *Successful Farming* **87**(12), 17 (1989).

294. A. Thelin, *Ann. Agric. Envir. Med.* **2**(1), 21–26 (1995).

CHAPTER SIXTY-ONE

Hazardous Wastes

Lisa K. Simkins, PE, CIH

1 INTRODUCTION

This chapter provides an introduction and overview of hazardous waste management issues. The topics include major regulatory requirements pertaining to hazardous waste, waste characterization, management options, and health and safety considerations. Owing to the vast number of local, state, and federal regulations pertaining to this subject, the practicing professional must always consult the latest regulations before embarking on a hazardous waste management program. This chapter can be viewed as a starting point that provides basic background information for the industrial hygienist or environmental professional involved in one or more facets of hazardous waste management.

1.1 Definition of Hazardous Waste

The standard definition of hazardous waste comes from the Resource Conservation and Recovery Act (RCRA) and pursuant regulations, which describes it as a solid waste, or combination of solid wastes that exhibits the "characteristics of hazardous waste," as define in Subpart C of **40** *CFR* Part 261 or is listed as a hazardous waste in Subpart D of **40** *CFR* Part 261 and is not otherwise excluded from regulation as a hazardous waste (1). A characteristic of hazardous waste is identified upon determining that a solid waste may:

1. Cause, or significantly contribute to, an increase in mortality or an increase in serious irreversible, or incapacitating reversible, illness, or
2. Pose a substantial present or potential hazard to human health or the environment when improperly treated, stored, transported or disposed of, or otherwise managed (2).

Patty's Industrial Hygiene, Fifth Edition, Volume 4. Edited by Robert L. Harris.
ISBN 0-471-29749-6

Criteria for listing a solid waste as a hazardous waste include:

1. The waste exhibits any of the characteristics of hazardous waste.
2. It has been found to be fatal to humans in low doses or in absence of data on human toxicity, it has been shown in studies to have an oral LD_{50} toxicity (rat) of less than 50 milligrams per liter, an inhalation LC_{50} toxicity (rat) of less than 2 milligram per liter, or a dermal LD_{50} toxicity (rabbit) of less than 200 milligrams per kilogram or is otherwise capable of causing or significantly contributing to an increase in serious irreversible, or incapacitating reversible, illness.
3. It contains any of the toxic constituents listed in appendix VIII of **40** *CFR* Part 261 and is capable of posing a substantial present or potential hazard to human health or the environment when improperly treated, stored, transported or disposed of, or otherwise managed after considering a number of factors (3).

Under the broad RCRA definition, solid waste could include semisolids, liquids, and contained gases.

Characteristics of hazardous wastes, including ignitability, corrosivity, toxicity, or reactivity, are discussed further in Section 3.2.2.

The following types of waste are excluded from RCRA's definition of solid waste and are therefore not regulated as hazardous waste under RCRA (4):

- Domestic sewage and any mixture of domestic sewage and other wastes that pass through a sewer system to a publicly owned treatment works for treatment.
- Industrial wastewater point source discharges subject to regulation under section 402 of the Clean Water Act (CWA).
- Irrigation return flows.
- Source, special nuclear, or by-product radioactive materials as defined by the Atomic Energy Act of 1954, as amended.
- Materials subject to *in-situ* mining wastes that are not removed from the ground as part of the extraction process.
- Nonspeculative, reused pulping liquors reclaimed in a pulping liquor recovery furnace.
- Nonspeculative spent sulfuric acid used to produce virgin sulfuric acid.
- Secondary materials that are reclaimed and returned to the original process or processes in which they were generated, where they are reused in the production process, under certain conditions.
- Spent wood preserving solutions that have been reclaimed and are reused for their original intended purpose.
- Certain iron and steel industry wastes and wastes from coke by-product processes, under certain reuse conditions.
- Nonwastewater splash condenser dross residue from certain high temperature metals recovery units.

- Recovered oil from petroleum refining, exploration, and production that will be reused in the refining process at a stage prior to removal of contaminants under certain conditions.
- Excluded scrap metal being recycled.
- Shredded circuit boards being recycled under certain conditions.
- Secondary materials generated within the primary mineral processing industry from which minerals, acids, cyanide, water or other values are recovered by mineral processing, under certain conditions.
- Comparable fuels or comparable syngas fuels under certain conditions.

Wastes that are RCRA solid wastes, but that are not hazardous wastes include the following:

- Household wastes (e.g., garbage, trash, septic tank wastes).
- Solid wastes generated by the growing and harvesting of agricultural crops or raising of animals when the wastes are returned to the soil as fertilizer.
- Mining overburden returned to the mine site.
- Fly ash waste, bottom ash waste, slag waste, and flue-gas emission control waste generated primarily from the combustion of coal or other fossil fuels except under certain conditions.
- Drilling fluids, produced waters, and other wastes associated with the exploration, development, or production of crude oil, natural gas or geothermal energy.
- Wastes that fail the test for the toxicity characteristic solely because of chromium, provided the chromium used and waste chromium produced is exclusively trivalent chromium and the wastes are typically and frequently managed in nonoxidizing environments.
- Specified solid waste from the extraction, beneficiation, and processing of ores and minerals, including coal when handled using certain methods.
- Cement kiln dust waste, except under certain conditions.
- Solid waste consisting of discarded arsenical-treated wood or wood products which fails the toxicity characteristic test only because of arsenic when the product is used for the materials' intended end use.
- Petroleum-contaminated media and debris that fail the toxicity characteristic test for hazardous waste codes D018 through D043 and are subject to corrective action regulations under RCRA underground storage tank regulations.
- Injected groundwater that is hazardous only because it fails the toxicity characteristic test for hazardous waste codes D018 through D043 under certain conditions.
- Used chlorofluorocarbon refrigerants from totally enclosed heat transfer equipment, provided the refrigerant is reclaimed for further use.
- Non-terne plated used oil filters that are not mixed with other hazardous wastes under certain conditions.

- Used oil re-refining distillation bottoms that are used as feedstock to manufacture asphalt products.

Other regulatory exclusions under RCRA include:

- Hazardous wastes which are generated in certain specified raw material processes are exempted from certain RCRA requirements until they exit the unit in which they are generated under specified conditions.
- Samples of water, soil, or air, which are collected for the sole purpose of testing to determine their characteristics or composition are exempted from certain RCRA requirements under specified circumstances.
- Samples collected for the purpose of treatability studies are exempted from certain RCRA requirements under specified circumstances.
- Samples undergoing treatability studies are exempted from certain RCRA requirements under specified circumstances.

Determination of whether a waste is defined as a "RCRA Hazardous Waste" is not always straightforward. Many more details pertaining to the above mentioned exclusions are included in the RCRA regulation. States and local governments often have more restrictive or different requirements. Therefore, the full text of the latest revisions of the applicable federal, state, and local regulations should be consulted to make a determination of whether a waste should be treated as hazardous.

1.2 Regulatory and Legal Liability

Enforcement of regulations and concern for liability have heightened industry's awareness and have resulted in more prudent practice relating to hazardous waste in recent years. The financial liabilities caused by environmental problems can be significant. Groundwater decontamination programs, involving extraction and treatment can be a significant cost and take many years to complete (i.e., obtain regulatory closure).

Pollution liabilities are typically strict (meaning that liability is imposed regardless of fault), and also joint and several (meaning that liability can be imposed on any or all responsible parties independent of their relevant contribution to the pollution). The United States Environmental Protection Agency (EPA) enforcement branch as well as state and local prosecutors may in some cases file criminal charges against company officers for causing environmental pollution and workplace injury or death.

It is the responsibility of generators and handlers to be familiar with regulations governing their operations, and to institute practices necessary to remain in compliance. Industrial activities operated without knowledge of waste management regulations create an unnecessary risk to human health and the environment.

2 REGULATORY REQUIREMENTS

Environmental laws and regulations originally focused on controlling air and water pollution. Subsequently, the focus expanded to include releases of toxic chemicals and disposal of hazardous waste. RCRA's definition of hazardous waste brought many industrial and commercial operations into the regulatory arena.

Environmental issues are a daily factor in industrial operations and business transactions. Businesses must consider federal, state, and local regulatory requirements in order to achieve compliance at each facility. These range from requirements for storing, treating and disposing of hazardous waste in accordance with RCRA and other hazardous waste regulations, to the recognition of due diligence standards that limit potential liabilities under the Superfund Amendments and Reauthorization Act (SARA).

A better understanding of the requirements can be gained with an understanding of the framework of the federal rulemaking system. Environmental laws are developed by Congress and signed into law by the president. These laws mandate that the EPA promulgate regulations to implement the law.

New regulations, as well as modifications to existing regulations, are printed in the Federal Register. The Code of Federal Regulations (CFR) is modified periodically to take into account changes promulgated in the Federal Register. The CFR is divided into 50 titles covering the areas subject to federal regulation; Title 40 deals with protection of the environment.

2.1 Resource Conservation and Recovery Act

The Resource Conservation and Recovery Act of 1976 (RCRA) marked the beginning of what has become the EPA's most far-reaching and complex regulatory program. RCRA tracks hazardous wastes from "cradle to grave." Generators of hazardous wastes are responsible for analyzing the material, maintaining appropriate records, submitting periodic reports of volumes, and submitting reports on offsite shipments for disposal. It took the EPA four years following the enactment of RCRA to promulgate the basic regulatory framework, which in turn has been amended many times since its initial publication in May 1980 (5). The 1980 regulations established permit requirements for treatment, storage, and disposal (TSD) facilities handling hazardous wastes but granted "interim status" to allow those facilities to operate, if appropriate applications and notices were submitted. Groundwater monitoring and financial responsibility are two of numerous other requirements imposed on such facilities by these regulations.

As discussed in the beginning of this chapter, RCRA provided a standard definition for hazardous waste in the United States. The EPA then had to determine the scope of the regulated community, and what areas of hazardous waste management were to be subject to these requirements. Table 61.1 provides the titles of the major regulatory sections found in Title 40 of the CFR that implement RCRA hazardous waste requirements.

RCRA's "cradle-to-grave" approach provides guidelines for generators, transporters, storage facilities, treatment operations, and disposal sites. Each of these waste management activities is subject to specific standards under these regulations.

2.1.1 Generators

A generator is any person, by site, whose act or process produces hazardous waste or whose act first causes a waste to be subject to regulation (2). All generators must determine

Table 61.1. Sections of 40 CFR Implementing RCRA Hazardous Waste Related Requirements

Part No.	Title
260	Hazardous Waste Management System: General
261	Identification and Listing of Hazardous Waste
262	Standards Applicable to Generators of Hazardous Waste
263	Standards Applicable to Transporters of Hazardous Waste
264	Standards For Owners and Operators of Hazardous Waste Treatment, Storage, and Disposal Facilities
265	Interim Status Standards For Owners and Operators of Hazardous Waste Treatment, Storage, and Disposal Facilities
266	Standards for the Management of Specific Hazardous Wastes and Specific Types of Hazardous Waste Management Facilities
268	Land Disposal Restrictions
270	EPA Administered Permit Program: The Hazardous Waste Permit Program
271	Requirements for Authorization of State Hazardous Waste Programs
272	Approved State Hazardous Waste Management Programs
273	Standards for Universal Waste Management
279	Standards for the Management of Used Oil
280	Technical Standards and Corrective Action Requirements for Owners and Operators of Underground Storage Tanks (UST)
281	Approval of State Underground Storage Tank Programs
282	Approved Underground Storage Tank Programs

if their waste is hazardous in accordance with the protocol described in **40** *CFR* 262.11. Other requirements depend on the amount of hazardous waste generated per calendar month. There are three categories of generators as described below.

- Large quantity generators (LQG) produce 1,000 kilograms (2,200 pounds) or more per month.
- Small quantity generators (SQG) produce greater then 100 kilograms (220 pounds) but less than 1,000 kilograms per month.
- Conditionally exempt small quantity generators (CESQG) produce no more than 100 kilograms per month.

The RCRA requirements that apply to LQG and SQG are detailed in **40** *CFR* 262. Requirements for conditionally exempt small quantity generators are described in 40 CFR 261.5. A summary of the requirements for each category generator is provided in Table 61.2.

2.1.2 Transporters

All persons transporting hazardous waste off the site where it was generated must comply with the requirements in **40** *CFR* 263. EPA adopted certain regulations of the Department

Table 61.2. Requirements for Hazardous Waste Generators

Requirements	Conditionally Exempt Small Quantity Generator (CESQG)	Small Quantity Generator (SQG)	Large Quantity Generator (LQG)
Quantity limits	<220 pounds	220 to 2,200 pounds	>2,200 pounds
EPA ID number	Not required	Required	Required
Uniform hazardous waste manifest	Not federally required[a]	Required	Required
Exception report	Not federally required[b]	Required within 60 days	Required within 45 days
Biennial report	Not federally required[b]	Not federally required[b]	Required
Personnel training	Not federally required[b]	Basic training required	Required
Contingency plan	Not federally required[b]	Basic plan required	Required
On-site accumulation quantity limits	<2,200 pounds	<13,200 pounds	No limit
On-site accumulation time limits	No limit	≤180 days or ≤270 days (if TSDF is over 200 miles away)	≤90 days unless EPA grants extension (+30 days)
Storage requirements	Not federally required[b]	Basic requirements with technical standards for tanks and containers	Full compliance with management of tanks and containers
Off-site management of waste	State approved facility	RCRA permitted facility	RCRA permitted facility

[a]Transporters or state may require manifest.
[b]State may require.

of Transportation (DOT) thus ensuring consistency between EPA and DOT requirements. The regulations apply to transport of hazardous waste within the United States if a manifest is required for transport under **40** *CFR* 262.20. Waste transporter requirements include:

- Obtain an EPA identification number.
- Comply with the manifest signature requirement.
- Deliver the waste in accordance with directions on the manifest.
- Maintain records of the waste shipment (e.g., manifest or shipping papers).
- Take appropriate immediate action in response to a hazardous waste discharge to protect human health and the environment.
- Clean up hazardous waste discharges.

2.1.3 *Waste Management Facilities*

Any person who owns or operates a facility which treats, stores, or disposes of hazardous waste is considered an owner or operator of a treatment, storage, and disposal (TSD) facility and is subject to the requirements of **40** *CFR* Part 264 or 265. General facility standards applicable to TSD facilities, address:

- EPA identification number.
- Required notices.
- Waste analysis.
- Security.
- Inspection requirements.
- Personnel training.
- General requirements for ignitable, reactive, or incompatible wastes.
- Location standards.
- Construction quality assurance program.

Other TSD facility standards contained in EPA regulations address the following issues:

- Preparedness and prevention.
- Contingency plan and emergency procedures.
- Manifest system, recordkeeping, and reporting.
- Releases from solid waste management units and required response actions.
- Closure and post-closure.
- Financial requirements.
- Air emission standards for various situations and processes.
- Containment buildings.
- Hazardous waste munitions and explosives storage.

Standards for owners and operators of TSD facilities also include technical standards applicable to waste management units, such as container storage units, tank systems, surface impoundments, waste piles, land treatment, landfills, and incinerators.

The basic difference between Part 264 (Final Operating Standards) and Part 265 (Interim Status Standards) is that the interim status standards were written for facilities treating, storing, or disposing of RCRA hazardous waste when the RCRA regulations first went into effect on November 19, 1980. The EPA was required to establish interim standards that would allow facilities to continue to operate as though they had a permit. A facility received interim status by filing a RCRA Section 3010 notification (Notification of Hazardous Waste Activity) and a Part A application. An owner or operator that qualified for and obtained interim status remains subject to the **40** *CFR* 265 standards until the final administrative disposition of the facility's permit application is made. At that point **40** *CFR* 264 (Final Operating Standards) goes into effect.

The general facility standards are applicable to all RCRA hazardous waste management facilities unless specifically excluded. The general compliance standards are essentially the same for both parts and include the following basic requirements:

- Every facility must have an EPA identification number.
- The owner or operator of a facility receiving hazardous waste from an offsite source must inform the waste generator that the facility has obtained appropriate permits.
- Before hazardous waste is treated, stored, or disposed of, the owner or operator must obtain a detailed chemical and physical analysis of a representative sample of the waste.
- The owner or operator must prevent inadvertent and, to the extent possible, deliberate entry of people or livestock into the active portion of the facility.
- The owner or operator must inspect his facility for malfunctions and deterioration, operator errors, and discharges, which may be causing or may lead to (*1*) release of hazardous waste constituents to the environment or (*2*) a threat to human health. These inspections must be done in accordance with a written schedule and must be recorded in an inspection log.
- Facility personnel must complete a program of training in procedures and emergency response to allow them to perform their duties in accordance with RCRA requirements. A written description of the training program, attendees, and instructors must be kept at the facility.
- The owner or operator must take precautions to prevent accidental ignition or reaction of ignitable or reactive waste. Precautions may include separating wastes and protecting them from sources of ignition or reaction.
- All permitted facilities must have equipment adequate to deal with unexpected fires, explosions, or any unplanned sudden or nonsudden spill or other release of hazardous waste.
- Emergency response procedures must be detailed in a contingency plan and must describe the personnel in charge of emergency, emergency equipment, and evacuation procedures.

In addition to these general requirements, other requirements apply to specific permitted facilities and types of equipment.

Once the facility no longer accepts wastes, a period of closure begins. Operators are required to develop a closure plan. The plan must include a schedule for closure, an estimate of the amount of wastes the facility handled at any time, and the necessary steps of closure. Closure will typically include removing hazardous wastes and decontaminating facility components, equipment, or structures. Testing of surrounding soils to determine the extent of decontamination needed is also done during closure. Other closure activities may include groundwater monitoring, leachate collection and precipitation control. Though the closure plan may be amended during the active life of the facility, once implemented, it must be strictly followed.

Following closure, postclosure requirements apply for units at which hazardous wastes or residues remain after closure. During the postclosure period, typically continuing for

30 years, the facility must continue groundwater monitoring and any maintenance activities necessary to preserve the integrity of the site. TSD facilities are also required to prepare cost estimates and provide financial assurance of ability to implement the closure and post-closure plans.

The full text of the regulation should be consulted by anyone responsible for complying with, auditing, or enforcing requirements applicable to TSD facilities.

2.3 Comprehensive Environmental Response, Compensation, and Liability Act

The enactment of the Comprehensive Environmental Response, Compensation, and Liability Act of 1980, better known as CERCLA or Superfund, brought environmental concerns to the forefront. By passing CERCLA, Congress authorized funds for the EPA to clean up abandoned dumps and other contaminated waste disposal sites.

The statute imposed strict, and joint and several liability for the cleanup costs on past and present site owners, past and present site operators, offsite generators who had arranged for disposal of hazardous wastes, and transporters who selected the site in question. In an effort to prevent future Superfund sites, Congress also imposed broad new reporting requirements for releases of hazardous substances into the environment.

Because the Superfund law is not fundamentally a regulatory statute, the primary obligation it imposes is liability for cleanup costs. Most Superfund sites require a full Remedial Investigation/Feasibility Study (RI/FS), which is typically very time consuming and costly. Even higher costs are encountered in the Remedial Design/Remedial Action (RD/RA) phase of site cleanup.

2.4 Superfund Amendments and Reauthorization Act

On October 17, 1986, when President Ronald Reagan signed the Superfund Amendments and Reauthorization Act of 1986 (SARA). It provided $8.5 billion over five years to the EPA and other federal agencies for the cleanup of abandoned and inoperative waste sites.

The amendments made major changes to the original Superfund law (CERCLA), including:

- Revisions that added strict cleanup standards strongly favoring permanent remedies at waste sites.
- Stronger EPA control over the process of reaching settlement with parties responsible for waste sites.
- A mandatory schedule for initiation of cleanup work and studies.
- Individual assessments of the potential threat to human health posed by each waste site.
- Increased state and public involvement in the cleanup decision-making process, including the right of citizens to file lawsuits for violation of the law.

The amended law retained the concept of strict, and joint and several liability, which EPA has found to be its most powerful enforcement tool for inducing responsible parties to clean up waste sites. During the reauthorization process, one of the key problems confronted was the lack of workable cleanup standards. SARA placed emphasis on remedial actions that permanently and significantly reduce the volume, toxicity, or mobility of the hazardous substances, pollutants, and contaminants. The discovery that disposal sites where wastes from Superfund sites had been dumped were also leaking led to language stating that offsite transport of wastes without treatment "should be the least favored alternative" where practical treatment technologies are available.

The revised law also contained two new separate sections not directly related to the task of cleaning up Superfund sites, but with important implications for the EPA's overall regulatory authority. A separate title within SARA requires industries that produce, use, or store hazardous chemicals or substances to report the presence of such substances to community authorities and to report routine and unauthorized releases of hazardous substances to the EPA. It also requires communities to improve emergency planning procedures for major chemical accidents.

A second section amended RCRA to require that owners of underground storage tanks take financial responsibility for cleaning up leaks and compensating third parties for property damage and bodily injury. A trust fund was established to pay for emergency cleanups where no responsible owner or operator of the tank could be found.

Finally, Superfund cleanup liability was extended to all owners of contaminated property, regardless of the circumstances of their ownership. CERCLA's liability standard had triggered several problems for property owners who took possession of land without knowledge that it was contaminated. Prompted by the perceived unfairness, Congress included a new defense in the amendments that was designed to protect innocent landowners.

Congress expanded the third-party exception by redefining the contractual relationship of property ownership. Innocent landowners who acquire property without having had reason to know that hazardous substances had ever been disposed of on the land are not liable as owners or operators. The landowner does not qualify as innocent unless he or she has completed a thorough examination of the land purchased. As a result, environmental site assessments are typically performed prior to the purchase of property. A standard for preliminary environmental site assessments was established by ASTM to help improve the consistency of environmental due diligence (6).

2.5 Other Regulations

Another major federal regulation pertains to health and safety when handling hazardous wastes. This regulation, the *Hazardous Waste Operations and Emergency Response Standard* (**29** *CFR* 1910.120), is discussed in Section 4.0.

In addition to the major federal regulations, there are numerous state and local regulations pertaining to hazardous wastes. State regulations vary in their application and are frequently updated. The federal regulations should therefore be considered to be minimum standards that will apply and additional regulations should be consulted in each state, county, and city in which a company operates a facility or disposes of waste.

3 WASTE MANAGEMENT

Numerous regulatory requirements govern what a generator of hazardous waste is allowed to do. As the preceding sections indicate, hazardous waste management can be a complex issue. This section provides a more detailed explanation of hazardous waste management programs.

The first step in the development of a sound hazardous waste management program is a comprehensive inventory of all solid and hazardous wastes generated at a facility. Following an inventory, waste must be characterized. At that point the various management and disposal options available to the generator can be evaluated.

3.1 Inventory

A complete list of all waste streams and sources (processes that result in the generation of waste) needs to be created. To the extent known, constituents of individual waste streams must be identified at the outset. This will be helpful in the characterization process, which can be time consuming and expensive.

The discussion in the following sections focuses on the hazardous waste component of solid wastes. However, all solid waste streams should be included in the inventory process. This should result in a clear distinction between hazardous and nonhazardous solid wastes.

In addition to identifying the chemical nature and source of all waste streams, a quantification must be made. The quantity may be computed on a daily basis (e.g., gallons per day), a monthly basis (e.g., kilograms per month), or an annual basis (e.g., tons per year). If the generator is a conditionally exempt small quantity generator (SQG) with monthly total hazardous waste generation of less than 100 kilograms (220 lb.) monthly quantities should be tracked to determine whether SQG status is maintained.

3.2 Characterization

Other than the obvious municipal garbage and household debris, prudent strategy is to assume a waste stream is hazardous until tests or specific exclusions prove otherwise. The characterization process is accomplished in one of several ways. A waste from a specific process may be designated hazardous by the EPA or a state environmental agency. These are known as listed wastes. A waste that is not listed by these agencies can be characterized through a series of tests.

3.2.1 Listed Wastes

EPA has listed some specific solid wastes as hazardous wastes based on criteria described in **40** *CFR* 261.11 (3). Listed hazardous wastes include the following categories of waste:

- Nonspecific sources (F codes).
- Specific sources (K codes).
- Commercial chemical products—acutely hazardous (P codes).
- Commercial chemical products—nonacutely hazardous (U codes).

Federally listed wastes are found in Subpart D of **40** *CFR* 261. Some states have generated their own lists. These lists are revised periodically and it is important to consult the latest listing of hazardous waste in a given state.

If a generator produces a listed waste but there is reason to believe that it is not hazardous, the generator or any other person may petition for a regulatory amendment. The petitioner must demonstrate that the waste does not exhibit any of the characteristics of hazardous waste by testing in accordance with prescribed analytical methods. A complete description of the information required in each petition is detailed in **40** *CFR* 260.20 and **40** *CFR* 260.22. If the petitioner successfully demonstrates that the waste is not hazardous, the EPA may issue a regulatory amendment to exclude the listed waste for that particular generating, storage, treatment, or disposal facility only.

3.2.2 Characteristics Criteria

A quantitative approach that can be applied directly to wastes that are not otherwise listed or excluded is the testing of characteristics. Characteristics are measured by standard testing protocol. When the characteristic exceeds a set threshold value, the material is designated hazardous.

When the hazardous status of a waste is being determined, tests are conducted for the characteristics of ignitability, corrosivity, reactivity, and toxicity. A solid waste that exhibits any of these characteristics is a hazardous waste, whether it is listed or not. For example, a waste could be subjected to a test to determine its flash point (ignitability characteristic). If the waste was determined to have a flash point of 140°F or less, it would be designated hazardous. These characteristics are described in detail in **40** *CFR* Parts 261.21 through 261.24 and are summarized below.

3.2.2.1 Ignitability. Waste meeting the ignitability characteristic is designated by an EPA hazard code "I" and has the EPA Hazardous Waste Number of D001. A solid waste exhibits the characteristic of ignitability if a representative sample of the waste is as follows (7):

1. A liquid other than an aqueous solution containing less than 24% alcohol by volume with a flash point less than 60°C (140°F).
2. A nonliquid capable under standard temperature and pressure of causing fire through friction, absorption of moisture or spontaneous chemical change, and, when ignited, burns so vigorously and persistently that it creates a hazard.
3. Ignitable compressed gas.
4. An oxidizer.

3.2.2.2 Corrosivity. Waste meeting the corrosivity characteristic is designated by an EPA Hazard Code "C" and has the EPA Hazardous Waste Number of D002. A solid waste exhibits the characteristic of corrosivity if a representative sample of the waste is (8):

1. A liquid with a pH less than or equal to 2 or greater than or equal to 12.5.
2. A liquid that corrodes steel at a rate greater than 6.35 mm (0.250 in.) per year at a test temperature of 55°C/130°F.

3.2.2.3 Reactivity. A solid waste exhibits the characteristic of reactivity if a representative sample of the waste has any of the following properties (9):

1. Normally is unstable and readily undergoes violent change without detonating.
2. Reacts violently with water.
3. Forms potentially explosive mixtures with water.
4. When mixed with water, it generates toxic gases, vapors, or fumes.
5. Contains cyanide or sulfide and generates toxic gases, vapors, or fumes between pH 2 and 12.5.
6. Capable of detonation or explosive reaction if it is heated under confinement or subjected to strong initiating source.
7. Readily capable of detonation or explosive decomposition or reaction at standard temperature and pressure.
8. Is defined as a forbidden explosive or a Class A or B explosive by the Department of Transportation (DOT).

Waste meeting the reactivity characteristic is designated by an EPA hazard code "R" and has the EPA Hazardous Waste Number of D003.

3.2.2.4 Toxicity. The characteristic of toxicity is determined using the Toxicity Characteristic Leaching Procedure (TCLP) Method 1311 in "Test Methods for Evaluating Solid Waste, Physical/Chemical Methods, EPA Publication SW 846 (10)." TCLP measures the leaching potential of toxic constituents, and is designed to determine the mobility of organic and inorganic contaminants in liquid, solid, and multiphasic wastes. If one or more of the constituents found in Table 61.3 can be leached from a waste in concentrations greater than the specified levels, the waste is regulated as hazardous. Waste meeting the toxicity characteristic is assigned a hazard code "E" and has EPA Hazardous Waste Numbers as shown in Table 61.3.

3.3 Management Options

Once components of the waste streams have been defined, the generator is in a position to evaluate treatment and management options. This section addresses some of the more commonly used management methods, including waste minimization, incineration, chemical and biological treatment, and land disposal.

3.3.1 Land Disposal

Solid waste, including hazardous wastes, have long been disposed of in landfills. Essentially, this amounts to long-term storage of the waste in a disposal site. Landfilling has

Table 61.3. Toxicity Characteristic Constituent, EPA Hazardous Waste Numbers, and Regulatory Levels

EPA HW No.[a]	Constituent	Regulatory Level (mg/L)
D004	Arsenic	5.0
D005	Barium	100.0
D018	Benzene	0.5
D006	Cadmium	1.0
D019	Carbon tetrachloride	0.5
D020	Chlordane	0.03
D021	Chlorobenzene	100.0
D022	Chloroform	6.0
D007	Chromium	5.0
D023	*o*-Cresol	200.0[b]
D024	*m*-Cresol	200.0[b]
D025	*p*-Cresol	200.0[b]
D016	2,4-D	10.0
D027	1,4-Dichlorobenzene	7.5
D028	1,2-Dichloroethane	0.5
D029	1,1-Dichloroethylene	0.7
D030	2,4-Dinitrotoluene	0.13[c]
D012	Endrin	0.02
D031	Heptachlor (and its hydroxide)	0.008
D032	Hexachlorobenzene	0.13[c]
D033	Hexachlobutadiene	0.5
D034	Hexachloroethane	3.0
D008	Lead	5.0
D013	Lindane	0.4
D009	Mercury	0.2
D014	Methoxychlor	10.0
D035	Methyl ethyl ketone	200.0
D036	Nitrobenzene	2.0
D037	Pentachlorophenol	100.0
D038	Pyridine	5.0[c]
D010	Selenium	1.0
D011	Silver	5.0
D039	Tetrachloroethylene	0.7
D015	Toxaphene	0.5
D040	Trichloroethylene	0.5
D041	2,4,5-Trichlorophenol	400.0
D042	2,4,6-Trichlorophenol	2.0
D017	2,4,5-TP (Silvex)	1.0
D043	Vinyl chloride	0.1

[a]Hazardous waste number

[b]If *o*-, *m*-, and *p*-Cresol concentrations cannot be differentiated, the total cresol (D026) concentration is used. The regulatory level for total cresol is 200 mg/L.

[c]Quantitation limit is greater than the calculated regulatory level. The quantitation limit therefore becomes the regulatory level.

been a favored method of disposal because of its relatively inexpensive cost. However, the available space in operating landfills is limited and approval for new sites is difficult to obtain. Real estate costs and strict federal and state regulations affect the cost of operation. In addition, under federal regulations, generators who contribute waste to a landfill can be held partially liable for any resulting contamination. RCRA also bans some hazardous wastes from land disposal. Finally, the landfill option does not solve the problem of hazardous waste, it only stores and, if designed and operated properly, contains it.

3.3.2 Waste Minimization

One way of reducing the hazardous waste that must be treated or stored is to reduce the amount of hazardous waste that is produced. Waste minimization techniques include inventory management, production process modification, volume reduction, and recovery of waste streams.

Waste minimization typically results in lower cost of raw materials and also reduces costs associated with waste disposal. It may also reduce a generator's liability for the hazardous waste produced and compliance costs for permits and monitoring. Waste minimization often cannot eliminate waste entirely though.

3.3.3 Incineration

Incineration, or the burning of hazardous wastes at high temperatures in the presence of oxygen, is another waste treatment option. The method is typically used for organic waste materials, for example, pesticides, solvents, and PCBs.

The heat produced by incineration of hazardous wastes has beneficial uses; for instance, it can be used to preheat combustion air. In addition to producing useful by-products, incineration detoxifies the hazardous waste by destroying the organic molecular structure, and reduces the volume of waste in the process. Another plus is that the incinerator operations do not require a large land area.

As with any method, incineration has its drawbacks. The process is one of the most expensive alternatives available; high capital and operating costs are associated with incinerators. The gaseous products and particulates are potential air pollutants that require expensive abatement equipment such as electrostatic precipitators, baghouse filters, wet air scrubbers, and activated carbon beds. Moreover, obtaining a permit for incinerator construction is often difficult due to community opposition.

3.3.4 Chemical Treatment

Certain hazardous wastes can be converted to a less hazardous form through chemical treatment, using reactions such as neutralization, precipitation, or oxidation and reduction. These processes often make resource recovery possible, and can produce useful byproducts and environmentally acceptable residual effluents. These processes are also used to reduce the volume of hazardous waste.

3.3.5 Biological Treatment

Another volume reduction process is biological treatment, or the use of microorganisms that feed on hazardous wastes to decompose them. The process is useful for treatment of organic wastes such as wastes from a petroleum refinery. Biological treatment can be used to lower the cost of downstream processes by reducing the organic load.

When using a biological treatment system, the proper conditions must be maintained. Considerations include the carbon and energy source, nutrients such as nitrogen, phosphorus, and trace metals, a source of oxygen (or a substitute in some cases), controlled temperature and pH, and removal of toxic organisms. Experimental studies are typically required to design the unit and the process.

These are just some of the more common treatment and disposal alternatives available for hazardous waste disposal. The inventory and characterization processes help to better define disposal needs. Only after completing these tasks can an informed decision be made about the best treatment options.

4 HEALTH AND SAFETY

The Occupational Safety and Health Administration (OSHA) regulates health and safety during hazardous waste operations and emergency response under **29** *CFR* 1910.120 (11). This OSHA standard applies to:

1. Cleanup operations required by a governmental body involving hazardous substances conducted at uncontrolled hazardous waste sites.
2. Corrective actions involving cleanup at sites covered by RCRA as amended.
3. Voluntary cleanup operations at sites recognized by federal, state, local, or other governmental bodies as uncontrolled hazardous waste sites.
4. Operations involving hazardous wastes that are conducted at treatment, storage, and disposal (TSD) facilities.
5. Emergency response operations for releases of, or substantial threats of releases of, hazardous substances.

This regulation includes numerous provisions that must be followed by employers involved in hazardous waste and emergency response operations. The following are the general categories of the requirements under this standard:

- Safety and health program.
- Site characterization and analysis.
- Site control.
- Training.
- Medical surveillance.
- Engineering controls, work practices, and personal protection for employee protection.
- Monitoring.
- Informational programs.
- Drum and container handling.

- Decontamination.
- Emergency response by employees at uncontrolled hazardous waste sites.
- Illumination.
- Sanitation at temporary workplaces.
- New technology programs.
- Operations at TSD facilities.
- Emergency response program to hazardous substance releases.

Some of the requirements of the OSHA standard and good health and safety practices are discussed here. Because regulations are constantly being updated, the latest version of the actual regulation should be consulted for further detail.

4.1 Program Development

The type and degree of hazard vary greatly depending on the site conditions and on variety and quantity of chemicals. Health and safety programs also vary, depending on the type and magnitude of the hazards present, the size of the site, the number of employees, type of operations, type of business, and the overall philosophy toward safety and health. The following are key elements to most successful health and safety programs (12):

1. A policy that is explained and made available to all employees in order to ensure a thorough understanding of program goals and individual responsibilities.
2. A clear definition of program objectives and a schedule for achievement.
3. An overall commitment which acknowledges management's responsibilities to the program.
4. A mechanism that will provide for mutual representation from all functional levels within the organization in the setting of priorities and the implementation of program objectives.
5. A clear definition of line and staff responsibilities and their reporting relationships. This is often best accomplished through the use of a functional organizational chart.
6. A means of periodically reviewing progress and accomplishments over the course of the program.

Major factors that should be considered in a health and safety program for dealing with hazardous wastes are discussed in the following sections.

4.1.1 Written Safety and Health Program

To meet the OSHA standard, **29** *CFR* 1910.120, the employer's safety and health program must include a written program designed to identify, evaluate, and control safety and health hazards, and provide for emergency response for hazardous waste operations. The written program must include:

- Organizational structure.
- Comprehensive work plan.
- Site-specific safety and health plan.
- Safety and health training program.
- Medical surveillance program.
- Standard operating procedures for safety and health.
- Interface between general program and site-specific activities.

4.1.2 Site Characterization

Site characterization and analysis of each individual site are necessary to identify specific site hazards and to determine appropriate safety and health control procedures for protection of employees. Based on the site characterization and hazard identification, personal protective equipment, engineering controls, and monitoring needs are determined, and employees are notified of the potential hazards and risks.

4.1.3 Air Monitoring

Airborne contaminants are one of the potential hazards that should be considered when developing a health and safety program. As with more typical industrial exposures, air monitoring to identify and quantify these contaminants is used to determine the need for personal protective equipment and medical surveillance. In the case of hazardous waste operations, the use of direct-reading instruments as well as laboratory analysis of air samples collected on sampling media are typically used.

Direct-reading instruments are widely used to evaluate airborne contaminants at hazardous waste sites due to the need for immediate information at sites with multiple contaminants and rapidly changing conditions. The instruments are one of the primary tools used during initial site characterization. Often direct-reading instrument readings are used to determine when various levels of personal protective equipment are required. These requirements are typically delineated in a site health and safety plan.

However, direct-reading instruments have a number of limitations that must be kept in mind, such as:

- They are selectively responsive, detecting only specific classes of chemicals.
- They have detection limits that may be above the concentration of concern for some chemicals.
- They are often subject to interferences from substances other than the substance of concern and to environmental conditions, such as humidity.
- They often have different responses to various chemicals.

The user must therefore be completely familiar with the instrument and its limitations in order to appropriately interpret the readings. Interpretation should be conservative, particularly when multiple or unknown contaminants may be present. These direct-reading instruments must also be carefully calibrated frequently, at least before and after each use.

Direct-reading instruments are often not adequate to determine personal exposures to specific contaminants, especially those with low exposure limits. Therefore, full-shift personal air sampling is also applicable to hazardous waste operations. These samples may be collected by drawing air through sampling media with personal sampling pumps or by using passive dosimeters. In both cases, sample analysis is done in a laboratory. Personal monitoring of workers with the highest potential exposure is recommended. The frequency of monitoring depends on a number of site-specific issues, such as the number and types of contaminants, the number of different operations, and changes in site conditions.

Although the primary purpose of air monitoring is typically worker protection, environmental concerns and regulations may require that sampling be performed to check for migration of airborne contaminants offsite. Sampling at the perimeter of the site is conducted for this purpose.

This sampling may be conducted using direct-reading and/or sampling media analyzed in the laboratory. When choosing the location of perimeter samples, wind speed and direction must be considered for an outdoor site.

4.1.4 Personal Protective Equipment

When working around hazardous waste, personal protective equipment (PPE) is often needed to protect individuals from chemical, physical, or biological hazards. Proper selection, use, and care of PPE are essential to adequate protection. PPE is selected based on the potential hazards identified during the site characterization and analysis.

The EPA has defined four levels of protection that are widely used for hazardous waste health and safety situations. Levels A, B, C, and D are summarized in Table 61.4. These levels are typically used as a starting point for choosing appropriate PPE; however, the PPE may need to be modified to meet the needs of the specific situation. In all cases, the limitations of PPE, such as respiratory protection factors and permeability of protective suits by various chemicals, must be taken into account.

On hazardous waste sites, the full extent of the hazard is often not defined at the beginning of a project. As more information becomes available, it may be necessary to modify PPE requirements by upgrading or downgrading the level of protection. Typical reasons for upgrading include (13):

- Known or suspected presence of dermal hazards.
- Occurrence or likely occurrence of gas or vapor emission.
- Change in work task that will increase contact or potential contact with hazardous materials.
- Request of the individual performing the task.

Typical reasons for downgrading include (13):

- New information indicating that the situation is less hazardous than was originally thought.
- Change in site conditions that decreases the hazard.
- Change in work task that will reduce contact with hazardous materials.

Table 61.4. Level of Protection[a]

Level of Protection	Equipment	Protection Provided
A	*Recommended* Pressure-demand, full-facepiece SCBA or pressure-demand supplied-air respiratory with escape SCBA Fully encapsulating, chemical resistant suit Inner chemical resistant gloves Chemical-resistant safety boots/shoes Two-way radio communications *Optional* Cooling unit Coveralls Long cotton underwear Hard hat Disposable gloves and boot covers	The highest available level of respiratory, skin, and eye protection
B	*Recommended* Pressure-demand, full-facepiece SCBA or pressure-demand supplied-air respirator with escape SCBA Chemical-resistant clothing (overalls and long-sleeved jacket; hooded, one- or two-piece chemical splash suit; disposable chemical-resistant one-piece suit) Inner and outer chemical-resistant gloves Chemical-resistant safety boots/shoes Hard hat Two-way radio communications *Optional* Coveralls Disposable boot covers Face shield Long cotton underwear	The same level of respiratory protection but less skin protection than Level A. It is the minimum level recommended for initial site entries until the hazards have been further identified.

(*Continued*)

Table 61.4. (continued)

Level of Protection	Equipment	Protection Provided
C	*Recommended* Full-facepiece, air-purifying, canister-equipped respirator Chemical-resistant clothing (overalls and long-sleeved jacket; hooded, one- or two-piece chemical splash suit) Inner and outer chemical-resistant gloves Chemical-resistant safety boots/shoes Hard hat Two-way radio communications *Optional* Coveralls Disposable boot covers Face shield Escape mask Long cotton underwear	The same level of skin protection as Level B, but a lower level of respiratory protection
D	*Recommended* Coveralls Safety boots/shoes Safety glasses or chemical splash goggles Hard hat *Optional* Gloves Escape mask Face shield	No respiratory protection. Minimal skin protection.

[a]Ref. 9

Training of the user is essential to the safety and effectiveness of PPE use. This is especially true at hazardous waste sites where the hazards can be high.

OSHA regulations for hazardous wastes sites require that a PPE program be prepared as part of the employer's written safety and health program. The written PPE program must address:

- PPE selection based upon site hazards.
- PPE use and limitations of the equipment.
- Work mission duration.
- PPE decontamination.
- PPE maintenance and storage.
- PPE training and proper fitting.
- PPE donning and doffing procedures.

- PPE inspection procedures prior to, during, and after use.
- Evaluation of the effectiveness of the PPE program.
- Limitations during temperature extremes, heat stress, and other appropriate medical considerations.

4.1.5 Training

All employees who engage in hazardous waste operations that could expose them to hazardous substances, health hazards, or safety hazards must receive health and safety training in accordance with requirements of **29** *CFR* 1910.120. Training requirements vary, depending on an employee's duties onsite. General site workers, such as equipment operators and general laborers, must receive 40 hours of instruction off the site and at least three days of actual field experience under the direct supervision of a trained, experienced supervisor. Workers who are on the site only occasionally for a specific limited task, such as groundwater monitoring, and who are unlikely to be exposed at levels greater than permissible exposure limits and published exposure limits, must receive at least 24 hours of instruction off the site and one day of actual field experience under the direct supervision of a trained, experienced supervisor.

In areas where a full characterization has indicated that (*1*) exposures are under permissible exposure limits and published exposure limits, (*2*) respirators are not necessary, and (*3*) there are no health hazards or the possibility of an emergency developing, workers regularly onsite must receive 24 hours of instruction offsite and one day of actual field experience under the direct supervision of a trained, experienced supervisor. Supervisors on hazardous waste sites must have the same training, plus an additional eight hours of specialized training. Additionally, employees must receive eight hours of refresher training each year.

Employees involved in emergency response to hazardous materials incidents require varying levels of training depending on their duties and functions. Most individuals that are actively involved in emergency response operations must receive at least 24 hours of training.

4.1.6 Engineering Controls and Work Practices

The use of engineering controls and work practices to reduce exposures to hazardous materials is preferred over the use of PPE, and is required by OSHA for reducing exposures to materials with a permissible exposure limit (PEL), except to the extent that such controls and practices are not feasible. At hazardous waste sites, available engineering controls and work practices are often limited.

Engineering controls that may be feasible for hazardous waste handling include the use of pressurized cabs or control booths on equipment, emergency alarm systems, forced air ventilation, and the use of remotely operated material handling equipment. Work practices that may be feasible at hazardous waste sites include controlling site access, minimizing the number of employees working in an area when a hazardous operation such as drum opening is performed, wetting down dusty operations, maintaining good housekeeping onsite, and locating employees upwind of potential sources of airborne hazards. Engineering controls and work practice design should take into account the potential offsite contamination/dispersion, as well as onsite employee protection.

4.1.7 Medical Program

A medical surveillance program is necessary to assess and monitor workers' health and fitness both prior to employment and during the course of work, and to provide emergency and other treatment as needed. The program also helps a company keep accurate records for future reference and meet OSHA requirements. OSHA requires a medical surveillance program to be instituted for (*1*) employees who may be exposed to hazardous substances at or above the permissible exposure limits (or other published exposure limits if permissible exposure limits do not exist) for 30 days or more a year; (*2*) employees who wear a respirator for 30 days or more a year; (*3*) employees who are injured, become or develop illness signs or symptoms due to possible overexposure involving hazardous substances or health hazards from an emergency response or hazardous waste operation; or (*4*) members of hazardous materials response teams.

Typically, a medical program includes the following components:

- Surveillance, including: pre-employment screening, periodic medical examinations, termination examination.
- Treatment (emergency and nonemergency).
- Record keeping.
- Program review.

A medical program should be developed for each site based on the specific needs at the location and potential exposures at the site. The medical program should be designed and directed by a medical doctor who has experience in managing occupational health services.

4.1.8 Site Layout and Control

Proper site layout and control is necessary to minimize potential contamination of workers, protect individuals offsite, and prevent unauthorized entry to the site.

To meet OSHA requirements, a site control program must include a site map; site work zones; the use of a "buddy system"; site communications, including alerting means for emergencies; standard operating procedures or safe work practices; and identification of the nearest medical assistance.

The work zones typically include an exclusion zone, a contamination reduction zone, and a support zone. This system provides an area for decontamination and reduces the likelihood of spreading hazardous substances from the contaminated area to clean areas. The exclusion zone is the contaminated area, where primary activities involving hazardous waste take place. The contamination reduction zone is the area where decontamination of personnel, clothing, and equipment takes place. The support zone is a clean area where administrative and other support functions are located. Access between the work zones is through access control points, usually one for personnel and one for equipment.

5 SUMMARY

Management of hazardous waste in accordance with EPA and OSHA regulations is essential not only for limitation of regulatory and tort liability, but for the protection of human health and the environment.

A logical approach for the generator faced with the task of developing and implementing a hazardous waste management program includes:

1. Identifying hazardous waste through an inventory and characterization process.
2. Minimizing the generation of hazardous waste.
3. Reviewing applicable federal, state, and local regulations.
4. Reviewing treatment and disposal options.
5. Tracking hazardous waste from "cradle to grave" with manifests.
6. Transporting wastes to registered treatment or disposal sites with registered transporters, within 90 days of generation.
7. Keeping adequate records of hazardous waste management practices.
8. Following the health and safety requirements of RCRA and OSHA.

Proper hazardous waste management is the responsibility of those who generate, transport, store, treat, and dispose of hazardous waste.

BIBLIOGRAPHY

1. *Code of Federal Regulations*, Title 40, Part 261.3.
2. *Code of Federal Regulations*, Title 40, Part 261.10.
3. *Code of Federal Regulations*, Title 40, Part 261.11.
4. *Code of Federal Regulations*, Title 40, Part 261.4.
5. **45** *FR* 33119 (May 19, 1980).
6. *ASTM E 1527-97, Standard Practice for Environmental Site Assessments: Phase I Environmental Site Assessment Process*, ASTM, Philadelphia, Pa, May 1997.
7. *Code of Federal Regulations*, Title 40, Part 261.21.
8. *Code of Federal Regulations*, Title 40, Part 261.22.
9. *Code of Federal Regulations*, Title 40, Part 261.23.
10. *Code of Federal Regulations*, Title 40, Part 261.24.
11. *Code of Federal Regulations*, Title 29, Part 1910.120.
12. S. P. Levine and W. I. Martin, *Protecting Personnel at Hazardous Waste Sites*, Butterworth, Stoneham, MA, 1985.
13. NIOSH, OSHA, USCG, EPA, *Occupational Safety and Health Guidance Manual for Hazardous Wastes Site Activities*, U.S. Department of Health and Human Services, Oct. 1985.

CHAPTER SIXTY-TWO

Industrial Hygiene Aspects of Hazardous Material Emergencies and Cleanup Operations

Ruth McIntyre-Birkner, MBA, Gary R. Rosenblum, MS, CIH
and Lawrence R. Birkner, MBA, MA, CIH

1 INTRODUCTION

Industrial hygiene issues have a significant impact on all aspects of responding to hazardous material emergencies and on the process of cleaning up a hazardous material spill. The health and safety of all response and cleanup personnel depend on effective evaluation of potential exposures and expert determination of the appropriate means to protect against those exposures.

This chapter covers how the professional judgment, expertise, and training of the industrial hygienist are counted on during all phases of a hazardous material emergency. The industrial hygienist is involved with contingency planning for potential incidents and recognizing, evaluating, and controlling health and safety risks during the incident. The industrial hygienist also helps develop the cleanup site safety plan and contributes to the incident termination process, where a critique of the response and cleanup is conducted, and final documentation is made.

Ideally, it is the industrial hygienist's responsibility as an expert and a teacher to help others identify, recognize, and understand the cause and effect of hazardous materials incidents so they can be prevented in the first place. However, whenever hazardous materials are transported or used, a spill or release could occur. One needs to be prepared for

Patty's Industrial Hygiene, Fifth Edition, Volume 4. Edited by Robert L. Harris.
ISBN 0-471-29749-6

unexpected incidents. Preparedness is as important as prevention; it is not possible to prevent all accidents, no matter what precautions are taken.

Hazardous material emergencies were etched into the public's consciousness by two major catastrophes in the 1980s, the Bhopal, India methyl isocyanate release, and the Exxon Valdez oil spill. Both disasters resulted in major changes in U.S. industry practices and have put a new emphasis on the industrial hygiene profession for all aspects of hazardous materials management.

In India, a catastrophic release of methyl isocyanate, a highly poisonous and irritating vapor, rapidly killed or injured thousands of people living near the Union Carbide facility. Images quickly flashed to the rest of the world showing the devastating results of an uncontrolled release from a chemical process that produced a deadly vapor cloud. As soon as it became clear how quickly so many lives could be lost, much of the industrialized world launched top-priority programs to examine its own chemical manufacturing processes to assess whether that kind of event could ever happen again, and to work to prevent it.

The Exxon Valdez oil spill took many weeks to develop fully, which gave the public time to develop a perception that the spill was of a magnitude that seemed beyond effective control, and the size of the area eventually touched by the spilled oil was astonishingly large. As a result of this incident, public attention became focused on the potential risks of hazardous materials transportation, from crude oil to just about any transported chemical. Prevention and control regulations can only go so far in preventing chemical release incidents. As long as transport vehicles are operated by human beings that must navigate around environmental factors out of their control, it is likely major incidents will occur for years to come.

2 REGULATORY FRAMEWORK

The immediate outcome of focusing public attention on the potential risks of catastrophic releases was to spur the U.S. government into action. Federal, state, and local governments quickly created many important new laws and regulations to change the way hazardous materials are handled in industrial processes and how they are transported, in order to reduce the risk of a hazardous materials emergency. Should a release or spill occur today, many additional requirements are in place to help protect the health and safety of emergency responders and the environment.

2.1 OSHA and EPA Regulations

Governmental standards dedicated to worker health and safety and environmental concerns have increased significantly since the early 1970s. The Department of Labor's OSH Act of 1970 established health and safety standards for the American workplace. Prior to 1970 the U.S. did not really have any type of environmental policy. With the establishment of the Environmental Protection Agency (EPA) numerous standards were put forth to protect the environment. The EPA's Superfund Amendment and Reauthorization Act of 1986(a) requires the Secretary of Labor to issue health and safety standards under Section 6 of the

OSH Act for the benefit of private sector employees engaged in hazardous waste operations and emergency response. SARA section 126(f) (1) requires the EPA to issue standards for hazardous waste operations and emergency response that are identical to OSHA standards. Although the two contain identical substantive provisions, EPA and OSHA address difference audiences. EPA's authority extends to state and local government employers conducting hazardous waste operations and emergency response in states that do not have an OSHA program.

SARA Title III covers emergencies arising from hazardous materials releases. This legislation established three major new requirements for all U.S. industry for emergency planning notification for hazardous materials, emergency release notification to the local community, and reporting on the hazards of the chemicals and releases to the local community. These requirements, described as community "right to know," are specifically designed to help inform the public of the potential risks of local industries handling hazardous materials.

The OSHA Hazardous Waste Operations and Emergency Response (HAZWOPER) Final Rule (**29** *CFR* 1910.120) was published on March 6, 1989 and became effective on March 6, 1990. It provides specific guidance on how to protect the safety and health of personnel responding to hazardous materials emergencies. Additionally, EPA's HAZWOPER standard (**40** *CFR* 311), which is virtually identical to the OSHA HAZWOPER standard, covers government employees such as firefighters, hazardous materials response teams, and volunteers in states where Federal OSHA has enforcement authority. The HAZWOPER standard was incorporated into the Construction Standards as **29** *CFR* 1926.65 on June 30, 1993.

Table 62.1 provides a summary of selected laws and regulations that can apply. HAZWOPER is referred to as a performance-oriented standard, which allows employers the flexibility to develop a safety and health program suitable for their particular facility. It is a collection of health and safety standards wrapped into one regulation. Since its original publication over 10 years ago, several new and/or revised OSHA standards now interface with some element of HAZWOPER. They include process safety management, bloodborne pathogens, confined spaces, respiratory protection, fire brigades, and others. If there is a conflict or overlap between OSHA standards, the provision that is more protective of employee safety and health applies. The level of training required is based on the responsibilities and duties expected of a worker during an emergency response operation.

Since the HAZWOPER standard was promulgated, many OSHA compliance directives have been put forth in the form of letters of interpretation. CLP 2-2.59 A (2) published in April 1998, provides clarification of emergency response sections of the standard and establishes policies for OSHA Compliance Officers to follow when conducting inspections where emergency response under **29** *CFR* 1910.120(q) is required. The update incorporates references to the recent emergency response plan requirements of other agencies (signed by OSHA, EPA, the Coast Guard, DOT's Office of Pipeline Safety, and Interior's Minerals Management Services). It includes the National Response Teams' guidance for employers on how to prepare an Integrated Contingency Plan that meets the criteria of multiple federal agencies.

Table 62.1 Standards and Acts Impacting Hazardous Materials Emergencies and Cleanup Operations[a]

Regulation	Summary	Agency
Clean Water Act; 1972 (PL 92-500) Amended in 1977 (PL 95-217) **40** *CFR* part 68	Protects our nation's waters, including lakes, rivers, acquifers, and coastal areas. Requires major industries to meet performance standards to ensure pollution control	EPA
Federal Insecticide, Fungicide, and Rodenticide Act (FIFRA) of 1972 PL 92-516	Gave EPA authority to suspend, cancel, or restrict existing pesticides to prevent unreasonable risk to humans or the environment	EPA
Hazardous Materials Transportation Uniform Safety Act of 1990 (HMTUSA) **49** *CFR* part 194; and 1992 **49** *CFR* 171–177, SubpartH—Hazardous Materials; Training for Safe Transportation	Regulations governing the transport of hazardous materials, which includes the design of transportation vehicles and vessels, and specifies shipping, packaging, labeling, and placarding requirements for both carriers and shippers; provided funding for health and safety and emergency planning training for public sector emergency response personnel	DOT
Toxic Substances Control Act, 1976 PL 94-469	Authorizes EPA to obtain data from industry regarding the production, use, and health effects of chemical substances and mixtures and regulates their manufacture and disposal	EPA
Resource Conservation and Recovery Act (RCRA), 1976: **40** *CFR* part 264-265, Subpart D; **40** *CFR* 279.52	Controls management and disposal of hazardous waste from "cradle to grave"	EPA
Comprehensive Environmental Response and Compensation Liability Act "CERCLA" 1980 PL 96-510	Addresses hazardous chemical releases into the environment. Those responsible for releases above "reportable quantities" are required to notify the National Response Center	EPA
Superfund Amendment and Reauthorization Act of 1986 (SARA) PL 99-499	Contains Title III, which established major requirements relating to Emergency Planning Notification, Emergency Release Notification, and Reporting on Chemicals for Community Right to Know (EPCRA)	EPA
Clean Air Act Amendments 1990—Reauthorized, extended and expanded the original act	Mandates new federal focus on prevention of chemical accidents. Its objectives are to prevent serious chemical accidents that have the potential to affect public health and the environment. Requires strict control of 189 air toxics, takes more stringent steps toward controlling acid rain and air toxins, strengthens enforcement provisions	EPA

Table 62.1 (continued)

Regulation	Summary	Agency
Pollution Prevention Act of 1990	Reviewed environmental regulations to determine their effect on source reduction, identify measurable goals and establish training. Act encourages initial prevention and voluntary actions	EPA
Oil Pollution Act of 1990 **40** *CFR* 112	Establishes requirements for facilities to prevent oil spills from reaching navigable waters of the U.S. or adjoining shorelines. Requires spill prevention, control and countermeasures	EPA
Risk Management Plan (**40** *CFR* 68, 1996)	Chemical accident prevention programs designed to protect workers (under **29** *CFR* 1910.119), the public and the environment. Joint effort between OSHA and EPA. Focus of the EPA rule is primarily on safety and health of the public and surrounding communities	EPA
National Response Team's Integrated Contingency Plan Guidance 1996 (61 FR 28641)	"One Plan" offers guidance on consolidation of various regulations into one functional emergency response plan or integrated contingency plan (ICP); Intended for facilities responding to releases of oil and nonradiological hazardous substances. Addresses **29** *CFR* 1910.38(a), 1910.119, and 1910.120. EPA and DOT/U.S. Coast Guard regulations are also covered in the one plan.	EPA
National Fire Protection Association Standards 1991–1993	Developed national voluntary concensus standards for Level A and Level B chemical protective suits for hazardous chemical operations	NFPA
Facility Response Plan **33** *CFR* part 154, Subpart F	Addresses Coast Guard specific issues in facility response planning	USCG
OSH Act of 1970 General Industry Safety and Health Standards PL 91-500	Established use and exposure limits for 22 hazardous substances; established exposure limits for 380 air contaminants	OSHA
29 *CFR* 1910	These standards cover workplace activities concerning hazardous substances, environmental controls, fire protection, hand and portable tools, and walking/working surfaces.	
OSHA Hazard Communication Standard **29** *CFR* 1910.1200	Regulates the determination of workplace hazards, and requires communication of the hazards to the affected employees, and the training to mitigate the hazards	OSHA
Employee Emergency Plans and Fire Prevention Plans **29** *CFR* 1910.38(a)	Describes the creation of emergency action plans and employees affected by them, includes emergency action plans and emergency response plans	OSHA
Hazard Communication Standard **29** *CFR* 1910.1200	Requires employers to train employees who may be exposed or potentially exposed to hazardous chemicals	OSHA

Table 62.1 (continued)

Regulation	Summary	Agency
Hazardous Waste Operations and Emergency Response Final Rule "HAZWOPER" **29** *CFR* 1910.120 March 6, 1989	Designed to protect the health and safety of persons involved with hazardous waste site cleanups, other hazardous waste activities, all persons responding to an uncontrolled release of a hazardous substance, and all persons involved in hazardous waste spill cleanup	OSHA
Occupational Exposure to Hazardous Chemicals in Laboratories, 1990 **29** *CFR* 1910.1450	Provides framework to educate laboratory workers about chemical hazards in the workplace and ensure protection from adverse chemical exposure	OSHA
Occupational Exposure to Bloodborne Pathogens **29** *CFR* 1910.1030 1991	Requires that employers and employees follow specific requirements when dealing with occupational exposure to blood or other potentially infectious agents	OSHA
HAZWOPER at Construction Sites **29** *CFR* 1926.65; 1993	HAZWOPER incorporated in construction standards; paragraph (q) applies to construction operations as well as general industry	OSHA
Process Safety Management of Highly Hazardous Chemicals, 1992 **29** *CFR* 1910.119	Geared toward preventing catastrophic releases, but they do not address specific procedures for responding to such releases. HAZWOPER's ER provisions apply to the actual emergency response effort at facilities covered by the PSM Standard.	OSHA
Permit-required Confined Spaces (PRCS) **29** *CFR* 1910.146	HAZWOPER does not address response to incidents involving PRCS with the detail provided in this standard. Targets work and emergency rescue related to confined spaces.	OSHA
Fire Brigades **29** *CFR* 1910.156	Industrial fire fighters must be aware of MSDSs, written procedures and training. Fire brigade employees must receive HAZWOPER	OSHA
Revised Respiratory Protection Standard **29** *CFR* 1910.134 1998	Employer must ensure site-specific procedures described in the health and safety plan (HASP) are consistent with the respiratory protection standard	OSHA

[a]Not an all-inclusive list; other federal, state and local laws and standards may apply.

HAZWOPER has dramatically changed the way emergency responders deal with incidents involving hazardous materials. Much of the remainder of this chapter is devoted to covering industrial hygiene activities and decisions required by the rule.

3 ADVANCE PLANNING AND COORDINATION

3.1 Preparing for Emergency Response

The first priority for any group anticipating an emergency response to a hazardous materials incident—an industrial corporation, government agency, local planning committee, or volunteer response—is planning and coordination with outside parties in advance of the incident. The planning and coordination procedures in the employer's written emergency response plan should state the conditions and circumstances under which outside responders will provide emergency response to the site or facility. Preplanning also includes determining the range of anticipated events, and then realistically preparing a response plan to ensure that the response group's efforts are timely, effective, and safe. Without advance planning and coordination, the emergency response to a hazardous materials incident is likely to result in confusion, delays, and possibly excess damage, unnecessary injuries, and even deaths.

When a major incident occurs in the U.S., as many as 16 federal agencies could respond with the goals of protecting worker health, public health, and the environment. This response system has been complex, confusing, costly, and many federal requirements overlap one another. In 1996, The National Response Team (NRT) published an "Integrated Contingency Plan (ICP or One Plan) (3) as a way for facility owners and operators to consolidate into one functional plan, the multiple oil and hazardous substances emergency response plans they may have prepared to comply with various federal regulations" (4). As previously mentioned, five key agencies signed the one-plan guidance. Its purpose is to:

- Provide a way to consolidate multiple facility response plans into one plan that can be used during an emergency.
- Improve coordination of planning and response activities within the facility and with the public and commercial responders.
- Minimize duplication of effort and unnecessary paperwork.

The format is based on the Incident Command System. Integrating the ICP format within the incident command system dovetails with established response management practices and should lesson the conflict between and among agencies, responders, employers and the public.

Some key elements for preplanning and coordination in the emergency response plan include:

- Site management and control.
- Anticipating the types of hazardous materials that may be involved.
- Making sure the appropriate exposure assessment equipment is available.
- Deciding the different levels of personal protective equipment needed.
- Assessing the control and containment parameters for the various scenarios.
- Determining the levels of decontamination that may be necessary.
- Anticipating the documentation needed throughout the response and at its conclusion.
- Training and drilling.

Each one of these elements comes from predicting the type, severity, and range of possible incidents at a facility or during the transportation of a hazardous substance.

Before response to a hazardous materials incident takes place, a charter should be developed for assigning key personnel roles and responsibilities. This includes who will staff the initial response, relief personnel, and so on. For example, the industrial hygienist needs to know in advance to whom to report, and his or her level of responsibility on scene, which could include becoming the safety officer for the event. Also, the plan should identify the "on-scene commander," the person with final authority for tactical response decisions, with the ultimate responsibility for the safety and health of all response and cleanup personnel. This incident command structure should be flexible and take into account various staffing contingencies, expanding and changing roles, should the incident grow or change in scope.

The site management plan should include the coordination of communications equipment. The amount and types needed, as well as anticipating the equipment used by other responding organizations and agencies, is important to ensure effective communication with all active response groups.

Another important element of the site management plan is determining the level of emergency that will require evacuation of any public nearby as well as evacuation of responding personnel. The conditions for evacuation do not have to be exactly laid out in advance. But it is possible to anticipate certain conditions, such as fire, involving certain materials or storage vessels that would lead the incident commander to call for evacuation. This element leads to another communication issue, which is the coordination of information with appropriate government agencies, responding teams, other authorities, and company management. Hazard potential and resource requirements must be clearly understood by all these groups, and the plan should be able to provide clear instructions to facilitate this process.

Other than site management, personnel coordination, and communications coordination, most of the other planning elements stem from anticipating the types of incidents and the materials involved. At first the sheer enormity of anticipating all possible hazardous materials incident scenarios can be daunting.

The hygienist should be aware of the importance of the identification of the materials likely to be involved in a hazardous materials incident, because all major safety and health decisions follow that determination. Preplanning of a hazardous materials incident can provide initial clues to the hazardous material identity, but it is also important to know and be able to utilize the labeling and placard systems in use.

One key factor for industrial hygienists conducting hazardous materials incident advance planning is anticipating exposure monitoring needs. Having the appropriate monitoring equipment immediately on site is of critical importance for the entire emergency response action. Without good knowledge of the exposure potential and the actual exposures faced by responding personnel, the level of personal protective equipment **must be maximized**. As a result of poor planning, if the proper direct-reading instruments are not available when the "on-scene commander" determines that the hazardous materials team must immediately go in to the danger zone, the team must be protected with maximum personal protective equipment. This may require the responders to wear the cumbersome fully encapsulating suit and self-contained breathing apparatus, which would limit their effectiveness and increase their risk of heat stress. HAZMAT incident planners and industrial hygienists should remember the HAZWOPER standard requires emergency responders to wear supplied air, unless air monitoring at the site documents the exposure level is low enough to use either air-purifying respirators or no respiratory protection.

The advance planning of personal protective equipment (PPE) is also of great importance. Again, response efforts cannot be performed without appropriate protection from exposure. Preplanning can help ensure that the proper level of personal protective equipment is available when it is needed.

Proper preplanning also estimates how much PPE will be needed. For instance, fully encapsulating suits are always stocked in pairs, because they are always used in groups of at least two for safety. The buddy system provides a safety net for getting someone out of danger if a problem arises. Again, the logistics of properly storing the equipment and getting it to the incident site should be addressed in the plan to increase the chances that the PPE will be there when it is needed, functioning, and ready to go without delay.

Advance planning should cover the amount and type of control and containment equipment and supplies needed. Control and containment equipment include all types of items, from granular absorbent to skimmers and booms to backhoes, boats, bulldozers, vacuum trucks, rakes, and shovels. On scene responders need access to functional equipment and supplies, of the right type and amount, when an incident arises. This requires anticipation of the potential emergencies and thoughtful preplanning. Proper planning of the logistics of storing and transporting the equipment will help the response proceed faster and smoother and will likely produce a more desirable result. Another preplanning element is proper disposal of contaminated material and decontamination of equipment and material that must be reused.

Decontamination (decon) is another important issue to consider during the advance planning stage. Based on the hazardous materials anticipated in the incident, a decon strategy can be developed. A major issue to consider in preplanning is whether to use reusable or disposable PPE and equipment. Effectiveness, cost, and utility of the PPE must be taken into consideration. If the plan calls for disposal as the decon procedure, steps must be taken to determine whether the material will be considered "hazardous waste" and thus require adherence to all applicable laws for transporting and disposing of hazardous waste from that location. Decon of reusable PPE, however, can be more labor intensive and require additional personnel for the multiple step cleaning and rinsing process.

Finally, the plan must cover the termination procedures to follow once the incident comes to a close. These procedures can consist of defining when the event has ended, when to demobilize the responders, and, if possible, how clean the cleanup should be. An additional element of the termination procedure is to critique the response in a constructive manner. This includes assessing the documentation of the event. The plan should anticipate and provide many of the generic documentation sheets that can be helpful, such as an incident commander log, material hazard summary sheet, lists of response personnel, and monitoring records. An example of this type of documentation is provided in Figure 62.1 (5).

3.2 Emergency Response Site Safety and Health Plan

The site safety and health plan is a critical element to the effective operation of a response and cleanup operation. It is a requirement for compliance with HAZWOPER. This plan is different from emergency response planning in that it addresses specific health and safety

I. PHYSICAL/CHEMICAL PROPERTIES

			SOURCE
Natural physical state: Gas ______ Liquid ______ Solid ______ (at ambient temperature of 20°–25°C)			______
Molecular weight	______	g/g-mole	______
Density	______	g/mL	______
Specific gravity	______ @ ______	°F/°C	______
Solubility: water	______ @ ______	°F/°C	______
Solubility ______	______ @ ______	°F/°C	______
Boiling point	______	°F/°C	______
Melting point	______	°F/°C	______
Vapor pressure	______ mmHg @ ______	°F/°C	______
Vapor density	______ @ ______	°F/°C	______
Flash point (open cup ______; Closed cup ______)	______	°F/°C	______
Other: ______	______		______

II. HAZARDOUS CHARACTERISTICS

A. TOXICOLOGICAL HAZARD	HAZARD?	CONCENTRATIONS (PEL, TLV, other)	SOURCE
Inhalation	YES NO	______	______
Ingestion	YES NO	______	______
Skin/eye absorption	YES NO	______	______
Skin/eye contact	YES NO	______	______
Carcinogenic	YES NO	______	______
Teratogenic	YES NO	______	______
Mutagenic	YES NO	______	______
Aquatic	YES NO	______	______
Other: ______	YES NO	______	______

B. TOXICOLOGICAL HAZARD	HAZARD?	CONCENTRATIONS	SOURCE
Combustibility	YES NO	______	______
Toxic by-product(s):	YES NO		
______		______	______
______		______	______
Flammability	YES NO		
LFL		______	______
UFL		______	______
Explosivity	YES NO		
LEL		______	______
UEL		______	______

Figure 62.1. Sample hazard summary sheet. For density/specific gravity, only one value is necessary. For solubility of organic components, recovery of spilled material by solvent extraction may require solubility data.

issues for the particular incident. Therefore it becomes an integral part of the emergency response planning process.

The emergency response plan is used to guide such items as interaction and coordination with outside agencies, training, and resource allocation, stockpiling of supplies, personnel requirements, and compliance issues. The site safety plan is created specifically to address the safety and health needs for the personnel responding to and cleaning up one specific incident. Because this is quite a large task, and it is done during an incident, it is highly recommended that a generic site safety and health plan be produced that contains blanks

C. REACTIVITY HAZARD	HAZARD? YES NO	CONCENTRATIONS	SOURCE
Reactivities:			
______________		______________	______________
______________		______________	______________

D. CORROSIVITY HAZARD	HAZARD? YES NO	CONCENTRATIONS	SOURCE
pH			
Neutralizing agent:			
______________		______________	______________
______________		______________	______________

E. RADIOACTIVE HAZARD	HAZARD?	EXPOSURE RATE	SOURCE
Background	YES NO	______________	______________
Alpha particles	YES NO	______________	______________
Beta particles	YES NO	______________	______________
Gamma radiation	YES NO	______________	______________

III. DESCRIPTION OF INCIDENT:

Quantity involved ______________________________

Release information ______________________________

____

Monitoring/sampling recommended ______________________________

____

IV. RECOMMENDED PROTECTION:

Worker ______________________________

____

Public ______________________________

____

V. RECOMMENDED SITE CONTROL:

Hotline ______________________________

____

Decontamination line ______________________________

____

Command Post location ______________________________

____

VI. REFERENCES FOR SOURCES:

____

____

____

____

____

____

____

____

Figure 62.1. Continued.

to be filled in as the information is developed. This fill-in-the-blanks approach has the advantage of reducing the chance that an important element of personnel safety is inadvertently forgotten in the heat of action, and it provides good documentation of the safety and health decision-making process when it is completed. An example of a model site safety plan blank is provided in Appendix A.

For guides and downloadable software to prepare HAZWOPER health and safety plans (e-HASP), visit the OSHA website (6). E-HASP is a software package that produces a site-specific health and safety plan (HASP) for work conducted under the HAZWOPER standard. The HASP is produced using site-specific input from the user. Organizations like the American Petroleum Institute, in conjunction with the American Society for Testing and Materials, have developed standards—site safety and health plans and emergency plans—for personnel responding to oil spills.

The minimum elements of a site safety plan include risk assessments for all site tasks and operations, frequency and type of air monitoring, personnel monitoring, environmental sampling techniques, specific personal protective equipment for all employees for each task and operation, medical surveillance requirements, site control measures, decontamination procedures, confined space entry procedures if applicable, task-oriented employee training requirements, and a spill containment program.

The risk assessments are based on an evaluation of the tasks to be performed by the emergency responders and cleanup crews. The specific hazards of the materials and environment to which the responders may be exposed are reviewed, and a task analysis is performed that includes characterizing exposure. A qualitative risk assessment is produced from these factors that enables the industrial hygienist to consider the need for monitoring and possible PPE strategies.

Next, the site safety plan considers the type and frequency of air monitoring as well as task-specific PPE strategies. Details of developing a monitoring strategy are covered in Section 5, Hazard Evaluation, but one key issue is identifying the types of direct-reading instruments for rapid airborne contaminant evaluation. The HAZWOPER final rule mandates exposure assessment (including air monitoring) before any reduction in respiratory protection below supplied air is considered. The site safety plan must provide for such exposure assessments.

The collection of air contaminant exposure level data plays a large role in determining the level of protection to be assigned to individuals who will be entering the danger zone in a hazardous material emergency response and to those personnel providing direct assistance to those in the danger zone. In some cases where quantitative assessments of exposures are not possible, the industrial hygienist must make a determination of the appropriate PPE based on qualitative factors. In almost all these cases, higher levels of protection are used to protect personnel against unknown risks. The model site safety plan should provide a format that allows the industrial hygienist to review all the exposure data that are available in the situation. It should also include areas to document the decision-making process for selecting the personal protection for the various tasks needed in the emergency response and cleanup.

Other site safety plan elements follow as the risk assessments, hazard determinations, and exposure characterizations are completed. The amount and type of HAZWOPER training responders need is based on their job functions, responsibilities and potential exposure to hazardous materials. This is noted in the site safety plan. If confined space entry procedures are needed; they are documented in the plan. Also included is a medical surveillance program based on the potential adverse effects that could be observed if overexposures were to occur. The need for decontamination or use of disposable equipment is also documented.

Site control procedures (zone control system) are always set up on a case-by-case basis. Zone control allows only protected, trained individuals near the danger zone and keeps the news media and other members of the public at a safe distance. During the course of an incident the site control plan could change several times as the material or wind moves, or the hazard is contained, and each change is accounted for and documented in the site safety plan.

In the scenario building process and through preplanning, the generic site safety plan can be constructed for a specific location. As far as the type of hazardous materials incident can be predicted, parameters of the site safety plan can be completed in advance. If the hazardous materials in an incident can be anticipated, then the hazards of that material, and the types of PPE that are most effective for that material, can be incorporated into the site safety and health plan in advance. When the hazardous material involved in an incident is unknown, the generic site safety and health plan is used, with blanks to be filled in on-site.

For example, in preplanning for an oil spill response, environmental concerns include any and/or all of the following: topography, climate of the spill area, wildlife/endangered species, marine environment, archaeological sites, earthquakes, avalanches, rock slides, tsunamis, commercial fishing, recreation and subsistence food areas. Further, the introduction of oil into the environment can affect water intake structures, tidal marshes, bird feeding grounds, oyster and coral reefs, river mouths and areas of high visibility (marinas, sea walls and beaches).

Factors affecting the severity of oil spills include:

- Type of oil spilled (chemical and physical nature).
- Amount and rate of oil spilled, length of time oil is allowed to weather.
- Geographic locality and size of the affected area.
- Meteorological and oceanographic weather conditions.
- Season—time of year.
- Type of biota present in the affected area.
- Cleanup technique used.
- Dispersant use and bioremediation strategies.

3.3 Emergency Response Training

In order to respond effectively to an incident, industrial hygienists who will be emergency responders must, in addition to their knowledge and experience in industrial hygiene, be trained in emergency response procedures, the incident command system and its management, the use of PPE, and decontamination procedures. More specific training requirements that include and go beyond these training requirements are covered in the HAZWOPER rule. Emergency response training mandated by the HAZWOPER rule is covered in Section q (6), and is outlined in Table 62.2.

Appendix E of HAZWOPER outlines nonmandatory training guidelines for assistance in developing site-specific training curricula used to meet the training requirements. Take note that OSHA State plan states may require more extensive training for HAZWOPER emergency responders. In addition, organizations like the U.S. Coast Guard and the American Petroleum Institute have developed standard guides for health and safety training of oil spill responders.

The industrial hygienist who will respond to hazardous materials emergencies must be trained in emergency response procedures, which are based on a simple premise: an emergency response means that all actions must be done as rapidly as possible, but always in an orderly and safe fashion. Training in emergency response procedures emphasizes that if an action cannot be done safely, it must not be done. At the same time, the training emphasizes that the determination of risk is being done in a situation that is far from optimum for risk data gathering. The best available information is gathered to reach the best available assessment, which is not the way an industrial hygienist normally gathers data.

The industrial hygienist may get only a few moments to sample only once, perhaps while wearing a fully encapsulating suit, and must recognize that this kind of pressure is not unusual.

A critical part of emergency response procedures is training in the incident command system. This system was developed by the fire service to coordinate effectively the activities of many different responding groups at a single incident. For example, when many fire companies from different cities within a county are called to fight a large fire, each arrives with a fire chief. Without a preplanned system for coordinating each chief's orders to his or her personnel, through the selection of one incident commander, each chief would not know what the other is doing and resources are sure to be wasted. Use of the incident command system at a hazardous materials emergency assures that a chain of command is established quickly, and the roles and responsibilities of all responding personnel are defined.

While the incident command system is beyond the scope of this chapter, it must be acknowledged that the incident commander is the single authority for making all strategic decisions concerning the activities of all responding personnel. This is accomplished through direct management, or indirectly through specific officers chosen to lead certain major categories of tasks, such as logistics, operations, finance and planning. The management system depends on the type of incident and the number of personnel needed and available. The incident commander has the final responsibility for the health and safety of all responding and cleanup personnel.

However, some of the decision-making authority for technical health and safety decisions may be delegated to a safety officer. The industrial hygienist may either be the site safety officer or may report directly to the site safety officer. The industrial hygienist may even be the incident commander in certain circumstances. This safety and health evaluation may take place in the advance planning stage to be carried out on-site by a nonindustrial hygienist, or may be done by the industrial hygienist under the pressure of instant analysis during a rapidly evolving uncontrolled release of a hazardous material. In all these cases,

Table 62.2 HAZWOPER Training Summary Under Paragraph Q[a]

First Responder Awareness	Determines that an uncontrolled release is occurring: Calls the HazMat Team Takes no action against the release Limited training
First Responder Operations	Initial spill control activity Defensive attack against spilled substance to protect human health, property, and the environment Peripheral spill containment; booms, dikes, absorbents, etc. Eight hours of training
HAZMAT Technician	Offensive attack on release Plugs, patches, stops the release At least 24 hours of training
HAZMAT Specialist	Experience with facility unit or hazardous material Assists technician in attacking and stopping the release Liaison with regulatory agencies At least 24 hours of training
On Scene Incident Commander	Ultimate authority/responsibility for worker safety, site security Implements company emergency response plan Also implements local, state, federal response plans when appropriate At least 24 hours of training

[a]These are minimal training requirements. Employers in OSHA state plan states must follow the specific HAZWOPER training requirements for the state(s) in which they reside.

the professional judgment needed to make decisions on PPE and other safety issues comes from the industrial hygienist(s) or other safety professional(s), who must know how this evaluation fits into the hazardous materials incident response process.

Although it may be taken for granted that an industrial hygienist knows how to select and use PPE, he/she should also be able to train others in how to use it safely and effectively.

Industrial hygienists must also understand PPE decontamination steps and procedures. The protocol for decontaminating nondisposable equipment is site-specific and requires on-site review. The industrial hygienist may be in charge of designing the decontamination facility, either in the emergency response plan or on-site, and may be responsible for training others in proper decon methods. See Section 6 for specific details on decontamination.

3.4 Training for Postemergency Response to Oil Spills

Guidelines have been published by OSHA (CPL 2-2.52, November 5, 1990) with regards to the training requirements for postemergency response to oil spills. Based on experience with the standard during oil spills off the coasts of Texas, Alaska and California, the hazards to employees vary widely in severity of potential injury or illness. For job duties and responsibilities with a low magnitude of risk, fewer than 24 hours of training may be appropriate for postemergency clean-up workers. OSHA recognizes that reduced training for certain categories of personnel involved in oil spill operations may be necessary, and may not constitute a serious violation (a *de minimis* violation). Basically, four hours of training is expected to be adequate to meet the *de minimis* criteria. This requirement could change based on specific state requirements, response circumstances, or both. Other requirements must also be met (adequate characterization of minimal hazards, and adequate supervision by fully trained personnel).

3.5 Emergency Response Drills

Classroom training plays a big role in preparing emergency responders for an incident. The most effective training includes practical, hands-on drills. The industrial hygienist and other response team members should participate in tabletop simulations of HAZMAT incidents and real-time drills that require donning and doffing PPE, use of air and other monitoring equipment, and practice in decon procedures. The drills should emphasize documentation of the decision-making process for all activities. The industrial hygienist has a role in planning and executing the drills, as well as participating as a responder. Without drills and related exercises, an emergency response team cannot be deemed ready for action. The drills are also an important part of annual refresher training to maintain readiness and competency.

4 HAZARD EVALUATION

After a hazardous material release has occurred, many decisions must be made based on the nature and severity of the incident. Accurate data determining the hazards faced by the responding personnel lead to effective personnel protection, which produces more efficient performance. It is very important for the industrial hygienist to be in a position to assess the hazards quickly and accurately.

This section covers some of the issues and methods affecting the rapid and accurate collection of data. It should be pointed out, however, that the industrial hygienist is frequently not the first responder on the scene. As a result, other first responder personnel will need to be able to make initial assessments of the scene and be able to perform some preliminary monitoring. Therefore this section is applicable for either the industrial hygienist on the scene or the first responder.

4.1 Developing the On-Scene Safety Plan

There are four basic steps to this process: (*1*) initial site assessment (*2*) identification of site hazards (*3*) evaluation of the site-specific safety plan, and (*4*) communication of the plan. Because managing hazardous materials incidents consists of caution in the face of the unknown, combined with aggressive management of known hazards, the person or group developing the on-scene safety plan needs to concentrate on separating verifiable facts from speculation and must be able to work quickly and effectively under pressure.

4.1.1 Initial Site Assessment

The initial site characterization consists of observing the general geographic features of the incident from a safe distance. If closer views are needed, then binoculars are recommended. This first stage of site characterization is essentially making a map of the incident with many important initial decisions based on the observations. Wind speed and direction, as well as pending weather conditions, such as rain, must be noted. Because approaches to a hazardous materials incident are from the upwind direction, the topography of the upwind side must be assessed. Roads, fields, hills, cliffs, streams, rivers, ditches, fences, buildings, and walls are the types of things that will determine many of the most critical initial site decisions such as setting up the command post (upwind and uphill if possible), access for heavy equipment and personnel, and determining zone control boundaries and potential escape routes. The topography is also a major factor in predicting where the spill material is moving, and where firefighting water or foam will collect or run off. This influences decisions on diking, booming, or otherwise impeding the flow of the spilled material.

Other atmospheric conditions besides winds are important. Temperature affects the vaporization of hazardous materials, but also has an impact on the responders, who may need to wear bulky PPE that could induce heat stress. High humidity can also play a role in inducing heat stress, whereas very low humidity can stress responders because of extra rapid depletion of fluids. If it is raining, or about to rain, many decisions may change concerning access to the site by crossing creek beds or streams, influence firefighting activities, and even create an increased hazard if the hazardous material is unstable or pyrophoric with water contact. Fog reduces visibility and lightning might inhibit the use of cranes or other heavy equipment.

Site geology is also assessed. The movement of the spilled material through the soils can range from rapid to virtually nil, depending on the local geology. Spilled material can soak through soils at one level and reappear below a cliff or seep down to groundwater. Hazardous materials soaked into soils can continue to expose workers throughout the cleanup phase of the operation as they remove the soil.

4.1.2 On-Scene Hazard Identification

The second aspect of on-scene risk evaluation deals with the identification of the hazards. These can be hazards to the responding personnel, the public, and/or the environment. The most critical aspect is to identify the hazardous material or materials involved in the incident.

Failing to identify the hazardous materials in an incident can lead to disastrous consequences for the responding personnel and restrict the incident commander's ability to manage the response. Without knowing the identity of the materials, it becomes impossible to send response personnel into hazardous material incident situations without taking maximum personal protection precautions. Maximum PPE, such as the level A totally encapsulated suit with self-contained breathing apparatus, is difficult to work in, is inefficient because it reduces the amount of time a person can work, and requires substantially increased levels of resources because it must be used in pairs, with two backup personnel standing by and with protected personnel assisting with donning, doffing, and decontamination. Therefore, when the hazards are not positively identified, it can make the emergency response effort much more dangerous, difficult, and more costly.

Materials are identified through either administrative means, such as examining shipping or transport information, including placards and labels, or through investigative means, such as observation of the location and function of the facility and of the appearance of the released material or its container, or use of direct-reading instruments that can identify the substances and evaluate the exposure.

The significant health and safety hazards, as well as the most significant routes of exposure to be faced by the emergency responders, can then be determined after a positive identification is made of the released substance(s). A review of the toxic properties is done using the best available information sources, such as Material Safety Data Sheets (MSDS), handbooks of properties of toxic materials, the Department of Transportation (DOT) *North American Emergency Response Guidebook* (7), computer data sources, or hotline phone sources such as CHEMTREC (Chemical Transportation Emergency Center). Identifying emergency data sources, listing appropriate phone numbers, and providing responders with handbooks with this information is also part of the advance planning process.

There are many regulatory and administrative identifiers of hazardous materials that can be seen from a safe distance that may aid in determining the identity of the substance and the risks faced by responders. The U.S. Department of Transportation requires placarding and labeling of bulk hazardous materials transported in the United States by rail or motor vehicle. Hazardous cargoes must be marked with a standard format placard that would enable an emergency responder to identify the type of hazard the cargo presents. The placard provides a pictogram of the hazard, a code number corresponding to the U.N. hazard classification system. The placard or label may also be used with a four-digit material-specific identification number that is coded to the DOT Emergency Response Guidebook. The number and guidebook provide emergency responders with general hazard and emergency response information for the material type identified by that code number. Both the *NIOSH Pocket Guide to Chemical Hazards* and the *North American Emergency Response Guidebook* are now available online in electronic formats (http://hazmat.dot.gov/gydebook.htm, http://www.cdc.gov/niosh/ngp/pgdstart.html).

Other label systems may also be used, and should be looked to for assistance in identifying the hazard. The National Fire Protection Association has created a placarding system for allowing the rapid identification of material hazards. This placard contains symbols for fire, health, reactivity, and other hazards in one diamond-shaped sign (see Fig. 62.2). Each section of the diamond contains a hazard summary number, with "0" indicating the minimum hazard and "4" indicating the maximum hazard. Petroleum companies may use a color code system of identification developed by the American Petroleum Institute (API). Each petroleum or petroleum product has a color or color combination associated with it. For instance, high-grade unleaded gasoline would be marked with an orange circle containing a white cross, and kerosene would be marked with a brown hexagon.

Shipping papers, bills of lading, and similar paper work often can confirm the identity of the hazardous material. Unfortunately, these papers often are carried directly with the material so when there is a spill, they are within the contaminated zone and are not readily available. In addition, if there is a fire situation, they could be destroyed. Finally, the shipping papers may not have been completed properly or indicate a mixed load, which

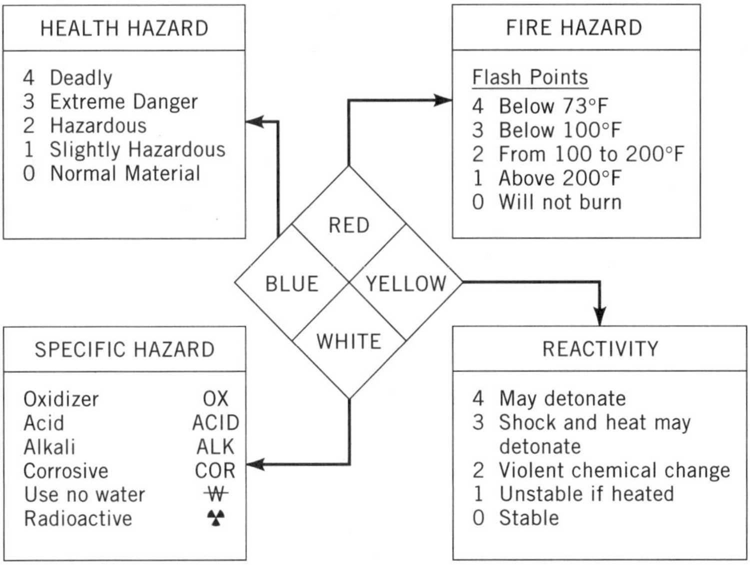

Figure 62.2. NFPA label.

could complicate an already difficult situation. As a result, it is often not possible to rely on these papers completely to identify the hazardous material.

In addition to relying on administrative identifiers, observation and deduction are also useful and important tools when sizing up a hazardous materials incident. The importance of preplanning cannot be overemphasized in helping to determine the identity of the hazardous material(s) through observation and deduction. An additional source of information is interviews with employees or other witnesses involved with the release or spill.

The type of facility or transport equipment involved is a key point to help determine the identity of an unknown hazardous material. The types of questions to answer would include whether the facility is a storage or a production facility, or whether or not the cargo is being shipped under pressure. For example, a round cylinder tank on a cargo truck would indicate it is pressurized, whereas an elliptical cylinder tank would indicate that the cargo is not pressurized. Quite different safety decisions would be made for some materials given a pressurized versus unpressurized situation. Besides shape, container color, and additional markings such as color, stripes can provide further clues. This is because some industries use standard colors and markings for transport of particular materials or material types. Observing the characteristics of the shipping container or the facility in combination with good advance planning data can go a long way toward identifying the hazardous material quickly and safely.

Observations should not be limited to the containers. The spilled or released material itself can present many identifiers when observed from a safe distance. Color, physical state, viscosity, and vaporization rate are some examples of properties that can help identify a hazardous material. In addition to the senses, using direct-reading instruments is important for many reasons. The direct-reading instrument can provide solid identification of a hazardous material as well as very important exposure level data.

4.1.3 Site-Specific Safety Plan

Federal regulations require that the decision to send emergency responders into an area that could potentially have a hazardous atmosphere in anything less protective than self-contained breathing apparatus must be based on documented exposure monitoring data. This means that on-scene measurements of airborne contaminants play a major role in emergency response decision making. Because these data are needed immediately on the scene, direct-reading instruments are the apparent choice. However, the use of direct-reading instruments also has some drawbacks to overcome.

First, the material has to be identified to the extent that the appropriate instrumentation can be used. Of course the monitoring equipment must be available on-scene, which suggests sufficient advance planning to anticipate equipment needs or sufficient flexibility of the equipment.

Another factor to consider is that the person performing monitoring in a contaminated area may need the maximum protection during the sampling procedure and must also be competent in field calibration and sampling. If direct-reading instruments do not exist for a particular hazardous material or mixture, then other sampling strategies may have to be developed that include laboratory analysis of the samples. Again, advance planning is important so that the right equipment will be rapidly available, and so that an analytical laboratory is identified and available when the need arises. Even the best advance planning, however, will not enable monitoring data developed from an analytical laboratory to have an impact on the initial emergency response process. In these cases the industrial hygienist will have to work within the constraints of these data gaps.

Ideally, once the identity of the hazardous material or material type has been established and the quantity of the released material has been sufficiently estimated, an evaluation of the situation follows. However, decisions are often made quickly under the time pressures inherent in an emergency response situation, and it is not always possible to wait until all the data are in. Thus, worst-case exposure assumptions may be required if the situation just cannot wait for monitoring equipment or laboratory results to arrive back on scene. Evaluation of the situation must be done when it is needed, with the existing data in hand; so much of the evaluation process may be based on professional judgment, knowledge, and experience within the context of emergency response.

During a hazardous materials incident, the conditions are always changing. Weather fluctuations, involvement of additional substances, fire situations, and effects of containment and control procedures are just a few of the factors that can produce changes in the health and safety situation. Data must be continually collected, and constant vigilance must be maintained so that the health and safety evaluation at the scene is maintained.

Environmental conditions can play a large role in the safety and health evaluation. Although the spilled material hazards determine the type and level of personal protective equipment, the relationship of that equipment, the personnel wearing the equipment and the environmental conditions is complex. If the weather is hot, then heat stress is a major concern. Personnel wearing totally encapsulated suits or other skin protection suits in hot weather may require reduced work time, careful maintenance of body fluids, additional rest time, and closer medical monitoring. In cold weather, or in marine situations where cold water is a concern, protecting personnel against hypothermia is important in the site evaluation. Additionally, the physical integrity of some protective suit materials could be compromised when the temperature drops below freezing.

Ergonomic factors must also be taken into consideration for hazardous material incident activities. The tools used must match the dexterity levels of personnel wearing various glove types and layerings. The impact of wearing PPE must be assessed for personnel operating heavy machinery. The proper handling methods for drums and other containers must also be evaluated.

Noise levels should be part of the hazard evaluation. If the type of equipment used and emergency response environment can produce excessive noise levels, then a hearing protection program to protect responders and cleanup personnel should be included in the response plan.

Sometimes the hazardous material incident is combined with a confined space entry. A confined space is a location that is large enough so that a person can enter, has a limited or restricted means of entry or exit, is not designed for continuous personnel occupancy, and may contain a hazardous atmosphere or lack of oxygen. Special procedures are needed to ensure the safety of personnel working with the additional hazards presented by the confined space. Newer OSHA requirements for confined space entry describe specifics for the entry team and back-up support personnel.

The evaluation of the hazardous materials incident combines an assessment of the chemical and physical hazards, the potential for hazardous exposures, the site and environmental conditions, and the risks to responding personnel. Many factors are weighed and balanced and interpreted. The industrial hygienist should be able to determine a site-specific health and safety strategy based on this evaluation. The interpretation of the hazard and exposure factors will lead to personal protective equipment recommendations that are appropriate for the level of involvement for each particular personnel group. Not all responders are performing the same task, and consequently each group of responders may have differing levels of hazardous exposures. The evaluations should therefore produce a PPE plan tailored to meet each group's objective.

The evaluation should also consider the time line of the incident. As the situation progresses, the release could be stabilized, and the risks could consequently become more defined. Conversely, it may become clear that the incident is progressing adversely and it will become worse as time progresses and the risks to responders' increases. In some cases, the evaluation indicates that nothing can be done, and responders back off to wait for the appropriate time to reenter the scene.

4.1.4 On-Scene Communications

The final step within the process for establishing the on-scene safety plan is communication with the incident commander. The incident commander has the final authority for the safety of every single person on the hazardous material incident scene, and must be fully apprised of the complete extent of the risks from both the hazardous material and the use of the PPE. There is no time for ineffective communication techniques. The key assumptions and variables must be covered efficiently and effectively. The incident commander must be made aware of the reasons for key decisions concerning PPE selection and other key factors in the on-scene plan.

The main factors of immediate concern to the incident commander include the nature of the hazard(s) faced by responding personnel, the levels of exposure, the PPE to be used to minimize exposure, recommended periods of exposure, decontamination procedures, and medical surveillance and/or biological monitoring requirements.

When the incident command system is operating as designed, all decision-critical information is brought to the incident commander, who may have received certain additional information that might change some of the evaluations made by the industrial hygienist. This must be communicated in an effective give-and-take process as the on-scene plan is made final and implemented. In addition, as the emergency response progresses, communication channels providing feedback from the field must be established so that any necessary adjustments to health and safety decisions can be made and quickly implemented.

5 PERSONAL PROTECTIVE EQUIPMENT (PPE) FOR EMERGENCY RESPONSE

One of the major tasks for the industrial hygienist who responds to hazardous material incidents is to select the appropriate level and type of personal protective clothing and equipment for all responding personnel. This decision is based on the evaluation discussed in the preceding section and also on an understanding of the PPE available, its protective features as well as its limitations.

PPE is to be used whenever there is a probability that response personnel will come in contact with a hazardous material. This potential exposure can include gases, liquids, or solids. It can occur through direct exposure activities, such as leak plugging, or indirectly through activities such as helping responders remove PPE during decontamination. If the potential exposure is unknown, the PPE selected must protect the wearer from the maximum credible exposure scenario.

This section first covers the major aspects of an emergency response PPE program, personal protective clothing and its limitations, and finally the different levels of protection defined by the U.S. Environmental Protection Agency (EPA). Respiratory protection is covered in Chapter 32.

5.1 Elements of Hazardous Materials Incidents PPE Program

The purpose of the PPE program is two-fold: (*1*) to protect emergency response personnel from site-specific hazards, and, (*2*) to protect the PPE wearer from injury resulting from incorrect use or malfunction of the PPE. The written program should detail the hazard identification and exposure characterization including any known exposure levels; cover PPE use, limitations, decontamination, maintenance and disposal; medical monitoring requirements, training and proper fitting; donning and doffing procedures; inspection procedures prior to, during, and after use; evaluation of PPE effectiveness; and limitations during temperature extremes; heat stress; and other appropriate medical considerations.

Respiratory protection is of primary importance because inhalation is a significant route of exposure, and hazardous materials in a gaseous state not only have a great potential for dispersion but are invisible in some cases. The principles for selecting respiratory protection for emergency responders are similar to those for the workplace, except that unless a site-specific exposure assessment including air monitoring has been performed, the maximum level of protection is warranted. Emergency responders should be prepared to wear supplied air respirators every time they arrive on scene. Only when the on-scene commander has reviewed on-scene exposure assessments, including monitoring data or other documented information, can lesser respiratory protection other than supplied air be considered.

Emergency responders must also always be prepared for the maximum level of skin and eye protection. Again, reduced levels of protective clothing may be worn when the on-scene commander deems it appropriate, based on site-specific exposure assessments including monitoring data or other documented information.

5.2 Personal Protective Clothing

The three main types of personal protective clothing are structural firefighting clothing, chemical protective clothing, and high-temperature protective clothing. The chemical protective clothing is of greatest importance for the hazardous material emergency responder. Structural firefighting clothing is usually worn to all calls by fire service responders—who may also be the HAZMAT responders in a specific locale. A brief summary of the types and differences of protective clothing follows:

Structural firefighting clothing, commonly called "turnout gear," is the standard PPE worn by firefighters. The gear generally consists of a helmet, fire-resistant hood, positive pressure self-contained breathing apparatus, turnout coat, turnout pants, gloves, and boots. The coats are generally made of three layers: the outer layer provides durability, tear resistance, and some thermal protection; the middle layer is usually made of a waterproof material designed as a moisture barrier; and the inner layer is designed for thermal protection (8). It is primarily designed to afford protection from burns, steam, hot particles, and falling debris resulting from structural fires. This ensemble is worn by fire service personnel to fight structural fires. Use of this clothing for HAZMAT emergency response may result in serious injury. This ensemble is not appropriate when contact with hazardous chemical liquids or vapors are possible. Although sealing of the arms and legs is possible to a limited extent, chemical splashes or vapors can easily penetrate the suit. In addition, the materials commonly used, such as leather, may be difficult if not impossible to decontaminate.

High-temperature protective clothing is designed to protect the wearer for short exposures to either heat or flame. A proximity suit protects the wearer from short exposures close proximity to heat and flame, and is made of a highly reflective aluminized outer surface over an inner shell of flame-retardant fabric. The fire entry suit offers protection for short-duration entries into a flame environment. It is composed of multiple layers of flame-retardant material. Both types of suits are of limited use in firefighting and have no significant chemical protection properties.

When any chemicals are present, selection of proper protective clothing can be complex and should be performed by a knowledgeable and experienced industrial hygienist. Choosing the most appropriate clothing depends on the chemicals present and the tasks to be performed. In the selection process, the industrial hygienist balances the performance of the protective clothing to protect against exposure, the physical limitations of using different types of protective clothing, and the site-specific factors of the incident.

Performance against chemical exposure depends on how well the protective clothing material resists permeation, degradation, and penetration. Permeation is the process whereby the chemical moves through protective clothing material by dissolving through the molecular structure of the clothing material. Degradation occurs when chemical exposure, use, or environmental conditions actually break down the protective clothing material through a change in the fabric's chemical composition or structure. Penetration is direct passage of a chemical through openings such as zippers, seams, or imperfections in the protective clothing material. Physical limitations that may influence the selection of protective clothing for response personnel include, but are not limited to, heat stress, excessive mobility or vision restrictions, and incompatibility with appropriate respiratory protection.

To make decisions on appropriate PPE, the industrial hygienist can consult a variety of sources that present the wide range of materials in a matrix with the chemicals to be protected against. A sample matrix is presented in Table 62.3. Bccausc no single material is completely impermeable to all chemicals, one key factor in the selection process is determining which materials will allow slowest permeation, and/or last the longest under the conditions of use. In some cases the industrial hygienist must estimate the maximum exposure times allowable before permeation or degradation becomes too great a risk and give the responders a time limit for exposure. Increasing thickness or doubling protection may be one appropriate remedy in some cases. In the case of mixtures or incidents involving multiple chemicals, a combination of protective clothing must be worn.

Table 62.3 Chemical Protection of Clothing Materials by Generic Class[a]

Generic Class	Butyl Rubber	Poly (vinyl chloride)	Neoprene	Natural Rubber
Alcohols	E	E	E	E
Aldehydes	E–G	G–F	E–G	E–F
Amines	E–F	G–F	E–G	G–F
Esters	G–F	P	G	F–P
Esthers	G–F	G	E–G	G–F
Fuels	F–P	G–P	E–G	F–P
Halogenated hydrocarbons	G–P	G–P	G–F	F–P
Hydrocarbons	F–P	F	G–F	F–P
Inorganic acids	G–F	E	E–G	F–P
Inorganic bases and salts	E	E	E	E
Ketones	E	P	G–F	E–F
Natural fats and oils	G–F	G	E–G	G–F
Organic acids	E	E	E	E

[a]Key: E, excellent; G, good; F, fair; P, poor.

Chemicals in mixtures can present a special challenge for industrial hygienists because determining appropriate protective clothing can be complex when one chemical can act as a carrier across the barrier, actually enhancing permeation of a second hazardous chemical. Environmental conditions of cold, rain, humidity, or sun and dryness may also change the rate of permeation or degradation. The industrial hygienist must carefully weigh many factors before making protective clothing selections. Material flexibility given the ambient conditions is another important factor, particularly for glove selection.

Materials used in personal protectives clothing regularly fall into three classes: elastomers, nonelastomers, and blends. Some examples of elastomers are rubber, butyl rubber, polyethylene, chlorinated polyethylene, poly(vinyl chloride), nitrile, polyurethane, poly(vinyl alcohol), and neoprene. Viton(TM) is a DuPont brand of fluoroelastomer. Nonelastomers include leather, Nomex(TM) and Tyvek(TM), DuPont brand products, and Goretex(TM). Blend materials often layer combinations of the different materials listed above to provide protection against a wider variety of hazards in a wider range of environmental conditions. The wide range of different PPE materials available highlights the need for proper advance planning and practice prior to an incident.

As an aid in selecting suitable chemical protective clothing, the National Fire Protection Association has developed standards on chemical protective clothing (9), including:

- NFPA 1991—Standards on Vapor-Protective Suits for Hazardous Chemical Emergencies (EPA Level A Protective Clothing)
- NFPA 1992—Standard on Liquid Splash-Protective Suits for Hazardous Chemical Emergencies (EPA Level B Protective Clothing)
- NFPA 1993—Standard on Support Function Protective Garments for Hazardous Chemical Operations (EPA Level B Protective Clothing)

These standards apply documentation and performance requirements to the manufacture of chemical protective suits.

5.3 PPE Ensembles for Hazardous Materials Incidents

Personal protective clothing when teamed with respiratory protection is often called an ensemble. The EPA has defined four levels of protection that combine the different protective values of respiratory PPE and protective clothing. These levels are termed Level A, B, C, and D. Each level is detailed below, but these groupings are considered the starting point for developing ensembles. Many site-specific circumstances will lead the industrial hygienist to recommend modifications for each of the basic levels of protection.

Level A provides maximum protection against chemical exposure. It fully protects the eyes, skin, and respiratory system from exposure to hazardous materials. This level is selected when the chemical that has been identified has a high degree of skin, eye, or respiratory hazard, when the chemical is suspected to be a high skin hazard, when operations are conducted in a poorly ventilated or confined space, perhaps with less than 19.5% oxygen, when there are unknown vapors or gases in the air, or when potential skin exposures to an unknown chemical may occur. One task that often requires Level A PPE at a hazardous material incident includes the initial walk-through or monitoring of a site where vapor, gas, or particulate skin hazards are or may be present. The key elements of the Level A protection are the fully encapsulating suit, and the pressure-demand supplied-air breathing apparatus or pressure-demand airline respirator. Table 62.4 provides a full list of the ensemble equipment.

If level A protection is used, a minimum of four persons should be equipped. This is because entry into an area with potential exposures warranting Level A protection requires use of a "buddy system" so that an individual who might be injured or accidentally exposed, or has problems with heat stress, has a partner in visual contact to provide immediate assistance. It is prudent practice to have two additional personnel suited up in a "ready" state, in case the first pair encounter unexpected trouble and cannot escape. The backups

Table 62.4 Level A and Personal Protective Equipment List[a]

Level A Protective Equipment	Level B Protective Equipment
To be selected when the greatest level of skin, respiratory, and eye protection is required. Pressure-demand, full-facepiece supplied air respirator approved by NIOSH: pressure-demand self-contained breathing apparatus (SCBA), or pressure-demand airline respirator (with an escape bottle for IDLH conditions) Totally encapsulating chemical resistant suit Chemical-resistant gloves (inner) Chemical-resistant gloves (outer) Chemical resistant boots with steel toe and shank Two-way radio communicator (intrinsically safe) *Optional* Hard hat (under suit) Long cotton underwear Disposable protective suit, gloves and boots (depending on suit construction, may be worn over totally-encapsulating suit.) Body cooling unit Coveralls	*Recommended when the highest level of respiratory protection is necessary but a lesser level of skin protection is needed. This is the minimum acceptable level for initial site entries until the hazards have been further identified.* Pressure-demand, full facepiece supplied-air respirator approved by NIOSH: pressure-demand self-contained breathing apparatus (SCBA), or pressure-demand, airline respirator (with escape bottle for IDLH conditions) Hooded chemical resistant clothing (overalls and long sleeved jacket; coveralls; one- or two-piece chemical splash suit, or disposable chemical-resistant overalls) Chemical-resistant inner and outer gloves Hard hat Chemical-resistant outer boots with steel toe and shank Two-way radio communicator (intrinsically safe) *Optional* Boot-covers, outer, chemical resistant (disposable) Long cotton underwear Coveralls Face shield

[a]The protective clothing must be compatible with the specific substance involved in the incident.

are needed because the exposure situation is clearly hazardous, and all rescuers need the maximum level of protection as well.

Because Level A PPE is impermeable to many substances, it also is very hot and claustrophobic to wear. Heat stress is a genuine hazard to personnel wearing Level A protection even in temperate weather conditions. Appropriate medical surveillance precautions must be taken. Also, unless an airline respirator is used, the combination of limited air, heat stress, equipment weight, suit bulkiness, and limited visibility severely limit the useful work time the personnel may have in the exposure zone.

Level B protection is similar to Level A protection with the exception of slightly reduced skin protection. Level B is worn when the skin hazard of chemical exposure is known not to be significantly hazardous. Level B can be worn in Immediately Dangerous to Life and Health (IDLH) atmospheres, if those concentrations do not pose a skin hazard or if the chemicals present do not meet the selection criteria to use air-purifying respirators. Essentially, supplied air is still needed, but the fully encapsulating suit is not. Although the protective clothing is not fully encapsulated against gases, vapors, or particulates, it still affords a very substantial level of skin protection. The protective clothing is chemical resistant, hooded, and one piece, with taped ankles and wrists over chemical-resistant boots and gloves.

This level may also be used for initial site entry and reconnaissance, if a lack of skin hazard is verified. Level B is slightly less bulky and cumbersome than Level A, but can be just as hot and stressful. Also, the supplied air canisters are on the outside of the protective clothing, as opposed to Level A where they are inside the encapsulated suit. When the air canisters are outside the protective clothing, they are subject to contamination and may therefore become additional equipment to decontaminate. The list of Level B equipment is in Table 62.4.

Level C protection maintains the chemical-resistant suit protection of Level B but allows for reduced respiratory protection. Air-purifying respirators are utilized in Level C protection in combination with the chemical-resistant suit, with hood, gloves, and boots. The chemical-resistant suit may also be downgraded one step to a two-piece outfit depending on anticipated exposure conditions. In order to utilize air-purifying respirators, all criteria for selection must be met. This requires identification not only of the chemicals but also their concentrations, and oxygen levels must be 19.5%. The site must not be likely to change exposure conditions or generate unknown compounds or excessive levels of known substances. Generally, the full-face air-purifying respirator is used. Level C has the advantages of providing a lighter, less bulky protection ensemble, and the personnel can usually work for longer periods of time than if they were wearing supplied air. The equipment list for Level C protection is given in Table 62.5.

Level D protection is primarily a work uniform to be worn where no respiratory hazard is known to exist. Respiratory protection is not needed, and skin protection is provided through coveralls. Wearing safety glasses is recommended as good safety practice in most situations, and splash goggles may be used in specific instances. Boots are also recommended. Gloves are optional depending on the chemicals involved and the task. Hazardous materials responders would not generally be outfitted in Level D, unless it is a very small, well-defined incident, such as a small container spill in a workplace, where the hazards of

Table 62.5 Level C and D Personal Protective Equipment List[a]

Level C Protective Equipment List	Level D Protective Equipment List
Recommended when the amount and type of airborne substance(s) is known and the criteria for using air purifying respirators are met. The same level of skin protection as Level B, but a lower level of respiratory protection.	*Work uniform offering minimal protection, used only for nuisance contamination. No respiratory protection and minimal skin protection.*
Air purifying respirator, full face or half-mask, canister-equipped, approved by NIOSH:	Safety boots/shoes, leather or chemical resistant, steel toe and shank
Hooded chemical resistant clothing (overalls or hooded, one-piece or two-piece chemical splash suit, or chemical-resistant hood and apron, or disposable chemical-resistant coveralls)	Coveralls
Chemical-resistant outer gloves	Hard hat
Chemical-resistant inner gloves	Safety glasses or chemical splash goggles
Chemical-resistant outer boots and steel toe and shank	*Optional*
Hard hat	Gloves
Two-way radio communicator (intrinsically safe)	Escape mask
Optional	Face shield
Coveralls	
Disposable boots covers	
Face shield	
Escape mask	
Long cotton underwear	

[a]The protective clothing must be compatible with the specific substance involved in the incident.

the material are well known and are of minimal risk to health and safety. Table 62.5 lists Level D equipment.

Intermediate ensembles combining elements of different levels are also used. For hazardous materials incidents where monitoring has established that respiratory protection is not needed, and the primary hazard of concern is skin exposure, personnel are usually outfitted in "modified Level C," where the chemical-resistant clothing of Level C is used, but without the respiratory protection. Glasses or goggles are used in place of full-face respirators.

In summary, the EPA levels of protection provide guidelines for PPE ensembles. The specifics of equipment such as gloves or double gloves, boots, and varieties of chemical protective suits are determined on a site-specific basis. PPE is also in a constant state of improvement, as manufacturers are now regularly modifying and improving the materials, blends, and designs, and introducing new products for the expanding hazardous materials response market. Much PPE can be purchased for disposable or multipurpose use. Contact manufacturers and local equipment suppliers regularly to learn of the latest specifications and ratings for new products.

6 EMERGENCY RESPONSE MONITORING

Monitoring the hazardous exposures at a hazardous materials incident is a critical function. Without accurate monitoring data, the on-scene incident commander will not have any basis to allow a reduction in PPE below maximum protection. Although it may be possible to determine that skin hazards may not be a factor, and reduce the level of protection to Level C, supplied air would still be required in a potentially hazardous atmosphere if exposure assessment and air monitoring did not show otherwise. Monitoring throughout the incident, even if not direct reading, will be valuable as documentation of exposures, which may be important if any injuries from chemical exposure are incurred, or if any litigation results. Sampling and analysis after the incident has stabilized or moved into the cleanup phase will also play an important role in the overall assessment of exposures and contamination caused by the hazardous materials spill.

Discussion of monitoring instrumentation can be found in Chapters 8 and 9.

7 DECONTAMINATION PRINCIPLES

7.1 Decontamination Planning and Implementation

Decontamination (decon) is a very critical process in a hazardous materials emergency. It is the process of removing or neutralizing contaminants from personnel and their equipment. All the extensive efforts to protect emergency responders from exposure to hazardous materials are wasted if decon is not effective. Materials transferred out of the hazardous material zone can produce adverse effects on the responders, support personnel, their families, and the next group of responders who use contaminated equipment. Very specific procedures must be set up and implemented in the decon plan. These procedures may range from a multi-step cleaning procedure to carefully disposing of all equipment and clothing in a hazardous waste disposal bin. Responders must also be concerned about exposure to bloodborne pathogens, particularly if they must deal with incident victims who have suffered bloody trauma, and must implement proper decon procedures to prevent contamination from blood or other potentially infectious agents.

The decon plan contains many elements that are site- and material-specific. However, based on advance planning, it may be possible to anticipate how the decon process will be handled. For instance, knowing that disposable protective clothing will be available and utilized enables the decon process to be structured one way, whereas knowing that expensive and high-maintenance reusable encapsulating suits will be utilized ensures that the decon setup will be handled differently.

Decontamination is one significant component of the site zone control system. When the PPE selection process and the site control zones are being established, the decon plan and setup must be established as well. Personnel must never enter a contamination zone without having the decon area fully prepared (see Figure 62.3).

The zone control system helps ensure that only properly protected personnel enter a "hot zone" where hazardous materials are present. The responders always enter the "hot zone" through controlled entry points from the "warm zone," and always exit the "hot zone" through a decon area located in a "warm zone," and then exit the decontamination area into the "cold zone." This system helps ensure that responders who are in the area of major contamination are wearing the proper protective equipment, and that residual contamination from the decon area is isolated from personnel who are not wearing any protection in the cold zone. Also, personnel performing decon procedures on the exiting responders may also be required to wear substantial PPE. Determining the proper level of protection for decon personnel is a major part of the decon plan.

The plan consists of six basic elements: (*1*) determining the number and layout of the decon stations, (*2*) determining the decon equipment needed, (*3*) determining the appropriate decon methods, (*4*) setting up procedures to prevent contamination of clean areas, (*5*) establishing work practices to minimize personnel contact with contaminants in the hot zone and while removing contaminated equipment, and (*6*) establishing the methods for properly disposing of materials that are not fully decontaminated. The plan should be flexible to change with changing conditions or equipment needs, or if the site hazards are reassessed after new information is received.

7.2 Contamination Prevention

Preventing contamination is a major aspect of the process. Work practices such as walking around contamination or not directly touching contamination must be stressed. Whenever

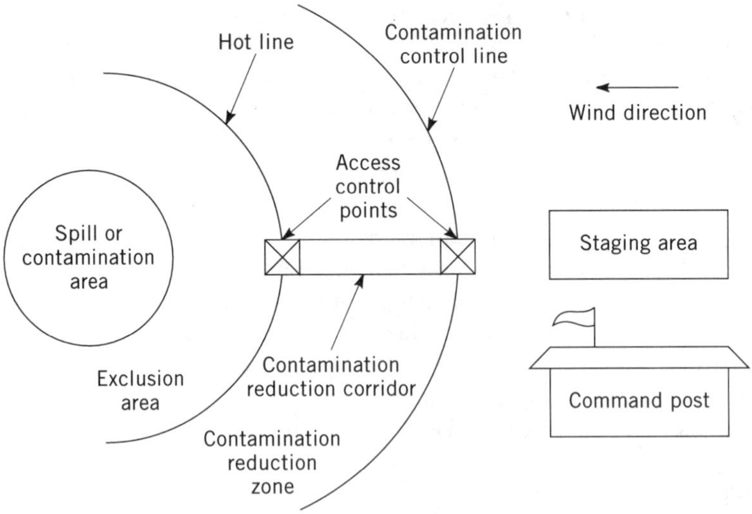

Figure 62.3. Site Zones and Decontamination

possible, remote handling equipment should be used. Monitoring and sampling instruments and equipment can be bagged. Sampling ports and sensors must be exposed through openings in the bag. Disposable outer garments and equipment are very effective in reducing the amount of decon to perform. Decontamination usually produces a certain amount of contaminated material such as wastewater or sorbent wipes. The purpose of reducing decon is partially to reduce contaminated waste.

7.3 Decontamination Methods

There are two basic decontamination methods: physical and chemical removal. Physical removal can consist of a variety of actions that include rinsing, scraping, and scrubbing, or using stream jets. The runoff from all these activities must be contained and disposed of as hazardous waste, or otherwise treated. It cannot be allowed to flow freely, spreading contamination.

Chemical removal can include dissolving contaminants with a solvent, utilizing specific surfactants, or solidifying the contamination through gelling or catalyzing agents, removal of water substrate with absorbents, or freezing. Again, the runoff or residue from these activities must be carefully collected and disposed or treated as a hazardous waste.

The effectiveness of the decon process must be monitored to demonstrate that the contaminants are indeed removed. If the decon methods prove ineffective, they must be revised. For example, if the hazardous material has permeated the protective clothing to the extent that it cannot be removed by scrubbing with the appropriate wash solution, disposal should be considered. Visual observation, wipe sampling, analysis of cleaning solution runoff, and even permeation testing should each be considered, as necessary. Consider whether any of the materials or processes in use could, by their nature, be hazardous. Compatibility of the contaminants and cleaning agents, compatibility of the cleaning agents and the protective clothing materials, and the toxicity of the cleaning agents should be assessed.

7.4 Decontamination Facility Design

The general design of the decon facility or area should be based on the principle that the area closest to the exit of the hot zone will receive the most contaminated personnel and equipment, and the decon area exits into the cold zone where emerging personnel should be free of contamination. The actual design is highly site specific. The first stages handle the outer garments and equipment, and later stages handle the inner garments. Each removal or cleaning is done stepwise to prevent cross contamination, and is often set up in a line.

The design must also take into account potential emergency decontamination for injured personnel. The decision whether or not to decontaminate a victim is based on the type and severity of the illness or injury and the nature of the contaminant. For some emergency victims, immediate decontamination may be an essential part of life-saving first aid. For others, decon may aggravate the injury or delay life-saving treatment. If decon does not interfere with essential treatment, it should be performed (11). It is highly desirable to avoid contaminating an ambulance, emergency room, etc. Some hospitals are not allowed

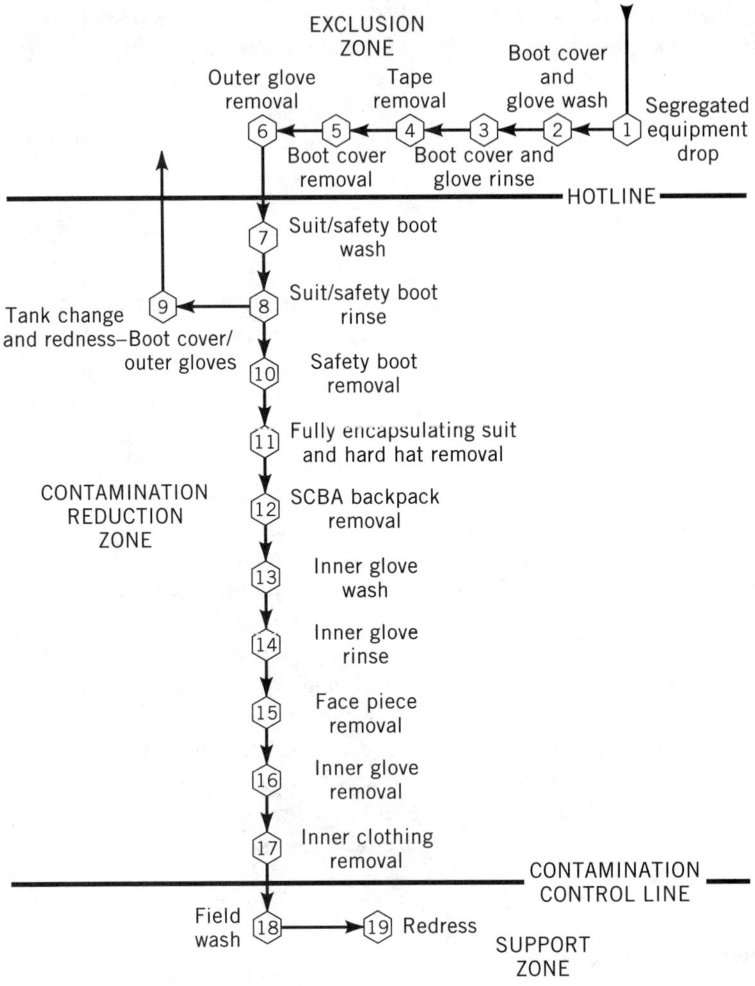

Figure 62.4. Decontamination Facility—Level A Protection

to accept contaminated personnel. The above situations should be addressed when planning for decon in medical emergencies.

A typical Level A maximum decontamination facility is outlined in Figure 62.4. Table 62.6 provides a general list of some of the materials needed for decon facilities. Of course, if disposable materials are being heavily relied upon, the decon facility does not have to be as extensive as one where multiple-step washes and rinses are used. When materials are collected for disposal, it must be done in a manner that complies with all local and other applicable regulations governing the generation and disposal of hazardous waste. Advance planning must address this issue.

Table 62.6 Decontamination Equipment and Methods

Typical Decontamination Equipment and Materials
Plastic tarps, heavy polyethylene sheeting
Portable plastic swimming pools
Soft brushes, long and short handles
Buckets
Water supply
Hoses, spray applicators, water pressure spray units
Salvage drums
Drum liners
Equipment hangers (for SCBA, boots, etc.)
Duct tape
Disposable wiping materials/absorbents
Liquid spill containment absorbent materials
Contaminant neutralization chemicals
Additional Equipment
Personal protective equipment for decon personnel
Contaminant test equipment

8 RESOURCES FOR ASSISTANCE

Many on-line resources are now available to assist in advance planning and on-scene management of hazardous materials incidents. Visit the website for OSHA's HAZWOPER emergency response at www.osha-slc.gov/SLTC/EmergencyResponse/index.html

Appendix B represents emergency response references and resources from the EPA/Labor Superfund Health and Safety Task Force, published by the OSHA Office of Health Compliance Advisor. References 5, 7, 10–13 are good general references.

APPENDIX A. MODEL SITE SAFETY PLAN

This appendix provides a generic plan based on a plan developed by the U.S. Coast Guard for responding to hazardous chemical releases (14). This generic plan can be adapted for designing a Site Safety Plan for hazardous waste site cleanup operations. It is not all inclusive and should only be used as a guide, not a standard.

A. SITE DESCRIPTION

Date ____________ Location ____________

Hazards ____________

Area affected ____________

____

Surrounding population ____________

Topography ____________

Weather conditions ____________

____

____

Additional information ____________

____

____

____

____

B. ENTRY OBJECTIVES—The objective of the initial entry to the contaminated area is to (describes actions, tasks to be accomplished; i.e., identify contaminated soil; monitor conditions, etc.) ____________

____

____

____

C. ONSITE ORGANIZATION AND COORDINATION—The following personnel are designated to carry out the stated job functions on site. (Note: One person may carry out more than one job function.)

PROJECT TEAM LEADER ____________

SCIENTIFIC ADVISOR ____________

SITE SAFETY OFFICER ____________

PUBLIC INFORMATION OFFICER ____________

SECURITY OFFICER ____________

RECORDKEEPER ____________

FINANCIAL OFFICER ____________

FIELD TEAM LEADER ____________

FIELD TEAM MEMBERS ____________

____

____

____

____

____

FEDERAL AGENCY REPS (i.e., EPA, NIOSH) ____________

____

____

____

APPENDIX A. (continued)

STATE AGENCY REPS ______________________________

LOCAL AGENCY REPS ______________________________

CONTRACTOR(S) ______________________________

All personnel arriving or departing the site should log in and out with the Recordkeeper. All activities on site must be cleared through the Project Team Leader.

D. ONSITE CONTROL

(Name of individual or agency) has been designated to coordinate access control and security on site. A safe perimeter has been established at (distance or description of controlled area)

No unauthorized person should be within this area.

The onsite Command Post and staging area have been established at ______________________________

The prevailing wind conditions are ______________________________. This location is upwind from the Exclusion Zone.

Control boundaries have been established, and the Exclusion Zone (the contaminated area), hotline, Contamination Reduction Zone, and Support Zone (clean area) have been identified and designated as follows: (describe boundaries and/or attach map of controlled area)

These boundaries are identified by: (marking of zones, i.e., red boundary tape—hotline; traffic cones—Support Zone; etc.)

E. HAZARD EVALUATION

The following substance(s) are known or suspected to be on site. The primary hazards of each are identified.

Substances Involved	Concentration (If Known)	Primary Hazards
(chemical name)		(e.g., toxic on inhalation)

APPENDIX A. (continued)

The following additional hazards are expected on site: (i.e., slippery ground, uneven terrain, etc.) ________

Hazardous substance information form(s) for the involved substance(s) have been completed and are attached.

F. PERSONAL PROTECTIVE EQUIPMENT

Based on evaluation of potential hazards, the following levels of personal protection have been designated for the applicable work areas or tasks:

Location	Job Function	Level of Protection
Exclusion Zone	________	A B C D Other
	________	A B C D Other
	________	A B C D Other
	________	A B C D Other
Contamination Reduction Zone	________	A B C D Other
	________	A B C D Other
	________	A B C D Other
	________	A B C D Other

Specific protective equipment for each level of protection is as follows:

Level A
- Fully encapsulating suit
- SCBA
- (disposable coveralls)

Level C
- Splash gear (type)
- Full-face canister respirator

Level B
- Splash gear (type)
- SCBA

Level D ________

Other ________

The following protective clothing materials are required for the involved substances:

Substance	Material
(chemical name)	(material name, e.g., Viton)

APPENDIX A. (continued)

If air-purifying respirators are authorized, __(filtering medium)__ is the appropriate canister for use with the involved substances and concentrations. A competent individual has determined that all criteria for using this type of respiratory protection have been met.

NO CHANGES TO THE SPECIFIED LEVELS OF PROTECTION SHALL BE MADE WITHOUT THE APPROVAL OF THE SITE SAFETY OFFICER AND THE PROJECT TEAM LEADER.

G. ONSITE WORK PLANS

Work party(s) consisting of _____ persons will perform the following tasks:

Project Team Leader	__(name)__	__(function)__
		________
		________
Work Party #1	________	________
		________
		________
		________
Work Party #2	________	________
		________
		________
		________
Rescue Team (required for entries to IDLH environments)	________	________
		________
		________
		________
		________
Decontamination Team	________	________
		________
		________
		________
		________

The work party(s) were briefed on the contents of this plan at____________________.

H. COMMUNICATION PROCEDURES

Channel______ has been designated as the radio frequency for personnel in the Exclusion Zone. All other onsite communications will use channel ______.

Personnel in the Exclusion Zone should remain in constant radio communication or within sight of the Project Team Leader. Any failure of radio communication requires an evaluation of whether personnel should leave the Exclusion Zone.

__(Horn blast, siren, etc.)__ is the emergency signal to indicate that all personnel should leave the Exclusion Zone. In addition, a loud hailer is available if required.

APPENDIX A. (continued)

The following standard hand signals will be used in case of failure of radio communications:

Hand gripping throat ———————————— Out of air, cannot breathe
Grip partner's wrist or ———————————— Leave area immediately
both hands around waist
Hands on top of head ———————————— Need assistance
Thumbs up ———————————————— OK, I am all right, I understand
Thumbs down ——————————————— No, negative

Telephone communication to the Command Post should be established as soon as practicable. The phone number is ____________________.

I. DECONTAMINATION PROCEDURES

Personnel and equipment leaving the Exclusion Zone shall be thoroughly decontaminated. The standard level ______ decontamination protocol shall be used with the following decontamination stations:

(1) ______________ (2) ______________ (3) ______________ (4) ______________
(5) ______________ (6) ______________ (7) ______________ (8) ______________
(9) ______________ (10) ______________ Other ______________________________

Emergency decontamination will include the following stations: ______________________________
____
____
____

The following decontamination equipment is required: ______________________________
____
____
____

(Normally detergent and water) will be used as the decontamination solution.

J. SITE SAFETY AND HEALTH PLAN

1. (name) is the designated Site Safety Officer and is directly responsible to the Project Team Leader for safety recommendations on site.

2. Emergency Medical Care

(names of qualified personnel) are the qualified EMTs on site.
(medical facility names) , at (address) ,
phone ______________ is located ______ minutes from this location.
(name of person) was contacted at (time) and briefed on the situation, the potential hazards, and the substances involved. A map of alternative routes to this facility is available at (normally Command Post) .

Local ambulance service is available from ______________________________ at phone ______________. Their response time is ______ minutes.
Whenever possible, arrangements should be made for onsite standby.

APPENDIX A. (continued)

First-aid equipment is available on site at the following locations:

First-aid kit ______________________
Emergency eye wash ______________________
Emergency shower ______________________
(other) ______________________

Emergency medical information for substances present:

Substance	Exposure Symptoms	First-Aid Instructions

List of emergency phone numbers:

Agency/Facility	Phone #	Contact
Police		
Fire		
Hospital		
Airport		
Public Health Advisor		

3. Environmental Monitoring

The following environmental monitoring instruments shall be used on site (cross out if not applicable) at the specified intervals.

Combustible Gas Indicator - continuous/hourly/daily/other ______________
O_2 Monitor - continuous/hourly/daily/other ______________
Colorimetric Tubes - continuous/hourly/daily/other ______________
(type) ______________
______________ ______________
______________ ______________
HNU/OVA - continuous/hourly/daily/other ______________
Other______________ - continuous/hourly/daily/other ______________
______________ - continuous/hourly/daily/other ______________

4. Emergency Procedures (should be modified as required for incident)

The following standard emergency procedures will be used by onsite personnel. The Site Safety Officer shall be notified of any onsite emergencies and be responsible for ensuring that the appropriate procedures are followed.

APPENDIX A. (continued)

Personnel Injury in the Exclusion Zone: Upon notification of an injury in the Exclusion Zone, the designated emergency signal ______________________ shall be sounded. All site personnel shall assemble at the decontamination line. The rescue team will enter the Exclusion Zone (if required) to remove the injured person to the hotline. The Site Safety Officer and Project Team Leader should evaluate the nature of the injury, and the affected person should be decontaminated to the extent possible prior to movement to the Support Zone. The onsite EMT shall initiate the appropriate first aid, and contact should be made for an ambulance and with the designated medical facility (if required). No persons shall reenter the Exclusion Zone until the cause of the injury or symptoms is determined.

Personnel Injury in the Support Zone: Upon notification of an injury in the Support Zone, the Project Team Leader and Site Safety Officer will assess the nature of the injury. If the cause of the injury or loss of the injured person does not affect the performance of site personnel, operations may continue, with the onsite EMT initiating the appropriate first aid and necessary follow-up as stated above. If the injury increases the risk to others, the designated emergency signal ______________________ shall be sounded and all site personnel shall move to the decontamination line for further instructions. Activities on site will stop until the added risk is removed or minimized.

Fire/Explosion: Upon notification of a fire or explosion on site, the designated emergency signal ______________________ shall be sounded and all site personnel assembled at the decontamination line. The Fire Department shall be alerted and all personnel moved to a safe distance from the involved area.

Personal Protective Equipment Failure: If any site worker experiences a failure or alteration of protective equipment that affects the protection factor, that person and his/her buddy shall immediately leave the Exclusion Zone. Reentry shall not be permitted until the equipment has been repaired or replaced.

Other Equipment Failure: If any other equipment on site fails to operate properly, the Project Team Leader and Site Safety Officer shall be notified and then determine the effect of this failure on continuing operations on site. If the failure affects the safety of personnel or prevents completion of the Work Plan tasks, all personnel shall leave the Exclusion Zone until the situation is evaluated and appropriate actions taken.

The following emergency escape routes are designated for use in those situations where egress from the Exclusion Zone cannot occur through the decontamination line: (describe alternative routes to leave area in emergencies)

____

____

____

In all situations, when an onsite emergency results in evacuation of the Exclusion Zone, personnel shall not reenter until:

1. The conditions resulting in the emergency have been corrected.
2. The hazards have been reassessed.
3. The Site Safety Plan has been reviewed.
4. Site personnel have been briefed on any changes in the Site Safety Plan.

APPENDIX A. (continued)

5. Personal Monitoring

The following personal monitoring will be in effect on site:

Personal exposure sampling: ____(describe any personal sampling programs being carried out on site personnel. This would include use of sampling pumps, air monitors, etc.)____

Medical monitoring: The expected air temperature will be (°F). If it is determined that heat stress monitoring is required (mandatory if over 70°F) the following procedures shall be followed: ____(describe procedures in effect, i.e., monitoring body temperature, body weight, pulse rate)____

All site personnel have read the above plan and are familiar with its provisions.

Site Safety Officer	(name)	(signature)
Project Team Leader		
Other Site Personnel		

APPENDIX B. LOCAL EMERGENCY RESPONSE COORDINATION: REFERENCES AND RESOURCES

Developed by OSHA's Office of Health Compliance Assistance, (202) 693-2190. Updates and/or additions to this worksheet are encouraged. Please contact MaryAnn Garrahan at *MaryAnn.Garrahan@osha-no.osha.gov*.

LISTSERVS

LEPC: Discussion of Local Emergency Planning Committee operations and local emergency response	posting address: *LEPC@list.uvm.edu* administrative address: *listproc@list.uvm.edu* human contact: Ralph Stuart, *rstuart@moose.uvm.edu*
California Emergency Medical Services Mail List	posting address: *CAEMS@ucdavis.edu* administrative address: *listproc@ucdavis.edu* human contact: *melochs@aol.com*

Internet Resources

Provider or Subject	Type of Resource	Internet Address (http://)
Hazardous Materials Emergency Preparedness (HMEP) Grants Program	Provides information about grant monies that can be used for emergency response hazmat training, and for LEPC HAZMAT planning	www.fema.gov/eml/hmep/bkground.htm
Local Government Reimbursement (LGR)	Information about obtaining federal funds for local governments for incidents involving the release of hazardous substances	www.epa.gov/oerrpage/superfund/programs/er/lgr/index.htm
National Fire Protection Association (NFPA)	Information about NFPA publications, standards, training. Key NFPA consensus standards for HAZMAT emergency response include NFPA 471, 472, and 473.	www.nfpa.org/home.html
NRT–One Plan Guidance	Integrated Contingency Plan (ICP) or One Plan Guidance–Intended to be used by facilities to prepare emergency response plans. Contains the suggested ICP outline as well as guidance on how to develop an ICP and demonstrate compliance with various regulatory requirements.	www.epa.gov/swercepp/pubs/one-plan.html
IAFF Hazardous Materials Training Dept.	International Association of Fire Fighters webpage for finding HAZMAT training across the country. Includes a distance learning option.	www.iaffhazmat.org
LEPC/SERC Net	Local Emergency Planning Committee/State Emergency Response Commission online network showcases several LEPC home pages, gives access to EPA information, and provides a discussion forum for current issues related to emergency planning and community right-to-know	www.rtk.net/www/lepc/webpage/mosaic.html
	LEPC/SERC Net—Documents and Resources that relate to emergency response and local coordination	www.rtk.net/lepc/webpage/documents.html

APPENDIX B. (continued)

Federal Emergency Management Agency (FEMA)	Emergency Management Institute Training Center—Provides links to the Emergency Management Institute, the National Fire Academy, and Independent Study Courses available through FEMA's Emergency Education Network, called EENET.	www.usfa.fema.gov/emi.training.htm
	Preparedness, Training and Exercises Directorate—Provides emergency planning fact sheets for local communities and families that can be used for public outreach. Links to Emergency Information Infrastructure Partnership (EIIP) Virtual Forum which provides links to discussion groups, training programs, conferences and workshops, research findings, on-line publications, and information about tools with application in the field of emergency management	www.fema.gov/pte/about.htm
Fire and EMS Information Network	Comprehensive on-line resource that provides links to community fire and EMS departments, publications, and training schools. A message board allows you to post questions/answers to other professionals about training, response planning, standards, etc. Includes the possibility to add your local department to the site.	www.fire-ems.net/
OSHA Emergency Response Page	The HAZWOPER emergency response web page provides training information, OSHA fact sheets, standards, and publications.	www.osh-slc.gov/SLTC/emergencyresponse/index.html
National Response Center (NRC)	Provides information about the basic reporting requirements that are necessary within the first 24 hours of a chemical spill.	www.nrc.useg.mil/index.html
NIOSH Pocket Guide to Chemical Hazards	Provides employers with chemical synonyms, exposure limits, incompatibilities and reactivities, respirator selection, signs and symptoms of exposure, and procedures for emergency treatment. Also available from NTIS at (800) 553-6847 or (703) 605-6000	www.edc.gov/niosh/npg.html
EPA Chemical Emergency Preparedness and Prevention Office (CEPPO)	Provides employers with access to laws, regulations, databases and software, publications, and Internet links that can be used for emergency prevention planning, emergency preparedness, emergency response, and the development of a Risk Management Program.	www.epa.gov/swercepp/index.html
EPA RCRA, Superfund, and EPCRA Hotline Webpage	Regulations developed under RCRA, Superfund, and EPCRA including reports, documents, state and local contracts, training modules, and regional contacts are found on-line to provide information on coordinating appropriate emergency response.	www.epa.gov/epaoswer/hotline

APPENDIX B. (continued)

National Response Team (NRT)	Offers relevant statutes and regulations, national guidance reports, accident databases, and lessons learned for HAZMAT and oil-related emergency response.	www.nrt.org
CHEMTREC (Chemical Transportation Emergency Center)	Emergency response information such as spill control and firefighting, emergency medical treatment, manufacturer contacts, and chemical information from a database of 1.5 million MSDSs. Lending library of audiovisual materials for training emergency response personnel.	www.cmahq.com/cmawebsite.nsf/pages/chemtrec
Agency for Toxic Substances and Disease Registry (ATSDR)	Offers toxicological profiles of hazardous substances that describe exposure routes, harmful exposure levels, health effects, and background information. Provides answers to frequently asked questions (FAQs) about exposure to hazardous substances and their effects on human health. Provides information on where to find Material Safety Data Sheets on the Internet	www.atsdr.ede.gov/atsdrhome.html
OSHA	Access to OSHA information including compliance guidance, publications, and standards.	www.osha.gov
OEM/Duke University	A valuable collection of health and safety websites including federal, private, and international agencies	oce-env-med.mc.duke.edu/oem/index2.htm

Publications

Many of the above publications can be ordered on-line or downloaded from the Internet home page of the originating Agency.

OSHA	*Field Directive for 1910.120(q), CPL 2-2.59A* (available on-line on OSHA website)	(202) 693-1888
OSHA	*Hospital and Community Emergency Response: What You Need to Know* (OSHA 3152, on-line)	(202) 693-1888
EPA	*NRT One Plan Guidance* (available on OSHA's website)	(800) 424-9346
EPA/NTIS	*An Overview of the Emergency Response Program* EPA 540/8-91/015	(800) 553-6847
DOT	*Emergency Response Guidebook* (NAERG96, online at http://hazmat.dot.gov/gydebook.htm)	(800) 327-6868
EPA	*Hazardous Materials Emergency Planning Guide NRT-1*	(800) 535-0202
EPA	*Criteria for Review of HAZMAT Emergency Plans NRT-1A*	(800) 535-0202
EPA	*Developing a HAZMAT Exercise Program, Handbook for State and Local Officials NRT-2*	(800) 535-0202
EPA	*Chemicals in Your Community: A Guide to the Emergency Planning and Community RTK Act*	(800) 535-0202
NPO	National Publications Catalog 200 B 96-001	(800) 490-9198

APPENDIX B. (continued)

CMA	Site Emergency Response Planning	(202) 741-5000
NIOSH	A Guide to Safety in Confined Spaces	(800) 356-4674
NIOSH	A Curriculum Guide for Public-Safety and Emergency-Response Workers 89-108	(800) 356-4674
AIHA	Emergency Response Planning Guidelines and Workplace Environmental Exposure Level Guides Handbook	(703) 849-8888
	Hotlines	
NRC	National Response Center (NRC) Chemical/Biological Hotline	(800) 424-8802 or (202) 426-2675
EPA	Emergency Planning and Community Right-to-Know Information Hotline	(800) 535-0202 or (202) 479-2449
EPA	RCRA, Superfund, and EPCRA Hotline	(800) 424-9346 or (703) 412-9810
EPA	Local Government Reimbursement Helpline	(800) 431-9209
DOT	Hazardous Material Information Center	(202) 366-4488
ASTDR	Agency for Toxic Substances and Disease Registry Helpline	(888) 422-8737
CHEMTREC	Chemical Transportation Emergency Center	(800) 262-8200
	24-Hour Emergency Hotline	(800) 424-9300
NIOSH	National Institute for Occupational Safety and Health	(888) 356-4674

BIBLIOGRAPHY

1. EPA Fact Sheet: *Hazardous Waste Operations and Emergency Response: General Information and Comparison*, Office of Emergency and Remedial Response Division, MS-101, April 1991.
2. OSHA updated "Q" Directive CPL 2-2.59A, April 1998.
3. "The National Response Team's Integrated Contingency Plan Guidance," EPA, *Federal Register* **61**, 109, (June 1996).
4. *NRT-RRT Factsheet*, National Response Team, May, 1998.
5. NIOSH, OSHA, USCG, and EPA, *Occupational Safety and Health Guidance Manual for Hazardous Waste Site Activities*, U.S. Government Printing Office, Washington, D.C. 1985.
6. EPA/OSHA Guide for Preparing HAZWOPER HASPs, Downloadable E-HASP software, www.osha-slc.gov/.
7. *1996 North American Emergency Response Guidebook*, U.S. Department of Transportation.
8. *Hazardous Materials Training for First Responders*, International Association of Fire Fighters Hazardous Materials Training for First Responders, Washington, D.C., 1991.
9. *National Fire Protection Association, Standard on Support Function Protective Garments for Hazardous Chemical Operations* (EPA Level B Protective Clothing), 1993.
10. G. G. Noll, M. S. Hildebrand, and J. G. Yvorra, *Hazardous Materials–Managing the Incident*, Fire Protection Publications, Stillwater, OK, 1988.
11. *Handbook on Hazardous Materials Management*, fifth ed., Institute of Hazardous Materials Management, 1995.

12. *Hazardous Waste and Emergency Response*, OSHA Publication #3114 NIOSH, Hazardous Waste Sites and Hazardous Substance Emergencies, Worker Bulletin, Publication #83-100.
13. U.S. E.P.A. Office of Emergency and Remedial Response, *Emergency Response Division, Standard Operating Safety Guides*, 1988.
14. U.S. Coast Guard, *Policy Guidance for Response to Hazardous Chemical Releases*, USCG Pollution Response, COMOTINST-M16465-30.

CHAPTER SIXTY-THREE

Health and Safety Factors in Designing an Industrial Hygiene Laboratory

Robert G. Lieckfield, Jr., CIH and Ronald C. Poore, CIH

1 INTRODUCTION

Health and safety are important concerns in any laboratory. Because the industrial hygiene laboratory, in particular, exists to help ensure worker protection, it should be the exemplary safe and healthful workplace. A well planned and well executed laboratory design, including the care, knowledge, and foresight that go into its planning, will do much to ensure the safety and health of its occupants and even the surrounding community. This chapter discusses health and safety factors that must be considered in designing the industrial hygiene laboratory and is intended to help management plan new laboratory construction or remodel existing laboratory spaces. The bibliography listed at the end of this chapter may be consulted for more information on the broad topic of laboratory design.

1.1 Nature and Scope of the Industrial Hygiene Laboratory

Industrial hygiene laboratories analyze samples collected to assess worker exposures or potential exposures to dusts, fibers, fumes, mists, gases, and vapors in the workplace. Typical industrial hygiene samples include airborne contaminants collected on various sampling media, such as particulate on filters, and organic solvent vapors on activated charcoal tubes. Additionally, the industrial hygiene laboratory typically analyzes bulk materials, such as asbestos-containing sprayed-on fireproofing or insulation; body fluids for

Patty's Industrial Hygiene, Fifth Edition, Volume 4. Edited by Robert L. Harris.
ISBN 0-471-29749-6

biological monitoring, such as lead in blood; or microbiological samples, such as microorganisms sampled to assess indoor air quality.

In almost all cases, samples are analyzed for trace quantities of materials. The industrial hygiene laboratory is not a laboratory for physical testing or for gross analytical work, such as is done in a quality-control laboratory. Properly designed and engineered laboratory facilities coupled with technically competent employees are the cornerstone for a safe workplace. This is bought about through effective leadership, written policies and procedures, employee training, regular quality and safety audits, a written Chemical Hygiene Plan (CHP), and is some cases, a routine medical surveillance program.

1.2 Laboratory Accidents

This chapter outlines steps to take to help ensure a building design that is safe and healthful to personnel, property, and the community. Poor laboratory building design is the culprit in many laboratory accidents. The following section lists common industrial hygiene laboratory accidents. Some of these may also result from poor organization and management. The reader should consult the references at the end of this and other chapters in this text for help in organizing and managing health and safety plans.

Industrial hygiene laboratory accidents can be classified by injury to personnel, property, and the community.

Injuries to personnel (1) can be caused by:

- Lack of or failure to wear protective equipment, such as safety glasses.
- Lack of hazard communication to workers, such as the use of material safety data sheets (MSDS) in providing information on precautions to be taken with specific chemicals.
- Lack of proper ventilation, such as inoperative or ineffective chemical fume hoods.
- Personal hygiene problems, such as eating in the laboratory or mouth pipetting.
- Electrical hazards, such as wiring that violates the National Electric Code.
- Storage problems, such as failure to secure compressed gas cylinders or to store incompatible chemicals in segregated storage locations.
- Inadequate emergency procedures and equipment, such as inoperative safety showers or fire extinguishers.
- Lack of proper management, such as failure to implement an industrial hygiene and safety plan or a hazard communication program.
- Lack of personal responsibility, such as a worker's failure to follow procedures.

The most common accidents resulting in property damage are fires and explosions. Water damage from firefighting also damages property. Several of the causes of injury to personnel listed above can also result in fire and explosion.

Fire and explosion can also harm people or property in the community. However, a very common and serious community concern is improper disposal of laboratory wastes. This is discussed further in Section 5.2.

2 GENERAL DESIGN CONSIDERATIONS

Building a laboratory (2) can be a challenge for the laboratory director. But for those who have worked in a less-than-ideal facility, it is exciting to be able to help plan and design a new one.

2.1 Design Elements

2.1.1 Location

Industrial hygiene laboratories exist as part of government agencies, academic institutions, corporations, and consulting groups. Most independent consulting laboratories are housed in one- or two-story buildings specially designed for the consulting firm or laboratory. Therefore, geographic location is a matter of choice of the owners of the facility. However, industrial hygiene laboratories with parent organizations are usually located within a major office complex. Location of the laboratory within such a complex is a major consideration because of the special ventilation, plumbing, electrical, storage, and access needs of the laboratory. The following should be considered when choosing a location for a laboratory within an existing building:

- The need to exhaust fume hoods to the external environment.
- The need for waterlines and acid-resistant drains.
- The need for special electrical installations.
- Convenient access for laboratory users.
- Convenient access for delivery of gases and supplies.
- Fire and safety considerations for laboratory staff and other building occupants.

2.1.2 Aesthetics and Environment

The laboratory director and the consulting firm or parent organization must provide a workplace that is not only free of health and safety hazards but that is comfortable and conducive to good employee morale and productivity.

Special consideration should be given to the needs of disabled laboratory staff as required by the Americans with Disabilities Act (ADA) of 1990. The ADA expands the scope and impact of laws and regulations on discrimination against individuals with disabilities. A key part of the ADA is the provision on "reasonable accommodation" that requires an employer to modify or adjust a job or work environment to enable a qualified individual with a disability to enjoy equal employment opportunity. Given the nature of the typical laboratory design, consideration of the needs of disabled persons is both good management practice and cost-effective. Design features, such as, bench height, aisle width, and hood configuration are important aspects of the laboratory accommodation under the ADA.

Aesthetics is important. Analysts and laboratory staff often spend their entire workday in the laboratory. Incorporating good laboratory design can assist in both recruiting and retaining good staff. The laboratory should be designed to be visually attractive as well as functional. Consideration should be given to modest amenities for staff, including lockers or areas designated for personal belongings; lunch- or break-rooms equipped with sink, refrigerator, and microwave; coat closet or rack; and convenient rest rooms.

Proper lighting, temperature, and humidity are necessary to keep both the employees and many of the instruments working well. Instruments produce a good deal of heat and contribute to the temperature of the laboratory environment. A separate thermostat control for areas with analytical instrumentation is desirable. Air exhausted by hoods and air contributed to makeup air also affect the temperature. These factors should be considered in planning for laboratory comfort and temperature control.

Designing the laboratory around the people and the technology used will help ensure the comfort, safety, and efficiency of the laboratory operation.

2.2 Health and Safety Considerations

2.2.1 Regulations

There are no federal regulations that apply to the design of industrial hygiene laboratories. However, laboratory activities related to chemical health and safety are governed by more than 100 separate regulations promulgated by the Occupational Safety and Health Administration (OSHA) and the Environmental Protection Agency (EPA). Table 63.1 lists several of these regulations and the areas of safety and health they address. These regulations must be consulted when constructing or remodeling laboratories.

Table 63.1. Health and Safety Regulations

Agency	Regulation	Pertains To
OSHA	**29** *CFR* 1910, 132–134, 1910.136, 1910.212	Personal protective equipment (eyes, face, head, extremities, respiratory tract)
	29 *CFR* 1910.106, 1910.157–164	Flammable and combustible liquids
	29 *CFR* 1910.101–105, 1910.166–167	Compressed gases
	29 *CFR* 1910.96–97	Radiation/Ionizing and nonionizing
	29 *CFR* 1910.137, 1910.301–308	Electrical hazards
	29 *CFR* 1910.176, 1910.141	Materials handling and storage, sanitation, and housekeeping
	29 *CFR* 1910.35–37	Means of egress
	29 *CFR* 1910.1000–1500	Air contaminants
	29 *CFR* 1910.20	Access to medical records
	29 *CFR* 1910.1200	"Right-to-Know"
	29 *CFR* 1910.1450	Occupational exposures to hazardous chemicals in laboratories
EPA	**40** *CFR* 720.36, 720.78	New chemicals
	40 *CFR* 355.10, 355.20, 355.30, 355.40, 355.50	Chemical emergency plans

Of the many regulations that apply to laboratories, the OSHA regulation for Occupational Exposures to Hazardous Chemicals in Laboratories (**29** *CFR*1910.1450) is particularly important when planning and managing a safe laboratory facility.

The OSHA laboratory standard applies to all laboratories that use small quantities of hazardous chemicals on a nonproduction basis. This includes laboratories in industrial, clinical, and academic settings. Laboratory operations excluded from the OSHA standard are laboratory operations that produce commercial quantities of materials, are part of the production process, or whose uses of hazardous chemicals provide no potential for employee exposure.

The main emphasis of the OSHA regulation is the implementation of a written Chemical Hygiene Plan (CHP) that documents procedures, equipment, and work practices that are capable of protecting employees from potential health hazards encountered in a particular laboratory. The CHP must outline procedures for keeping chemical exposures below either permissible exposure limits (PELs), as outlined in Subpart Z, **29** *CFR* 1910.1000 of the OSHA standard, or recommended exposure limits for hazardous chemicals where there is no applicable OSHA standard.

A CHP must include the following eight elements:

- Safety and health standard operating procedures for working with hazardous chemicals.
- Criteria to determine and implement control measures to reduce employee exposure to hazardous chemicals.
- Procedures to ensure the proper functioning of fume hoods and other protective equipment.
- Provisions for employee information and training.
- Designation of circumstances in which a particular laboratory operation, procedure, or activity will require prior approval from the employer or supervisor.
- Provisions for medical consultation, surveillance, and examination.
- Designation of a responsible individual (a chemical hygiene officer or chemical hygiene committee) to oversee implementation of the CHP.
- Provisions and procedures for employee protection when hazardous chemicals (e.g., carcinogens, reproductive toxins, teratogens, or acutely toxic chemicals) are used in a laboratory.

The OSHA Hazard Communication standard (**29** *CFR*1910.1200) should be consulted for guidance in determining the scope of health hazards and determining the hazard clarification of a particular chemical.

2.2.2 Ventilation

Laboratory ventilation is needed to provide a safe and comfortable environment. Balanced amounts of supply and exhaust air are needed as well as temperature and humidity control. Good laboratory exhaust ventilation contains or captures toxic contaminants and transports them out of the building. Exhaust ventilation must be designed to prevent contamination of other areas of the building that can occur by re-entrainment of contaminants from discharge points into outside air inlets, or by negative pressure inside the building that can cause downdrafts in fume hoods. The same supply ventilation system may be used to provide makeup air for exhaust air systems as well as a comfortable and safe work environment, or a separate supply system may be used for each function.

The design of a ventilation system for the industrial hygiene laboratory should consider air balance, pressure relationships among areas, supply and exhaust air criteria, fume hoods and local exhaust, and criteria for outside air intakes and hood exhaust discharges. These items are discussed in Section 3.4.2 and Section 3.4.3.

2.2.3 Egress and Handicapped Person Access

Safe egress (1), as defined by the National Fire Codes 45 and 101 prepared by the National Fire Protection Association (NFPA) and adopted as the building code in many localities, is an essential part of laboratory planning. Two means of egress are usually required. Exceptions are based on limitations of square footage and in the types and quantities of chemicals housed within the laboratory. The primary exit door should be no more than 75 ft in travel distance from the farthest occupied space of the laboratory. The recommended minimum width of exit doors is 36 in. clear. Exit doors must swing in the direction of egress. Doors opening into corridors should not reduce the required width of the egress passage. Design of building egress outside of individual laboratories is governed by NFPA 101 provisions.

Arrangement of laboratory benches should facilitate egress. Benches arranged in parallel rows form regularly spaced working aisles. The recommended minimum width of working aisles is five feet, with a maximum of seven feet. Working aisles should join directly to at least one egress aisle of equal or greater width. The egress aisle should lead directly to a fire-protected exit.

Hazardous operations affect planning for safe egress. The most hazardous operations, such as those using flammable solvents, should be located farthest from the primary exit. Less hazardous activities such as paperwork, should be next to the primary exit.

State and local regulations specify laboratory design for access by persons in wheelchairs or those otherwise handicapped. The prime concern is safe egress. The regulations specify minimum door widths, minimum clearances on the latch side of in-swinging doors, the directions of egress, elimination of threshold conditions, ramp access and allowable slopes, handrail heights, and appropriately designed personal hygiene facilities. In general, slightly more floor space is required to comply with most of these regulations.

2.2.4 Emergency Facilities

Clustering safety equipment into "Safety Niches" along major corridors with emergency showers, fire blankets, fire extinguishers, burn and acid treatment kits, and general medical supplies is a good design feature. Emergency facilities and supplies include emergency deluge showers, eyewash fountains, spill kits, fire extinguishers, fire blankets, telephones and communications devices, control panels, and building fire protection systems. These are discussed in Sections 5.3, 5.4, 5.5, and 5.6.

2.2.5 Laboratory Furnishings

Selection of wood or metal casework is largely a matter of economy or aesthetics, not safety. Wood rots under wet conditions and it can burn; it can be treated to minimize these disadvantages but not eliminate them. Metal corrodes, especially in typical laboratory atmospheres. Although it can burn, this is not likely in promptly extinguished fires. Metal can be coated to resist corrosion, but can not be totally protected. A good quality finish and construction are primary considerations in selecting laboratory casework.

All wall-hung storage units, shelves, and equipment should be securely attached with fittings of sufficient strength to support units under maximum loading conditions. Walls that support storage units should be reinforced to bear the maximum load. In normal construction of gypsum wallboard on metal stud partitions, lateral bracing is not included. Even lightweight concrete masonry unit wall construction cannot hold the weight of typical laboratory storage. Reinforcing for interior partitions must be directly called for in the specifications and drawings.

2.2.6 Laboratory Features

Lighting should be designed and installed so there is low glare at the laboratory bench height and no shadows. Normal lighting levels range from 50 to 100 footcandles (500 to 1000 lux) at countertop height. The intensity of light depends on the nature of the work. All emergency lighting and exit signs should be provided as required by local building codes.

If compressed gas cylinders are used in the laboratory, sturdy supports must be provided for securing cylinders in use and empty cylinders or those awaiting use.

An area out of the flow of traffic should be provided for lab staff to store personal articles such as coats, hats, raingear, and briefcases. Recessed locations out of the flow of traffic are desirable for waste containers, including receptacles not only for ordinary trash but also for temporary storage of different chemical wastes.

Walls between adjacent laboratories and between laboratories and corridors should have a ¾-hr or more fire-resistant construction rating according to local building codes.

3 FACILITY DESIGN

3.1 Introduction

The basic layout of laboratories, with fixed benches and utilities, has changed little since the mid-1800s. The laboratories of this period consisted of fixed benches along two walls with cupboards and drawers underneath and shelves above. Center rows of parallel benches with cupboards underneath were spaced about 6 ft. apart. The side benches were 2 ft deep, and the center benches approximately 3 ft wide (3). As is true in today's laboratories, every surface of the laboratory was covered with apparatus. This basic layout has been used in almost every laboratory built in the last 100 years.

Laboratory designers are moving away from the stoic designs of the past and embarking on a fresh look; making laboratories a place where employees want to work rather than have to work. There appears to be a relationship between the building design and aesthetics and the resulting efficiency of the operation. The concept of parallel benches was indeed practical, because most laboratory analyses have been traditionally performed on the bench surface. However, the industrial hygiene laboratory has evolved from classical wet chemistries, such as titrations and colorimetric and gravimetric methods, to analyses using sophisticated electronic instrumentation with computer-based data reduction. The continuing trends in laboratory equipment, instrumentation, and personal computers have made the traditional laboratory design a thing of the past (4). These trends require much more flexibility in design than that provided by fixed benches. Design features allowing easy access to both the front and back of instrumentation for repair and maintenance is now an important consideration.

In any case, the design of the physical space is the biggest challenge in building or renovating a laboratory. The most important and effective tool is a sound master plan. This plan must cover all aspects of the design process and reflect a clear understanding of both aesthetic and functional requirements by the designers and the users of the facility.

Key points in planning and executing a laboratory design project are

- Hire competent architects that have specific experience in laboratory design and understand the activities and science that will be conducted in the space.
- Design and size utilities and ventilation systems to accommodate growth and flexibility.
- Provide generous space for offices, conference rooms, and break areas.
- Avoid creating identical spaces for all laboratories within the facility. Design each laboratory area for its intended use and individuality.
- Design adaptation into the original plan, since laboratory requirements and technology change rapidly.

The National Research Council (NRC) plans on publishing a report on the design, construction, and renovation of chemical and biochemical laboratory facilities. This report is planned for Summer 1999.

The difficulty in laboratory design is predicting the growth of both the individual components of a laboratory and that of the whole organization. Even so, it is imperative that the design team develop a strategy to address future requirements. Benches, services, and partitions must be planned with anticipation of changes in work patterns, scientific instrumentation, staffing, and work demands, which all affect space requirements. If such growth is not planned for in the original design, the result will be a misuse of space or the need for major redesign. Ignoring future laboratory requirements for space, services, and function is extremely costly in terms of both additional capital investment and loss of analyst productivity caused by inefficient space design.

The capital investment in a laboratory expansion is generally not just the cost of constructing new walls or expanding into new areas. Laboratory expansion often requires extensive redesign of the ventilation and heating and cooling systems, as well as other utilities. Because the costs of ventilation and heating and cooling systems are linear, upgrading and expansion is often two to three times the cost of systems that are designed for expansion (3).

With all the traps associated with laboratory design, how can anyone get it right? First, it must be acknowledged that some aspect of the laboratory design will end up insufficient for the work being performed. But this should not deter the designer from covering all aspects of the design in minute detail with a goal of a functional, practical, and efficient work environment. The design will have succeeded if the facility can accommodate change easily at a reasonable cost. Successful laboratory design requires a careful and well-thought-out plan and design process involving the architects, engineers, and users of the laboratory. Please note the term "users" refers to both laboratory management and analysts.

Anderson, deBartolo, Pan, Inc., an architectural engineering firm, offers the following recommendations based on its design of the Sandia National Laboratory in Albuquerque, New Mexico (5):

- A multidisciplinary architectural and engineering team working intimately with the client, particularly the user, is the best combination for ensuring successful architectural and engineering solution to design problems.
- The architect and engineer and the user must all consciously push for innovation, to guarantee that new concepts will be explored.
- Sufficient time must be available to test the proposed innovations thoroughly.
- The client's management and decision-makers should be involved in the details of the project at the earliest stages of design. This is an excellent way to minimize red tape and complications in interactions between the architect and engineer and laboratory management.
- The architect and engineer and the laboratory must achieve a balance between solution concepts and follow-through on the details.
- The architect and engineer and laboratory must share an absolute commitment to satisfying the analyst's requirements and achieving design goals.

3.2 Space Planning

Space planning is conducted for the entire laboratory facility and includes planning the physical layout, individual laboratory space, work patterns and adjacencies, as well as planning for future expansion. There is no perfect design and layout for an industrial hygiene laboratory because each laboratory has unique space requirements and work habits. Therefore, space planning must be project-specific as identified by laboratory management and personnel.

3.2.1 Physical Layout

The physical layout of the laboratory depends on the particular constraints imposed by the actual building structure. Although these constraints are certainly minimized in construction of a new building, land use and economics are always a consideration.

There are a number of basic design plans that can be used as a starting point. However, because function and space requirements will certainly change over time, flexibility must be built into the planning process.

The first step is to develop a building design program. The building design program is a written document that provides details on the construction site and the various functions and requirements of the laboratory. Because laboratories require extensive ventilation and safety features, the building design program must address the relationship between function and safety and health aspects.

The building design program should provide a description of all rooms, and include information on the type of work performed, instrumentation and equipment used, and physical activities in each area. There are five general areas within every laboratory: laboratory areas; support areas, such as computer and data calculation facilities; glassware washing and storage; offices; and personnel support facilities, such as lunchroom, study and library areas, meeting rooms, and mechanical rooms.

The building design program must also consider the issue of shared versus single use of the laboratory areas described above. Depending on the size and function of the particular laboratory, areas such as sample and chemical storage, data calculation, breakrooms, and shipping and receiving can have either centralized or decentralized functions. Although it is generally more economical to build laboratories around centralized support facilities rather than to duplicate support areas in each laboratory or department, the long-term costs of centralized areas must be calculated in terms of work flow inefficiencies and loss of productivity.

A major consideration in the physical design of a laboratory is the circulation of staff and work patterns. This can be greatly enhanced by a carefully thought-out floor plan. Aside from the more obvious adjacencies, such as locating a sample receipt area next to a sample storage area, and locating the chemical storeroom central to all work areas, there are six principal patterns of use in a laboratory:

- Circulation of staff and samples.
- Individual laboratory function.
- Distribution of mechanical equipment and services.
- Structural system.
- Site regulations.
- Building enclosure.

For each laboratory area, the building design program must detail the placement of the extensive mechanical service areas needed for laboratory ventilation, electricity, plumbing, and compressed gas. Figures 63.1, 63.2, 63.3, and 63.4 show various physical layout plans and the advantages and disadvantages of each compiled by DiBerardinis et al. (6). The purpose of illustrating these basic plans is to show the variety of approaches that can be taken. As noted above, the end product must be developed as a consensus of the architect, engineer, and laboratory management and analysts.

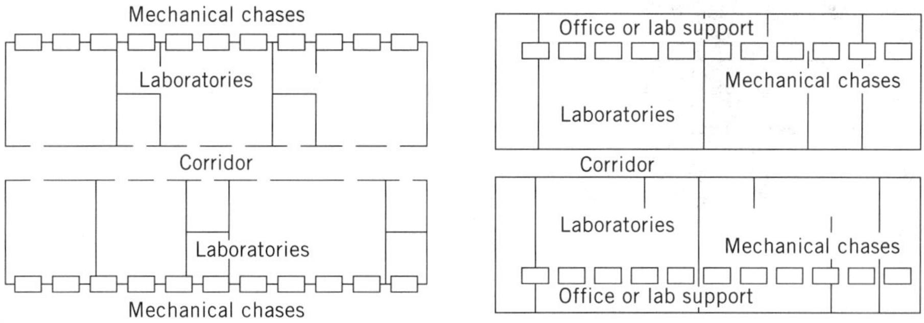

Figure 63.1. Service chase at exterior walls. Advantages: short horizontal exhaust duct from fume hood; fume hood is at the end of an aisle; major inner lab traffic near corridor egress; and equipment and desk at corridor wall. Disadvantages: pipes and ducts only accessible from laboratory; secondary egress near corridor exit; restricted window area; and outswinging doors temporarily obstruct corridor, wide corridors are recommended.

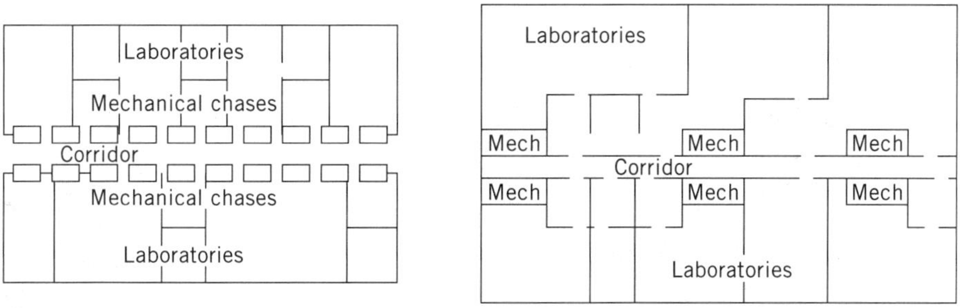

Figure 63.2. Service chase at interior walls. Advantages: pipes and ducts accessible from corridor; secondary egress opposite corridor; expansion window area possible on perimeter wall; outswinging door shielded in alcove; and equipment and desk at corridor wall. Disadvantages: long horizontal exhaust duct from fume hood and persons pass in front of fume hood, causing turbulence.

3.2.2 Room Size Planning

Determination of space requirements for each laboratory area is critical to the overall design plan. The existing space design can be used as a starting point. If a new laboratory is being built, area requirements can be approximated from review of similar laboratories.

Starting from the space occupied by the existing laboratory areas, the design plan focuses on areas needed for instrumentation, benches, hoods, data reduction and research, and on the mechanical and storage requirements of each laboratory area, factoring in future needs.

Space requirements can be determined by construction of a model of the laboratory. This is done using scale models (two- or three-dimensional) of laboratory benches, instrumentation, equipment, hoods, sinks, workstations, and computer terminals. The design

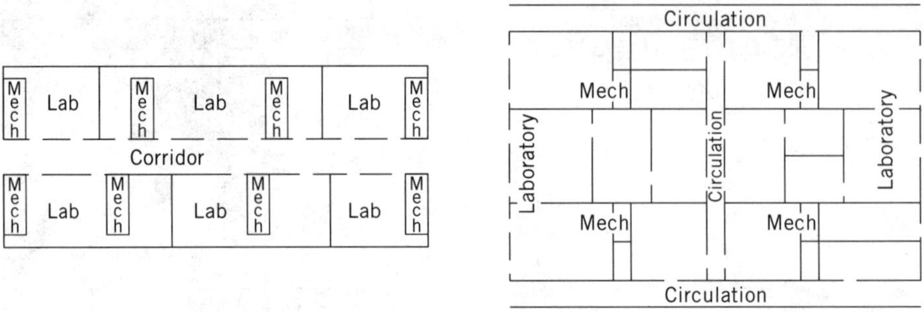

Figure 63.3. Service chase between modules. Advantages: short horizontal exhaust duct from fume hood; fume hood is at the end of an aisle; major inner lab traffic near corridor egress; expansion window area possible on perimeter wall; and applicable to conversion of building to laboratory use. Disadvantages: single module laboratory has only one wet wall; outswinging doors temporarily obstruct corridor, wide corridors are recommended; pipes and drains have long horizontal runs to peninsula benches; and drains require additional venting.

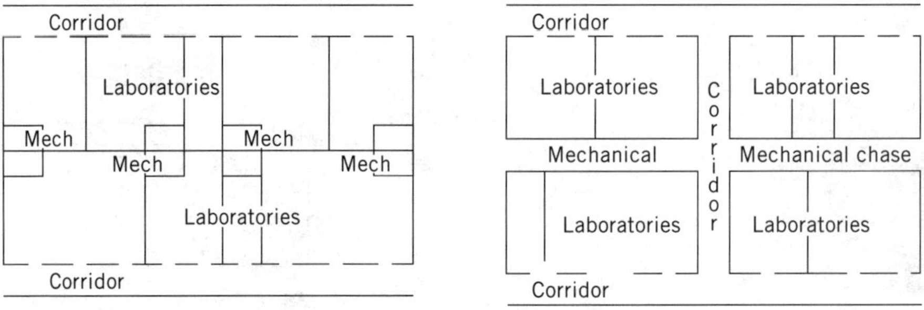

Figure 63.4. Central service chase. Advantages: minimum mechanical area; efficient for low buildings; fume hood is at the end of an aisle; short horizontal exhaust duct from fume hood; pipes and ducts for cluster of four laboratories accessible from each; and maximum variation in laboratory size possible. Disadvantages: limited chase area not suitable for multi-storied buildings; no window area in laboratory; and outswinging doors temporarily obstruct corridor, wide corridors are recommended.

team then constructs a scale laboratory showing the bench plan, instrument layout, and hoods. Using the existing laboratory as a guide, the design team can physically determine current and future laboratory space needs. Areas for data review and report writing should not be forgotten in this process. Scale models should also be constructed for storage, computer, and library areas.

The results of the scale model study will then define the individual laboratory area space requirements. The total facility space needs are then determined by adding up the individual laboratory area space needs and the space needed for corridors, offices, lunchrooms, and mechanical rooms.

As discussed previously, the mechanical and service requirements of a laboratory are extensive and must be carefully planned. Adjacencies for efficient distribution of services should be planned as practical to reduce the overall design and construction costs. The design plan must include access to the mechanical or utility service by both installation and maintenance personnel.

In addition to the mechanical and service design, special design and construction are required for certain instrumentation, that which is sensitive to vibration, such as mass spectrometers, electron microscopes, or inductively coupled argon plasma spectrophotometers.

Local government regulations on land use and construction methods have to be consulted. Local zoning ordinances govern the building use classification, easements, building height, allowable floor area ratio, number of parking spaces required, and use of sewer and water utilities. Some state regulations also require permits for exhausting air from laboratory hoods.

The actual building enclosure can be determined after all the above criteria are established.

3.2.3 Expansion Planning

Once constructed, a laboratory is difficult to expand. It is extremely important to build in a well thought-out expansion plan to meet the ever-changing requirements of the laboratory. Expansion of laboratory facilities into laboratory support areas such as offices and storage rooms can be planned. Because these areas are usually converted to laboratory use anyway, up-front planning of the inevitable will save considerable renovation expense. Locating mechanical and utility services in expansion areas during initial construction will certainly provide the necessary flexibility. As discussed, expansion planning must consider the capacity of ventilation systems, utilities, and other services. It is more cost-effective to oversize these services in initial construction than to upgrade them later.

3.3 Laboratory Planning

Once the physical space is determined, the design should be tested using the layout model of actual bench and instrument placement discussed in Section 3.2.1. The purpose of this portion of the design phase is to test the physical space of the entire facility both for current requirements and for future expansion. In this stage of planning, it is important to establish (if not already done) the location of doorways, corridors, and hoods. In general, the bench and instrument layout must first be planned for easy egress and efficient travel within the laboratory.

3.3.1 Bench Layout

The laboratory bench layout is unique to a particular laboratory's needs and use. Typical bench configurations are parallel to a partition containing utility hookups, extending from a partition at right angles, or free-standing. Any or all of these design concepts can be incorporated into a laboratory bench layout. In the typical industrial hygiene laboratory, the use of instrumentation is extensive and requires careful planning for easy access to the back of the instrument for maintenance.

In deciding configuration, the ease of utility hookups (gas, water, electric) should be considered. Services can easily be extended when benches are parallel or perpendicular to a partition. Services to free-standing or island benches are somewhat more difficult, but can be extended from the ceiling or from underground through mechanical chases.

The minimum clearance between benches should be five feet. This allows for free traffic flow with chemists working at benches. Bench surfaces and storage units should not block utility outlets. Storage units placed above the benches should be easily accessible. The standard work surface is 24 in. deep. Deeper surfaces may be required depending on the type of instruments being placed on the surface. If wall-mounted cabinets are placed over benches, a minimum clearance of 2 feet should be provided to accommodate taller instruments, allow easy access to top and back of instrument, and minimize heat buildup under the wall cabinet.

The work surface height of the bench depends on the use intended for the bench. If workers are seated, the bench can be designed at a height of 30 to 32 in., typical of normal desk height. If workers stand at the bench, the bench height should be 35 to 37 in. and knee spaces should be provided. A drawback of the standing height knee space is its awkwardness and need for laboratory stools.

The lower-height bench is recommended in laboratories doing microscopy and microbiological analyses.

The remainder of the benches should be the standard 35 to 37 in. high. This height has been found to be ergonomically efficient, allowing easy access to the rear of the bench surface, and a comfortable height at which to perform analytical tasks in a standing position.

Incorporated into the bench plan may be desk surfaces for calculations and related activities. Laboratory desks must be positioned away from potentially dangerous laboratory operations and not impede the aisles when the desk is occupied.

An integral aspect of the bench layout plan is the placement of fume hoods. In general, hoods should be located away from entrances and exits and major corridors. This will reduce the traffic flow in front of the hood, thus minimizing airflow disturbance, which affects hood performance. Hoods handling highly hazardous material should be isolated to minimize safety hazards and potential for explosion and fire damage.

3.3.2 Instrument Layout

Once the basic bench and hood layout is complete, the location of each piece of instrumentation and equipment should be planned. A complete list of instrumentation should be drawn up to include dimensions, operating characteristics and specifications, clearances, utilities required, related equipment such as glassware, and ventilation requirements. A checklist is shown in Figure 63.5 to aid in defining these needs.

Considerations in planning for instrumentation placement include:

- A centralized equipment room to service multiple laboratories with utilities and layout designed specifically for instrumentation.

1. List all major pieces of equipment in each laboratory according to the following characteristics:

   (a) Size and weight
   (b) Whether bench mounted, free standing, or other mounting method
   (c) Fixed to wall or floor with permanent utility connections
   (d) Moveable with quick disconnects to utilities
   (e) Required utilities and services
      (1) Water: city, purified, recirculated process
      (2) Drain: floor, piped
      (3) Gas: piped, manifolded tanks
      (4) Compressed air
      (5) Vacuum
      (6) Electricity: voltage, amperage, phases, location and number of standard and special receptacles
      (7) Local exhaust services: type, volume, filters, other treatment
      (8) Air supply for local exhaust services: tempered, filtered, humidified
      (9) Stream: pressure, volume, flowrate
   (f) Maintenance clearances
   (g) Operator's position and clearances
   (h) Equipment attachments
   (i) Heavy rotary components

2. Determine the location of and area required for supplies associated with each piece of equipment:

   (a) Glassware
   (b) Instruments, manipulators
   (c) Chemicals
   (d) Disposables: paper goods, plastic ware

Figure 63.5. Checklist for selection and location of laboratory furnishings and equipment.

- A review of lighting requirements for areas using video display terminals and personal computers.
- The need for vibration-free and temperature- and humidity-controlled locations (for balances, electron microscopes, mass spectrometers).
- Clearances for instrument delivery and service by maintenance staff.

Because remote terminals control many of the newer instruments, placement of instrumentation adjacent to offices may be considered for efficient use of analyst's time.

3.4 Services

A good laboratory design requires accurate specifications for ventilation, electrical, and plumbing services. Concentration in this chapter has been on space planning and physical layout of equipment and benches with little mention of the services required to make the laboratory functional. Ideally, all of these aspects are considered together, with the best possible solutions for each design component addressed at the same time.

Laboratory services can be provided by either a centralized or decentralized laboratory distribution system. Factors and constraints that affect this decision are

- Ceilings/access space.
- Number of floors.
- Structural system.
- Types of laboratories and services.
- Flexibility desired.
- Expansion plans.
- Budget limitations.

3.4.1 Distribution Schemes

Typical distribution schemes include continuous wall service corridors, vertical distribution, and horizontal distribution (7).

3.4.1.1 Continuous Wall Service Corridors. The mechanical and electrical services in continuous wall service corridor design are distributed horizontally along a central service corridor. The services are distributed via a main vertical service feed. In this plan, each laboratory area has direct access to the utilities. The continuous wall service corridor has many advantages, including easy maintenance and the flexibility to rearrange and modify services without affecting the entire operation.

3.4.1.2 Vertical Distribution. The vertical distribution design introduces the necessary services to each laboratory area through a series of vertical utility plenums located to service one or more laboratories. Services are accessed by tapping into the vertical service plenums and extending services to each laboratory area horizontally either through the ceiling or below the floor. This type of design has a lower initial construction cost. However, it has limited flexibility unless the service plenums are closely spaced.

3.4.1.3 Horizontal Distribution. The horizontal distribution design organizes the laboratory services through ceiling plenums. The individual services are then accessed through vertical service chases at the particular bench area. Horizontal distribution offers flexibility and comparable simplicity of design.

Planning the most effective distribution scheme requires knowledge of present and future design needs. The type of distribution system selected should be determined in conjunction with the laboratory space planning and layout phases of the design process, because layout and services access are mutually dependent.

3.4.2 Heating, Ventilation, and Air Conditioning

Laboratory heating, ventilation, and air conditioning (HVAC) systems must be capable of maintaining a uniform temperature throughout the laboratory. Much of the analytical equipment in use at modern laboratories is extremely temperature sensitive. Wide fluctuation in temperature conditions is detrimental to instrument performance and typically results in higher-than-normal maintenance expense.

Laboratory internal heat generation is an important design consideration for the heating, ventilating, and air conditioning system. The laboratory heat load is a significant contributor of internal heat gains through the heat output of laboratory instrumentation and equipment, such as, furnaces and ovens. The magnitude of heat load varies widely between laboratory operations. The objective of the laboratory control and ventilation system is to maintain both temperatures and building pressures within operating specifications. A discussion on the modeling of heat loads and selection of design parameters are described in a recent ASHRAE Symposia, Volume 102, 1996.

In many cases, simultaneous heating and cooling are required for different laboratory areas. Both local cooling units and central heating and cooling systems may be necessary. For example, laboratories equipped with gas chromatographs/mass spectrometers, inductively coupled argon plasma spectrophotometers and inductively coupled argon plasma spectrophotometers/mass spectrometers, and electron microscopes need individualized cooling units because of the larger heat loads (150,000 to 200,000 BTUs) and extremely temperature-sensitive equipment.

If the laboratory is located within office space, the design team should specify separate heating and cooling units for the laboratory and office areas. In areas where office HVAC allows for 20 to 25 percent recirculation of exhaust air, separate systems may be the only viable alternative because recirculation of laboratory exhaust air is generally discouraged. Generally, 6 to 10 air changes per hour for laboratory areas are recommended.

Discharge and supply air volumes must be carefully balanced in the laboratory. Laboratories handling hazardous materials must always be maintained at a negative pressure relative to corridors and other access areas such as offices, lunchrooms, and public access areas.

There are three types of laboratory ventilation systems. Comfort ventilation considers the supply and removal of air for breathing and temperature control. Exhaust ventilation, through fume hoods, is designed principally for protection of health and safety. Makeup air supply is designed to replenish the amount of air discharged through the fume hoods.

Laboratory ventilation design is affected by the following factors:

- Fume hood design
- Minimum exhaust air required by fume hoods
- Location of the fume hoods within the laboratory
- Pressure relationships with respect to surrounding areas
- Removal of chemical vapors, gases, and odors

3.4.3 Fume Hoods

3.4.3.1 General. The purpose of a fume hood is to remove toxic and harmful fumes, gases, and vapors from the laboratory environment by exhausting air at a sufficient velocity to capture and remove the substances. The fume hood must be designed and located to allow the analyst to perform tasks or observe processes both easily and safely. The laboratory fume hood accomplishes this by providing both a work area and a sufficient face velocity (airflow across the hood opening). Examples of typical laboratory hood configurations are shown in Figures 63.6, 63.7, and 63.8.

The fume hood face velocity is crucial for protection of the analyst's health and safety. The face velocity and, therefore, hood performance are adversely affected by drafts across

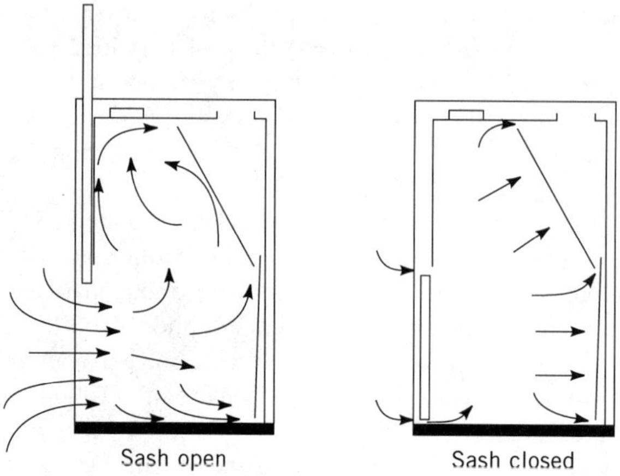

Figure 63.6. Conventional fume hood.

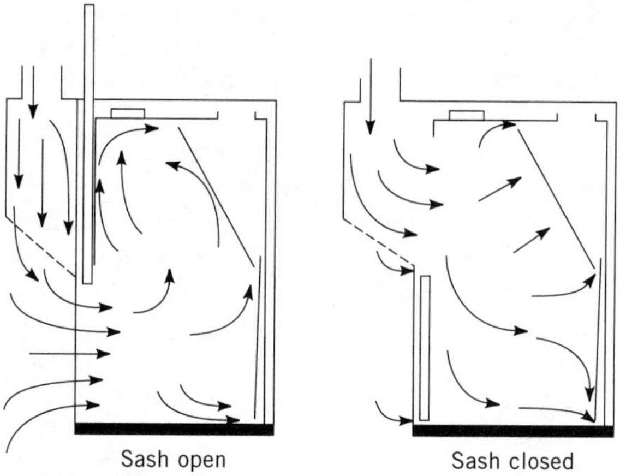

Figure 63.7. Auxiliary air fume hood.

the hood face, large temperature differences between the hood and surrounding air, equipment contained within the hood cabinet, and analyst's work habits.

The upper limit on face velocity should be no more than 150 feet per minute (fpm). Flow rates above this level develop air disturbances, created by the analyst's position in front of the hood, resulting in a negative pressure in front of the analyst, which actually pulls vapors and gases back toward the analyst.

Until recently, it was generally accepted that high face velocities offered better protection. However, most data support a maximum face velocity of 100 fpm with face velocities of 60 to 80 fpm, provided proper hood safety procedures are followed (8).

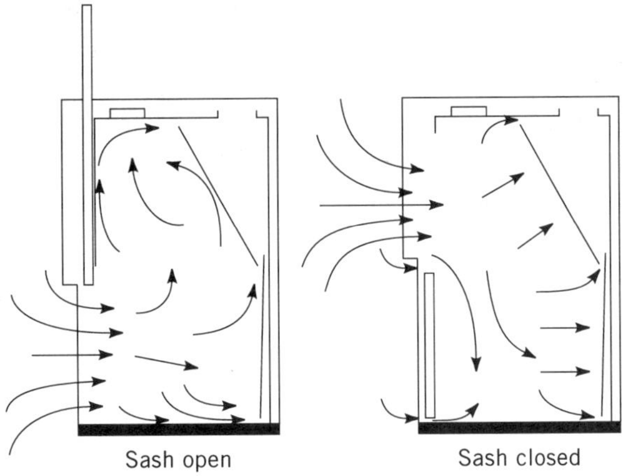

Figure 63.8. Bypass fume hood.

Ultimately, the specific use and practices in the laboratory should determine hood face velocities. The American Society of Heating, Refrigerating, and Air Conditioning Engineers (ASHRAE) has published guidelines that should be consulted during the design phase (9).

A properly designed laboratory hood should have the following characteristics:

- Uniform face velocity (vary no more than 20 percent over the access opening).
- Smooth, rounded, and tapered openings to minimize turbulence at the hood face.
- Constant face velocities maintained regardless of sash opening height.
- Airflow monitors incorporated into each hood such as liquid-filled draft gauges and magnehelic gauges.

Most laboratory furniture suppliers have a selection of models that meet these performance standards.

The supply of makeup air is critical to the optimum performance of laboratory hoods. The design of the room air supply system is as important to lab hood performance as face velocity. The volume of air exhausted from fume hoods must be replaced with an equal amount of incoming air. The location of the makeup air outlets and temperature of the supply air are also important design factors.

Supply air should be introduced at low velocity (approximately ½ to ⅔ of the hood face velocity) and in a direction that does not cause disruptive cross-drafts at the hood opening. Supply air can be introduced through diffusers, wall grilles, or perforated ceiling panels located adjacent or opposite to the fume hood. Introducing supply air through perforated ceiling panels is the most efficient means because these panels are of simple design and easy to apply, and do not require precise adjustment. The ceiling panels should be located at least one meter from the hood face.

Proper location of fume hoods within the laboratory is another important aspect of effective and safe hood performance. Fume hoods should be located away from:

- Doorways, for a fire or explosion in the hood could block an exit.
- Major traffic corridors to minimize cross-drafts and turbulence.
- Room corners and openable windows.

In addition to the standard laboratory fume hood, special hoods may be needed to handle specific chemical hazards. These include perchloric acid hoods, glove boxes for highly toxic materials, high-efficiency particulate air (HEPA)-filtered hoods, and clean bench hoods. Each of these hoods is discussed briefly below.

3.4.3.2 Perchloric Acid Hoods. Perchloric acid hoods must meet the same face velocity and other design criteria noted for general fume hoods. They do, however, have several specific design features. The perchloric acid hood should be constructed of stainless steel and have welded seams. Taped seams, putties, or sealers can be used in the fabrication of the hood or duct system. An internal water wash capability is required to eliminate the buildup of perchlorate material. Extremely explosive organic perchlorate vapors can condense in the hood exhaust system and perchlorate residue can detonate on contact during cleaning and repair. The water system must be capable of rinsing all portions of the hood system including ductwork. A vertical duct system should be designed for ease in cleaning and visual inspection.

3.4.3.3 Glove Boxes. The typical laboratory fume hood is very effective in protecting the analyst from volatile solvent. However, when highly toxic or carcinogenic materials are being handled, it is prudent to use a glove box enclosure. The glove box is the ultimate enclosure. It is a type of hood contained in a sealed box fitted with flexible airtight gloves for performing manual operations inside the box.

The exhaust airflow is calculated to effectively remove contaminants generated from work performed inside the enclosure, maintain the glove box interior at a negative pressure, and prevent contaminant escape if a glove fails. Generally, the airflow exhaust is 50 cubic feet per minute (cfm) per square foot of open door area.

The glove box must be an isolated system to protect the analyst from exposure to toxic and carcinogenic materials. The isolated glove box includes an air lock system. The airlock allows materials to be transferred into and out of the glove box while preventing the escape of contaminants from inside the box.

Air filtering equipment is used in glove box enclosures to remove contaminants within the glove box before they are exhausted to the outside. The typical glove box contains three filters. The first is a particulate filter. The second filter is generally a charcoal bed, which is effective in removing volatile solvents. The third filter is a HEPA filter.

3.4.3.4 HEPA Filters. The HEPA filter is constructed of pleated paper or other filter material bonded to the filter frame, and is at least 99.97% efficient in collecting a 0.3-μm aerosol. Because HEPA filters are highly efficient in removing small particles from the air stream, the prefilter must be kept in proper working condition. The effectiveness of the prefilter will extend the life of the HEPA filter.

There are a number of products on the market designed specifically for use in preparation of asbestos samples. The HEPA filter hood is one of two ventilation designs that must be considered when a laboratory is handling asbestos. The second, the clean bench hood, is discussed in the next section.

It is important to prevent analyst exposure and release of asbestos to the outside air during asbestos fiber counts and analysis of asbestos-containing materials. Use of HEPA filters in a standard hood design can be extremely complicated because of pressure differences upstream and downstream of the filter and fan speed required to maintain adequate face velocity.

The advantages of commercial specialty hood systems are their small size and self-contained HEPA filter and fan assemblies. At a relatively low cost of $600 to $800 per hood, the design team should consider this purchase before designing a HEPA ventilation system for a standard fume hood. Most of these units allow for exhaust into the general room environment. While there may be data demonstrating the capture efficiency of the HEPA filtration system, it is desirable to exhaust the hood effluent to the outside of the building.

3.4.3.5 Clean Bench Hoods. The second specialty unit for asbestos is the clean bench hood, which is used in preparing samples for transmission electron microscopy. The purpose of the clean bench is to ensure that samples do not become contaminated with laboratory air. Laboratory air is filtered through a HEPA filter before being exhausted across a work surface. Clean benches are commercially available, and purchase, rather than design, of a separate system is recommended.

3.4.4 Electrical

The electrical power requirements of laboratories are significantly greater than the typical office building. Careful study must be done to plan for current and future requirements. The design of electrical systems must allow for growth and change with reasonable ease and without major disruption to operations during renovation. Recommended electrical design features include 50% excess capacity in electrical service and reserve for additional equipment and distribution panels.

The first step in designing the electrical system is to assemble information on the use of electrical power by the laboratory. The electrical requirements should be listed by area according to laboratory equipment in use and the voltage and operating amperage of each instrument. All laboratory equipment should be listed, including such items as hot plates, centrifuges, and meters.

Current and anticipated use rates should also be provided to the electrical system design team. Use rates will allow calculation of continuous power required and peak power.

The initial space plan model should be used to place electrical equipment together with electrical requirements. This allows the design team to determine optimum placement and sizing of distribution panels within the laboratory. Noting the growth potential of individual laboratory areas will also allow the design team to plan future growth in areas of probable need, rather than oversizing all electrical systems in all areas of the laboratory.

Because laboratories rely on computers to operate instruments and analyze data, it is important to consider the special electrical requirements of computer operations. Because most computers require minimal line disturbances, an isolated ground should be installed that reduces the electrical fluctuations to the equipment. Isolated grounds should be installed where necessary and as specified by the equipment manufacturer. Installation of line filters, static voltage regulators, or shielded transformers may also be necessary. Electrical service should be located so that it can be easily and quickly disconnected. This is an important safety consideration and should not be overlooked during the design phase.

3.4.5 Plumbing

Laboratories require complete sanitary and storm water drainage systems. Complicating water supply design are special requirements for neutralizing sumps before contents are discharged to the public sanitary sewer. A comprehensive report on each laboratory function should be developed to determine the types of chemicals to be discharged. Waste piping can then be selected to handle the particular chemicals. Piping materials for these systems include high silicon iron, borosilicate glass, polypropylene, and poly(vinyl chloride).

Acid waste must be neutralized by chemical reaction. A typical application is a neutralization sump filled with limestone chips to raise the pH of the discharge to suitable levels. A centrally located neutralizing sump can be used to handle wastes from a number of sources.

The water system design for a laboratory is more complicated than that for typical office buildings. Systems usually encountered in laboratories are potable water, both hot and cold; cooling water for mass spectrometers, electron microscopes, and X-ray diffraction instruments; and high-purity distilled water.

All laboratory areas should have access to the water supply. Again, an inventory of needs will help the design team in locating water outlets. Laboratory water supplies should be designed to prevent the drinking water supply from being contaminated. All plumbing fixtures should be equipped with antisiphon devices or backflow preventers.

Most laboratories require a source of pure water. Pure water systems consist of distilled water, deionized water, or demineralized water. There are four basic methods of producing pure water-distillation, demineralization, reverse osmosis, and filtration. Depending on the type of pure water that is required, one or more of these methods will be needed.

The use of mass spectrometers, electron microscopes, inductively coupled argon plasma spectrophotometers, and X-ray diffraction units, for example, requires a continuous cooling water system. Cooling water systems can be designed to use either a cold water supply and discharge or a recirculating cooling unit depending on the frequency of use, economics, and environmental considerations.

3.4.6 Compressed Gas System

A wide variety of compressed gases are needed to operate laboratory instrumentation, including air, nitrogen, hydrogen, helium, acetylene, and argon. The design of compressed gas distribution depends on the overall requirements of the laboratory. In many cases, compressed gas is delivered to the instruments by compressed gas tanks located within the particular laboratory.

In other cases, compressed gases can be delivered via gas manifold systems. When designing a gas manifold system, the design team must pay particular attention to local building and fire protection codes and pressure requirements. Safety features that should be considered are backflow preventers, spark arrestors, and automatic shutoffs.

If the manifold method is chosen, it is imperative that ultraclean piping (oxygen supply grade) be used. Joints must not be soldered. Fittings such as those marketed by Swagelok must be used.

4 CONSTRUCTION

4.1 The Construction Team

The efforts of the exhaustive design planning and successful building of the laboratory hinge on the quality of the general contractor. The general contractor should be hired early in the design phase. Employing the contractor at this time allows the general contractor to be part of the design process and provide valuable insight into specifics on the actual construction details.

Selection of a general contractor should be approached in much the same manner as choosing a doctor, lawyer, or accountant. All aspects of a contractor's experience and qualifications must be thoroughly investigated, especially as it pertains to the construction of laboratory facilities. All general contractors will claim to be able to build buildings, and most could do a fine job. However, the building of laboratory facilities requires special expertise and qualifications. In many cases, the general contractor can also arrange services of architects and engineers.

The construction team includes the client, the architect, the mechanical engineer, and the general contractor. To ensure success, each representative of the construction team must be qualified with respect to experience specific to laboratory design.

In addition to reviewing the qualifications of the general contractor, each trade participating in the construction should be investigated. Trades such as HVAC, plumbing, electrical, and carpentry should submit a list of previous laboratory projects completed.

Visits to laboratories designed and constructed by the construction team should be conducted to help assess qualifications. Interviews with laboratory staff at these locations will be helpful. Staff should be asked about satisfaction with design and workmanship. As many visits and interviews should be made as are necessary to satisfy questions concerning qualifications and experience completely. The time spent in qualifying the construction team is easily justified when put in terms of the useful life of the laboratory facility.

The construction of a laboratory is a complicated process subject to many changes during the building process. To aid communication between the design team and the construction team, a construction manager should be appointed. The construction manager should handle questions, changes, and problems that are inevitable in the building process. The construction manager must be intimately involved in the project and knowledgeable in laboratory operation and construction techniques. Above all, the manager must be an effective communicator and must be detail-oriented and diplomatic.

An independent expert in building construction should also be employed to serve as the "tenant's representative." The tenant's representative is in fact the client's building inspector. The principal purpose of this position is to ensure the building owner that the contractor and trades are following the construction blueprints. In other words, the tenant's representative is a project overseer.

Although it may appear to be unnecessary to employ a tenant's representative, because the qualifications of the contractor and trades have been thoroughly investigated, it does provide another means of ensuring the quality of the work. As noted before, the construction of a laboratory is extremely complex, involving extensive HVAC, electrical, and plumbing construction. An outside expert can help prevent, and quickly correct, building errors not uncovered by the general contractors, site supervisors, or local building inspectors.

4.2 Review of Final Blueprints

With the construction team of construction manager, general contractor, architect, and tenant's representative in place, the next phase of the construction process is to perform a critical review of the final blueprints. As noted before, laboratory design and engineering are extremely complicated. Neglecting a final exhaustive review may result in a less-than-optimal working design and can prove costly in future changes. Depending on the construction team's overall confidence in the design and engineering phases, the blueprint review process may need to include outside consultants and engineers to give a third-party unbiased evaluation of the final blueprints.

Each building blueprint should be reviewed in the context of the original design and specifications. The following are questions to ask during the review phase:

- Are the overall plans consistent with the original design concept?
- Will the ventilation and hood systems operate according to desired specifications?
- Are the electrical and plumbing plans consistent with instrument placement and service requirements, such as voltage and amperage necessary, hot and cold water, and sanitary and chemical drains?

During this phase of the construction program, the laboratory staff or at least senior level staff should critically review the final plans. It is important for the construction team to be open-minded, at this point, to small changes in space requirements and bench layout.

Typically, the process from early design to final blueprint is 9 to 12 months. Depending on the degree of change in the laboratory organization over time, it is sometimes necessary to rethink the original design in terms of space allocations and services. This is not to imply that wholesale changes should and can be made, but the design team should not rule out current thinking that would be best addressed before construction. Types of changes could include varying interior room sizes or bench layout or modifying service requirements.

The decision to make changes during final blueprint review should rest with the construction team, after consideration of up-front costs versus after-construction costs. Much depends on the overall flexibility and expansion built into the original designs.

The review phase should also include inspection and definition of the actual construction materials specified in the plan. This does not necessarily mean reviewing the color scheme or wall coverings, although these may be considered. Because laboratories use and store corrosive chemicals, and background contamination must be minimized or eliminated, the construction team must determine that all construction materials meet laboratory specifications. For instance, the acid and solvent resistance of the floor covering and plumbing and ventilation ductwork should be determined. Adhesives used in construction, such as flooring and drywall adhesives, can cause long-term off-gassing problems with background solvents interfering in laboratory analysis of samples. Because of typical solvent off-gassing interfering with future analysis, we recommend the use of latex rather than solvent-based paints.

4.3 Communication During Construction

Once the design and construction teams are satisfied with the plans, specifications, and materials, the physical construction of the facility can begin. The construction manager and tenant's representative must take an active role in overseeing the facility construction. Constant communication and sometimes arbitration and mediation are required to ensure the timely completion of the project. Written communication is highly recommended for all exchanges of information between the contractor, trades, and construction manager. All conversations involving problems or change orders must be summarized in writing and copied to the design and construction team. This will provide some assurance that changes are appropriate and implemented.

4.4 Building Checklist

The design and construction teams must guard against a hasty move-in and start-up of the new facility. When construction is complete, a general excitement at completion of the project can overcome even the most stoic of laboratory managers. However, prior to move-in and start-up, the entire facility must be carefully checked to ensure proper operation of services and utilities.

To facilitate this process, a checklist should be created listing all electrical outlets, hot and cold water supplies, drains, hoods, and compressed gases if manifolds are used. The next step is to test each individual point for proper operation. This entails verifying that electrical outlets are functioning by using a voltmeter, hot and cold water supplies and drains are functioning by allowing sufficient flow to verify the hot water supply, and hood and ventilation system operations are functioning using flowmeters, such as thermoanenometers and velometers. In addition to the fume hood and ventilation systems check, a certified air-balancing contractor should conduct a complete ventilation balancing audit.

The gas lines and vacuum system, if used, should be leak tested with soap solution. If a gas manifold is used for compressed gases such as air, helium, hydrogen, argon, and acetylene, it may be appropriate, depending on the complexity of the manifold, to verify that the correct gas is being delivered to the exit point and the lines are free of contamination. This can be done by performing an operations check using the appropriate laboratory instrument. For instance, on a manifold designed for multiple gas chromatographs, a single unit could be used to verify all exit points, rather than testing the operation after moving all the instrumentation.

Conducting a thorough check of all systems before move-in will help ensure a smooth start-up and minimize costly downtime.

5 SAFETY DESIGN AND PLANNING

5.1 Chemical Storage

Space for storing chemicals in a storeroom and within areas of use in the laboratory must be planned for in the design of the laboratory. Lack of sufficient storage space can create hazards because of overcrowding, storage of incompatible chemicals together, and poor housekeeping. Adequate, properly designed and ventilated storage facilities must be provided to ensure personnel safety and property protection.

5.1.1 Chemicals Used in the Industrial Hygiene Laboratory

The industrial hygiene laboratory analyzes trace amounts of contaminants and typically does not store large volumes of chemicals. However, chemicals used include flammable liquids, high chronic toxicity substances, and incompatible chemicals.

Chemicals that may be stored in quantities of one gallon or more include:

- Carbon disulfide for desorption of organic compounds from activated charcoal and analysis by gas chromatography (GC).
- Solvents for use in high-pressure liquid chromatography (HPLC), including methanol, acetonitrile, and methylene chloride.
- Solvents, including chloroform and dimethyl formamide, for use in preparing air sample filters for analysis of asbestos by transmission electron microscopy (TEM).
- Acids such as nitric and hydrochloric for digestion of air sample filter materials for analysis of metals.

Because of the large number of analytes within the laboratory's scope of work, the laboratory may store small quantities of several hundred materials to be used as standards. Some of the chemicals may be known human carcinogens or suspect carcinogens. A small number of reagents may be stored typically in quantities of 500 g or less.

5.1.2 Chemical Storeroom Design

Storeroom size must be large enough to meet current chemical storage needs and to accommodate future growth. Areas must be planned for segregation of incompatible chemicals, storage of high chronic toxicity substances in an identified area, and storage of certain chemicals in an explosion-proof refrigerator. More than one room may be required.

All chemical storerooms should be under negative pressure with respect to the surrounding area. Air supplied to the room should not be returned to the building HVAC system but should be exhausted to the roof.

5.1.2.1 Flammable Liquids. The most effective way to minimize the impact of a hazard such as fire and explosion is to isolate it. A flammable liquid storeroom is best located in a special building separated from the main building. If the room must be located within a main building, the preferred location is a cut-off area on the first floor (at grade) level with at least one exterior wall. In any case, storage rooms for flammable liquids should not be placed on the roof, on a below-grade level, an upper floor, or in the center of the building. All of these locations are undesirable because they are less accessible for firefighting and are potentially dangerous to personnel in the building.

The walls, ceiling, and floor of an inside storage room for flammable liquids should be constructed of materials having at least a 2-hr fire resistance. The room should have self-closing Class B fire doors [see the OSHA standard (**29** *CFR* 1910.106 (d) (4) (i)) or the National Fire Protection Association (NFPA) standard (No. 30.4310)]. All storage rooms should have adequate mechanical ventilation controlled by a switch outside the door and explosion-proof lighting and switches. Other potential sources of ignition, such as cigarettes and open flames, should be forbidden.

When space is limited or renovation out of the question, a flammable liquid storage cabinet exhausted to the outside is an alternative for safely storing small quantities of flammable liquids.

5.1.2.2 High Chronic Toxicity Substances Including Carcinogens. The industrial hygiene laboratory stores and uses small quantities of high chronic toxicity substances including known and suspected human carcinogens. Examples are benzo[*a*]pyrene, bis(chloromethyl) ether, and nitrosamines.

High chronic toxicity substances should be segregated from other substances and stored in a well-defined or identified area that is cool, well ventilated, and away from light, heat, acids, oxidizing agents, and moisture. For protection of laboratory and nonlaboratory personnel, such chemicals should be stored in a locked cabinet or in an area accessible only by designated personnel. The area should be identified with warning signs such as: WARNING! HIGH CHRONIC TOXICITY OR CANCER-SUSPECT AGENT. All containers of substances in this category should have labels that identify the contents and include a warning as above.

Procedures for working with substances of known high chronic toxicity should be adopted and followed (10).

5.1.2.3 Incompatible Chemicals. Incompatible chemicals should not be stored together. Such contact could result in a serious explosion or the formation of highly toxic and/or flammable substances.

Table 63.2 gives examples of incompatible chemicals. Several of the chemicals listed are not in common use in industrial hygiene laboratories. However, information on these

Table 63.2. Examples of Incompatible Chemicals[a]

Chemical	Is Incompatible With
Acetic acid	Chromic acid, nitric acid, hydroxyl compounds, ethylene glycol, perchloric acid, peroxides, permanganates
Acetylene	Chlorine, bromine, copper, flucrine, silver, mercury
Acetone	Concentrated nitric and sulfuric acid mixtures
Alkali and alkaline earth metals (such as powdered aluminum, magnesium, calcium, lithium, sodium, potassium)	Water, carbon tetrachloride or other chlorinated hydrocarbons, carbon dioxide, and halogens
Ammonia (anhydrous)	Mercury (in manometers, for example), chlorine, calcium hypochlorite, iodine, bromine, hydrofluoric acid (anhydrous)
Ammonium nitrate	Acids, powdered metals, flammable liquids, chlorates, nitrites, sulfur, finely divided organic or combustible materials
Aniline	Nitric acid, hydrogen peroxide
Arsenical materials	Any reducing agent
Azides	Acids
Bromine	See Chlorine
Calcium oxide	Water
Carbon (activated)	Calcium hypochlorite, all oxidizing agents
Carbon tetrachloride	Sodium
Chlorates	Ammonium salts, acids, powdered metals, sulfur, finely divided organic or combustible materials
Chromic acid and chromium trioxide	Acetic acid, naphthalene, camphor, glycerol, alcohol, flammable liquids in general
Chlorine	Ammonia, acetylene, butadiene, butane, methane, propane (or other petroleum gases), hydrogen, sodium carbide, benzene, finely divided metals, turpentine
Chlorine dioxide	Ammonia, methane, phosphine, hydrogen sulfide
Copper	Acetylene, hydrogen peroxide
Cumene hydroperoxide	Acids (organic or inorganic)
Cyanides	Acids
Flammable liquids	Ammonium nitrate, chromic acid, hydrogen peroxide, nitric acid, sodium peroxide, halogens
Fluorine	Everything
Hydrocarbons (such as butane, propane, benzene)	Fluorine, chlorine, bromine, chromic acid, and sodium peroxide
Hydrocyanic acid	Nitric acid, alkali
Hydrofluoric acid (anhydrous)	Ammonia (aqueous or anhydrous)
Hydrogen peroxide	Copper, chromium, iron, most metals or their salts, alcohols, acetone, organic materials, aniline, nitromethane, combustible materials

Table 63.2. (continued)

Chemical	Is Incompatible With
Perchloric acid	Acetic anhydride, bismuth and its alloys, alcohol, paper, wood, grease, oils
Peroxides, organic	Acids (organic or mineral), avoid friction, store cold
Phosphorus (white)	Air, oxygen, alkalis, reducing agents
Potassium	Carbon tetrachloride, carbon dioxide, water
Potassium chlorate	Sulfuric and other acids
Potassium perchlorate (See also chlorates)	Sulfuric and other mineral acids
Potassium permanganate	Glycerol, ethylene glycol, benzaldehyde, sulfuric acid
Selenides	Reducing agents
Silver	Acetylene, oxalic acid, tartartic acid, ammonium compounds, fulminic acid
Sodium	Carbon tetrachloride, carbon dioxide, water
Sodium nitrite	Ammonium nitrate and other ammonium salts
Sodium peroxide	Ethyl or methyl alcohol, glacial acetic acid, acetic anhydride, benzaldehyde, carbon disulfide, glycerin, ethylene glycol, ethyl acetate, methyl acetate, furfural
Sulfides	Acids
Sulfuric acid	Potassium chlorate, potassium perchlorate, potassium permanganate (similar compounds of light metals, such as sodium, lithium)
Tellurides	Reducing agents
Hydrogen sulfide	Fuming nitric acid, oxidizing gases
Hypochlorites	Acids, activated carbon
Iodine	Acetylene, ammonia (aqueous or anhydrous), hydrogen
Mercury	Acetylene, fulminic acid, ammonia
Nitrates	Sulfuric acid
Nitric acid (concentrated)	Acetic acid, aniline, chromic acid, hydrocyanic acid, hydrogen sulfide, flammable liquids, flammable gases, copper, brass, any heavy metals
Nitrites	Acids
Nitroparaffins	Inorganic bases, amines
Oxalic acid	Silver, mercury
Oxygen	Oils, grease, hydrogen, flammable liquids, solids, or gases

[a]Ref. 10.

chemicals is provided since many laboratories change their scope of work over time or inherit chemicals from previous operations.

5.1.2.4 Water-Sensitive Chemicals. If the laboratory uses any water-sensitive chemicals, such as sodium metal, it must store them in an area with no sources of water. The area should not have an automatic sprinkler system. Storage areas for such chemicals should be of fire-resistant construction, and other combustible materials should not be stored in the same area.

5.1.2.5 Chemicals Requiring Refrigeration. Some chemicals that are very volatile or unstable at room temperature must be refrigerated. Chemical storage refrigerators must be explosion-proof and wired to an explosion-proof outlet. Such refrigerators should be located in a designated storeroom that is supplied with single-pass air exhausted to the outside and that is under negative pressure with respect to the surrounding area.

5.1.2.6 Compressed Gases. Storage and use of compressed gases are covered in Section 3.4.6 and Section 5.1.3.3.

5.1.3 Chemical Storage in the Laboratory

Every chemical in the laboratory should have a definite storage place and should be returned to that location after each use. Chemicals should not be stored on benchtops because of the potential exposure to fire and the possibility of being knocked over. Chemicals should not be stored in hoods because they will interfere with the airflow, clutter up the working space, and increase the amount of materials that could become involved in a hood fire.

Volatile solvents, concentrated acids, and toxic substances should be handled in a chemical fume hood. Design of the industrial hygiene laboratory should include an adequate number of hoods of appropriate size for staff to conduct their work safely.

5.1.3.1 Beneath-Hood Storage. Because most laboratory workers tend to store chemicals in the cabinet space under the hood, ventilated cabinets designed by the manufacturer for safe storage of acids or solvents should be provided. Cabinets located directly under the hood also allow for the safe practice of making transfers of hazardous materials in the hood. The use of each hood should be planned and the appropriate base cabinet purchased for beneath-hood chemical storage.

5.1.3.2 High Chronic Toxicity Substances. Only minimum quantities of toxic materials should be present in the work area. Such substances should be handled in glove boxes or hoods according to procedures established by the laboratory.

5.1.3.3 Compressed Gases. In the industrial hygiene laboratory, compressed gases are used for atomic absorption, gas chromatography, and other instrumentation. If the laboratory is small, these specialty gas needs may be best handled by the use of gas cylinders stored and tied down at the bench or point of use. However, if the laboratory is large or its needs extensive, a central piping system may be needed. Gases may be provided from banks of high-pressure cylinders located in a convenient and safe place from which gas is piped through reducing valves to desired locations. This system is also an asset to laboratory housekeeping and safety, because gas cylinders are removed from the immediate work area.

5.2 Disposal of Waste

Typical waste generated daily from the analysis of industrial hygiene samples include:

- Carbon disulfide from analysis of solvents on charcoal tubes.
- Dilute nitric acid from analysis of metal fumes on filters.
- Bulk samples of building materials containing asbestos.
- Aqueous mixtures of reagents from colorimetric analyses.

All waste generated by the laboratory must be disposed of in accordance with applicable federal, state, or local waste disposal regulations. Proper disposal methods vary with the type of waste, the quantities generated, and regulatory requirements.

5.2.1 Regulations

Laboratories (11) have a moral and legal obligation to see that chemical waste is handled and disposed of in ways that pose minimum potential harm, both short-term and long-term, to health and the environment. The legal obligations began on November 19, 1980, when the U.S. Environmental Protection Agency (EPA) put into effect federal regulations on a Hazardous Waste Management System (**40** *CFR* Parts 260 through 266), under the authority of the Resource Conservation and Recovery Act (RCRA) of 1976, as amended. These regulations were designed to establish a "cradle-to-grave" system for the management of hazardous waste from all sources.

EPA's November 19, 1980 RCRA regulations focused on large generators who produce most of the hazardous waste in the country. As a result, the initial regulations exempted small-quantity generators—those producing less than 1000 kg per calendar month (kg/mo) of hazardous waste—from manifesting and record-keeping requirements applicable to large generators. More recently, however, the U.S. Congress also has directed EPA to regulate small-quantity generators of hazardous waste. Specifically, Congress lowered from 1000 kg/mo to 100 kg/mo the quantity at which generators of hazardous waste are exempt from compliance with certain RCRA regulations. This means that generators of more than 100 kg/mo of hazardous waste, which include many small laboratories, are now subject to specific RCRA regulations.

Waste generated in laboratories can differ significantly from waste generated at industrial production sites. Laboratories, which typically produce small quantities of many different wastes, face problems that set them apart from production facilities that generate large quantities of homogeneous waste. Proper handling and disposal of laboratory waste may involve procedures different from those used in a typical manufacturing environment. Therefore, when following the RCRA regulations, the unique nature of a laboratory environment must be taken into consideration.

The EPA can authorize state hazardous waste programs to operate in lieu of the federal program (12). Such state programs must be at least equivalent to and consistent with the federal program. The RCRA regulations do not preclude states from adopting more stringent hazardous waste regulations, and many have done so. All waste generators should check with state and local authorities to determine their local responsibilities.

5.2.2 Hazardous Waste Management System

The selection of a hazardous waste manager or coordinator for each laboratory or facility is important because laboratory managers are obligated to understand and to comply with the applicable elements of RCRA and the corresponding U.S. Department of Transportation (DOT) regulations. The laboratory hazardous waste coordinator should be familiar with the regulations and be able to organize a program that is practical to implement and that addresses both environmental and economic concerns.

As mentioned earlier, typical waste generated by industrial hygiene laboratories include spent solvents, unused chemicals, samples, and various contaminated containers, supplies, and equipment. Although many wastes generated by laboratories are not wastes appearing on EPA lists, many will be regulated because of hazardous characteristics (i.e., ignitability, corrosivity, reactivity, and EP toxicity). A hazardous waste management program should take into account both regulations and a common sense approach to hazardous material safety.

5.2.3 Waste Management Options

Until recently, the majority of laboratory waste was placed in secure chemical landfills. Because of long-term concerns with liability, landfill space, and regulations that may restrict land disposal, alternative disposal methods are now being considered. Landfilling of waste still may be the most economical disposal method in some geographic areas, but long-range concerns suggest that the waste generator should give careful consideration to alternative disposal methods. The most common alternative methods are discussed briefly below.

5.2.3.1 Waste Reduction. The most effective method of waste management is minimization of waste generation. The control of chemical purchasing practices and sharing of chemicals among departments, labs, or facilities offer two ways of doing this. By reducing the volume of chemicals entering the user system, the generator ultimately reduces the waste. Increasing costs of treatment or disposal will provide an incentive for waste reduction.

5.2.3.2 Recycling. In-house recycling is an environmentally sound and cost-effective way to manage certain waste chemicals, particularly solvents. However, the wastes accumulated for recycling are regulated under RCRA from the time of accumulation until they are recycled. The waste materials must be labeled, segregated, and stored according to the regulations applicable to the particular class of hazardous waste generator performing the recycling. The recycling activities and resulting products are not regulated under RCRA, but the residues from recycling are generally hazardous wastes.

5.2.3.3 Thermal Treatment. Thermal treatment of hazardous waste involves using high temperatures to change the chemical, physical, or biological character or composition of a waste. The objective of this process is to convert hazardous materials into nonhazardous by-products.

Incineration is the type of thermal treatment of most use to laboratories. Incineration can be used to destroy organic hazardous wastes safely by reducing waste materials into nontoxic gases and small amounts of ash and other residues (note, however, that the ash itself is considered a hazardous waste unless proved otherwise to EPA).

Many materials, however, are not suitable for incineration because of their high toxicity or reactivity. Mercury and mercury compounds, for instance, are not accepted for disposal by most incinerators. Highly reactive or explosive materials should not be incinerated.

Incineration may involve a different choice of outer packaging for off-site transportation from that of other disposal methods. Check with the incineration facility for packaging procedures before preparing a shipment. Because of a shortage of available facilities, lead times prior to shipment may vary considerably. This variation could affect the length of on-site storage. In the short term, incineration may be an expensive option; however, in the long term, if may offer the lowest liability.

5.2.3.4 Chemical Treatment. By altering the character or composition of a material by chemical treatment, a hazardous waste can be rendered less hazardous. Chemical treatment includes solidification, neutralization, oxidation, reduction, hydrolysis, and precipitation. Chemical treatment often can be done in the laboratory by the chemist who generated the material and knows its properties best. Information on reducing or destroying chemical wastes can be found in the National Research Council's Prudent Practices for Disposal of Chemicals from Laboratories (11).

5.2.3.5 Recovery. Sometimes it is economically feasible to regenerate chemical wastes so that the original products are reusable, especially large volumes of solvents. To date, laboratory recovery practices have been limited. However, the recovery of laboratory chemicals such as valuable metals (e.g., mercury, silver, platinum) and common water-insoluble solvents (e.g., chloroform) is increasing. As disposal costs rise, other recoveries will become more economical.

5.2.3.6 Sewage System. Under certain circumstances it is possible to flush certain waste materials into a municipal sewage system. Neutralized acids and bases and organic chemicals with 3% or greater solubility in water that are not highly toxic, flammable, or foul smelling are possible candidates if local regulations permit this type of disposal. Local regulations must be consulted and complied with.

5.2.4 Sources of Information on Chemical Waste Disposal

The most widely used reference on laboratory waste disposal is Prudent Practices for Disposal of Chemicals from Laboratories (11). Other references on waste disposal are also listed at the end of this chapter.

Questions on EPA regulations can be directed to EPA's RCRA "hotline" at (800) 424-9346 (toll-free) and (202) 382-3000 (metropolitan Washington, DC area).

5.3 Safety Showers and Eyewash Stations

Each laboratory should be equipped with at least one safety shower and one eyewash station. Safety showers and eyewash stations should be placed near operations with the greatest risk of splashes, spills, or fire.

The American National Standards Institute (ANSI) standard on emergency deluge showers (13) provides the following design criteria for safety showers and eyewashes:

- Both must be within 30 steps walking distance from the farthest occupied space in the laboratory.
- Shower and fountain should be at least 5 ft. apart so that two people could use them simultaneously.
- Both should be equipped with tempered water (the necessary 15-min minimum flushing period is difficult to achieve when the water is cold).
- Showers should have approximately 50 gal/min flow.
- Eyewash fountains should have approximately 2.5 gal/min flow.

An adequately sized floor drain is useful to speed cleanup.

If possible, deluge showers and eyewash fountains should be positioned so that laboratory staff can quickly and reliably locate them in an emergency. Bright distinctive floor markings and signs aid in safety shower and eyewash fountain identification.

5.4 Fire Extinguishers and Fire Suppression Systems

Fire extinguishers should be of the type and size appropriate to the work conducted in the laboratory. They should be mounted on a wall at a comfortable height in a convenient location. They should be placed where coats or equipment cannot impede their use.

The use of fire blankets in laboratories is controversial. While a tightly wrapped blanket may extinguish burning clothing, it may also press molten synthetic fabric deeply into burned skin and tissue. If mounted in a laboratory, a fire blanket should be within 30 steps walking distance from any part of the lab and well marked.

Typically, laboratory fire suppression systems are water sprinklers. Water systems are hazardous in high-voltage locations and in locations where chemicals that are incompatible with water are used or stored. Also, a water system may not be the best choice in rooms where expensive equipment would be ruined by water. Alternative, but more expensive, fire suppression systems are available for these applications. Fire detection devices should be installed in all areas not protected by fire suppression systems. Local fire codes dictate requirements for fire suppression systems, when used, and should be consulted during the design phase. In buildings in which water sprinkler pipes run through unheated plenum space, pipes may freeze and burst during cold weather. The design phase should include consideration of pipe insulation or a compressed-air dry system to prevent water damage from frozen pipes.

5.5 First Aid Planning

Most industrial hygiene laboratories do not have medical personnel on staff or within the building. The lab should plan to have personnel trained in first aid, including cardiopulmonary resuscitation (CPR), available during working hours to respond to emergencies until medical help can be obtained. Emergency phone numbers should be displayed on all laboratory phones.

An emergency room staffed with medical personnel specifically trained in proper treatment of chemical exposures should be identified and readily accessible. Prior arrangement with a nearby hospital or emergency room may be necessary to ensure treatment will be available promptly. The services of an ophthalmologist especially alerted to and familiar with chemical injury treatment should also be available to minimize the damage to eyes that may result from many types of laboratory accidents. Proper and speedy transportation of the injured to the medical treatment facility should be available. In addition to the normal facilities found in an emergency room, there should be specific standing orders for emergency treatment of chemical accidents. Emergencies that should be anticipated include:

- Thermal and chemical burns, including those caused by hydrofluoric acid.
- Cuts and puncture wounds from glass or metal, including possible chemical contamination.
- Skin irritation by chemicals.
- Poisoning by ingestion, inhalation, or skin absorption.
- Asphyxiation (chemical or electrical).
- Injuries to the eyes from splashed chemicals.

5.6 Emergency Evacuation Design

The laboratory should establish an emergency plan that includes evacuation routes, shelter areas, medical facilities, and procedures for reporting all accidents and emergencies.

Evacuation procedures (10) should be established and communicated to all personnel. Evacuation routes and alternatives may have to be established and, if so, communicated to all personnel. An outside assembly area for evacuated personnel should be designated.

An emergency alarm system should be available to alert personnel in an emergency that may require evacuation. Laboratory personnel should be familiar with the location and operation of this equipment. A system should be established to relay telephone alert messages; the names and telephone numbers of personnel responsible for each laboratory or department should be prominently posted in case of emergencies outside regular working hours.

Isolated areas (e.g., microscopy dark rooms) should be equipped with alarm or telephone systems that can be used to alert outsiders to the presence of a worker trapped inside or to warn workers inside of the existence of an emergency outside that requires their evacuation. Where unusually toxic substances are handled, it may be desirable to have a monitoring and alarm system that is activated if the concentration of the substances in the work environment exceeds a set limit.

Brief guidelines for shutting down operations during an emergency or evacuation should be communicated to all personnel. Return procedures to ensure that personnel do not return to the laboratory until the emergency is ended, as well as start-up procedures that may be required for some operations, should be prominently displayed and regularly reviewed.

All aspects of the emergency procedure should be tested regularly (e.g., every 6 months to a year), and trials of evacuations (if there are such procedures) should be held periodically.

5.7 Facilities Maintenance

5.7.1 Housekeeping

There is a definite relationship between safety performance and orderliness in the laboratory. When housekeeping standards fall, safety performance inevitably deteriorates. The work area should be kept clean, and chemicals and equipment should be properly labeled and stored. The following guidelines should be established.

- Work areas should be kept clean and free from obstructions. Cleanup should follow the completion of any operation or be done at the end of each day.
- Wastes should be deposited in appropriate receptacles.
- Spilled chemicals should be cleaned up immediately and disposed of properly. Disposal procedures should be established, and all laboratory personnel should be informed of them.
- Unlabeled containers and chemical wastes should be disposed of promptly using appropriate procedures. Such materials, as well as chemicals that are no longer needed, should not accumulate in the laboratory.
- Floors should be cleaned regularly for removal of accumulated dust, chromatography adsorbents, and other assorted chemicals some of which may pose respiratory hazards.
- Stairways and hallways should not be used as storage areas.
- Access to exits, emergency equipment, and controls should never be blocked.
- Equipment and chemicals should be stored properly; clutter should be minimized.

5.7.2 Equipment Maintenance

Good equipment maintenance is important for safe, efficient operations. Equipment should be inspected and maintained regularly. Servicing schedules will depend on both the possibilities and the consequences of failure. Maintenance plans should include a procedure to ensure that a device that is out of service cannot be restarted.

5.7.3 Security

Laboratory security is important in protecting the safety of employees and those who might enter the laboratory, and in protecting laboratory work and capital assets. The laboratory should be in an area that has limited access. All individuals entering the laboratory should come through a reception area, provide identification, and sign in. During hours when the laboratory is not staffed, areas housing equipment, samples, or laboratory records should be locked. Only laboratory, custodial, and maintenance staff should have access to the laboratory during nonoffice hours.

BIBLIOGRAPHY

1. J. A. Young, ed., *Improving Safety in the Chemical Laboratory*, John Wiley & Sons, Inc, New York, 1987.
2. A. C. Farrar, and H. A. Hurley, "Industrial Hygiene Laboratory," in J. Garrett, L. J. Cralley, and L. V. Cralley, eds., *Industrial Hygiene Management*, John Wiley & Sons, Inc., New York, 1988.
3. S. Braybrooke, ed., *Design for Research, Principles of Laboratory Architecture*, John Wiley & Sons, Inc., New York, 1986.
4. S. L. Wilkinsin, *Chem. Eng. News*, February 15, 1999, 53–64. (Feb. 15, 1999).
5. AIA, C. Linn, ed., "Interactive Approach Produces Innovations," *Lab. Planning Design*, **1**(1) (Spring 1989).
6. L. DiBerardinis et al., *A Guide to Laboratory Design: Health and Safety Considerations*, Wiley-Interscience, New York, 1st ed., 1987; 2nd ed., 1993.
7. D. Guise, *Design and Technology in Architecture*, John Wiley & Sons, Inc., New York, 1985.
8. American Conference of Governmental Industrial Hygienists, *Industrial Ventilation, A Manual of Recommended Practice*, 20th ed., 1988.
9. American Society of Heating, Refrigerating, and Air-conditioning Engineers, Inc., "1985 Fundamentals," *ASHRAE Handbook*, Atlanta, GA, 1985.
10. National Research Council, *Prudent Practices for Handling Hazardous Chemicals in Laboratories*, National Academy Press, Washington, DC, 1981.
11. National Research Council, *Prudent Practices for Disposal of Chemicals from Laboratories*, National Academy Press, Washington, DC, 1983.
12. American Chemical Society, Task Force on RCRA, *RCRA and Laboratories*, American Chemical Society, Washington, DC, 1986.
13. American National Standards Institute, *ANSI Z358.1*, New York, 1985.

GENERAL REFERENCES

S. L. Wilkinsin, "Building for Success, Lab Designers Nurture Research with Views, Quiet Nooks, Coffee, and a Dash of Whimsy," *Chem. & Eng. News*, February 15, 1999, pp 53–64.

L. DiBerardinis, et al, *Guidelines for Laboratory Design: Health and Safety Considerations*, Second Edition, John Wiley & Sons, New York, 1993.

E. Crawley Cooper, *Laboratory Design Handbook*, CRC Press, 1994.

J. C. Rock and S. A. Anderson, "Benefits of Designing for Ventilation Diversity in a Large Industrial Research Laboratory—A Case Study," *Appl. Occ. Env. Hyg. J.* 1233–1237 (Oct. 1996).

T. C. Smith and S. M. Crooke, "Maximizing the Effectiveness of your Laboratory Ventilation Management Program," *Today's Chemist at Work*, 30–34 (May 1996).

R. Newill, "Designing Labs for People," *Environmental Testing & Analysis*, 1996, 19–20, 35 (March/April 1996).

K. M. Magon, "How to Prevent Your Lab from Becoming a Frankenstein," *Safety and Health* (July 1990).

L. DiBerardinis, "Laboratory Ventilation Standards: Revision of ANSI Z9.5: Why, What, How"? *The Synergist*, December 1998, 28–29 (Dec. 1998).

"Influence of Heat Load on Selection of Laboratory Design Parameters and Dynamic Performance of Laboratory Environment." *ASHRAE Transaction: Symposia*, **102**(1), 723–731 (1996).

M. J. Pitt and E. Pitt, *Handbook of Laboratory Waste Disposal*, John Wiley & Sons, Inc., New York, 1987.

M. J. Lafevre, *First Aid Manual for Chemical Accidents*, Van Nostrand Reinhold, New York, 1989.

N. H. Proctor and J. P. Hughes, *Chemical Hazards of the Workplace*, J. B. Lippincott, Philadelphia, 1978.

R. S. Stricoff and D. B. Walters, *Laboratory Health and Safety Handbook*, John Wiley & Sons, Inc., New York, 1990.

Industrial Indemnity Company, *Safety Engineering Standard Handbook*, San Francisco, 1990.

CHAPTER SIXTY-FOUR

Occupational Epidemiology: Some Guideposts

John F. Gamble, Ph.D.

1 INTRODUCTION

It is the nature of modern human beings to speculate on the causes of disease. Often this speculation centers on factors associated with work, and the recognition of disease-producing agents in occupational settings has resulted in striking health improvements in some groups of workers. Such conditions as lead intoxication, mercury poisoning, and benzol poisoning have largely disappeared because they could be easily identified with a particular substance, and industrial hygiene measures were relatively easy to apply (1). For some occupational disease, such as bladder cancer in the dye industry and phossy jaw in the match industry, the use of certain substances was simply stopped in order to protect the health of the worker.

As the more obvious industrial hazards have been identified and brought under control, the task of identifying disease produced by occupational exposure has become increasingly difficult. In many instances environmental or occupational exposures may be low so that our ability to distinguish response from background is limited. Nevertheless, studying human populations to assess the determinants of disease has several advantages over other methods. In particular, humans comprise the population of interest, and they are studied in the real world where the exposure actually occurs.

What is the nature of epidemiology? Epidemiology may be defined as the study of health and illness in human populations. It is the study of the distribution of determinants of health-related states and events in populations, and the application of this study to the control of health problems. It is a method of reasoning about disease that deals with

Patty's Industrial Hygiene, Fifth Edition, Volume 4. Edited by Robert L. Harris.
ISBN 0-471-29749-6

biological inferences derived from observations of disease phenomenon in population groups (2).

Although epidemiologists do not always agree on a definition of epidemiology, or on the scope of activities included in this scientific discipline, there are features of epidemiology that can serve to characterize it as a field of study. The statements and viewpoints listed below should help to orient one as to what epidemiology is about.

1. Findings should relate to a defined population, a population at risk. A population is either a finite or infinite universe of people about whose health a statement is to be made. Conventionally, in a study of a population of concern, three groups are considered in succession: target population, available study population, and study sample.

In general, one evaluates the status of the study sample and then proceeds to make statistically justifiable generalizations based on findings, provided that the sample is representative. To meet this requirement there should be an unambiguous definition of the study population composition; for example, it is composed of all factory employees hired between 1950–1960. There should be an enumeration of all the members of the study population. And there should be an adequate definition of the sampling process to ensure an even probability for all members of the study population to be selected into the sample. The study sample should provide, in other words, an adequate insight on the target population (3).

2. Epidemiology deals with biological inferences about the cause and/or natural history of a disease. Clues to etiology come from comparing disease rates among groups with different levels of exposure that is defining the relationships in the form of exposure–response curves. Note that exposure–response (E–R) is often called dose–response. However, exposure (e.g., air concentration of dust) is what is being measured and not the amount of dust deposited in the lung, which is dose. An example of an indirect measure of dose is the severity of pneumoconiosis. Rules of causation generally also include criteria such as consistency of the association, strength of association (both magnitude and gradient), temporal association (cause must precede or coincide with effect), and biological plausibility (see Refs. 4–6 for discussion).

3. Association is a term meaning the quantitative relationship between two variables such as, for example, smoking and lung cancer. In epidemiology there are two main measures of association, and these can also be measures of effect. One compares disease frequency (e.g., in exposed and control groups) by assessing their differences (attributable risk) or their ratio (risk ratio, or RR). The best measures of the RR are incidence and prevalence in the exposed group divided by occurrence in the nonexposed or control group. Incidence rate is the proportion of new cases of disease in a population at risk during a specified period of time. Prevalence rate is the total number of cases divided by the total population at a given point in time or over a short time interval.

4. Often in epidemiology, one investigates associations between exposure (the independent or effect variable) and disease (the dependent or response variable). When the comparison is between exposed and nonexposed populations, the paradigm is what is called a natural experiment. In the occupational experience, epidemiology deals with the real world and cannot control the composition of the exposure groups. Unlike traditional ex-

periments, comparability between exposed and referent groups is often affected by factors other than mere exposure. Therefore, in all epidemiology studies one must be concerned about bias, that is, an effect tending to produce results that depart systematically from the results of a given exposure.

5. Epidemiology deals with the evaluation of scientific hypotheses. A common practice is to assume that hypothesis testing involves simply accepting the null hypothesis or rejecting the hypothesis in favor of an alternative. Statistical testing quantifies the likelihood of falsely rejecting the null hypothesis. It is not uncommon to think of the p value, or the confidence interval around the RR, as the decision points for the truth or falsity of the hypothesis.

In epidemiology, many, if not most, so-called hypothesis-testing situations are actually estimation exercises. Every epidemiological study is an exercise in measurement for the purpose of obtaining estimates of the disease occurrence (such as prevalence and incidence rates) or some derivatives of these measures (i.e., means). It is an attempt to quantify the association between exposure and effect.

6. Epidemiology can be defined by what epidemiologists do. What are the hypotheses or questions they ask? In general terms these questions relate to the occurrence of disease by time, place, and person, such as the following:

- Has there been an increase or decrease in disease over the years?
- Does one geographic area (industry, plant, work area, job title) have a higher frequency of disease than another?
- What are the characteristics of persons with a particular disease or condition that distinguish them from those without the disease or condition? These characteristics may be risk factors such as age, gender, race, physiological function, biochemical status, socioeconomic status, occupation, and personal habits.

In occupational epidemiology, studies often address specific questions such as, what is the association between low-level ionizing radiation and leukemia, or what is the association between asbestos and cancer (or more specifically, chrysotile exposure and mesothelioma or non-asbestiform amphibole exposure and lung cancer).

In occupational epidemiology the primary purpose is to determine if there are cause–effect relationships between disease and exposure encountered at work. The purpose of this chapter is to outline some of the methods and pitfalls encountered in determining cause–effect. Or as Hill (4) put it in 1965, "How in the first place do we detect these relationships between sickness, injury and conditions of work? How do we determine what are physical, chemical and psychological hazards of occupation, and in particular those that are rare and not easily recognized?"

Section 2 contrasts experimental research and epidemiological research, which is observational in nature. The form and presentation of results in an experimental animal study or clinical trial appear to be similar in the comparison of exposed and control groups. However, the experimental study has control over study subjects and exposures, while the epidemiology study is observational and cannot determine *a priori* the composition of the study population or the intensity and duration of exposure. The study designs that are unique to epidemiology are discussed with consideration of their strengths and weaknesses. They differ fundamentally from experimental research in that cause–effect cannot be assumed, even if the results are statistically significant.

The problems of inference and interpretation in observational studies are different than those in experimental studies. Not all studies and designs have equal merit for determining cause–effect. Some studies may be useful for generating a hypothesis or be descriptive in nature. Some studies are designed for testing a hypothesis, but because of bias or confounding are unsuitable for assessing cause–effect or exposure–response (E–R) trends. Thus there is a continuum in the quality or usefulness of observational studies for determining cause–effect. Part of the science is to design useful studies, and to determine which are, and which are not, useful for testing a hypothesis to determine cause–effect. The criteria for selecting the appropriate and relevant studies for assessing cause–effect, and therefore risk, are outlined in Section 3.

Cause–effect is based on the weight of all the relevant evidence. Much of the evidence is obtained from the relevant studies following the criteria discussed in Section 3. Evidence is also obtained from other sources such as toxicology (experimental animal studies), levels of organization (e.g., individual/organ/cell levels), and biological theory. The criteria used in epidemiology for weighing all this information in determining cause–effect relationships are discussed in Section 4.

The observational nature of epidemiology means that one must always be wary of biases that can lead to spurious interpretations. Descriptions of studies will be presented to help explain the concepts presented and provide examples of the kind of skeptical consideration required to find a true cause. Or to avoid the "wish bias," the tendency to make causal conclusions (or not) on the basis of an inherent tendency to "wish for" a desired result (7, 8).

2 TYPES OF EPIDEMIOLOGICAL RESEARCH

There are basically two types of research: (*1*) experimental and (*2*) nonexperimental or observational.

2.1 Experimental Research

Experiments in epidemiology consist of clinical trials or field trials where the objective is to test etiologic hypotheses and to detect "exposure" effects. The questions are, for example, of the form: does drug A work better than drug B? If it does work, at what dose? The paradigm of this scientific process is an experiment or controlled observation, for example, an animal experiment. In a clinical epidemiological study (or animal experiment) which is designed to estimate the effect of a particular exposure, the procedure is the following.

Out of a population available for study one might select subjects of one gender and age range. Subjects would then be randomly assigned, respectively, to exposed and nonexposed groups and followed over a specified time period adequate to determine the incidence of a particular condition. The measure of effect in this case is a relative risk (risk ratio, rate ratio) which is the ratio of the incidence rate of the condition among exposed to the incidence rate among the nonexposed. The variables need not be dichotomous variables. In both experimental and observational sciences they may also be continuous. Continuous semiquantifiable response variables such as fibrosis may be observed, instead of a discrete variable such as the occurrence of a tumor. The outcome can be measured by severity ranks and the contrast still made between the exposed and unexposed groups. Several exposure groups (none, low, medium, high) may allow the computation of an exposure–response relationship. Thus one gains evidence not only of etiology but also of potency and risk at various levels of exposure or dose.

2.2 Nonexperimental or Observational ("Real World") Research

The goal of an observational study is to simulate an experiment had one been possible. This is the common kind of research in occupational epidemiology because one cannot assign workers to industries and jobs, nor require that they stay on the job. Because of the observational nature of occupational epidemiological research, there are a number of potential problems one must be aware of in the conduct and interpretation of epidemiological studies that are not characteristic of experimental research.

2.2.1 Potential Problems

Experimental and observational research often appears to have a similar structure, and their analysis is parallel. There are, however, scientific standards in experimental research that cannot be met in observational studies. These are (a) random assignment to an exposure group and (b) control of exposure. Other scientific standards that are routinely applied in laboratory experiments but may be absent in observational research include (c) a stipulated research hypothesis; (d) a well-specified cohort or population at risk (PAR); (e) high-quality data, and (f) analysis of attributable actions (9).

Conclusions about cause–effect relationships drawn from observational studies may be fraught with difficulties introduced by unknown or undetected biases. Some potential consequences of the inability to randomize and control exposure are discussed. Also discussed are standards that can be controlled in epidemiology. If not controlled the result can be spurious findings.

2.2.1.1 Randomization. When the members of a population selected for an experiment are similar in relevant and measurable characteristics (such as age, gender, or health status), each individual is randomly assigned to an exposed or unexposed group. In this way one hopes that the random exposure assignments will reduce if not prevent bias, or that differences between individuals will be evenly distributed between groups. If there are undetermined characteristics that can influence the outcome, one hopes that such characteristics will be more or less equally distributed in all exposure groups and therefore have no significant effect on the ultimate analysis. Indeed, if characteristics are equally distributed, there will be no bias, and the presence or absence of the characteristic(s) becomes irrelevant for the results of interest.

It is a major function of randomization to achieve "approximate comparability with all variables, whether known or not" (10). If exposure and nonexposure groups differ with respect to the outcome, that difference may be attributed to the exposure. Obviously, if the exposed and comparison (referent) groups differ with respect to variables other than the exposure, a cause–effect relationship is weakened or denied.

An often overlooked aspect of the randomization principle is the fact that "in order to insure generalizability it is necessary that samples be random. Random samples imply that each member of the population should have the same probability of being included in the sample" (11). This idea has been stated most emphatically with the warning to the observer that "in the absence of random sampling, the whole apparatus of inference from sample to population falls to the ground, leaving the sample without a scientific basis for the inferences which he wishes to make" (quoted in Ref. 10).

However, randomization is not a cure-all because it does not always assure comparability and prevent bias. Randomization is neither necessary nor sufficient for rigorous statistical inferences. However, with or without randomization one should try to control for base-line inequalities or biases, for example, by stratification (10).

2.2.1.2 Control of Exposure. A second important characteristic of an experiment is the ability to control and measure exposure with adequate precision. In a bench experiment one can and does precisely control both the duration and intensity of the exposure of interest and eliminate all unwanted or confounding exposures. In epidemiological research, comparability must be assured by more complex arrangements. A common procedure, for instance, is to exclude from the experiment all subjects presenting the outcome at the beginning of the study period. Each study group hopefully begins with the same base-line demographic characteristics and health status. One can further control the experiment by allowing sufficient time (latency) for the exposure to have an effect.

Weiss (12) discusses two biases that can arise because of differences in the onset of exposure. In his examples these biases may account for some apparent elevated risks observed in occupational cohorts. A chronology bias can occur if cohort members have a different length of exposure at the beginning of the study period. Because exposure intensity is generally greater in earlier eras, workers with a longer duration of exposure and who had worked during earlier time periods will probably also have a greater intensity of exposure than workers hired more recently. This bias can be avoided if comparisons are made between subjects whose exposure begins at about the same time.

Selection bias may occur when exposure begins before the study period begins. The more remote the onset of exposure before the beginning of the study period, the greater the selection pressures and therefore the greater the reduction in the study population to a fraction of the original. The remedy is to include all subjects beginning work at about the same time.

Occupational exposures are generally erratic in nature because of changes in the work schedule and in the intensity of the environmental characteristics. Workers are often exposed to a variety of nonoccupational agents, such as tobacco, dietary, and other risks of everyday life, that can have effects at least as significant as the occupational exposure singled out for study. In such situations, problems in attributing the outcome to a given exposure may depend on the quantitative characteristics of the exposure, the eligibility of workers for inclusion in the "exposed" category, the identification of independent secular trends as a factor of pathology, and so on. Thus indicating an exposure history as causing an outcome is a somewhat subjective judgment, based upon considerations of the intensity and duration of exposure, classification of exposure status, and whether there is a progressive increase in the outcome variable that is in parallel with increased exposure.

2.2.1.3 A Priori Hypothesis. In an experiment or clinical trial there is a research hypothesis that identifies the cause–effect comparison that will be tested. In an epidemiological study there may or may not be a clearly stated *a priori* hypothesis. Because of the wide availability of computing equipment, the ability to collect large amounts of diverse information has vastly increased, and consequently many different associations in addition to an *a priori* hypothesis can be explored in "data dredging" activities. In this process a new hypothesis may be suggested by newly collected data. Setting statistical significance at $p < 0.05$, one might expect a positive computer-generated conjecture by chance alone for every 20 associations explored. And voila, a new cause–effect relationship is discovered. Beware!

2.2.1.4 A Well-Specified Cohort. In experiments, the characteristics of the cohort members are determined at the beginning of the experiment, that is, before exposure has occurred. During the course of the experiment periodic checks must monitor the occurrence of the outcome. This process assures that (a) only eligible subjects are included in the cohort, and (b) all subjects are accounted for statistically.

In epidemiology the actual base-line conditions are often unknown and unknowable. In retrospective studies one relies on data collected for other purposes. In retrospective cohort studies and case-control studies, for example, work history and study populations are determined from personnel records and from exposure estimates collected for other purposes; often smoking information is not available at all.

The "healthy worker effect" is a phenomenon where the overall health experience (based on both mortality and morbidity) of the employed population is usually more favorable than that of the general population. The employed population usually appears healthier than the general population because of two main factors, selection bias and higher socioeconomic status (13).

The most important factor is probably selection bias, that is, selection by the employer for healthy "employable" workers, and self-selection by the employee out of the work force. The selecting out process does not necessarily lead to a healthier work force, because short-term employees are often less healthy than longer-term workers (14, 15). And employability is not necessarily related to susceptibility or occurrence of cancer.

Wang and Miettinen (16) describe the healthy worker effect as a matter of confounding. Owing to different requirements and incentives for job entry and exit between the study population and the reference population, these two populations may not be comparable for many risk factors. Therefore there will be interference (confounding) when their results are compared. Thus the importance of having an "internal" reference group from the same population base as the exposed population. Expected values derived from an "external" reference group such as a national population can be made comparable on a few risk factors such as age, gender, and race. But since the external comparison population is not from the same population-base as the exposed population, the two populations must be thought of as potentially, if not probably, different. In an experimental study one hopes that randomization will result in groups that are comparable regarding confounding risk factors that are difficult or impossible to measure. In an analogous fashion one hopes that use of an internal comparison group will result in exposed and referent groups that are roughly comparable regarding potentially confounding risk factors for which information cannot be obtained. One expects some degree of comparability when the exposed and referent populations are from the same population-base.

2.2.1.5 High-Quality Data. In a bench experiment base-line data are obtained directly with calibrated methods worked out before the onset of the experiment. These measurements are specific for the outcome defined in the hypothesis. In epidemiology, outcome information often must be obtained from second-hand sources and therefore errors are less readily screened out. For instance, in mortality studies the outcome variable, death, is determined from death certificates. The determination of cause of death leaves room for inaccuracies and omissions, even deception (17). Patients may die at home of chronic illness in which the immediate cause of death may be an intervening complication rather than the underlying pathological process. Even hospital deaths with high autopsy rates show discrepancies between the autopsy results and the cause of death (COD) originally recorded on the death certificate. The current trend for fewer autopsies and the differential rates of autopsies for ethnic and racial groups increases further the potential for misclassification. An example, taken from asbestos workers, showed that when comparing the COD recorded from autopsy and/or hospital records with the respective COD recorded on the death certificate, there were underestimates of mesothelioma and overestimates of gastrointestinal and other tumors (18).

The accuracy of the COD varies depending on disease category (>90% for ischemic heart disease, for example). Mortality studies are quite useful for some causes of death (e.g., lung cancer), when the COD used is broad and when very elderly decedents are not used (17).

Exposure information can come from a variety of sources. In retrospective cohort studies they may come from personnel records that are accurate in terms of time spent (tenure) in given jobs at that facility. Actual environmental conditions are often unmeasured or sparsely documented. As a result, duration of exposure is often used as a surrogate of dose in the determination of exposure–response relationship. To minimize this source of error, exposed subjects included in the exposure–response analysis should belong to uniform exposure categories, with a similar duration of exposure (19). Unfortunately such remedies are not always feasible.

In case-control studies information on exposure for both cases and controls may come from surrogate sources (such as friends and relatives). Often the information is obtained long after the occurrence of the event(s) in question, and problems therefore exist in verifying the accuracy of such information. For example, in attempting to control confounding from tobacco smoking the only source of information may be the spouse and/or children. In general, measurements of work history, job exposure, and exposure to toxic substances are less accurate than smoking. Table 64.1 reports examples of these discrepancies.

2.2.1.6 Analysis of Attributable Actions. An ideal experiment allows one to determine if there is a statistical difference between the exposed and nonexposed groups. If there is

Table 64.1. Percent Agreement for Smoking Status and Work History

Reference	Respondent	Units	Agreement (%)
20	Husband and wife	Smoking—yes/no	96
		Age started	28
		Cigarettes/day	36
21	Surrogates	Smoking—yes/no	
		Smoker	91
		Nonsmoker	85
22	Self-report	Work History	82
23	Spouse	Exposure to solvents—yes/no	58
24	Self-report	Date of Hire	95
		Exact	53
		± One year	73
		Tenure—exact	48
25	Spouse	Number of work area assignments	48
		Usual work area	72
		Type of chemical or physical exposure	3
26	Next of kin	Smoking status	87
27	Wife	Industry	51
		Occupation	48
		Exposure to asbestos	50
		Exposure to radiation	86
		Exposure to arsenic	70
		Exposure to formaldehyde	70
		Smoking—yes/no	100

such a difference (or no difference) one may conclude the exposure is (or is not) the responsible causative agent. Such a conclusion is the accepted outcome of an experiment, assuming those scientific standards have been maintained and the number of observations is adequate to achieve sufficient power.

In epidemiology the exposure is often erratic, variable in intensity and duration, uneven among different subjects, and often contaminated by exposures to agents other than those characteristic of the workplace, such as personal habits (hobbies, smoking), diet, and similar risks of life. These risk factors for the disease of interest are often unmeasured or are often not amenable to statistical tests or adequate analysis.

To reiterate, scientific standards common in experimental science are difficult to achieve in all instances in observational science. Therefore bias is a constant concern in cause–effect research. In the conduct of an epidemiological study biases can occur at all stages from selecting the study population through to the collection and entry of data (28) as described below.

Susceptibility bias occurs when the base-line states of the contrasted groups are not comparable. The differences can be quite varied and might include factors such as smoking, alcohol consumption, socioeconomic status, and health status. Perhaps the best-known bias at this stage is the healthy-worker effect or selection bias.

Performance bias occurs when there are differences in the acquisition of information on the "maneuvers" or exposures of the compared groups. Enterline (29) discusses two pitfalls that fit in this category. The first case occurs when an occupational cohort experiences competing causes of death so the probability of death from the cause(s) of interest may be underestimated. A second pitfall can occur when there are variations in latency periods. Because susceptibility to disease is on a continuum and is influenced by the intensity of exposure, observations should be collected by time since first exposure. For an effect having a long latency, the inclusion of large numbers of workers with a short time since first exposure will reduce the computed relative risk for that effect. Similarly, the evaluation of an exposure–response relationship for a chronic disease with a long latency should be done on workers observed through adequate length of time. For example, to evaluate lung cancer in its relationships with asbestos exposure, the effect should be evaluated in workers with 20 or more years since exposure inception.

Detection bias occurs in the determination of outcome and arises because of inconsistencies of diagnostic criteria and/or exposure. Examples include the following:

- A greater effort to get smoking or other exposure information from members of a case than from those of a control group can bias estimates of exposure.
- Knowing the origin of the study population may result in a radiological interpretation of nonspecific changes on a chest X-ray as being consistent with a specific pneumoconiosis. The entire study population should be read "blind," without distinction between "exposed" and "referent" group. In addition, more than one qualified reader should be used.

Correcting the cause of death recorded on death certificates based on additional information from hospital and/or autopsy records for exposed but not the referent population invalidates the between-group comparisons and the calculated relative risks (29). A cause of death based on "best evidence" for the exposed population should not be used to calculate an SMR since the external referent cause of death is based only on information from the death certificate, and the degree of diagnostic misclassification will not be comparable.

Transfer bias is an incomplete collection of data or an uneven loss of data when transferred for analysis. This bias occurs in cohort studies when subjects are lost to follow-up because of unknown status and/or incomplete information such as gender, race, and date of birth, or work history. In a cross-sectional morbidity study this bias occurs when study subjects do not participate for reasons unknown, which may include being late, or absent, or coincidence of vacation schedule with examination time.

In summary, occupational epidemiological studies may appear to have the same structure as experiments. But unlike experiments the exposed and comparison groups are often not perfectly comparable, random apportionment to exposures and control groups is not possible, and exposure cannot be controlled. In order to reduce bias and increase accuracy the sources of possible errors must be diligently reviewed and when possible controlled, lest the results and interpretations be distorted.

2.3 Study Designs

Epidemiological studies are designed to detect diseases produced by occupational or other exposures. That is, they are meant to determine one or more of the following objectives:

1. Fishing expedition to look for possible relationships of a large number of variables to a specific disease. The best that can come of this exercise is discovering a hypothesis that is plausible and should be tested.
2. Determine the characteristics of the individuals that are experiencing ill health. This is a descriptive study.
3. Determine if the occurrence of disease in an exposed population exceeds that in an unexposed or general population. Is there a hazard?
4. Determine risk factors or etiologic variables that are associated with, or cause, a specific disease, including occupational risk factors such as asbestos.
5. Test *a priori* hypotheses and determine whether there is a causal relationship between a putative etiologic agent and an adverse health outcome (30). This can include a quantitative assessment of health risks associated with specific levels of exposure, i.e., assessment of exposure-response relationships.

Epidemiological studies can be classified into about five categories, each providing a different level of inferential knowledge toward the definition of a causal agent or agents.

2.3.1 Ecological Study Design

These are studies assessing the correlation between some measure of morbidity or mortality in a population and an exposure variable. In this type of study either the independent or dependent variables, or both, are based on aggregate or group data, and the characteristics of an individual's health status and/or individual exposure status cannot be determined. Individuals cannot be classified as to their health status or their exposure. As a result the association between exposure and disease cannot be determined for any individual. Ecological studies are unable to test a hypothesis or determine with reliability an exposure–response relationship. To infer that the associations between variables based on aggregate data will be the same as the association between variables based on individual-level data is likely to result in a false inference called the ecological fallacy. A correlation based on group (or ecological) characteristics may disappear, or even be reversed, when assessed at the individual level.

Data for ecological studies are easy to obtain. The results are useful for generating a hypothesis. They should not be used to test a hypothesis or determine cause-effect or estimate exposure-response.

2.3.1.1 Example of Ecological Study: Cancer Death Rates by County. The NCI published cancer death rates by county for the years 1950–1969, accompanied by an atlas of cancer mortality (31). One hypothesis generated from these data was that the presence of chemical and petroleum industries may be related to cancer death rates (32).

A large number of epidemiological studies of workers in the petrochemical industry have been published since then using individual-level data. Wong and Raabe (33) reviewed close to 100 such studies of petroleum industry employees potentially exposed to petroleum products and including gasoline-like hydrocarbons. The bulk of these were cohort studies reporting SMRs (observed deaths/expected deaths from external national populations and adjusted for age and gender) for major causes of death. The combined results of these studies showed a significantly lower mortality than the general population for cancers of the digestive system, stomach, lung, and all cancers combined. The mortality experience for skin and brain cancers was similar to the general population. Prostate and kidney cancer mortality overall were similar to the general population, but some small groups of workers were at increased risk. For example, one study showed truck drivers were at increased risk of kidney cancer. Some refinery workers, particularly those employed prior to 1940, were at increased risk of leukemia. And some lymphatic tissue cancers were elevated. However these results should for the most part still be considered as associations that need further testing. The presence, or absence, of an association may be due to only a small number of exposed persons (dilution), confounding (the reduced lung cancer might be due to reduced smoking), noncomparability of populations (petroleum workers may have better health care or socioeconomic (SES) than the general population).

More analytic studies were considered necessary to define possible risk of kidney cancer because of excess risk in some cohorts of petroleum workers and because of increased kidney tumors in male rats exposed to gasoline. Three nested case-control studies within the petroleum industry have assessed the association of petroleum hydrocarbons and kidney cancer (34–36). None found any apparent increased risk associated with tenure, or with semiquantitative estimates of exposure to petroleum hydrocarbons.

In these examples the hypothesis generated by the ecological studies was not confirmed by individual-level studies.

2.3.1.2 Example of Ecological Study: Particulate Matter (PM) Air Pollution. A number of air pollution studies have been conducted to evaluate the association of ambient air particulate matter (PM) with citywide mortality and morbidity (37). The design was a hybrid ecological study design where individual health characteristics were known, but exposure to PM was a group variable based on one or a few ambient monitors for an entire metropolitan area. In time-series studies, the association of daily mortality in a geographic area such as a city is correlated with a daily PM ambient air concentration from one or a limited number of area samplers. Exposure is an ecological or group variable as all persons in the area are assumed to have the same exposure and there are no data on personal exposure. Mortality is based on individual-level data, but is used as an ecological variable, i.e., the number of daily deaths. Based on a limited number of studies of this design the EPA concluded that a 24-hour increase of 50 $\mu g/m^3$ $PM_{2.5}$ is associated with a relative risk (RR) of 1.025–1.05 or a 2.5–5% increase in mortality in the general population. Somewhat higher risks were said to exist for the elderly and those with preexisting cardiopulmonary conditions.

The EPA's interpretation of these data resulted in a new 24-hour air quality standard of 50 $\mu g/m^3$ $PM_{2.5}$. Whether the observed associations are causal, confounded, biased, or statistical remains a controversial issue as seen in various reviews (38–40).

A similar type of hybrid ecological study was conducted to assess the association of chronic mortality with long-term exposure to $PM_{2.5}$. Two major studies have been used to set an annual air quality standard (AQS) of 15 $\mu g/m^3$ for $PM_{2.5}$ (41, 42). In these studies the predicted effect of $PM_{2.5}$ in ambient air could be tested against the known effects of particulate matter in cigarette smoke, using the individual-level risk of mortality assessed in these same studies. The ecological estimate of the risk of ambient air $PM_{2.5}$ overestimated the risk of cardiopulmonary mortality by several orders of magnitude. By this analysis the association is not causal and the ecological fallacy has been committed (43).

2.3.1.3 Summary. Ecological studies may be useful in describing differences between populations, and such differences at least signal the presence of possible effects that may be worthy of investigation at the level of the individual. However, the results should be validated by individual-level studies such as the type discussed further below, and should not be viewed as a valid estimate of E–R relationships or as providing support for causality. When tested the ecologic risk estimates may not be corroborated by individual-level studies that are a more appropriate design for determining risk (44). One should beware of the ecological fallacy.

Thus far ecological correlations have not provided much useful information with regard to occupation epidemiology. Until the right variables can be measured and identified, progress will be slow (45). It is interesting that a recent book on ecological correlation suggests methods for estimating individual voting preferences at the level of the precinct (46). Whether this is possible in epidemiology remains to be seen.

2.3.2 Cohort (Follow-up, Incidence, Prospective) Study Design

A prospective or cohort study analyzes a group of persons (cohort) defined by some exposure, and the cohort is followed forward from the initial time of exposure. The basic design of a cohort study includes the following characteristics:

1. Enumeration of an exposed cohort (population at risk).
2. Identification of a comparison nonexposed population.
3. Follow-up of each individual in the cohort to determine the incidence of disease.
4. Comparison of rates of disease between cohort and reference population .

Some examples of typical occupational cohorts include all workers ever employed at a single location (mine, plant, company, etc.), or all workers from several plants, factories, or mines from the same or different companies. The important distinction is that the industrial process (exposure) is similar. A cohort might also comprise the members of a particular union or association with common occupational exposures. A historical cohort mortality study is often based on company employment or union records. The minimum amount of information required is name, social security number, date of birth, date of hire, and date of termination with enough detail to identify where and for how long each individual worked. The records must not only contain this information but must also be complete; that is, they must include all persons ever employed (47).

There are basically two kinds of reference or comparison populations: external and internal. The most common is an external reference population, usually provided by the national population experience. Sometimes, state or even county populations are similarly used to compare the rates of disease. Disease rates from these sources are available; can be stratified by gender, race, year, and age at death; and are large enough for the expected disease rates to be stable. The standardized mortality rate or ratio (SMR) or standardized incidence rate (SIR) are measures of effect and are the ratio of observed number of deaths (or incidence of disease) to expected number of deaths (or incidence of disease). Observed deaths are from the exposed population and are stratified by age, gender, and race. Expected deaths are derived from the reference population and are the number of deaths one would expect in a nonexposed population with the same age, race, and gender distribution as the exposed population.

The choice of a comparison population can produce different results and conclusions, so the actual SMR in any given study has no absolute meaning. Although an increased SMR is often used as the driving force for deriving interpretations of risk, there are dangers and pitfalls in using this information in an unqualified and absolute fashion (see Ref. 48 for a discussion of the use of the SMR).

The use of the national population in these comparisons has a variety of justifications:

1. Rates of disease are stable for even rare conditions.
2. These rates are available on computer programs, making the related calculation convenient and prompt.
3. The national population is the reference group in common use and makes comparisons with other studies possible. However to use for comparison, either with other studies or in an exposure–response analysis, the age distributions should be similar in the groups being compared.

An internal reference group comprises all eligible workers in the cohort with little or no exposure. The internal comparison group is advantageous in that selective forces and quality of data are similar to the exposed workers. Except for exposure, the internal controls should be quite similar to the exposed group, thus reducing the bias originating from the healthy worker effect. The internal comparison group also has obvious disadvantages, because it is usually small and often presents less stable disease rates. Sometimes there is no “nonexposed” group but only a minimal or less exposed group for reference.

The base for cohort studies is the person–time experience of their members. Stratification according to time-related factors may be necessary in addition to the specification of age at first exposure and calendar year of first exposure. Risk of disease varies with age and often calendar year. Other time-related variables such as latency (time since exposure onset), cumulative exposure, and tenure (a surrogate assessment of exposure) contribute to the estimates of exposure.

The essence of a cohort study consists of comparing rates between the exposed and reference populations. When the comparison is obtained with an external reference population the results may indicate exposure effects but should not be considered, in any way, the final proof of adverse effects (49, 50, 51). To establish causation, rather than just association, the demonstration of a graded exposure–response relationship cannot be overemphasized (4, 51, 52).

Measures of exposure can be quite varied. The most common is tenure (length of exposure), a surrogate of cumulative exposure (53). Estimates of exposure are subject to error, and the quality and quantity of past exposure estimates vary widely. Important factors to consider are duration of exposure (with or without gaps), intensity of exposure, age at first exposure, and time since last exposure. Concomitant stratification by cumulative exposure and latency is also important in the analysis of chronic diseases of gradual evolution. Separate analyses by latency and then by exposure can be misleading unless essentially all cases have sufficient latency to develop the disease.

The method of evaluating exposure-response trends across multiple levels of exposure is an extension of the process of comparing incidence rates between exposed and referent populations. In this type of analysis the low or nonexposed category is set at unity, and the higher exposure categories are gauged relative to the lowest exposure category.

Table 64.2 provides a checklist of factors to be considered in interpreting the results of individual cohort studies. (See also Ref. 54–57.)

2.3.2.1 Example of Cohort Study: U.K. Pottery Workers. Cherry et al. (60) studied mortality in a cohort of male pottery, refractory, and sandstone workers born 1916–1945 and identified from U.K. medical center records at Stoke-on-Trent. Since the exposure of interest was silica, workers with exposure to asbestos or in foundries or one or more years of work with other dusts were excluded. Because of possible loss of records prior to 1985, only the 470 deaths occurring from 1985–1992 were included in the analysis. Vital status was determined on 98.9% of the cohort, and death certificates obtained on 96.6% of the known deaths. Two external referent populations were utilized for expected deaths; national rates from England and Wales, and local rates from Stoke-on-Trent. Primary interest

Table 64.2. Criteria of Quality in Evaluating Cohort Studies

Enumeration and Verification: Is 100% of the population at risk enumerated and has the enumeration been verified, such as through Social Security records?[a]

Follow-up: Is the loss to follow-up less than 5%?

Latency: Is the period between first exposure and end of follow-up longer than the latency of the disease of interest?

A priori hypothesis: Is the hypothesis being investigated clearly stated and hypothesized prior to data analysis? (Other significant findings should be tested in an independent study; otherwise there is the possibility of "data-dredging bias")[b]

Control Populations: The national population is a minimum requirement. State or local reference populations may be more desirable. An internal reference population provides greater comparability than an external population.

Stratification: Are results stratified by latency and exposure, or is this adjustment accounted for in another manner (e.g., regression techniques)?

Exposure: Is there an exposure-response analysis? Is it stratified by latency? Are there adequate data to assess intensity and duration of exposure? Are cumulative exposure and/or tenure used as exposure variables? Is there a trend analysis? Is exposure time long enough and intensity great enough to evaluate an exposure-response relationship?

Confidence Intervals: Are there confidence intervals on SMRs and on the slopes of the exposure–response regression?

Bias: Are confounding biases such as selection, smoking, and other occupational exposures considered?

Interpretation: Is there a discussion of possible sources of error and their potential impact? Are the descriptions of methods, material, and data in sufficient detail that the reader has the opportunity to reach his own conclusions? Are criteria relating to causality considered?

was on lung cancer mortality, but other causes of death were also assessed. The choice of referent population is critical in evaluating risk of lung cancer as well as other causes of death as seen when comparing SMRs by referent group (Table 64.3).

In the U.K. there are wide differences in health from region to region, and Stoke-on-Trent has a poor record. Except for accidental deaths, SMRs were substantially reduced when local rates were used. All cancers and heart disease SMRs were no longer significantly elevated using local expected deaths and 90% confidence intervals. If 95% C.I. were used lung cancer and all causes would not be significantly elevated using local rates. The SMR analyses provide ambiguous results, and are not adequate to determine if exposure to quartz (or cristobalite for some workers) caused an increase in lung cancer. This is essentially a hypothesis-generating study, and there are several reasons no convincing interpretation of possible lung cancer risk is possible.

- There could be confounding from smoking, which could explain the excess risk.
- The possibility of bias cannot be completely eliminated. Destruction of medical records resulted in some loss of past work histories so some person-years at risk may have been reduced in some workers; thereby overestimating the true SMR. Destruction of records forced the investigators to exclude some workers, thereby reducing statistical power. Selection bias was not considered probable. Finally the use of medical records may have resulted in a disproportionate number of workers with long or heavy exposure, also causing possible overestimate of risk.

Because of these uncertainties in the cohort study, two nested lung cancer case-control analyses were conducted. One was part of the cohort study report and evaluated E–R trends using tenure as a surrogate for exposure and adjusting for smoking and silicosis (60). This study showed no apparent association of lung cancer risk with tenure or silicosis.

The other nested case-control analysis constructed a quantitative job–silica exposure matrix (61), which was utilized to assess E–R trends. In this analysis there was no association with exposure to quartz. There was however an excess risk of lung cancer associated with ever being exposed to cristobalite as estimated by ever working in firing or postfiring jobs (62).

Table 64.3. Comparison of SMRs by Referent Group

		SMRs (90% CI)	
Cause of Death	Observed	England and Wales	Stoke-on-Trent
All Cancers	150	1.44 (1.25, 1.65)	1.12 (0.97, 1.28)
Lung cancer	68	1.91 (1.55, 2.34)	1.28 (1.04, 1.57)
NMRD	57	2.87 (2.28, 3.58)	2.04 (1.62, 2.55)
Heart disease	171	1.36 (1.19, 1.54)	0.98 (0.86, 1.11)
Cerobrovascular disease	17	0.91 (0.58, 1.37)	0.80 (0.51, 1.19)
Accidental injury	9	0.58 (0.30, 1.0)	0.65 (0.34, 1.13)
All causes	470	1.46 (1.35, 1.58)	1.15 (1.06, 1.24)

These series of studies were initiated to specifically address the risk of lung cancer in men exposed to crystalline silica, and involved first a proportional mortality study (63), then an SMR study, and then case-control studies using tenure and then dichotomous exposure estimates in E–R analyses (60–64). A final report (65) more fully described these results. Differences included use of 95% confidence intervals instead of 90% confidence intervals, and analyses of E–R by continuous measures of cumulative and mean exposure, and duration rather than dichotomous analyses. The results appear to be similar as shown in comparisons of odds ratios (ORs) that adjusted for smoking (Table 64.4).

This example is somewhat unusual in several ways. The cohort was derived from medical rather than company personnel records, and destruction of some of these records at prescribed intervals made reconstruction of the cohort somewhat problematic. Two external referent populations were utilized; only one referent population is more common. There was no latency by tenure analysis, which should be the norm for causes of death of interest. And a nested case-control study was a concomitant feature of the cohort study, assessing and adjusting for potential confounders such as smoking (and silicosis in this case) while evaluating E–R using tenure. Analysis of lung cancer risk in the case-control study was restricted to cases with 10 or more years latency. The cohort portion of the report is basically a hypothesis-generating study for many causes of death, while the case-control portion of the analysis is testing the hypothesis that silica exposure increases the risk of lung cancer.

This series shows a natural progression from PMR to SMR to nested case-control with increasing analysis, power, and control of confounding to test the hypothesis. These results appear similar. What is not clear is why the 95% confidence intervals are narrower than the 90% confidence intervals when they should be wider; and why the ORs adjusted for smoking often increased compared to unadjusted ORs when cases had a higher prevalence of heavier smokers than did the controls.

2.3.2.2 Example of Cohort Study: Diatomaceous Earth Workers. A series of historical cohort mortality studies were reported on a group of diatomaceous earth workers to assess the potential carcinogenicity of crystalline silica to cause lung cancer (66–68). Exposure was primarily to cristobalite, and it was initially thought exposure to other potential occupational carcinogens was minimal. These studies provide a number of methodological examples of cohort analyses and how some decisions can result in controversial results.

Table 64.4 Comparisons of Odd Ratios Adjusted to Smoking

	Ref. 62 Dichotomous OR (90% CI)	Ref. 65 Continuous OR (95% CI)
Cumulative exposure	0.60 (0.26, 1.41)	1.01 (0.85, 1.19)
Duration	0.48 (0.21, 1.09)	0.79 (0.56, 1.13)
Mean	1.68 (0.93, 3.03)	1.67 (1.13, 2.77)
Post-firing (cristobalite)	2.17 (1.16, 4.07)	2.19 (1.06, 4.51)
Conclusions	"the only risk factor so far identified . . . was work in firing or postfiring occupations."	"crystalline silica may well be a human carcinogen."

After the 1993 report it was discovered that there was considerable exposure to asbestos. An independent analysis of asbestos use estimated about 50% of the study population had been exposed to asbestos (69). Smoking data were available on about 50% of the cohort and was collected in the 1980s. However some of the cohort began working in the early 1900s. Prevalence of smoking was 64% in the low dust exposure category and 81–84% in the higher dust exposure categories. The authors' concluded it was very unlikely that confounding by smoking caused the high risk of lung cancer in the high silica exposure category (67, 68). The exposure–response trend of lung cancer and cristobalite provided support for the hypothesis that cristobalite is a human lung carcinogen, although not an overwhelmingly potent one.

This conclusion has been criticized on several grounds, which are briefly summarized:

Confounding from asbestos: The time periods when asbestos exposure was clearly significant was prior to 1930 and from 1951 to 1977, with over 50% of the cohort definitely or probably exposed (69). The most problematic exposures were for the workers hired before 1930 when asbestos was commonly used but individual exposures could not be determined. These workers were excluded in the 1996 study, but were added back in the 1997 study where exposure estimates were extrapolated backward from 1930 levels. However no asbestos was used in 1930, so there is misclassification of asbestos exposure for workers hired before 1930.

Thus the critics argue that there is confounding from asbestos, that because of exposure misclassification the adjustment for asbestos is not meaningful, and the pre-1930 hires should have been excluded. The highest silica exposed group classified as having no asbestos exposure had a nonsignificant RR of 2.03 (95% CI = 0.93, 4.45) for lung cancer. But this group probably includes some pre-1930 hires that may have significant asbestos exposure, and the excess risk may be due to asbestos and not cristobalite. This is the only group of workers that show a significant excess risk.

It is also possible there is misclassification of cristobalite in this same group. Flux calcining of DE at high temperatures is required to transform the amorphous silica in DE to cristobalite. Prior to 1930 flux calcining was not widely used, but it was used in 1930 (70). As a result, although exposure to amorphous silica may be high, exposure to cristobalite may be low. Thus it is possible that some persons in the high silica exposure groups actually belong in a lower exposure group.

Referent population: These studies appropriately conducted internal E–R analyses. However, for external comparisons two referent populations were used, namely U.S. males and males from three counties surrounding the plant (68). But the county in which the plant was located, compared to the three surrounding counties, had a 37% higher incidence of lung cancer. This does not impact the internal E–R analysis, but it does raise the academic question of what is the most appropriate referent population for calculating an SMR. Is the most appropriate reference based on national rates (as in Refs. 66, 67), or local rates that include the study population (68), and/or local rates surrounding but not including the facility being studied? Analysis of E–R trends using an internal comparison avoids most of the problems caused by the hypothesis-generating analysis that assesses only a point estimate risk using an external comparison group.

Confounding from smoking: When workers classified as exposed to asbestos are excluded, there is no monotonically increasing E–R trend of lung cancer and exposure. The

RR of the high exposure category was elevated about twofold, and with the lower 95% confidence interval less than one. The RRs of groups with lower exposures were around the null value of one.

Now the question of whether there is an E–R trend revolves around the 13 lung cancer deaths in the high exposure category. If an Axelson (71) adjustment is made in an attempt to account for the potential confounding effect of smoking, the RR is reduced to a point estimate of 1.86. Should the excess risk of lung cancer be attributed to chance, confounding from asbestos/smoking, or bias due to misclassification of pre-1930 workers. The risk does not appear to be great in this group, and the cause of the nonsignificant increase cannot be determined at this point by these data alone.

2.3.2.2 Summary. The question of silica carcinogenicity is the hypothesis being tested in examples 1 and 2. Neither alone can answer the question, as the total weight of the evidence from all relevant epidemiology silica studies must be evaluated to determine the consistency of strength of association and E–R trends, coherence, and mechanistic studies to evaluate biological plausibility (72).

2.3.3 Case-Control or Case-Referent Study Design

The case-control study is, like the cohort design, a prospective or longitudinal study. The basic difference is that the cohort or prospective study identifies exposure and looks for disease, whereas the case-control study identifies disease and looks for exposure. As a result the case-control study involves only a sample and not the total of the population at risk (73). In a "nested" case-control (or cohort-based) study, for example, the cases would be all those in the cohort who had the disease of interest and the controls would be a sample of those in the cohort without the disease.

In the case-control design, cases are selected on the basis of disease. The measure of association is the odds ratio, that is, the ratio of the odds of exposure among the cases to the odds of exposure among the controls or $(a/c)/(b/d)$, or ad/bc (Table 64.5). For example, suppose there were 100 lung cancer cases, with 60% exposed to a particular chemical, and 400 nonlung cancer controls with 40% exposed. The controls are selected from within the cohort but without lung cancer. This computation indicates that the relative risk or odds of exposure is 2.25 times greater among workers with lung cancer than workers not exposed to the chemical.

To control for possible confounding from smoking, one could also obtain smoking histories, stratify on smoking, and thus calculate odds ratios for smokers and nonsmokers.

If this were a nested case-control study, the cohort might comprise 10,000 workers. It would be costly and time-consuming to examine such a large study and to code each work history and determine who was exposed. It is essentially impossible to get smoking histories on this many workers unless collected during their working lifetimes. It would be necessary to use an internal analysis, as there are no individual smoking data for external referent population. It is relatively cheap and quick to collect smoking histories for 500 workers, thus making case-control studies attractive alternatives.

A disadvantage of a case-control design is said to be that often there is less familiarity with this design than the cohort method, making findings more difficult to qualify and

Table 64.5. Sample Calculations of Odds Ratio (ORs) in Case-Control Study

	Cases	Controls
Exposed	*a*	*b*
Non-exposed	*c*	*c*

OR = *ad*/*bc*

Example 1

	Lung Cancer	Controls	Odds Ratio
Exposed	60	160	2.25
Nonexposed	40	240	1.0

OR = (60 × 240)/(40 × 160) = 2.25

Example 2

	Smokers			Nonsmokers		
	Cases	Controls	OR	Cases	Controls	OR
Exposed	60	80	4.5	10	80	1.5
Nonexposed	20	120	1.0	10	120	1.0
	OR = (60 × 120)/(20 × 80) = 4.5			OR = (10 × 120)/(80 × 10) = 1.5		

interpret. The problem of unfamiliarity is compounded by the incorrect idea that a case-control study is a fundamentally different type of study. However, the cohort and case-control study designs are architecturally the same in that both are prospective, following workers from some base-line state forward in time to some end point (28). In cohort studies the population studied is defined on the basis of exposure to a particular substance. Data on risk factors and incidence are then collected on all cohort members. The association is a comparison of the incidence of disease in exposed and nonexposed members. In a case-control study the cases may be obtained from the same population base, but only a subsample of the population at risk comprises the controls. The contrast is obtained by the comparison of the exposure parameter in diseased and nondiseased persons (74).

Recall bias can be a problem if the subject (or surrogate) attributes the disease to a particular exposure, and/or if the exposure is rare. Using diseased controls that are apt to answer questions in a manner similar to cases may alleviate the former problem. This bias is more likely to occur in a community-based case-control study where the primary source of exposure history is from the cases and controls.

Selection of controls can be a knotty problem in case-control studies. If diseased controls are selected, the diagnosis should be unrelated to exposure. If the disease in controls is related to the exposure, the frequency of exposure among controls may be falsely elevated and the effects of exposure underestimated. Two important principles are involved in selecting controls. Controls (*1*) should be comparable to the cases in all respects except the presence of disease; and (*2*) controls should be as representative as possible of the nondiseased population.

Thus in the literature there are discussions such as that initiated by Gordis (75) about using dead controls matched to dead cases. McLaughlin et al. (76, 77) argue against using cancer patients as controls in case-control studies, but Smith et al. (78) argue in favor of the procedure. Similarly Linet and Brookmeyer (79) are generally in favor. The discussion about controls will go on indefinitely, for "it is unlikely any formula for control selection in studies involving occupational diseases will hold for all circumstances" (80).

2.3.4 Some Biases in Cohort and Case-Control Studies

Bias in the selection of subjects might occur when a particular exposure requires medical surveillance and the diagnosis of symptoms, signs, or cause of death are influenced by the knowledge of that exposure. This has been called the *diagnostic suspicion* bias by Sackett (81) and *publicity bias* by Feinstein (28). For example, working in the rubber industry might encourage the diagnosis of bladder cancer (82). Such a suspicion could result in a *detection bias* as well. Wells and Feinstein (83) showed that male smokers with a chronic cough were 22 times more likely to be a given a sputum test in the diagnostic pursuit of lung cancer than nonsmoking women without a cough.

Information on a surrogate of exposure such as job tenure should not be a problem when company personnel records are used to confirm work history. An advantage of case-control studies is the feasibility of going further than is possible with records to obtain more complete exposure information. If interviews are used to ascertain work exposure and/or nonoccupational exposure, an *exposure suspicion bias* could result (81). This could occur if the interviewer has knowledge of a subject's disease status, and such knowledge may in turn influence the intensity and outcome of the search for exposure. For example upon routine questioning about exposure to irradiation, the evidence of exposure was, respectively, 28 and 0% in a study of thyroid cancer (81). Upon intensive questioning and search of records, the evidence of previous irradiation was 47 and 50%, respectively. Using standardized methods and employing blinded interviewers can reduce such biases.

If there is less than 95% follow-up, an analysis should include an appropriate scrutiny of the nonresponders, in order to ascertain potential *selection or ascertainment bias.*

If diagnosis of disease (or death) is incorrect, the inclusion of a misclassified case into either a cohort or case-control study may bias the SMR and odds ratio toward the null. Collection of additional diagnostic data may create an ascertainment bias if this addition is not applied equally among exposed and controls in a cohort study or among cases and controls in a case-control study.

Confounding is a potential problem in all study designs involving nonexperimental research. The confounding factor is important when (a) it is predictive of the response, and (b) it is associated with the exposure. For example, smoking is often said to be a confounder in a lung cancer study because smoking is predictive of the response, lung cancer. However, if the proportion of smokers in the case and control groups is the same, smoking cannot be a confounder because it is not associated with exposure.

Confounding factors must be identified in an epidemiological study to avoid misinterpretation of the results. Attempts to control for confounding are more common in case-control than cohort studies. This is because of the more focused *a priori* hypothesis and because fewer subjects are required on which more information regarding confounding factors must be obtained. Some methods of controlling confounding include:

Matching: Select one or more controls with the same confounding factor(s) as the cases; that is, match a smoking case with a smoking control.

Stratification: Group cases and controls in similar categories of the confounding factors; that is, compare the risk of cases who smoke using controls who have the same smoking status.

Restriction: Include only subjects within a specified range of the confounder.

Internal control group: Fellow workers are likely to be comparable on many lifestyle factors so confounders are unlikely to be associated with exposure, and the potential for significant confounding is reduced.

2.3.5 Cross-Sectional or Survey or Prevalence Study Design

Cross-sectional studies are among the most common of epidemiological studies (at least when studying morbidity). The distinguishing characteristic of a cross-sectional study is that each person in the study population is examined only once. At the time of the examination both exposure status and disease status are determined, as well as the presence or absence of potential confounding variables. When quotients are calculated to summarize the prevalence of disease, the conditions in the numerator and denominator are determined at the time of the examination. Ideally, these parameters should be determined at a midpoint of the time interval studied.

For the longitudinal study design, in contrast to the cross-sectional design, each person is examined on at least two separate occasions. When quotients are calculated to summarize the incidence of disease, the first examination delineates the base-line conditions (denominator), the second examination delineates the change in state (occurrence of new cases), and the latter appears in the numerator.

A cross-sectional study is distinctive in that the data collected at one point can be analyzed and interpreted (*1*) for the current situation and (*2*) looking backward or retrospectively, as well as (*3*) forward or prospectively. For example, suppose one examines all underground coal miners in 10 bituminous coal mines and a control group of surface workers at these same mines, collecting the following observations on each person: chest X-ray for determination of pneumoconiosis, pulmonary function, age, height, smoking history, and work history.

The data are analyzed retrospectively, by comparing the prevalence of pneumoconiosis and reduced pulmonary function in both the underground miners and controls. One assesses the current situation by describing the prevalence of pneumoconiosis (cases/total), and the related pulmonary function indices. A forward look at the data occurs when the data are stratified by smoking category, grouping the miners into discrete exposure categories (such as <5, 5–10, 10–20, and >20 years tenure), and reporting the prevalence of pneumoconiosis and indices of pulmonary function (adjusted for age, race, gender, and height). Again it should be stressed that observations obtained within the context of cross-sectional studies may not be used as definite estimates of pertinent risks. The presentation of these results for all temporal directions can be arranged in a single table. Table 64.6 is an example for the computation of prevalence of pneumoconiosis in men 40 to 50 years of age. These hypothetical data show that the prevalence of pneumoconiosis among smokers is four times higher among underground miners, compared to surface workers. The current overall prev-

Table 64.6. Sample Data Layout for Cross-Sectional Study: Prevalence of Pneumoconiosis

	Underground Miners (%)		Controls (Surface Workers) (%)	
Tenure (years)	Smokers	Nonsmokers	Smokers	Nonsmokers
<5	1	0	1	0
5–15	5	2	2	0
>15	18	6	3	1
Total	8	2.5	2	0.3

alence of pneumoconiosis is 8% among smokers and 2.5% among nonsmokers. The prevalence of pneumoconiosis is increased in both smokers and nonsmokers.

A similar format could be used if the outcome variable was lung function. In that instance, average values could be used.

The interpretation becomes more difficult in regard to a reduced pulmonary function, because this is a nonspecific effect, often related to a number of different exposures.

According to hypothetical data in Table 64.7 both smoking and exposure to coal mine dust appear to affect lung performance. However, in a study of this type, it may not be possible to differentiate completely the causes of the reduced lung function. One reason is the intercorrelation existing between age, smoking, and exposure, thereby making separate analysis of each variable difficult if not impossible. A second problem has to do with the study design. Everything that has happened to the exposed and control group occurred before the data collection began. Did the reduced pulmonary function happen as a result of exposure, had it already occurred before exposure, or is the relationship observed in the data largely the result of healthier workers selectively leaving employment? The use of cross-sectional study design does not allow the observer to address these questions directly. The problem is reduced somewhat for radiological effects. Coal workers pneumoconiosis (CWP), silicosis, and asbestosis as well as some other pneumoconioses have distinct features that are specific and characteristic of particular exposures, e.g., rounded opacities in the upper lobes and a history of silica exposure; or linear opacities in the lower lobes and

Table 64.7. Sample Data Layout for Cross-Sectional Study Expressed as Percent of Predicted Pulmonary Function

	Underground miners (%)[a]		Controls (Surface Workers) (%)[a]	
Tenure (years)	Smokers	Nonsmokers	Smokers	Nonsmokers
<5	100 (3.5)	95 (3.4)	97 (3.2)	96 (3.3)
5–15	90 (3.3)	93 (3.3)	92 (3.0)	94 (3.2)
>15	80 (3.0)	90 (3.2)	85 (2.9)	92 (3.2)

[a]Figures in parentheses are mean FEV1 in liters.

a history of asbestos exposure. Some abnormalities may occur because of poor technique, smoking, age, variability in interpretation, at categories below 1/1 or so. The correct diagnostic category can usually be distinguished from background noise for some specific dusts above category 1/1 (84).

Cross-sectional studies are relatively easy to conduct and are useful for conditions that are measured quantitatively and vary over time, such as lung function or pneumoconiosis. They are also useful for relatively frequent conditions of a chronic nature (i.e., chronic bronchitis). It is quite easy in cross-sectional studies to collect information for many different hypotheses. Because of the usually abundant data, cross-sectional studies can result in new etiologic hypotheses regarding risk factors and/or disease. A major value of the cross-sectional study is descriptive, that is, the simple characterization of a target population.

A survey is also the initial stage of a cohort study. If a cross-sectional study is used to substitute for a longitudinal study, the occurrence rates in the two types of studies will probably be different. For example, Glindmeyer et al. (85) found the age coefficients for lung function derived from a five-year longitudinal study were lower than the age coefficients from cross-sectional analysis. Weiss (12) discusses three examples that show differences in lung cancer mortality between cohort and cross-sectional analysis. These results are summarized in Table 64.8. All workers included in these analyses had 20 or more years of latency. Clearly the cross-sectional design shows the higher risk.

Both kinds of studies must consider errors due to differential losses or selection. In a cross-sectional study one is investigating a residual cohort consisting of the remains of an original cohort affected by attrition. Often the causes of attrition, and most important, the health status of those leaving are unknown. Even the magnitude of attrition is not known. Alternatively, in a cohort study one knows the number and original health status of all members and so can better estimate the effort of attrition.

2.3.5.1 Example of Cross-Sectional Study: U.S. Coal Miners. The National Study of Coal Workers Pneumoconiosis (NSCWP) is a continuing epidemiological study of respiratory health of U.S. coal miners. The first round took place between 1969 and 1971 in 31 underground mines in all of the major coal fields. About 91% of the miners were examined. The examination consisted of a chest X-ray, spirometry, and a questionnaire that included work history, smoking, and respiratory symptoms. The Bureau of Mines conducted intensive dust sampling surveys in 17 of the 31 mines that were part of the

Table 64.8. Lung Cancer Mortality—Comparison by Study Design[a]

	SMR for Lung Cancer	
Occupation	Cohort	Cross-sectional
New Jersey insulation workers	4.44	8.67
Asbestos manufacturing workers	1.11	1.51
Homestake gold miners	1.24	3.21

[a]Ref. 12.

NSCWP. These data plus compliance samples from the mine operators were used to estimate past exposure levels of each miner.

The dependent or response variables in the two examples discussed below were lung function (Forced Expiratory Volume in 1 second or FEV1), and coal workers pneumoconiosis (CWP). The confounding variables included age, height, pack-years of smoking, tenure underground, and rank of coal (or geographic area where the coal was mined). The purpose of the study was to assess E–R relationships between the independent variables and cumulative exposure to coal mine dust. These were the first studies of U.S. coal miners to correlate lung function and pneumoconiosis with dust levels, as previous analyses used years underground as a surrogate for exposure.

2.3.5.1.1 Pulmonary Function. The multiple regression of FEV1 against age, height, region, smoking and cumulative dust exposure for 7,139 coal miners >25 years of age explained 47% of the total variability in observed FEV1 values. Cumulative exposure is expressed as gram-hours per cubic meter (gh/m^3) using a factor of 1740 hours per year/1000 (mg per g). (Data from Ref. 86). The decrements of FEV1 associated with each variable are summarized in Table 64.9.

The results indicate that coal miners working at about 6 mg/m^3 could experience effects similar to that of a one pack year smoker over a 40-year working life. The results are generally consistent with those of British coal miners.

2.3.5.1.2 Coal Workers Pneumoconiosis (CWP). Prevalences of CWP were associated with cumulative dust exposure, as well as with age and coal rank (% carbon). Logistic regression models predicted the following prevalences of CWP and progressive massive fibrosis (PMF) for a 40-year exposure to 1 mg/m^3 (Table 64.10). Data from Ref. 87.

These results show an association between severity of pneumoconiosis with dust exposure, coal rank, and age and are in general agreement with recent findings for British coal miners.

Table 64.9. Decrements FEV1

Independent Variable	Decrement in FEV1 (mL)	*p* Value
1 year	31	<0.0001
1 packyear	5	<0.0001
Smoker	200	
Ex-smoker	36	<0.0001
Regional Effects		<0.0001
Anthracite mines	242	
Central Pennsylvania	158	
Northern Appalachia	225	
Southern Appalachia	224	
South	204	
Midwest	219	
Cumulative exposure-gh/m^3[a]	0.69	<0.0001

[a]gh/m^3 = gram-hours per cubic meter.

Table 64.10. Predicted Prevalence of CWP and PMF, %[a]

	Category 1	Category >2	PMF
Anthracite	13	5	3
Bituminous			
Medium/low volatile (89% carbon)	12	4	3
High volatile A (83% carbon)	4	2	2
High volatile-Midwest	6	2	1
High volatile-West	6	1	1

[a]At 1 mg/m^3 for 40 years.

The findings "should be interpreted with caution" for several reasons. There are no exposure data prior to 1969, so back extrapolation to earlier mine exposure levels is problematic. It is not clear that the mine-specific adjustments are reliable, so the predicted prevalences are not as reliable as the small standard errors suggest.

The pneumoconiosis diagnoses are based on only one reader because of problems associated with the interpretations of the other two readers.

The statistical model was unable to determine if there was a threshold in the concentration range of 0 to 2 mg/m^3 (where most exposures occur today). Age was a significant variable and it is possible that noncoal related factors could affect the radiographic interpretation at low exposures and low categories of CWP. For example, the effects of age and smoking may mimic radiographic signs suggestive of CWP.

The results to some extent are dependent on the statistical model used, and also only apply to miners who work for 40 years in a 2 mg/m^3 environment. What happens at lower exposures and shorter working lifetimes remains to be determined.

2.3.6 Summary

There are four major epidemiological study designs. Each has its own advantages and disadvantages in the continuum of developing a hypothesis to testing the hypothesis and determining an etiologic agent causing a particular disease.

The ecological study is easy to conduct, and is primarily useful for generating a hypothesis. Because group-level data, and not individual-level, data are used, it is subject to the ecological fallacy and should not be used to test a hypothesis.

The cross-sectional or survey study is also relatively easy to conduct, and is useful for determining prevalence of disease and describing the characteristics of the population under study. It can address the association of risk factors and disease. For responses that are characteristic of a specific exposure it can be quite useful in assessing risk and exposure–response relationships. For nonspecific responses caused by multiple risk factors the etiologic relationship is less certain because of uncertainty as to the time sequence of events.

The prospective study design is the most appropriate for testing a hypothesis and determining etiologic agents. In practice the cohort study often generates a hypothesis because of the capability of assessing multiple responses (e.g., all causes of death). Enumeration of all causes of death, for example, is basically a hypothesis generating type of study except when testing an *a priori* hypothesis. As noted in section 3, the cohort study is limited in its ability to test fully the hypothesis unless further analyses are conducted, including internal referents to help control for confounding and analysis of exposure–response trends. The nested case-control study within a cohort is the most appropriate design for testing a hypothesis. It is focused on a particular hypothesis, and designed to control for specific confounders and to determine whether a specific exposure is an etiologic agent. As a result it is ill equipped to generate hypotheses, but is well equipped to test a hypothesis. The nested case-control design is the most appropriate for investigating occupational causes of disease. It is less appropriate for investigating lifestyle risk factors.

The community-based or hospital-based case-control study is useful for determining the role of life-style risk factors for disease such as cholesterol, diet, and smoking. It is less useful for occupational studies because the job histories are usually poorly known, the range of jobs is both broad and heterogeneous, and the number of cases in any homogeneous job or occupation is small. Thus the community-based case-control study lacks both statistical power and well-defined quantitative or qualitative exposures to test a hypothesis regarding occupational exposures. The community-based case-control study assessing occupations has been used to generate a hypothesis, but generally cannot test a hypothesis relating to occupational exposures.

3 CRITERIA FOR INCLUSION OF INDIVIDUAL STUDIES FOR EVALUATION OF WEIGHT OF EVIDENCE IN DETERMINING CAUSALITY

A primary goal of occupational epidemiology is to determine what exposure(s) in the workplace are associated with disease or injury, and to determine exposure–response (E–R) associations to determine exposures low enough to prevent disease. The ultimate objective is to determine as clearly as possible whether or not the association between a putative etiological agent (such as asbestos, silica, radon, nonfibrous amphiboles, talc, asphalt fumes, coal tar) and adverse health outcomes (such as lung cancer or pneumoconiosis or restrictive/obstructive lung disease) is causal or only statistical. The determination of this objective has importance in policy-making and in setting regulatory workplace standards or recommendations. Often the evidence is not overwhelming for a causal relationship, and so it is usually necessary to assess the weight of the evidence using criteria well known in epidemiology. These criteria were used in the Surgeon General's 1964 Report on Smoking (88), and are commonly discussed in the literature (4, 52, 89, 90).

Gordis (30) suggested some of the difficulties in deriving causal inferences from epidemiological data arise from at least two factors. One is the tendency to disregard or de-emphasize biological plausibility, which is related to the coherence of the data with the biology and natural history of disease. The second is an over-emphasis on "positive" studies where the results suggest a significant association of a putative cause with an adverse health outcome. Gordis (30) suggests many studies are positive because of highly sophisticated statistical techniques. Positive results may also be emphasized while negative studies may be neglected or overlooked for other reasons. These include the fact that they are less likely to get published, or selective emphasis may be placed on the positive studies in the mistaken belief that a causal associations is determined on the strength of a few positive studies. As a result only part of the evidence may be reviewed, without due consideration of the whole body of evidence. The potential effect of bias and confounding on the interpretation of results is also often overlooked.

In making inferences about causality there are two distinct processes. One has to do with the validity and usefulness of each individual study; a determination of the design and conduct of each piece of evidence relevant to the inquiry. The general requirements commonly cited are that the results should not be due to chance, bias, or confounding. Other important considerations include an appropriate referent population and evaluation of E–R in attempting to test a clearly articulated and biologically meaningful hypothesis. Although there may not be unanimity on each study, these criteria should be applied *a priori* and are necessary to establish credibility of a particular study, and in the larger sense, credibility of epidemiology itself. These are the criteria discussed in this Section.

The second process is to assess the weight of evidence from credible studies with regard to Hill's (4) causal criteria of temporality, strength of association, biological gradient, consistency of the associations, coherence, biological plausibility. The only criterion that must be met without equivocation is that the cause must precede the effect; the time sequence must be correct. Fulfillment of the other criteria will never be complete, and will to some extent also be a matter of judgment. These criteria are discussed in the next section.

All of the putative agents in the first paragraph of this section have at one time or another been suggested as risk factors that increase the risk of lung cancer. And some are generally accepted as lung carcinogens at some exposure level (e.g. asbestos, coal tar, and radon). But a number of unconfirmed hazards have been linked to adverse health effects for a variety of reasons, which often boil down to a too quick judgment without due consideration of the criteria for causation. For example the American Council on Science and Health has published a review of the greatest unfounded health scares over the last 30 years or so, scares due to a quick, but false, rush to judgment (91). The judgment that silica is a probable human carcinogen (92) appears to be based on a limited view of confounding and bias with incomplete assessment of relevant studies, which may have resulted in a spurious conclusion (72). The alleged role of PM air pollution in increased mortality and morbidity is based on studies of an inappropriate study design (38, 43). Neglect of the first process in assessing causality, namely that of inclusion of appropriate studies, may result in spurious conclusions even if the second process of evaluating the weight of evidence is valid.

But first the role of chance is reviewed. Statistical significance should not be a reason for including or excluding a study in a weight of evidence review. But statistical significance is not useful in estimating strength of association, consistency, coherence, or biological plausibility. And statistical significance is meaningless and irrelevant if there is significant bias.

3.1 Chance or Statistical Significance

A significance test assesses the probability p (or plausibility) that there is no difference between cases and controls, or no difference between high and low exposed groups. The p value (or 95% confidence interval) is the probability that the observed outcome would have occurred by chance. The 95% CI says one is 95% confident that the true value of the risk ratio lies between the upper and lower bounds of the interval. One chance out of 20 is generally accepted as unlikely to be a random occurrence ($p = 0.05$).

A nonsignificant result ($p > 0.05$) means the data are compatible with the effect being due to chance; it does not mean there is no effect, or even that there is no causal association. It means that a causal relationship is not the most likely explanation. Associations that are highly significant ($p < 0.001$) are rarely to be doubted. Findings of less significance ($p < 0.05$) may be due to chance (93). In any event a test of significance is inappropriate unless there is a specific hypothesis is being tested.

Note that p values are commonly reported for such tests as the departure of a risk ratio from unity or a test of an E–R trend. The p-value is a confounded mix of the magnitude of the effect measured and the precision of the estimate, which is a function of sample size (94). A more informative approach is to present the two pieces of information separately. The two pieces are the size of the effect (the risk ratio or strength of the association), and the precision of the estimate (the confidence interval and related to the sample size).

In some studies there may be so many comparisons that false positive results are likely, raising the question of what is an appropriate level of significance. For example in some studies, such as cross-sectional studies assessing neurobehavioral performance in solvent-exposed workers, there may be multiple tests and exposure variables. Bleecker et al. (95) used a battery with 23 elements and two exposure variables resulting in a minimum of 46 E–R comparisons with two to three positive results expected by chance alone. Orbaek et al. (96) administered nearly 500 clinical, psychiatric, clinical chemistry, nervous system function, and psychometric tests in a group of solvent-exposed workers. They commented that "even when no real difference exists, it is likely that certain variables show a statistically significant result by pure chance." A suggestion for reducing the likelihood false positives from multiple comparisons is to divide the level of significance equally across each test (97). Thus, in a study with 10 comparisons the correction would require differences at the $p = 0.005$ level of significance ($0.05/10 = 0.005$). Rothman (98) argues against adjustment for multiple comparisons because that increases the probability of rejecting true positive associations and does not aid in interpretation. Thompson (99) argues that if p values are used then adjustment should be made to avoid, in part, exaggerated confidence in weak results.

Hill (4) notes that for there to be an association it must not be due to chance. However statistical significance is not the only, or even the most important, criterion for making judgments. Formal statistical tests are framed to give a precise mathematical answer to a structured question. The validity of a study and the question of causality are based on answers to unstructured questions and are based on judgment (100).

3.2 Criteria for Inclusion of Individual Studies

As Gordis (30) points out, the "essence of epidemiology is the design and conduct of studies appropriate for testing the validity of biologically meaningful hypotheses in human populations." Some suggested criteria by which studies should be judged for inclusion when weighing the evidence regarding causality are discussed below.

3.2.1 Comparability

Susser (101) notes that comparison is at the heart of the epidemiological method, and that a valid comparison requires "situations that differ only in respect either of the determinant or the presence of the disease." Epidemiology studies are similar in construction to experiments; however unlike experiments which can be fully controlled and replicated, studies of humans are observational with limited ability to control exposures, confounding risk factors, and composition of the study population. Even in studies of pathogenesis and clinical issues, rigor and replication are not under complete experimental control. In epidemiology studies comparability is achieved for some risk factors by stratification (statistical adjustment), but can not be complete for all factors. One can also limit the confounding effect of noncomparability by using referent groups from the same population base as the study group, i.e., internal comparisons as in nested case-control designs or internal cohort nalyses, and to some extent by using local rather than national referents in calculations of SMRs.

Comparability and credibility are further increased in a study when internal comparisons of disease frequency is made among individuals at different levels of exposure (92).

Cherry et al. (60) in a cohort study of pottery workers in the U.K. found significantly increased SMRs for the workers when expected values were based on national rates (England and Wales); SMRs were reduced, in most cases to nonsignificance, when the expected values were based on local rates (Stoke-on-Trent). The authors then conducted two case-control studies using internal referents from the same study population to ascertain the true association between silica exposure and lung cancer. They note that "evidence for an increase in risk of lung cancer . . . lies critically . . . in the choice of an appropriate (comparison) population (60)."

Sometimes the criteria of lack of confounding and comparability overlap. In tests of cognitive ability on neurobehavioral test performance, intelligence is related to performance and even has a surprisingly large impact on motor function performance (102). Years of education is the usual surrogate measure used to adjust for intelligence in referent groups. Sometimes "hold' tests such as tests of vocabulary may be used for adjustment of pre-exposure intellectual capacity since they are thought to be resistant to the effect of neurotoxic chemicals. Gade et al. (103) retested solvent-exposed workers diagnosed with toxic encephalopathy and significant intellectual impairment. When compared with nonexposed control subjects closely matched on age, education, and intelligence, the diagnosis of brain damage and painters' syndrome could not be confirmed. Cherry et al. (104) reported that differences in psychometric test performances were virtually eliminated when proper allowances were made for differences in intellectual capacity measured as reading score.

3.2.2 Relative Lack of Bias

Bias is a consistent nonrandom error, such as systematic selection of the study population so that they differ in important ways from a second group of referents with which they are being compared. An example is participation, volunteer, or selection bias—as might occur if there is consistently more participation of smokers or nonhealthy subjects in a study or voluntary surveillance program.

For example, it is generally accepted that silicotics are at greatly increased risk of high mortality from lung cancer and other diseases. However the vast majority of these studies are based on registries of compensated silicotics and usually consist of volunteers from an unknown population base with inadequate information on exposure to silica and other potential carcinogens. Compensated silicotics are almost always smokers, and so registries have a higher than expected proportion of smokers. Smokers are at greater risk of lung cancer, and are more likely to have respiratory impairment and other symptoms, and for the same degree of fibrosis are more likely to seek and obtain compensation. In addition the diagnosis is frequently based on degree of disability rather than the kind of pneumoconiosis and the specific kind of dust that caused it. Studies of compensated silicotics provide a model of biased studies that should in general be excluded from consideration in determining causation (92). Because of the many shortcomings of silicosis registries they provide useful examples of various biases (including confounding biases). McDonald (105) grouped them under headings, two of which are briefly summarized here as examples of biases to be aware of and avoid in all studies.

3.2.2.1 Ascertainment Bias. Severe selection or incomplete ascertainment bias can often characterize registry studies, as the silicotics are volunteer applicants for compensation with smoking-related disabling symptoms. They are more likely to have social, psychological, and industry-related differences in addition to having more severe symptoms, radiographic changes, and lung function impairment. In short they are unlikely to be representative of all silicotics. There may in fact be no appropriate referent group.

The potential effects of incomplete ascertainment or volunteer bias can be quite severe even in nonregistry studies. Gillam et al. (106) studied gold miners with five or more years tenure from a cross-sectional survey of volunteers. When compared to the results from two other studies of these miners (107, 108) which had complete enumeration, up to four to five times longer duration of exposure and longer follow-up, the volunteers had more than three times the lung cancer risk than the more exposed miners. Incomplete ascertainment bias can produce statistically significant but incorrect estimations of excess risk, so statistical significance becomes irrelevant.

3.2.2.2 Diagnosis Bias. Compensation boards often decide between various pneumoconioses in workers exposed to mixed dusts and where the degree of disability is more important than the kind of diagnosis. Prior to 1960 the radiographic categories did not differentiate between types of pneumoconiosis. In addition, heavy smoking may result in a dirty lung that may be diagnosed for a low category of some pneumoconioses. Detection bias is likely to occur if ill health from incipient lung cancer and/or smoking-related symptoms triggered an examination to get compensation for silicosis. In either case the apparent risk of lung cancer would be spuriously elevated.

In a study of Finnish silicotics (109) misclassification of mixed dust pneumoconiosis may have biased results upward by an estimated 40% or so. Misclassification of silicosis was minimized in a study of dusty trade workers in NC because diagnosis was not based on smoking-related health problems. Nevertheless, there was still a 28% error rate of false-positives on re-evaluation of full-size radiographs (110). The conclusion of de Klerk and Musk (111) is that smoking-related symptoms and disability and radiographic changes might have increased the probability of an Australian gold miner making a successful compensation claim for silicosis.

3.2.2.3 Misclassification of Exposure Bias. This bias may occur when earlier exposure conditions are higher than recent exposures, but can only be assessed indirectly such as by interview. Subjects with symptoms may overestimate exposures that are the same as asymptomatic workers. While random misclassification is likely to reduce the ability to detect an association, consistent overestimates of exposures by symptomatic workers in cross-sectional studies will spuriously overestimate the risk.

3.2.3 Freedom from Significant Confounding

Confounding is a particular kind of bias where a factor can cause the disease and is also unequally distributed among the exposed and unexposed populations. For example, smoking and radon at some exposure level increase the risk of lung cancer. If radon is correlated with dust exposure in a mine, it may confound the lung cancer/dust exposure association, so that an increased lung cancer risk may be falsely attributed to dust exposure. Similarly, if smoking prevalence is higher in high-dust exposed groups than in low-dust exposed groups, a higher risk of lung cancer in high-dust exposed groups may be falsely attributed to dust exposure when, in fact, it is smoking that causes it. If a group of workers exposed to solvents is older and/or has less education or intellectual ability than the referents (less exposed group), then lower performances on neurobehavioral tests may be due to differences in age and education and not to differences in solvent exposure. Thus, confounding risk factors will vary and depend on the health effect of interest. When lung cancer is the health effect of concern and silica the occupation exposure of interest, at least three major confounders of concern come to mind—smoking, occupational carcinogens, and dust exposure. When considering the association of solvent exposure and effects on the nervous system important potential confounders include age, education and SES, health and drug history, alcohol, caffeine, fatigue, motivation, sleep the previous night (112). Age, race, and gender are generally considered to be potential confounders, and are easy to control.

The relative degree of confounding that is unlikely to significantly affect the results and interpretation will also vary. When the effect is weak, i.e., when there is a weak association, then the concern is increased because then the presence of only a weak confounder (or confounders) are required to produce a similar weak effect. If the effect of the exposure of interest is strong, then the effect of the confounder must be even stronger and the distribution even more unequal between exposed and controls (or cases and controls) to produce a measurable and significant effect.

3.2.3.1 Examples of Potential Confounders. Three major potential confounders that potentially mask the true association between silica and lung cancer are discussed as examples—namely smoking, occupational carcinogens, and dust. These examples apply to some extent to lung disease and dust exposure or chemical exposure in a more general sense.

3.2.3.1.1 Smoking. Potential confounding from smoking is a constant concern in assessing the risk of lung cancer as well as other diseases such as obstructive lung disease which are caused by cigarette smoke. The likely bias of overestimating risk increases among those with long latency as both smoking-related and occupationally-related diseases have long latencies so the two outcomes may be correlated. Thus an analysis that only shows increasing risk with increasing time from first exposure (date of hire) is likely to be confounded by smoking since both smoking and date of hire may be similar.

If the strength of association is weak (RR < 2), differences in the prevalence and intensity of smoking between exposed and controls, or high exposed and low exposed groups, could be, at least in part the cause of the difference in risk. The 1985 Surgeon General's report on smoking (113) comments that adjustment for differences in smoking patterns and depth of inhalation may cause a two-fold excess risk attributable to smoking, and the use of high-tar cigarettes might increase the risk even more.

Various methods have been used to try and estimate if smoking is a likely cause of an increased risk of lung cancer. In the complete absence of smoking data the risks from other smoking-related diseases have been used to estimate if smoking prevalence in the exposed group is likely to be high (114). This is at best a crude indicator. One reason it is unreliable is because the risks of smoking-related diseases vary from one study to another. In eight studies summarized in the 1989 report on smoking and lung cancer (115), the RR of laryngeal cancer ranged from 6.1 to 13.6, from 0.7 to 3 for bladder cancer, from 1.3 to 1.9 for cardiovascular disease. In four studies summarized by Steenland et al (114) the risks of emphysema ranged from 0.6 to 12.3, bronchitis from 2.9 to 24.7. Another reason is that smoking attributable risk differs from one disease to another. For example, the percent of deaths attributed to smoking were 90% for lung cancer, 81% for laryngeal cancer, 48% for bladder and kidney cancer, and 29% for pancreatic cancer in one study (116).

The Axelson method (71) uses the differences in smoking prevalence between study and control populations to estimate whether smoking is likely to cause an observed increase in risk. Using RRs of 10 and 60 as extreme risks of smokers compared to nonsmokers and the indicated differences in prevalence of smoking between exposed and control, the proportion of lung cancers attributable to smoking range from 18% to 57% in the examples in Table 64.11.

The "adjusted" lung cancer risk attributed to smoking should lie below the lower 95% confidence interval of the exposed workers' RR before one can exclude the role of smoking (113). Thus if the lower 95% confidence interval is less than unity after subtracting the estimated effect from smoking, then smoking is a reasonable alternative hypothesis for the

Table 64.11. Prevalence of Smoking and Cancer

Prevalence of Smoking		RR of Smoking and Cancer Attributable to Smoking[a]	
Exposed (%)	Controls (%)	RR = 10	RR = 60
85	70	1.18	1.21
90	65	1.33	1.38
95	60	1.49	1.57

[a]RR = relative risk.

increased risk. For example, smoking data were available on about 50% of a cohort of diatomaceous earth workers. Smoking prevalence was 64% in the lowest silica exposure group and 84% in the highest exposure group (68). Using the Axelson (71) method and RR of 20, about 29% of the RR is attributable to smoking. Subtracting this estimated effect of smoking from the highest RR of 2.15 in the high exposure category gives a RR of 1.86 with a lower 95% confidence interval of 0.79. Therefore smoking must be considered a possible alternative hypothesis to help explain the apparent E–R trend.

3.2.3.1.2 Occupational Carcinogens. In the IARC (92) assessment of silica as a lung carcinogen studies were not considered if there was thought to be potential confounding from occupational carcinogens. The IARC panel selected 12 of the "least confounded" studies as the most important ones in making a decision that silica is a probable human carcinogen. However, there was potential confounding from occupational carcinogens in several of these studies; for example asbestos among Finnish silicotics and DE workers and slight confounding from polyaromatic hydrocarbons (PAHs) among Chinese pottery workers (68, 109, 117). Radon was of particular concern in mining cohorts and so none of the cohorts of South African gold miners were included. However, the potential for confounding need not always be assumed. Radon is likely to be correlated with dust exposure in mining, so that if radon is a confounder it is likely that there would be an increasing risk as dust exposure increased, and a possible increase in risk with increasing tenure. However, if no association is found between lung cancer and dust exposure, confounding by radon is unlikely to be an issue, since it is counterintuitive to believe radon is protective against lung cancer. If radon, arsenic, PAHs, asbestos, or other lung carcinogens are correlated with dust exposure, they are unlikely to be confounders in the absence of an E–R trend. Thus several studies should have been included among the "least confounded" studies (72). These include several South African gold miner cohorts (118, 119, 120), tungsten and iron–copper miners in China (117), and foundry workers in the U.S. (121) where there was no apparent increased risk of lung cancer associated with exposure to silica. Significant coal tar exposure is common in foundry environments where coal tar pitch is used as an additive resulting in high concentrations of benzo(*a*)pyrene (B(*a*)P). However, in this study of gray foundry workers coal was used instead of coal tar, formaldehyde was present but not associated with exposure, and potential respiratory irritants such as acrolein, bentonite clay, and ammonia were not considered confounders for lung cancer. The lack of a trend for risk to increase with increasing silica exposure is confirmatory evidence that confounding is unlikely.

3.2.3.1.3 Dust Exposures. Silica and dust exposures are generally highly correlated, as is silica exposure and silicosis. Therefore workers with high silica exposure will also generally have high dust exposure and a relatively high prevalence of silicosis. Impaired lung function can be caused by high dust exposure, and also may be an independent risk factor for lung cancer (122–124). In studies of silica exposed workers and silicotics impaired lung function attributable to dust exposure may confound associations of lung cancer with silica and silicosis. Several investigators (125–127) separated the effects on lung function of silicosis and cumulative dust exposure by closely matching silicotics with non-silicotics on age, smoking, and dust exposure. When controlling for dust exposure they found no difference in lung function between silicotics and nonsilicotics. To ameliorate the potential confounding effect of dust exposure in assessing the risk of lung cancer for silicotics it would be helpful to adjust for silicosis or adjust for differences in dust exposure in the analysis. This has been rarely done.

3.2.4 Measure of Exposure–Response Trends

The presence of a biological gradient is a major criterion in assessing causality and provides strong evidence for a causal association. An analytic study that is attempting to test a clearly articulated hypothesis should conduct an E–R analysis as an integral part of that test. Assessment of E–R is a kind of internal comparison that reduces the potential for confounding and inappropriate referent populations while evaluating coherence and sensitivity by determining if the association becomes stronger in higher exposure groups. The use of point estimates of risk using external referent populations is sometimes useful for generating a hypothesis if the association is very strong. But in these days of relatively weak association, a determination of only point estimates should, in general, be considered uninformative on the issue of whether there is an association between an occupational exposure and adverse health effect.

The use of duration of exposure is seldom a reliable surrogate for exposure unless all jobs have similar concentrations over the time period covered by the study. Such is rarely, if ever, the case.

3.2.5 Summary

It is important to consider what studies are relevant for a weight of evidence assessment to determine causality. Study design and conduct are important, and what confounders are important will change depending on the particular hypothesis. No epidemiology study will perfectly meet every criterion, and is not expected to in an observational science. That is why it is important to consider the weight of evidence. By selecting available studies that have different biases of greater or lesser magnitude it may be possible to make an informed and valid interpretation of whether there is or is not a causal association between a putative agent and an adverse health effect. The criteria for evaluating the weight of evidence from individual studies that meet the criteria for inclusion is now discussed.

4 CRITERIA FOR ASSESSING CAUSALITY CONSIDERING THE WEIGHT OF ALL THE EVIDENCE

Experiments are an attempt to determine a cause–effect relationship and provide the model for epidemiological studies. In animal experiments, for example, the question of whether the rate of disease in the exposed group differs from the rate of disease in the nonexposed group is answered (generally) by either chance or identifiable cause:

	Disease		
Exposure	Yes	No	
Yes	a	b	$a + b$
No	c	d	$c + d$

Is the difference between $a/a + b$ and $c/c + d$ due to chance or to exposure? The answer to this question is based largely on tests of statistical significance that provide reasonable assurance of the probability that the results are (or are not) due to chance alone. Statistics play an important role because the exposed and nonexposed groups are comparable in essentially every way but their exposure. This is so because (a) the assignment to exposure groups is determined by random selection among similar subjects; and (b) the investigator has continuous control and observation during the course of the entire experiment.

Epidemiological studies are also cause–effect studies. They are investigations of the association between "exposure" and "disease." Because they are based on observations rather than experiments, noncomparability between exposed and nonexposed groups is practically guaranteed. The investigator has no control over the content of these groups and no control of exposure. Therefore, in epidemiology, one must be concerned about bias, or an effect tending to produce results that depart systematically from the true values. As a result the decision about whether an association is causal, or not causal, cannot be a mere statistical judgment. A partial list of biases is presented in Appendix A.

Epidemiology is concerned with determining causation and estimating the magnitude of the effect of the causative factor. Because epidemiology is largely observational, rarely can a single study or critical point decide or determine the etiologic significance of an association. Biases can occur at all stages of a study. There are, however, well-known and widely used criteria or rules of evidence that are applied when determining whether an association is causal or not. One of the best-known applications is recognized in the association of smoking and lung cancer presented by the first Surgeon General's report on smoking (88) and discussed by others (4, 5, 8, 52, 89, 90, 100, 101).

The following guidelines provide a framework for the most difficult part of epidemiology, the interpretation of the meaning of the data.

4.1 Temporality

A criterion reliable under all circumstances is that the cause must precede the effect. If the reverse occurs, then causality is refuted and one need proceed no further. Not only must the cause precede the effect, but for chronic diseases that require a long latent period for the disease to develop, there must also be adequate time from first exposure for causation to be affirmed. For chronic diseases such as most cancers and pneumoconiosis, the length of time between first exposure and diagnosis of the disease (called the latency) is generally considered to be at least 15 years, and more often 20 or more years. The latency for lung cancer from smoking is closer to 30 years, and for mesothelioma from asbestos is more than 40 years. If exposure is relatively low, it may be that average latency may be extended.

To fulfill the temporal criterion only workers who have had an adequate time to develop the disease of interest should be included in the analysis of risk. E–R and strength of association should only be assessed among this group of workers. A problem with this approach is that for some diseases, such as lymphohaematopoietic cancers, the appropriate latency period is not precisely known. Nevertheless, when known, a valid E–R analysis should be stratified by latency, or as a second choice, include only subjects with adequate latency.

4.1.1 Example of Temporality: Diesel Exhaust and Lung Cancer

There have been many studies of the association of diesel exhaust and lung cancer. Diesel engines began to be introduced in any quantity in the late 1940s. By about 1959 the railroad and trucking industries were using almost exclusively diesel engines. In the absence of information on individual worker exposures to diesel exhaust, which is characteristic of the epidemiological studies of diesel-exposed workers, the beginning of diesel exposure is set at 1959. The earliest date for a 20-year latency is 1979, assuming the worker was employed in 1959. Follow-up needs to be longer than 20 years, for if less than say 25 years the ability to detect an association is reduced. The strength of an association will increase as latency and exposure increase beyond the minimum 20-year period.

Most of the mortality studies on diesel exhaust did not include enough workers with adequate latency to meet the temporality criterion. Latency has consistently not been considered, so any excess risk that was observed is likely to be inappropriately attributed to diesel exhaust. In fact, latency and temporality were not discussed in these early studies. A brief examination of a retrospective cohort mortality study of railroad workers is an example of the problems. This study is one of the more important studies because it is one of the largest and has some estimate of E–R.

Garshick et al. (128) conducted a retrospective cohort study of U.S. male railroad workers aged 40–64 years in 1959 and with 10–20 years railroad service. Follow-up was through 1980, and beginning of diesel exhaust exposure was considered to be 1959. The exposure variable was years in a diesel-exposed job. Persons dying before 1979 would have less than 20-years latency. The maximum latency is 22 years for those dying in 1980, the last year of follow-up when the oldest subjects would be 86 and the youngest 62. No data were provided on distribution by latency. Latency was not considered when the analysis by diesel-years was conducted. Crump et al. (129) determined that for the last four years of follow-up data on deaths were missing, about 25% in 1980 and somewhat less in 1977–1979. Because of the missing data and the possibility of under-ascertainment bias, data from these years should not be used. Now the maximum latency period is reduced to 17 years, which Crump et al. (129) considered insufficient for the development of lung cancer due to diesel exhaust.

4.2 Strength of Association

The strength of association is the magnitude of the outcome, usually measured in the exposed group compared to the unexposed or referent group. The strength refers to the size of the RR (SMR, SIR, or OR). The stronger the association the more likely there is a causative link, because strong associations are less likely than weak association to be due to bias or confounding. As explained by Hill (4) the risk of lung cancer is 10 or so times greater among smokers than nonsmokers, while coronary thrombosis is no more than two times greater, perhaps less. "Though there is good evidence to support causation it is surely much easier in this case to think of some features of life that may go hand-in-hand with smoking-features that might conceivably be the real underlying cause or, at the least, an important contributor, whether it be lack of exercise, nature of diet or other factors. But to explain the pronounced excess in cancer of the lung in any other environmental terms requires some feature of life so intimately linked with cigarette smoking and with the amount of smoking that such a feature should be easily detectable." If one cannot detect it or reasonably infer a specific one, then in such circumstances, it is reasonable to reject the vague contention of the armchair critic that since you cannot prove there is such a confounder, there may in fact be one present.

Weak associations detract from the determination of a causal association as the presence of relatively weak confounding or bias could explain the association. A weak association is commonly considered to be a RR of less than 1.5 to about 2 (7). When the association is weak the other criteria become more important. Exposure–response trends, use of internal referent population for comparisons, and consistency of the findings become more important, as they are needed to counteract the weak association that could be due to biases.

Definition of the association is sometimes ambiguous. It could be the SMR for a specific cause of death in a cohort. This is the least powerful, as the total cohort potentially includes a proportion of workers with little or no exposure and short latencies. The strength of the association is assessed more appropriately for workers with adequate latency and highest exposure. For example, the risks of lung cancer among current smokers compared to nonsmokers ranges in various studies from 4 to 16, but 60 to 80-fold for heavy smokers (115, 116). For other causes the associations are less strong: 0.7 to 3.0 for bladder cancer; 1.1 to 1.6 for kidney cancer; 1.3 to 2.1 for coronary heart disease; 1.3 to 1.9 for CVD (115); 2.9 to 25 for bronchitis; 0.6 to 12.3 for emphysema (114). Thus a risk ratio for a total cohort may present a weak or even no association because of dilution or other reasons even if the evidence is strong for a causal association.

4.2.1 *Example of Confounding or Bias as an Alternative Explanation for a Weak Association*

Blue-collar workers frequently smoke more than the general population. Using a RR for lung cancer of 10 for smokers, a difference in the prevalence of smoking of 15% between exposed and referents could explain about 18% (or a RR of 1.18) of an excess risk of lung cancer. For a RR of 60 and a 35% difference in prevalence, about 57% of lung cancer is attributable to smoking using the Axelson (71) adjustment for smoking. If other factors of smoking such as tar content, length of smoking, number of cigarettes, use of filters, depth of inhalation are taken into account then smoking-attributable risk may be increased even more. If the lower confidence of a smoking adjusted RR must be below one before the role of smoking can be excluded (113), then smoking is often as likely an explanation for a weak association as work exposure. This may sometimes be the case even when there is an internal E–R analysis where smoking is more likely to be similar in high exposed compared to low exposed. In a cohort of diatomaceous earth workers smoking data were available on about 50% of the cohort. Prevalence of smoking was 64% in the lowest silica exposed group and 84% in the highest exposed. Using the Axelson (71) method for indirect adjustment of smoking and a RR of 20 for smokers versus nonsmokers, about 29% of the RR is attributable to smoking. Subtracting this estimated effect of smoking from the highest RR (2.15 in the highest exposure category) gives a nonsignificant RR of 1.86 with a lower confidence of 0.79. Thus, the increased risk of lung cancer in the high exposure group could be due to chance, exposure to cristobalite, or smoking.

4.3 Biological Gradient

The presence of an exposure–response gradient is generally considered strong evidence of a causal association. Hill (4) notes that the fact that mortality from lung cancer rises linearly with the daily number of cigarettes smoked adds considerably to the simpler evidence that smokers have a higher death rate than nonsmokers do. The comparison would be weakened and call for a more complex explanation if heavier smokers had lower mortality than light smokers, as now there would be no biological gradient. The same holds true for dust or other hazards in the occupational environment. Although it may be difficult to explore exposure–response and dose–response gradients, "we should invariably seek it."

An E–R gradient could be noncausal if there is confounding. For example, one might observe an association of increased SMRs for lung cancer with increasing tenure. In fact, increasing age, increasing years of smoking, and increasing dust exposure may also be associated with lung cancer risk and tenure. And so the gradient could be in part or totally due to these confounding factors. The stronger the gradient, the less likely the effect is caused by a confounder. The use of a cumulative exposure index sometimes reduces the interference of age on exposure.

The lack of a biological gradient does not by itself disprove a causative association. Misclassification of exposure or too narrow an exposure range to show an effect may reduce a gradient toward the null. Variations on the idea of a narrow exposure range include exposure of such magnitude that it produces a maximum effect so there is no gradient, as well as exposures too low to produce a detectable effect.

A variety of measures for exposure and dose have been used to evaluate biological gradients. Tenure or duration of exposure is relatively easy to determine from questionnaires or personnel records. Tenure is an adequate measure in the unlikely event that workplace concentrations are similar for all jobs and over the entire time period of the study population. If not, exposure misclassification occurs, and the risk is most likely to be biased toward the null. An improvement on simple tenure is to estimate qualitatively exposure for each job in the study population over the time of the study, and summate the product of job rank and time in job. The disadvantage is that the numerical rank given for each job may not be correct. For example, should low, medium, high ranks be 1, 2, 3 or 1, 3, 9 or what?

Cumulative exposure is estimated as the summation of the products of (air) concentration for each job and time worked in that job. The disadvantage of cumulative exposure is the difficulty of estimating exposure for jobs where there is little or no industrial hygiene sample data to quantify the relevant concentrations, and often little is known of processes and work environment when the exposures occured decades earlier. Attempting to reconstruct past exposures is a difficult and labor-intensive job. Although also subject to misclassification, the probability is less than for the less quantitative estimates such as tenure.

The rewards for estimating E–R are potentially great. Assessing E–R gradients and not relying simply on point estimates of risk increases the possibility of determining causality, of determining an effect level or level of risk. Often the referents in an E–R analysis are from the same work force (internal controls) so they are likely to be similar with regard to the major risk factors (e.g. smoking, diet, SES, and other features of lifestyle) that are likely to confound the association. The potential for confounding by time-related elements is also reduced. By matching or statistical adjustment the effects of exposure may be largely disassociated from potential biases such as the different levels of exposure and selection pressures that occurred in the different decades of a study. Sometimes the work histories of a study population may span 50 or more years, making it important to include referents that are from the same birth cohort. Analysis of biological gradients converts a largely hypothesis-generating study to a hypothesis-testing study.

Some references that discuss methodological issues of exposure reconstruction or have analyzed E–R trends include: Armstrong et al. (130) for benzene; Burgess et al. (61), Dosemeci et al. (131), Rice et al. (132), Seixas et al. (133) for crystalline silica; Gibbs and Christensen (69) for asbestos.

The ideal measure for assessing biological gradient is dose, or the actual concentration or mass of a substance that is deposited in or on the study population members. When the lung is the route of exposure and dust is the exposure of interest, dose may be estimated by categories of pneumoconiosis read from a chest X-ray, or autopsy. For the radiographic diagnoses it is preferable to use the ILO 12-point scale for pneumoconiosis developed for epidemiological studies. The diagnosis should be made using several readers comparing with standard films and blind to other factors not related to the occupational dust exposure.

4.3.1 Example of Biological Gradient: Exposure-Response (E–R)

Cherry et al. (60) conducted a cohort mortality study of U.K. pottery workers and found significantly increased SMRs for lung cancer using expected deaths from either national rates (England and Wales) or local rates (Stoke-on-Trent). To help determine whether the excess mortality was due to quartz exposure in this industry they conducted a nested case-control study and analyzed first by years worked and then by cumulative exposure and peak exposures to quartz. They matched on date of birth and date of first exposure and adjusted for smoking (they also included only controls who also smoked since all cases were smokers) and silicosis. Study subjects were restricted to those without exposure to asbestos or foundry work A job exposure matrix was constructed from 1300 air samples collected since 1930. The estimates of cumulative and average exposure appeared to be valid based on strong associations between exposure and radiographic evidence of silicosis (61). Maximum exposure was added to the analysis. There were no significant gradients between lung cancer and tenure, or cumulative exposure to quartz. Maximum exposure to quartz was associated with increased risk of lung cancer (62, 65).

4.3.2 Example of Biological Gradient: Dose–Response and Registry Studies

Since 1980 there have been a number of studies designed to consider the question of whether silicotics are at increased risk of lung cancer. This is a worst case situation where the dose of quartz in the lung is measured by severity of pneumoconiosis on the chest radiograph.

Most of these studies were based on registries of compensated silicotics, which, in general, have shown strong associations with lung cancer. For example in the IARC (92) evaluation of the carcinogenicity of crystalline silica, a study of compensated silicotics in Finland was considered among the most important and least confounded studies. There was a statistically significant three-fold excess of lung cancer in this study (109). Smoking was known for about 25% of the silicotics, and the incidence of lung cancer was increased nearly seven times compared to the Finnish population. The authors concluded this study supports an association between silicosis (dose) and lung cancer.

However this and most registry studies suffer from a number of significant biases that severely limit their value in determining a valid dose–response gradient. These shortcomings have been grouped under six headings (105). *Ascertainment bias* occurs as there is incomplete enumeration of the silica exposed population, and the silicotics enumerated are often self-selected for compensation and are unlikely to be representative of all silicotics. *Confounding from smoking* is likely to occur as registered silicotics are more likely to be smokers, at greater risk of lung cancer, and not comparable to external referent populations or even silicotics without smoking-related symptoms. *Confounding from occupational exposures* such as radon, PAHs, arsenic, nickel, and chromium may occur before and after silica exposure and be undocumented because of incomplete work histories. The RR in the study of Finnish silicotics was estimated to have been increased 40% by asbestos exposure (108). *Diagnosis* is based on compensation criteria that may be related to smoking symptoms and disability that may be more important than the kind of pneumoconiosis and characteristics of the dust exposure. *Timing* may bias the findings as lung cancer may contribute to the symptoms that resulted in discovery and diagnosis of silicosis. Only a few such incidents can result in an over-estimate of perceived risk. *Expectation bias* occurs because compensated silicotics comprise a highly selected case series of sick workers that will result in elevated risk ratios, in part because there is no appropriate referent group.

It is possible to evaluate dose–response trends in cohorts where both occupational and smoking histories are known, and where there is complete enumeration and objective diagnosis of silicosis. Among white South African gold miners, the autopsy rate is about 85%. Based on autopsy and histological diagnoses the degree of silicosis is determined. In a case-control study of these miners, Hessel et al. (119) found no increased risk of lung cancer for miners with parenchymal silicosis compared to nonsilicotics, and no trend for lung cancer risk to increase with increasing severity of silicosis. Hessel et al. (118) conducted a similar study among a different group of miners, but also determined radiographic silicosis status. Based on autopsy diagnoses there was no significant increased risk of lung cancer for parenchymal silicotics compared to nonsilicotics (OR $= 1.49$, $p = 0.11$), and a nonsignificant ($p = 0.08$), but suggestive trend for risk to increase with increasing severity. The risk of lung cancer was not increased among the radiologically diagnosed silicotics compared to nonsilicotics (OR $= 1.08$, $p = 0.92$). Several studies had radiological diagnoses of pneumoconiosis based on the ILO classification system of workers within a cohort, and could assess the risk of lung cancer among those with and without silicosis (62, 117, 122: see 72 for review of these studies and their use in assessing dose–response trends).

4.4 Consistency

Consistency is repeated observations of an association in different populations by different investigators under different circumstances in different places and at different times. Exact replication is impossible in epidemiology, but even though separate situations differ, a consistent health effect related to similar exposures provides strong evidence of a causal association. An association that is shown repeatedly by different investigators in different places under different circumstances and at different times and using different study designs, is unlikely to be due to a constant bias. Hill (4) gave an example from the 1964 Report of the Surgeon General (88) where the association of smoking with lung cancer was observed in 29 retrospective and 7 prospective studies. The lesson is that the same answer has been reached in a wide variety of situations and techniques so it is reasonable to infer that the association is not due to a constant error or fallacy that would be present in every inquiry. Since then the variety of situations and techniques has increased.

Several components of the consistency criterion need to be kept in mind. Consistency of a strong association and the consistent presence of biological gradients provide strong evidence of a causal association. So an aspect of both of these criteria is whether there is consistency. An isolated finding of a strong association (and many weak or no associations) and isolated findings of E–R or D–R trends does not instill confidence that bias or error could not have caused a result that is in effect an outlier or spurious or biased finding.

Susser (89) defines consistency as the persistence of an association upon repeated testing. It is not just the number of tests, but also the variety and the rigor of the tests. He defines these subclasses of consistency as replication and survivability respectively. Survivability is the more important.

Replication is the number and, specifically, variety and diversity of the tests of the hypothesis. Replication of a result under widely varying conditions enhances the credibility of the hypothesis and the case for causality. Examples of diverse replications of the silica/lung cancer hypothesis are the many different working populations studied (e.g., metal miners, potters, refractory brick, dusty trades, foundries) on five continents and dozens of different investigators.

Survivability is defined by the number and, specifically, the rigor and severity of tests of association. Persistence of results under increasingly severe conditions enhances credibility of the hypothesis, and contributes most to the verification, or refutation, of the hypothesis. Severity of the test is determined by the study design, and whether further worst case situations are addressed and whether an alternative hypothesis can be refuted. For example, the most severe test of the silica/lung cancer hypothesis is to evaluate risk among a representative and unbiased sample of silicotics.

Study design. The rigor of study design can be roughly ranked by the confidence it generates about the definitive attributes of cause. To assure confidence the study design should have individual-level data for both exposure and response. Thus an ecological study with group-level data is subject to the ecological fallacy, does not provide confidence that the results apply to individuals, and is not a rigorous test of the hypothesis. Some would say an ecological study can generate a hypothesis, but cannot test a hypothesis. Clinical trials (controlled experiments) are perhaps the most rigorous tests of a hypothesis (89), but play little or no role in occupational epidemiology. Longitudinal cohort design follows forward from exposure to effect, and nested case-control within the cohort have similar rigor and the quality of the data are similar. Cross-sectional studies may have trouble speaking to questions of time order and direction, and so is generally a less rigorous design. Since it is not prospective in nature, the information on exposure prior to the survey is more subject to error. Study designs ranked by severity or rigor in testing a hypothesis in the occupational setting are: cohort = nested case-control > cross-sectional > population-based case-control, with ecological studies considered to not be valid for testing a hypothesis.

Refutation of Alternative Hypotheses by E–R and/or D–R. The most rigorous test is to determine the risk among the most susceptible and those with the highest dose. If there is a consistent finding of little or no increased risk in the most susceptible population defined by highest dose and highest response to exposure, then the hypothesis is refuted. For example, a test of the alternative to the silica/lung cancer hypothesis is to assess the risk of lung cancer among silicotics. If silicotics are not at increased risk of lung cancer, then the credibility of the silica/lung cancer hypothesis is severely eroded; likewise if the risk does not increase with increasing severity of silicosis. If the risk does increase as severity of silicosis increases (presence of a D–R trend), then the silica/lung cancer hypothesis is supported. However, the hypothesis about silica exposure (not silicosis) must be further tested by a consistent finding of E–R trends among an unbiased and representative sample of exposed workers. There are about a dozen studies with a minimum of bias that have evaluated E–R between the risk of lung cancer and silica exposure, and these results also show no consistent gradients. This combination of a consistent lack of E–R and D–R gradients provides strong evidence against the silica/lung cancer hypothesis (72).

4.4.1 Example of Consistency: Particulate Matter (PM) Air Pollution

The EPA (37) suggests there is a causal association between increased mortality and particulate matter air pollution less than 2.5 μm ($PM_{2.5}$) and less than 10 μm (PM_{10}) in size. This assertion is based, in part, on the argument that there are consistent findings from dozens of ecological studies of populations in dozens of cities in different parts of the world. It is their contention that the consistency criterion for replication is met because of the number (for PM_{10}, but not $PM_{2.5}$) and variety of studies. However it does not meet the survivability portion of this criterion, as ecological studies are not a rigorous or valid test of the hypothesis. Consistency is based on ecological studies, so the validity of the E–R cannot be accepted because there is no individual-level exposure data, and exposure misclassification is biasing the results to an unknown degree and direction. In addition, the consistency criterion was not met as several investigators analyzed the data from five different communities, and arrived at different results and conclusions. The analysis of similar data from the same location by different investigators is perhaps as close as one is likely to get in epidemiology to replicating an experiment, a common procedure in experimental science (38). The EPA's (37) use of the consistency criterion regarding PM air pollution pertains only to the replication portion. The Surgeon General's (88) use of the consistency criterion for lung cancer and cigarette smoking pertains both to replication and survivability, as it is not the mere repetition of evidence of the same kind (100).

4.5 Coherence

Coherence is the idea that all the pieces should fit together and be compatible with the hypothetical relationship being tested. A cause-effect interpretation should have coherence. That is the data should not seriously conflict with the generally known facts of the natural history and biology of the disease. Hill (4) considered the histopathological evidence from the bronchial epithelium of smokers and the isolation of substances from tobacco smoke that produce tumors on the skin of laboratory animals as evidence contributing to coherence. Coherence is another nonquantitative criterion of judgment regarding the reasonableness of the association in biologic terms. Susser (89) considers analogy, biological gradient, and biological plausibility to be subsumed under the criterion of coherence. It is an ultimate, but not a necessary criterion, as existing theory may be incorrect (100). By Susser's (100) definition, the increasing risk of lung cancer is coherent with the observed association with the number of cigarettes smoked, tar content, and depth of inhalation.

Results that are plausible in terms of current theory affirm the hypothesis; results that contradict the theory suggest that either the study results are in error or the theory needs revision. For example, if cigarette smokers who inhale are at no greater risk of lung cancer then those who do not inhale (or women are at less risk than men) and the results appear to be true, then one should consider modifying the hypothesis that smoking/lung cancer hypothesis.

4.5.1 Example of Coherence: Silica/Lung Cancer Hypothesis

IARC (92) recently declared that silica is a probable human lung carcinogen. This was based primarily on 11 cohort and case-control studies that were judged to be the least confounded (for occupational carcinogens) studies published. IARC acknowledged that increased risk was not always observed, but the results were considered consistent enough to change the classification for silica from a possible to probable carcinogen. Only the replication portion of consistency was considered. In a weight-of-evidence review of this hypothesis, a different set of studies was selected to test the replication and survivability portions of the consistency criterion. Only studies that had minimal bias and confounding and evaluated E–R were used to evaluate the silica hypothesis. In addition the alternative hypothesis that the most susceptible workers (silicotics) should show the strongest risk was tested in cohort studies where enumeration bias and compensations diagnoses were not biasing the results. There were no consistent findings of increased risk among silicotics and there were no consistent D–R or E–R trends. These data provided a significant portion of the information that led to the conclusion that the silica/lung cancer hypothesis was refuted (72).

To be coherent, one would expect other outcomes if the silica/lung cancer hypothesis were correct:

The predicted increased risk for silicotics, called theoretical coherence (89) is not supported by the data. In addition a series of necropsy studies conducted prior to the 1950s going back as far as the 1920s show no "demonstrable causal connection . . . with the prior existence of pneumoconiosis" (134). In these case series of pottery workers, sandstone workers, and gold miners with and without silicosis, those with silicosis had either less or the same prevalence of lung cancer as those without silicosis. If the hypothesis is correct, one would predict the quartz content of the lung should increase in the progression going from persons without silicosis or lung cancer, to persons with silicosis but not lung cancer, to persons with both silicosis and lung cancer. In a small series of such cases, silicotics without lung cancer had higher amounts of quartz in the lung than did silicotics with lung cancer and victims of accidental deaths without either silicosis or lung cancer (135). These data are not compatible with preexisting theory.

4.6 Biological Plausibility

Are the findings plausible in terms of biological knowledge? Do the facts fit current theory? Do findings from experiments in non-human species verify the hypothesis? Plausibility is not a required criterion to demonstrate causality. However, if the evidence supporting the other criteria is weak, then plausibility may take on added importance.

For example, the hypothesis that PM air pollution at low concentrations increases mortality and morbidity (such as hospital admissions) is exclusively based on ecological studies. An increased level of proof is required because such studies are subject to the ecological fallacy, and because the results are not coherent when comparing group-level risk with individual-level risk estimated from smoking (43).

4.6.1 *Examples of Biological Plausibility: Diesel Exhaust, PM, and Mortality*

The PM/mortality hypothesis predicts that long-term exposure to fine particulate increases mortality. Experimental data of lifetime exposure of animals to fine particulate matter (in this case diesel exhaust that consists of submicron particulate) show no increased mortality despite exposures so high as to produce lung overload. Overload occurs when exposures are so high that there is reduced clearance, increased retention of particulate matter, and increased lung burden (136). In fact at the lower concentrations rat mortality is reduced slightly (see Ref. (43). These data suggest the plausibility criterion of the PM/mortality hypothesis is not met.

These same experiments show a trend for the incidence of lung cancer to increase in rats as exposure to diesel exhaust increases. At first glance these data appear to provide plausibility for the diesel exhaust/lung cancer hypothesis. However, other data suggest this interpretation is unlikely to be correct, and that the rat is not an appropriate model as the rat is not relevant to the human experience. When the rat is exposed to overload concentrations of many substances including inert matter, lung tumors also develop. The tumors develop late in the life of the rat, and do not reduce survival. The list of such substances is long, and includes non-carcinogens, such as titanium dioxide, silica, and carbon black (136). In this instance the rat data neither support nor detract from the plausibility criterion as the rat data are not relevant to humans. If the rat were an appropriate model, then the question that would need to be addressed is whether concentrations sufficient to cause overloading meets the criterion, since humans are not usually exposed to overload concentrations. Pritchard (137) estimates human overload occurs at similar lung burdens seen in the rat experiments. For 40 hour/week exposures this is on the order of 4 mg/m^3 concentrations. Overload is predicted to occur in smokers of 25 middle tar cigarettes/day who are breathing daily concentrations of 25 mg/m^3 fine particulate in mainstream smoke. (137).

4.6.2 *Example of Biological Plausibility: Petroleum Hydrocarbons and Kidney Cancer*

The male rat exposed to gasoline develops tumors in the kidney. These findings stimulated an extensive research program to discover the mechanism. As it turned out the tumors are seen only in male rats that produce an excess of β-globulin, and not in the female rat, a variety of male rat without excess of β-globulin, or in other species. Thus the male rat is not a relevant species for determining biological plausibility for some potential kidney carcinogens. Several epidemiology studies are suggestive of limited or no risk for workers (34, 35, 36).

4.7 Summary

Conclusiveness in inferring causality is a goal, but it is not always an accomplishment. The rules of evidence must be applied to the question of causality. They overlap with each other, can be mutually reinforcing, and have different levels of conclusiveness (138).

It seems appropriate to conclude this discussion with a quote from Feinstein (28).

> If industrial workers and the public are to receive suitable protection against occupational and environmental hazards, and if everyone is to be protected against irrational fears that may lead to a needless loss of income, jobs, and homes, the first step in the process is to obtain accurate and unbiased information about what the hazards are, where they occur, and how relatively great they may be. The task of acquiring credible information is difficult but not impossible. Until this task is approached with rigorous scientific methods, workers and the public may be victimized far more often by the products of defective research than by the toxins of occupation or the pollutions of environment.

5 CONCLUSION

Hopefully these guideposts and points of view will assist the reader in the interpretation of epidemiological studies. To conclude, some of the viewpoints discussed herein are summarized in the words of others.

> Although there is long historical precedent for using "tests of significance" as a primary means with which to interpret epidemiological data, there is no compelling reason to do so. To describe relations between exposure and disease, there is no need for a commitment to reject or to fail to reject null hypotheses. Furthermore, dichotomization of the p value range into regions labeled "significant" and "nonsignificant" is highly arbitrary and can result in misleading interpretation (139).

> Formal statistical tests are framed to give mathematical answers to structural questions leading to judgments, whereas in . . . [epidemiology one] must give answers to unstructured questions leading from judgment to decision implementation. These questions of decision generally hinge around judgments about causality and prediction (100).

> Full epidemiologic analysis assesses bias, confounding, causation and chance. Of these, chance is least important but still receives the most attention (140).

> The results of epidemiological studies are often difficult to interpret, and claims and counterclaims of the existence, or absence, of a risk are sometimes made on the basis of fragmentary and contradictory data. It is understandable, therefore, that those responsible for protecting the health of individual workers have sometimes sought to lay down rules that could ease

> their task. Unfortunately, epidemiology cannot be regulated in the same way as laboratory science. A laboratory investigator may properly be told to use so many animals, test so many species, test at so many levels of dose, and observe for an optimum length of time, as all these are under his personal control-subject only to the constraints of finance. It is unproductive, however, to lay down similar rules for the epidemiologist, as unintended experiments on Man cannot be repeated at will and the conditions of the experiment are not under the observer's control. All we can do is to require that the observations that are made should be relevant and not biased and that consideration should be given to the possibility of confounding as an explanation for them. So far as numbers are concerned, we can deal with this only by combining the results from different series, for this purpose, all data that are relevant and unbiased must be grist to the mill. Indeed, if there is one general rule in the assessment of epidemiological evidence, it is that no conclusion can be reached until the totality of the evidence is taken into account (141).

Sound epidemiological evidence offers a robust and often unique resource in shaping our beliefs and driving our conclusions. By the same token, erroneous epidemiological data and observations may be the prescription for waste and unjustifiable decisions. The task for any responsible practitioner is to examine the evidence with patience, care, and an open mind. This should be an admonition and a rule.

APPENDIX A. BIASES THAT CAN OCCUR IN EPIDEMIOLOGIC RESEARCH BY STAGE OF RESEARCH

See Sackett (81) for additional references and discussion

1. In reading-up on the field
   a. Positive results bias (publication or file drawer bias): Positive results are more likely to get published than are negative results (142).
   b. One-sided reference bias: References may be restricted to those supporting the authors' results. Or only the results that support the authors' view or a governmental or other policy are presented.
2. In specifying and selecting the study sample (selection bias)
   a. Diagnostic suspicion bias: Knowledge of occupation or exposure may influence the diagnosis and/or the search for a putative cause. The examiners being blind to the other information can accomplish the elimination of this bias.
   b. Sample size bias: Too small a sample may prove nothing; too large a sample can make any difference statistically significant (143).
   c. Nonrespondent bias: Nonrespondents (or "latecomers") in a survey may have different exposures or health status from respondents or "early comers." A mortality follow-up of gold miners examined in a cross-sectional morbidity study showed a positive association of lung cancer to exposure (106). Because this was not a complete cohort and subsequent mortality studies of the complete cohort (107, 108) were negative, this original result may be incorrect because of nonrespondent bias. Williams (144) suggests that in surveys the effects of nonrespondent bias can be subtle, can change relationships, do not react to increasing

response rates in a desirable way, and can be serious in magnitude. For example, a 4% bias in estimating unemployment rates can occur with a nearly 98% response.

d. Volunteer bias: Volunteers or "early comers" may have different exposures or health status from nonvolunteers or "latecomers." In fact, volunteers are in many ways different from nonvolunteers.

e. Healthy worker bias: An employed population of active workers is generally healthier than the general population, often with a mortality risk 60 to 90% of that of the general population. There is a large literature on the healthy worker effect. Early and recent comments can be found in References 145 and 146.

f. Chronology bias: This is a selection bias caused by variation in exposure over time. The effects of this bias were observed in a cross-sectional survey where the risk of respiratory cancer was overestimated by about 30% compared to a cohort study (147).

3. In measuring and classifying exposures and outcomes (information bias)

a. Detection bias: This occurs if the outcome (response) event is detected unequally in the exposed and nonexposed groups. For example, Enterline (148) showed that correction in the cause of death of asbestos-exposed workers in three studies incorrectly increased respiratory cancer risk by about 13% and mesothelioma risk by about 67% (149; see also Ref. 28 for other examples).

b. Instrumental error: Faulty calibration and inaccurate measuring instruments result in bad data and wrong results. Graham et al. (150) reasoned that a leaky spirometer resulted in a reported decrease in pulmonary function in a cohort of granite shed workers. The decrease was incorrectly attributed to exposure (151) because the instrument error was undetected during its use.

c. Interviewer bias: This is a systematic error due to interviewers gathering selective data. The data may be biased if the interviewers are aware of the research hypothesis and the health status of the individual. For example, an interviewer who believes asbestos causes lung cancer may readily accept no exposure from a control but may inquire more intensively of a case. Such knowledge may influence both an interviewer's attitude and mode of interrogation. Blinded interviewers and a standard questionnaire (28) can eliminate this bias. Gould (149) discusses an interesting example of "unconscious manipulation of data," which argues for the necessity of blindness as a minimum, and perhaps even supports the suggestion (28) that the investigator should not be the one both to collect the data and later to test the hypothesis.

d. Recall bias: This bias is the underreporting of true exposures in controls or overreporting of true exposures in cases. The potential for its occurrence exists whenever self-reported information is required. Although this bias is given a lot of attention in epidemiological textbooks, the research literature on this subject is sparse (152).

e. Measurement bias: This is the inaccurate or misclassification of subjects on a study variable such as smoking category or exposure. Bias due to misclassification of exposure is generally considered to reduce risk toward the null (153).

Greenberg et al. (154) showed that misclassification occurring in a nuclear worker group produced a substantial bias away from the null hypothesis of no association between cancer and nuclear work.

4. In analyzing the data
   a. Data dredging bias: If there is no hypothesis, significant associations are suitable for hypothesis-forming activities but should be confirmed in specific studies of different cohorts to test the hypothesis. So-called "hypothesis-generating research" presents a hazard of interpreting a statistical association as causative when it is not. Such an association may be due to factors overlooked during the data dredging or may be simply accidental because of the numerous associations examined. (See Refs. 28 and 59 for more discussion of this bias.)
   b. Managing unknown data: Unknown data should not be used as evidence of nonexposure, and the report should indicate how unknown data are handled. The treatment of lost-to-follow-up in a cohort study can affect the apparent forces of mortality. Methods of dealing with workers lost-to-follow-up include counting them as lost at the time of loss, assuming they are alive till the end of follow-up, and assuming them dead at loss or at the end of follow-up. The last two methods result in abnormally high SMRs. The second method is considered the most conservative and is a common method in many cohort studies. However, considering subjects lost to follow-up at the time of loss may be the least biased estimate of expected mortality (155, 156).
   c. Retroactive demarcation of exposure: Odds ratios may show an eight-fold change in the same data set by changing the exposure criteria. To avoid this bias, criteria for establishing exposure categories should be set before analysis of the data, and there should be no changes in exposure categories after collection of the data (28).
5. In interpreting the analysis
   a. Significance bias: This bias occurs when there is confusion about the statistical (or stochastic) significance of a result and the quantitative (or biological) significance. If n is too small, a biologically important difference may not be statistically significant. If n is quite large, a biologically insignificant difference may appear to be statistically significant. For whatever reason, statistical significance is commonly set at $p < 0.05$. There are no accepted standards for biologically significant differences (28).
   b. Correlation bias: Correlation should not be equated with causation. One might avoid being misled by correlation coefficients (r) by examining a scattergraph of the data, considering r^2, which indicates the proportion of the variance that has been reduced by the linear model, and remembering that correlations from relatively small sample sizes can easily achieve statistical significance (28).
   c. Bias of interpretation: This error arises from failure to consider all interpretations consistent with the facts and failure to assess the credentials of each. One of the major advances of scientific method in the nineteenth and twentieth centuries was the obligation of investigators to rule out alternative hypotheses that might explain the observed results.

APPENDIX B. GLOSSARY

Association: An association is when events occur together more frequently than would be expected to occur by chance. For example, a *statistical association* exists when the occurrence of disease in two groups is statistically significant. A statistical association does not necessarily imply a causal relationship.

An *artificial* or *spurious association* occurs when a statistically significant relationship between two events is due to chance, creating a type I error. One implication of probability is that a certain proportion of outcomes will be declared statistically significant even though due to random fluctuation. Artificial associations can also be a result of bias, flaws in methods, flaw in study design, and/or nonrepresentative selection of study groups.

A noncausal (indirect) association is when there is a statistically significant relationship between an exposure and health outcome, but the association occurs because of a common relationship to an underlying condition. For example, the miasma theory hypothesized that cholera death was due to bad air and was supported by the inverse relationship between altitude and cholera death. However, bad air and low altitudes were also associated with contaminated water, and the true association with contaminated water was missed.

A *causal association* is when A causes B and can occur only if (*1*) A occurs before B, (*2*) a change in A is correlated with a change in B, and (*3*) the correlation is not due to confounding; that is, the correlation is not due to both A and B being correlated to some factor C. A number of criteria are widely used to help evaluate the likelihood of a causal association. These include strength of association, exposure-response relationship, a consistent association in different studies, a correct temporal sequence of events (A occurs before B), and biological plausibility.

Attributable Risk or Risk Difference: This is a measure to estimate the magnitude of a given cause (referred to in epidemiology as an exposure). This is an absolute effect, namely, differences in incidence or prevalence rates between exposed and nonexposed populations. It should be used only when there is a cause–effect relationship between exposure and disease. It is a useful measure to estimate the magnitude of the public health problem caused by the exposure.

Bias: A bias is an effect that can occur at any stage of an investigation or inference. The effect of the bias is to produce results that depart systematically from the truth. We have described some biases in Appendix A.

Confounding: Confounding is a central issue in nonexperimental research and is a mixing of the effect of exposure with the effect of some extraneous factor, thereby distorting the effect one is trying to estimate. For example, if lung cancer is the effect or response and asbestos the exposure, a confounder could be smoking if the exposed asbestos workers have a higher (or lower) prevalence of smokers than the nonexposed control group. For an extraneous factor to be a confounder, two criteria must be met:

1. The extraneous factor must be associated with the exposure under study. That is, the occurrence of the factor must be "different" in exposed and control groups.
2. The extraneous factor must be associated with the disease under study. The factor must be predictive of the occurrence of the disease but not necessarily causal. Thus

a lower social class is predictive of increased cancer risk, not because it causes cancer, but because it is correlated with other factors that do cause cancer.

One could remove the confounding effect of smoking by stratifying exposed and non-exposed groups, that is, comparing asbestos workers who smoked with controls who smoked. Then smoking would not be associated with exposure.

Dose–Response Relationship: More correctly termed exposure–response, this is a relationship in which a change in exposure is associated with a change in the risk of a specified outcome (response).

Exposure is an estimate of dose. In occupational epidemiology a common surrogate of dose is years worked (tenure). If quantitative estimates of exposure are available (such as mg/m^3, dust-years, $fibers/cm^3$-years, million particles per cubic foot-years or mppcf-years, and ppm-years) then cumulative exposure can be estimated. Cumulative exposure is defined as (concentration $\times$ time) and so cumulative exposure is estimated by multiplying the estimated exposure in each job times the time worked in that job (such as $mg/m^3 \times$ years exposed) and adding the results over the life work history of each worker. This is not a measure of "dose" because the relationship between the external level of contact with the exposure agent(s) and contact at the critical site for each individual may be quite different from one individual to another. "Dose" also varies depending on properties of the agent, circumstance of exposure, performance characteristic of portal of entry, pathways of transport to thc critical sitc, and sampling mcthod (157).

The main point of an epidemiological analysis is to estimate the magnitude of the response as a function of exposure status (52). This cannot be accomplished by determining if average exposures of, say, cases and controls are statistically different. The most desirable analysis is to determine response rates at several levels of exposure. Estimation of effects at every level is possible, but such a separate analysis at each exposure level results in lost information because the continuity of the underlying variables is not considered and because of a potential loss in precision due to fewer data in each exposure strata. If there is an exposure–response relationship, the estimates of effect in bordering exposure categories provide additional information such that the effect may be characterized in an exposure category that has few data.

Exposure–response relationships are among the most important criteria for inferring causality. Statistical hypothesis testing using central measures of exposure are not the preferred measure of exposure-response because this method suffers from loss of information and is more susceptible to confounding. It is more desirable to estimate an overall measure of trend in response at various levels of exposure rather than a separate estimate of overall effect.

Latent Period (Latency, Induction Period): This is the time between exposure to a disease-causing agent and the appearance of the disease. For chronic disease this can be years. For example, the latent period for lung cancer due to occupational exposure is generally considered to be at least 20 years.

Null Hypothesis: This is a statement about causation phrased in the negative, such as "smoking is not a cause of lung cancer." If the p value is statistically significant (say, $p < 0.05$), then the hypothesis that exposure does not cause the effect is repeated.

Odds Ratio: Commonly used in connection with case-control data, the odds ratio (OR) is the ratio of the odds in favor of exposure among cases (a/b) to the odds in favor of exposure among controls or noncases (c/d). See section 2.3.3 for examples and sample calculations.

Proportional Mortality Ratio (PMR): Number of deaths from a given cause during a given time period divided by the total number of deaths during the same time period times 100; [(number of deaths)/(total deaths)] $\times$ 100.

Significance Testing (Statistical Inference versus Scientific Inference): Feinstein (9) has called "significance" the "single greatest intellectual pathogen in both biological and statistical domains today." There are two types of significance; the one used almost exclusively is "statistical significance." This is a measure of the probability that an observed difference is due to chance. It is an assessment of random variability and involves primarily a test to reject a null hypothesis of no association between two variables and, by exclusion, accept the alternative hypothesis. The p value is a quantification of error in the decision process. If p is set by the investigator at 0.05, then when $p < 0.05$ the observation is "significant" and the null hypothesis is rejected; about 5% of the time the hypothesis will be rejected when true. This kind of "decision" is both misleading and counterproductive because:

- The results of a single study virtually never provide the only basis for a decision in epidemiology.
- Statistical testing is derived from experiments set up to provide a choice between two or more alternatives of action. Observational studies are not experiments formed to give "mathematical answers to structured questions" (100).
- Statistical testing does not account for biases, quality of data, lack of follow-up, and all other inherent problems in observational studies. The p value addresses only the likelihood of falsely rejecting the null hypothesis, and has nothing to do with the chain of reasoning that must precede a conclusion.
- Statistical significance is a function of the number of observations and the variability of the observations. One can make any difference "statistically significant" by making n large enough. It is only when "scientific significance" (or "biological significance") is apparent that statistical significance becomes important (158, 159).
- "Chance as an explanation of an apparent hazard is most effectively eliminated by repetition of the study on another population or over a further period. It can seldom be eliminated on the basis of a test of significance alone, for such tests are only guides to interpretation and never absolute arbiters" (141).

Scientific (or Biological) Significance refers to the quantitative difference in the observations. For example, a mean difference of 1 inch in height between smokers and nonsmokers may be statistically significant, but its quantitative significance is trivial (see also Ref. 52).

Standardization is the weighted averaging of the stratum-specific rates according to an artificial or standard distribution for the factor of interest. The artificial distribution (known as the standard) might be the exposed population (indirect method), comparison population (direct method), or the combined populations. For example, the standardized mortality ratio (SMR) is the ratio of observed deaths in the study population (stratified by

Table 64.12. Circumstances when Type I and Type II Errors Occur[a]

True Biological Significance	Observed Biological Significance	Calculated Statistical Significance ($p < 0.5$)	Agreement of Biological and Statistical Significance	Agreement of Statistical Conclusion and Truth
1 Yes	Yes	No	No	Type II Error
2 Yes	No	No	Yes	Type II Error
3 No	Yes	Yes	Yes	Type I Error
4 No	No	Yes	No	Type I Error

[a]See also Last (160).

age, race, and sex) to the expected number of deaths based on mortality rates in the comparison population stratified in the same manner.

Stratification: To control for possible effects of confounding, the comparison is between samples of exposed and controls separated into subsamples according to specified criteria such as age, race, sex, and smoking status. By stratifying in the analysis of results one compares "like with like." After stratifying the data into selected strata to compare strata-specific rates and thereby reducing the effects of unfair comparisons, a standardized rate can be calculated.

TYPE I: Alpha errors occur when the null hypothesis is rejected even though it is true. When the p value is set at 0.05, the likelihood of this error will be about 5%.

TYPE II: Beta errors occur when the null hypothesis is erroneously accepted. As the criteria for statistical significance is made more stringent (i.e., as the "significant" p value is made smaller) the probability of making a type II error increases. The circumstances when type I and type II errors occur can be visualized as shown in Table 64.12. In the first instance there is a true biologically significant effect but not a statistically significant effect. The reason for this error may be that the sample size is too small.

The fear in the second instance is that one may miss a truly significant difference. The finding of no statistical and no biological significance could be due to a small sample size. The lack of quantitative difference could be due to a flaw in the research design or to a small sample size that did not detect a true effect because of the low precision.

In the third instance sample size is not the problem for there is statistical significance. The error is caused by distorted results such as bias or confounding, or because 1 out of 20 times the probabilities are wrong.

In the fourth instance statistical significance is achieved despite the lack of biological significance. This may occur because of the huge sample size. (See Refs. 28 and 52 for further discussion.)

BIBLIOGRAPHY

1. D. Hunter, *The Diseases of Occupations*, 6th ed., Hodder and Stoughton, London, 1978.
2. D. E. Lilienfeld, *Am. J. Epidemiol.* **107**, 87–90 (1978).

3. G. Rose and D. J. P. Barker, *Br. Med. J.* **2**, 803–804 (1978).
4. A. B. Hill, *Proc. Roy. Soc. Med.* **58**, 295–300 (1965).
5. K. J. Rothman, ed., *Causal Inference*, Epidemiology Resources, Inc., Chestnut Hill, MA, 1988.
6. M. J. Gardner and D. G. Altman, *Br. Med. J.* **292**, 746–750 (1986).
7. E. J. Wynder, I. T. Higgins, R. E. Harris, *J. Clin. Epidemiol.* **43**, 619–621 (1990).
8. D. L. Weed, *Int. J. Epid.* **26**, 1137–1141 (1997).
9. A. R. Feinstein, *Science* **242**, 1257–1263 (1988).
10. R. M. Royall, *Am. J. Epidemiol.* **104**, 463–474 (1976).
11. J. R. Jamison, *S. Afr. Med. J.* **57**, 783–785 (1980).
12. W. Weiss, *J. Occup. Med.* **25**, 290–294 (1983).
13. C. P. Wen and S. P. Tsai, *Scand. J. Work. Environ. Health*, **8** (Suppl. 1), 48–52 (1982).
14. E. S. Gilbert, *Am. J. Epidemiol.* **116**, 177–188 (1982).
15. J. C. McDonald, F. D. K. Liddell, G. W. Gibbs, G. E. Eyssen, and A. C. McDonald, *Br. J. Ind. Med.* **7**, 11–24 (1980).
16. J. D. Wang and O. S. Miettinen, *Scand. J. Work Environ. Health* **8**, 153–158 (1982).
17. J. M. Harrington, *Scand. J. Work. Environ. Health* **10**, 347–352 (1984).
18. M. L. Newhouse and J. C. Wagner, *Br. J. Ind. Med.* **26**, 302–307 (1969).
19. E. S. Johnson, *Br. J. Ind. Med.* **43**, 427–429 (1986).
20. L. N. Kolonel, T. Hirohata, and A. M. Nomura, *Am. J. Epidemiol.* **106**, 476–484 (1977).
21. J. K. McLaughlin, M. S. Dietz, E. S. Mehl, and N. J. Blot, *Am. J. Epidemiol.* **126**, 144–146 (1987).
22. M. Baumgarten, J. Siemiatycki, and G. W. Gibbs, *Am. J. Epidemiol.* **118**, 583–591 (1983).
23. S. L. Shalat, D. C. Christiani, and E. L. Baker, *Scand. J. Work. Environ. Health* **13**, 67–69 (1987).
24. W. F. Stewart, J. A. Tonascia, and G. M. Matanoski, *J. Occup. Med.* **29**, 795–800 (1987).
25. G. G. Bond, K. M. Bodner, W. Sobel, R. J. Shellenberger, and G. H. Flores, *Am. J. Epidemiol.* **128**, 343–351 (1988).
26. K. Steenland and T. Schnorr, *J. Occup. Med.* **30**, 348–353 (1988).
27. M. L. Lerchen and J. M. Samet, *Am. J. Epidemiol.* **123**, 481–489 (1986).
28. A. R. Feinstein, *Clinical Epidemiology*, Saunders, Philadelphia, 1985.
29. P. E. Enterline, *J. Occup. Med.* **18**, 150–156 (1976).
30. Gordis L, *Am. J. Epidemiol.* **128**, 1–9 (1988).
31. *Atlas of Cancer Mortality for U.S. Counties: 1950–1960*, U.S. Department of Health, Education and Welfare, DHEW Publication No. (NIH) 75-780, 1975.
32. W. J. Blot, L. A. Brinton, J. F. Fraumeni, and B. J. Stone, *Science* **198**, 51–53 (1977).
33. O. Wong and G. K. Raabe, *Am. J. Ind. Med.* **15**, 283–310 (1989).
34. C. P. Wen, *Adv. Mod. Env. Toxicol.* **7**, 245–258 (1984).
35. C. Poole, N. A. Dreyer, M. H. Satterfield, L. Levin, and K. J. Rothman, *Environ. Health Perspect.* **101**(suppl 6), 53–62 (1993).
36. J. F. Gamble, E. D. Pearlman, and M. J. Nicolich, *Environ. Health Perspect.* **104**, 642–650 (1996).
37. U.S. EPA. *Air Quality Criteria for Particulate Matter*, EPA-600-AP-95-001C, U.S. Environmental Protection Agency, Washington, DC.

38. J. F. Gamble and R. J. Lewis, *Environ. Health Perspect.* **104**, 838–850 (1996).
39. S. Vedal, *J. Air Waste Manage. Assoc.* **47**, 551–581 (1997).
40. J. G. Watson, *J. Air Waste Manage Assoc.* **47**, 995–1008 (1997).
41. D. W. Dockery, C. A. Pope, X. Xu, J. D. Spengler, J. H. Ware, et al., *NEJM* **329**, 1753–1759 (1993).
42. C. A. Pope, M. J. Thun, M. M. Namboodiri, D. W. Dockery, J. S. Evans, et al. *Am. J. Respir. Crit. Care Med.* **151**, 669–674 (1995).
43. J. F. Gamble, *Environ. Health Perspect.* **106**, 535–549 (1998).
44. S. Richardson, I. Stucker, and D. Hsemon, *Int. J. Epidemiol.* **16**, 111–120 (1987).
45. P. E. Enterline, "Detecting Disease Produced by Occupational Exposure, in L. J. and L. V. Cralley, eds., *Patty's Industrial Hygiene and Toxicology*, 2nd ed., Vol. IIIA, John Wiley & Sons, Inc., New York, 1985, pp. 73–92.
46. G. King, *A Solution to the Ecological Inference Problem: Reconstructing Individual Behavior from Aggregate Data*, Princeton University Press, Princeton, NJ, 1997.
47. K. Steenland, L. Stayner, and A. Griefe, *Am. J. Ind. Med.* **12**, 419–430 (1987).
48. S. P. Tsai and C. P. Wen, *Int. J. Epidemiol.* **15**, 8–21 (1986).
49. H. Checkoway, N. Pearce, and J. M. Dement, *Am. J. Ind. Med.* **15**, 363–373 (1989).
50. H. Checkoway, N. Pearce, and J. M. Dement, *Am. J. Ind. Med.* **15**, 375–394 (1989).
51. F. D. K. Liddell, *J. Clin. Epidemiol.* **41**, 1217–1237 (1988).
52. K. J. Rothman, *Modern Epidemiology*, Little, Brown, Boston, 1986.
53. E. S. Johnson, *Br. J. Ind. Med.* **43**, 427–429 (1986).
54. G. M. H. Swaen and J. M. M. Meijers, *Br. J. Ind. Med.* **45**, 624–629 (1988).
55. J. Wang and O. Miettinen, *Scand. J. Work. Environ. Health* **8**, 153–158 (1982).
56. C. M. J. Bell and D. A. Coleman, *Stat. Med.* **2**, 363–371 (1983).
57. S. Hernberg, *Scand. J. Work. Environ. Health*, **7** (Suppl. 4), 121–126 (1981).
58. G. M. Marsh and P. E. Enterline, *J. Occup. Med.* **21**, 665–670 (1979).
59. D. C. Thomas, J. Siemiatycki, R. Dewar, J. Robins, M. Goldberg, and B. G. Armstrong, *Am. J. Epidemiol.* **122**, 1080–1095 (1985).
60. N. Cherry, G. Burgess, R. McNamee, S. Turner, and C. McDonald, *Appl. Occup. Environ. Hyg.* **10**, 1042–1045 (1995).
61. G. L. Burgess, J. Turner, J. C. McDonald, and N. M. Cherry, *Ann. Occup. Hyg.* **41**(suppl 1) 403–407 (1997).
62. N. Cherry, N. Burgess, S. Turner, and J. C. McDonald, *Ann. Occup. Hyg.* **41**(suppl. 1), 408–411 (1997).
63. J. C. McDonald, N. M. Cherry, R. McNamee, G. Burgess, and S. Turner, *Scand. J. Work Environ. Health* **21**, 63–65 (1995).
64. J. C. McDonald, G. L. Burgess, S. Turner, et al., *Ann. Occup. Hyg.* **41**(suppl. 1), 412–414 (1997).
65. N. M. Cherry, G. L. Burgess, S. Turner, and J. C. McDonald, *Occup. Environ. Med.* **55**, 779–785 (1998).
66. H. Checkoway, N. J. Heyer, P. A. Demers, and N. E. Breslow, *Br. J. Ind. Med.* **50**, 586–597, (1993).

67. H. Checkoway, J. N. Heyer, P. A. Demers, and G. W. Gibbs, *Occup. Environ. Med.* **53**, 645–647 (1996).

68. H. Checkoway, J. N. Heyer, N. S. Seixas, E. A. E. Welp, P. A. Demers, J. M. Hughes, and H. Weill, *Am. J. Epidemiol.* **145**, 680–688 (1997).

69. G. W. Gibbs and D. R. Christensen, *The Asbestos Exposure of Workers in the Manville Diatomaceous Earth Plant*, Final Report to the International Diatomite Producers Association, IDPA, Lompoc, CA, 1994.

70. M. Merliss, *Am. J. Epidemiol.* **148**, 307–308 (1998).

71. O. Axelson, *Scand. J. Work Env. Health* **4**, 85–88 (1978).

72. P. Hessel, J. F. Gamble, J. B. L. Gee, G. W. Gibbs, F. H. Y. Green, W. K. C. Morgan, and B. T. Mossman, "Silica, Silicosis, and Lung Cancer: A Critical Review," submitted for publication to *J. Occup. Env. Hlth.*, 1998.

73. N. Pearce, H. Checkoway, and J. Dement, *Am. J. Ind. Med.* **15**, 395–402, 403–416 (1989).

74. S. Greenland and H. Morgenstern, *J. Clin. Epidemiol.* **41**, 715–716 (1988).

75. L. Gordis, *Am. J. Epidemiol.* **115**, 1–5 (1982).

76. J. K. McLaughlin, W. J. Blot, E. S. Mehl, et al., *Am. J. Epidemiol.* **121**, 131–139 (1985).

77. J. K. McLaughlin, W. J. Blot, E. S. Mehl, and J. S. Mandel, *Am. J. Epidemiol.* **122**, 485–494 (1985).

78. A. H. Smith, N. E. Pearce, and P. W. Callas, *Am. J. Epidemiol.* **17**, 298–306 (1988).

79. M. S. Linet and R. Brookmeyer, *Am. J. Epidemiol.* **125**, 1–11 (1987).

80. P. A. Hessel and G. K. Sluis-Cremer, *Int. Arch. Occup. Environ. Health* **59**, 97–105 (1987).

81. P. L. Sackett, *J. Chronic Dis.* **32**, 51–63 (1979).

82. A. J. Fox and G. C. White, *Lancet* **1**, 1009–1010 (1976).

83. C. K. Wells and A. R. Feinstein, *Am. J. Epidemiol.* **128**, 1016–1026 (1988).

84. J. A. Dick, W. K. C. Morgan, D. F. C. Muir, R. B. Reger, and N. Sargent, *Chest* **102**, 251–260 (1992).

85. H. W. Glindmeyer, J. E. Diem, R. N. Jones, et al., *Am. Rev. Resp. Dis.* **125**, 544–548 (1982).

86. M. D. Attfield and T. K. Hodous, *Am. Rev. Respir. Dis.* **145**, 605–609 (1992).

87. M. D. Attfield and K. Morring, *Ind. Hyg. Assoc. J* **53**, 486–492 (1992).

88. *Smoking and Health*, Report of the Advisory Committee to the Surgeon General of the Public Health Service, U.S. Department of Health, Education, and Welfare/PHS, USGPO, Washington, DC, 1964.

89. M. Susser, *Am. J. Epidemiol.* **133**, 635–648 (1991).

90. M. Susser, *Causal Thinking in the Health Sciences: Concepts and Strategies in Epidemiology*, Oxford University Press, New York, 1973.

91. A. J. Lieberman and S. C. Kwon, *Facts Versus Fears: a Review of the Greatest Unfounded Health Scares of Recent Times*, 3rd ed., American Council on Science and Health, New York, 1998.

92. International Agency for Research on Cancer, *Monograph on the Evaluation of Carcinogenic Risks to Humans*, Vol. 68, *Silica, Some Silicates, Coal dust, and Para-aramid Fibrils*, IARC Press, Lyon, France, 1997.

93. R. Peto and R. Doll, *Br. Med. J.*, 259 (July 23, 1977).

94. J. M. Lang, K. J. Rothman, and C. I. Cann, *Epidemiology* **9**, 7–8 (1998).

95. M. L. Bleecker, K. I. Bolla, J. Agnew, B. S. Schwartz, and D. P. Ford, *Am. J. Ind. Med.* **19**, 715–728 (1991).

96. P. Orbaek, J. Risberg, I. Rosen, B. Haeager-Aronsen, S. Hagstadius, U. Horsberg, G. Regnell, et al., *Scand. J. Environ. Health* **11** (Suppl. 2), 1–28 (1985).

97. W. K. Anger, O. F. Sizemore, S. J. Grossmann, J. A. Glasser, and C. A. Kovera, "Human Neurobehavioral Study Methods: Effects of Subject Variables on Research Results," API manuscript #4648, The American Petroleum Institute, Washington, DC, 1996.

98. K. J. Rothman, *Epidemiology* **1**, 43–46 (1990).

99. J. R. Thompson, *Am. J. Epidemiol.* **147**, 801–806 (1998).

100. M. Susser, *Am. J. Epidemiol.* **105**, 1–15 (1977).

101. M. Susser, *Epidemiol. Prev.* **21**, 160–168 (1997).

102. W. K. Anger, O. J. Sizemore, S. J. Grossmann, J. A. Glasser, R. Letz, and R. Bowler, *Environ. Res.* **73**, 18–43 (1997).

103. A. Gade, E. Mortensen, and Brijm, *Acta Neurol. Scand.* **77**, 293–306 (1988).

104. N. Cherry, H. Hutchins, T. Pace, and H. A. Waldron, *Br. J. Ind. Med.* **42**, 291–300 (1985).

105. C. McDonald, *Appl. Occup. Env. Hyg.* **10**, 1056–1063 (1995).

106. J. Gillam, J. Dement, R. Lemen, J. Wagoner, V. Archer, and H. Blejer, *Ann. NY Acad. Sci.* **271**, 336–344 (1976).

107. J. C. McDonald, G. W. Gibbs, F. D. K. Liddell, and A. D. McDonald, *Am. Rev. Resp. Dis.* **118**, 271 277 (1978).

108. D. Brown, S. Kaplan, R. D. Zumwalde, M. Kaplowitz, and V. E. Archer, "Retrospective Cohort Mortality Study of Underground Gold Miner Workers, in D. Goldsmith Winn, and C. Shy, eds., *Silica, Silicosis, and Lung Cancer*, Praeger, New York, 1986, pp. 311–336.

109. T. Partanen, E. Pukkala, H. Vainio, K. Kurppa, and H. Koskinen, *J. Occup. Med.* **36**, 616–622 (1994).

110. H. E. Amandus, R. M. Castellan, C. Shy, E. F. Heineman, and A. Blair, *Am. J. Ind. Med.* **22**, 147–153 (1992).

111. D. H. de Klerk and A. W. Musk, *Occup. Environ. Med.* **55**, 243–248 (1998).

112. J. F. Gamble, "Low-level Hydrocarbon Solvent Exposure and Neurobehavioral Effects," accepted for publication in *Occupational Medicine*, 1999.

113. *The Health Consequences of Smoking: Cancer and Chronic Lung Disease in the Workplace, a report of the Surgeon General*, U.S. Department of Health and Human Services. PHS/Office on Smoking and Health, GPO, Rockville, MD, 1985.

114. K. Steenland, J. Beaumont, and W. Halperin, *Scand. J. Work Environ. Health* **10**, 143–149 (1984).

115. *Reducing the Health Consequences of Smoking: 25 Years of Progress*, a report of the Surgeon General, U.S. Department of Health and Human Services, PHS, CDC, Center for Chronic Disease Prevention and Health Promotion, Office of Smoking and Health, GPO, Rockville, MD, (1989).

116. D. R. Shopland, and H. G. Egre, *JNCI* **83**, 1142–1148 (1991).

117. J. K. McLaughlin, C. Jing-Qiong, et al., *Br. J. Ind. Med.* **49**, 167–171 (1992).

118. P. A. Hessel, G. K. Sluis-Cremer, and E. Hnizdo, *Am. J. Ind. Med.* **10**, 57–62 (1986).

119. P. A. Hessel, G. K. Sluis-Cremer, and E. Hnizdo, *Br. J. Ind. Med.* **47**, 4–9 (1990).

120. P. G. Reid and G. K. Sluis-Cremer, *Occup. Environ. Med.* **53**, 11–16 (1996).

121. D. A. Andjelkovich, C. M. Shy, M. H. Brown, D. A. B. Janszen, R. F. Levine, and R. B. Richardson, *J. Occup. Med.* **36**, 1301–1309 (1994).

122. P. Carta, P. L. Cocco, and D. Casula, *Br. J. Ind. Med.* **48**, 122–129 (1991).

123. M. S. Tockman, N. R. Anthonisen, E. C. Wright, and M. G. Donithan, *Ann. Internal. Med.* **106**, 512–518 (1987).

124. D. M. Skillrud, K. P. Offord, and R. D. Miller RD, *Ann. Internal. Med.* **105**, 503–507 (1986).

125. F. J. Wiles, E. Baskind, P. A. Hessel, B. Bezuidenhout, and E. Hnizdo, *Int. Arch. Occup. Env. Health* **63**, 387–391 (1992).

126. C. Bucca, W. Arossa, M. Bugiana, G. Rolla, and M. Cacciabne, *Med. Lav.* **76**, 466–470 (1985).

127. L. M. Irwig and P. Rocks, *Am. Rev. Respir. Dis.* **117**, 429–435 (1978).

128. E. Garshick, M. B. Schenker, A. Munoz, M. Segal, T. J. Smith, S. R. Woskie, K. Hammond, and F. E. Speizer, *Am. Rev. Resp. Dis.* **137**, 820–825 (1988).

129. K. S. Crump, T. Lambert and C. Chen, "Assessment of Risk from Exposure to Diesel Engine Emissions," prepared for U.S. Environmental Protection Agency 68-02-4601, Work Assignment #182, July 1991.

130. T. W. Armstrong, E. D. Pearlman, A. R. Schnatter, S. M. Bowes, III, N. Murray, and M. J. Nicolich, *Am. Ind. Hyg. Assoc. J.* **57**, 333–343 (1996).

131. M. Dosemeci, J-Q. Chen, F. Hearl, R-G. Chen, M. McCawley, et al., *Am. J. Ind. Med.* **24**, 55–66 (1993).

132. C. H. Rice, R. L. Harris, H. Checkoway, and M. J. Symons, "Dose–Response Relationships for Silicosis from a Case-Control Study of North Carolina Dusty Trades Workers," in D. Goldsmith, D. Winn, and C. Shy, eds., *Silica, Silicosis, and Lung Cancer*, Praeger, New York, 1986, pp. 311–336.

133. N. S. Seixas, N. J. Heyer, E. A. E. Welp, and H. Checkoway, *Ann. Occup. Hyg.* **41**, 591–604 (1997).

134. A. G. Heppleston, *Am. J. Ind. Med.* **1**, 285–294 (1985).

135. P. Loosereewanich, A. C. Ritchie, B. Armstrong, R. Begin, D. C. F. Muir, and A. Dufresne, *Appl. Occup. Environ. Hyg.* **10**, 1104–1106 (1995).

136. P. E. Morrow, *Toxicol. Appl. Pharmacol.* **113**, 1–12 (1992).

137. J. N. Pritchard, *Exp. Pathol.* **37**, 39–42 (1989).

138. M. Susser, "Falsification, Verification and Casual Inference in Epidemiology: Reconsiderations in the Light of Sir Karl Popper's Philosophy," in Ref. 5.

139. S. F. Lanes and C. Poole, *J. Occup. Med.* **26**, 571–574 (1984).

140. P. Cole, *J. Chronic Dis.* **32**, 15–28 (1979).

141. R. Doll, *Ann. Occup. Hyg.* **28**, 291–305 (1984).

142. C. B. Begg and J. A. Berlin, *J. Roy. Stat. Soc., Ser. A* **151**, 419–463 (1988).

143. J. A. Freiman, T. C. Chalmers, H. Smith, and R. R. Kuebler, *N. Engl. J. Med.* **299**, 690–694 (1978).

144. B. Williams, "How Bad Can "Good" Data Really Be?," presented at American Statistical Association Meeting, Atlanta, GA, August, 1975.

145. A. J. McMichael, *J. Occup. Med.* **18**, 165–168 (1976).

146. L. M. Carpenter, *Br. J. Ind. Med.* **44**, 289–291 (1987).

147. W. Weiss, *J. Occup. Med.* **31**, 102–105 (1989).

148. P. E. Enterline, *Am. Rev. Resp. Dis.* **113**, 175–180 (1976).

149. S. J. Gould, *Science* **200**, 503–509 (1978).
150. W. G. B. Graham, R. V. O'Grady, and B. Dubuc, *Am. Rev. Resp. Dis.* **123**, 25–28 (1981).
151. G. P. Theriault, J. M. Peters, and W. M. Johnson, *Arch. Environ. Health*, **28**, 23–27 (1974).
152. K. Raphael, *Int. J. Epidemiol.* **16**, 167–170 (1987).
153. K. T. Copeland, H. Checkoway, and A. J. McMichael, *Am. J. Epidemiol.* **105**, 488–495 (1977).
154. E. R. Greenberg, B. Rosner, C. Hennekens, R. Rinsky, and T. Colton, *Am. J. Epidemiol.* **121**, 301–308 (1985).
155. J. E. Vena, H. A. Sultz, G. L. Carlo, R. C. Fiedler, and R. E. Barnes, *J. Occup. Med.* **29**, 256–261 (1988).
156. E. S. Johnson, *J. Occup. Med.* **30**, 60–62 (1988).
157. F. Hatch, *Arch. Env. Health* **16**, 571–578 (1968).
158. C. Poole, *Am. J. Public Health* **77**, 195–199 (1987).
159. D. R. Cox, *Scand. J. Stat.* **4**, 49–70 (1977).
160. J. M. Last, *A Dictionary of Epidemiology*, Oxford University Press, New York, 1983.

CHAPTER SIXTY-FIVE

Indoor Air Quality in Nonindustrial Occupational Environments

Philip R. Morey, Ph.D., CIH, Elliott Horner, Ph.D., Barbara L. Epstien, MPH, CIH, Anthony G. Worthan, MPH and Marilyn S. Black, Ph.D.

1 HISTORY

Indoor air in occupied buildings is always more polluted from human-sourced contaminants than the air outside the building. This is true for modern buildings constructed during the past decade as well as for primitive shelters erected centuries or millennia ago (1). It follows that a continuous source of outdoor air is required to prevent the degradation of indoor air from contaminants arising from people and their activities. This was realized historically by the inventors of roof vents for exhaust of fire smoke and by early designers of openable windows for introduction of make-up (outdoor) air.

Historically, the concept of indoor air quality (IAQ) has included view points that outdoor ventilation air is required both to prevent adverse health effects and to provide for comfort of occupants. Thus, over two centuries ago, Benjamin Franklin wrote that " . . . I am persuaded that no common air from without is so unwholesome as the air within a closed room that has been often breathed and not changed" (2). He further stated that outdoor air and "cool air does good to persons in the smallpox and other fevers. It is hoped, that in another century or two we may find out that it is not bad even for people in health."

By the mid-nineteenth century it was realized that tuberculosis and other airborne contagious diseases were more readily contracted in crowded places with deficient ventilation.

Patty's Industrial Hygiene, Fifth Edition, Volume 4. Edited by Robert L. Harris.
ISBN 0-471-29749-6

Accordingly, ventilation codes by the late nineteenth century recommended the provision of large amounts of outdoor air to lower the risk of disease from airborne infectious agents. By the beginning of the twentieth century the pioneering work of Pettenkofer and Flugge had shown (1) that the staleness of room air was associated with a build-up of carbon dioxide (CO_2). Carbon dioxide itself, however was not recognized as an air contaminant but rather as a surrogate indicator of human body emissions.

Studies in the 1930s (3) showed that in order to prevent malodor complaints among most building occupants, a minimum of 10 to 30 cubic feet per minute (cfm); or 5 to 15 liters per second (L/s) of ventilation by outdoor air was required. The dual concern over both the health and comfort effects of indoor air is summarized in the definition of acceptable IAQ found in the American Society of Heating, Refrigerating, and Air-conditioning Engineers (ASHRAE) Standard 62-1989 (Ventilation for Acceptable Indoor Air Quality) namely (4):

> Air in which there are no known contaminants at harmful concentrations as determined by cognizant authorities and with which a substantial majority (80% or more) of the people exposed do not express dissatisfaction.

1.1 Building Related Symptoms and Building Related Illness

In a paper entitled “The Sick Building Syndrome” (SBS) Stolwijk (5) described a constellation of nonspecific complaints (e.g., mucous membrane irritation, headache) that occurred at higher prevalence rates in some buildings. Although these complaints seem to be present at some level in all buildings, their prevalence appears highest in a subset of “sick” buildings. In the decade that followed Stolwijk’s 1984 publication, most work-related occupant discomfort and annoyance complaints were referred to as SBS. Documentable clinical disease associated with building occupancy was defined as building related illness (BRI).

Menzies and Bourbeau (6) recommended the term “Nonspecific Building Related Illness” to be used to refer to the nonspecific symptoms or complaints commonly present in nonindustrial office buildings. The term “Specific Building Related Illness” was suggested to refer to clinically demonstrable diseases that can be associated with building occupancy. Use of the terms “nonspecific building related illness” or “building related symptoms” (BRS) has been suggested (7) to be preferable to SBS because the former focus attention on people rather than the building as the repository of symptom complaints. In this review the term BRS is used for simplicity (7) to refer both to the nonspecific building related illness proposed by Menzies and Bourbeau (6) and the historic concept of the SBS.

The terms BRS and BRI are compatible with the artificial but useful terms “problem” or “healthy” buildings. A healthy building is one in which contaminants and factors associated with BRS and BRI are minimal. By contrast, in a problem building, defects occur which are associated with the manifestation of BRS in a large number of occupants and possibly with the presence of BRI in one or more occupants.

The prevention of both BRS and BRI associated with occupancy of nonindustrial buildings has been the subject of much research during the past decade. The occurrence of BRIs such as a lung cancer from radon and environmental tobacco smoke (ETS) exposure, asthma from house dust mite allergens, and allergic and nonallergic respiratory diseases from microbial agents is a major public health concern (8). In like manner the disruptive consequences of BRS in workplaces is of considerable economic concern (9). The historic concern for minimizing both the occurrence of BRS and BRI is seen in the suggestion by Fanger (10) that the perception of unacceptable odors and sensory irritation may provide an early warning of the presence of some air contaminants that affect health.

1.2 Buildings, Pollutants, and Definitions

A number of factors collectively make IAQ a more important issue now compared to 50 or 60 years ago. Construction practices today are vastly different than those employed in the 1930s or 1940s. Windows in modern buildings generally do not open, whereas in the 1940s natural ventilation (open windows) was common. In modern buildings, reliance is placed on the heating ventilation and air-conditioning (HVAC) system to transport outdoor air to the breathing zone. Insufficient quantities of outdoor air or incompletely conditioned air are often transported to the occupant breathing zone as a result of energy conservation measures that reduce HVAC system operational costs during the cooling and heating seasons. Poor maintenance and cleaning of today's complex building systems are other reasons cited for increased attention to IAQ issues.

Construction materials have markedly changed over the past 50 years. Stone, wood, and other "natural" construction/finishing materials have largely been replaced by synthetic, pressed wood, and amorphous cellulose products. Some of these are less resistant to microbial growth that the materials used earlier. The ceiling tiles, wall and floor coverings, and even desks and chairs in modern buildings are primarily synthetic in nature. Volatile organic compounds (VOCs) are emitted into the indoor air from modern finishing and construction materials. Volatile agents from cleaning and graphics materials, office machines, pesticides, hydraulic elevator fluid, and personal care products are additional sources of indoor air pollution.

Table 65.1 provides a list of some representative kinds of air contaminants that have been associated with BRS and BRI.

BRS in nonindustrial workplaces have been associated with factors such as overcrowding, the use of fleecy finishing materials, photocopier usage, and ventilation rates less than 10 L/s or 20 cfm per occupant (11). Generally, there is a higher prevalence of BRS in mechanically ventilated buildings. In other words the prevalence of BRS has been found to be less in naturally ventilated buildings.

Objective physiological or clinical abnormalities that may cause BRS are unknown (9). Most investigators consider BRS to be multifactorial in origin (12). Physical factors (e.g., presence of dusts), chemical agents (e.g., VOCs), microbial agents (e.g., endotoxins and β1,3 glucans), and psychosocial factors have all been considered in the etiology of BRS.

BRI occurs when occupant health problems are clearly recognizable as disease (with overt symptoms or abnormal physical signs) upon medical examination and are associated with indoor environmental exposure. Perhaps the most well-known case of BRI was the outbreak of pneumonia (later known as Legionnaire's disease) that occurred in a Philadelphia hotel in 1976 as a result of indoor exposure of guests to the bacterium *Legionella pneumophila*. This case of BRI is still remembered by the public because of the 29 fatalities

Table 65.1. Representative Kinds of Indoor Air Pollutants and Their Sources

Pollutant	Sources
2-Butoxyethanol	Glass cleaners
4-Phenylcyclohexene	New carpet with styrene–butadiene backing
C_5–C_{16} aliphatic hydrocarbons	Adhesives, waxes, cleaning chemicals
C_{15}–C_{18} aliphatic hydrocarbons	Hydraulic elevator fluid
Chlorpyrifos	Pesticide used to control fleas and cockroaches
Dust/particles	Copy machines; unfiltered air
Formaldehyde	Composite wood products, textiles
Fungi	Moist or damp finishing materials or humid air
Legionella	Hot water service systems and cooling towers
Man-made fibers	Some ceiling tiles
Nicotine	Environmental tobacco smoke
Sulfur dioxide	Diesel fuel combustion

that occurred during the disease outbreak. Examples of BRIs and other air contaminants that are involved in disease etiology include:

- Cancer caused by gaseous and particulate components of ETS, some VOCs, and radon.
- Dermatitis caused by fibers from manmade insulation and some irritant molds.
- Hypersensitivity pneumonitis, humidifier fever, allergic rhinitis, asthma, and nonallergic respiratory disease caused by microorganisms or their toxins.

It is important to realize that the diagnosis of a BRI is based on medical examination of one or more occupants confirming the presence of a specific disease plus an environmental assessment showing the likelihood of a contaminant exposure that would cause the disease. Thus, the isolation of *Legionella pneumophila* from a pneumonia patient, and the recovery of the same serotype of *L. pneumophila* from the water system in a building where that patient worked, or otherwise spent time is the basis for a diagnosis of building-related Legionnaire's disease. The diagnosis of BRS (SBS) on the other hand, is a group diagnosis based on review of the nonspecific discomfort symptoms. Since BRS remit upon exiting the building, medical examination of individuals at the physician's office is usually of little use in making BRS diagnosis. Similarly, diagnosing the building as the cause without an environmental assessment is not reliable. The same environmental agents may cause both BRS and BRI. For example, exposure to VOCs, ETS, or combustion products may cause BRS. Chronic exposure to the same air contaminants may also elicit BRI. The following definitions are used in this chapter:

Building related symptoms (BRS) is defined as the constellation of nonspecific work-related symptoms such as irritation of the mucous membranes of the eyes, nose, and throat plus headache, fatigue, skin irritation, and difficulty concentrating. BRS is used in a manner similar to Menzies and Bourbeau's (6) nonspecific building related illness (1997) and BRS replaces the term SBS. Occupants affected by BRS generally report relief when they breathe outdoor air even for a short time. Because BRS are nonspecific and can be potentially associated with many different contaminant sources, an exact etiology is not readily apparent even in buildings that have been thoroughly studied.

Building related illness (BRI) is defined as clinically demonstrable disease (with overt symptoms or physical signs) that can be associated with building occupancy (similar to Menzies and Bourbeau's (6) specific BRI).

Healthy buildings are characterized by minimal occurrence of contaminants and factors associated with BRS and BRI.

Problem buildings are characterized by defects associated with the occurrence of BRS in large numbers of occupants and possible with BRI in one or more persons. Problem buildings are also associated with impaired productivity because of poor IAQ (13).

2 INDOOR AIR POLLUTANTS

In the following discussion pollutants that are important in IAQ evaluations are reviewed. The reader is also referred to the proceedings of international conferences on indoor air quality and climate (14–16), where numerous papers are presented on each type of pollutant covered in this section.

2.1 Microbials

Microbial contaminants in indoor environments include viruses, bacteria, fungi, protista, and cellular or toxic components such as endotoxins and mycotoxins. It should be understood that surfaces of construction and finishing materials in buildings are characterized by dusts and deposits which almost always contain culturable and nonculturable fungi and bacteria. The indoor air in most nonproblem buildings normally contains fungi and bacteria similar to those found in the atmosphere outside the building. Thus, background concentrations of various kinds of microorganisms are normally present in indoor environments. Microbial contaminants and some of the diseases they can cause (Table 65.2) are often

Table 65.2. Diseases Caused by Microbial Contaminants in Buildings

Disease	Symptoms	Agent and Sources	Refs.
Legionnaire's disease	Pneumonia	Hot water service system; cooling towers	17
Humidifier fever	Acute influenza-like symptoms	Endotoxin in humidifier water	18
Hypersensitivity pneumonitis	Acute fever and cough, fibrosis of lung	Fungi and bacteria growing in the HVAC system	19
Asthma	Constriction of the airways	Fungi, bacteria, house dust mites	20
Pulmonary hemosiderosis	Bleeding in lungs	*Stachybotrys* and other fungi	21, 22

associated with moisture and dampness in building infrastructure and poor maintenance of plumbing and HVAC system components, or inadequate humidity control.

2.1.1 Fungi

Most of the fungal or mold spores commonly found indoors in nonproblem buildings originate from outdoor sources. The air outdoors during clear weather is dominated by common fungal spores such as *Cladosporium* or *Alternaria* species, which grow on the leaves of plants [phylloplane (i.e. leaf surface) fungi]. Other fungi such as *Penicillium* and *Aspergillus* species occur in topsoil and decay litter. When air enters a building such as through the HVAC system outdoor air inlet, through an open door, or by infiltration through cracks in the building envelope, phylloplane spores are carried into the building. Since the source of spores is normally the outdoor air the mix of fungal species found indoors and outdoors is similar in buildings without mold-growth problems.

In buildings with highly efficient HVAC system filters, the total concentration of airborne fungi indoors is generally much lower (exception, during winter snow cover) than that found outdoors. Tracking of soil indoors often leads to indoor accumulation of *Penicillium* and *Aspergillus* species, especially in poorly maintained carpet. While a diversity of fungi are normally present in the air in healthy buildings, the occurrence of only one or two dominating kinds of fungi indoors (especially if the same kind is not present outdoors at the time of sampling) is an indication of a moisture problem in the building infrastructure or the HVAC system.

The growth of fungi in buildings is primarily dependent on the availability of moisture in the capillary spaces and on the presence of moisture on the surfaces of construction and finishing materials. Since fungi grow on surfaces and in materials the relative humidity in room air is less important than the available moisture in the substrate. Water activity (a_w) which is the vapor pressure of water in the substrate divided by the vapor pressure of pure water at the same temperature and pressure (23, 24) is used as a measure of moisture that is available for fungal growth. The a_w of a material is important in governing the kinds of fungi that grow on or in the substrate. At an a_w of 0.65 virtually no fungus of any significance will grow on even the most highly biodegradable building material (24, 25). Damp materials (a_w 0.65 to 0.85) subject to biodeterioration can support the growth of xerophilic (dry-loving) fungi such as *Eurotium* spp. (*Aspergillus glaucus* group) or *Wallemia sebi*. Construction materials that are chronically wet ($a_w > 0.9$) are dominated by hydrophilic fungi such as *Ulocladium, Stachybotrys*, and *Fusarium* spp.

There are many areas in buildings where moisture problems can lead to fungal growth. These include:

- *The building envelope*. In cold climates, fungi may grow on the interior (occupied) surface of poorly insulated envelope walls because moisture in room air condenses on the cold wall surface. In air-conditioned buildings in hot, humid climates, fungi may grow on the inner surfaces of the envelope wall because moisture in infiltrating humid air from outdoors condenses on cool (air-conditioned) surfaces in contact with the interior conditioned space.
- *Porous materials in damp locations*. Porous materials collect dirt, and molds grow in damp or moist dirt; thus porous materials must be protected from moisture. Fungal

growth may occur on damp HVAC system insulation especially in locations downstream of operating cooling coils when the airstream surface remains damp or wet. Insulation that is dirty (poor HVAC system filtration) is most susceptible to fungal growth because dirt is hydrophilic (attracts moisture). In like manner, carpet that is placed in building locations that are damp or chronically flooded is susceptible to fungal growth after dirt has accumulated in the carpet.

- *Gypsum board.* Signature flood molds such as *Stachybotrys* and *Chaetomium* grow on amorphous cellulose (paper) covered gypsum wall board that is chronically wet or that is wetted and not dried promptly. Other molds such as *Aspergillus versicolor* may grow on other areas of the same wallboard that are damp due to capillary movement of moisture.

In Finland approximately 55% of buildings including residences are moisture damaged. Growth of fungi has been estimated to occur on interior surfaces or in construction materials in approximately one-fifth of the moisture damaged buildings (26). BRS such as irritation of the eyes, respiratory tract, and skin as well as BRIs such as asthma, hypersensitivity pneumonitis, and organic dust toxicity syndrome have been reported in some of the water damaged Finnish buildings.

Changes in available moisture in dusts and on surfaces of materials affects fungal growth and the production of fungal spores. When house dust was maintained for 25 days at a relative humidity of 84 to 86% (a_w = 0.84 to 0.86), the concentration of fungal spores (mainly *Aspergillus*, *Eurotium*, and *Penicillium* spp.) increased 45-fold (27). In laboratory experiments when the relative humidity was maintained at 95% there was minimal spore release from air stream surfaces of ductwork colonized by *Penicillium chrysogenum* (28). Fungal spore emissions from the same colonized surface increased when the relative humidity was suddenly lowered to 64%.

A considerable amount of current research involves fungal materials such as β1,3 glucan and ergosterol and secondary metabolites of fungi such as mycotoxins (29–31). β1,3 glucans are polyglucose molecules found in the cell walls of fungi and some bacteria (32). β1,3 glucans are thus present in fungal components (spores, spore and hyphal fragments) in settled dusts (33, 34). More β1,3 glucan would likely be present in dusts in those buildings affected by water damage and fungal growth. Inhalation of β1,3 glucans is thought to be associated with the occurrence of some BRS such as fatigue, nasal and throat irritation, and headache (32).

Ergosterol is a sterol present in the cell membrane of filamentous fungi and is used as an indicator of fungal biomass in air and dust samples (35). A correlation occurs between the ergosterol level and the total concentration of culturable fungi in settled dust samples (34). Ergosterol content of biocontaminated gypsum board and wood chip insulation correlated well with the level of culturable filamentous fungi in these materials (36). Measurement of ergosterol and β1,3 glucan levels in air or dust samples provides a chemical means of quantifying fungal biomass without the necessity of using culture or direct microscope analytical methods. However, these assays do not provide information on what fungal species are present.

Mycotoxins are secondary metabolites produced by fungi such as *Penicillium*, *Aspergillus*, *Fusarium*, and *Stachybotrys* species that can be toxic to animals or humans (2).

These metabolites are found not only in spores but also in the mycelium as well as in substrates on which toxigenic fungi grow. Aerosolization of particulates from moldy building materials provides a route of potential exposure to mycotoxins. Because mycotoxins are chemicals, the presence of mycotoxins in dusts from moldy materials is unrelated to the culturability of fungi or to whether or not the cell has been killed. Mycotoxins have been detected in laboratory experiments in moldy construction materials (37, 38) as well as in moldy construction materials in buildings where BRS or BRIs have occurred (39, 40).

Fungi and bacteria growing in building materials can produce a wide range of volatile compounds known as microbial volatile organic compounds (MVOCs). MVOCs produced by fungi include alcohols and ketones such as ethanol and acetone, respectively which can also be present from nonmicrobial sources in construction materials as well as in common cleaning agents. Some MVOCs such as 1-octen-3-ol and 2-octen-1-ol are uniquely produced by fungi (41, 42) and are almost never emitted from construction and finishing materials unaffected by biodeterioration. These chemicals often are useful indicators of microbial growth.

Higher levels of MVOCs have been found in indoor air in Swedish buildings with microbial problems than in control buildings or in outdoor air (43). MVOCs can diffuse through construction materials such as plastic sheeting, vapor barriers, and wallpaper (44). Biodeterioration of water based paint has been associated with emission of MVOCs such as 1-octen-3-ol, 2-octen-1-ol, and 3-methyl furan (45). MVOCs are emitted by fungi growing on HVAC system porous insulation (46).

MVOCs adsorbed on and emitted from settled dust in a room have been used as an indicator of past fungal growth on surfaces in the room (47). It is also known that different MVOCs are emitted by fungi at various stages in the growth cycle (48). In general, the presence of MVOCs in indoor air over and above the levels and kinds of MVOCs found in control buildings and the outdoor air indicates moisture in building infrastructure and consequential growth of microorganisms (49, 50).

2.1.2 Bacteria

Lukewarm water (30–40°C) in cooling towers, evaporative condensers, and potable water service systems provides niches for the growth of *Legionella*, which are gram-negative bacteria. At present there are approximately 35 species known in the genus Legionella (51). At least 17 species have been associated with human illness (51). Some *Legionella* species can be distinguished into serogroups, for example *L. pneumophilia* serogroups 1,4, and 6. *Legionella pneumophila* is the species that caused the outbreak of Legionnaire's disease in Philadelphia in 1976. Legionnaire's disease is an infection (the lung of a susceptible person is the target organ) that results in pneumonia, which may be fatal. *Legionella* may cause a milder form of illness known as Pontiac Fever, which is nonpneumonic and similar to a severe flu. Legionnaire's disease and Pontiac Fever differ in the clinical reaction, incubation time, and attack rate (51, 52). Legionnaire's disease is a pneumonia with a 2–10 day incubation, and only 5% of exposed subjects are affected. Further, there are specific risk factors for Legionnaire's disease, such as immune deficiencies, underlying illnesses, diabetes mellitus, smoking, and the elderly and young children are at greater risk. Pontiac fever is nonpneumonic, with symptoms that resemble influenza and start about

36 hours after exposure and affect over 90% of those exposed. The reason for the differing health effects of Legionella exposure is unknown.

Legionella pneumophila is the species that most frequently causes legionellosis. Several other species are also important. Further, there are several serotypes of *L. pneumophila* which are distinguished by their reactivity to diagnostic antibodies (52, 53). This means that cultures from infected subjects and *Legionella* isolates from the environmental source will match, and sources that do not match are not the infective source. So, it is important to determine the species and the serotype of both clinical and environmental isolates when attempting to identify the environmental source of an infection or outbreak.

Several documents provide a review of sources of *Legionella* in cooling towers, hot water systems, and other water reservoirs (17, 51–55). Protocols for collection of *Legionella* from these sources are well known (35, 51). Analysis of water samples for culturable *Legionella* is currently the preferred method used in field studies. Polymerase chain reaction (PCR) methods wherein a predetermined sequence of *Legionella* nucleic acid is isolated, amplified, and made detectable by addition of a marker may become the analytical method of choice in the future (31, 51). The advantage of PCR is that it is quick (one day). A disadvantage is that PCR can not distinguish between dead and living *Legionella*, but only living *Legionella* is infectious.

Gram-negative bacteria such as species of *Pseudomonas*, *Flavobacterium*, and *Blastobacter* grow in stagnant water (e.g., humidifier sumps) or on wet surfaces of HVAC system cooling coils or drain pans (56–58). A biofilm (slime) on a wet cooling coil surface or on a wet surface in a humidifier sump is a certain indication of microbial growth that likely includes these gram-negative bacteria. Humidifier fever, a disease characterized by influenza-like symptoms that remit after cessation of exposure, has in some cases been associated with endotoxin (from the toxic outer membrane of gram-negative bacteria) present in water droplets emitted from some humidifiers and water spray systems with a recirculation system or sump. These systems are prone to microbial contamination especially if maintenance is poor (59–62).

Endotoxins have been causally associated with disease (for example, byssinosis from cotton dust; organic dust toxicity syndrome in silos and barns) in industrial and agricultural settings. The endotoxin in cotton dust is derived from gram-negative bacteria that grew on the cotton plant prior to harvest (63). Gram-negative bacteria growing on moist silage and hay are sources of endotoxin in some agricultural settings. Metalworking fluids and the water in humidifiers and air washers are sources of endotoxin in industrial environments. In general, airborne endotoxin concentrations that are two or three orders of magnitude greater than background (outdoor) levels are considered significant exposures (64).

It has recently been suggested that in the presence of respiratory symptoms a relative limit value (RLV) action level for endotoxin of 10 times background is appropriate (65). In the absence of respiratory symptoms a RLV of 30 times background has been suggested (65). Studies in Danish buildings have shown that the concentration of endotoxin in settled dusts was correlated with the prevalence of BRS (66).

Other kinds of bacteria may be found in buildings. These include gram-positive bacteria such as *Staphylococcus* and *Micrococcus* species present on human skin scales and *Streptococcus* species emitted as aerosols from the nasal/pharynx when a person is talking. Other than being an indictor of the occurrence of people in the indoor environment, adverse health effects have not been directly ascribed to these gram-positive bacteria in nonhospital settings (24).

The reader is referred to the ACGIH publication on *Bioaerosols: Assessment and Control* for detailed information on sources of specific microbial agents (Ref. 7, Chapt 23, 24). Additional information on sources of microbial agents is found in several specialized publications (35, 67–69).

2.1.3 Allergens

Regular exposure to airborne allergens can induce allergic sensitization and in some cases respiratory allergic symptoms (20). These exposures can be occupational, recreational, or domestic and they may be perennial or seasonal. The allergens may be from plant, animal, or mold sources. Since the exposure is more straightforward to identify, and often to measure, seasonal and occupational allergens have been more thoroughly studied. Seasonal allergens include tree and ragweed pollen, as well as caddisfly and mayfly hatches, and symptoms closely follow the presence of the allergens. Occupational allergens are usually perennial, but the symptoms still closely follow exposure, i.e., symptoms remit when away from exposure. Examples of occupational allergens are microbial enzymes that are used in detergent manufacture or in baking, allergens from laboratory insects or rodents, and crustacean allergens from crabmeat being processed (70).

Common indoor allergens are produced mostly from animals and molds, although some plant pollen allergens accumulate indoors. A number of molds can amplify in indoor environments that have adequate moisture. Molds may be allergenic depending on the exposure; many likely have nonallergic effects as well (71). Important animal sources of indoor allergens are dust mites, cockroaches and mammals (pets or pests) (72–74). Indoor allergens are now recognized as an important cause of allergy and among some segments of the population are significant risk factors for the development of asthma (20, 75).

Exposure to common environmental antigens normally elicits an IgG antibody response in most individuals. Antigens that induce an allergic sensitization, i.e., elicit an IgE rather than an IgG antibody response are termed allergens. IgE antibodies participate in the allergic reaction, and individuals that are genetically prone to sensitization, i.e., produce IgE against common environmental antigens, are said to be atopic. In the last decade, it was recognized that T lymphocytes respond to antigens in one of two ways, and the type of response governs sensitization. The control of this response is not fully known.

Dust mites (*Dematophagoides pteronyssinus* and *D. farinae*) are found in dust from mattresses, bedding, upholstered furniture and floors. Mites consume human skin scales which are abundant in settled dusts in most occupied spaces. Mites grow and reproduce optimally when the relative humidity is in the 75 to 80% range (76). They can still survive when the relative humidity is above 50% for only part of each day (77).

Two major allergens Der p 1 and Der f 1 are found in mite fecal pellets. These allergens are found at high levels in dusts in residences especially in bedroom bedding, mattresses, and carpet. Considerable variation exists between allergen content in different areas of the same room and between dusts in different rooms (78). Average concentrations of Der p 1 in dusts in offices have been reported to be as high as 1.5 μg/g (79). In Florida hotels Der p 1 concentrations as high as 40 μg/g have been found in carpet dust (unpublished data).

Dust mite allergens become airborne in large amounts during cleaning activities, or during other disturbances, and presumably this is one of the principal exposures (72). However, the pellets are aerodynamically large (>10 μm) and hence usually settle out quickly from room air. This makes airborne measurement of dust mite allergen difficult and generally unreliable which is why mite allergens are usually measured in settled dust.

Measuring allergen in dust does not measure inhalation exposure, but rather is an index of exposure. The concentration in dust has been used widely to estimate exposures for epidemiological studies of risk factors for asthma. Exposure to dust mite allergens at a concentration of 2 μg allergen/g dust is associated with an increased risk of sensitization to mite allergens. At 10 μg/g, there is an increased risk of developing asthma symptoms in sensitized individuals (20). Surveys of residential and other buildings indicate that residences often have greater concentrations of mite allergen in settled dust than other buildings.

To reduce a person's exposure, control measures are likely justified only if the principal reservoirs have been identified. In other words, strict control efforts in an office or school will have little effect if allergen concentrations in those areas are already lower than the concentration in mattress dust. It is often overlooked that the typical person spends 8 hours per day asleep with their mattress surface, with its dust, in their breathing zone. Fortunately, the ELISA tests for mite allergens are affordable and available, so that identifying allergen reservoirs is now feasible and reliable, only requiring samples of settled dust.

Cockroach infestations are commonly thought of as only occurring in buildings that are poorly kept or maintained. Most species live apart from humans though, (in the "wild"); some live in and around human dwellings, as well as elsewhere, and a few "domestic" species essentially cohabit with humans (73). The domestic German cockroach and brown-banded cockroach typically are associated with poor housekeeping, but other species, such as the American, Oriental and smokybrown cockroaches may only transiently enter buildings, regardless of the housekeeping or maintenance. American cockroaches are also known locally as Palmetto bugs and Oriental cockroaches as waterbugs.

Cockroaches have become recognized over the last decade or so as a very important source of indoor allergens, probably second only to dust mites. The allergens can be associated with various secretions, body parts, egg casings, and fecal material (73). As with dust mite allergens, environmental reservoirs are detected by sampling settled dust rather than air. Monoclonal and polyclonal antibody based ELISA tests are commercially available to detect and quantify the allergens in dust (80). Cockroaches produce a number of allergens. The group 1 allergens, Bla g 1 from German cockroach and Per a 1 from American cockroach are cross-reactive, whereas Bla g 2 does not cross-react with American cockroach allergens. At present, evidence suggests that some sensitization occurs from exposure to Bla g 1 levels below 1 unit/g dust, but that the prevalence of sensitization increases dramatically in the range of 2 U/g, and is maximal (>40%) above 4 U/g (81). Individuals that are sensitized to cockroach allergens may develop symptoms of asthma if they are subsequently exposed to significant (>8 U/g) concentrations of allergen (82).

Allergens from various species of mammals almost invariably contaminate human environments. The most prevalent are the major allergens from pets such as dogs, cats, and rabbits. Pest species such as mice and rats, unfortunately, are also important in some buildings (74). Specialized facilities may have other, less commonly encountered allergens from horses (stables), cows (barns), big cats (zoos), and guinea pigs, rats and mice (laboratories). The rodent allergens have been studied as a source of occupational allergy among laboratory workers, but far more people worldwide are probably exposed to these allergens due to building infestations. Several common features apply to the mammalian allergens characterized to date. They are all proteins and the sources are generally from dander, except for rodents which excrete allergens in urine (74).

Dog and cat allergens are present in many buildings where these animals are not usually kept (83). It has been shown, particularly with cats, that allergen bearing particles can readily be transported on owners' clothing into these areas, and Scandinavian studies have shown that cat allergen levels in schools can exceed the provisional threshold for sensitization (84). Further, the clothing of children without pets can become sufficiently contaminated at school to carry measurable amounts of cat allergen into their homes. Special day care facilities for allergic children in Scandinavia are open only to children from pet-free homes because of this "second-hand" pet exposure.

Cat allergen occurs on particles with a broad size range, from more than 10 μm to less than 2.5 μm (85). Cat allergen bearing particles are also adherent, and thus many surfaces and materials can become contaminated and successful cleaning to reduce cat allergen levels is often very difficult. Removal of the source (the cat) will ultimately reduce allergen levels, but declines in allergens may not be detectable until at least 4 or 5 months. Commercial assays are available for the major cat allergen Fel d 1 and for the major dog allergen, Can f 1. At this time, a provisional threshold level of 2 μg Fel d 1 allergen/g dust is recognized as being associated with an increased risk of sensitization.

2.1.4 Specific Health Effects

Asthma has been defined in several different ways. Partly, this resulted from changing concepts over the last half century of the "cause" of asthma, but also from the orientations of different medical subspecialties (86). Asthma was recently defined as a "usually chronic condition characterized by intermittant episodes of wheezing, coughing and difficulty in breathing, sometimes caused by an allergy to inhaled substances" (20). Another recent expert panel stated that "asthma, is a chronic inflammatory disorder of the airways in which many cells and cellular elements play a role, in particular, mast cells, eosinophils, T lymphocytes, neutrophils, and epithelial cells" (75).

Asthma apparently may result from more than one factor. A major portion of the dramatic rise in asthma in industrialized countries in the last 30 years is due to asthma in children and young adults (20). Among those less than 20 years old, allergic sensitization is a significant risk factor for asthma, and familial correlations occur. However, in adults the importance of common environmental allergens decreases with age, there is no familial correlation, and asthma often occurs with other chronic lung diseases, such as chronic bronchitis and emphysema (86). Finally, occupational asthma can develop among adults exposed to sensitizers. Some occupational sensitizers clearly induce specific IgE reactions, which indicates that allergy is involved. Other occupational sensitizers induce asthma through immunologic mechanisms other than allergy, and some induce allergy though nonimmunologic mechanisms (87).

Asthma that is associated with indoor, nonindustrial environments often has an allergic basis. A hallmark of building-related asthma is that symptoms are associated with being in the building. This association often aids in the diagnosis, but does not necessarily aid in identifying the particular sensitizer and hence confirming the cause. Allergic symptoms may develop within an hour of exposure (early phase reaction) or after 4–12 hours (late phase reactions). Very common indoor allergens are dust mites, cockroaches, mammal danders, and molds. With any allergic sensitization, it is crucial to identify the relevant sensitizer and reduce or prevent further exposure (75). The indoor allergens, except for mold, are often present in greater concentrations in residential than in other buildings. Molds though flourish in any building with inadequate moisture control, regardless of the use. Mammal danders are typically from pets, except for certain occupations, such as laboratory exposure to mice. Case reports have associated asthma with indoor exposures to bioaerosol sources (7, 40). These bioaerosol sources include humidifiers contaminated with mold and/or bacteria, moldy building materials, and moisture damaged walls.

Humidifier fever. Episodes of fever, muscle aches, and malaise with only minor pulmonary function changes have been associated with inhalation of aerosols from humidifiers contaminated with gram-negative bacteria and protozoa (7). Symptoms of this flu-like illness generally subside within a day after exposure without any long-term adverse effects.

An example of this type of BRI occurred in an office when three of the seven occupants reported attacks of fever and chills that started late during the workday and lasted well into the night (88). Illness was associated with the use of a humidifier on occasions when the air was considered too dry for comfort. Analysis of water from the humidifier reservoir showed that *Flavobacterium* was present at a concentration of about 8×10^4/mL. The concentration of airborne *Flavobacterium* increased from nondetectable when the humidifier was not running to about 3000/m^3 within 15 min. of operation. Endotoxin from this gram-negative bacterium was thought to be the etiologic agent of this outbreak of humidifier fever (18). Subsequent work by Rylander and Haglind (88) showed that endotoxin from humidifiers at a concentration of about 130 to 390 ng/m^3 from a *Pseudomonas* species caused another outbreak of BRI.

Hypersensitivity pneumonitis is an immunologic lung disease that occurs in some individuals after inhalation of organic dusts (89, 90). Hypersensitivity pneumonitis is suspected when symptoms such as fever, cough, and chest tightness occur several hours after exposure. The diagnosis of the disease is made on the basis of a physician's review of patient symptomology plus a battery of tests (91) including restrictive pulmonary function measurements, precipitins (IgG antibodies) to extracts of microbial agents collected in the building, and in rare instances, inhalation exposure of the patient to suspect antigens in a clinical setting (19).

A wide variety of specific microbial agents have been implicated in outbreaks of hypersensitivity pneumonitis. *Cladosporium* (92), thermophilic actinomycetes (93), and *Bacillus subtilis* (94) have each been implicated as etiologic agents in indoor residential outbreaks of this BRI. In small buildings or portions of buildings, *Penicillium* species have been shown to be causative agents of disease (95, 96).

In large buildings with complex HVAC systems and with many potential sites of microbial amplification, outbreaks of hypersensitivity pneumonitis often cannot be ascribed to a single agent (19, 97, 98). For example, an outbreak of acute hypersensitivity pneumonitis associated with cafeteria flooding in a large office building could not be etiologically related to potential agents such as thermophilic actinomycetes and *Acanthamoeba polyphaga* even though they were abundant in environmental samples (97). Changing patterns of microbial amplifications and dissemination as well as loss of viability and antigenicity of disseminated organic dusts may account for the difficulty in establishing specific etiology in these instances of BRI.

Acute Febrile Illness. Outbreaks of infective illness in the indoor environment may be caused by airborne exposure to specific human-shed microorganisms such as the influenza virus and the viruses and bacteria that cause the common cold. Provision of adequate amounts of outdoor air or highly filtered recirculated air to occupants in buildings is thought to reduce the risk of infection due to the common cold. It has been reported (99) that in military barracks with very low outdoor air ventilation rates [<2 cfm/occupant (100)] the incidence of febrile respiratory infection (common cold) was higher than that found in more adequately ventilated (approximately 14 cfm/occupant) barracks. A similar scenario was reported by Hoge et al. (101) in an overcrowded jail where inadequate ventilation was associated with an outbreak of pneumoccocal pneumonia. Providing outdoor ventilation air to reduce the risk of "contagion" by dilution has been recognized in ventilation codes since the late nineteenth century.

Tuberculosis (TB) is caused by several species of mycobacteria in the *Mycobacterium tuberculosis* complex. The mycobacteria are spread from infectious individuals through airborne droplets released by coughs and sneezes. The risk of contracting tuberculosis is thus greater in indoor environments or enclosed spaces where exposure to droplets is prolonged because droplet dispersal is reduced. This has been documented in passenger aircraft, hospital wards and other buildings. It also explains why tuberculosis only became an important disease after humans began living in crowded cities (102).

Indoor levels of *M. tuberculosis* are generally low, but since TB is infectious, the clinically relevant exposure may be only a few propagules. The dose–response relationship is not known, and there are no established exposure limits. Culturable sampling techniques do not always reliably recover mycobacteria. The infective mycobacteria are very slow growing and even when recovered successfully, the colonies may not attain detectable size before being overgrown by other microbes. Sampling is thus used principally in research projects, epidemiological studies or in other specialized situations. Culture media and sampling precautions are available though (103). *M. tuberculosis* detection is an area where the great sensitivity of PCR detection will be very useful (104). Indeed, PCR procedures to detect mycobacteria have already been developed (105).

Cancer. Certain molds that occur in buildings with moisture problems can produce carcinogenic mycotoxins. These molds include *Aspergillus versicolor* which is especially common in buildings with poor humidity control, and *A. flavus*. Aflatoxin (produced by *A. flavus*) is a potent carcinogen, and sterigmatocystin (*A. versicolor*) is a precursor in the biosynthesis of aflatoxin. Retrospective studies have detected elevated cancer prevalences among grain handlers that were exposed to aflatoxin containing grain in industrial settings (106). Other mycotoxins produced by molds, although not directly carcinogenic, have been shown in the laboratory to be immunosuppressive to selected cell types (7, 29). Hence, it is plausible although certainly not proven, that the effects of carcinogen exposure might be increased if the individual's immune system has been suppressed by exposure to certain molds.

It is important to note that using industrial exposures to infer a response from a nonindustrial exposure is a significant extrapolation. Likewise, inferring augmented effects due to an unproven, even though very plausible mechanism is also an extrapolation. However, the scope of exposure and the potential magnitude of any effect is quite large, since mold growth is present in many buildings. This means that understating the risk could have serious consequences. Also, there is anyway a general consensus that regardless of cancer issues, sustained, active mold growth is not acceptable in occupied spaces (7, 67, 69). Hence, there are a number of valid reasons to avoid moldy buildings. Realistically though, exposure in moldy buildings continues. The best means of understanding the true risk of mold exposures is by following cohorts of subjects that have been exposed to molds in buildings that were well characterized mycologically and toxicologically. Unfortunately, even though a number of such cases occurs each year, no apparent effort is being made in the United States to follow these subjects and determine any long-term health effects such as incidence of cancer.

2.1.5 Sampling and Data Interpretation

The principles of sampling for microorganisms and microbial products (e.g. endotoxins) are reviewed elsewhere (7, 35, 107). An enormous diversity of microbial contaminants may occur in buildings, including hundreds of different culturable fungi and bacteria, nonculturable (nonviable) microorganisms, other cellular components (β1,3 glucans), and microbial toxins (e.g., mycotoxins, endotoxins). Procedures used to collect microbial contaminants are also varied, including air and source (bulk) sampling methods (7, 35). Because moisture problems and fungal growth frequently occur in buildings, the following discussion is restricted primarily to the collection of culturable fungi.

A plan for data interpretation must be in place prior to the start of sample collection. The objective of the sampling plan as well as appropriate controls must be clearly defined. Thus, source samples should be collected from moisture problem areas as well as from background (reference control) areas not affected by moisture or dampness. Air sampling should include outdoor controls, generally obtained from high-quality air on the roof of the building. Chapters 5 (developing a sampling plan) and 6 (sample analysis) of the ACGIH *Bioaerosols Assessment and Control* (7) should be understood prior to onset of sampling. Sampling objectives are usually best assessed when the objectives are specifically stated. For example, sampling with a general intent to "see if there is a problem" often yields no useful information. Conversely, sampling for specific types of fungi in different areas can provide data to answer a specific question such as "are the fungi in these areas similar enough to represent the same population?"

Regardless of the kind of sampling methodology used, enough replicates must be collected to allow for data interpretation because of the natural order-of-magnitude variation in fungal populations normally encountered in air or surface samples. At least three kinds of culture media (35) as follows, should be considered for the collection and enumeration of the varied kinds of fungi that may occur as a result of moisture problems in buildings:

- DG-18 or malt extract agar plus 20 or 40% sucrose for xerophilic fungi.
- Malt extract agar for hydrophilic fungi (see discussion on various kinds of malt extract agar in Ref. 108.
- Cellulose agar for cellulosic fungi such as *Stachybotrys* and *Chaetomium* species.

Sampling data for culturable fungi may be interpreted as follows: (*1*) Building surfaces are not sterile. However, the occurrence of extensive visible fungal growth in a building is unacceptable. Visible mold should be physically removed (see section 6.1) in such a manner that spores are not dispersed into occupied or clean areas; (*2*) in healthy buildings the diversity of airborne fungi indoors and outdoors should be similar; (*3*) the dominating presence of one to two fungal species indoors and the absence of the same species outdoors indicates a moisture problem and degraded air quality; and (*4*) the consistent presence of fungi such as *Stachybotrys chartarum*, and *Aspergillus versicolor, Aspergillus fumigatus* or various *Penicillium* species over and beyond background conditions indicates the occurrence of a moisture problem and an atypical exposure.

The rank order assessment of culturable fungi in Table 65.3 for indoor and outdoor samples illustrates the kind of comparisons useful in data interpretation. *Cladosporium cladosporioides* is clearly the dominant culturable fungus in the outdoor air. *Aspergillus versicolor* and *Wallemia sebi*, two xero-tolerant fungi are abundant in the indoor but not the outdoor air. The different rank order of fungi in the outdoor and indoor samples plus the presence of two xerotolerant species indoors suggests the occurrence of a dampness problem in building infrastructure.

The reader is referred to extensive discussions on interpretation of fungal sampling data in the *Bioaerosols: Assessment and Control* (see Chaps. 7 and 19 in ref. 7) as well as in other publications (35, 67, 69).

2.2 Volatile Organic Compounds

VOCs are commonly associated with indoor air pollution and BRS. VOCs are characterized by boiling points ranging from about 50–260°C and include alcohols, aldehydes, straight chain and cyclic alkanes, aromatic hydrocarbons, halogenated hydrocarbons, terpenes, ketones, and esters. Those VOCs with boiling points less than 50–100°C are referred to as very volatile organic compounds (VVOC). Those with boiling points above 240–

Table 65.3. Airborne Fungi Present in Indoor and Outdoor Air

Location	CFU/m^{3a}	Rank Order of Taxa	%
Outdoors	700	*Cladosporium cladosporioides*	84
		Penicillium brevicompactum	4
		Penicillium implicatum	3
		Penicillium citrinum	2
		Aspergillus niger	2
Indoors	380	*Aspergillus versicolor*	38
		Cladosporium cladosporioides	18
		Penicillium citrinum	10
		Penicillium sclerotiorum	7
		Wallemia sebi	7

[a]CFU/m^3 means colony forming units per cubic meter of air. Culture medium was DG-18. Top five taxa only presented in table.

260°C are called semivolatile organic compounds (SVOC). SVOCs include pesticides, polynuclear aromatic hydrocarbons (PAH), and certain plasticizers such as phthalate esters (109).

Even in those geographic areas where outdoor air pollution from sources such as heavy vehicle traffic and petroleum refineries is significant, indoor total VOC (TVOC) concentrations have been shown to be two to ten times higher than those outdoors (110). In new office buildings, the TVOC concentration at the time of initial occupancy is often 50 to 100 times than that present in outdoor air. The variety of VOCs found in indoor air is almost always greater than that in outdoor air, mainly because of the very large number of possible sources in indoor environments.

TVOC cannot be directly correlated with the incidence of BRS (111). However, the presence of elevated TVOC (generally, greater than 1,000 to 3,000 micrograms per cubic meter of air [$\mu g/m^3$] or the dominating presence of specific classes of VOCs indoors (for example, terpenes such as limonene; aromatic hydrocarbons such as benzene; halocarbons such as methylene chloride) indicates strong contaminant sources in the building. Elevated TVOC or the dominating presence of a single class of VOC indoors is not characteristic of well operated and maintained buildings (112, 113).

2.2.1 Sources and Health Effects

Major VOC sources indoors include construction and finishing materials, furnishings, activities performed by people including cleaning and photocopying, and occupants themselves. Alkanes and aromatic hydrocarbons from building construction materials are commonly present in indoor air in most buildings. Alkanes such as *n*-decane and *n*-undecane may be present indoors in new buildings (and during renovation of portions of older buildings) at concentrations of 100 to 1000 times that present outdoors. Toluene, xylene, and ethylbenzene are almost always found in indoor air because of the extensive use of these aromatic hydrocarbon solvents in interior finishes (for example, linoleum and paints). In new buildings the substantial majority of VOCs in indoor air are emitted from construction materials and furnishings such as vinyl composition tile, carpet, office furniture, insulation, adhesives, and paints (114). The most common VOCs are toluene and formaldehyde.

Activities such as cleaning and photocopying may be associated with strong VOC emissions. Styrene emissions from freshly photocopied paper averaged 5 $\mu g/m^2 \cdot hr$ (115). Emission rates varied among copiers from different manufacturers. Halogenated hydrocarbons (e.g., methylene chloride and tetrachloroethylene) may be released into indoor air from solvents used to clean workstation panels and to strip paint from finishes. Dry process photocopiers, laser printers, and computers have been found to be sources of VOCs in addition to ozone and particles (116). Common VOCs emitted include styrene, butanol, formaldehyde, benzaldehyde, and acetone. Computers often emit acrylates, phenol, and plastizers (116). Liquid process photocopiers and photocopied paper itself are strong sources of branched C_{10} and C_{11} alkanes (117). Alcohols and alkanes from cleaning solvents and soaps can become predominant VOCs in buildings where original interior finishes are no longer significant emission sources. In the example in Table 65.4, isopropanol and various alkanes present in cleaning agents accounted for almost 90% of the TVOC found in indoor air.

Table 65.4. Concentrations of Alcohols and Alkanes in a 5-Year Old Building with and without Cleaning Activities

	Concentration ($\mu g/m^3$)	
Cleaning Status	Alcohols	Alkanes
No cleaning[a]	3	250
Routine cleaning	1,030[b]	1,300[c]

[a]Zero cleaning was performed during a six-month period.
[b]Mostly isopropanol.
[c]Mostly C_9 to C_{11} alkanes.

The occupants themselves are major sources of indoor VOCs. Benzene is present in ETS and higher concentrations of this aromatic hydrocarbon are expected in designated smoking areas. Limonene (a terpene) and various siloxane compounds (e.g., decamethylcyclopentasiloxane) may be present in many personal care products including antiperspirants and deodorants (117). Tetrachloroethylene is emitted from clothing that has been recently dry-cleaned. Hand and body lotion, moisturizing soaps, and cosmetics can be strong sources of C_{12} to C_{16} alkanes (117). Finishing materials and products used in newer buildings are characterized by reduced emissions of traditional solvents (e.g., toluene, xylene) and the slow emission of higher boiling point VOCs such as Texanol and various monoterpenes (117, 118). Wolkoff (119) has comprehensively reviewed the sources of VOCs in the indoor environment.

Among the VOCs, formaldehyde is often associated by the public with indoor air pollution. Although formaldehyde may be emitted from a number of sources such as from gas stoves and from smoking, its major source indoors is from construction materials such as particleboard, fiberboard, and plywood. Concentrations of formaldehyde in residential buildings are higher than in office buildings because of the relatively large ratio of pressed wood products to air volume in the former as compared to the latter type of buildings (120). Molhave et al. (121) has estimated potential exposures associated with formaldehyde emissions from manufactured products and concluded that there is no significant cancer risk to the consumer in the home environment.

Adverse health responses potentially caused by VOCs in nonindustrial indoor environments fall into three categories, namely, (*1*) irritant effects including the perception of unpleasant odors and mucous membrane irritation, (*2*) systemic effects such as fatigue and difficulty concentrating, and (*3*) toxic effects such as carcinogenicity (120). Many of the VOCs emitted from new furnishings and building materials are mucous membrane irritants. Molhave et al. (122) exposed 62 people with a history of difficulty with poor air quality to a mixture of 22 irritant VOCs (TVOC 5 or 25 mg/m^3) in a chamber setting. Subjective tests showed that the perception of poor air quality and odor intensity increased with increase in the TVOC. Similarly, the perception of mucous membrane irritation was elevated at both 5 and 25 mg/m^3 concentrations.

Studies have shown that human physiological responses to sensory irritants are similar to some BRS. Some studies have been undertaken with bioassays to determine the degree of sensory irritation associated with VOC releases from certain products (123, 124). Results showed that sensory irritation did not generally occur if products were kept at normal ambient conditions. However, sensory irritation could occur with products heated to 70°C and this was partly attributable to increased VOC exposure. Products containing formaldehyde were most likely to produce sensory irritation.

It has been hypothesized that complaints of eye, nose, and throat irritation may be due to exposure to VOC mixtures that are often found in new or renovated buildings (125). Occupant complaints are almost always encountered when the TVOC is about 3 mg/m^3 or higher (113). At TVOC levels in the range of 0.2 to 3.0 mg/m^3 occupant discomfort and irritation complaints are manifested if other pollutant exposures occur simultaneously (113). At TVOC levels below 0.2 mg/m^3 discomfort and irritation complaints due to VOCs should be minimal.

The dose–response relationship of TVOC to adverse health responses of occupants is negated when highly reactive VOCs (e.g., aldehydes or amines) or highly odorous VOCs are present in indoor air. In other words, even low concentrations of these VOCs can elicit complaints. Highly odorous aldehydes, esters, and alkoxy alcohols are known to degrade the human perception of air quality even at minute concentrations. Thus, the quality of the air perceived by building occupants is not likely governed by TVOC but by specific VOCs having high odor indices (126). A relationship has been proposed between the irritancy of some specific VOCs and BRS (127).

It is known that ozone and possibly other oxidants (e.g., nitrogen oxides) react with unsaturated carbon to carbon bonds of some VOCs resulting in the production of aldehydes such as formaldehyde and acetaldehyde (126, 128, 129). Ozone also reacts with nonvolatile components of carpet resulting in VOC emissions (128). The interaction of oxidants and organic compounds in buildings should be considered in the etiology of BRS (126).

Studies in the Nordic countries have suggested that health effects of VOCs should be viewed in a multifaceted manner including effects of specific VOCs on odor perception, sensory irritation and noncarcinogenic and carcinogenic health endpoints (130–132). Odor and sensory irritation thresholds for acetic, propionic, and butyric acid have been proposed (133). Toxicological guidelines for exposure to various carboxylic acids, phenols, glycol ethers, aldehydes, terpenes, and alcohols have also been proposed (130). The TVOC concept was considered not to be relevant from a health viewpoint (131).

2.2.2 Sampling and Interpretation

Existing methods of sampling established for VOCs in industrial workplaces are not readily adaptable to nonindustrial indoor studies. Sampling methods developed for industrial workplaces are often bulky, noisy, and validated for concentrations about an order of magnitude below the applicable industrial threshold limit value (TLV). Concentrations of VOCs found in nonindustrial indoor air are usually two or more orders of magnitude lower than industrial TLVs (109). More sensitive sampling and analytical methods are therefore required for characterization of VOCs in indoor air.

Most IAQ sampling evaluations for VOCs utilize in-field trapping of pollutants on various sorbents and subsequent laboratory analysis of specific VOCs. A variety of sorbents (often two or more sorbents per tube), including Tenax, graphitized carbon black, or carbon molecular sieves, are often used to trap VOCs (134). After thermal desorption in the laboratory, gas chromatography and mass spectrometry (GC/MS) are used to provide the sensitivity (generally 1 μg/m^3 for specific VOCs) required for IAQ studies (135). The use of passive sampling devices based on the principle of molecular diffusion into charcoal combined with GC/MS provides another method of measuring VOCs in indoor air at levels of 1 μg/m^3 or less (117).

Interpreting VOC sampling results is difficult because of the absence of dose–response data, the complexity and changing nature of VOC mixtures, and the variation caused by changes in ventilation and occupant activities indoors. In general, indoor air may contain 50 to 200 different individual VOCs at low concentrations. The TVOC can be used as an indication of the presence of strong VOC emission sources. Most buildings have TVOC levels in the 50 to 200 μg/m^3 range. However, there may be very low levels of individual VOCs that may be odorants and irritants (126, 127, 130, 133). Speciation or identification of individual VOCs including odorants and irritants should be made during sample analysis.

Sampling is appropriate when trying to demonstrate that activities in one zone (for example, interior renovation, paint spraying, or dry cleaning) may be degrading air quality elsewhere in the building. Sampling should be performed in close proximity to the suspected source of the VOCs, in zones where air quality may be degraded by ingress of VOCs, and in control locations such as in the outdoor air on the roof in an area distant from building vents and exhausts. Analytical methods used in the laboratory (usually GC/MS) must be able to distinguish individual compounds representative of each VOC class (112, 117, 135). Indoor/outdoor TVOC ratios or VOC class ratios consistently exceeding 10 or 20 suggest the presence of strong indoor sources of VOCs which are not expected in well operated and maintained buildings. In a review of the VOC literature, Brown et al. (136) concluded that concentrations of individual VOCs in indoor air seldom exceed 50 μg/m^3. Most individual VOCs in buildings occur at concentrations less than 5 μg/m^3.

To determine whether or not a mixture of VOCs is typical or atypical, it is useful to establish both the TVOC as well as the rank order concentration of specific VOCs both indoors and outdoors. For example, sampling performed shortly after the opening of Building 1 (Table 65.5) showed that the TVOC indoors was almost 30 times higher than that in the outdoor air. Further, the presence of elevated concentrations of four specific VOC indoors (Table 65.5) suggested the presence of strong emission sources such as from cleaning chemicals and from solvents present in new construction materials.

The equivalent TVOC indoors and outdoors for Building 2 (Table 65.5) might at first suggest an absence of unusual VOC sources. However, examination of the concentration data for three specific VOCs indicates that methylene chloride from a roofing repair near the outdoor air inlet was being entrained in the building. In addition, there is indication of indoor emission sources for alkanes and aromatic hydrocarbons. The presence of alkanes and aromatic hydrocarbons indoors was due to use of solvents on a laboratory bench without proper local exhaust ventilation. Sampling in Building 2 thus showed that quantification of concentrations of specific VOCs is required for interpretation. The determination of TVOC alone was inadequate for data interpretation. See Section 6.7 for other examples of interpretation of VOC data.

Table 65.5. Volatile Organic Compounds in Two Office Buildings[a]

	VOC Concentration In	
Compound	Outdoor air (μg/m^3)	Indoor air (μg/m^3)
	Building 1	
Total VOCs	62	2200
Tetrachloroethylene	6	700
Cyclohexanes	1	140
Dodecane	4	200
Chloroform	1	40
	Building 2	
Total VOCs	542	448
Methylene chloride	400	40
Decane	6	200
Ethylbenzene	2	80

[a]Sample collection on Tenax sorbent; analysis by gas chromatography/mass spectrometry.

Variables such as building age and outdoor air ventilation must also be considered when interpreting VOC sampling results. In new office buildings or in portions of older buildings undergoing renovation, the indoor/outdoor concentration ratio of TVOC may be 50 or 10 to 1 (137, 138). With adequate outdoor air ventilation, these ratios fall to less than 5 to 1 after 4 or 5 months of aging. In older buildings, an indoor/outdoor concentration ratio of TVOC may vary from nearly 1 when maximum amounts of outdoor air are being used in HVAC systems to greater than 10 during winter and summer months when minimum amounts of outdoor air are being used (139).

2.2.3 VOC Emissions From Materials

Chamber testing is used to measure VOC emissions from finishing materials and a wide variety of products used in buildings. Environmental chamber technology involves the use of stainless steel (or other inert material) chambers which are operated under precisely controlled environmental conditions. The actual measurement specifications require that the chamber simulate conditions of temperature, humidity, and air flow as typically found indoors. The chambers vary in size to accommodate products of virtually any size, ranging from carpet, to office furniture and business machines.

Chemicals emitting from the product being tested are identified and results are expressed in units of measurement called emission rates. The emission rate is product specific and refers to the amount of chemical being emitted per product per unit time such as μg/hr (micrograms of chemical per hour) or μg/m^2·hr which represents the micrograms of chemicals being emitted per each square area (square meter, m^2) of product per hour. Emission rate measurements require an extended test period, typically one week. If emissions are constant over time, a single point measurement can be made to predict emission characteristics. This single point measurement is called an emission factor and is expressed as $mg/m^2 \cdot hr$ for most products. Computer modeling techniques can subsequently be used to combine the specific product emission data with building use parameters. Predictions can then be made of pollutant concentration levels that will result in a building from use of a specific product.

The use of large scale chambers was pioneered by the wood products industry for the measurement of formaldehyde from pressed wood products. The technology was successfully implemented into a voluntary industry wide certification program in the United States for controlling formaldehyde emissions from pressedwood products used in premanufactured housing (140). The U.S. EPA initiated the development and use of small-scale environmental chambers for the study of VOC emissions from consumer products (141). The EPA Carpet Policy Dialogue Group subsequently accepted environmental chamber technology for use in industry related studies of carpet, cushions, and adhesives (142). American Society for Testing and Materials (ASTM) D5116-90 provides guidance for the use of environmental chambers to measure organic emissions from products (143). Similar guidance is provided by a committee from the European Community (144).

Environmental chamber studies of products have shown significant variations in VOC emissions among generic types of products. Common chemicals frequently identified with certain products are listed in Table 65.6. Additionally, product specific emission rates found from low-emitting products are given in Table 65.7. A study of paint for the State of

Table 65.6. VOCs Frequently Identified with Product Groups

Product	Common VOCs
Wall coverings	TXIB, naphthalene, toluene
Floor coverings	Styrene, 2-ethyl-1-hexanol, trimethylbenzenes
Paints	Propylene glycol, butyl propionate, Texanol®, acetates
Textiles	Formaldehyde, hexanol, nonanal
Ceiling tiles	Formaldehyde, acetic acid, hexanol
Office furniture	Formaldehyde, acetone, cyclohexanes

Table 65.7. Some Emission Levels For Products Categorized As "Low-Emitting" For State of Washington's East Campus Plus Program

Product	TVOC Emission Factor (24 hr) $mg/m^2 \cdot hr$
Office furniture	12–17 (mg/workstation·hr)
Office chairs	0.2–7.2 (mg/chair·hr)
Carpet	0.04–12
Floor adhesive	0.8–23
Latex paint	2.1
Vinyl composition tile (VCT)	0.03–1.4
Air duct liner	<1.0–4.2
Ceiling tile	0.02–0.03

Washington East Campus Plus Program (145) showed TVOC emission factors ranging from 1,500 $\mu g/m^2 \cdot hr$ to 825,000 $\mu g/m^2 \cdot hr$. Adhesives used in carpet installations have been found to vary in emission factors from 2,783 $\mu g/m^2 \cdot hr$ for one classified as "low VOC" adhesive, to 38,406 $\mu g/m^2 \cdot hr$ for a general purpose adhesive. In efforts to control product emissions and to encourage the use of low-emitting products, numerous voluntary and compliance programs have been established. These programs are described in Section 6.2.

2.3 Pesticides

Pesticides including insecticides, termiticides, and fungicides are often used in interior spaces to control a wide variety of organisms including wood boring insects, moths, and fungi. Although pesticides are by definition poisons, their toxicity varies toward different types of organisms. Several million pounds of naphthalene and paradichlorobenzene are used annually in U.S. homes as moth repellants (146). Similar quantities of pentachlorophenol are used annually in U.S. homes to protect wood and paints from degradation by insects and fungi. Pesticides such as chlordane, heptachlor, aldrin, dieldrin, and chlorpyrifos have been used or are still used in the soil or under the foundations of buildings (147).

In 1985, the U.S. EPA initiated studies on pesticide exposures in the general population. The Non-Occupational Pesticide Exposure Study (NOPES) has facilitated the development of improved methodologies for sampling for pesticides in indoor environments (148, 149). Using techniques described in these studies, it is now possible to sample and analyze for over 50 specific organochlorine, organophosphate, and organonitrogen, and pyrethroid pesticides using polyurethane foam sorbent with sensitivities as low as 0.01 $\mu g/m^3$. For example, among 50 residences monitored in the Jacksonville, Florida, area, 46 contained detectable levels of the pesticide chlorpyrifos (mean concentration 0.47 $\mu g/m^3$). Chlorpyrifos was detectable in outdoor air around only 9 of the 50 residences, with a mean concentration of 0.059 $\mu g/m^3$. Concentrations of chlorpyrifos as high as 37 $\mu g/m^3$ have been found in other residences (109). A chlorpyrifos concentration of 0.52 $\mu g/m^3$ was found in one house four years after pest control treatment (150). After application, pesticides may be present not only in indoor air, but also in settled dusts (which may become airborne) and as secondary contaminants in interior finishes such as flooring and door frames (151). Thus, following application of the pesticide lindane to ethnological objects in museum store rooms, it could be found in the air (0.2 to 3.7 $\mu g/m^3$), in dust (up to 128 mg/kg), and on room finishes (4 to 32 mg/kg) (151). The pyrethroid insecticides permethrin and deltamethrin used for cockroach control have been shown to be very persistent (for periods longer than 70 weeks) in settled dust (152, 153) on interior surfaces. Japanese studies in an experimental room showed that air concentrations of *d*-tetramethrin and *d*-resmethrin following spraying declined quickly as a function of ventilation rate (154).

A combination of air and dust (surface) sampling is probably best for characterization of pesticide residues that may be present in indoor environments. If sampling is performed, the analytical methods used must be sufficiently sensitive (for example, 0.01 $\mu g/m^3$ for air samples) so as to detect background concentrations of specific pesticides that may be present in the outdoor air. The objective of sampling is often to determine if pesticide concentrations in areas of suspect contamination exceed background outdoor levels and average concentrations that have been previously reported in the literature (148).

Health effect data available for pesticides are based primarily on oral or dermal exposures from animal studies. Little information is available on adverse health effects on inhalation exposures. It is significant, however, that most pesticide related injuries (including those from organophosphate insecticides) occur in the home (146). Of the accidents that were not related to ingestion, inhalation and dermal exposures were found to be equally important. Because of the potential chronic health effects from exposure to termiticides, the National Research Council has recommended that airborne concentrations in indoor air be limited as follows: chlorpyrifos (10 $\mu g/m^3$), chlordane (5 $\mu g/m^3$), heptachlor (2 $\mu g/m^3$), and dieldrin and aldrin (1 $\mu g/m^3$) (155). It should be noted that these exposure levels are more than an order of magnitude lower than concentrations considered acceptable in industrial workplaces. One report showed that industrial exposure limits can be exceeded following residential application of pesticides. Methyl bromide concentrations in homes following fumigation to control drywood termites were 5 ppm even after several aerations (156). Since the TLV for methyl bromide is 1 ppm (157) this study showed that better ventilation procedures following fumigation are needed.

2.4 Combustion Products

A variety of combustion sources contribute to indoor air pollution. In residential environments, these include fuel burning appliances such as gas ranges, unventilated kerosene or gas heaters, and wood or coal-burning stoves. Important combustion sources in commercial buildings are vehicle exhaust and other external sources which may be entrained into the outdoor air inlets of HVAC systems. Carbon dioxide and water vapor are major products of combustion. However, other combustion by-products such as CO, NO_2, sulfur dioxide (SO_2), and respirable particulates are significant causes of indoor air pollution. The type and amount of combustion by-products in indoor air of commercial buildings depend upon the type of fuel consumed (for example, diesel oil used by trucks contains significant amounts of sulfur, and SO_2 is an important combustion by-product) and the location of outdoor air inlets and emission sources. Combustion processes that are oxygen-starved and characterized by yellow-colored flames are characterized by elevated CO emissions (158). Combustion under oxygen rich conditions results in higher flame temperatures thus emitting greater amounts of oxides of nitrogen. Wood burning combustion sources can contribute respirable particulates which include PAHs. PAHs are produced as a result of incomplete combustion and are of concern because of their carcinogenic potential (159). Considerable literature is available describing characteristics of combustion devices found in indoor environments (primarily residences) and factors affecting the emission rates of various combustion by-products (160, 161). The monitoring of combustion by-products such as CO, NO_2 and SO_2 is reviewed elsewhere (149, 162).

Indoor air pollutants such as CO and NO_2 originate from multiple sources and concentrations indoors depend on parameters such as source emission rates, the volume of air indoors, outdoor air ventilation rates, and HVAC system characteristics. Pollutants such as CO and NO_2 are emitted intermittently and are usually concentrated only in certain areas of the building. Outdoor conditions such as ambient concentrations of NO_2 have a strong influence on indoor pollutant concentrations (163).

Considerable literature exists on exposure to combustion by-products in residential environments. For example, use of an unvented gas stove adds about 25 ppb of NO_2 to the background concentration in indoor air (164). Peak concentrations of NO_2 in kitchens during cooking with a gas range may be elevated by 200 to 400 ppb (165). In a study carried out at the Newark International Airport in 1985, 24-hour time weighted average (TWA) indoor concentrations of NO_2 indoors in gate areas ranged from 19 to 116 ppb (166). Average outdoor concentrations were higher and varied from 41 to 233 ppb. During cooking with a gas range, CO concentrations in residences typically range from 2 to 6 ppm. Carbon monoxide levels in vehicles during urban commuting may become elevated by as much as five times that present in ambient air (164).

It is well known that CO binds with hemoglobin to form carboxyhemoglobin (COHb), thereby decreasing the oxygen-carrying capacity of the blood. COHb levels higher than 4 to 5% are known to exacerbate symptoms of individuals with preexisting cardiovascular disease. Limiting average CO exposures to 9 ppm (maximum) for 8 hr or to 35 ppm for 1 hr, as specified in the National Ambient Air Quality Standards (NAAQS), is intended to provide a margin of safety with regard to COHb buildup in individuals with cardiovascular disease. It is worth noting that even in environments with no detectable CO levels, normal COHb levels in blood are not zero. Nonsmoking adults breathing CO-free air have COHb levels between 0.3 and 0.5%; among tobacco smokers' COHb levels are between 3 and 10% (167). For these and other reasons, the effects of chronic exposure to lower levels of CO that may occur in buildings are still considered controversial.

Most studies of the health effects of NO_2 exposure at concentrations commonly found in nonindustrial environments have focused on respiratory symptoms and illnesses. NO_2 is a pulmonary irritant and health effects of concern range from increased risk of respiratory infections to exacerbation of asthma (164). Some (but not all) epidemiologic studies have shown a higher prevalence of respiratory symptoms in children exposed to NO_2 at levels below 1 ppm. Airways reactivity may be increased in some asthmatics by exposure levels that occur in homes. Subjective complaints have been reported at 0.5 ppm and asthmatics may be responsive at that level. Experimental animal studies indicate an elevated incidence of respiratory infections from exposures to NO_2 concentrations above 1 ppm.

Although HVAC systems' outdoor air intakes located near a loading dock or garage are obvious pathways for entry of combustion products in commercial buildings, the intermittent nature of pollutant generation and HVAC and building operational variables make it difficult to establish combustion by-products as the cause of air quality complaints in a building. In one large building where occupants had been evacuated because of suspected exposure to combustion products, round-the-clock sampling for NO_2 for six days was necessary before it could be demonstrated that combustion products from a loading dock were being entrained in the HVAC system outdoor air inlet serving the affected office (139). The concentrations of NO_2 in the outdoor air inlet, in the vacated office, and in the outdoor air on the roof far removed from emission sources were 2.0, 0.7, and 0.08 ppm, respectively.

Table 1 of ASHRAE Standard 62-1989 (4) lists maximum concentrations of certain combustion by-products that are considered acceptable for introduction with outdoor air into HVAC systems. For example, NO_2 concentrations in outdoor air used for ventilation should not exceed 55 ppb (on a yearly basis). An addendum to ASHRAE Standard 62-1989 (168) recommends that respirable particulate levels (PM_{10}) not exceed 50 $\mu g/m^3$. In spite of considerable research, there is an absence of consensus on what constitutes "safe" exposure levels for various combustion products. For example, although it appears that adverse pulmonary function response to NO_2 occurs at exposures above 300 ppb, there is considerable diversity of opinion on the range of the lowest-observed-effect level. It is often argued that the quality of indoor air should be at least as good or acceptable as that outdoors. However, obtaining "safe" indoor target levels of combustion by-products such as NO_2 is difficult, owing to the poor ambient air quality in some urban areas where NAAQS are frequently exceeded.

2.5 Environmental Tobacco Smoke

Environmental tobacco smoke (ETS) is composed of a complex mixture of chemicals, including combustion gases and respirable particles. Because the tobacco leaf does not burn completely, ETS contains more than 4700 chemical compounds (169), including nicotine, tars (containing a variety of PAHs), vinyl chloride, formaldehyde, benzene, styrene, ammonia, NO_2, SO_2, CO, hydrogen cyanide, and arsenic. Many carcinogenic compounds are found in ETS.

ETS is derived from mainstream smoke that is drawn through the cigarette by the smoker and from side-stream smoke that is emitted from the cigarette itself directly into room air between puffs. Individual chemical components of ETS may be found in both the gaseous and the particulate phases. During smoking, particulate ETS may adsorb on surfaces in the occupied space and HVAC system. Gaseous components can then be reemitted into the indoor air from adsorbed particulate. Smoking characteristically results in a significant rise in the concentration of respirable particulate with a mass median diameter of about 0.2 to 0.4 μm (170). Although many chemical components of ETS can also be attributed to other sources, nicotine is considered a specific marker for this indoor pollutant.

Exposure to ETS is associated with numerous and significant health risks including lung cancer, coronary disease, and in children an increased risk of acute lower respiratory tract illness (171). Eye, nose, and throat irritation are frequent complaints in people exposed to ETS. Odor annoyance due to ETS is perceived at concentrations much lower than those that cause irritation (172); consequently much greater amounts of outdoor air are required to control odor annoyance as opposed to the acute irritation effects of ETS (173). Both odor annoyance and irritational effects are thought to be caused by the gaseous phase of ETS. In studies in a series of Danish buildings, about 25% of the perception of stale, unacceptable air is due to smoking (10).

The major chronic health effect from ETS is lung cancer. In 1985, major study panels were convened by the U.S. Public Health Service, the National Research Council, and the Federal Interagency Task Force on Environmental Cancer, Heart and Lung Disease, to consider the risk associated with breathing ETS in indoor air (passive smoking). All groups arrived at the same conclusion that passive smoking significantly increased the risk of lung cancer in adults (169, 174).

The presence of ETS in indoor air is associated with increased concentrations of respirable particulate, nicotine, and a variety of other contaminants. One smoker consuming a pack of cigarettes daily in a residence contributes approximately 20 μg/m^3 of respirable particulate to the (24-h) particle concentration (164). Concentrations of respirable particulates less than 2.5 μm in aerodynamic diameter in indoor environments where smoking is permitted may rise up to 500 μg/m^3 (170).

Nicotine, which is the most specific marker for ETS, can be measured at concentrations less than 1 μg/m^3 with personal or area monitors (175). Miesner et al. (170) measured nicotine concentrations in a number of office environments with varying smoking policies and found an increase in nicotine levels as respirable particulate concentrations increased. Nicotine and respirable particulate levels in a designated smoking room in one office were 26.5 and 520.8 μg/m^3, respectively. In a nonsmoking office located directly above a floor where smoking was permitted, the concentration of nicotine was 2.0 μg/m^3. Nicotine concentrations in nightclubs, taverns, and bars range from about 10 to 100 μg/m^3 (176). Nonsmoking nightclub musicians are exposed to nicotine concentrations in the range of 28 to 50 μg/m^3 (177).

ASHRAE Standard 62-1989 (4) recommends that a minimum of 20 cfm of outdoor air per occupant is needed to provide acceptable IAQ in indoor environments with a moderate amount of smoking. In smoking lounges, a minimum ventilation rate of 60 cfm/occupant (including transfer air) is required for acceptable IAQ.

Because there is no acceptable level of ETS due to its carcinogenicity, the 1996 public review of ASHRAE Standard 62-1989 (178) stated that acceptable IAQ can only be achieved in environments that are smoke free. For indoor environments with smoking allowed, acceptable "perceived" air quality can be achieved by providing adequate volumes of dilution air (based on the number of cigarettes smoked) for adapted or nonadapted (nonsmoking visitors) persons. The concept that acceptable IAQ can only be achieved in the absence of ETS is retained in the continuous maintenance process for ASHRAE Standard 62 (179).

In commercial buildings in the United States, ETS is less of an issue because of the increasing number of smoke-free facilities. For those buildings where smoking lounges are provided, ETS-IAQ problems can be minimized by complete exhaustion of smoke outdoors. The measurement of airborne concentrations of nicotine provides a sensitive analytical measure of ETS in cases where smoke may not be completed exhausted from the building. Because ETS is carcinogenic and because nicotine is a specific marker for ETS, the presence of airborne nicotine in a building is an indication of degraded IAQ.

2.6 Radon

Radon-222 (radon) is a noble gas and a decay product of radium-226, which in turn is a product of the decay of the uranium-238 series. Uranium-238 and radium-226 are found in very small amounts in most rocks and soils. Because the half-life of radium-226 is about 1600 years, radon gas is released at an almost constant rate into the soil atmosphere (180). Because it is not chemically bound or attached to other materials, radon migrates through soil porous and is released into ambient air or it diffuses through cracks and pores in building foundations and enters the indoor environment. Radon may also enter the indoor environment as a solution in well water or by diffusion from construction materials containing radium or uranium. Infiltration of soil gas into buildings is by far the largest contributor to indoor radon concentrations. The amount of radon that enters indoor air is influenced by a number of factors including the amount of radium in the soil or rock around the building foundation, the permeability of the soil around the foundation, the air pressure inside the building (for example, air pressure indoors may be lowered by the operation of exhaust fans), and the extent of cracks or openings in the building foundation. In general, radon is more of a problem in buildings with a high foundation surface area and low internal volume (residences) as opposed to those structures with a relatively low foundation surface area and large internal volume (large commercial office buildings).

Radon has a half-life of 3.8 days and its decay progeny (polonium-218, lead-214, bismuth-214, and polonium-214) have half lives of less than 30 min. Unlike radon, which is almost chemically inert, the decay products tend to adhere to dust particles and surfaces including those in the lung. Thus, although an inhaled radon atom is likely to be exhaled before it decays, inhaled radon progeny will likely decay (emit alpha or beta particles or gamma rays) before they can be removed by normal lung clearance mechanisms. Polonium-218 and polonium-214 both emit alpha particles, and consequently are the radon progeny of major concern in terms of potential adverse health effects (149).

There may be a significant distinction in health risk from radon progeny that attach to dust particles (attached decay products) and those progeny that do not attach (unattached decay products). The latter agglomerate to form very small particles (ranging from 0.002 to 0.02 μm), which have a higher probability of being deposited deep in the lung. Radon decay products attached to dust particles are more likely deposited in the moist epithelial lining of the bronchi. Some mathematical models indicate that a higher lung dose of alpha emissions is associated with deposition of unattached radon decay products (149).

Amounts of radioactive materials are specified in terms of their activity or the rate at which atoms decay. One picocurie (pCi), the term traditionally used in the United States, corresponds to 0.037 decays per second; the international unit for activity, the Becquerel (Bq), corresponds to one decay per second (181).

Radon decay product concentration has traditionally been measured in terms of potential alpha energy concentrations expressed in the working level (WL) unit. A radon level of 1.0 pCi/l (37 Bq/m^3) in dynamic equilibrium with its decay products is equivalent to a decay product concentration of about 0.005 WL (149, 181). Extensive studies on radon exposure in U.S. residences indicate that the yearly average exposure is about 1.5 pCi/l with approximately 1 to 3 percent of homes exceeding 8 pCi/l (181). In commercial buildings and health care facilities, radon levels are generally less than 2 pCi/l (182).

Inhalation of radon decay products is a significant health risk (lung cancer). Estimates of health risk due to radon exposure are based on extrapolation from epidemiologic studies on underground miners whose exposures were higher than those characteristic of indoor environments. Although the extrapolation calculations are somewhat uncertain, it is estimated that aggregate health effects from exposure to radon decay products range from 7,000 to 30,000 lung cancer deaths per year in the United States (183, 184). In Swedish studies, radon levels were measured in over 8,000 dwellings occupied by persons with and without lung cancer (185). Residential exposure to radon was shown to be an important cause of lung cancer and the risk is consistent with estimates based on radon exposure in miners (185).

Short-term and long term measurement methods are available for the quantification of indoor radon (181). Short-term screening measurements are generally carried out over a period of 1 to 7 days using canisters containing activated charcoal. Gamma radiation from the decay products collected in the activated charcoal is measured to quantify radon levels. Long term radon measurements are commonly carried out over period of up to a year using alpha track detectors. Ionization tracks in the plastic detector film produced by the emission of alpha particles from radon and its decay products are enlarged by etching and then visually counted. Other detection methods and extensive discussions on measurement protocols are reviewed elsewhere (181, 186).

Measurements of indoor radon concentration can vary on a hourly, daily, and seasonal basis. Thus measurements made under condition of maximum outdoor air ventilation (open windows) can underestimate the average annual radon level. Indoor measurements made when a building is subject to minimum outdoor air ventilation (for example, during the winter) tend to overestimate average concentrations. The most reliable estimates of radon concentration are those made over periods of 6 months to 1 year.

The U.S. EPA recommends a two phase strategy for measuring indoor radon concentrations in residences. A short term measurement of radon is carried out under closed-house conditions in a living area where elevated concentrations might be expected (ground level). If measured concentrations indoors are between 4 and 20 pCi/l, follow-up measurements made in the living space over a 12 month period are recommended. If short term measurements exceed 20 pCi/l, intensive follow-up tests are recommended because these concentrations can significantly increase health risk (187). Follow-up measurement is probably unnecessary when screening measurements show that radon levels are less than 4 pCi/l. Interpretation of indoor radon measurements should take into consideration that naturally occurring radon concentrations in outdoor air range from 0.1 to 1 pCi/l. Various kinds of actions to lower radon levels such as subslab depressurization and ventilation have been extensively reviewed (182, 188, 189).

2.7 Particles

A variety of particles, most solid (for example, dusts, smoke, and microorganisms) and a few liquid (for example, mist from humidifiers) are present in indoor environments. Particles present in indoor air may originate in the outdoor air, in the HVAC system, or in occupied spaces. Particles in the outdoor air, for example, from heavy vehicular traffic, may enter HVAC air intakes and especially if filtration is poor, be transported to the breathing zone in ventilation supply air. Particulates from soil are tracked into buildings and can accumulate in porous finishes, especially carpet. The HVAC system becomes a major source of particles when fibers are eroded from damaged or improperly installed porous insulation or when fungal spores growing on damp surfaces are released.

A major source of particles in most buildings is occupants (skin scales and fiber from clothing) and their activities (settled dust aerosolized by foot traffic, particulates from copiers). Dusts from interior renovations, if not properly contained, may be dispersed throughout the entire building. Activities such as vacuuming carpets and upholstered surfaces may disperse settled dust into the indoor air. Maintenance activities such as moving ceiling tiles or replacing dirty filters may inadvertently lead to dispersal of particles if not carefully performed.

In general, airborne particles of a size from about 0.1 to 10 μm are of concern for human health. Particles less than 0.1 μm are exhaled, and those greater than 10 μm do not enter the lower regions of the lung. ASHRAE Standard 62-1989 provides guidance on conditioning of outdoor for use in HVAC systems in terms of acceptable maximum concentrations of airborne particles. When the yearly average concentration of particles less than 10 μm (PM_{10}) exceeds 50 μg/m^3, filtration is required (168).

A number of studies have associated dust and indoor surface pollution with BRS (66, 190, 191). Dust components that may cause BRS include allergens from mites, endotoxin, macromolecular organic or microbial materials (192), manmade mineral fibers (193), and adsorbed ETS and VOCs. Cleaning of surface dusts has been shown to lower airborne particulate concentrations and surface particulate concentrations (194) and to reduce the prevalence of BRS.

Large differences in particulate emissions have been reported between different brands of vacuum cleaners (195). While HEPA filtered instruments in general emit fewer aerosols than other types of cleaners, a poorly fitted HEPA filter can allow polluted air from the motor to pass directly into room air (195).

In general, the particle concentrations in healthy buildings in developed countries seldom exceed 50 μg/m^3. It has been shown in some buildings that the outdoor air is the main contributor to indoor particulate and that air cleaning by the HVAC system filter deck is the critical factor which reduces particle levels in room air (196). In developing countries where wood, crop residues, and coal are major fuels, particle levels in buildings often exceed 200 μg/m^3 (197; see also Section 3.6).

2.8 Quick Reference Guide for Sampling

Table 65.8 provides a quick reference guide on data interpretation for major categories of indoor air pollutants as well as references useful for sampling and analysis methodology.

3 VARIED APPROACHES TO IAQ STUDIES

The different approaches that can be used during IAQ evaluations reflect the varied kinds of problems that can occur in buildings. Complaints may be related epidemiologically to ventilation or to defects in building performance. Thermal environmental parameters can interact strongly with IAQ problems in buildings. In some buildings where environmental conditions are deteriorated, BRI occurs in addition to BRS. Problem buildings can also be viewed in terms of the economic losses associated with lost productivity of occupants. Ideally, in order to understand the etiology of BRS, the prevalence of symptoms in occupants should be studied, preferably in problem and nonproblem buildings. Objectives of such case-control studies should include an understanding of whether the percentage of dissatisfied occupants exceeds some unacceptable level, and if dissatisfaction can be related to building, air contaminant, work-practice, and other variables.

Table 65.8. Quick Reference Guide for Major Categories of Indoor Air Pollutants and Appropriate Sampling Methodology

Pollutant	Guideline Consideration	Sampling Guidance	Refs.
Fungi	The kinds of fungi found indoors should be similar to those found outdoors[a]	AIHA ACGIH	35 7
VOCs	Indoor/outdoor TVOC ratios >10 indicate strong indoor sources; a single class of VOC should not dominate indoor air[b] odor perception, sensory irritation, genotoxic and nongenotoxic effects should be considered[c]	NKB EPA	130 135
Formaldehyde	≤0.05 ppm (60 μg/m^3) maximum target level[d]; 0.1 ppm (120 μg/m^3) long term exposure level of concern[e]; labeling requirement for products that may release >0.1 ppm[f]	EPA	135, 201, 202
Pesticides	Following post-treatment ventilation, pesticides should not be detectable in air above background outdoor levels	Polyurethane foam	148
Combustion products	Avoid entrainment from external sources; remove internal sources by local exhaust ventilation	EPA	149
Particulate	PM10 seldom exceeds 50 μg/m^3; avoid dust raising activities	Morawska	196
Nicotine	Presence indicates degraded IAQ	ASTM	203
Radon	Avoid entry of soil gas into buildings[g] see remedial recommendations when radon occurs at 4 and 20 pCi/l[h]	EUR	205

[a]Ref. 69
[b]Ref. 112
[c]Ref. 132
[d]Ref. 198
[e]Ref. 199
[f]Ref. 200
[g]Ref. 189, 204
[h]Ref. 183

3.1 North American Studies

By 1990, the U.S. National Institute for Occupational Safety and Health (NIOSH) had investigated over 500 problem buildings (206). Factors crudely associated with complaints were in descending order of frequency: inadequate ventilation (53%), indoor pollutant sources (15%), entrainment of outdoor contaminants (10%), microbial problems (5%), building fabric contamination (4%), and unknown causation (13%). It should be understood that this early estimate on likely causes of IAQ problems was not based on systematic studies. Microbial or chemical pollutants, defects in building infrastructure and in HVAC system design, operation, and maintenance were not investigated in every building.

Subsequently NIOSH evaluated 104 problem buildings where complaints and building defects were systematically studied (206). Problems of the following type were identified: (*1*) 63% had defects such as water damage, water intrusion, or poor housekeeping; (*2*) 60% had defective HVAC operation such as insufficient outdoor air ventilation, poor air distribution to the breathing zone, and flooded drain pans; (*3*) 58% had defective HVAC maintenance such as systems that were dirty and in disrepair or suffered from an absence of written operation and maintenance plans; (*4*) 51% had defective HVAC design problems such as outdoor air inlets that needed to be moved, inefficient or poorly fitting filters, or absence of minimum stops on VAV terminals; (*5*) 30% had thermal environment problems; (*6*) 24% had indoor contaminant sources such as VOCs, cooking odors, renovation dusts, and chemicals in mechanical equipment rooms; (*7*) 22% had combustion gas and restroom entrainment problems; and (*8*) 12% had ergonomic (workstation design) or physical agent (lightning/glare) problems (206). Although NIOSH studies do not provide information on the cause of complaints, they do provide practical information on the frequency of building defects that give rise to IAQ problems.

NIOSH found that levels of common BRS as follow occurred in problem buildings: strained eyes (33%), eye irritation (30%), fatigue (26%), irritability (23%), and stuffy nose (22%) (207). A review of the literature indicated that prevalence rates for some symptoms were similar in problem and nonproblem buildings (208). NIOSH studies also indicated that BRS of some occupants tended to become more debilitating over time and no longer remitted after the individual left the workplace each day (207).

An additional study by NIOSH (209) correlated BRS and BRI with building factors and defects. Symptoms such as shortness of breath, cough, chest tightness, and wheezing were associated with the presence of debris inside outdoor air inlets and poor drainage of drain pans. Multiple atopic symptoms such as sneezing, eye irritation, stuffy nose–nasal congestion were related to the occurrence of suspended ceiling tiles. Asthma diagnosed after beginning to work in the building was correlated with the occurrence of dirty filters, debris in the outdoor air inlet, and renovation including the installation of new drywall during the past three weeks.

In 1994 the EPA initiated a Building Assessment and Survey Evaluation (BASE) study in a wide cross section of public and private buildings in the United States. An objective of the BASE study was to provide baseline data on chemical, physical, microbiological, ventilation, and subjective parameters often measured in IAQ evaluations. Preliminary VOC sampling data from a subset of 41 building was reported in 1997 (210). VOCs with the highest median indoor concentrations were, in descending order: ethanol, acetone, 2 propanol, toluene, 1,1,1-trichloroethane, dichlordifluoromethane, and *m*- and *p*-xylenes. Forty-seven VOCs found indoors had indoor to outdoor median ratios greater than one (210). Questionnaire data obtained from the randomly selected buildings in the BASE study has been compared to that from the NIOSH database on complaint buildings (211). Preliminary indications suggest that perceived air quality and thermal environmental conditions are better in the BASE than in the NIOSH buildings. This is not surprising since NIOSH-studied buildings were known to have considerable building and HVAC defects (206) whereas the BASE protocol only included buildings with no history of complaints.

Hedge et al. (212) performed questionnaire studies and made measurements of physical and chemical factors in five air-conditioned offices where smoking was prohibited. Levels of CO_2, formaldehyde and particulate as well as temperature, relative humidity and illuminance did not correlate with BRS. BRS were however correlated with perceived indoor air quality, job stress, job satisfaction and allergies.

3.2 European Studies

Several systematic studies on complaints related to building characteristics were carried out in British office buildings in the 1980s (213, 214). In the largest study, complainant symptoms were recorded using a common protocol in 47 different groups of occupants in 42 buildings. The questionnaire used in these studies elicited information on whether the following 10 symptoms occurred during the past year and whether symptoms disappeared during periods when occupants were away from the building; dryness of eyes, itching of eyes, stuffy nose, runny nose, dry throat, lethargy, headache, fever, breathing difficulty, and chest tightness. Ventilation systems serving the 47 different occupant groups were categorized as *natural* (open windows, no forced air), *mechanical* (forced air system without cooling or humidification), *local induction units* (IUs), *central induction/fan coil unit* (FCU), and *variable or constant-volume system*. The later three ventilation categories were characterized by cooling of air, and in some cases, also by humidification. Questionnaire results showed that the lowest prevalence of work-related symptoms was found in the naturally or mechanically ventilated categories. Although there were considerable variations between buildings of each ventilation type, the highest symptom prevalence rates were found in ventilation systems with IUs or FCUs. A somewhat intermediate symptom prevalence occurred in variable/constant-air systems.

These British studies show that naturally ventilated and non-air conditioned, mechanically ventilated buildings had the lowest prevalence of work related symptoms. Because the other ventilation types examined were characterized by air-conditioning, and, in some cases, also by humidification, it has been hypothesized that microbiological air contaminants may be responsible for some of the higher prevalence rates of work-related symptoms found in these studies (213).

Additional epidemiological studies relating to BRS etiology have been carried out in 44 British buildings in the 1990s (215). These studies have shown a relation between symptom prevalence and some building factors such as exposure to passive ETS, inadequate thermal environmental conditions, and high TVOC levels. However, the study also suggested that psychosocial factors were more strongly related to symptom prevalence (215).

Systematic questionnaire studies of complaints were carried out in the 1980s in 14 town halls and 14 affiliated buildings in the Copenhagen area (216). The buildings chosen in this study were not previously categorized as being problem or healthy facilities (217). Environmental measurements such as the concentration of TVOC and CO_2, the microbial content in floor dusts, and thermal/air moisture parameters were made in one representative office in each building. Two parameters, the fleece and shelf factor, were measured in each study office. The fleece factor is a measure of the surface area of all porous room furnishings (e.g., carpets, drapes, upholstery) divided by the room volume. The shelf factor is the length of open shelves in the study room divided by total volume of the room studied (216).

Work-related symptoms reported most frequently in the Danish town hall study included eye, nose, and throat irritation, fatigue, and headache. A great variation in the prevalence of work-related symptoms was found between buildings. Workplace environmental factors such as total amount of floor dusts and the fleece and shelf factors of studied offices were related to symptom prevalence. Certain indoor environmental factors such as CO_2 concentrations, however, could not be related to the prevalence of work-related symptoms. The Danish town hall study is important because the amount of adsorptive and absorptive surface (fleece and shelf factors) for pollutants was postulated to be important in IAQ. In addition BRS were found not to be related to CO_2 concentration.

In the late 1980s a study in 61 office buildings was conducted in the Netherlands to evaluate BRS and building factors such as the presence or absence of air-conditioning (218). The presence of BRS was higher in air-conditioned as compared to naturally ventilated buildings. Factors such as gender, work satisfaction, presence of allergies and personal control of thermal environmental conditions were related to prevalence of symptoms.

Comprehensive IAQ studies were performed in 56 office buildings in nine European countries during the heating season of 1993–1994 (111). Measurements were made of physical factors such as temperature, relative humidity, and air velocity. Chemical measurements included CO_2, CO, and TVOC concentration. HVAC parameters such as the rate of outdoor air ventilation were determined. Perceived air quality was determined by using trained and untrained panels. A number of important findings were reported in the 56 building European study (111). Outdoor air ventilation rates were on average about 25 L/s and yet 30% of the occupants (50% of visitors) were dissatisfied with the air quality. This finding shows that compliance with ventilation standards (see Section 4) does not guarantee acceptable IAQ. Source control is relatively more important than dilution ventilation in achieving acceptable IAQ (see Section 6). The measurement of TVOC did not correlate with the subjective evaluation of perceived air quality (111). The 56 building European study suggested that occupants were a less significant source of pollutants than building materials, activities in offices, and the HVAC system itself.

3.3 Perceived Air Quality

A prominent complaint in buildings is the perception of odors and sensory irritants. Studies have shown that about 2.5 L/s (5 cfm) per person of high quality (clean) outdoor air is required to satisfy the substantial majority of adapted persons in the occupied space (3, 219) (see also Appendix A.1. of Ref. 178). A higher rate of outdoor air ventilation 7.5

L/s (15 cfm) is required to provide acceptable perceived air quality to visitors to the occupied space.

An approach toward understanding and quantifying the perception of air quality uses the bioeffluents emitted by a "standard" person. One *olf* is defined as the bioeffluent emitted by a person 18 to 30 years old, with a skin area of 1.8 m^2 and hygiene characterized by a daily change of underwear and 0.7 baths per day (10). The unacceptability of the indoor air caused by other pollutants sources such as building furnishings or the HVAC system itself is defined in terms of the number of olfs needed to cause the same degree of dissatisfaction as the standard person. Thus, moist dirt in a HVAC system might be the source of enough MVOCs to cause the same equivalent dissatisfaction as 14 standard persons or 14 olfs. Perceived air quality can not be measured by the collection and analysis of pollutants by instrumentation. The measurement of olf values as well as the determination of perceived air quality requires a trained panel of judges.

The *decipol* is used to quantify the perception of odors and irritants. One decipol is equivalent to the perceived air pollution that would cause the same dissatisfaction as the bioeffluents from a standard person (one olf) diluted by 20 cfm or 10 L/s of unpolluted or outdoor air. In the panel studies carried out by Fanger and colleagues (10), one decipol is the perceived air pollution that results in causing about 15% of judges to find the air unacceptable when entering the occupied space. Three decipols is the perceived air pollution that causes slightly more than 30% of judges to be dissatisfied with the air quality.

The olf and decipol concepts (widely used in research studies in the European community) were used to quantify air pollution sources in 15 office buildings and five assembly halls in the Copenhagen area (10). Panels of judges visited each building; and immediately assessed air quality during periods when the building was unoccupied and unventilated, unoccupied and ventilated, and occupied and ventilated. Although the ventilation rate in offices was on average 25 L/s (50 cfm) per occupant, the percentage of dissatisfied judges was in excess of 30%. For each olf associated with human occupancy (primarily bioeffluents), 6 or 7 olfs were estimated to be emitted from other sources in the buildings. The perception of poor air quality in these studies was assigned to the following pollutant sources: the HVAC system (42%), ETS (25 percent), furnishings and construction materials (20%), and occupants themselves (13%). The Copenhagen study showed that people themselves (body odors) were not the primary source of indoor air pollutants. Contaminants from the HVAC system and interior furnishings plus ETS were major contributors to the perception of poor air quality. In addition, the Copenhagen study showed that excessive amounts of outdoor air ventilation (25 L/s) were insufficient to dilute strong indoor contaminant sources that affect perceived air quality.

The outdoor air ventilation rates of 7.5 to 10 L/s (15 to 20 cfm) prescribed in the ventilation rate table of ASHRAE Standard 62-1989 (see section 4.2) are sufficient to provide acceptable perceived air quality as long as pollutant sources from the building itself are minimal (11). European studies have found that panels of judges can detect odors and sensory irritants from building sources that might not easily be detectable by instrumentation. For example, in a naturally ventilated building with different kinds of floor coverings, the perceived air quality (% dissatisfied) varied by as much 15% per floor (220). The sensory pollution load varied from 0.02 to 0.10 olf/m^2 depending on the pollutants emitted from the various floor coverings (220).

3.4 Crisis Buildings, Mass Psychogenic Illness, and Other Conditions

Baker (221) has reviewed the psychological and social factors that are important in office environments and may be important as predisposing factors in building problems. He points out that cues from the environment such as unusual odors or stuffy air over a period of time may be interpreted by occupants as an indicator that a toxic agent exists in the indoor air. This phenomenon may lead to the development of a "crisis building". In a very different situation, an anxiety reaction known as mass psychogenic illness may occur over a matter of a few hours or days. Mass psychogenic illness is characterized by hyperventilation and often by visits to hospital emergency rooms.

Stress and workplace organizational factors are considered to be predisposing factors for the development of crisis buildings and mass psychogenic illness. Many stresses exist in the modern office, such as routine and boring work, low pay, low status, little chance for advancement, lack of control over rigid work patterns, and lack of control over environmental conditions in the building (see Refs. 212 and 215).

Baker (221) has described how a toxic-hysteria crisis (crisis building) can develop in a stressful office environment. One or more occupants report nonspecific symptoms such as odor annoyance, headache, or irritational complaints. Over a period of weeks or months occupants come to believe that a toxic agent is present in their office environment. Management on the other hand believes that this is merely hysteria. Environmental investigations to detect toxic agents are inconclusive. Deep and sincere concern by occupants about their safety leads to heightened anxiety. An indoor environment in which occupants feel that it is no longer safe to work is a crisis building (221).

A number of intervention strategies have been suggested for dealing with crisis buildings (221). Most important is an understanding of the underlying stress factors and organizational characteristics of the office. The likelihood that the employee symptoms have a multifactorial basis should be considered. Effective communications between employees and management should be encouraged (see Section 6.8). This might involve a joint environmental committee to monitor symptoms and to measure parameters such as thermal environmental conditions (221). Changes in rigid workplace organizations and involvement of employees in the control of their workplace environment should be considered. Reynolds et al. (222) provide an example of a case which in part resembled a crisis-like building.

In contrast to the phenomenon of a crisis building, mass psychogenic illness is recognized by the acuteness of the anxiety reaction and the characteristic occurrence of hyperventilation, often followed by symptoms such as headache, dizziness, faintness, nausea, and weakness (221, 223). Gender-specific attack rates may occur and, most importantly, illness is transmitted both along sight lines and/or verbally. The triggering event for mass psychogenic illness is often a strong index case(s) such as an individual who precipitates an epidemic by announcing to colleagues that a toxic agent in the air caused illness (224) for example, the development of uncontrollable itching in view of colleagues (225). Mass psychogenic illness has been reviewed elsewhere (226–228). It should be realized that although instances of mass psychogenic illness have occurred in nonindustrial buildings, the vast majority of IAQ complaints do have an objective basis (221, 223).

A set of subjective symptoms including fatigue, depression, mood changes and headaches, among others, have been attributed to various environmental causes for most of the

20th century. At various times these symptoms have been called environmental illness, total allergy syndrome, environmental hypersensitivity disorder and twentieth century disease (229). More recently, a very similar suite of symptoms has been referred to as multiple chemical sensitivities (MCS) or ecologic illness, especially when the condition is acquired after exposure to low doses of unrelated chemicals. This concept was originally proposed in the 1950s by Randolph, a founder of the Society for Clinical Ecology, later known as the American Academy of Environmental Medicine (230). Clinical ecologists define MCS as a " . . . polysymptomatic, multisystem chronic disorder manifested by adverse reactions to environmental excitants . . ." (230). More recently, this condition has been referred to as idiopathic environmental intolerance (IEI) (231).

This concept has proven controversial and some consider the "diagnostic and therapeutic approaches to be inadequately supported by published studies" (230, 231). Others in the medical community have noted that there is no precise definition or case criteria of IEI; IEI symptoms are subjective, hence there are no physical nor measurable physiological abnormalities. The American Medical Association, because of these points, recommended in 1992 that MCS not be recognized as a medical syndrome (232). Finally, there is lack of agreement on pathogenic mechanisms and no experimental models are available for testing hypotheses pertaining to IEI. Some studies have claimed that subjects with IEI may have co-existing psychiatric disorders which will need to be addressed (233). Research protocols were suggested in 1992 by an expert panel from the National Research Council in 1992 (230), but the controversy remains essentially unchanged as of 1999 (231).

Since IEI or MCS cannot be defined clinically and information is lacking on specific agents in the indoor environment which cause the condition, a presumptive diagnosis of this condition in a building is more difficult to resolve than BRS or BRI. Several studies suggest that this condition is primarily a psychologic disorder (234–236).

3.5 Economic Issues

The exact number of occupants affected by BRS in nonresidential, nonindustrial buildings is unknown. However, estimates suggest that about 20 to 30% of the nonindustrial building stock may be classified as problem buildings; that is, those where BRS and impaired productivity occur because of poor IAQ (13). Since the total number of nonindustrial commercial buildings in the United States is about 4,000,000, this means that about 1,000,000 buildings have occupants with BRS.

Some estimates suggest that as many as one-third of problem building are also characterized by BRI (13). Although this estimate may be too high, it is clear that the costs associated with each instance of BRI can be significant. Several instances where BRI has led to the evacuation and restoration of a building or portion of a building are found in the literature. The building studied by Arnow et al. (19) originally housed about 1000 employees and was vacated for about three years during renovation. Renovations in the buildings studied by Hodgson et al. (97, 98) required 1 to 3 years. Costs associated with renovations of these buildings varied from $150,000 to $10,000,000. Direct health care and litigation costs in these types of problem buildings were also significant. Costs of renovating two newly constructed Florida courthouses that had to be vacated primarily because of fungal growth have exceeded $20,000,000 in one case and $40,000,000 in the second case.

It has been estimated that the indirect cost to U.S. employers for three sick days per year or a loss of six minutes of concentration ability per day is about $10 billion annually (13). Costs for improvement in operation and maintenance of existing buildings and better design and construction for new buildings to improve IAQ should be balanced against potential gains in occupant productivity. Various costs associated with the construction and operation of nonindustrial buildings are estimated in the 1980s as follows (13, 187):

- Salary cost: \$100 to \$300/(ft^2) (year).
- Lease cost: \$15 to \$50/(ft^2) (year).
- Building construction cost: \$50 to \$125/gross ft^2.
- Capital assets cost (furnishing, equipment): \$20 to \$100/ft^2 of floor area.
- Maintenance and operation costs: \$2 to \$4/(ft^2) (year).
- Environmental control costs: \$2 to \$10/(ft^2) (year).
- Utility cost: \$2 to \$4/(ft^2) (year).
- Costs to improve operation and maintenance programs: \$0.25 to \$1.0/(ft^2) (year).

The costs associated with upgrading the performance of buildings is predicted to be more than offset by savings achieved by decline in transmission of infectious aerosols (e.g., viruses and bacteria that cause the common cold), decline in allergic disease (e.g., asthma), and reduction in prevalence of BRS (237). The kinds of upgrades in building performance necessary to decrease morbidity and improve productivity include increased outdoor air ventilation (or upgrading the cleaning of recirculated air) (238) to dilute human sourced bioaerosols, prevention of moisture damage to lessen impacts of fungal and mite growth, and emphasis on source (pollutant) control in the design of building components including HVAC systems (see Section 6). Fisk and Rosenfeld (237) have calculated that these kinds of indoor environmental quality upgrades, if used in buildings in the United States would result in the following economic benefits:

- Savings of about 6 to 19 billion dollars because of reduced respiratory disease.
- Savings of about 1 to 4 billion dollars from reduction in asthma and allergies.
- Savings of 10 to 20 billion dollars from reduction in BRS.
- Gain of 12 to 125 billion dollars from increased productivity unrelated to health effects.

The costs associated with improved building performance are potentially more than offset through improved productivity of building occupants. It thus makes good economic as well as environmental sense to transform a problem building into a healthy building, and also to prevent a healthy building from degrading into a problem building.

Flatheim (239) has calculated the investment costs in Norway associated with the design operation and maintenance of buildings constructed according to the "clean building philosophy." Wages and wage related expenses totally dominate costs associated with workstation construction. The annual cost to upgrade a workstation over the standard cost is about $3,641. Total cost per work station as a unit of total building cost is about $70,000. The increase in occupant productivity and well being associated with enactment of the clean building philosophy should according to the calculations of Fisk and Rosenfeld (237) more than offset the investment costs of building upgrades.

3.6 Developing Countries

Higher morbidity and mortality in developing countries have multifactorial causes, with contaminated food, water, and air as major risk factors. The importance of sanitation is unquestionable; however, still in question is the extent to which indoor air pollution contributes to ill health in these nations. From the limited studies that have been conducted in developing countries, disease patterns appear consistent with a substantial effect of indoor pollution, although causal relationships are unclear.

Globally, particulates are one of the most important classes of air pollutants (see Section 2.7). This is especially true in developing countries, where there are significant combustion sources contributing to indoor air pollution. While air pollution is often associated with industrialization and urbanization, use of unprocessed or dirty solid fuels for cooking and heating, both in rural and urban households, contributes substantially to indoor pollutant emissions. Compared to modern kerosene and gas as cooking fuels, for example, unprocessed solid fuels produce 10 to 100 times more respirable particulates (240). In addition to respirable particulates, high emissions of CO, PAHs, and numerous VOCs are produced by open combustion of wood, coal, and biomass. Biomass (a term for organic matter such as plant material and animal dung) is widely used as a cooking and heating fuel in developing countries (240).

Chronic lung disease in women has been found to be strongly associated with cooking exposures on open biomass stoves. A study characterizing indoor air pollution in underground dwellings in a rural area of China demonstrated relatively high levels of combustion products, including combustion gases, particles, and PAHs resulting from cooking and heating with biomass or coal (241, 242). Other studies in China, where coal is widely used as a household fuel, provide evidence of higher lung cancer rates among women, implicating exposure to indoor coal smoke as a major risk factor. Other stove pollutants, such as wood smoke, have been shown to increase the risk of acute respiratory infections in young children living in homes with wood stoves. Such concerns led the World Bank, in 1992, to class indoor air pollution in developing countries as one of the four most critical global environmental problems (197).

4 VENTILATION SYSTEMS IN COMMERCIAL BUILDINGS

A HVAC system should provide the occupied space with acceptable IAQ and acceptable thermal environmental conditions. In addition, the HVAC system should remove stale contaminated air from the building and maintain appropriate pressure relationships between clean areas, contaminated areas (e.g., toilets), and the outside atmosphere.

The HVAC system should provide the occupied zone with an appropriate quantity of clean outdoor air and remove (dilute) air contaminants to the extent that the vast majority of people perceive the air quality as being acceptable. While dilution ventilation offers the only practical method to control odorous bioeffluents from occupants, other potential building contaminants such as VOCs, ETS, combustion products, and particulate (dusts) should be controlled at their source. This section describes the relationship between ventilation and IAQ, a topic discussed extensively in other reviews (243–247).

4.1 General Description of a HVAC System

The HVAC system of a large building contains one or more air handling units (AHUs). In the mixed air plenum of an AHU, outdoor air is usually mixed with a portion of the HVAC system's return air. The mixture of outdoor and return air enters a plenum which may contain low efficiency prefilters and more highly efficient filters located downstream of the prefilters. The prefilters trap larger particles and are replaced frequently so as to extend the life of the more costly downstream filters.

Filters (air cleaners) are generally rated in terms of weight arrestance and dust spot efficiency (ASHRAE 52.1) (178, 248). Dust spot efficiency is used to classify air cleaners according to their ability to remove fine dusts that can visually soil interior surfaces of occupied spaces. Bag filters used in some AHUs have a dust spot efficiency of 60 to 90%. Unfortunately, in many buildings only low dust spot efficiency (20% or less) filters are used to remove particulate from the ventilation air. Weight arrestance refers to the ability of an air cleaner to capture a coarse synthetic mixture of Arizona road dust, carbon black and cotton linters. The large particles whose capture enhances an arrestance rating will settle out quickly, whereas the small particles that are likely to cause BRS or BRI will have minimal effect on an arrestance rating. A prefilter with no rated (<20%) dust spot efficiency may have an arrestance of 50 to 90%. A new ASHRAE Standard (52.2) is pending to measure the efficiency of air cleaners in terms of ability to remove particles as a function of size (249, 250). For example, a 25 to 30% dust spot efficiency rating by ASHRAE Standard 52.1-1992 would be equivalent to removing at least 60% of 3μm particles by ASHRAE Standard 52.2-1999.

The air mixture after filtration enters the heat exchanger section, where heat is either added to or removed from the airstream as required to maintain the thermal comfort of occupants in the building. During the summer air-conditioning season, moisture is removed from the airstream as it passes over the dehumidification cooling coils. These coils are maintained at a temperature of about 42 to 52°F by a refrigeration or chiller system. Moisture from the airstream passing through the heat exchanger condenses on the outside coil surface when the dew point temperature of the air is greater than the coil temperature. This moisture collects in drain pans beneath the heat exchanger and should exit the AHU through drain lines with deep sealed traps. Water should not be allowed to stagnate in drain pans or in other portions of the AHU.

After passing through the heat exchanger and supply fan, conditioned air is distributed to occupied spaces through a system of ducts. The main air supply duct is usually constructed of sheet metal, and it, as well as the plenum housing the fan and heat exchanger, may be internally insulated with a porous liner for both thermal and acoustic control.

Internal fiber glass liners in plenums and ducts must be undamaged and have a structural integrity that does not allow for loose fibers to be entrained in the airstream.

Air from main supply ducts enters rigid branch ducts which in modern buildings (designed since the 1970s) often contain variable air volume (VAV) terminals. In a VAV system, the air volume delivered to the occupant zone is varied to maintain the interior temperature, and this is accomplished by a control system including a thermostat in the occupied space. In a constant air volume system that is often found in older buildings, the interior space temperature is maintained by varying the temperature of the conditioned air.

Air from rigid branch ducts usually enters a flexible duct and then passes into the occupied space through a ceiling-mounted diffuser. A portion of room air is then entrained into the supply airstream being discharged into the occupied space. If air supply inlets and return air outlets are properly sized and selected, outdoor air can be effectively distributed to the occupied zone.

In many buildings a considerable amount of the outdoor air that enters the HVAC system does not reach the breathing zone of occupants (244). Thus, air may leak out of unsealed portions of supply ducts into wall, ceiling, and other unoccupied spaces. Supply air may also short circuit into a closely positioned return air vent without benefiting persons in the occupied zone. Outdoor air may fail to reach the occupied spaces because of a malfunction in a VAV terminal (Table 65.9). VAV terminals should always provide a certain minimum supply of outdoor air per occupant (4, 247) to the occupied zone even when the thermostat setting is satisfied.

In an office (Table 65.9) of 1000 square feet with 7 occupants, the measured maximum supply air flow (approximately 50% outdoor air) was 800 cfm or 57 cfm outdoor air per person. Under minimum design flow rate conditions (see Table 65.9) 14 cfm of outdoor air per person would be provided, which is somewhat less than the minimum recommended by ASHRAE Standard 62 (4) for offices. Under actual minimum conditions (thermostat thermally satisfied) no outdoor air was being provided to office occupants by the defective VAV terminal. ASHRAE Standard 62 requires that when the total supply of conditioned

Table 65.9. Supply Air Flow Rates from Variable Air Volume Terminal Serving Office Zone with Seven Occupatns (Approximately 1000 sq. ft. Floor Space)[a]

Supply Air Measurements Condition	Total cfm from Supply Diffusers	Outdoor Air per Occupant[b]
Design maximum	750	54
Actual maximum[c]	800	57
Design minimum	200	14[d]
Actual minimum[e]	0	0[d]

[a]Ref. 251.

[b]Outdoor air comprises approximately 50% of supply air.

[c]Maximum flow rate measured when VAV terminal thermostat set to minimum temperature (55°F) setting, calling for maximum cool air.

[d]Does not comply with ASHRAE Standard 62-1989.

[e]Minimum flow rate measured when VAV terminal thermostat set to maximum temperature (85°F) setting, calling for minimum cool air.

air to an office space is reduced a minimum amount of outdoor air (20 cfm per person) must still be provided for acceptable IAQ.

The fraction of the outdoor air delivered to the occupied space (via the diffuser) that actually reaches the occupied zone (generally 3 to 72 in. above the floor) is defined as ventilation effectiveness. The minimum outdoor air ventilation rates recommended in ASHRAE Standard 62-1989 assume a ventilation effectiveness that approaches 100%. The ventilation effectiveness in most buildings is only 60 to 90%. It is implicit in ASHRAE Standard 62-1989 that more than the minimum amount of outdoor air as recommended by the ventilation rate procedure is required to compensate for imperfect mixing of outdoor air in the occupied space.

In many modern buildings, the above-ceiling cavity is used for the passage of return air from the occupied space back to the AHU. The use of a ceiling plenum instead of return ducts is associated with a number of problems such as entrainment of fibrous fire-proofing into return air and blockage of airflow by walls that extend to the underside of the floor above. The air in the return air plenum may travel back to the AHU through a main return duct or riser. Some of the return air enters the AHU mixed air plenum and some of the return air is discharged directly outdoors. Outdoor air inlets and relief (exhaust) air vents must be spacially separated so as to prevent reentrainment.

The amount of outdoor air brought into the HVAC system should be slightly more than the combined total of relief air plus air exhausted from the building by toilet and local exhaust fans. The building as a whole is thus maintained slightly positive (0.05 inches water column) (243, 244) with respect to the atmosphere so as to prevent the infiltration of unfiltered and unconditioned air through loose construction in the envelope and through other building openings.

In many buildings, FCUs and IUs located along exterior walls are used to condition the air in perimeter zones. FCUs contain small fans, low-efficiency filters, and small heat exchangers. These units condition and recirculate supply air (usually without any outdoor air supply) in peripheral zones. IUs are generally supplied with conditioned outdoor air from a central AHU. Conditioned air from the unit passes through nozzles and mixes with a portion of the room air. Because a large building may contain several hundred FCUs or IUs, maintenance of this part of the HVAC system is often neglected.

ASHRAE Standard 62-1989 (4) and the European Guidelines for Ventilation in Buildings (247) provide guidance on how much outdoor air ventilation is necessary in buildings. A minimum outdoor air ventilation rate of 7.5 l/s (15 cfm) and 10 l/s (20 cfm) per occupant is recommended for classrooms and offices, respectively by ASHRAE Standard 62-1989. This is in keeping with Mendell's (11) finding that BRS are more likely to occur in buildings where the ventilation rate drops below 10 l/s per occupant. In the 1992 European guideline (247) three different air quality levels (A, B, C) are proposed based on the % of occupants dissatisfied with the perceived air quality. The required ventilation rate to achieve quality level A (10% dissatisfied) is about 4× higher (16 l/s olf) than level C (30% dissatisfied; 4 L/s olf).

A definitive relationship between BRS and outdoor air ventilation rate has not been demonstrated (252). Even when the outdoor air ventilation is very high (e.g., 25 L/s per occupant) the percentage of occupants perceiving that air quality is unacceptable may be high (30%) (111). Thus, compliance with a ventilation rate of 10 L/s (Standard 62-1989) does not preclude poor air quality. The occurrence of a high incidence of BRS or a high number of occupants who perceive the air to be unsatisfactory even when outdoor air ventilation exceeds recommended guidelines is likely caused by factors such as strong emission sources of contaminants in the occupied space or the HVAC system. Thus, providing outdoor air ventilation through a dirty HVAC system is unlikely to promote the perception that the indoor air is acceptable.

Ventilation rates in a building are affected by pressurization. Temperature driven pressure differentials (stack effect) result from buoyancy of heated air (243). During the winter, warm air rises to upper portions of the building causing positive pressure and cool air enters the negatively pressurized lower floors. When stack effect is extreme (e.g., negative pressure of 0.05 to 0.10 water column on lower floors of a tall building) enormous volumes of outdoor air can enter openings in the envelope (253). Contaminants such as combustion products from loading docks can enter the occupied space with unconditioned outdoor air. Stack effect can be reduced by designing compartmentalized HVAC systems with one AHU on each floor or by providing more return air volume for lower floors (254).

Negative pressurization associated with HVAC operation can also occur for other reasons. The leakage of conditioned air into the return air plenum can cause depressurization in the occupied space. When the occupied space becomes depressurized in air conditioned buildings in warm humid climates, moisture infiltrates the envelope wall potentially resulting in condensation or near condensation conditions that allow fungal growth on biodegradable construction and finishing materials. Mechanical equipment rooms can be depressurized because of leakage of air into AHUs on the suction side of the fan. VOCs from cleaning chemicals, particulates from welding, and other contaminants generated in and around the mechanical equipment room then enter the ventilation air and occupied spaces. It is essential therefore that mechanical equipment rooms be fastidiously clean, not serve as work areas, and be kept free of stored materials that can contaminate ventilation air.

In several early studies CO_2 concentrations above 1000 ppm have been associated with an increase in occupant complaints in buildings (255, 256). The normal concentration of CO_2 in the outdoor air is about 325 ppm. In the indoor environment the CO_2 concentration is always somewhat elevated relative to the outdoor level because a CO_2 concentration of about 38,000 ppm is present in the air exhaled from the lung.

ASHRAE Standard 62-1989 recommends that steady-state levels of CO_2 in indoor air not exceed 1000 ppm, which is equivalent to the provision of about 15 cfm of outdoor air per occupant (100). Panel studies (219) (see Section 3.3) have shown that approximately 15 cfm outdoor air per occupant is needed to dilute human body odor adequately to a concentration where 20% or less of visitors are dissatisfied (80 percent acceptability) with the IAQ.

Although CO_2 concentration remains a good surrogate for human-generated contaminants such as body odor (257) and a good predictor of over occupancy conditions, the consensus of epidemiologic studies shows that prevalence of BRS symptoms is not related to CO_2 levels (91). Studies summarized by Bluyssen (111) indicate that air contaminants derived from nonhuman sources such as HVAC systems and interior construction and finishing materials are major sources of perceived dissatisfaction with indoor air. The concentration of CO_2 in indoor air would, of course, have no relation to these emission sources.

CO_2 can be useful in determining the quantity of outdoor air delivered to the occupied zone but only if the number of people remain constant and the CO_2 builds up to an equilibrium concentration (243, 244). Other methods such as decay of sulfur hexafluoride tracer gas and measuring supply air volumes and estimating the percentage of outdoor air in supply air are available to estimate the amount of outdoor air delivered to occupied spaces (244).

4.2 ASHRAE Standard 62

An ASHRAE ventilation standard at the end of the nineteenth century recommended the provision of outdoor air a minimum rate of 15 L/s per person to lower the risk of tuberculosis in occupied spaces (258). In the 1930s Yaglou et al. (3) showed that from 5 to 15 L/s of outdoor air per person was required to provide acceptable dilution of ETS and bioeffluents. The ASHRAE Standard for Natural and Mechanical Ventilation (259) subsequently prescribed a "minimum" outdoor air ventilation rate of 2.5 L/s (5 cfm) per person to provide acceptable but not odor-free conditions and a "recommended" outdoor air ventilation rate [generally a minimum of 5 L/s (10 cfm) outdoor air per occupant or more] to control odor annoyance. The minimum outdoor air ventilation rate recommended in ASHRAE Standard 62-1973 (259) (5 cfm per occupant) was intended to prevent indoor CO_2 levels from rising above 2500 ppm, or half the recognized occupational threshold limit for this gas.

As a result of the energy crisis of the mid 1970s, ASHRAE Standard 90 (260) published in 1975 recommended that only the "minimum" outdoor air ventilation rate of ASHRAE Standard 62-1973 (259) (2.5 L/s outdoor air per occupant) be used for ventilation (260). Concerns about providing enough outdoor air to dilute ETS and bioeffluents to acceptable levels were overlooked. Many of the HVAC systems designed for buildings constructed or retrofitted in the 1970s and 1980s have been designed on the basis of the "minimum" outdoor ventilation rate recommended in ASHRAE Standard 90 (260). Thus it is not surprising that air quality complaints have occurred in these buildings.

ASHRAE Standard 62-1981 (261) partially rectified the error of Standard 90 by recommending a minimum outdoor air ventilation rate of 10 L/s (20 cfm) per occupant for zones where smoking was permitted. The 2.5 L/s (5 cfm) of outdoor air per occupant value recommended for nonsmoking zones was intended to prevent CO_2 concentrations from rising above 2500 ppm. In practice, many building designers throughout the 1980s continued to specify outdoor air ventilation rates between 2.5 and 5 L/s (5 and 10 cfm) per occupant, by misinterpreting ASHRAE Standard 62-1981 (261) to imply that the nonsmoking and smoking ventilation rates could be averaged according to the anticipated number of smokers in the building.

ASHRAE Standard 62-1989 (4) was probably the most important document in the IAQ literature in the early 1990s. A key feature of Standard 62-1989 and its ventilation rate procedure is the increase in the minimum outdoor ventilation rate compared to its predecessor standard from 2.5 to 7.5 L/s (5 to 15 cfm) per person. Outdoor air requirements recommended by the ventilation rate procedure make no distinction between "smoking-allowed" and "smoking prohibited" areas. A minimum of 7.5 L/s (15 cfm) of outdoor air per person as specified in the ventilation rate procedure is recommended because it was believed that this is the minimum amount of outdoor air needed to dilute body and tobacco smoke odors to acceptable levels (100). The outdoor air requirements specified by ASHRAE Standard 62-1989 (4) must be delivered to the occupant breathing zone.

Standard 62-1989 also required that the design documentation for a HVAC system state clearly which assumptions are used in design. This allows others to estimate the limits of the HVAC system in removing air contaminants prior to commissioning and prior to the introduction of new contaminant sources into the occupied space.

A key provision in Standard 62-1989 (4) requires that when the supply of air to the occupied zone is reduced (for example, in VAV systems), provision be made to maintain a minimum flow rate of outdoor air to the occupied zone. Building maintenance is recognized as a key important factor in providing acceptable IAQ. Thus AHUs and FCUs should be easily accessible for both inspection and maintenance. Furthermore, inspection and maintenance should occur regularly. Specific mention is made of avoiding stagnant water in HVAC system. Caution is urged in the use of recirculated water spray systems that are invariably prone to microbial contamination. Special care is urged to prevent entrainment of moisture drift from cooling towers into outdoor air inlets. However, a minimum distance between makeup air inlets and external contaminant sources such as cooling towers, sanitary vents, and loading docks is not specified.

The outdoor air used in HVAC systems must meet ambient air quality standards for priority pollutants before it is introduced into the occupied zone. According to this criterion, the average long-term concentration of NO_2 and SO_2 for example, in outdoor air introduced into HVAC systems, shall not exceed 0.055 and 0.03 ppm, respectively. Ozone shall not exceed a short-term concentration of 0.12 ppm. An addendum to Standard 62-1989 (168) recommends that in accordance with the 1987 revision of the National Ambient Air Quality Standards, the long-term respirable particulate concentration (PM10) in outdoor air used in HVAC systems shall not exceed 50 $\mu g/m^3$. When outdoor air is unacceptable, contaminant levels must be reduced to acceptable limits by cleaning.

An indoor air quality procedure for achieving acceptable IAQ is present in Standard 62-1989 (4). In this procedure it is required that the concentration of air contaminants in the occupied zone be held below acceptable limits. The outdoor air ventilation rate is left unspecified. Guidance in Standard 62-1989 (4) as to what concentrations of indoor air contaminants are acceptable is furnished for outdoor contaminants covered in the ambient air quality standards and for four additional air contaminants of indoor origin, namely, CO_2 (1000 ppm), chlordane (5 $\mu g/m^3$), ozone (0.05 ppm), and radon (0.027 WL). Because acceptable limits and health effects data are currently unavailable for the vast majority of other indoor air contaminants, most designers do not use the air quality procedures of the Standard. The Standard does, however, make provision for the subjective evaluation of perceived air quality (see Section 3.3) as one means of implementing the IAQ procedure.

ASHRAE standards are periodically reviewed and updated, generally every 5 to 10 years. The first attempt at revising ASHRAE Standard 62-1989 culminated with the release of the public review draft ASHRAE Standard 62-1989R in August 1996 (178). The 1996 public review draft, which was subsequently withdrawn by ASHRAE, is a substantial milestone in the development of ideas concerning ventilation in buildings. The 1996 draft document contains 10 sections and 15 appendices and is useful because it represents an attempt to reach consensus on difficult topics ranging from the ventilation rates necessary to achieve acceptable perceived air quality to guidance on the use of porous materials on the airstream surfaces of AHUs and supply air ducts.

Some highlights of the 1996 public review draft of ASHRAE Standard 62-1989R (178) follow (see also Ref. 262):

- The draft standard was formatted in code (shall) and advisory (should) language so as to make it easier for portions of the standard to be incorporated into building codes.
- The provision of outdoor air ventilation to achieve acceptable perceived air quality was based on the necessity to dilute contaminants from people (e.g., bioeffluents) *plus* contaminants from the building itself (e.g., VOC emissions).
- The intent of the standard was to achieve acceptable perceived air quality for a substantial majority of occupants. The ventilation rates listed is in Table 6.1 of the Standard (178) were based on no smoking. A separate appendix (Appendix E) described the ventilation necessary to achieve acceptable perceived air quality in the presence of smoking. Acceptable air quality, according to the draft Standard, can not be achieved in the presence of ETS because of the associated health risks.
- CO_2 is not mentioned as a pollutant in the body of the draft standard (178). The notion that acceptable perceived air quality can be achieved if the CO_2 level is less than 1000 ppm was dropped. It was recognized that many pollutants such as those from building furnishings can effect sensory irritation and the perception of acceptable perceived air quality. Appendix F of the draft standard (178) describes the relation of CO_2 to perceived air quality and the use of CO_2 in demand controlled ventilation.
- Cleaning of the outdoor air introduced into HVAC systems to meet NAAQS standards was deleted.
- Section 5 on "general requirements" of the draft standard contained detailed design principles for HVAC systems. Thus, buildings in humid climates that were air-conditioned must be designed to maintain a net positive or neutral pressure relative to ambient air. In buildings in humid climates, the mechanical cooling system was to be operated to achieve a relative humidity no greater than 60% during occupied periods and no greater than 70% during unoccupied periods in order to limit fungal growth.
- Sections 7 and 8 of the draft standard contained requirements for "construction and system start-up" and "operating and maintenance procedures," respectively. Regular inspection of HVAC components such as outdoor air dampers, plenums, dehumidifying cooling coils, drain pans, and outdoor air intake louvers was sensibly required.

In October, 1997 ASHRAE Standard 62-1989 Ventilation for Acceptable Indoor Air Quality was placed under "continuous maintenance," meaning that various components of the existing standard (4) will be revised on a section-by-section basis. In the continuous maintenance process various sections of the current standard such as the IAQ procedure, operation and maintenance, ventilation rate procedure, etc., will be revised in the form of "addenda." At the time of writing there were over 15 addenda under various stages of development. Another major change in the Standard 62 revision process has occurred in the scope of buildings covered by the standard. The standard setting committee has been divided into ASHRAE Standard 62.1 Ventilation and Acceptable Indoor Air Quality in Commercial, Institutional, and High Rise Residential Buildings and a separate committee (ASHRAE Standard 62.2) on smaller residential buildings.

Although the outcome of the continuous maintenance process will evolve over several years, these general aspects of the revision process are evident:

- The Standard will apply only to new buildings or additions to existing buildings.
- The costs associated with changes relative to benefits derived will be considered. The influence of various economic interest groups on the ASHRAE Standard 62 development process has been reviewed by Vogt (263).
- Concentration limits for pollutants will be specified only if recognized by a nationally known authority and only if standardized test procedures are available over the concentration range of interest.
- Informative language on how to improve indoor environmental quality beyond the requirements of the current standard (4) and future standard will be deleted.

Comparison of the technical content of some of the continuous maintenance addenda (ASHRAE addendum I, reference 264; ASHRAE addendum M, reference 265) with the 1996 public review draft (178) offer insight into the future direction of the standard development process.

Section 7.1.3 of the 1996 public review draft (178) contained guidance on isolation of major construction areas. This section required that when an occupied building undergoes a major renovation, the construction area was to be isolated from directly adjacent non-construction areas by . . . "temporary walls, plastic sheeting, or other vapor retarding barriers." These construction areas were to be maintained at a negative pressure relative to the adjacent non-construction areas by either exhausting construction areas and/or pressurization of adjacent areas. In Addendum I (264) Section 7.1.4.2 describes protective measures when alterations occur in existing buildings as follows: "Measures shall be employed to limit the migration of contaminants to occupied areas during construction. Examples of acceptable measures include, but are not limited to, sealing the construction area using temporary walls or plastic sheeting, exhausting the construction area, and/or pressurizing contiguous occupied areas." Careful reading of the 1996 (178) and 1999 (264) versions indicates that the requirement for negative pressurization in the construction area was deleted according to the 1999 addendum.

Section 8.3.8 on drain pans in the 1996 public review draft (178) stated that . . . "In humid climates, drain pans and other adjacent surfaces that are subject to wetting shall be visually inspected for cleanliness and microbial growth at a minimum of once every six months and cleaned if needed." Furthermore, " . . . Biofilms, slimes, and fungal growth found in drain pans and other adjacent surfaces shall be physically removed." Section 8.4.6 on drain pans in the 1999 Addendum M (265) states that " . . . Drain pans shall be visually inspected for cleanliness and microbial growth at a minimum of once per year during the cooling season . . . and "cleaned if necessary . . ." A later section (8.4.13) states that . . .

"visible microbiological contamination shall be investigated and rectified." The 1999 (265) addendum thus lessened the frequency of inspection and deleted the requirement for physical removal of biofilms, slimes, and fungal growth.

The proposed ventilation rate procedure in addendum N (179) of the continuous maintenance of ASHRAE Standard 62 recognizes that outdoor air is required to dilute occupant generated and building sourced pollutants. The minimum requirement for offices is 3.0 L/s (6.0 cfm) per person *plus* a building component of 0.3 L/s m^2 (0.06 cfm/ft^2). For general classrooms the minimum outdoor ventilation requirement is 3.0 L/s (6.0 cfm) per person plus a higher building component of 0.5 L/s m^2 (0.1 cfm/ft^2). These requirements are based on no smoking and an adjustment to compensate for the ventilation efficiency of the occupied space (179). The impact of the proposed ventilation rate procedure in Addendum N (179) is to generally lower the outdoor air ventilation rates for densely occupied spaces while raising the ventilation rates in some sparsely occupied spaces.

4.3 Pollutant Sources and Design Issues in HVAC

In mechanically ventilated buildings, both the quantity and the quality of air supplied to occupants is highly dependent upon the HVAC system. In most studies where the quantity of outdoor air supplied to occupants has been measured, the risk of BRS increased as the ventilation decreased below 10 L/s (20 cfm) (266). The occurrence of BRS symptoms has also been associated with air-conditioning (267). In studies where the perception of acceptability of conditioned air from a building's HVAC system has been compared to that of the outdoor air, the former was generally found to be worse or poorer in quality. The perception of poor quality conditioned air was thought to be caused by pollutants originating from the HVAC system itself.

HVAC systems can become major sources of pollutants that lead to degraded air quality for the following reasons:

- External, often malodorous air contaminants such as water droplets from cooling towers (may contain *Legionella* and water treatment chemicals) and combustion products from garages or loading docks may enter the building through the HVAC system outdoor air intakes.
- Dirt on airstream surfaces in HVAC systems is a reservoir for various kinds of pollutants (e.g., VOCs, ETS, fungal spores, pollen), which can be subsequently emitted into supply air (268). The accumulation of dirt in HVAC systems is especially a problem where the filter deck efficiency is poor or where porous materials line airstream surfaces and collect dirt.
- Microbial contaminants including MVOCs originate in moist niches in HVAC system mechanical equipment. Fungi grow on dirty surfaces such as in moist filters and damp insulation. Standing water in drain pans and humidifiers and water droplets emitted from these sources as well as from the surfaces of dehumidifying cooling coils may contain yeasts, gram-negative bacteria, and endotoxins.
- Disinfectants and biocides used in humidifiers and drain pans may be emitted into supply air.

- VOCs from paints, solvents, and other chemicals often stored in mechanical equipment rooms can enter the HVAC system through openings in negatively pressurized plenums (e.g., fan plenum of AHU; see Section 4.1).

Several recent studies have shown that fungi can colonize AHU filters (269) primarily as a function of the increasing moisture content which results from the hygrophilic property of the dirt which becomes trapped in the filter meshwork (270). Fungi can grow in unexpected niches in HVAC systems such as the buried piping system where outdoor air enters a heat recovery unit (271). During the summer, condensation can occur as moist outdoor air flows through the relatively cool piping system leading to the growth of fungi and actinomycetes. Nonmicrobial pollutants can also enter the breathing zone from unexpected niches in HVAC systems such as the airpack unit supplying cabin air in a commercial airplane (272). Methylated siloxane derivatives, methylated propane and butane derivatives, and combustion products can enter cabin air when the plane's outdoor ventilation air passes through a malfunctioning catalytic converter.

Perceived air quality can be degraded by use of aged HVAC system filters (273–275). As filters become progressively dirty up to about 1/3 of occupants become dissatisfied with air quality (274). Perceived air quality does not improve with increased ventilation (275) presumably because of contaminant emissions associated with oily and dusty airstream surfaces in AHUs and duct work. The amount of dust in supply ducts in Finnish buildings has been reported to range from about 3 to 58 g/m^2 (276). Dust on airstream surfaces of HVAC systems presumably acts as an adsorptive sink for varied pollutants including VOCs, ETS, and MVOCs (276, 277). Sampling for airborne contaminants before and after air duct cleaning in the United States failed to demonstrate a decrease in particulate or bioaerosol levels in indoor air following cleaning (278). However, supply air flow rates generally increased following cleaning (278).

HVAC systems can be designed so as to minimize possible sources of contaminants from or in components such as outdoor air inlets, AHUs, and supply duct work (279). Construction guidelines for Minnesota schools recommend (*1*) location of outdoor air inlets remote from possible outdoor pollutant sources, (*2*) nonporous airstream surfaces in duct work, (*3*) avoidance of elbows in duct work, and (*4*) use of low VOC emitting materials (280). New concepts such as supplying 100% outdoor air delivered low in classrooms have been successfully introduced (281). In Norway, Flatheim (239) has introduced a proactive clean building design including features as follows: (*1*) The intent of design is to reduce BRS as much as possible, (*2*) dust and debris accumulation are reduced as much as possible, (*3*) IAQ concerns override energy conservation issues, (*4*) the outdoor air ventilation rate for offices is 4.2 L/s m^2 (0.84 cfm/ft^2) and (*5*) the extra costs associated with this proactive design are marginal compared to increase in productivity and occupant well being (see Section 3.5).

4.4 Thermal Environmental Conditions

Thermal environmental complaints are common in office buildings. Often IAQ problems that are initially perceived to be due to chemical or ventilation parameters are found upon detailed evaluation to be caused by unsatisfactory air temperature and air movement parameters.

In the 1970s, the imposition in buildings of temperature restrictions as a means of reducing energy costs led to widespread thermal environmental problems characterized by complaints that the air was too warm or cold, too humid or "stuffy" (282). Human acceptance of a thermal environment is related to a number of variables such as metabolic heat production, the transfer of heat between the occupant and the environment, and physiological and body temperature adjustment. Heat transfer is influenced by variables such as dry bulb temperature, thermal radiation, moisture levels, air velocity, clothing insulation, and metabolic activity. The body is in thermal equilibrium when the net heat gain or loss is zero (i.e, when heat gain, including metabolic heat, exactly equals body heat loss). Thermal comfort is presumed to be optimal when the body is close to or at thermal equilibrium with its indoor environment. A number of factors in offices may adversely affect the body's thermal equilibrium and lead to thermal dissatisfaction. Examples include asymmetric thermal radiation that affects people sitting near sun-facing windows, vertical temperature differences in a room, cold drafts, and cold floors and walls. Performance criteria for environmental conditions that 80 percent or more of sedentary occupants in indoor environments will find acceptable have been defined in ASHRAE Standard 55-1992 (283). This standard recommends ranges of temperature and air moisture levels as well as limits on thermal radiation, air velocity, and other variables that should collectively be acceptable to the majority of occupants carrying out light sedentary work when attired in typical summer or winter clothing. Some highlights of this standard follow.

- Winter: an operative temperature (average of mean radiant temperature and dry bulb temperature) range of 68 to 76°F (20 to 24.5°C) is acceptable in a relative humidity range of 30 to 60%.
- Summer: an operative temperature range of 73 to 80°F (22.5 to 27°C) is acceptable in the relative humidity range of 30 to 60%.
- In order to prevent local thermal discomfort the vertical air temperature gradient measured 4 inches and 76 inches above the floor should not exceed 3°C (5°F).
- In order to minimize foot discomfort the surface temperature of the floor should be in the 18 to 29°C (65 to 84°F) range.

ASHRAE Standard 55-1992 (283) provided for an upper limit of humidity of 60% to promote both comfort and to minimize microbial growth and other moisture related problems. An addendum to ASHRAE Standard 55-1992 (284) eliminated the upper acceptable relative humidity limit of 60%. Consequently, according to Standard 55-1992 (283, 284), it is acceptable in terms of thermal comfort for the relative humidity to exceed 70 and even 75%; conditions which are known to be conducive to the growth of a number of molds.

Recent work has shown that perceived air quality is influenced by physical factors such as temperature, relative humidity, and air motion (285). In general, air at a given pollution level is perceived as more acceptable when the temperature and relative humidity are lowered (286, 287). Air at a temperature of 28°C and a relative humidity of 70% is perceived as unsatisfactory by 95% of people regardless of whether the air is clean or polluted (286). Thus, temperature and relative humidity have more of an effect on recommended outdoor air ventilation rates than is currently recognized in ventilation standards such as ASHRAE 62 (4). Cool dry air is generally perceived as being freer of contaminants than warmer or humid air (285). It should be recognized however, that the potential for draft complaints increases at lower air temperatures (283).

5 IAQ EVALUATION PROTOCOLS AND GUIDELINES

Approaches to solving IAQ problems using traditional industrial hygiene techniques are generally inadequate. The traditional industrial hygiene approach to solving occupational health problems involves recognition of the hazard (because input and output materials are known, the identity of air pollutants is generally preestablished), measuring the contaminant(s) by NIOSH- or Occupational Safety and Health Administration (OSHA)-approved procedures, interpretation of analytical data in terms of TWA, TLVs, and suggestion of corrective action such as containment of the contaminant at its source. A common result of this approach when applied to IAQ evaluations is that compliance with industrial workplace standards is demonstrated but building-associated complaints from occupants persist (288).

For IAQ evaluations, more sensitive protocols are needed because comfort, occupant well-being, and general population susceptibilities are not addressed by industrial workplace TLVs (288). In IAQ evaluations, sources of contaminants such as bioeffluents and VOCs are diverse and diffuse. Concentrations of air contaminants are often one to four orders of magnitude less than the TLVs. Contaminant control strategy in mechanically ventilated buildings is based almost universally on dilution ventilation by a HVAC system that is designed primarily to provide acceptable thermal conditions in the occupied space.

The evaluation approach described here is basically that of building diagnostics, a process wherein an experienced investigator uses available knowledge, techniques, and instruments to predict the likely performance of a building over a period of time (288, 289). The net result of building diagnostics is the development of recommendations to improve the HVAC system and occupant well being. Diagnostics as used in the context of IAQ evaluations is usually divided into a qualitative and a quantitative phase. Considerable literature is available on protocols dealing with IAQ investigations and the reader is referred to these publications for additional guidance (7, 290–294).

5.1 Qualitative Evaluation—General Principles

The qualitative IAQ evaluation begins with a consultation phase that may be accomplished through an extensive telephone interview or by a site visit. During this phase of the evaluation, objectives and scope are defined and a preliminary hypothesis as to the likely reason for occupant complaints is formulated. At the building site, the nature of the health complaints is reviewed, environmental factors that may be responsible for complaints are visually evaluated and a qualitative engineering assessment of the HVAC system is conducted. An objective of the qualitative evaluation is to provide recommendations for remedial actions in a manner that avoids the necessity of costly objective measurement of air contaminant concentrations or HVAC system performance parameters.

Instrumentation used during the qualitative IAQ evaluation is limited, and intended primarily to guide the investigator in verifying the hypothesis as to the nature of the building-associated problem. Direct-reading instruments to measure parameters such as relative humidity, operative temperature, and CO_2 and respirable particulate concentrations are often useful.

A key aspect of the qualitative evaluation is the analysis of a building's HVAC system. HVAC system mechanical components are visually examined for deficiencies with regard to design, operation, and maintenance parameters. The control strategies that govern HVAC system operation must be thoroughly understood. Essential components of the qualitative evaluation of the HVAC system include a smoke pencil for visualization of airflow patterns, instruments to measure air velocity and building pressurization, a complete set of mechanical plans and specifications, and the on-site availability of the building facility engineer.

Care must be taken at this phase of the evaluation to exclude the possibility that occupant complaints are due to lighting, acoustic, ergonomic, or psychosocial problems. Thus the possibility that complaints are correlated with labor management difficulties or with occupant perceptions that a toxic agent is present in the air (see Section 3.4) should be explored.

During the qualitative evaluation consideration should be given to the many objective factors that may cause IAQ problems including poor design, operation and maintenance of the HVAC system as well as to strong pollution sources in the occupied space and in the HVAC system:

When health complaints are reviewed, it is essential to determine if the problem is one of occupant discomfort and annoyance (thermal discomfort) or if one or more occupants have a BRI. When a BRI is apparent, immediate medical attention for the affected occupants is required concomitant with the initiation of appropriate remedial actions (288).

Performance criteria for maintaining an indoor environment in which the substantial majority of occupants do not express annoyance or dissatisfaction should be reviewed with building management/occupants during the qualitative evaluation (288). Some performance criteria for comfort are found in ASHRAE Standards 55 1992 (283), ASHRAE Standard 62-1989 (4), European Guidelines for Ventilation Requirements in buildings (247) and the Nordtest Report in Indoor Climate Problems (246). Exact performance criteria for the vast majority of contaminants found in indoor air are unknown. However, the prevention of discomfort and annoyance in nonindustrial environments, such as offices, implies control of air contaminants to concentrations several orders of magnitude below industrial TLVs. An in-depth discussion of the qualitative IAQ evaluation is found elsewhere (288).

5.2 The Qualitative Evaluation—Checklist

Review all telephone conversations and written information relative to existence of thermal environmental problems, BRS, BRI, ergonomic issues, and potential psychosocial issues. Define a hypothesis. What is the likely problem based on pre-assessment telephone interviews and review of available documentation? Be prepared to discuss your hypothesis and your plan for the qualitative evaluation. Also be prepared to conduct an exit interview (later in day) at which time major observations, conclusions, and preliminary recommendations should be discussed with building management and employee representatives.

Carry out interviews with office managers and employees or their representatives in order to form a conclusion as to the presence or absence of BRS. Determine by interview if higher attack rates of BRS occur in certain occupied zones. Elicit information on whether or not prevalence of BRS symptoms follow a daily, weekly, or seasonal pattern.

Obtain information on the possible existence of BRI among one or more occupants during interviews with management and employee representatives. What evidence suggests that occupant illnesses are building-related? For example, is there evidence to suggest that acute febrile respiratory illness among building occupants follows a nonseasonal pattern and occurs at a higher prevalence rate than in the general community population?

Determine by the interview process if occupant complaints have a thermal environmental basis. Is there a seasonal or a zonal aspect (for example, thermal complaints in interior zones only) to thermal problems? Finally, are complaints in the building multifactorial, suggesting that BRS, thermal environmental problems, and possibly BRI all coexist (13, 246, 295). It is important during the qualitative evaluation to examine not only the complainant work area but also noncomplainant zones and the HVAC system serving both areas. At the conclusion of the qualitative evaluation the investigator should clearly understand how the building operates (293).

5.2.1 Microbial Contaminants

Is there evidence of current of past water damage? How extensive or localized is the water damage and what are the likely sources of the water? (Table 65.10).

Is visible mold present on interior finishes and construction materials? How extensive is the surface area covered by mold? (Table 65.10).

Table 65.10. Corrective Actions for Microbial Contamination

Type of Problem	Corrective Action	Refs.
Floor and roof leaks	Remove moisture from building infrastructure so as to prevent mold growth	296
Visually moldy materials	Remove in a manner that prevents dispersion of fungi spores	7, 69
Condensation in building envelope	Prevent condensation by proper use of air barriers and vapor diffusion retarders; eliminate locally cool surfaces	69
Materials susceptible to biodeterioration	Avoid their use in wet niches in occupied spaces or the HVAC system	69
Water stagnation	Design drain pans for self drainage	178
	Continuous bleedoff required for sump humidifiers	295
Biofilm in HVAC	Physical removal required	69, 178
Biocides and disinfectants	Must not be aerosolized in functioning HVAC systems	69, 178

Are moisture problems evident in the building envelope? Is there evidence of biodeterioration (condensation, mold growth) on the room (occupied) side of the envelope? Are vapor diffusion barriers correctly positioned for the building's climatic location? (Table 65.10).

Is there evidence of hidden microbial growth such as musty odors (MVOCs)?

Do records indicate that relative humidity consistently exceeds 60%? Measure the temperature and relative humidity of room air plus the surface temperature and relative humidity of finishes and construction material that may be moldy or show evidence of mold growth.

Are porous materials such as carpet and insulation present in damp niches in the building? (Table 65.10).

Are dead botanical materials such as bark chips used in moist locations such as in an atrium?

Are the HVAC outdoor air inlet and other building openings positioned such that bioaerosols from cooling towers are likely to be entrained? Consider horizontal and vertical separation distances as well as airflow patterns around the building and prevailing wind direction.

Is there standing water in the HVAC (e.g., drain pans, humidifier sumps)? Are biofilms (slime) present on wet surfaces? (Table 65.10).

Are airstream surfaces in the outdoor air intake, mixed air, filter, fan, and supply air (and supply ducts) plenums damp and dirty? Is visible mold present? How much of the airstream surface is lined with porous insulation? (Table 65.10).

Are HVAC filters wet? Check both visually and by touching. Discard wet filters.

Does moisture carry over from cooling coils to downstream HVAC surfaces?

What protocols are used to clean HVAC components? Is cleaning done during off-hour periods? Are biocides used during cleaning? Are biocides and odor-maskers (fragrances) used in active HVAC systems? (Table 65.10).

What protocols are used to clean and to control moisture in unit ventilators, FCUs, and IUs that may be present?

Has a sewage backflow occurred? If so, consult IICRC guidelines (297).

5.2.2 *Volatile Organic Compounds and Odors*

Are malodors present? The presence of body odors indicates inadequate ventilation. Chemical odors might be associated with cleaning compounds, restroom deodorants, personal care products, or building activities. Control is achieved by identifying and eliminating the odor source. Environmentally preferred products with "low or no odors" should be substituted (Table 65.11).

Has renovation recently occurred? If yes, were VOCs exhausted directly out of the building or recirculated by the HVAC system? Determine if high solvent materials were used in the renovation. Recommend use of low solvent emitting materials (Table 65.11).

Table 65.11. Corrective Actions for VOCs and Other Contaminants

Event	Preventive or Corrective Action	Refs.
Renovations including painting, new carpet, ceiling system installation	Allow wet products to dry or cure as much as possible before installing dry, porous materials which may become secondary sources (sinks).	116, 120, 144, 298
Strong odor during normal operation of building	Identify chemical(s) and trace to its source; eliminate source and replace with alternative product. If source is an activity such as photocopying or graphic production, provide local exhaust ventilation.	130, 201
Sewer-like odor in building	Fill all plumbing traps with water. Make sure there are no cracks or leaks in plumbing lines.	201
Musty or moldy odors	Confirm presence of MVOCs; find moisture and mold sources and remediate.	50
Odors from nearby activities in HVAC mechanical equipment room	Adjust pressurization.	244
Eye irritation in new remoldeled office	Select and use low-formaldehyde and/or VOC emitting products such as wood and upholstered furnishing.	130, 198
Solvent-like odors in office areas during roof repair	Schedule these activities during non-business hours; maintain renovation and construction work areas under negative pressure relative to adjacent spaces.	201

Are specialized processes such as printing, graphics production, or darkroom work occurring in occupied spaces? Are chemicals from these processes exhausted directly out of the building? If not, install local exhaust ventilation. Determine if office equipment has scrubbers for ozone.

Have pressedwood products that contain or emit formaldehyde been used during renovation? If chemical odors are present increase outdoor air ventilation. Use low VOC (including formaldehyde) emitting wood products for next renovation. Consider alternative wood products that do not emit formaldehyde (Table 65.11).

Determine who is responsible for monitoring the use of cleaning chemicals and solvents by occupants and cleaning personnel. Are protocols in place to restrict usage of toxic, irritative, and odorous chemicals in the building and its HVAC system?

5.2.3 Other Indoor Pollutants

Are HVAC outdoor air inlets located within three stories of the ground (or at grade or below-grade levels) where they may be contaminated by combustion products from garages, loading docks, or vehicular traffic? Does the time of peak occupant complaint correlate with vehicular activity if combustion product entrainment occurs.

Are intermittently operated sources of combustion products near HVAC outdoor air intakes (e.g., gas fired heaters in rooftop AHUs or diesel powered emergency generators)?

Does stack effect induce infiltration of combustion gases during winter months?

Are pesticides applied in the building, and what protocol is followed in their application?

Is there a pesticide management program?

Are routes present by which soil gas including radon can enter building sealed? A subfloor depressurization system may be needed to prevent radon entry (189, 204, 205).

Can ETS be transported to nonsmoking areas? This is important in multiple-tenant buildings where ETS can be transported from a smoking floor to a nonsmoking floor. ETS can be transported by negative pressurization and stack effect into smoke-free buildings from smokers standing just outside a negatively pressurized building lobby?

What kinds of practices are used by facilities and housekeeping staff to remove settled dusts from floors and walls and from airstream surfaces in the HVAC system? When are these cleaning activities performed? Are settled dusts dispersed into the air during the cleaning process? Initiate cleaning with a HEPA filter vacuum if dust control is a problem.

5.2.4 HVAC System Design

Are HVAC system outdoor air inlets and other building openings inappropriately located near external contaminant sources?

Are original HVAC plans available? Do available plans incorporate remodeling that has occurred or changes in occupant and/or thermal loads?

Is porous liner located on airstream surfaces in moist niches in the HVAC system? Are liners moldy? Remove moldy liners.

Was the building designed to meet the minimum outdoor air ventilation rates recommended by ASHRAE Standard 62-1989?

What is the design efficiency of the HVAC system filter deck? Can filtration efficiency be increased from 60% for 3 μm particles to 60% for 1 μm particles?

Do VAV terminal boxes have minimum set points that provide at least 15 to 20 cfm of outdoor air per occupant when the zone thermostat is satisfied? If not, replace or fix the equipment and/or controls.

5.2.5 HVAC System Operation

Are control system plans available that describe how much outdoor air ventilation is provided during the heating and cooling season and during transitional climatic periods?

Do outdoor air inlet dampers open and close (check for minimum stops) according to design criteria?

Is the occupied space (especially complaint area) stuffy and warm? Do VAV terminal boxes close when the thermostat is thermally satisfied? Does supply air short circuit to return air vents? Are return air vents present?

Are thermostats properly positioned to control heating and cooling requirements in perimeter (envelope) and core (interior) areas? Are temperature recommendations of ASHRAE Standard 55-1992 (283) being met for all occupied zones?

Are exhaust systems properly removing combustion products from garage and dock areas and keeping these areas negatively pressurized relative to occupied spaces?

Are unexpected air flows likely (243, 244)? Do observations indicate that unconditioned outdoor air can enter the return air plenum through openings in the envelope? Is the occupied space negatively pressurized relative to the outdoor air (see Section 4.1)?

5.2.6 HVAC System Maintenance

Are written preventive maintenance protocols available for the HVAC system including FCUs and IUs? Are the protocols being followed?

Is a written preventive maintenance plan available for the cooling tower including bacterial, scale, and corrosion control?

Are HVAC facilities staff trained with regard to recognition and control of potential IAQ problems?

Are mechanical equipment rooms and HVAC plenums used for storage of chemicals, solvents, paints, fertilizers, pesticides, and other potential indoor pollutants?

Are air airstream surfaces in the HVAC system clean and dry? This includes dampers and pressure sensor stations, the filter plenum, plenums downstream of the heat exchanger, and air supply ductwork.

5.3 Quantitative Evaluation

In some building investigations, objective measurements of contaminant concentrations, subjective measurements of occupant perception of acceptability, and measurement of HVAC system operational parameters are required to document the hypothesis, to justify recommendations (288, 299), or for litigation purposes. Clear pathways should be predetermined on data interpretation when sampling is performed (7, 246).

In order to make appropriate interpretations, sampling for indoor air pollutants must be carried out with a complete understanding of HVAC system dynamics. For example, concentrations of total and specific VOCs in an office building may vary by an order of magnitude depending on whether the occupied zone is being provided with maximum (economizer) or minimum amounts of outdoor air (139).

The qualitative evaluation should have indicated potential pollutant sources and pathways, and sampling should include contaminant collection in worse case locations. In the case of VOCs, the building itself (emissions from finishing materials), activities of people (use of a photocopier), or outdoor activities (vehicular traffic) are all potential sources. The sampling strategy used should include contaminant collection in a worst case location. If building finishes are suspected to be strong sources of VOC, the worst case samples should be collected early in the workday before ventilation has diluted contaminant concentrations. If VOCs are likely being emitted from photocopied paper, sampling should occur in the copy room or where printed paper is stored. In the case of entrainment of vehicular emissions, sampling at the outdoor air inlet is required. In addition to sampling in worse-case locations, quantitative assessment of contaminants should occur in control locations (e.g., in "control" locations where minimal pollutants are anticipated), as well as in the zone of highest-quality outdoor air (e.g., facing into the wind on the roof at a site remote from external contaminants). Data interpretation often involves a comparison of worst case with indoor and outdoor control locations.

Sampling is performed differently depending upon the objective of the quantitative evaluation. If the objective is to determine occupant exposure, the samples are collected in breathing zone locations under normal building operation conditions (HVAC system is on, routine activities occur in the occupied space). However, if the objective is to demonstrate that a contaminant source is present, then the air sample, for example, for VOCs may be collected adjacent to the finishing material (or determine emission characteristics of the finishing material directly in the laboratory; see Section 2.2.3). In the case of fungi, source sampling may include collection of dust, sticky tape and bulk (piece of material) samples for analysis (7).

Quantitative measurements are highly affected by the operation of the HVAC system. Contaminants sourced in the building will likely build up rapidly in occupied zones if a VAV terminal completely closes. Since less outdoor air is used in HVAC systems of many buildings during periods when it is very warm or very cold outdoors, the season of the year is important with regard to dilution of building sourced contaminants like VOCs. Season of the year is an important consideration when air samples of culturable fungi are obtained. Different concentrations and viability of fungi are expected in the cooling season when the building and its HVAC system are humid and moist as compared to the heating season when materials are dry.

Quantitative evaluations may involve subjective measurements made by questionnaire or by panels of judges. These procedures have been reviewed elsewhere (247, 300, 301; see also Section 3.3). Procedures for quantitatively assessing the performance of HVAC systems are reviewed by Bearg (243).

5.4 IAQ Guidelines

No federal agency has authority to regulate BRI. The Clean Air Act of 1970 (U.S. Public Law 90-148) cannot be used to regulate IAQ (302). According to the EPA (187, 302), Congress in 1970 conceived of air pollution as an outdoor phenomenon and contemplated regulation only for outdoor air. The only U.S. federal agency to implement comprehensive air quality regulations for indoor environments is OSHA (187) (see section 5.4.1), but these regulatory efforts are directed only to the industrial workplace. Although U.S. federal agencies do not have any comprehensive regulatory authority for IAQ, several agencies (for example, the Consumer Products Safety Commission and Housing and Urban Development) have been involved in efforts to limit emissions of pollutants from products or materials used indoors and to proactively promote the concept of "healthy homes."

Performance criteria that could form the basis of IAQ guidelines in the United States have been influenced by both health and comfort concerns. Concern for both health and comfort is perhaps best embodied in the preface of the ACGIH TLVs for chemical substances in the work environment (157):

> The basis on which the values are established may differ from substance to substance: protection against impairment of health may be a guiding factor for some, whereas reasonable freedom from irritation, narcosis, nuisance or other forms of stress may form the basis for others (157). Although it is clear that the ACGIH TLVs deal with exposure to chemical substances in the industrial workplace, it is evident that the portion of the TLV rationale dealing with "reasonable freedom from irritation, narcosis, nuisance or other forms of stress" is similar to the World Health Organization definition of health, namely, "a state of complete physical, mental, and social well-being, and not merely the absence of disease or infirmity (303).

In the United States there are numerous building codes that provide guidance on how buildings should be constructed. While ASHRAE Standards are referenced in some but not all building codes, IAQ in general is given only limited consideration in most building codes. Turner et al. (244) point out that currently in the United States there are no federal requirements to ensure that buildings are operated and perform as designed. A number of guidance documents and professional society consensus publications do provide a standard-of-care performance criteria for the design, operation, and maintenance of a healthy building. Some of these guidelines are discussed in the sections that follow:

5.4.1 Proposed OSHA Rule

In 1994, U.S. Occupational Safety and Health Administration (OSHA) proposed a comprehensive, performance-oriented standard for IAQ in general industry under **29** *CFR* 1910.1033, with identical standards also proposed for the construction industry and maritime and agricultural workplaces (304). The proposed rule focuses not on permissible exposure limits (PELs) but instead on requiring written compliance plans with a means of developing, implementing, and evaluating the effectiveness of applicable control measures (305). The scope of the proposed rule is broad, applying to all nonindustrial workplaces, and was initiated in response to petitions by three U.S. public health groups in the late 1980s (304). A number of key components have been considered controversial, resulting in significant delay of progress toward establishing the proposed regulation as final rule. Perhaps the most important and widely disputed of these is the restriction of tobacco smoking to designated areas which are posted, enclosed, exhausted directly to the outdoors, and negatively pressurized relative to surrounding areas. It has also been argued that private sector initiatives could achieve the same objectives as the regulation. For example, ASHRAE's proposed revision to Standard 62 (section 4.2) contained many provisions similar to those in the OSHA rule.

While the requirements of the proposed OSHA IAQ rule are not currently enforceable by U.S. federal regulatory authority, a number of its components have been used as a standard of care throughout the IAQ industry (305). In addition to the designated smoking areas requirement described above, key elements of the proposed rule included a written IAQ compliance program with:

- Description of building systems and performance criteria.
- Pertinent HVAC drawings and as-built construction documents.
- Proactive and standard operation and maintenance procedures.
- Description of occupancy patterns.
- Information on any known air contaminants.
- Checklist for periodic qualitative inspections.
- Maintenance of records of employee complaints.
- Designation of a responsible person to oversee the IAQ program.

Some specific elements of program implementation as required by the OSHA proposed rule included:

- Inspection and maintenance of HVAC systems to ensure the functioning of equipment as originally designed and based on actual occupancy.
- Assurance of HVAC operation at all times during occupied hours, except during emergency repairs and scheduled maintenance.
- Use of general or local exhaust ventilation during housekeeping and maintenance activities, to minimize associated exposures.
- Maintenance of relative humidity below 60% to inhibit microbial growth.
- Monitoring CO_2 levels as a surrogate indicator of adequacy of outdoor air ventilation; concentrations greater than 800 ppm would trigger a HVAC system evaluation (see discussion on CO_2 in Section 4.1).
- Restriction of hazardous substances in HVAC air distribution systems, e.g., assuring that components such as plenums are clean; avoidance of storage of hazardous materials in mechanical rooms.

Other air contaminants were to be controlled by:

- Restriction of entry of outdoor air contaminants such as vehicle exhaust.
- Control measures such as local source capture exhaust or source substitution where general dilution ventilation is not adequate to control point sources within work spaces.
- Application of pesticides, cleaning, and maintenance chemicals in accordance with manufacturers' specifications, and prior notification of occupants regarding these activities.

- Preventive and corrective measures to minimize microbial growth, such as prompt repair of water leaks; drying, replacing, and/or cleaning of damp or wet materials; removal of visible mold growth.

During renovation or remodeling, a work plan was to be required for controlling air contaminants generated by these activities to minimize exposures both for employees performing these activities and for occupants elsewhere in the building. In addition, the proposed OSHA rule required:

- Maintaining adequate ventilation during building cleaning and maintenance.
- Minimizing adverse effects on IAQ resulting from the use and disposal of chemicals.
- Use of appropriate personal protective equipment for building operations and maintenance employees.
- Retention of inspection and maintenance records.
- Maintaining a record of employee/occupant complaints and remedial actions taken.

5.4.2 Canadian Guides

In 1993 Health Canada published "Indoor Air Quality in Office Buildings: A Technical Guide" (295). The intent of this document was to provide guidance for persons responsible for conducting IAQ evaluations in office buildings including maintenance personnel and health and safety officials. The qualitative (initial assessment) and quantitative (detailed assessment) evaluations are described. Checklists are provided with regard to sampling considerations, thermal environmental measurements, and various potential pollutants such as CO, formaldehyde, particulates, VOCs and microbials. Another publication from Public Works and Governmental Services Canada provides detailed information on the operational and maintenance requirements of humidification systems commonly used in buildings (306).

Two publications provide guidance on microbial problems in Canadian buildings (307, 308). Because of a fungal growth problem in a Prince Edward Island school, the Federal-Provincial Committee on Environmental and Occupational Health produced a consensus document on mycological air quality in buildings titled *Fungal Contamination in Public Buildings: A Guide to Recognition and Management* (307). The Health Canada publication contains background information on health effects associated with fungal exposure and an extensive protocol for investigating fungi in buildings. Although the collection and analysis of *Stachybotrys* is emphasized, the general principles in the document are applicable to all fungi. Other sections of the Health Canada guide cover the prevention of moisture problems in buildings as well as the procedures to be used when fungal growth is physically removed from buildings. The second publication (308) describes microbial problems in office buildings including sampling protocols and remedial actions that have been used by Canadian investigators.

Canada Mortgage and Housing corporation published a guide (309) on clean-up procedures recommended for homes with mold growth problems. This very practical document contains information such as procedures useful in cleaning bedding, clothing, carpet, drapes, upholstered furniture, walls, ceilings and floors affected by mold growth. The concept that people should not live in moldy homes is emphasized. The use of personal protective equipment including respirators is recommended for persons involved in clean-up of moldy materials. A second Canada Mortgage and Housing Corporation Guide (310) on “Toxic Mold Cleanup Procedures” is being prepared. This document when released will provide detailed protocols on restoration recommended for contractors involved in cleaning of mold in housing and other small buildings.

5.4.3 NATO and European Commission Guides

The North Atlantic Treaty Organization (NATO) Committee on the Challenges of Modern Society (CCMS) has during the period from 1989 to 1993 published guidelines on approaches to solving IAQ issues in NATO countries (311, 312). Guidelines for pollutants found in indoor air, for ventilation, and for control of emissions from materials are discussed in CCMS documents (313). Topics covered in one of these guides include chemical and microbial pollutants in indoor air, the role of the HVAC system at both a source of pollutants and a means to control pollutants, material selection for low pollution emitting materials, perceived air quality, and approaches that can be used for regulating indoor air (311).

A series of 19 (as of 1997) publications from the European Community known as the “European Collaborative Action—Indoor Air Quality and its Impact on Man” (ECA-IAQ) covers a wide range of topics including BRS, microbial contaminants, radon, VOCs, and formaldehyde. Several of these reports are reviewed in this section. Report No. 11 (247) describes ventilation guidelines required to achieve different comfort levels in occupied spaces. Ventilation rates prescribed are a function of the pollution loads from occupants and the building itself. The ventilation rates required for both health and comfort are the highest values used in this guide. Report No. 11 (247) also includes several appendices on control of pollutants in HVAC system components, perceived air quality, health effects of carcinogenic and noncarcinogenic materials.

ECA-IAQ Report No. 14 (298) provides guidance on sampling strategies for VOCs. The variable nature of VOCs found in indoor environments and the interpretation of VOC measurements in relation to comfort and health are considered. Variables such as position of the sampler, sampling duration, and quality control and quality assurance are also described. An appendix describes solid adsorbents that can be used to collect VOCs.

ECA-IAQ Report No. 16 (144) describes an interlaboratory comparison of VOCs emitted from water-based paint samples in small test chambers. Factors considered in the report include chamber air velocity, control of the source layer thickness, and dilution and sink mathematical models. ECA-IAQ Report No. 18 (301) outlines general procedures applicable to evaluation of VOC emissions from solid flooring materials. The procedures are intended to form the basis for classification and/or labeling of flooring (and other) materials that could serve both voluntary and regulatory purposes. Factors considered include the selection and handling of test specimens, the determination of VOC concentrations, and measurement of perceived air quality resulting from the emissions.

5.4.4 Nordic Guidelines

A guideline prepared by the Nordic Committee on Building Regulations (314) provides general principles on how buildings should be constructed, ventilated, and operated. Section 3.7 of this guide states that building surfaces in contact with supply or room air shall be cleaned before the building is occupied. In addition, surfaces that are likely to be heavily soiled shall be readily accessible and easily cleanable. Surfaces of finishing materials used in the building shall be selected so that dirt is not unnecessarily concealed or deposited. Section 4.6.3 of the guide (314) recommends that accumulated dirt in HVAC ductwork be cleaned so that neither the quantity of airflow nor the quality of supply air is effected.

A 1993 publication by the Nordic Ventilation group (246) provides a comprehensive review of indoor climate conditions in offices, schools, and dwellings. The multifactorial causes of BRS are described including the view that in the early history of a building (just after construction) IAQ complaints are often due to emissions from building materials. With time the building deteriorates and IAQ complaints arise primarily due to defects such as dirty finishing materials, and a dirty and often out of balance HVAC system. This guide contains an inspection checklist for the qualitative evaluation. A comprehensive section on indoor climate measurements useful for the quantitative IAQ evaluation is included. Topics covered under indoor climate measurements include physical factors such as air velocity, lighting, sound, static electricity, electromagnetic fields, ions, and dusts and fibers.

5.4.5 EPA Guides

In 1991 U.S. EPA published *Building Air Quality: A Guide for Building Owners and Facility Managers* (201). Developed jointly with NIOSH, this guide aimed to provide practical information and advice to building operators and managers on actions to prevent IAQ problems. The guide emphasizes communication between facility managers, staff, contractors, and occupants, developing a description of IAQ in each building (an "IAQ profile"), and designation of an IAQ manager responsible for coordinating actions affecting IAQ. Other sections outline steps for taking proactive actions, IAQ diagnostics, and mitigation of IAQ issues. An overview of HVAC system design is provided, as well as general guidance on identifying and correcting fungal and moisture problems. Suggestions are made regarding pertinent information to be gathered during a qualitative inspection, including description of pollutant sources and HVAC system operations. General guidance is provided on common IAQ measurements. Examples of various forms and checklists are included which are useful to the building IAQ coordinator and those conducting evaluations. These include: an IAQ management/tracking checklist; various HVAC checklists and ventilation worksheets; IAQ complaint and incident logs; occupant interview forms; and forms to help develop a building IAQ profile such as checklists for recording pollutant pathways and pollutant source inventories.

Guidance to school officials involved in IAQ is provided in a second document entitled "Indoor Air Quality Tools for Schools" (315). This guide is presented as an "action kit," directed primarily to kindergarten through grade 12 schools, and was designed to help schools proactively address IAQ problems with minimal cost. The intent is to educate and enable existing school staff to perform qualitative IAQ evaluations, to upgrade building operation and maintenance activities, and to promote the concept of an educated consumer. Emphasis is on proactive actions, advising the user that expenses and efforts necessary for preventing most IAQ problems are less than those required to resolve problems after they develop. The guide includes substantial background information, primarily addressing IAQ issues pertinent to schools. Much guidance is presented as checklists for various categories of school staff. In addition to addressing general items such as cleanliness and ventilation, some key items included in these checklists are

- Teachers—regular classroom vacuuming and dusting, restrictions on animals in the classroom, careful selection and safe use and storage of art supplies and science supplies.
- Administrative staff—proper operation of printing/duplicating equipment.
- Renovation and repairs—selection of low-emitting paints and flooring materials, scheduling of pollutant-producing activities such as roofing work.

Checklists are also provided for maintenance activities, food service and waste management staff, and health officers.

In addition to these guidance documents, U.S. EPA has taken proactive IAQ measures in the construction of new buildings including its new Headquarters facility in Washington, DC. These measures include specifications to ensure pollutant source control, with guidelines established for emissions criteria for products such as furniture (202).

6 CONTROL OF INDOOR AIR POLLUTION

A major objective of both the qualitative and quantitative IAQ evaluation is to provide recommendations to remediate building problems, including the environmental and physical causes of occupant complaints that may lead to BRS, BRI or thermal discomfort. Two very different approaches can be taken when performing remedial actions to upgrade building IAQ. Remedial actions can be directed at source control, which involves removal of the pollutant (253). Alternatively, remedial actions may be directed at exposure control, where the concentrations of pollutants that may be present in indoor air are diluted by provision of relatively clean outdoor air or by cleaning (filtration) of indoor air.

Source control is usually pollutant specific and the more effective remediation strategy because occupant exposure can be eliminated or greatly reduced. For example, if occupant complaints are being caused from emission of 4-phenylcyclohexene from newly installed carpet, replacement of the carpet with a brand that does not contain this chemical (replacement carpet should also have a low emission rate of other VOCs) is an effective remediation strategy. Alternatively, the exposure control approach using the same example would be to increase the ventilation rate to the complainant space to dilute 4-phenylcyclohexene to levels that hopefully would not be perceived by occupants.

In general, recommendations for both source and exposure control in a building are often made simultaneously. For example, the minimum outdoor air ventilation rates recommended by ASHRAE Standard 62-1989 (4) are adequate for dilution of VOCs emitted from interior finishes with low source strengths. For intense sources of VOCs (e.g., liquid process photocopier) the appropriate recommendation is removal of the source (the photocopier) or effective removal of the contaminant by local exhaust ventilation.

IAQ evaluations may be reactive or proactive. The reactive evaluation is usually accompanied by recommendations to lessen the concentration of pollutants in occupied spaces (exposure control) rather than to eliminate exposure (source control). For example, in an existing building with a tuck-under loading dock, recommendations such as negative pressurization of the dock area, installation of double doors and air curtains, and limiting vehicle idling time are made to limit entry of combustion products into occupied spaces (exposure control). Moving the loading dock (source control) would eliminate exposure, but the cost associated with this form of remediation often inhibits action. Proactive approaches toward control of indoor air pollutants through source control are almost always easiest to implement during design of new buildings or during major retrofit of existing buildings.

IAQ problems can occur in some buildings because operators and managers fail to comprehend the complexity of building systems and the potential effects of pollutants on occupants. For example, significant morbidity and mortality occurred among infants in the neonatal intensive care unit in a major medical center. Air being supplied to respirators used by infants originated in a compressor in a mechanical equipment room which in addition to AHUs also housed shop activities such as paint spraying and welding (without local exhaust ventilation). The air entering the neonatal respirator system was passed through high efficiency particulate air (HEPA) filters to remove particulate, but zero consideration had been given to the likelihood that VOCs and combustion products would pass unfiltered to infants on respirators. In this case, source control by relocation of the air inlets for the compressor to a location of high quality air was the appropriate control option. Exposure control by attempting to limit shop activities in the mechanical equipment room was not an option. That this kind of case would even occur in a medical center indicates that risks associated with pollutant exposure are often unappreciated by building operators. In the case of medical centers the Joint Commission on Accreditation of Health care Organizations recently published *A Guide to Managing Indoor Air Quality in Health Care Organizations* (316). The intent of this book is to educate operation and maintenance personal of hospitals on IAQ issues. The US EPA's *Building Air Quality* guide (201) and *Tools for Schools* (315) are also intended to educate operators of other kinds buildings on control and prevention of IAQ problems.

6.1 Microbials

Growth of microorganisms in buildings is caused by moisture and dampness in building infrastructure. The primary approach toward control is achieved by eliminating sources of moisture that support growth. Since dirt, dust, and soiling of building materials provide nutrient for microbial growth when moisture is nonlimiting, the proper cleaning of the building and its equipment is also essential for microbial control. While biocides may have some role in controlling microbial growth, the fundamental aspect of control should always be exercised through elimination of excess moisture and nutrients (69). The control of moisture problems in buildings has been extensively reviewed (7, 317, 318).

The following general actions should be considered to prevent the growth of microbials in buildings.

- Building components should be thoroughly inspected for moisture damage and presence of visible fungal growth. A recent survey in Finland found that moisture damage occurred in 80% of buildings inspected (319). Moisture damage and fungal growth are often hidden in building infrastructure (320).

- Identify the reasons why water damage and dampness occur in building infrastructure (317, 318) and prevent future damage.
- Keep the surface relative humidity of interior construction and finishing materials from consistently exceeding the 65–70% range.
- Avoid the use of fleecy, extended surface finishing material such as porous insulation and carpet in portions of the building or building equipment where conditions of dampness occur.
- Dry building infrastructure as rapidly as possible following a leak or a flood so as to prevent microbial growth on susceptible materials. Porous finishes such as gypsum wall board, ceiling tiles, and carpet that remain wet for periods of 24 hours or more are likely to be colonized by gram-negative bacteria, yeasts, and fungi.
- Develop a plan of action for dealing with moisture problems such as floods and water spills that may occur in buildings. Removing moisture hidden within building infrastructure (e.g., water in wall cavities) is an important aspect of the plan.
- Incorporate a highly efficient cleaning program into facilities maintenance so that dust and dirt are physically removed from surfaces, especially those niches that may be affected by dampness.

Condensation occurs in the envelope of air-conditioned buildings in warm humid climates when moist air infiltrates into the wall and encounters a cool surface on the side of the wall facing the occupied space. Condensation or near condensation conditions (surface relative humidity of 65% or greater) in the building envelope can be controlled or prevented by the following actions:

- If vapor diffusion and air retarders are used they should, typically be located toward the exterior of the envelope. Avoid the use of low-permeance materials on the interior surface of external walls.
- Maintain a net positive pressurization in the building so that infiltration of humid air through the envelope is prevented.
- Prevent rain from entering the envelope. If water accidentally enters the envelope provision should be made during construction to have water drain to the outside.
- Avoid cooling the interior space below the average monthly outdoor dew-point temperature.
- Outdoor air that enters the HVAC system even during unoccupied periods must be dehumidified.

In cold climates condensation and consequent fungal growth occur on the inner surface of the envelope on windows, in corners of walls, at the junction of the ceiling and the external wall, and at thermal bridges or at locations where there is an absence of protective insulation (320). Condensation also occurs within the envelope when relatively moist indoor air flows outward and encounters a cold surface near the outside of the building. In cold climates, the following actions are intended to reduce the likelihood of moisture and microbial problems in the building envelope:

- Ensure that the envelope has adequate thermal resistance to prevent interior surfaces from becoming too cold.
- Prevent room moisture from entering the envelope. This may be accomplished by placing the vapor diffusion retarder toward the warm (interior) side of the envelope. The movement of moist room air into the envelope can also be reduced by provision of an air barrier system typically toward the warm (interior) side.
- Lower the relative humidity in indoor air by source control (e.g., locally exhaust water vapor from simmering or boiling of foods) or by ventilation with relatively dry outdoor air.

Moisture and dirt in HVAC systems result in microbial growth, and the mechanical equipment then becomes a source of microbial pollutants for the rest of the building. The following actions are useful for controlling moisture and microbial problems in HVAC systems:

- Locate outdoor air inlets sufficiently away from (preferably above-grade) standing water, leaves, soil, dead vegetation, and bird droppings. The outdoor air inlet should be designed so as to prevent the intake of rain, snow, and in coastal areas, fog.
- Keep the floors, walls, and ceilings of HVAC system plenums clean and dry. Airstream surfaces in wet or damp niches in HVAC system should not accumulate or retain moisture, should be smooth or easily cleanable, and should be resistant to biodeterioration. Avoid use of uncovered fiberglass insulation in the HVAC system, since it cannot be effectively cleaned and cannot be decontaminated once it is colonized.
- Upgrade the efficiency of HVAC filtration to 65% for particles in the 1 to 3 μm size range (approximately 85% of dust spot efficiency). This will remove fungal spores from the airstream and reduce dust accumulation in portions of the system that are difficult to clean.
- Prevent filters from becoming wetted by rain, snow, fog, or water droplets from dehumidifying cooling coils or the humidification system.
- Conditions of water stagnation (accumulation) must not occur in operating HVAC systems. Prevent the buildup of biofilm on wet surfaces of drain pans and dehumidifying cooling coils by frequent thorough cleaning. Microbiocidal chemicals and disinfectants, if used, must be physically removed after cleaning of pans and coils. These chemicals must not be aerosolized into occupied spaces.
- Humidifiers and dehumidifying cooling coils in HVAC systems should not wet downstream components. When humidifiers are used, the types that emit water vapor rather than droplets are desirable. Control of microbial growth in recirculating water humidifiers is achieved by keeping the unit clean, using high-quality water, and continuous blowdown or replacement of some of the water from the sump (306).

- Avoid use of unit ventilators, FCUs and IUs unless these HVAC components can be kept clean and dry through high-quality maintenance.

Controlling the growth of *Legionella* involves design, operation, and maintenance of cooling towers and potable water systems where this organism can potentially amplify. Some general guidelines follow:

- Design hot water tanks and piping so that the water system can be disinfected by superheating throughout to 60°C in the event that *Legionella* contamination occurs.
- Avoid conditions in hot water tanks and piping that are conducive to *Legionella* growth such as poor mixing, lukewarm water and water stagnation.
- Locate HVAC outdoor air inlets and other building openings, at a minimum horizontal separation distance of 25 feet (preferably 50 feet) from cooling towers and evaporative condensers. Air flow patterns around buildings and prevailing wind direction are important considerations with regard to protecting inlets and openings from cooling tower drift.
- Cooling tower water systems should be treated to control the growth of microbial contaminants. Microbial control involves the use of biocides and/or physical methods to kill bacteria. A testing program to verify the effectiveness of microbiocidal treatment may be considered (51).

The occurrence of visible fungal growth on interior surfaces of buildings is unacceptable (7, 307, 321). When the surface area covered by visible fungal growth in a room is extensive, large scale engineering containment controls should be used during clean-up (7, 307, 322). General principles to be followed during removal of visible mold include the following actions:

- Fix the moisture problem in building infrastructure that led to microbial growth.
- When extensive visible fungal growth occurs on interior surfaces, remove visually moldy materials under negative pressure containment in such a manner that dusts and spores are not dispersed into adjacent clean or occupied areas. Biocide treatment or encapsulation of the moldy surface does not substitute for physical removal of the contaminant.
- Remove all fine dusts (particulate) from the formerly moldy area through damp wiping and/or HEPA vacuuming prior to installation of new finishes in the occupied space.

6.2 Volatile Organic Compounds

VOC emissions from finishing and construction materials used in new buildings or major renovations in existing buildings are best controlled by source reduction or elimination. Voluntary emission rate guidelines for VOCs have been established for carpet, carpet cushion, and carpet adhesive, as follows (116).

For carpet:

- TVOC: <0.5 $mg/m^2 \cdot hr$
- Formaldehyde: <0.05 $mg/m^2 \cdot hr$
- 4-Phenylcyclohexene: <0.05 $mg/m^2 \cdot hr$
- Styrene: <0.4 $mg/m^2 \cdot hr$

For carpet cushion:

- TVOC: <1.0 $mg/m^2 \cdot hr$
- Formaldehyde: <0.05 ppm
- Butylated hydroxytoluene: <0.3 $mg/m^2 \cdot hr$

For adhesive:

- TVOC: <10 $mg/m^2 \cdot hr$
- Formaldehyde: <0.05 ppm
- 2-Ethyl-1-hexanol: <3.0 $mg/m^2 \cdot hr$

These emission values are based on 24-hour exposure concentrations. It is expected that these emission rate values will result in low indoor air concentrations of the contaminants within a one week period.

According to the guidelines established by the State of Washington, all finishes and furnishings installed in new buildings should not increase certain VOC concentrations above the following levels (116).

- TVOC: 0.5 mg/m^3
- Formaldehyde: 0.05 ppm
- Particles: 0.05 mg/m^3
- 4-Phenylcyclohexene: 1 ppb

These concentration guidelines for VOCs (above) should be achievable for low-source strength materials provided that at least 35 days of continuous preoccupancy outdoor air ventilation occurs at a rate of 20 cfm per 140 square feet of floor space.

ASHRAE Standard 62-1989R (178) required no less than the design rate of outdoor air ventilation be provided continuously for 48 hours to zones following completion of major construction in order to purge VOCs and other contaminants prior to occupancy. Under the continuous maintenance revision process for Standard 62, the requirement for preoccupancy ventilation to flush out VOCs after major construction or renovation has been deleted (264).

Specifications from the Environmental Protection Agency's Headquarters project require that office furniture (workstations) meet the following concentration criteria within a seven day period (202). Office chairs must meet one-half of these concentration criteria:

- TVOC: <0.5 mg/m^3
- Total aldehydes: <0.1 mg/m^3
- Formaldehyde: <0.1 ppm

Guidelines for VOC emissions from building finishing and construction materials have been established by European countries. European Collaborative Report No. 18 (301) on *Evaluation of VOC Emissions from Building Products* provides background information useful in selecting low emitting construction and finishing materials. In Finland the following criteria have been established (323).

- TVOC: <0.2 $mg/m^2 \cdot hr$
- Formaldehyde: <0.05 $mg/m^2 \cdot hr$
- Carcinogenic compounds as defined by IARC <0.005 $mg/m^2 \cdot hr$
- Dissatisfaction because of odorous emission is less than 15%

VOCs may be controlled during new construction and major renovation by following general principles such as:

- Avoid use of finishes and construction materials that contain carcinogens, toxins, or teratogens.
- Avoid use of materials containing highly odorous VOCs that are known to degrade perceived air quality at minute concentration levels (119, 126).
- Avoid the use of finishing and construction materials that increase VOC concentrations above the guidelines described by Black et al., (116).
- Minimize the use of solvents, sealants, caulks, paints, and other products that have high emission rates for VOCs (324) or that are highly odorous.
- Provide continuous pre-occupancy outdoor air ventilation with conditioned air at the design rate preferably for at least 48 hours to flush out VOCs emitted from new finishes (178). Increased outdoor air ventilation will probably be needed during the first 3 months of occupancy to dilute VOCs from strong sources (295).
- Isolate areas of existing buildings undergoing major renovation by construction of critical barriers (temporary walls, plastic sheeting, or other vapor retarding layer). The construction area should be maintained at a negative pressure relative to adjacent occupied areas. Air from the construction areas should be locally exhausted outdoors in order to prevent VOC adsorption into finishes elsewhere in the building.
- If high solvent/high VOC emitting products such as adhesives and paints are used allow them to air out (offgas) prior to installation of porous materials such as ceiling tile, fiberglass ductwork, and carpet. This will minimize adsorption of the VOCs and prevent secondary emissions.

Many diverse sources of VOCs potentially occur in existing buildings not undergoing major renovation. Sources include cleaning chemicals and products, minor touch-up and painting, equipment operated by occupants, and occupants' personal care products. Principles useful in controlling VOCs from diverse sources in occupied buildings include the following:

- Monitor and reduce to the lowest feasible level of usage of the volatile chemicals in cleaning agents and work processes in the building. Choose low VOC or environmentally preferred products.
- Use local exhaust ventilation to remove VOCs and other air contaminants from printing machines, graphics operations, darkrooms, closets or rooms where paints, solvents, and cleaning agents are stored, or from any other strong VOC sources.
- Prohibit the storage of cleaning chemicals, solvents, paints, etc. in HVAC mechanical equipment rooms or in HVAC plenums.
- Carpets, modular partitions, or other finishes brought into the building during minor touch-up can be previously off-gassed in a clean, well-ventilated warehouse.
- Determine, based on emission data provided by manufacturers, that new computers, furnishings, printers, copiers, etc., are not significant sources of VOCs.

6.3 Other Indoor Pollutants

The following principles should be considered for control of air pollutants including pesticides, combustion products, ETS, and particles:

- *Pesticides*. Limit pesticide usage to building areas where its application is required. Avoid environmental conditions (e.g., food in offices) that cause pest problems. Apply pesticides only during unoccupied periods and accurately follow manufacturer's directions. Thoroughly ventilate the space prior to occupancy. Never use or store pesticides in active HVAC equipment or plenums (e.g., in FCUs or mechanical equipment rooms).
- *Combustion products*. Avoid building designs where sources of combustion products such as garages or loading docks are located beneath or within buildings or where combustion sources are located near HVAC outdoor air inlets or other building openings.
- Keep garages and loading docks that may be present within or adjacent to buildings negatively pressurized relative to office (occupied) areas. Block paths (air curtains, doors) through which air contaminants may travel from the garage or dock into occupied spaces. An alarm should be triggered at the building central security desk when CO sensors in the garage or dock exceed high limits so that prompt action can be taken to control combustion product emissions and protect occupants.
- Eliminate stack effect as a mean through which combustion products (and other pollutants) may enter tall buildings. Prevent idling of vehicles in dock or garage areas.
- *ETS*. Contaminants from smoking lounges must be totally exhausted outdoors. Acceptable IAQ is not compatible with the recirculation of ETS from any source (178, 179).
- *Particulates*. Avoid activities that generate particulate contaminants. These activities may include dry mopping that disturbs settled dust and use of inefficient vacuum

cleaners that allow fine particles to pass through the instrument and back into the indoor environment.

- Prevent dusts generated during major renovation from contaminating occupied spaces by use of containment barriers and exhausting particulate laden air outdoors through HEPA filters.
- Avoid the use of extended surface, fleecy finishes in building areas where dust control is critical or where soiling is difficult to prevent.
- Provide occupied spaces with highly filtered conditioned air at a sufficient air exchange rate so as to dilute particles that may be aerosolized by occupant activities. Air cleaning is almost always more effective when performed in the central HVAC system rather than by portable air cleaners.

6.4 Thermal Environmental Conditions

Thermal environmental conditions strongly affect occupant perception of acceptable IAQ (285–287). Control of thermal comfort is achieved by following the recommendations in ASHRAE Standard 55-1992 (283) especially those dealing with temperature, temperature gradients, and air motion in occupied spaces. Provision of relative humidity in the 30 to 50% range is best for maintaining acceptable perceived air quality (286).

Control of thermal environmental problems involves an evaluation of the HVAC system's capacity to adequately heat or cool air in the occupied spaces. Changes in thermal loads (e.g., more occupants or heat generating equipment) necessitates additions to capacity perhaps by upgrading fan capacity or addition of new AHUs. Other considerations for controlling thermal environmental problems include the following (295):

- Rebalance the HVAC system to reflect current loads. Verify that conditioned air being supplied to and returned from zones is in balance.
- Verify that HVAC controls are operating correctly. For example, cooling may be provided to core (interior) zones while heating is occurring in envelope area.
- Insulate envelope surfaces to control thermal loads and to reduce temperature gradients in perimeter zones.

6.5 HVAC Systems

HVAC systems can be the primary source of pollutants that are associated with unacceptable perceived in air quality, BRS, and BRI. The presence of contaminants in HVAC systems is almost always associated with inadequate maintenance, operation, and design. The following discussion reviews actions that maximize the likelihood that the HVAC system will provide high quality air to occupied spaces (253, 325).

Outdoor air intakes. At the time of design, outdoor air intakes should be positioned in locations with the highest quality outdoor air. This means that outdoor air intakes should be located as far as possible from pollutant sources such as cooling towers, sanitary vents, loading docks, and laboratory exhaust vents (239). The placement of outdoor air inlets within an architectural fence together with contaminant sources such as relief and exhaust vents, cooling towers, etc. almost certainly results in entrainment of pollutants into ventilation air. The outdoor air inlet itself needs to be protected by a bird screen and the outdoor air intake and connecting passages should be kept dry and clean.

Mixed air plenum and other AHU plenums. These AHU components should be kept clean and dry. Airstream surfaces that are smooth and cleanable are preferred. The storage of pesticides, cleaning materials, paints, HVAC filters, maintenance equipment and supplies in AHU plenums should be avoided. Avoid the presence of open floor drains in plenums unless traps are protected by automatic trap primers. Access into plenums should be made through hinged panels or doors so that maintenance personnel can physically enter plenums for routine cleaning.

HVAC filters. European studies have shown that dirty filters in some buildings can be associated with unacceptable perceived air quality (see Section 4.3). Thus, fabric roughing filters should be changed approximately six times per year. High efficiency filters should be changed several times per year to minimize pollutant emissions from the trapped dust cake. The use of filters that remove >65% of 1 to 3 μm particles (178) will not only improve IAQ with respect to particulate contamination but will also reduce the accumulation of dust and dirt in supply ductwork where cleaning is usually difficult. Filters should fit tightly in their holding frames and be kept dry.

Dehumidification cooling coils and other sources of water. Air-conditioning has been associated with BRS (267). Water spray systems in older office buildings have been associated with BRI (19). Humidifiers can be sources of microbial contaminants (59). In order to minimize pollutants from wet portions of the HVAC system the following actions should be considered (253):

- Drain pans should self drain. See design language in ASHRAE Standard 62-1989R (178).
- Biofilms should be removed by frequent cleaning. Microbiocidal chemicals that may be used in cleaning should not be aerosolized into occupied spaces.
- Water spray systems in the heat exchanger should not be used in office buildings (326).
- Odor maskers, pesticides and rodenticides should not be used in heat exchangers on other HVAC components.
- When humidification is required, use devices that emit water molecules and not water droplets (314).

Air supply ductwork and plenum. Dirt and debris that accumulated in air supply ductwork can cause unacceptable perceived air quality (see sections 3.3 and 4.3). Protocols are available (327) for the cleaning of air conveyance systems involving physical removal of dusts with HEPA vacuum systems. It is important to determine why dust and debris has accumulated in ductwork and take corrective actions such as enhancing the filtration capability of the HVAC system. The use of biocides in place of physical removal of dust or the use of encapsulants to cover up dust and debris should be avoided (328).

Porous insulation in HVAC. Porous insulation on airstream surfaces of HVAC components will in most systems become soiled and subject to fungal growth if moisture is adequate (329). A porous surface that is colonized by fungi is difficult if not impossible to clean because hyphae have penetrated into the substrate. Appendix M of the 1996 public review draft of ASHRAE Standard 62-1989 (178) offered the following advice to HVAC designers . . ." materials that are used to line the airstream surface near moisture producing equipment (e.g., dehumidifying cooling coils, humidifiers, etc.) should not contribute to biodeterioration, should minimize the accumulation of dirt, and should not absorb and retain moisture".

Peripheral heating, ventilation and air-conditioning units. In many buildings FCUs and IUs are used to heat and cool peripheral (envelope) zones of a building. These kinds of HVAC components have been associated with a high index of BRS (213). Because hundreds of these peripheral units may be used in large buildings, it should be determined during initial design if resources will be available to maintain this "distributed" HVAC system. If maintenance of FCUs and IUs cannot be guaranteed, it is best to design a HVAC system that relies on a few large easily accessible central AHUs for purposes of providing conditioned air (330). If peripheral HVAC components are used in HVAC systems, these units should be designed so that there is access into and beneath units for purposes of cleaning. Occupants of buildings should be educated not to block access panels, air inlets and air outlets of peripheral units with office furnishings.

Return air systems. It is generally cost effective to design office buildings with common return plenums as opposed to ducted returns. As a consequence of using common return plenums the designer should realize that any contaminant emitted into this space actually enters the ventilation air stream. Therefore low VOC emitting, nonfleecy materials should be used wherever possible in common return plenums. Because common return plenums are negatively pressurized it is important to protect these areas from infiltration of unconditioned outdoor air (through the envelope) or from high contaminant zones in the occupied space beneath the dropped ceiling. Use of ducted returns is preferable in those buildings where provision of air of the highest quality is desirable.

6.6 Pollutants Generated in Occupied Spaces

Human activities including cleaning are important in contaminant control in occupied spaces. Devices such as hot tubs and saunas and the boiling of foods, showering, and even the application of water-based latex paints can be important sources of moisture especially in indoor environments where low air change rates exist (92, 320). Ventilation provides the only effective means for removing moisture that may be present in an indoor environment. In order to prevent fungal growth on biodegradable interior surfaces it most important to keep the water activity (a_w) or surface relative humidity of the material from consistently exceeding 65% (69).

Repair work such as replacing windows and walls generally results in order of magnitude increases in indoor particulate including fungal spores (331). It is well known in medical centers that fungal aerosols generated during interior renovations can be a cause of nosocomial infection in highly susceptible patients (332). Fleecy or porous materials can be reservoirs of fungi and various allergens such as those from mites and cats. The process of vacuum cleaning of carpet with an instrument devoid of a HEPA filter can result in an order of magnitude increase in fungal aerosols (331) and cat allergen (85). A publication by the Canada Mortgage and Housing Corporation (309) provide practical advice on clean-up of molds including instruction on proper use of vacuum cleaners.

The sources of chemical pollutants in occupied spaces are numerous and diverse, ranging from emissions of cleaning materials to periodic interior renovations. Reduction in VOCs emitted from finishing materials in an occupied building often rely on dilution ventilation. Some principles useful in controlling VOCs in buildings include:

- During planning and construction of new buildings or during major renovation of an existing building, control VOCs by source elimination or source reduction. Qualified low-emitting (environmentally friendly) products should be used in construction and furnishing.
- Strong VOC sources (e.g., photocopiers using large amounts of organic solvents) should be controlled by local exhaust ventilation.
- Strong emissions of VOCs associated with major renovation should be exhausted directly outdoors and not circulated to occupied spaces by an active HVAC system.

Portable air cleaners are often used in indoor environments because occupants believe that these instruments are effective in controlling indoor pollutants. There is general agreement however, that these devices should not be the primary means for improving IAQ (333). Devices that contain HEPA or highly efficient fabric filters are more effective than air cleaners which rely on ozone or ionization (334). Some air cleaners can degrade IAQ. Ozone generating devices are not only ineffective in terms of particle removal (335) but they also cause chemical reactions that increase indoor levels of aldehydes and potentially can elevate room ozone levels to unsafe concentrations (336). Portable air cleaning devices are generally ineffective in most occupied environments because the rate of air filtration through the device is too low relative to the emission rates of various indoor pollutants. Consider the following when portable devices are suggested to enhance IAQ:

- Contaminant source avoidance or source removal is more effective than the exposure control benefits derived from portable air cleaners.
- Air purification devices should be incorporated into a HVAC system in order to provide maximum exposure control.
- Air purifiers that generate ozone should be avoided (336). These devices are ineffective in killing fungal spores in building materials even when used at levels above those considered safe for human exposure (337).

6.7 Case Studies

6.7.1 Case Study #1: Evaluation of IAQ Pollutant Concentrations in a Newly Constructed Building

The Labor and Industries Building was the first state office building constructed under the State of Washington's East Campus Plus IAQ Program (145). This program was developed with the overall goal of creating a healthy, productive work environment for state employees. A significant portion of this program outlined specifications designed to minimize exposures to common IAQ contaminants through pollutant source control and effective building ventilation.

One significant specification was the requirement of a 90 day, 100% outdoor air "flush-out" procedure prior to building occupancy. The flush-out requirement was an emission control specification designed to lower indoor pollutant levels originating from processes and materials used in the construction, finishing, and furnishing of the building. During the flush-out, the building's two primary air handling units (AHUs) were set to operate on the economizer cycle for 100% outdoor air. AHU 1 was set to operate at approximately 31,300 cfm of outdoor air while AHU 2 was set to operate at approximately 29,100 cfm of outdoor air (total square footage = 275,000). VAV boxes throughout the building were set for 0.20 cfm per square foot (28 cfm/140 sq ft.) through the entire flush-out period.

Prior to commencement of the flush-out period, certain construction procedures were required to be completed, including installation of all wet material applications (such as paints and adhesives) and certain dry product installations, such as carpet and ceiling systems. In an effort to determine the overall effectiveness of the flush-out, chemical and particle monitoring were conducted immediately prior to and 90 days following start of the flush-out period.

Pollutant measurements were made using methodologies adapted specifically for non-industrial indoor environments. VOCs were collected on multi-sorbent cartridges and subsequently analyzed by thermal desorption/gas chromatography/mass spectrometry (135). The TVOC concentration was determined along with identification and quantitation of specific VOCs. This technique is generally applicable for compounds in the C_5–C_{16} range and has a method detection limit of 1 $\mu g/m^3$ for TVOC and most individual VOCs. Sorbent collection followed by thermal desorption/gas chromatography/mass spectrometry detection is the desired technique for analyzing low-level VOCs in nonindustrial indoor environments (135). Airborne particles were collected onto preweighed, 37 millimeter membrane filters housed in a plastic cassette. The net collection weight was determined using methodology similar to that outlined in NIOSH Method 0500 (338). Large collection volumes were obtained so that a detection limit of 1 $\mu g/m^3$ was achieved. Formaldehyde was collected using passive monitors. Monitors were analyzed using a chromotropic acid technique with a detection limit of 2 $\mu g/m^3$ (198, 339).

Pollutant concentrations measured prior to start of the flush-out period averaged 10,400 $\mu g/m^3$ for TVOC, 32 $\mu g/m^3$ for formaldehyde, and 76 $\mu g/m^3$ for total particles. The types of VOCs detected and elevated concentrations measured during preflush monitoring were consistent with recent construction activities in the building including painting, ceiling system installation, and vinyl flooring installation. Pollutant concentrations measured upon completion of the 90 day flush-out period were significantly lower than preflush levels, averaging 77 $\mu g/m^3$ for TVOC, 13 $\mu g/m^3$ for formaldehyde, 4 $\mu g/m^3$ for total particles. Some additional details on concentrations of specific VOCs prior to and after flush-out are provided in Table 65.12.

The study indicated that occupancy immediately following completion of new construction would result in elevated exposure to airborne contaminants. However, the study demonstrated that these exposures can be reduced by implementing an effective building flush-out period prior to occupancy. It should also be realized that the subject building was located in a geographic location where ventilation with large amounts of relatively dry

Table 65.12. Concentrations of Specific VOCs Before and After Flush-Out[a]

Specific VOC	Before Flush-Out	After Flush-Out
Decane	451 μg/m³	2 μg/m³
2-Methylnonane	430 μg/m³	1 μg/m³
Toluene	263 μg/m³	7 μg/m³
Mixed xylenes	479 μg/m³	9 μg/m³
c-Decahydronaphthalene	240 μg/m³	<1 μg/m³
Propyl cyclohexane	346 μg/m³	<1 μg/m³

[a]U.S. EPA Method IP-1B (135).

outdoor air was possible. See section 6.2 for a discussion of the requirement for preoccupancy outdoor air ventilation in the public review draft of ASHRAE Standard 62-1989 (178) and the deletion of the preoccupancy ventilation requirement in the ASHRAE continuous maintenance process (264).

6.7.2 Case Study #2: Indoor VOCs and VOCs from Subsurface Remediation Site

A restaurant was located adjacent to a gasoline service station where an underground storage tank had leaked petroleum products into the soil. A remediation system (vacuum extraction with air sparging) was removing subsurface contamination; however, the restaurant employees were concerned that VOCs from the subsurface conditions at the service station were migrating from the soil into the restaurant building.

Table 65.13 presents a comparison of VOC sampling data from the indoor air of the restaurant and the exhaust of the vacuum extraction system of the service station. A TVOC

Table 65.13. Comparison of VOCs from Building Air and Vacuum Extraction System Exhaust.[a]

Specific VOC	Indoors, μg/m³	Exhaust, μg/m³
TVOC	2,740	86,041
Isopropanol	451	<1.0
Ethanol	338	<1.0
1,4-Dichlorobenzene	268	<1.0
Acetone	262	10.0
Limonene	126	<1.0
% of TVOC	53	<0.1
Xylenes	9.9	4,027
2, 3, 4-Trimethylpentane	2.5	3,437
2, 2, 4-Trimethylpentane (isooctane)	<0.1	3,037
2, 4-Dimethylhexane	2.9	2,726
1-Ethyl-4-methylbenzene	12.7	2,417
% of TVOC	1	18

[a]U.S. EPA Method IP-1B (135).

of approximately 86,000 μg/m^3 was found in the vacuum extraction system exhaust, while an average TVOC of 2,740 μg/m^3 was found in the restaurant. The primary individual VOCs from the indoor air and from the vacuum extraction system exhaust are rank-order listed in Table 65.13. The typical sources of VOCs identified in the restaurant (isopropanol, ethanol, etc.) are cleaning materials, deodorizers, etc., commonly found in buildings. The VOC profile from the vacuum extraction system exhaust was very different from the indoor VOCs. These data suggested that the petroleum release at this service station was not contaminating the indoor air of the nearby building.

6.7.3 *Case Study #3: Evaluation of VOC Cross-Contamination in a Mixed-Use Facility*

Buildings which house both office space and industrial process areas are referred to as mixed-use facilities. The threat of cross-contamination from emissions generated by industrial processes into the attached office space is always present in these facilities. Mismanagement of appropriate pressure relationships in mixed use facilities (i.e., negative pressurization of the office space relative to the industrial space) often results in BRS among office personnel. An evaluation was conducted in a mixed-use facility housing a retail space (where occupants reported BRS) located adjacent to an industrial paint shop. The objective of the evaluation was to determine if VOC emissions from the industrial paint shop were entering the attached retail space. The TVOC level in the paint shop was 1,648 μg/m^3. TVOC concentrations in the retail space ranged from 1,716 μg/m^3 to 2,433 μg/m^3. Individual VOCs identified in the retail space were numerous and included aromatic hydrocarbons; cyclic substituted and normal alkanes; ketones; and esters. Those VOCs in the retail space were similar to those identified in paint shop air (Table 65.14). Primary VOCs, including butyl acetate, toluene, mixed xylenes, and substituted benzenes were common to both areas. These VOCs were consistent with those commonly associated with solvent-containing paints and associated materials such as cleaners and thinners. The data strongly suggested that VOC emissions generated in the paint shop were migrating into the adjacent retail area.

6.7.4 *Case Study #4: Evaluation of VOC Cross-Contamination in a Retail Shopping Center*

A study was conducted in response to complaints of BRS in a retail tenant space (subject space). Tenants in the subject space believed the source of this IAQ problem was caused

Table 65.14. Comparison of Individual VOCs in the Paint Shop and Adjacent Retail Space[a]

Chemical	Paint Shop	Retail Space
Butyl acetate	191 μg/m^3	99 μg/m^3
Toluene	197 μg/m^3	299 μg/m^3
Mixed Xylenes	108 μg/m^3	143 μg/m^3
Substituted benzenes	331 μg/m^3	295 μg/m^3
TVOC	1,648 μg/m^3	2,075 μg/m^3

[a]U.S. EPA Method IP-1B (135).

by an adjacent nail salon. The objective of the evaluation was to determine if VOC contaminants from the nail salon were entering the adjacent subject space. The TVOC level in the nail salon was 5,768 μg/m^3. TVOC concentrations in the subject space ranged from 1,146 μg/m^3 to 1,404 μg/m^3. Individual VOCs identified in the subject space were numerous and included acrylates, acetates, aromatic hydrocarbons; cyclic substituted and normal alkanes; ketones, and alcohols. Those VOCs in the subject space were similar to those identified in nail salon air (Table 65.15). Primary VOCs, including methyl methacrylate, butyl acetate, isobutyl acetate, ethyl acetate, isopropanol, and acetone were common to both areas. These VOCs were consistent with those commonly associated with nail polishes, hardeners, and nail polish remover. The data strongly suggested that VOC emissions generated in the nail salon were migrating into the adjacent retail area.

6.8 Communications

Open communications between building management and occupants are necessary for the resolution of IAQ complaints (295). Successful resolution of IAQ complaints requires communication and participation of occupants and/or their representatives, housekeeping and HVAC facility staff, health and safety committee representatives, and the building manager (or owner). Cooperation and early action to investigate and solve problems usually results in a successful resolution of complaints. Denial of the existence of a problem in the face of continued occupant complaints often results in a situation where problem solving is difficult and delayed.

When IAQ complaints occur in a building, the following communication actions should occur: (*1*) openly discuss the complaints (or problem) with all concerned parties; solicit occupant views on reasons for the complaints; establish a policy through which all parties will be appraised of progress during complaint investigation and resolution; (*2*) develop a written procedure for recording the time, location, and nature of complaints as they occur; housekeeping and facilities staff should record building operational parameters that may coincide with the complaint (e.g., occurrence of renovation or cleaning activities); a team approach is often needed to resolve the problem; for example, the building manager may

Table 65.15. Comparison of Individual VOCs in the Nail Salon and Adjacent Retail Space[a]

Chemical	Nail Salon	Retail Space
Acetone	19,000 μg/m^3	2,700 μg/m^3
Methyl methacrylate	3,026 μg/m^3	297 μg/m^3
Butyl acetate	241 μg/m^3	29 μg/m^3
Ethyl acetate	197 μg/m^3	21 μg/m^3
Isobutyl acetate	204 μg/m^3	13 μg/m^3
Isopropanol	223 μg/m^3	122 μg/m^3
Toluene	218 μg/m^3	46 μg/m^3
TVOC	5,768 μg/m^3	1,275 μg/m^3

[a]U.S. EPA Method IP-1B (135).

be unaware that a renovation or cleaning activity is degrading IAQ, but occupants may readily recognize the source of the problem; (*3*) all parties, including occupants, should be provided with information on compliance with remedial recommendations; highly technical reports dealing with problem resolution should be accurately summarized in layman's terms so as to be understandable by the general public. Abend (293) has provided a protocol for the kind of communication system especially useful to resolve IAQ problems and schools.

BIBLIOGRAPHY

1. J. Sundell, *Indoor Air. Suppl. No. 2*, 1–49 (1994).
2. P. Morey and J. Woods, *Occup. Med: State of the Art Rev.* **2**, 547–563 (1987).
3. C. Yaglou, C. Riley, and D. Coggins. *ASHRAE Trans.* **42**, 133–163 (1936).
4. American Society of Heating, Refrigerating and Air-conditioning Engineers, *Ventilation for Acceptable Indoor Air Quality*, Standard 62-1989. Atlanta, GA, 1989.
5. J. Stolwijk, in *Proceedings of the 3rd International Conference on Indoor Air Quality and Climate*, Vol. 1, Stockholm, Sweden, 1984, pp. 23–29.
6. D. Menzies and J. Bourbeau, *New England J. Med.* **337**, 1624–1631 (1997).
7. American Conference of Governmental Industrial Hygienists, *Bioaerosols Assessment and Control*, Cincinnati, OH, 1999.
8. J. Samet, *Indoor Air* **3**, 219–226 (1993).
9. C. Redlich, J. Sparer, and M. Cullen, *Lancet* **349**, 1013–1016 (1997).
10. P. Fanger, J. Lauridsen, P. Bluyssen, and G. Clausen, *Energy and Buildings*, **12**, 7–19 (1988).
11. M. Mendell, *Indoor Air* **3** 227–236 (1993).
12. M. Lahtinen, P. Huuhtanen, and K. Reijula, *Indoor Air, Suppl. No. 4*, 71–80 (1998).
13. J. Woods, *Occupational Medicine: State of the Art Reviews* (4), 753–770 (1989).
14. O. Seppänen, *Proceedings of the 6th International Conference on Indoor Air Quality and Climate*, Helsinki, Finland (6 volumes), 1993.
15. S. Yoshizawa, *Proceedings of the 7th International Conference on Indoor Air Quality and Climate*, Nagoya, Japan (4 volumes), 1996.
16. G. Raw, *Proceedings of the 8th International Conference on Indoor Air Quality and Climate*, Edinburgh, U.K., (5 volumes) 1999.
17. Health and Safety Executive, *The Control of Legionellosis Including Legionnaires' Disease*, Health and Safety Series Booklet HS (G) 70, Sudbury, U.K., 1994.
18. R. Rylander, P. Haglind, M. Lundholm, I. Mattsby, and K. Stengvist, *Clinical Allergy* **8**, 511–516 (1978).
19. P. Arnow, J. Fink, D. Schlueter, J. Barboriak, G. Mallison, S. Said, S. Martin, G. Unger, G. Scanlon, and V. Kurup, *Am. J. Med.* **64**, 236–242 (1978).
20. Institute of Medicine, *Indoor Fungal Allergens*, National Academy Press, National Academy of Sciences, Washington, DC, 1993.
21. R. Etzel, E. Montana, W. Sorenson, G. Kullman, T. Allen, and D. Dearborn, *Arch. Pediatr. Adolesc. Med.* **152**, 757–762 (1998).

22. Health Effects of Toxin-Producing Indoor Molds in California, *California Morbidity*, Dept. of Health Services, Sacramento, CA, 1998, 3 pp.
23. C. Hunter, and C. Sanders, "Mould, Energy and Condensation in Buildings and Community Systems Programme: Final Report" Vol. 1, of *Condensation and Energy Sourcebook*, International Energy Agency, Annex 14, Leuven University Belgium/IEA 1991, pp. 2.1–2.30.
24. B. Flannigan, in *IAQ '92, Environments for People*, American Society of Heating, Refrigeration and Air-Conditioning Engineers, Inc. Atlanta, GA, 1992, pp. 139–145.
25. C. Grant, C. Hunter, B. Flannigan, and A. Bravery, *Intern. Biodeterioration and Biodegradation*, **25** 259–289 (1989).
26. K. Reijula, *Indoor Air Suppl. 4*, 40–44 (1998).
27. A. Korpi, A-L. Pasanen, P. Pasanen, and P. Kalliokoski, *Intern. Biodeterioration and Biodegradation* **40**, 19–27 (1997).
28. K. Foarde, D. Van Osdell, and J. Chang, in *Proceedings of Engineering Solutions to Indoor Air Quality Problems*, Air and Waste Management Association, Pittsburgh, PA, 1997, pp. 325–333.
29. D. Miller, in *Indoor Air an Integrated Approach*, Elsevier, Amsterdam 1995, pp. 159–168.
30. B. Flannigan, and D. Miller, "Health Implications of Fungi in Indoor Environments An Overview", in R. Samson, B. Flannigan, M. Flannigan, et al., eds., *Health Implications of Fungi in Indoor Environments*, Elsevier, Amsterdam, 1994, pp. 3–28.
31. R. Samson, *Eur. J. Epidemiol.* **1**, 54–61, (1985).
32. R. Rylander, *Arch. Environ. Health* **52**, 281–285 (1997).
33. J. Douwes, G. Doekes, J. Heinrich, A. Koch, W. Bischof, and B. Brunekreef, *Indoor Air* **8**, 255–263 (1998).
34. A. Saraf, L. Larsson, H. Burge, and D. Milton, *Appl. Environ. Microbiol.* **63**, 2554–2559 (1997).
35. American Industrial Hygiene Association, *Field Guide for the Determination of Biological Contaminants in Environmental Samples*, Fairfax, VA, 1996.
36. A-L. Pasanen, K. Yli-Pietila, P. Pasanen, P. Kalliokoski, and J. Tarhanen, *Appl. Environ. Microbial.* **65**, 138–142 (1999).
37. M. Nikulin, A-L. Pasanen, S. Berg, and E-L. Hintikka, *Appl. Environ. Microbiol.* **60**, 3421–3424 (1994).
38. M. Andersson, M. Nikulin, U. Koljalg, M.C. Andersson, F. Rainey, K. Reijula, E-L. Hintikka, and M. Salkinoja-Salonen, *Appl. Environ. Microbiol.* **63**, 387–393 (1997).
39. E. Johanning, P. Morey, and B. Jarvis, in *Proceedings of the Sixth International Conference on Indoor Air Quality and Climate*, Helsinki, Vol. 4, 1993, pp. 311–316.
40. M. Hodgson, P. Morey, W-Y, Leung, L. Morrow, D. Miller, B. Jarvis, H. Robbins, J. Halsey, and E. Storey, *J. Occup. Environ. Med.* **40**, 241–249 (1998).
41. E. Kaminski, L. Libbey, S. Stawicki, and E. Wasowicz, *Appl. Microbial.* **24**, 721–726 (1972).
42. R. Tressl, D. Bahri, and K-H. Engel, *J. Agric. Food Chem.* **30**, 89–93 (1982).
43. G. Strom, D. Norback, J. West, B. Wessen, and U. Palmgren, in *Proceedings Building Design, Technology, and Occupant Well-Being in Temperate Climates*, ASHRAE, Atlanta, GA, 1993, pp. 351–357.
44. G. Strom, J. West, B. Wessen, and U. Palmgren, in R. Samson, B. Flannigan, M. Flannigan, A. Verhoeff, O. Adan, and E. Hoeskstra, eds., *Health Implications of Fungi in Indoor Environments*, Elsevier, Amsterdam 1994, pp. 291–305.
45. D. Norback, G. Wieslander, G. Strom, and C. Edling, *Indoor Air* **5**, 166–170 (1995).

46. I. Ezeonu, D. Price, R. Simmons, S. Crow, and D. Ahearn, *Appl. Environ. Microbiol.* **60**, 4172–4173 (1994).
47. K. Wilkins, E. Nielsen, and P. Wolkoff, *Indoor Air* **7**, 128–134 (1997).
48. J. Bjurman, E. Nordstrand, and J. Kristensson, *Indoor Air* **7**, 2–7 (1997).
49. B. Wessen, G. Strom, and K-O, Schoeps, in *Indoor Air an Integrated Approach*, Elsevier, Amsterdam 1995, pp. 67–70.
50. P. Morey and E. Horner, in *Design, Construction and Operation of Healthy Buildings*, ASHRAE, Atlanta, GA, 1998, pp. 123–129.
51. American Society for Testing and Materials, *1998 Annual Book of ASTM Standards 11.03*, D5952-96, ASTM, Philadelphia, PA, 1998, pp. 600–615.
52. B. Fields, "Legionellae and Legionnaires' Disease", in *Manual of Environmental Microbiology*, American Society of Microbiology, Washington, DC, 1997, pp. 667–675.
53. C. Fliermans, "Legionella Ecology" in H. Burge, ed., *Bioaerosols*, Lewis Publishers, Boca Raton, FL, 1995, pp. 49–70.
54. Wisconsin Division of Health 1987, *Control of Legionella in Cooling Towers, Summary Guidelines*. Wisconsin Division of Health, Madison, WI, 1987.
55. Standards Australia, *Control of Microbial Growth in Air-Handling and Water Systems in Buildings*, HB 32-1992, North Sydney, NSW, 1992.
56. K. Anderson, *Med. J. Australia*, 529 (April 18, 1959).
57. P. Burge, M. Finnegan, N. Horsfield, D. Emery, P. Austwick, P. Davies, and C. Pickering, *Thorax*, **40**, 248–254 (1985).
58. P. Hugenholtz, and J. Fuerst, *Appl. Environ. Microbiol.* **58**, 3914–3920 (1992).
59. B. Ager and J. Tickner, *Ann. Occup. Hyg.* **27**, 341–358 (1983).
60. G. Brundrett, *Maintenance of Spray Humidifiers*, Electricity Council, Capenhurst, Chester, U.K., 1979.
61. U.S. Environmental Protection Agency, Indoor Air Facts No. 8, *Use and Care of Home Humidifiers*, Washington, DC, 1991.
62. A. Mohan, C. Feigley, and C. Macera, *Appl. Occup. Environ. Hyg.* **13**, 782–787 (1998).
63. P. Morey, in *Cotton and Microorganisms*, U.S. Dept. Agriculture, ARS-138, Washington, DC, 1997, pp. 10–17.
64. D. Milton, Bacterial Endotoxins: A Review of Health Effects and Potential Impact in the Indoor Environment, in R. Gammage and B. Berven, eds., *Indoor Air and Human Health*, Lewis Publishers, Boca Raton, FL, 1995, pp. 179–95.
65. D. Milton, "Endotoxin and Other Bacterial Cell-Wall Components" in *Bioaerosols Assessment and Control*, ACGIH, Cincinnati, OH 1999, Chapt. 23.
66. F. Gyntelberg, P. Suadicani, J. Nielsen, P. Skov, O. Valbjorn, P. Nielsen, T. Schneider, O. Jorgensen, P. Wolkoff, C. Wilkins, S. Gravesen, and S. Norm, *Indoor Air* **4**, 223–228 (1994).
67. R. Samson, B. Flannigan, M. Flannigan, A. Verhoeff, O. Adan, E. Hoekstra, eds., *Health Implications of Fungi in Indoor Environments*, Elsevier, Amsterdam 1994.
68. J. Singh, *Building Mycology*, E & FN Spon, London, 1994.
69. International Society Indoor Air Quality, *Guidelines for Control of Moisture Problems Affecting Biological Indoor Air Quality*, Task Force 1, Milano, Italy, 1996.
70. M. Chan-Yeung, *Environmental Health Perspectives*, **103** (Suppl 6); 249–262 (1995).
71. E. Horner, S. Lehrer, and J. B. Salvaggio, "Fungi" in *Immunology and Allergy Clinics of North America* **14**(3), 551–556 (1994).

72. E. Fernandez-Caldas, L. Puerta, and R. Lockey, "Mite Allergens" in R. Lockey and S. Burkantz, eds., *Allergens and Allergen Immunotherapy*, Marcel Dekker, New York, 1999, pp. 181–201.
73. R. Helm, "Cockroach and Other Inhalant Insect Allergens" in R. Lockey and S. Burkantz, eds., *Allergens and Allergen Immunotherapy*, Marcel Dekker, New York, 1999, pp. 203–223.
74. C. Schou, "Mammaliean Allergens" in R. Lockey and S. Burkantz, eds., *Allergens and Allergen Immunotherapy*, Marcel Dekker, New York, 1999, pp. 225–235.
75. U.S. Department of Health and Human Services, Public Health Service, National Heart Lung and Blood Institute, *Expert Panel Report II, Guidelines for the Diagnosis and Management of Asthma*, NIH Publications No. 97-4051, 1997.
76. I. Andersen, and J. Korsgaard, *Environmental International* **12**, 121–127 (1986).
77. L. Arlian, "House Dust Mites", in *Bioaerosols Assessment and Control*, ACGIH, Cincinnati, OH 1999, Chapt. 22.
78. A. Simpson, R. Hassall, A. Custovic, and A. Woodcock, *Allergy* **53**, 602–607 (1998).
79. J. Janko, D. Gould, L. Vance, C. Stengel and J. Flack, *Amer. Industrial Hyg. Assoc. J.* **56**, 1133–1140 (1995).
80. S. Pollart, D. Mullins, L. Vailes, M. Hayden, T. Platts-Mills, W. Sutherland, and M. Chapman, *J. Allergy Clin. Immunol*, **87**, 511–521 (1991).
81. P. Eggleston, D. Rosenstreich, H. Lynn, P. Gergen, D. Baker, M. Kattan, K. Mortimer, H. Mitchell, D. Ownby, R. Slavin, and F. Malveaux, *J. Allergy Clin. Immunol.* **102**, 563–570 (1998).
82. D. Rosenstreich, P. Eggleston, M. Kattan, D. Baker, R. Slavin, P. Gergen, et al., *N. Eng. J. Med.* **336**, 1356–1363 (1997).
83. M. Perzanowski, E. Ronmark, B. Nold, B. Lundback, and T. Platts-Mills, *J. Allergy Clin. Immunol.* **103**, 1118–1024 (1999).
84. C. Aimqvist, P. Larsson, A-C. Egmar, M. Hadren, P. Malmberg, and M. Wickman, *J. Allergy Clin. Immunol.* **103**, 1012–1017 (1999).
85. C. Luczynska, Y. Li, M. Chapman, and T. Platts-Mills, *Am. Rev. Respir. Dis.* **141**, 361–367 (1990).
86. C. Reed, *J. Allergy Clin. Immunol.* **103**, 539–547 (1999).
87. M. Chan-Yeung, "Occupational Asthma," in R. Rylander and R. Jacobs, eds., *Organic Dusts*, Lewis Publishers, Boca Raton, FL, 1994, pp. 161–174.
88. R. Rylander and P. Haglind, *J. Clin. Allergy* **14**, 109 (1984).
89. H. Richardson, "Hypersensitivity Pneumonitis," in R. Rylander and R. Jacobs, eds., *Organic Dusts*, Lewis Publishers, Boca Raton, FL, 1994, pp. 139–160.
90. J. Salvaggio, "Extrinsic Allergic Alveolitis," in S. Holgate and M. Church, eds., *Allergy-1993*, Gower Medical Publishing, London, 1993, pp. 1–11.
91. K. Kreiss, *Occupational Medicine: State of the Art Reviews* **4**, 575–592 (1989).
92. R. Jacobs, R. Thorner, J. Holcomb, et al., *Ann. Intern. Med.* **105**, 204–206, (1986).
93. J. Fink, E. Banaszak, W. Thiede, and J. Barboriak, *Ann. Internal Med.* **74**, 80 (1971).
94. C. Johnson, J. Bernstern, J. Gallagher, P. Bonventre, and S. Brooks, *Am. Rev. Respir. Dis.* **122**, 339 (1980).
95. R. Bernstein, W. Sorenson, D. Garabrant, C. Reaux, and R. Treitman, *Am. Ind. Hyg. Assoc. J.* **44**, 161 (1983).
96. G. Solley and R. Hyatt, *J. Allergy Clin. Immunol.* **65**, 65 (1980).

97. M. J. Hodgson, P. R. Morey, M. Attfield, W. Sorenson, J. N. Fink, W. W. Rhodes, G. S. Visvesvara, *Arch. Environ. Health*, **40**, 96–101 (1985).

98. M. J. Hodgson, P. R. Morey, J. S. Simon, T. D. Waters, and J. N. Fink, *Am. J. Epidemiol.* **125**, 631–638 (1987).

99. J. Brundage, R. Scott, W. Lednar, D. Smith, and R. Miller, *J. Am. Med. Assoc.* **259**, 2108–2112 (1988).

100. J. E. Janssen, *ASHRAE J.* **31**, 40 (1989).

101. C. Hodge, M. Reichler, E. Dominguez, J. Bremer, T. Mastro, K. Hendricks, D. Musner, J. Elliott, R. Facklam and R. Breman, et al., *N. Engl. J. Med*, **331**, 643–648 (1994).

102. E. Nardell and J. Macher, "Respiratory Infections-Transmissions and Environmental Control", in *Bioaerosols Assessment and Control*, ACGIH, Cincinnati, OH (1999), Chapt 9.

103. P. Jensen, "Airborne Mycobacterium" spp., in *Manual of Environmental Microbiology*, Washington, DC, 1997, pp. 676–681.

104. A. Alvarez, M. Buttner, and L. Stetzenback, et al. *Appl. Environ. Microbiol.* **61**, 3610–3614, (1995).

105. K. Stark, J. Nicolet, and J. Frey, *Appl. Environ. Microbiol.* **64**, 543–548, 1998.

106. J. Autrup, J. Schmidt, and T. Seremet, et al. *Scand. Environ. Health Perspectives* **99**, 195–197 (1993).

107. B. Flannigan, Guidelines for Evaluation of Airborne Microbial Contamination of Buildings, in E. Johanning, and C. Yang, eds., *Fungi and Bacteria in Indoor Environments*, Eastern New York Occupational Health Program, Latham, NY, 1995, pp. 123–130.

108. European Collaboration Action, *Biological Particles in Indoor Environments*. Report No. 12, Commission of the European Communities, Brussels, Eur 14988EN, 1993.

109. R. Lewis and L. Wallace, *ASTM Standardization News* **16**, 40 (1988).

110. L. Wallace, T. Hartwell, K. Perritt, L. Sheldon, L. Michael, and E. D. Pellizzari, in *Proceedings of the 4th International Conference on Indoor Air Quality and Climate*, Vol. 1, Berlin (West), 1987, pp. 117–121.

111. P. Bluyssen, E. DeOliveiro Fernandes, L. Groes, G. Clausen, P. Fanger, O. Valbjorn, C. Bernhard, and C. Roulet, *Indoor Air* **6**, 221–238 (1996).

112. B. Seifert, in *Proceedings of the 5th Intern Conference on Indoor Air Quality and Climate*, Toronto, Vol. 5, 1990, pp. 35–49.

113. L. Molhave and G. Clausen, "*Proceedings of the 7th Intern Conference, on Indoor Air Quality and Climate* Vol. 2, Nagoya, Japan, 1996, pp. 37–46.

114. M. Black, L. Work, A. Worthan, and W. Pearson, in *Proceedings of the 6th Intern Conference on Indoor Air Quality and Climate* Vol. 2, Helsinki, Finland, 1993, pp. 401–405.

115. P. Wolkoff, C. Wilkins, P. Clausen and K. Larsen, *Indoor Air*, **3**, 113–123, (1993).

116. M. Black, in *Proceedings The Institute of Electrical and Electronics Engineers, Intern Symposium on Electronics & the Environment*, Piscataway, NJ, 1998, pp. 75–78.

117. H. Shields, D. Fleisher, and C. Weschler, *Indoor Air* **6**, 2–17 (1996).

118. M. Reitzig, S. Mohr, B. Heinzow, and H. Knoppel, *Indoor Air* **8**, 91–102 (1998).

119. P. Wolkoff, *Indoor Air Suppl.* **3**, 73 (1995).

120. J. Girman, *Occupational Medicine: State of the Art Reviews* **4**, 695–712 (1989).

121. L. Molhave, S. Dueholm, and L. Jensen, *Indoor Air* **5**, 104–119 (1995).

122. L. Molhave, B. Bach, and O. Pedersen, *Environmental International* **12**, 167–175 (1986).

123. W. Muller and M. Black, *Am. Ind. Hyg. Assoc. J.* **56**, 794–803 (1995).
124. L. Hansen, G. Nielsen, J. Tottup, A. Abildgaard, O. Jensen, M. Hanson, and O. Nielsen, *Indoor Air* **2**, 95–110 (1991).
125. S. Kjaergaard, L. Mohave, and O. Pedersen, in *Proceedings of the 4th Intern Conference on Indoor Air Quality and Climate*, Vol. 1, Berlin (West), 1987, pp. 97–101.
126. P. Wolkoff, P. Clausen, B. Jensen, G. Nielsen, and C. Wilkins, *Indoor Air* **7**, 92–106 (1997).
127. J. Brinke, S. Selvin, A. Hodgson, W. Fisk, M. Mendell, C. Koshland, and J. Daisey, *Indoor Air* **8**, 140–152, (1998).
128. C. Weschler, A. Hodgson, and J. Wooley, *Environ. Sci. Technol.* **26**, 2371–2377 (1992).
129. N. Dunston, and S. Spivak, *J. Applied Fire Science*, **6**, 231–242 (1996–1997).
130. Nordic Committee on Building Regulations, *Toxicological Based Air Quality Guidelines for Substances in Indoor Air*, NKB 11E, Helsinki, Finland, 1996.
131. K. Andersson, J. Bakke, O. Bjorseth, C-G., Bornehag, G. Clausen, J. Hongslo, M. Kjellman, S. Kjaergaard, F. Levy, L. Molhave, S. Skerfving, and J. Sundell, *Indoor Air* **7**, 78–91 (1997).
132. H. Levin, *Indoor Air Suppl.* **5**, 5–7 (1998).
133. G. Nielsen, L. Hansen, B. Nexo, and O. Poulsen, *Indoor Air Suppl.* **5**, 8–24 (1998).
134. A. Hodgson, *Indoor Air*, **5**, 247–257, (1995).
135. W. Winberry, L. Forehand, N. Murphey, A. Ceroli, B. Phinney, and A. Evans, *Compendium of Methods for the Determination of Air Pollutants in Indoor Air*, U.S. EPA, EPA/600/4-90/010, Washington, DC, (1990).
136. S. Brown, M. Sim, M. Abramson, and C. Gray, *Indoor Air* **4**, 123–134 (1994).
137. L. Sheldon, R. Handy, T. Hartwell, R. Whitmore, H. Zelon, and E. Pellizzari, *Indoor air Quality in Public Buildings*, Vol. I, Washington, DC, EPA/600/S6-88/009A, 1998.
138. L. Sheldon, H. Zelon, J. Sickles, C. Easton, T. Hartwell, and L. Wallace, *Indoor Air Quality in Public Buildings*, Vol. II. Research Triangle Park, NC, EPA/600/009b, 1988.
139. P. Morey and B. Jenkins, in *Proceedings of IAQ '89, The Human Equation: Health and Comfort*, ASHRAE, Atlanta, GA, 1989, pp. 67–71.
140. U.S. Housing and Urban Development, *The Manufactured Home Construction and Safety Standard*, Washington, DC, 1984.
141. B. Tichenor, *Indoor Air Sources: Using Small Environmental Test Chambers to Characterize Organic Emissions from Indoor Materials and Sources*, U.S. Environmental Protection Agency, Research Triangle Park, NC, 1989.
142. U.S. Environmental Protection Agency, *Carpet Policy Dialogue: Compendium Report*, EPA/560/2-91-002, U.S. EPA, Washington, DC, 1991.
143. American Society of Testing and Materials, *Guide D5116-90 for small scale Environmental Chamber Determination of Organic Materials/Products*, West Conshohocken, PA, ASTM, 1990.
144. European Collaborative Action, *Determination of VOCs Emitted for Indoor Materials and Products*, Report No. 16, Commission of the European Communities, Brussels, EUR 16284EN, (1995).
145. State of Washington, *East Campus Plus, Indoor Air Quality Specification*, Tacoma, WA, 1990.
146. J. Reinert, in *Proceedings of the 3rd Intern Conference on Indoor Air Quality and Climate*, Vol. 1, Stockholm, Sweden, 1984, pp. 233–238.
147. P. Dingle, and P. Tapsell, in *Proceedings of the 7th Intern Conference on Indoor Air Quality and Climate* Vol. 1, Nagoya, Japan, 1996, pp. 595–598.

148. R. Lewis and A. Bond, in *Proceedings of the 4th Intern Conference on Indoor Air Quality and Climate*, Vol. 1, Berlin (West), pp. 195–199, 1987.

149. U.S. Environmental Protection Agency *Indoor Air Quality Implementation Plan, Appendix A. Preliminary Indoor Pollution Information Assessment*, Washington, DC, EPA/600/8-87/014, 1987.

150. G. Schenk, H. Rothweiler, and C. Schlatter, *Indoor Air* **7**, 135–142 (1997).

151. J. Kroos, and P. Stolz, *Proceedings of the 7th International Conference on Indoor Air Quality and Climate*, Vol. 2, Nagoya, Japan, 1996, pp. 249–252.

152. E. Berger-Preiss, A. Preiss, and K. Levsen, *Proceedings of the 7th International Conference on Indoor Air Quality and Climate*, Vol. 1, Nagoya, Japan, 1996, pp. 507–512.

153. E. Berger-Preiss, A. Preiss, K. Sielaff, M. Raabe, B. Ilgen, and K. Levsen, *Indoor Air* **7**, 248–261 (1997).

154. Y. Motoba, Y. Takimoto, and T. Kato, *Amer. Indust. Hyg. Assoc.* **59**, 181–190 (1998).

155. National Research Council, *An Assessment of the Health Risks of Seven Pesticides Used for Termite Control*, Committee on Toxicology, Washington, DC, 1982.

156. R. Scheffrahn, C. Bloomcamp, and N.-Y. Su, *Indoor Air* **2**, 78–83 (1992).

157. American Conference of Governmental Industrial Hygienists, *Threshold Limit Values for Chemical Substances and Physical Agents*, Biological Exposure Indices, Cincinnati, OH, 1999.

158. J. Spengler and M. Cohen, In: R.B. Gammage and S. V. Kaye, Eds., *Indoor Air and Human Health*, Lewis Publishers, Boca Raton, FL, 1985, pp. 261–278.

159. U.S. Environmental Protection Agency, *Introduction to Indoor Air Quality; A Reference Manual*, Washington, DC, EPA/400/3-91/003, 1991.

160. U.S. Department of Energy, *Indoor Air Quality Environmental Information Handbook: Combustion Sources*, Washington, DC, DOE/EV10450-1, 1985.

161. J. Woodring, T. Duffy, J. Davis and R. Bechtold, *Am. Ind. Hyg. Assoc. J.* **46**, 350 (1985).

162. N. Nagda, H. Rector, and M. Koontz, *Guidelines for Monitoring Indoor Air Quality*, Hemisphere Publishing Corporation, Washington, DC, 1987, 270 pp.

163. I. Billick, J. Spengler, P. Ryan, P. Baker and S. Colome, *J. Air Pollut. Control Assoc* **39**, 1169 (1985).

164. J. Samet, M. Marbury, J. Spengler. *Am Rev. Respir. Dis.* **136**, 1486–1508, (1987).

165. J. Spengler and K. Sexton, *Science* **221**, 9 (1983).

166. G. Thurston, in *Proceedings of the 4th Intern Conference on Indoor Air Quality and Climate*, Vol. 1, Berlin (West), 1987, pp. 451–455.

167. S. Hall, "Toxic Responses of the Blood", in *Chemical Exposure and Toxic Responses*, S. Hall, J. Chakraborty, and R. Ruch, (eds.), CRC Lewis Publishers, Boca Raton, FL, 1997.

168. American Society of Heating, Refrigerating, and Air-Conditioning Engineers, *Addendum 62A-1990 to Ventilation for Acceptable Indoor Air Quality*, Atlanta, 1990.

169. U. S. Environmental Protection Agency, *Indoor Air Facts, No.*, 5, Washington, DC, ANR-445, 1989.

170. E. Miesner, S. Rudnich, F. Hu, J. Spengler, L. Preller, H. Ozkaynak, and W. Nelson, Paper 88-76.4, *Air Pollution Control Association*, 16 pp. 1988.

171. J. Hanrahan and S. Weiss, "Environmental Tobacco Smoke", in P. Harber, M. Schenker, and J. Balmes, eds., *Occupational and Environmental Respiratory Disease*, Mosby, St. Louis, MO, 1995, pp. 767–783.

172. G. Clausen, in *Proceedings of IAQ '88, Engineering Solutions to Indoor Air Problems*, ASHRAE, Atlanta, GA, pp. 267–274, 1988.

173. W. Cain, B. Leaderer, R. Isseroff, L. Berglund, R. Huey, E. Lipsitt and D. Periman, *Atm. Environ.* **17**, 1183–1197, (1983).

174. U.S. Public Health Service, *Surgeon Generals' Report: The Health Consequences of Involuntary Smoking*, USDHHS, CDC, 87-8398, 1987.

175. K. Phillips, D. Howard, M. Bentley, and G. Alvan, *Proceedings of the 7th Intern Conference on Indoor Air Quality and Climate*, Nagoya, Japan, Vol. 1, 1996, pp. 495–500.

176. C. Collett, J. Ross, and K. Levine, *Environmental Intern.* **18**, 347–352 (1992).

177. T. Bergman, D. Johnson, D. Boatright, K. Smallwood, and R. Rando, *Amer. Ind. Hyg. Assoc. J.* **57**, 746–752 (1996).

178. American Society of Heating, Refrigerating, and Air-Conditioning Engineers, *Public Review Draft, Ventilation for Acceptable Indoor Air Quality*, ASHRAE Standard 62-1989R, Atlanta, GA, August 1996.

179. American Society of Heating Refrigerating and Air-Conditioning Engineers, *Addendum N, Ventilation for Acceptable Indoor Air Quality*, Atlanta, GA, June (1999).

180. R. Wadden and P. Scheff, *Indoor Air Pollution*, John Wiley & Sons, Inc., New York, 1983.

181. R. Fortmann, *ASTM Standardization News*, **16**, 50 (1988).

182. L. Sheldon, H. Zelon, J. Sickles, C. Eaton, T. Hartwell, and R. Jungers, Paper 88-109-8, Air Pollution Control Association, 1988, 9 pp.

183. U.S. Environmental Protection Agency, *A Citizen's Guide to Radon*, 2nd Ed., Washington, DC, 1992.

184. J. Samet, "Radon and Lung Cancer Revisited" in R. Gamage and B. Berven, eds., *Indoor Air and Human Health*, 2nd ed., Lewis Publishers, Bacon Raton, FL, 1996, pp. 325–339.

185. G. Pershagen, G. Akerblom, O. Axelson, B. Clavensjo, L. Damber, G. Desai, A. Enflo, F. Lagarde, H. Meilander, M. Svartengren, and G. Swedjemark, *N. Engl. J. Med.* **330**, 159–164 (1994).

186. D. Harrje, I. Hubbard, and D. Sanchez, in *Healthy Buildings '88*, Vol. 2, Stockholm, Sweden, 1988, pp. 143–152.

187. U.S. Environmental Protection Agency, *Report to Congress on Indoor Air Quality; Volume II: Assessment and Control of Indoor Air Pollution*, Washington, DC, EPA/400/1-89/001C, 1989.

188. P. Korhonen, H. Kokotti, and P. Kalliokoski, *Amer. Ind. Hyg. Assoc. J.* **58**, 366–369, 1997.

189. Canada Mortgage and Housing Corporation, *Guide to Radon Control*, NHA 618, Ottawa, Canada, 1990.

190. L. Wallace, C. Nelson, R. Highsmith, and G. Dunteman, *Indoor Air* **3**, 193–205 (1993).

191. G. Raw, "The Importance of Indoor Surface Pollution in Sick Building Syndrome", *BRE Information paper IP/94*, Garston, Watford, U.K., 1994, 3 pp.

192. S. Gravesen, H. Ipsen, and P. Skov. in *Proceedings of the 6th Intern Conference on Indoor Air Quality and Climate* **4**, 33–35 (1993).

193. A. Hedge, W. Erickson, and G. Rubin, in *Proceedings of the 6th Intern Conference on Indoor Air Quality and Climate*, **1**, 291–96 (1993).

194. D. Franke, E. Cole, K. Leese, K. Foarde, and M. Berry, *Indoor Air* **7**, 41–54 (1997).

195. O. Bjorseth and B. Malvik, in *Proceedings of Healthy Buildings '97*, Vol 1, Washington, DC, 1997; pp. 429–432.

196. L. Morawska, M. Jamriska, and D. Francis, *Indoor Air*, **8**, 285–294 (1998).

197. K. Smith, in *Proceedings Healthy Buildings 97*, vol 3, Washington, DC, 1997, pp. 13–27.

198. State of California, California Air Resources Board, *Indoor Air Quality Guideline for Formaldehyde in the Home*, Sacramento, CA, 1991.

199. World Health Organization, *Indoor Air Quality Research, EURO Reports and Studies* 103, Copenhagen, Denmark, 1984.

200. U.S. Occupational Safety and Health Administration, *Occupational Exposure to Formaldehyde*; Final Rule, **29** *CFR* Part 1910, *Federal Register* Vol. 57, No. 102, 22310–22328.

201. U.S. Environmental Protection Agency, *Building Air Quality, A Guide for Owners and Facilities Managers*, EPA/400/1-91/033, Washington, DC, 1991.

202. U.S. Environmental Protection Agency, *New Headquarters Project Furniture Project Furniture Procurement: Office Seating*, Washington, DC, 1996.

203. American Society for Testing & Materials, "Standard Test Method for Nicotine in Indoor Air, D5075-90a," *Annual Book of ASTM Standards*. 11.03, West Conshohocken, PA, 1995, pp. 418–424.

204. T. Godish, *Indoor Air Pollution Control*, Lewis Publishers, Boca Raton, FL 1990, 401pp.

205. European Collaborative Action, *Radon In Indoor Air*, Report No., 15, Commission of the European Communities, EUR 16123 EN, 1995.

206. M. Crandall, and W. Sieber, *Appl. Occup. Environ Hyg.* **11**, 533–539 (1996).

207. R. Malkin, T. Wilcox, and W. Sieber, *Appl. Occup. Environ. Hyg.* **11**, 540–545 (1996).

208. N. Nelson, J. Kaufman, J. Burt, J. Karr, *Scand. J. Work Environ Health* **21**, 51–59 (1995).

209. W. Sieber, L. Stayner, R. Malkin, M. Petersen, M. Mendell, K. Wallingford, M. Crandall, T. Wilcox, and L. Reed, *Appl. Occup. Environ Hyg.* **11**, 1387–1392 (1996).

210. G. Hadwen, J. McCarthy, S. Womble, J. Girman, and H. Brightman, in *Healthy Buildings '97*, Vol. 2, Washington, DC, 1997, pp. 465–470.

211. H. Brightman, S. Womble, J. Girman, W. Sieber, J. McCarthy, and R. Buck, in *Healthy Buildings '97*, Vol 2, Washington, DC, 1997, pp. 453–458.

212. A. Hedge, W. Erickson, and G. Rubin, *Indoor Air* **5**, 10–21 (1995).

213. S. Burge, A. Hedge, S. Wilson, J. Bass, and A. Robertson, *Ann. Occup. Hyg.* **31**, 493–504 (1987).

214. J. Harrison, A. Pickering, M. Finnegan, P. Austwick, in *Proceedings of the 4th International Conference on Indoor Air Quality and Climate*, Vol. 2, Berlin (West), 1987, pp. 487–491.

215. A. Marmot, J. Eley, M. Nguyen, E. Warwick, and M. Marmot, in *Healthy Buildings '97*, Vol. 2, Washington, DC, 1997, pp. 483–488.

216. P. Skov, O. Valbjorn, and Danish Indoor Climate Study Group, *Environ. Int.* **13**, 339–349 (1987).

217. O. Valbjorn and P. Skov, in *Proceedings of the 4th Intern Conference on Indoor Air Quality and Climate*, Vol. 2, Berlin (West), 1987, pp. 593–597.

218. T. Zweers, L. Preller, B. Brunekreef, and J. Boleij, *Indoor Air* **2**, 127–136 (1992).

219. B. Berg-Munch, G. Clausen, and P. Fanger, *Environ. Intern.* **12**, 195–200 (1986).

220. P. Wargocki and P. Fanger, in *Proceedings of Healthy Buildings*, Vol. 2, Washington DC, 1997, pp. 243–248.

221. D. Baker, *Occup. Med. State of Art Reviews* **4**, 607–624, 1989.

222. S. Reynolds, P. Morey, J. Gifford, and S. Li, *Indoor Air* **6**, 168–180 (1996).

223. M. Hodgson, and P. Morey, in *Immunology and Allergy Clinics of North America*, Vol. 9. W. Solomon, ed., W. Saunders, 1989, pp. 399–412.

224. R. Alexander and J. Fedoruk, *J. Occup. Med.* **28**, 42 (1986).
225. R. Levine, D. Sexton, F. Romm, B. Wood, and J. Kaiser, *Lancet*, 1500 (1974).
226. P. Boxer, *J. Occup. Med.* **27**, 867 (1985).
227. G. Magarian, *Medicine* **61**, 219 (1982).
228. M. Colligan and M. Smith, *J. Occup. Med.* **20**, 401 (1978).
229. J. Salvaggio, *J. Allergy Clin. Immunol.* **85**, 689–699 (1990).
230. National Research Council, *Multiple Chemical Sensitivities*, National Academy Press, Washington, DC, 1992.
231. Position Statement, *J. Allergy Clin. Immunol.* **113**, 36–40, 1999.
232. American Medical Association, *JAMA* **268**, 3465–3467, 1992.
233. D. Black, *Reg. Tox. Pharamacol.* **18**, 23–31 (1993).
234. A. Leznoff, *J. Allergy Clin. Immunol.* **99**, 438–442 (1997).
235. K. Binkley and S. Kutcher, *J. Allergy Clin. Immunol.* **99**, 570–574 (1997).
236. H. Staudenmayer, *J. Allergy Clin. Immunol.* **99**, 434–437 (1997).
237. W. Fisk and A. Rosenfeld, *Indoor Air* **7**, 158–172 (1997).
238. A. Wheeler, in *IAQ '93, Operating and Maintaining Buildings for Health, Comfort, and Productivity*, ASHRAE, Atlanta, GA, 1993, pp. 131–136.
239. G. Flatheim, in *Design, Construction, and Operation of Healthy Buildings*, ASHRAE, Atlanta, GA, pp. 231–239 (1998).
240. K. Smith, in *Proceedings of the 7th Intern Conference on Indoor Air Quality and Climate*, Vol. 3, Climate, Nagoya, Japan, 1996, pp. 33–44.
241. B. Ligman, R. Shaughnessy, Z. Wang, R. Kleinerman, J. Lubin, E. Fisher, S. Zhang, and L. Wang, in *Healthy Buildings '97*, Vol. 3, Washington, DC, 1997, pp. 51–56.
242. R. Shaughnessy, B. Ligman, E. Fisher, Z. Wang, and R. Kleinerman, in *Healthy Buildings '97*, Vol. 3, Washington, DC 1997, pp. 57–62.
243. D. Bearg, *Indoor Air Quality and HVAC Systems*, Lewis Publishers, Boca Raton, FL, 1992, 220 p.
244. W. Turner, D. Bearg, and T. Brennan, *Occupational Medicine: State of the Art Reviews*, **10**, 41–57 (1995).
245. W. Fisk, R. Spencer, D. Grimsrund, F. Offermann, B. Pedersen, and R. Setro, *Indoor Air Quality Control Technologies*, Noyes Data Corporation, Park Ridge, NJ, 1987, 245 pp.
246. Nordtest Report, *Indoor Climate Problems*, NT Technical Report 204, Espoo, Finland (1993).
247. European Concerted Action, *Guidelines for Ventilation Requirements in Buildings*, Report No. 11, Commission of European Communities, Eur 14449 EN, Brussels, 1992.
248. American Society of Heating, Refrigerating and Air-Conditioning Engineers, ASHRAE Standard 52.1-1992, *Gravimetric and Dust Spot Procedures for Testing Air Cleaning Devices Used in General Ventilation for Removing Particulate Matter*, ASHRAE, Atlanta, GA, 1992.
249. American Society of Heating, Refrigerating, and Air-Conditioning Engineers, ASHRAE Standard 52.2, *Method of Testing General Ventilation Air Cleaning Devices for Removal Efficiency by Particle Size*, ASHRAE, Atlanta, GA, 1999.
250. A. Veeck, *Invironment Prof.* **4**, 1 and 4–5 (1998).
251. P. Morey, "Indoor Air Quality in Non-industrial Occupational Environments" in L. DiBerardinis, ed., *Handbook of Occupational Safety and Health*, 2nd ed., John Wiley & Sons, Inc., New York, 1999, pp. 743–789.

252. T. Godish and J. Spengler, *Indoor Air*, **6**, 135–145 (1996).

253. P. Morey, "Control of Indoor Air Pollution" in P. Harber, M. Schenker, and J. Balmes, eds., *Occupational and Environmental Respiratory Disease*, Mosby, St. Louis, 1995, pp. 981–1003.

254. R. Tamblyn, *ASHRAE Transactions, Paper* DE-93-10-1 (1993).

255. G. Raijhans, *Occup. Health Ontario* **4**, 160 (1983).

256. K. Ikeda, S. Yoshizawa, T. Irie, et al., *ASHRAE Trans.*, Paper No. 2955, 1986.

257. M. Narasak, in *Proceedings of the 4th Intern Conference on Indoor Air Quality and Climate*, Vol. 3, Berlin (West), 1987, pp. 277–282.

258. P. Morey and J. Singh, "Indoor Air Quality in Nonindustrial Occupational Environments", in G. Clayton and F. Clayton, eds., Vol 1, Part A., *Pattys Industrial Hygiene and Toxicology*, 4th ed., John Wiley & Sons, New York 1991, pp. 531–594.

259. American Society of Heating, Refrigerating, and Air-Conditioning Engineers, *Standards for Natural and Mechanical Ventilation*, Standard 62-1973 (ANSI 3 194-1-1977), Atlanta, GA, 1977.

260. American Society of Heating Refrigerating, and Air-Conditioning Engineers, *Energy Conservation in New Building Design*, ASHRAE Standard 90-75, Atlanta, GA, 1975.

261. American Society of Heating, Refrigerating, and Air-Conditioning Engineers, *Ventilation for Acceptable Indoor Air Quality*, ASHRAE Standard 62-1981, Atlanta, GA, 1981.

262. H. Levin, *Indoor Air Bulletin* **3**(8), 1–9 (1996).

263. C. Vogt, *IEQ Strategies* **12**(6), 1–8 (1999).

264. American Society of Heating, Refrigerating, and Air-Conditioning Engineers, *Addendum L, Ventilation for Acceptable Indoor Air Quality*, Atlanta, GA June 1999.

265. American Society of Heating Refrigerating and Air-Conditioning Engineers, *Addendum M, Ventilation for Acceptable Indoor Air Quality*, Atlanta, GA May 1999.

266. O. Seppanen, in *Proceedings of the 7th Intern Conference on Indoor Air Quality and Climate*, Nagoya, Japan, Vol. 3, 1996, pp. 15–32.

267. M. Mendell and A. Smith, *Amer. J. Public Health* **80**, 1193–1199 (1990).

268. P. Pasanen, *Emissions from Filters and Hygiene of Air Ducts in the Ventilation Systems of Office Buildings*, Doctoral Dissertation, University of Kuopio, Finland, 1998.

269. R. Simmons, D. Price, J. Noble, S. Crow, and D. Ahearn, *Amer. Industr. Hyg. Assoc. J.* **58**, 900–904 (1997).

270. R. Bock, H. Schleibinger, and H. Ruden, in *Healthy Buildings 97*, Vol. 1, Washington, DC, 1997, pp. 593–598.

271. B. Fluckiger, C. Monn, P. Luthy, and H.-U. Wanner, *Indoor Air* **8**, 197–202 (1998).

272. C. van Netten, *Appl. Occup. Environ. Hyg.* **13**, 733–739 (1998).

273. M. Hujanen, O. Seppanen, and P. Pasanen, in *Healthy Buildings 91*, ASHRAE, Atlanta, GA, 1991, pp. 329–333.

274. P. Pasanen, J. Teijonsalo, O. Seppanen, J. Ruuskanen, and P. Kalliokoski, *Indoor Air* **4**, 106–113 (1994).

275. M. Bjorkroth, A. Torkki and O. Seppanen, in *Healthy Buildings 97*, Vol. 1, Washington, DC, 1997, pp. 599–603.

276. M. Bjorkroth, O. Seppanen, and A. Torkki, in *Design Construction and Operation of Healthy Buildings*, ASHRAE, Atlanta, GA, 1998, pp. 47–55.

277. J. Pejtersen, *Indoor Air* **6**, 239–248 (1996).

278. R. Kulp, R. Fortmann, C. Gentry, D. VanOsdell, K. Foarde, T. Hebert, R. Krell, and C. Cochrane, in *Healthy Buildings '97*, Vol. 1, Washington, DC, 1997, pp. 605–610.

279. P. Morey, "Internal HVAC Pollution", NATO, *CCMS Pilot Study on Indoor Air Quality*, Sainte-Adele, Quebec, 1991, pp. 47–55.

280. K. Boone, P. Ellringer, J. Sawyer, and A. Streifel, in *Healthy Buildings '97*, Vol. 1, Washington, DC, 1997 pp. 75–79.

281. L. Belinda, W. Turner, S. Martel, and W. Johnson, in *Healthy Buildings '97*, Vol. 1, Washington, DC, 1997 pp. 123–128.

282. J. Carlton-Foss, *Ann. Am. Conf. Gov. Ind. Hyg.* **10**, 93 (1984).

283. American Society of Heating, Refrigerating, and Air-Conditioning Engineers, *Thermal Environmental Conditions for Human Occupancy*, Standard 55-1992, ASHRAE, Atlanta, GA, 1992.

284. American Society of Heating, Refrigerating, and Air-Conditioning Engineers, *Addendum to Thermal Environmental Conditions for Human Occupancy*, Atlanta, GA, ASHRAE 55a-1995, 1995.

285. L. Berglund, *ASHRAE J.*, 35–41 (Aug. 1998).

286. P. Fanger, *Indoor Air Suppl* **4**, 81–86 (1998).

287. L. Fang, G. Clausen, and P. Fanger, *Indoor Air*, **8**, 80–90 (1998).

288. J. Woods, P. Morey, and D. Rask, in N. Nagda and J. Harper, eds., *Design and Protocol for Monitoring Indoor Air Quality*, American Society for Testing and Materials, Philadelphia, PA, 1989, pp. 80–98.

289. National Research Council, *Building Diagnostics, A Conceptual Framework*, Building Research Board, National Academy Press, Washington, DC, 1985.

290. J. Samet and J. Spengler, *Indoor Air Pollution, A Health Perspective*, John Hopkins University Press, Baltimore, MD, 1991.

291. H. Levin, *Occupational Medicine: State of the Art Reviews* **10**, 59–94, (1995).

292. B. Samimi, *Occupational Medicine: State of the Art Reviews* **10**, 95–118 (1995).

293. A. Abend, *Occupational Medicine: State of the Art Reviews* **10**, 195–204 (1995).

294. M. Lytton, in *Design Construction and Operation of Healthy Buildings*, ASHRAE, Atlanta, 1998, pp. 15–22.

295. Health Canada, *Indoor Air Quality in Office Buildings: A Technical Guide*, H46-3/3-1993E, Ottawa, 1993, pp. 55.

296. P. Morey, in *Proceeding of the 7th Intern Conference on Indoor Air Quality and Climate*, Nagoya, Japan, Vol 2, 27–36, 1996.

297. Institute of Inspection, Cleaning, and Restoration Certification, *Standard and Reference Guide for Professional Water Damage Restoration*, 5500-94, IICRC, Vancouver, WA, 1994, 75 pp.

298. European Collaborative Action, *Sampling Strategies for Volatile Organic Compounds (VOCs) in Indoor Air*, Report No. 14, Commission of European Communities, Brussels, Eur 16051 EN, 1994, 41 pp.

299. A. Persily, W. Turner, H. Burge, and R. Grot, in N. Nagda and J. Harper, eds., *Design and Protocol for Monitoring Indoor Air Quality*, ASTM STP 1002, American Society for Testing and Materials, Philadelphia, PA, 1989, pp. 35–50.

300. M. Hodgson, *Occupational Medicine: State of the Art Reviews* **10**, 167–175 (1995).

301. European Collaborative Action, *Evaluation of VOC Emissions from Building Products*, Report No. 18, Commission of European Communities, Brussels, 1997, Eur 17334 EN, 108 pp.

302. U.S. Environmental Protection Agency, *Report to Congress on Indoor Air Quality, Executive Summary and Recommendations*, 1989, Washington, DC, EPA/400/1-89/001A.

303. *Constitution of the World Health Organization*, 1946, Official Record of the World Health Organization, Vol. 2, p. 100.

304. U.S. Occupational Safety and Health Administration, *Indoor Air Quality; Proposed Rule*, **29** *CFR* Parts 1910, 1915, 1926, 1928 *Federal Register* **59** (65), 15967–16039 (1994).

305. B. Epstien, *EIA Technical Jour.* **3**(2), 16–18 (1995).

306. T. Nathanson, *Humidification System: Function, Operation, and Maintenance*, Public Works and Government Services Canada, Ottawa, 1995, 11 pp.

307. Health Canada, *Fungal Contamination in Public Buildings: A Guide to Recognition and Management*, Environmental Health Directorate, Ottawa, 1995, 76 pp.

308. T. Nathanson, *Microbials in Office Buildings*, Public Works and Government Services Canada, Ottawa, 1995, 7 pp.

309. Canada Mortgage and Housing Corporation, *Clean-up Procedure for Molds in Houses*, NH15-91/1993E, Ottawa, 1993, 32 pp.

310. Canada Mortgage and Housing Corporation, *Toxic Mold Clean-up Procedures*, Ottawa, 1999 (draft).

311. Committee on Challenges of Modern Society, *Managing Indoor Air Quality Risks*, NATO, EPA/400/7-90/005, 1990, 200 pp.

312. Committee on Challenges of Modern Society, *Energy and Building Sciences in Indoor Air Quality*, NATO, Institut de Recherche en Santé et en Sécurité du Travail du Québec, Montreal IRSST, 1991, pp. 136.

313. M. Maroni, R. Axelrad, and A. Bacaloni, *Am. Ind. Hyg. Assoc. J.* **56**, 499–508, 1995.

314. Nordic Committee on Building Regulations, *Indoor Climate-Air Quality*, NKB Publication No 61E, 1991, 36 pp.

315. U. S. Environmental Protection Agency, *Tools For Schools Action Kit*, Washington, DC, EPA 402-K-95-001, September 1995.

316. Joint Commission on Accreditation of Healthcare Organizations, W. Hansen, ed., *A Guide to Managing Indoor Air Quality*, Oakbrook Terrace, IL, 1997.

317. American Society of Testing and Materials, H. Trechsel, ed., *Moisture Control in Buildings*, ASTM Manual Series MNL 18, West Conshohocken, PA, 1994.

318. American Society of Testing and Materials, R. Kudder and J. Erdly, eds., *Water Leakage through Building Facades*, ASTM STP 1314, West Conshohocken, PA, 1998.

319. A. Nevalainen, P. Partanen, E. Jaaskelainen, A. Hyvarinen, O. Koshinen, T. Mekin, M. Vahteristo, J. Koivisto, and T. Husman, *Indoor Air Suppl.* **4**, 45–49 (1998).

320. M. Lawton, R. Dales, and J. White, *Indoor Air* **8**, 2–11, (1998).

321. New York City Department of Health, New York City Human Resources Administration, and Mount Sinai-Irving, J. Salikoff, Occupational Health Clinical Center, *Guidelines on Assessment and Remediation of Stachybotrys atra in Indoor Environments*, New York, New York, 1993.

322. P. Morey and D. Sawyer "Mitigation of Visible Fungal Contamination in Buildings: Experience from 1993–1998," in E. Johanning, ed., *Third International Conference on Bioaerosols, Fungi, and Mycotoxins: Health Effects Assessment, Prevention, and Control*, Saratoga Springs, NY, 2000, pp. 255–265.

323. O. Seppanen, "Finnish Criteria for Materials Emissions," *Finnish Society of Indoor Air Quality and Climate*, Helsinki, Finland, 1998.

324. H. Levin, *Occupational Medicine: State of the Art Reviews*, **4**, 667–693 (1989).
325. P. Morey, and D. Shattuck, *Occupational Medicine: State of the Art Reviews* **4**, 625–642 (1989).
326. P. Morey, M. Hodgson, W. Sorenson, G. Kullman, W. Rhodes, and G. Visvesvara, *ASHRAE Trans.* **93**, 399 (1986).
327. National Air Duct Cleaners Association, *Mechanical Cleaning of Non-porous Air Conveyance System Components*, NADCA-01, Washington, DC, 1992.
328. A. Nevalainen, in *Proceedings of the Sixth Intern Conference on Indoor Air Quality and Climate*, Helsinki, Finland, Vol. 4, 1993, pp. 3–13.
329. P. Morey in *IAQ '94, Engineering Indoor Environments*, ASHRAE, Atlanta, 1994, pp. 79–88.
330. P. Morey, H. Eisenstein, J. Girman, H. Levin, J. Woods, in *Healthy Buildings '91 Postconference Proceedings*, ASHRAE, Atlanta, GA, 1991, pp. 54.
331. C. Hunter, C. Grant, B. Flannigan, et al., *Intern. Biodeterioration and Biodegradation* **24**, 81–101 (1988).
332. T. Walsh and D. Dixon, *Europ. J. Epidemiol.* **5**, 131–142 (1989).
333. R. Fox, *J. Allergy Clin. Immunology* **94**, 413–416 (1994).
334. R. Shaughnessy, E. Levetin, J. Blocker, and K. Sublette, *Indoor Air* **4**, 179–188 (1994).
335. R. Shaughnessy, and L. Oatman in *IAQ 91 Healthy Buildings*, ASHRAE, Atlanta, 1991, pp. 318–324.
336. M. Boeniger, *Am. Ind. Hyg. Assoc. J.* **56**, 590–598 (1995).
337. K. Foarde, D. Van Osdell, and R. Steiber, *Appl. Occup. Environ. Hyg.* **12**, 535–542 (1997)
338. National Institute for Occupational Safety and Health, Method 0500, *NIOSH Manual of Analytical Methods*, USDHHS, 1994.
339. P. Eller and M. Cassinelli, eds., Method 3500, in *NIOSH Manual of Analytical Methods*, 4th ed., US DHHS, 1994.

CHAPTER SIXTY-SIX

Role of the Industrial Hygiene Consultant

Henry J. Muranko, MPH, CIH, CSP

Editorial Note: Concurrent with the development of the industrial hygiene profession has been the increase in the number of consultants specializing in industrial hygiene services. Although the functions of the individual consultant and that of the organizational consultant may differ because the latter's capacity may be broader in scope of available specialties, they are similar in their objectives: that of helping the client and of achieving, in their modus operandi, the highest degree of technical expertise and professional ethics. Therefore the reader will find some duplication and overlapping in the two sections that follow. This is deliberate.

A THE NONORGANIZATIONAL INDUSTRIAL HYGIENE CONSULTANT

1 CONSULTATION BY INDIVIDUALS

1.1 History: Growth of Industrial Hygiene Consultation

Every professional endeavor and, indeed, many activities that might not qualify as professional endeavors appear to have individuals and organizations who offer their specialized services to society as consultants. The profession of industrial hygiene is no exception. The reasons for the proliferation of consultants in this field are not difficult to identify. Industrial hygiene is a specialty requiring expertise not immediately available within many

Patty's Industrial Hygiene, Fifth Edition, Volume 4. Edited by Robert L. Harris.
ISBN 0-471-29749-6

organizations; in addition, specialized equipment for various types of monitoring the industrial environment is also required.

There is no record of the number of industrial hygiene consultants or organizations that have existed at any time in the past or, for that matter, even today. In all probability, though, the number of consultants has increased in proportion to the growth of industrial hygiene as a profession, although it seems likely that the percentage of professional industrial hygienists engaged in consulting activities today is greater than in past years. There has always existed an enormous need for consulting industrial hygienists to provide services to the many hundreds of thousands of medium sized and small industrial operations in the United States that may require professional assistance, but do not have professional industrial hygienists on their staffs.

When the American Industrial Hygiene Association (AIHA) was founded in 1939, its list of founding members numbered about 160 individuals. It is probable that not more than 10% of these individuals offered industrial hygiene consultation services, so although no records exist to support any estimates, it is likely that in the years between World War I and World War II, only a handful of individuals were engaged in industrial hygiene consulting, and very few organizations existed for this purpose. Subsequent to World War II, steady, sustained growth of industrial hygiene as a profession began, until today the AIHA membership numbers in excess of 13,000 persons, and the American Academy of Industrial Hygiene (AAIH) consists of more than 6,000 active certified individuals in the U.S., and more than 300 Canadians. In demographics surveys in 1994 and 1996 the AIHA reported that approximately 25% of its members were employed by consultants or were self-employed.

Matching the growth of individuals pursuing private consulting during the post-World War II period, industrial hygiene consulting organizations began to form, and such organizations have grown greatly both in number and in size during the past 20 years. The many reasons for this rapid growth, as well as the kinds of industrial hygiene services presently available, are detailed later in this chapter. The remainder of this discussion addresses matters related to consultation offered by individuals, or small groups of individuals, operating with little or no laboratory or field analysis equipment.

1.2 Backgrounds of Industrial Hygiene Consultants

There are many individuals offering services related to industrial hygiene matters who are not actually considered industrial hygiene consultants, although at times the distinction may be difficult to define. There are, for example, individuals whose principal specialty is measuring or controlling noise, or ionizing radiation, or perhaps they consider themselves to be industrial ventilation engineers. This discussion is limited, however, to individuals offering full-service industrial hygiene assistance, which is to say that they are prepared to consider or study any workplace, and make recommendations related to potential or actual health hazards arising from work operations. They may, of course, call upon other specialists, such as noted, to be of further assistance, but in general the industrial hygiene consultant is capable of taking appropriate actions related to any risk of adverse health effects arising from the occupational environment.

All industrial hygiene consultants, as just defined, should have a common background, namely, proficiency in industrial hygiene, and they should be professionals who are primarily dedicated to the practice of industrial hygiene. It is reasonable to expect such individuals to be members of the AIHA, possibly the American Conference of Governmental Industrial (ACGIH), and also to be Certified Industrial Hygienists (CIHs). Many also hold certification as a safety professional and frequently in environmental management. Today, with the rapid growth in the number of CIHs, and the growing requirement by many employers that job candidates and consultants be certified, it is likely that few, if any, future industrial hygiene consultants will lack this qualification.

Although it is true that all consultants are or should be industrial hygienists, their individual backgrounds may differ greatly, and it is useful to list some of the many backgrounds of individuals actively consulting today. Table 66.1 lists the most probable prior employment of individuals who choose to become consultants. Prior to becoming industrial hygienists, of course, their backgrounds differed to an even greater extent, with the more common educational backgrounds of industrial hygienists listed in Table 66.2. To some extent, the capabilities and preferences of individual consultants vary with the particular educational background that preceded their employment in industrial hygiene, even though they may be qualified to offer broad-spectrum industrial hygiene consultation. Thus the analytical chemist who became an industrial hygienist will in all likelihood tend to emphasize sampling and analytical activities, whereas a physicist may feel more comfortable dealing with physical phenomena such as radiation and noise, and seek to become more knowledgeable in these areas.

Table 66.1. Professional Backgrounds of Practicing Industrial Hygienists

1. Industrial hygienists presently employed by industry, government, insurance companies, etc.
2. Industrial hygienists formerly employed by industry, government, insurance companies, consulting organizations, etc.
3. University faculty and research associates, primarily from undergraduate and graduate programs in industrial hygiene.
4. Retired industrial hygienists, all sectors.

Table 66.2. Educational Backgrounds of Industrial Hygiene Consultants

1. Undergraduate and graduate degree in one of the physical sciences, e.g., chemistry, physics, geology, etc.
2. Undergraduate in one of the physical sciences, graduate degree in industrial hygiene.
3. Undergraduate and graduate degrees in one of the biological sciences, e.g., biology, biochemistry, etc.
4. Undergraduate degree in one of the biological sciences, graduate degree in industrial hygiene.
5. Undergraduate and graduate degrees in engineering (chemical industrial, other).
6. Undergraduate degree in engineering, graduate degree in industrial hygiene.
7. Undergraduate degree in a variety of specialties not included above, graduate degree in industrial hygiene.

It is apparent that industrial hygienists who have retired or resigned their positions with industry, government, labor, and so on are perhaps the most logical prospects for becoming consultants, but it is not so apparent that individuals presently employed by the private sector, government, and so forth may also be consultants. In fact, a limited number of corporations, government agencies, and other organizations do permit their employees to engage in consulting, provided generally that they do it on their own time and meet certain other requirements. Clearly they must be careful to avoid conflicts of interest with their employers, and the amount of time they can devote to consulting will be limited. It is important that consultants who do have corporate or other employers inform their clients of this relationship so that there can be no misunderstandings and possible embarrassments owing to potential or actual conflicts of interest. It is reasonable to assume that industry trends over the last several years resulting in mergers, down-sizing, restructuring and outsourcing have caused this category of private consultant to become increasingly rare as a result of the added demands placed on the available time of such individuals by their full-time employer.

Industrial hygienists from every sector tend to consider becoming consultants once they retire, on either a full- or part-time basis. Such individuals are usually very well qualified, and frequently have assurance from previous employers and closely related organizations that they will become clients of the new consultant. There are pitfalls to be considered, however, and not infrequently they are unanticipated by prospective retirees, who may be disappointed at their lack of success as consultants. More often than not, individual consultants tend to rely on word-of-mouth advertising among their professional associates and clients, and in this regard, some individuals who retire tend to be somewhat disadvantaged. Unlike retired university staff, for example, who may have been consulting on a part-time basis for many years, the corporate industrial hygienist who retires may suddenly learn that although well known within his ordinary circle of professional acquaintances, he is not well known to the larger community that may require consulting assistance. Direct advertising is of limited usefulness, because the universe of persons or organizations who may need assistance is very large, and essentially unknown to most would-be consultants.

Most, if not all, industrial hygienists on university staffs do some limited amount of consulting, and it is logical and even essential that they do so. In a dynamic field such as industrial hygiene, a university professor who does no consulting runs the risk of being poorly prepared to familiarize students with important current problems and the challenges of the ever-changing workplace. Consulting activities, particularly when they involve on-site visits and workplace studies, are extremely valuable to university professors, and spawn many research projects both large and small for the professor and the students. The benefits of consulting by university staff are twofold, of course, including in addition to the benefits just noted, a reverse benefit to the client in retaining an individual who may possess unique skills in the area of expertise required. It may be argued that the more practical reason for encouraging university staff to consult is the supplementation of income that is often required so that university persons will not succumb to the inevitable economic pressure to accept more lucrative positions elsewhere. Given the inevitability of inadequate university budgets, this factor is an important one that helps ensure the maintenance of competent staff in our universities.

1.3 Ideal Qualifications of an Industrial Hygiene Consultant

In view of the varied backgrounds of industrial hygiene consultants, it is reasonable to ask what would be considered the ideal or optimal qualifications for such an individual? There is probably no single answer to which all professionals would agree, but certainly the single qualification that would be of the greatest importance is proficiency in the field of industrial hygiene. Today this goal is best realized by taking formal graduate training in a good university program. As of 1998, 28 graduate schools and five undergraduate programs had achieved accreditation by the Accreditation Board for Engineering and Technology (ABET) for their industrial hygiene curriculum.

In years past, the probable answer to the question concerning the ideal educational background of an industrial hygienist would have been an engineering degree, preferably chemical engineering, or perhaps a degree in chemistry. These backgrounds presuppose primary concern with the effects of substances in the environment rather than physical phenomena, whereas today it would be difficult to argue that the general subject areas of noise, radiation, or ergonomics are not equally deserving of special educational backgrounds in the appropriate engineering or physical sciences. Also, given the close relationship between the effects of substances and various forms of energy in the workplace, and the importance of toxicologic studies, a case can certainly be made for the desirability of a strong background in the biological sciences, and, in fact, a substantial percentage of incoming students to graduate industrial hygiene programs have been graduates of biological science programs.

In the final analysis it can be argued that both undergraduate and graduate training are secondary to the field experience of the individual who contemplates industrial hygiene consulting as a career. There is probably little disagreement that inexperienced persons who have just completed graduate studies should not begin their professional careers as consultants. The reasons are self-evident, of course, but perhaps in industrial hygiene more than many professions, a period of learning on the job and solving real-life problems is of the greatest importance in developing the degree of proficiency required of a consultant. No minimum time of employment prior to becoming a consultant can be defended, but certainly someone entering the field at present should be a CIH prior to becoming a consultant, and for this and other reasons it is not recommended that anyone with less than five years experience attempt to become established as a consultant.

Brief mention may also be made of personal qualifications, and here, of course, it is difficult to defend any particular set of attributes. As just noted, proficiency and technical competence are of the utmost importance, but second only to these attributes are those personal qualities that enable an individual to interact with other persons in a natural and easy manner. Consultants, by definition, are constantly meeting new persons, and must always attempt to sell both their own personal capabilities and the importance of carrying out their recommendations. Individuals who find it difficult to interact with strangers on short notice will probably not fare well as consultants. Consultants are also required to live a somewhat "unstructured" life, which is to say that they may travel considerably and on short notice, and must be prepared to offer their services when needed, and not necessarily at their own convenience. Those individuals who are most comfortable with a five-day week, nine-to-five schedule are probably not suited to the irregular hours and working habits frequently required of consultants.

1.4 Relationships Between Consultants and Client's Professional Staff

Consultants are frequently identified and retained by industrial hygienists, safely engineers, or other health professionals, or attorneys, if in fact the clients have such persons on their staffs. Not infrequently, however, a consultant may be retained by someone else in the client organization who is either unaware of the industrial hygiene staff, or has not felt it necessary to consult with it. Regardless of by whom retained, it is important that the industrial hygiene consultant make every effort to meet with the other staff members concerned with employee health and safety. These persons are obviously familiar with the principal problems within their own organization, and in all probability have ongoing programs that may be attempting to address problems that have resulted in calling in a consultant in the first place. Generally their insights will be invaluable and, in addition, they may be expected to have records that can be of value to the consultant. If the organizational industrial hygienist has not been involved in the particular problem that necessitated the services of a consultant, then it is incumbent upon the consultant to attempt to see that that person is properly informed and, if possible, becomes part of the effort to address the problem. Transcending these considerations, of course, is the propriety of exercising ordinary professional courtesy toward one's professional associates.

Not infrequently a given organization may request the services of a consultant, while unknown to the consultant that organization may be part of a larger corporation or entity that does have professional industrial hygiene staff, who may be located at a central headquarters elsewhere. In such instances, the consultant should make a special effort at least to discuss the problems with the staff industrial hygienist, if permitted to do so by the organization retaining him or her.

Ideally, an industrial hygiene consultant would report to a counterpart in the client's organization, but frequently this will not be possible, and instead a consultant may report to some other individual, in all probability the person who retained the consultant initially. Such individuals may be legal counsel, department heads, or other administrative persons with special interests in the matter that prompted the need for consultation assistance. The consultant may be unable to insist that the report be submitted to the corporate industrial hygienist, but certainly every prudent effort should be made to ensure that that person is fully informed and is given a copy of any reports that may be generated.

1.5 Advantages and Limitations of Being an Industrial Hygiene Consultant

An individual may choose to become an industrial hygiene consultant, and may eventually be sufficiently successful that pressure builds to form an organization to handle the increasing work load. This expansion may take several forms, ranging from a loose federation with other professionals, available on a part-time basis, to the formation of a consulting organization requiring larger quarters, secretarial staff, more equipment, and perhaps a laboratory. Each individual consultant must weigh the advantages and shortcomings of expanding and make whatever decision seems best suited to his or her needs. There are readily definable advantages and shortcomings to any course of action selected. For example, it is obviously attractive to form an organization and begin an expansion that may lead to becoming a full-service organization. The opportunity for substantially greater income is probably the greatest motivation, but the opportunity to offer services not otherwise possible may also be an important factor. Additionally, acquiring field sampling equipment and other devices certainly is desirable, and the convenience of a laboratory serving not only the in-house consulting activity but also offering such services to others is readily apparent. Offsetting these advantages are all of the problems related to the formation of any organization, such as finding and keeping qualified personnel, greatly increased overhead costs, concern with attracting sufficient professional activity to ensure the payment of staff salaries, benefits administration, the large capital investment required to acquire modern equipment and to form a modern laboratory, and other matters.

If an individual consultant chooses, on the other hand, to remain a one-person organization or at most to form a loose relationship with others, to be available on a part-time basis, most of the problems associated with large organizations will be avoided, but the consultant must continue to deal with the pressures of trying to accommodate clients' needs within limited time, and will in all likelihood find it difficult or impossible to acquire as much equipment as might be needed. It will also be necessary to rely upon one or more existing consulting laboratories if analysis of collected samples is required.

There is obviously no single best choice, and it remains for each individual consultant to practice the profession in the manner that seems best suited to his or her needs. There are many examples today of individual consultants continuing to practice more or less on their own, and there are many examples of individuals who have started out as one-person organizations but have built large and successful consulting organizations.

2 PRINCIPAL INDUSTRIAL HYGIENE CONSULTING ACTIVITIES

All industrial hygiene consulting activities will in all probability bear some clear relationship to industrial hygiene-related problems, but there are a number of different kinds of activities that deserve classification and discussion. Table 66.3 lists the principal kinds of industrial hygiene consulting activities that may be anticipated, although occasional consulting assignments will not fit into any of the suggested categories.

Table 66.3. Principal Industrial Hygiene Consulting Activities

Problem solving; industry, labor, government
Performing industrial hygiene surveys of plants, industries, specific operations, etc.
Industrial hygiene, safety, or occupational health program auditing.
Industrial hygiene training, workers, management
Assisting industry or other organizations with regulatory compliance.
Litigation support, including serving as expert witness.
Designing and conducting, or assisting in the conduct of scientific studies related to health problems in occupational or environmental settings.
Appearing at hearings, or meeting with regulatory officials *re* proposed legislation or regulation of interest to a client.
Serving as an advisor to Trade Organizations, Labor Unions, Professional Organizations or Other Special Interest Group.

2.1 Problem Solving: Industry, Labor, Government

Consulting for the purpose of problem solving, or problem-stimulated consultation, is probably the most common consulting activity performed. Typical problems may arise because of management concerns, or concerns expressed by individual workers, often identified by labor organizations. Citations by the Occupational Safety and Health Administration (OSHA) or local regulatory agencies are also a common basis for identifying problems requiring solutions, but no matter what the source of the original concern may have been, it has subsequently been determined that outside assistance by a qualified consultant is required. There are several reasons why such a decision may have been made, and it is well for the consultant to be aware of the real reasons for seeking outside assistance. The more common reasons for requesting consultation assistance are genuine lack of adequate expertise within the requesting organization; "political" considerations, wherein the opinions of a disinterested expert are required; insufficient time and staff of the requesting organization to address the problem adequately, even though they possess the expertise; and perhaps the need for someone to act as an arbitrator in an attempt to reconcile opposing points of view. The consultant will normally approach the problem to be solved in the same manner, regardless of the reasons that stimulated the request for services, but he or she may stand a better chance of satisfying the client if fully informed as to the circumstances prompting the investigation. It goes without saying that no matter by whom retained, or why, the consultant must make every effort to be completely neutral with respect to the political aspects of the problems involved, and totally objective, regardless of the opinions of the particular individual who retained him or her initially.

2.2 Performing Industrial Hygiene Surveys of Plants, Industries, and Specific Operations

For a variety of reasons, industrial hygiene consultants may be requested to perform "wall-to-wall" industrial hygiene surveys of entire plants, or specific buildings, units, or operations. Such surveys may be related to problems that initially required some problem-solving activity, but after which it became apparent that a complete survey was needed. In some instances, complete surveys may be completed in one or two days, whereas others may require weeks of effort involving several persons. Regardless of the size or complexity of the area to be surveyed, each survey should be based on the classic industrial hygiene principles of recognition, evaluation, and control.

The recognition phase of the survey is perhaps the most important, and should be accomplished by extensive discussions with management, labor, and other concerned parties; detailed studies of all of the processes, raw materials, by-products, and finished products involved; and careful examination of all existing records, including plant layout prints, ventilation diagrams, and of course, all existing industrial hygiene records. One or more thorough walk-through inspections should also be made, accompanied by knowledgeable persons who can explain everything that is of possible interest. Consultations with the plant medical department, including interviews with both the physician and nurse, are also of great value and may provide insights into the existence of problem areas or operations. In cases where medical services are provided by a part-time physician or industrial medicine clinic it is often useful to consult with them, to the extent that is possible.

The evaluation phase of the survey may consume the greatest amount of time, and may require the acquisition of additional personnel and equipment. Basic industrial hygiene principles should be followed, of course, with particular attention directed both to obvious problem areas and to jobs or operations that are subject to regulation by specific OSHA standards or regulations. Whenever applicable, efforts should be made to supplement an air sampling program with biospecimen analyses, whenever these analyses are recognized as being useful in evaluating the extent of exposures.

Recommendations for controlling proven overexposures, or potential overexposures, must be made with the realization that a consultant is not a compliance officer for a regulatory agency, and must instead present options in a clear and lucid manner, emphasizing those actions that are most urgently required but also presenting those that may be optional. Individual clients may posses the required expertise and may be capable of providing their own controls, but in many instances it will be the responsibility of the consultant to make specific recommendations, or to recommend expert organizations capable of designing and installing good industrial ventilation, for example.

2.3 Industrial Hygiene or Occupational Health Program Auditing

Most good corporate industrial hygiene programs periodically audit their programs to ensure the level of performance that has been deemed necessary. Many corporations have internal auditing procedures and units for just this purpose, and are quite capable of doing an excellent job. Other organizations may not elect to have such self-auditing capability, and may choose instead to retain an industrial hygiene consultant for this purpose. Such audits are by their nature very demanding, for the consultant is required to evaluate the performance of peers, who may be well informed and quite capable of evaluating their own performance level. The requested audits may be quite limited in scope, or may be so broad as to include not only the performance of the industrial hygiene group but also its interrelationships with other members of the occupational health team. In the latter event, it is necessary to meet with and interview members of the other groups, in order to make an appraisal of the relationships that are essential to the successful operation of the entire program. Typically, audits require extensive interviews with all staff personnel, meeting with key individuals in private and being sensitive to the need for respecting confidences. Additionally, interviews should be held with the administrative head of the department of which industrial hygiene is a part and with other administrative personnel, up to and including the corporate president if possible. Ideally, interviews should also be held with selected members of the work force, including union representatives, and perhaps a random sampling of workers who are most concerned with the quality of industrial hygiene programs. Not infrequently, however, corporate policy will not permit the consultant to interview workers, unlike OSHA compliance officers, who may do so. Nonetheless, every reasonable effort should be made to convince the client that access to employees often provides valuable input to important aspects of the audit in such areas as effectiveness of training, management safety and health communications, and employee perceptions of management commitment.

In addition to extensive discussions and interviews with staff persons, all relevant records should be made available to the consultant, as well as all documents such as the current industrial hygiene manual, safety publications, policies, and standard procedures. The consultant should note whether these are current and complete and whether they are consistent with good practices in industrial hygiene and, if applicable, in industrial hygiene laboratories. Samples of reports generated by the staff should be reviewed, as well as records indicating the process for the operation's response to a survey and the extent of follow-up activities by the industrial hygiene function.

It is important that the consultant draw up a comprehensive outline of the audit to be performed, or else acquire such an outline from organizations that have developed them over a number of years. As noted earlier, many corporations perform their own audits and generally have formalized and very complete outlines and operating procedures. Similarly, other consultants who have performed such audits in the past may likewise have prepared reliable operating procedures, and the consultant who is performing an audit for the first time would do well to seek out such preexisting outlines, if possible.

Prior to performing the audit there should be agreement concerning the form in which the results will be summarized. Some formalized audit systems have devised numerical rating systems, whereas others rely simply on word descriptions and listings of deficiencies, strengths, and so on. Either approach may be satisfactory, but there should be agreement prior to conducting the audit concerning how the results will be expressed.

2.4 Industrial Hygiene Training: Workers and Management

Specialized training of groups of workers or management personnel is frequently required, sometimes in relation to special projects, or else as a part of an ongoing routine training program. Companies or other organizations may elect to request that such training be given by an industrial hygiene consultant, or that the consultant organize a training session or sessions that assemble a staff of qualified plant personnel and others with expertise in selected areas. Consultants who are associated with a university are frequently asked to organize such activities and are generally well qualified and prepared to do so. Subject areas to be addressed may be as general as a survey of the principles of industrial hygiene, or may be very specific, treating relatively advanced subjects, such as industrial ventilation design. Training activities are sometimes requested in order to satisfy contractual agreements between management and labor, and in such instances, the required subject matter is ordinarily a matter of contractual agreement.

In preparing a training activity of any kind, it is extremely important that the consultant design the activity for the particular students expected to attend the course. There is little value in presenting complex subjects using language that is unfamiliar to the students, and care must be taken to avoid, whenever possible, technical jargon that may impress but not educate the students. It is equally wasteful of time to underestimate the level of knowledge of the class and explain fundamentals that may be well known to them.

Whenever possible, training sessions should use visual aids to the maximum extent, and take advantage of modern computer programs and interactive techniques, as well as hands-on laboratory sessions whenever these are appropriate. For example, if a research program requires that air samples be taken in various locations by persons not familiar with air sampling techniques, it is useful to have a training session at some central location, in order to explain carefully the purpose of the entire effort, its goals, and the techniques to be used. The actual sampling equipment to be used should be available, and there should be enough units to permit each student to use the unit in the manner required for the study. In addition, printed material should be prepared and distributed. This material should describe the procedures in as much detail as required, discussing the importance of calibration procedures, methods of handling the samples, and instructions concerning recording the required sampling data.

Occasionally, a consultant may decide that for various reasons it may not be possible to provide the depth of coverage required by the client and instead may elect to make the client aware of training courses offered by the AIHA, National Institute for Occupational Safety and Health (NIOSH), or some other well established training organization. In the long run, a client's needs may better be met by such a decision, and the consultant will be spared the trauma of conducting a training activity that did not work.

2.5 Activities Related to Litigation, Including Serving as Expert Witness

All industrial hygiene consultants may anticipate being asked to become involved in matters related to litigation, including offering their expertise to attorneys, and ultimately appearing as an expert witness. There are a number of individual consultants for whom litigation-related activities are their most important single source of income, and there are some who do nothing else. Typically, such litigation involves matters related to worker compensation cases, toxic torts, or even environmentally related problems. Much has been written about this kind of litigation, and it is beyond the scope of this chapter to address in detail the complexities of actual lawsuits and the particular roles of the industrial hygiene consultant. A few general observations may be useful, however.

There is no single activity of an industrial hygiene consultant that requires more adherence to a strict code of ethics than does litigation. All industrial hygiene activities, of course, must be conducted in accordance with ethical principles, but it is more likely that ethical dilemmas may arise in matters related to litigation than in most other consulting activities. By definition, litigation always involves two legal teams with opposing views, and each seeks to do the best possible job for its client. It is important that the industrial hygienist avoid taking an advocacy position, if at all possible. The role of the industrial hygiene consultant is to provide information based upon expertise, and to offer opinions that, to the best of his or her knowledge, arise from an objective investigation of all of the facts and circumstances involved. An industrial hygiene consultant should be able to state that no matter whether retained by counsel for the plaintiff or counsel for the defense, the opinions, given the same body of facts, will be the same. It may well be that in some instances the opinions will not be those that the retaining attorney finds useful, in which case the relationship in all probability will terminate.

It is extremely important that the industrial hygiene consultant recognize the boundaries of his or her expertise, and not be persuaded to go beyond them. Obviously, an industrial hygienist is knowledgeable concerning toxicologic, medical, epidemiologic, and environmental matters, and it is quite proper that such knowledge be factored into the formation of his or her opinions. The industrial hygienist must never, however, assume the role of a physician, for example, and offer opinions that he or she is unqualified to make. Conversely, other experts should not assume the role of an industrial hygienist if they are not qualified to do so, but unfortunately in practice it is difficult to determine whether an individual is qualified as an industrial hygienist, and not infrequently poorly qualified individuals are in fact rendering opinions requiring industrial hygiene expertise.

A consultant may be approached by counsel for the plaintiff or the defense, and there is no reason why any consultant should refuse to offer assistance to one or the other. Ideally, in fact, it could be argued that all consultants should be equally involved with litigation for the plaintiff, as well as for the defense, but in practice such is frequently not the case. Certain individuals apparently choose, or are chosen, repeatedly to assist counsel for the plaintiff, while others routinely offer their services to counsel for the defense. The single, most important principle that should guide the activities of any consultant in this regard is the avoidance of any semblance of conflict of interest. Obviously, the consultant cannot offer advice to both plaintiff and defense in the same case, but perhaps not so obvious are the potential conflicts of interest that may arise with other cases, which may have some connection with either the organizations involved in previous litigation or the central issues. Common sense should serve to prevent most conflicts of interest from developing, but the consultant is well advised to give considerable thought to each request for assistance in relation to previous activities for the same clients or other clients.

The code of ethics (Appendix A) developed and used by the AAIH and adopted subsequently by the AIHA and ACGIH is totally adequate both for its intended purpose of defining ethical practices for the professional practice of industrial hygiene and more particularly for the practice of industrial hygiene, in relation to litigation.

Note particularly the statement in the code, "Avoid circumstances where compromise of professional judgment or conflict may arise." Also of particular concern in litigation is the admonition to "not distort, alter or hide facts in rendering professional opinions . . .," and finally the warning to "not knowingly make statements that misrepresent or omit facts."

2.6 Scientific Studies Related to Health Problems in Industry

On occasion, a consultant may be fortunate enough to be asked to conduct or to participate in a study designed to provide required information regarding a health problem in industry. These studies may be of several kinds, perhaps the most common being limited to specific work operations or perceived problems in a given department or building. Such studies will generally require that the industrial hygiene consultant perform all the industrial hygiene activities normally associated with plant surveys, problem solving, and so on, but with a different focus. It is essential that such studies, or any research endeavors for that matter, be carefully designed so that they are capable of answering the questions that initiated them. The investigators should also anticipate preparing the comprehensive report that will ultimately be required, as well as a publication, preferably in a refereed journal in which findings of such studies are ordinarily published.

The most comprehensive studies likely to be carried out in the workplace are epidemiologic studies of large worker populations believed to be at risk for one or more environmental factors. Such studies generally require the cooperative efforts of a team consisting of an epidemiologist, a physician, who may also be the epidemiologist, an industrial hygienist or industrial hygiene consultant, and other persons technically proficient in areas of importance to the study. All such studies are generally expensive, and require a substantial period of time for completion. The industrial hygiene consultant will generally not be the principal investigator of a study, but may share responsibility with the epidemiologist who will be responsible for the overall design of the study.

The principal concern of the industrial hygiene consultant will ordinarily be the collection of required quantities of all environmental data. Under ideal conditions the air sampling study, if one is required, will be carried out under the direction of the consultant utilizing an appropriate number of industrial hygienists or industrial hygiene technicians, probably from a cooperating industrial hygiene consulting organization. Frequently, however, the realities are that budgetary limitations and other concerns require that sampling be carried out by plant personnel, thus entailing training activities, quality control procedures, and other necessary steps to ensure the objectivity and validity of the data.

One of the most common motivations for conducting such large-scale studies is the pressure exerted by pending or existing governmental standards or regulations. Thus a study may be intended to address specifically the question of an appropriate standard for a substance, and the investigators must take care to ensure that the findings will have the required degree of credibility with the regulatory agencies concerned. In all such instances, it is strongly recommended that the study be planned very carefully, and that prior to commencing the study, meetings be held with the regulatory agency personnel or perhaps with the Threshold Limit Values Committee if appropriate, with their inputs welcomed and used to modify the studies where required. These same principles, of course, apply to any studies conducted for the purpose of providing information to regulatory agencies relative to proposed or actual standards. As noted earlier, all large-scale epidemiologic studies can be very time consuming and demanding, but can also be very rewarding professionally to the industrial hygiene consultant who finds himself engaged in a research activity that may result in significant information affecting the health of the working populations under study.

2.7 Activities Related to Proposed Legislation and Open Hearings

Given the fact of ever-increasing regulatory pressures and activities by federal agencies such as OSHA and the Environmental Protection Agency, as well as numerous state and local government agencies, it is almost a certainty that an industrial hygiene consultant will be asked to assist some organization having a direct interest in such matters. Most commonly, given a new proposed standard, for example, special-interest groups such as industrial establishments, industrial associations, labor groups or public interest groups may have sharply different opinions regarding the standards being considered, and will request the opportunity to make statements at the public hearings that are usually held in connection with such standard-setting activities. Any or all of these special-interest groups may send persons to testify or give opinions at hearings, and on occasion the most suitable person is an independent industrial hygiene consultant. The consultant selected will ordinarily have had some prior familiarity with the problem area, and may have consulted for the company or other organization requesting the consultant's services. Regardless of the extent of prior involvement or lack of it, the consultant must become thoroughly familiar with the issues involved and make every effort to acquire all available information. Then opinions must be formulated with which he or she is completely comfortable, and which he or she strongly believes. Should these opinions be consistent with those of the client, the consultant will logically then express them in open hearings and ordinarily will prepare a written document for submission. As in activities related to litigation, the consultant should be guided by the code of ethics that govern all of his or her activities, for there may well be suspicions of bias favoring the group represented. Such charges are frequently leveled at consultants who have been retained by industry, and who thereafter make appearances at hearings on behalf of industry. Even though the consultant may have been selected because of not being an employee of the concerned industry, the simple fact of the consulting relationship with the organization will not infrequently be used as a basis for charges of industry-slanted bias. Similar charges, of course, may be made against consultants representing labor, public interest groups, or other organizations, but they are not as probable as those made against industry-retained consultants. It is unfortunate when such charges are made without cause, but in anticipation of this possibility, the conduct of the consultant should always be consistent with the highest principles required by the AAIH code of ethics. It is obviously very important that the consultant not be tempted to make statements favorable to the client's position that cannot be supported by objective consideration of all available information.

2.8 Advisory Activities

Industrial hygiene consultants are frequently asked to become members of advisory groups established by industry, labor, professional organizations, and others. Providing such services should be considered a particularly desirable activity, for it allows the consultant to express opinions without the pressures associated with litigation, regulatory activities, and so on, while being presented with the opportunity to make significant contributions to the organization that formed the advisory group. Typical advisory groups include those established by specific industrial or trade organizations sharing common problems, governmental agencies, and somewhat less frequently, combined labor union and management groups. For example, the union-management contracts of the tire and rubber industry have for some time provided for jointly funded research programs requiring advisory groups, and more recently the United Auto Workers union and the plants of the General Motors Corporation, Chrysler Corporation, and Ford Motor Company that employ UAW workers have signed contracts involving the expenditure of substantial sums on employee health and safety related matters. The advisory committees formed are charged with the very important task of assisting the joint union-management committees in training efforts, research directed at potential problems, and related matters.

In contrast to the usual "problem-solving" activities of a consultant, service on such advisory committees provides unique opportunities to help identify present and future problems and design research programs or other measures that can significantly affect the health and safety of many thousands of employees. As might be anticipated, it is frequently the practice in forming such committees to tend to select consultants who are affiliated with university programs.

3 ETHICAL CONSIDERATIONS

The importance of ethical considerations in relation to ligitation related activities has been stressed. In fact, however, strict adherence to a code of ethics is an essential activity for anyone practicing industrial hygiene, whether the person be a consultant or otherwise employed. Because of the great variety of situations in which an industrial hygiene consultant may be involved, it is perhaps even more important, or at least potentially difficult, to maintain the highest degree of ethical conduct. The AAIH code of ethics discussed earlier is a well conceived document that has withstood the test of time. Every industrial hygiene consultant should review it carefully, and make constant effort to comply with its various provisions. All the clients of a consultant should be made aware that this code of ethics exists and that the consultant does indeed conduct himself or herself accordingly. Perhaps the central theme of the code is that all industrial hygienists must constantly be aware that their primary mission is to protect the health of people who work. It does not matter whether the employer or the client, in the case of a consultant, is a corporate entity, a labor organization, or an insurance company; the industrial hygienist is committed to recommending whatever actions are best for the working population involved. It will be recognized, of course, that industrial hygiene consultants cannot implement all of the recommendations they may make, but they should never fail to make such recommendations because of the costs involved, or because of other practical considerations if they believe that in doing so adverse health consequences will follow. There may be many reasons why management will not implement recommendations made by a consultant, and may select alternative strategies that are better suited to their overall needs, at least in their own opinion. Nevertheless, the consultant should make such recommendations as are believed necessary to whatever extent possible.

3.1 Confidentiality Concerns

Another provision of the code of ethics is perhaps uniquely important to the industrial hygiene consultant. The code calls for the industrial hygienist to maintain the confidentiality of personal and business matters of the client but not at the expense of the duty to protect the health and safety of workers and the community. Virtually all industrial hygienists may have access to confidential information, but in the specific instance of consultants, it is probably true that not only will the consultant have access to considerable confidential information, but the opportunity also exists to violate this simple rule more frequently. A consultant must always exercise the greatest caution in transmitting to one client information obtained from another. Certainly there are instances when some kinds of information may be freely exchanged, but whenever trade secrets or proprietary matters are involved, exchanging such information with others is a clear breach of trust. It is generally wise for the consultant to assume the latter circumstances and avoid discussing information concerning one client with another.

Practical difficulties may arise within an organization when information of a confidential nature is provided by an individual who does not wish it to be known to others in the organization, or when employees confide in a consultant with the understanding that such information will be kept confidential. In such instances, common sense and the exercise of good judgment are obviously indicated, but a consultant is urged to attempt to avoid such situations whenever possible. It may even be advisable for the consultant to advise everyone with whom he deals that all information provided to him or her will not be kept confidential, at least insofar as the client is concerned.

4 BUSINESS ASPECTS OF CONSULTING

Consulting is ordinarily conducted on a fee-for-service basis, and the client has no obligation to pay the consultant or provide fringe benefits except when requested services are being rendered. On occasion, however, it may suit the needs of both the consultant and the client to establish a continuing relationship to provide service throughout a stated period of time, frequently one year. In such instances, contracts may be drawn up by the client and will probably be found acceptable by the consultant. Unfortunately for consultants, there does not exist a document entitled "Code of Business Practices" comparable to the code of ethics to use as a reliable guideline. There does not exist a generally accepted fee schedule for services rendered by industrial hygiene consultants, for example, nor are there universally accepted practices concerning such matters as charging for travel time, or differential fee rates for certain activities such as serving as an expert witness. Nonetheless, there does exist at any given time a general awareness of the fees being charged by consultants, and the individual consultant is advised to be aware of such information. All of the larger consulting organizations have fee schedules, of course, and individuals acting as consultants are well advised to obtain a sufficient number of such fee schedules to enable them to arrive at an appropriate fee for their services. Certain generalities concerning the fee schedules are possible: a young, relatively inexperienced, and noncertified industrial hygienist is logically going to command the lowest fee, whereas individuals with increasing amounts of experience, who are certified, will generally earn more. Other things being equal, an individual with a Ph.D. will in all probability command a higher fee than one without such a degree. In general, however, an individual considering consulting as a profession will in all probability find it necessary to resort to the time-honored practice of asking friends, associates, and those already familiar with consulting practices for guidance in arriving at a reasonable basis for charging for services rendered. It is perhaps self-evident also that the individual contemplating consultation as a career should have adequate resources to be able to weather periods of decreased activity and, more commonly, periods during which payments are slow. A consultant cannot count on receiving payments in a regular and predictable manner, as do corporate employees, for example, and in addition, a consultant frequently must make expenditures of substantial size for travel, secretarial assistance, or assistance by other professional associates prior to billing his client.

The basic principles of successfully running a business are no different for a consultant than for others, but many industrial hygienists who consult may not have strong business backgrounds and are well advised to seek advice and guidance from a certified public accountant or other persons qualified to give such advice. Keeping good records is extremely important, and particularly so in relation to income taxes. A consultant receives payments from a large number of sources, and may spend a great deal on travel, entertainment, and so on, all of which are of special interest to the Internal Revenue Service. Any number of computer software programs are available to aid in the financial management of small businesses, as well as for maintaining invoicing records, and project management.

5 AIHA INDUSTRIAL HYGIENE CONSULTANT LISTING

For a number of years, the AIHA has assembled and published a list of industrial hygiene consultants. Presently, this listing is updated twice annually, and printed in the January/

February and July/August issues of the *American Industrial Hygiene Association Journal* as well as in a separate publication list of consultants distributed bi-annually to a variety of potentially interested parties. The listing has become a very useful source of information concerning industrial hygiene consultants to those seeking assistance, but it is important to recognize its limitations. The AIHA makes the following statement concerning this listing: "The American Industrial Hygiene Association provides these listings as an informational service, accepting no responsibility for the performance of services by the consultants, their claims of specialization and their competence as consultants." The only requirements for being placed on the list are that the consultants must be full members of the AIHA, and must pay the required fees. The January 1999 listing contains the names of approximately 280 individuals and organizations who offer a spectrum of industrial hygiene consulting services. Obviously, there are many more industrial hygiene consultants than choose to be listed, for a simple inspection of the list will reveal the absence of a number of well-known consultants, and many, if not most, part-time consultants. The potential user of consulting services is well advised to keep in mind that the listing, as stated by AIHA, is noncritical, and there is no assurance of competence. At the same time, it is probable that most of the individuals and organizations listed are competent, and the listing provides a very useful service to those seeking assistance. For convenience, the listing is presented first according to the state in which a consultant does business and, then all names are presented alphabetically. The speciality that each consultant has identified competence in is identified as part of each listing. Although as noted, many consultants have apparently not found it necessary to be included in this listing, it is potentially useful both to consultants and to those seeking assistance.

6 RELATIONSHIP WITH CONSULTING ORGANIZATIONS

Individual industrial hygiene consultants will rarely be capable of providing all requested services without the assistance of others specialized help. For example, to help perform field studies, the industrial hygiene consultant may use the services of an industrial hygiene technician or work closely with an accredited industrial hygiene laboratory to prepare and analyze samples collected in the field. Such assistance is readily found today, and simple reference to the consulting listing just described should be ample to identify consulting organizations who are geographically convenient to the consultant. In some cases, geographical convenience may not be important, and the consultant should then try to identify the organization best suited to meet his or her needs. It is reasonable to require that the consulting laboratory be accredited by the AIHA, and in all probability most, if not all, of the consulting laboratories in the AIHA consultant listing are accredited. It is recommended that consultants identify consulting laboratories and organizations upon whom they plan to rely, and learn in a general way the spectrum of services available and the prevailing fee schedules. Given the number of competent laboratories and organizations that presently exist, the individual consultant should have little difficulty in satisfying every need which may arise.

7 CONCLUSIONS

The profession of industrial hygiene is relatively young, and the process of maturing requires that attitudes and practices gradually change to meet contemporary needs. The substantial growth in the number of industrial hygiene consultants and their use by those requiring such assistance is part of the maturing process, and these trends are likely to continue indefinitely. Industrial hygienists who retain consultants should not be viewed as professionals who lack basic skills, but rather as enlightened professionals who choose to supplement their skills as required to meet their employer's/client's needs in the best manner possible. In this regard, industrial hygiene follows traditions established many years ago by such long established professions as engineering, medicine, and accounting, whose practitioners routinely retain required consultative assistance as soon as the need is perceived.

B THE CONSULTING ORGANIZATION

1 INDUSTRIAL HYGIENE CONSULTATION FIRMS

1.1 Introduction

Emergence of the industrial hygiene consulting firm in the United States is a relatively recent phenomenon. With few exceptions most of the industrial hygiene firms in business today have their beginnings in the passage of the Occupational Safety and Health Act (OSH) of 1970. A 1966–1967 survey of the industry conducted by the American Industrial Hygiene Association (AIHA) showed that as recently as 1967, only about 10 to 15 percent of the workforce ever came in contact with an industrial hygienist. Dr. Morton Corn, in this keynote address at the opening session of the 1976 American Industrial Hygiene Conference, said, "In the pre-OSHA (Occupational Safety and Health Administration) days, larger, enlightened firms can be pointed to as the main consumers of our knowledge and our primary employers" (2).

2 WHAT NEEDS DO THE CONSULTANTS SERVE?

Industrial hygiene consulting firms fill a critical need. For companies that are small or medium size, independent consultants are the main source of industrial hygiene expertise because many smaller firms cannot justify hiring a full-time professional. In addition, larger companies such as service organizations and academic and religious institutions may not always need a full-time in-house industrial hygiene staff. For these organizations, engaging consulting firms provides a cost-effective means of procuring industrial hygiene services because consultants are used only on an as-needed basis.

Companies with in-house industrial hygiene capabilities may also retain consultants from time to time in areas these companies may lack in-house. These areas may include toxicological evaluation, developing material safety data sheets (MSDS) and industrial hygiene programs, and performing exhaust ventilation engineering or epidemiologic studies.

Consultants are also called upon in situations where independent third-party investigations are required because of legal considerations or because of certain labor contract agreements.

3 SERVICES OFFERED

Services provided by consulting firms range from asbestos management only to comprehensive industrial hygiene-related services. Increasingly, larger consulting firms combine industrial hygiene services with related environmental activities in the areas of air pollution, water pollution, hazardous waste management, and subsurface environmental investigations (to assess and remediate contaminated soil and groundwater problems). Although the depth and range of services offered by industrial hygiene consulting firms vary, consulting firms offer the following typical services.

3.1 Field Surveys

Field surveys are a major activity for many industrial hygiene consulting firms; they include sampling to develop baseline exposure data and/or conduct routine monitoring to measure worker exposures. Such measurements are often made to satisfy mandatory periodic monitoring required by federal or state regulatory agencies. For example, OSHA mandates periodic monitoring for substances such as asbestos, lead, arsenic, acrylonitrile, and benzene. Periodic monitoring is also required to measure noise exposures using personal dosimeters and/or sound level meters. Consulting firms often perform sampling surveys for companies that have in-house industrial hygiene functions but may lack sufficient manpower resources and field equipment.

3.2 Program Development and Program Auditing

Motivated by concerns for worker health and safety and/or as a result of requirements imposed by law, many firms in the United States are confronting industrial hygiene issues for the first time and turning to consulting firms for assistance. Because of the passage of the OSHA's hazard communication regulations (1910.1200), firms in both the manufacturing and the nonmanufacturing segments of U.S. industry have had to develop specific written hazard communication programs. Such programs require generating chemical inventories, preparing MSDS, training employees, and instituting personal protection maintenance programs. Many companies have found it advantageous to engage consultants to develop computerized data bases for chemical inventories, MSDS, employee exposure data, and medical records.

3.3 Asbestos Management

More and more, consulting firms are being called upon to provide asbestos management services. Assessment of asbestos hazards has become particularly important during property transfers, lease negotiations, mergers and acquisitions, and financing of commercial properties. Asbestos consulting services are also in high demand by educational institutions and owners of commercial and industrial buildings concerned about employee exposure to asbestos and resultant legal liabilities. As a result, a major asbestos consulting industry has developed in the United States and overseas. In the United States, a large number of consulting firms are almost exclusively engaged in asbestos consulting services including:

Building inspections to identify suspect asbestos-containing materials (ACM).

Laboratory analyses of ACM for asbestos content.

Risk assessments.

Identification of areas needing abatement.

Development of plans and specifications for the safe removal and disposal of asbestos.

Surveillance of the asbestos abatement projects to provide safe work practices, air sampling, and on-site microscopic analysis of samples to determine exposure of abatement workers and building occupants as well as to verify the integrity of fiber-containment barriers.

Final inspections and "clearance" monitoring to determine if the abated area has been adequately cleaned up and whether airborne fiber concentrations are within the specified "acceptable" limits.

3.4 Legal Testimony

In our increasingly litigious society, employers are facing mounting environmental liabilities. In the United States, litigation related to environmental and safety and health matters has proliferated at an accelerated pace. Such litigation is often complex, and the prosecution or defense of such cases sometimes requires extensive literature searches and gathering of field data. Consulting firms are often called upon to generate such data in support of litigation. More experienced consultants, particularly those who specialize in certain aspects of industrial hygiene and have attained recognition in their field, are generally called upon as expert witnesses.

3.5 Indoor Air Quality

Indoor air quality (IAQ) is a fast developing area of endeavor for industrial hygiene consultants. Energy conservation measures along with the rising use of synthetic building materials and interior furnishing have resulted in poor air quality in many buildings. Moreover, employers and building occupants have become more cognizant of IAQ problems and their resultant effects on employee health, productivity, and employee relations. In the United States, legislation has been proposed to address IAQ issues. This legislation is likely to result in further demand for industrial hygiene consulting services.

Indoor air quality projects can be complex and often require a team of experts including an industrial hygienist, a heating, ventilating, and air-conditioning (HVAC) engineer, and a microbiologist. In addition, an epidemiologist may be needed, unless the industrial hygiene professional is well versed in epidemiologic techniques. Very few consulting firms have such in-house multidisciplinary expertise and are therefore not equipped to perform comprehensive IAQ investigations. In some IAQ projects, an occupational physician may be needed to interview building occupants or examine them physically. Services of an occupational physician are either subcontracted by the consulting firm or engaged directly by the employer.

3.6 Industrial Hygiene Engineering

Full-service industrial hygiene consulting firms provide local exhaust ventilation engineering services including conceptual design as well as detailed drawings and specification, field supervision during installation, and system performance testing. To a lesser extent, industrial hygiene consulting firms provide general building ventilation design services. Such services, however, are generally limited to a review of the ventilation design performed by mechanical engineering firms specializing in HVAC system design.

Industrial hygiene engineering services also include noise control engineering, including machine and process enclosure design and, occasionally, process modifications. Unfortunately, fewer industrial hygiene firms provide engineering services today, for the number of engineering graduates entering the practice of industrial hygiene has gradually dwindled.

3.7 Laboratory Services

Several industrial hygiene consulting firms maintain analytical laboratories. Some firms specialize only in laboratory services. The number of consulting industrial hygiene laboratories has steadily grown since the AIHA laboratory accreditation program began (4).

Capabilities and resources of accredited industrial hygiene laboratories vary considerably. Services offered range from limited microscopic analyses of asbestos samples to comprehensive analyses covering the entire range of contaminants encountered in the workplace. Most industrial hygiene laboratories now use modern and automated analytical instrumentation such as gas and liquid chromatography, mass spectrometry, atomic absorption spectrometry, and optical and electron microscopy in addition to conventional wet chemical analyses.

Consulting industrial hygiene laboratories serve a vital function. Even large industrial firms with in-house industrial hygiene programs rely heavily on consulting laboratories for chemical analyses because commercial laboratories, owing to the large volume of analyses they may perform, conduct certain tests more cost effectively than in-house laboratories that perform such specialized tests only occasionally.

3.8 Other Services

An increasing trend among larger consulting firms has been to provide multidisciplinary environmental services including industrial hygiene, hazardous waste management, and air pollution and water pollution control. In particular, environmental engineering firms specializing in assessment and remediation of hazardous waste sites have added industrial hygiene expertise to their capabilities because safety and health issues are prominent in hazardous waste site cleanup programs.

The broad range of industrial hygiene related services offered by consulting firms is reflected in the directory of industrial hygiene consultants (5) published in the AIHA journal. The recent listing included the following 21 areas of specialization:

Asbestos
Biological monitoring
Comprehensive industrial hygiene practice
Computer software and information services
Emergency management/disaster planning
Environmental practice
Environmental and occupational medicine
Ergonomics
Expert witness
Hearing conservation and noise control
Indoor air quality
Industrial hygiene chemistry
Lead
Management/audits/inspection
Radiological control
Respiratory protection/personal protective equipment
Safety specialist
Toxicology
Training-instruction
Ventilation
Vibration

4 INDUSTRIAL HYGIENE CONSULTING AS A CAREER

Consulting firms now employ significant numbers of industrial hygienists. Moreover, the industrial hygiene consulting business appears to be poised for considerable growth as governmental regulations become more complex and stringent, and workers and employers become more aware of and concerned about industrial hygiene issues. Another reason for the growth of the industrial hygiene consulting business is a rising trend among large U.S. corporations to contract out work that is not directly related to the manufacture of their basic product. Similar trends are also apparent for environmental, legal, accounting, medical, and design engineering services.

Industrial hygiene consulting can be a rewarding career for professionals who strive for diversity, desire to solve problems, enjoy interaction with others, and derive satisfaction and pride in helping clients to maintain safer workplaces. If the consulting firm is multidisciplinary, there is opportunity to interact with other specialists and develop skills in related areas. Because industrial hygiene service is the end product of a consulting firm, an industrial hygiene professional is the most valuable commodity in that business, for the success and growth of the business depends directly on the industrial hygienist's productivity and competence.

To succeed in a consulting career, an industrial hygienist should:

Possess good communication (oral and written) skills.

Have a relentless drive to please clients.

Be adaptable to highly variable situations.

Possess appropriate certifications (such as certification by the ABIH).

Be service-motivated, flexible, and patient.

Be willing and able to travel.

A typical industrial hygiene consultant handles several projects simultaneously, faces critical deadlines, travels on short notice, and sometimes works long and odd hours.

Frequently consulting services are requested at a moment's notice due to emergency situations. Report deadlines and frequent emergencies can sometimes strain certain individuals. Consulting careers may therefore prove stressful to those whose life styles are not amenable to such conditions. Another discouraging factor for some consultants sometimes can be their inability or helplessness to implement recommendations. If a client chooses not to act on a consultant's recommendations, the industrial hygienist generally has little influence or recourse.

Successful industrial hygiene consultants, however, derive considerable satisfaction from helping clients solve problems, making workplaces safer, and preventing liability for the client. Professional and financial awards for the consultant can be substantial. A competent and energetic industrial hygienist can advance quickly in a growing consulting firm and can attain a managerial rank. In some firms, an aspiring industrial hygienist can attain the status of a partner.

5 ETHICS AND RESPONSIBILITIES

The primary responsibility of every industrial hygienist is to protect the health of the worker (6). Industrial hygiene consultants are responsible for fulfilling the requirements of the client who engages their services to identify problems and develop cost-effective solutions. Consultants are also responsible for providing documentation and accountability to the client in support of their findings.

Industrial hygiene consultants of course subscribe to the code of ethics imposed by their profession and as defined by the AIHA (7). The code of ethics has been described earlier in this chapter. For the industrial hygiene consultant, one of the most important ethical issues is to avoid conflict of interest. For example, a consultant who is affiliated with an equipment manufacturer or a contractor who installs the recommended control system has an obvious conflict of interest. In the asbestos abatement field, conflict of interest also occurs when the abatement contractor acts as the industrial hygiene investigator and project consultant, thus specifying procedures and then certifying his or her own work.

6 LIABILITIES

Failure to perform adequately entails certain legal risks for all industrial hygienists. An industrial hygiene consultant could face additional potential liability from an unhappy client not satisfied with the consultant's performance. Some legal experts feel (8, 9) that developments in the area of toxic tort liability can engulf industrial hygiene consultants for negligence or breach of duty if the conduct of the industrial hygienist is below the standard normally accepted in the profession (see also Chapter 39). To avoid liability, industrial hygiene consultants should use procedures and test methods generally recognized in the profession as reliable and accurate.

To guard against such liabilities, most industrial hygiene consulting firms carry insurance coverage, generally in the form of errors and omissions. Although in recent years, such insurance has become more affordable to the individual consultant, coverage exclusions for certain high risk work and deductibles should be carefully studied and understood by both the consulting firm or individual consultant.

7 SELECTING INDUSTRIAL HYGIENE FIRMS

7.1 Consulting Firms

Listings of industrial hygiene consulting firms can be obtained from several sources. Many firms advertise their services through professional journals. The AIHA publishes a list of practitioners representing individual consultants as well as those affiliated with larger firms in various association publications (5). The *Air and Waste Management Association Journal* (9) also publishes a guide to consultants that includes practitioners in industrial hygiene.

Selection of a qualified industrial hygiene firm is a crucial task and can prove to be difficult and frustrating because, currently, any person can claim to be an industrial hygiene expert and legally offer services as an industrial hygiene consultant. To avoid consultants who are inexperienced, incompetent, or lacking suitable training and equipment, the following criteria should be used in the selection:

Certification by the ABIH.

References from past clients.

Adequacy of staff, equipment, and laboratory facilities (it is not uncommon for smaller firms to rent specialized equipment and use outside laboratory services).

Insurance coverage for professional liabilities, including errors and omissions.

Successful completion of AHERA training programs (for asbestos projects).

AIHA-accredited laboratory facilities (in-house, or assurance that an outside laboratory will be accredited).

Financial stability.

Competitive fees.

No conflict of interest.

7.2 Analytical Laboratories

Equal care should to be exercised in selecting an industrial hygiene laboratory. Prospective clients should take the following steps before selecting a laboratory:

Talk with the key laboratory personnel about their capabilities and credentials.

Visit the laboratory and examine sample handling facilities, chain-of-custody procedures, instrumentation, written procedures, and quality control logs.

Look for evidence of competence such as laboratory accreditation and proficiency in external testing programs.

Ask for references from colleagues and professionals.

The best source for finding a qualified laboratory is the listing of AIHA-accredited laboratories published biannually in the AIHA Journal and the AIHA website (4). The AIHA accreditation program requires participating laboratories to demonstrate proficiency in the following five categories of analyses:

- Metals (lead, cadmium, zinc and chromium)
- Asbestos and man-made fibers
- Silica
- Organic solvents
- All proficiency analytical testing materials

The AIHA listing indicates the proficiency of the listed laboratories based on the laboratories' successful participation in the AIHA's Proficiency Analytical Testing (PAT) program. Over 1,160 laboratories participate in the PAT program for one or more of the above analyte groups. In the 1999 listing (4), 296 laboratories were listed as accredited as part of the Industrial Hygiene Laboratory Accreditation Program; a number of these laboratories participate in all of the above analyte proficiency groups.

8 CONCLUSIONS

Industrial hygiene consulting firms serve a vital function by providing services to employers who lack in-house expertise or need backup from time to time. As employers and the public become increasingly aware of workplace hazards, and regulatory requirements become more complex and stringent, demand for services of consulting firms will continue to rise. Consulting firms will provide challenging careers to many who strive for diversity, who desire to help others, and who seek rapid career advancement.

APPENDIX A. THE AMERICAN ACADEMY OF INDUSTRIAL HYGIENE CODE OF ETHICS FOR THE PRACTICE OF INDUSTRIAL HYGIENE

OBJECTIVE

These canons provide standards of ethical conduct for Industrial Hygienists as they practice their profession and exercise their primary mission, to protect the health and well-being of working people and the public from chemical, microbiological and physical health hazards present at, or emanating from, the workplace.

CANONS OF ETHICAL CONDUCT AND INTERPRETIVE GUIDELINES

Canon 1

Industrial Hygienists shall practice their profession following recognized scientific principles with the realization that the lives, health and well-being of people may depend upon their professional judgment and that they are obligated to protect the health and well-being of people.

Interpretive Guidelines

- Industrial Hygienists should base their professional opinions, judgments, interpretations of findings and recommendations upon recognized scientific principles and practices which preserve and protect the health and well-being of people.
- Industrial Hygienists shall not distort, alter or hide facts in rendering professional opinions or recommendations.
- Industrial Hygienists shall not knowingly make statements that misrepresent or omit facts.

Canon 2

Industrial Hygienists shall counsel affected parties factually regarding potential health risks and precautions necessary to avoid adverse health effects.

Interpretive Guidelines

- Industrial Hygienists should obtain information regarding potential health risks from reliable sources.
- Industrial Hygienists should review the pertinent, readily available information to factually inform the affected parties.

- Industrial Hygienists should initiate appropriate measures to see that the health risks are effectively communicated to the affected parties.
- Parties may include management, clients, employees, contractor employees, or others dependent on circumstances at the time.

Canon 3

Industrial Hygienists shall keep confidential personal and business information obtained during the exercise of industrial hygiene activities, except when required by law or overriding health and safety considerations.

Interpretive Guidelines

- Industrial Hygienists should report and communicate information which is necessary to protect the health and safety of workers and the community.
- If their professional judgment is overruled under circumstances where the health and lives of people are endangered, industrial hygienists shall notify their employer or client or other such authority, as may be appropriate.
- Industrial Hygienists should release confidential personal or business information only with the information owners' express authorization, except when there is a duty to disclose information as required by law or regulation.

Canon 4

Industrial Hygienists shall avoid circumstances where a compromise of professional judgment or conflict of interest may arise.

Interpretive Guidelines

- Industrial Hygienists should promptly disclose known or potential conflicts of interest to parties that may be affected.
- Industrial Hygienists shall not solicit or accept financial or other valuable consideration from any party, directly or indirectly, which is intended to influence professional judgment.
- Industrial Hygienists shall not offer any substantial gift, or other valuable consideration, in order to secure work.
- Industrial Hygienists should advise their clients or employer when they initially believe a project to improve industrial hygiene conditions will not be successful.
- Industrial Hygienists should not accept work that negatively impacts the ability to fulfill existing commitments.
- In the event that this Code of Ethics appears to conflict with another professional code to which industrial hygienists are bound, they will resolve the conflict in the manner that protects the health of affected parties.

Canon 5

Industrial Hygienists shall perform services only in the areas of their competence.

Interpretive Guidelines

- Industrial Hygienists should undertake to perform services only when qualified by education, training or experience in the specific technical fields involved, unless sufficient assistance is provided by qualified associates, consultants or employees.
- Industrial Hygienists shall obtain appropriate certifications, registrations and/or licenses as required by federal, state and/or local regulatory agencies prior to providing industrial hygiene services, where such credentials are required.
- Industrial Hygienists shall affix or authorize the use of their seal, stamp or signature only when the document is prepared by the Industrial Hygienist or someone under their direction and control.

Canon 6

Industrial Hygienists shall act responsibly to uphold the integrity of the profession.

Interpretive Guidelines

- Industrial Hygienists shall avoid conduct or practice which is likely to discredit the profession or deceive the public.
- Industrial Hygienists shall not permit the use of their name or firm name by any person or firm which they have reason to believe is engaging in fraudulent or dishonest industrial hygiene practices.
- Industrial Hygienists shall not use statements in advertising their expertise or services containing a material misrepresentation of fact or omitting a material fact necessary to keep statements from being misleading.
- Industrial Hygienists shall not knowingly permit their employees, their employers or others to misrepresent the individuals professional background, expertise or services which are misrepresentations of fact.
- Industrial Hygienists shall not misrepresent their professional education, experience or credentials.

ACKNOWLEDGMENTS

Other than updating and minor revisions, the vast majority of the A and B sections of this chapter remain the work product of my former mentor Ralph G. Smith, PhD, CIH (deceased) and my colleague, Jaswant Singh, PhD, CIH.

BIBLIOGRAPHY

1. "Code of Ethics for the Practice of Industrial Hygiene," American Academy of Industrial Hygiene, Roster of Diplomates of the American Board of Industrial Hygiene and Members of the American Academy of Industrial Hygiene, 1998.
2. M. Corn, "Role of the American Conference of Governmental Industrial Hygienists and the American Industrial Hygiene Association in OSHA Affairs," *Am. Ind. Hyg. Assoc. J.* 391–394 (1976).
3. M. Corn, "Influence of Legal Standards on the Practice of Industrial Hygiene," *Am. Ind. Hyg. Assoc. J.*, (6), 353–356 (1976).
4. "IH Accredited Laboratories" *Am. Ind. Hyg. Assoc. J.* **60** (2), 266–279 (1999). AIHA website, http://www.aiha.org.
5. "Industrial Hygiene Consultants," *Am. Ind. Hyg. Assoc. J.* **50**, A562–A577 (1989).
6. P. D. Halley, "Industrial Hygiene—Responsibility and Accountability (Cummings Memorial Lecture—1980)," *Am. Ind. Hyg. Assoc. J.* **41**, 609–615 (1980).
7. R. L. Harris, "Information, Risk and Professional Ethics," *Am. Ind. Hyg. Assoc. J.* **47**, 67–71 (1986).
8. M. E. Alexander, "Professional Hazards for the Industrial Hygienist," *Econ. Environ. Contractor*, 85–87 (1989).
9. Consultant Guide, *J. Air Pollut. Control Assoc.* **38**, 1605–1627 (1998).
10. M. Nash, "A Question of Balance," *OH & S Canada*, **4**, 92–98 (1988).

CHAPTER SIXTY-SEVEN

Industrial Hygiene Abroad: Occupational Hygiene

Thomas A. Hethmon, CIH and Henry J. Muranko, CIH, CSP

1 INTRODUCTION

The industrial hygiene profession has grown and matured steadily since the first edition of this chapter was published in 1991. In that period, the scope of practice has been debated and expanded and new occupational and environmental hazards are recognized, evaluated and controlled. In addition, a broader diversity of practitioners have entered the profession and are enhancing the experience base established by the pioneers of the profession in the preceding decades and there is a greater recognition of the social and economic benefits of industrial hygiene. However, as the twentieth century draws to a close, it can be reasonably argued there has been no greater change in the profession during the 1990s than its recognition and growth outside the United States. This chapter describes the current status of the profession of industrial hygiene abroad.

2 DEVELOPMENTAL FACTORS

The rate of development of occupational hygiene outside the U.S. has been affected by a number of factors, including, but not limited to: (*1*) the influence of multinational corporations; (*2*) the Internet as a global information tool; (*3*) nongovernmental organizations such as the World Health Organization (WHO), the International Labour Organization (ILO), and the International Occupational Hygiene Association (IOHA); (*4*) improved economic conditions in many developing countries with a concurrent recognition of the

Patty's Industrial Hygiene, Fifth Edition, Volume 4. Edited by Robert L. Harris.
ISBN 0-471-29749-6

moral, social and economic costs of occupational and environmental illness; and (5) international trade agreements with binding and nonbinding requirements for environmental, health and safety standards, among others. Also of indisputable importance has been the vision and determination of many occupational hygienists whose efforts have greatly facilitated development of the profession in their respective countries.

Globalization and trade liberalization present new challenges to developing countries both economically and from a public health perspective. The North American Free Trade Agreement (NAFTA), the General Agreement on Tariffs and Trade (GATT) (1), and the European Union (EU) are three of the most notable examples of economic and trade agreements which have influenced occupational health and safety by altering environmental, health and safety legislation and increasing the international distribution of the workers. The "spillover effect" due to the export of hazardous industries from developed countries has resulted in increased financial and social costs for occupational health, and treatment of industrial wastes in developing countries. This occurrence can also be seen in the proliferation of commercial joint ventures financed with foreign capital in the 1980s wherein the management of worker health and safety is shared with an organization in the host country, and, in some instances, the host government itself. Much of this investment comes from multinational corporations, many of whom integrated their corporate values and systems for responsible health and safety management. However, others have not exercised the same foresight.

Nongovernmental organizations such as the ILO and WHO have been instrumental in promoting the development of the profession through workshops, conferences, and consensus documents summarizing needed resources, as in WHO's "Global Strategy for Occupational Health for All" (2) and "Development of a Profession—Occupational Hygiene in Europe" (3); providing financial assistance for education and training programs, and developing collaborative programs with liaison organizations such as the Pan American Health Organization (PAHO) under WHO, and the ILO's International Occupational Safety and Health Information Centers (CIS). ILO's conventions (guidance documents) have been influential throughout the world, e.g., Conventions No. 155 (1981) Occupational Safety and Health, No. 161 (1985) Occupational Health Services, No. 167 (1988) Safety and Health in Construction, No. 170 (1990) Chemicals Convention, No. 174 (1993) Prevention of Major Industrial Accidents, No. 176 (1995) Safety and Health in Mines (4).

In Europe, the European Union (EU) directives set minimum standards for health and safety requirements in all member states. By treaty, these directives have resulted in a higher level of parity between European countries in their health and safety legislation and provided for a wide coverage of issues; however, this has not necessarily resulted in statutory recognition of the profession in all cases. The EU is currently developing a list of chemical exposure limits (Indicative Limit Values) which will have strong influence on national requirements. The EU also requires harmonization of standards applied to the marketing and use of products including new chemicals, machinery, and personal protective equipment. The work of the Comité Européen de Normalisation (CEN) has helped to harmonize occupational hygiene practices in the EU.

Formed in 1987, the International Occupational Hygiene Association (IOHA) is currently an affiliation of 22 professional associations from 20 countries representing more than 25,000 professionals worldwide. The IOHA has contributed significantly to the development of a variety of technical and organizational partnerships aimed at improving worker health and advancing the profession internationally (5, 6). IOHA is also responsible for disseminating an international code of ethics among the member organization for the betterment of the profession. In the U.S., The International Affairs Committee of the American Industrial Hygiene Association (AIHA) has provided a forum for exchange and collaboration aimed at the development of the profession internationally.

All of these developments, and significantly more unseen, have been facilitated by the advent of the Internet with its limitless power of accessible information. It has resulted in an improved global understanding of environmental and occupational hazards for anyone with curiosity in the issues and access to the medium, while promoting interaction among allied occupational health professionals worldwide, and the sharing of resources, educational opportunities, and experiences.

But the most significant influence remains the undeniable human spirit and the expressed will to improve the quality of life. While the right to a safe and healthy workplace remains the privilege of too few, the view that diminished occupational health is a cost to be borne for economic progress is being replaced by the moral and economic realization that individuals and nations must share a mutual interest in protecting occupational health.

3 OCCUPATIONAL HYGIENE OR INDUSTRIAL HYGIENE

Whereas the title *industrial hygiene* has been used consistently in the United States for more than 60 years, with only limited exceptions, the rest of the world utilizes the term *occupational hygiene*. The prevailing use of occupational hygiene outside the United States is not motivated by a need to differentiate the profession internationally from that practiced domestically, but in the perceived limitations of the word industrial which is viewed by many practitioners outside the U.S. as not accurately reflecting the full scope of the profession in nonindustrial environments. They consider the term occupational hygiene to have a more inclusive connotation than industrial hygiene. In recognition of this broader vision of the profession, and in light of its customary use worldwide, the term occupational hygiene is used throughout the remainder this chapter.

4 COUNTRY PROFILES

The following country profiles provide an overview of some of the functional elements of the occupational hygiene profession in 22 countries: Australia, Brazil, Canada, China excluding Hong Kong, Egypt, Finland, Hong Kong (special administrative region of China), Germany, India, Ireland, Italy, Japan, Mexico, Norway, the Netherlands, Poland, Saudi Arabia, South Africa, South Korea, Sweden, Switzerland, Thailand, and the United Kingdom. Each profile includes a review of the country's economic status as it may relate to the utilization and/or rationalization of occupational hygiene, population and professional demographics, the historical development of the profession where it is notable, educational and organizational resources, information on professional certification and/or registration, ethics, and an overview of occupational health and safety legislation and regulations, including occupational exposure limits. While these individual elements do not fully represent the breadth of the profession, they provide general points of reference for the comparison of the relative rate of progress in the global development of occupational hygiene. The profiles also contain sources of occupational health information for readers to facilitate research and the exchange of information that is essential to the continuing growth of the profession worldwide.

Some profiles are more comprehensive than others which is a reflection of both the varying degrees of maturity of the profession in different countries, as well as the limited availability of information for some countries represented in this chapter. Regrettably, many countries with developing professions could not be included in this edition due to a lack of available information on which to adequately characterize the country, or space for publication. Some data used to characterize occupational hygiene practice in the countries profiled was acquired through the generous contributions of practitioners in the host country. The authors have relied upon those contributors for the accuracy of that information.

4.1 Australia

4.1.1 Economics and Demographics

The Commonwealth of Australia consists of six states: Queensland, New South Wales, Tasmania, South Australia, Victoria and Western Australia, and two self-governing territories: the Australian Capital Territory and the Northern territory. English is the official language for the cumulative population of approximately 19 million with an adult literacy rate of 99%. Provided with vast natural resources, the Australian economy is comprised of agricultural and mining sectors coupled with diversified manufacturing elements and a growing services sector. The distribution of the Australian workforce of 9.2 million reflects a shift toward a service-based economy (7). The percentage of work force per economic sector is as follows: Services, 69%; Mining, manufacturing and utilities, 22%; Agriculture, 5%; and Public administration and defense 4%.

By estimation of the Australian Institute of Occupational Hygiene (AIOH), there are about 350 practicing occupational hygienists and/or technicians working in the country (8). Their distribution relative to the four primary occupational hygiene employment sectors is weighted toward industry with an estimated occupational hygienist to workforce ratio of 1:26,300. The percentage distribution of AIOH members per employment sector is as follows: Industry, 40%; Consulting, 25%; Government, 25%; and Academia, 10%.

4.1.2 Historical Development

The beginnings of the occupational hygiene profession in Australia are traced to the early 1920s when social awareness and concern for various occupational illnesses such as silicosis, lead poisoning and coal dust pneumoconiosis resulted in the development of the first Division of Industrial Hygiene in New South Wales under the Ministry of Labor. This focus on industrial disease control was subsequently repeated in Victoria and the other Australian states. These organizations were established within state health departments and state analytical laboratories.

The profession continued to grow as a result of natural resource-based commercial projects and major infrastructure projects after World War II. In 1949, the federal government established the Commonwealth Institute of Health at Sydney University to conduct research and to provide an educational framework for occupational medicine. In the late 1950s, the federal government established the National Health and Medical Research Council of Australia (NHMRC). Through its Occupational Health Committee and the Occupational Hygiene Subcommittee, the council has played an advisory role on occupational hygiene issues through the states (8).

4.1.3 Occupational Health and Safety Legislation

In occupational health and safety matters, the federal government has direct responsibility for its employees throughout the country, as well as for the population of the territories. The six states have their own autonomous governments but rely on the federal government for federally financed projects. The states have legislated occupational health and safety matters since the late nineteenth and early twentieth centuries through the promulgation of various acts, such as the Factories, Construction Safety, and Mining Acts. More recently, the federal government has taken the lead in regulating health and safety matters.

The Robens Report of Great Britain also had a major impact on the Australian approach to occupational health and safety. Several states have passed acts based on the tripartite approach recommended in the report and adopted by the British Health and Safety Com mission wherein employers, labor organizations, and governmental bodies are brought together to develop and draft occupational health and safety regulations. Today, employers and employees are jointly responsible for implementing such regulations. This is one of the hallmarks of the Australian approach. The ultimate responsibility for providing a safe and healthful workplace, however, rests with the employer while governmental agencies are charged with enforcement. Jones (9) adds:

> To ensure that there was "expert" contribution to these deliberations, which was not universally the case in the initial stages, a peak council of professional organizations was formed to lobby for the most appropriate person, nominated by it, to be appointed to the committees. This peak council consists of the presidents of the associations of occupational physicians, hygienists, nurses, safety engineers, ergonomists, and it has brought a balance to the development of regulations and standards at the national level. Acts promulgated by the states call for tripartite councils or commissions responsible for occupational health and safety in mining and industry.

In 1985 the federal government established the tripartite National Occupational Health and Safety Commission (NOHSC) known as Worksafe Australia. Under WorkSafe, the legislative focus is on the common law principle of duty of care for occupational health and safety. Worksafe develops general legislation that is adopted or modified by the states and territories, but it is the responsibility of the employer to recognize the hazard and apply appropriate controls as feasible. The emphasis is placed on industry self-regulation rather than prescriptive enforcement by government regulatory agencies. Tripartite participation supports this system by giving equal influence to the participating parties in the development and enforcement of health and safety regulations.

As the result of a change in the National Government in 1996, the role of Worksafe has changed to one of coordination, with the States once again assuming a greater responsibility for occupational health and safety regulation. This has resulted in less emphasis on the promulgation of health and safety legislation including occupational exposure standards. The AIOH has sought to assist where possible to expedite the process. The degree of resource dilution for WorkSafe over the past few years as it relates to occupational hygiene is reflected in the fact that there is currently only one professional occupational hygienist employed by Worksafe. Similar downsizing has occurred in major companies, as a result of recent economic changes in 1997–1998, with many organizations contracting expertise only when required. This in turn has impacted the mentoring system that had traditionally been based on the relationship between senior occupational hygienists and their counterparts in industries employing occupational hygienists (10). This trend is likely to increase as Australia moves away from its traditional economic roots in heavy industry to a more service-based economy. As the service sector has developed in the past ten years the level of consultation to small industry has rapidly increased.

4.1.4 Occupational Exposure Limits

The current process for establishing occupational exposure limits is based on the 1985 NOHSC Act which follows the tripartite model with state, territory and commonwealth government representatives working in conjunction with trade unions, employers and employer representatives, e.g., trade associations (11). The NOHSC sponsors the activities of the Standards Development Standing Committee (SDSC) which in turn coordinates with the Exposure Standards Expert Working Group (ESEWG), yet another tripartite body whose function is to finalize the occupational exposure limits. Like the U.S. ACGIH TLV's, Australian occupational exposure limits are defined as levels of exposure that are anticipated to neither cause impairment of health nor dysfunction in the majority of individuals who are exposed over a 30-year period. They are promulgated as either an eight-hour time-weighted average or a short-term (15-min.) exposure value.

ESEWG recommends potential new occupational exposure limits based on the activities of other international standards and guidelines development organizations and the input of industry and state or territorial government agencies involved in occupational health and safety. Organizations that are monitored closely include the U.S. ACGIH TLV Committee, the Nordic Expert Group on Limit Value Documentation (NEG), the British Health and Safety Executive (HSE), and the German Commission for the Investigation of Health Hazards of Chemical Compounds in the Work Area, among others. Relevant documentation is reviewed, and based on its scientific merit, a proposal for a new standard is submitted to the SDSC. The SDSC considers the economic and technical feasibility and either adopts the ESEWG's recommendation or modifies it before it is released by the NOHSC for public comment. The ESEWG amends the proposed standard based on the public comments, which leads to formal adoption by the NOHSC. Once adopted, the new standards do not have the force of law until subsequently integrated into the regulations of each state and territory. The history of regulatory development in Australia suggests that most new occupational exposure standards are adopted and enforced by the Commonwealth governments. In most instances when a standard has not been adopted it is the result of a perceived need to lower the standard (8).

In recent years a new group, the Hazardous Substances Subcommittee, has been organized to replace the work of the existing work groups addressing the classification, health surveillance and occupational exposure standards. This new framework has resulted in a slower rate of development for new standards as this new system prioritizes standard setting development.

4.1.5 Education and Training

The majority of Australian occupational hygienists have an undergraduate degree in the sciences or engineering with many having graduate degrees, although there are a number of practitioners who developed their knowledge of occupational hygiene strictly through on the job experience. Mentors, the majority of whom have been employed by the government, have played a significant role in developing young professionals (8). However, many of these opportunities are disappearing as a result of government and industry downsizing and outsourcing pressures. This is likely to have a significant impact on the profession in the next decade, when the current generation of senior occupational hygienists leaves the profession.

The general trend in Australia for formal academic training in occupational hygiene is at the post-graduate level, however, there are undergraduate and graduate programs offered by academic institutions in several states in Australia on a residency and nonresidency basis. The University of Sydney offers a Masters of Public Health (M.P.H., Occupational Health) degree within its School of Medicine. The curricula include occupational hygiene topics. In addition, Ballarat College of Advanced Education in Victoria, the South Australia Institute of Technology, and the Western Australian Institute of Technology offer graduate courses with occupational hygiene content as part of their diploma programs in occupational health and/or safety. The Footscary Institute of Technology in Victoria also offers hygiene courses as part of a program leading to an associate diploma in occupational health and safety. Since about 1980, the School of Biological and Chemical Sciences at Deakin University in Victoria has offered a graduate certificate, diploma and masters degree in occupational hygiene using a distance learning curriculum. A small number of Australian occupational hygienists have earned postgraduate degrees at universities abroad.

4.1.6 Certification and Registration

There is no independent certification or registration process for occupational hygiene in Australia; however, the AIOH operates as a pseudo-certifying organization through defined categories of membership based on level of competencies. There are five grades: Fellow, Honorary Fellow, Full, Provisional, and Associate (8). Fellow members must be a full member for at least five years, have worked for more than 15 years in a professional capacity as an occupational hygienist or one of its specialty branches, have made a substantial contribution to the advancement of the profession, and be nominated by a full member of the Institute. Individuals obtain honorary fellow status by invitation from the AIOH Council in recognition of distinguished contribution to the advancement of the profession. A candidate for admission as a full member is required to have a baccalaureate degree or diploma in science or engineering, or an equivalent qualification, five years of experience in the field of occupational hygiene, and must demonstrate a satisfactory level of professional competence to the AIOH Council. Provisional members must meet the same qualifications as full members but are required to have only one year of relevant experience in addition to demonstrating an acceptable level of understanding of the basic principles of occupational hygiene. An associate member is required to be working as an occupational hygienist or in a closely aligned field of scientific endeavor and hold a certificate issued by the Australian Department of Technical and Further Education in either chemistry, physics, biology, medical technology or a similar field. For the purposes of qualifications for membership, "work in the field of occupational hygiene" is defined by the AIOH to mean at least 50% of the candidate's time directly involved in occupational hygiene.

In addition to the professional designations defined by the AIOH, there are approximately 50 Australians who have achieved certification through the American Board of Industrial Hygiene (ABIH) and others listed by the British Examination and Registration Board in Occupational Hygiene (BERBOH). The AIOH has a formal administrative agreement with ABIH to administer the core, comprehensive and specialty examinations in Australia under the supervision of Australian diplomates of ABIH functioning as examination proctors.

4.1.7 Professional Organizations

The AIOH was organized in 1978 to provide a professional network for occupational hygienists. The early focus of the Institute was membership qualifications and professional development. There was a conscientious effort in the 1980s to maintain an increasingly high level of education among members, which continues today through professional conferences and short course development. AIOH is a member organization of IOHA and continues to promote the profession through advocacy with the Federal and State governments, interaction with allied health professions including the Australian Ergonomics Society, and is involved in the accreditation of undergraduate and graduate courses at various academic institutions.

As with many professional occupational hygiene organizations worldwide, AIOH maintains a code of ethics for its members. The code places obligations on its members to practice their profession in an objective manner recognizing that the lives, health and welfare of individuals may be dependent on their professional judgment. The Council of the Institute recognizes the code as criteria to be applied when determining whether the conduct of a member renders them unfit to remain a member pursuant to the Rules and Statement of Purpose of the Institute.

According to the AIOH Code of Conduct and Ethics (12):

Primary Responsibility: In providing advice to employers, clients or employees, members shall give paramount consideration to safeguarding the health of the workforce.

Professional Conduct: Members shall conduct their affairs so as to promote and improve the professional practice of occupational hygiene, and shall so order their conduct as to uphold the dignity, standing and reputation of the profession. Members shall base the advice they give on the best available scientific evidence.

Responsibility to Employers: (*1*) Advise the employer, responsibly and competently so that healthy working conditions may be achieved and maintained without unnecessary expense. (*2*) Keep confidential all information relating to the employer's business operation or manufacturing processes, which is not common knowledge. (*3*) Advise the employer so that unwitting contraventions of any relevant legislation or professionally accepted standard can be avoided; in particular, to inform the employer when he has a statutory duty to disclose findings to workers or their representatives. (*4*) Report findings clearly and factually to the employer directly and to no other body without the permission of the employer, unless there is not way other than disclosure, of averting a high risk of death or serious injury. Where disclosure is to occur, the relevant member should notify the employer.

Responsibilities to the Workforce: (*1*) Adopt an objective attitude towards the recognition, evaluation, and control of environmental factors adverse to health. (*2*) Report clearly and factually; ensure that matters of opinion are founded on adequate knowledge and are within the member's expertise. (*3*) Ensure that all information obtained is used solely for the purpose of promoting occupational health.

Responsibilities to the General Public: Make public statements claiming professional knowledge in an area of public interest only if competent to do so, and only if such statements are not inconsistent with other responsibilities set out in this Code.

Responsibilities of Consultants: In addition to conforming to the above standards of ethical conduct, a member acting as a consultant shall: (*1*) Ensure that work performed by other persons at the member's behest is competently performed and honestly and reliably reported. (*2*) Inform the client of any interest or employment such as might compromise the exercise of independent professional judgement or conduct. (*3*) Work for one client only on the same matter unless the consent of all relevant clients is obtained. (*4*) Not solicit for work either by calling into question ability or integrity of another member or by offering or paying to a prospective client financial or material inducements. (*5*) Not disclose to any third person any finding on behalf of the client without the client's permission, unless there is no way other than disclosure, of averting an immediate risk of death or serious injury. Where disclosure is to occur, the relevant member should notify the client.

4.1.8 Status of the Profession

The profession in Australia has grown substantially in the last 20 years with a current estimate of more than 350 practitioners, a strong professional association with a system for differentiating qualifications among the membership, and a viable code of ethics. There are practitioners in a wide variety of industries including manufacturing, chemical and petroleum companies, mining, shipping, telecommunications, health care, transport, academic institutions and the government including the military. The profession has representation in the deliberation on regulatory developments and enjoys a growing number of educational opportunities. Australian occupational hygienists have also impacting the international scene through the early work in forming IOHA and have subsequently held leadership positions in the association. However, recent economic change has resulted in the outsourcing of some occupational hygienists and an increase in new consultants, which has affected on-the-job training and mentoring opportunities. Despite the optimistic outlook, public recognition of the profession remains relatively weak, a realization that is by no means unique to Australia.

4.1.9 Information Resources

The Australian Institute of Occupational Hygiene (AIOH, Inc.)
PO Box 1205, Tullamarine
Victoria 3043, Australia
Tel: 03 9335 2577, Fax: 03 9335 3454
Internet: www.curtin.educ.au/org/aioh/who/htm
Status: Professional Association.

Worksafe Western Australia
Westcentre, 1260 Hay Street
West Perth WA 6005, Australia
Or, PO Box 294, West Perth WA 6872, Australia
Tel: 619 327 8777, Fax: 619 481 8427
Internet: www.wt.com.au/safetyline/sl_info.htm
Status: Regulatory Agency.

National Occ. Health and Safety Commission (WorkSafe Australia)
92 Parramatta Road, Camperdown NSW 1460, Australia
or, GPO Box 58, Sydney NSW 2001, Australia
Tel: 61 2 9577 9555, Fax: 61 2 9577 9202
Internet: www.worksafe.gov.au
Status: Regulatory Agency

National Safety Council of Australia Ltd
PO Box 810, Mascot NSW 1460, Australia
Level 2, 8 Lord Street, Botany, Sydney NSW 2019, Australia
Tel: 61 2 9666 4899, Fax: 61 2 9666 4811
Internet: www.safetynews.com/nsca/nsca.html
Status: Nonprofit organization with industry and government representatives.

VIOSH Australia
Asia-Pacific Center for Teaching & Research in Occupational Health & Safety
University of Bellarat
PO Box 663, Ballarat VIC 3353, Australia
Tel: 63 3 5327 9160, Fax: 63 3 5327 9151
Internet: www.ballaret.edu.au/viosh
Status: Academic Institution.

Deakin University
Geelong, VIC, 3217, Australia
Tel: 61 352 47111, Fax: 61 352 472001
Internet: www.deakin.oz.au
Status: Academic Institution.

4.2 Brazil

4.2.1 Demographics and Economics

Brazil has one of the largest economies in the world with a gross domestic product of $862 billion U.S. dollars in 1998 (7). The country has substantial natural resources including iron ore, manganese, bauxite, nickel, uranium, gemstones, and petroleum and is a primary exporter of agricultural products including coffee, soybeans, sugar cane, cocoa, rice, beef, corn, cotton, citrus and wheat. In addition, raw, manufactured and durable goods such as steel, chemicals including petrochemicals, machinery, motor vehicles, computers hardware and software, cement and lumber make Brazil's industrial sector one of the most advanced in South America. The official language in this country with 156 million people is Portuguese and the literacy rate is approximately 81% in the adult population. There is a recognized work force of more than 65 million people with the number of unofficial employees estimated at 30 million (7). As of 1995, there were more than two million companies in Brazil with at least 50 or more employees. Percentage of distribution of workers in the economic sector is as follows: Services, 40%; Mining, manufacturing and utilities, 25%; and Agriculture, 35%.

Brazil has approximately 300 full-time practicing occupational hygienists with significantly more individuals performing occupational hygiene-related activities whose primary job title is something other than occupational hygienist (13). The majority of occupational hygienists are employed with either multinational corporations, or the Brazilian government. An accurate estimation of the distribution of professional occupational hygienists within the primary employment sectors is not available. The ratio of professional occupational hygienists to the total Brazilian workforce is estimated to be 1:216,700.

4.2.2 Historical Development

The first occupational health and safety inspectorate in Brazil was established in 1923 under the Ministry of Labor. In 1943, Brazil's Minister of Labor formally noted the importance of occupational health and safety to Brazilian society: "There is no doubt that a human life has economic value. It is capital with a production capacity that can be appraised by actuaries and mathematicians. But the affective and spiritual value of human life is so great that all the money in the world could not buy it. This, above all, is why it is so important to prevent occupational hazards and, thus, avoid the irreplaceable loss of a father, a husband, a son, and, especially, of a breadwinner or head of any family. Prevention is very similar to enjoying good health. It is only taken seriously after the accident occurs or the disease sets in." As part of the 1946 Constitution, modern Brazilian labor laws were developed including those relating to accident prevention. When the new Federal Constitution was written in 1988 having "occupational hazards reduced through the establishment of health, hygiene and safety standards" reinforced the social rights of all Brazilian citizens.

In 1966, the Fundacentro (Fundacao Jorge Figureiredo de Seguranca e Medicina do Trabalho) was established as a research organization charged with addressing occupational health and safety issues. It is funded by workers compensation premiums assessed to all private sector employers. Fundacentro is functionally analogous to the U.S. National Institute for Occupational Safety and Health (NIOSH). In 1969, the "Serviço Especializado em Engenharia de Segurança e Medicina do Trabalho" or Specialized Services for Occupational Safety and Medicine (SESMT) "obligation" was developed which required private employers to retain or employ the services of safety technicians, engineers, occupational health nurses and physicians depending on the level of risk associated with the work environment and the number of employees present. However, from its inception, SESMT does not specifically recognize the role of occupational hygienists. This is a significant political and economic barrier to the development of the profession in Brazil. Approximately 1.6% of all companies in Brazil today have SESMT occupational health and safety support groups in place (8). By 1972, Act 3237 set forth requirements for the prevention of occupational disease and accidents and began to be enforced in 1976.

While occupational hygiene-related activities have been practiced in Brazil for decades, the primary practitioners have been occupational medicine physicians, safety engineers and other engineering professionals. It is only in the last decade that progress has been made in the development of an autonomous occupational hygiene profession. Much of that progress has been facilitated by a small and dedicated group of occupational hygienists supplemented by the introduction of comprehensive health and safety legislation.

4.2.3 Occupational Health and Safety Legislation

Despite the highly visible recognition of occupational health in the 1948 Brazilian Constitution, it was not until 1974 that significant regulatory requirements were established. Fundacentro and the Ministry of Labor adopted WHO, ILO, and U.S. Department of Labor Occupational Safety and Health Administration (OSHA) regulations in Brazil (Reg. 3214/78) which became law in 1978. For example, Brazil ratified ILO agreements Nos. 12, 42, 115, 119, 127, 139, 148, 155, 161, 162 and 170 (13, 14).

Laws addressing occupational safety and health come under the authority of the federal government and specifically the Ministry of Labor which made significant changes to occupational safety and health regulations in 1978 with the publication of Ordinance No. 3214/78. The ordinance contained all the rules governing issues relating to occupational hazard prevention.

The Labor Ministry recently created the "Work Committee" composed of occupational hygienists from different organizations within the country including Fundacentro. Other representatives on the Work Committee included in the regional labor agencies (DRT), academic institutions and representatives from the private sector. The purpose of the Committee is to develop a new regulatory standard to be incorporated into Ordinance No. 3214/78 to create an "efficient environmental hazard prevention work methodology which is to be complied with by employers across the country, regardless of how many employees they have working for them."

The final version of Standard No. 9 was promulgated by the Ministry of Labor in December 1994 and serves as the basic framework for the control of occupational health hazards in Brazil (13):

> This standard (NR-9), establishes that all employers and organizations who hire employees are required to prepare and implement an Environmental Hazard Prevention Program (PPRA). The program must be aimed at preserving the health of workers through the anticipation, recognition, evaluation and control of existing or potential hazards in the workplace, with due regard for the protection of the environment and of the natural resources.

Employers are legally authorized to either use their own employees or outside specialized professionals, whichever they deem most appropriate to fulfill the goals and objectives. Mandatory provisions of all PPRA's include the anticipation and recognition of hazards, the establishment of evaluation and control priorities and goals, evaluation of hazards and worker exposure, the implementation of control measures and evaluation of their efficiency, and exposure and risk assessment; and, data registration and disclosure.

The PPRA must also include a written program which outlines the steps taken to address the fundamental elements of the regulation, including: (*1*) an annual plan with an outline of the main action goals, priorities, and schedule for each year; (*2*) a defined strategy and accompanying methodology to be used in the implementation of the program for planning, development, and execution; (*3*) the form in which data is to be collected, managed, and communicated; and, (*4*) the frequency and manner, in which the program's development is to be evaluated, which must be at least annually (14).

Exposure controls must be adopted, independent of quantitative analysis, in the following instances: (*1*) when a potential risk to health is identified during the anticipating stage; (*2*) when an obvious risk to health is detected during the recognition stage; and, (*3*) when medical controls show that a cause and effect relationship has been established between exposure and health effects, as well as exposures above the Brazilian TLV's. The use of personal protective equipment is only permitted in emergencies or when it is obvious that control measures are technically unfeasible or insufficient, or in instances when the controls are still being evaluated, designed or implemented.

The employer's PPRA must be consistent and compatible with the Occupational Health Medical Control Program companies are required to develop under Standard No. 7 (NR-7), which requires certain medical services be made available for employees based on the number of employees affected. Brazilian workers are entitled to information on the content of their workplace PPRA, and are encouraged to make recommendations that will assure protection against health risks resulting from identified environmental hazards. Workers are also empowered with the right of work refusal for situations that pose a serious or imminent risk to health.

Although NR-9 has resulted in considerable new opportunities for occupational hygiene professionals in Brazil, it does not restrict any other allied health professional or engineer from performing the predominantly occupational hygiene-related activities mandated in the standard. In addition, and not uncommon for countries with developing private and government health and safety systems, the scope and depth of comprehensive regulations like NR-9 are not always matched by effective, voluntary implementation and enforcement efforts.

4.2.4 Occupational Exposure Limits

There is no independent process for developing occupational exposure limits in Brazil. The first significant application of occupational exposure limits occurred in 1978 when Act 3214 was promulgated which codified the 1978 ACGIH TLVs with a 22% reduction to adjust for differences in work shift lengths between the U.S. and Brazil (15). Act 3214 has not been updated since 1978 and as such the 1978 TLVs are still enforced. With the promulgation of NR-9 and subsequently NR-15, if a substance of concern is not covered by a 1978 TLV, a current TLV or another exposure limit that is mutually agreed upon between employer and employees as stipulated in collective bargaining agreements may be applied. Lastly, NR-9 also introduces the concept of action levels. When an action level is exceeded, employers are statutorily obligated to implement preventive action such as exposure monitoring and/or medical surveillance and inform workers of the need for these actions. The ABHO has recently translated the 1996 ACGIH TLV booklet into Portuguese (13).

4.2.5 *Education and Training*

There are currently two Brazilian universities that offer post-graduate courses in occupational hygiene, and universities in Rio de Janeiro, São Paulo, and Minas Gerias provide training in occupational health and safety (16). A few professionals have acquired graduate degrees in the U.S. or the U.K. However, most opportunities for professional development come in the form of short courses, usually one week in duration, that are often developed and taught by either members of "Associação Brasileira dè Higienistas Ocupacionais" (Brazilian Association of Occupational Hygienists, or ABHO), corporations with occupational hygiene staffs, or other nonacademic institutions.

4.2.6 *Professional Organizations*

The ABHO was founded in August, 1994 in São Paulo at the first national meeting of Brazilian occupational hygienists with about 100 hundred attendees who formed the core of the association. The ABHO has grown rapidly and in 1998 had approximately 450 members with another 75 applications under review (13, 14).

4.2.7 *Certification and Registration*

There is no certification or accreditation system in place in Brazil. The ABHO is currently working to develop such a framework modeled after the American Board of Industrial Hygiene (ABIH) and the British Examination and Registration Board in Occupational Hygiene (BERBOH) (13). Although there is no certification system in Brazil, the ABHO has established a code of ethics for occupational hygiene practitioners that are similar to that of the ABIH with minor modifications.

4.2.8 *Status of the Profession*

Brazil's occupational hygienists struggle for broader recognition as a result of a lack of statutory recognition in the primary health and safety regulatory framework. This is compounded by the concurrent statutory recognition of other allied professions in the safety, engineering, and medical fields to address the same issues. However, Brazilian industry is beginning to recognize and accept the profession, although the allied professions have been less accepting of the growth of the profession due primarily to socioeconomic pressure (13).

This notwithstanding, occupational hygiene is growing in Brazil as a result of public sector requirements for services and government recognition of the contribution of professional occupational hygienists. Role delineation and public and statutory recognition of the different capabilities of the allied health professionals have not been fully realized in

Brazil and will continue to passively restrict the growth of the profession until fully addressed.

4.2.9 *Information Resources*

Brazilian Association of Occupational Hygienists (ABHO)
Caixa Postal 3066
Campinas, SP, Brazil 13033-990
Tel: 55 19 242-6946, Fax: 55 19 242-6946
Status: Professional association

Fundacentro (Fundacao Jorge Figureiredo de Seguranca e Medicina do Trabalho)
Caixa Postel 11.484
05422-970, Sao Paulo (SP), Brazil
Tel: 551 1 3066 6000, Fax: 551 1 3066 6234
Status: Governmental research agency

4.3 Canada

4.3.1 *Demographics and Economics*

Canada's 10 provinces (Alberta, British Columbia, Manitoba, New Brunswick, Newfoundland, Nova Scotia, Ontario, Prince Edward Island, Quebec and Saskatchewan) and two territories (Northwest Territories, Yukon Territory) are endowed with a wide variety of natural resources including nickel, zinc, copper, gold, lead, molybdenum, potash, silver, fish, timber, coal, petroleum and natural gas (7). The Northwest Territories is currently being subdivided into two distinct territories: the eastern section, which is now self-governing and renamed Nunavut, and the west section which is as yet unnamed. This affluent, industrial society began a substantial program of development after World War II with advances in its mining, manufacturing and services sectors which transformed this country of 30 million from a largely rural society into a primarily industrial and urban society. The distribution of the 15.3 million work force is as follows: Services, 75%; Mining, manufacturing and utilities, 16%; Construction, 5%; Agriculture, 3%; and other 1%.

The growth of the Canadian economy has been mirrored by the growth of the occupational hygiene profession. Today, there are more than 750 people practicing occupational hygiene in Canada (17) representing an occupational hygienist to work force ratio of approximately 1:20,400.

4.3.2 *Historical Development*

The history of occupational hygiene as such can be traced back to the 1930s, to the era of industrial expansion and of growing recognition within industrialized nations of the need to prevent occupational diseases. Canadian occupational health scientists in industry and government established professional links with their colleagues abroad, particularly in the United States and in Great Britain. Such collaboration originated from their common cultural, political, and industrial heritage. Over the years, links between Canadian occupational hygienists and their American counterparts led to many Canadians becoming members of, and working with, the American Industrial Hygiene Association (AIHA), ACGIH, the American Academy of Industrial Hygiene (AAIH), and ABIH. Similar links were established in Great Britain with the London School of Hygiene and Tropical Medicine, the British Occupational Hygiene Society (BOHS), the Institute of Occupational Hygiene (IOH), and BERBOH. Only three times in the 60-year history has the American Industrial Hygiene Conference and Exposition meet outside the United States. All three times the event was held in Canada, in Toronto in 1971 and 1999, and in Montreal in 1987.

Among Canadian Provinces, Ontario has been one of the leaders in occupational health. In 1976, in addition to safety, the Ontario Ministry of Labor also assumed the responsibility for occupational health. In the ensuing administrative reorganization, existing safety branches in the Ministry of Labor combined with the Occupational Health Protection Branch, formerly with the Ministry of Health, and the Mines Engineering Branch, previously under the Ministry of Mines and Natural Resources, thus giving birth to the Occupational Health and Safety Division.

This reorganization changed the status of governmental occupational hygienists. Until this time, occupational hygienists had participated in workplace inspections as technical advisers to safety inspectors of the Ministry of Labor that were responsible for enforcing health regulations. Once occupational hygienists were brought under the Ministry of Labor, they assumed the full responsibilities of inspectors. The Occupational Health and Safety Act of 1978 further changed the way occupational hygiene was practiced in Ontario. It made occupational hygienists responsible for periodic inspections of workplaces in addition to answering requests for occupational hygiene investigations made by other inspectorate branches and by the Workplace Safety and Insurance Board.

The Act requires that health and safety committees be formed in places of employment with 20 or more workers. Such committees consist of equal numbers of representatives of management and of labor, the latter being selected by the workers. This concept of a cooperative effort between management and labor is based on the recognition that the two parties are best qualified to anticipate and control work-related hazards. The minister of labor may order that such joint management-labor committees be formed in smaller places of employment depending on the use of hazardous materials.

An important feature of the Occupational Health and Safety Act of Ontario is the Internal Responsibility System (IRS) that places joint responsibility for health and safety on both employers and employees. It creates an interlocking set of rights and obligations on the part of management as well as of workers. Both parties are expected to work in concert rather than in conflict, to help and support each other in important health and safety functions in the workplace, with government standing by to see that these responsibilities are fulfilled. The IRS thus represents an effort to promote in-plant problem solving and decrease reliance on government inspectors.

The Canadian Centre for Occupational Health and Safety (CCOSH) was established as a Crown Corporation by the Federal Government to promote the right of Canadians to a healthful and safe work environment. The Center's primary funding comes through cost-recovery of services provided, and is aided by a grant from the federal government. It operates under the direction of a council of governors representing federal, provincial, and territorial agencies, industry, and labor and reports to Parliament through the federal Minister of Labor. The CCOSH provides occupational health and safety information to anyone requesting it. It publishes technical information and provides computerized information through access to data banks and databases.

Other developments have taken place in the broad area of occupational medicine including occupational hygiene. Canada and its provinces promulgated a set of acts and regulations in 1987 to create the requirements for Material Safety Data Sheets (MSDSs), container labels, and effective worker training in the safe use and handling of hazardous materials. This system is known as the Workplace Hazardous Materials Information System (WHMIS).

Several provinces have instituted new ways of providing occupational health services to workers and, in some instances, to communities. The Province of Quebec has granted workers the right to participate in choosing occupational health personnel through joint worker-management committees. Ontario, Manitoba, and Alberta have fostered the establishment of worker-controlled occupational health centers. Such centers differ in several respects from one another.

Yassee (18) reports that the Ontario Workers' Health Clinic has a board of directors composed of union officials, community leaders, representatives of Canadian occupational safety and health groups, and others. It has established clinics in Hamilton, Toronto, Sudbury, and Windsor. All funding originally came from union contributions and from the Clinic's outside services and contracts. It is now largely funded by grants from the Ontario government.

The Manitoba Federation of Labor Occupational Health Centre was founded with seed money provided by organized labor (18). All members of the board of directors are from the Manitoba Federation of Labor. Presently, the Social-Democratic government of Manitoba provides the operating funds for the center. Continued financial support by organized labor has enabled the center to acquire an extensive occupational health library and a computer link to international databases. The center provides occupational health information as well as medical and occupational hygiene services to employers and employees alike, irrespective of their union affiliation or lack thereof.

4.3.3 Occupational Health and Safety Legislation

The first significant Canadian legislation dealing with occupational health and safety was the 1884 Act for the Protection of Persons Employed in Factories. The act was intended to protect children, young girls, and women working in industrial workplaces. Among others, the act stated (19):

> Every factory shall be ventilated in such a manner as to render harmless, so far as reasonably practicable, all gases, vapours, dust or other impurities generated in the course of the manufacturing process or handicraft carried on therein that may be injurious to health.

It established penalties if workers suffered permanent loss of health.

The Workmen's Compensation Act of 1914 set guidelines for worker health education and for control measures intended to prevent accidents and work-related diseases while the Factory, Shop and Office Buildings Act, promulgated in World War I, was expanded in the late 1920s to require compulsory medical supervision of workers handling hazardous substances. Amendments to the act in 1932 required employers to label containers of benzene and to report any cases of industrial diseases directly to the Director of Industrial Hygiene.

In 1926 a study among workers at the Porcupine Mining Camp indicated that only 98 of the 236 miners employed did not suffer from tuberculosis, silicosis, or both diseases. Thus in 1929, the Workmen's Compensation Board set up clinics for periodic health examinations of underground miners and established the Silicosis Referee Board.

The Canadian Constitution grants jurisdiction for the regulation of occupational health and safety issues to the provincial and territorial governments across Canada, except for limited jurisdiction over Federal government employees and certain industry sectors such as airlines, railroads, communications, and uranium mining. As such, each of the 14 jurisdictions has separate legislation and enforcement structures.

A review of Nova Scotia Health and Safety legislation, with the promulgation of the 1996 Occupational Health and Safety Act which heavily stresses the principle of internal responsibility. Several draft pieces of legislation are also on the books to be passed within the next 12 months, which should have a significant impact on Occupational Hygiene practice in Nova Scotia.

In 1998, changes to the British Columbian Occupational Health and Safety Regulation administered by the Workers' Compensation Board (not the Ministry of Labor, as in other Canadian provinces) requires employers to conduct exposure risk assessments, and includes specific requirements for monitoring and control implementation. This has increased the need for smaller employers to use the services of occupational hygienists (17).

The provincial and federal Health and Safety Acts provide for the formation of joint management–labor health and safety committees and the right for workers to refuse unsafe work.

WHMIS is functionally equivalent to OSHAs Hazard Communication Standard. It requires suppliers of hazardous materials to provide information in MSDSs and descriptive container labels to purchasers and it requires employers to provide this information to workers with appropriate training and education. The Hazardous Products Act, the federal legislation for WHMIS covering suppliers and the equivalent provincial requirements for employers were developed for the first time in Canadian history as a cooperative, consensus effort by representatives of federal and provincial governments, organized labor and employer associations. This model legislative development by multipartite consultation has now become the norm in Canada.

4.3.4 Occupational Exposure Limits

The federal Department of Human Resources and Development (Labor Branch) in Ottawa has jurisdiction over matters affecting the occupational safety and health of workers employed in federal establishments across Canada, such as transportation and civil service workers. The department exercises its authority through the Canada Dangerous Substances Regulations that recognize the most recent ACGIH TLVs as permissible exposure limits.

Because Canada has 10 provinces, applicable rules and exposure limits may vary from province to province and substance to substance. Proposals have been made among regulators in recent years to harmonize the various occupational exposure limits across Canada (17), but no progress to this end has been made. In Ontario, occupational exposure standards have been developed on a substance-by-substance basis including coke oven emissions, isocyanates, asbestos, silica, benzene, arsenic, vinyl chloride, mercury, lead, ethylene oxide and acrylonitrile. Each standard has an accompanying code specifying exposure assessment methods and controls which must be conducted in consultation with the joint health and safety committee (20). In addition to designated substances, the Ministry of Labor applies the ACGIH TLVs for other substances. All occupational exposure limits carry the weight of law including the potential for criminal and civil penalties.

4.3.5 Professional Organizations

There is no single national association representing occupational hygienists in Canada. There are, however, occupational hygiene associations in several provinces, and three AIHA local sections (Alberta, British Columbia-Yukon, and Manitoba). The two largest associations are the Occupational Hygiene Association of Ontario (OHAO) and l'Association Québeçoise pour l'Hygiene, la Santé et la Securité du Travail (AQHSST) of Quebec.

A third organization, the Canadian Registration Board of Occupational Hygienists (CRBOH), provides the dual function of administering the national registration (certification) system and serves as the primary professional association in the country. In its latter role, the CRBOH promotes the profession nationally, represents Canada as a member organization of the IOHA, sponsors "Canadian Issues" functions at the American Industrial Hygiene Conference and Exposition, and in 1999 launched an internet-based discussion group (OccHygPro) for accredited occupational hygienists around the world (17).

The OHAO is dedicated to the practice of occupational hygiene in the province and governed by an elected board of directors with several categories of membership based on professional responsibilities (17). The OHAO organizes symposia and other technical meetings, publishes a newsletter, and sponsors a provincial award to recognize lifetime achievement in the profession.

The AQHSST has approximately 800 members representing various fields in occupational health and safety such as industrial hygiene, health and safety, ergonomics and environment. The AQHSST is not a restricted professional association but offers membership to any individual with an interest in the area of occupational health and safety. Its mission is to provide technical information and to offer quality training that will enable inquirers and trainees to be knowledgeable and competent in their field of practice. The AQHSST also fosters interaction between its members to share experiences and problem-solving in various industries. Also, the association offers its opinion regarding legislation in the area of occupational health and safety.

Nova Scotia also has an association, the Health and Safety Associations of Nova Scotia (HSANS) which is a loose affiliation of the Canadian Society of Safety Engineers (CSSE), Occupational Health Nurses of Nova Scotia, the Human Factors Association of Canada (HFAC), the Occupational and Environmental Medical Association of Nova Scotia (OEMANS), and local sections of the AIHA. It functions as an information and advocacy organization with its primary focus on worker protection.

In 1993, members of the professional groups which constitute the broad field of worker health and safety in Ontario formed the Alliance of Environment, Health, and Safety Professionals to leverage resources and to increase the mutual influence of the independent groups to better promote professionalism in all aspects of worker health and safety (17). The alliance also lobbies for legal recognition of the allied professions, and advises other organizations, and the government, of the consequences of their policies and regulatory decisions.

4.3.6 *Education and Training*

The University of Toronto first offered physicians a diploma in industrial hygiene in 1942. Canadian universities began offering courses leading to Master of Science degrees in occupational hygiene in the 1950s. The first two institutions to offer such courses were the School of Hygiene at the University of Toronto and McGill University in Montreal, Quebec. Subsequently, both universities also established doctoral programs in occupational hygiene. In 1978, the Ontario Ministry of Labor established manpower training and made funds available for the establishment of a Diploma course in occupational hygiene at McMaster University in Hamilton, Ontario. More recently, Quebec's Laval University has offered courses leading to a certificate in occupational health and the University of Quebec offers both a masters of science (M.Sc.) and an undergraduate certificate in occupational hygiene. There are currently no Canadian undergraduate degree programs in occupational hygiene. Three other Canadian universities also offer a masters degree in occupational hygiene: the University of Toronto, McGill University, and the University of British Columbia (21).

Canada also benefits from a network of community colleges offering diploma programs in occupational health and safety that produces a steady supply of technologist-level professionals (21). In the opinion of Smith (22), "The large number of occupational health and safety technology programs offered at community colleges in Ontario suggests that there is probably an oversupply of safety and hygiene trained technologists in Ontario, and hence in Canada." Also, "It would appear . . . that the present production rate of 30 to 40 hygienists per year is adequate to meet the national need."

4.3.7 *Registration and Certification*

Certification (referred to as registration in Canada) in occupational hygiene is administered by the CRBOH. Two levels of registration are available (23): the Registered Occupational Hygienist (ROH) and the Registered Occupational Hygiene Technologist (ROHT).

The CRBOH is unique in North America for the utilization of both oral and written examinations to evaluate registration applicants. While CRBOH recognizes that an oral examination presents unique challenges in the maintaining of objectivity and need for stringent record keeping, the CRBOH also holds that the oral examination measures skills such as problem solving skills and communication abilities unavailable through other testing formats.

In 1998, there were 350 ROHs and 73 ROHTs of which 254 were Canadian residents. Most of the other 96 ROHs were based in the U.S. with occupational hygiene responsibilities in Canada. There are 312 Canadians certified by ABIH with 191 of these in Toronto (23).

The CRBOH defines Registered Occupational Hygienists (ROHs) as qualified professionals with a university degree in science and a minimum of five years of professional experience. The examination is comprised of three components: (*1*) a multiple-choice examination, (*2*) a short answer and essay examination; and (*3*) an oral examination. Prospective candidates for the examination are expected to be familiar with a broad range of occupational hygiene topics including basic science, legislation and other health, safety and environment-related standards, recognition of hazards and their effects, sampling and evaluation of hazards, control of hazards, and related topics such as training and communication.

Registered Occupational Hygiene Technologists (ROHTs) are qualified technologists who meet a minimum experience requirement and successfully complete a written examination (multiple choice and essay) covering all aspects of occupational hygiene technology. Eligibility requirements for admission to the ROHT examination are five years experience in occupational hygiene or related field subsequent to receipt of a high school diploma. Completion of a community college program in occupational hygiene technology; in a related science or engineering field, or of two years of a university undergraduate program in related sciences/engineering may be recognized as an equivalent to field experience of two years, subject to approval by the CRBOH. More than 75% of each year for which credit is claimed must have been spent in occupational hygiene or closely related activities.

The code of ethics of the CRBOH contains ten tenets for ROHs and ROHTs (24): (*1*) Place the health and safety of workers above all other interests in the performance of their professional work; (*2*) Direct professional activities toward the protection and improvement of the health, safety, and well being of all persons; (*3*) Make every reasonable effort to protect the environment from adverse effects resulting from the performance of their work; (*4*) Perform their work honestly, objectively, and in accordance with currently accepted professional standards; (*5*) Respect the privacy of confidential personal, professional, and business information; (*6*) Participate only in projects or situations that do not place them in personal or business conflicts of interest. This provision is waived if the principal parties to the ROHs or ROHTs conflict of interest have given their informed, specifically expressed consent; (*7*) Conduct themselves with integrity; (*8*) Maintain a working knowledge of current developments in the profession and a detailed knowledge of areas in which they claim expertise; (*9*) Promote activities that advance and disseminate occupational hygiene knowledge; and, (*10*) Cooperate with the directors of the Canadian Registration Board of Occupational Hygienists in administering this Code of Ethics.

4.3.8 Status of the Profession

Over half of the occupational hygienists in Canada practice their profession in large industries such as the petrochemical, automotive, and foundry industries. Smaller industries handling highly toxic materials also have occupational hygienists on their staffs. About one-fourth of occupational hygienists work for regulatory agencies with the remainder employed in academia or in other capacities. Occupational health regulations and the need of small industries for occupational hygiene assistance have led to the establishment of a number of consulting services. Recently, private consulting firms have become available that offer both field and analytical services in occupational hygiene.

Many small establishments continue to assign occupational health responsibilities to safety personnel, chemists, metallurgists, or other individuals familiar with the chemicals and processes used at the plant. There is good recognition in industry for the value of protecting the workforce and the subsequent role of occupational hygienists to contribute to this issue. This is particularly true in major companies in the petrochemical, forestry, mining, and health service sectors. Occupational hygiene consulting services are primarily available from private firms although some universities, community colleges and industry health and safety associations also provide services.

Verma summarized the factors that have influenced occupational health and safety in Ontario and Canada as: "(*1*) Increased recognition of occupational health and safety through regulatory activities; (*2*) increased involvement in occupational health and safety by sectorial, industrial and trade associations; (*3*) empowerment of workers; (*4*) workers' and communities' right-to-know (*5*) standard development through bi-partite (management and labor) and tri-partite (management, labour and government) consensus; (*6*) emphasis on occupational health and safety education and training at various levels; (*7*) nationwide access to information through the Canadian Centre for Occupational Health and Safety; and (*8*) increased numbers of occupational health and safety professionals" (25).

A reflection of the growing prominence of Canadian practitioners is the number with international stature including Dr. Ernest Mastrometteo, emeritus professor at the University of Toronto who has made significant contributions to the literature on the toxicology of metals and other materials, past-Chair of the ACGIH TLV Committee and 1986 AIHA Yant Memorial Award recipient; John H. Johnston, who passed away in 1998 and was the first full-time occupational hygienist hired by a Canadian company (Imperial Oil Ltd.) in 1953, the first non-American member of the AIHA Board of Directors, general conference chair of the 1971 AIHCE in Toronto and who was instrumental in forming a number of occupational health and safety committees within major trade associations such as the Canadian Chemical Producers Association, the Canadian Petroleum Products Institute, and the American Petroleum Institute; Gyan S. Rajhans of the Ontario Ministry of Labor and an expert in industrial ventilation who among other accomplishments was the 1997–1998 Chair of the ACGIH; and Dr. Dave K. Verma of McMasters University winner of the 1999 ACGIH Meritorious Service Award.

4.3.9 Information Resources

Canadian Registration Board of Occupational Hygienists (CRBOH)
224 Parkside Court
Port Moody, BC, Canada V3H 4Z8
Tel: 604 878 3040, Fax: 604 949 8601
Internet: www.crboh.ca
Status: Professional accreditation organization

l'Association Québéçoise pour l'Hygiène, la Santé et la Sécurité du Travail
7400, boul. Les Galeries d'Anjou, bureau 410

Ville d'Anjou, Québec, Canada H1M 3M2
Tel: 514 355 3830, Fax: 514 355 4159
Internet: www.aqhsst.qc.ca
Status: Professional association

McGill University
Purvis Hall, 1020 Pine Avenue West
Montreal, Canada H3A 1A2
Tel: 514 398 6258, Fax: 514 398 4503
Internet: www.mcgill.ca/occh
Status: Academic institution

University of Toronto
Occupational & Environmental Health Unit
150 College Street, Suite 140
Toronto, Ontario, Canada M5T 1R4
Tel: 416 978 4353, Fax: 416 978 7262
Internet: www.utoronto.ca/occmed
Status: Academic Institution

Canadian Centre for Occupational Health and Safety (CCOHS)
250 Main Street East, Hamilton, Ontario, Canada
L8N 1H6
Tel: 905 572 4400, Fax: 905 572 4500
Internet: www.ccohs.ca
Status: Agency governed by a tripartite council.

Occupational Hygiene Association of Ontario
6519B Mississauga Road
Mississauga, Canada L5N 1A6
Tel: 905 567 7196, Fax: 905 567 7191
Status: Professional association

McMaster University
1200 Main Street West, HSC 3H58
Hamilton, Ontario, Canada L8N 3Z5
Tel: 905 525 9140, Fax: 905 528 8860
Internet: www-fhs.mcmaster.ca/oehl
Status: Academic institution

l'Association Québéçoise pour l'Hygiène, la Santé et la Sécurité du Travail (AQHSST)
7400, boul. Les Galaries d'Anjou, bureau 410,
Ville d'Anjou, Québec, H1M 3M2, Canada
Tel: 514 355 3840, Fax: 514 355 4159
Internet: www.aqhsst.qc.ca
Status: Professional Association

University of British Columbia
Occupational Hygiene Program
3rd Floor, LPC, 2206 East Mall

Vancouver BC, Canada V6T 1Z3
Tel: 604 822 9595
Internet: www.interchange.ubc.ca/occhyg/occhhome.html
Status: Academic Institution

4.4 China (Excluding Hong Kong)

4.4.1 Demographics and Economics

With a population of nearly 1.23 billion, China is the most populous country in the world and the largest remaining socialist state (7). Following the decline of the "collective" economic policy in the 1970's, the Chinese leadership developed a hybrid system of socialism and free-market capitalism with an element of central control, particularly of financial institutions and key state enterprises termed, "a socialist market economy." This strategy has been effective in that the gross domestic product of China has quadrupled since 1978 (7), with the greatest contribution coming from industry sector (49%) followed by services (31%) and agriculture (20%). The industrial sector consists of iron and steel production, the manufacture of motor vehicles, armaments, industrial and telecommunications equipment and machinery, textiles and apparel, and a wide variety of consumer products such as toys and electronics; as well as the production of coal, petroleum, cement and chemical fertilizers. The economy is supported by a work force of more than 620 million with a general population literacy rate of 82% (7).

The Chinese planning authorities modified the old collective agriculture system to one focused on family responsibility for agriculture and increased authority for local officials and plant managers in industry. A wide variety of small-scale enterprises in services and light manufacturing developed, and opened the economy to increased foreign trade and investment. However, as a result of the economic decentralization, more than 75 million rural workers travel between the villages and the cities subsisting on part-time, low-wage work. There has also been a weakening of population control. Pollution threatens both the population and the economic growth. Economic growth is expected to continue, but at a declining rate (7). The Percent Distribution of the Chinese work force is as follows: Agriculture and forestry, 53%; Industry and commerce, 26%; Construction and mining, 7%; Social services, 4%; and Other, 10%.

China is a classic example of the fine balance between occupational, public health, and economic development. Rapid development has resulted in considerable occupational health and safety problems based on the results of morbidity and mortality studies. According to statistics, occupational lung diseases rank at the top of occupational health problems, accounting for more than two-thirds of all verified occupational diseases. They are followed by chemical poisonings, occupational skin damage, physical agent-related disorders, and occupational cancers (26).

4.4.2 Historical Development

Occupational health problems in China have been documented as far back as the Song (1000 B.C.) and Ming (14th–17th Century) Dynasties. For example, descriptions of mining-related "lung damage" and symptoms consistent with occupational lead exposure as well as preventative measures for asphyxiant intoxication are recorded in early Chinese writings (27).

As a distinct discipline, occupational health started as a preventative medicine in the early 1950s, soon after the founding of the People's Republic of China. In the 1950s–1960s occupational health services consisted of occupational medicine for workers suffering from occupational illnesses and work-related diseases, occupational hygiene for work environmental monitoring and assessment of control engineering, and industrial toxicology assaying for chemical toxicity. By 1979, following the beginning of economic reform and the policy of opening the country, the structures and activities of occupational health were further developed to include identification, evaluation, prevention, and control of occupational hazards at work in the same way as they have been adopted in most of the industrialized countries. Gradually, the concept of occupational health has been expanded to the prevention of traditional occupational diseases and the work-related disorders using occupational hygiene, occupational medicine and industrial toxicology, but also the ergonomic aspects of the work environment. It also includes primary health care for general health problems (27).

4.4.3 Occupational Health and Safety Legislation

The Chinese government began focusing attention on occupational health and safety issues in the 1950's based on Mao Zedong's philosophy "Health and Safety for people at work, and work performed along with health and safety." However, China does not currently have comprehensive national legislation addressing occupational health and safety issues. Limited legislation pertaining to occupational health and safety is developed by the National People's Congress. These legislative actions are further developed by the national government ministries such as the Ministry of Health, which are enforced by "labor bureaus" at the provincial and city level. With rapid national industrialization beginning in the 1970's, a variety of regulations for occupational health and safety have been developed at both the provincial and municipal levels. Examples include (28): the Regulation on Prevention of Silica Dust in Enterprises (1956), the Health Standards for the Design of Industrial Premises (Standards TJ 36-79) (1979), the Regulation of Work Safety for Mining Industries (1982), the Rules of the Management of Verified Occupational Diseases (1984), the Act on Silicosis Prevention (1987), and the Rules on Occupational Health Services for Town- and Village-Owned Enterprises (1987).

Comprehensive legislation is also adopted at the municipal or provincial level without involving the People's Congress such as the "Occupational Diseases Control Act" adopted by the Shanghai Municipal People's Congress in 1996. This legislation defines general requirements for monitoring the work environment, health protection and promotion, and legislative responsibilities designed to control occupational disease. Effective in May 1996, the Act addresses: (*1*) the control of chemical, physical (noise, heat) and biological hazards; (*2*) requirements for personal protective equipment; (*3*) establishment of occupational hygiene programs with the goal of minimizing exposures to airborne toxic chemicals; (*4*) worker "right to know" provisions; (*5*) worker training requirements; (*6*) the right of refusal for unsafe environment; (*7*) right for health care coverage for occupational injuries and illnesses; (*8*) worker responsibilities for rules (implied worker liability); (*9*) prohibition on the transfer of hazardous work processes to enterprises with no protective measures is prohibited; (*10*) all new construction, (and) the importation of technology to be accompanied by an occupational health and safety program and of controls; (*11*) requirements for the disclosure of toxic substances in the registration of new chemicals; (*12*) the segregation of hazardous processes from nonhazardous operations; (*13*) emergency response and alarm requirements; (*14*) reporting of occupational disease; (*15*) implementation of abatement plans based on the reporting of occupational disease; (*16*) disclosure of relevant information to the government; a requirement that only licensed companies may perform workplace monitoring; (*17*) mandatory medical examinations for occupational health; (*18*) requirement that only licensed persons may perform medical examinations; (*19*) a requirement that only licensed hospitals can give long-term care (municipal governments are responsible for licensing); (*20*) the reporting of periodic monitoring results to the government; (*21*) a requirement to reassign workers suffering an occupational illness or injury; (*22*) the requirement that treatment of ill and injured workers is paid by the employer; (*23*) fines for violations established as between Y500 and Y10,000 (Y8 is approximately equal to $1) with the provision of confiscation of income; (*24*) as the establishment of a dispute resolution mechanism; and (*25*) unethical behavior by an inspector is a criminal offense. The Act represents a very significant potential for change in the practice of occupational hygiene in China due to its comprehensiveness and inclusion of provisions for controls and sanctions for failing to protect workers.

One of the most interesting aspects of China's health and safety legislation is the definition of a worker as an individual who is covered by a collective agreement between a labor union and an employer at a public or private enterprise. The consequence of the requirement for representation is far-reaching in that any professional not covered is also not covered under any relating health and safety legislation.

Chapter VI of the 1995 Labor Law of the People's Republic of China provides the most general framework upon which lower level regulations are now based (29). It requires that employers establish a system for occupational safety and health and implement the rules and standards of the State on occupational safety and health. The system must include worker education, prevention of accidents through work process control, and reduce occupational hazards. New processes and facilities must incorporate these requirements. Regular health examinations must be provided for workers engaged in "work with occupational hazards." The 1995 Labor Law also requires all workers to abide by the rules of safe operations established by management and the government. Conversely, workers have the right to refuse unsafe or unhealthy work as well as the right to criticize, report or file charges against those whose actions endanger or disregard life. Article 57 of the Labor Law requires that private and public employers establish a system for reporting injuries, deaths and cases of occupational disease.

Enforcement of regulations through periodic workplace health inspections and monitoring are conducted by the Health and Anti-epidemic Station at the provincial and municipal levels.

4.4.5 Occupational Exposure Limits

Chinese occupational exposure limits were first defined in the 1950s including chemicals, dusts and physical agents. The Chinese system is based on information contained in *Health*

Standards for the Design of Industrial Premises (Standard TJ 36-79) published in 1979. In 1981, the National Committee of Health Standards Setting (NTC-HSS) was established under the Ministry of Health. Within the NTC-HSS is the Subcommittee of Occupational Health Standards which is responsible for developing proposed standards for promulgation by the Ministry of Health. Occupational exposure limits established using this structure are recognized as government standard with the weight of Chinese law (30).

The process involves two general steps: (*1*) the collection of toxicological information, epidemiology research, risk assessment; and the development of a proposed standard; and (*2*) an analysis of the technical and economic feasibility of the proposed standard. The basis of most Chinese occupational exposure limits is the detection of adverse effects rather than the expression of occupational disease or dysfunction. As such, they are generally lower than most occupational exposure standard promulgated worldwide. For example, of the 120 agents with current limits, 50 are lower than the respective ACGIH TLV®. The Chinese occupational exposure limits are expressed as Maximum Allowable Concentrations (MAC) and are defined as the concentration of a chemical substance or compound in air to which nearly all workers could be exposed repeatedly without observed adverse health effects. The MAC values are ceiling limits for which no representative compliance air sampling should exceed. However, the majority of representative air sampling in China is conducted using area or static sampling techniques (30). The government's Health and Anti-epidemic stations charged with verifying compliance with the MAC values are not currently equipped to perform personal sampling. There has been discussion among the NTC-HSS members regarding the precision of this type of sampling. It has been proposed that a "bi-track" system be developed for occupational exposure limits which would require time-weighted average sampling for chemicals with cumulative or chronic effects, and grab sampling for substances with acute effects (30).

Once promulgated by the Ministry of Health, the occupational exposure limits must be adopted by local authorities. In practice, this is not accomplished with any degree of consistency throughout the country. The result is a patchwork of different standards in different regions of the country with the highest degree of compliance occurring in regions with strong academic programs in occupational medicine. Compliance assurance for occupational exposure limits is the responsibility of the Bureau of Technical Inspection. There are about 225 occupational exposure standards promulgated by the Ministry of Health.

Though many of the MACs are lower than their TLV counterparts, it is not clear if these standards are technologically or economically feasible. Enforcement poses another potential problem. According to the 1996 Occupational Diseases Control Act, the fine for failure to utilize personal protective equipment or develop an occupational hygiene plan to minimize airborne exposures (Item 8.1 of the Act) ranges from 2,000 and 20,000 yuan (U.S. $250–$2,500) (31).

Stimulated by the relevant regulations, the compliance with national occupational exposure limits increased in State-owned enterprises from 51.4%–63.0% in 1986–1989 to 65.6%–68.3% in 1991–1995. However, the coverage rates of environmental monitoring and health surveillance were not high enough and often varied widely (27).

4.4.6 Education and Training

Occupational health curriculum has been developed to address the needs of undergraduate and graduate students and on-the-job refresher training. There are 142 medical school programs in China and 34 academic institutions with schools of public health offering courses in occupational health and industrial toxicology. About 20 medical universities and the Institute of Occupational Medicine, Chinese Academy of Preventive Medicine in Beijing offer specialized training leading to a master's and/or doctorate degree in occupational health, occupational medicine or industrial toxicology. Among the most prominent universities for occupational health training is the Shanghai Medical University, which houses a WHO Collaborating Center for Occupational Health. Research at this institution is diverse but emphasizes applications for small-scale enterprises and township and village owned enterprises (TVEs) which represent a growing proportion of the Chinese economy.

Professional development opportunities for those individuals who perform occupational hygiene activities frequently are offered in the form of short courses of one to 14 day duration. Topics have included occupational epidemiology, occupational health services, chemical safety and regulatory issues, and industrial toxicology, among others. There are several universities with programs in environmental and occupational air sampling as part of schools of public health. All graduates of a school of public health receive a bachelor of medicine degree. Despite the very wide range of occupational health educational options available in the country, there are no academic programs devoted to the development of occupational hygienists and it is the physicians and engineers that conduct most occupational hygiene work.

To meet broader needs, televised courses on occupational health, broadcast once a week on the Chinese TV, have been offered by the Ministry of Health since the 1980s. In addition, a second WHO Collaborating Center for Occupational Health has been designated at the Institute of Occupational Medicine, Chinese Academy of Preventive Medicine in Beijing. WHO centers have been developed throughout the world to provide education and training, to promote scientific research, and to provide information in community and occupational health (27).

4.4.7 Certification and Accreditation

China does not have a certification system for occupational hygiene.

4.4.8 Professional Organizations

The Chinese Occupational Health Association in Beijing was developed under the auspices of the Chinese Medical Association's Division of Preventive Medicine. In turn, the Chinese Medical Association is supported by the Ministry of Health. There are no organizations dedicated specifically to occupational hygiene.

4.4.9 Status of the Profession

In China, the practice of occupational hygiene is largely the concern of safety engineers and physicians and is primarily an adjunct activity to these safety and public health disciplines. The majority of these two groups of professionals are civil servants while some work in the research institutes and others for provincial governments. Training and education in occupational hygiene, as known in the west, is relatively recent. China has a number of occupational hygiene-related components including occupational health regulations, exposure standards, research institutions, personnel conducting workplace exposure evaluations, to name just a few, but lacks a specific occupational hygiene profession as is known in the U.S. and Western Europe.

There is no accurate estimate of the number of professional occupational hygienists in China nor of the number of allied professionals who conduct occupational hygiene-related activities; however, according to Chinese law, there should be one safety or health professional for every 100 workers. If true, this would suggest there are more than 600,000 health and safety personnel in China. The distribution of those who report working in occupational hygiene functions is primarily in government, industry, academics, and consulting.

The primary emphasis in occupational health in China is the occupational health service. There is a nation-wide network consisting of seven regional centers and more than 200 institutes of occupational health at the municipal and provincial level. There are 1789 Health and Anti-epidemic Stations at the county level. These institutes reportedly provide medical surveillance services, conduct workplace monitoring and consult with industry regarding occupational health and safety management. Services may also include emergency and routine medical care. While the government-sponsored occupational health services to small private and public enterprises, similar services have been established in most large industrial businesses.

The occupational health services may consult on safety issues, but the administrative system of occupational accident prevention and control (safety) in China is independent from the system responsible for occupational health.

Chinese health center workers may decide if a workplace is out of compliance, but they can not make or recommend engineering changes that for American industrial hygienists would be considered part of a routine evaluation (31). Management is obligated to take responsibility by asking for suggestions or seeking other help to make necessary engineering changes that can improve working conditions and the quality of the workers' lives.

There is a mature system of health surveillance information collection and dissemination in China including a central database on occupational diseases and injuries managed by the Ministry of Health and the Ministry of Labor. There are four scientific journals focusing on occupational health and medicine that are published in China including the *Chinese Journal of Preventive Medicine, Chinese Journal of Industrial Hygiene and Occupational Medicine*, the *China Journal of Industrial Medicine*, and *China Safety Sciences and Technology.*

4.4.10 Information Resources

All-China Federation of Trade Unions
Occupational Safety and Health Information and Training Center
10 Fuxingmemwai Street, Beijing, P.R. China
Tel: 86 10 6859 3629; Fax: 86 10 6859 3616
Status: Labor Union Institute

Shanghai Medial University
College of Preventative Medicine
Department of Occupational Health

138 Yi Xue Yuan Road, Shanghai, 200032,
P.R. China
Tel: 86 021 64041900, Fax: 86 021 64178160
Internet: www.shmu.edu.cn
Status: Academic Institution

4.5 Egypt

4.5.1 Demographics and Economics

In the 1990s, Egypt has developed a more decentralized, market-oriented economy in the face of low productivity compounded by the adverse social impacts of high population growth and the accompanying urban overcrowding, combined with high inflation. As the population reaches 66 million, natural resources include petroleum, natural gas, iron ore, phosphates, manganese, limestone, gypsum, talc, asbestos, lead, and zinc are being capitalized (7). The work force of 17.4 million is employed in a variety of industries such as textiles, food processing, chemicals, petroleum, construction, cement and metals. However, the primary source of employment is agriculture that accounts for as much as 40% of the total workforcc. Pcrcentages of distribution of the Egyptian work force is as follows: Agriculture, 40%; Services including government, 38%; and Industry 22%.

While an accurate count of the number of full-time professional occupational hygienists is not available, the estimate of no more than 25 suggests an occupational hygienist to work force ratio of approximately 1:696,000.

4.5.2 Historical Development

Egypt has made great efforts and achieved considerable success in modernizing its agriculture and its industry in the last 50 years. Ancient tools and advanced technologies are coming together in a society known for the splendor of its ancient culture and for its rich documentation reaching back thousands of years. Temples and tombs in Upper Egypt display drawings showing farmers using plows similar to those still used in some parts of the world. In addition, those murals provide graphic proof that over 2000 years ago, Egyptians processed flax, tanned hides, made pottery, and mined, refined, and cast metals.

Rapid modernization of Egyptian industry began after World War I. It has brought about a higher standard of living, better education, and improved health care for certain segments of the population. Such improvements, however, have also exacted a price. According to Noweir (32), recipient of the 1979 Yant Memorial Award,

> The process of industrialization in most developing countries has not been accompanied by a parallel process in establishing occupational health services to counteract the hazards associated with industrialization, which are usually manifested in the form of work-related diseases. In many instances, both management and workers are not even aware of the risks in their industries.

In that same article Professor Noweir also notes that "the rapid introduction of complex work methods in developing countries has been associated with considerably higher rates of industrial accidents and occupational diseases than in industrial countries." In a presentation made at the International Symposium on the Biomedical Impact of Technology Transfer (33), Professor Noweir identified several contributing factors such as the hot climate, malnutrition, and unsanitary conditions in crowded urban areas.

4.5.3 Occupational Health and Safety Legislation

The Minister of Manpower first set minimum health standards for workers in industry in 1967 by means of Decree No. 48. This decree was based on Law No. 91 promulgated in 1959. This decree was superseded by Decree No. 55 in 1983, issued on the basis of the new Labor Law of 1981. This latter decree set standards for heat exposure, noise, lighting, radiation, and ventilation in an effort to reduce or possibly even prevent occupational hazards (34). Today, medium-sized companies of 50 to 499 employees and large-sized companies of more than 500 employees must utilize the services of a "safety officer." There is no explicit requirement for the services of an occupational hygienist based on either worker populations or specific workplace hazards.

4.5.4 Occupational Exposure Limits

In 1987, Cook (15) listed 40 occupational exposure limits for Egypt in his Table "Occupational Exposure Limits of Countries other than USA and Canada." New exposure limits are developed through the Division of Occupational Health in the Ministry of Health and the Industry, and representatives of the Department of Labor in the Ministry of Manpower enforce applicable health and safety standards and permissible exposure limits.

4.5.5 Education and Training

In order to train occupational hygienists locally, Egypt sent a number of prospective scientific educators to the United States to pursue doctoral studies in the early 1960s. The University of Alexandria in Egypt established in 1955 the Department of Occupational Health as part of the High Institute of Public Health. Today the department offers courses leading to a Diploma of Science (D.Sc.), a Master of Science (M.Sc.), and a Doctor of Public Health Sciences (Dr.P.H.Sc.) (34).

Developing countries face a number of problems even when they are able to modernize industry as rapidly as local resources permit. Thus, even though Egypt has been able to train a number of capable occupational hygiene professionals, it has lost many to prosperous Middle Eastern countries or to the West because of higher salaries and/or better research opportunities. Despite considerable difficulties, Egypt has been able to lay the foundation for an occupational hygiene profession.

4.5.6 Certification and Registration

There is currently no formal certification or accreditation system for occupational hygiene in Egypt.

4.5.7 Status of the Profession

The number of professional occupational hygienist employed in Egypt is estimated to be between 10 and 25. Many of these are employed in academic establishments and particularly the University of Alexandria, a few with regulatory agencies such as the Division of Occupational Health, and the remainder in industry. However, due to a number of factors including more diverse economic opportunities in other northern African and neighboring OPEC countries, the actual number of Egyptian occupational hygienists is closer to 100.

4.5.8 *Information Resources*

The University of Alexandria
The High Institute of Public Health
Department of Occupational Health
Alexandria, Egypt
Status: Academic Institution

National Institute of Occupational Safety and Health
P.O. Box 2208, El-Horreya
Heliopolis, Cairo, Egypt
Tel: 20 2245 2630, Fax: 20 2242 4355
Status: National Research Institute

4.6 Finland

4.6.1 *Demographics and Economics*

Finland is a highly industrialized country of 5.1 million with one quarter of its landmass located above the Arctic Circle. Although the country has a high per capita output, it lacks substantial natural resources and must import much of its energy and raw materials for manufactured goods. Manufacturing is the central component of the economy with forest products (pulp and paper), steel and copper, chemicals, shipbuilding, foodstuffs, textiles, and clothing representing the primary industry sectors (7). The Finnish work force contains approximately 2.5 million and is a member of the EU. Percentage distribution of the Finnish work force is as follows: Public services, 32%; Industry, 22%; Commerce, 14%; Finance, insurance and business services, 10%; Agriculture and forestry, 8%; Transportation and communications, 8%; Construction, 6%.

Based on the membership of the Finnish Occupational Hygiene Society, there are approximately 175 professional occupational hygienists in Finland (35). The members are employed across the public and private spectrum, and based on the size of the Finnish workforce, constitute an occupational hygiene to work force ratio of 1:14,300. Percentage of Employment Distribution of FOHS Members is as follows: Institute of Occupational Health, 39%; Industry, 30%; Regulatory agencies, 8%; and Manufacturing, consultants, etc. 6%; Other research institutions, 6%; Academic institutions, 6%; and Insurance companies, 3%.

4.6.2 *Historical Development*

The first instances of occupational health-related activity in Finland date back to 1754 when Carl Fredrik Zandt was appointed medical officer for the Suomenlinna Fortress outside Helsinki with the task of providing medical services for the military personnel and for the building workers "for as long as the ongoing building of the fortress continues." A second physician (barber-surgeon) Brosell was engaged by Fiskars in 1755 to provide health care for the 35 person work force at their blast furnace (35). During the 1800s "large scale" industries emerged in Finland and many of the big industrial enterprises had their own hospitals to care for the health of the workers and their families. By the 1940s there were already about 20 full-time occupational health physicians in Finland.

In May 1941, Leo Noro M.D., Ph.D., publicly defended his doctoral thesis on the morbidity associated with explosives poisoning in the munitions manufacturing industry and subsequently wrote a guidebook on the "prevention of poisoning in the munitions industry." During his assignment as a World War II surgeon in the Finnish tank battalion he began to investigate carbon monoxide poisoning caused by exhaust-gas exposure and from the wood-gas generators used on cars (35). In 1945 a department of occupational disease was established in conjunction with the first Department of Internal Medicine at the General Hospital in Helsinki. This was the beginning of The Institute of Occupational Health, which was founded as a private foundation in 1951. Dr. Noro was the Institute's director from the inception until 1970. The Institute was nationalized in 1978. To recognize his contributions in the field of occupational health, the AIHA honored him with the 1968 AIHA Yant Memorial Award. Today, Leo Noro is recognized as one of the pioneers of occupational hygiene in Finland.

Since 1926 information and official statistics have been compiled on occupational diseases diagnosed in Finland. The Finnish Registry of Occupational Diseases was established in 1964 by the Institute of Occupational Health. Since 1979 the Institute has also maintained a registry of employees exposed to carcinogenic substances.

4.6.3 Occupational Health and Safety Legislation

Finland is a member of the EU and much of its contemporary occupational safety legislation is based on EU community directives. The Occupational Safety Act (OSA, 299/58), enacted by the Finnish parliament in 1958 is the cornerstone of workplace protection in Finland. In the latest amendments, community directives have been transposed into the Act (35). The OSA is a general document and contains no specific regulations but has been developed through amendments passed under the Act. The Ministry of Social Affairs and Health is responsible for implementing the Labor Safety Act throughout the country. The country is divided into 11 districts with each district having its own occupational health and safety authorities that supervise the implementation of regulatory requirements within the Act.

Under the OSA, the employer is ultimately responsible for protecting employees against injury and health hazards. Employees, in turn, are required to observe all safety and health regulations such as those dealing with chemical agents, illumination, ventilation, noise, vibration, radiation, fire and explosion and major hazards.

The Occupational Health Care Act (OHCA, 743/78) mandates every employer to organize occupational health services at the workplace. In addition to preventative health care, surveys of working conditions are an important responsibility of occupational health services. External experts, like occupational hygienists, are also used where special expertise is needed (35).

Since 1974, the Act on the Supervision of Labor Protection and Appeal in Labor Protection Matters (131/73) has mandated cooperation between employers and employees on matters relating to health and safety. For workplaces with more than 10 employees, the employees elect one part-time safety representative. If the number of employees exceeds 20, management and labor elect a health and safety committee with equal representation. Where the work force exceeds 500 employees, safety representatives work full-time (35). Occupational health personnel, mainly occupational physicians and nurses, can attend committee meetings as advisors. The employer appoints a health and safety manager (safety officer), who is responsible for coordination of labor protection issues in the workplace.

Occupational safety delegates in Finland are entitled to stop production for safety reasons. If the employer does not accept the interruption he must submit the case to the district occupational safety authorities. Also, a single worker has right to refuse work if there is evidence of an imminent and serious risk for injury or exposure leading to illness.

4.6.4 Occupational Exposure Limits

The National Board of Labor Protection is responsible for developing and publishing occupational exposure limits for airborne toxic substances in Finland. The board is the central administrative agency for occupational safety and health under the Ministry of Social Affairs and Health. Finnish occupational exposure limits are called "Haitalliseksi Tunnettu Pitoisuus" (HTP), with an approximate English translation of "concentrations known to be hazardous" (35). By definition, long-term exposures to toxic substances at levels above Finnish occupational exposure limits may cause deleterious health effects. These limits are not meant to protect hypersensitive individuals but to protect healthy workers adequately as long as the exposure limits are not exceeded. The employer must take these limits into consideration when assessing the workers exposure to chemicals at work.

The National Board of Labor Protection is also responsible for developing biological exposure limits for airborne toxic substances of which are currently four addressing ethylbenzene, phenol, carbon disulfide, and toluene.

4.6.5 Education and Training

In Finland most occupational hygienists have a masters of science in chemistry, physics, biology or engineering. An increasing number of occupational hygienists have doctoral degrees in their field. The University of Kuopio offers full-time basic and advanced education (Ph.Lic and Ph.D.) in environmental and occupational hygiene. Professor Pentti Kalliokoski, recipient of the 1999 AIHA Yant award, heads the program.

The Institute of Occupational Health also provides training in occupational hygiene for those with different science education. The Institute in Helsinki and the regional institutes also provide occupational health and safety training both for their own personnel and for others, i.e., occupational health physicians and nurses, occupational hygienists, occupational psychologists, occupational physiotherapists, occupational safety officers and delegates, design engineers, etc.

The National Institute of Occupational Health has a certificate system with a written examination for individuals who fulfill certain prerequisites. Candidates must have a master's degrees in chemistry, physics, industrial hygiene or engineering and they must have worked as industrial hygienists full time for a minimum of three years. They also must have attended occupational hygiene courses offered by the Institute of Occupational Health. Upon successful completion of the written examination they are given an unofficial certificate. Thus far about 40 occupational hygienists have taken the examination.

4.6.6 Certification and Registration

There is no official certification mechanism for occupational hygiene in Finland. The Finnish Occupational Hygiene Society (FOHS) has begun work to develop a certification system, but additional development on the structure and administration is required before it can be finalized. However, individuals with certain qualifications can sit for a written examination offered by the Institute of Occupational Health. See Section 4.6.5.

4.6.7 Professional Organizations

The Finnish Occupational Hygiene Society (FOHS) was founded in 1975 as a "learned society" with the purpose of furthering the knowledge of occupational hygiene. The FOHS represents Finland as a member of the IOHA and administers a code of ethics. During the early 1990s there was a drop in the number of members due to the deep economic recession in Finland, but during the last several years the number of members has increased and has reached about 180 members. Some of the members are also members of ACGIH, AIHA, and BOHS as well as other national organizations in occupational health and safety.

Among other activities, the FOHS arranges an annual two-day conference and is one of the collaborating parties in the yearly two-day "Occupational Health and Safety" conference jointly arranged with the Finnish Association of Industrial Medicine (1000 + members), the Finnish Association of Occupational Health Nurses (1750 members), the Finnish Association of Occupational Physiotherapists (600 members), and the Safety Officers of Finland (450 members).

The FOHS also has regional activities, with regional members meeting (typically twice a year for about half a day) to exchange information on occupational hygiene matters of interest. The FOHS also provides input on government regulations as appropriate.

While not a membership organization, the Institute of Occupational Health acts as an expert body in matters of occupational health and safety, but it does not have any enforcement authority. Its activities are focused on the different fields of occupational hygiene, medicine and safety. Through broad and multidisciplinary research, training and international co-operation, it is constantly broadening and deepening its national and international role in its field. In addition to the central institute in Helsinki, it now has five regional institutes (R.I.): the Uusimaa R.I. (in Helsinki) specializing in indoor air quality problems and small and medium size enterprises, the Turku R.I. specializing in the shipping industry and welding, the Tampere R.I. specializing in training of safety personnel, field-clinic activities and occupational safety and health in the building industry, the Lappeenranta R.I. specializing in sawmill-industry, pulp and paper, the Kuopio R.I. specializing in Agriculture and Forestry, and the Oulu R.I. specializing in cold climate hazards and electronics industry. In addition to pursuing research in their special fields, they also provide extensive and comprehensive occupational hygiene services to the work places in their region.

4.6.8 Status of the Profession

Occupational hygiene has established itself as a profession in Finland although as of 1998 there is no legal definition of occupational hygiene or occupational hygienist in Finland. To address this matter there is an official task force attempting to define the scope of education and experience required for occupational hygienists working the occupational health services in Finland.

Many large companies with more than 1000 employees have their own full- or part-time occupational hygienist as a member of their occupational health service. Most often these comprehensive occupational health services are provided through in-house staff. As a member of the occupational health service, the occupational hygienist works in close cooperation with the other occupational health professionals (occupational doctors, nurses and physiotherapists) as well as with the research and design, facilities and line-organizations. Occupational hygienists may also work in the environmental or quality departments.

Smaller companies may provide some of the occupational health/occupational hygiene services as in-house services and/or use external consultants. The Institute of Occupational Health is one of the major providers of occupational hygiene consultant services in the country, but there are also private consultants providing services on a commercial basis. As stated by the Institute (36):

> Finland has ratified the ILO Convention no. 161 on occupational health services which encourages member countries to develop further their systems both at the national and particularly at the level of undertakings. On the other hand, in 1986 the Finnish Government adopted a national level programme of "Health for All by the Year 2000" in which the development of occupational health services plays an important role. These international challenges will certainly guide the development of the Finnish health service systems and help in the achievement of the ambitious goal: Health for All by the Year 2000.

In addition to these resources there are close to 100,000 persons (including elected safety ombudsmen) dealing full- or part-time with occupational safety and health issues at work. Manpower surveys conducted in 1972 and 1981 determined that there were an adequate number of occupational hygienists in the country to address the then-current need (35).

The peer-reviewed occupational hygiene periodical, the *Scandinavian Journal of Work, Environment and Health* is published in Finland.

4.6.9 Information Resources

The Finnish Occupational Hygiene Society (FOHS)
Helsinki, FIN-00250, Finland
Tel: 358 9 474 7345, Fax: 358 9 474 7548
Status: Professional Association

Ministry of Labor
Occupational Health and Safety Department
Tyoministerio, Tyosuojeluosasto

Etelaesplanadi 4, P.O. Box 524
Helsinki 00101, Finland
Status: Government Agency

The Finnish Institute of Occ. Health (FIOH)
Tpoeliuksenkatu 41a1, FIN-00250, Helsinki, Finland
Tel: 358 9 47471, Fax: 358 9 241 4634
Internet: www.occuphealth
Status: Government Research Institute

Association of European Toxicologists and Societies if Toxicology (Eurotox)
Secretary-General EUROTOX
% University of Turku
Department of Clinical Physiology
FIN-20520, Turko, Finland
Tel: 358 2 261 2664, Fax: 358 2 261 1666
Status: Association of Toxicologists in Europe.

4.7 Germany

4.7.1 Demographics and Economics

Germany has a population of some 82 million and one of the most powerful economics in the world (7). The industrial sector encompasses the production of iron, steel, coal, cement, chemicals, machinery, automobiles and trucks, ships, fabricated metals, machine tools, a wide variety of commercial and consumer electronics, food and beverages, textiles and petroleum products. The German work force involves 38 million individuals working in more than 2.5 million workplaces, primarily in the services sector (37).

To prepare for the start of the European Monetary Union, the government in Bonn made significant efforts in 1996–1997 to control the fiscal deficit which was made more difficult by rising unemployment, a decrease in the tax base, and transfer of $100 billion a year to eastern Germany to refurbish that region (7). In recent years, German manufacturers have begun to move production and manufacturing facilities outside Germany to be closer to markets and to avoid Germany's high tax rates, rigid labor structures and high costs, and extensive regulations (7). This has had the effect of decreasing foreign investment in Germany. Percentage of distribution of the German work force is as follows: Services, 56%; Industry, 41%; and Agriculture 3%.

There is no accurate estimate of the number of professional occupational hygienists working in Germany (38). Similarly, the distribution of occupational hygienists in industry, government, academia and consulting is not known.

4.7.2 Occupational Health and Safety Legislation

It is the right of all German citizens to enjoy life and freedom from physical injury as mandated in Article 2 of the German Basic Law (37). To exercise this right as it applies to the workplace, Germany has three levels of occupational health and safety protection: the Federal government, the independent laws and regulations written by mutual indemnity institutions, and collaboration of worker and employers in the workplace.

The 1996 Occupational Health and Safety Law was established to integrate EC Directive 89/391/EEC ("On the introduction of measures to encourage improvements in the safety and health of workers at work.") into German legislation and is the primary legislative structure for occupational health and safety along with the Die gewerblichen Berufsgenossenschaften (BG).

The BG is an umbrella organization of the Statutory Accident Prevention and Insurance Institutions in Industry legislation representing the 35 insurance institutions mandated by legislation and empowered to write and enforce regulations. It is one component of the overall German social insurance system. Companies pay into the BG system based on a premium of 1.4 percent of all wages and salaries. In 1995, 37 million workers were insured under one or more of the Employers' Liability Insurance organizations (39).

The BG plays an important role in the implementation of German and EC legislation, including the Occupational Health and Safety Law, by promulgating regulations and guidance documents addressing occupational health and safety management which are legally binding. These include defining the safety and health requirements to be applied to equipment and operating procedures, identify codes of practice, and occupational health and safety management.

Under the Occupational Health and Safety Law and BGs, employers are responsible for establishing the systems and administrative procedures to ensure the safety and health of their workers, compliance with existing regulations, and for providing adequate emergency response capabilities. To this end, management can request assistance from the BGs consultants or maintain staff health and safety professionals. Companies with more than 20 employees must have designated safety delegates and develop an occupational health and safety committee with employees and management representatives to address health and safety matters. The employer also must appoint a safety professional to support implementation of occupational health and safety regulations. The professional can be employed by the company or by a consultant.

In accordance with the German Occupational Safety Act, a company physician must also be retained to conduct medical examinations and consult with the employer on matters relating to occupational health (39). Some BGs also offer the services of a company physician in the framework of their overall occupational medical service scheme. However, occupational hygienists are not a statutory requirement in Germany. This element of exclusion represents the primary institutional barrier to the development and acceptance of the profession as most occupational hygiene activities are most commonly conducted by safety professionals, engineers, physicians and chemists.

Under the BG, the Executive Division is responsible for coordinating all prevention activities carried out by the federation of statutory accident insurance institutions for the industrial sector. The division includes the BG Central office for health and safety at work (BGZ); the BG Institute for Occupational Safety (BIA); the BG Institute for Occupational Medicine (BGFA); the Commission for Occupational Health and Safety Standardization; the Academy Dresden Cooperative Program for Work and Health (KOPAG); and the Technical Inspection Service (TAD). The principal functions of the TAD are to inform and advise business managers on health and safety issues, and to monitor performance of occupational health and safety measures. The TAD employs "health and safety inspectors" for this purpose, as well as other experts including specialists in occupational medicine and education and engineering.

The Berufsgenossenschaftliches Institut für Arbeitssichicherheit (BIA) is the central research institute for occupational health and safety under the BG. It is the main department within the Executive Division Prevention of the Central Federation of the German Berufsgenossenschaften. BIA's activities focus primarily on research involving chemical and physical hazards, the BG-prescribed system for hazard assessment (BGMG), exposure databasing (MEGA, GESTIS and GISBAU), analyses of air and bulk samples, toxicology, biological hazards, noise, vibration, radiation, ergonomics, epidemiology, safety engineering, control technology, computer technology, material science, structural engineering, transportation and traffic, and personal protective equipment, among others.

The Hazardous Substances Ordinance addresses many issues relating to occupational health in the workplace including those governing the distribution and handling of chemicals. Its scope includes regulation of (*1*) conditions for packaging and marking of hazardous substances (*2*) handling of hazardous substances (*3*) hierarchical controls of chemical exposures (*4*) user instructions, (*5*) storage; and (*6*) preventive health monitoring required. It requires that no hazardous gases, vapors or aerosols be released in the workplace and that employees are protected from dermal exposure involving hazardous substances. Companies are required to evaluate the risks in their work environments. Such risk assessments provide them with scope for action other than mere adherence to the regulations or reaction to accidents after the event. They can eliminate specific risks and take suitable safety measures in advance and are supported in this activity by the BG health and safety inspectors and by guidance documents developed by the BGs.

Health and safety inspectors are actually supervisory personnel specified in Part VII of the German Social Code. They are generally safety engineers but may also be chemists or physicians. Their other functions include: (*1*) consulting with employers on all issues relating to occupational health and safety; (*2*) monitoring the occupational health and safety performance measures in the companies; (*3*) participating in the development of health and safety regulations at the national and the European level; (*4*) providing training within the scope of the BG's training activities and at various universities; (*5*) working with equipment manufacturers in the safety and ergonomic design of equipment; and (*6*) participating in safety testing of equipment. Inspectors are authorized to visit businesses without advanced notice, to request necessary information or data, and, should they recognize an imminent hazard, issue citations for violation of regulations. Violations can carry fines of up to DM 20,000 (Part VII of the German Social Code—§ 209 Clause 1 No. 1 SGB VII).

The Federal Institute of Occupational Safety and Health (FIOSH, BAuA) was founded in 1996 through the merger of the Federal Institute for Occupational Safety and Health (Bundesanstalt für Arbeitsshutz, BAU) and the Federal Institute for Occupational Medicine (Bundesanstalt für Arbeitsmedizin, BAfAM). It is responsible for developing occupational health systems, conducting medical surveillance and epidemiological studies in industry, performing toxicological studies of chemicals and developing guidelines relating to occupational health. It is also "unification unit" under the German Chemicals Act with responsibility for implementation of regulations pertaining to human and environmental protection from hazardous substances.

The German Chemicals Act requires that a new chemical substance may only be placed on the market after prior notification and approval. Manufacturers or importers must submit data on the physical–chemical, and toxicological properties of the substance to be introduced. After assessment of the information by the authorities, the data, together with a risk assessment, are exchanged between the EC Member States through the European Chemicals Bureau (ECB). The Notification Unit subsequently informs the German federal states about the data on the substances and the results of the assessment.

A risk assessment posed by any substance (from production through to disposal) is performed on the basis of information on the properties and uses of the substance. Recognized dangers are to be avoided by means of targeted measures—classification, labeling, occupational exposure limits, and restrictions. The notification unit receives and manages the relevant chemical data. The notification unit is also responsible for the notification procedure for new substances.

4.7.3 Occupational Exposure Limits

The German occupational exposure limits are developed by a Commission of the German Research Society [Deutsche Forschungs-gemeinschaft (DFG)]. The principal exposure limits are the "maximum allowable concentration" (MAK) in the workplace which are defined as the maximum permissible concentration of a chemical compound present in air within a working area which, according to current knowledge, generally does not impair the health of the employee nor result in undue annoyance. Under these conditions, exposure can be repeated up to eight hours per day, based on four working days per week. MAK's are recommendations.

Technical reference concentration (TRK) represents the concentration of a carcinogenic substance (in gas, vapor, or aerosol form) at which specified control measures must be implemented. Lastly, biological tolerance values (BAT) describe the concentration of a substance in the body at which the health of the employee is generally not affected.

The Federal Ministry of Labor and Social Affairs is the regulatory agency responsible for determining compliance with the MAKs which also be enforced through BG inspectors working at the "Laender" or state level.

4.7.4 Education and Training

Some German universities and technical colleges offer course work associated with occupational hygiene; however, there are currently no undergraduate or graduate degrees in the country. German professionals who hold a degree in occupational hygiene have acquired their degrees outside Germany. Most professional development opportunities are offered through the insurance companies, federal research institutes and large companies with occupational hygiene staffs.

4.7.5 Certification and Accreditation

While there is no current certification process for occupational hygiene in Germany, the German Occupational Hygiene Society is developing a system based on WHO criteria with qualifications similar to those of the ABIH. The Society currently has no formal code of ethics.

4.7.6 Professional Organizations

The primary professional organization concerned with occupational hygiene is the German Occupational Hygiene Society (Deutsche Gesellscaft für Arbeithygiene DGAH). A member organization of IOHA, there are about 100 individuals members in the society representing the core of professional occupational hygienists in the country and interested physicians, chemists and engineers.

The Federal Association for Occupational Safety and Health, or BASI, is a federal association of 50 occupational safety and health institutions in Germany including government departments, social agencies, BGs, and professional occupational safety and health associations. The BASI supports cooperation and exchange of information among its members through the acquisition and sharing of knowledge and new developments in occupational safety and health management. It also manages the A + A Congress of Occupational Safety and Health, the largest event of its kind held annually in Germany.

4.7.7 Status of the Profession

The profession in Germany is anchored by a small group of practitioners, whose precise number is unknown (38), based on a social insurance and regulatory structure that integrates the activities traditionally associated with occupational hygiene, but does not formally recognize the profession. Like many European countries, Germany emphasizes the occupational health service which integrates occupational medicine with physicians and nurses, safety engineers, and other allied professionals as a the model for the delivery of worker health resources. Recent legislative advances including the Occupational Health and Safety Law of 1996 which details the comprehensive EC Directive 89/391/EEC specifying the inclusion of occupational hygienists as a standard component of occupational health services. Ironically, this is one of the aspects of the 89/391/EEC that is not reflected in German law.

Although German occupational hygienists strive for recognition at home, they wield influence in Europe through participation in various agency committees, associations, and consensus organizations.

4.7.8 Information Resources

Federal Association for Occupational Safety and Health (Basi)
Alte Heerstr. 111, D 53754 Sankt Augustin, Germany
Tel: 49 2241 231 1162, Fax: 49 2241 231 1391
Internet: www.basi.de
Status: Government Association

Institute of Hygiene and Environmental Medicine
Rheinisch-Westfälische Technische Hochschule, Aachen, Pauwelsstrasse 30, Klinikum, D-52057, Germany
Tel: 49 241 808 8385, Fax: 49 241 888 8477
Status: Government Research Institute

German Occupational Hygiene Society (DGAH)
Am Waldrand 42, D-23627 Gross Groneau, Germany

Tel: 49 4509 1655, Fax: 49 4509 2295
Status: Professional Association

Central Federation of the German Berufsgenossenschaften (BIA)
Office for Safety and Health
D-53754, Sankt Augustin, Germany
Tel: 49 2241 23101, Fax: 49 2241 231 1333
Status: Organization of Regulatory and Insurance Institutions

Deutsche Forschungsgemeinschaft (DFG)
Committee of the German Research Society
Kennedyallee 40, D-5300 Bonn-Bad Godesberg, Germany
Status: Regulatory Development Organization

Academy of Occupational Medicine and Health Protection
Lorenzweg 51 D-12099, Berlin, Germany
Tel: 49 30 757953-11
Status: Professional Association

4.8 Hong Kong

4.8.1 Demographics and Economics

Hong Kong became a "special administrative region" of China on July 1, 1997 following an agreement between the United Kingdom and China in December 1984. The terms of the agreement call for China to provide Hong Kong with a high degree of autonomy in all matters except defense and foreign relations. Autonomy is an important factor to Hong Kong's economic viability and to the continuing development of the occupational hygiene profession in the region.

Hong Kong has a dynamic free market economy based almost exclusively on international trade. With very limited natural resources, Hong Kong must import much of the raw material and energy needed to sustain its economy which consists of textiles, clothing, electronics, plastics, toys, watches, clocks and other consumer goods manufacturing as well as tourism and financial services. The Hong Kong labor force consists of just under 3.2 million workers (7), including nonresidents to supplement the short labor market. Percentage of distribution of the Hong Kong workforce is as follows: Wholesale/retail trade, restaurants and hotels, 32%; Other, 26%; Finance, insurance, and real estate, 13%; Social services, 10%; Manufacturing, 10%; Transportation and communications, 6%; and Construction, 3%.

There are about 40 occupational hygienists and 20 technicians doing occupational hygiene work in Hong Kong representing an occupational hygienist to work force ratio of 1:53,000 (40). Hong Kong maintains it own administrative system and functions independent of those in China including those relating to occupational health and hygiene. Employment Distribution of Hong Kong Occupational Hygienists is as follows: Industry, 20%; Government, 40%; Academia, 25%; and Consulting, 15%.

4.8.2 Historical Development

In Hong Kong, the practice of occupational hygiene began in the mid-1950s with the development of the Occupational Health and Safety Branch of the Labor Department.

Almost all of the early work in occupational health in Hong Kong was a reflection of the British system of government and social influence for this colony state at the time. It was upgraded to a profession in the mid-1970s with an increase in number of qualified professionals, most of who were educated and trained in Great Britain or the U.S. The profession was known in the government as advisors to the industry and government authorities. At the time, practitioners were referred to as industrial hygienists because they primarily addressed health issues in industrial settings. In about 1983, the profession was re-named occupational hygiene to reflect the gradual extension of the roles of occupational hygienists into non-industrial sectors of the Hong Kong economy (40). This coincided with the establishment of organizations devoted to occupational health and local universities providing educational opportunities in occupational hygiene at about the same time.

4.8.3 Occupational Health and Safety Legislation

The Occupational Safety and Health Ordinance was enacted in 1997 and provides basic protection of all workers in both the private and public sectors. This law addresses a wide range of health, safety, and welfare issues such as housekeeping, control measures including ventilation and other means of engineering controls, lighting, prevention of injuries, first aid, fire safety, manual handling, and sanitary facilities, among others. However, while the Ordinance defines only the minimum requirement for all workplaces, it provides a framework of general principles that future specific regulations can be built upon ("subsidiary regulations").

The enactment of the Ordinance and its accompanying regulations in 1997 and 1998 has brought the profession into a new era in Hong Kong. The new regulations include many relating to occupational health and hygiene such as chemical and manual handling, the use of video display terminals. As the regulatory enforcement agency over health and safety matters, the Department of Labor emphasizes protection of all employees through self-regulation and the establishment of a safety and health culture among the citizens. Since 1997–1998, the number of the governmental occupational hygienists has tripled with nearly 50% of all practicing professionals being employed by the Hong Kong government (40).

4.8.4 Occupational Exposure Limits

While no independent occupational exposure limit development mechanism is currently in place in Hong Kong (40), the Department of Labor codifies and enforces the U.S. ACGIH TLVs for airborne contaminants and physical agents.

4.8.5 Education and Training

The Chinese University of Hong Kong, School of Medicine administers a Master of Public Health degree (M.P.H.) in occupational hygiene; however, there are no undergraduate programs in the area. The majority of current senior practitioners who have been academically trained in occupational hygiene received their education in either Great Britain or the U.S. The Hong Kong University of Science and Technology located in Kowloon, has a master degree in environmental engineering and one of the largest staffs of professional occupational hygienists in Hong Kong (40).

4.8.6 Certification and Accreditation

There is no national certifying organization for occupational hygiene in Hong Kong.

4.8.7 Professional Organizations

The Hong Kong Institute of Occupational and Environmental Hygiene is the focal point for organized activities associated with occupational hygiene in the region and is a member organization of the IOHA. The institute conducts research, provides professional development opportunities for practitioners, and develops educational and public information materials. The institute offers "fellow" status to qualified candidates in the occupational hygiene community.

4.8.8 Status of the Profession

In Hong Kong, occupational hygiene is a recognized profession like other engineering disciplines. In general, the allied health professions work well together as they have clearly defined roles wherein physicians and nurses have responsibilities for the clinical care of workers while occupational hygienists address the prevention and field work aspects of occupational health. The allied health professions work together to formulate administrative programs and provide comment on pending legislation of mutual interest. The other safety professions recognize occupational hygiene as a specialized field and make referrals as needed.

Until the early 1990s, there were only a few occupational hygienists working in Hong Kong, primarily in the government. Since the early 1990s, the field has undergone a marked expansion. There were three CIH's in 1991, there are about 20 today. This growth is due to an increased awareness of occupational health and safety issues in society at large and government's increased effort to address occupational hygiene issues through the promulgation of new laws and an increased investment in resources and educational opportunities.

The primary factor behind the development of the profession in Hong Kong in recent years has been the statutory recognition of occupational health in the form of the 1997/1998 Occupational Safety and Health Ordinance. This has facilitated the introduction of occupational hygiene in areas of the Hong Kong economy that had previously been relatively unaccustomed to the profession. For future development, the profession will require the continuing autonomy of the Hong Kong economy, more specific statutory roles for occupational hygienists, enhancement of existing academic opportunities and continued education for practicing occupational hygienists (40).

4.8.9 Information Resources

Hong Kong Institute of Occupational and Environmental Hygiene
P.O. Box 25645, Harbour Building Post Office
Central, Hong Kong
Status: Professional Association

Occupational Safety and Health Council (OSHC)
19/F, China United Centre, 28 Marble Road, North

Point, Hong Kong
Tel: 852 2739 9377, Fax: 852 2739 9779
Internet: www.oshc.org.hk
Status: Government Consultancy

The Chinese University of Hong Kong
School of Medicine
Department of Community and Family Medicine
Shatin, Hong Kong
Internet: www.cuhk.edu.hk
Status: Academic Institution

The Hong Kong University of Science and Technology
Clear Water Bay Road, Kowloon, Hong Kong
Tel: 852 2358 6000, Fax: 852 2358 0537
Status: Academic Institution

Industrial Safety Training Centre
Labour Department
Occupational Safety and Health Branch
13th Floor, Harbour Building
38 Pier Road, Central, Hong Kong
Tel: 852 2852 3561, Fax: 852 2544 3497
Status: Regulatory Agency

4.9 India

4.9.1 Demographics and Economics

With one of the largest populations on Earth, India's 990 million citizens represent a huge economic challenge. India's current economy comprises a commercial agriculture, traditional village farming, broad scale manufacturing and handicrafts, and an abundance of support services. However, agriculture remains the primary employer in the country with 67% of India's 400 million workers (7). Production, trade, and investment reforms since 1991 have provided new opportunities for Indian businesses including foreign investment. Forty percent of the Indian population lives below a level providing an adequate diet and the national literacy rate is less than 55% (7). In contrast, India, per capita, also has one the highest number of scientists and engineers in the world. Distribution of the Indian work force is as follows; Agriculture, 67%; Services, 18%; and Industry, 15%.

In recent years economic gains have been substantial but less than expected due in part to shortages of fuels and inadequate telecommunication and transportation systems. Industries in the Indian economy include textiles, chemicals, food processing, steel, transportation equipment, cement, mining, petroleum and machinery. The dichotomy between the modern and the antiquated is further highlighted by the percentage of India's population that rely on manually powered vehicles while it develops and manages a sophisticated space program.

An accurate count of the number of practicing occupational hygienists in India is not available; however, an estimate of 100 would include both full-time professional practitioners and technicians (41). In light of the nearly 400 million people in the Indian work-force, even if the estimate of active hygienists was off by a factor of 100, India would still have one of the highest ratios of occupational hygienists to work force ranging from 1:4,000,000 to 1:40,000.

4.9.2 Historical Development

No Information available.

4.9.3 Occupational Health and Safety Legislation

The Factories Act of 1948 and subsequent amendments in 1987 are the key framework for the regulation of occupational health and safety in India. Promulgated by the Federal government and adopted by each state under the title of Factories Rules, the regulations are predominantly oriented toward safety with a minimum of occupational hygiene substance. This law identifies the basic requirements for employers and the responsibilities of the state to enforce the regulation. In general, it requires employers to maintain a safe and healthy work environment making provisions for the control of harmful atmospheres, confined space entry, permissible exposure limits, medical surveillance and record keeping, handling hazardous materials, ventilation and other control measures, worker right to know and hazard communication, maximum lifting limits, noise, and occupational health centers, among others.

Along with the relatively progressive subject matter like worker "right to know," the 1948 law also contains regulations addressing the working conditions of children which is still part of the Indian economic reality.

Changes made to the Factory Act in 1987 conferred significant power on Factory Inspectors with the intent of furthering the implementation of the safety and health provisions of the Act and creating health and safety awareness among employers and employees (41). The Act permits the inspectors to make interpretive decisions about important and complex health and safety issues while the level of occupational health and safety education among inspectors.

Chemical accidents, and particularly the Bhopal disaster in December 1984, brought public safety and chemical manufacturing in India into sharp focus and had collateral effects on the public's perception of the importance of occupational safety and health (41). Considered in the context of Bhopal, the 1987 Amendment Act is a milestone in safety consciousness in India as it has imposed a number of conditions and limitations on manufacturers with the aim of ensuring health and safety of workers as well as the public at large living in adjacent areas.

4.9.4 Occupational Exposure Limits

Provisions of the Factories Act of 1948 call for the adoption of exposure standards which are developed by the Department of Environment based on occupational exposure limits from other countries as well as an independent review of toxicity and volume of chemical use in India. Proposed standards are developed as time-weighted averages over an entire work shift without specific reference to the number of hours, and short-term exposure limits not to exceed 15 minutes, and no more than four times per shift. The short-term exposure limits are also ceiling values not to be exceeded. Limits proposed by the Department of Environment are referred to the Ministry of Labor and Rehabilitation for promulgation and enforcement.

4.9.5 Education and Training

The most significant development in occupational hygiene education in India was the recent development of a graduate degree program at the Center for Environmental and Industrial Hygiene Studies, Birla Vishvakarma Mahavidyalaya (BVM), Saradar Patel University, Vallabh Vidya Nagar, Gujarat, northwest of Bombay (41). Beginning with eight students selected from 30 applicants, the three-semester program includes a six-month internship in industry. The curriculum involves 16 courses including industrial hygiene and safety technology, air sampling and analysis, safety and law, toxicology and physiology, industrial ventilation, noise, etc. The program is a collaboration with the University of Cincinnati's Division of Environmental and Industrial Hygiene.

Training in occupational hygiene is also available in the form of seminars, short courses and lectures through the National Institute on Occupational Health, the Central Labor Institute, the All India Institute of Hygiene and Public Health in Calcutta, and the Indian Toxicology Research Center. General occupational health topics have been part of the preventative and social medicine curricula in Indian medical schools for many years, albeit on a limited basis (42). Also, occupational hygiene subject matter is taught as part of the graduate diploma program in occupational safety and occupational health at several Indian universities.

4.9.6 Certification and Registration

There is no certification system for occupational hygiene in India. A number of Indian practitioners have achieved certification and/or registration in the U.S., Canada, Australia and the U.K. while working abroad and in India.

4.9.7 Professional Organizations

There is no one association dedicated to occupational hygiene although there are several with ties to the profession including: Industrial Toxicology Research Center (ITRC), the National Institute of Occupational Health in Meghaninagar, Ahmedabad, India; the National Safety Council of India, and the Indian Occupational Health Association.

4.9.8 Status of the Profession

While progress has been made in prevention and control of occupational safety hazards in India, less advancement has occurred in the area of occupational health hazards. There is neither a clear understanding of the number of full-time, professional occupational hygienists practicing in India, nor the number of individuals who perform occupational hygiene-related activities under the auspices of the allied professions. However, the best estimate of occupational hygienists is less than 100 at both professional and technician levels. It has been estimated that there is an unrealized demand for as many as 7,000 qualified professionals based on the current degree of industrialization in India (41). Strategically, professionals in India point to the need for additional education as the key driver in the long-term development of occupational hygiene in the country. The new Masters of Industrial Hygiene (MIH) Program at BVM Saradar Patel University is expected to be duplicated at four other universities in India in the next five years (41).

Despite notable progress, reports continue to be published in India of significant cases of occupational diseases in industry, the most recent being perforation of nasal septum affecting 30 chemical manufacturing workers working with hexavalent chromium at a facility in Baroda, Gujarat. This has impacted the public perception of the need for occupational health intervention. Press coverage of recent decisions from state courts and India's Supreme Court involving requirements to reduce production capacity and shut down operations as ordered by the courts with significant economic impact has also raised public awareness.

The lack of a reliable occupational disease registry has affected the perceived magnitude of occupational illness in the country. As such, the need for occupational hygiene has not been driven by public demand, as has been the case in other countries. A disease registry would be facilitated by the promulgation of legislation requiring all physicians to report every case of potential or confirmed occupational illness. The need for a network of poison control centers would provide needed public access to important toxicology information and referral services.

Another important influence in the development of the profession in India is the need for improved occupational hygiene qualifications for Factory Inspectorate personnel with exclusive responsibility for occupational health issues. An increase in competency would result in more attention regarding workplace conditions and thereby catalyze the need for qualified practitioners to address issues of concern identified during compliance inspections. One of the reliable sources of occupational hygiene resources is the Employee State Insurance Corporation (ESIC) which has central and regional departments of occupational hygiene staffed with competent and committed professionals. An ESIC physician is often the first level of identification of potential cases of occupational disease. Hence, ESIC provides not only diagnosis and treatment of occupational diseases, but also can assist in prevention and control.

Other influences include multinational corporations bringing their corporate occupational health and safety systems and culture to the region. One example of this influence is Pfizer Corporation, which contributed to both the development of the BVM Saradar Patel University graduate program and has provided other resources for professional development (41).

4.9.9 Information Resources

National Institute of Occupational Health
Meghaninagar, Ahmedabad, India
Tel: 079-2867351, 2867352
Status: Government Research Institution

Industrial Toxicology Research Center (ITRC)
P.B. No-80, Mahatma Gandhi Marg

Lucknow 226001
Telex 0535-456
Status: Private Research Institute

National Safety Council of India
C.L.I. Bldg. N.S. Mankikar Marg
Sion, Mumbai, India 400 022
Tel: 407 3694, 407 3283, Fax: 91 22 4075937
Status: Non-governmental Organization

Indian Occupational Health Association
Mumbai, India
Status: Professional Association

Central Labour Institute
Mumbai, India
Tel: 022-4092203
Status: Government Agency

Directorate General Factory Advice Service and Labor Institutes
Ministry of Labor,
N. S. Mankikar Marg, Sion, Mumbai, India, 400 022
Status: Government Agency

Indian Journal of Industrial Medicine
Hindustan Lever Limited,
Hindustan Lever House,
165/166 Backbay Reclamation,
Mumbai, India 400 020
Status: Scientific Journal

Centre for Occ. and Environmental Health
42 Tughlakabad Institutional Area
New Delhi, India 110 062
Tel: 91 011 6981908, Fax: 91 011 6980183
Internet: pria@da.tool.nl
Status: Non-governmental Organization

4.10 Ireland

4.10.1 Demographics and Economics

Ireland's population of nearly 3.7 million has until recently lived with high unemployment and a relatively weak economy based on agriculture and manufacturing. In the early 1990s, the Irish economy began to expand with the influence of foreign investment, export gains and a rise in consumer spending. Today more than 50% of the gross domestic product results from the services sector while food products, brewing, textiles, clothing, chemicals, pharmaceuticals, machinery, transportation equipment, glass and crystal round out the industries that represent more than 38% of the economic driving force (7). The labor force of 1.52 million is employed primarily in the services sector. Distribution of the Irish work force is as follows: Services (including government), 62%; Manufacturing and construction, 27%; Agriculture, forestry and fisheries, 10%; and Utilities, 1%.

There are fewer than 20 occupational hygienists in Ireland (43). There are an uncounted number of chemists and engineers who have responsibilities for occupational hygiene-related activities in their companies and organization. For the most part, occupational hygiene in Ireland is relatively new and poorly understood in the general public (43). There are less than a half dozen academically trained occupational hygienists working in Ireland across a range of industry as independent consultants. The estimated ratio of occupational hygienists to the total work force is 1:80,000.

4.10.2 Occupational Health and Safety Legislation

The primary instruments for the regulation of workplace health and safety in Ireland are the Safety, Health, and Welfare at Work Act of 1989, (No. 7 of 1989) and the Safety, Health and Welfare at Work (General Application) Regulations, 1993 (S.I. 44 of 1993). These laws set out the full scope of workplace health and safety, the statutory systems necessary to achieve it, the responsibilities and roles of employers, the self-employed and employees, as well as enforcement procedures.

Under these two acts, employers (including self-employed persons) are primarily responsible for creating and maintaining a safe and healthy workplace. This includes: (*1*) the management of issues through consultation with the work force; (*2*) the development of a written "Safety Statement" identifying hazards and outlining measures to protect employees; (*3*) the periodic evaluation of health and safety risks and a written record of these assessments kept as part of the Safety Statement; (*4*) having available competent advice on health and safety matters; (*5*) the inclusion of all employees (full or part-time, permanent or temporary) as participants in the management system; (*6*) medical surveillance made available where justified by the risk assessment; and (*7*) proper use of all machines, tools, substances, and personal protective equipment, among other provisions (44).

The National Authority for Occupational Health and Safety, formally the Health and Safety Authority (HSA), within the Ministry of Labor is the government agency responsible for enforcing health and safety in the workplace. The National Authority inspectors provide consultation or issue citations and fines for violations of the Acts. Prosecutions under the Acts can lead to imprisonment.

4.10.3 Occupational Exposure Limits

Exposure to workplace chemical and physical hazards in Ireland is regulated by the Safety in Industry Act of 1980, Section 20. This Act requires employers to provide a safe workplace and accordingly to verify the control of airborne contaminants. As a member of the EU, the occupational exposure limits specified by the EU Council of Directives are incorporated into the Irish health and safety legislation by reference. Where EC Council of Directive guidelines are not available, the ACGIH TLVs are codified and enforced under Section 20 of the Act. Enforcement of occupational exposure limits is the responsibility of the National Authority industry inspectorate.

4.10.4 Education and Training

No undergraduate or graduate degrees in occupational hygiene are currently available in Ireland (43). The Dublin City University (UDC) has courses addressing aspects of occupational health and safety; however, the majority of professional development opportunities in Ireland originate from the National Authority, the Occupational Hygiene Society of Ireland or other outside sources. By its statutory mandate, the National Authority promotes education and training in occupational hygiene and conducts research. The majority of professional occupational hygienists currently practicing in Ireland obtained their professional education in the United Kingdom or elsewhere. The development of a graduate program in occupational hygiene remains an important prerequisite for the further development of the profession on a national level.

4.10.5 Certification & Registration

There is no certification or registration process for occupational hygiene in Ireland at this time.

4.10.6 Professional Organizations

The Occupational Hygiene Society of Ireland (OHSI) is member of the IOHA and forms the nucleus of the occupational hygiene profession in Ireland. Membership in the society is open to anyone with an interest in the profession. The OHSI conducts seminars, interacts with the National Authority and other government agencies to promote the profession, and to provide technical advice to interested parties (43). There is interaction with allied professional associations including the Irish Ergonomics Society.

4.10.7 Status of the Profession

There is a relatively small number of trained practitioners in Ireland, but occupational hygiene is still at the relatively early stage of development. In the last twenty years, environmental pollution and ecological issues have attracted public interest and government resources not unlike many other countries with substantial economic growth. The primary social concern has been chemical contaminants in the environment and the food chain rather than the workplace. There is a growing regulatory requirement for consideration of occupational hygiene issues. However, public awareness for the existence of or need for occupational hygiene is low.

Litigation associated with the asbestos industry and the construction sector as well as noise-induced hearing loss in the military has also resulted in an increase in the profile of the profession in Ireland based on technical expertise provided by the Irish practitioners (43). Workers' compensation insurance premiums are also becoming a significant source of pressure for companies to improve health and safety in the workplace including occupational hygiene. Investment by foreign multinational companies with established occupational hygiene policies and systems provide yet another developmental influence.

Common to many countries with developing economies discussed in this chapter, Ireland lacks certain infrastructure elements that can facilitate the development of the profession, including available graduate education in occupational hygiene, social awareness of occupational illness, government and industry recognition of the profession, adequate enforcement of existing occupational health legislation through improved technical standards for inspectors, and a national occupational health research institute with a designated focus on occupational hygiene.

4.10.8 Information Resources

National Irish Safety Organization
Temple Court, Hogan Place
Dublin 2, Ireland
Tel: 353 1 662 0399, Fax: 353 1 662 0397
Status: Professional Association

National Authority for Occ. Health and Safety
10 Hogan Place, Dublin 2 Ireland
Tel: 353 1 614 7000, Fax: 353 1 614 7020
Internet: www.HSA.ie/osh

Occupational Hygiene Society of Ireland
℅ IOHA, Georgian House, Suite 2, Great Northern Road, Derby De1 1LT, United Kingdom
Status: Professional Association

4.11 Italy

4.11.1 Demographics and Economics

In the fifty years since the end of World War II, Italy has modified its economy from an agricultural base to a ranking industrial economy. Geographically, the economy is divided into the more industrialized north and the more rural and agricultural southern region. The total and per capita output of the Italian economy is similar to that of the United Kingdom and France (7). The majority of the raw materials required by industry and more than three quarters of the energy requirements are imported. The population of 56.7 million enjoys a very high literacy rate. The resident work force represents 40% of the total population (7). The primary industries in Italy include the production of machinery, iron and steel, chemicals, food processing, textiles, motor vehicles, clothing, footwear, ceramics, and tourism. Distribution of the Italian work force is as follows: Services (including government), 61%; Manufacturing and construction, 32%; and Agriculture and fisheries, 7%.

The estimated ratio of occupational hygienist to the total work force in Italy is approximately 1:9,600 (45). Italy is one of the countries outside the U.S., which continues to use the title industrial hygiene in reference to the profession.

4.11.2 Historical Development

Italy has benefited from a long history of attention to occupational health matters beginning with Bernardino Ramazzini (1633–1714) who is recognized as one of the founding fathers of occupational medicine. Ramazzini is known for many contributions to the field including delivery of a brilliant series of lectures at the University of Padova linking occupational diseases to various trades.

Before World War II, Italy was among the highly industrialized nations of Europe, known for both its light and its heavy industries. Such industries ranged from agricultural products to footwear and textiles, from passenger vehicles and trucks to aircraft and heavy weapons. In June 1940, Italy entered World War II as a partner of the Rome-Berlin Axis. Five years later, by the end of the war, Italy lay in ruins, from the tip of Sicily to its northern borders.

The development of modern occupational hygiene in Italy coincides with the massive effort of industrial reconstruction and expansion. In the words of Dr. Danilo Sordelli (46), past-president of the Italian Association of Industrial Hygienists (AIDII), "The early 1950s were characterized by post-war reconstruction and industrial recovery; the available products and their use were not yet a problem for the environment. Social awareness was directed towards problems connected with in-plant conditions."

The rapid growth of industry, an awakening social awareness toward occupational and environmental problems, and the desire to control and prevent such problems as early as feasible are reflected by regulatory efforts by central and local governmental agencies. Beginning in 1955, safety and occupational hygiene regulations followed one another in steady succession (47). Although the early regulations were somewhat generic in nature, they were important because they opened the way for more stringent and specific regulatory efforts. Major industries began to review their operations and seek better workplace controls while occupational physicians were striving to offer their employees better health care. Organized labor and government inspectors participated in such efforts adding to the momentum.

4.11.3 Occupational Health and Safety Legislation

Regulatory requirements for occupational health and safety protection in Italy are based on Article 32 of the Constitution of the Italian Republic, which holds human health as a "fundamental right of the individual and the goal of society."

Italian legislation involving health and safety encompasses air quality standards, hazard control in industry, substance-specific rules for lead, noise, asbestos and vinyl chloride; material handling, personal protective equipment, use of video display terminals, controlling carcinogens, and a wide a variety of other general health and safety guidelines and standards. The Workers Health Act was promulgated to address health-specific issues and is enforced by the National Health Service that administers occupational hygiene laboratories and conducts workplace surveys.

As a member of the EC, Italy has adopted the general EC Directives for occupational health and safety including the Framework Directives 80/1007 and 89/391 for the prevention of work hazards.

The first piece of significant legislation relating to the health and well being of workers was passed in 1970. This law assumed particular significance. It required, among other things, the formation of plant environmental committees and made them responsible for implementing this law. A law limiting industrial air pollution was passed in 1971. It was followed in 1976 by one addressing water pollution, and in 1978, by another establishing the National Health Service. Through this law the Italian Parliament made the Ministry of Public Health responsible for occupational hygiene and specifically charged the Prime Minister's office with setting and periodically updating occupational exposure limits for places of employment.

Large industries led the way in controlling in-plant and outdoor exposures to contaminants such as vinyl chloride, acrylonitrile, and benzene. These industries also sought to identify and produce biodegradable detergents, selective photobiodegradable pesticides, and other environmentally acceptable products.

Italy experienced several major industrial accidents in the 1970s and early 1980s that threatened the health of workers and segments of the population living near the plants. In addition, a growing concern for environmental issues further affected the increasingly tense relations between various industries and the public at large. Overall, public opinion came to see the chemical industry as the major polluter and culprit behind the broad environmental threat.

Environmental regulations were passed with increasing frequency, often in reaction to specific events or social pressures rather than because of new hazards or developments. Italy, like other members of the European Community, began to adopt EC health standards and regulations in an effort to update its own and to achieve conformity within the EC. Thus Italy replaced its own regulations on packaging and labeling promulgated in the 1970s with corresponding EC regulations.

The first European law for the protection of workers against chemical, physical, and biological hazards in the workplace appeared in 1980. Shortly thereafter, Italy adopted the EC directive regulating workplace exposures to vinyl chloride. Laws regulating air quality and air pollution from industrial sources saw the light in the 1980s, followed by laws on waste disposal, on prohibiting smoking in public places, and on abatement of asbestos in schools, hospitals, and other public buildings (45).

4.11.4 Occupational Exposure Limits

The first Italian regulation limiting air contaminants in the workplace appeared in 1966 (48). Industry came to recognize the need to limit employee exposures to air contaminants in factories and began to incorporate such controls into the initial process design. At first most enterprises emitted the captured air contaminants directly outdoors. Some went beyond indoor controls and installed the first dust collectors and water treatment plants at about this time.

Italian environmental experts soon recognized that early in-plant controls were still inadequate and jointly with organized labor continued their search for better ways to control pollutants. As a result of such efforts, the chemical industry and organized labor incorporated the ACGIH TLVs into their 1969 national contract. The ACGIH TLV booklet was translated into Italian. Updated translations of the booklet have appeared regularly ever since.

With regard to Italian occupational exposure limits, Cook (15) writes:

> In 1975 a list of occupational exposure limits, termed "Valori Limite Ponderati (VLP)" was prepared jointly by the Italian Society of Occupational Medicine and the Italian Society of

Industrial Hygienists. At about the same time the Technical Committee prepared a more nearly complete list with a few differences for Maximum Allowable Concentrations instituted by the ENPI for the Minister of Labor.

4.11.5 Education and Training

While it is estimated that Italy has over 2000 practicing industrial hygienists and various universities and other institutions of higher learning conduct extensive and sophisticated occupational hygiene research, presently there are no university programs offering degrees in occupational hygiene (45). About 20 medical colleges and occupational health services offer occupational hygiene courses, primarily as part of their curricula for occupational physicians. About half offer occupational physicians advanced occupational hygiene training beyond their regular course work.

Relatively few Italian occupational hygienists have occupational hygiene degrees from universities abroad. Most begin their professional careers as chemists, physicists, occupational physicians, or graduates of other physical sciences. Once they move toward occupational hygiene, they receive their training on the job under the direction of experienced occupational hygienists and by attending courses offered by AIDII, by universities, and by other institutions. Such courses are available to university graduates, be they in industry or in government. They may cover basic as well as advanced topics and they last from 70 to 80 hours.

The universities of Pavia and Padova in northern Italy, and one at the southern seaport of Bari on the Adriatic, offer two-year courses for occupational hygiene technicians.

4.11.6 Certification and Registration

Certification in occupational hygiene is administered by the Institute for Certification of Industrial Hygienists (Instituto per la Certificazone Degli Igienisti Industriali, ICCII). There are two levels of certification: the Certified Industrial Hygienist (IIC), and Certified Industrial Hygiene Technician (TIIC). The qualifications for the ICC include three years of comprehensive experience or equivalent part-time experience and at least an undergraduate degree from a university in chemistry, physics, engineering, biology, or engineering. The examination consists of multiple choice questions. Under the Italian system, the ICC differentiates those occupational hygienists that by possession of the requisite experience and successful completion of an examination are competent in the comprehensive practice of occupational hygiene at an advanced level. The TIIC examination requires a technical diploma in chemistry or engineering, three years of full-time practice in the field of air and/or water pollution monitoring and control, and the successful completion of a multiple choice examination. The ICCII mandates certification maintenance through continuing education on a five-year cycle as well as adherence to a code of ethical conduct.

4.11.7 Professional Organizations

The AIDII Associazione Italiana Degli Igienisti Industriali was founded in 1969 and currently has a membership of over 1200. The number of practicing occupational hygienists in Italy is conservatively estimated at over 2500. A 1992 membership survey by the association revealed that 90% were male, 38.1% under the age of 40 and 23.5% over the age of 51, 69.9% had college degrees of which 41.5% were degrees in chemistry, 22% in medicine, 19.9% in engineering; 7.7% in biology, less than 2.0% in geology or physics; and less than one percent in occupational hygiene (45).

There are some statutory limitations in the practice of occupational hygiene including ionizing radiation where only those individuals with specialized licensure are permitted to work. However, in 1985 the AIDII amended its bylaws to expand the scope of practice to include responsibilities for environmental matters—a growing trend among occupational hygiene organizations in Europe.

AIDII has held annual conferences for several years. Such annual conferences offer professional development courses and exhibits of industrial hygiene instruments. The AIDII has several organizational members from allied associations devoted to safety, health physics, occupational medicine, ergonomics, and chemistry, among others. Such associations plan and coordinate future activities with AIDII through regularly scheduled joint meetings.

In 1969, AIHA awarded Enrico Vigliani, M.D. its prestigious Yant Memorial Award in recognition of his work regarding the toxicity of chemicals and particularly of heavy metals. In 1993, Dr. Danilo Sordelli was recognized with the Yant Award for his significant contributions to the profession.

4.11.8 Information Resources

Higher Institute for Worker Security and Protection (ISPESL)
Via Urbana 167, 00184 Rome, Italy
Tel: 47141, Fax: 4818493
Internet: www.ispesl.it
Status: Government Research Institute

4.12 Japan

4.12.1 Demographics and Economics

With a high degree of control over technology, a national ethic for work and strong industry-government cooperation, Japan's 125.9 million citizens enjoy the second most powerful economy in the world behind the U.S. This despite having to import much of its industry's need for raw materials and energy. More than 41% of Japan's gross domestic product comes from industry, 56% from the services sector (7). Leading industry segments include: commercial electrical equipment, steel and nonferrous metallurgy construction and mining equipment, motor vehicles and parts, electronic and telecommunication equipment, machine tools, automated production systems, locomotives and railroad rolling stock, ships, chemicals, textiles and processed foods (7). Japan has one of the largest fishing fleets in the world and captures more than 10 million metric tons of fish per year. Distribution of the Japanese work force is as follows: Trade and services, 50%; Manufacturing, mining, and construction, 33%; Utilities and communication, 7%; Agriculture, forestry, and fishing, 6%; and Government, 3%. The work force in Japan is about 67.2 million strong and there are approximately 2025 professional occupational hygienists giving Japan an occupational hygienist to total work force ratio of 1:33,200 (49).

4.12.2 Historical Development

The first legislative action taken in Japan on behalf of the protection of worker health and safety was the Mining Act of 1890, which defined compensation for mine injuries and fatalities, but did not include a provision for occupational illnesses. Prior to 1890, there were occupational hygiene activities associated with the protection of military personnel and the health of government workers. Subsequent to the establishment of local legislation designed to address worker health, the Factories Act was promulgated in 1911 (50). In the 1910s, government officials conducted the first formal occupational hygiene research focusing on such issues as illumination, dust, chemical exposure, tuberculosis, occupational disease, and working conditions in mines and factories. In 1921, the Institute of Labor was founded by Mr. M. Ohara, a cotton mill owner, and in 1929 facilitated the organization of the first occupational hygiene association, although it was short-lived. By the end of World War II, the Ministry of Labor was established and occupational hygiene was formally introduced into legislation through the 1947 Labor Standards Law and the 1947 Workers Accident Compensation Act (50). Much of the early work was directed at controlling tuberculosis. The first committee for the development of occupational exposure limits was established in the Japan Society for Industrial Health in 1952.

4.12.3 Occupational Health and Safety Legislation

Under the Industrial Safety and Health Law, Japanese employers are required to take certain measures to address process and environment controls and worker protection. In addition, an occupational health management system must be established at each workplace administered by an occupational health team consisting of a senior manager, a health manager, an industrial physician, and a safety and health committee. A management system is required for employers with more than 50 employees (50).

Typical environment controls that must be considered include: (*1*) manufacturing process enclosures; (*2*) local exhaust ventilation and/or general ventilating system; (*3*) installing exhaust equipment such as gas disposition systems; (*4*) hygiene facilities; (*5*) conducting exposure assessment surveys and conducting regular voluntary inspections of the workplace; (*6*) employee education; (*7*) safe work procedures; (*8*) physical segregation of work areas; (*9*) personal protective equipment; (*10*) warning signage and labeling; and (*11*) the prohibition of meals and drink consumption except in designated areas of the workplace.

Provisions for health care must include conducting medical surveillance examinations and maintaining personal health records. To the benefit of the profession, this also requires the appointment of a health manager who is responsible for ensuring that government regulations are complied with and to ensure a healthy workplace.

Article 65 of the Industrial Safety and Health Law obligates employers to conduct workplace exposure assessment for airborne contaminants, for which "Working Environment Measurement Standards" detail analytical methods depending upon the type of substances subject to measurement. In addition, the Ministry of Labor's "Working Environment Appraisal Standards" describes analytical methods to be used in conducting workplace exposure assessments. Measurement results are statistically processed and compared to occupational exposure limits (called "control level") established by the Minister of Labor for each substance to determine if there is a need for additional environmental controls.

A system for toxicity testing of new chemical substances was introduced by the revision of the Industrial Safety and Health Law in 1979. This system requires companies which plan to manufacture or use new chemical substances, other than those announced by the Minister of Labor as "existing chemical substances," to conduct specified toxicity testing and submit the results to the Ministry of Labor before a chemical can be marketed. The Ames mutagenicity test is required as the screening test for carcinogenicity.

Subsequent to the adoption of the ILO Convention No. 170 which addresses the need to communicate the hazards and toxicity of chemical substances used at workplaces, the Ministry of Labor announced guidelines in 1992 requiring Japanese employers to develop MSDSs to be affixed at the time of chemical transfer, and encouraging users of chemicals to address MSDS as educational tools to prevent health impairment due to misuse of chemicals.

As an indication of one negative outcome of the very strong Japanese work ethic, a revision to the Industrial Safety and Health Law was promulgated in June 1996 (Law No. 89) with the aim of promoting measures to protect workers' health and minimize disabling work anxieties. This was done in response to a rise in the incidence of cerebral and cardiovascular diseases, occupational stress and anxiety as well as "karoshi," or death caused by overwork (50).

4.12.4 Occupational Exposure Limits

In Japan there are two sets of occupational exposure limits available to address worker protection. One set was developed by the Japan Society for Occupational Health (JSOH), and the second by the Ministry of Labor.

The JSOH annually recommends occupational exposure limits ("permissible exposure limits") for chemicals and physical agents based on a review of the international toxicology literature and health hazards associated with physicals hazards using a deliberative process similar to the ACGIH TLVs. The JSOH Committee on Occupational Exposure Limits consists of members from academia and government research institutes, the majority being physicians and the minority occupational hygienists (51, 52). The "Recommendations for Occupational Exposure Limits" are published annually in the *Journal of Occupational Health* along with biological permissible limits (analogous to ACGIH BEIs) (53).

Since 1984, the Ministry of Labor has developed its own "control limit indices" (CLI) that are used as guidelines. The control limits are used in conjunction with area air sampling that is conducted by licensed Working Environment Measurement Experts (52). The resulting data are analyzed using prescribed mathematical models to determine compliance. The degree of risk associated with a working environment contaminated with a toxic substance is defined or categorized as Control Class I, II or III with reference to the "administrative control level." The control class dictates what actions will be required of the employer to improve the environmental conditions. Class I indicates acceptable air quality, ordinary occupational hygiene measures are advisable; Class II refers to borderline conditions, certain additional occupational hygiene steps are advisable; and Class III indicates unacceptable conditions, specific and stringent controls are required.

The Japan Association publishes the control limits for Working Environment Measurement. The Ministry of Labor itself issues documents titled "Working Environment Evaluation Criteria" which include referrals to the ACGIH TLVs, JSOH's limits (as time-weighted averages and ceiling limits), epidemiological data and information addressing the technical feasibility of the standards (50).

4.12.5 Education and Training

In Japan, occupational health physicians are given primary responsibility for the prevention of occupational disease. The Occupational Health Training Center (OHTC) was established as an institute of the University of Occupational and Environmental Health (UOEH) in Kitakyushu, Fukuoka for the purpose of educating qualified physicians in occupational health. The major focus of UOEH is undergraduate students and medical residents engaged in occupational health training, but also the continuing education program for occupational health personnel throughout Japan. To fulfill these purposes, UOEH also functions as an occupational health service center in addition to an information service center for the whole country. However, UOEH's role is gradually expanding with a much larger objective of providing resources for health promotion for all Japanese workers. Specifically, the UOEH's mandate is: (*1*) to "address the health status of the workers based on the sound understanding of the nature of the work, management, and industrial structure; (*2*) to seek comprehensive and lifelong health; (*3*) to maintain good relations with the community health care system; (*4*) to address not only the regional but also the global environment; (*5*) to act on the basis of international health; (*6*) to establish an independent specialty through unique theoretical groundwork; and (*7*) to create an educational system based on ethical guidelines for Active Occupational Health."

The National Institute of Industrial Health (Sangyo Igaku Sogo Kenkyusho) was established in the 1970s under the Ministry of Labor to conduct interdisciplinary research to promote workers' health and the prevention of occupational diseases. It also provides scientific and technical information to the Ministry's Labor Standards Administration (50). The institute focuses on six areas of research: (*1*) work management and human factor engineering in response to changes in working conditions; (*2*) work capacity and fitness of women and the elderly; (*3*) prevention of illness caused by toxic substances; (*4*) biological monitoring; assessment of chemical and physical hazards; and (*5*) work environment measurement, evaluation and countermeasures. The institute also conducts performance testing for gas masks and dust respirators, and conducts national approval tests for the all new respirators.

4.12.6 Certification and Registration

There are four registration groups pertaining to occupational health and safety matters in Japan (49): (*1*) Labor inspectors, subcategorized as either "legal experts" or "engineers." (*2*) The consultant group includes physicians, occupational hygienists and safety engineers. The number of registered occupational hygienists is 305. (*3*) Health managers comprise the third group and which includes managers and occupational hygienists. There are some 800,000 registered in this group of which it is estimated that about 123,000 part-time and 1,850 full-time health managers are active in the field and another 1,713 occupational hygienists working in this category. The last group is (*4*) Working Environment Measurement Experts for which there are five subspecialties: particles, radioisotopes, metals, specified chemicals and organic vapors. It is believed that about 6,000 measurement experts are active. Many of the 6,000 hold multiple licenses. It is not known what percentage of occupational hygienist consultants, occupational hygiene health managers, and working environment measurement experts overlap. The registration system is administered by the Institute for Safety and Health Qualifying Examination.

4.12.7 Professional Organizations

The Japan Industrial Safety and Health Association (JISHA) was established in 1964 under the Industrial Accident Prevention Organizations Law for the purpose of upgrading the standards of occupational safety and health through the promotion of voluntary activities by private employers. It has more than 6,000 members. The JISHA works closely with the government and other organizations concerned with the prevention of injuries and illnesses in occupational settings. It is functionally analogous to the U.S. National Safety Counsel.

The Japan Association of Occupational Hygiene (JAOH) has approximately 450 members including researchers from institutions and academia, industrial health managers, engineers, consultants, and workplace measurement experts. The JAOH is involved in traditional functions associated with the development of the profession including membership services, interaction with the government relative to new legislation, and professional development opportunities. The association has codes of ethics for both consultants and measurement experts (49).

4.12.8 Status of the Profession

The profession continues to grow in Japan, benefiting from a long history of social policy relating to occupational health issues, university and government research programs (Ministry of Labor, Institute of Public Health, Institute of Labor Science, etc.), and the integration of occupational hygiene activities into the occupational health services system.

The JAOH has about 450 members and advocates on behalf of the membership; however, it is not known how many practitioners consider themselves to be occupational hygienists by profession (50).

Future growth will occur as a result of an increase in the understanding for the profession among the general public and growth of educational and employment opportunities (49).

4.12.9 Information Resources

Japan Occupational Hygiene Association
4-4-5 Shiba, Minato-ku, Tokyo 108-0014, Japan

Tel: 81 3 3456 5851, Fax: 81 3 3456 5854
Status: Professional Association

Japan Industrial Safety and Health Association
5-35-1 Shiba, Minato-Ku, Tokyo, 108-0014, Japan
Tel: 81-3-3452-6841
Internet: www.jisha.or.jp
Status: Professional Association

National Institute of Industrial Safety Ministry of Labor (NIIS)
1-4-6, Umezono, Kiyose, Tokyo 204-0024, Japan
Tel: 00 81 424 91 4512, Fax: 00 81 242 91 7846
Internet: www.anken.go.jp
Status: Government Research Institute

Technology Institute of Industrial Safety
Tokuei Building 3F, 5-33-7 Shiba, Minato-ku,
Tokyo, 108-0014, Japan
Tel: 81 3 3455 3957, Fax: 81 3 345 3957
Status: Academic Institution

National Institute of Industrial Health
Ministry of Labour
Tel: 81 44 865 6111, Fax: 81 44 865 6116
Internet: www.niih.go.jp
Status: Government Research Institute

Japan Association of the Working Environment
4-4-5, Shiba, Minato-Ku, Tokyo, 108-0014, Japan
Tel: 81 3 3456 5851, Fax: 81 3 3456 5854
Status: Professional Association

Japan Association of Safety and Health Consultants
4-4-5, Shiba, Minato-Ku, Tokyo, 108-0014, Japan
Tel: 81 3 3453 7935
Status: Professional Association

Institute for Safety and Health Qualifying Examination
Suidbashi Building 8F, 1-3-12, Misaki-cho,
Chiyodaku, Tokyo, 101-0061, Japan
Tel: 81 3 3295 1088
Status: Government Licensure Agency

Japan Society for Occupational Health
Koshueisei Building 4F, 1-29-8, Shinjuku,
Shinjuku-ku, Tokyo, 160-0022, Japan
Tel: 81 3 3356 1536, Fax: 81 3 5362 3746
Status: Professional Association

University of Occ. and Environmental Health
1-1, Iseigaoka, Yahata-nishi-ku, Kitakyushu,
Fukuoka, 807-8555, Japan

Tel: 81 93 603 1611
Internet: www.noeh_u.ac.jp
Status: Academic Institution

4.13 Korea (South)

4.13.1 Demographics and Economics

A country in transition, the Republic of Korea's population and population density has increased steadily since the early 1970s, although the annual rate of increase has slowed from 1.7% in 1975 to 0.9% in 1996. This change correlates to a shift in South Korea's economic base from agriculture, mining, fisheries, and forestry in the 1960s, to the current emphasis on manufacturing, construction, and service industries. There was a drop of 34% in the working population between these industry sectors in that same time period and the gross national product (GNP) grew 7.5 times from $60 billion U.S. dollars to $455 billion U.S. dollars (7). South Korea has rapidly industrialized during the past 30 years with a heavy focus on steel production and automotive manufacturing, shipbuilding, mining, electronics manufacturing, and construction. However, in that time little attention was paid to the conditions in the work environment. The work force of 20 million represents approximately 42% of the population of 46 million (7). Distribution of the South Korean work force is as follows: Services, 52%; Mining and manufacturing, 27%; and Agriculture, fisheries, forestry, 21%.

4.13.2 Historical Development

The occupational health movement began in 1958 by Dr. Y. T. Choi, a physician with the Korean Coal Company. Dr. Choi introduced medical surveillance for workers and began tracking disease incidence rates such as pneumoconiosis. The company also developed the first occupational hygiene laboratory in South Korea. In 1962, the Institute of Industrial Medicine was founded in the Catholic Medical School where the country's first occupational hygienist, K. M. Lee, a chemist, worked as an instructor and developed sampling and analytical methods for airborne contaminants (54). The Institute also published a quarterly journal, and conducted occupational hygiene surveys and medical surveillance examinations. By 1992, the Institute had performed more than 4,200 surveys in the coal and cement industries, among many others.

In 1963, the nonprofit Korean Industrial Health Association (KISA) was founded with the goal of educating industrial health personnel, conducting occupational hygiene surveys and performing medical examinations for workers. It is now the largest industrial health organization in South Korea with 600 occupational hygienists (Korea continues to use the term industrial hygiene), occupational physicians and nurses. Annually, KISA performs occupational hygiene surveys for 10,000 industrial facilities and as well as more than 300,000 medical examinations (54).

4.13.3 Occupational Health and Safety Legislation

The most significant South Korean legislation for worker protection is the Occupational Safety and Health (OSH) Act promulgated in 1981 and revised in 1990. The Act requires employers to provide adequate occupational health and safety resources for worker protection through programs, training, hazard evaluation and control, and health and safety committees, among other provisions (54). Private sector employers are also required to employ a health manager (Article 16), occupational hygienists and nurses in specific ratios relating to the size and type of industry. Mechanisms for evaluation and control of health hazards are also stipulated, e.g., radiation, temperature extremes, ultrasonic waves, noise, vibration and abnormal atmospheric pressure.

Despite this regulatory recognition of the profession, enforcement of the occupational hygiene quota system is not widely enforced due primarily to the lack of qualified professionals. The Act also requires periodic evaluation of worker exposures to chemical and physical agents, which resulted in substantial efforts to conduct baseline surveys throughout industry as historically only area sampling had been conducted prior to 1985. By 1993, the use of personal breathing zone samples were accepted as a standard method of exposure assessment by the Ministry of Labor. Currently, exposure assessment surveys must be conducted once every six months for facilities using toxic materials such as benzene, asbestos, lead, and carbon disulfide while general occupational hygiene surveys are required once per year at most industrial plants.

Enforcement of the Act is the focus of the Ministry of Labor. The Ministry contains the Bureau of Industrial Safety, which in turn contains the Division of Industrial Hygiene, the Division of Industrial Health (medicine) and the Division of Industrial Safety. There are regional offices that conduct work under the Ministry of Labor including the Division of Work Standards.

The existing legislation supports the concept of occupational health services. Employers with more than 50 workers must retain the services of a physician to conduct medical surveillance and education. This is done in cooperation with health managers whose qualifications vary widely. Almost two-thirds of all occupational health services are managed by group service providers, which usually includes occupational hygiene capabilities (55). Occupational health personnel must visit work facilities under their responsibility at least 12 times per year, i.e., twice by a physician, four times by an occupational hygienist and six times by an occupational health nurse.

4.13.4 Occupational Exposure Limits

The Ministry of Labor has adopted approximately 650 occupational exposure standards for chemical and physical agents. Historically, exposure standards were adopted from the ACGIH TLVs, although today there is growing interest in developing an autonomous system for exposure standard and control limits specific to South Korea. The NIOSH sampling and analytical methods have been generally adopted and sanctioned for use in Korea. Enforcement of the South Korean standards in the public and private sectors is the responsibility of the Division of Industrial Hygiene under the Bureau of Industrial Safety in the Ministry of Labor. South Korean exposure limits carry the weight of law.

4.13.5 Education and Training

In 1985, the Seoul National University, School of Public Health began teaching occupational health at the graduate level. This was the first formal academic training available in occupational hygiene in South Korea and today remains the primary source of graduates to industry and government producing some 125 graduates since 1985. Other institutions offering some level of training in occupational health are the Catholic University School of Occupational Health in Seoul and In-Je University in Pusan.

4.13.6 Certification and Registration

Occupational hygienists in South Korea are certified by the Korean Manpower Agency as Professional Engineers in Industrial Safety Management. Achieving certification requires an acceptable baccalaureate degree in engineering or a science and a minimum of seven years of professional experience, or a diploma from a two-year junior college in engineering or science and a minimum of nine years of professional experience, or a high school diploma in engineering or science and a minimum of 11 years of experience (54). Each candidate for certification must also pass a two-part examination containing both written and oral components.

As of 1996, there were 54 Professional Engineers certified in occupational hygiene and six Koreans who had obtained the CIH designation through the ABIH. Under a similar system, there were 479 Professional Engineers in safety management (54).

4.13.7 Professional Organizations

The primary association dedicated to occupational hygiene is the Korean Industrial Hygiene Association (KIHA). Membership in KIHA has risen steadily since its inception in 1990 to 530 in 1996 (54). The association holds biannual technical conferences and has been highly influential in the development of the profession in South Korea. Among its many advocacy activities, the KIHA aided the Ministry of Labor to begin an occupational hygiene laboratory quality control program in 1992 that today is jointly managed by KIHA and the Industrial Health Research Institute (IHR), the research arm of the Korean Industrial Safety Corporation (KISCO). There are approximately 45 consulting laboratories in South Korea and many more private laboratories in large companies. The number of laboratories participating in the quality control program is close to 100 (56).

Prior to its introduction by the Ministry of Labor in 1982, a small number of occupational hygiene laboratories participated in AIHA's Proficiency Analytical Testing Program (PAT), and the British Asbestos Fiber Regular Informal Counting Arrangement (AFRICA). The first round of testing was conducted in 1996 with less than 28% of the participating laboratories being rated as proficient in metals and organic analyses (56); however, by the second round of sequential testing, the number of laboratories rated as proficient had more than doubled to 63%.

Other professional associations in South Korea include the Korean Industrial Health Association and the Korean Society of Occupational Medicine. Formed in 1963, the Korean Industrial Health Association has a current membership of more than 600 occupational hygienists, physicians and nurses and is involved in the education of occupational health professionals through the development of short courses and seminars as well as offering medical surveillance and occupational hygiene services. Yet another association is the government-sponsored Korean Industrial Safety Corporation (KISCO), founded in the early 1990s, and now employs more than 1,000 occupational hygienists, occupational physicians and safety engineers who provide occupational health and safety services throughout the country.

Occupational hygiene and medicine services are also available through the Korea University Institute of Environmental Health in Seoul, Sun-Chun-Hyang University College of Medicine in Chun-an, and Yonsei University College of Medicine in Seoul. The Institute of Industrial Medicine at the Catholic University in very active in the field providing consulting services to industry, publishing journals and conducting research.

4.13.8 Status of the Profession

The number of professional engineers and researchers in South Korea has rapidly increased in the last five years. In 1995, there were 479 safety engineers among the 12,534 professional engineers. Of these, 54 were certified in "industrial hygiene management." The field continues to gain exposure in the public and credibility with the government. There are at least 150 practicing occupational hygienists which good academic training and field experience (54).

The 1970s and 1980s were difficult times for proactive occupational health in South Korea as the country was heavily focused on production and growth. As in many countries, the popular focus on occupational health issues is often catalyzed by the reports of occupational disease in the news media. Such is the case for South Korea where an outbreak of carbon disulfide-related illness among 450 workers at a large rayon manufacturing facility in the late 1980s resulted in an increased demand for worker protection and facilitated additional government oversight of industry with regard to occupational health issues. As a result of mass media reports addressing the cases of rayon-related disease, government, labor and the general public increased their emphasis on occupational health. Today, the Ministry of Labor continues to monitor occupational disease, although the accuracy of the scope and breadth of such surveillance is debated.

4.13.9 Information Resources

Korean Industrial Hygiene Association (KIHA)
℅ Seoul National University
School of Public Health
Seoul, South Korea
Tel: 02 740 8883, Fax: 02 745 9104
Status: Professional Association

Catholic University
School of Occupational Health & Institute of Industrial Medicine
Seoul, South Korea
Tel: 02 590 1114
Status: Academic Institution

Korea University, College of Medicine, Institute of Environmental Medicine
Seoul, Korea
Tel: 02 926 2641, Fax: 02 920 1114
Status: Academic Research Institute

Yonsei University, College of Medicine
Seoul, Korea
Tel: 02 361 2114
Status: Academic Institution

Seoul National University
School of Public Health, 28 Yunkeun-Dong,
Chongro-Ku, Seoul, Korea
Tel: 02 740 8883
Status: Academic Institution

Korean Industrial Health Association
#1022-1 Bangbee-3-Dond
Sucho-Ku Seoul, South Korea
Tel: 02 586-2412, Fax: 02 585-1584
Status: Quasi-governmental Association

In-Je University
School of Public Health
Pusan, South Korea
Tel: 051 890 6654
Status: Academic Institution

Korea Industrial Safety Corporation (KISCO)
34-4 Gusandong Bupyong-ku
Inchon, South Korea 403-711
Tel: 82 32 5100 749, Fax: 82 32 512 8311
Status: Quasi-governmental Association

Sun-Chun-Hyang University, College of Medicine
Seoul, Korea
Tel: 0418 530 1389
Status: Academic Institution

4.14 Mexico

4.14.1 Demographics and Economics

Following decades with an economy based on outmoded industry and agriculture, Mexico is emerging as a major power in the Western Hemisphere with a population approaching 100 million people. Mexico has had difficulty capitalizing on its relatively abundant natural resources while most of the country's industrial structure was nationalized. In the period from 1982 to 1998 the government privatized more than 800 state-owned enterprises (7). Today the Mexican economy is based on free market reforms and an industrial sector comprised of chemicals, iron and steel, petroleum, mining, textiles, clothing, motor vehicles, consumer durable goods, food and beverages, and tobacco. Tourism has become a major source of external currency and capital.

Today, the agriculture sector employs more than 20% of the labor force but produces less than 10% of Mexico's gross domestic product (7). Trade with Canada and the U.S.

has doubled since the 1994 implementation of the North American Free Trade Agreement (NAFTA). However, Mexico is pursuing additional trade agreements with other Latin American countries and with the European Union to reduce its dependence on the United States, which accounts for 80% of Mexico's total trade.

The Mexican work force consists of about 36.6 million people. Distribution of the Mexican work force is as follows: Services, 28.8%; Agriculture, forestry, hunting, and fishing, 21.8%; Commerce, 17.1%; Manufacturing, 16.1%; Construction, 5.2%; Public administration and national defense, 4.4%; and Transportation and communications, 4.1%. There are less than 50 full-time occupational hygienists in the country (57), which translates to an occupational hygienist to work force ratio of approximately 1:732,000. There are reportedly many hundreds of individuals who perform occupational hygiene-related activities in other disciplines such as safety, medicine and engineering (57). Practitioners in the country predominantly refer to themselves as industrial hygienists.

4.14.2 Occupational Health and Safety Legislation

The primary occupational health and safety legislation in Mexico includes Article 123 of the Mexican Constitution; the Federal Labor Law (Ley Federal del Trabajo); and the Federal Regulation on Safety, Hygiene and the Workplace Environment (Reglamento Federal de Seguridad, Higiene y Medio Ambiente de Trabajo).

Article 123 of the Mexican Constitution is the primary authority for laws pertaining to health and safety. It provides a framework on which all subsequent laws have been based and enables the federal government to exercise its authority to enforce matters relating to health and safety, Article 123 also establishes certain social principles that form the basis for many of Mexico's labor and social policies such as minimum wage, maternity leave, a ban on children in the workplace less than age 14, a maximum eight-hour work day, employers obligation to follow all health and safety laws, and their responsibility for workplace injuries and illnesses. Employers are obligated to provide employees with a safe workplace and to be responsible for injuries and illnesses related to normal work practices or emergencies.

The Federal Labor Law was first ratified in 1931 and revised in 1972 thereby codifying the labor principles outlined in the Constitution. It addresses the general responsibilities of employers, the reporting of injuries, sanctions for failure to comply with the law, and established the National and State Advisory Commissions on Workplace Health and Safety (57). When the law was revised again in 1978, the General Regulation on Workplace Safety and Hygiene was adopted which established programs to facilitate cooperation between employers and workers in addressing workplace preventive measures. The regulation requires employers to establish joint management-labor committees, control hazardous materials and conditions, and set provisions for fire prevention, among others. The members of these safety committees ("comisiones mixtas") have responsibility for inspecting the workplace on a monthly basis and formulating recommendations to improve working conditions. They also verify the implementation of both the safety and health measures mandated by law and the technical improvements dictated by Federal health inspectors (58).

In 1987, the Secretariat of Health issued a series of 22 technical requirements for health and safety. However, it is the Secretariat of Labor and Social Welfare that has authority for the enforcement of health and safety regulations in Mexico.

Official Mexican Norms ("Normas Oficiales Mexicana"—NOM) are developed by various government agencies to establish specific requirements for health and safety management in the workplace. Examples of these NOMs include: NOM-017-STPS-1994, relating to personal protection equipment for workers in workplaces; NOM-003-SSA1-1993, Environmental Health and Sanitary requirements for labeling of paints, inks, varnishes, lacquers and enamels; NOM-009–STPS-1993, relating to safety and hygiene conditions for the storage, transportation and handling of corrosive, irritant and toxic substances in work centers; NOM-005-STPS-1993, relating to safety conditions for storage, transportation and handling of flammable and combustible substances in work centers; NOM-010-STPS-1994, relating to safety and hygiene conditions in workplaces where chemical substances capable of generating contamination in the work environment are produced, stored or handled; NOM-013-STPS-1993, relating to safety and hygiene conditions in work centers in which non-ionizing electromagnetic radiation is generated; and NOM-048-SSA1-1993, establishing the standardized method for the assessment of health risks due to environmental agents, to list just a few.

The Federal Regulation on Safety, Hygiene and the Workplace Environment was promulgated on January 21, 1997. It repealed six existing regulations and reduced the number of articles in the law from 1353 to 168. However, the most notable change is the privatization of government enforcement and regulation of occupational health and safety in Mexico—the most unique aspect of the Mexican regulatory scheme. This act permits privately owned "verification units" to conduct workplace inspections and, upon "verification of compliance" with regulations, to exempt the inspected companies from further inspections and fines by the Department of Labor and Social Welfare. As a result, enforcement can now occur in one of three different ways: (*1*) traditional safety and health workplace inspections by both federal and state officers; (*2*) voluntary safety and health at work programs; and (*3*) third-party verification units (57, 58).

The Secretariat of Labor and Social Welfare (Secretaria del Trabajo y Prevision Social), through the General Direction on Safety and Hygiene at Work, is responsible for developing, reviewing, updating and publication of the Mexican occupational health and safety regulations. Most of this work is done through several technical committees in which health and safety professionals from private industry, government agencies, professional associations, worker unions, and consultants provide input and expertise. Another important organization in the Mexican occupational health and safety system is the Social Security Institute that employs groups of health and safety professionals and conducts workplace surveys including occupational hygiene.

4.14.3 Occupational Exposure Limits

Mexico has established occupational exposure limits for 562 toxic substances and airborne contaminants through the adoption of the most recent ACGIH TLVs (57). The Department of Medicine and Occupational Security under the Secretariat of Labor and Social Welfare (STPS) has traditionally had responsibility for establishing these limits which are included in substance-specific NOM which may also include a specified analytical method for the substance or hazard in question (15). For example, NOM-010-STPS-1993, addressing chemical agents; NOM-015-STPS-1994, relating to occupational exposure to elevated or reduced thermal conditions in the workplace; NOM-073-STPS-1993, Determination of isobutyl alcohol in the air—Gas chromatography method; NOM-031-STPS-1993, Determination of Vinyl Chloride in the Air. Gas Chromatographic Method; NOM-036-STPS-1993, Determination of Formaldehyde in the Air—Spectrophotometric method; NOM-035-STPS-1993, Determination of Carbon Monoxide in Air—Electrochemical Method; NOM-072-STPS-1993, Determination of Free Silica Dust (quartz, cristobalite and trydamite) Airborne in the Work Environment—X-Ray Diffraction Method; NOM-079-ECOL-1994, which establishes maximum permissible limits for noise emission from inplant new motor vehicles and the method for measurement thereof, among many others.

4.14.4 Education and Training

No undergraduate or graduate university programs have been established in Mexico that are specific to occupational hygiene (57). There have been efforts to establish a masters degree program in occupational hygiene at several large universities. Curricula has been developed several times and reviewed for both a 160-hour course work and for a two-year masters degree program. Unfortunately, none has been sustained longer than one class (57). Acquisition of a graduate degree in occupational hygiene from an U.S. institution is an increasing trend.

Most occupational hygienists in Mexico obtain professional development through one to five-day continuing education courses sponsored by various organizations including the National Federation of Occupational Health-Related Societies and Associations (Federacion Nacional de Asociaciones y Sociedades Relacioandas con la Salud en el Trabjo, A.C.—FeNASSTAC). Most courses are offered in Mexico City.

4.14.5 Certification and Registration

The Mexican Industrial Hygiene Association (AMHI) is leading an effort to establish a national certification program in occupational hygiene. An alliance is being formed with the national safety certification body to facilitate the development of the process. Certification for occupational hygienists is expected in the next two to three years. There are five Mexican occupational hygienists who are certified by the ABIH.

4.14.6 Professional Organizations

The AMHI (Association Mexicana de Higiene Industrial) was founded in Mexico City in July 1995, by a small group of six enthusiastic Mexican occupational hygienists (57). The formation of AMHI was the culmination of several other unsuccessful efforts at establishing long-term local or national occupational hygiene associations. The mission of the AMHI is to "promote the development of occupational hygiene in the country" through three primary functions: membership recruitment, professional development, and advocacy within the government on behalf of the profession. AMHI has organized technical seminars with the participation of speakers representing well-known organizations, such as NIOSH,

ACGIH, WHO, and ILO. Three national occupational hygiene conferences have been jointly organized with the Mexican Hygiene and Safety Association (Asociacion Mexicana de Higiene y Seguridad, A.C.—AMHSAC) in Mexico City in 1996, 1997, and 1998. The conferences have attracted more than 200 attendees who heard representatives from various national and international organizations speak on a wide range of issues. As of March 1999, AMHI had more than 150 members (57). The association was recently accepted for membership in the IOHA. No code of ethics has been developed yet; however, this is expected in conjunction with the development of a certification system.

The AMHSAC was founded in 1982, and provides training in occupational safety and occupational hygiene, environmental compliance, hazard materials handling, emergency response, and publishes a monthly journal. Membership is open to private companies.

A third significant private organization in Mexico is the FeNASSTAC, an umbrella association consisting of approximately 40 local, state, and national organizations with an interest in occupational health. AMHI is a member of FeNASSTAC.

4.14.7 Status of the Profession

Occupational hygiene is a young profession in Mexico and generally is misunderstood among the general public. Although occupational health and safety regulations have existed in the country since the promulgation of the 1917 Mexican Constitution, occupational health requirements are not always correctly followed, enforced, and/or understood by employers, employees, government agencies, or health and safety professionals. Few full-time occupational hygiene practitioners or mature occupational hygiene programs exist within companies with the general exception of multinational companies who have imported programs (57).

According to AMHI membership, safety professionals and occupational medicine physicians do most occupational hygiene work (57). Most members hold a bachelor's degree, so there are very few occupational hygiene technicians in the country and most full-time practitioners have a chemistry or chemical engineering background. There are some occupational hygienists working for private companies, government agencies, nongovernment organizations, consulting, and academia. There is a very close association between most full-time and part-time occupational hygienists and safety engineers, occupational physicians and occupational health nurses.

There are some occupational hygiene laboratories in Mexico of which the majority are governmental laboratories—none currently AIHA accredited. Some laboratories have been accredited by the Mexican national accreditation program, which differs in philosophy from AIHA's system. e.g., in Mexico there is no proficiency analytical testing (PAT) program, so accreditation is based only on a system review and audit rather than a combination of performance and system review.

4.14.8 Information Resources

Mexican Industrial Hygiene Association
Asociacion Mexicana de Higiene Industrial, A.C.
Tecualiapan #36-II-6, Mexico City, D.F. 04000 Mexico

Tel & Fax: 52 55 54 9676
Status: Professional Association

National Chemical Industry Association
Occupational Safety and Health Committee
Asociacion Nacional de la Industria Quimica, A.C.
Providencia # 1118, Mexico City, D.F. 03100 Mexico
Tel: 52 52 30 5100, Fax: 52 55 59 2208
Status: Trade Association

Secretariat of Labor and Social Welfare
Secretaria del Trabajo y Prevision Social
Av. Azcapotzalco-La Villa #209, Edif C
Mexico City, 02020 Mexico
Tel: 52 53 94 5166, Fax: 52 53 94 264
Status: Regulatory Agency

Mexican Hygiene and Safety Association
Asociacion Mexicana de Higiene y Seguridad, A.C.
Lirio #7, Mexico City, D.F. 06400 Mexico
Tel: 52 55 47 8782, Fax: 52 55 41 1566
Status: Professional Association

Mexican Social Security Institute
Instituto Mexicano del Seguro Social
Centro Medico Siglo XXI, Av. Cuauhtemoc #330
Mexico City, 06725, Mexico
Tel: 52 55 19 1999, Fax: 52 55 38 7739
Status: Government Medical Institute

National Federation of Occ. Health-Related Societies and Association
Federacion Nacional de Asociaciones y Sociedades
Relacioandas con la Salud en el Trabjo, A.C.
Cd. Juarez, Chiuhahua, Mexico
Status: Professional Association

4.15 The Netherlands

4.15.1 Demographics and Economics

The Netherlands has a highly developed economy centered on private enterprise. Industrial activity features food processing, petroleum refining, and the manufacture of electrical machinery and equipment, agrochemicals and microelectronics, metalworking and construction with a work force of 6.6 million within an overall population of 15.7 million (7). The agricultural sector employs only 2% of the labor force, but provides large surpluses for export and the domestic food-processing industry. The Netherlands is a member state of the European Monetary Union. Distribution of the Dutch work force is as follows: Services, 75%; Manufacturing and construction, 23%; and Agriculture and fisheries, 2%.

It is estimated that there are about 530 occupational hygienists in the Netherlands (59). About 70 percent work in occupational health services organizations. Another 20% work for governmental agencies or research institutions, or hold teaching positions. Less than 10% percent are with the Labor Inspectorate, and the remainder work as consultants either or in related services such as safety and environmental departments. The ratio of occupational hygienists to the Dutch work force 1:12,400. Distribution of Dutch Occupational Hygienists is as follows: Occupational health services, 70%; Research and academia, 13%; Government and other, 8%; and Consulting, 6%.

4.15.2 Historical Development

In the Netherlands, occupational hygiene is a young discipline that dates from the early 1970s. Until then, hazardous working conditions were addressed by safety engineers, who focused on short-term risks and by sanitary engineers, whose approach emphasized appropriate design of machinery and equipment, occupational physicians were also involved responsible for all human-related activities, including risk assessment. Beginning in the early 1970s, the increasing importance of identifying and controlling hazardous chemical, physical, mechanical and biological agents at the workplace was recognized. Occupational physicians, as the primary occupational health care providers, acknowledged that occupational hygiene activities had grown beyond their scope of practice. Universities and polytechnical colleges were called upon to educate professionals who were able to assess hazardous chemical and physical agents in the workplace, perform risk assessments, and interface with designers, architects, engineers, and ergonomists.

A hallmark meeting was convened in 1978 when the national association for occupational physicians officially acknowledged that occupational hygiene was an indispensable discipline in managing long-term risks to the health of workers. It was advised that each occupational health service should employ an occupational hygienist and that the Agricultural University of Wageningen should develop an educational program leading to a Master of Science (M.Sc.) in occupational hygiene.

Soon after this meeting educational programs were launched at Wageningen University and two polytechnical colleges. The latter offered concise one-year courses for chemists to meet the strong demand for occupational hygienists. As a result of these educational opportunities, the occupational hygiene discipline began to take shape. In 1983, the Dutch Occupational Hygiene Society was launched with about 30 members. By 1985 there were approximately 50 occupational hygienists employed nationally in occupational health services. Their numbers grew steadily to 75 in 1988, 140 in 1992, and 375 in 1997 (59). These professionals make up about 70% of the estimated number of occupational hygienists in the Netherlands.

4.15.3 Occupational Health and Safety Legislation

The first government regulation of working conditions in Dutch factories took place in 1874. Such early legislation focused on improving the lot of children working in industry. The law limited the number of hours children were permitted to work and addressed some recognized hazards linked to working conditions characteristic of the times. Traditionally, such laws were intended to protect persons unable to protect themselves. The law of 1874 covered children and in 1889 the law was broadened to include minors and women. Six years later, in 1895, the Labor Act extended its protection to cover all workers, including mine workers. The Act also enabled the Dutch government to set standards for the proper use of hazardous materials and of industrial equipment.

In the 1960s, the Occupational Safety Act broadened occupational health care. Initially, medical examinations sought to determine the physical fitness of job seekers and their ability to work in factories. In addition, physicians examined workers to determine whether they were able to return to work after an absence due to illness or an accident.

Over the years, the existing network of safety and health laws and regulations had grown inadequate. Traditionally, legislators had focused exclusively on employers and their responsibilities addressing the workers' health and safety issues and their firsthand knowledge of work-related hazards. To correct this situation, Her Majesty's Government passed the Working Environment Act on November 8, 1980, which has been partially in force since January 1, 1982 (60). It consists of a number of nationally and internationally recognized regulations on occupational health, safety, and welfare. It enables legislators further to strengthen such regulations through subsequent amendments. The act provides for the creation of worker Health-Safety-Welfare Committees as partners to management on such matters. It specifies management and worker responsibilities in matters of health and safety and enables employees to participate in the formulation of joint policies.

The Act requires the establishment of health services for places of employment with more than 500 employees and for certain categories of factories such as those producing lead batteries and lead pigments. Such health services are charged with promoting and protecting the health and welfare of workers at their place of employment. These services may exist either within larger companies or as regional centers. The latter provide their customers with comprehensive occupational health services for a fee and include assistance in occupational hygiene. Private health services may operate as consultants to their customers, but cannot enforce the law. They may, however, notify official agencies of the existence of occupational hazards in the workplace.

Traditionally, occupational physicians have provided medical services to workers and safety engineers have focused on eliminating physical hazards and preventing accidents (61). What occupational hygiene may have lacked in tradition in the Netherlands, it has made up with the concerted and coordinated efforts to protect the work force and its active role in promoting health and safety matters within the European Economic Community (61).

The 1989 implementing order on toxic substances of the 1980 Working Environment Act stipulated that knowledge of occupational hygiene is essential in addressing risks due to the use of toxic substances. This implementing order further stipulated that the approach of anticipation, recognition, evaluation, and control had to be adopted companies dealing with all toxic substances. Today, companies with 15 employees or more are obliged to conduct a hazard identification and risk evaluation equivalent to the implementation of the EU's Workplace Framework Directive 89/391/eec. This also includes work environments such as offices, hospitals, etc.

In the past three years, additional legislative measures were introduced to reduce disability benefit claimants, the rate of sick leave, and to inventory and evaluate occupational hazards. In the latter amendment, all employers are obliged to perform hazard identification and risk evaluation in accordance with the EU Workplace Framework Directive. Companies with more than 15 employees must engage a certified occupational health and safety service when developing an inventory of occupational hazards and evaluating the potential hazards of their use, or at least have their risk assessment approved by the certified occupational health and safety service. This requirement was instituted as a quality control mechanism for risk assessments.

As a result of the Workplace Framework Directive, enforced in all Dutch companies as of 1998, there has been a national dialogue on the qualifications and necessary experience of personnel and management structures within certified occupational health and safety services. Various professional disciplines lobbied for recognition based on their qualifications and competence. The occupational hygienist profession was able to demonstrate its relevancy to any occupational health and safety service based on their expertise in hazard identification and assessment, and advisory skills in planning, designing and implementing of control measures. Currently, the Dutch government sets the minimum requirements for a certified occupational health and safety service. These include, but are not limited to, the requirement that such a service must employ an occupational hygienist that is responsible for the occupational hygiene program, including supervision of the mandatory risk assessments performed by employers. Other professionals required in a certified occupational health and safety service are occupational physicians, safety engineers, and a labor and management consultant. The occupational hygienist must be certified in accordance with the approved standards of thc Dutch ccrtification scheme (62). This system has provided a legal structure for occupational hygiene activities and the profession.

The authority to enforce health and safety laws and regulations rests with the Labor Inspectorate under the Director General of Labor.

4.15.4 Occupational Exposure Limits

The Labor Inspectorate, under the Director General of Labor, use the "Nationale MAC-list Arbeidsinspectie P No. 145" as a guide in enforcing regulatory requirements relating to occupational exposures. The Dutch MAC values are derived primarily from the ACGIH TLV's. Some are based on limits listed by the Senate Commission for the Investigation of Health Hazards of Work Materials of the Federal Republic of Germany. Others draw from recommendations published by NIOSH in the United States (15).

Chemical Abstracts Service (CAS) numbers also identify substances on the MAC list. Permissible exposure levels fall into two categories: maximal accepted concentrations (Maximale Aanvaarde Concentratie-Tijdgewogen Gemiddelde, or 'MAC-TGG'), averaged over a period of up to eight hours per day, 40 hours per week; and, ceiling concentrations (Maximale Aanvaarde Concentratie-Ceiling, or MAC-C), which may not be exceeded at any time because of their acute toxic effects.

4.15.5 Education and Training

The Agricultural University of Wageningen is the only Dutch educational institution conferring M.Sc. and Ph.D. degrees in occupational hygiene. The program was established in the early 1980s at this institution based on its long-standing research focus on the effects of air and water pollution on soil contamination as well as on flora, fauna, and human life. According to Drown (63), the program began in 1978 within the Department of Air Pollution and soon branched out to include the Departments of Public Health and Toxicology.

The program is a full-time course of study modeled after the Harvard University occupational hygiene program (61). Students focus on the sciences and public health and receive a substantial amount of practical experience as a part of ongoing industrial research projects carried on by the three departments involved.

Graduates of the master's program usually occupy positions of responsibility in occupational health services or in academia. The Post-HBO Hogeschool West-Brabant and the Polytechnic Ijsselland and Polytechnic Breda offer a two-year training program in occupational hygiene. Courses are taught in the evenings to accommodate the needs of working students, many of whom are employed in occupational health services as technicians, occupational health nurses, and safety specialists. Graduates receive a certificate of completion and find such training helpful, for some are called upon to assist with occupational hygiene-related problems even though fully trained occupational hygienists are becoming more numerous.

4.15.6 Certification and Registration

In 1993, a third party certification process was introduced by the Foundation for Certifying Occupational Hygienists (Stichting Ter Certificering Van Arbeidshygiënisten). The process is based on the belief that regulation of the profession was an important step in the quality assurance of occupational hygiene care delivered to industry. The development of a certification scheme was accelerated by the introduction of the governmental licensing scheme for occupational health services.

The Dutch certification scheme awards eligible candidates with the designation registered occupational hygienist (RAH in Dutch). Requirements for certification are related to appropriate education in occupational hygiene at an academic degree level, and to full-time work experience acquired while working for an accredited service or consultant in the field of occupational health and safety (62). The certifying body has established detailed guidelines and standards for an appropriate occupational hygiene curriculum. However, the Dutch system is unique in that it does not require the completion of an examination specific to the certification process, but as part of the candidate's prerequisite course work.

A certification maintenance program is being set up based on a three-year cycle of certification renewal with emphasis on active participation in the profession and continuous education and training. Although discussed extensively when the scheme was first established, a separate level of qualification pertaining to occupational hygiene technicians had not been approved as of 1998 (61). The Dutch Council has formally approved the certification scheme for accreditation, which acts as a national organization auditing certifying bodies. As of the first quarter of 1999, there were 225 RAH's in The Netherlands (59).

4.15.7 Professional Organizations

In the early 1980s, the DOHS (Nederlandse Vereniging Voor Arbeidshygiene) was formed and today has a membership of 532. It is estimated that approximately 95% of all occupational hygienists are members of the DOHS (59). Members come from industry, academia, the occupational health services, and the Dutch government. Applicants must be actively engaged in the field of occupational hygiene and have a degree from an accredited institution. The DOHS is a member of the IOHA from which the DOHS code of ethics has been adapted.

The Society is very active in national and international affairs (61). In addition to sponsoring professional conferences at home and participating in conferences abroad such as the American Industrial Hygiene Conference and Exposition, it has proposed a limit for hand-arm vibrations (Voorstel voor een Grenswaarde voor Hand-Arm Trillingen, January 1988) as a guide for practicing occupational hygienists and researchers. It maintains close liaison with BOHS, ACGIH, AIHA, and other societies. It has a number of active technical committees. Among the most active is the Education Committee, which has compiled recommendations for training programs in occupational hygiene. The committee is presently considering the need for professional refresher courses and for certification in line with the needs and conditions prevailing in the Netherlands.

The DOHS is also considering ways to influence EEC occupational health regulations to represent specific Dutch interests in cooperation with BOHS and other European sister societies. Other recent developments include initiating an annual technical symposium, launching a scientific journal (Tijdschrift Voor Toegepaste Arbowetenschappen), establishing special interest groups for the building, metals, health care industries and the office environment. The DOHS has also established an award for the best project work, the Bob van Beek Award.

4.15.8 Status of the Profession

Occupational hygiene has legal status in the Netherlands and has been defined as the "applied science concerned with recognition, evaluation and control of environmental stresses especially chemical, physical, and biological stresses arising from work which might adversely affect the health and/or well-being of people at work and/or their posterity" (64).

The definition is interesting because of both its wording and its omissions. According to the authors of the definition, the word "especially" was inserted to emphasize the preeminent nature of chemical, physical, and biological stresses in the workplace while leaving room for physiological, psychological, and other factors. The definition specifically mentions adverse reproductive effects and excludes effects beyond the workplace because these are considered the domain of environmental health professionals.

There are four primary factors that have impacted the current standard of occupational hygiene practice in the Netherlands: (*1*) The DOHS has concentrated its efforts on professional recognition with the national government, industry, and allied professions, but has not addressed the clear lack of professional recognition among the general public. (*2*) Education and training programs for occupational hygienists are well established and the academic degree programs have been favorably evaluated with regard to course content and quality. (*3*) The majority of the occupational hygienists are bound to restrict themselves to the practice of occupational hygiene. This has created a strong profession with a primary focus on competency. (*4*) The timely introduction of a certification process has been essential in achieving legal recognition of professional occupational hygienists in the Netherlands. The profession was able to set the required qualifications and offer the government a well-designed system of certification (65).

Four important trends will shape the future of occupational hygiene in The Netherlands: (*1*) Education and training programs as well as certification in-line with international developments; (*2*) A clear shift towards developing an integrated policy at the company level, covering health and safety, environment, and quality assurance in one management system; (*3*) A shift in Dutch legislative approach to hazard control from specific standards to guidelines. Occupational hygienists will have to adjust to this new regulatory paradigm. Less focus on compliance will increase the contribution of professional judgment; and (*4*) The revived focus on prevention and engineering control among occupational hygienists in the Netherlands (61). New issues like life-cycle analysis and product stewardship have been introduced in industry. Companies are increasingly concerned about the environmental impact of their products, and research and development programs are directed towards recycling of products. This presents another challenges for occupational hygienists in combining environmental issues such as factory emissions with the control of exposures during production.

Although the number people entering the discipline in The Netherlands has leveled off, occupational hygiene is firmly rooted in the occupational health care system that is catering to almost 85% of the Dutch work force (61).

4.15.9 Information Resources

Wageningen Agricultural University
Department of Epidemiology and Public Health
P. O. Box 238, 6700 Ae Wagenigen, The Netherlands
Tel: 0 317 482080, Fax: 0 317 482782
Status: Academic Institution

Foundation for Certifying Occupational Hygienists
Stichting Ter Certificering Van Arbeidshygiënisten
Sutton 7, 7327 Ab Apeldoorn, The Netherlands
Tel: 0 55 534 2328, Fax: 0 55 534 2329
Internet: www.sch-rah.nl
Status: Professional certification organization

The Dutch Occupational Hygiene Society (DOHS)
P.O. Box 1762, 5602 Bt Eindhoven, The Netherlands
Tel: 0 40 292 6575, Fax: 0 40 248 2328
Status: Professional Association

4.16 Norway

4.16.1 Demographics and Economic

Endowed with natural resources including petroleum, hydroelectric power, fish, forests and minerals. Norway's economic structure is based on a combination of free markets and government controls. The government maintains tight controls on the vital petroleum sector and provides extensive subsidizes to agriculture, fishing, and areas with sparse resources. Norway also maintains an extensive welfare system that results in public sector expenditures of more than 50% of GDP (7). As a result, Norway has one of the highest average tax levels in the world.

A small country in population (4.4 million citizens), Norway is an exporter of raw materials and converted goods, is highly dependent on international trade and has an abundance of small- and medium-sized companies (7). It is one of the world's major shipping nations. The country is also highly dependent on its oil sector that is second only to Saudi Arabia in oil exports (7). In a November 1994 referendum, Norway's population voted to remain separate from the European Economic Community. The Norwegian work force is 2.13 million. In addition to oil and gas, Norway's industries include: food processing, metals processing, shipbuilding, pulp and paper products, chemicals, timber, mining, textiles and fishing. Distribution of the Norwegian work force is as follows: Services, 71%; Industry, 23%; and Agriculture, forestry, and fishing, 6%.

There are 300 members in the Norwegian Occupational Hygiene Association (NYF), but not all occupational hygienists and/or technicians are members (66). The distribution of occupational hygienists in industry, government, academia and consulting is dominated by Occupational Health Services. Norway enjoys one of the lowest occupational hygienists to work force ratios in the world at approximately 1:7,100. Distribution of the Norwegian Occupational Hygiene Association membership is as follows: Occupational Health Services, 32%; Others, 23%; Industry, 17%; Government, 12%; Academia (STAMI and hospitals), 9%; and Consulting, 8%.

4.16.2 Historical Development

Occupational hygiene was formally recognized in Norway in 1974 beginning with the development of the Labor Inspectorate Authority. At the time, both the Labor Inspectorate and private companies required access to knowledge relating to complex health problems and began looking for and promoting an institutional resource for occupational hygiene competence. By the 1980s, the need for occupational hygiene in occupational health services had become acute and a number of engineers began to do occupational hygiene work without formal education. This resulted in a government review of the issue and the subsequent development of a national certification system to verify occupational hygiene competence for consumers.

4.16.3 Occupational Health and Safety Legislation

The key piece of occupational health and safety legislation in Norway is the Working Environment Act of 1977. The Act was developed and is enforced by the Norwegian Labour Inspection Authority and codifies regulations pertaining to the systematic control of health and safety activities in private and government enterprises.

4.16.4 Occupational Exposure Limits

The original set of occupational exposure limits was issued by the Directorate for Labor Inspection in 1978 and were based on the Danish "hygienic limit values" (15). Norwegian exposure limits are treated as guidelines unless there is formal notification from the government that the status has changed to legally binding. The limits are updated every two years and developed on the basis of disease and disorders associated with the chemical substance, but also justified on the basis of economic and technical feasibility. Skin, mixture and ceiling limit designation are included where appropriate.

4.16.5 Education and Training

The increasing concern about chemical, physical and biological hazards in the environment and the health and safety of workers has prompted the formation of a new graduate program in occupational hygiene at the University of Bergen (66). This is the only university program in occupational hygiene in Norway, but also attracts students from other countries.

The primary objective of the Bergen program is to provide students and working professionals with the expertise to identify, evaluate, and control the risks associated with the workplace. It is an interdisciplinary degree coordinated between the Health Sciences program at the Centre for International Health, and the Department of Public Health and Primary Health Care, Division of Occupational Medicine within the Faculty of Medicine. Successful completion of the program results in a Master of Public Health degree (M.P.H.).

Admission requirements include a Bachelor of Science (B.Sc.) from a university, technical college or equivalent qualifications with prerequisite coursework in chemistry, biology and/or physics. The program curriculum includes epidemiology, disease prevention and health promotion, introduction to occupational hygiene, introduction to occupational medicine, monitoring chemical factors in the working environment, preventive measures for chemical factors, biological factors in the working environment, physical factors in the working environment, noise and vibrations, ionizing and non-ionizing radiation, electromagnetic fields, the indoor climate, safety and risk assessment, and environmental pollution. Completion of a thesis is required.

4.16.6 Certification and Accreditation

Norway introduced a new occupational hygiene certification in 1996 which is administered by the Norwegian Occupational Hygiene Certification Association. To become certified, candidates must have at least four to five years of academic experience at a technical school or a bachelors degree from a university, and at least three years of relevant work experience in an occupational health service, industry or as a consultant in occupational health and safety. The certification carries certification maintenance requirements.

4.16.7 Professional Organizations

The Norwegian Occupational Hygiene Association (Norsk Yrkeshygienisk Forening—NYF) has membership of about 300 and meets semi-annually to discuss association matters and to review relevant activities in the profession and among other associations. Once every three to four years there is a conference between all the Norwegian associations related to health and safety. The NYF is a member of the IOHA and enforces the IOHA Code of Ethics.

4.16.8 Status of the Profession

Occupational hygiene is relatively new in Norway having been introduced by the Labour Inspectorate in 1974. Occupational hygienists in Norway are generally university graduates with training in the sciences or engineering. They are part of the occupational health services (teams) within industry and now have a national certification mechanism in effect. Norway enjoys one of the highest ratios of professional occupational hygienists to the total work force in the world. Further growth of the profession in Norway will require government recognition and academic opportunities, including a guaranteed and suitable level of training and competence for practitioners (66).

4.16.9 Information Resources

Norwegian Occupational Hygiene Association
Norsk Yrkeshygienisk Forening (NYF)
Arbeidsmedisin Avdeling—RiT
Olav Kyrresgt. 3, N-7006 Trondheim, Norway
Internet: www.takvam.no/nyf
Status: Professional Association

NTN University
N-7034 Trondheim, Norway
Tel: 47 73 59 50 00
Status: Academic Institution

University of Bergen
Division for Occupational Medicine
Ulriksdal 8C, N-5009 Bergen, Norway
Tel: 47 55 58 61 00
Status: Academic Institution

National Institute of Occupational Health (STAMI)
Pb. 8149 Dep.
N-0033 Oslo, Norway
Tel: 47 23 19 51 00
Status: Governmental Research Institute

Høyskolen Bø
N-3800 Bø, Norway
Tel: 47 35 95 25 48
Status: Private Research Organization

Department of Labour Inspection
Direktoratet for Arbeidstilsynet
P.O. Box 8103 Dep, Oslo 0032, Norway
Tel: 47 22 95 70 00, Fax: 47 22 46 62 14
Internet: www.arbeidstilsynet.no
Status: Government Regulatory Agency

4.17 Poland

4.17.1 Demographics and Economics

Poland's population of 38.5 million occupies 16 separate (provincial) administrative regions called voivodships (7). Each voivodship maintains its own administrative system tied to the federal government that is structured as a parliamentary democracy. With the transition from a socialist system under the influence of the Soviet Union in 1990–1991 to the current parliamentary democracy, most of Poland's government institutions and economic policies have gone through a rapid and pronounced transformation.

The Polish economy is based on manufacturing and agriculture with nine primary manufacturing sectors: machine building, iron and steel, coal mining, chemicals, shipbuilding, food processing, glass, beverages, and textile (7). There are a growing number of small and large businesses that are creating an increasing need for occupational hygiene legislation and consultants to address the demands.

Poland's development under central planning and its subsequent transition to a market economy beginning in 1990–1991 has helped the country stand out as one of the most successful and open transition economies. The privatization of small- and medium-sized state-owned companies and a liberal law on establishing new companies has characterized much of the development of the private sector that is currently responsible for two-thirds of overall economic activity in Poland. Restructuring and privatization of remaining state sectors, so called "sensitive sectors" (7), including coal and steel has been delayed while privatization efforts in the aviation, energy, and telecommunications sectors is ongoing.

The distribution of total labor force of some 17.7 million workers in the Polish economy is highlighted as follows: Agriculture and forestry, 28.1%; Manufacturing, 24.3%; Other (including government), 22.6%; Wholesale and retail sales and other services, 12.3%; Construction, 5.6%; Transportation and telecommunication, 5.3%; and Coal mining, 1.8%. The number of professional occupational hygienists in Poland is not available; however, an estimate of less than 300 is based on membership in the Polish Industrial Hygiene Association (67). Assuming that this estimate is accurate, the ratio of occupational hygienists to the working population would be 1:59,000.

4.17.2 Historical Development

While the first significant occupational hygiene activity in Poland occurred with the first national legislation to address occupational health, the Decree of the President of 1927. Pioneers of the field including Dr. Kasprzak and Dr. Nowakowski are credited with bringing focus to the issues of occupational health protection of workers between 1918 and 1939. After World War II, Poland, like many Central and Eastern European countries modified the Soviet system of occupational health which resulted in a primary focus on the diagnosis and treatment of occupational disease, and less on prevention (68). This trend coincided with extensive industrialization throughout the country in what had traditionally been an agricultural economy. Heavy industry was advanced with output the key goal and with less emphasis on the protection of resources, including worker health.

In the early 1950s, four independent research institutes of occupational medicine were established: Nofer Institute of Occupational Medicine (NIOM) in Lodz; the Institute of Occupational Medicine and Environmental Health (IOMEH) in Sosnowiec; the Institute of Marine and Tropical Diseases (IMMTD) in Gdynia; and the Institute of Rural Medicine (IRM) in Lublin (69). Since their development, these institutes have become the focal point for occupational health including occupational hygiene in Poland. The establishment of the institutes coincided with the establishment of the State Sanitary Inspectorate in 1954 under the Ministry of Health and Social Welfare that continues today as the primary compliance mechanism for occupational hygiene legislation.

4.17.3 Occupational Health and Safety Legislation

Poland is divided administratively into 16 voivodships (provinces) and 2,121 gminas (municipalities). Much of the implementation of occupational health and safety policy is addressed at the voivodship level including the primary instrument, the Labor Code Act of 1974. The initial version of the Labor Code defined the basic responsibilities of state-owned enterprises, but was completely revised in 1996 to integrate the rapidly developing private sector as well as EU and International Labour Organization (ILO) directives. It continues to be enforced through the State Sanitary Inspectorate and the State Labor Inspectorate through separate acts of Parliament. The State Labor Inspectorate focuses on compliance with labor laws and other regulations affecting occupational health and safety whereas the State Sanitary Inspectorate deals primarily with health-related issues including the Polish occupational exposure limits. The basic enforcement unit of the State Sanitary Inspectorate is the Sanitary Epidemiological Stations under the Department of Occupational Hygiene which maintains more than 2,000 inspectors and laboratory analysts across all 16 voivodships and in 193 smaller administrative units. More than 1,305,000 field measurements and laboratory analyses of chemicals and physical agents were conducted relating to 250,000 workplace in 1995 alone (68).

Employers are obligated to provide access to health and safety services if they employ more than 10 people and maintain a full-time health and safety officer for every 600 workers. These services can be acquired through external providers for companies with less than 50 employees. Under the 1996 Ordinance for Carcinogenic Agents, employers are also obligated to register the use or production of any of 55 designated carcinogens and 48 probable carcinogens, review working conditions with regard to these agents, assess exposures at least twice a year, provide hazard communication information including safe work practices, and maintain records of exposed employees for 40 years.

State-owned industries have also maintained occupational hygiene laboratories, generally for those manufacturing plants with more than 500 employees as recommended by the Ministry for Health and Social Welfare. However, with the economic transition of the early 1990s, the number of industry laboratories dropped from 777 in 1989 to 481 in 1995. New private laboratories have compensated for this decline, but until 1998 only a few had obtained accreditation from the Polish Centre for Testing and Certification in accordance with ISO Guide 25. To address this quality assurance issue, in 1996 the Ministry of Health and Social Welfare published an ordinance requiring that the analysis on occupational hygiene samples be conducted only by State Sanitary Inspectorate, scientific research institutes dedicated to occupational hygiene and medicine measurements, or laboratories authorized by the regional sanitary inspector. To improve analytical quality control, the Nofer Institute of Occupational Medicine in Lodz developed an interlaboratory proficiency-testing program. As of 1999, there are 280 laboratories participating in chemical analysis (metals and organic solvents), 230 involved in the analysis of crystalline silica and 10 involved in mineral fibers.

The Central Institute for Labor Protection was established in 1950 by the Minister of Labor and Social Policy to conduct research in developing working conditions that would conform to the requirements of the ILO and EU, develop the Polish maximum allowable concentrations (MSC) and maximum allowable intensities (MAI), develop standards for occupational safety and ergonomics, test and certify personal protective equipment, and provide education in occupational health and safety. The institute is also the coordinating center for the national strategic program: "Occupational Safety and Health Protection in the Working Environment," launched in 1995 by the Minister of Health and Social Welfare and the State Committee for Scientific Research.

4.17.4 Occupational Exposure Limits

Polish occupational exposure limits, referred to as Maximum Allowable Concentrations (MAC) for chemical agents, and Maximum Allowable Intensities (MAI) for physical agents, were first published in 1956 through the codification of the 1956 ACGIH TLVs. The Ministry of Labor and Social Policy began regulating MACs and MAIs in 1982 and established the Intersectoral Commission for Updating the Register of MAC and MAI in the Working Environment in 1983 to review existing exposure limits and develop new limits. This permanent commission is composed of representatives of industry, labor, and various occupational medicine and hygiene research institutes, such as the NIOM and IOMEH (70). The commission submits documentation for proposed standards to the Ministry of Labor and Social Policy, whereby after approval, introduce the new exposure limits into legislation. However, new exposure limits can not be legislated and enforced without the development and publication of an accompanying exposure assessment and analytical method that is published in the form of a Polish Standard. MACs are developed on the basis of eight-hour time-weighted averages, 30-minute short-term exposure limits, and ceiling concentrations. There are currently 321 chemical and dust MACs, and separate MAIs for noise, vibration, and heat stress (71). Approximately 40 new standard proposals are developed by the Intersectoral Commission for Updating the Register of MAC and MAI in the Working Environment each year. In addition to the MACs and MAIs, biological exposure indices are also regulated by the Ministry of Health and Social Welfare.

4.17.5 Education and Training

Today, most practicing occupational hygienists in the country have a nonmedical undergraduate or graduate degree and have attended advanced professional development courses offered through the Institutes, the units of the State Sanitary Inspectorate or State Labor Inspectorate, and programs and courses in Europe and the United States. There are no undergraduate programs in occupational hygiene in Poland. In 1990, the School of Industrial and Environmental Hygiene was established at the Nofer Institute of Occupational Medicine to develop occupational hygienists. Candidates for this program must be under 35 years of age and possess an undergraduate degree in chemistry, physics, biology, medicine, pharmacy, or engineering. This postgraduate certificate program consists of three semesters of curriculum totaling 400 hours of instruction covering the fundamental of occupational hygiene, physiology and ergonomics, occupational psychology and environmental toxicology, epidemiology and biostatistics, health risk analysis, legislation, occupational safety and occupational pathology, among other topics.

There are two universities that offer undergraduate programs in environmental sciences that incorporate some occupational hygiene principles, including: Silesian University in Katowice and Warsaw Polytechnic in Warsaw. The Silesian University of Technology in Gliwice, in cooperation with the Institute of Occupational Medicine and Environmental Health in Sosnowiec is organizing a five year graduate engineering degree in occupational hygiene.

4.17.6 Certification and Registration

There is presently no governmental or professional certifying body for occupational hygiene in Poland. The development of such a scheme is one of the objectives of the Polish Industrial Hygiene Association.

4.17.7 Professional Organizations

In 1992 the Polish Association of Industrial Hygienists (PAIH) was registered as a professional and scientific organization in Poland with the objective of promoting worker protection and upgrading the stature and recognition of the profession to a scientific discipline (67, 68). As of 1998, the PAIH had in excess of 300 members from local branches of the State Inspectorates, occupational hygiene laboratories, research institutes, academic institutions and industry. There are three local sections of the PAIH throughout Poland.

Other objectives of the PAIH include the development of graduate training programs and other forms of professional development, the organization of relevant scientific meetings and conferences, upgrading the qualifications of the association's members, supporting relevant research, and cooperating with occupational hygiene associations and institutions abroad. To this end, PAIH has been working with AIHA and ACGIH to better characterize the profile of occupational hygienists internationally, define the demand for occupational hygiene services, and to determine the optimal method to obtain legal recognition for the profession. The association applied for membership to IOHA in 1999.

4.17.8 Status of the Profession

As a profession, occupational hygiene is relatively new in Poland and not fully understood or formally recognized as a scientific discipline within the country. As with many European countries, it has developed in Poland under the influence of physicians and engineers. Overall, the medical profession stills controls much of the practice and while the profession remains active and continues to grow, it lacks official recognition by the Polish government. However, various sectors of the Polish government have expressed interest in sanctioning the profession.

An accurate estimate of the population of professional occupational hygienists in Poland is difficult to ascertain because individuals from various disciplines conduct occupational hygiene-related activities including safety engineers, physicians, and "sanitarians", e.g., government employees who collect and analyze air samples. There are 700 such sanitarians, 1350 researchers at the national institutes of occupational medicine and labor protection. There are also 700 safety inspectors who are classified as government employees who measure noise exposures, assess exposure risks and conduct safety audits. However, it is estimated that there are about 100 private consultants with varying qualifications and experience.

Several factors have affected the development of the profession in Poland including the statutory control of occupational hygiene activities by safety specialists who may not have adequate training in the discipline resulting from an underestimate in the differences in knowledge and experience. As such, occupational safety issues dominate the allocation of available resources. In addition, very limited training is required to qualify as a member of an occupational health service in Poland. This has lead to a dilution of qualified staff and a systematic underestimation of the seriousness of certain occupational hazards.

Advances in professional development in Poland have centered on the network of medical institutes that carry out research in occupational hygiene and medicine, and provide education and training. Representatives of the institutes serve as advisors to the Ministry of Health and Social Welfare, other governmental ministries, industry and labor unions. They are a point of dissemination for occupational health information through the publications of peer-reviewed and non-peer-reviewed journals, including: Medycyna Pracy (Occupational Medicine), Higiena Pracy (Occupational Hygiene), Bezpieczenstwo Pracy (Safety at Work), and Atest-Ochrona Pracy (Atest-Labour Protection); In English: the *International Journal of Occupational Medicine and Environmental Health*, and the *International Journal of Occupational Safety and Ergonomics*.

The prevailing view of those dedicated to the advancement of occupational hygiene in Poland is that more focused government legislation specific to the improvement of occupational health and safety in Polish industry, and Polish membership in the EU where formal recognition of the profession has occurred, are both necessary for the continued growth of the profession (67). The next step in the development of occupational hygiene in Poland is the promulgation of the Occupational Health Services Act based on the EU Directive 89/391, which addresses measures necessary to improve worker health and safety. It is hoped this will harmonize Polish legislation with that of the EU and provide a defined role for occupational hygienists within the occupational health services model becoming more common in Europe and elsewhere (68).

4.17.9 Information Resources

Polish Association of Industrial Hygienists
Nofer Institute of Occupational Medicine
Ul Sw. Teresy 8, P.O. Box 199, Lodz 90-950, Poland
Tel: 48 42 63148 30, Fax: 48 42 634 83 31
Status: Professional Association

Nofer Institute of Occupational Medicine
Ul Sw. Teresy 8, P.O. Box 199, Lodz 90-950, Poland
Tel: 48 42 63148 29, Fax: 48 42 634 83 31
Status: Government Research Institute

Central Institute for Labor Protection
Ul. Czermiakowska 16, Warszawa 00-71, Poland
Tel: 48 22 623 73, Fax: 48 22 623 36 95

Internet: www.ciop.waw.pl
Status: Regulatory Agency

4.18 Saudi Arabia

4.18.1 Demographics and Economics

Saudi Arabia's population of 20.7 million is ruled by an institutional monarchy that achieved independence in 1932. The constitution is based on Islamic Law. With a total labor force of seven million, roughly four million foreign workers play an important role in the Saudi economy and primarily in the oil and service sectors (7). The literacy rate is approximately 63% for adults. Distribution of the Saudi Arabian Work force is as follows: Government, 40%; Industry, construction and oil, 25%; Services, 30%; and Agriculture, 5%.

Saudi Arabian industries include crude oil production, petroleum refining, a variety petrochemical products, cement, steel-rolling products, construction, fertilizer and plastics; however, the petroleum sector accounts for as much as 75% of total revenues, and 90% of export earnings. Saudi Arabia has the largest reserves of petroleum in the world with 26% of the proven total and is the largest exporter in the world (7). The majority of businesses in Saudi Arabia are government-controlled.

There are approximately 25 occupational hygienists in Saudi Arabia (73), equivalent to an occupational hygienist to work force ratio of 1:280,000.

4.18.2 Historical Development

Occupational hygiene is a relatively recent development in Saudi Arabia. In 1937, the U.S. petroleum company, SOCAL, began oil exploration and in 1939 deposits were discovered. World War II slowed the development of the newly identified deposits, but post-war demands sped development and eventually became the economic mainstay of the country. Aramco was a joint venture of American petroleum companies that was purchased in 1990 by the Saudi government and renamed Saudi Aramco. Since oil discovery, the development of occupational hygiene has mirrored the development of this commodity, but today there is only limited professional infrastructure and most of the full-time practitioners are foreign nationals including expatriate Americans who are independent contractors or full-time employees of Saudi companies (73). Today, almost all occupational hygienists work for Aramco.

4.18.3 Occupational Health and Safety Legislation

The Saudi Arabian Standards Organization (SASO) has published a series of mandates requiring "safe and healthy" work environments. Similar in structure to the U.S. OSHA regulations, they have been in effect for only a few years. These mandates are supplemented by ministerial or royal decrees such as the decree that banned the importation of asbestos. The Ministry of Labor is responsible for enforcing health and safety-related government mandates.

The SASO is also responsible for the development and approval of national standards for all commodities and products as well as standards concerned with metrology, calibration, marking and identification of commodities and products, methods of sampling, inspection and testing.

4.18.4 Occupational Exposure Limits

Practitioners in the country generally rely on OSHA PELs and ACGIH TLVs as defacto occupational exposure limits, but there is no formal mechanism for establishing standards in the country (73).

4.18.5 Education and Training

There are no undergraduate or graduate programs specific to occupational hygiene in Saudi Arabia, although the King Abdul Aziz University in Jeddah offers an overview course in occupational hygiene. All practicing Saudi professionals have been trained outside the country.

4.18.6 Certification and Accreditation

There are currently no certifying or accrediting organizations for occupational hygiene in Saudi Arabia. There are six CIH's (American Board of Industrial Hygiene), one ROH (Canadian Registration Board of Occupational Hygiene), and one MIOH (UK Institute of Occupational Hygiene) working in the country. All eight are expatriate employees.

4.18.7 Professional Organizations

There are no dedicated national occupational hygiene organizations in Saudi Arabia. However, there are local chapters of the AIHA and the American Society of Safety Engineer each with roughly 20 members (73).

4.18.8 Status of the Profession

There are presently about 20 full-time occupational hygienists in the country, and a hand full of technicians. Most of the occupational hygienists that work in Saudi Arabia work for Saudi Aramco, the national petroleum company. The remainder are employed in other parts of the petrochemical industry. There are a fewer than 10 government officials who have responsibility for occupational hygiene related compliance activities. These individuals generally do not have formal training or education in occupational hygiene (73). The primary factor in development of occupational hygiene in the country has been U.S. and other foreign multinational companies. Many of these organizations are accustomed to utilizing occupational hygienists which has led to a degree of familiarity and recognition among management in Saudi Arabian companies and to a lesser extent in the government, but there is little recognition in the general public regarding the profession (73).

There are also several allied health professions who work in the country including health physicists, occupational physicians and nurses, and there are numerous safety professionals who may have some responsibility for issues related to occupational hygiene, primarily for construction projects.

4.18.9 Information Resources

Saudi Arabian Standards Organization
Imam Saud Ibn Abdul-Aziz Ibn Mohammad Street
Mohammadia Quarter, Riyadh
P.O. Box 3437, Riyadh 11471
Tel: 966 1 452 0000, Fax: 966 1 452 0086
Status: Governmental Standards Organization

King Abdul Aziz University
Jeddah, Saudi Arabia
Status: Academic Institution

4.19 South Africa

4.19.1 Demographics and Economics

The South African industry has historically been dominated by mining as the world's leading producer of gold, platinum and chromium. Mining plays a less important role in the current economy, but there are diverse industry sectors including automobile assembly, metalworking, machinery, textile, iron and steel, chemical, fertilizer, food products. Following the dismantling of the apartheid government beginning in 1994, the South African economy has struggled to attract foreign investment but has seen positive growth (7). Poverty and economic empowerment for all South African citizens remains a challenge despite an abundant supply of natural resources, and well-developed legal, financial, energy, communications and transportation sectors. Unemployment in the population of 43 million is 30%. Segments of the workforce of 14.2 million are under-employed as well, i.e., part-time or temporary (7). Distribution of the South African work force is as follows: Services, 35%; Industry and mining, 20%; Mining, 9%; and Other, 6%.

There are approximately 150 full-time occupational hygienists and/or technicians employed in South Africa, although this estimate is believed to be on the low side (74). With a work force of 14.3 million, this results in occupational hygienist to work force ratio of 1:95,000. The majority of occupational hygienists are employed in services industries. Distribution of South African Occupational Hygienists is as follows: Industry, 46%; Consulting, 26%; Government, 16%; and Academia, 12%.

4.19.2 Historical Development

The history of occupational health development in South Africa parallels the history of the country itself. Industrialization in South Africa began with the development of the mining industry and the creation of secondary, supportive manufacturing industries. Improved standards for occupational health, safety and working conditions have resulted mainly from pressure from workers and advocacy by health and safety professionals.

The first step toward the development of occupational health standards was Lord Milner's Commission of Inquiry into Pthisis in the mining industry in 1902, which resulted in the Miners' Pthisis Act in 1907. A Factories Act followed in 1918 for the nonmining sector, and the first Workman's Compensation Act in 1941. By 1951, the National Occupational Safety Association (NOSA) was founded which today is a significant nongovernmental consultancy in occupational health and safety management systems, third party verification and education. Much of the development at this point in time was directed at addressing mining-related illnesses such as pneumoconiosis and silicosis based on South Africa's heavy economic reliance on gold, platinum and diamond mining (75). The Erasmus Commission of Inquiry into Occupational Health was founded and determined that the existing provisions for occupational health services in South Africa were inadequate and that standard for occupational health and safety were lower than those in other countries (76). The commission also determined that there was inadequate surveillance and statistics to track occupational morbidity and mortality. As a result, a single consolidated piece of legislation was recommended and subsequently promulgated in the Machinery and Occupational Safety Act of 1983 (MOS Act) (76). The MOS Act repealed the Factories Act of 1918.

In the last five years, the occupational health legislative structure has been revised. However, an enduring legacy of this history is the division between mining and non-mining sectors, which still have separate and, in several aspects very different, treatments of occupational health and safety (75).

4.19.3 Occupational Health and Safety Legislation

Based on the principle of self-regulation, the 1993 Occupational Health and Safety Act (OSH Act) contains the majority of South Africa's current occupational health and safety regulatory structure for all work excluding mining. It replaced the 1974 Machinery and Occupational Safety Act as the primary preventative legislation in the country. The Act specifically references occupational hygiene and defines it as the anticipation, recognition and control of conditions arising in or from the workplace, which may cause illness or adverse health effects to persons. Where risk assessment indicates that workers are exposed to the relevant hazard, medical surveillance must be instituted by a qualified person—usually an occupational health nurse or occupational medicine physician.

The OSH Act requires private and public employers to establish and maintain systems for hazard identification and assessment including occupational hygiene practices to prevent or eliminate exposure. There are also provisions for health and safety representatives and committees. It contains a wide range of regulations within the Act which address specific subjects such as Hazardous Chemical Substances, Asbestos, Noise, Lead, Electrical Installations, General Safety, Major Hazardous Installations and Environmental Regulations for Workplaces. Occupational exposure limits are also defined in the OSH Act. Specifics regarding the implementation of the general provisions of the Act are addressed in a series of guidelines and codes. For example, the guidelines for occupational noise measurement was developed by the South African Bureau of Standards (SABS). The OSH Act is enforced by the Department of Labor.

The Advisory Council for Occupational Health and Safety is an independent tripartite body consisting of 20 experts supported under the provisions of the OSH Act which is responsible for advising the Minister of Labor regarding appropriate occupational health and safety practices and regulation. Included in these recommendations are occupational exposure limits. The constitution of the Council requires that one of the 20 members be a qualified occupational hygienist.

Compensation for occupational injuries and illness diseases in the non-mining sector is controlled by the Compensation for Occupational Injuries and Diseases Act of 1993 that provides for temporary disability leave, compensation for medical expenses and permanent disability, and death benefits. The Act recognizes a broad variety of compensable occupational illnesses.

The Mines Health and Safety Act of 1996 is the mining sector equivalent to the OSH Act with the two sharing many similarities; however, the Mines Health and Safety Act includes a mechanism requiring occupational medicine practitioners to report annually on the health status of personnel under their review to the Department of Mineral and Energy Affairs as well as to a health and safety committee with management and worker representatives. The Occupational Diseases in Mines and Works Act, provides compensation for occupational illness in South Africa's mines and quarries. It is the function of the Medical Bureau for Occupational Diseases to provide benefit examinations for miners wishing to claim compensation in terms of this Act.

Occupational health service remains the primary vehicle for the dispensing of occupational medicine and hygiene activities in the country. The primary impetus for these services comes from a network of legislation addressing occupational health services. While modeled after Article 105 of the ILO Convention on Occupational Health Services, South Africa lacks a central national policy on the utilization of occupational health services. To date, South Africa has not ratified any of the primary ILO directives on occupational health and safety.

Occupational health services may be staffed by a combination of allied health professional including: (*1*) occupational medicine practitioners including physicians with a two-year postgraduate diploma in occupational health or medicine; (*2*) occupational health practitioners including qualified nurses with postgraduate training in occupational health nursing; (*3*) occupational hygienist; (*4*) safety manager with a diploma in safety management; (*5*) and support staff that includes administrators, and medical technicians. Ironically, while occupational hygiene and safety management are closely aligned with the occupational medicine service model, the degree of interaction is relatively low between the different specialties. Occupational medicine is not currently recognized as a medical specialty, although it is expected to be included as a subspecialty of Community Health.

Occupational health services are delivered by one of several models: Regional occupational health units centered around at least one referral center per province, with facilities for the diagnosis and management of occupational illnesses, rehabilitation and disability assessment, and occupational hygiene and toxicology services. These systems are in the very early stages of development in most regions of South Africa. In general, there are very few occupational health services available for public sector employees. What little public sector services that are available are supplemented by nongovernmental organizations which rely on private funding and offer occupational health services primarily to trade unions. These include the Industrial Health Research Group (IHRG) in Cape Town, the Industrial Health Unit (IHU) in Durban and the Industrial Health and Safety Education Project (IHSEP) in the Eastern Cape. A third option is the local general practitioner. Because there is no statutory requirements for advanced training in occupational health in order to deliver occupational health services, these outsourced services often lack technical expertise in occupational health and safety (74).

4.19.4 Occupational Exposure Limits

Recommendations are developed by the Advisory Council on Occupational Health and Safety which are approved and made law through adoption by the chief inspector of the Division of Manpower. The Council relies on both the ACGIH TLVs and the UK COSHH regulations as source material in the development of the recommendations. There are two types of occupational exposure limits in South Africa: Control Limits (OEL-CL) and Recommended Limits (OEL-RL) (74).

4.19.5 Education and Training

While there are no undergraduate or graduate programs specific to occupational hygiene in South Africa, the University of Potchefstroom offers both a bachelor of science and masters of science degrees in industrial physiology which is the closest approximation to specific occupational hygiene degrees in the country.

The Erasmus Commission Report of 1976 reported the lack of occupational hygiene education and training in South Africa resulting in a slow but consistent increase in available opportunities in the country (75).

The system of higher education in South African includes both technical colleges (technikons) and universities. A variety of individual courses and diploma, certificate and degree programs are available throughout South Africa for subjects such as occupational hygiene measurement techniques, risk assessment, acoustics, noise and hearing conservation, safety management, epidemiology, occupational toxicology, ergonomics, occupational health and medicine, and environmental and public health. Institutions offering these educational opportunities include the University of Pretoria, Technikon Witwatersrand, Rand Afrikaans University, Vaal Trainangle Technikon, Technikon Northern Gauteng, University of the Orange Free State, University of Natal, Mangosuthu Technikon, Rhodes University, University of Port Elizabeth, University of Cape Town, Cape Technikon, University of Stellenbosch and the Peninsula Technikon.

Some South African practitioners have elected to go overseas to advance their education and have acquired graduate degrees at such institutions as the University of Cincinnati, Harvard University, the University of Michigan, the University of London, and the University of Newcastle Upon Tyne (75).

4.19.6 Certification and Accreditation

The Institute for Occupational Hygienists of Southern Africa (IOSHA) was formed in 1993 and manages the country's occupational hygiene certification system (75). The initial categories of membership in the Institute are assistant, technologist, occupational hygienist and professional occupational hygienist. Registered Occupational Hygienist (M + 4 qualification) requires at least a bachelors degree in an appropriate field (chemistry, physics, biology, engineering), five years of comprehensive experience, and the successful completion of written and oral examinations. The Registered Occupational Hygiene Technologist (M + 3 qualification) designation also requires at least a bachelors degree in an appropriate discipline, two years of comprehensive experience, and the successful completion of written and oral examination. The Registered Occupational Hygiene Assistant must work under the direct supervision of a registered occupational hygienist or be engaged in occupational hygiene studies and successful complete an oral examination to earn this designation. The interim requirements for professional certification in occupational hygiene include graduation from a university or technikon with five years of study (two years of relevant experience can substitute for each year of formal education); plus two years of occupational hygiene experience; plus successful completion of a written examination with multiple choice and short definition questions, and successful completion of an oral examination. However, the IOSHA deleted the professional occupational hygienist designation in 1998 and the British Institute of Occupational Hygiene (BIOH) designation is now recognized as the highest level of certification in the country. These include the Certificate of Operational Competence in Comprehensive Occupational Hygiene and the Diploma of Professional Competence in Comprehensive Occupational Hygiene. BIOH certificate and diploma holders are exempt from the IOSHA written examination, at the discretion of the IOSHA examination board. Training for the BIOH core modules are available at the Pretoria Technikon. Demand for the BIOH qualifications has been significant with more than 500 registering for the module training. There are currently less than 10 BIOH and ABIH diplomates working permanently in South Africa.

It is the intent of the IOSHA to incorporate their certification system into the Board of Registration of Occupational Health, Safety & Associated Professionals (OHSAP), the national registration body for occupational hygiene, safety and associated professionals jointly formed by the IOSHA and the Institute of Safety Management (IoSM). This will enable IOHSA to comply with the country's legislation on national vocational qualifications.

Of regional significance, the Southern Africa Workshop on OHS education and Training for OHS Professionals was convened in Johannesburg in October 1997. The conference was attended by representatives of most member states of the Southern Africa Development Community (SADC) and concluded with various recommendations for collaborative development of improved education and training standards between educational institutions, governments, and nongovernmental organizations.

In 1998, the Department of Labor began requiring the BIOH modules and examination as statutory qualifications [Approved Inspection Authorities (AIA)] for any consultant wishing to perform occupational hygiene measurement work under the OHS Act. Also, draft regulations introduced under the Mine Health and Safety Act requires that mine operators with more than 300 employees who may be occupationally exposed must appoint at least one full-time occupational hygiene practitioner. A part-time practitioner is required if the number is less than 300.

4.19.7 Professional Organizations

The Occupational Hygiene Association of Southern Africa (OHASA) was established in 1983 and currently has a membership of more than 150. The association's charter includes improving awareness in the public and government for occupational hygiene and to enhance the profession. OHASA is a member of the IOHA and has adopted a code of ethics based on IOHA's code.

Code of Ethics for the Professional Practice of Occupational Hygiene

(*Objective*): To enunciate standards of professional and ethical conduct for members of the Institute of Occupational Hygienists of Southern Africa in order for them to act professionally and with integrity at all times for the benefit of workers, the public, employers, clients and the environment.

(*Professional Conduct*): (*1*) To promote occupational hygiene as a professional discipline; (*2*) To advance the professional practice of occupational hygiene; (*3*) To maintain the highest level of personal integrity and professional competence; (*4*) To be responsible in the application of recognized scientific methods and objective in the interpretation of findings; (*5*) To disseminate scientific knowledge for the benefit of the profession and society in general; (*6*) To protect confidential information, insofar as this confidence does not endanger the well being of people; (*7*) To confine themselves to matters on which they can speak with authority based on knowledge and relevant experience.

(*Duties and Responsibilities to the Workers, the Public and the Environment*): (*1*) To recognize that the primary responsibility of the occupational hygienist is to protect the health and well being of workers, also accounting for the impact on the surrounding communities and the general environment; (*2*) To adopt and maintain an objective attitude towards the recognition, identification, evaluation and control of health hazards regardless of external influences, realizing that the health and welfare of people may depend upon the occupational hygienist's professional judgment; (*3*) To inform people objectively and factually regarding health hazards and the precautions necessary to avoid adverse health effects; (*4*) To act responsibly in the application of occupational hygiene principles so as to contribute towards the attainment of environmentally sound practices on a national level.

(*Duties and Responsibilities to Employers and Clients*): (*1*) To advise the employer or client honestly, responsibly and competently so that recognition, identification, evaluation and control of health hazards in workplaces may be attained in accordance with the recognized principles of the profession; (*2*) To advise the employer or client so that contravention of any legislation, or professionally accepted standard or guideline, can be avoided; (*3*) To respect confidences regarding all information relating to the employer's/client's business operations or manufacturing processes, insofar as this confidence does not compromise the health and well being of workers; (*4*) To report findings and recommendations factually and to ensure those matters of opinion are founded on adequate knowledge and are within the member's expertise; (*5*) To hold responsibilities to the employer or client subservient to the ultimate responsibility to protect the health of workers, surrounding communities and the general environment.

Contact with allied health professions generally occurs through an umbrella organization, the Associated Societies in Occupational Safety and Health (ASOSH), and through the OHSAP. Other society members of the ASOSH include the Ergonomics Society of South Africa, the South African Society of Occupational Medicine, the South African Society of Occupational Nurses, the Mine Ventilation Society of South Africa, the National Occupational Safety Association, and the South African Institute of Environmental Health, among others.

4.19.8 Status of the Profession

The isolation incurred during the apartheid (sanction) era between the 1960s to 1994 forced South African occupational hygienists to become self-reliant which in turn has led to a very dynamic and action-oriented professional community. Today, the profession of occupational hygiene in South Africa is the most advanced on the African continent.

The rate of development has been influenced by many factors including the needs of large multinational companies with international links. It is the large corporations that employ staff occupational hygienists in South Africa (74). Recent general industry and mining legislation has had a significant impact by requiring registered occupational hygienists where there are occupational health risks. However, occupational hygienists are still fairly unknown outside the major industries. Demand from legislators for reliable results for use in the prosecution of irresponsible employers and the demand from occupational hygienist practitioners to protect their profession and the general public from poorly qualified practitioners has also been influential (74). The continued development of the profession will require improved credibility with the South African workforce, demand for services in industry and greater international acceptance.

4.19.9 Information Resources

Institute of Occupational Hygienists of Southern Africa (IOHSA)
P.O. Box 14402, Clubview, Pretoria 0014, South Africa
Tel and Fax: 012 654 8349
Status: Professional Association

Association of Societies for Occupational Safety and Health (ASOSH), and Board of Registration for Occupational Hygiene, Safety and Associated Professionals (OHSAP)
PO Box 14402, Clubview 0014, South Africa
Tel and Fax: 012 654 8349
Status: Professional Certification Organization

Department of Labor
Occupational Health and Safety Directorate
Private Bag X117, Pretoria 0001, South Africa
Tel: 012 309 4000, Fax: 012 322 0413
Internet: www.gcis.gov.za
Status: Government Ministry

National Centre for Occupational Health (NCOH)
25 Hospital Street, Hillbrow, Johannesburg 2001, South Africa
Tel: 27 11 720 5734, Fax: 27 11 720 6103
Status: Government Research Institute

Potchefstroom University of CHE
Environmental Management Unit
Potchefstroom, 2520 South Africa
Tel: 018 299 1579, Fax: 018 299 1581
Status: Academic Institution

Department of Minerals and Energy
Mine Health and Safety Directorate
Private Bag X59, Pretoria, 0001, South Africa
Tel: 012 317 9000, Fax: 012 322 3416
Status: Government Ministry

National Occupational Safety Association (NOSA)
508 Proes Street, P.O. Box 27434
Arcadia 0007, Pretoria, South Africa
Tel: 27 12 321 7736, Fax: 27 12 325 6056
Internet: www.nosa.co.za
Status: Quasi-governmental Consultancy

Industrial Health Unit
Department of Community Medicine
Faculty of Medicine
Private Bag 7, Congella 4013, South Africa
Tel: 27 31 260 4528, Fax: 27 31 260 4211
Status: Government Research Agency

4.20 Sweden

4.20.1 Demographics and Economics

With a skilled labor force and a history of peace and neutrality, Sweden has developed a strong economy based on a mixed system of high-tech capitalism and extensive welfare systems (7). The economy is centered on foreign trade and relies on timber and wood products, hydroelectric power, and iron ore as the resource base. Other industry sectors include precision equipment (bearings, radio and telephone parts, armaments), wood pulp and paper products, processed foods, motor vehicles and aircraft. Private companies account for 90% of industrial output of which engineering accounts for 50% of output and exports (7). Agriculture plays a minor role in the gross domestic product and accounts for only 2% of workforce employment. The work force is comprised of approximately 4.5 million workers. Labor unions represent 85% of the workforce (7).

In the mid-1990s, Sweden's strong economic outlook began to experience difficulties due to budgetary issues, rising inflation, and unemployment. Sweden has harmonized its economic policies with the EU, which it joined in 1995. Distribution of Swedish work force is as follows: Community, social and personal services, 38%; Mining and manufacturing, 21%; Commerce, hotels and restaurants, 14%; Banking, insurance, 9%; Communications, 7%; Construction, 7%; and Agriculture, fishing and forestry, 4%.

There are approximately 145 full-time occupational hygiene practitioners in Sweden (77, 78) giving the country an occupational hygienist to work force ratio of about 1:30,000. The current economic depression is anticipated to lead to a reduction in the number of occupational hygienists working in industry. Distribution of Swedish Occupational Hygienists by employment sector is as follows: Industry, 37%; Departments of Occupational Medicine, 33%; National Institute of Occupational Health, National Board of Occupational Safety and Health, Labor Inspectorate, 21%; and Universities and research institutes, 9%.

4.20.2 Historical Development

Sweden passed its first industrial safety law in 1889 intended mainly to protect workers against fire hazards. The Labor Inspectorate was established in 1890 as the regulatory enforcement mechanism for the 1889 law. A more comprehensive act, the Workers' Protection Act, was promulgated in 1913, then amended in 1949 and again in 1974. In 1942 the employer's federation and national labor unions reached an agreement that established safety committees in private companies and began the long-term process of consensus in health and safety matters that is the hallmark of the Swedish system today.

The 1949 version of the Act compares in many ways to the Occupational Safety and Health Act of 1970 (OSHAct) in the United States. The amended Swedish Workers' Protection Act applies nationwide and establishes a central regulatory agency while holding employers responsible for providing safe and healthful working conditions in their places of employment. The governing board is known as the National Board of Occupational Safety and Health (NBOSH) and was housed in the Ministry of Labor. It consists of 11 members: a director-general, an assistant to the director-general, four representatives from the Confederation of Trade Unions (LO), three from the Swedish Employers' Confederation (SAF), and two members of the Swedish Parliament. It was during the late 1940s that occupational hygiene began to emerge as a recognized discipline (78).

The board is responsible for issuing "directions" or "codes of practice," comparable to U.S. OSHA standards. Employer and labor representatives on the board participate directly in drafting such "directions" and "codes of practice." Other employers, equipment manufacturers, labor representatives, and any other interested parties also have the opportunity to comment on such documents before promulgation.

Another important function of the board and its six technical departments is to explain how such standards apply to specific workplaces and to review plans of new processes and industrial plants for adequacy of health and safety controls (78). The 1974 version of the Workers' Protection Act makes it mandatory that experts employed by the board review such plans and blueprints before construction begins. The Act further requires that local joint safety committees also approve such plans and drawings, unless they participated in development.

4.20.3 Occupational Health and Safety Legislation

At its core, the occupational health and safety system in Sweden is based on occupational health services, safety delegates, occupational health and safety engineers and a high degree of cooperation between employers, employees, government and labor.

The Work Environment Act of 1977 superseded the 1949 Workers' Protection Act. The Work Environment Act sets out the basic framework for legislation addressing occupational health and safety in the Sweden. The rules of the Work Environment Act set the framework for Provisions issued by the National Board of Occupational Safety and Health which in turn set out more detailed obligations affecting the workplace. The Board's provisions may, for example, refer to mental and physical capacities, hazardous substances or equipment and the product of collaboration with labor unions. Subsequent amendments to the Work Environment Act since 1978 include the 1985 rule regarding occupational health services and job health and safety controls. In 1992 certain amendments were made to the

Work Environment Act to bring it in line with the Agreement on the European Economic Area (EEA).

The Swedish Occupational Safety and Health Administration (SOSHA) consists of the National Board of Occupational Safety and Health (NBOSH) and the Labor Inspectorate. The NBOSH consists of six departments or functions (Engineering, Occupational Hygiene, Medical and Social, Analysis and Planning, Administration, and Information). There is also a Legal Affairs Secretariat and a Secretariat of the Director-General. The Labor Inspectorate reports to the NBOSH and is divided into 11 districts (Borås, Falun, Gothenburg, Härnösand, Linköping, Luleå, Malmö, Stockholm, Umeå, Växjö and Örebro). Each district has a labor inspection committee consisting of a maximum of eight members representing different industries and economic sectors in the region. There are approximately 400 labor inspectors who are responsible for inspecting approximately 300,000 workplaces across Sweden.

If a workplace injury or exposure causes death, severe injury, affects several employees simultaneously, or seriously endangers life or health, the employer must immediately notify the Labor Inspectorate. Swedish physicians are also legally required to notify the National Board of Occupational Safety and Health or the Labor Inspectorate of occupational illness that may be connected with employment.

The Board also supervises the Labor Inspectorate, which is responsible for enforcing board rules in 19 districts. Over half of the annual health and safety inspections are conducted by local municipal inspectors who generally visit small firms with simpler operations and fewer than 10 employees. Swedish inspectors function as technical consultants rather than as enforcement officers; they have the authority, but normally issue "warnings." Employers who ignore such warnings may be fined or even sent to prison for up to one year. Employers can appeal warnings directly to the NBOSH as there is no review commission such as stipulated by the OSHAct in the United States.

The Swedish approach to occupational health and safety stresses cooperation over contention between industry and labor unions. Traditionally, industrial firms participate in and act through the Swedish Employer's Confederation, SAF. On the labor side, over 85% of workers are represented by unions that are members of the Confederation of Trade Unions, LO. As early as 1938, SAF and LO agreed to settle important issues through negotiations. Over the years, as the strength of SAF and LO remained fairly evenly matched, most local issues have been resolved locally. Swedish industry and unions both want to maintain and further the economic health of the country. According to Clack (79), Birger Viklund, former Secretary of the Swedish Metal Workers' Union and Labor Attaché at a Swedish Embassy, explains: "Swedish industry is geared to export, and the workers realize that strikes could destroy foreign trade built up over the years."

Workers receive extensive health and safety training. They are expected to observe applicable regulations, use chemical substances properly, and operate mechanical equipment as intended. They are required to use proper personal protective equipment and must work safely using proper precautions to protect themselves and the health and safety of their co-workers.

A central party in Swedish occupational health and safety system is the safety delegate(s). These individuals and their deputies are appointed for a period of three years and must have knowledge of and interest in the working environment and must complete a

40-hour course in basic health and safety management and legislation. The number of safety delegates appointed at any one work site depends on the size of the operation, the number of employees at the site, and the nature of the work conditions. Delegates are also appointed for each work shift within the same facility. Delegates are involved in all activities associated with workplace safety and health including internal audits or inspections and participating on safety committees. There are more than 100,000 safety delegates in Sweden.

Management members of joint safety committees are principally responsible for providing safe working conditions in their plants. They must train workers to work safely with equipment and chemical substances and must also provide health services for workers in establishments with over 1000 employees. Such health services are on-site and include occupational physicians and occupational hygienists. Smaller plants have access to health services provided by over 100 regional health centers built by the SAF. Mobile medical surveillance stations staffed with nurses and health technicians service temporary work sites, such as construction projects. Factories with more than five employees are required to have a safety steward. Factories with more than 50 workers must have a joint labor-management safety committee.

Safety committee membership includes a management representative, a member of the executive committee of the local union, one or more safety delegates, and representatives of the occupational safety and health service. The health services provide services to more than 3.5 million employees in over 80% of all Swedish workplaces. In large companies with more than 1,000 employees, the health service will be staff whereas in smaller companies it may be provided by contract services.

A second basic structure in the Swedish system is the occupational health service, which is staffed by occupational physicians, occupational hygienists, and safety engineers. The service reports to management and communicates with in-house safety committees. Unlike safety that is dictated primarily by legal policy and legislation, occupational health is managed under a set of agreements between the labor unions and employer's federations (77). This is the framework law for health and safety in Sweden. The National Board of Occupational Health and Safety issues guidelines and procedures most of which are mandatory. The employer is responsible for a safe and healthy working environment for his workers (80).

Yet because of extensive cooperation between unions and employers in health and safety, duties of occupational hygienists in industry are determined mostly by agreements among SAF, LO, and the Negotiation Cartel for Salaried Employees in the Private Business Sector (PTK). According to Gerhardsson (81), three such agreements are of fundamental importance:

1. The Development Agreement of 1982, which strives to foster industrial efficiency and preserve employment as areas of common interest among the three signatories.

2. The Equal Employment Agreement, also signed in 1982, providing for equal employment opportunities for men and women with regard to promotions and equal pay for equal work. This agreement applies to salaried and hourly employees alike. Both agreements foster a spirit of cooperation between management and labor so necessary to promote occupational health and safety.

3. The Work Environment Agreement, reached in 1983, and its impact on occupational health. It sets health and safety rules for the work environment. It establishes guidelines for industrial health programs and defines details of occupational health training for employees. The Agreement specifies that corporate health programs must include occupational hygiene and medical services with special emphasis on psychosocial factors. It also calls for a joint safety committee or a company health program committee to supervise the occupational health staff and thus ensure that its activities are based on sound science and experience. This approach is intended to maintain strict impartiality and fairness on the part of company health programs and strengthen their commitment to prevent potential health problems. Disagreements over scientific or other important issues are brought before a national scientific committee for review and resolution.

The three signatories of these agreements contribute funding and staff in support of joint activities. All three parties are also free to pursue individual initiatives within the framework of the agreed-upon programs and guidelines.

Under such tripartite agreements, occupational hygienists are responsible for a number of activities. Such activities include preparing for and working toward ever more stringent permissible exposure levels, considering indoor as well as outdoor pollution problems, and maintaining effective industrial emission controls. Swedish occupational hygienists are also expected to strive for energy and cost savings while reducing and controlling environmental problems.

4.20.4 Occupational Exposure Limits

The NBOSH may issue provisions concerning occupational exposure limit values to be observed in the planning and control of the working environment (78). The NBOSH order these documents and collect reports and epidemiological studies on exposure level and adverse effects. Economic feasibility is also considered when a new or lower occupational exposure limit is being considered. Swedish exposure limits are revised and updated every two to three years (80).

The Swedish Institute of Occupational Medicine published the first list of permissible exposure limits in 1969. NBOSH has had the responsibility for issuing lists of occupational exposure limits (TWA) for chemical substances since 1974. According to Edling and Lundberg (80),

> That list, which included values for about 70 substances, was based on the ACGIH TLVs and Swedish experiences but did not recognize carcinogens specifically. In the second list, published in 1974, background data from NIOSH, OSHA, ANSI, BRD [the German Federal Republic], Czechoslovakia and ILO were considered. That list included an appendix of recognized carcinogens. The list has been revised every few years with the current list comprising some 400 substances. These lists have gradually been expanded and legally strengthened.

The Work Environment Act, 1977:1166, provides that substances liable to cause ill health or accidents may only be used in conditions affording adequate security. Authorized by this Act, the NBOSH issued its Hygienic Limits Values Ordinance in May 1981, effective January 1982. The ordinance established two types of occupational exposure limits for airborne contaminants. "Level limit values" apply to daily eight-hour exposures and "ceiling limit values" represent maximum permissible exposure levels measured over 15-minute periods, unless otherwise specified in Appendix I of the Ordinance (15).

In the development process, the Occupational Health Department of the NBOSH develops criteria documents using a committee of occupational hygienists, representatives of the national institute, labor unions and employer associations taking into consideration dose–response and dose–effect data. However, the committee does not recommend a numerical value. That process is left to the Supervision Department that establishes the magnitude of the values based on the criteria document, public comments on the criteria document and additional documentation.

In recent years, Swedish exposure limits have become increasingly stringent to avoid or minimize biological effects (78). Some limits were lowered based on worker complaints rather than on demonstrable dose–effect relationships. Others were made more stringent because of demonstrable effects on the central and/or peripheral nervous system(s). Thus Swedish occupational exposure limits for many chemicals, and particularly for some industrial solvents, are lower than corresponding limits in several industrialized countries.

4.20.5 Education and Training

While Swedish legislation does require occupational hygiene services as part of an integrated occupational health approach, it does not define competence in occupational hygiene.

In 1966, the Swedish Parliament combined four major organizations devoted to occupational health and safety research into the National Institute of Occupational Health in Stockholm. Six years later, in 1972, it placed the Institute under NBOSH. At this point, the board assumed the important role of providing technical and practical training for occupational safety and health engineers.

But the Swedish system for developing occupational hygienists is varied and broad in scope, while the Swedish government through the NBOSH provides training in "occupational health and hygiene engineering." Such individuals have an engineering degree from a university or from an equivalent institution and several years of field experience (81). Practicing occupational health and safety engineers attend 15 weeks of theoretical training and are required to complete a special project. Prospective engineers receive 20 weeks of classroom instruction and work for another 20 weeks at an occupational health and safety service unit under the supervision of an experienced OSH engineer. There are more than 1,500 trained occupational health and hygiene engineers (OSH Engineers) in the country (78).

Early on, some Swedish governmental occupational hygienists received scholarships from the Rockefeller Foundation and earned master's degrees in industrial hygiene at universities in the United States. For a number of years, no formal university courses leading to degrees in occupational hygiene were available in Sweden. The Swedish Institute of Occupational Health offered class work and practical training in occupational hygiene patterned after courses offered at American universities and in particular at Harvard. Such courses taught students how to measure and evaluate chemical and physical stresses in the plant environment and emphasized control methods to reduce or eliminate such stresses (15).

Since 1952, SAF has offered occupational hygiene courses and on-the-job training to members of their occupational health staffs, including physicians and other technical personnel. Beginning in 1959, SAF offered a one-year course in occupational health for occupational hygienists and safety engineers with particular emphasis on prevention and control of hazards.

The University of Lund is the only university offering a graduate degree in occupational hygiene (M.Sc.), while Uppsala University offers post-graduate courses in occupational medicine relating to occupational hygiene (77, 78). The program at the University of Lund requires candidates to have a master's degree in chemistry or be a professional safety engineer with a strong chemistry background. The program is geared mainly toward control of chemical hazards.

In the last ten years approximately 12 doctorate degrees have been conferred on Sweden occupational hygienists in Sweden. Further, there are occupational hygiene-related short courses offered at the Nordic Institute for Advanced Training in Occupational Health, NIOH, and the University of Lund.

4.20.6 Certification and Registration

In Sweden there is no official certification or registration mechanism for occupational hygienists (77, 78). However, the Swedish Occupational and Environmental Hygiene Association is developing a third party certification process that is expected to begin in 1999. Requirements will include the completion of a degree or course of study from a university including chemistry, physics and/or biology for at least three years. Candidates will also be required to study occupational hygiene and related courses like toxicology, epidemiology and risk analysis for another one year and a half years.

4.20.7 Professional Organizations

Until the early 1990s, occupational hygienists working at occupational medicine clinics belonged to the Swedish Occupational Hygiene Association, which, in addition to pursuing technical goals, represented its members in collective bargaining negotiations for wages and in other contractual matters. All other occupational hygienists tended to join the Swedish Occupational Hygiene Society. These two organizations have merged to form the Swedish Occupational and Environmental Hygiene Association (SOEHA).

Membership in the SOEHA is limited to those with active professional interest in occupational or environmental hygiene and who have a degree in chemistry, physics or biology. About 50% of members have a master's degree in occupational hygiene acquired in Sweden or other countries. Activities of the association include representing the profession in rulemaking, development of occupational exposure limits, providing professional development opportunities and serving as a mechanism for information exchange among Swedish occupational hygienists and other allied health professionals in Sweden and Europe. The SOEHA is a member organization of the IOHA and has adopted the IOHA code of ethics for its own members.

The National Institute for Occupational Health (NIOH) was created in 1966 through a consolidation of the four major organizations devoted to occupational health research. By

1987, the NIOH became independent and currently has a staff of more than 400 including, 325 researchers with a primary research focus on chemical risk assessment.

4.20.8 Status of the Profession

Occupational hygiene in Sweden is characterized by a strong research focus and specialization within the occupational health service model, and by cooperation between management and labor (78). Such cooperation is not limited to health and safety, but extends into other important areas of labor relations. Broad implementation of occupational hygiene principles and practices in industry, particularly with regard to control of chemical hazards and the implementation of ergonomic controls also characterize Swedish occupational hygiene. Occupational hygiene has been codified in Swedish law as a reflection of the societal concern for working conditions; however, owing to commensurate statutory definition of occupational hygiene competence(s), and the strong emphasis on research in Sweden, there are no efforts underway to develop a certification system.

The number of practicing professionals in Sweden is approximately 145 according to the SOEHA (77, 78). Most practitioners work in the area of occupational medicine or in government or industry health departments. Their primary emphasis is research. Within the field of occupational health, occupational hygienists are well accepted, but outside this area they are less well known. There are also 1,500 individuals working as occupational safety engineers who have primary responsibility for occupational health and safety issues in the workplace.

In 1981, a governmental advisory committee estimated that by the year 2010 the number of occupational hygienists with governmental and municipal affiliations would more than double, and that their numbers in industry would increase fourfold. Gerhardsson (81) disagrees:

> From the present terms of reference—which are too narrow—this assessment has been shown to be an over-estimation of the numbers needed. In industry managerial skills are upgraded; enforcement tasks are increasingly transferred to line functions. In governmental and municipal agencies, systems strategies require fewer occupational hygienists but more specialized ones.

The scope of practice in Sweden is generally reflected in three types of activities: (*1*) conducting field investigations of the work sites to define health risks, measure and evaluate exposures and make recommendations regarding effective methods to minimize risk; (*2*) participating in research programs to investigate changes in the work conditions to identify early signs of poor or inappropriate conditions and propose preventive measures. These practitioners may also be involved in developing models for retrospective exposure assessment as well as studying the pharmacokinetics of different substances used at the workplaces; and (*3*) providing education, training and information in a university setting, in industry to safety engineers, occupational physicians and other occupational professionals, as well as to the public. The work of occupational hygienists in risk assessment has lead to a greater overall involvement in environmental issues in general, a change that is becoming more common throughout Europe (78).

However, occupational hygienists working in the occupational medicine arena or within health departments also play a role in the evaluation of the medical patients. In cases when occupational disease is suspected of developing over an extended period of time, the occupational hygienist conducts a retrospective exposure assessment for the patient's entire work life. This assessment may be conducted as part of a compensation claim. The implementation of recommendations to minimize exposure in the workplace is often performed in co-operation with the safety engineer responsible for workplace health and safety.

As a reflection of Sweden's contributions to the occupational hygiene profession, the AIHA has conferred the Yant Award on three distinguished Swedish scientists: Sven Forssman, M.D. in 1967 (Lecture: Occupational Health Institutes: An International Survey); Harry G. Ohman in 1978 (Lecture: Prevention of Silica Exposure and Elimination of Silicosis); and Lars T. Friberg, M.D. in 1985 (Lecture: The Rationale of Biological Monitoring of Chemicals—With Special Reference to Metals).

4.20.9 Information Resources

Swedish Occ. and Environ. Hygiene Association (SOEHA)
Department of Occupational Medicine
St Sigridsgatan 85, SE- 412 66 Gothenburg, Sweden
Tel: 46 31 335 48 77, Fax: 46 31 40 97 28
Status: Professional Association

National Institute for Working Life
SE-171 84 Solna, Sweden
Tel: 46 87 30 91 00, Fax: 46 87 30 98 88
Internet: www.niwl.se
Status: Government Research Institute

Institute for Occupational Medicine
Uppsala University
Department of Occ. And Envir. Medicine
University Hospital, S-751 85 Uppsala, Sweden
Tel: 46 18 66 30 00, Fax: 46 19 51 99 78
Internet: www.medfak.bmc.uu.se
Status: Academic Research Institute

National Board of Occ. Safety and Health
S-171 84 Solna, Sweden
Tel: 46 08 730 90 00, Fax: 46 08 83 35 10
Status: Government Regulatory Agency

4.21 Switzerland

4.21.1 Demographics and Economics

Switzerland has a stable modern economy based on financial services, biotechnology, pharmaceuticals, and machinery, but has experienced difficulties in the mid-1990s. The service sector accounts for the majority of the gross domestic product followed by manufacturing (machinery, chemicals, watches, textiles, precision instruments) and agriculture (7). The work force of 3.8 million includes approximately 850,000 foreign nationals, primarily from Italy and tracks the gross domestic productivity in terms of the distribution of workers. The current rate of unemployment is 5%. Currently, Switzerland is not a member of the European Economic Community. Distribution of the Swiss work force is as follows: Services (government, financial, tourism), 67%; Manufacturing and construction, 29%; and Agriculture and forestry, 4%.

There are approximately 85 practicing occupational hygienists in Switzerland (82) giving the country an occupational hygienist to work force ratio of 1:44,700. Additionally, another 20 to 40 technicians in industry (mainly chemical industry) may be charged with occupational hygiene obligations. Distribution of Swiss Occupational Hygienists by employment sector is as follows: Industry, 40%; Government, 25%; Academia, 20%; and Consulting, 15%.

4.21.2 Historical Development

While Switzerland, was one of the first European countries to issue legislation addressing occupational health and safety, with the Factories Act in 1877, occupational hygiene as a profession developed slowly beginning in the late 1960s due to the efforts of a few motivated chemists and progressive physicians.

4.21.3 Occupational Health and Safety Legislation

Today, Swiss occupational health and safety legislation is addressed in the Work Act of 1964 and the Accident Insurance Act (AIA) of 1981. The Work Act addresses general work conditions including working hours, shift work, special regulations for women and young people, occupational health and hygiene with the exception of occupational illness as stipulated in the AIA (83). Employers are required to maintain a safe and healthy work environment, but there is no requirement for facility health and safety committees. However, some industries must provide occupational health services to their employees including physicians, occupational hygienists and safety engineers. The AIA relates to the prevention of occupational injuries and illnesses and their compensation.

Swiss workers are insured against occupational injury and illness through employer compensation premiums. The Swiss National Accident Insurance Organization (SUVA) is the sole organization charged with tracking and reporting occupational injury and illness data. Concern regarding the accuracy of national occupational disease reporting is based on the reliance of an insurance organization and not a public health agency or service (83). If underreporting were occurring, this would have an impact on the development of the profession owing to a reduced social concern and less legislation directed at preventing occupational disease.

Enforcement of the Work Act is the responsibility of the cantonal (or state) and Federal labor inspectorates while enforcement of the AIA is the joint responsibility of both cantonal and federal labor inspectors and inspectors of the SUVA. Occupational health issues are the exclusive responsibility of the SUVA. The SUVA delegates part of the enforcement activities associated with the AIA to recognized associations such as the Advisory Office for Accident Prevention in Agriculture and the Swiss Welding Association. Further complicating this situation is the Federal Commission of Coordination for Work Safety that was established to coordinate the enforcement of the AIA and to provide interpretive material for it. The commission's members include the SUVA, representatives of the cantonal and federal inspectorates, private insurance companies, but not employers or employees. This enforcement scheme is among the most complicated in Europe because it subdivides responsibility for health and safety issues between so many parties. This problem is illustrated by Guillemin (83).

> The dichotomy between these two main laws to protect the workers implies an arbitrary division of hazards according to their intensity. For example, a noise below 87 dB(A) belongs to the field of the Work Act because this level is not considered able to produce an occupational disease but only discomfort or annoyance; however, above this level, noise comes into the competence of the SUVA through the AIA and is a recognized source of potential occupational disease.

A second unusual aspect of the Swiss system is the potential conflict of interest that arises from the SUVA's regulation of Swiss law in preventing occupational disease and their simultaneous obligation to provide compensation for occupational disease that develops.

4.21.4 Occupational Exposure Limits

The SUVA is also responsible for developing the Swiss occupational exposure levels, referred to as tolerance limit values, or MAC values, which were first published in 1945 by the Insurance Institute. Development is conducted cooperatively with the Commission on MAC of the Swiss Society of Occupational Medicine, Hygiene and Safety, an interdisciplinary group of scientists and occupational hygiene professionals (84). The MAC values are promulgated as time-weighted averages for which there are no health effects with up to 45 hours of exposure per week to a pure substance. Mixtures are not addressed by MACs because of the Swiss contention that mixtures can only be addressed through toxicological testing of the mixture. The development of the Swiss MAC values closely follows the process used to develop the German MAK-values and the values themselves are often adopted from the German MAK list (84). The ACGIH TLVs may also be used in the event there is a substance under study that is not addressed by a German MAK value.

Biological exposure indices are not published as they are not yet recognized as legally enforceable standards (83). The SUVA enforces compliance with Swiss MAC values.

4.21.5 Education and Training

The University of Lausanne has the only structured program in occupational hygiene education in the country (83). The two-year program is a joint project of the Institute of Occupational Health Science of the University of Lausanne and the Institute of Hygiene and Work Physiology of the Federal Institute of Technology in Zurich. The curriculum is split between occupational medicine and occupational hygiene with 560 hours of common topics and 240 hours of instruction in each specialty. Candidates to the occupational hygiene program include chemists, physicists, biologists, and engineers with a Master of Science (M.S.) degree. The occupational medicine track is available to licensed physicians.

There are no governmental institutes that conduct research in occupational health and safety. Research that is conducted takes place primarily in universities and private industry. The main research center is the Institute of Occupational Health Sciences at the University of Lausanne and integrates occupational medicine and hygiene. The institute has developed an international reputation for its contribution to the occupational hygiene literature and the participation of its principle scientists in occupational health policy issues in Europe and elsewhere.

4.21.6 Certification and Registration

In 1996, a federal regulation (RS 822.116) was promulgated that required certain qualifications to be recognized as an occupational hygienist by the Federal Office of Social Security and the Labor Inspectorate (82). The qualifications include a technical or scientific diploma issued by a university; a sufficient number of years of professional experience; and, completion of the University of Lausanne postgraduate course in occupational hygiene. While this title recognition is not a formal certification system, one is being developed by the Swiss Society of Occupational Hygiene (Industrial Hygienist SSOH) based on the same qualification requirements and the accumulation of 80 "credit points" over a three year period based on various professional activities such as attending conferences and authoring publications (82).

4.21.7 Professional Organizations

In 1984, there were a sufficient number of practitioners to establish the first professional association in Switzerland with the support of the AIHA. The initial organization was formed as a local section of the AIHA with about 40 members. This group evolved into the Swiss Society for Occupational Hygiene (SSOH) and subsequently joined the Swiss Association of Occupational Medicine, Occupational Hygiene and Safety, thus consolidating the three national societies for medicine, occupational hygiene, and safety. Membership in SSOH requires active part- or full-time status in occupational hygiene and two sponsors who are full members of the society. In 1999, SSOH's membership reached 100 with members ranging from full-time, academically trained professionals to part-time technicians with on-the-job skills. Like many associations, the SSOH conducts short courses, advocates on behalf of the profession, interacts with other allied health professions and societies, interacts in official activities within the Swiss legislative structure for occupational health and safety, and is the main resource for the dissemination of relevant information about and within the profession. There is good cooperation between occupational hygienists and occupational physicians in Switzerland due to the good collaboration between the two above-mentioned academic institutions.

The SSOH is a founding member of IOHA. The SSOH follows the code of ethics of the IOHA (82).

4.21.8 Status of the Profession

The profession is not widely known or accepted in Switzerland except among certain sectors of the government and in private industry. There is not substantial social or labor pressure driving development and among some groups critical of government intervention and the current economic conditions in the country, occupational hygiene is viewed as not being cost-effective (83). This perception may be due in part to an unusual occupational illness reporting scheme and a workforce with historically low rates of union participation.

The current status of the profession in Switzerland is the result of a combination of factors (82, 83), including: (*1*) the recognition of the importance of occupational hygiene activities by physicians in the prevention of occupational illnesses and the subsequent collaboration with the medical establishment; (*2*) several large domestic chemical companies have also recognized the need for occupational hygiene services and added staff accordingly; (*3*) a base of academically trained occupational hygienists with broad experience influenced by the occupational hygiene systems in the U.K. and the U.S.; and (*4*) the formation of a professional society at an early stage in the profession's development with permanent contacts with related safety and medical societies. The present situation is also supported by legal requirements for occupational hygienists in industry where they are defined and recognized as full partners in the prevention of occupational diseases.

Based the opinion of SSOH members, additional employment and academic opportunities must be developed and certification and additional government recognition must occur to facilitate the continuing development of the profession in Switzerland (82).

4.21.9 Information Resources

Swiss Society for Occupational Hygiene (SSOH)
Arbeitshygiene BWA, Kreuzstr. 26, 8008 Zürich, Switzerland
Tel. 41 1 261 77 78, Fax: 41 1 251 65 02
Internet: www.sgah.ch
Status: Professional Association

Institute of Hygiene and Occupational Physiology
Swiss Federal Institute of Technology,
ETH-Zentrum, CH 8092 Zurich, Switzerland.
Tel and Fax: 41 1 632 11 73
Internet: www.iha.bepr.ethz.ch
Status: Academic Research Institute

University of Lausanne
Institute for Occupational Health Sciences
19 Rue du Bugnon, CH 1005 Lausanne, Switzerland
Tel and Fax: 41 21 314 74 20
Internet: www.hospvd.ch/public/instituts/ist
Status: Academic Institution

4.22 Thailand

4.22.1 Demographics and Economics

While the Thai economy has been in a deep recession for several years, the country has a wide variety of natural resources including tin, rubber, natural gas, tungsten, tantalum, timber, lead, fish, gypsum, lignite and fluorite (7). It has become one of the key manufacturing centers in Southeast Asia. In 1997, the Thai government signed a $14 billion agreement with the International Monetary Fund (IMF) to help restore financial market stability. The population of 60 million enjoys an adult literacy rate of 94% of which approximately 32.6 million are part of the Thai work force (7). In terms of gross domestic product, the Thai economy is centered on the service sector, although a majority of the work force is occupied in agriculture (7). Industry sectors include tourism; textiles and garments, agricultural processing, beverages, tobacco, cement, light manufacturing, such as jewelry; electric appliances and components, computers and parts, integrated circuits, furniture, and plastics; Thailand is the world's second-largest tungsten producer and third-largest tin producer. Distribution of the Thai work force is as follows: Agriculture, 54%; Industry, 15%; and Services (including government), 31%.

There is no registration system for occupational hygienists or occupational hygiene technicians in Thailand. However, a survey of 658 graduates of Mahidol University with degrees in occupational hygiene suggests that a majority of those working in the occupational hygiene profession in Thailand are employed in private industry (85). Distribution of Thai Occupational Hygienists is as follows: Private industry, 57% (petrochemicals/chemicals; petroleum and natural gas; electronics and semiconductors; Portland cement; plastic products; metal processing; automobile assembly; and food and other agricultural products) (86); Government, 32%; Academia, 5%; and Consulting, 4%.

4.22.2 *Historical Development*

Occupational hygiene first gained recognition in Thailand as part of the national occupational health project under the Ministry of Public Health in 1966, when an outbreak of manganese poisoning occurred in the dry cell battery manufacturing industry. Successful implementation of an occupational hygiene program as a part of a larger occupational health project led the country to successful control of manganese poisoning. From that point until today, WHO and ILO have contributed significantly to the development of the occupational health system in the country through technical and financial assistance. In 1972, the project evolved into the Division of Occupational Health in the Ministry of Public Health. At the same time the Department of Labor was also established and subsequently became the Ministry of Labor and Social Welfare with responsibility for occupational safety and health of workers and employers throughout Thailand.

4.22.3 *Occupational Health and Safety Legislation*

The primary national legislation relating to occupational health and safety is the Labor Protection Act (1998) which addresses health, safety, and welfare of workers including safety standards, occupational health organization and management, occupational exposure standards, related health standards and labor laws for women and children. The Factory Acts of 1992 covers safety of factory buildings, machines and equipment and the control of plant emissions. The 1992 Hazardous Substances Act was promulgated to control hazardous substances including product registration, and risk assessment and management. The Public Health Act of 1992 addresses occupational health standards and the control of hazardous activities. Another piece of Thai legislation that impacts occupational health and safety is the Promotion and Preservation of Environmental Quality Act which contains provisions for employers to conduct environmental impact assessments including other pertaining to occupational health impacts of industrial operations.

According to the Safety at Work Law, the professional safety officers employed in industry have to perform the following duties (87): (*1*) examine and recommend actions for employers to take in relations to the Safety at Work Law; (*2*) submit safety-at-work programs, projects and safety measures to the employers; (*3*) follow up on safety-at-work programs, projects and safety measures; (*4*) enforce health and safety rules and regulations; (*5*) conduct safety training; (*6*) investigate occupational injuries and illnesses and provide mitigation and prevention measures; and, (*7*) compile, analyze and report occupational injuries and illnesses to the government.

The Ministry of Public Health has developed multidisciplinary occupational health service teams including occupational hygienist, physicians, nurses, chemists, toxicologists and epidemiologists to undertake the primary roles of preventative medicine and medical surveillance, including: (*1*) investigation of occupational health problems: (*2*) administering occupational diseases surveillance programs; and (*3*) developing legal standards for occupational health and safety. In industries where physicians, nurses and other professions are scarce, occupational hygienists often seek help from the government sector, consultants, or the professional organization. The quality, coverage, accessibility, affordability and sustainability of occupational health teams continues to be a challenge in Thailand.

4.22.4 Occupational Exposure Limits

Thai occupational exposure limits are adopted from the ACGIH TLVs and enforced through the Labor Protection Act and the Public Health Act. The Ministry of Public Health determines compliance verification relative to occupational exposure limits (15).

4.22.5 Education and Training

There are both undergraduate and graduate programs in occupational hygiene in Thailand Mahidol University has been offering occupational health as a bachelor degree for more than 25 years (B.S., Occupational Health and Safety), as well as a graduate program in occupational hygiene (M.S., Occupational Hygiene and Safety), and a postgraduate certificate in occupational nursing. Other institutions with academic programs in occupational health include Khonkaen University (B.S., Public Health with an Occupational Health major), Burapa University (B.S., Occupational Health and Safety), Hua Chiew University (B.S., Occupational Health), and Sukhothai University (B.A., Education; Occupational Health and Safety with certificate program in Occupational Health and Safety). Some 1,600 undergraduate and graduate degree have been conferred at these institutions since 1975 (85).

There is also a 180-hour curriculum in occupational health provided by the Ministry of Labor and Social Welfare to qualify as a safety officer which is required to conduct certain types of health and safety work (85). Individuals with a bachelor degree in occupational health or equivalent can become the safety officers without having to take the safety officer training course and examination.

4.22.6 Certification and Accreditation

While there are professional certification processes for physicians, nurses, dentists, pharmacist, architects and engineers in Thailand, there is currently no certification or registration mechanism for occupational hygienists. Discussions between government officials and members of the profession regarding the development of a certification system have begun.

4.22.7 Professional Organizations

There are several professional organizations dedicated to occupational health in Thailand; however, none that are specific to occupational hygiene. The Occupational Health and Safety at Work Association, Occupational Medicine Association of Thailand and the Occupational and Environmental Medicine Association of Thailand are private associations while the Safety and Health at Work Promotion Association is sponsored by the Thai government.

The National Institute for the Improvement of Working Conditions and Environment (NICE) was established by the Thai Government with the assistance of the United Nations Development Program (UNDP) and the ILO. NICE is responsible for worker protection through research and enforcement of the Labour Protection Acts and is an agency under the Department of Labour Protection and Welfare in the Ministry of Labour and Social Welfare.

4.22.8 Status of the Profession

In Thailand, occupational hygiene enjoys general acceptance as a recognized scientific discipline (85). Hygienists are acknowledged by the government, employers, and workers and the demand for their services in both the public and private sectors is generally greater than the supply of qualified practitioners (85). However, occupational hygiene is not statutorily recognized to the same extent as the safety profession. Occupational physicians, nurses and engineers continue to play a significant role in the practice in industry. In the past, occupational hygiene and safety programs were mainly the responsibility of the government sector but today there are a number of industries which have well-developed occupational health and safety programs. There are a small number of private consultants providing occupational hygiene services and training.

Factors that have impacted the development of the profession include: (*1*) demands by labor for greater worker protection relative to occupational illness and injury; (*2*) stringent enforcement of occupation health and safety laws; (*3*) a lack of occupational hygiene data relating to compensible occupational diseases; (*4*) Thailand's ISO 18000 (BS 8800) which had an impact on the improvement of occupational health and safety management for exporting industries; and (*5*) the lack of occupational health and safety professionals.

For the profession to continue to grow in Thailand, a certification system for occupational hygienists with a code of ethics and professional standards must be established, the system of professional development through continuing education and training must be addressed and expanded, and alliances with allied health professions must be strengthened (85). But as in many countries around the world, the general public must be better educated regarding the social value of the work of occupational hygiene.

Other issues of concern to occupational hygienists in Thailand involve the development of appropriate occupational health service system, the development of appropriate technology for prevention and control of occupational health problems, legislative measures (appropriate and effective enforcement), appropriate approaches for small and medium sized enterprises, and the integration of occupational health programs with other programs leading to achieving total quality of life for workers, e.g., WHO's Healthy Work Approach.

4.22.9 Information Resources

Mahidol University
Dept. of Occ. Health, Faculty of Public Health
420/1 Rajthevi Road, Bangkok, Thailand 10400
Tel: 066 02 245 7793, Fax: 066 02 247-9458
Status: Academic Institute

Sukhothai Thammathirat Open University
Department of Health Science
Bang Pood, Pak Kred, Nonthaburi, Thailand 11120
Tel: 066 02 5033576, Fax: 066 02 5033607
Status: Academic Institute

Safety and Health at Work Promotion Association
The Department of Labor Protection and Welfare
22/3 Boromrach Chonnanee Road, Talingchun
Bangkok, 10170 Thailand
Tel: 066 02 880-4803, Fax: 066 02 884-1853
Status: Government Association

Occ. Health and Safety at Work Association
420/1 Building 2, Faculty of Public Health,
Rajthevi Road, Bangkok, 10400 Thailand
Tel: 066 02 245-7793, Fax: 066 02 247 9458
Status: Professional Association

Burapa University
Department of Industrial Hygiene and Safety,
Faculty of Public Health
Chonburi, Thailand 20131
Tel 066 038 745900, Fax: 066 038 745900
Status: Academic Institution

Occupational Medicine Association of Thailand
Department of Family Medicine, Faculty of Medicine, Ramatibodi Hospital
Praram 6 Road
Bangkok, 10400 Thailand
Tel: 066 02 246-5102, Fax: 066 02 201-148
Status: Professional Association

Occupational and Environmental Medicine Association of Thailand
P.O. Box 11, Rajvithi Post Office,

Bangkok, 10408 Thailand
Status: Professional Association

Ministry of Labour and Social Welfare
Department of Labour Protection and Welfare
Thanon Fuang Nakhon
Bangkok, Thailand, 10200
Tel: 0662 5140 4
Status: Government Agency

National Institute for the Improvement of Working Conditions and Environment (NICE)
22/3 Baromrachachonnanee Hwy, Thaling-chan,
Bangkok, Thailand, 10170
Tel: 066 2 448 1727, Fax: 066 2 448 6509
Internet: www.inet.th
Status: Government Research Institution

Huachiew Chalerm Prakiet University
Faculty of Public Health and Environment
Bangna-Trad Road, Bangplee, Samudprakran,
Thailand 10540
Tel: 066 02 312-6300 ext. 1174, Fax: 066 02 312-2415
Status: Academic Institution

Ministry of Public Health
Department of Health, Division of Occ. Health
Tivanond Road, Nonthaburi, Thailand, 11000
Tel: 0662 5918149
Internet: www.anami.moph.go.th
Status: Government Agency

Khon Kaen University
Department of Occupational Health and Safety, Faculty of Public Health
123 Thanon Friendship Highway, Amphoe
Muang Khonkaen, Thailand 40002
Tel: 066 043 236906, 2423319 ext. 2480
Status: Academic Institution

4.23 The United Kingdom

4.23.1 Demographics and Economics

The 58.9 million people that comprise the population of the United Kingdom benefit from one of the most powerful economies in Western Europe. The U.K. contains one of the world's leading financial centers and is a major trading entity. Banking, insurance, and business services are the largest proportion of the U.K.'s gross domestic product and percentage of the work force (7). Unemployment has fallen gradually in recent years and inflation has been held at moderate levels. The U.K. is a member of the EU but continues to struggle with its role in the monetary and economic integration with Europe (7).

Through a long-term program of privatization, the government has relinquished control of a number of industries in the last 20 years. The industries sector which accounts for the second largest percentage of the workforce, consists of the production of machine tools, electric power equipment, electronics and communications equipment, automation equipment, railroad equipment, ships, aircraft, motor vehicles and parts, paper and paper products, textiles, clothing, and other consumer goods, as well as metals, chemicals, coal, and petroleum (7). The U.K. is a major producer of energy in the form of coal, natural gas, and petroleum. However, the percentage of the population employed in these industries that have been traditionally associated with the utilization of occupational hygiene services is declining. Industrial production growth rate in 1997 was approximately 2%. Agriculture remains a significant economic force producing almost 60% of food needs with only about 1% of the labor force (7). Distribution of the British work force is as follows: Services, 68.9%; Manufacturing and construction, 17.5%; Government, 11.3%; Energy, 1.2%; and Agriculture, 1.1.

The population of 23 million people is served by some 210 occupational hygienists (diploma professionals), and another 240 certified occupational hygienists (88). This estimate yields an estimated occupational hygienist to work force ratio of 1:51,000. There are many more individuals who may perform occupational hygiene-related duties.

4.23.2 Historical Development

In 1833, the United Kingdom established the first Factory Inspectorate. Its charge was to deal with the problem of the long hours that children and men spent working in factories, among others. The Factory Inspectorate broke new ground in several respects. The four original inspectors divided the country into four districts. They recognized early on that they could not visit all establishments in their districts and thus could not play a true policing role. They could, however, promote good will among employers and elicit their cooperation. Inspectors had the opportunity to see problems in the factories they visited and to seek solutions that would benefit employers as well as those working for them. As the inspectors traveled from workplace to workplace, they appealed to common sense and to the good will of factory owners and left enforcement as a matter of last resort (89).

Despite the Factory Inspectorate's successes, serious risks continued to exist in British factories as industry continued to grow and expand. Such risks were serious enough to claim the attention of the British public. Luxon (90) writes:

> Contemporary writers, including Charles Dickens who drew attention forcibly to the dangers, highlighted the risks of the new industrial processes. In the latter part of this period there were several disastrous fires that paved the way for additional legislation and industrial hygiene, as we know it today, was born. The British Factories Act of 1864 contained the first requirements. Every factory was to be ventilated to render harmless any gases, dusts or impurities that may be damaging to health, i.e., what we would call now dilution ventilation. An Act of 1878 took a further step and required exhaust ventilation by means of fans for the removal of dust likely to be injurious to health. It is interesting to note that wording almost identical to that setting out these two concepts appears in present day UK legislation.

The Factories Act of 1901 marked another major step forward for it also regulated hazardous trades. The Act led to in-depth investigations and control of work-related hazards. Early surveys focused on the hazards of silicosis in the pottery industry. The public of the time had read about the shakes of the Mad Hatter in Alice in Wonderland. A study revealed that Lewis Carroll had given a good description of occupational poisoning among hatters owing to their using mercury to change fur into felt. Another study linked phossy jaw to the use of yellow phosphorus in the production of matches. Based on this study, Great Britain was the first country to outlaw the use of yellow phosphorus.

By 1914, World War I had engulfed Europe. Once again, long working hours became a matter of concern in Great Britain. In 1915 the British government appointed the Health of Munitions Workers' Committee under the chairmanship of the Right Honorable David Lloyd George, later Prime Minister of Great Britain from 1916 to 1922. The Committee was asked to investigate and advise on questions regarding working hours and matters that may affect the health and efficiency of munitions workers. As in other wars, women had entered the work force in large numbers.

The need for people trained in industrial physiology and psychology prompted the London School of Hygiene and Tropical Medicine to offer several related courses in 1938. By 1942 the responsibilities of the Industrial Health Research Board increased to include responsibilities for studying occupational and environmental factors that could cause illness and disease (89). Following the end of World War II, a number of research centers, mostly at universities, began to specialize in various areas of knowledge based mainly on staff interest and existing facilities. Thus Cambridge specialized in applied psychology. The Research Unit at Penarth in South Wales conducted studies on pneumoconiosis whereas groups at Oxford and London concentrated on the effects of heat stress. An atmospheric pollution research unit was set up in London, and the Atomic Research Establishment at Harwell conducted investigations into the hazards of ionizing radiation.

At about that time, about 20% of British workers employed by large industries came to enjoy the benefits of in-house occupational hygiene and medical services. The Slough Industrial Health Service was set up in 1947 and was intended to provide first medical services to small industrial establishments with a total work force of some 20,000 workers. Two years later the service was expanded to include occupational hygiene support, the sole provider of such services in Great Britain at the time (89). Unfortunately, the intended beneficiaries lacked the financial resources and may have failed to grasp the need for such services. The Slough Industrial Health Service was available to its customers for about 18 years. By 1964 it had to close its doors for lack of financial support.

By the mid-1960s, occupational hygiene was receiving strong support and increased recognition in Great Britain. The Factory Inspectorate added an occupational hygiene unit to advise industry on occupational hygiene problems and provide field as well as analytical services. The Trades Union Congress expressed strong interest in occupational hygiene and what it could do for British workers. Although the congress did not add hygienists to its staff, on the occasion of the one-hundredth anniversary of its founding, it did endow an Institute of Occupational Health at the London School of Hygiene and Tropical Medicine to provide consultation and field and laboratory services to various organizations, industries, and individuals.

Until the early 1970s, British industry was subject to many safety statutes, mostly under some earlier parent statute and usually specifying duties of factory occupiers (typically the employers). The inspectorate was fragmented by industry, and there was no national body to oversee the systematic development of safety legislation. A government committee, under Lord Robens, was appointed in 1970 to review this situation (89). The committee published its report two years later and identified a series of shortcomings that existed in industry. It recommended increased involvement and self-regulation on the part of industry. It also stressed the need for greater cooperation in matters of health and safety between management and workers, and for a more effective role on the part of the government.

Two major events, unrelated to each other, followed and had a major impact on occupational medicine and hygiene in the United Kingdom. In 1973 Great Britain joined the EU. The original signatories, Belgium, France, the Federal Republic of Germany, Holland, Italy, and Luxembourg had signed the treaty of Rome committing themselves mainly to promoting economic cooperation among the members.

The other major event resulted from the findings of the Robens report published in 1972 (89). Two years later, in 1974, the British Parliament passed the Health and Safety at Work Act that placed all occupational health and safety agencies under one umbrella and forms the basis of the current regulatory structure in Britain. The Act covers areas such as occupational health and safety, explosive and flammable materials, noxious contaminants, radioactive materials, and nuclear installations. It imposes a number of obligations on manufacturers, designers, importers, and suppliers of substances and articles. It requires such individuals to test items covered by the act. They either have to carry out, or have others carry out, research to ensure that their products can be used safely or with minimum risk. They also have to provide customers with adequate information for the proper and safe use of such products. If, however, customers use these substances or articles improperly and in violation of instructions provided, the act places the responsibility for any harm on the user.

4.23.3 Occupational Health and Safety Legislation

The 1974 Health and Safety at Work Act provides the primary framework of duties for employers and employees, within which specific legislation on hazardous substances, noise, radiation, asbestos, work shift length, electrical safety and a host of other issues are addressed (88, 89). The primary regulatory agency for matters pertaining to occupational health and safety is the Health and Safety Executive (HSE).

The Act imposes duties on employees as well as employers. Employees are expected to exercise due care and caution to protect their own health and safety and that of their co-workers. Otherwise, they can be held accountable if they knowingly violate health and safety instructions.

The Act makes the operators of industry responsible for the health and safety of the public if work-related hazards threaten the well-being of the latter. It replaces overly complex regulations with more flexible codes of practices. Such codes carry weight in certain legal proceedings but allow industry to satisfy the standards by equivalent means, i.e., they are performance-based. There are mandatory requirements and others that are applicable "so far as is reasonably practicable." Although the act does not clearly define this expression, the courts have interpreted it to mean that the risks of a given process or activity must be weighed against the cost and effort involved in controlling such risks. Though cost is clearly a factor, British courts have not required that such costs be compared against or linked to an employer's ability to pay. It is equally clear, however, that without the moderating expression "so far as reasonably practicable," agencies of the government might tend to enforce provisions of the act more strictly (88).

The Health and Safety Commission (HSC) and the HSE were created to implement the provisions of the Act. The HSC develops strategies, formulate policies, and acts as an advisor on occupational health and safety to other governmental agencies. It consists of a chairman and eight members, three appointed by the National Employers Organisation, three by the National Trade Union Organisation, and two representing the public at large. The HSC operates through the HSE, which consists of three members supported by six inspectorates, a research laboratory, and an Employment Medical Advisory Service. By the tripartite nature of its membership, the HSC is the principal arena where technical and economic issues are raised and major interests voiced. It is a system wherein interested parties have the opportunity to review and scrutinize proposed regulations and codes of practice and it is subject to political debate and exposed to public scrutiny (88). Only after all parties have had the opportunity to be heard does the HSE present a proposal to the HSC for ratification. By the time a regulation or code of practice is promulgated, all concerned parties have been heard. Although the process may be slow and laborious, it is time-tested and works to the satisfaction of the participants for the purpose and within the framework for which it was created.

The commission also set up an Advisory Committee on Toxic Substances (ACTS) with representatives from employers, employees, and professional organizations. They were charged with drawing up a document assessing occupational risks, recommending control measures, and establishing requirements for health surveillance. They provided information to affected parties and training to minimize or avoid ill health among those exposed in the work place or among the public. Under the ACTS is the Working Group on the Assessment of Toxic Chemicals (WATCH) which is responsible for developing the UK's occupational exposure limits.

By virtue of section 16(1) of the Health and Safety at Work Act 1974, and with the consent of the Secretary of State for Employment, the HSC approved the Code of Practice entitled Control of Substances Hazardous to Health (COSHH). This code of practice, which took effect in 1988, gives practical guidance with respect to the Control of Substances Hazardous to Health Regulations 1988 (SI 1988 No 1657) including exposure assessment and control, and medical surveillance.

U.K. health and safety legislation is also substantially driven by the EU in Brussels with a high degree of agreement with the EU Directives (88). The HSE coordinates with the EU to ensure that directives adopted under the HSE are based on good science and practice. HSE is beginning to write into Codes of Practice that certain tasks, such as those conducted by asbestos technicians, are conducted by a person with BIOH accreditation in the subject.

4.23.4 Occupational Exposure Limits

The establishment of the HSC gave a major impetus to the initiative of developing British occupational exposure limits. In 1979, the commission adopted a new system of setting control limits intended to protect industrial workers from airborne substances at work. Subsequently the HSE published a list of "control limits" and of "recommended limits" in a publication entitled Guidance Note Environmental Hygiene (EH 40 April 1984).

The current system for the development of occupational exposure limits begins with the authority granted to the COSHH regulations to develop limits based on the employers obligation to assess the risk of exposure to their workers and to prevent or control the risk (91). Under the auspices of the HSC, the WATCH determines the need and scientific feasibility for an occupational exposure limit for a particular substance or compound and the appropriate numerical value.

Two types of limits are available: Occupational Exposure Standards (OESs) and Maximum Exposure Limits (MELs). Substances considered for review are derived from the EU's Existing Substances Regulations (ESR) and an internal priority system prioritized by carcinogens, mutagens, respiratory sensitizers, reproductive toxins and neurotoxins. To recommend an OES, WATCH must meet three qualifying criteria: (*1*) the available scientific information for the chemical in question is sufficient to identify a "level at which there is no indication that exposure is likely to be injurious" (91); (*2*) temporary exposures above the OES will not result in an increase risk of injury; and (*3*) compliance with the OES is feasible. Numerical values are derived by dividing the no adverse effects level (NOAEL), if it can be identified, by a safety factor that generally ranges from one to 50. Higher safety factors are generally applied to substances with chronic effects such as teratogens. If no NOAEL is identified, WATCH may recommend a MEL. This is normally the case for genotoxic carcinogens, although for some non-genotoxic carcinogens an OES may be feasible. The MEL value is based on what is judged to be achievable in the work environments in which the substance is encountered as well as risk assessment information at the proposed level. However, extrapolation of high dose animal toxiciology data is not applied in determining a MEL value in Britain due to the variability in risk associated with different animal systems (91). MELs may not be exceeded.

Once these criteria are met the recommendation is forwarded to the ACTS for cost benefit analysis and consideration of the social acceptability of risk prior to endorsement and recommendation to the HSC for final review. Because violations of MELs represent a potential criminal offense for failure to control the work environment under the COSHH, promulgation requires the approval of the Secretary of State (91). There are currently about 500 occupational exposure limits with 350 being adopted from the 1989 ACGIH TLV list, and an additional 150 developed through the COSHH system. Of these, 100 are OESs and 50 are MELs.

The British authorities recognize the high cost and relatively slow rate of development of new limits resulting from the deliberative process working on a substance-by-substance basis. As such, the British authorities are now considering the concept of "risk banding" wherein control of exposure is the priority, and where the method of control is determined by a combination of toxicity, bioavailability and chemical usage. It is anticipated that this modelling of exposures for risk assessment purposes and linkage of toxicological endpoints to fundamental practical controls will greatly affect the UK occupational hygiene community in future.

HSE inspectors determine whether an industrial establishment complies with applicable regulations based on its implementation of recommended limits and on overall working conditions.

4.23.5 *Education and Training*

The education and training of occupational hygienists in Great Britain is similar to the U.S. with some candidates obtaining an undergraduate degree in the natural sciences (biology, chemistry, physics, etc.) and engineering, followed by the pursuit of a graduate degree in occupational hygiene, while some candidates elect to acquire an undergraduate degree in occupational hygiene first (92). With either option there are a variety of undergraduate and graduate degree programs in occupational hygiene in the UK. The British Institute of Occupational Hygiene (BIOH), the certification entity in UK, presently recognizes seven degree programs (88): Bachelor of Science (B.Sc.) in Occupational Hygiene from the University of the South Bank in London; Bachelor of Science (B.Sc.) in Environmental Science with an option in occupational hygiene from the University of Bradford; the Masters of Science (M.Sc.) in Occupational Hygiene from University of Newcastle, University of Manchester, University of Aberdeen, University of Surrey and the University of Birmingham. Doctoral programs are available at the London School of Hygiene and Tropical Medicine (Ph.D.), and University of Newcastle-upon-Tyne (Ph.D.). Other universities in the UK offer degree programs with an emphasis in occupational hygiene or related disciplines.

To the extent that the British academic system can sustain the current demand for qualified occupational hygiene candidates without maximizing capacity, British universities will continue to play an increasingly important role in the development of occupational hygienist internationally, but particularly among EU member countries. Another trend that has been noted is the lack of parity between the development of academic programs and the rate of development in occupational hygiene research in the U.K. (92).

Yet another trend in the U.K. is the movement toward the use of module academic programs in the development of core occupational hygiene competencies. This is best seen in the BIOH module training and certification system.

4.23.6 *Certification and Registration*

In 1967, the British Occupational Hygiene Society (BOHS) formed the British Examining Board in Occupational Hygiene (BEBOH) with members nominated by the society. The examining board set professional criteria and granted diplomas to those qualified and able to pass appropriate oral and written examinations. Over the years, BEBOH grew in both strength and influence. To serve its award holders better, under a constitution agreed to jointly by BOHS and IOH, the functions of the board were expanded in 1978 to include registration of qualified individuals. Thus 11 years after its founding, BEBOH became the British Examining and Registration Board in Industrial Hygiene (BERBOH). Initially, BOHS and IOH each appointed four board members. By 1985 BERBOH amended its constitution. Henceforth, award holders, rather than BOHS and IOH members, would elect candidates to BERBOH.

The 1978 official BERBOH booklet explains the board's aims and functions as follows: "The aims of the Board are to certify to the attainment of recognised standards of competence in the practice of Occupational Hygiene, and to establish and maintain a public Register of those people who have achieved such recognised standards." To that end, BERBOH defined two levels of professional competence in the initial stages of the certification process: a senior practitioner who could work at a professional level without direct supervision and a junior practitioner who would normally work under the direction of a senior professional. By 1975 the need for a more junior level practitioner was apparent and the Board responded by introducing a system of 13 occupational hygiene topic modules. The subjects included general principles of workplace control, noise and vibration, toxic metals, asbestos sampling and analysis, to name a few. This system has proved to successful and has been subsequently enhanced to permit certification for those practitioners who acquire six appropriately grouped Preliminary Certificates to be exempt from sitting for the Certificate Core examination.

At present, the BIOH recognizes four levels of qualification (93). The most fundamental of these is the Occupational Hygiene Modules. This consists of a series of eight occupational hygiene modules including a requirement of 60 hours of pre-class preparation, a one-week individual subject course and an examination. The available subject areas include risk assessment, measurement of hazardous substances, asbestos, noise and vibration, thermal environment, testing of ventilation systems, COSHH and basic ergonomics. The module can be used to gain competence in the subject areas or grouped together to gain exemption from the Certificate or Diploma Core examination.

The second level of qualification is the Certificate of Competence in an Individual Subject. A two-part examination including written and oral components in an individual subject area. The oral examination is usually 30 minutes in duration and includes a requirement to submit an example of written work also in the subject area.

The third level is the Certificate of Operational Competence in Occupational Hygiene. This examination is available to practitioners with responsibility for conducting surveys of the work environment. This requires a "good standard of general education," three years of comprehensive experience, and either successfully complete the two-part written/oral examination or qualify for an examination exemption through the completion of five of the specified modules (general principles of occupational hygiene, instrumentation, sampling methods, standards, principles of control), having an approved undergraduate or graduate degree in occupational hygiene. Individuals acquiring the qualification must have a minimum of three years of comprehensive field experience but are not required to make the interpretations and judgments expected of a diplomate.

The fourth and highest level of qualification is the Diploma of Professional Competence in Occupational Hygiene. To become a Diplomate of BIOH, the candidate must have a qualified undergraduate or graduate degree in science, engineering, medicine, or "other equivalent qualifications or experience," have a minimum of five years of comprehensive experience, be able to demonstrate their knowledge of comprehensive, advanced occupational hygiene through risk assessment in a variety of environments and appropriate control measures. A two-part written/oral examination is required.

The occupational hygiene system in the U.K. also recognizes four general categories of professionals; (*1*) Professional Hygienist: "A person who is recognized as being capable of practicing at a professional level in all aspects of occupational hygiene." (*2*) Limited Professional Hygienist: "A person who is recognized as being capable of practicing at a professional level in a specified area of occupational hygiene practice, as indicated in the Register." (*3*) Operational Hygienist: "A person who is recognized as being capable of making environmental surveys in the comprehensive field of occupational hygiene." (*4*)

Competent Person: "A person recognized as being competent to carry out a specified range of environmental surveys, as indicated in the Register, to the same standard as an Operational Hygienist."

A voluntary continuous professional development program that has been in place for some years become mandatory in 1999 (88).

4.23.7 Professional Organizations

British professionals from diverse occupational health-related areas founded the British Occupational Hygiene Society (BOHS) as a Learned Society in 1953 (94). The aim of the society is to promote awareness of occupational hygiene through education, information, research and professional networking. Over the years, the society has grown rapidly both in the field of comprehensive practice as well as in specialized areas. In 1958 the Society began publishing the technical journal, *Annals of Occupational Hygiene* which today enjoys an international reputation. The BOHS also maintains several special interest groups (SIG) dealing with chemical hazard and risk, the electronics and telecommunications industries, the national health service, occupational hygiene information systems, offshore industry, radiation, and sampling and analytical methods. These SIGs operate as a forum for information exchange and as a repository of subject matter knowledge for the benefit of other practitioners.

By 1975 the ranks of British occupational hygienists had grown to the point where their own professional organization was needed, distinct and separate from BOHS. Thus the Institute of Occupational Hygienists (IOH) was founded but continued to work closely with BOHS. In fact, today many members hold dual membership. In 1997, the IOH and BEBOH merged with a subsequent name change to the British Institute of Occupational Hygiene (BTOH) (88).

Membership in the BIOH includes eight grades: (*1*) "Associate: Open to those who have attained qualifications with an element of occupational hygiene and can demonstrate an active interest in occupational hygiene practice. (*2*) Student: Open to those undertaking full- or part-time study leading to any of the qualifications specified as acceptable for application to graduate membership. Student: Open to those undertaking full- or part-time study leading to any of the qualifications specified as acceptable for application to graduate membership. (*3*) Graduate: Open to those with an honours or post-graduate degree in occupational hygiene or an equivalent. (*4*) Licentiate: Open to holders of the 'Certificate of Operational Competence in Occupational Hygiene'. (*5*) Specialist Member: Open to those who have experience at an advanced level in specialist aspects of occupational hygiene; such as, occupational health or safety, toxicology, epidemiology, etc. (*6*) Member: Open to those who are holders of the 'Diploma of Professional Competence in Occupational Hygiene'. (*7* Fellow: Awarded to applicants who can demonstrate seniority in the field of occupational hygiene and have made a distinct contribution to the advancement of the profession. (*8*) Honorary Fellow: Bestowed at the discretion of the board on those who have made an outstanding contribution to the profession." The Institute currently has some 60 Fellows, 150 Members, 240 Licentiates, 50 Graduates and 60 Associates for a total membership of 560, including those in training.

BIOH has five working committees addressing different aspects of the organization's charter: Examinations and Membership, Conference and Publications, Membership Services, Marketing and External Relations, and Technical Affairs. The BIOH inherited the code of ethics from the merger of the Institute of Occupational Hygienists (IOH) with the British Examining Board in Occupational Hygiene (BEBOH) in 1997. It is currently being revised.

To enhance alignment between the allied occupational health professions on important issues, the occupational health related organizations meet informally with the HSE and representatives of other professions such as the Faculty of Occupational Medicine, the Society for Occupational Medicine, the Institution for Occupational Safety and Health, and the Occupational Health Nursing division of the Royal College of Nursing.

4.23.8 Status of the Profession

In Great Britain as in many other countries, including the United States, recognition of the occupational hygiene profession in the general public has been slow. Despite growth of occupational hygiene as an accepted science among the allied health professions and the government, there is no statutory requirement for certification or registration required to practice in the U.K. Nevertheless, it is increasingly common to find employers in the industrial sector requiring evidence of competence on the basis of academic or professional certification, and consequently there is much greater reliance on the BIOH's standards than was the case only a few years ago. Integration of the occupational hygienist into a occupational health services team with allied professions and the perceived change in the definition of the occupational hygienist from one focusing on the methodologies to one focused on the objectives within the UK are both important issues at this point in time (95).

But recognition is not the most significant driver in the development of the profession in the last decade or so. Instead, the most important change in the development of occupational hygiene was an evolution in the longstanding governmental practice of identifying a specific disease or impairment associated with a specific industry and implementing regulations to address the issue (88). Recognizing the inadequacy of this approach in the modern industry work environment has been significant. The change to a more system-based, proactive, preventative posture has required a substantial reliance on the occupational hygienist to evaluate the work environment and exercise profession judgment regarding its control. This change really began with the 1972 "Robens Committee" which was particularly concerned with the pace of change in chemical industries. The resulting Health and Safety at Work Act of 1974 is what is regarded as "performance based" legislation, setting general guidelines for duties which are further defined in the specific regulations aided by codes of practice. In many cases the action that is required is based on determination of risk requiring the use of skilled professionals, such as the occupational hygienist, though in many cases the nature of the problem and the available guidance means that effective controls can be introduced without this level of input.

With regard to future developmental needs, the U.K. is very similar to the U.S. in its need for a coordinated approach among the different stakeholders in the profession, e.g., government, academia, industry, etc. With the relatively small number of professionals in the U.K., and the relatively large working population with a potential need for occupational hygiene services, it is reasonable to estimate that many thousands of engineers, safety professionals, nurses, and chemists are conducting work involving some element of occupational hygiene (88). Role delineation and increased public recognition will be important in this regard (95). However, occupational hygienists in the U.K. will continue to yield significant influence on the profession regionally (EC) and globally.

Great Britain can lay claim to being among the first to recognize the need for occupational hygiene. It has offered the world outstanding examples of scientific advances and legislative controls in this field. It is rich in occupational health organizations, academic institutions, and publications. Among the many who have made significant contributions to the profession are eight British scientists have been honored with the AIHA Yant Memorial Award: Henry L. Green in 1965, W. H. Walton in 1971, C. G. Warner in 1974, S. G. Luxon, in 1984, R. J. Sherwood in 1987, P. L. Bidstrup in 1989, S. A. Roach in 1994, and C. Veys in 1997.

4.23.9 Information Resources

British Occupational Hygiene Society (BOHS)
The Occupational Hygiene Secretariat,
Suite 2, Georgian House, Great Northern Road,
Derby DE1 1LT, United Kingdom
Tel: 44 0 1332 298101, Fax: 44 01332 298099
Internet: www.bohs.org
Status: Professional Association

British Standards Institution
British Standards House
389 Chiswick High Road
London, W4 4 AL, United Kingdom
Tel: 44 0 181 996 9000, Fax: 44 0 181 996 7400
Internet: www.bsi.org.uk
Status: Consensus Standards Organization

The Institute of Occupational Health
The University of Birmingham.
Edgbaston, Birmingham, B15 2TT UK
Tel: 44 0 121 414 6030, Fax: 44 0 121 414 6217
Internet: www.bham.ac.uk/IOH
Status: Academic Research Institute

The Society of Occupational Medicine
6 St Andrew's Place, London, NW1 4LB, United Kingdom
Tel: 44 171 486 2641, Fax: 44 171 486 0028
Internet: www.med.ed.ac.uk/hew/som
Status: Professional Association

Health and Safety Executive
London Information Center
Rose Court, Ground Floor North, 2 Southwark
Bridge, London, SE1 9HS, United Kingdom
Tel: 44 054 154 5500

Internet: www.open.gov.uk/hse
Status: Regulatory Agency

British Institute of Occupational Hygiene (BIOH)
Suite 2, Georgian House, Great Northern Road,
Derby DE1 1LT, United Kingdom
Tel: 4401332 298087, Fax: 4401332 298099
Internet: www.ed.ac.uk
Status: Professional Association

Institution of Occupational Safety and Health
The Grange, Highfield Drive, Wigston,
Leicestershire, LE18 1NN, United Kingdom
Tel: 44 0116 257 3100, Fax: 44 0116 257 3101
Internet: www.iosh.co.uk
Status: Professional Association

Health and Safety Executive
Health and Safety Laboratory
Broad Lane, Sheffield, S3 7HQ, United Kingdom
Tel: 44 0114 289 2920
Status: Regulatory Research Laboratory

5 THE FUTURE

The future possibilities and prospects for the occupational hygiene profession are significant. As the world economy grows and more opportunities emerge for developing countries, the demand for qualified occupational hygienists will continue to grow. However, there are many issues that must be addressed to make this a sustainable trend. Among these are: (*1*) need to define a consensus of education for occupational hygiene and provide adequate access to universities and other institutions offering education appropriate to the needs of the country or region; (*2*) accurate and consistent morbidity and mortality surveillance mechanisms; (*3*) harmonized legislative structures for occupational health and safety that support the direct involvement of occupational hygienists in the protection of workers and the community; (*4*) greater recognition of the profession in the general public; (*5*) reciprocity between countries with regard to professional certification; (*6*) equitable distribution of resources necessary to promote the development of the profession; (*7*) greater worker advocacy where there is none; and (*8*) increased synergism between allied occupational health professionals with the purpose of applying a multidisciplinary approach to occupational health.

In the future, occupational hygienists around the world will need to work in the context of new issues and paradigms in order to remain an effective and viable profession. These may include: (*1*) the principle of sustainable development; (*2*) the application of the precautionary principle to occupational risk assessment; (*3*) the integration of occupational safety, environmental management, public health and sanitation principles as necessary to address comprehensive protection; (*4*) economic rationalization and justification of occupational hygiene services; (*5*) the application of health and safety management systems such as ISO, BSE, etc., and (*6*) the systematic application of primary control measures over traditional exposure assessment in regions with chronically poor resources (96–102).

It is entirely likely that the very concept of a healthy workplace will change in the future through the sanction of a more holistic approach to "occupational health promotion"—whether the occupational environment is a factory, farm, office, home, or hybrid environment—linking occupational health with sustainable development of the economy conducive to harmonization of work, health and safety, an improved ecological system, and economic growth.

As stated by Berenice Goelzer principal occupational hygienist of the WHO (103):

> Looking at the world at large, one can observe the whole history of occupational health and occupational hygiene happening at the same time. In the past, each step, each achievement, took a long time, sometimes decades, sometimes centuries! The great difference, at present, is that these steps can now be conquered at a much faster rate, because the barriers of ignorance are being increasingly replaced by bridges that are more easily crossed. How fast these bridges will be crossed depends largely on us because passwords for crossing them include sharing and collaboration among occupational hygienists everywhere.

ACKNOWLEDGMENTS

This chapter would not have been possible without the conscientious assistance and input of occupational hygienists from the countries represented herein. The authors express their gratitude to all those who contributed to this chapter with special appreciation to the following individuals: Brian Davies and Heather Jackson (Australia), Irene F. Souza D. Saad, Marcos Domingos da Silva, and Saeed Pervaiz (Brazil), Ugis Bickis, Dave Verma, G. Peter Robson, Mark A. Nazar, Kay Teschke, Helen Mersereau, and Dhanani Nasrin, (Canada), Nick Yin, Liu Juan Luang, Jiong Liang Zhou, Steven P. Levine, Roxanne Present, and Y. Cheng (China), Riitta Viinanen, Rauno Hanhela, and Ben I. Bjorkqvist (Finland), Kurt Leichnitz (Germany), Joseph Kwan and Tai Wa Tsin (Hong Kong), Maharshi Mehta (India), Joe Kearney (Ireland), Yoshimi Matsumura (Japan), Bjorg Eli Holland (Norway), Rafael Echavarria (Mexico), Paul Swuste (The Netherlands), Janusz Indulski, Jan Grzesik, Jan Gromiec, Edward Wiecek, and Alexsandra Nawakowski, (Poland), Nam Won Paik (Republic of Korea), David R. Hicks (Saudi Arabia), Robert Ferrie, and Raymond Strydom (South Africa), Gun Nice (Sweden), Alfred Steinegger and Rudolf Knutti (Switzerland), Twisuk Punpeng and Harry A. Trout (Thailand), and David Asker-Browne and Paul J. Oldershaw (United Kingdom). The authors also recognize the efforts of Gene Kortsha in developing the first edition of this chapter upon which this current work is based.

BIBLIOGRAPHY

1. D. Markey, S. Levine, and C. Redinger, *Am. Ind. Hyg. Assoc.* **57**, 936 (1996).
2. *WHO Global Strategy on Occupational Health for All*, WHO/OCH/95.1, World Health Organization, Geneva, 1995.

3. WHO, *Occupational Hygiene in Europe, Development of the Profession*, European Occupational Health Series No. 3, WHO Regional Office for Europe, Copenhagen, 1992.
4. N. T. Watfa, "ILO Programme in Occupational Hygiene," ILO African Newsletter, International Labout Organization, Geneva, Switzerland, Vol. 3, (1996).
5. IOHA, *IOHA Newsletter*, International Occupational Hygiene Association, Derby, United Kingdom, Volume 6, 1998.
6. J. S. Barnett and J. S. Lee, *Appl. Occup. Environ. Hyg.* **6**, 98 (1991).
7. CIA, *CIA World Factbook*, U.S. Central Intelligence Agency, National Technical Information Service, Springfield, Virginia, 1998.
8. B. Davies and H. Jackson, personal communication, March 1999.
9. T. Jones, *Am. Ind. Hyg. Assoc. J.* **49**, 593 (1988).
10. V. Vasak, *Am. Ind. Hyg. Assoc. J.* **57**, 119 (1996).
11. J. H. Vincent, *Am. Ind. Hyg. Assoc. J.* **59**, 729 (1998).
12. AIOH, AIOH Bylaws and Code of Ethics, Australian Industrial Hygiene Association, 1998.
13. I. Saad, M. Domingos Da Silva, and S. Pervaiz, personal communication, Feb. 1999.
14. M. Domingos da Silva: *La Higiene Ocupacional en Brasil*, 3rd Congresso Nacional de Higiene Industrial, Queretaro, Mexico, Sept. 23, 1998.
15. W. A. Cook, *Occupational Exposure Limits—Worldwide*, American Industrial Hygiene Association, Fairfax, VA 1987.
16. H. Frumkin, V. de M. Camara, *Am. J. Pub. Hlth.* **81**, 1619 (1991).
17. U. Bickis, D. Verma, G. P. Robson, K. Teschke, H. Mersereau, and D. Nasrin, personal communication, Feb.–March 1999.
18. A. Yassee, *Am. J. Public Hlth.* **78**, 689 (1988).
19. The Ontario Factories Act of 1884, c. 39, s. 11(3), as quoted in *Occup. Health Ontario*, **9**(1), 20 (Winter 1988).
20. Ontario Ministry of Labour, *Designated Substances in the Workplace: A General Guide to the Regulations, Ontario Government Publications*, 0-7729-6964-7, 1985.
21. D. K. Verma, A. M. Sass-Kortsak, and D. H. Gaylor, *Am. Ind. Hyg. Assoc. J.* **55**, 364 (1994).
22. J. W. Smith, "Training and Education in Occupational Hygiene: An International Perspective," in *Ann. Am. Conf. Gov. Ind. Hyg.* **15**, 133, 1988.
23. CRBOH, *Newletter*, Canadian Registration Board of Occupational Hygiene, 1998.
24. CRBOH, *Code of Ethics*, Canadian Registration Board of Occupational Hygiene, 1998.
25. D. K. Verma, *Ann. Occup. Hyg.* **40**, 477 (1996).
26. S. Y. Chen, D. Pan, Y. Gao, F. F. Wang, and X. Liu, *Chinese J. Health Inspection* **3**(Suppl), 21 (1996).
27. Y. X. Liang, H. Fu, and X. Q. Gu: "Provision of Occupational Health Services in China," *ILO-CIS Asian-Pacific Newsletter on Occupational Health and Safety*, ILO/Finnida, Asian-Pacific Regional Programme on Occupational Safety & Health (ASIA-OSH) (1997).
28. Y. X. Liang; "Occupational Health in China: The Current Status and Future Perspectives," in: *Symposium Proceedings From Research to Reality: Improving Workplace Safety and Health in China*, co-sponsored by Liberty International, Shanghai Medical University, and Harvard School of Public Health, held on 3–4 Dec. 1997, Shanghai.

29. Order of the President of the People's Republic of China—No. 28: Labour Law of the People's Republic of China.
30. Y. X. Liang, B. Q. Gang, and X. Q. Gu, *Reg. Tox. Pharm.* **22**, 162 (1995).
31. R. Present, S. P. Levine, and Y. X. Liang, personal communications, May 1999.
32. M. H. Noweir, *Am. J. Ind. Med.* **9**, 125 (1986).
33. M. H. Noweir, "Overview of Occupational Health Research in Egypt," *International Symposium on the Biomedical Impact of Technology Transfer, Faculty of Medicine Ain Shams University, Cairo, Egypt*, National Institute of Environmental Sciences, North Carolina, 1986.
34. M. H. Noweir, "Training and Education in Occupational Hygiene in Egypt, Training and Education in Occupational Hygiene: An International Perspective," *Ann. Am. Conf. Gov. Ind. Hyg.* **15**, 143 (1988).
35. R. Viinanen, R. Hanhela, and B. I. Bjorkqvist, personal communications, Feb. 1999.
36. *Occupational Health Services in Finland*, 2nd rev. ed., Institute of Occupational Health, Helsinki, Finland, 1988.
37. K. Leichnitz, personal communication, Feb. 1999.
38. D. Walters, ed., "The Identification and Assessment of Occupational Health and Safety Strategies in Europe," Volume I: *The National Situations—Germany European Foundation for the Improvement of Living and Working Conditions*, Dublin, Ireland, 1995, pp. 75–89.
39. G. Lehnert, and R. Wrbitzy, *Int. J. Occ. Med. Env. Hlth.* **11**, 9 (1998).
40. J. Kwan and T. W. Tsin, personal communication, Feb. 1999.
41. M. Mehta, personal communication, April 1999.
42. D. J. Parikh, and S. K. Kashyap, *Ann. Am. Conf. Gov. Ind. Hyg.* **15**, 153 (1988).
43. J. Kearney, personal communication, Jan. 1999.
44. HSE, "Guide to the Health and Welfare at Work Act 1989 and The Safety Health and Welfare at Work (General Application) Regulations, 1993 (S.I. 44 of 1993)," Health and Safety Executive, Dublin, Ireland, 1995.
45. D. Sordelli, *Am. Ind. Hyg. Assoc. J.* **55**, 491 (1993).
46. From a speech by Dr. Danilo Sordelli at the II Conferencia Nacional de Higiene Industrial, November 17, 1998, Valencia, Spain.
47. DPR April 27, 1955, No. 547, Norme per la prevenzione degli infortuni sul lavoro e successive integrazioni. DPR of March 19, 1956, No. 303, Norme generali per l'igiene del lavoro, followed by 320/1956, 321/1956, and 128/1959.
48. Legge No. 615, July 13, 1966, Provvedinenti control l'inquinamento atmosferico.
49. Y. Matsumura, *Ann. Occup. Hyg.* **39**, 261 (1995).
50. Y. Matsumura, personal communication, Feb. 1999.
51. T. Toyama, *Am. J. Industrial Med.* **8**, 87 (1985).
52. J. H. Vincent, *Am. Ind. Hyg. Assoc. J.* **59**, 729 (1998).
53. T. Takebayashi, K. Omae, and R. J. Sherwood, *Appl. Occup. Environ. Hyg.* **11**, 457 (1996).
54. N. W. Paik, "Industrial Hygiene in Korea" AIHA International Affairs Committee Country Profile, American Industrial Hygiene Association, Fairfax, VA, 1997.
55. S-H. Lee, "Occupational Health Services in Korea," *ILO Asian Pacific Newletter*, (1998).
56. N. W. Paik, *Appl. Occup. Environ. Hyg.* **11**, 215 (1997).
57. R. Echavarria, personal communication, May 1999.

58. J. A. Ortega, "*la Normatividad de la salud en el trabajo en Mexico*" (working paper) Instituto Nacional de Salud Publica, Cuernavaca, Mexico.
59. Dutch Occupational Hygiene Association, *1997 DOHS Annual Report*, Nvva Eindhoven, The Netherlands.
60. P. B. Meyer, "Training and Education in Occupational Hygiene: An International Perspective," *Ann. Am. Conf. Gov. Ind. Hyg.* **15**, 165 (1988).
61. P. Swuste, personal communication, Jan. 1999.
62. A. Burdorf, *Certification of Occupational Hygiene, A Survey of Existing Schemes Throughout the World*, The International Occupational Hygiene Association, Derby, U.K., 1995.
63. D. B. Brown, *Occupational Hygiene Education in the EEC: A Survey of Existing Programs* Wageningen University, The Netherlands, 1988, pp. 105–110.
64. From "Note 'Occupational Hygiene' NvvA," English Translation, Utrecht, The Netherlands, May 1987.
65. A. Burdorf, *Ann. Occup. Hyg.* **38**, 939 (1994).
66. B. E. Holland, personal communication, Feb. 1999.
67. J. Indulski, J. Janusz, J. Grzesik, J. Gromiec, E. Wiecek, and A. Nawakowski, personal communications, March–April 1999.
68. J. Indulski, J. Gromiec, and E. Wiecek, *Occupational Hygiene in Poland*, The Nofer Institute of Occupational Medicine, Lodz, Poland, 1998.
69. J. Indulski, L. Dawydzik, and M. Jakubowski, *Int. Arch. Occup. Environ. Hlth.* **70**, 289 (1997).
70. M. Jakubowski and S. Czerczak, *Principles for Setting Hygienic Standards in Poland*, Nofer Institute of Occupational Medicine, Lodz, Poland, 1994.
71. J. Indulski, Z. Kowlaski, J. Majka, et al, *Pol. J. Occ. Med.* **1**, 111 (1988).
72. J. P. Gromiec, *Ann. Conf. Gov. Ind. Hyg.* **15**, 169 (1988).
73. D. R. Hicks, personal communication, Jan. 1999.
74. R. Ferrie and R. Strydom, personal communication, Feb. 1999.
75. D. Truter and R. Ferrie, *Occ. Health S. Africa* **3**, 12 (1997).
76. J. R. Johnston, *Ann. Occ. Hyg.* **37**, 237 (1993).
77. G. Nice, personal communication, March 1999.
78. L. Lillienberg, A. Burdorf, and G. Nise, *Appl. Occup. Environ. Hyg.* **8**, 1005 (1997).
79. G. Clack, *Job Safety Health* **2**, 11 (1974).
80. C. Edling and P. Lundberg, "The Procedure of Occupational Standard Setting in Sweden" *Proceedings of the Seventh Annual Conference*, Australian Institute of Occupational Hygiene, Dec. 1988.
81. G. Gerhardsson, "Training and Education in Occupational Hygiene: An International Perspective," *Ann. Am. Conf. Gov. Ind. Hyg.*, **15**, 69 (1988).
82. A. Steinegger and R. Knutti, personal communications, Jan.–Feb. 1999.
83. M. P. Guillemin, *Appl. Occup. Environ. Hyg.* **8**, 921 (1993).
84. M. A. Boillat, M. P. Guillemin, and H. Savolainen, *Int. Arch. Occup. Environ. Hlth.* **70**, 361 (1997).
85. T. Punpeng and H. A. Trout, personal communications, Feb. 1999.
86. Anonymous, "Review of Occupation Health and Safety Conditions to Promote Workers' Health and Safety in Factories," Thailand Research Fund, 1998.

87. Anonymous, The Announcement of Ministry of Labour and Social Welfare on Safety at Work of Employee, Thailand, 31 March 1997.
88. T. Punpeng, and H. A. Trout, personal communications, Feb. 1999.
89. C. Asker-Browne and P. J. Oldershaw, personal communications, March–April 1999.
90. S. G. Luxon, *Am. Ind. Hyg. Assoc. J.* **45**, 731 (1984).
91. T. L. Ogden, M. D. Topping, *Appl. Occup. Environ. Hyg.* **12**, 302 (1997).
92. P. J. Hewitt, *Ann. Occup. Hyg.* **17**, 321 (1993).
93. BIOH, BIOH Membership & Qualifications, British Institute of Occupational Hygiene, 1998.
94. E. Hickish, *The First Forty Years—An Account of the Formation and Development of the British Occupational Hygiene Society*, The British Occupational Hygiene Society, Derby, UK, 1993.
95. J. W. Cherrie and B. S. Bord, *Ann. Occup. Hyg.* **35**, 665 (1991).
96. G. Gerhardsson, *Ann. Occ. Hyg.* **32**, 1 (1988).
97. A. Burdorf and D. Heederik, *Am. Ind. Hyg. Assoc. J.* **53**, A-484 (1990).
98. D. W. Purchon, *Ann. Occ. Hyg.* **33**, 645 (1989).
99. K. Elgstrand, *Am. J. Ind. Med.* **8**, 91 (1985).
100. R. J. Sherwood, *Am. Ind. Hyg. Assoc. J.* **53**, 398 (1992).
101. A. Burdorf, *Safety Sci.* **20**, 191 (1995).
102. M. P. Guillemin, *Ann. Occup. Hyg.* **36**, 669 (1992).
103. B. F. Goelzer, *Am. Ind. Hyg. Assoc. J.* **57**, 984 (1996).

Index

Cumulative Index, Volumes 1–4